DATE DUE

DE 10 '96			
MR 3 '97			
AP 7 '97			
RENEW MY 12 '97 DE 4 '00			

DEMCO 38-296

Handbook of

INORGANIC COMPOUNDS

Edited by
Dale L. Perry
Lawrence Berkeley Laboratory
University of California
Berkeley, California

Sidney L. Phillips
CAMATX Chemistry
Orinda, California

CRC Press
Boca Raton   New York   London   Tokyo

Library of Congress Cataloging-in-Publication Data

Catalog record available from Library of Congress.

PREFACE

Over the past five decades a large number of new compounds have been synthesized and their basic properties obtained by laboratory measurements. At the same time, new uses for older products have emerged and new methods have been developed to synthesize and manufacture inorganic compounds. The objective of this work is to provide an information resource to the reader who uses numerical values of the basic chemistry properties of inorganic compounds for various applications.

The compounds in the *Handbook* were selected for several reasons. For example, many are now commercially important, and the data can be used to assist those who are looking at potential uses for new compounds, and wish to compare key data of the current commercial product with their new compound. Other compounds have been included because they are or can be precursors in processes for commercially preparing important materials. Examples are use of the precursors in sol-gel, vacuum-deposition and hydrothermal crystallization as steps in a larger preparative procedure. A third reason for inclusion is to provide completeness to a family of compounds such as oxides and halides.

It is not possible to include either all inorganic compounds or all the important data. The *Handbook of Inorganic Compounds* consists of data for 3326 selected gas, liquid and solid compounds. The selection was based on considerations such as inclusion of the compounds in various handbooks of laboratory chemicals, discussion in recent research publications, and from comments of the Advisory Committee for this work.

The material in this work includes mainly the chemical elements, binary compounds of the elements with anions such as sulfate and chloride, and metal salts of some simple organic acids. If a compound has more than one form, then each form may be listed individually. One example is separate listings for an anhydrous compound and its hydrates. A second example is the separate listing of the three calcium carbonates. With some exceptions, minerals, organometallic compounds, the metal alloys, noncrystalline materials, and nonstoichiometric materials are not included in this *Handbook*.

The format for presenting information has both numerical data and descriptive information. The data are solubility, melting point, boiling point, density, thermal conductivity and thermal expansion coefficient. Other data may also be included, e.g., vapor pressure, viscosity, hardness, lattice parameters, electrical resistivity, Poisson's ratio, and dielectric constant. There may also be thermodynamic values, mainly enthalpy of vaporization, fusion and sublimation. However, thermodynamic values for the individual compounds such as enthalpy of formation are not covered by this work. Descriptive information for the various compounds are organized into the three categories: form, e.g., color and particle size; preparation or manufacturing procedure; and commercial or other uses.

A significant effort has been made to tabulate numerical values for each compound. Thus, the reader is saved considerable time in looking for the basic properties from many sources. The *Handbook* is intended to be useful to chemists, chemical engineers, and materials scientists who need:

1. Property data for compounds which they wish to use in their research, development, and applications work.

2. CAS RN numbers for computer and other searches. An effort has been made to include CAS RN numbers for hydrates, as well as for compounds with no waters of crystallization.

3. A consistent tabulation of molecular weights. In this work, molecular weights have been calculated to three decimal places in all cases.

4. To synthesize inorganic materials on a laboratory scale.

5. Information on commercial and other uses for many compounds, which can be used for teaching purposes in a chemistry course.

Additional numerical data for inorganic compounds are found in the 75th edition of the *Handbook of Chemistry and Physics*, the *Encyclopedia of Chemical Technology*, the *Merck Index*, *Comprehensive Inorganic Chemistry*, *Gmelin*, *Lange's Handbook of Chemistry*, and *Hawley's Comprehensive Chemical Dictionary*. Some of these sources are also available by computer access. More recent information and data can be found in research journals such as the *Journal of Material Research*, *Journal of the American Ceramic Society*, *Chemistry of Materials*, *Chemical Reviews*, *Material Research Bulletin*, *Journal of the Electrochemical Society,* and *Solid-State Ionics*.

Sidney L. Phillips
Dale L. Perry

Handbook of Inorganic Compounds

Editors

Dale L. Perry
Lawrence Berkeley Laboratory
University of California
Berkeley, CA 94720
Phone: (510) 486-4819
FAX: (510) 486-5799

Sidney L. Phillips
CAMATX Chemistry
171 El Toyonal
Orinda, CA 94563
Phone: (510) 254-1144
FAX: (510) 253-1358

Editorial Advisory Board

THE EDITORS

Dale L. Perry received his B.S. in chemistry from Midwestern State University in 1969, his M.S. in inorganic chemistry from Lamar University in 1972, and his Ph.D. in inorganic chemistry from the University of Houston in 1974. He was a Welch Postdoctoral Fellow from 1975 to 1976 and a National Science Postdoctoral Fellow from 1976 to 1977 at Rice University. From 1977 to 1979, he was a Miller Fellow in chemistry at the University of California, Berkeley. He has been on the scientific staff in chemistry at Lawrence Berkeley Laboratory, University of California, since 1979, being appointed a Senior Scientist in chemistry at the same institution in 1987.

His research interests are in solid state inorganic synthesis and spectroscopy, inorganic systems which include those of transition, main group, lanthanide, and actinide metal ions. The classes of compounds and materials on which his research has focused include metal ion-organic complexes, inorganic thin films, semiconductors, superconductors, mixed metal ion oxide catalysts, inorganic crystals, inorganic scintillation materials, and inorganic polymers. He is the author and co-author of over 200 contributed and invited scientific presentations, refereed journal publications, and numerous invited seminars at universities, national laboratories, and industry. He is the editor and author of several books, including *Instrumental Surface Analysis of Geologic Materials, Applications of Analytical Techniques to the Characterization of Materials*, and *Applications of Synchrotron Radiation Techniques to Materials Science*. He has conducted workshops related to the characterization of inorganic materials and analysis using x-ray photoelectron, Auger, infrared, Raman, nuclear magnetic resonance, and Mossbauer spectroscopy.

His honors include a Sigma Xi National Research Award and Traineeship, a Miller Fellowship, and a National Science Foundation Fellowship. He is a member of the American Chemical Society, Materials Research Society, and the Society for Applied Spectroscopy; he is also a Fellow of the Royal Society of Chemistry (London). He is a member of the Committee for Corporate Participation in the Materials Research Society and both a member and chairman of the Chemistry and Engineering Materials Subdivision in the Industrial & Engineering Division of the American Chemical Society.

In addition to research, he has been a member of several *ad hoc* panels for the U.S. Department of Energy related to instrumentation needs in both heavy metal chemical research and research as it pertains to heavy metals in the environment. He is an organizer of symposia concerning the application of spectroscopy to materials research, synthesis and characterization of inorganic materials, and the application of surface spectroscopy to materials studies. He is also very active as a principal investigator and mentor to Hispanic, Native American, and African American students in the Center for Science and Engineering Education at Lawrence Berkeley Laboratory.

Sidney L. Phillips has been consulting in electrochemistry and computerized databases since 1991. He received his A.B. in Chemistry from Boston University in 1953, A.M. in Chemistry from Dartmouth College in 1958, and Ph.D. in Chemistry from the University of Wisconsin in 1964. He was a Staff Scientist at Lawrence Berkeley Laboratory of the University of California from 1972 to 1991 where his research centered on electrochemistry and thermochemical databases for basic energy research. From 1958 to 1972 he was a chemist with IBM Research and Advanced Materials Laboratories. He was a member of the Chemistry Department at Vassar College 1967 to 1968, and a Research Associate at the National Bureau of Standards (now National Institute of Standards and Technology) 1968 to 1969.

He has served as Chairman of the Mid-Hudson Section of the American Chemical Society in 1967, and has been a member of the Electrochemical Society. Membership on Advisory Committees includes: Atomic Energy Commission's Division of Physical Research for Collection, Evaluation and Dissemination of Physical and Chemical Data in 1974, International Geothermal Information Exchange from 1974 to 1975, Instrumentation Advisory Panel for *Analytical Chemistry* from 1977 to 1979, ASTM Materials Task Group for Utilization of Geothermal Chemistry in 1979, U.S. Department of Energy Advisory Committee on Drill Pipe and Well Casing for Geothermal Energy in 1980, Doctoral Promotion Committee for Delft Technical University in 1992. He was awarded an IBM Creative Development Award in 1965, and an IBM Patent Award in 1968.

Besides Editor of the *Handbook of Inorganic Compounds*, he is a Senior Research Associate at Herguth Laboratories, and has co-edited the *Actinide Nitrates* volume of the Solubility Data Series for the International Union of Pure and Applied Chemistry published in 1994. He has published over 80 research publications and technical reports.

ACKNOWLEDGMENTS

The Editors express their thanks to the following individuals for their contributions to this work:

Solubility Data:

Boris Krumgalz
Israel Oceanographic & Limnological Research
Tel Shikmona
P.O.B. 8030
Haifa, Israel 31080
Phone: 04-515202
FAX: 04-511911

Computer Work:

Daniel J. Phillips
Camatx/Basic Chemistry
171 El Toyonal
Orinda, CA 94563
Phone: 510-254-7717
FAX: 510-253-1358

Barium and Cerium Data:

Mariska Scholten
Delft University of Technology
P.O.B. 5045
Julianalaan 136
2628 BL Delft
The Netherlands
Phone: 31 (0)15-782615
FAX: 31 (0)15-782655

Barium and Cerium Data:

Joop Schoonman
Delft University of Technology
P.O.B. 5045
Julianalaan 136
2628 BL Delft
The Netherlands
Phone: 31 (0)15-782615
FAX: 31 (0)15-782655

Sodium Silicates Data:

Barry Schenker
Occidental Chemical Corporation
P.O.B. 344
Niagara Falls, NY 14302-0344
Phone: 1-800-733-1165

Consistency of Data:

David R. Lide
CRC Press, Inc.
13901 Riding Loop Dr.
Gaithersburg, MD 20878
Phone: 301-738-7147
FAX: 301-738-7147

Lanthanide and Ruthenium Data:

Joseph A. Rard
Lawrence Livermore National Laboratory
University of California
Livermore, CA 94550
Phone: 415-422-1100

Oxide Solubility Data:

Steven E. Ziemniak
Knolls Atomic Power Laboratory
P.O. Box 1072
Schnectady, NY 12301
Phone: 518-395-4000
FAX: 518-395-4422

ORGANIZATION OF DATA FOR THE COMPOUNDS

References to sources of the data are given in the form [XXXYY].

Compound: Commonly used name of inorganic compound.

Formula: Commonly used chemical formula.

Molecular Formula: Modified Hill system in which carbon is always listed first, followed by hydrogen (if any), then other elements in alphabetical order. If there is no carbon, then the elements are given in alphabetical order. Stoichiometry is always shown in the usual subscript form.

Molecular Weight: Consistently calculated from the stoichiometry of the formula to three decimal places, using atomic weights from *Pure & Applied Chemistry,* **1992,** *64,* pages 1522—1523. Significant figures are not taken into account.

CAS RN: Chemical Abstracts Service Registry Number. Where possible, the CAS RN for the compound with hydrated waters is given, as well as that for the anhydrous compound.

Properties: Consists of all or some of the following basic chemical data: crystalline form with lattice parameters; color; gas, liquid, or solid; vapor pressure; hardness; viscosity; dielectric constant; electrical resistivity; Poisson's ratio; enthalpy of vaporization; enthalpy of fusion; preparation; uses.

Solubility: Concentration of compound in solvent under the stated conditions. The solvent is usually water; effort has been made to include the equilibrium solid phase.

Density: Density of the solid, liquid, or gas.

Melting Point: Temperature at which the pure solid becomes liquid.

Boiling Point: Temperature at which the pure liquid becomes a gas.

Reactions: Limited generally to phase changes, decomposition, and hydrolytic reactions.

Thermal Conductivity: Property of the compound which attributes a numerical value to its capability to transmit heat.

Thermal Expansion Coefficient: Change in volume or length per degree change in temperature. This is an important property for ceramics.

LITERATURE CITED

[AES93]. AESAR Catalog, Johnson Matthey, Ward Hill, MA 01835-0747 (**1992-1993**).

[AIR87]. Specialty Gases, Air Products and Chemicals, Inc., Allentown, PA 18105 (**1987**).

[ALD93]. Aldrich Catalog Handbook of Fine Chemicals, Aldrich Chemical Co., St. Louis, MO 63178-9916 (**1992-93**).

[ALD94]. Aldrich Catalog Inorganics, Aldrich Chemical Co., Milwaukee, WI 53233 (**1994**).

[ALF93]. Alfa Catalog of Research Chemicals and Accessories, Johnson Matthey Catalog Co., Inc., Ward Hill, MA 01835-0747 (**1993-94**).

[ALF95]. Alfa Aesar Chemicals Catalog, Johnson Matthey Catalog Co., Inc., Ward Hill, MA 01835-0747 (**1995-96**).

[ASM93]. W. Assmus and W. Schmidbauer, "Crystal Growth of HTSC Materials", *Supercond. Sci. Technol.,* **1993**, *6*, 555-566.

[AST92]. "Standard Practice for Use of the SI International System of Units", ASTM E 380-92, American Society for Testing and Materials, Philadelphia, PA 19103 (**1992**).

[BAB85]. V.I. Babushkin, G.M. Matveyev, and O.P. Mchedlov-Petrossyan, "Thermodynamics of Silicates", transl. by B.N. Frenkel and V.A. Terentyev, Springer-Verlag, New York (**1985**).

[BAN90]. N.P. Bansal, "Influence of Several Metal Ions on the Gelation Activation Energy of Silicon Tetraethoxide", *J. Am. Ceram. Soc.,* **1990**, *73*, 2647-2652.

[BHA78]. H. Bhattacharya and B.N. Sammaddar, "Formation of Nonstoichiometric Spinel on Heating Hydrous Magnesium Aluminate", *J. Am. Ceram. Soc.,* **1978**, *61*, 279-280.

[BOU93]. W.L. Bourcier, K.G. Knauss, and K.J. Jackson, "Aluminum Hydrolysis Constants to 250°C from Boehmite Solubility Measurements", *Geochim. Cosmochim. Acta,* **1993**, *57*, 747-762.

[BRO73]. D. Brown, "The Actinides", in *Comprehensive Inorganic Chemistry*, Pergamon Press, New York, (**1973**), p. 278.

[CAB85]. KBI Electronic Materials, Potassium and Rubidium, Cabot Corp., Revere, PA 18953 (**1985**).

[CAB93]. Cabot Performance Materials, "Tantalum", Cabot Corp., Revere, PA 18953 (**1993**).

[CEN92]. F.J. Adrian and D.O. Cowan, "The New Superconductors", *Chem. & Eng. News*, Vol. *70*, Dec. 21, **1992**.

[CEN94]. R. Dagani, "New Opportunities in Materials Science Draw Eager Organometallic Chemists", *Chem. & Eng. News*, Vol. *72*, Oct. 3, **1994**.

[CER91]. Cerac, Inc., Milwaukee, WI 53201 (**1991**).

[CHA90]. F. Chaput and J.-P. Boilot, "Alkoxide-Hydroxide Route to Synthesize $BaTiO_3$-Based Powders", *J. Am. Ceram. Soc.,* **1990**, *73*, 942-948.

[CIC73].   J.C. Bailar, H.J. Emeleus, R. Nyholm, and A.F. Trotman-Dickenson, Eds., *Comprehensive Inorganic Chemistry*, Pergamon Press, New York (**1973**).

[CLA66].   S.P. Clark, Ed., Handbook of Physical Constants, Memoir 97, Geological Society of America, New York (**1966**).

[CON87].   L.E. Conroy, A.N. Christensen, and J. Bottiger, "Preparation and Characterization of Hi-T_c Oxides. $YBa_2Cu_3O_7$ and $REBa_2Cu_3O_7$", *Acta Chim. Scand.*, **1987**, *A41*, 501-505.

[COT88].   F.A. Cotton and G. Wilkinson, *Advanced Inorganic Chemistry*, 5th ed., John Wiley & Sons, New York (**1988**).

[CRC77], [CRC86], [CRC92], [CRC93].   *CRC Handbook of Chemistry and Physics*, various editions, D.R. Lide, Ed., CRC Press, Inc., Boca Raton, FL.

[CRC94].   Ibid., 75th ed. (**1994**).

[DAH90].   J.R. Dahn, U. von Sacken, and C.A. Michal, "Structure and Electrochemistry of $Li_{1\pm y}NiO_2$ and a New Li_2NiO_2 Phase with the $Ni(OH)_2$ Structure", *Solid State Ionics*, **1990**, *44*, 87-97.

[DES91].   P. Descamps, S. Sakaguchi, M. Poorteman, and F. Cambier, "High Temperature Characterization of Reaction Sintered Mullite-Zirconia Composites", *J. Am. Ceram. Soc.*, **1991**, *74*, 2476-2481.

[DOU83].   B. Douglas, D.H. McDaniel, and J.J. Alexander, *Concepts and Models of Inorganic Chemistry*, 2nd ed., John Wiley & Sons, Inc., New York (**1983**).

[DRE93].   M.S. Dresselhaus, C. Dresselhaus, and P.C. Eklund, "Fullerenes", *J. Mater. Res.*, **1993**, *8*, 2054-2097.

[ERI92].   T.E. Eriksen, P. Ndalamba, J. Bruno, and M. Caceci, "The Solubility of $TcO_2.nH_2O$ in Neutral to Alkaline Solutions Under Constant pCO_2", *Radiochim. Acta*, **1992**, *58/59*, 67-70.

[FIE87].   P.E. Fielding and T.J. White, "Crystal Chemical Incorporation of High Level Waste Species in Aluminotitanate-Based Ceramics", *J. Mater. Res.*, **1987**, *2*, 387-414.

[FMC93].   FMC Corporation, Lithium Division, "Lithco Products. Your Source of Quality Lithium Chemicals", Gastonia, NC 28053-3925 (**1993**).

[FRE87].   B. Freudenberg and A. Mocellin, "Aluminum Titanate Formation by Solid-State Reaction of Fine Al_2O_3 and TiO_2 Powders", *J. Am. Ceram. Soc.*, **1987**, *70*, 33-38.

[FRI87].   J.J. Fritz and E. Luzik, "Solubility of Copper(I) Bromide in Aqueous Solutions of Potassium Bromide", *J. Solution Chem.*, **1987**, *16*, 79-85.

[GEI92].   G. Geiger, "Ceramic Coatings", *Bull. Am. Ceram. Soc.*, **1992**, 1470-1481.

[GME76].   *Gmelin Handbuch der Anorganischen Chemie. Seltenerdelemente Teil C3: Sc, Y, La und Lanthanide, Fluoride, Oxidifluoride und zugehorige Alkilidoppelverbindungen*, H. Bergmann, Springer, Berlin (**1976**), in [SCH93]: J. Scholten and J. Schoonman, Delft University of Technology, The Netherlands, (October 18, **1993**).

[GME77].   *Gmelin Handbuch der Anorganischen Chemie*, Springer-Verlag, New York (**1977**).

[GUM92].   R.J. Gummow, M.M. Thackeray, W.I.F. David, and S. Hull, "Structure and Electrochemistry of Lithium Cobalt Oxide Synthesised at 400°C", *Mat. Res. Bull.*, **1992**, *27*, 327-337.

[HAW93]. *Hawley's Condensed Chemical Dictionary*, 12th ed., R.J. Lewis, Van Nostrand Reinhold Co., New York (**1993**).

[HIR87]. S.-I. Hirano and K. Kato, "Synthesis of $LiNbO_3$ by Hydrolysis of Metal Alkoxides", *Adv. Ceram. Mater.,* **1987**, *2*, 142-145.

[HIR89]. Y. Hirata, K. Sakeda, Y. Matsushita, K. Shimada, and Y. Ishihara, "Characterization and Sintering Behavior of Alkoxide Derived Aluminosilicate Powders", *J. Am. Ceram. Soc.,* **1989**, *72*, 995-1002.

[HO 72]. C.Y. Ho, R.W. Powell, and P.E. Liley, *J. Phys. Chem. Ref. Data,* **1972**, *1*, 279-421.

[HOL73]. C.E. Holcombe and A.L Coffey, "Calculated X-Ray Powder Diffraction Data for $\beta\text{-}Al_2TiO_5$", *J. Am. Ceram. Soc.,* **1973**, *56*, 220-221.

[HOU82]. V. Houlding, T. Geiger; U. Kölle, and M. Grätzel, "Electrochemical and Photochemical Investigations of Two Novel Electron Relays for Hydrogen Generation from Water", *J. Chem. Soc., Chem. Commun.,* **1982**, 681-683.

[HUA91]. G. Huan, J.W. Johnson, A.J. Jacobson, and D.P. Goshorn, "Hydrothermal Synthesis, Single-Crystal Structure, and Magnetic Properties of $VOSeO_3.H_2O$", *Chem. Mater.,* **1991**, *3*, 539-541.

[IUP92]. IUPAC "Atomic Weights of the Elements 1991", *Pure & Applied Chem.,* **1992**, *64*, 1519-1534.

[IUP93]. PAC Review, "Fullerenes", *Chem. Int.,* **1993**, *15*, 94.

[JAN71]. D.R. Stull and H. Prophet, "JANAF Thermochemical Tables", Nat. Stand. Ref. Data Ser., NIST, 37 (**1971**).

[JAN82]. M.W. Chase, J.L. Curnutt, J.R. Downey, R.A. McDonald, A.N. Syverud, and E.A. Valenzuela, "JANAF Thermochemical Tables, 1982 Supplement", *J. Phys. Chem. Ref. Data,* **1982**, *11*, 695.

[JAN85]. M.W. Chase, C.A. Davies, J.R. Downey, D.J. Frurip, R.A. McDonald, and A.N. Syverud, "JANAF Thermochemical Tables", 3rd ed., *J. Phys. Chem. Ref. Data,* **1985**, Vol. *14*, Suppl. 1.

[KAT86]. J.J. Katz, G.T. Seaborg, and L.R. Morss, Eds., *The Chemistry of the Actinide Elements*, 2nd ed., Vol. 1 and Vol. 2, Chapman and Hall, New York (**1986**).

[KAZ90]. A.M. Kazakos, S. Komarneni, and R. Roy, "Sol-Gel Processing of Cordierite: Effect of Seeding and Optimization of Heat Treatment", *J. Mater. Res.,* **1990**, *5*, 1095-1103.

[KIR78], [KIR79], [KIR80], [KIR81], [KIR82], [KIR83], [KIR84]. Kirk-Othmer, *Encyclopedia of Chemical Technology*, 3rd ed., various volumes, John Wiley & Sons, New York (**1978** to **1984**).

[KIR91]. Kirk-Othmer, Ibid., 4th ed., (**1991**).

[KLE93]. J.D. Klein, R.D. Herrick, D. Palmer, M.J. Sailor, C.J. Brumlik, and C.R. Martin, "Electrochemical Fabrication of Cadmium Chalcogenide Microdiode Arrays", *Chem. Mater.,* **1993**, *5*, 902-904.

[KNA91]. O. Knacke, O. Kubascheski, and K. Hesselmann, *Thermochemical Properties of Inorganic Substances*, Springer-Verlag, Berlin (**1991**).

[KOH88]. S.C.H. Koh and R. McPherson, "Sintering of Metastable $ZrO_2\text{-}Al_2O_3$", "Ceramic Developments", in *Materials Science Forum*, 1988, Vols. 34-36, p. 117-121, Trans. Tech. Publications, Ltd., Switzerland (**1988**).

[KOR91]. N.E. Korte and Q. Fernando, "A Review of Arsenic(III) in Groundwater", *Crit. Rev. in Environ. Control,* **1991**, *21*, 1-39.

[KRE91]. A.M. Kressin, V.V. Doan, J.D. Klein, and M.J. Sailor, "Synthesis of Stoichiometric Cadmium Selenide Films via Sequential Monolayer Electrodeposition", *Chem. Mater.,* **1991**, *3*, 1015-1020.

[KRU93]. B.S. Krumgalz, "Mineral Solubility in Water at Various Temperatures", Israel Oceanographic & Limnological Research, Tel-Shikmona, P.O.B. 8030, Haifa 31080 Israel (**1993**).

[LAN52]. *Lange's Handbook of Chemistry*, N.A. Lange, Ed., 8th ed., Handbook Publisher's, Inc., Sandusky, OH (**1952**).

[LAN85]. *Lange's Handbook of Chemistry*, J.A. Dean, Ed., 13th ed., McGraw-Hill Book Co., New York (**1985**).

[LAU73]. R.A. Laudise, "Hydrothermal Growth", in *An Introduction to Crystal Growth*, P. Hartman, Ed., North-Holland (**1973**).

[LAU87]. R.A. Laudise, "Hydrothermal Crystallization", *Chem. Eng. News*, September 28, **1987**, p. 30.

[LID94]. D.R. Lide, "Physical Constants of Inorganic Compounds", CRC Press, Inc., Private Communication (Nov. **1994**).

[LOP84]. M.A. Lopez-Quintela, W. Knoche, and J. Veith, "Kinetics and Thermodynamics of Complex Formation between Aluminum(III) and Citric Acid in Aqueous Solutions", *J. Chem. Soc., Faraday Trans., 1*, **1984**, *80*, 2313-2321.

[MAE90]. K. Maeda, F. Mizukami, S. Miyashita, S.-I. Niwa, and M. Toba, "Synthesis of Cordierite by Complexing Agent Assisted Sol-Gel Procedure", *J. Chem. Soc., Chem. Commun.,* **1990**, 1268-1269.

[MAK90]. A. Makishima, M. Asami, and K. Wada, "Preparation and Properties of TiO_2-CeO_2 Coatings by the Sol-Gel Process", *J. Non Crystalline Solids,* **1990**, *121*, 310-314.

[MAR58]. W.L. Marshall, F.J. Loprest, and C.H. Secoy, "The Equilibrium $Li_2CO_3 + CO_2 + H_2O = 2Li^+ + 2HCO_3^-$ at High Temperature and Pressure", *J. Am. Chem. Soc.,* **1958**, *80*, 5646.

[MER52]. *The Merck Index*, 6th ed., Merck & Co., Inc., Rahway, NJ (**1952**).

[MER89]. *The Merck Index*, 11th ed., Merck & Co., Inc., Rahway, NJ (**1989**).

[MIT87]. M. Mitomo and Y. Yoshioka, "Preparation of Si_3N_4 and AlN Powders from Alkoxide-Derived Oxides by Carbothermal Reduction and Nitridation", *Adv. Ceram. Mater.,* **1987**, *2*, 253-256.

[MIT72]. P.W.D. Mitchell, "Chemical Method for Preparing $MgAl_2O_4$ Spinel", *J. Am. Ceram. Soc.,* **1972**, *55*, 484.

[MIZ89]. M. Mizuno and H. Saito, "Preparation of Highly Pure Fine Mullite Powder", *J. Am. Ceram. Soc.,* **1989**, *72*, 377-382.

[MOI86]. A. Moini, R. Peascoe, P.R. Rudolf, and A. Clearfield, "Hydrothermal Synthesis of Copper Molybdates", *Inorg. Chem.,* **1986**, *25*, 3782-3785.

[MOY86]. J.R. Moyer, A.R. Prunier, N.N. Hughes, and R.C. Winterton, "Synthesis of Oxide Ceramic Powders by Aqueous Coprecipitation", *Mater. Res. Soc. Symp. Proc.*, Vol. *73*, Material Research Society (**1986**).

[OGU88]. Y. Oguri, R.E. Riman, and H.K. Bowen, "Processing of Anatase Prepared from Hydrothermally Treated Alkoxy-Derived Hydrous Titania", *J. Mater. Sci.,* **1988**, *23,* 2897-2904.

[OKA91]. K. Okada, N. Otsuka, and S. Somiya, "Review of Mullite Synthesis Routes in Japan", *Ceram. Bull.,* **1991**, *70,* 1633-1640.

[OXY93]. B. Schenker, "Sodium Metasilicate", Occidental Chemical Corp., Niagara Falls, NY 14302. Private Communication Nov. 10, **1993**.

[OZB80]. H. Ozbek and S.L. Phillips, "Thermal Conductivity of Aqueous Sodium Chloride Solutions from 20 to 330°C", *J. Chem. Eng. Data,* **1980**, *25,* 263-267.

[PAR90]. F.J. Parker, "Al_2TiO_5-$ZrTiO_4$-ZrO_2 Composites: A New Family of Low-Thermal-Expansion Ceramics", *J. Am. Ceram. Soc.,* **1990**, *73,* 929-932.

[PAS88]. J.A. Pask, "Phase Equilibria in the Al_2O_3-SiO_2 System with Emphasis on Mullite", *Mater. Sci. Forum,* **1988**, *34-36,* 1-8.

[PER89]. D.S. Perera, "Reaction-Sintered Aluminum Titanate", *J. Mater. Sci. Lett.,* **1989**, *8,* 1057-1059.

[PFA93]. Pfaltz & Bauer Chemicals Catalog, Aceto Corp., Waterbury, CT 06708 (**1993**).

[PHI93]. S.L. Phillips and F.V. Hale, "Hydrolysis Constants for $AlOH^{++}$ in High Temperature Water from Thermodynamic Calculations", *Proc. 1991 Symp. High Temperature Chemistry,* Provo, UT (August **1993**).

[PHU89]. P.P. Phule and S.H. Risbud, "Low Temperature Synthesis and Dielectric Properties of Ceramics Derived from Amorphous Barium Titanate Gels and Crystalline Powders", *Mater. Sci. Eng.,* **1989**, *B3,* 241-247.

[PLU82]. L.N. Plummer and E. Busenberg, "The Solubilities of Calcite, Aragonite and Vaterite in CO_2-H_2O Solutions Between 0 and 90°C, and an Evaluation of the Aqueous Model for the System $CaCO_3$-CO_2-H_2O", *Geochim. Cosmochim. Acta,* **1982**, *46,* 1011-1040.

[POT78]. R.W. Potter and M.A. Clynne, "The Solubility of the Noble Gases He, Ne, Ar, Kr, and Xe in Water up to the Critical Point", *J. Solution Chem.,* **1978**, *7,* 837-844.

[PRA92]. A.V. Prasadarao, U. Selvaraj, S. Komarneni, A.S. Bhalla, and R. Roy, "Enhanced Densification by Seeding of Sol-Gel-Derived Aluminum Titanate", *J. Am. Ceram. Soc.,* **1992**, *75,* 1529-1533.

[RAR83]. J.A. Rard, "Critical Review of the Chemistry and Thermodynamics of Technetium and Some of Its Inorganic Compounds and Aqueous Species", UCRL-53440, Lawrence Livermore National Laboratory, Livermore, CA (**1983**).

[RAR84]. J.A. Rard, "Solubility of $Eu(NO_3)_3.6H_2O$ in Water at 298.15 K", *J. Chem. Thermodynamics,* **1984**, *16,* 921-925.

[RAR85]. J.A. Rard, "Chemistry and Thermodynamics of Ruthenium and Some of Its Inorganic Compounds and Aqueous Species", *Chem. Rev.,* **1985**, *85,* 1-39.

[RAR85a]. J.R. Rard, "Chemistry and Thermodynamics of Europium and Some of Its Simpler Inorganic Compounds and Aqueous Species", *Chem. Rev.,* **1985**, *85,* 555-582.

[RAR85b]. J.A. Rard, "Solubility Determinations by the Isopiestic Method and Application to Aqueous Lanthanide Nitrates at 25°C", *J. Solution Chem.,* **1985**, *14,* 457-471.

[RAR87a]. J.A. Rard, "Osmotic and Activity Coefficients of Aqueous La(NO$_3$)$_3$ and Densities and Apparent Molal Volumes of Aqueous Eu(NO$_3$)$_3$ at 25°C", *J. Chem. Eng. Data*, **1987**, *32*, 92-98.

[RAR87b]. J.A. Rard, "Isopiestic Determination of the Osmotic and Activity Coefficients of Aqueous NiCl$_2$, Pr(NO$_3$)$_3$, and Lu(NO$_3$)$_3$ and Solubility of NiCl$_2$ at 25°C", *J. Chem. Eng. Data*, **1987**, *32*, 334-341.

[RAR88]. J.A. Rard, "Aqueous Solubilities of Praseodymium, Europium and Lutetium Sulfates", *J. Solution Chem.*, **1988**, *17*, 499-517.

[RAR92]. J.A. Rard, "Isopiestic Investigation of Water Activities of Aqueous NiCl$_2$ and CuCl$_2$ Solutions and the Thermodynamic Solubility Product of NiCl$_2$-H$_2$O at 298.15K", *J. Chem. Eng. Data*, **1992**, *37*, 433-442.

[RIO92]. A. Riou, A. Lecerf, Y. Gerault, and Y. Cudennec, "Etude Structurale de Li$_2$MnO$_3$", *Mat. Res. Bull.*, **1992**, *27*, 269-275.

[RIT86]. J.J. Ritter, R.S. Roth, and J.E. Blendell, "Alkoxide Precursor Synthesis and Characterization of Phases in the Barium-Titanium Oxide System", *J. Am. Ceram. Soc.*, **1986**, *69*, 155-162.

[ROB67]. R.A. Robie, P.M. Bethke, and K.M. Beardsley, "Selected X-Ray Crystallographic Data, Molar Volume, and Densities of Minerals and Related Substances", U.S. Geol. Surv. Bull. 1248, 87p. (**1967**). In [CRC92].

[ROB78]. R.A. Robie, B.S. Hemingway, and J.R. Fisher, "Thermodynamic Properties of Minerals and Related Substances at 298.15K and 1 Bar (10^5 Pascals) Pressure and at Higher Temperatures", Geol. Surv. Bull. 1452, U.S. Geological Survey, U.S. Government Printing Office, Washington, D.C. (**1978**).

[ROS91]. M.H. Rossouw and M.M. Thackeray, "Lithium Manganese Oxides from Li$_2$MnO$_3$ for Rechargeable Lithium Battery Applications", *Mat. Res. Bull.*, **1991**, *26*, 463-473.

[ROS92]. M.H. Rossouw, D.C. Liles, and M.M. Thackeray, "Alpha Manganese Dioxide for Lithium Batteries: A Structural and Electrochemical Study", *Mat. Res. Bull.*, **1992**, *27*, 221-230.

[SAF87]. A. Safari, Y.H. Lee, A. Halliyal, and R.E. Newnham, "O-3 Piezoelectric Composites Prepared by Coprecipitated PbTiO$_3$ Powder", *Am. Ceram. Soc. Bull.*, **1987**, *66*, 668-670.

[SCH88]. G. Schwartz, R. Bennett, and S.A. Prokopovich, "Sol-Gel Processing in the ZrO$_2$-SiO$_2$ System", *Ceramics Developments*, C.C. Sorrell and B. Ben-Nissen, Eds., *Materials Science Forum*, Vols. *34-36* (**1988**), pp. 841-843.

[SCH93]. M. Scholten and J. Schoonman, "Barium and Cerium", Delft University of Technology, 2600GA Delft, The Netherlands, October 18, **1993**.

[SIE94]. S. Siekierski and S.L. Phillips, Eds., Actinide Nitrates, Solubility Data Series, Vol. 55, Oxford University Press, UK (**1994**).

[SOM91]. S. Somiya and Y. Hirata, "Mullite Powder Technology and Applications in Japan", *Ceram. Bull.*, **1991**, *70*, 1624-1632.

[STR93]. Strem Catalog No. 15, **1993-1994**, Strem Chemicals, Inc., 7 Mulliken Way, Newburyport, MA 01950-4098.

[STR94]. The Strem Chemiker, April 1994, vol. XV, No. 1, Strem Chemicals, Inc. 7 Mulliken Way, Newburyport, MA 01950-4098.

[SUB90]. E.C. Subbarao, D.K. Agrawal, H.A. McKinstry, C.W. Sallese, and R. Roy, "Thermal Expansion of Compounds of Zircon Structure", *J. Am. Ceram. Soc.*, **1990**, *73*, 1246-1252.

[TAY84a]. D. Taylor, "Thermal Expansion Data. I. Binary Oxides with the Sodium Chloride and Wurtzite Structures, MO", *Br. Ceram. Trans. J.*, **1984**, *83*, 5-9.

[TAY85]. D. Taylor, "Thermal Expansion Data. VIII. Complex Oxides, ABO_3, the Perovskites", *Br. Ceram. Trans. J.*, **1985**, *84*, 181-188.

[TAY86]. D. Taylor, "Thermal Expansion Data. X. Complex Oxides, ABO_4", *Br. Ceram. Trans. J.*, **1986**, *85*, 146-155.

[TAY87]. D. Taylor, "Thermal Expansion Data: XI. Complex Oxides, A_2BO_5, and the Garnets", *Br. Ceram. Trans. J.*, **1987**, *86*, 1-6.

[TAY88a]. D. Taylor, "Thermal Expansion Data. XII. Complex Oxides: AB_2O_6, AB_2O_7, $A_2B_2O_7$, Plus Complex Aluminates, Silicates and Analogous Compounds", *Br. Ceram. Trans. J.*, **1988**, *87*, 39-45.

[TAY88b]. D. Taylor, "Thermal Expansion Data. XIII. Complex Oxides with Chain, Ring and Layer Structures and the Apatites", *Br. Ceram. Trans. J.*, **1988**, *87*, 87-95.

[TAY91a]. D. Taylor, "Thermal Expansion Data. XIV. Complex Oxides with the Sodalite and Nasicon Framework Structure", *Br. Ceram. Trans. J.*, **1991**, *90*, 64-69.

[TAY91b]. D. Taylor, "Thermal Expansion Data. XV. Complex Oxides with the Leucite Structure and Frameworks based on the Six-Membered Ring of Tetrahedra", *Br. Ceram. Trans. J.*, **1991**, *90*, 197-204.

[THA92]. M.M. Thackery, A. deKock, M.H. Rossouw, D. Liles, R. Bittihn, and D. Hoge, "Spinel Electrodes from the Li-Mn-O System for Rechargeable Lithium Battery Applications", *J. Electrochem. Soc.*, **1992**, *139*, 363-366.

[TOU77]. Y.S. Touloukian, R.K. Kirby, R.E. Taylor, and T.Y.R. Lee, *Thermophysical Properties of Matter, Volume 13: Thermal Expansion Nonmetallic Solids*, IFI/Plenum, New York (**1977**).

[WU 88]. J.-M. Wu and H.-W. Wang, "Factors Affecting the Formation of $Ba_2Ti_9O_{20}$", *J. Am. Ceram. Soc.* **1988**, *71*, 869-875.

[VIE91]. D.J. Viechnicki, M.J. Slavin, and M.I. Kliman, "Development and Current Status of Armor Ceramics", *Ceram. Bull.*, **1991**, *70*, 1035-1039.

[YAM87]. O. Yamaguchi, H. Taguchi, and K. Shimizu, "Formation of Spinel from Metal Organic Compounds", *Polyhedron*, **1987**, *6*, 1791-1796.

[YAM89]. O. Yamaguchi and Y. Mukaida, "Formation and Transformation of TiO_2 (Anatase) Solid Solution in the System TiO_2-Al_2O_3", *J. Am. Ceram. Soc.*, **1989**, *72*, 330-333.

[YIN92]. Y. Ying and Y. Rudong, "Study on the Thermal Decomposition of Tetrahydrated Ceric Sulfate", *Thermochim. Acta*, **1992**, *202*, 301-306.

[ZIE89]. S.E. Ziemniak, M.E. Jones, and K.E.S. Combs, "Solubility and Phase Behavior of Nickel Oxide in Aqueous Phosphate Solutions at Elevated Temperatures", *J. Solution Chem.*, **1989**, *18*, 1133-1152.

[ZIE92a]. Ibid., "Copper(II) Oxide Solubility Behavior in Aqueous Phosphate Solutions at Elevated Temperatures", *J. Solution Chem.*, **1992**, *21*, 179-200.

[ZIE92b]. Ibid., "Zinc(II) Oxide Solubility and Phase Behavior in Aqueous Phosphate Solutions at Elevated Temperatures", *J. Solution Chem.*, **1992**, *21*, 1153-1176.

[ZIE93]. Ibid., "Solubility Behavior of Titanium(IV) Oxide in Alkaline Media at Elevated Temperatures", *J. Solution Chem.*, **1993**, *22*, 601-623.

GLOSSARY OF TERMS

Reference: "Abbreviated List of Quantities, Units and Symbols in Physical Chemistry", IUPAC, Blackwell Scientific Publications, Oxford, England (1987). "Standard Practice for Use of the International System of Units", ASTM E 380-92, Philadelphia, PA 19103 (1992). *Handbook of Chemistry and Physics*, 74th edition, CRC Press, Inc., Boca Raton, FL 33431 (1993).

, (comma): and, and also, in
$\rightarrow$: giving, yielding
<: below
>: above
~: approximately
μ: micro, E -06
10^{-n}: exponent to number, n
a, b, c: lattice parameters
aq: aqueous
atm: atmosphere, atmospheric
bcc: body-center(ed) cubic
bluish: having a tinge of blue; somewhat blue
bp: boiling point
Btu: British thermal unit
cal: calorie
CAS RN: Chemical Abstracts Service Registry Number
cm: centimeter
conc: concentrated
cp: centipoise
cryst: crystal(s), crystalline
cub: cubic
deliq: deliquescent
dil: dilute
E + or -: exponent (10^{+} or $^{-}$)
eV: electron volt
fcc: face-center(ed) cubic
fp: freezing point
GPa: gigapascal
h: hour(s)
hex: hexagonal
hygr: hygroscopic
i: insoluble
J: joule
K: Kelvin
k: kilo
kgf: kilogram force
kJ: kilojoule

liq: liquid
m: meter
micro: E -06
min: minute(s)
mL: milliliter
mm: millimeter
mol: mole
monocl: monoclinic
mp: melting point
MPa: megapascal
mW: milliwatt
nm: nanometer
off-white: a yellowish or grayish white
ohm·cm: ohm centimeter
ortho-rhomb: ortho-rhombic
ortho: ortho-rhombic
Pa: pascal
powd: powder(s)
reddish: having a tinge of red
rhomb: rhombic
sec: second(s)
s: soluble
sl: slightly
soln: solution
$t_{1/2}$: half-life
T_c, T_K: superconducting transition temperature
temp: temperature
tetr: tetragonal
tric: triclinic
trig: trigonal
v: very
V: volt
W: watt
W/(m·K): $W\ m^{-1}\ K^{-1}$
[XXXYY]: Reference to the source publication; XXX denotes first three letters of last name of senior author, YY denotes last two digits of year of publication

CONVERSION OF UNITS

References: "Abbreviated List of Quantities, Units and Symbols in Physical Chemistry", IUPAC, Blackwell Scientific Publications, Oxford, England (1987). "Standard Practice for Use of the International System of Units", ASTM E 380-92, Philadelphia, PA 19103 (1992).

Å (angstrom) $\times$ E -01 = nm

atm $\times$ 1.013 250 E $+05$ = Pa

bar $\times$ 1.000 000 E $+05$ = Pa

Btu in/(h ft^2 °F) $\times$ 1.441 314 E -01 = W/(m·K)

Btu (thermochemical) $\times$ 1.054 35 E $+03$ = J

cal/(cm·s·°C) $\times$ 4.184 000 E $+02$ = W/(m·K)

cal (thermochemical) $\times$ 4.184 = J

cp $\times$ 1.000 000 E -03 = Pa·s

g/cm^3 $\times$ 1.000 000 E $+03$ = kg/m^3

g $\times$ 1.000 000 E $+03$ = kg

J $\times$ E $+03$ = kJ

kgf/mm^2 $\times$ 9.806 650 E $+06$ = Pa

kcal $\times$ 4.184 = kJ

m $\times$ E -09 = nm

mm Hg (0°C) $\times$ 1.333 22 E $+02$ = Pa

ohm·cm $\times$ 1.000 000 E -02 = ohm·m

Pa $\times$ E $+09$ = GPa

W/(m·K) = 0.1 mW/(cm·K)

W/(m·K) = 100 W/(cm·K)

°C = (°F $-$ 32)/(1.8)

°C = K $-$ 273.15

Table of atomic weights to five significant figures

The following 1991 Table of Standard Atomic Weights Abridged to Five Significant Figures was prepared for publication by J. Cesario, N. N. Greenwood and H. S. Peiser (Commission on Atomic Weights and Isotopic Abundances, Inorganic Chemistry Division).

Introduction

The detail and the number of significant figures in the IUPAC Table of Standard Atomic Weights exceed the needs and interests of most users, who are more concerned with the length of time during which a given table has validity to the precision limit of their interests. The Commission on Atomic Weights and Isotopic Abundances in 1987 therefore decided to prepare for publication a revised and updated version of the 1981 Table of Atomic Weights abridged to Five Significant Figures, or fewer where uncertainties do not warrant even five-figure accuracy (this currently applied to 9 elements). When an atomic weight is known to more than five significant figures, it is abridged in this Table to the five-figure value closest to the unabridged best value. When the sixth digit of the unabridged value is five exactly, it is rounded up or down to make the fifth digit in this abridged Table even. The single-digit uncertainty in the tabulated atomic weight is held to be symmetric, that is, it is applicable with either a positive or a negative sign.

The abridged table is given here with the reasonable expectation that the quoted values for most elements will never need to be changed – a desirable attribute for textbooks and numerical tables derived from atomic-weight data. However, it should be understood that the atomic-weight values for 29 elements are still uncertain by more than one unit in the fifth significant figure. Moreover, for 21 additional elements the indicated uncertainty range in the unabridged table includes values which when rounded to five significant figures would show a change in the fifth figure. For some of these elements minor changes in their best standard atomic weight to five significant figures could occasionally be required as more accurate values become available as a result of the biennial revision of the unabridged table. Most annotated warnings of anomalous geological occurrences, isotopically altered materials, and variability of radioactive elements are relevant even in the abridged table.

This Table may be freely reprinted provided it includes the annotations and the rubric at the head of the Table, and provided the IUPAC source is acknowledged.

Atomic weights are here quoted to five significant figures unless the dependable accuracy is more limited by either the combined uncertainties of the best published atomic-weight determinations, or by the variability of isotopic composition in normal terrestrial occurrences (the latter applied to elements **annotated r**). The last significant figure of each tabulated value is considered reliable to ± 1 except when a larger single-digit uncertainty is inserted in parentheses following the atomic weight. Neither the highest nor the lowest actual atomic weight of any normal sample is thought likely to differ from the tabulated value by more than the assigned uncertainty. However, the tabulated values do not apply either to samples of highly exceptional isotopic composition arising from most unusual geological occurrences (for elements **annotated g**) or to those who isotopic composition has been artificially altered. Such might even be found in commerce without disclosure of that modification (for elements **annotated m**). Elements annotated by an asterisk (*) have no stable isotope and are generally represented in this Table by just one of the element's commonly known radioisotopes, with a corresponding relative atomic mass in the atomic-weight column. However, three such elements (Th, Pa and U) do have a characteristic terrestrial isotopic composition, and for these an atomic weight is tabulated. For more detailed information users should refer to the full IUPAC Table of Standard Atomic Weights, as is found in the biennial reports of the Commission on Atomic Weights and Isotopic Abundances. The most recent table was published in *Pure Appl. Chem.*, 64, 1519–1534, (1992).

Photocopied and reproduced with permission from *Chemistry International*, 1993, Vol. 15, No. 4.

1991 Table of standard atomic weights abridged to five significant figures. Scaled to the relative atomic mass, $A_r(^{12}C) = 12$.

	Name	Symbol	Atomic weight	Annotations		Name	Symbol	Atomic weight	Annotations
1	Hydrogen	H	1.0079	g m	53	Iodine	I	126.90	
2	Helium	He	4.0026		54	Xenon	Xe	131.29(2)	g m
3	Lithium	Li	6.941(2)	g m r	55	Caesium	Cs	132.91	
4	Beryllium	Be	9.0122		56	Barium	Ba	137.33	
5	Boron	B	10.811(5)	g m r	57	Lanthanum	La	138.91	
6	Carbon	C	12.011	g r	58	Cerium	Ce	140.12	g
7	Nitrogen	N	14.007		59	Praseodymium	Pr	140.91	
8	Oxygen	O	15.999		60	Neodymium	Nd	144.24(3)	g
9	Fluorine	F	18.998		61	Promethium *	^{147}Pm	146.92	
10	Neon	Ne	20.180	m	62	Samarium	Sm	150.36(3)	g
11	Sodium (Natrium)	Na	22.990		63	Europium	Eu	151.96	g
12	Magnesium	Mg	24.305		64	Gadolinium	Gd	157.25(3)	g
13	Aluminium	Al	26.982		65	Terbium	Tb	158.93	
14	Silicon	Si	28.086		66	Dysprosium	Dy	162.50(3)	g
15	Phosphorus	P	30.974		67	Holmium	Ho	164.93	
16	Sulfur	S	32.066(6)	g r	68	Erbium	Er	167.26(3)	g
17	Chlorine	Cl	35.453	m	69	Thulium	Tm	168.93	
18	Argon	Ar	39.948	g r	70	Ytterbium	Yb	173.04(3)	g
19	Potassium (Kalium)	K	39.098	g	71	Lutetium	Lu	174.97	g
20	Calcium	Ca	40.078(4)	g	72	Hafnium	Hf	178.49(2)	
21	Scandium	Sc	44.956		73	Tantalum	Ta	180.95	
22	Titanium	Ti	47.88(3)		74	Tungsten (Wolfram)	W	183.84	
23	Vanadium	V	50.942		75	Rhenium	Re	186.21	
24	Chromium	Cr	51.996		76	Osmium	Os	190.23(3)	g
25	Manganese	Mn	54.938		77	Iridium	Ir	192.22(3)	
26	Iron	Fe	55.847(3)		78	Platinum	Pt	195.08(3)	
27	Cobalt	Co	58.933		79	Gold	Au	196.97	
28	Nickel	Ni	58.693		80	Mercury	Hg	200.59(2)	
29	Copper	Cu	63.546(3)	r	81	Thallium	Tl	204.38	
30	Zinc	Zn	65.39(2)		82	Lead	Pb	207.2	g r
31	Gallium	Ga	69.723		83	Bismuth	Bi	208.98	
32	Germanium	Ge	72.61(2)		84	Polonium *	^{210}Po	209.98	
33	Arsenic	As	74.922		85	Astatine *	^{210}At	209.99	
34	Selenium	Se	78.96(3)		86	Radon *	^{222}Rn	222.02	
35	Bromine	Br	79.904		87	Francium *	^{223}Fr	223.02	
36	Krypton	Kr	83.80	g m	88	Radium *	^{226}Ra	226.03	
37	Rubidium	Rb	85.468		89	Actinium *	^{227}Ac	227.03	
38	Strontium	Sr	87.62	g r	90	Thorium *	Th	232.04	g
39	Yttrium	Y	88.906		91	Protactinium *	Pa	231.04	
40	Zirconium	Zr	91.224(2)	g	92	Uranium *	U	238.03	g m
41	Niobium	Nb	92.906		93	Neptunium *	^{237}Np	237.05	
42	Molybdenum	Mo	95.94	g	94	Plutonium *	^{239}Pu	239.05	
43	Technetium *	^{99}Tc	98.906		95	Americium *	^{241}Am	241.06	
44	Ruthenium	Ru	101.07(2)	g	96	Curium *	^{244}Cm	244.06	
45	Rhodium	Rh	102.91		97	Berkelium *	^{249}Bk	249.08	
46	Palladium	Pd	106.42	g	98	Californium *	^{252}Cf	252.08	
47	Silver	Ag	107.87		99	Einsteinium *	^{252}Es	252.08	
48	Cadmium	Cd	112.41		100	Fermium *	^{257}Fm	257.10	
49	Indium	In	114.82		101	Mendelevium *	^{258}Md	258.10	
50	Tin	Sn	118.71		102	Nobelium *	^{259}No	259.10	
51	Antimony (Stibium)	Sb	121.76(3)	g	103	Lawrencium *	^{262}Lr	262.11	
52	Tellurium	Te	127.60	g					

TABLE OF CONTENTS

Inorganic Compound Data

1

Compound: Acetylferrocene
Formula: $CH_3COC_5H_4FeC_5H_5$
Molecular Formula: $C_{12}H_{12}FeO$
Molecular Weight: 228.074-
CAS RN: 1271-55-2
Properties: orange cryst [STR93]

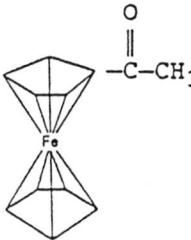

Melting Point, °C: 83 [STR93]

2

Compound: Actinium
Formula: Ac
Molecular Formula: Ac
Molecular Weight: 227
CAS RN: 7440-34-8
Properties: silvery white metal; fcc; a = 0.5311 nm; $t_{1/2}$ of ^{227}Ac is 21.8 years; a decay product of ^{235}U; stable, colorless solution for Ac^{+++}; ionic radius Ac^{+++}, 0.1119 nm; enthalpy of vaporization 293 kJ/mol; chemistry closely follows that of lanthanum; first discovered in 1899 by Diebierne; preparation: transmutation of radium: $^{226}Ra + n \rightarrow ^{227}Ra + \gamma \; ^{227}Ra \rightarrow ^{227}Ac$ [KIR78] [KAT86] [HAW93]
Density, g/cm³: 10.07 (25°C) [KIR91]
Melting Point, °C: 1100 [KIR91]
Boiling Point, °C: ~3300 [MER89]

3

Compound: Actinium bromide
Formula: $AcBr_3$
Molecular Formula: $AcBr_3$
Molecular Weight: 467
CAS RN: 33689-81-5
Properties: white; hex, a = 0.806 nm, c = 0.468 nm; preparation: reacting Ac_2O_3 with $AlBr_3$ at 750°C [CIC73] [KAT86]
Solubility: s H_2O [CRC93]
Density, g/cm³: 5.85 [KAT86]
Melting Point, °C: sublimes 800 [CRC93]

4

Compound: Actinium chloride

Formula: $AcCl_3$
Molecular Formula: $AcCl_3$
Molecular Weight: 333
CAS RN: 22986-54-5
Properties: white cryst; hex, a = 0.762 nm, c = 0.455 nm; preparation: reacting $Ac(OH)_3$ with CCl_4 at 500°C [KAT86] [KIR78]
Density, g/cm³: 4.81 [KIR78]
Melting Point, °C: sublimes 900 [CRC93]

5

Compound: Actinium fluoride
Formula: AcF_3
Molecular Formula: AcF_3
Molecular Weight: 284
CAS RN: 33689-80-4
Properties: white cryst; hex, a = 0.741 nm, c = 0.755 nm; preparation: from the reaction $Ac^{+++} + 3F^- = AcF_3$ at 25°C [KIR78] [KAT86]
Solubility: i H_2O [CRC93]
Density, g/cm³: 7.88 [KIR78]

6

Compound: Actinium hydride
Formula: AcH_2
Molecular Formula: AcH_2
Molecular Weight: 229
CAS RN: 60936-81-4
Properties: black; cub, fluorite structure, a = 0.5670 nm [CIC73]
Density, g/cm³: 8.35 [CIC73]

7

Compound: Actinium hydroxide
Formula: $Ac(OH)_3$
Molecular Formula: AcH_3O_3
Molecular Weight: 278
CAS RN: 12249-30-8
Properties: white [CRC93]
Solubility: i H_2O [CRC93]

8

Compound: Actinium iodide
Formula: AcI_3
Molecular Formula: AcI_3
Molecular Weight: 608
CAS RN: 33689-82-6
Properties: white [CRC93]
Solubility: s H_2O [CRC93]
Melting Point, °C: sublimes 700-800 [CRC93]

9
Compound: Actinium oxalate decahydrate
Formula: $Ac_2(C_2O_4)_3 \cdot 10H_2O$
Molecular Formula: $C_6H_{20}Ac_2O_{22}$
Molecular Weight: 898
Properties: monocl, a = 1.126 nm, b = 0.997 nm, c = 1.065 nm; obtained by adding soluble oxalate to soluble Ac^{+++} solution at 25°C [KAT86]
Solubility: i H_2O [CRC93]
Density, g/cm³: 2.68 [KAT86]

10
Compound: Actinium oxide
Formula: Ac_2O_3
Molecular Formula: Ac_2O_3
Molecular Weight: 502
CAS RN: 12002-61-8
Properties: white cryst; hex, a = 0.407 nm, c = 0.629 nm; preparation: from decomposition of actinium oxalate at 1100°C [KIR78] [KAT86]
Solubility: i H_2O [CRC93]
Density, g/cm³: 9.19 [KIR78]

11
Compound: Actinium oxybromide
Formula: AcOBr
Molecular Formula: AcBrO
Molecular Weight: 323
CAS RN: 49848-33-1
Properties: tetr, a = 0.427 nm, c = 0.740 nm; preparation: reacting $AcBr_3$ with NH_3 and H_2O at 1000°C [KAT86]
Density, g/cm³: 7.89 [KAT86]

12
Compound: Actinium oxychloride
Formula: AcOCl
Molecular Formula: AcClO
Molecular Weight: 278
CAS RN: 49848-29-5
Properties: white, tetr, a = 0.424 nm, c = 0.708 nm; preparation: reacting $AcCl_3$ with H_2O at 1000°C [KAT86] [CIC73]
Density, g/cm³: 7.23 [KAT86]

13
Compound: Actinium oxyfluoride
Formula: AcOF
Molecular Formula: AcFO
Molecular Weight: 262
CAS RN: 49848-24-0

Properties: white; cub, a = 0.5931 nm; preparation: by reaction of AcF_3 with NH_3 and H_2O at 900-1000°C [KAT86]
Density, g/cm³: 8.28 [KAT86]

14
Compound: Actinium phosphate hemihydrate
Formula: $AcPO_4 \cdot 1/2\text{-}H_2O$
Molecular Formula: $AcHO_{4.5}P$
Molecular Weight: 331
Properties: hex, a = 0.721 nm, c = 0.664 nm; preparation: precipitation of a soluble Ac^{+++} salt with a solution of PO_4^- [KAT86]
Density, g/cm³: 5.48 [KAT86]

15
Compound: Actinium sulfide
Formula: Ac_2S_3
Molecular Formula: Ac_2S_3
Molecular Weight: 550
CAS RN: 50647-18-2
Properties: bcc, a = 0.897 nm; preparation: reaction between Ac_2O_3 and H_2S at 1400°C [KAT86] [CIC73]
Density, g/cm³: 6.75 [KAT86]

16
Compound: Aluminum
Formula: Al
Molecular Formula: Al
Molecular Weight: 26.981539
CAS RN: 7429-90-5
Properties: silvery white, ductile, metal; fcc, a = 0.40496 nm; forms corrosion resistant oxide film ~5 nm thickness in moist air; enthalpy of fusion 10.71 kJ/mol; enthalpy of sublimation 314.0 kJ/mol; enthalpy of vaporization 294 kJ/mol; electrical resistivity 2.6548 μohm·cm; tensile strength 6800 psi; hardness 2.9 Mohs; ionic radius of Al^{+++} 0.050 nm; used in mirrors, beverage cans, buildings and construction [CIC73] [KIR78] [HAW93] [CER91]
Solubility: i H_2O and conc HNO_3 [HAW93]
Density, g/cm³: 2.70 [MER89]
Melting Point, °C: 660.37 [ALD94]
Boiling Point, °C: 2517.66 [JAN85]
Reactions: reacts with dil HCl, H_2SO_4, KOH, NaOH to evolve hydrogen [MER89]
Thermal Conductivity, W/(m·K): 236 (0°C), 237 (25°C), 240 (100°C) [HO 72]
Thermal Expansion Coefficient: linear coefficient (30-300°C) 24.9 x 10^{-6}/°C [CIC73]

17

Compound: Aluminum acetate

Synonyms: aluminum triacetate

Formula: $Al(CH_3COO)_3$

Molecular Formula: $C_6H_9AlO_6$

Molecular Weight: 204.115

CAS RN: 139-12-8

Properties: white powd; preparation: by heating aluminum or $AlCl_3$ with an acetic acid solution containing acetic anhydride; uses: an antiseptic, astringent and in antiperspirant applications; there is a hydroxyaluminum diacetate, CAS RN 142-03-0 [CIC73] [HAW93] [ALD94]

Solubility: s H_2O [HAW93]

Melting Point, °C: decomposes [CRC94]

Reactions: minus acetic anhydride at 120 to 140°C, forming basic acetates [CIC73]

18

Compound: Aluminum acetylacetonate

Synonyms: 2,4-pentanedione, aluminum(III) derivative

Formula: $Al[CH_3COCH=C(O)CH_3]_3$

Molecular Formula: $C_{15}H_{21}AlO_6$

Molecular Weight: 324.310

CAS RN: 13963-57-0

Properties: white powd or monocl cryst; preparation: by reacting $AlCl_3$ and acetylacetone; uses: to vapor deposit aluminum, and as a catalyst [HAW93] [CIC73] [STR93]

$$O-\quad O$$
$$|\quad\ \ ||$$
$$[CH_3-C=CH-C-CH_3]_3Al$$

Solubility: i H_2O; s benzene, alcohol [CIC73] [HAW93]

Density, g/cm³: 1.27 [CRC94]

Melting Point, °C: 189 [HAW93]

Boiling Point, °C: 315 [HAW93]

Reactions: decomposes at 320°C; sublimes at 150°C (1 mm Hg) [STR93]

19

Compound: Aluminum antimonide

Formula: AlSb

Molecular Formula: AlSb

Molecular Weight: 148.739

CAS RN: 25152-52-7

Properties: electronic dielectric constant 10.2; enthalpy of fusion 58.6 kJ/mol; cryst, lattice constant 0.61361 nm; band gap 1.68 eV at 0 K and 1.58 eV at 300 K; mobility (300 K) 200 cm²/(V·s) electrons, 420 cm²/(V·s) holes; effective mass 0.12 electrons and 0.98 holes; preparation: by fusion of Al and Sb, followed by purification using zone melting; uses: semiconductor research [CIC73] [MER89] [KIR82]

Density, g/cm³: 4.15 [CIC73]

Melting Point, °C: 1050 [MER89]

Thermal Conductivity, W/(m·K): 60 (25°C) [CRC93]

Thermal Expansion Coefficient: 4.2×10^{-6}/K [CRC93]

20

Compound: Aluminum arsenide

Formula: AlAs

Molecular Formula: AlAs

Molecular Weight: 101.903

CAS RN: 22831-42-1

Properties: -20 mesh powd with 99.5% purity; semiconductor; band gap, 2.13 eV (22°C); electronic dielectric constant, 10.3; lattice constant a = 0.5662 nm; enthalpy of fusion 24.5 kJ/mol; uses: in rectifiers, transistors and thermistors [CIC73] [ALF93] [HAW93]

Density, g/cm³: 3.81 [CIC73]

Melting Point, °C: 1740 [CIC73]

21

Compound: Aluminum borate

Synonyms: eremeyevite, jeremejevite

Formula: $2Al_2O_3 \cdot B_2O_3$

Molecular Formula: $Al_4B_2O_9$

Molecular Weight: 273.543

CAS RN: 11121-16-7

Properties: white granular powd, or cryst needles; prepared by heating Al_2O_3 with B_2O_3; used in glass and ceramics [HAW93] [MER89]

Solubility: i H_2O with decomposition [HAW93] [MER89]

Melting Point, °C: ~1050 [MER89] [CIC73]

Reactions: forms $2Al_2O_3 \cdot B_2O_3$ at 1000°C, and $9Al_2O_3 \cdot 2B_2O_3$ at 1100° [MER89]

22

Compound: Aluminum borohydride

Synonyms: aluminum tetrahydroborate

Formula: $Al(BH_4)_3$

Molecular Formula: AlB_3H_{12}

Molecular Weight: 71.510
CAS RN: 16962-07-5
Properties: volatile; liq; enthalpy of vaporization 30 kJ/mol; ignites spontaneously in air; can be formed by reacting sodium borohydride and aluminum chloride in the presence of small quantity of tributyl phosphate; used as a reducing agent and as a fuel for jet engines and rockets [HAW93] [MER89] [CRC93]
Solubility: reacts vigorously with H_2O and HCl evolving H_2 [MER89]
Melting Point, °C: -64.5 [MER89]
Boiling Point, °C: 44.5 [MER89]

23

Compound: Aluminum bromate nonahydrate
Formula: $Al(BrO_3)_3 \cdot 9H_2O$
Molecular Formula: $AlBr_3H_{18}O_{18}$
Molecular Weight: 572.826
CAS RN: 11126-81-1
Properties: white cryst; hygr; can be obtained from mixing aq solutions of $Al_2(SO_4)_3$ and $Ba(BrO_3)_2$, followed by crystallization [CIC73] [CRC93]
Solubility: s H_2O; s sl acids [CRC94]
Melting Point, °C: 62 [CIC73]
Boiling Point, °C: decomposes >100 [CIC73]

24

Compound: Aluminum bromide
Formula: $AlBr_3$
Molecular Formula: $AlBr_3$
Molecular Weight: 266.694
CAS RN: 7727-15-3
Properties: white to yellow-red; trig cryst or powd; very hygr; fumes strongly in air; enthalpy of sublimation 35.9 kJ/mol; enthalpy of fusion 11.25 kJ/mol; enthalpy of vaporization 23.5 kJ/mol; can be prepared by heating Al and Br_2; used as an acid catalyst for organic syntheses, similarly to $AlCl_3$, but is more reactive and more soluble in organic solvents [CIC73] [MER89] [KIR78] [CRC93]
Solubility: reacts with H_2O violently [KIR78]
Density, g/cm³: 3.01 [KIR78]; 2.64, 100°C (liq) [STR93]
Melting Point, °C: 97.45 [KIR78]
Boiling Point, °C: sublimes 256 [CIC73]

25

Compound: Aluminum bromide hexahydrate
Formula: $AlBr_3 \cdot 6H_2O$
Molecular Formula: $AlBr_3H_{12}O_6$
Molecular Weight: 374.785

CAS RN: 7784-11-4
Properties: colorless to sl yellow; deliq; may be prepared by dissolution of Al or aluminum hydroxide in HBr, followed by precipitation; used as an acid catalyst [MER89] [KIR78]
Solubility: s H_2O, alcohol [MER89]
Density, g/cm³: 2.54 [HAW93]
Melting Point, °C: 93 [MER89]
Boiling Point, °C: decomposes 135 [CRC94]

26

Compound: Aluminum carbide
Formula: Al_4C_3
Molecular Formula: C_3Al_4
Molecular Weight: 143.959
CAS RN: 1299-86-1
Properties: yellow hex cryst, or olive green powd; can be prepared by reacting stoichiometric amounts of Al and C in the absence of both oxygen and nitrogen at ~1000°C; used to generate methane and to manufacture AlN [MER89] [ALF93] [CIC73]
Solubility: decomposes with evolution of CH_4 in H_2O [MER89]
Density, g/cm³: 2.36 [MER89]
Melting Point, °C: 2100 [MER89]
Boiling Point, °C: decomposes >2200 [MER89]

27

Compound: Aluminum chlorate
Formula: $Al(ClO_3)_3$
Molecular Formula: $AlCl_3O_9$
Molecular Weight: 277.332
CAS RN: 15477-33-5
Properties: colorless cryst; deliq; occurs as hexahydrate and nonahydrate; evaporation of an aq solution yields the nonahydrate; used as a disinfectant and to prevent yellowing of acrylic fibers [CIC73] [MER89] [HAW93]
Solubility: s H_2O, alcohol [HAW93]
Melting Point, °C: decomposes [CRC94]

28

Compound: Aluminum chlorate nonahydrate
Synonyms: mallebrin
Formula: $Al(ClO_3)_3 \cdot 9H_2O$
Molecular Formula: $AlCl_3H_{18}O_{18}$
Molecular Weight: 439.472
CAS RN: 15477-33-5
Properties: deliq cryst; obtained by mixing $Ba(ClO_4)_2$ and $Al_2(SO_4)_3$ solution, followed by evaporation [CIC73] [MER89]

Solubility: s H$_2$O, alcohol [MER89]

29
Compound: Aluminum chloride
Formula: AlCl$_3$
Molecular Formula: AlCl$_3$
Molecular Weight: 133.340
CAS RN: 7446-70-0
Properties: white, or light yellow; cryst or powd; hygr; hex; odor of HCl; exists as dimer <327°C; triple point 192.5°C (233 kPa); enthalpy of sublimation of dimer Al$_2$Cl$_6$ (25°C) 115.73 kJ/mol; enthalpy of solution at 20°C is -325.1 kJ/mol; enthalpy of fusion 35.40 kJ/mol; can be made by reacting HCl and Al at ~150°C; used as an acid catalyst, in cracking petroleum and in the manufacture of rubbers and lubricants [CIC73] [KIR78] [ALF93]
Solubility: dissolves violently in H$_2$O evolving HCl [MER89]; g/100g soln H$_2$O: 30.84±0.25 (0°C), 31.10 (25°C), 33.23 (98°C), equilibrium solid phase, AlCl$_3$·6H$_2$O [KRU93]; s HCl, ether, ethanol [KIR78]
Density, g/cm^3: 2.44 [KIR78]
Boiling Point, °C: sublimes 181.2 [KIR78]
Reactions: Al$_2$Cl$_6$ (dimer) → 2AlCl$_3$ >327°C [KIR78]

30
Compound: Aluminum chloride hexahydrate
Formula: AlCl$_3$·6H$_2$O
Molecular Formula: AlCl$_3$H$_{12}$O$_6$
Molecular Weight: 241.431
CAS RN: 7784-13-6
Properties: cryst powd; white or yellow; deliq; preparation: dissolution of Al(OH)$_3$ in conc HCl, followed by cooling to 0°C and addition of gaseous HCl; uses: preserve wood, disinfect stables, and in deodorants and antiperspirants [MER89] [HAW93] [KIR78]
Solubility: 1g/0.9mL H$_2$O; s alcohol [MER89]; gAl$_2$O$_3$/100mL at 35°C in each of following solvents: 15.91 H$_2$O, 9.43 methanol, 4.77 ethanol, 6.04 ethylene glycol [OKA91]
Density, g/cm^3: 2.398 [STR93]
Melting Point, °C: 100, decomposes [ALF93]

31
Compound: Aluminum chromate
Synonyms: aluminum oxide-chromium oxide
Formula: Al$_2$O$_3$·Cr$_2$O$_3$
Molecular Formula: Al$_2$Cr$_2$O$_6$
Molecular Weight: 253.952

CAS RN: 57921-51-4
Properties: yellow amorphous solid; fused 98 wt % Al$_2$O$_3$, 2 wt % Cr$_2$O$_3$; -200, +325 mesh and -325 mesh, +10 microns of 99 % purity; used in ceramics [KIR78] [CER91]

32
Compound: Aluminum citrate
Formula: AlC$_6$H$_5$O$_7$
Molecular Formula: C$_6$H$_5$AlO$_7$
Molecular Weight: 216.084
CAS RN: 31142-56-0
Properties: white powd or scales; study of complex formation in [LOP84] [MER52]

```
        CH2COO        O2H
         |         \  /
        HOC-COO — Al — O2H
         |          /  \
        CH2COO        O2H
```

Solubility: dissolves slowly in cold H$_2$O, s hot H$_2$O; s ammonia [MER52]

33
Compound: Aluminum diacetate
Synonyms: aluminum subacetate
Formula: Al(CH$_3$COO)$_2$(OH)
Molecular Formula: C$_4$H$_7$AlO$_5$
Molecular Weight: 162.079
CAS RN: 142-03-0
Properties: white amorphous powd or curdy precipitate; can be produced by reacting sodium aluminate solution with acetic acid; used as a mordant in dyeing, to manufacture color lakes, to waterproof and fireproof fabrics, in antiperspirant formulations and as a disinfectant by embalmers [MER89] [CIC73]
Solubility: i H$_2$O when dried at 100°C [MER89]

34
Compound: Aluminum diboride
Formula: AlB$_2$
Molecular Formula: AlB$_2$
Molecular Weight: 48.604
CAS RN: 12041-50-8
Properties: powd; made by reaction of the elements above 600°C; high neutron absorption; used as a nuclear shielding material [HAW93] [CIC73]
Solubility: s dil HCl [CIC73]
Density, g/cm^3: 3.19 [ALF93]
Melting Point, °C: decomposes to AlB$_{12}$ >920 [CIC73]

35

Compound: Aluminum distearate
Formula: $Al(OH)[CH_3(CH_2)_{16}COO]_2$
Molecular Formula: $C_{36}H_{71}AlO_5$
Molecular Weight: 610.939
CAS RN: 637-12-7
Properties: white powd; preparation in [KIR78]; used as a thickener for paints, inks and greases, and as a water repellent and lubricant [HAW93]
Solubility: i H_2O, alcohol, ether; forms gel with aliphatic and aromatic hydrocarbons [HAW93]
Density, g/cm³: 1.009 [HAW93]
Melting Point, °C: 145 [HAW93]

36

Compound: Aluminum dodecaboride
Formula: AlB_{12}
Molecular Formula: AlB_{12}
Molecular Weight: 156.714
CAS RN: 12041-54-2
Properties: 3-8 micron powd; high neutron absorption [ALF93] [HAW93]
Solubility: s hot HNO_3; i acid, alkalies [CRC92] [CIC73]
Density, g/cm³: 2.55 [CRC92]
Melting Point, °C: decomposes to boron and carbon at 1900 [CIC73]

37

Compound: Aluminum ethoxide
Synonyms: aluminum ethylate
Formula: $Al(C_2H_5O)_3$
Molecular Formula: $C_6H_{15}AlO_3$
Molecular Weight: 162.165
CAS RN: 555-75-9
Properties: liq which slowly solidifies to a white powd; sensitive to moisture; prepared from a reaction of Al with ethanol in the presence of catalytic amounts of I_2 and $HgCl_2$; used as polymerization catalyst, and to reduce aldehydes and ketones [MER89] [STR93] [HAW93]
Solubility: s sl in high boiling organic solvents [HAW93]
Density, g/cm³: 1.142 (20°C) [CRC94]
Melting Point, °C: 130 [STR93]
Boiling Point, °C: 210 (10 mm Hg) [STR93]

38

Compound: Aluminum fluoride
Synonyms: aluminum trifluoride
Formula: AlF_3
Molecular Formula: AlF_3

Molecular Weight: 83.977
CAS RN: 7784-18-1
Properties: hex white powd, or 99.5% pure highly dense 3-6 mm sintered pieces; enthalpy of fusion 98.0 kJ/mol; dielectric constant 6; formed by heating $(NH_4)_3AlF_6$ in nitrogen; used in electrolyte for production of Al, in ceramics as a flux in metallurgy and to inhibit fermentation, and as an evaporation material and sputtering target for preparation of low index films [CIC73] [KIR78] [STR93] [MER89] [CER91] [CRC93]
Solubility: g/100g soln H_2O: 0.25 (0°C), 0.50 (25°C), 1.64 (100°C), equilibrium solid phase $AlF_3 \cdot 3H_2O$ [KRU93]
Density, g/cm³: 3.10 [KIR78]
Melting Point, °C: 1290 [COT88]
Reactions: transition from α to β at 455°C; dissociates at 776°C [ROB78]

39

Compound: Aluminum fluoride monohydrate
Synonyms: fluellite
Formula: $AlF_3 \cdot H_2O$
Molecular Formula: AlF_3H_2O
Molecular Weight: 101.992
CAS RN: 32287-65-3
Properties: ortho-rhomb [MER89]
Solubility: sl s H_2O [CRC93]
Density, g/cm³: 2.17 [MER89]

40

Compound: Aluminum fluoride trihydrate
Formula: $AlF_3 \cdot 3H_2O$
Molecular Formula: $AlF_3H_6O_3$
Molecular Weight: 138.023
CAS RN: 15098-87-0
Properties: white hygr; cryst powd [HAW93] [STR93]
Solubility: sl s H_2O [HAW93]
Density, g/cm³: 1.914 [STR93]
Reactions: minus H_2O at both 100°C and 200°C [MER89]

41

Compound: Aluminum hexafluorosilicate nonahydrate
Synonyms: aluminum silicofluoride
Formula: $Al_2(SiF_6)_3 \cdot 9H_2O$
Molecular Formula: $Al_2F_{18}H_{18}O_9Si_3$
Molecular Weight: 642.329
CAS RN: 17099-70-6

Properties: hex prisms; occurs naturally as topaz; used to protect and preserve construction materials and in the manufacture of glass [MER89] [HAW93]

Solubility: s H_2O, decomposes in hot H_2O [MER89]

Melting Point, °C: decomposes ~1000 [MER89]

Reactions: minus H_2O <500°C [MER89]

42

Compound: Aluminum hydride

Formula: AlH_3

Molecular Formula: AlH_3

Molecular Weight: 30.005

CAS RN: 7784-21-6

Properties: colorless nonvolatile solid; can be obtained by reacting an ether solution of $AlCl_3$ with LiH; used as a catalyst for organic polymerization processes [MER89]

Solubility: evolves H_2 in H_2O [HAW93]

Melting Point, °C: decomposes, 160 [HAW93]

43

Compound: Aluminum hydroxide

Formula: $Al(OH)_3$

Molecular Formula: AlH_3O_3

Molecular Weight: 78.004

CAS RN: 21645-51-2

Properties: white bulky amorphous powd; forms gels if in prolonged contact with H_2O; absorbs CO_2; many uses such as an absorbent and emulsifier, in ion-exchange chromatography, as a mordant in dyeing [MER89] [ALF93]

Solubility: i H_2O; s acids, alkalies [MER89]

Density, g/cm³: 2.42 [HAW93]

Reactions: minus H_2O at 300°C [CRC94]; forms gel on contact with H_2O [MER89]

44

Compound: Aluminum hydroxide (β')

Synonyms: nordstrandite

Formula: β'-$Al(OH)_3$

Molecular Formula: AlH_3O_3

Molecular Weight: 78.004

CAS RN: 12752-71-0

Properties: tricl, a = 0.875 nm, b = 0.507 nm, c = 1.024 nm [KIR78]

45

Compound: Aluminum hydroxide (α)

Synonyms: gibbsite

Formula: α-$Al(OH)_3$

Molecular Formula: AlH_3O_3

Molecular Weight: 78.004

CAS RN: 14762-49-3

Properties: white, pearly vitreous; hardness 2.5-3.5 Mohs; monocl: a = 0.868 nm, b = 0.507 nm, c = 0.972 nm; tricl: a = 1.733 nm, b = 1.008 nm, c = 0.973 nm [KIR78]

Density, g/cm³: monocl: 2.441; tricl: 2.42 [KIR78] [ROB78]

Reactions: transition to boehmite at 103°C [ROB78]

46

Compound: Aluminum hydroxide (β)

Synonyms: bayerite

Formula: β-$Al(OH)_3$

Molecular Formula: AlH_3O_3

Molecular Weight: 78.004

CAS RN: 20257-20-9

Properties: monocl, a = 0.506 nm, b = 0.867 nm, c = 0.471 nm; used to make η-alumina catalyst [KIR78]

Density, g/cm³: 2.53 [KIR78]

47

Compound: Aluminum hydroxychloride

Synonyms: aluminum chlorohydroxide

Formula: $Al_2(OH)_5Cl \cdot 2H_2O$

Molecular Formula: $Al_2ClH_9O_7$

Molecular Weight: 210.483

CAS RN: 1327-41-9

Properties: glassy solid; prepared by electrolysis of Al solutions; used as an antiperspirant and in medicine [MER89]

Solubility: s H_2O, forms sl turbid colloidal solution [MER89]

48

Compound: Aluminum hydroxystearate

Formula: $Al(OH)[OOC(CH_2)_{10}CH_2O(CH_2)_5CH_3]_2$

Molecular Formula: $C_{36}H_{71}AlO_7$

Molecular Weight: 642.938

CAS RN: 637-12-7

Properties: white powd; preparation in [KIR78]; used to weatherproof leather and cement, to lubricant plastics and rope, and in paints and inks [HAW93]

Density, g/cm³: 1.045 [HAW93]

Melting Point, °C: 155 [HAW93]

49

Compound: Aluminum hypophosphite

Formula: $Al(H_2PO_2)_3$
Molecular Formula: $AlH_6O_6P_3$
Molecular Weight: 221.948
CAS RN: 7784-22-7
Properties: cryst powd; can be made by precipitation with slow heating from a solution of an aluminum salt with 50% hypophosphorus acid at 80-90°C; used in finishes for acrylonitrile polymer fibers [MER89] [CIC73]
Solubility: i H_2O; s warm NaOH; decomposes in H_2SO_4 and HCl [MER89]
Reactions: decomposes ~220°C evolving phosphine [MER89]

50
Compound: Aluminum iodide
Formula: AlI_3
Molecular Formula: AlI_3
Molecular Weight: 407.695
CAS RN: 7784-23-8
Properties: solid; white leaflets, if pure; yellowish to brownish black lumps; fumes in moist air; strong exothermic reaction with H_2O; enthalpy of sublimation 112.1 kJ/mol; enthalpy of fusion 15.90 kJ/mol; enthalpy of vaporization 32.2 kJ/mol; can be prepared by heating of Al and I_2 in a sealed tube; used as a catalyst for organic reactions [CIC73] [MER89] [HAW93] [CRC93]
Solubility: reacts violently with H_2O [KIR78]
Density, g/cm³: 3.98 [KIR78]
Melting Point, °C: 191 [KIR78]
Boiling Point, °C: 381-382, sublimes [CIC73]

51
Compound: Aluminum iodide hexahydrate
Formula: $AlI_3 \cdot 6H_2O$
Molecular Formula: $AlH_{12}I_3O_6$
Molecular Weight: 515.786
CAS RN: 10090-53-6
Properties: yellowish cryst powd; deliq; can be obtained from a reaction between Al or $Al(OH)_3$ and HI; there is a $AlI_3 \cdot 15H_2O$ [MER89] [KIR78]
Solubility: s H_2O, alcohol, ether [MER89]
Density, g/cm³: 2.63 [CRC92]
Melting Point, °C: decomposes at 185 [CRC94]

52
Compound: Aluminum isopropoxide
Synonyms: aluminum isopropylate
Formula: $Al([OCH(CH_3)_2]_3$
Molecular Formula: $C_9H_{21}AlO_3$
Molecular Weight: 204.245

CAS RN: 555-31-7
Properties: hygr white solid; prepared from a reaction between aluminum and isopropyl alcohol with $HgCl_2$ catalyst; used to prepare aluminum soaps, as a waterproofing finish for textiles, and to synthesize aluminum titanate [MER89] [YAM89]
Solubility: decomposed by H_2O; s ethanol, isopropanol, benzene, toluene, chloroform, carbon tetrachloride [MER89]
Density, g/cm³: 1.0346 (20°C) [CRC94]
Melting Point, °C: 119 [MER89]
Boiling Point, °C: 141 [CRC94]

53
Compound: Aluminum lactate
Synonyms: aluctyl
Formula: $Al(C_3H_5O_3)_3$
Molecular Formula: $C_9H_{15}AlO_9$
Molecular Weight: 294.194
CAS RN: 18917-91-4
Properties: powd; preparation: lactic acid and $AlCl_3$; used in foam fire extinguishers and in dental impression materials [MER89] [ALF95]

$$[CH_3-\overset{\overset{\displaystyle OH}{|}}{\underset{\underset{\displaystyle H}{|}}{C}}-COO]_3Al$$

Solubility: v s H_2O [MER89]

54
Compound: Aluminum metaphosphate
Formula: $Al(PO_3)_3$
Molecular Formula: AlO_9P_3
Molecular Weight: 263.898
CAS RN: 32823-06-6
Properties: colorless powd; tetr; used as a component of glazes, enamels and glasses, and in high temp insulating cement [CRC92] [HAW93]
Solubility: i H_2O [HAW93]
Density, g/cm³: 2.780 [ALD94]
Melting Point, °C: ~1527 [HAW93]

55
Compound: Aluminum molybdate
Formula: $Al_2(MoO_4)_3$
Molecular Formula: $Al_2Mo_3O_{12}$
Molecular Weight: 533.776
CAS RN: 15123-80-5

Properties: -325 mesh powd with 99% purity [ALF93]

56
Compound: Aluminum monopalmitate
Formula: $Al(OH)_2C_{16}H_{31}O_2$
Molecular Formula: $C_{16}H_{33}AlO_4$
Molecular Weight: 316.418
CAS RN: 555-35-1
Properties: white powd; made by heating aluminum hydroxide with palmitic acid and H_2O, followed by filtration and drying; used to water proof leather, paper and textiles, and to thicken lubricating oils, also used in varnishes and as a food additive [HAW93]
Solubility: i H_2O, alcohol; gels in hydrocarbons [HAW93]
Density, g/cm³: 1.072 [HAW93]
Melting Point, °C: 200 [HAW93]

57
Compound: Aluminum monostearate
Formula: $Al(OH)_2[CH_3(CH_2)_{16}COO]$
Molecular Formula: $C_{18}H_{37}AlO_4$
Molecular Weight: 344.472
CAS RN: 7047-84-9
Properties: faint odor; white to yellowish white powd; prepared by mixing solutions of sodium stearate and a soluble aluminum salt; used in paints, inks, greases, waxes, to thicken lubricating oils, for waterproofing [HAW93]
Solubility: i H_2O; forms gel with aliphatic and aromatic hydrocarbons [HAW93]
Density, g/cm³: 1.020 [HAW93]
Melting Point, °C: 155 [PFA93]

58
Compound: Aluminum nitrate nonahydrate
Formula: $Al(NO_3)_3 \cdot 9H_2O$
Molecular Formula: $AlH_{18}N_3O_{18}$
Molecular Weight: 375.134
CAS RN: 7784-27-2
Properties: white hygr cryst; monocl, a = 1.086 nm, b = 0.959 nm, c = 1.383 nm; prepared by adding lead nitrate solution to aluminum sulfate solution; used in leather tanning, as a corrosion inhibitor and as an antiperspirant [MER89] [CIC73] [STR93]
Solubility: 67.3 g/100mL H_2O (25°C) [CIC73]; g Al_2O_3/100mL, 35°C, in the following solvents: methanol 14.45, ethanol 8.63, ethylene glycol 18.32 [OKA91]
Density, g/cm³: 1.72 [CIC73]

Melting Point, °C: 73 [MER89]
Boiling Point, °C: decomposes 135 [MER89]
Reactions: decomposes to oxides of Al and N_2 at 500°C [KIR78]

59
Compound: Aluminum nitride
Formula: AlN
Molecular Formula: AlN
Molecular Weight: 40.989
CAS RN: 24304-00-5
Properties: powd or bluish white cryst; hex or ortho-rhomb; hex, a = 0.311 nm, c = 0.4975 nm; has odor of NH_3 in moist air; hardness 9 to 10 Mohs; band gap 4.26 eV; manufactured by heating bauxite in flowing nitrogen at 1500°C for 8 h; semiconductor material, used in steel manufacturing, as a crucible to grow cryst of gallium arsenide, and as a 99.8% sputtering target to prepare diodes and integrated circuits [MER89] [CIC73] [ALF93] [MIT87] [CER91]
Solubility: decomposes in H_2O to $Al(OH)_3$ + NH_3; decomposes in acids and alkalies [MER89] [CRC93]
Density, g/cm³: 3.05 [MER89]
Melting Point, °C: 2150-2200 [MER89]
Reactions: sublimes 2000°C; decomposes to Al gas and N_2 from 1340°C to 1654°C [CRC92] [JAN85]
Thermal Conductivity, W/(m·K): 30 [KIR81]
Thermal Expansion Coefficient: coefficient is 4.03 x 10^{-6}/°C [KIR81]

60
Compound: Aluminum oleate
Synonyms: 9-octadecanoic acid, aluminum(III) salt
Formula: $Al[CH_3(CH_2)_7CH=CH(CH_2)_7COO]_3$
Molecular Formula: $C_{54}H_{99}AlO_6$
Molecular Weight: 871.358
CAS RN: 688-37-9
Properties: yellowish mass; formed from freshly precipitated aluminum hydroxide and oleic acid; used as a lacquer for metals in oil or turpentine solutions and as drier and waterproofing agent for paints [MER89]
Solubility: i H_2O; s alcohol, benzene, ethanol, oil, turpentine [MER89]
Density, g/cm³: 1.01 [KIR78]
Melting Point, °C: 120 [KIR78]

61
Compound: Aluminum oxalate monohydrate
Formula: $Al_2(C_2O_4)_3 \cdot H_2O$

Molecular Formula: $C_6H_2Al_2O_{13}$
Molecular Weight: 336.037
CAS RN: 814-87-9
Properties: white powd; used as a mordant to print textiles and to dye cotton [MER89]
Solubility: i H_2O, alcohol; s acids [MER89]

62

Compound: Aluminum oxide
Synonyms: native aluminum oxide, bauxite
Formula: Al_2O_3
Molecular Formula: Al_2O_3
Molecular Weight: 101.961
CAS RN: 1344-28-1
Properties: white powd; hex; hardness 8.8 Mohs; electrical resistivity at 300°C ~1.2 x 10^{+13} ohm cm; used as an adsorbent (see corundum); evaporated material of 99.99% purity is used as a high temp dielectric to protect aluminum mirrors and as a support for specimens in electron diffraction [MER89] [CER91]
Solubility: i H_2O; s sl alkalies [MER89]
Density, g/cm³: 3.965 [ALF93]
Melting Point, °C: 2045 [ALF93]
Boiling Point, °C: 2980 [ALF93]
Thermal Conductivity, W/(m·K): 28.9 (100°C), 21.2 (200°C), 12.5 (400°C), 8.70 (600°C), 6.86 (800°C), 5.86 (1000°C), 5.27 (1200°C) [HO 72]

63

Compound: Aluminum oxide (α)
Synonyms: corundum
Formula: α-Al_2O_3
Molecular Formula: Al_2O_3
Molecular Weight: 101.961
CAS RN: 1302-74-5
Properties: naturally occurring; white rhomb cryst, a = 0.47591 nm, c = 1.2894 nm; impedance 384-410 MPa·s/m; hardness 9 Mohs; enthalpy of fusion 111.1 kJ/mol; used as an abrasive powd, in grinding wheels and in crucible form to melt metals; some precious stones are forms of corundum which contain traces of other metals include ruby (chromium) and sapphire (cobalt) [JAN85] [ROB67] [VIE91] [KIR78] [CER91]
Solubility: i H_2O; v sl s acids, alkalies [HAW93]
Density, g/cm³: 3.987 [ROB78]
Melting Point, °C: 2054 [CRC93]
Boiling Point, °C: 2980 [CRC93]
Thermal Expansion Coefficient: ⊥ c-axis: 100°C (0.044), 200°C (0.112), 400°C (0.278), 600°C (0.455), 800°C (0.632), 1000°C (0.815), 1200°C (0.998) [CLA66]

64

Compound: Aluminum oxide (γ)
Formula: γ-Al_2O_3
Molecular Formula: Al_2O_3
Molecular Weight: 101.961
CAS RN: 1344-28-1
Properties: -60 mesh with 96% purity; white powd; a = 0.562 nm, b = 0.780 nm; enthalpy of fusion 78.49 kJ/mol [JAN85] [STR93] [BHA78] [KIR78] [ALF93]
Solubility: i H_2O; s sl acids, alkalies [CRC92]
Density, g/cm³: 3.97 [STR93]
Melting Point, °C: 2018 [ROB78]
Boiling Point, °C: 2980 [STR93]

65

Compound: Aluminum oxide (δ)
Formula: δ-Al_2O_3
Molecular Formula: Al_2O_3
Molecular Weight: 101.961
CAS RN: 1344-28-1
Properties: ortho-rhomb: a = 0.425 nm, b = 1.275 nm, c = 1.021 nm; tetr: a = 0.796 nm, c = 2.34 nm; enthalpy of fusion 93.3 kJ/mol [JAN85] [KIR78]
Density, g/cm³: 3.2 [KIR78]
Melting Point, °C: 2035 [JAN85]
Reactions: transition from δ to α in two steps: from 800°C to 1100°C and at 1200°C [JAN85]

66

Compound: Aluminum oxide (κ)
Formula: κ-Al_2O_3
Molecular Formula: Al_2O_3
Molecular Weight: 101.961
CAS RN: 1344-28-1
Properties: hex; a = 0.971 nm, c = 1.786 nm; enthalpy of fusion 91.2 kJ/mol [JAN85] [KIR78]
Density, g/cm³: 3.1-3.3 [KIR78]
Melting Point, °C: 2040 (fusion) [JAN85]
Reactions: transition from κ to α ~1200°C [JAN85]

67

Compound: Aluminum oxyhydroxide (α)
Synonyms: boehmite
Formula: α-AlO(OH)
Molecular Formula: $AlHO_2$
Molecular Weight: 59.989
CAS RN: 1318-23-6

Properties: white; ortho-rhomb; a = 0.286 nm, b = 1.2227 nm, c = 0.380 nm; hardness is 3.5-4 Mohs; obtained by hydrothermal reaction of hydroxide slurries at 200-250°C [KIR78] [ROB67]; solubility data are in [BOU93] and [PHI93]
Solubility: i H_2O; s hot acids, hot alkalies [CRC92]
Density, g/cm³: 3.07 [ROB78]
Reactions: transforms to diaspore at 227°C, dehydrates to corundum at 400°C [LAU73]

68
Compound: Aluminum oxyhydroxide (β)
Synonyms: diaspore
Formula: β-AlO(OH)
Molecular Formula: $AlHO_2$
Molecular Weight: 59.989
CAS RN: 14457-84-2
Properties: ortho-rhomb; a = 0.439 nm, b = 0.942 nm, c = 0.284 nm; hardness is 6.5-7 Mohs; stable from 275 to 425°C [KIR78]
Solubility: i H_2O; s hot acids, hot alkalies [CRC92]
Density, g/cm³: 3.44 [KIR78]
Reactions: dehydrates to corundum at 400°C [LAU73]

69
Compound: Aluminum palmitate
Synonyms: hexadecanoic acid, aluminum(III) salt
Formula: $Al[CH_3(CH_2)_{14}COO]_3$
Molecular Formula: $C_{48}H_{93}AlO_6$
Molecular Weight: 793.244
CAS RN: 555-35-1
Properties: white to yellow powd; made by heating aluminum hydroxide, palmitic acid and water; used to thicken petroleum and lubricants and for water proofing fabrics [MER89] [HAW93]
Solubility: i H_2O, alcohol; s petroleum ether [MER89]
Density, g/cm³: 1.095 (dihydroxy monopalmitate) [CRC94]
Melting Point, °C: 200 (dihydroxy monopalmitate) [CRC94]

70
Compound: Aluminum perchlorate
Formula: $Al(ClO_4)_3$
Molecular Formula: $AlCl_3O_{12}$
Molecular Weight: 325.329
CAS RN: 14452-39-2

Properties: can be prepared by evaporation of solutions of $AlCl_3$ and $AgClO_4$ in methanol or benzene, with subsequent evaporation at 150°C; forms hydrates with 3,6,9 and 15 waters of hydration; finds use in studies of cation-ligand interactions, e.g. hydrolysis products [CIC73]
Solubility: g/100g soln, H_2O: 54.87 (0°C), 64.62 (91.5°C); equilibrium solid phase $Al(ClO_4)_3·9H_2O$ [KRU93]

71
Compound: Aluminum perchlorate nonahydrate
Formula: $Al(ClO_4)_3·9H_2O$
Molecular Formula: $AlCl_3H_{18}O_{21}$
Molecular Weight: 487.470
CAS RN: 81029-06-3
Properties: white cryst [STR93]
Density, g/cm³: 2.0 [STR93]
Melting Point, °C: 82 [ALF93]

72
Compound: Aluminum phosphate
Synonyms: berlinite, aluminum orthophosphate
Formula: $AlPO_4$
Molecular Formula: AlO_4P
Molecular Weight: 121.953
CAS RN: 7784-30-7
Properties: white; rhomb plates; cryst, a = 0.4942 nm, c = 1.097 nm, isomorphous with quartz; naturally occurring; can be prepared by mixing a solution of aluminum sulfate and sodium phosphate; used as a flux for ceramics, in dental cements, for special glasses [MER89] [CRC92] [HAW93]
Solubility: i H_2O; v sl s HCl, HNO_3 [MER89]
Density, g/cm³: 2.56 [MER89]
Melting Point, °C: >1460 [MER89]
Reactions: α-$AlPO_4$ form is stable below 584°C [LAU87]

73
Compound: Aluminum phosphate dihydrate
Synonyms: variscite
Formula: $AlPO_4·2H_2O$
Molecular Formula: AlH_4O_6P
Molecular Weight: 157.984
CAS RN: 7784-30-7
Properties: rhomb; green; hardness 3.5-4.5 Mohs; $AlPO_4·xH_2O$ can be prepared as a gelatinous precipitate by adding a neutral solution of an Al salt to a solution of an alkali metal phosphate [CIC73] [CRC93]

Solubility: i H_2O; sl s HNO_3, HCl [CIC73]
Density, g/cm³: 2.57 [CIC73]
Melting Point, °C: 1850 [CIC73]

74

Compound: Aluminum phosphate trihydroxide
Synonyms: angelite
Formula: $Al_2(PO_4)(OH)_3$
Molecular Formula: $Al_2H_3O_7P$
Molecular Weight: 199.954
CAS RN: 12004-29-4
Properties: naturally occurring mineral; colorless, white, yellowish-white or rose; monocl; hardness 4.5-5 Mohs [CRC93] [MER89]
Density, g/cm³: 2.696 [CRC93]

75

Compound: Aluminum phosphide
Synonyms: celphos,detia,phostoxin
Formula: AlP
Molecular Formula: AlP
Molecular Weight: 57.956
CAS RN: 20859-73-8
Properties: dark gray or yellow powd; cub, a = 0.5467 nm; band gap 2.42 eV; reacts readily in moist air to produce phosphine; electronic dielectric constant 8.5; can be prepared from red phosphorus and aluminum powd; used as a fumigant, as a source of phosphine and in semiconductor work [CIC73] [MER89]
Solubility: reacts with H_2O to produce phosphine [MER89]
Density, g/cm³: 2.40 [MER89]
Melting Point, °C: >1000 [MER89]
Thermal Conductivity, W/(m·K): 92.0 [CRC93]

76

Compound: Aluminum selenide
Formula: Al_2Se_3
Molecular Formula: Al_2Se_3
Molecular Weight: 290.843
CAS RN: 1302-82-5
Properties: 6.5 mm and down black pieces or yellowish to light brown powd; unstable in air; formed by reaction of stoichiometric amounts of Al and Se at 1000°C; used to prepare H_2Se for semiconductor work; a = 0.389 nm, c = 0.630 nm [STR93] [MER89] [CIC73]
Solubility: decomposed by H_2O, acids [MER89]
Density, g/cm³: 3.437 [MER89]

77

Compound: Aluminum silicate
Synonyms: metakaolinite
Formula: $Al_2O_3 \cdot 2SiO_2$
Molecular Formula: $Al_2O_7Si_2$
Molecular Weight: 222.128
CAS RN: 1302-76-7
Properties: white powd [STR93]
Density, g/cm³: 2.60 (dickite) [ROB78]
Reactions: forms mullite and SiO_2 at 1200°C [BAB85]; forms amorphous aluminosilicate at 980°C [CHA90]

78

Compound: Aluminum silicate
Synonyms: sillimanite
Formula: $Al_2O_3 \cdot SiO_2$
Molecular Formula: Al_2O_5Si
Molecular Weight: 162.041
CAS RN: 12141-45-6
Properties: ortho, a = 0.78483 nm, b = 0.7673 nm, c = 0.57711 nm [ROB67]
Density, g/cm³: 3.247 [ROB78]
Reactions: forms mullite and SiO_2 from 1345°C to 1550°C [CLA66]
Thermal Expansion Coefficient: 100°C (0.088), 200°C (0.215), 400°C (0.531), 1000°C (1.979) [CLA66]

79

Compound: Aluminum silicate
Synonyms: andalusite
Formula: $Al_2O_3 \cdot SiO_2$
Molecular Formula: Al_2O_5Si
Molecular Weight: 162.041
CAS RN: 12183-80-1
Properties: gray, greenish, reddish or bluish; hardness 7-7.5; used in dental cements and the glass industry, in enamels, ceramics and as a paint filler [MER89] [HAW93]
Density, g/cm³: 3.145 [ROB78]
Reactions: forms mullite and SiO_2 from 1325°C to 1410°C [CLA66]
Thermal Expansion Coefficient: 100°C (0.151), 200°C (0.417), 1000°C (3.606) [CLA66]

80

Compound: Aluminum silicate
Synonyms: mullite
Formula: $3Al_2O_3 \cdot 2SiO_2$
Molecular Formula: $Al_6O_{13}Si_2$
Molecular Weight: 426.048

CAS RN: 1302-93-8

Properties: colorless; rhomb, a = 0.7557 nm, b = 0.76876 nm, c = 0.28842 nm; hardness: hot pressed 13.6 GPa, sintered 12.7 GPa; indentation microfracture 2.02 MPa·m$^{1/2}$; submicrometer powd can be prepared by hydrolysis of mixed alkoxides, followed by drying and calcining up to 1600°C; sinter, microstructure [SOM91]; phases equilibria in mullite [PAS88]; other data in [HIR89] [ROB67] [MIZ89]

Solubility: i H$_2$O, acids, HF [CRC92]

Density, g/cm^3: theoretical 3.17 [MIZ89]

Melting Point, °C: 1750 [JAN85]

Thermal Conductivity, W/(m·K): 100°C (5.39), 200°C (4.89), 400°C (4.18), 600°C (3.81), 800°C (3.59), 1000°C (3.43) [HO 72]

Thermal Expansion Coefficient: 100°C (0.070), 200°C (0.188), 400°C (0.471), 600°C (0.786), 800°C (1.121), 1000°C (1.439) [CLA66]

81

Compound: Aluminum silicate

Synonyms: kyanite

Formula: Al$_2$O$_3$·SiO$_2$

Molecular Formula: Al$_2$O$_5$Si

Molecular Weight: 162.041

CAS RN: 12141-46-7

Properties: mineral; tric, a = 0.7123 nm, b = 0.7848 nm, c = 0.5564 nm [ROB67] [HAW93]

Density, g/cm^3: 3.247 [CRC94]

Reactions: forms mullite and SiO$_2$ from 1000°C to 1325°C [CLA66]

Thermal Expansion Coefficient: 100°C (0.127), 200°C (0.360), 400°C (0.890), 600°C (1.478), 800°C (2.081), 1000°C (2.687) [CLA66]

82

Compound: Aluminum silicate dihydrate

Synonyms: kaolin, China clay

Formula: Al$_2$O$_3$·2SiO$_2$·2H$_2$O

Molecular Formula: Al$_2$H$_4$O$_9$Si$_2$

Molecular Weight: 258.161

CAS RN: 1332-58-7

Properties: white to yellowish or grayish fine powd; high lubricity, feels slippery to touch; tricl, a = 0.5055 nm, b = 0.8959 nm, c = 1.4736 nm; used as a filler and coating for paper and rubber, in paint [STR93] [HAW93]

Solubility: i H$_2$O, dil acids and alkali hydroxides [HAW93]

Density, g/cm^3: 2.594 [ROB78]

Reactions: transforms to metakaolinite about 525°C [BAB85]

83

Compound: Aluminum sulfate

Synonyms: alum, pearl alum

Formula: Al$_2$(SO$_4$)$_3$

Molecular Formula: Al$_2$O$_{12}$S$_3$

Molecular Weight: 342.154

CAS RN: 10043-01-3

Properties: white, lustrous cryst; can be prepared by treating kaolin, aluminum hydroxide or bauxite with sulfuric acid, followed by filtration and crystallization; used in tanning leather as a mordant for dyeing, to purify water and to waterproof and fireproof cloth [MER89] [HAW93]

Solubility: g/100g soln, H$_2$O: 27.50 (0°C), 27.82 (25°C), 43.9 (99.2°C); equilibrium solid phase, Al$_2$(SO$_4$)$_3$·16H$_2$O [KRU93]

Density, g/cm^3: 1.61 [MER89]

Melting Point, °C: 770, decomposes [ALD94]

Reactions: decomposes to γ-Al$_2$O$_3$ and SO$_3$ from 580°C to 900°C [KIR78]

84

Compound: Aluminum sulfate octadecahydrate

Synonyms: alunogen, cake alum

Formula: Al$_2$(SO$_4$)$_3$·18H$_2$O

Molecular Formula: Al$_2$H$_{36}$O$_{30}$S$_3$

Molecular Weight: 666.429

CAS RN: 7784-31-8

Properties: colorless; monocl; used in paper industry and in water treatment [KIR78] [STR93]

Solubility: anhyd/100g, H$_2$O: 27.5 (0°C), 27.8 (25°C), 46.9 (103.2°C) [KIR78]; gAl$_2$O$_3$/100mL at 35°C in the following solvents: methanol 2.89, ethanol 0.45, ethylene glycol 8.76 [OKA91]

Density, g/cm^3: 1.69 [KIR78]

Melting Point, °C: decomposes 86.5 [STR93]

Reactions: minus 15 H$_2$O from 40°C to 250°C; minus 3 H$_2$O from 250°C to 400°C [KIR78]

85

Compound: Aluminum sulfide

Formula: Al$_2$S$_3$

Molecular Formula: Al$_2$S$_3$

Molecular Weight: 150.161

CAS RN: 1302-81-4

Properties: yellowish gray powd; hex, a = 0.642 nm, c = 1.783 nm; H$_2$S odor; decomposes in moist air; formed by heating stoichiometric amounts of Al and S at 700-100°C; used as a semiconductor [CIC73] [MER89]

Solubility: decomposes in H_2O [MER89]
Density, g/cm³: 2.32 [CIC73]
Melting Point, °C: 1100 [MER89]
Reactions: sublimes at 1500°C in N_2 atm; $\alpha \rightarrow \gamma$ transition at 1000°C [CRC92] [JAN85]

86
Compound: Aluminum tartrate
Synonyms: 2,3-dihydroxybutanedioic acid, aluminum(III) salt
Formula: $Al_2(C_4H_4O_6)_3$
Molecular Formula: $C_{12}H_{12}Al_2O_{18}$
Molecular Weight: 498.179
CAS RN: 815-78-1
Properties: odorless granules; used in dyeing textiles [MER89]

$$
\begin{array}{c}
\text{HO} \quad \text{H} \\
| \quad | \\
[\text{OOC–C–C–COO}]_3\text{Al}_2 \\
| \quad | \\
\text{H} \quad \text{OH}
\end{array}
$$

Solubility: s H_2O, dissolves faster in hot H_2O; s ammonia [MER89]

87
Compound: Aluminum telluride
Formula: Al_2Te_3
Molecular Formula: Al_2Te_3
Molecular Weight: 436.763
CAS RN: 12043-29-7
Properties: dark gray or black cryst; a = 0.407 nm, c = 0.693 nm; electrical resistivity (27°C) 0.0054 ohm·cm; can be formed by reacting Al and Te at 1000°C; used in semiconductor work [CIC73] [STR93]
Density, g/cm³: 4.5 [CIC73]
Melting Point, °C: decomposes [ALF93]

88
Compound: Aluminum tellurite
Formula: $Al_2(TeO_3)_3$
Molecular Formula: $Al_2O_9Te_3$
Molecular Weight: 580.758
CAS RN: 58500-12-2
Properties: reacted product, -80 mesh particle size; 99% purity [CER91]

89
Compound: Aluminum thiocyanate

Formula: $Al(CNS)_3$
Molecular Formula: $C_3AlN_3S_3$
Molecular Weight: 201.233
CAS RN: 538-17-0
Properties: yellowish powd; aq solution used as mordant in the dye industry and in pottery manufacturing [MER89] [HAW93]
Solubility: s H_2O; i alcohol, ether [HAW93]

90
Compound: Aluminum titanate
Synonyms: tielite
Formula: Al_2TiO_5
Molecular Formula: Al_2O_5Ti
Molecular Weight: 181.827
CAS RN: 12004-39-6
Properties: -100 mesh with 99.5% purity; ortho-rhomb, pseudobrookite; preparation: sol-gel [PRA92], by reaction of equimolar amounts of Al_2O_3 and TiO_2 at 1300°C, then sintering at 1400°C for 4 h [PER89]; cryst structure in [FIE87]; other references [ALF93] [FRE87]
Density, g/cm³: 3.73 [ROB78]
Melting Point, °C: 1860 [HOL73]
Reactions: decomposes at 800°C to 1300°C to corundum and rutile, which recombine to tielite >1300°C [PAR90]
Thermal Expansion Coefficient: $-2 \times 10^{-6}/°C$ (24 to 1000°C) [PAR90]

91
Compound: Aluminum tristearate
Synonyms: aluminum stearate
Formula: $Al[CH_3(CH_2)_{16}COO]_3$
Molecular Formula: $C_{54}H_{105}AlO_6$
Molecular Weight: 877.406
CAS RN: 637-12-7
Properties: white powd; prepared by reacting stearic acid with aluminum salts; used in waterproofing fabrics and ropes, in paint and varnish driers [MER89] [HAW93]
Solubility: i H_2O, alcohol and ether; s alkali; forms gel with aliphatic and aromatic hydrocarbons [HAW93]
Density, g/cm³: 1.070 [HAW93]
Melting Point, °C: 115 [HAW93]

92
Compound: Aluminum tungstate
Formula: $Al_2(WO_4)_3$
Molecular Formula: $Al_2O_{12}W_3$
Molecular Weight: 797.476

CAS RN: 15123-82-7
Properties: -325 mesh with 99% purity; white powd [ALF93] [STR93]

93

Compound: Aluminum zirconate
Formula: $Al_2O_3 \cdot 3ZrO_2$
Molecular Formula: $Al_2O_9Zr_3$
Molecular Weight: 471.630
CAS RN: 70692-95-4
Properties: high fracture toughened ceramic material at room temp due to a dispersion of tetr ZrO_2 particles in an alumina matrix; can be prepared by plasma synthesis from the vapor phase by injecting mixtures of $AlCl_3$ and $ZrCl_4$ vapor in an argon-oxygen plasma, resulting in powd consisting of a mixture of δ-Al_2O_3 and tetr ZrO_2; reacted product, -100 mesh; 99% purity; reaction sintering with mullite in [DES91]; other references [CER91] [KOH88], wear coating [GEI92]

94

Compound: Aluminum zirconium
Formula: Al_2Zr
Molecular Formula: Al_2Zr
Molecular Weight: 145.187
CAS RN: 12004-50-1
Properties: -100 mesh with 99% purity; powd [ALF93]
Melting Point, °C: 1645 [ALF93]

95

Compound: Americium
Formula: Am
Molecular Formula: Am
Molecular Weight: 243
CAS RN: 7440-35-9
Properties: silvery metal with two forms; α: hex; a = 0.3468 nm, c = 1.1241 nm; β-Am: cub; a = 0.4894 nm; $t_{1/2}$ ^{241}Am = 433 year, $t_{1/2}$ ^{242}Am = 152 h, $t_{1/2}$ ^{243}Am = 7400 year; Am^{+++} stable in aq solution; enthalpy of vaporization 230 kJ/mol; enthalpy of fusion 14.4 kJ/mol; enthalpy of sublimation 276 kJ/mol; ionic radius of Am^{+++} is 0.0982 nm; can be prepared from ^{241}Pu; used to diagnose thyroid disorders [KIR78] [CIC73]
Solubility: s dil acids [CRC92]
Density, g/cm³: α-Am: 13.67; β-Am: 13.65 [CIC73]
Melting Point, °C: 1173 [KIR91]
Boiling Point, °C: 2011 [KIR91]

Reactions: α transforms to β at 1079°C [CIC73]

96

Compound: Americium bromide
Formula: $AmBr_3$
Molecular Formula: $AmBr_3$
Molecular Weight: 483
CAS RN: 14933-38-1
Properties: white; ortho-rhomb; a = 0.4064 nm, b = 1.2661 nm, c = 0.9144 nm [CIC73]
Solubility: s H_2O [CRC92]
Density, g/cm³: 6.85 [KIR78]
Melting Point, °C: sublimes [CRC92]

97

Compound: Americium carbonate dihydrate
Formula: $Am_2(CO_3)_3 \cdot 2H_2O$
Molecular Formula: $C_3H_4Am_2O_{11}$
Molecular Weight: 702
Properties: a = 1.409 nm [CIC73]

98

Compound: Americium chloride
Formula: $AmCl_3$
Molecular Formula: $AmCl_3$
Molecular Weight: 349
CAS RN: 13464-46-5
Properties: pink; hex, a = 0.7382 nm, c = 0.4214 nm [KIR78]
Density, g/cm³: 5.87 [KIR78]
Melting Point, °C: 715 [KIR91]

99

Compound: Americium fluoride
Synonyms: americium trifluoride
Formula: AmF_3
Molecular Formula: AmF_3
Molecular Weight: 300
CAS RN: 13708-80-0
Properties: pink; hex, a = 0.7044 nm, c = 0.7225 nm [CIC73]
Solubility: i H_2O [CRC92]
Density, g/cm³: 9.53 [KIR78]
Melting Point, °C: 1393 [KIR91]

100

Compound: Americium hydride
Formula: AmH_3
Molecular Formula: AmH_3
Molecular Weight: 246

CAS RN: 13774-24-8
Properties: black; hex, a = 0.377 nm, c = 0.675 nm
[CIC73]
Density, g/cm³: 9.76 [CIC73]

101
Compound: Americium hydroxide
Formula: Am(OH)₃
Molecular Formula: AmH₃O₃
Molecular Weight: 294
CAS RN: 23323-79-7
Properties: hex, a = 0.6426 nm, c = 0.3745 nm
[CIC73]

102
Compound: Americium iodide
Formula: AmI₃
Molecular Formula: AmI₃
Molecular Weight: 624
CAS RN: 13813-47-3
Properties: yellow; ortho-rhomb, a = 0.742 nm, c =
2.055 nm [KIR78]
Solubility: s H₂O [CRC92]
Density, g/cm³: 6.9 [CRC92]
Melting Point, °C: ~950 [KIR91]

103
Compound: Americium oxide(α)
Formula: α-Am₂O₃
Molecular Formula: Am₂O₃
Molecular Weight: 534
CAS RN: 12254-64-7
Properties: tan; hex, a = 0.3805 nm, c = 0.696 nm
[KIR78]
Solubility: s mineral acids [CRC92]
Density, g/cm³: 11.77 [KIR78]

104
Compound: Americium oxide(β)
Formula: β-Am₂O₃
Molecular Formula: Am₂O₃
Molecular Weight: 534
CAS RN: 12254-64-7
Properties: reddish brown; cub, a = 1.103 nm
[KIR78]
Density, g/cm³: 10.57 [KIR78]

105
Compound: Americium oxychloride
Formula: AmOCl

Molecular Formula: AmClO
Molecular Weight: 294
CAS RN: 37961-19-6
Properties: cryst; a = 0.400 nm, b = 0.678 nm
[BRO73]

106
Compound: Americium phosphate
Formula: AmPO₄
Molecular Formula: AmO₄P
Molecular Weight: 338
Properties: monocl, a = 0.673 nm, b = 0.693 nm, c
= 0.641 nm [CIC73]

107
Compound: Americium sulfide
Formula: Am₂S₃
Molecular Formula: Am₂S₃
Molecular Weight: 582
CAS RN: 12446-46-7
Properties: bcc, a = 0.845 nm [CIC73]

108
Compound: Americium(IV) fluoride
Formula: AmF₄
Molecular Formula: AmF₄
Molecular Weight: 319
CAS RN: 15947-41-8
Properties: tan; monocl, a = 1.254 nm, b = 1.052
nm, c = 0.820 nm [KIR78]
Density, g/cm³: 7.23 [KIR78]

109
Compound: Americium(IV) oxide
Formula: AmO₂
Molecular Formula: AmO₂
Molecular Weight: 275
CAS RN: 12005-67-3
Properties: black; cub, a = 0.5374 nm [KIR78]
Solubility: s mineral acids [CRC92]
Density, g/cm³: 11.68 [KIR78]

110
Compound: Ammonia
Formula: NH₃
Molecular Formula: H₃N
Molecular Weight: 17.031
CAS RN: 7664-41-7

Properties: colorless gas; very pungent odor; critical pressure 111.5 atm; critical temp 132.4°C; vapor pressure of liq 8.5 atm (20°C); enthalpy of vaporization 23.33 kJ/mol; enthalpy of fusion 5.66 kJ/mol; autoignition temp 690°C; produced by reaction of steam forced through incandescent coke; used in the manufacture of nitric acid, explosives, synthetic fibers, nitrides, fertilizers, and in electronics and in refrigeration [MER89] [HAW93] [AIR87]

Solubility: H_2O: 47% (0°C), 31% (25°C), 28% (50°C) [MER89]

Density, g/cm³: 0.5967 (air = 1.0000) [MER89]

Melting Point, °C: -77.7 [MER89]

Boiling Point, °C: -33.35 at 1 atm [MER89]

Reactions: mixtures of NH_3 and air can explode when ignited [MER89]

111

Compound: Ammonium 12-molybdophosphate hydrate

Synonyms: ammonium phosphomolybdate

Formula: $(NH_4)_3PMo_{12}O_{40} \cdot xH_2O$

Molecular Formula: $H_{12}Mo_{12}N_3O_{40}P$ (anhydrous)

Molecular Weight: 1834.345 (anhydrous)

CAS RN: 12026-66-3

Properties: yellow, cryst powd; prepared by reacting ammonium molybdate with phosphoric and nitric acids; used as a reagent, in ion-exchange columns and as a photographic additive [HAW93]

Solubility: v sl s H_2O; s alkali; i alcohol, acids [HAW93]

Melting Point, °C: decomposes [CRC94]

112

Compound: Ammonium acetate

Synonyms: acetic acid, ammonium salt

Formula: CH_3COONH_4

Molecular Formula: $C_2H_7NO_2$

Molecular Weight: 77.084

CAS RN: 631-61-8

Properties: white cryst; deliq; prepared from acetic acid and ammonia by exact neutralization to pH7.0; finds use in analytical chemistry, drugs and in textile dyeing; specific gravity of aq solutions, in % of CH_3COONH_4: 10% (1.022), 20% (1.042), 30% (1.062), 40% (1.077), 50% (1.092) [MER89] [KIR78]

Solubility: 148 g/100g H_2O (4°C) [KIR78]

Density, g/cm³: 1.073 [KIR78]

Melting Point, °C: 114 [MER89]

Boiling Point, °C: decomposes [KIR78]

113

Compound: Ammonium aluminum sulfate

Synonyms: burnt ammonium alum

Formula: $NH_4Al(SO_4)_2$

Molecular Formula: $AlH_4NO_8S_2$

Molecular Weight: 237.148

CAS RN: 7784-25-0

Properties: white powd [MER89]

Solubility: g $NH_4Al(SO_4)_2$/100 g H_2O: 2.10 (0°C), 5.00 (10°), 7.74 (20°), 10.9 (30°), 14.9 (40°C), 26.7 (60°C) [LAN85]

Density, g/cm³: 2.45 [CRC94]

Melting Point, °C: 280 (decomposes) [KIR78]

Reactions: forms δ-Al_2O_3 at 1000-1250°C [KIR78]

114

Compound: Ammonium aluminum sulfate dodecahydrate

Synonyms: ammonium alum

Formula: $NH_4Al(SO_4)_2 \cdot 12H_2O$

Molecular Formula: $AlH_{28}NO_{20}S_2$

Molecular Weight: 453.331

CAS RN: 7784-26-1

Properties: colorless cryst, white granules or powd; obtained by crystallization from a mixture of ammonium and aluminum sulfates; used to purify drinking water, in baking powd, for dyeing fabrics and fireproofing [MER89] [HAW93] [KIR78]

Solubility: 1g/7mL H_2O; 1g/0.5mL hot H_2O [MER89]; s glycerol; i alcohol [HAW93]

Density, g/cm³: 1.65 [MER89]

Melting Point, °C: 94.5 [MER89]

Boiling Point, °C: decomposes above 280 [MER89]

Reactions: minus $10H_2O$ (250°C) forms γ-Al_2O_3 (1000-1250°C) [MER89] [KIR78]

115

Compound: Ammonium arsenate hydrate

Synonyms: ammonium orthoarsenate

Formula: $(NH_4)_3AsO_4 \cdot xH_2O$

Molecular Formula: $AsH_{12}N_3O_4$ (anhydrous)

Molecular Weight: 193.035 (anhydrous)

CAS RN: 13462-93-6

Properties: -6 mesh with 99.9% purity; x=3: ortho [CER91] [CRC94]

Reactions: minus NH_3 on heating [CRC94]

116

Compound: Ammonium azide

Formula: NH_4N_3

Molecular Formula: H_4N_4

Molecular Weight: 60.059
CAS RN: 12164-94-2
Properties: ortho-rhomb, a = 0.893 nm, b = 0.864 nm, c = 0.380 nm [CIC73]
Solubility: g NH_4N_3/100g H_2O: 16.0 (0°C), 25.3 (20°C), 37.1 (40°C) [LAN85]
Density, g/cm³: 1.346 [CIC73]
Melting Point, °C: 160 [CIC73]
Boiling Point, °C: 134 (explodes) [CIC73]

117

Compound: Ammonium benzoate
Synonyms: benzoic acid, ammonium salt
Formula: $C_6H_5COONH_4$
Molecular Formula: $C_7H_9NO_2$
Molecular Weight: 139.154
CAS RN: 1863-63-4
Properties: white cryst or powd; manufactured from benzoic acid and ammonia; used in medicine and as a preservative for latex [HAW93] [MER89]
Solubility: 1g/4.7mL H_2O, 1g/1.2mL hot H_2O [MER89]; s alcohol, glycerol [HAW93]
Density, g/cm³: 1.26 [MER89]
Melting Point, °C: 198 [MER89]
Boiling Point, °C: sublimes at 160 [HAW93]
Reactions: gradually loses NH_3 in air [MER89]

118

Compound: Ammonium bimalate
Synonyms: hydroxybutanedioic acid, monoammonium salt
Formula: $NH_4OOCCH_2CH(OH)COOH$
Molecular Formula: $C_4H_9NO_5$
Molecular Weight: 151.119
CAS RN: 5972-71-4
Properties: ortho-rhomb cryst [MER89]
Solubility: s 3 parts H_2O, sl s alcohol [MER89]
Density, g/cm³: 1.15 [MER89]
Melting Point, °C: 160-161 [MER89]

119

Compound: Ammonium bromide
Formula: NH_4Br
Molecular Formula: BrH_4N
Molecular Weight: 97.943
CAS RN: 12124-97-9

Properties: colorless cryst or yellowish white powd; tetr, a = 0.4034 nm; vapor pressure, kPa: 7.3 (300°C), 13.3 (320°C), 41.2 (360°C), 73.4 (380°C), 115.4 (400°C); prepared by reacting HBr and NH_4OH; used to make AgBr salts for photography, in medicine, engraving, textile finishing and as a fire retardant material [CIC73] [HAW93] [KIR78]
Solubility: g/100g soln, H_2O: 37.5 (0°C), 43.9 (25°C), 57.4 (100°C) [KRU93]; g/100g H_2O: 60.6 (0°C); 75.5 (20°C), 145.6 (100°C) [KIR78]; s alcohol [HAW93]
Density, g/cm³: 2.429 [CIC73]
Melting Point, °C: sublimes 452 [CIC73]
Boiling Point, °C: 235 in vacuum [CIC73]

120

Compound: Ammonium caprylate
Synonyms: octanoic acid ammonium salt
Formula: $C_7H_{15}COONH_4$
Molecular Formula: $C_8H_{19}NO_2$
Molecular Weight: 161.245
CAS RN: 5972-76-9
Properties: hygr; monocl cryst; prepared by reacting caprylic acid and ammonia; used in photographic emulsions, and as an insecticide [MER89]
Solubility: s acetic acid, ethanol [MER89]
Melting Point, °C: 70-85 [MER898]
Reactions: easily hydrolyzed in H_2O [MER89]

121

Compound: Ammonium carbamate
Formula: NH_2COONH_4
Molecular Formula: $CH_6N_2O_2$
Molecular Weight: 78.071
CAS RN: 1111-78-0
Properties: cryst powd; ammonia odor; white; rhomb; very volatile; forms urea when heated; made from liq NH_3 and solid CO_2; used as a fertilizer [HAW93] [MER89]
Solubility: v s H_2O; s alcohol [MER89]
Melting Point, °C: volatilizes ~60 [MER89]
Reactions: gradually evolves NH_3 in air [MER89]

122

Compound: Ammonium carbonate
Synonyms: normal ammonium carbonate
Formula: $(NH_4)_2CO_3$
Molecular Formula: $CH_8N_2O_3$
Molecular Weight: 96.086
CAS RN: 506-87-6

Properties: powd or lumps; colorless cryst; prepared by passing CO_2 gas through NH_4OH solution, to crystallize the carbonate [ALF93] [ALD94] [KIR78]

Solubility: 20g/100g saturated solution in water (25°C) [MER89]

Melting Point, °C: decomposes at 58 [CIC73]

123

Compound: Ammonium cerium(III) nitrate tetrahydrate

Synonyms: cerous ammonium nitrate

Formula: $(NH_4)_2Ce(NO_3)_5 \cdot 4H_2O$

Molecular Formula: $CeH_{16}N_7O_{19}$

Molecular Weight: 558.278

CAS RN: 15318-60-2

Properties: large, colorless transparent monocl cryst [CRC94] [MER52]

Solubility: g/100g anhydrous, H_2O: 242 (10°C), 276 (20°C), 318 (30°), 376 (40°), 681 (60°C) [LAN85]

Melting Point, °C: 74 [CRC94]

124

Compound: Ammonium cerium(III) sulfate tetrahydrate

Synonyms: cerous ammonium sulfate

Formula: $(NH_4)Ce(SO_4)_2 \cdot 4H_2O$

Molecular Formula: $CeH_{12}NO_{12}S_2$

Molecular Weight: 422.342

CAS RN: 21995-38-0

Properties: prepared by slow evaporation of a solution containing stoichiometric amounts of cerous and ammonium sulfates; monocl cryst [MER89]

Solubility: g/100g anhydrous, H_2O: 5.53 (20°C), 4.49 (30°C), 3.48 (40°C), 2.02 (60°C), 1.33 (80°C) [LAN85]

Density, g/cm³: 2.523 (octahydrate) [CRC94]

Reactions: anhydrous at 150°C [CRC94]

125

Compound: Ammonium cerium(IV) nitrate

Synonyms: ceric ammonium nitrate

Formula: $(NH_4)_2Ce(NO_3)_6$

Molecular Formula: $CeH_8N_8O_{18}$

Molecular Weight: 548.222

CAS RN: 16774-21-3

Properties: small reddish orange cryst; precipitates from ceric nitrate solution containing excess nitric acid, on addition of NH_4NO_3; has been used as an oxidizing agent in analytical chemistry, and as a catalyst for polymerization of olefins [ALF93] [MER89] [KIR78]

Solubility: g/100g H_2O: 135 (20°C), 150 (30°C), 169 (40°C), 213 (60°C) [LAN85]; s alcohol, i conc HNO_3 [HAW93]

126

Compound: Ammonium cerium(IV) sulfate dihydrate

Synonyms: ceric ammonium sulfate dihydrate

Formula: $(NH_4)_4Ce(SO_4)_4 \cdot 2H_2O$

Molecular Formula: $CeH_{20}N_4O_{18}S_4$

Molecular Weight: 632.556

CAS RN: 10378-47-9

Properties: cryst powd; oxidizing agent; precipitates from a solution of ceric sulfate and ammonium sulfate; anhydrous compound, 13840-04-5, exists [ALF93] [ALD93] [KIR78]

127

Compound: Ammonium chlorate

Formula: NH_4ClO_3

Molecular Formula: ClH_4NO_3

Molecular Weight: 101.490

CAS RN: 10192-29-7

Properties: colorless or white cryst; oxidizing agent; obtained as a product of the reaction of solutions of ammonium chloride and sodium chlorate; can be used as an explosive [HAW93]

Solubility: g/100g H_2O: 28.7 (0°C), 115 (75°C) [CIC73]

Density, g/cm³: 1.80 [CIC73]

Melting Point, °C: 102, explodes [CIC73]

128

Compound: Ammonium chloride

Synonyms: ammonium muriate, sal ammoniac

Formula: NH_4Cl

Molecular Formula: ClH_4N

Molecular Weight: 53.492

CAS RN: 12125-02-9

Properties: white; cub cryst, a = 0.3866 nm; enthalpy of formation 317 kJ/mol; enthalpy of sublimation 165.7 kJ/mol; vapor pressure, kPa, at temp shown: 6.5 (250°C), 17.9 (280°C), 33.5 (300°C), 60.9 (320°C), 101.1 (338°C); prepared by mixing ammonium sulfate and sodium chloride solutions; used in batteries, as a soldering flux, in electroplating [CIC73] [HAW93] [KIR78]

Solubility: g/100g H_2O: 29.4 (0°C), 39.3 (25°C), 77.3 (100°C); equilibrium solid phase, NH_4Cl [KRU93]

Density, g/cm³: 1.527 [CIC73]

Melting Point, °C: sublimes without melting [MER89]

Reactions: decomposes at 520°C [CIC73]

129

Compound: Ammonium chromate(VI)

Synonyms: ammonium chromate

Formula: $(NH_4)_2CrO_4$

Molecular Formula: $CrH_8N_2O_4$

Molecular Weight: 152.071

CAS RN: 7788-98-9

Properties: yellow cryst; obtained by adding NH_4OH to ammonium dichromate solution, followed by crystallization; used as a mordant in dyeing and as a corrosion inhibitor [HAW93] [KIR78]

Solubility: g/100g soln, H_2O: 19.9 (0°C), 27.02 (25°C), 41.20 (75°C) [KRU93] [MER89]; i alcohol [HAW93]

Density, g/cm³: 1.90 [KIR78]

Melting Point, °C: decomposes at 185 [MER89]

Reactions: minus some NH_3 in air [MER89]

130

Compound: Ammonium chromic sulfate dodecahydrate

Synonyms: ammonium chrome alum

Formula: $(NH_4)Cr(SO_4)_2 \cdot 12H_2O$

Molecular Formula: $CrH_{28}NO_{20}S_2$

Molecular Weight: 478.345

CAS RN: 10022-47-6

Properties: green powd or deep violet cryst; can be crystallized from a solution of chromic sulfate and ammonium sulfate; used as a mordant for dyeing, and in tanning [HAW93] [MER89]

Solubility: g/100g anhyd, H_2O: 3.95 (0°C), 18.8 (30°C), 32.6 (40°C) [LAN85]; sl s alcohol [HAW93]

Density, g/cm³: hydrated form: 1.72 [HAW93]

Melting Point, °C: hydrated form: 94 [HAW93]

Reactions: minus $9H_2O$ on melting, minus $12H_2O$ by 300°C [MER89]

131

Compound: Ammonium citrate tribasic

Formula:
$H_4NOOCCH_2C(OH)(COONH_4)CH_2COONH_4$

Molecular Formula: $C_6H_{17}N_3O_7$

Molecular Weight: 243.217

CAS RN: 3458-72-8

Properties: structure:

$$CH_2COONH_4$$
$$|$$
$$HO-C-COONH_4$$
$$|$$
$$CH_2COONH_4$$

Melting Point, °C: 185, decomposes [ALD94]

132

Compound: Ammonium cobalt(II) phosphate monohydrate

Synonyms: cobaltous ammonium phosphate

Formula: $NH_4CoPO_4 \cdot H_2O$

Molecular Formula: CoH_6NO_5P

Molecular Weight: 189.959

CAS RN: 14590-13-7

Properties: red to violet powd or monocl; results from a reaction between a cobalt(II) salt and ammonium phosphate; used as a pigment in ceramic glass, and to indicate temp in textile industry [MER89]

Solubility: i H_2O; s acids [MER89]

133

Compound: Ammonium cobalt(II) sulfate hexahydrate

Synonyms: cobaltous ammonium sulfate

Formula: $(NH_4)_2Co(SO_4)_2 \cdot 6H_2O$

Molecular Formula: $CoH_{20}N_2O_{14}S_2$

Molecular Weight: 395.229

CAS RN: 13586-38-4

Properties: red; monocl prismatic cryst [MER89]

Solubility: g $(NH_4)_2Co(SO_4)_2$/100 g H_2O: 6.0 (0°C), 9.5 (10°C), 13.0 (20°C), 17.0 (30°C), 22.0 (40°C), 33.5 (60°C), 49.0 (80°C), 58.0 (90°C), 75.1 (100°C) [LAN85]; i alcohol [MER89]

Density, g/cm³: 1.902 [HAW93]

134

Compound: Ammonium copper(II) chloride dihydrate

Synonyms: cupric ammonium chloride dihydrate

Formula: $2NH_4Cl \cdot CuCl_2 \cdot 2H_2O$

Molecular Formula: $Cl_4CuH_{12}N_2O_2$
Molecular Weight: 277.464
CAS RN: 10060-13-6
Properties: blue to bluish green cryst; tetr; preparation: by evaporating a solution containing a stoichiometric amount of NH_4Cl and $CuCl_2$; has been used as an analytical reagent; anhydrous material has yellow, hygr rhombohedral cryst [MER89]
Solubility: g $2NH_4Cl \cdot CuCl_2$/100g H_2O: 28.2 (0°C), 32.0 (10°C), 35.0 (20°C), 38.3 (30°C), 43.8 (40°C), 56.6 (60°C), 76.5 (80°C), 76.5 (90°C) [LAN85]; s alcohol [MER89]
Density, g/cm³: 1.993 [STR93]
Melting Point, °C: decomposes >120 [MER89] [STR93]
Reactions: minus $2H_2O$ over range 110-120°C [MER89]

135
Compound: Ammonium cyanide
Formula: NH_4CN
Molecular Formula: CH_4N_2
Molecular Weight: 44.056
CAS RN: 12211-52-8
Properties: colorless cryst solid; readily decomposes; can be formed by mixing ammonium sulfate solution with barium cyanide solution [KIR78]
Solubility: v s H_2O [KIR78]
Density, g/cm³: 1.02 (100°C) [CRC94]
Melting Point, °C: decomposes 36 [CRC94]
Reactions: decomposes to NH_3 and HCN at 36°C [KIR78]

136
Compound: Ammonium dichromate(VI)
Synonyms: ammonium dichromate
Formula: $(NH_4)_2Cr_2O_7$
Molecular Formula: $Cr_2H_8N_2O_7$
Molecular Weight: 252.065
CAS RN: 7789-09-5
Properties: reddish orange cryst; monocl; can be prepared from ammonium sulfate and sodium dichromate, followed by crystallization from the solution; used as a mordant for dyeing, in leather tanning, oil purification, photography [HAW93] [KIR78]
Solubility: g/100g H_2O: 18.2 (0°C), 25.5 (10°C), 35.6 (20°C), 46.5 (30°C), 58.5 (40°C), 156 (100°C) [LAN85]; g/100g soln, H_2O: 15.37 (0°C), 28.615 (25°C), 60.89 (100°C) [KRU93]; s alcohol [HAW93]
Density, g/cm³: 2.155 [KIR78]

Melting Point, °C: decomposes at 180 [KIR78]
Reactions: decomposes with swelling and evolution of heat and N_2 [MER89]

137
Compound: Ammonium dihydrogen arsenate
Formula: $NH_4H_2AsO_4$
Molecular Formula: AsH_6NO_4
Molecular Weight: 158.975
CAS RN: 13462-93-6
Properties: cryst [ALF93]
Solubility: g/100g H_2O: 33.74 (0°C), 48.67 (20°C), 122.4 (90°C); equilibrium solid phase, $NH_4H_2AsO_4$ [KRU93]
Density, g/cm³: 2.311 [ALF93]
Melting Point, °C: 300, with decomposition [ALF93]

138
Compound: Ammonium dihydrogen phosphate
Synonyms: ammonium phosphate monobasic
Formula: $(NH_4)H_2PO_4$
Molecular Formula: H_6NO_4P
Molecular Weight: 115.026
CAS RN: 7722-76-1
Properties: odorless cryst or white powd; made from ammonia and phosphoric acid; used with sodium bicarbonate as baking powd, and for fireproofing materials such as paper, wood and fiberboard [MER89] [HAW93]
Solubility: g/100g soln, H_2O: 17.8±0.8 (0°C), 28.8±0.4 (25°C), 72.4±7.0 (100°C) [KRU93]; i alcohol [HAW93]
Density, g/cm³: 1.803 [HAW93]
Melting Point, °C: 190 [STR93]

139
Compound: Ammonium dimolybdate
Synonyms: ammonium molybdenum dioxide
Formula: $(NH_4)_2Mo_2O_7$
Molecular Formula: $H_8Mo_2N_2O_7$
Molecular Weight: 339.953
CAS RN: 27546-07-2
Properties: white powd; obtained by crystallization from a solution of MoO_3 containing excess NH_3; used as a high purity source for the preparation of Mo metal [KIR81] [ALF93]
Density, g/cm³: 3.1 [ALF95]

140
Compound: Ammonium dithiocarbamate

Synonyms: ammonium sulfocarbamate
Formula: $NH_2CSS(NH_4)$
Molecular Formula: $CH_6N_2S_2$
Molecular Weight: 110.204
CAS RN: 513-74-6
Properties: yellow, lustrous; ortho-rhomb; decomposes in air, odor of H_2S; prepared from carbon disulfide and ammonia; used to precipitate metals and metal sulfides, and in the synthesis of heterocyclic compounds [MER89]
Solubility: s H_2O [MER89]
Density, g/cm³: 1.451 [MER89]
Melting Point, °C: decomposes 99 [MER89]
Reactions: reversible exothermic transition at 63°C [MER89]

141
Compound: Ammonium ferric chromate
Synonyms: ferric ammonium chromate
Formula: $(NH_4)Fe(CrO_4)_2$
Molecular Formula: $Cr_2FeH_4NO_8$
Molecular Weight: 305.871
CAS RN: 7789-08-4
Properties: carmine red microcryst powd; can be prepared by adding ammonia to a solution of CrO_3 and ferric nitrate hexahydrate [MER89]
Solubility: i H_2O [MER89]

142
Compound: Ammonium ferric citrate
Synonyms: ferric ammonium citrate
CAS RN: 1185-57-5
Properties: undetermined structure; reddish-brown granules, red scales, or brownish-yellow powd, and green form; deliq; light sensitive; preparation: $Fe(OH)_3$ addition to aq solution of citric acid and ammonia; uses: blueprints, photography, for iron deficiency [MER89]
Solubility: v s H_2O, i alcohol [MER89]
Reactions: light causes reduction to ferrous salt [MER89]

143
Compound: Ammonium ferric oxalate trihydrate
Synonyms: ferric ammonium oxalate trihydrate
Formula: $(NH_4)_3Fe(C_2O_4)_3 \cdot 3H_2O$
Molecular Formula: $C_6H_{18}FeN_3O_{15}$
Molecular Weight: 428.065
CAS RN: 13268-42-3

Properties: bright green; monocl hygr prismatic cryst; sensitive to light; prepared by addition of ammonium binoxalate and ferric hydroxide; used in blueprint photography, and to color aluminum and aluminum alloys [ALD93] [MER89] [HAW93]
Solubility: v s H_2O; i alcohol [MER89]
Density, g/cm³: 1.780 [ALD93]
Melting Point, °C: decomposes 160-170 [MER89]
Reactions: minus $3H_2O$ by 100°C [MER89]

144
Compound: Ammonium ferric sulfate dodecahydrate
Synonyms: ferric alum
Formula: $(NH_4)Fe(SO_4)_2 \cdot 12H_2O$
Molecular Formula: $FeH_{28}NO_{20}S_2$
Molecular Weight: 482.194
CAS RN: 10138-04-2
Properties: colorless to pale violet; effloresces; octahedral cryst; prepared by mixing solutions of ferric sulfate and ammonium sulfate with subsequent evaporation and crystallization; used in medicine, as a mordant to dye textiles, and as an astringent [HAW93] [MER89] [STR93]
Solubility: v s H_2O; i alcohol [MER89]
Density, g/cm³: 1.71 [MER89]
Melting Point, °C: 39-41 [ALD94]
Reactions: minus $12H_2O$ at 230°C [HAW93]

145
Compound: Ammonium ferricyanide trihydrate
Synonyms: ammonium hexacyanoferrate(III) trihydrate
Formula: $(NH_4)_3Fe(CN)_6 \cdot 3H_2O$
Molecular Formula: $C_6H_{18}FeN_9O_3$
Molecular Weight: 320.113
CAS RN: 14221-48-8
Properties: red cryst; sensitive to light [MER89]
Solubility: s H_2O [MER89]

146
Compound: Ammonium ferrous sulfate hexahydrate
Synonyms: Mohr's salt, ferrous ammonium sulfate
Formula: $(NH_4)_2Fe(SO_4)_2 \cdot 6H_2O$
Molecular Formula: $FeH_{20}N_2O_{14}S_2$
Molecular Weight: 392.141
CAS RN: 10045-89-3

Properties: pale bluish green cryst powd; slowly oxidizes and effloresces in air; sensitive to light; prepared from a mixture of ferrous sulfate and ammonium sulfate solutions, with subsequent evaporation and crystallization; used in analytical chemistry and in metallurgy [HAW93] [MER89] [STR93] [ALF93]

Solubility: g/100g H_2O: 12.5 (0°C), 17.2 (10°C), 26.4 (20°C), 33 (30°C), 46 (40°C); i alcohol [MER89] [LAN85]

Density, g/cm³: 1.865 [HAW93]

Melting Point, °C: decomposes at 100-110 [HAW93]

147

Compound: Ammonium fluoride

Formula: NH_4F

Molecular Formula: FH_4N

Molecular Weight: 37.037

CAS RN: 12125-01-8

Properties: white deliq cryst; hex, a = 0.439 nm, c = 0.702 nm; tends to lose NH_3 to form the more stable $NH_4F \cdot HF$; can be obtained by adding ammonia to an ice cold 40% HF solution; NH_4F is used as a laboratory reagent [KIR78] [CIC73]

Solubility: g/100g soln, H_2O: 41.72 (0°C), 45.5±0.3 (25°C), 54.05 (80°C) [KRU93]

Density, g/cm³: 1.009 [CIC73]

Melting Point, °C: decomposes [CIC73]

Reactions: decomposed by hot water to NH_3 and HF [MER89]

148

Compound: Ammonium fluoroborate

Synonyms: ammonium tetrafluoroborate

Formula: NH_4BF_4

Molecular Formula: BF_4H_4N

Molecular Weight: 104.844

CAS RN: 13826-83-0

Properties: white powd; ortho-rhomb below 205°C, a = 0.7278 nm, b = 0.9072 nm, c = 0.5678 nm; cub above 205°C; can be prepared by reacting ammonia gas with fluoroboric acid [KIR78] [ALF93]

Solubility: g/100mL H_2O: 3.09 (-1.0°C), 5.26 (-1.5°C), 10.85 (-2.7°C), 12.20 (0°C), 25 (16°C), 25.83 (25°C), 44.09 (50°C), 67.50 (75°C), 98.93 (100°C), 113.7 (108.5°C); s HF 19.89% (0°C) [KIR78]

Density, g/cm³: 1.871 [HAW93]

Melting Point, °C: 487 decomposes [KIR78]; sublimes 220 [ALF93]

149

Compound: Ammonium fluorosulfonate

Formula: NH_4SO_3F

Molecular Formula: FH_4NO_3S

Molecular Weight: 117.101

CAS RN: 13446-08-7

Properties: long colorless needles [KIR78]

Solubility: s H_2O, alcohol, methanol [KIR78]

Melting Point, °C: 245 [KIR78]

150

Compound: Ammonium formate

Synonyms: formic acid, ammonium salt

Formula: $HCOONH_4$

Molecular Formula: CH_5NO_2

Molecular Weight: 63.056

CAS RN: 540-69-2

Properties: deliq cryst, or white powd; formed by reaction of ammonia and formic acid; has been used to precipitate metals [HAW93] [STR93]

Solubility: g/100g H_2O: 102 (0°C), 143 (20°C), 204 (40°C), 311 (60°C), 533 (80°C) [LAN85]; s alcohol [HAW93]

Density, g/cm³: 1.26 [HAW93]

Melting Point, °C: 119-121 [STR93]

Boiling Point, °C: decomposes 180 [CRC94]

151

Compound: Ammonium germanium oxalate hydrate

Synonyms: ammonium tris(oxalato)germanate

Formula: $(NH_4)_2Ge(C_2O_4)_3 \cdot xH_2O$

Molecular Formula: $C_6H_8GeN_2O_{12}$ (anhydrous)

Molecular Weight: 372.745 (anhydrous)

Properties: hygr [ALD94]

152

Compound: Ammonium heptafluorotantalate

Formula: $(NH_4)_2TaF_7$

Molecular Formula: $F_7H_8N_2Ta$

Molecular Weight: 350.014

CAS RN: 12022-02-5

Properties: hygr [ALD93]

153

Compound: Ammonium hexabromoosmiate(IV)

Formula: $(NH_4)_2OsBr_6$

Molecular Formula: $Br_6H_8N_2Os$

Molecular Weight: 705.731

CAS RN: 24598-62-7

Properties: black powd [ALF93]

154
Compound: Ammonium hexabromoplatinate(IV)
Formula: $(NH_4)_2PtBr_6$
Molecular Formula: $Br_6H_8N_2Pt$
Molecular Weight: 710.581
CAS RN: 17363-02-9
Properties: reddish brown powd [ALF93]
Density, g/cm³: 4.26 [ALD94]
Melting Point, °C: 145, decomposes [ALF93]

155
Compound: Ammonium hexachloroiridate(III)
Formula: $(NH_4)_3IrCl_6$
Molecular Formula: $Cl_6H_{12}IrN_3$
Molecular Weight: 459.048
CAS RN: 15752-05-3
Properties: olive green powd [ALF93]

156
Compound: Ammonium hexachloroiridate(III)
    monohydrate
Formula: $(NH_4)_3IrCl_6 \cdot H_2O$
Molecular Formula: $Cl_6H_{14}IrN_3O$
Molecular Weight: 477.063
CAS RN: 29796-57-4
Properties: hygr [ALD93]

157
Compound: Ammonium hexachloroiridate(IV)
Formula: $(NH_4)_2IrCl_6$
Molecular Formula: $Cl_6H_8IrN_2$
Molecular Weight: 441.010
CAS RN: 16940-92-4
Properties: black cryst powd [ALD93] [STR93]
Solubility: g/100g H_2O: 0.556 (0°C), 0.706 (10°C),
    0.77 (20°C), 1.21 (30°C), 1.57 (40°C), 2.46
    (60°C), 4.38 (80°C), decomposes (90°C)
    [LAN85]
Density, g/cm³: 2.856 [ALD93]
Melting Point, °C: decomposes [STR93]

158
Compound: Ammonium hexachloroosmiate(IV)
Synonyms: Ammonium osmium chloride
Formula: $(NH_4)_2OsCl_6$
Molecular Formula: $Cl_6H_8N_2Os$
Molecular Weight: 439.023
CAS RN: 12125-08-5
Properties: red powd, or dark red octahedral cryst
    [MER89]
Solubility: s H_2O, alcohol [MER89]

Density, g/cm³: 2.930 [ALD93]
Melting Point, °C: 170, sublimes [ALF93]

159
Compound: Ammonium hexachloropalladate(IV)
Formula: $(NH_4)_2PdCl_6$
Molecular Formula: $Cl_6H_8N_2Pd$
Molecular Weight: 355.213
CAS RN: 19168-23-1
Properties: hygr; reddish brown cryst [ALF93]
    [STR93]
Density, g/cm³: 2.418 [STR93]
Melting Point, °C: decomposes [STR93]

160
Compound: Ammonium hexachloroplatinate(IV)
Synonyms: ammonium chloroplatinate
Formula: $(NH_4)_2PtCl_6$
Molecular Formula: $Cl_6H_8N_2Pt$
Molecular Weight: 443.873
CAS RN: 16919-58-7
Properties: cub, reddish orange cryst or yellow
    powd [KIR82] [MER89]
Solubility: g/100g H_2O: 0.289 (0°C), 0.374 (10°C),
    0.499 (20°C), 0.637 (30°C), 0.815 (40°C), 1.44
    (60°C), 2.16 (80°C), 2.61 (90°C), 3.36 (100°C)
    [LAN85]; i alcohol [HAW93]
Density, g/cm³: 3.065 [ALD93]
Melting Point, °C: decomposes >380 [KIR81]

161
Compound: Ammonium hexachlororhodate(III)
    monohydrate
Formula: $(NH_4)_3RhCl_6 \cdot H_2O$
Molecular Formula: $Cl_6H_{14}N_3ORh$
Molecular Weight: 387.752
CAS RN: 15336-18-2
Properties: red hygr cryst [STR93] [ALF95]

162
Compound: Ammonium hexachlororuthenate(IV)
Formula: $(NH_4)_2RuCl_6$
Molecular Formula: $Cl_6H_8N_2Ru$
Molecular Weight: 349.863
CAS RN: 18746-63-9
Properties: red cryst [ALF93]

163
Compound: Ammonium hexacyanoferrate(II)
    monohydrate

Synonyms: ammonium ferrocyanide
Formula: $(NH_4)_4Fe(CN)_6 \cdot H_2O$
Molecular Formula: $C_6H_{18}FeN_{10}O$
Molecular Weight: 302.120
CAS RN: 14481-29-9
Properties: yellowish green powd; hygr; light sensitive [STR93] [ALD94]; a trihydrate with similar properties is listed in [CRC94] and [MER89]
Solubility: trihydrate s H_2O [MER89]
Melting Point, °C: decomposes [MER89]
Reactions: minus NH_3 on exposure to air and light [MER89]

164
Compound: Ammonium hexafluoroaluminate
Synonyms: ammonium aluminum fluoride
Formula: $(NH_4)_3AlF_6$
Molecular Formula: $AlF_6H_{12}N_3$
Molecular Weight: 195.087
CAS RN: 7784-19-2
Properties: white powd; cub; does not attack glass; preparation: NH_4F and $Al(OH)_3$; uses: preparation of pure NH_4F [STR93] [MER89]
Solubility: s H_2O [MER89]
Density, g/cm³: 1.78 [MER89]
Melting Point, °C: thermally stable to above 100 [MER89]

165
Compound: Ammonium hexafluorogallate
Formula: $(NH_4)_3GaF_6$
Molecular Formula: $F_6GaH_{12}N_3$
Molecular Weight: 237.828
CAS RN: 14639-94-2
Properties: octahedra; preparation: reaction of $Ga(OH)_3$, HF, NH_4F; uses: preparation of GaF_3 [MER89]
Reactions: heating in air forms Ga_2O_3, forms GaN if heated in a vacuum at 200°C [MER89]

166
Compound: Ammonium hexafluorogermanate
Formula: $(NH_4)_2GeF_6$
Molecular Formula: $F_6GeH_8N_2$
Molecular Weight: 222.677
CAS RN: 16962-47-3
Properties: white cryst [HAW93]
Solubility: s H_2O; i alcohol [HAW93]
Density, g/cm³: 2.564 [HAW93]
Melting Point, °C: 380 (sublimes) [HAW93]

167
Compound: Ammonium hexafluorophosphate
Formula: $(NH_4)PF_6$
Molecular Formula: F_6H_4NP
Molecular Weight: 163.003
CAS RN: 16941-11-0
Properties: white cryst; square leaflets or tables; cub [MER89] [STR93]
Solubility: 74.8g/100mL H_2O (20°C) [MER89]
Density, g/cm³: 2.180 [MER89]
Melting Point, °C: decomposes 58 [CIC73]

168
Compound: Ammonium hexafluorosilicate
Synonyms: cryptohalite
Formula: $(NH_4)_2SiF_6$
Molecular Formula: $F_6H_8N_2Si$
Molecular Weight: 178.153
CAS RN: 16919-19-0
Properties: white odorless cryst powd; cub or trig; used in soldering flux, to etch glass, and in pesticides [MER89]
Solubility: s H_2O; i alcohol [MER89] [HAW93]
Density, g/cm³: 2.011 [ALD93]
Melting Point, °C: decomposes [ALF93]

169
Compound: Ammonium hexafluorotitanate dihydrate
Formula: $(NH_4)_2TiF_6 \cdot 2H_2O$
Molecular Formula: $F_6H_{12}N_2O_2Ti$
Molecular Weight: 233.964
CAS RN: 16962-40-6
Properties: white cryst [STR93]
Melting Point, °C: decomposes [CRC94]

170
Compound: Ammonium hydrogen acetate
Synonyms: ammonium acetate double salt
Formula: $(NH_4)H(CH_3COO)_2$
Molecular Formula: $C_4H_{11}NO_4$
Molecular Weight: 137.136
CAS RN: 631-61-8
Properties: prepared by dissolving ammonium acetate in hot acetic acid; the product crystallizes as long, deliq needles [KIR78]
Solubility: v s H_2O [KIR78]
Melting Point, °C: 66 [KIR78]

171
Compound: Ammonium hydrogen arsenate

Formula: $(NH_4)_2HAsO_4$
Molecular Formula: $AsH_9N_2O_4$
Molecular Weight: 176.004
CAS RN: 7784-44-3
Properties: white powd, efflorescing in air with loss of NH_3 [HAW93]
Solubility: s H_2O, decomposed by hot H_2O [HAW93]
Density, g/cm³: 1.99 [HAW93]
Melting Point, °C: decomposes [CRC77]

172

Compound: Ammonium hydrogen borate trihydrate
Formula: $(NH_4)HB_4O_7 \cdot 3H_2O$
Molecular Formula: $B_4H_{11}NO_{10}$
Molecular Weight: 228.332
CAS RN: 10135-84-9
Properties: colorless cryst; effloresces evolving NH_3; obtained by reaction of NH_4OH and boric acid, then cryst of product; used to fireproof materials [HAW93]
Solubility: s H_2O [HAW93]
Density, g/cm³: 2.38-2.95 [HAW93]

173

Compound: Ammonium hydrogen carbonate
Synonyms: ammonium bicarbonate
Formula: NH_4HCO_3
Molecular Formula: CH_5NO_3
Molecular Weight: 79.056
CAS RN: 1066-33-7
Properties: white cryst; vapor pressure, kPa: 59 (25.4°C), 122 (34.2°C), 201 (40.7°C), 278 (45.0°C), 395 (50.0°C), 72.1 (54.0°C), 108.4 (59.2°C); manufactured by passing CO_2 gas through NH_4OH solution, which evolves heat, followed by crystallization of ammonium bicarbonate; used as a leavening agent for cookies, crackers, and in fire-extinguisher materials [HAW93] [KIR78]
Solubility: g/100g soln, H_2O: 10.6 (0°C); 19.9 (25°C); 78.0 (100°C); equilibrium solid phase, NH_4HCO_3 [KRU93]; i alcohol [HAW93]
Density, g/cm³: 1.586 [KIR78]
Melting Point, °C: 107.5 (v rapid heating) [MER89]
Boiling Point, °C: sublimes ~60 with decomposition [MER89]
Reactions: decomposed by hot H_2O [MER89]

174

Compound: Ammonium hydrogen citrate

Synonyms: diammonium citrate
Formula: $(NH_4)_2HC_6H_5O_7$
Molecular Formula: $C_6H_{14}N_2O_7$
Molecular Weight: 226.186
CAS RN: 3012-65-5
Properties: granules or cryst; white; stable in air; used in metal cleaning applications [KIR78] [MER89]

$$CH_2COONH_4$$
$$|$$
$$HO-C-COOH$$
$$|$$
$$CH_2COONH_4$$

Solubility: 100 g/100 mL H_2O at 25°C; sl s alcohol; i ether [KIR78]
Density, g/cm³: 1.48 [MER89]

175

Compound: Ammonium hydrogen fluoride
Synonyms: ammonium bifluoride
Formula: NH_4HF_2
Molecular Formula: F_2H_5N
Molecular Weight: 57.044
CAS RN: 1341-49-7
Properties: white deliq flakes; ortho-rhomb cryst; enthalpy of fusion 19.1 kJ/mol; enthalpy of vaporization 65.3 kJ/mol; enthalpy of solution 20.3 kJ/mol; enthalpy of dissociation to form NH_3 and HF 141.4 kJ/mol; can be prepared by dehydration of NH_4F solutions; can be used as a less hazardous substitute for HF [KIR78] [MER89] [HAW93]
Solubility: 41.5% in H_2O at 25°C; 1.73% in 90% alcohol at 25°C [KIR78] [HAW93]
Density, g/cm³: 1.50 [KIR78]
Melting Point, °C: 126.1 [KIR78]
Boiling Point, °C: 239.5 [KIR78]

176

Compound: Ammonium hydrogen oxalate hemihydrate
Synonyms: ammonium binoxalate
Formula: $NH_4HC_2O_4 \cdot 1/2H_2O$
Molecular Formula: $C_2H_6NO_{4.5}$
Molecular Weight: 116.022
CAS RN: 37541-72-3
Properties: monohydrate: rhomb; uses: remove ink stains [MER89]
Solubility: monohydrate: s 25 parts H_2O; s sl alcohol [MER89]
Melting Point, °C: 220, decomposes [ALD93]

177

Compound: Ammonium hydrogen oxalate monohydrate
Synonyms: ammonium binoxalate monohydrate
Formula: $NH_4OOCCOOH \cdot H_2O$
Molecular Formula: $C_2H_7NO_5$
Molecular Weight: 125.081
CAS RN: 5972-72-5
Properties: colorless rhomb cryst; obtained from solution of NH_4OH and oxalic acid, then cryst; used to remove ink from fabrics [HAW93] [MER89]
Solubility: s 25 parts H_2O [MER89]
Density, g/cm³: 1.56 [MER89]
Melting Point, °C: decomposed by heating [HAW93]

178

Compound: Ammonium hydrogen phosphate
Synonyms: ammonium phosphate dibasic
Formula: $(NH_4)_2HPO_4$
Molecular Formula: $H_9N_2O_4P$
Molecular Weight: 132.055
CAS RN: 7783-28-0
Properties: cryst or powd; gradually loses about 8% NH_3 when exposed to air [MER89]
Solubility: g/100g soln, H_2O: 36.4 (0°C), 41.0 (25°C), 58.6 (100°C); equilibrium solid phase $(NH_4)_2HPO_4$ [KRU93]
Density, g/cm³: 1.619 [ALF93]
Melting Point, °C: 155, decomposes [ALF93]

179

Compound: Ammonium hydrogen phosphite monohydrate
Formula: $(NH_4)_2HPO_3 \cdot H_2O$
Molecular Formula: $H_{11}N_2O_4P$
Molecular Weight: 134.072
CAS RN: 51503-61-8
Properties: deliq cryst [MER89]
Solubility: s H_2O [MER89]
Density, g/cm³: 1.619 (anhydrous) [CRC94]
Melting Point, °C: decomposes 155 (anhydrous) [CRC94]

180

Compound: Ammonium hydrogen sulfate
Synonyms: ammonium bisulfate
Formula: $(NH_4)HSO_4$
Molecular Formula: H_5NO_4S
Molecular Weight: 115.111
CAS RN: 7803-63-6

Properties: white powd; cryst, deliq; used as catalyst in organic syntheses and in hair wave formulations [HAW93] [STR93] [MER89]
Solubility: 100g/100g H_2O (0°C) [CIC73]; i acetone, alcohol [HAW93]
Density, g/cm³: 1.78 [CIC73]
Melting Point, °C: 146.9 [CIC73]

181

Compound: Ammonium hydrogen sulfide
Synonyms: ammonium hydrosulfide, ammonium bisulfide
Formula: NH_4HS
Molecular Formula: H_5NS
Molecular Weight: 51.113
CAS RN: 12124-99-1
Properties: white; tetr or ortho-rhomb; vapor pressure 99.7 kPa at 32.1°C; readily sublimes; produced from stoichiometric amounts of NH_3 and H_2S at 0°C [MER89] [KIR78]
Solubility: 128.1g/100g H_2O (0°C), decomposes in hot H_2O [CIC73]; i ether, benzene [KIR78]
Density, g/cm³: 1.17 [CIC73]
Melting Point, °C: 118 [CIC73]
Reactions: decomposes to H_2S and NH_3 at room temp [MER89]

182

Compound: Ammonium hydrogen sulfite
Synonyms: ammonium bisulfite
Formula: $(NH_4)HSO_3$
Molecular Formula: H_5NO_3S
Molecular Weight: 99.111
CAS RN: 10192-30-0
Properties: cryst; available commercially only in solution form; uses: preservative [MER89]
Solubility: g/100g soln, H_2O: 72.2±0.4 (0°C), 78.41 (25°C), 85.4±0.7 (60°C); equilibrium solid phase, $(NH_4)_2S_2O_5$ [KRU93]
Density, g/cm³: 2.03 [CIC73]
Melting Point, °C: sublimes 150 [CIC73]

183

Compound: Ammonium hydrogen tartrate
Synonyms: ammonium bitartrate
Formula: $(NH_4)OOCCH(OH)CH(OH)COOH$
Molecular Formula: $C_4H_9NO_6$
Molecular Weight: 167.118
CAS RN: 3095-65-6
Properties: white odorless cryst; used in baking powd [HAW93] [MER89]

Solubility: g/100g H_2O: 1.00 (0°C), 1.88 (10°C), 2.70 (20°C) [LAN85]; s acids, alkalies; i alcohol [HAW93]
Density, g/cm³: 1.68 [MER89]
Melting Point, °C: decomposes [CRC94]

184
Compound: Ammonium hydrogen tetraborate dihydrate
Formula: $(NH_4)HB_4O_7 \cdot 2H_2O$
Molecular Formula: $B_4H_9NO_9$
Molecular Weight: 210.317
CAS RN: 12228-86-3
Properties: cryst [ALF93]
Melting Point, °C: decomposes (tetraborate) [CRC94]

185
Compound: Ammonium hydroxide
Synonyms: ammonia solution
Formula: NH_4OH
Molecular Formula: H_5NO
Molecular Weight: 35.046
CAS RN: 1336-21-6
Properties: colorless liq; strong odor of ammonia; dissolved NH_3 concentration ranges up to 30% [HAW93]
Density, g/cm³: 0.900 [ALD94]

186
Compound: Ammonium hypophosphite
Formula: $NH_4H_2PO_2$
Molecular Formula: H_6NO_2P
Molecular Weight: 83.028
CAS RN: 7803-65-8
Properties: hygr; deliq cryst or white granules; uses: catalyst in manufacture of polyamide [MER89]
Solubility: g/100mL H_2O: 83 (room temp) [KRU93]; s alcohol [HAW93]
Density, g/cm³: 1.634 [CRC94]
Melting Point, °C: on heating, decomposes evolving phosphine gas [MER89]

187
Compound: Ammonium iodate
Formula: NH_4IO_3
Molecular Formula: H_4INO_3
Molecular Weight: 192.941
CAS RN: 13446-09-8
Properties: white, granular powd; oxidizing agent [HAW93]

Solubility: g/100mL H_2O: 2.06 (15°C), 14.5 (101°C) [CRC94]
Density, g/cm³: 3.309 [CRC94]
Melting Point, °C: 150, decomposes [CRC94] [ALF93]

188
Compound: Ammonium iodide
Formula: NH_4I
Molecular Formula: H_4IN
Molecular Weight: 144.943
CAS RN: 12027-06-4
Properties: white, odorless, very hygr; tetr; yellow to brown on exposure to air due to liberation of some iodine; vapor pressure, kPa: 31.3 (360°C), 54.1 (380°C), 89.8 (400°C), 101.1 (405°C); enthalpy of fusion 21.00 kJ/mol; formed by reacting NH_3 with I_2; finds some use in photography [KIR78] [MER89] [CRC93]
Solubility: g/100g soln, H_2O: 60.55 (0°C), 64.65 (25°C), 71.3 (100°C); equilibrium solid phase NH_4I [KRU93]; g/100g H_2O: 154.2 (0°C), 172.3 (20°C), 250.3 (100°C) [KIR78]; s alcohol [HAW93]
Density, g/cm³: 2.514 [CIC73]
Melting Point, °C: partly decomposes and sublimes if heated [MER89]
Boiling Point, °C: 220 in vacuum [CIC73]

189
Compound: Ammonium magnesium chloride hexahydrate
Synonyms: carnallite
Formula: $NH_4Cl \cdot MgCl_2 \cdot 6H_2O$
Molecular Formula: $Cl_3H_{16}MgNO_6$
Molecular Weight: 256.793
CAS RN: 39733-35-2
Properties: deliq cryst [MER89]
Solubility: s in 6 parts H_2O [MER89]
Density, g/cm³: 1.456 [CRC77]
Reactions: minus $2H_2O$, 100°C [CRC94]

190
Compound: Ammonium mercuric chloride dihydrate
Synonyms: mercuric ammonium chloride
Formula: $(NH_4)_2HgCl_4 \cdot 2H_2O$
Molecular Formula: $Cl_4H_{12}HgN_2O_2$
Molecular Weight: 440.852
CAS RN: 33445-15-7
Properties: powd; uses: ointment for chronic eczema, antifungal [MER89]

Solubility: s H_2O [MER89]

191

Compound: Ammonium metatungstate hexahydrate
Formula: $(NH_4)_6W_7O_{24} \cdot 6H_2O$
Molecular Formula: $H_{36}N_6O_{30}W_7$
Molecular Weight: 1887.188
CAS RN: 12028-48-7
Properties: white cryst; formula also given as $(NH_4)_6H_2W_{12}O_{40}$ and $(NH_4)_6W_{12}O_{39} \cdot 4H_2O$; preparation: reaction of NH_4OH with tungstic acid; uses: to prepare tungsten alloys and ammonium phosphotungstate [HAW93] [STR93] [ALD94]
Solubility: s H_2O; i alcohol [HAW93]

192

Compound: Ammonium metavanadate
Synonyms: ammonium vanadate
Formula: NH_4VO_3
Molecular Formula: H_4NO_3V
Molecular Weight: 116.979
CAS RN: 7803-55-6
Properties: white or sl yellow cryst powd; loses H_2O and NH_3 on heating; obtained by precipitation with NH_4Cl from alkaline V_2O_5 solutions; used as a catalyst in dyes and varnishes [HAW93] [MER89]
Solubility: g/100g H_2O: 0.48 (20°C), 0.84 (30°C), 1.32 (40°C), 2.42 (60°C) [LAN85]
Density, g/cm³: 2.326 [HAW93]
Melting Point, °C: decomposes, 210 [HAW93]

193

Compound: Ammonium molybdate tetrahydrate
Synonyms: ammonium molybdate(VI)
Formula: $(NH_4)_6Mo_7O_{24} \cdot 4H_2O$
Molecular Formula: $H_{32}Mo_7N_6O_{28}$
Molecular Weight: 1235.857
CAS RN: 12054-85-2
Properties: colorless, sl greenish or yellowish cryst; obtained by crystallization from a solution of MoO_3 with excess NH_3; used to prepare specialty catalysts [KIR81] [MER89]
Solubility: s 2.3 parts H_2O [MER89]; i alcohol [HAW93]
Density, g/cm³: 2.498 [ALD93]
Melting Point, °C: decomposes [HAW93]

194

Compound: Ammonium nickel chloride hexahydrate
Synonyms: nickel ammonium chloride
Formula: $NH_4Cl \cdot NiCl_2 \cdot 6H_2O$
Molecular Formula: $Cl_3H_{16}NNiO_6$
Molecular Weight: 291.181
CAS RN: 16122-03-5
Properties: green cryst; deliq; obtained from salt solutions by crystallization; used as a dye mordant, and for metal finishing [KIR81] [HAW93]
Solubility: s H_2O [HAW93]
Density, g/cm³: 1.65 [HAW93]

195

Compound: Ammonium nickel sulfate hexahydrate
Synonyms: nickel ammonium sulfate
Formula: $(NH_4)_2SO_4 \cdot NiSO_4 \cdot 6H_2O$
Molecular Formula: $H_{20}N_2NiO_{14}S_2$
Molecular Weight: 394.989
CAS RN: 7785-20-8
Properties: bluish green cryst; decomposes on heating; obtained from an aq solution by crystallization; used as a dye mordant and in metal finishing [KIR81] [MER89] [HAW93] [ALF93]
Solubility: g$(NH_4)_2Ni(SO_4)_2$/100g H_2O: 1.00 (0°C), 4.00 (10°C), 6.50 (20°C), 9.20 (30°C), 12.0 (40°C), 17.0 (60°C) [LAN85]; i alcohol [HAW93]
Density, g/cm³: 1.923 [MER89]

196

Compound: Ammonium nitrate
Formula: NH_4NO_3
Molecular Formula: $H_4N_2O_3$
Molecular Weight: 80.043
CAS RN: 6484-52-2
Properties: white transparent hygr cryst; vapor pressure of saturated NH_4NO_3 solutions, kPa: 0.85 (10°C), 1.5 (20°C), 2.5 (30°C), 3.9 (40°C); 5 cryst forms, stable at: α, < -18°C; β, -18 to 32.1°C; γ, 32.1 to 84.2°C; δ, 84.2 to 125.2°C; ϵ, 125.7 to 169.6 °C; enthalpy of fusion 6.40 kJ/mol; enthalpy of neutralization 51.8 kJ/mol; manufactured by neutralization of HNO_3 solutions with NH_3; used as a fertilizer and in explosives [MER89] [KIR78] [CRC93]
Solubility: g/100g soln, H_2O: 54.2 (0°C), 68.2 (25°C), 90.3 (100°C); equilibrium solid phase NH_4NO_3 [KRU93]
Density, g/cm³: 1.725 [CIC73]

Melting Point, °C: 169.6 [CRC93]
Boiling Point, °C: 210 (11mm Hg) [CIC73]
Reactions: decomposes ~210°C to $H_2O + N_2O$ [MER89]
Thermal Expansion Coefficient: coefficient of expansion is 0.000920 (20°C), 0.001113 (100°C) [KIR78]

197

Compound: Ammonium nitrite
Formula: NH_4NO_2
Molecular Formula: $H_4N_2O_2$
Molecular Weight: 64.044
CAS RN: 13446-48-5
Properties: white-yellowish cryst; uncertain stability; can be made by adding solution of barium nitrite to ammonium sulfate solution [KIR78] [CRC77]
Solubility: g/100g soln, H_2O: 56.0 (1.4°C), 64.3 (19.15°C); equilibrium solid phase NH_4NO_2 [KRU93]
Density, g/cm³: 1.69 [CIC73]
Melting Point, °C: explodes 60-70, producing N_2, H_2O and other products [KIR78] [CIC73]

198

Compound: Ammonium nitroferricyanide
Synonyms: ammonium nitroprusside
Formula: $(NH_4)_2Fe(CN)_5NO$
Molecular Formula: $C_5H_8FeN_8O$
Molecular Weight: 252.017
CAS RN: 14402-70-1
Properties: red to brownish red cryst [MER89]
Solubility: s H_2O, alcohol [MER89]

199

Compound: Ammonium O,O-diethyldithiophosphate
Formula: $(C_2H_5O)_2P(S)SNH_4$
Molecular Formula: $C_4H_{14}NO_2PS_2$
Molecular Weight: 203.267
CAS RN: 1068-22-0
Properties: cryst [ALF95]
Melting Point, °C: 164-165 [ALF95]

200

Compound: Ammonium oleate
Synonyms: ammonium soap
Formula: $CH_3(CH_2)_7CH=CH(CH_2)_7COONH_4$
Molecular Formula: $C_{18}H_{37}NO_2$
Molecular Weight: 299.498

CAS RN: 544-60-5
Properties: yellowish brown paste, softens at 10-13°C; used to emulsify products, and in cosmetics [MER89] [HAW93]
Solubility: s H_2O (27°C); s sl acetone [MER89]
Melting Point, °C: 21-22 [MER89]

201

Compound: Ammonium oxalate
Synonyms: ethanedioic acid, diammonium salt
Formula: $(NH_4)_2C_2O_4$
Molecular Formula: $C_2H_8N_2O_4$
Molecular Weight: 124.097
CAS RN: 1113-38-8
Properties: colorless cryst; used as an analytical chemistry reagent, to manufacture oxalates, for removing rust and scale [HAW93]
Solubility: g/100g soln, H_2O: 2.31 (0°C), 4.95 (25°C), 25.73 (100°C) [KRU93]
Density, g/cm³: 1.5 [ALF93]

202

Compound: Ammonium oxalate monohydrate
Synonyms: ethanedioic acid, diammonium salt monohydrate
Formula: $(NH_4)_2C_2O_4 \cdot H_2O$
Molecular Formula: $C_2H_{10}N_2O_5$
Molecular Weight: 142.111
CAS RN: 6009-70-7
Properties: white granular odorless cryst; ortho-rhomb; used in analytical chemistry, and to remove rust [MER89] [ALF93] [HAW93]
Solubility: 1g/20mL H_2O, 2.6mL boiling H_2O [MER89]
Density, g/cm³: 1.502 [HAW93]
Melting Point, °C: decomposes on heating [HAW93]

203

Compound: Ammonium palmitate
Synonyms: hexadecanoic acid, ammonium salt
Formula: $CH_3(CH_2)_{14}COONH_4$
Molecular Formula: $C_{16}H_{35}NO_2$
Molecular Weight: 273.460
CAS RN: 593-26-0
Properties: yellowish white powd, softens at 3-4°C [MER89]
Solubility: s H_2O [MER89]; s hot alcohols, benzene [HAW93]
Melting Point, °C: 21-23 [MER89]

204
Compound: Ammonium pentaborate tetrahydrate
Formula: $(NH_4)B_5O_8 \cdot 4H_2O$
Molecular Formula: $B_5H_{12}NO_{12}$
Molecular Weight: 272.150
CAS RN: 12229-12-8
Properties: two cryst forms: ortho-rhomb (α) and monocl (β); very stable with respect to loss of NH_3, losing less than 1% NH_3 at 50°C and only 2% at 200°C; prepared from an aq solution of NH_3 and boric acid; used in flameproofing formulations [HAW93] [KIR78]
Solubility: % anhydrous by weight, H_2O: 4.00 (0°C), 8.03 (25°C), 14.4 (50°C), 30.3 (90°C) [KIR78]
Density, g/cm³: 1.58 [KIR78]
Reactions: minus 75% of the H_2O content at 50°C [KIR78]

205
Compound: Ammonium pentachlororhodate(III) monohydrate
Formula: $(NH_4)_2RhCl_5 \cdot H_2O$
Molecular Formula: $Cl_5H_{10}N_2ORh$
Molecular Weight: 334.261
CAS RN: 63771-33-5
Properties: red cryst [ALF93]
Melting Point, °C: 210-230, decomposes [ALF93]

206
Compound: Ammonium pentachlororuthenate(III) monohydrate
Formula: $(NH_4)_2RuCl_5 \cdot H_2O$
Molecular Formula: $Cl_5H_{10}N_2ORu$
Molecular Weight: 332.425
CAS RN: 68133-88-0
Properties: cryst powd [ALF93]

207
Compound: Ammonium pentachlorozincate
Synonyms: zinc ammonium chloride
Formula: $(NH_4)_3ZnCl_5$
Molecular Formula: $Cl_5H_{12}N_3Zn$
Molecular Weight: 296.769
CAS RN: 14639-98-6
Properties: ortho-rhomb cryst; hygr; uses: manufacture of dry cell batteries, welding flux, soldering, galvanizing [MER89]
Solubility: v s H_2O [MER89]
Density, g/cm³: 1.81 [MER89]
Melting Point, °C: sublimes 340 [MER89]

208
Compound: Ammonium perchlorate
Formula: NH_4ClO_4
Molecular Formula: ClH_4NO_4
Molecular Weight: 117.490
CAS RN: 7790-98-9
Properties: white cryst; oxidizing agent; ortho-rhomb, a = 0.9202 nm, b = 0.5816 nm, c = 0.7449 nm; prepared from NH_4OH, HCl and sodium chlorate with subsequent cryst; cub >240°C, a = 0.763 nm; used as an oxidizer in rocket propellants [CIC73] [HAW93] [ALD93] [KIR79] [ALF93]
Solubility: g/100g soln, H_2O: 10.8 (0°C), 19.8 (25°C), 46.9 (100°C); mol/kg H_2O: 1.019 (0°C), 2.122 (25°C) [KRU93]
Density, g/cm³: 1.95 [CIC73]
Melting Point, °C: can explode; decomposed by heating [MER89] [ALD93]
Reactions: transition from ortho-rhomb to cub at 513K [KIR79]

209
Compound: Ammonium permanganate
Formula: NH_4MnO_4
Molecular Formula: H_4MnNO_4
Molecular Weight: 136.975
CAS RN: 13446-10-1
Properties: dark purple; rhomb; oxidizing agent [KIR78]
Solubility: g/100g H_2O: 8 (15°C), 86 (25°C) [KIR78]
Density, g/cm³: 2.22 [KIR78]
Melting Point, °C: decomposes above 70 [KIR78]

210
Compound: Ammonium peroxydisulfate
Synonyms: ammonium persulfate
Formula: $(NH_4)_2S_2O_8$
Molecular Formula: $H_8N_2O_8S_2$
Molecular Weight: 228.204
CAS RN: 7727-54-0
Properties: odorless plate-like or prismatic cryst; monocl; strong oxidizing agent [MER89]
Solubility: g/100g soln, H_2O: 37.0 (0°C), 45.5 (25°C), 62.0 (80°C) [KRU93]
Density, g/cm³: 1.982 [ALF93]
Melting Point, °C: decomposes on heating forming O_2 and $(NH_4)_2S_2O_7$ [MER89]

211
Compound: Ammonium perrhenate

Synonyms: ammonium perrhenate(VII)
Formula: NH_4ReO_4
Molecular Formula: H_4NO_4Re
Molecular Weight: 268.244
CAS RN: 13598-65-7
Properties: colorless powd; weak oxidizing agent [HAW93] [ALF93]
Solubility: sl s in cold H_2O; s hot H_2O [HAW93]
Density, g/cm^3: 3.97 [ALD93]
Melting Point, °C: decomposes 365 [HAW93]

212
Compound: Ammonium phosphate dibasic
Synonyms: diammonium hydrogen phosphate
Formula: $(NH_4)_2HPO_4$
Molecular Formula: $H_9N_2O_4P$
Molecular Weight: 132.07
CAS RN: 7883-28-0
Properties: odorless cryst or powd; used to fireproof textiles, paper, wood and vegetable fibers, to impregnate lamp wicks and in soldering fluxes [MER89]
Solubility: 1g/1.7 mL H_2O, 1g in 0.5 mL boiling H_2O [MER89]
Density, g/cm^3: 1.619 [ALD94]
Melting Point, °C: decomposes 155 [CRC94]

213
Compound: Ammonium phosphomolybdate
Synonyms: ammonium molybdophosphate
Formula: $(NH_4)_3PO_4 \cdot 12MoO_3$
Molecular Formula: $H_{12}Mo_{12}N_3O_{40}P$
Molecular Weight: 1876.345
CAS RN: 54723-94-3
Properties: heavy yellow cryst powd [MER89]
Solubility: 0.2g/L H_2O (20°C) [MER89]
Melting Point, °C: decomposes [ALF93]

214
Compound: Ammonium phosphotungstate dihydrate
Synonyms: ammonium tungstophosphate
Formula: $(NH_4)_3PO_4 \cdot 12WO_3 \cdot 2H_2O$
Molecular Formula: $H_{16}N_3O_{42}PW_{12}$
Molecular Weight: 2967.176
CAS RN: 1311-90-6
Properties: microcryst powd [MER89]
Solubility: 0.15g/L H_2O (20°C) [MER89]

215
Compound: Ammonium picrate
Synonyms: ammonium carbazoate

Formula: $(NH_4)C_6H_2N_3O_7$
Molecular Formula: $C_6H_6N_4O_7$
Molecular Weight: 246.137
CAS RN: 131-74-8
Properties: bright yellow; ortho-rhomb; explodes easily from heat or shock [MER89]
Solubility: 1g/100mL H_2O (20°C) [MER89]
Density, g/cm^3: 1.72 [MER89]
Reactions: explodes at 423 [CRC94]

216
Compound: Ammonium polysulfide
Formula: $(NH_4)_2S_x$
CAS RN: 9080-17-5
Properties: exists only in solution; yellow unstable solution with an odor of H_2S; prepared by dissolving H_2S gas in 28% NH_4OH solution, then dissolving excess sulfur; used as an analytical chemistry reagent, and as an insect spray [HAW93]
Reactions: decomposed by acids, evolving H_2S [HAW93]

217
Compound: Ammonium salicylate
Synonyms: salicylic acid, monoammonium salt
Formula: $(NH_4)C_7H_5O_3$
Molecular Formula: $C_7H_9NO_3$
Molecular Weight: 155.153
CAS RN: 528-94-9
Properties: odorless lustrous cryst, or white cryst powd; discolors on exposure to light; loses some NH_3 on long exposure to air [MER89]
Solubility: 1g/1mL H_2O [MER89]
Boiling Point, °C: sublimes [CRC94]

218
Compound: Ammonium selenate
Formula: $(NH_4)_2SeO_4$
Molecular Formula: $H_8N_2O_4Se$
Molecular Weight: 179.035
CAS RN: 7783-21-3
Properties: white powd; colorless; monocl cryst; used to mothproof [HAW93] [MER89] [STR93]
Solubility: g/100g soln, H_2O: 54.02 (25°C) [KRU93]; i alcohol [HAW93]
Density, g/cm^3: 2.194 [MER89]
Melting Point, °C: decomposed by heating [MER89]

219

Compound: Ammonium selenite

Formula: $(NH_4)_2SeO_3$

Molecular Formula: $H_8N_2O_3Se$

Molecular Weight: 163.035

CAS RN: 7783-19-9

Properties: white or sl reddish cryst; deliq; used to color glass [HAW93] [MER89]

Solubility: g/100g soln, H_2O: 49.21 (1°C), 54.70 (25°C), 69.08 (70°C); equilibrium solid phase, $(NH_4)_2SeO_3 \cdot H_2O$ [KRU93]

Melting Point, °C: decomposes [MER89]

220

Compound: Ammonium sesquicarbonate

Synonyms: hartshorn

Formula: $NH_2COONH_4 \cdot NH_4HCO_3$

Molecular Formula: $C_2H_{11}N_3O_5$

Molecular Weight: 157.126

CAS RN: 10361-29-2

Properties: mixture of ammonium bicarbonate and ammonium carbamate; colorless; hard translucent cryst with ammonia odor; obtained from a mixture of ammonium sulfate and calcium carbonate by sublimation; used in baking powd, as a mordant in dyeing and as an expectorant; changes to bicarbonate in air [MER89] [HAW93]

Solubility: slowly s in 4 parts H_2O, decomposes in hot H_2O, evolving NH_3 and CO_2 [MER89] [HAW93]

Melting Point, °C: volatilizes, ~60 [MER89]

221

Compound: Ammonium stearate

Synonyms: octadecanoic acid, ammonium salt

Formula: $CH_3(CH_2)_{16}COONH_4$

Molecular Formula: $C_{18}H_{39}NO_2$

Molecular Weight: 301.514

CAS RN: 1002-89-7

Properties: yellowish white powd, softens at 1.7-4.4°C; used in vanishing creams, brushless shaving and other cosmetics [MER89] [HAW93]

Solubility: s H_2O [MER89]; s hot toluene [HAW93]

Density, g/cm³: 0.89 [HAW93]

Melting Point, °C: 21-24 [MER89]

222

Compound: Ammonium sulfamate

Synonyms: sulfamic acid, monoammonium salt

Formula: $(NH_4)NH_2SO_3$

Molecular Formula: $H_6N_2O_3S$

Molecular Weight: 114.125

CAS RN: 7773-06-0

Properties: white hygr cryst; made by reacting urea with fuming sulfuric acid; used to flameproof textiles, and in metal finishing [HAW93] [MER89]

Solubility: v s H_2O [MER89]

Melting Point, °C: 131 [MER89]

Boiling Point, °C: decomposes, 160 [HAW93]

223

Compound: Ammonium sulfate

Synonyms: mascagnite

Formula: $(NH_4)_2SO_4$

Molecular Formula: $H_8N_2O_4S$

Molecular Weight: 132.141

CAS RN: 7783-20-2

Properties: brownish gray to white odorless orthorhomb cryst; manufactured from NH_3 and sulfuric acid; used as a nitrogen fertilizer, for water treatment, and as a food additive [MER89] [HAW93]

Solubility: g/100g soln, H_2O: 41.35 (0°C); 43.30 (25°C), 50.61 (100°C); equilibrium solid phase, $(NH_4)_2SO_4$ [KRU93]; 70.6g/100g H_2O (0°C), 103.8g/100g H_2O (100°C); i alcohol, acetone [KIR78]

Density, g/cm³: 1.77 [MER89]

Melting Point, °C: decomposes >280 [MER89]

224

Compound: Ammonium sulfide

Formula: $(NH_4)_2S$

Molecular Formula: H_8N_2S

Molecular Weight: 68.143

CAS RN: 12135-76-1

Properties: yellowish orange liq; cryst below -18°C; stable only below -18°C, at higher temperatures loses NH_3 to form NH_4HS and polysulfides; made from NH_3 and H_2S; used in textile industry and photography [STR93] [MER89] [KIR78] [HAW93]

Solubility: s H_2O, alcohol, alkalies [HAW93]

Density, g/cm³: 0.997 [ALD94]

Melting Point, °C: decomposes [HAW93]

Reactions: evolves NH_3 to form hydrosulfide above -18°C [KIR78]

225

Compound: Ammonium sulfite monohydrate

Synonyms: sulfurous acid, diammonium salt

Formula: $(NH_4)_2SO_3 \cdot H_2O$
Molecular Formula: $H_{10}N_2O_4S$
Molecular Weight: 134.156
CAS RN: 7783-11-1
Properties: colorless cryst; loses H_2O and gradually oxidized to $(NH_4)_2SO_4$ when heated in air; hygr; used in medicine, metal lubricants, as a chemical reducing agent [MER89] [HAW93]
Solubility: g anhydrous/100g soln, H_2O: 32.40 (0°C), 39.29 (25°C), 60.44 (100°C); equilibrium phase: monohydrate (0,25°C), anhydr (100°C) [KRU93]
Density, g/cm³: 1.41 [HAW93]
Melting Point, °C: sublimes with decomposition, 150 [HAW93]

226
Compound: Ammonium tartrate
Synonyms: tartaric acid, diammonium salt
Formula: $(NH_4)_2C_4H_4O_6$
Molecular Formula: $C_4H_{12}N_2O_6$
Molecular Weight: 184.148
CAS RN: 3164-29-2
Properties: white cryst; decomposes when heated; used in medicine and in industry [HAW93]
Solubility: g/100g H_2O: 45.0 (0°C), 55.0 (10°C), 63.0 (20°C), 70.5 (30°C), 76.5 (40°C), 86.9 (60°C) [LAN85]; s alcohol [HAW93]
Density, g/cm³: 1.601 [HAW93]
Melting Point, °C: decomposes [CRC94]

227
Compound: Ammonium tellurate
Synonyms: ammonium tellurate(VI)
Formula: $(NH_4)_2TeO_4$
Molecular Formula: $H_8N_2O_4Te$
Molecular Weight: 227.675
CAS RN: 13453-06-0
Properties: -60 mesh with 99.5% purity; white powd [ALF93] [STR93]
Density, g/cm³: 3.024 [STR93]
Melting Point, °C: decomposes [STR93]

228
Compound: Ammonium tetraborate tetrahydrate
Synonyms: ammonium borate
Formula: $(NH_4)_2B_4O_7 \cdot 4H_2O$
Molecular Formula: $B_4H_{16}N_2O_{11}$
Molecular Weight: 263.377
CAS RN: 12228-87-4

Properties: tetr colorless cryst; unstable, has an appreciable ammonia vapor pressure; uses: fireproofing wood and textiles [MER89] [STR93] [KIR78]
Solubility: % anhydrous by weight (H_2O): 3.75 (0°C), 9.00 (25°C), 21.2 (50°C), 52.7 (90°C) [KIR78]
Density, g/cm³: 1.58 [KIR78]
Melting Point, °C: decomposes [CRC94]

229
Compound: Ammonium tetrachloroaluminate
Synonyms: aluminum ammonium chloride
Formula: NH_4AlCl_4
Molecular Formula: $AlCl_4H_4N$
Molecular Weight: 186.831
CAS RN: 7784-14-7
Properties: white cryst; preparation: from $AlCl_3$ and NH_4Cl; finds use in treating furs [MER89] [HAW93]
Solubility: s H_2O, ether [MER89]
Melting Point, °C: 304 [MER89]

230
Compound: Ammonium tetrachloroaurate(III) hydrate
Formula: $(NH_4)AuCl_4 \cdot xH_2O$
Molecular Formula: $AuCl_4H_4N$ (anhydrous)
Molecular Weight: 356.816 (anhydrous)
CAS RN: 13874-04-9
Properties: yellow cryst [STR93]
Melting Point, °C: 520 [ALD93]

231
Compound: Ammonium tetrachloropalladate(II)
Formula: $(NH_4)_2PdCl_4$
Molecular Formula: $Cl_4H_8N_2Pd$
Molecular Weight: 284.308
CAS RN: 13820-40-1
Properties: reddish brown powd; olive green cryst [STR93] [ALF93]
Density, g/cm³: 2.17 [ALD93]
Melting Point, °C: decomposes [CRC77]

232
Compound: Ammonium tetrachloroplatinate(II)
Synonyms: ammonium platinous chloride
Formula: $(NH_4)_2PtCl_4$
Molecular Formula: $Cl_4H_8N_2Pt$
Molecular Weight: 372.968
CAS RN: 13820-41-2

Properties: dark ruby red cryst; used in photography [HAW93] [MER89]
Solubility: s H_2O [MER89]; i alcohol [HAW93]
Density, g/cm³: 2.936 [ALD93]
Melting Point, °C: decomposes, 140-150 [HAW93]

233
Compound: Ammonium tetrafluoroantimonate(III)
Formula: NH_4SbF_4
Molecular Formula: F_4H_4NSb
Molecular Weight: 215.789
CAS RN: 14972-90-8
Properties: white powd [STR93]

234
Compound: Ammonium tetranitrodiamminecobaltate(III)
Synonyms: Erdmann's salt
Formula: $NH_4[Co(NH_3)_2(NO_2)_4]$
Molecular Formula: $CoH_{10}N_7O_8$
Molecular Weight: 295.054
CAS RN: 13600-89-0
Properties: reddish, pale brown rhomb [KIR79] [MER52]
Solubility: s H_2O [LID94]
Density, g/cm³: 1.876 [KIR79]

235
Compound: Ammonium tetrathiocyanodiammonochromate(III) monohydrate
Synonyms: Reinecke salt
Formula: $NH_4[Cr(NH_3)_2(SCN)_4] \cdot H_2O$
Molecular Formula: $C_4H_{12}CrN_7OS_4$
Molecular Weight: 354.446
CAS RN: 13573-16-5
Properties: dark red cryst or red powd; can be produced by fusion of ammonium thiocyanate with ammonium dichromate; used to precipitate primary and secondary amines and as a reagent for mercury [MER89]
Solubility: sl s cold H_2O, s hot H_2O; can decompose in aq solutions [MER89]
Melting Point, °C: 268-272, decomposes [ALD94]

236
Compound: Ammonium tetrathiomolybdate
Formula: $(NH_4)_2MoS_4$
Molecular Formula: $H_8MoN_2S_4$
Molecular Weight: 260.281
CAS RN: 15060-55-6

Properties: dark red cryst powd; a preparation is passing H_2S through a solution of ammonium molybdate [STR93] [ALF93] [KIR81]

237
Compound: Ammonium tetrathiotungstate
Formula: $(NH_4)_2WS_4$
Molecular Formula: $H_8N_2S_4W$
Molecular Weight: 348.181
CAS RN: 13862-78-7
Properties: orange cryst powd; H_2S odor; sensitive to heat; commonly made by adding NH_3 to a solution of tungstic acid, followed by saturation with H_2S; can be used as a source of WS_2 by decomposition in a non-oxidizing atm [KIR83] [HAW93]
Solubility: s H_2O, ammonia solutions [HAW93]
Density, g/cm³: 2.71 [ALD93]
Melting Point, °C: decomposes [HAW93]

238
Compound: Ammonium tetrathiovandate(IV)
Formula: $(NH_4)_3VS_4$
Molecular Formula: $H_{12}N_3S_4V$
Molecular Weight: 233.321
CAS RN: 14693-56-2
Properties: dark violet cryst [STR93]

239
Compound: Ammonium thiocyanate
Synonyms: ammonium rhodanide
Formula: NH_4SCN
Molecular Formula: CH_4N_2S
Molecular Weight: 76.122
CAS RN: 1762-95-4
Properties: colorless; deliq cryst; formed from solution of NH_4CN with sulfur on boiling; used as fertilizer, in chemicals and for dyeing fabrics [MER89] [HAW93]
Solubility: g/100g soln, H_2O: 64.95 (26.33°C), 81.73 (71.53°C); equilibrium solid phase, NH_4SCN [KRU93]; s alcohol, acetone, ammonia solutions [HAW93]
Density, g/cm³: 1.3057 [HAW93]
Melting Point, °C: ~149 [MER89]
Boiling Point, °C: decomposes 170 [HAW93]

240
Compound: Ammonium thiosulfate
Synonyms: ammonium hyposulfite
Formula: $(NH_4)_2S_2O_3$

Molecular Formula: $H_8N_2O_3S_2$
Molecular Weight: 148.207
CAS RN: 7783-18-8
Properties: white cryst; used in photography, fungicides, silver plating and in hair wave preparations [MER89] [HAW93]
Solubility: 103.3 g/100g H_2O at 100°C [CIC73]
Density, g/cm^3: 1.679 [ALD93]
Melting Point, °C: decomposes 150 [MER89]

241
Compound: Ammonium titanium oxalate monohydrate
Synonyms: ammonium bis(oxalato)oxotitanate(IV)
Formula: $(NH_4)_2TiO(C_2O_4)_2 \cdot H_2O$
Molecular Formula: $C_4H_{10}N_2O_{10}Ti$
Molecular Weight: 293.997
CAS RN: 10580-03-7
Properties: hygr cryst; used as a mordant to dye leather and cellulosic fabrics [HAW93] [MER89] [ALD93]
Solubility: v s H_2O [MER89]

242
Compound: Ammonium tungstate pentahydrate
Synonyms: ammonium tungstate(VI)
Formula: $(NH_4)_{10}W_{12}O_{41} \cdot 5H_2O$
Molecular Formula: $H_{50}N_{10}O_{46}W_{12}$
Molecular Weight: 3132.516
CAS RN: 1311-93-9
Properties: 99.999% pure plates or cryst powd; usually prepared by crystallization from a boiling solution [ALF93] [MER89]
Solubility: v s H_2O [MER89]
Density, g/cm^3: 2.3 [ALF93]

243
Compound: Ammonium uranate(VI)
Synonyms: ammonium diuranate
Formula: $(NH_4)_2U_2O_7$
Molecular Formula: $H_8N_2O_7U_2$
Molecular Weight: 624.131
CAS RN: 7783-22-4
Properties: -80 mesh with 99.5% purity; reddish yellow amorphous powd [MER89] [CER91]
Solubility: i H_2O [MER89]

244
Compound: Ammonium valerate
Synonyms: pentanoic acid, ammonium salt
Formula: $CH_3(CH_2)_3COONH_4$

Molecular Formula: $C_5H_{13}NO_2$
Molecular Weight: 119.164
CAS RN: 42739-38-8
Properties: very hygr cryst; used to flavor foods [HAW93] [MER89]
Solubility: v s H_2O [MER89]
Melting Point, °C: 108 [MER89]

245
Compound: Ammonium zirconyl carbonate dihydrate
Synonyms: zirconium ammonium carbonate
Formula: $(NH_4)_3ZrOH(CO_3)_3 \cdot 2H_2O$
Molecular Formula: $C_3H_{17}N_3O_{12}Zr$
Molecular Weight: 378.404
CAS RN: 12616-24-9
Properties: large prisms from H_2O; unstable in air, gradually evolving CO_2 and NH_3; aq solution decomposes rapidly above 60°C; used in paper and textile water repellents [HAW93] [MER89]
Solubility: s H_2O [MER89]; decomposed by dil acid, alkalies [HAW93]
Density, g/cm^3: aq solution: 1.238 [MER89]

246
Compound: Antimony
Synonyms: stibium
Formula: Sb
Molecular Formula: Sb
Molecular Weight: 121.757
CAS RN: 7440-36-0
Properties: silver-white metal available with 99.999% purity; hex, a = 0.4307 nm, c = 1.1273 nm; hardness 3.0-3.5 Mohs; enthalpy of fusion 19.866 kJ/mol; enthalpy of vaporization 195.1 kJ/mol; electronegativity 1.82; electrical resistivity (0°C) 37 μohm·cm; uses: in alloys such as solder, type metal and bearings, and as an evaporated semiconductor film [KIR78] [COT88] [CER91]
Solubility: oxidized by HNO_3 [HAW93]
Density, g/cm^3: 6.697 [KIR78]
Melting Point, °C: 630.7 [KIR78]
Boiling Point, °C: 1587 [KIR78]; 1635, 1440 [MER89]
Reactions: forms $SbCl_3$ and $SbCl_5$ from reaction of Sb and Cl_2 [KIR78]
Thermal Conductivity, W/(m·K): 25.5 (0°C), 24.4 (25°C), 21.9 (100°C) [KIR78] [HO 72]
Thermal Expansion Coefficient: coefficient of linear expansion at 20°C is 8 to 11 x 10^{-6} m/(m·°C) [KIR78]

247

Compound: Antimony arsenide
Formula: Sb₃As
Molecular Formula: AsSb₃
Molecular Weight: 440.193
CAS RN: 12255-36-6
Properties: 6 mm pieces and smaller of 99.999% purity [CER91]

248

Compound: Antimony iodide sulfide
Formula: SbIS
Molecular Formula: ISSb
Molecular Weight: 280.727
CAS RN: 13816-38-1
Properties: dark red; -20 mesh [CRC94] [ALF95]
Solubility: i H₂O [CRC94]
Melting Point, °C: 392 [CRC94]
Boiling Point, °C: decomposes [CRC94]

249

Compound: Antimony phosphide
Formula: SbP
Molecular Formula: PSb
Molecular Weight: 152.731
CAS RN: 53120-23-3
Properties: black powd; -100 mesh with 99.5% purity [CER91] [AES93]

250

Compound: Antimony(III) acetate
Synonyms: antimony triacetate
Formula: Sb(CH₃COO)₃
Molecular Formula: C₆H₉O₆Sb
Molecular Weight: 298.893
CAS RN: 3643-76-3
Properties: off-white powd; can be prepared by dissolution of Sb(III) salt in acetic acid, followed by crystallization [KIR78] [STR93]

251

Compound: Antimony(III) bromide
Synonyms: antimony tribromide
Formula: SbBr₃
Molecular Formula: Br₃Sb
Molecular Weight: 361.472
CAS RN: 7789-61-9

Properties: yellow deliq cryst; enthalpy of vaporization 59 kJ/mol; entropy of vaporization at 560°C is 94.9 J/(mol·K); prepared by reacting Sb and Br₂; used as a mordant [HAW93] [KIR78] [CRC93]
Solubility: decomposed by H₂O; s dil HCl, HBr [HAW93] [MER89]
Density, g/cm³: 4.148 [HAW93]
Melting Point, °C: 96.0 [KIR78]
Boiling Point, °C: 280 [CRC93]

252

Compound: Antimony(III) chloride
Synonyms: antimony trichloride
Formula: SbCl₃
Molecular Formula: Cl₃Sb
Molecular Weight: 228.118
CAS RN: 10025-91-9
Properties: colorless ortho-rhomb cryst; very hygr; enthalpy of vaporization 45.19 kJ/mol; entropy of vaporization at 496°C is 93.3 J/(mol·K); enthalpy of fusion 12.70 kJ/mol; can be prepared from Sb or Sb₂O₃ and conc HCl; used as a catalyst to chlorinate olefins, and to polymerize hydrocarbons [HAW93] [KIR78] [CRC93]
Solubility: g/100g soln, H₂O: 601.6 (0°C), 988.1 (25°C), infinite (72°C); Solid phase at equilibrium: SbCl₃ [KRU93]; s CHCl3 (22%), CCl₄ (13%), CS2 and benzene [KIR78]
Density, g/cm³: 3.14 [MER89]
Melting Point, °C: 73.4 [KIR78]
Boiling Point, °C: 220.3 [CRC93]
Reactions: gradual hydrolysis to SbOCl in H₂O [MER89]

253

Compound: Antimony(III) fluoride
Synonyms: antimony trifluoride
Formula: SbF₃
Molecular Formula: F₃Sb
Molecular Weight: 178.755
CAS RN: 7783-56-4
Properties: white to gray powd; hygr; ortho-rhomb; enthalpy of fusion 21.4 kJ/mol; entropy of fusion 38.2 J/(mol·K); enthalpy of vaporization at 298°C is 102.8 kJ/mol; entropy of vaporization at 25°C is 175.8 kJ/(mol·K); vapor pressure at mp is 26.34 kPa; slowly hydrolyzes in H₂O; can be prepared by dissolution of Sb₂O₃ in anhydrous HF; used as a fluorinating agent for organic compounds [HAW93] [KIR78]
Solubility: g/100g H₂O: 384.7 (0°C), 492.4 (25°C); 154 g/100mL in methanol; i benzene, chlorobenzene, heptane [KRU93] [KIR78]

Density, g/cm³: 4.379 [MER89]
Melting Point, °C: 292 [COT88]
Boiling Point, °C: 319 [STR93]
Reactions: transforms from SbF_3 to $Sb_3O_2(OH)_2F_3$ then SbOF at 100°C [KIR78]

254
Compound: Antimony(III) hydride
Synonyms: stibine
Formula: SbH_3
Molecular Formula: H_3Sb
Molecular Weight: 124.784
CAS RN: 7803-52-3
Properties: colorless gas; slowly decomposes at room temp, readily at 200°C; flammable material; enthalpy of vaporization 21.3 kJ/mol; distibine, Sb_2H_4, 14939-42-5, has been reported; a preparation is to add acid to a metal antimonide; used as n-type gas-dopant in Si semiconductors [KIR78] [CRC93]
Solubility: sl s H_2O; s CS_2, ethanol [KIR78]
Density, g/cm³: 4.344 (air = 1.000), 15°C; liq at bp 2.204 [KIR78]
Melting Point, °C: -88 [KIR78]
Boiling Point, °C: -17 [KIR78]
Reactions: decomposes at 200°C to $Sb + H_2$ [KIR78]

255
Compound: Antimony(III) iodide
Synonyms: antimony triiodide
Formula: SbI_3
Molecular Formula: I_3Sb
Molecular Weight: 502.473
CAS RN: 7790-44-5
Properties: ruby red trig cryst; volatile at high temperatures; enthalpy of fusion at 444°C is 22.7 kJ/mol; entropy of fusion at 444°C is 51.5 J/(mol·K); enthalpy of sublimation 101.6 kJ/mol at 298°C; enthalpy of vaporization 68.6 kJ/mol; can be obtained by reacting Sb and I_2 [MER89] [HAW93] [KIR78] [CRC93]
Solubility: decomposed in H_2O; s CS_2, HCl; i alcohol, $CHCl_3$ [HAW93]
Density, g/cm³: 4.921 [MER89]
Melting Point, °C: 170.5 [KIR78]
Boiling Point, °C: 401 [CRC93]
Reactions: decomposed by water and air to SbOI [MER89]

256
Compound: Antimony(III) nitrate
Formula: $Sb(NO_3)_3$

Molecular Formula: N_3O_9Sb
Molecular Weight: 307.775
CAS RN: 20328-96-5
Properties: can be obtained by dissolution of Sb(III) salt in HNO_3 solution, followed by crystallization [KIR78]
Solubility: hydrolyzes [KIR78]

257
Compound: Antimony(III) oxide
Synonyms: valentinite
Formula: Sb_2O_3
Molecular Formula: O_3Sb_2
Molecular Weight: 291.518
CAS RN: 1317-98-2
Properties: colorless; ortho-rhomb; stable above 570°C [KIR78]
Density, g/cm³: 5.67 [KIR78]
Melting Point, °C: 656 [KIR78]
Boiling Point, °C: 1425 [KIR78]

258
Compound: Antimony(III) oxide
Synonyms: antimony trioxide, antimony white
Formula: Sb_2O_3
Molecular Formula: O_3Sb_2
Molecular Weight: 291.518
CAS RN: 1309-64-4
Properties: white, odorless cryst powd; obtained by igniting Sb in air; used to flameproof materials and in paints; evaporated material of 99.9% purity used in dielectric interference filter for ultraviolet radiation [HAW93] [CER91]
Solubility: s sl H_2O; i organic solvents [KIR78]; s conc HCl, H_2SO_4, alkalies [HAW93]
Density, g/cm³: 5.67 [HAW93]
Melting Point, °C: 656 [KIR78]
Boiling Point, °C: 1425 [KIR78]; sublimes 1550 [STR93]

259
Compound: Antimony(III) oxide
Synonyms: senarmontite
Formula: Sb_2O_3
Molecular Formula: O_3Sb_2
Molecular Weight: 291.518
CAS RN: 12412-52-1
Properties: colorless; cub; stable below 570°C; can be prepared by heating Sb in air; used as a flame retardant for fabrics, and as a catalyst [KIR78]
Solubility: v sl s H_2O; i organic solvents [KIR78]
Density, g/cm³: 5.2 [KIR78]

Melting Point, °C: 656, in absence of O_2 [KIR78]
Boiling Point, °C: 1425, partial sublimation [KIR78]

260
Compound: Antimony(III) perchlorate trihydrate
Formula: $Sb(ClO_4)_3 \cdot 3H_2O$
Molecular Formula: $Cl_3H_6O_{15}Sb$
Molecular Weight: 474.157
CAS RN: 65277-48-7
Properties: can be prepared by dissolution of Sb(III) salt in perchloric acid, followed by crystallization [KIR78]
Solubility: hydrolyzes [KIR78]

261
Compound: Antimony(III) phosphate
Formula: $SbPO_4$
Molecular Formula: O_4PSb
Molecular Weight: 216.731
CAS RN: 12036-46-3
Properties: can be prepared by dissolving Sb(III) compound in H_3PO_4, followed by crystallization [KIR78]
Solubility: hydrolyzes [KIR78]

262
Compound: Antimony(III) selenide
Synonyms: antimony triselenide
Formula: Sb_2Se_3
Molecular Formula: Sb_2Se_3
Molecular Weight: 480.400
CAS RN: 1315-05-5
Properties: gray powd; obtained by passing H_2Se through a solution of potassium antimonyl tartrate [MER89]
Solubility: v sl s H_2O [MER89]
Melting Point, °C: 611 [MER89]

263
Compound: Antimony(III) sulfate
Synonyms: antimonous sulfate
Formula: $Sb_2(SO_4)_3$
Molecular Formula: $O_{12}S_3Sb_2$
Molecular Weight: 531.711
CAS RN: 7446-32-4
Properties: white cryst powd; deliq; can be obtained by dissolution of Sb(III) salt in H_2SO_4, followed by crystallization; used in matches, and pyrotechnics [MER89] [KIR78] [HAW93]

Solubility: s H_2O, but can form insol basic salt [MER89]
Density, g/cm³: 3.62 [HAW93]
Melting Point, °C: decomposes [ALF93]

264
Compound: Antimony(III) sulfide
Synonyms: antimony orange
Formula: Sb_2S_3
Molecular Formula: S_3Sb_2
Molecular Weight: 339.718
CAS RN: 1345-04-6
Properties: black cryst (stibnite) or amorphous reddish orange powd; amorphous material prepared by passing H_2S through $SbCl_3$ solution; used in the form of 99.9% pure material as a sputtering target to produce infrared filter with high-index in red part of visible spectrum, and used in pyrotechnics, certain matches, in manufacturing ruby glass; the nonahydrate is a lemon-yellow cryst [KIR78] [HAW93] [CER91]
Solubility: i H_2O, s conc HCl [MER89]
Density, g/cm³: 4.562 [HAW93]
Melting Point, °C: 550 [KIR78]
Boiling Point, °C: ~1150 [STR93]

265
Compound: Antimony(III) telluride
Synonyms: antimony tritelluride
Formula: Sb_2Te_3
Molecular Formula: Sb_2Te_3
Molecular Weight: 626.314
CAS RN: 1327-50-0
Properties: gray; 3-12 mm fused pieces of 99.999% purity [CER91] [CRC94]
Density, g/cm³: 6.5 [ALD94]
Melting Point, °C: 629 [CRC94]

266
Compound: Antimony(IV) oxide
Synonyms: antimony tetroxide
Formula: β-Sb_2O_4
Molecular Formula: O_4Sb_2
Molecular Weight: 307.518
CAS RN: 1332-81-6
Properties: monocl; formed by heating valentinite in dry air at 1130°C; composed of half Sb(III) and half Sb(V); used as an oxidation catalyst [STR93] [KIR78]
Density, g/cm³: 5.82 [CRC94]
Melting Point, °C: vaporizes [KIR78]
Reactions: minus O at 930 [CRC94]

267

Compound: Antimony(IV) oxide
Synonyms: cervantite
Formula: α-Sb_2O_4
Molecular Formula: O_4Sb_2
Molecular Weight: 307.518
CAS RN: 1332-81-6
Properties: colorless; ortho-rhomb; composition half Sb(III), half Sb(V); formed by heating valentinite in air at 460 to 540°C; used as an oxidation catalyst [STR93] [KIR78]
Density, g/cm³: 4.07 [KIR78]
Melting Point, °C: vaporizes [KIR78]

268

Compound: Antimony(V) chloride
Synonyms: antimony pentachloride
Formula: $SbCl_5$
Molecular Formula: Cl_5Sb
Molecular Weight: 299.024
CAS RN: 7647-18-9
Properties: reddish yellow or colorless (if pure), hygr oily liq; fumes in air; enthalpy of vaporization at 449°C is 43.45 kJ/mol; entropy of vaporization at 449°C is 95.44 J/(mol·K); made by action of chlorine on molten $SbCl_3$; useful for providing chlorine for reactions such as formation of ICl from I_2; decomposes if distilled [HAW93] [KIR78] [MER89]
Solubility: hydrolyzes in H_2O; s HCl [MER89]
Density, g/cm³: 2.34 [HAW93]
Melting Point, °C: 3.2 [KIR78]
Boiling Point, °C: 68 (1.82 kPa), 176 (extrapolated) [KIR78]

269

Compound: Antimony(V) dichlorotrifluoride
Formula: $SbCl_2F_3$
Molecular Formula: Cl_2F_3Sb
Molecular Weight: 249.660
CAS RN: 7791-16-4
Properties: viscous liq; made by reacting SbF_3 and Cl_2; used as a catalyst in fluorocarbon manufacturing [MER89] [HAW93]

270

Compound: Antimony(V) fluoride
Synonyms: antimony pentafluoride
Formula: SbF_5
Molecular Formula: F_5Sb
Molecular Weight: 216.752
CAS RN: 7783-70-2

Properties: colorless hygr viscous liq; viscosity 460 mPa·s at 20°C; tendency to polymerize can be prevented by addition of 1% anhydrous HF; can be prepared by direct fluorination of SbF_3 or Sb powd; used as a catalyst in fluorinating reactions [HAW93] [KIR78]
Solubility: reacts vigorously with H_2O, becoming hydrolyzed [HAW93] [KIR78]
Density, g/cm³: 3.145 (15.5°C) [KIR78]
Melting Point, °C: 7 [KIR78]
Boiling Point, °C: 142.7 [KIR78]
Reactions: reacts with I_2, S, NO_2, graphite [KIR78]

271

Compound: Antimony(V) oxide
Synonyms: antimony pentoxide
Formula: Sb_2O_5
Molecular Formula: O_5Sb_2
Molecular Weight: 323.517
CAS RN: 1314-60-9
Properties: yellowish powd; cub; always somewhat hydrated; prepared by reacting Sb or Sb_2O_3 with conc HNO_3; used as a flame retardant for textiles [HAW93] [MER89]
Solubility: sl s H_2O, i HNO_3; dissolves slowly in warm HCl, KOH [MER89]
Density, g/cm³: 3.78 [MER89]
Melting Point, °C: decomposes [MER89]
Reactions: loses oxygen at 300°C [MER89]

272

Compound: Antimony(V) oxide hydrate
Synonyms: antimonic acid
Formula: $Sb_2O_5 \cdot xH_2O$
Molecular Formula: O_5Sb_2 (anhydrous)
Molecular Weight: 323.517 (anhydrous)
CAS RN: 12712-36-6
Properties: cub yellowish powd; material with approximate composition $Sb_2O_5 \cdot 3$-$1/2H_2O$ is prepared by hydrolysis of $SbCl_5$ [KIR78] [MER89]
Solubility: s sl H_2O; i HNO_3; s KOH [KIR78]
Density, g/cm³: 3.78 [MER89]
Reactions: forms cub white Sb_6O_{13} ~700°C [KIR78]

273

Compound: Antimony(V) oxychloride
Synonyms: basic antimony chloride
Formula: SbOCl
Molecular Formula: ClOSb
Molecular Weight: 173.212
CAS RN: 7791-08-4

Properties: white monocl cryst or powd; can be prepared by adding $SbCl_3$ to water; used in flameproofing textiles [HAW93] [MER89] [KIR78]

Solubility: hydrolyzed by H_2O; s HCl [MER89]; i alcohol, ether [HAW93]

Melting Point, °C: 170, decomposes [HAW93]

Reactions: heating to 250°C gives $Sb_2O_5Cl_2$, to >320°C gives Sb_2O_3 [MER89]

274

Compound: Antimony(V) sulfide

Synonyms: golden sulfide of antimony

Formula: Sb_2S_5

Molecular Formula: S_5Sb_2

Molecular Weight: 403.850

CAS RN: 1315-04-4

Properties: yellow to orange to red solid; amorphous; odorless; can be formed by boiling Sb_2S_3 and sulfur in alkaline media, followed by precipitation with HCl; finds use as a red pigment, and in the vulcanization of rubber [HAW93] [KIR78]

Solubility: i H_2O; s HCl to evolve H_2S [MER89]

Density, g/cm³: 4.12 [STR93]

Melting Point, °C: 75, decomposes [STR93]

275

Compound: Argon

Formula: Ar

Molecular Formula: Ar

Molecular Weight: 39.948

CAS RN: 7440-37-1

Properties: colorless, odorless tasteless inert gas; air contains 9.340 µL/L of argon; enthalpy of vaporization 6.469 kJ/mol; enthalpy of fusion 1.12 kJ/mol; sonic velocity (101.32 kPa, 0°C) 307.8 m/s; viscosity (101.32 kPa, 25°C) 22.64 Pa s; critical temp -122.29°C; critical pressure 48.3 atm; crystallizes as fcc; triple point, -189.37°C; used as a carrier gas and for sputtering/VLSI [KIR78] [MER89] [CRC93]

Solubility: 33.6 mL/1000g H_2O (20°C) [KIR78]; Henry's law constants, k x 10^{-4}: 3.974 (25.0°C), 5.359 (65.1°C), 5.342 (91.1°C), 3.812 (222.7°C), 2.541 (267.3°C), 1.870 (287.9°C) [POT78]

Density, g/cm³: gas, 101.3 kPa, 0°C, 0.0017838 [KIR78]; solid, 1.623 at triple point [MER89]

Melting Point, °C: -189.35 [CRC93]

Boiling Point, °C: -185.87 [KIR78]

Thermal Conductivity, W/(m·K): gas (101.32 kPa, 0°C): 1.694 [KIR78]

276

Compound: Argon fluoride

Formula: ArF

Molecular Formula: ArF

Molecular Weight: 58.946

CAS RN: 56617-31-3

Properties: unstable gas; used as a light emitting source in lasers [KIR78]

277

Compound: Arsenic (α)

Formula: α-As

Molecular Formula: As

Molecular Weight: 74.92159

CAS RN: 7440-38-2

Properties: gray, shiny, brittle, metallic looking; rhomb, a = 0.376 nm, c = 1.0548 nm; oxidizes to As_2O_3 in air; hardness 3.5 Mohs; enthalpy of fusion 27.44 kJ/mol; enthalpy of sublimation 31.974 kJ/mol; specific heat (25°C) 24.6 J/(mol·K); electrical resistivity (0°C) 26 µohm·cm; electronegativity 2.20; used in semiconductors [KIR78] [MER89] [COT88] [CER91] [CRC93]

Solubility: i H_2O; s conc HNO_3 [KIR78]

Density, g/cm³: 5.778 [KIR78]

Melting Point, °C: 817.1, 28 atm [ALD94]

Boiling Point, °C: 615, sublimes [KIR78]

Reactions: reacts with conc HNO_3 → H_3AsO_4 [KIR78]

Thermal Conductivity, W/(m·K): 50.2 (25°C) [ALD94]

Thermal Expansion Coefficient: 20°C, linear coefficient of thermal expansion is 5.6 micrometer/(m·°C) [KIR78]

278

Compound: Arsenic (β)

Formula: β-As

Molecular Formula: As

Molecular Weight: 74.92159

CAS RN: 7440-38-2

Properties: dark gray amorphous solid; electrical resistivity 107 ohm·cm [KIR78]

Density, g/cm³: 4.700 [KIR78]

Melting Point, °C: sublimes [KIR78]

Reactions: transformation from amorphous to cryst at 280°C [KIR78]

279

Compound: Arsenic disulfide

Synonyms: realgar

Formula: As_4S_4
Molecular Formula: As_4S_4
Molecular Weight: 427.950
CAS RN: 12279-90-2
Properties: red or orange solid; naturally occurring mineral; can be manufactured by heating iron pyrites and arsenopyrite; used in pyrotechnics [KIR78]
Solubility: i H_2O, hot HCl; s warm alkali [KIR78]
Density, g/cm^3: 3.5 [MER89]
Melting Point, °C: 307 [KIR78]
Boiling Point, °C: 565 [KIR78]
Reactions: transforms to black allotropic modification at 267°C [KIR78]

280
Compound: Arsenic hemiselenide
Formula: As_2Se
Molecular Formula: As_2Se
Molecular Weight: 228.803
CAS RN: 1303-35-1
Properties: black cryst, with metallic luster; formed by melting stoichiometric amounts of As and Se in nitrogen atm; used in glass manufacturing [MER89]
Solubility: i most solvents; decomposed by boiling alkali hydroxides [MER89]

281
Compound: Arsenic(II) iodide
Synonyms: arsenic diiodide
Formula: AsI_2
Molecular Formula: AsI_2
Molecular Weight: 328.731
CAS RN: 13770-56-4
Properties: red solid; formula also given as As_2I_4 [KIR78]
Solubility: s organic solvents [KIR78]
Melting Point, °C: 130 [KIR78]
Reactions: with $H_2O \rightarrow AsI_3$ and As [KIR78]

282
Compound: Arsenic(II) sulfide
Synonyms: realgar
Formula: As_2S_2
Molecular Formula: As_2S_2
Molecular Weight: 213.975
CAS RN: 1303-32-8
Properties: reddish brown monocl powd; α and β forms [CRC94] [ALF95]
Density, g/cm^3: α: 3.506; β: 3.254 [CRC94]
Melting Point, °C: 360 [ALF95]
Boiling Point, °C: 565 [ALF95]
Reactions: $\alpha \rightarrow \beta$ at 267°C [CRC94]

283
Compound: Arsenic(III) bromide
Synonyms: arsenic tribromide
Formula: $AsBr_3$
Molecular Formula: $AsBr_3$
Molecular Weight: 314.634
CAS RN: 7784-33-0
Properties: colorless to yellow lumps; deliq; orthorhomb; fumes in moist air; enthalpy of vaporization 41.8 kJ/mol; enthalpy of fusion 11.70 kJ/mol; dielectric constant 8.33 (35°C); can be formed from As and Br_2 dissolved in CS_2; used in analytical chemistry and in medicine [STR93] [KIR78] [MER89] [CRC93]
Solubility: decomposed in H_2O, forming HBr, As_2O_3; miscible with ether, benzene [MER89]
Density, g/cm^3: 3.66 [KIR78]
Melting Point, °C: 31.1 [CRC93]
Boiling Point, °C: 221 [KIR78]

284
Compound: Arsenic(III) chloride
Synonyms: arsenic trichloride
Formula: $AsCl_3$
Molecular Formula: $AsCl_3$
Molecular Weight: 181.280
CAS RN: 7784-34-1
Properties: colorless or pale yellow oily liq; fumes in air; enthalpy of vaporization 35.01 kJ/mol; enthalpy of fusion 10.10 kJ/mol; decomposed by UV light; may be obtained by reaction of As and Cl_2; used as an intermediate in organic preparations, and in ceramics [MER89] [KIR78] [HAW93] [CRC93]
Solubility: decomposed by H_2O, giving $As(OH)_3$ and HCl products [MER89]; s conc HCl and most organic solvents [HAW93]
Density, g/cm^3: 2.205 [KIR78]
Melting Point, °C: -16 [MER89]
Boiling Point, °C: 130.2 [MER89]

285
Compound: Arsenic(III) fluoride
Synonyms: arsenic trifluoride
Formula: AsF_3
Molecular Formula: AsF_3
Molecular Weight: 131.917
CAS RN: 7784-35-2

Properties: colorless liq; fumes in air; enthalpy of
vaporization 29.7 kJ/mol; enthalpy of fusion
10.40 kJ/mol; can be prepared by fluorinating
As_2O_3 with H_2SO_4 and CaF_2; used as a
fluorinating agent and to synthesize AsF_5
[MER89] [KIR78] [CRC93]
Solubility: hydrolyzed by H_2O; s alcohol, ether,
benzene [MER89]
Density, g/cm³: 2.666 [KIR78]
Melting Point, °C: -5.9 [CRC93]
Boiling Point, °C: 57.8 [CRC93]

286
Compound: Arsenic(III) iodide
Synonyms: arsenic triiodide
Formula: AsI_3
Molecular Formula: AsI_3
Molecular Weight: 455.635
CAS RN: 7784-45-4
Properties: red powd; enthalpy of formation, -58.2
kJ/mol; entropy, 213.0 J/(mol·K); enthalpy of
vaporization 59.3 kJ/mol; decomposes slowly in
air at 100°C, rapidly at 200°C, to give As, I_2,
As_2O_3; made by precipitation from a hot $AsCl_3$-
HCl solution by the addition of KI [KIR78]
[STR93]
Solubility: 1g/12mL H_2O; not easily hydrolyzed
[MER89] [KIR78]
Density, g/cm³: 4.39 (15°C) [KIR78]
Melting Point, °C: 140 [COT88]
Boiling Point, °C: 424 [CRC93]

287
Compound: Arsenic(III) oxide
Synonyms: arsenolite
Formula: As_2O_3
Molecular Formula: As_2O_3
Molecular Weight: 197.841
CAS RN: 1327-53-3
Properties: white odorless and tasteless powd; cub;
may be obtained by strongly heating As in air,
or by roasting arsenopyrite, FeAsS; used in
pigments, ceramic enamels, insecticide
[HAW93] [KIR78]
Solubility: 1.7g/100g H_2O (25°C); s acids and
alkalies [KIR78]; s glycerol [HAW93]
Density, g/cm³: 3.865 [HAW93]
Melting Point, °C: 275 [KIR78]
Reactions: sublimes freely above 135°C [KIR78]

288
Compound: Arsenic(III) oxide
Synonyms: claudetite

Formula: As_2O_3
Molecular Formula: As_2O_3
Molecular Weight: 197.841
CAS RN: 1327-53-3
Properties: white powd; monocl;
thermodynamically stable form; can be prepared
by ignition of As in air; used as a pigment in
ceramics, as a decolorizing agent in glass
[HAW93] [STR93] [KIR78]
Solubility: g/100g H_2O: 1.20 (0°C), 1.49 (10°C),
1.82 (20°C), 2.31 (30°C), 2.93 (40°C), 4.31
(60°C), 6.11 (80°C), 8.2 (100°C) [LAN85]; s dil
HCl [MER89]
Density, g/cm³: 3.738 [STR93]
Melting Point, °C: 313 [MER89]
Boiling Point, °C: 465 [MER89]
Reactions: sublimes when slowly heated [MER89]

289
Compound: Arsenic(III) selenide
Synonyms: arsenic triselenide
Formula: As_2Se_3
Molecular Formula: As_2Se_3
Molecular Weight: 386.723
CAS RN: 1303-36-2
Properties: black cryst; dark brown solid;
preparation: from melted As and Se; uses:
vacuum deposition [CER91] [STR93] [MER89]
Solubility: i H_2O; s HNO_3 [MER89]
Density, g/cm³: 4.75 [MER89]
Melting Point, °C: 260 [MER89]; ~360 [STR93]

290
Compound: Arsenic(III) sulfide
Synonyms: orpiment, arsenic trisulfide
Formula: As_2S_3
Molecular Formula: As_2S_3
Molecular Weight: 246.041
CAS RN: 1303-33-9
Properties: yellow or orange powd; forms when
As_2O_3 is heated with sulfur; used as a pigment,
reducing agent, and in the form of 99.9% or
99.99% material as a sputtering target to
produce adherent, stable, non-hygr, anti-
reflection films on germanium and silicon
[HAW93] [KIR78] [MER89] [CER91]
Solubility: i H_2O; s alkalies, slowly s HCl;
decomposes in HNO_3 [MER89]
Density, g/cm³: 3.46 [MER89]
Melting Point, °C: 320 [KIR78]
Boiling Point, °C: 707 [KIR78]
Reactions: transition to red form at 170°C
[HAW93]

291

Compound: Arsenic(III) telluride
Synonyms: arsenic tritelluride
Formula: As_2Te_3
Molecular Formula: As_2Te_3
Molecular Weight: 532.643
CAS RN: 12044-54-1
Properties: black cryst; uses: vacuum deposition [CER91] [STR93]
Density, g/cm³: 6.50 [STR93]
Melting Point, °C: 621 [STR93]

292

Compound: Arsenic(V) acid hemihydrate
Formula: $H_3AsO_4 \cdot 1/2H_2O$
Molecular Formula: $AsH_4O_{4.5}$
Molecular Weight: 150.951
CAS RN: 7778-39-4
Properties: white translucent; hygr cryst; acid, $K_1 = 5.6 \times 10^{-3}$, $K_2 = 1.7 \times 10^{-7}$, $K_3 = 3.0 \times 10^{-12}$; loses water above 300°C forming the anhydrous As_2O_3; can be obtained by treating As_2O_3 with conc HNO_3; used in glassmaking, wood-treatment [HAW93] [KIR78] [MER89]
Solubility: v s H_2O, alcohol, glycerol [MER89]
Density, g/cm³: 2-2.5 [HAW93]
Melting Point, °C: 35.5 [HAW93]
Reactions: minus H_2O forming H_4AsO_7 at 100°C; forms $HAsO_3$ >100°C [KIR78]

293

Compound: Arsenic(V) fluoride
Synonyms: arsenic pentafluoride
Formula: AsF_5
Molecular Formula: AsF_5
Molecular Weight: 169.914
CAS RN: 7784-36-3
Properties: colorless gas; condenses to yellow liq; forms white clouds in moist air; enthalpy of vaporization 20.8 kJ/mol; dielectric constant, 12.8 (20°C); can be formed by reacting AsF_3 with fluorine; used as doping agent for electroconductive polymers [HAW93] [KIR78] [MER89] [CRC93]
Solubility: hydrolyzed quickly in H_2O; s alcohol, benzene, ether [MER89]
Density, g/cm³: liq: 2.33 at bp [KIR78]
Melting Point, °C: -88.7 [KIR78]
Boiling Point, °C: -53.2 [CRC93]

294

Compound: Arsenic(V) oxide

Synonyms: arsenic pentoxide
Formula: As_2O_5
Molecular Formula: As_2O_5
Molecular Weight: 229.840
CAS RN: 1303-28-2
Properties: white amorphous lumps or powd; uncertain structure; oxidizing agent, can liberate Cl_2 from HCl; deliq; obtained by reaction of As or As_2O_3 with O_2 under pressure; used as an insecticide, in the manufacture of colored glass, and for weed control [HAW93] [STR93] [KIR78] [MER89]
Solubility: g/100g H_2O: 59.5 (0°), 62.1 (10°C), 65.8 (20°C), 69.8 (30°C), 71.2 (40°C), 73.0 (60°C), 75.1 (80°C), 76.7 (100°C) [LAN85]; s alcohol [MER89]
Density, g/cm³: 4.32 [STR93]
Melting Point, °C: 315 decomposes [STR93]

295

Compound: Arsenic(V) selenide
Synonyms: arsenic pentaselenide
Formula: As_2Se_5
Molecular Formula: As_2Se_5
Molecular Weight: 544.643
CAS RN: 1303-37-3
Properties: black, brittle solid; metallic luster; obtained when stoichiometric amounts of As and Se are melted in a nitrogen atm [MER89]
Solubility: i H_2O, dil acids; s alkali hydroxides [MER89]
Melting Point, °C: decomposes on heating in air [MER89]

296

Compound: Arsenic(V) sulfide
Synonyms: arsenic pentasulfide
Formula: As_2S_5
Molecular Formula: As_2S_5
Molecular Weight: 310.173
CAS RN: 1303-34-0
Properties: yellow or orange powd; stable in air up to 95°C; can be prepared by fusing As and S, or by passing H_2S through HCl solution of arsenic acid; used as a paint pigment, in light filters [HAW93] [KIR78]
Solubility: g/L soln, H_2O: 0.00136 (0°C) [KRU93]
Melting Point, °C: decomposes [KIR78]
Reactions: decomposes to As_2S_3 and S >95°C [KIR78]

297

Compound: Arsenious acid

Formula: H_3AsO_3
Molecular Formula: AsH_3O_3
Molecular Weight: 125.944
CAS RN: 13464-58-9
Properties: exists only in solution; is a weak acid, K = 8 x 10^{-16}; structure: HO(OH)AsOH; preparation: 1 g As_2O_3, 5 mL dil HCl, dil to 100 mL with H_2O; uses: for skin and blood disorders in animals [MER89] [KIR78]

298
Compound: Arsine
Synonyms: arsenic trihydride
Formula: AsH_3
Molecular Formula: AsH_3
Molecular Weight: 77.946
CAS RN: 7784-42-1
Properties: colorless gas; garlic like odor; highly toxic; critical temp 105.4°C; critical pressure 6.60 MPa; enthalpy of vaporization 16.69 kJ/mol; decomposes 230°C; formed by reaction of Zn, HCl and As compound, and by hydride reduction, e.g. $NaBH_4$ in NaOH solution; used in organic synthesis, some use in electronics industry [HAW93] [KIR78] [KIR80] [AIR87] [KOR91] [CRC93].
Solubility: mL/100g H_2O (760 mm): 42 (0°C), 30 (10°C), 28 (20°C) [LAN85]
Density, g/cm^3: liq: (-64.3°C) 1.640; gas: 2.695 g/L [KIR78] [KIR80]
Melting Point, °C: -116.3 [KIR78]
Boiling Point, °C: -62.4 [KIR78]
Reactions: becomes hydrated to $AsH_3 \cdot 6H_2O$ at -10°C [KIR78]

299
Compound: Astatine
Formula: At
Molecular Formula: At
Molecular Weight: 210
CAS RN: 7440-68-8
Properties: radioactive cryst halogen with 20 isotopes; heaviest of the halogens; ^{209}At, $t_{1/2}$ = 5.5 h; ^{210}At, $t_{1/2}$ = 8.3 h; more metallic than iodine; preparation: from Bi by α-particle bombardment; possible medical uses, concentrates in thyroid gland [HAW93] [MER89]
Solubility: s organic solvents [MER89]
Melting Point, °C: 302 [CRC94]
Boiling Point, °C: 337 (estimated) [CRC94]

300
Compound: Barium
Formula: Ba
Molecular Formula: Ba
Molecular Weight: 137.327
CAS RN: 7440-39-3
Properties: yellow-silver soft metal; bcc; a = 0.5025 nm; enthalpy of fusion 7.66 kJ/mol; enthalpy of vaporization 149.20 kJ/mol; vapor pressure, kPa: 0.00133 (629°C), 1.33 (1050°C), 101.3 (1640°C); easily air oxidized; gruneisen parameter -0.2; electrical resistivity 29.4 μohm·cm for the pure element; electron work function 2.11 eV; Ba^{++} radius 0.143 nm; electronegativity 1.02 [CIC73] [KIR91] [MER89]
Solubility: s with H_2 evolution in cold H_2O and hot H_2O; sl s alcohol; i benzene [CRC92]
Density, g/cm^3: 3.62 [CIC73]
Melting Point, °C: 729 [KNA91]
Boiling Point, °C: 1640 [KIR91]
Thermal Conductivity, W/(m·K): 18.4 (25°C) [CRC93]
Thermal Expansion Coefficient: coefficient of linear expansion 1.85 x 10^{-5} m/(m·°C) [KIR91]

301
Compound: Barium 2-ethylhexanoate
Formula: $[CH_3(CH_2)_3CHC_2H_5COO]_2Ba$
Molecular Formula: $C_{16}H_{30}BaO_4$
Molecular Weight: 423.739
CAS RN: 2457-01-4
Properties: precursor used in the preparation of thin-film superconductors [ALD94]
Melting Point, °C: >300 [ALD93]

302
Compound: Barium acetate
Synonyms: acetic acid, barium salt
Formula: $Ba(CH_3COO)_2$
Molecular Formula: $C_4H_6BaO_4$
Molecular Weight: 255.417
CAS RN: 543-80-6
Properties: white powd; crystallizes from H_2O as the trihydrate below 24.7°C, as a monohydrate from 24.7 to 41°C, and as the anhydrous material above 41°C; can be prepared from acetic acid and either $BaCO_3$ or BaS, followed by crystallization and dehydration [KIR78] [STR93]
Solubility: g/100g H_2O: 58.8 (0.3°C), 78.1 (24.1°C), 74.8 (99.2°C); solid equilibrium phase, $Ba(CH_3COO)_2 \cdot 3H_2O$ (0.3 and 24.1°C), $Ba(CH_3COO)_2$ (99.2°C) [KRU93]

Density, g/cm³: 2.47 [KIR78]

303

Compound: Barium acetate monohydrate
Synonyms: acetic acid, barium salt monohydrate
Formula: Ba(CH₃COO)₂·H₂O
Molecular Formula: C₄H₈BaO₅
Molecular Weight: 273.432
CAS RN: 5908-64-5
Properties: white cryst; made by addition of acetic acid to barium sulfide solution, followed by evaporation and crystallization; used as a chemical reagent and as a textile mordant, used in paint and varnish driers [HAW93]
Solubility: 1 g/1.5mL cold or boiling H₂O; 1 g/700mL alcohol [MER89]
Density, g/cm³: 2.19 [KIR78]
Melting Point, °C: decomposes [HAW93]
Reactions: minus H₂O 110°C [MER89]

304

Compound: Barium acetylacetonate octahydrate
Synonyms: 2,4-pentanedione, barium derivative octahydrate
Formula: Ba[CH₃COCH=C(O)CH₃]₂·8H₂O
Molecular Formula: C₁₀H₃₀BaO₁₂
Molecular Weight: 479.668
CAS RN: 12084-29-6
Properties: hygr powd [STR93] [ALD93]
Melting Point, °C: 123, decomposes [ALD93]

305

Compound: Barium aluminate
Formula: BaO·Al₂O₃
Molecular Formula: Al₂BaO₄
Molecular Weight: 255.288
CAS RN: 12004-04-5
Properties: nepheline type structure; -100 mesh, 99.5% purity; a = 0.5224 nm c = 0.8792 nm [TAY91b] [CER91] [CRC77]
Melting Point, °C: 1827 [KNA91]
Thermal Expansion Coefficient: from 25°C to 100°C (0.18), 200°C (0.45), 400°C (0.96), 600°C (1.50), 800°C (2.01), 1000°C (2.52), 1200°C (3.06) [TAY91b]

306

Compound: Barium aluminate
Formula: 3BaO·Al₂O₃
Molecular Formula: Al₂Ba₃O₆
Molecular Weight: 561.940

Properties: gray mass [HAW93]
Solubility: s H₂O, acid [HAW93]
Melting Point, °C: 1750 [KNA91]

307

Compound: Barium aluminide
Formula: BaAl₄
Molecular Formula: Al₄Ba
Molecular Weight: 245.253
CAS RN: 12672-79-6
Properties: 6 mm pieces and smaller [CER91]

308

Compound: Barium antimonide
Formula: Ba₃Sb₂
Molecular Formula: Ba₃Sb₂
Molecular Weight: 655.501
CAS RN: 55576-04-0
Properties: 6 mm pieces and smaller [CER91]

309

Compound: Barium arsenide
Formula: Ba₃As₂
Molecular Formula: As₂Ba₃
Molecular Weight: 561.824
CAS RN: 12255-50-4
Properties: brown; 6 mm pieces and smaller [CRC94] [CER91]
Density, g/cm³: 4.1 [CRC94]

310

Compound: Barium azide
Formula: Ba(N₃)₂
Molecular Formula: BaN₆
Molecular Weight: 221.367
CAS RN: 18810-58-7
Properties: cryst solid; monocl, a = 0.622 nm, b = 2.929 nm, c = 0.702 nm; M-N₃ bond length 0.2937 nm; unstable and can explode when heated or on impact; used in high explosives [CIC73] [HAW93] [CRC93]
Solubility: g/100g H₂O: 12.5 (0°C), 16.1 (10°C), 17.4 (20°C) [LAN85]; alcohol 0.17 (16°C), i ether [CRC92]
Density, g/cm³: 2.936 [HAW93]
Boiling Point, °C: explodes [CRC94]
Reactions: evolves N₂ at 120°C [HAW93]

311

Compound: Barium bis(2,2,6,6-tetramethyl-3,5-heptanedionate) hydrate

Formula: $[(CH_3)_3CCOCH=C(O)C(CH_3)_3]_2Ba \cdot xH_2O$

Molecular Formula: $C_{22}H_{38}BaO_4$ (anhydrous)

Molecular Weight: 503.866 (anhydrous)

CAS RN: 17594-47-7

Properties: used in the preparation of superconducting thin films [ALD94]

Melting Point, °C: 175-180 [ALD94]

312

Compound: Barium bismuth oxide

Formula: $BaBi(III)_{0.5}Bi(V)_{0.5}O_3$

Molecular Formula: $BaBiO_3$

Molecular Weight: 394.305

CAS RN: 12785-50-1

Reactions: transitions: monocl to hex at 132°C; hex to cub at 440°C [TAY85]

Thermal Expansion Coefficient: from 25°C to: 100°C (0.15); 200°C (0.45); 400°C (1.26); 600°C (2.10); 800°C (3.96) [TAY85]

313

Compound: Barium bromate

Formula: $Ba(BrO_3)_2$

Molecular Formula: $BaBr_2O_6$

Molecular Weight: 393.131

CAS RN: 13967-90-3

Properties: can be prepared from potassium bromate and barium chloride [MER89]

Solubility: g/100g soln, H_2O: 0.286 (0°C), 0.788 (25°C); 5.39 (99.65°C); equilibrium solid phase $Ba(BrO_3)_2$ [KRU93]

314

Compound: Barium bromate monohydrate

Formula: $Ba(BrO_3)_2 \cdot H_2O$

Molecular Formula: $BaBr_2H_2O_7$

Molecular Weight: 411.147

CAS RN: 10326-26-8

Properties: white monocl cryst or powd; obtained by addition of bromine to hot barium hydroxide solution, followed by crystallization; used as an oxidizing agent and corrosion inhibitor [HAW93] [MER89]

Solubility: g/100mL: 0.44 (10°C), 0.96 (30°C), 5.39 (100°C) [MER89]; i alcohol [HAW93]

Density, g/cm³: 3.99 [MER89]

Melting Point, °C: decomposes 260 [MER89]

315

Compound: Barium bromide

Formula: $BaBr_2$

Molecular Formula: $BaBr_2$

Molecular Weight: 297.135

CAS RN: 10553-31-8

Properties: white powd; hygr; -20 mesh 99.9% and 99.995% purity; enthalpy of fusion 31.96 kJ/mol; made by reacting barium carbonate and hydrobromic acid [KIR78] [ALD94] [CIC73] [STR93] [CER91]

Solubility: g/100g soln, H_2O: 47.5(0°C), 50.0 (25°C), 57.8 (100°C); equilibrium solid phase, $BaBr_2 \cdot 2H_2O$ [KRU93]

Density, g/cm³: 4.781 [KIR78]

Melting Point, °C: 857 [KNA91]

Boiling Point, °C: 1835 [KNA91]

316

Compound: Barium bromide dihydrate

Formula: $BaBr_2 \cdot 2H_2O$

Molecular Formula: $BaBr_2H_4O_2$

Molecular Weight: 333.166

CAS RN: 7791-28-8

Properties: white cryst; can be obtained from HBr and BaS solutions, followed by crystallization; used in manufacturing bromides [HAW93] [STR93]

Solubility: g/100g H_2O: 98 (0°C), 101 (10°C), 104 (20°C), 109 (30°C), 114 (40°C), 123 (60°C), 135 (80°C), 149 (100°C) [LAN85]

Density, g/cm³: 3.58 [KIR78]

Melting Point, °C: see anhydrous $BaBr_2$

Boiling Point, °C: decomposes [CIC73]

Reactions: minus H_2O at 75°C; minus $2H_2O$ at 100°C [KIR78]

317

Compound: Barium calcium tungstate

Synonyms: barium calcium tungsten oxide

Formula: Ba_2CaWO_6

Molecular Formula: Ba_2CaO_6W

Molecular Weight: 594.568

CAS RN: 15552-14-4

Properties: -325 mesh 99.9% purity [ALF93]

Melting Point, °C: 1420 decomposes [ALF93]

318

Compound: Barium carbide

Formula: BaC_2

Molecular Formula: C_2Ba

Molecular Weight: 161.349

CAS RN: 50813-65-5
Properties: gray tetr; -8 mesh 99.5% purity [CER91] [CRC92]
Solubility: decomposes in H_2O yielding acetylene, $HC{\equiv}CH$; decomposed in acids [CRC92]
Density, g/cm³: 3.74 [CRC92]
Melting Point, °C: decomposes [KNA91]

319
Compound: Barium carbonate
Synonyms: witherite
Formula: $\alpha\text{-}BaCO_3$
Molecular Formula: $CBaO_3$
Molecular Weight: 197.336
CAS RN: 513-77-9
Properties: white heavy powd; rhomb; witherite is naturally occurring mineral; hardness 3.0-3.75 Mohs; manufactured by precipitation from a solution of BaS by Na_2CO_3 at 60-70°C; used to remove sulfates from chlor-alkali cells, in brick-making, in oil well industry, to produce barium titanate, as a ceramic flux and in radiation resistant glass for color television [HAW93] [CIC73] [KIR78]
Solubility: g/1000g H_2O: 0.0180 (25°C) [KRU93]; s acid, NH4Cl; i alcohol [CRC92]
Density, g/cm³: 4.2865 [MER89]
Melting Point, °C: 174 (90 atm), 811 (1 atm) [HAW93]
Reactions: ~ 1300°C decomposes into BaO and CO_2 [MER89]

320
Compound: Barium chlorate
Formula: $Ba(ClO_3)_2$
Molecular Formula: $BaCl_2O_6$
Molecular Weight: 304.228
CAS RN: 13477-00-4
Properties: -80 mesh 99.9% purity; prepared by electrolysis of $BaCl_2$ solutions; monohydrate: colorless monocl [CER91] [CRC94] [MER89]
Solubility: 27.5g/100g soln (25°C); 67g/100g soln (100°C) [CIC73]
Density, g/cm³: monohydrate: 3.18 [CRC94]
Reactions: monohydrate: minus H_2O at 120°C, minus O at 250°C [CRC94]

321
Compound: Barium chlorate monohydrate
Formula: $Ba(ClO_3)_2 \cdot H_2O$
Molecular Formula: $BaCl_2H_2O_7$
Molecular Weight: 322.244

CAS RN: 10294-38-9
Properties: monocl prismatic white cryst; combustible, used in fireworks, explosives and as a textile mordant [MER89]
Solubility: g/100g H_2O: 20.3 (0°C), 26.9 (10°C), 33.9 (20°C), 41.6 (30°C), 49.7 (40°C), 66.7 (60°C), 84.8 (80°C), 105 (100°C); s HCl [LAN85] [MER89]
Density, g/cm³: 3.179 [MER89]
Melting Point, °C: 414 [MER89]
Reactions: minus H_2O at 120°C; evolves oxygen at 250°C [MER89]

322
Compound: Barium chloride
Formula: $\alpha\text{-}BaCl_2$
Molecular Formula: $BaCl_2$
Molecular Weight: 208.232
CAS RN: 10361-37-2
Properties: white powd; two forms: monocl, cub; enthalpy of fusion 16.00 kJ/mol [KIR78] [STR93] [CIC73] [CRC93]
Solubility: g/100g soln, H_2O: 23.8 (0°C), 27.1 (25°C), 37.0±0.3 (100°C); Solid phase, $BaCl_2 \cdot 2H_2O$ (100°C) [KRU93]
Density, g/cm³: 3.856 [KIR78]
Melting Point, °C: 960 [KNA91]
Boiling Point, °C: 1560 [KNA91]
Reactions: transition α (monocl) to β (cub) at 925°C [SCH93] [KIR78]

323
Compound: Barium chloride dihydrate
Formula: $BaCl_2 \cdot 2H_2O$
Molecular Formula: $BaCl_2H_4O_2$
Molecular Weight: 244.263
CAS RN: 10326-27-9
Properties: white monocl; manufactured from BaS and HCl, followed by evaporation; used to make barium pigments and as a flux for Mg metal [KIR78]
Solubility: 31.7g/100g H_2O (0°C), 35.8 g/100g H_2O (20°C), 58.7g/100g H_2O (100°C); i alcohol [LAN85] [KIR78] [HAW93]
Density, g/cm³: 3.097 [KIR78]
Reactions: minus $2H_2O$ at 113°C [KIR78]

324
Compound: Barium chromate
Synonyms: lemon chrome, baryta yellow
Formula: $BaCrO_4$
Molecular Formula: $BaCrO_4$

Molecular Weight: 253.321
CAS RN: 10294-40-3
Properties: heavy yellow powd; rhomb, monocl; prepared from $BaCl_2$ and Na_2CrO_4 solutions, followed by filtering resulting precipitate; used in safety matches, as a corrosion inhibitor [MER89] [HAW93]
Solubility: g/L soln, H_2O: 0.002 (0°C), 0.00291 (25°C) [KRU93]
Density, g/cm³: 4.50 [MER89]
Melting Point, °C: decomposes [KIR78]

325
Compound: Barium chromate(V)
Formula: $Ba_3(CrO_4)_2$
Molecular Formula: $Ba_3Cr_2O_8$
Molecular Weight: 643.968
CAS RN: 12345-14-1
Properties: greenish black cryst [KIR78]
Solubility: s, decomposing in H_2O; s dil acids [KIR78]

326
Compound: Barium cyanide
Formula: $Ba(CN)_2$
Molecular Formula: C_2BaN_2
Molecular Weight: 189.362
CAS RN: 542-62-1
Properties: white cryst powd; slowly decomposes in air; obtained by reaction of HCN and barium hydroxide, followed by crystallization; used in metallurgy and electroplating [HAW93] [MER89]
Solubility: 80.0 g/100mL H_2O (14°C) [CRC93]; s alcohol [HAW93]

327
Compound: Barium dichromate dihydrate
Formula: $BaCr_2O_7 \cdot 2H_2O$
Molecular Formula: $BaCr_2H_4O_9$
Molecular Weight: 389.346
CAS RN: 10031-16-0
Properties: brownish red needles; used in ceramics [KIR78]
Solubility: decomposed in H_2O [KIR78]
Melting Point, °C: decomposes [CRC92]
Reactions: minus $2H_2O$ at 120°C [CRC92]

328
Compound: Barium diphenylamine-4-sulfonate

Synonyms: diphenylamine-4-sulfonic acid, barium salt
Formula: $(C_6H_5NHC_6H_4SO_3)_2Ba$
Molecular Formula: $C_{24}H_{20}BaO_6S_2$
Molecular Weight: 633.878
CAS RN: 6211-24-1
Properties: white cryst leaflets; prepared by acetylation and subsequent sulfonation of diphenylamine; used as an oxidation/reduction indicator [ALD94] [HAW93] [MER89]
Solubility: sl s H_2O [MER89]

329
Compound: Barium dithionate dihydrate
Synonyms: barium hyposulfate dihydrate
Formula: $Ba(SO_3)_2 \cdot 2H_2O$
Molecular Formula: $BaH_4O_8S_2$
Molecular Weight: 333.486
CAS RN: 13845-17-5
Properties: colorless cryst; prepared from barium hydroxide and manganese dithionate [HAW93]
Solubility: s in 4 parts H_2O; sl s alcohol [MER89]
Density, g/cm³: 4.54 [MER89]
Reactions: minus SO_2 >150°C forming $BaSO_4$ [MER89]

330
Compound: Barium ferrite
Formula: $BaFe_{12}O_{19}$
Molecular Formula: $BaFe_{12}O_{19}$
Molecular Weight: 1111.456
CAS RN: 11138-11-7
Properties: powd, -325 mesh 98% purity; used as a permanent magnet material [HAW93] [CER91]

331
Compound: Barium ferrocyanide hexahydrate
Synonyms: barium hexacyanoferrate(II)
Formula: $Ba_2Fe(CN)_6 \cdot 6H_2O$
Molecular Formula: $C_6H_{12}Ba_2FeN_6O_6$
Molecular Weight: 594.696
CAS RN: 13821-06-2
Properties: yellow, becomes colorless with loss of H_2O; rectangular monocl [MER89]
Solubility: 0.17 g/100mL (288 K) H_2O; i alcohol [CRC92]
Density, g/cm³: 2.666 [CRC94]
Reactions: minus H_2O (40°C); decomposes, losing HCN, at 80°C [MER89]

332

Compound: Barium fluoride
Formula: BaF_2
Molecular Formula: BaF_2
Molecular Weight: 175.324
CAS RN: 7787-32-8
Properties: white cub cryst, or 99.9% pure 3-6 mm melted pieces; enthalpy of fusion 23.36 kJ/mol; enthalpy of vaporization 347.3 kJ/mol; may be prepared by reacting barium carbonate with HF solution; finds used as a component in welding flux, and to produce infrared transparent films [KIR78] [CIC73] [HAW93] [MER89] [CER91]
Solubility: g/L soln, H_2O: 1.586 (10°), 1.617±0.003 (25°C); s HCl, HNO_3 [KRU93] [MER89]
Density, g/cm³: 4.83 [MER89]
Melting Point, °C: 1354 [HAW93]
Boiling Point, °C: 2260 [CIC73]

333

Compound: Barium hexaboride
Synonyms: barium boride
Formula: BaB_6
Molecular Formula: B_6Ba
Molecular Weight: 202.193
CAS RN: 12046-08-1
Properties: metallic black cub; -100 mesh 99.5% purity [CER91] [CRC93]
Solubility: i H_2O; s HNO_3; i HCl [CRC93]
Density, g/cm³: 4.36 [CRC93]
Melting Point, °C: 2070 [KIR78]

334

Compound: Barium hexafluorogermanate
Formula: $BaGeF_6$
Molecular Formula: BaF_6Ge
Molecular Weight: 323.927
Properties: white cryst solid [HAW93]
Density, g/cm³: 4.56 [HAW93]
Melting Point, °C: ~665 [HAW93]
Reactions: decomposes to BaF_2 and GeF_4 [HAW93]

335

Compound: Barium hexafluorosilicate
Formula: $BaSiF_6$
Molecular Formula: BaF_6Si
Molecular Weight: 279.403
CAS RN: 17125-80-3

Properties: white ortho-rhomb needles; prolonged contact with water induces hydrolysis, which is accelerated by alkali; formed from $BaCl_2$ and H_2SiF_6; used in ceramics and insecticides [HAW93] [MER89]
Solubility: 0.015 g/100mL H_2O (0°C), 0.0235 (25°C), 0.091 (100°C) [MER89]
Density, g/cm³: 4.29 [MER89]
Melting Point, °C: decomposes 300 [MER89]

336

Compound: Barium hydride
Formula: BaH_2
Molecular Formula: BaH_2
Molecular Weight: 139.343
CAS RN: 13477-09-3
Properties: gray cryst; -60 mesh with 99.7% purity; sensitive to moisture; resembles CaH_2 in properties [KIR80] [CER91] [STR93]
Solubility: decomposes in water to $Ba(OH)_2 + H_2$; decomposes in acids [CRC92]
Density, g/cm³: 4.16 [KIR80]
Melting Point, °C: decomposes 675 [STR93]
Boiling Point, °C: ~1673 [CRC92]

337

Compound: Barium hydrogen phosphate
Synonyms: barium phosphate, dibasic
Formula: $BaHPO_4$
Molecular Formula: $BaHO_4P$
Molecular Weight: 233.306
CAS RN: 10048-98-3
Properties: cryst white powd; used as a flame retardant and in phosphors [STR93] [MER89] [HAW93]
Solubility: i H_2O; s dil HCl or HNO_3 [MER89]
Density, g/cm³: 4.16 [MER89]
Melting Point, °C: decomposes 410 [CRC94]

338

Compound: Barium hydrosulfide
Formula: $Ba(HS)_2$
Molecular Formula: BaH_2S_2
Molecular Weight: 203.475
CAS RN: 25417-81-6
Properties: yellow cyst; hygr; preparation: reaction of H_2S with BaS, followed by precipitation of $Ba(HS)_2 \cdot 4H_2O$ by alcohol and dehydration [KIR78] [HAW93]
Solubility: g/100g H_2O soln: 0°C (32.6); 20°C (32.8); 100°C (43.7) [KIR78]

339

Compound: Barium hydrosulfide tetrahydrate
Formula: $Ba(HS)_2 \cdot 4H_2O$
Molecular Formula: $BaH_{10}O_4S_2$
Molecular Weight: 275.536
CAS RN: 12230-74-9
Properties: yellow rhomb; obtained by passing H_2S through BaS solution, followed by addition of alcohol and subsequent crystallization [KIR78]
Solubility: g/100g, H_2O: 32.6 (0°C), 32.8 (20°C), 43.7 (100°C) [KIR78]
Melting Point, °C: decomposes 50 [KIR78]

340

Compound: Barium hydroxide
Synonyms: caustic baryta, anhydrous barium hydroxide
Formula: $Ba(OH)_2$
Molecular Formula: BaH_2O_2
Molecular Weight: 171.342
CAS RN: 17194-00-2
Properties: white powd, hygr; enthalpy of fusion 16.70 kJ/mol [CRC93] [STR93]
Solubility: g/100 g soln, H_2O: 1.67 (0°C), 4.68 (25°C), 101.4 (80°C); Solid phase, $Ba(OH)_2 \cdot 8H_2O$ [KRU93]
Melting Point, °C: 408 [KNA91]
Boiling Point, °C: decomposes 1032 (calculated) [KNA91]

341

Compound: Barium hydroxide monohydrate
Formula: $Ba(OH)_2 \cdot H_2O$
Molecular Formula: BaH_4O_3
Molecular Weight: 189.357
CAS RN: 22326-55-2
Properties: white powd; formed when barium hydroxide octahydrate is 'boiled dry' in CO_2 free atm; used in manufacturing oil and grease additives, soaps and refining beet sugar [KIR78] [MER83] [HAW93]
Solubility: sl s H_2O; s acids [HAW93]
Density, g/cm³: 3.743 [MER83]
Reactions: minus H_2O <407°C, dehydrates to BaO ~800°C [KIR78]

342

Compound: Barium hydroxide octahydrate
Formula: $Ba(OH)_2 \cdot 8H_2O$
Molecular Formula: $BaH_{18}O_{10}$
Molecular Weight: 315.464
CAS RN: 12230-71-6

Properties: white monocl cryst; rapidly absorbs CO_2 from air; vapor pressure at mp 30.3 kPa; prepared by dissolution of BaO in hot water, followed by crystallization; used as plastic stabilizer, an additive in papermaking, a pigment dispersant, and to protect limestone materials from deterioration [KIR78] [STR93] [MER89]
Solubility: $gBa(OH)_2$/100g soln: 1.65 (0°C), 3.76 (20°C), 48.5 (78°C) [KIR78]
Density, g/cm³: 2.18 [KIR78]
Melting Point, °C: 77.9, melts in waters of crystallization [KIR78]

343

Compound: Barium hypophosphite monohydrate
Formula: $Ba(H_2PO_2)_2 \cdot H_2O$
Molecular Formula: $BaH_6O_5P_2$
Molecular Weight: 285.320
CAS RN: 14871-79-5
Properties: monocl platelets; can be prepared by reacting white phosphorus and barium hydroxide; used in nickel plating; anhydrous material, $Ba(H_2PO_2)_2$, is white odorless cryst powd [MER89] [HAW93]
Solubility: g/100mL: 28.6 (17°C), 33.3 (100°C); i alcohol [MER89]
Density, g/cm³: 2.90 [MER89]
Melting Point, °C: decomposes 100-150 [CRC94]

344

Compound: Barium iodate
Formula: $Ba(IO_3)_2$
Molecular Formula: BaI_2O_6
Molecular Weight: 487.132
CAS RN: 10567-69-8
Properties: white cryst powd [HAW93]
Solubility: g/L soln, H_2O: 0.395 (25°C) [KRU93]; g/100g H_2O: 0.035 (20°C), 0.046 (30°C), 0.057 (40°C) [LAN85]
Density, g/cm³: 5.23 [HAW93]
Melting Point, °C: 476, decomposes [HAW93]

345

Compound: Barium iodate monohydrate
Formula: $Ba(IO_3)_2 \cdot H_2O$
Molecular Formula: $BaH_2I_2O_7$
Molecular Weight: 505.148
CAS RN: 7787-34-0
Properties: cryst [MER89]
Solubility: s in 3350 parts H_2O (25°C), 625 parts boiling H_2O; s HCl, HNO_3; i alcohol [MER89]

Density, g/cm³: 5.00 [MER89]
Reactions: minus H_2O at 130°C [MER89]

346
Compound: Barium iodide
Formula: BaI_2
Molecular Formula: BaI_2
Molecular Weight: 391.136
CAS RN: 13718-50-8
Properties: off-white powd; hygr; enthalpy of fusion 26.53 kJ/mol [CIC73] [STR93] [CRC93]
Solubility: g/100g soln, H_2O: 62.5 (0°C), 68.8 (25°C), 73.35 (98.9°C); solid phase, $2BaI_2 \cdot 15H_2O$ (0°C, 25°C), $BaI_2 \cdot 2H_2O$ + $BaI_2 \cdot H_2O$ (98.9°C) [KRU93]
Density, g/cm³: 5.15 [KIR78]
Melting Point, °C: 711 [CIC73]
Boiling Point, °C: decomposes [CIC73]

347
Compound: Barium iodide dihydrate
Formula: $BaI_2 \cdot 2H_2O$
Molecular Formula: $BaH_4I_2O_2$
Molecular Weight: 427.167
CAS RN: 7787-33-9
Properties: colorless, odorless cryst; rapidly becomes reddish in air due to liberation of iodine; prepared from HI and barium hydroxide solutions with subsequent crystallization from hot solutions; used to prepare other iodides [MER89] [HAW93]
Solubility: g BaI_2/100g soln: 169.4 (0°C), 271.0 (100°C) [KIR78]; g/100g H_2O: 182 (0°C), 223 (20°C), 301 (100°C) [LAN85]
Density, g/cm³: 4.917 [KIR78]
Melting Point, °C: 740 [HAW93]
Reactions: minus $2H_2O$ at 150°C [KIR78]

348
Compound: Barium lead oxide
Formula: $BaPbO_3$
Molecular Formula: BaO_3Pb
Molecular Weight: 392.525
CAS RN: 12047-25-5
Properties: monocl [TAY85]
Reactions: transition monocl to hex (127°C); hex to cub (423°C) [TAY85]
Thermal Expansion Coefficient: from 25°C to: 500°C (1.29); 600°C (1.62); 800°C (2.22) [TAY85]

349
Compound: Barium manganate(VI)
Synonyms: manganese green
Formula: $BaMnO_4$
Molecular Formula: $BaMnO_4$
Molecular Weight: 256.263
CAS RN: 7787-35-1
Properties: greenish gray cryst; sensitive to moisture; uses: oxidizes primary and secondary alcohols to carbonyl compounds, paint pigment [HAW93] [STR93] [ALD93]
Solubility: disproportionates in H_2O to $Ba(MnO_4)_2$ + MnO_2 [MER89]
Density, g/cm³: 4.85 [MER89]

350
Compound: Barium metaborate dihydrate
Synonyms: barium borate dihydrate
Formula: $Ba(BO_2)_2 \cdot 2H_2O$
Molecular Formula: $B_2BaH_4O_6$
Molecular Weight: 258.977
CAS RN: 23436-05-7
Properties: prepared by precipitation by adding sodium metaborate solution to a solution of barium chloride at 90-95°C; at room temp the tetrahydrate precipitates; finds use as a fire retardant for paints, plastics textiles and paper products [KIR78]
Solubility: 12.5g/L of $BaO \cdot B_2O_3 \cdot 4H_2O$ in H_2O (25°C) [KIR78]
Reactions: dehydrates above 140°C [KIR78]

351
Compound: Barium metaborate monohydrate
Synonyms: barium borate monohydrate
Formula: $Ba(BO_2)_2 \cdot H_2O$
Molecular Formula: $B_2BaH_2O_5$
Molecular Weight: 240.962
CAS RN: 26124-86-7
Properties: white powd; manufactured from a solution of BaS and sodium tetraborate; used to add mold, corrosion and fire resistance to paint [KIR78]
Solubility: 0.3% H_2O [KIR78]
Density, g/cm³: 3.25-3.35 [KIR78]
Melting Point, °C: >900 [KIR78]

352
Compound: Barium metaphosphate
Formula: $Ba(PO_3)_2$
Molecular Formula: BaO_6P_2
Molecular Weight: 295.271

CAS RN: 13466-20-1
Properties: white powd; used in glasses, porcelain and enamel [HAW93]
Solubility: i H_2O; slowly dissolves in acids [HAW93]
Melting Point, °C: 1560 [ALF93]

353

Compound: Barium metasilicate
Synonyms: monobarium silicate
Formula: $BaSiO_3$
Molecular Formula: BaO_3Si
Molecular Weight: 213.411
CAS RN: 13255-26-0
Properties: colorless rhomb powd; can be formed by heating BaO, $BaCO_3$ or $BaSO_4$ to white heat with SiO_2, which also forms $3BaO \cdot SiO_2$. The tribarium silicate hydrolyzes to form $BaSiO_3$ and $Ba(OH)_2$, which is the basis for the Deguide process; used in ceramics [KIR78] [HAW93]
Solubility: i H_2O; s acids [HAW93]
Density, g/cm^3: 4.4 [STR93]
Melting Point, °C: 1605 [KNA91]

354

Compound: Barium molybdate
Formula: $BaMoO_4$
Molecular Formula: $BaMoO_4$
Molecular Weight: 297.265
CAS RN: 7787-37-3
Properties: white powd; scheelite structure, c/a = 2.29; used in electronic and optical equipment, and in paint pigments for protective coatings [HAW93] [MER52] [KIR81]
Solubility: 0.0055 g/100g H_2O [KIR81]
Density, g/cm^3: 4.975 [KIR81]
Melting Point, °C: 1450 [KNA91]

355

Compound: Barium niobate
Synonyms: barium niobate(V)
Formula: $Ba(NbO_3)_2$
Molecular Formula: $BaNb_2O_6$
Molecular Weight: 419.136
CAS RN: 12009-14-2
Properties: yellow hex or ortho; -100 mesh, 99.9% purity [CER91] [LID94]
Density, g/cm^3: 2.8 [LID94]
Melting Point, °C: 1450 [LID94]

356

Compound: Barium nitrate
Synonyms: nitrobarite
Formula: $Ba(NO_3)_2$
Molecular Formula: BaN_2O_6
Molecular Weight: 261.336
CAS RN: 10022-31-8
Properties: white cryst powd; prepared from $BaCO_3$ suspension and HNO_3, followed by crystallization; used in pyrotechnics, green flares, tracer bullets and detonators [STR93] [MER89] [KIR78]
Solubility: g/100g soln, H_2O: 4.72 (0°C), 9.27 (25°C), 25.6 (100°C); Solid phase, $Ba(NO_3)_2$ [KRU93]
Density, g/cm^3: 3.24 [KIR78]
Melting Point, °C: 592 [KIR78]
Reactions: decomposes above 590°C [MER89]

357

Compound: Barium nitride
Formula: Ba_3N_2
Molecular Formula: Ba_3N_2
Molecular Weight: 439.994
CAS RN: 12047-79-9
Properties: yellowish brown; -20 mesh, 99.7% purity [CER91] [CIC73]
Solubility: decomposed in H_2O [CRC92]
Density, g/cm^3: 4.78 [ALF93]
Boiling Point, °C: 1000, vacuum [ALF93]

358

Compound: Barium nitrite
Formula: $Ba(NO_2)_2$
Molecular Formula: BaN_2O_4
Molecular Weight: 229.338
CAS RN: 13465-94-6
Properties: colorless, hex [CRC92]
Solubility: 67.5 g/100mL (20°C) H_2O; sl s alcohol [CRC92]
Density, g/cm^3: 3.234 [KIR78]
Melting Point, °C: 267 [KIR78]
Reactions: decomposes at 270°C to BaO, NO and N_2 [KIR78]

359

Compound: Barium nitrite monohydrate
Formula: $Ba(NO_2)_2 \cdot H_2O$
Molecular Formula: $BaH_2N_2O_5$
Molecular Weight: 247.353
CAS RN: 7787-38-4

Properties: white to yellowish hex cryst powd; crystallized from a stoichiometric solution of $BaCl_2$ and $NaNO_2$; used as a corrosion inhibitor, in explosives, and for diazotization [HAW93] [KIR78]
Solubility: g $Ba(NO_2)_2$/100g H_2O: 54.8 (0°C), 319 (100°C) [KIR78]; s alcohol [HAW93]
Density, g/cm³: 3.173 [KIR78]
Melting Point, °C: 217 decomposes [HAW93]
Reactions: minus H_2O at 116°C [KIR78]

360

Compound: Barium oxalate
Synonyms: ethanedioic acid, barium salt
Formula: BaC_2O_4
Molecular Formula: C_2BaO_4
Molecular Weight: 225.347
CAS RN: 516-02-9
Properties: white powd, 99.999% purity [ALF93]
Solubility: g/1000 g soln, H_2O: 0.053 (0°C), 0.1087 (25°C), 0.285 (73°C); Solid phase, $BaC_2O_4 \cdot 2H_2O$ [KRU93]
Density, g/cm³: 2.658 [STR93]
Melting Point, °C: 400, decomposes [STR93]

361

Compound: Barium oxalate monohydrate
Synonyms: ethanedioic acid, barium salt monohydrate
Formula: $BaC_2O_4 \cdot H_2O$
Molecular Formula: $C_2H_2BaO_5$
Molecular Weight: 243.362
CAS RN: 13463-22-4
Properties: white cryst powd; used in pyrotechnics, as an analytical reagent [HAW93]
Solubility: s in 10,000 parts cold H_2O, 5000 parts boiling H_2O; s dil HCl, HNO_3 [MER89]
Density, g/cm³: 2.66 [MER89]

362

Compound: Barium oxide
Synonyms: barium monoxide
Formula: BaO
Molecular Formula: BaO
Molecular Weight: 153.326
CAS RN: 1304-28-5

Properties: white to yellowish white powd; reacts with atm CO_2 and H_2O forming the hydroxide and carbonate, with evolution of heat; two forms: cub, hex; a = 0.55391 nm; enthalpy of fusion 59.00 kJ/mol; made by heating $BaCO_3$ and carbon; uses: dehydrate solvents, additive in detergents and lubricating oils [CIC73] [HAW93] [KIR78] [CRC93]
Solubility: 3.48 g/100mL (0°C) H_2O; s dil a, alcohol; i acetone, NH_3 [CRC92]
Density, g/cm³: 5.72(cub), 5.32(hex) [KIR78]
Melting Point, °C: 2013 [CRC93]
Boiling Point, °C: ~2000 [KIR78]
Reactions: $BaO + O_2(g) = BaO_2$ at 500°C [KIR78]
Thermal Expansion Coefficient: from 25°C to 100°C (0.33); 200°C (0.78); 400°C (1.77); 600°C (2.91); 800°C (4.08); 1000°C (5.25); 1200°C (6.51) [TAY84a]

363

Compound: Barium perchlorate
Formula: $Ba(ClO_4)_2$
Molecular Formula: $BaCl_2O_8$
Molecular Weight: 336.227
CAS RN: 13465-95-7
Properties: colorless hex cryst; uses: efficient desiccant [ALD93] [CRC92] [ALF93]
Solubility: g/100 g soln, H_2O: 67.3 (0°C), 74.3 (20°C), 84.9 (100°C); Solid phase, $Ba(ClO_4)_2 \cdot 3H_2O$ [KRU93]; s 125g/100g ethanol (25°C) [CIC73]
Density, g/cm³: 3.20 [ALF93]
Melting Point, °C: 505 [ALD93]

364

Compound: Barium perchlorate trihydrate
Formula: $Ba(ClO_4)_2 \cdot 3H_2O$
Molecular Formula: $BaCl_2H_6O_{11}$
Molecular Weight: 390.273
CAS RN: 10294-39-0
Properties: colorless cryst; oxidizing agent; used in the manufacture of explosives and in rocket fuels [HAW93] [MER89] [STR93]
Solubility: g/100g H_2O: 239 (0°C), 336 (20°C), 653 (100°C); s methanol [HAW93] [LAN85]
Density, g/cm³: 2.74 [HAW93]

365

Compound: Barium permanganate
Formula: $Ba(MnO_4)_2$
Molecular Formula: $BaMn_2O_8$
Molecular Weight: 375.198
CAS RN: 7787-36-2

Properties: brownish violet to black cryst; oxidizing agent; used as a disinfectant, in dry cell batteries [HAW93]

Solubility: 62.5 g/100mL (10°C) H_2O; decomposed by alcohol [MER89] [CRC92]

Density, g/cm³: 3.77 [MER89]

Melting Point, °C: decomposes 200 [CRC92]

366

Compound: Barium peroxide

Synonyms: barium dioxide

Formula: BaO_2

Molecular Formula: BaO_2

Molecular Weight: 169.326

CAS RN: 1304-29-6

Properties: white or grayish white heavy powd, -80 mesh 99% pure; decomposes slowly in air; oxidizing agent; can be prepared by heating BaO in oxygen or air at 500°C; used to bleach materials and to decolorize glass [HAW93] [MER89] [KIR78] [CER91]

Solubility: i H_2O, but slowly decomposed by contact with H_2O [MER89]

Density, g/cm³: 4.96 [MER83]

Melting Point, °C: 450, decomposes [STR93]

Reactions: decomposes at 700°C by reaction: BaO_2 to $BaO + O_2$ [KIR78]

367

Compound: Barium potassium chromate

Synonyms: Pigment E

Formula: $BaK_2(CrO_4)_2$

Molecular Formula: $BaCr_2K_2O_8$

Molecular Weight: 447.511

CAS RN: 27133-66-0

Properties: pale yellow solid; has lower chloride and sulfate content than other chromate pigments; prepared by reacting $K_2Cr_2O_7$ and $BaCO_3$ at 500°C; used in paints to protect Fe and steel from corrosion, and to form strong and elastic paint films [HAW93] [KIR78]

Solubility: g/100g H_2O: 57.2 (0°C), 57.5 (30°C), 82.7 (100°C) [LAN85]

Density, g/cm³: 3.65 [KIR78]

368

Compound: Barium pyrophosphate

Formula: $Ba_2P_2O_7$

Molecular Formula: $Ba_2O_7P_2$

Molecular Weight: 448.597

CAS RN: 13466-21-2

Properties: white powd; rhomb [CRC77] [HAW93]

Solubility: 0.01 g/100 mL H_2O; s acids, NH_4 salts [CRC77] [HAW93]

Density, g/cm³: 3.9 [CRC93]

369

Compound: Barium selenate

Formula: $BaSeO_4$

Molecular Formula: BaO_4Se

Molecular Weight: 280.285

CAS RN: 7787-41-9

Properties: ortho-rhomb cryst; preparation: heating $BaCO_3$ and Se [MER52]

Solubility: g/L soln; H_2O; 0.081 (25°C); s HCl, i HNO_3 [MER52] [KRU93]

Density, g/cm³: 4.75 [MER52]

Reactions: heating causes decomposition [MER52]

370

Compound: Barium selenide

Formula: BaSe

Molecular Formula: BaSe

Molecular Weight: 216.287

CAS RN: 1304-39-8

Properties: cub microcryst powd, -20 mesh 99.5% purity; turns red in air; used in semiconductors and photocells [HAW93] [CER91] [MER89]

Solubility: decomposed by H_2O [MER89]

Density, g/cm³: 5.02 [MER89]

371

Compound: Barium selenite

Formula: $BaSeO_3$

Molecular Formula: BaO_3Se

Molecular Weight: 264.285

CAS RN: 13718-59-7

Properties: solid [ALD93]

Solubility: g/100 g soln, H_2O: 0.005 (0°C), 0.005 (25°); Solid phase, $BaSeO_3$ [KRU93]

372

Compound: Barium silicate

Synonyms: pentabarium octasilicate

Formula: $5BaO \cdot 8SiO_2$

Molecular Formula: $Ba_5O_{21}Si_8$

Molecular Weight: 1247.306

Properties: a = 3.365 nm, b = 0.4697 nm, c = 1.3896 nm [TAY88a]

Melting Point, °C: 1445 [TAY88a]

Thermal Expansion Coefficient: from 25°C to:
100°C (0.21), 200°C (0.51), 400°C (1.29),
600°C (2.19), 800°C (3.24), 1000°C (4.47),
1200°C (5.85) [TAY88a]

373
Compound: Barium silicate
Synonyms: sanbornite
Formula: BaO·2SiO$_2$
Molecular Formula: BaO$_5$Si$_2$
Molecular Weight: 273.495
CAS RN: 12650-28-1
Properties: monocl, a = 2.3206 nm, b = 0.4661 nm,
c = 1.3613 nm [TAY87]
Density, g/cm^3: 3.70 [LID94]
Melting Point, °C: 1420 [TAY87]
Thermal Expansion Coefficient: from 25°C to:
100°C (0.27), 200°C (0.66), 400°C (1.44),
600°C (2.22), 800°C (3.06), 1000°C (3.90),
1200°C (4.74) [TAY87]

374
Compound: Barium silicate
Synonyms: dibarium trisilicate
Formula: 2BaO·3SiO$_2$
Molecular Formula: Ba$_2$O$_8$Si$_3$
Molecular Weight: 486.906
CAS RN: 14871-82-0
Properties: a = 1.246 nm, b = 0.4687 nm, c =
1.3950 nm [TAY88a]
Melting Point, °C: 1446 [TAY88a]
Thermal Expansion Coefficient: from 25°C to:
100°C (0.11), 200°C (0.25), 400°C (0.54),
600°C (0.84), 800°C (1.13), 1000°C (1.43)
[TAY88a]

375
Compound: Barium silicide
Formula: BaSi$_2$
Molecular Formula: BaSi$_2$
Molecular Weight: 193.498
CAS RN: 1304-40-1
Properties: metallic gray lumps, 6 mm pieces and
smaller, 98% pure; quite permanent in dry air,
but decomposed by moisture to evolve H$_2$;
metallurgic use to deoxidize steel [HAW93]
[MER89] [CER91]
Melting Point, °C: 1180 [STR93]

376
Compound: Barium sodium niobium oxide

Formula: Ba$_2$NaNb$_5$O$_{15}$
Molecular Formula: Ba$_2$NaNb$_5$O$_{15}$
Molecular Weight: 1002.167
CAS RN: 12323-03-4
Properties: white powd of 99.999% purity; electro-
optical cryst; used to produce coherent green
light in lasers [HAW93] [ALF93]
Melting Point, °C: 1483 [ALD94]

377
Compound: Barium stannate trihydrate
Formula: BaSnO$_3$·3H$_2$O
Molecular Formula: BaH$_6$O$_6$Sn
Molecular Weight: 358.081
CAS RN: 12009-18-6
Properties: anhydrous, 51404-76-3, -325 mesh 99%
pure; cub, a = 0.4117 nm; trihydrate is white
cryst powd; used in the production of special
ceramic insulations requiring dielectric
properties [HAW93] [TAY85] [CER91]
Solubility: sl s H$_2$O, s HCl [HAW93]
Thermal Expansion Coefficient: from 25°C to
100°C (0.18), 200°C (0.45), 400°C (1.02),
600°C (1.62), 800°C (2.31), 1000°C (3.00),
1200°C (3.78) [TAY85]

378
Compound: Barium stearate
Formula: Ba[CH$_3$(CH$_2$)$_{16}$COO]$_2$
Molecular Formula: C$_{36}$H$_{70}$BaO$_4$
Molecular Weight: 704.277
CAS RN: 6865-35-6
Properties: white powd; used as a waterproofing
agent, lubricant in metal working and in wax
compounding [HAW93] [STR93]
Solubility: i H$_2$O, alcohol [HAW93]
Density, g/cm^3: 1.145 [HAW93]
Melting Point, °C: 160 [HAW93]

379
Compound: Barium strontium niobium oxide
Formula: BaSr(NbO$_3$)$_4$
Molecular Formula: BaNb$_4$O$_{12}$Sr
Molecular Weight: 788.566
CAS RN: 37185-09-4
Properties: -325 mesh white powd [ALF93]

380
Compound: Barium strontium tungsten oxide
Formula: Ba$_2$SrWO$_6$
Molecular Formula: Ba$_2$O$_6$SrW

Molecular Weight: 642.110
CAS RN: 14871-56-8
Properties: -325 mesh powd, 99.9% purity; sensitive to moisture [ALD94] [ALF93]
Melting Point, °C: 1400 [ALD94]

381
Compound: Barium sulfate
Synonyms: barite
Formula: $BaSO_4$
Molecular Formula: BaO_4S
Molecular Weight: 233.391
CAS RN: 7727-43-7
Properties: white or yellowish, odorless and tasteless rhomb powd; hardness 3-3.5 Mohs; enthalpy of fusion 40.60 kJ/mol; obtained from mining; used in drilling muds in the form of an aq suspension, to lubricate and cool drill bits, and to plaster walls of drill holes [HAW93] [KIR78] [CRC93]
Solubility: g/L soln, H_2O: 0.00115 (0°C), 0.00223 (25°C), 0.0039 (100°C); s conc H_2SO_4 [KRU93] [KIR78]
Density, g/cm³: 4.50 [KIR78]
Melting Point, °C: 1350 [CRC93]
Boiling Point, °C: decomposes 1580 [KIR78]
Reactions: transition rhomb to monocl at 1150°C [KIR78]
Thermal Expansion Coefficient: (volume) 100°C (0.434), 200°C (1.023), 400°C (2.381) [CLA66]

382
Compound: Barium sulfide
Formula: BaS
Molecular Formula: BaS
Molecular Weight: 169.393
CAS RN: 21109-95-5
Properties: heavy, grayish-white or pale yellow powd; -100 mesh, 99% purity; used as depilatory, in luminous paints and in vulcanization of rubber [MER89] [CER91]
Solubility: g/100g soln, H_2O: 2.88 (0°), 8.95 (25°C), 60.29 (100°C) [KRU93]
Density, g/cm³: 4.36 [MER89]
Melting Point, °C: >2000 [MER89]
Reactions: slowly oxidizes in air [MER89]

383
Compound: Barium sulfide
Synonyms: black ash
Formula: BaS
Molecular Formula: BaS

Molecular Weight: 169.393
CAS RN: 21109-95-5
Properties: black powd; colorless cub if pure; oxidizes in air; black ash is commercial product produced by reduction of $BaSO_4$ with carbon at 1000 to 1250°C; used as a precursor to produce $BaCO_3$, $BaCl_2$ and other Ba compounds, as a flame retardant, to dehair hides; other sulfides are: Ba_2S_3, 5311-28-7, BaS_2, 12230-99-8, BaS_3, 12231-01-5, $BaS_4 \cdot H_2O$, 12248-67-8, BaS_5 and $BaS \cdot 6H_2O$ [HAW93] [STR93] [MER89] [KIR78]
Density, g/cm³: 4.25 [KIR78]
Melting Point, °C: 1200 [STR93]
Reactions: in H_2O: $2BaS + 2H_2O = Ba(HS)_2 + Ba(OH)_2$ [KIR78]

384
Compound: Barium sulfite
Formula: $BaSO_3$
Molecular Formula: BaO_3S
Molecular Weight: 217.391
CAS RN: 7787-39-5
Properties: white powd, cub (hex) cryst; oxidizes gradually in air to $BaSO_4$; formed by reacting a soluble sulfite and soluble barium salt; used in paper manufacturing and in analysis [MER89] [KIR78] [HAW93]
Solubility: 0.0197g/100g H_2O (20°C); 0.0018g/100g (80°C); s dil HCl [KIR78] [HAW93]
Density, g/cm³: 4.44 [LID94]
Melting Point, °C: decomposed by heat [HAW93]

385
Compound: Barium tantalate
Formula: $Ba(TaO_3)_2$
Molecular Formula: BaO_6Ta_2
Molecular Weight: 595.219
CAS RN: 12047-34-6
Properties: -100 mesh 99% pure solid [CER91]

386
Compound: Barium tartrate
Formula: $BaC_4H_4O_6$
Molecular Formula: $C_4H_4BaO_6$
Molecular Weight: 285.399
CAS RN: 5908-81-6
Properties: white cryst; used in pyrotechnics [HAW93]
Solubility: s 3400 parts H_2O; i alcohol [MER52]
Density, g/cm³: 2.98 [HAW93]

387
Compound: Barium telluride
Formula: BaTe
Molecular Formula: BaTe
Molecular Weight: 264.927
CAS RN: 12009-36-8
Properties: -20 and -28 mesh yellow powd, 99.5% pure; cub [CER91] [ALF93] [CRC94]
Density, g/cm³: 5.13 [CRC94]

388
Compound: Barium tetracyanoplatinate(II) tetrahydrate
Formula: BaPt(CN)$_4$·4H$_2$O
Molecular Formula: C$_4$H$_8$BaN$_4$O$_4$Pt
Molecular Weight: 508.540
CAS RN: 13755-32-3
Properties: yellow powd; large dichroic cryst; yellowish green by transmitted light, bluish violet by reflected light; used in x-ray screens [HAW93] [MER89] [STR93]
Solubility: s in about 35 parts H$_2$O, more in hot H$_2$O; i alcohol [HAW93] [MER89]
Density, g/cm³: 2.076 [STR93]; 3.05 [MER89]
Reactions: minus 2H$_2$O at 100°C [HAW93]

389
Compound: Barium tetraiodomercurate(II)
Synonyms: mercuric barium iodide
Formula: BaHgI$_4$
Molecular Formula: BaHgI$_4$
Molecular Weight: 845.535
CAS RN: 10048-99-4
Properties: yellow or reddish, deliq cryst [MER89]
Solubility: v s H$_2$O, alcohol [MER89]

390
Compound: Barium thiocyanate
Formula: Ba(SCN)$_2$
Molecular Formula: C$_2$BaN$_2$S$_2$
Molecular Weight: 253.494
CAS RN: 2092-17-3
Properties: deliq cryst [MER89]
Solubility: v s H$_2$O, s acetone, methanol, ethanol [MER89]; g/100 g soln, H$_2$O: 62.63 (25°C); Solid phase, Ba(SCN)$_2$·3H$_2$O [KRU93]

391
Compound: Barium thiocyanate trihydrate
Formula: Ba(SCN)$_2$·3H$_2$O
Molecular Formula: C$_2$H$_6$BaN$_2$O$_3$S$_2$

Molecular Weight: 307.540
CAS RN: 68016-36-4
Properties: white cryst; needles from H$_2$O, deliq [STR93] [MER89]
Solubility: g/100mL: 4.3 (20°C) H$_2$O; 35.0 (20°C) alcohol [CRC92]
Density, g/cm³: 2.286 [CRC92]
Reactions: loses H$_2$O at 160°C [CRC92]

392
Compound: Barium thiosulfate monohydrate
Synonyms: barium hyposulfite monohydrate
Formula: BaS$_2$O$_3$·H$_2$O
Molecular Formula: BaH$_2$O$_4$S$_2$
Molecular Weight: 267.473
CAS RN: 7787-40-8
Properties: white cryst powd; used in explosives, luminous paints, matches, varnishes and in photography [MER89] [HAW93]
Solubility: v sl s H$_2$O; i alcohol [HAW93] [MER89]
Density, g/cm³: 3.5 [HAW93]
Melting Point, °C: decomposed by heat [HAW93]

393
Compound: Barium titanate
Formula: BaO·2TiO$_2$
Molecular Formula: BaO$_5$Ti$_2$
Molecular Weight: 313.084
CAS RN: 12009-27-5
Thermal Expansion Coefficient: from 25°C to: 100°C (0.21), 200°C (0.48), 400°C (1.20), 600°C (1.95) [TOU77]

394
Compound: Barium titanate
Formula: BaO·4TiO$_2$
Molecular Formula: BaO$_9$Ti$_4$
Molecular Weight: 472.842
CAS RN: 12009-31-3
Properties: ortho-rhomb [WU 88]
Density, g/cm³: 4.55 [WU 88]
Thermal Expansion Coefficient: 100°C (0.18), 200°C (0.42), 400°C (0.96), 600°C (1.53) [TOU77]

395
Compound: Barium titanate
Synonyms: barium metatitanate
Formula: BaTiO$_3$
Molecular Formula: BaO$_3$Ti
Molecular Weight: 233.192

CAS RN: 12047-27-7

Properties: white powd or sintered lumps; two forms: tetr, a = 0.39932 nm, c = 0.40347 nm; dielectric constant ~4000; preparation: by calcining and sintering barium carbonate and anatase powd at 1300 to 1450°C, by hydrothermal synthesis, and sol-gel process using Ti(IV) isopropylate; used in ferroelectric ceramic materials, as an evaporated ceramic at 99.995% purity in dielectric films and thin film capacitors [HAW93] [STR93] [KIR83] [CER91] [CHA90] [PHU89]

Solubility: i H_2O [CRC92]

Density, g/cm³: tetr, 6.017; hex, 5.806 [KNA91]

Melting Point, °C: ~1625 [KIR78]

Reactions: transition: hex to tetr (-5°C); tetr to cub (120°C) [TAY85]

Thermal Expansion Coefficient: from 25°C to: 100°C (0.12), 200°C (0.36), 400°C (1.02), 600°C (1.80), 800°C (2.61), 1000°C (3.48), 1200°C (4.44) [TAY85]

396

Compound: Barium titanium silicate

Synonyms: benitoite

Formula: $BaO \cdot TiO_2 \cdot 3SiO_2$

Molecular Formula: BaO_9Si_3Ti

Molecular Weight: 413.446

CAS RN: 15491-35-7

Properties: gemstone; a = 0.6643 nm, c = 0.9757 nm [TAY88b] [LAN52]

Solubility: s HF [LAN52]

Density, g/cm³: 3.6 [LAN52]

Thermal Expansion Coefficient: from 25°C to: 100°C (0.09), 200°C (0.21), 400°C (0.48), 600°C (0.72), 800°C (0.99), 1000°C (1.23) [TAY88b]

397

Compound: Barium tungstate

Synonyms: barium white

Formula: $BaWO_4$

Molecular Formula: BaO_4W

Molecular Weight: 385.165

CAS RN: 7787-42-0

Properties: white powd, -200 mesh, 99.9% purity; tetr, a = 0.5614 nm, c = 1.2715 nm; used in x-ray photography, and as a pigment [STR93] [HAW93] [TAY86] [CER91]

Solubility: i H_2O [HAW93]

Density, g/cm³: 5.04 [HAW93]

Thermal Expansion Coefficient: from 25°C to: 100°C (0.27), 200°C (0.63), 400°C (1.35), 600°C (2.04), 800°C (2.76), 1000°C (3.45), 1200°C (4.14) [TAY86]

398

Compound: Barium uranium oxide

Synonyms: barium uranate

Formula: BaU_2O_7

Molecular Formula: BaO_7U_2

Molecular Weight: 725.381

CAS RN: 10380-31-1

Properties: orange or yellow powd; used in painting on porcelain [MER89]

Solubility: i H_2O, s acids [MER89]

399

Compound: Barium vanadate

Formula: $Ba_3(VO_4)_2$

Molecular Formula: $Ba_3O_8V_2$

Molecular Weight: 641.859

CAS RN: 39416-30-3

Properties: -200 mesh, 99.9% pure [CER91]

Melting Point, °C: 707 [KNA91]

400

Compound: Barium yttrium tungsten oxide

Formula: $Ba_3Y_3WO_9$

Molecular Formula: $Ba_3O_9WY_3$

Molecular Weight: 1006.534

CAS RN: 37265-86-4

Properties: -325 mesh, 99.9% purity [ALF93]

Melting Point, °C: 1470, decomposes [ALF93]

401

Compound: Barium zirconate

Formula: $BaZrO_3$

Molecular Formula: BaO_3Zr

Molecular Weight: 276.549

CAS RN: 12009-21-1

Properties: light gray or white powd, -100, +200 mesh 99% purity; a = 0.4193 nm; used in the manufacture of white silicone rubber compounds [HAW93] [TAY85] [CER91]

Solubility: i H_2O, alkalies; sl s acids [HAW93]

Density, g/cm³: 5.52 [HAW93]

Melting Point, °C: 2500 [ALF93]

Thermal Expansion Coefficient: from 25°C to: 100°C (0.18), 200°C (0.42), 400°C (0.90), 600°C (1.38), 800°C (1.86), 1000°C (2.34) [TAY85]

402

Compound: Barium zirconium phosphate
Formula: $Ba_{0.5}Zr_2(PO_4)_3$
Molecular Formula: $Ba_{0.5}O_{12}P_3Zr_2$
Molecular Weight: 536.026
Properties: NASICON structure, a = 0.8642 nm, c = 2.398 nm [TAY91a]
Thermal Expansion Coefficient: from 25°C to: 100°C (0.03), 200°C (0.06), 400°C (0.13), 600°C (0.20), 800°C (0.27), 1000°C (0.34) [TAY91a]

403

Compound: Barium zirconium silicate
Formula: $BaO \cdot ZrO_2 \cdot SiO_2$
Molecular Formula: BaO_5SiZr
Molecular Weight: 336.634β
Properties: white powd; uses: producing electrical resistor ceramics, and in glass opacifiers [HAW93]
Solubility: i H_2O, alkalies; sl s acids; s HF [HAW93]

404

Compound: Berkelium (α)
Formula: α-Bk
Molecular Formula: Bk
Molecular Weight: 247
CAS RN: 7440-40-6
Properties: metal; hex, a = 0.3416 nm, c = 1.1069 nm; in trivalent state, its properties are close to that of Ce^{+++}; ionic radius of Bk^{+++} is 0.096 nm, of Bk^{++++} is 0.0860 nm; enthalpy of vaporization 382 kJ/mol; enthalpy of fusion 7.92 kJ/mol; first discovered in 1949 [KIR78] [KAT86] [MER89]
Density, g/cm³: 14.78 (25°C) [KIR78]
Melting Point, °C: 1050 [KIR91]
Boiling Point, °C: ~2630 [KAT86]
Reactions: transfroms from hex to cub ~930°C [KAT86]

405

Compound: Berkelium (β)
Formula: β-Bk
Molecular Formula: Bk
Molecular Weight: 247
CAS RN: 7440-40-6
Properties: discovered in 1949; fcc, a = 0.4997 nm; stable <986°C [KIR78]
Density, g/cm³: 13.25 (25°C) [KIR78]
Melting Point, °C: 986 [MER89]
Boiling Point, °C: ~2630 [KAT86]

406

Compound: Beryllium
Synonyms: glucinium
Formula: Be
Molecular Formula: Be
Molecular Weight: 9.012182
CAS RN: 7440-41-7
Properties: metal; two forms, α: gray; hex, a = 0.22856 nm, b = 0.35832 nm, c/a = 0.15677 nm; β: bcc, a = 0.2551 nm; enthalpy of fusion 7.90 kJ/mol; enthalpy of sublimation ~320 kJ/mol; enthalpy of vaporization, 230 to 310 kJ/mol; electrical resistivity at 25°C is 4.266 x 10^{-8} ohm·m; velocity of sound 12,600 m/s; reflectivity, white light, 50-55%; used in semiconductor junctions [CIC73] KIR78] [CER91] [CRC93]
Solubility: s acids except for HNO_3; s alkalies [HAW93]
Density, g/cm³: 1.8477 [CIC73]
Melting Point, °C: 1278 [COT88]
Boiling Point, °C: 2970 [KIR78]
Reactions: transformation α to β at 1250°C [CIC73]
Thermal Conductivity, W/(m·K): 190 (25°C), 170 (100°C), 150 (200°C), 130 (400°C), 75 (800°C) [KIR78]
Thermal Expansion Coefficient: coefficient of linear expansion, K^{-1}: 25-100°C, 11.5 x 10^{-6}; 25-200°C, 12.7 x 10^{-6}; 25-400°C, 14.8 x 10^{-6} [KIR78]

407

Compound: Beryllium acetate
Synonyms: acetic acid, beryllium salt
Formula: $Be(CH_3COO)_2$
Molecular Formula: $C_4H_6BeO_4$
Molecular Weight: 127.101
CAS RN: 543-81-7
Properties: white cryst; preparation: can be crystallized from hot acetic acid in pure form; uses: source of pure beryllium salts; formula also given as $Be_4O(CH_3COO)_6$ [MER89] [HAW93]
Solubility: s hot H_2O, with hydrolysis; i alcohol [MER89]
Reactions: decomposes at 60-100°C [MER89]

408

Compound: Beryllium acetylacetonate
Synonyms: 2,4-pentanedione, beryllium derivative
Formula: $Be[CH_3COCH=C(O)CH_3]_2$
Molecular Formula: $C_{10}H_{14}BeO_4$
Molecular Weight: 207.231
CAS RN: 10210-64-7

Properties: monocl cryst powd [MER89] [STR93]

$$[CH_3-\overset{\displaystyle O^-}{\underset{\displaystyle |}{C}}=\overset{\displaystyle O}{\underset{\displaystyle \|}{CHC}}-CH_3]_2Be$$

Solubility: i H_2O, hydrolyzed in boiling H_2O [MER89]; v s alcohol, ether [HAW93]
Density, g/cm³: 1.168 [MER89]
Melting Point, °C: 108 [MER89]
Boiling Point, °C: 270 [MER89]

409
Compound: Beryllium aluminate
Synonyms: chrysoberyl
Formula: $BeAl_2O_4$
Molecular Formula: Al_2BeO_4
Molecular Weight: 126.973
CAS RN: 12004-06-7
Properties: ortho-rhomb; enthalpy of fusion 176 kJ/mol [JAN71] [CIC73]
Density, g/cm³: 3.76 [CRC94]
Melting Point, °C: 1870 [JAN71]

410
Compound: Beryllium aluminum silicate
Synonyms: beryl
Formula: $3BeO\cdot Al_2O_3\cdot 6SiO_2$
Molecular Formula: $Al_2Be_3O_{18}Si_6$
Molecular Weight: 537.502
CAS RN: 1302-52-9
Properties: gemstone; green or blue; chief ore for beryllium; hex, a = 0.9188 nm, c = 0.9189 nm [CIC73] [KIR78]
Density, g/cm³: 2.66 [CRC94]
Melting Point, °C: 1410 [CRC94]
Thermal Expansion Coefficient: (volume) 100°C (0.061), 200°C (0.079) [CLA66]

411
Compound: Beryllium basic acetate
Formula: $Be_4O(CH_3COO)_6$
Molecular Formula: $C_{12}H_{18}Be_4O_{13}$
Molecular Weight: 406.316
CAS RN: 1332-52-1
Properties: white cryst; can be crystallized in very pure form from acetic acid; tetrahedra when crystallized from chloroform; used as a source of pure Be salts [HAW93] [MER89]

Solubility: i H_2O, hydrolyzed by hot H_2O and dil acids; s glacial acetic acid, chloroform and other organic solvents except alcohol and ether [MER89] [HAW93]
Density, g/cm³: 1.25 [MER89]; 1.360 [ALD93]
Melting Point, °C: 285-286 [MER89]
Boiling Point, °C: 330-331 [MER89]

412
Compound: Beryllium borides
Formula: Be_4B
Molecular Formula: BBe_4
Molecular Weight: 46.859
CAS RN: 12536-52-6
Properties: refractory materials; -80 mesh with 98% purity; there are four borides: Be_4B [12536-52-6], Be_2B [12536-51-5], BeB_2 [12228-40-9], BeB_6 [12429-94-6]; borides in general are used in wear-resistant films, and to produce semiconductor films [CER91] [KIR78]
Melting Point, °C: Be_4B: 1160; Be_2B: 1520; BeB_2: >1970; BeB_6: 2070 [KIR78]

413
Compound: Beryllium borohydride
Synonyms: beryllium tetrahydroborate
Formula: $Be(BH_4)_2$
Molecular Formula: B_2BeH_8
Molecular Weight: 38.698
CAS RN: 17440-85-6
Properties: spontaneously flammable; obtained by reaction of diborane with dimethylberyllium [MER89]
Solubility: vigorous reaction with H_2O, HCl, evolving H_2 [MER89]
Melting Point, °C: sublimes, 91.3 [MER89]
Boiling Point, °C: decomposes >123 [MER89]

414
Compound: Beryllium bromide
Formula: $BeBr_2$
Molecular Formula: $BeBr_2$
Molecular Weight: 168.820
CAS RN: 7787-46-4
Properties: ortho-rhomb; very hygr; -80 mesh, 99% purity; enthalpy of fusion 9.80 kJ/mol; made by reaction of Be and Br_2 at 500-700°C [KIR78] [CER91] [MER89] [CRC93]
Solubility: v s H_2O [MER89]
Density, g/cm³: 3.465 [MER89]
Melting Point, °C: 508 [CRC93]
Boiling Point, °C: 520 [MER89]

Reactions: sublimes at 473°C [MER89]

415
Compound: Beryllium carbide
Formula: Be_2C
Molecular Formula: CBe_2
Molecular Weight: 30.035
CAS RN: 506-66-1
Properties: brick red or yellowish red octahedra; hard, refractory material; -200 mesh, 98% purity; prepared by hot pressing mixture of Be and C to 900°C; used in nuclear-reactor cores [CER91] [HAW93] [KIR78] [MER89]
Solubility: decomposed very slowly by H_2O; hydrolysis yields methane and beryllium hydroxide [KIR78] [MER89]
Density, g/cm^3: 1.90 [MER89]
Melting Point, °C: decomposes >2100 [MER89]

416
Compound: Beryllium carbonate tetrahydrate
Formula: $BeCO_3 \cdot 4H_2O$
Molecular Formula: CH_8BeO_7
Molecular Weight: 141.083
CAS RN: 60883-64-9
Properties: unstable unless kept under CO_2 atm; obtained by passing CO_2 through aq suspension of $Be(OH)_2$ [KIR78]

417
Compound: Beryllium chloride
Formula: $BeCl_2$
Molecular Formula: $BeCl_2$
Molecular Weight: 79.917
CAS RN: 7787-47-5
Properties: white to faint yellow powd, and sublimed fibers and clumps of 99.5% purity; very deliq; ortho-rhomb cryst; hydrolyzed by water vapor; enthalpy of vaporization 105 kJ/mol; enthalpy of fusion 8.66 kJ/mol; prepared by heating BeO, Cl_2 and C at 600-800°C [KIR78] [MER89] [CRC93]
Solubility: v s H_2O with evolution of heat [MER89]: g/100 g soln, H_2O: 40.35 (0°C), 41.72 (25°C); Solid phase, $BeCl_2 \cdot 4H_2O$ [KRU93]; s alcohol, benzene, ether [HAW93]
Density, g/cm^3: 1.90 [MER89]
Melting Point, °C: 405 [DOU83]
Boiling Point, °C: ~550 [DOU83]
Reactions: sublimes in vacuum at 300°C [MER89]

418
Compound: Beryllium fluoride
Formula: BeF_2
Molecular Formula: BeF_2
Molecular Weight: 47.009
CAS RN: 7787-49-7
Properties: 0.25 in. pieces and down; two forms; glassy, hygr; tetr when heated above 230°C; glassy form crystallizes spontaneously to quartz modification; enthalpy of fusion 4.76 kJ/mol; made by thermal decomposition of $(NH_4)_2BeF_6$ [STR93] [KIR78] [MER89]
Solubility: v s H_2O [MER89]; sl s alcohol [HAW93]
Density, g/cm^3: 1.986 [MER89]
Melting Point, °C: 552 [CRC93]
Boiling Point, °C: sublimes at 1036 under 1 mm Hg [MER89]
Reactions: transition from α to β form at 227°C [KIR78]

419
Compound: Beryllium formate
Synonyms: formic acid, beryllium salt
Formula: $Be(OOCH)_2$
Molecular Formula: $C_2H_2BeO_4$
Molecular Weight: 99.048
CAS RN: 1111-71-3
Properties: powd [MER89]
Solubility: very slowly hydrolyzed by H_2O [MER89]
Reactions: forms $Be_4O(HCOO)_6$ >250°C, which sublimes ~320°C [MER89]

420
Compound: Beryllium hydride
Formula: BeH_2
Molecular Formula: BeH_2
Molecular Weight: 11.028
CAS RN: 7787-52-2
Properties: white amorphous solid; inert to laboratory air; can be prepared by continuous thermal decomposition of a di-t-butylberyllium ethyl ether complex in a boiling hydrocarbon; has found use as rocket fuel and as a moderator for nuclear reactors [KIR80] [MER89]
Solubility: reacts slowly with H_2O, rapidly with dil acids evolving $H_2(g)$ [MER89]
Density, g/cm^3: 0.65 [LID94]
Reactions: rapid $H_2(g)$ evolution at 220°C [MER89]

421
Compound: Beryllium hydroxide (α)
Formula: $Be(OH)_2$

Molecular Formula: BeH_2O_2
Molecular Weight: 43.027
CAS RN: 13327-32-7
Properties: amorphous powd or cryst; ortho-rhomb; prepared by precipitation from beryllium acetate solution with alkali [HAW93] [MER89] [KIR78]
Solubility: v sl s H_2O, dil alkali; s hot NaOH, acids [MER89]
Density, g/cm^3: 1.92 [MER89]
Reactions: minus H_2O at >950°C [KIR78]

422
Compound: Beryllium hydroxide (β)
Formula: $Be(OH)_2$
Molecular Formula: BeH_2O_2
Molecular Weight: 43.027
CAS RN: 13327-32-7
Properties: white powd; tetr; metastable; decomposes to the oxide at 138°C [KIR78] [HAW93]
Solubility: i H_2O; s acids, alkalies [HAW93]
Reactions: transition from β to α after months of standing [KIR78]

423
Compound: Beryllium iodide
Formula: BeI_2
Molecular Formula: BeI_2
Molecular Weight: 262.821
CAS RN: 7787-53-3
Properties: -60 mesh, 99.5% pure and needles; very hygr; enthalpy of vaporization 70.5 kJ/mol; enthalpy of fusion 21.00 kJ/mol; obtained by reaction of Be and I_2 at 500-700°C [CER91] [KIR78] [MER89] [CRC93]
Solubility: reacts violently with H_2O giving off HI [MER89]
Density, g/cm^3: 4.325 [CRC94]
Melting Point, °C: 510 [CRC94]
Boiling Point, °C: 590 [CRC94]

424
Compound: Beryllium nitrate trihydrate
Formula: $Be(NO_3)_2 \cdot 3H_2O$
Molecular Formula: $BeH_6N_2O_9$
Molecular Weight: 187.068
CAS RN: 13597-99-4
Properties: white to sl yellow; deliq; prepared from BeO and HNO_3 solution, followed by evaporation and crystallization; used as a reagent [HAW93] [MER89]

Solubility: g $Ba(NO_3)_2$/100g H_2O: 97 (0°C), 108 (20°C), 178 (60°C); s alcohol [LAN85] [HAW93]
Density, g/cm^3: 1.557 [CRC94]
Melting Point, °C: ~60 [MER89]
Boiling Point, °C: decomposes 100-200 [HAW93]

425
Compound: Beryllium nitride
Formula: Be_3N_2
Molecular Formula: Be_3N_2
Molecular Weight: 55.050
CAS RN: 1304-54-7
Properties: hard white to grayish white; refractory; cub; a = 0.814 nm; obtained from a reaction of Be and NH_3 at 1100°C [KIR78] [HAW93] [CIC73]
Solubility: decomposes slowly in H_2O, quickly in acids and alkalies to evolve NH_3 [MER89]
Density, g/cm^3: 2.71 [LID94]
Melting Point, °C: 2200 decomposes [KIR78]
Boiling Point, °C: volatile [MER89]
Reactions: oxidized in air at 600°C [MER89]

426
Compound: Beryllium oxalate trihydrate
Formula: $BeC_2O_4 \cdot 3H_2O$
Molecular Formula: $C_2H_6BeO_7$
Molecular Weight: 151.078
CAS RN: 15771-43-4
Properties: rhomb; obtained by evaporating a solution of $Be(OH)_2$ in excess oxalic acid; used to prepare ultra pure BeO [CRC94] [KIR78]
Solubility: g/100 g soln, H_2O: 63.2 (25°C) [KRU93]
Melting Point, °C: 2200 [CRC94]
Boiling Point, °C: 2240 decomposes [CRC94]
Reactions: decomposes to BeO above 320°C [KIR78]

427
Compound: Beryllium oxide
Synonyms: beryllia
Formula: BeO
Molecular Formula: BeO
Molecular Weight: 25.011
CAS RN: 1304-56-9

Properties: light, amorphous white powd; insulates electrically like a ceramic, conducts heat like a metal; electrical resistivity, >1 x 10^{+16} ohm m; hardness 9 Mohs; tensile strength 150 MPa; compressive strength 1400 MPa; Poisson's ratio 0.164-0.380; modulus of rupture 250 MPa; modulus of elasticity 345 GPa; made from $Be(OH)_2$ and H_2SO_4; enthalpy of fusion 85.00 kJ/mol; used in electron tubes, resistor cores [HAW93] [MER89] [STR93] [KIR78] [CRC93]

Solubility: v sl s H_2O; slowly s conc acid, alkali [MER89]

Density, g/cm³: 3.01 [STR93]

Melting Point, °C: 2507 [CRC93]

Boiling Point, °C: 4300 [STR93]

Thermal Conductivity, W/(m·K): 25°C (290-330); 100°C (190-220); 500°C (65.4); 1000°C (20.3) [KIR80] [KIR78]

Thermal Expansion Coefficient: coefficient of thermal expansion at 100°C is 9.7 x 10^{-6}/K; at 500°C is 13.3 x 10^{-6}/K [KIR78]

428

Compound: Beryllium perchlorate tetrahydrate

Formula: $Be(ClO_4)_2 \cdot 4H_2O$

Molecular Formula: $BeCl_2H_8O_{12}$

Molecular Weight: 279.974

CAS RN: 7787-48-6

Properties: very hygr cryst; retains waters of crystallization tenaciously [MER89]

Solubility: g/100 g soln, H_2O: 59.5 (25°C); Solid phase, $Be(ClO_4)_2 \cdot 4H_2O$ [KRU93]

429

Compound: Beryllium selenate tetrahydrate

Formula: $BeSeO_4 \cdot 4H_2O$

Molecular Formula: BeH_8SeO_8

Molecular Weight: 224.031

CAS RN: 10039-31-3

Properties: colorless ortho-rhomb cryst [CRC94] [MER89]

Solubility: v s H_2O [MER89]

Density, g/cm³: 2.03 [MER89]

Reactions: minus $2H_2O$ at 100°C; minus $4H_2O$ at 300°C [MER89]

430

Compound: Beryllium sulfate

Formula: $BeSO_4$

Molecular Formula: BeO_4S

Molecular Weight: 105.076

CAS RN: 13510-49-1

Properties: colorless cryst [HAW93]

Solubility: g/100 g soln, H_2O: 26.69 (0°C), 29.22 (25°C), 45.28 (100°C); Solid phase, $BeSO_4 \cdot 4H_2O$ [KRU93]; i alcohol [HAW93]

Density, g/cm³: 2.443 [CRC94]

Melting Point, °C: decomposes 550-600 [CRC94]

431

Compound: Beryllium sulfate dihydrate

Formula: $BeSO_4 \cdot 2H_2O$

Molecular Formula: BeH_4O_6S

Molecular Weight: 141.106

CAS RN: 14215-00-0

Properties: forms when the tetrahydrate is heated at 92°C [KIR78]

Reactions: minus $2H_2O$ at 400°C; decomposes to BeO ~650°C [KIR78]

432

Compound: Beryllium sulfate tetrahydrate

Formula: $BeSO_4 \cdot 4H_2O$

Molecular Formula: BeH_8O_8S

Molecular Weight: 177.137

CAS RN: 7787-56-6

Properties: colorless cryst, lump 99.9% purity; produced from $Be(OH)_2$ and H_2SO_4 solution, followed by fractional crystallization; used to produce BeO for ceramics [KIR78] [HAW93]

Solubility: v s H_2O [MER89]

Density, g/cm³: 1.71 [MER89]

Reactions: minus $2H_2O$ ~100°C [MER89]

433

Compound: Beryllium sulfide

Formula: BeS

Molecular Formula: BeS

Molecular Weight: 41.079

CAS RN: 13598-22-6

Properties: regular; -100 mesh with 99% purity [CRC94] [CER91] [ALF93]

Solubility: decomposes in H_2O [CRC94]

Density, g/cm³: 2.36 [CRC94]

434

Compound: Bis(cyclopentadienyl)ruthenium

Synonyms: ruthenocene

Formula: $(C_5H_5)_2Ru$

Molecular Formula: $C_{10}H_{10}Ru$

Molecular Weight: 231.259

CAS RN: 1287-13-4

Properties: light yellow cryst [STR93]

Melting Point, °C: 194-198 [STR93]

435

Compound: Bis(diethylamino)chlorophosphine
Formula: [(C₂H₅)₂N]₂PCl
Molecular Formula: C₈H₂₀ClN₂P
Molecular Weight: 210.687
CAS RN: 685-83-6
Properties: liq [ALF95]
Boiling Point, °C: 124-125 (15 mm Hg) [ALF95]

436

Compound: Bismuth
Formula: Bi
Molecular Formula: Bi
Molecular Weight: 208.980373
CAS RN: 7440-69-9
Properties: grayish white soft brittle metal, 99.999% vacuum deposition grade; rhomb, a = 0.47457 nm; enthalpy of fusion 11.30 kJ/mol; enthalpy of vaporization 151 kJ/mol; Poisson's ratio 0.33; electrical resistivity (20°C) 129 μohm·cm; Brinell hardness 7; electronegativity 1.67; used in ferromagnetic and resistive films, in pharmaceuticals and medicine [KIR78] [MER89] [CRC93] [COT88] [CER91] [ALD94]
Solubility: s dil HNO₃, conc HCl [MER89]
Density, g/cm³: 9.808 (25°C) [KIR78]
Melting Point, °C: 271.4 [KIR78]
Boiling Point, °C: 1564 [KIR78]
Thermal Conductivity, W/(m·K): 7.92 (25°C) [ALD94]
Thermal Expansion Coefficient: (volume) 100°C (0.37) [CLA66]

437

Compound: Bismuth acetate
Formula: Bi(CH₃COO)₃
Molecular Formula: C₆H₉BiO₆
Molecular Weight: 386.113
CAS RN: 22306-37-2
Properties: white cryst 99.999% pure; sensitive to moisture [ALF93] [STR93] [ALD93]
Solubility: i H₂O [CRC94]
Melting Point, °C: decomposes [CRC94]

438

Compound: Bismuth antimonide
Formula: BiSb
Molecular Formula: BiSb
Molecular Weight: 330.740
CAS RN: 12323-19-2

Properties: 99.99% pure cryst; used as a semiconductor material in the form of a single cryst [HAW93] [ALF93]
Melting Point, °C: 475 [ALF93]

439

Compound: Bismuth basic carbonate hemihydrate
Synonyms: bismuth subcarbonate
Formula: (BiO)₂CO₃·1/2H₂O
Molecular Formula: CHBi₂O₅.₅
Molecular Weight: 518.976
CAS RN: 5892-10-4
Properties: odorless, tasteless white powd; light sensitive; used in a mixture with other compounds in glazes for ceramics and to give a pearly surface for plastics [ALD93] [ALF93] [CRC94] [MER89]
Solubility: i H₂O; s mineral acids, glacial acetic acid [MER89]
Density, g/cm³: 6.86 [ALF93]
Boiling Point, °C: decomposes [CRC94]

440

Compound: Bismuth basic dichromate
Formula: Bi₂O₃·2CrO₃
Molecular Formula: Bi₂Cr₂O₉
Molecular Weight: 665.948
Properties: reddish orange amorphous powd; preparation: reaction between Bi(NO₃)₃ and potassium chromate [HAW93]
Solubility: i H₂O; s acids, alkalies [HAW93]

441

Compound: Bismuth bromide
Synonyms: bismuth tribromide
Formula: BiBr₃
Molecular Formula: BiBr₃
Molecular Weight: 448.692
CAS RN: 7787-58-8
Properties: yellowish cryst, -60 mesh 99.999% purity; odor of HBr; hygr; enthalpy of fusion 21.8 kJ/mol; enthalpy of sublimation 115 kJ/mol; enthalpy of vaporization 75.4 kJ/mol; can be prepared by dissolution of Bi₂O₃ in conc HBr solution, followed by dewatering in gentle stream of N₂ and distillation [CER91] [KIR78] [CRC93] [HAW93] [MER89]
Solubility: decomposed by H₂O forming BiOBr; s dil HCl, acetone, ether; i alcohol [HAW93] [KIR78] [MER89]
Density, g/cm³: solid: 5.72; liq 4.572 at 271.5°C [KIR78]
Melting Point, °C: 218 [MER89]

Boiling Point, °C: 453 [CRC93]

442

Compound: Bismuth chloride
Synonyms: bismuth trichloride
Formula: $BiCl_3$
Molecular Formula: $BiCl_3$
Molecular Weight: 315.338
CAS RN: 7787-60-2
Properties: white to yellowish very deliq cryst, -60 mesh 99.99% and 99.9% purity; HCl odor; enthalpy of fusion 10.90 kJ/mol; enthalpy of sublimation 114 kJ/mol; enthalpy of vaporization at bp 72.61 kJ/mol; a preparation is by chlorination of molten metal; used as a catalyst for organic reactions [KIR78] [HAW93] [MER89] [CER91] [CRC93]
Solubility: decomposed by H_2O forming BiOCl [HAW93] [MER89]; s acids; i alcohol [HAW93]
Density, g/cm³: 4.75 [MER89]
Melting Point, °C: 230 [CRC93]
Boiling Point, °C: 447 [CRC93]
Reactions: sublimes ~430°C [MER89]

443

Compound: Bismuth chloride monohydrate
Synonyms: bismuth trichloride monohydrate
Formula: $BiCl_3 \cdot H_2O$
Molecular Formula: $BiCl_3H_2O$
Molecular Weight: 333.353
CAS RN: 39483-74-4
Properties: 99.99% pure cryst [ALF93]
Density, g/cm³: 4.75 [ALF93]
Melting Point, °C: 230-232 [ALF93]
Boiling Point, °C: 447 [ALF93]

444

Compound: Bismuth citrate
Formula: $BiC_6H_5O_7$
Molecular Formula: $C_6H_5BiO_7$
Molecular Weight: 398.082
CAS RN: 813-93-4
Properties: white powd; preparation: boiling bismuth subnitrate and citric acid solution; used in medicine [HAW93]
Solubility: i H_2O; s ammonia; sl s alcohol [HAW93]
Density, g/cm³: 3.458 [HAW93]
Melting Point, °C: decomposes [CRC77]

445

Compound: Bismuth fluoride

Synonyms: bismuth trifluoride
Formula: BiF_3
Molecular Formula: BiF_3
Molecular Weight: 265.975
CAS RN: 7787-61-3
Properties: white to gray powd, -60 mesh of 99.9% and 99.999% purity; moisture sensitive; can be obtained by dissolving either bismuth oxide or oxyfluoride in HF, followed by careful evaporation; enthalpy of fusion 21.6 kJ/mol; enthalpy of sublimation 201 kJ/mol; used to prepare BiF_5 [KIR78] [MER89] [CER91] [ALD93] [CRC93]
Solubility: 0.00503 mol/L in H_2O [KIR78]
Density, g/cm³: 8.3 [MER89]
Melting Point, °C: 725 [COT88]
Boiling Point, °C: 900 [KIR78]
Reactions: volatilizes >730°C, slowly, without decomposition [MER89]

446

Compound: Bismuth germanium oxide
Formula: $2Bi_2O_3 \cdot 3GeO_2$
Molecular Formula: $Bi_4Ge_3O_{12}$
Molecular Weight: 1245.744
CAS RN: 12233-56-6
Properties: white chips or powd, 99.99995% purity [ALF93]

447

Compound: Bismuth hydride
Synonyms: bismuthine
Formula: BiH_3
Molecular Formula: BiH_3
Molecular Weight: 212.004
CAS RN: 18288-22-7
Properties: colorless gas, unstable at room temp and rapidly decomposes to Bi and H_2; can be prepared by disproportionation of either methyl- or-ethyl bismuthine; enthalpy of vaporization about 25.15 kJ/mol; finds use in manufacturing Ge or Si semiconductors [MER89] [KIR78]
Density, g/cm³: 9.303 g/L [LID94]
Melting Point, °C: -67 [LID94]
Boiling Point, °C: ~16.8 [MER89]

448

Compound: Bismuth hydroxide
Formula: $Bi(OH)_3$
Molecular Formula: BiH_3O_3
Molecular Weight: 260.002
CAS RN: 10361-43-0

Properties: white to yellowish white amorphous powd; used in plutonium separation, as an absorbent for rutin and quercetin [HAW93] [MER89]
Solubility: i H_2O [MER89]; s acids [HAW93]
Density, g/cm³: 4.962 [MER89]
Reactions: readily loses one H_2O to form metahydroxide [MER89]

449
Compound: Bismuth hydroxide nitrate oxide
Synonyms: bismuth subnitrate
Formula: $4BiNO_3(OH)_2 \cdot BiO(OH)$
Molecular Formula: $Bi_5H_9N_4O_{22}$
Molecular Weight: 1461.987
CAS RN: 1304-85-4
Properties: odorless, tasteless, heavy, sl hygr microcryst powd, 98+% [MER89] [ALF93]
Solubility: i H_2O; s dil HNO_3, HCl [MER89]
Reactions: decomposes at 260°C [ALF93]

450
Compound: Bismuth iodide
Synonyms: bismuth triiodide
Formula: BiI_3
Molecular Formula: BiI_3
Molecular Weight: 589.693
CAS RN: 7787-64-6
Properties: black minute hex cryst or gray powd, -20 mesh and -40 mesh 99.999% and 99.9% purity; metallic sheen; enthalpy of fusion 39.1 kJ/mol; enthalpy of sublimation 134.3 kJ/mol; sensitive to moisture; prepared from Bi and I_2; used in analytical chemistry [HAW93] [MER89] [STR93] [ALF93]
Solubility: i H_2O, decomposed by hot H_2O; s alcohol and HI, KI solutions [HAW93]
Density, g/cm³: 5.778 [MER89]
Melting Point, °C: 408.6 [KIR78]
Boiling Point, °C: 542 [KIR78]
Reactions: sublimes at 439°C; decomposes at 500°C [MER89]

451
Compound: Bismuth iron molybdenum oxide
Formula: $Bi_3FeMo_2O_{12}$
Molecular Formula: $Bi_3FeMo_2O_{12}$
Molecular Weight: 1066.659
CAS RN: 59393-06-5
Properties: -325 mesh powd; used as an oxidation catalyst [ALF93]

452
Compound: Bismuth molybdate
Synonyms: bismuth molybdenum oxide
Formula: $Bi_2(MoO_4)_3$
Molecular Formula: $Bi_2Mo_3O_{12}$
Molecular Weight: 897.774
CAS RN: 51898-99-8
Properties: colorless trig; hygr; -200 mesh and -325 mesh powd; used as an oxidation catalyst [CRC94] [CER91] [ALF93]
Density, g/cm³: 5.95 [LID94]
Melting Point, °C: decomposes 30 [CRC94]

453
Compound: Bismuth molybdenum oxide
Formula: Bi_2MoO_6
Molecular Formula: Bi_2MoO_6
Molecular Weight: 609.897
CAS RN: 13565-96-3
Properties: -325 mesh powd; used as an oxidation catalyst [ALF93]
Density, g/cm³: 9.32 [KIR78]

454
Compound: Bismuth nitrate pentahydrate
Formula: $Bi(NO_3)_3 \cdot 5H_2O$
Molecular Formula: $BiH_{10}N_3O_{14}$
Molecular Weight: 485.071
CAS RN: 10035-06-0
Properties: lustrous clear colorless hygr cryst, 99.999% purity; acid taste; odor of HNO_3; used to prepare other bismuth salts, in luminous paints and enamels [HAW93] [MER89] [STR93] [ALF93]
Solubility: decomposed by H_2O to the subnitrate; s in dil HNO_3 and glycerol, acetone [HAW93]
Density, g/cm³: 2.83 [MER89]
Boiling Point, °C: 75-80 (decomposes) [HAW93]

455
Compound: Bismuth oleate
Synonyms: oleic acid, bismuth salt
Formula: $[CH_3(CH_2)_7CH=CH(CH_2)_7COO]_3Bi$
Molecular Formula: $C_{54}H_{99}BiO_6$
Molecular Weight: 1053.356
CAS RN: 52951-38-9
Properties: yellowish brown, soft granular mass; obtained from Bi_2O_3, oleic acid and acetic anhydride; used in catalysts for manufacturing aldehydes and alcohols by oxo process [MER89] [HAW93]

Solubility: i H₂O; s ether, s in about 1500 parts benzene [MER89] [HAW93]

456

Compound: Bismuth oxide
Formula: Bi₂O₃
Molecular Formula: Bi₂O₃
Molecular Weight: 465.959
CAS RN: 1304-76-3
Properties: yellow heavy odorless powd or monocl cryst, of various grades: -30 mesh of 99.9999% purity, 3-12 mm pieces (sintered); used to enamel cast iron, in ceramic coloring and as an evaporated material and sputtering target for beam splitting and as a base coating for gold films which are used as transparent heating elements on glass [HAW93] [MER89] [CER91]
Solubility: i H₂O [MER89]; s acids [HAW93]
Density, g/cm³: 8.9 [STR93]
Melting Point, °C: 817 [STR93]
Boiling Point, °C: 1890 [STR93]

457

Compound: Bismuth oxybromide
Synonyms: bismuth bromide oxide
Formula: BiOBr
Molecular Formula: BiBrO
Molecular Weight: 304.883
CAS RN: 7787-57-7
Properties: colorless cryst or amorphous powd; used in manufacture of dry cell cathodes [MER89] [CRC94]
Solubility: i H₂O, alcohol; s HCl, HBr and HNO₃ [MER89]
Density, g/cm³: 8.082 (15°C) [CRC94]
Melting Point, °C: melts at red heat [MER89]

458

Compound: Bismuth oxychloride
Synonyms: bismuth chloride oxide
Formula: BiOCl
Molecular Formula: BiClO
Molecular Weight: 260.432
CAS RN: 7787-59-9
Properties: white 99.999% fine powd or tetr cryst; used in face powd, to manufacture artificial pearls, in dry cell cathodes [MER89] [ALF93]
Solubility: i H₂O [MER89]
Density, g/cm³: 7.72 [MER89]
Melting Point, °C: melts at low red heat [MER89]

459

Compound: Bismuth oxyiodide
Synonyms: bismuth iodide oxide
Formula: BiOI
Molecular Formula: BiIO
Molecular Weight: 351.883
CAS RN: 7787-63-5
Properties: brick red, heavy, odorless powd; or, copper-colored cryst; used in manufacturing dry cell cathodes and as an anti-infective [MER89]
Solubility: i H₂O, alcohol, chloroform; s HCl; decomposed by HNO₃ or alkali [MER89]
Density, g/cm³: 7.92 [MER89]
Melting Point, °C: fuses at red heat with partial decomposition [MER89]

460

Compound: Bismuth oxynitrate
Synonyms: bismuth subnitrate
Formula: BiONO₃
Molecular Formula: BiNO₄
Molecular Weight: 286.985
CAS RN: 10361-46-3
Properties: heavy white powd; sl hygr; uses: cosmetics, ceramic glasses and enamel fluxes; there is a 99.99+ pure monohydrate, CAS RN 13595-83-0 [HAW93] [ALD94]
Solubility: i H₂O, alcohol; s acids [HAW93]
Density, g/cm³: 4.928 [HAW93]
Melting Point, °C: 260, decomposing [HAW93]

461

Compound: Bismuth oxyperchlorate monohydrate
Formula: BiOClO₄·H₂O
Molecular Formula: BiClH₂O₆
Molecular Weight: 342.445
CAS RN: 44584-78-3
Properties: white cryst powd [STR93] [ALF93]

462

Compound: Bismuth pentafluoride
Synonyms: bismuth(V) fluoride
Formula: BiF₅
Molecular Formula: BiF₅
Molecular Weight: 303.972
CAS RN: 7787-62-4
Properties: long white needles; body-centered tetr cryst; very sensitive to moisture; discolors quickly in moist air; can be formed by fluorinating BiF₃ or Bi metal at 120°C; used as a fluorinating agent [KIR78] [MER89]

Solubility: violent reaction with H_2O to form BiF_3 and ozone [MER89]
Density, g/cm³: 5.55 [MER89]
Melting Point, °C: 151 [KIR78]
Boiling Point, °C: 230 [KIR78]

463
Compound: Bismuth phosphate
Formula: $BiPO_4$
Molecular Formula: BiO_4P
Molecular Weight: 303.951
CAS RN: 10049-01-1
Properties: odorless powd or monocl cryst; used as an antacid, to recover plutonium, and in optical glass [HAW93] [MER89]
Solubility: sl s H_2O; s conc HNO_3, HCl [MER89]
Density, g/cm³: 6.323 [MER89]
Melting Point, °C: does not melt on heating [MER89]

464
Compound: Bismuth selenide
Synonyms: guanajuatite
Formula: Bi_2Se_3
Molecular Formula: Bi_2Se_3
Molecular Weight: 654.841
CAS RN: 12068-69-8
Properties: black cryst, 1-6 mm pieces (fused) of 99.999% purity; rhomb and hex; decomposes when heated in air, by conc HNO_3, and by aqua regia; used in semiconductors, and in the form of a 99 or 99.999% pure material used as a sputtering target to produce multilayer thin films and magneto-resistant films [HAW93] [MER89] [CER91]
Solubility: i H_2O [MER89]
Density, g/cm³: 7.70 [MER89]
Melting Point, °C: 710 [MER89]
Thermal Conductivity, W/(m·K): 2.4 [CRC93]

465
Compound: Bismuth stannate
Formula: $Bi_2O_3 \cdot 2SnO_2$
Molecular Formula: $Bi_2O_7Sn_2$
Molecular Weight: 767.377
CAS RN: 12338-09-9
Properties: -200 mesh, 99.9% purity [CER91]
Reactions: pentahydrate: $-5H_2O$ at ~140°C [HAW93]

466
Compound: Bismuth stannate pentahydrate
Formula: $Bi_2O_3 \cdot 3SnO_2 \cdot 5H_2O$
Molecular Formula: $Bi_2H_{10}O_{14}Sn_3$
Molecular Weight: 1008.162
CAS RN: 12777-45-6
Properties: light colored cryst; used in ceramic capacitor such as barium titanate [HAW93]
Solubility: i H_2O [HAW93]
Reactions: minus $5H_2O$ ~140°C [HAW93]

467
Compound: Bismuth strontium calcium copper oxide (1112)
Synonyms: supercon 1112
Formula: $BiSrCaCu_2O_x$
CAS RN: 114901-61-0
Properties: 99.999% and 99.99% pure 20 micron powd; dry processed from 99.999% pure oxides and carbonates [STR93]

468
Compound: Bismuth strontium calcium copper oxide (2212)
Synonyms: supercon 2212
Formula: $Bi_2Sr_2CaCu_2O_8$
Molecular Formula: $Bi_2CaCu_2O_8Sr_2$
Molecular Weight: 888.366
CAS RN: 114901-61-0
Properties: O_8 can be $O_{8.15}$ to $O_{8.20}$; superconductor; 99·999% pure, 20 micron powd; dry processed from 99·999% oxides and carbonates; intermediate precursor available as fine agglomerate and ballmilled powd; T_c 85 K [STR93] [ASM93] [ALF93]
Density, g/cm³: 6.40 [ALD93]

469
Compound: Bismuth strontium calcium copper oxide (2223)
Synonyms: supercon 2223
Formula: $Bi_2Sr_2Ca_2Cu_3O_{10}$
Molecular Formula: $Bi_2Ca_2Cu_3O_{10}Sr_2$
Molecular Weight: 1023.989
CAS RN: 114901-61-0
Properties: superconductor; 99.999% and 99.9% pure 20 micron powd; dry processed from 99.999% and ACS grades, respectively, of oxides and carbonates; T_c 110 K; for $Bi_2Sr_2Ca_3Cu_4O_{12}$, T_c is <120 K [STR93] [ASM93]

470
Compound: Bismuth subacetate
Synonyms: bismuth acetate oxide
Formula: $CH_3COOBiO$
Molecular Formula: $C_2H_3BiO_3$
Molecular Weight: 284.024
CAS RN: 5142-76-7
Properties: thin cryst plates; slight acetic acid odor [MER89]
Solubility: i H_2O; s glacial acetic acid [MER89]

471
Compound: Bismuth sulfate
Formula: $Bi_2(SO_4)_3$
Molecular Formula: $Bi_2O_{12}S_3$
Molecular Weight: 706.152
CAS RN: 7787-68-0
Properties: white needles or powd; used in the analysis of other metal sulfates [HAW93]
Solubility: i H_2O, alcohol; s dil HCl, HNO_3 [HAW93]
Density, g/cm³: 5.08 [HAW93]
Melting Point, °C: 405, decomposes [HAW93]

472
Compound: Bismuth sulfide
Synonyms: bismuth glance, stibnite
Formula: Bi_2S_3
Molecular Formula: Bi_2S_3
Molecular Weight: 514.159
CAS RN: 1345-07-9
Properties: blackish brown, -200 mesh 99.9% and 99.999% purity; ortho-rhomb bipyramidal cryst; hardness 2 Mohs [HAW93] [MER89] [CER91]
Solubility: i H_2O; s HNO_3, HCl [MER89]
Density, g/cm³: 7.6-7.8 [HAW93]
Melting Point, °C: 685 (decomposes) [STR93] [ALF93]

473
Compound: Bismuth telluride
Synonyms: tetradymite
Formula: Bi_2Te_3
Molecular Formula: Bi_2Te_3
Molecular Weight: 800.761
CAS RN: 1304-82-1

Properties: gray hex platelets, various grades: 1-6 mm pieces (fused) -325 mesh powd, 99.999% purity; preparation: by heating stoichiometric quantities of of Bi and Te at 475°C in a vacuum for several days; used as a semiconductor; resistivity 0.00033 ohm·cm; energy gap 0.15 eV [MER89] [CER91] [AlF93]
Solubility: i H_2O, alcohol [HAW93]
Density, g/cm³: 7.642 [MER89]
Melting Point, °C: 585 [MER89]
Thermal Conductivity, W/(m·K): 3.0 [CRC93]

474
Compound: Bismuth tetroxide
Synonyms: bismuth peroxide
Formula: Bi_2O_4
Molecular Formula: Bi_2O_4
Molecular Weight: 481.959
CAS RN: 12048-50-9
Properties: reddish orange to yellowish brown heavy powd, -200 mesh as the dehydrate; used as a lubricant for metal-extrusion dies [HAW93] [MER89] [CER91]
Solubility: slowly decomposed by H_2O [MER89]
Density, g/cm³: 5.6 [HAW93]
Melting Point, °C: 305 [HAW93]

475
Compound: Bismuth titanate
Formula: $Bi_2O_3 \cdot 4TiO_2$
Molecular Formula: $Bi_2O_{11}Ti_4$
Molecular Weight: 785.422
CAS RN: 12233-34-0
Properties: -325 mesh, 10 micron average, or less or 99.9% purity; used as sputtering target for beam splitter, and base coating for gold films to prepare heating elements on glass [CER91]

476
Compound: Bismuth titanate
Formula: $Bi_2O_3 \cdot 2TiO_2$
Molecular Formula: $Bi_2O_7Ti_2$
Molecular Weight: 625.723
CAS RN: 12048-51-0
Properties: white powd, -325 mesh 99.9% purity; used as sputtering target for beam splitter, and base coating for gold films as heating elements on glass; there are two other compounds: $Bi_4Ti_3O_{12}$ (12010-77-4) and $Bi_2Ti_4O_{11}$ (12233-34-0) [CER91] [STR93]

477
Compound: Bismuth tungstate
Formula: $Bi_2(WO_4)_3$
Molecular Formula: $Bi_2O_{12}W_3$
Molecular Weight: 1161.479
CAS RN: 13595-87-4
Properties: off-white powd, -200 mesh of 99.9% purity [CER91] [STR93]

478
Compound: Bismuth vanadate
Synonyms: pucherite
Formula: $Bi_2O_3 \cdot V_2O_5$ or $BiVO_4$
Molecular Formula: $Bi_2O_8V_2$
Molecular Weight: 647.839
CAS RN: 14059-33-7
Properties: reddish-green; rhomb; -200 mesh, 99.9% purity [CRC94] [CER91]
Density, g/cm^3: 6.25 [CRC94]

479
Compound: Bismuth zirconate
Formula: $2Bi_2O_3 \cdot 3ZrO_2$
Molecular Formula: $Bi_4O_{12}Zr_3$
Molecular Weight: 1301.586
CAS RN: 37306-42-6
Properties: reacted powd, -325 mesh, 5 micron or less with 99% purity [CER91]

480
Compound: Borazole
Synonyms: borazine,s-triazoborzane
Formula: $B_3N_3H_6$
Molecular Formula: $B_3H_6N_3$
Molecular Weight: 80.501
CAS RN: 6569-51-3
Properties: inorganic analog of benzene; colorless liq; preparation: heating equimolar mixture of ammonia and BH_3 at 250-300°C for 30 min [MER89] [HAW93]
Solubility: hydrolyzes, evolving boron hydrides [HAW93]
Density, g/cm^3: 0.824 (0°C) [HAW93]
Melting Point, °C: -58 [HAW93]
Boiling Point, °C: 53 [HAW93]

481
Compound: Boric acid
Synonyms: orthoboric acid,sassolite
Formula: $B(OH)_3$
Molecular Formula: BH_3O_3

Molecular Weight: 61.833
CAS RN: 10043-35-3
Properties: colorless, odorless cryst or white powd; tricl; used in borosilicate glass, as an ointment and eyewash [MER89] [HAW93] [KIR78]
Solubility: g/100g soln in H_2O: 2.52 (0°C); 4.72 (20°C); 27.53 (100°C) [KIR78]
Density, g/cm^3: 1.5172 [KIR78]; 1.435 [STR93]
Melting Point, °C: 171 [JAN85]

482
Compound: Boron
Formula: B
Molecular Formula: B
Molecular Weight: 10.811
CAS RN: 7440-42-8
Properties: black hard solid; polymorphic; α: rhombohedral; clear red cryst, almost as hard as diamond; β: rhomb; black; α': tetr; black; opaque cryst; enthalpy of vaporization 480 kJ/mol; enthalpy of fusion 50.20 kJ/mol; Young's modulus of filamentary boron 3040 to 3330 MPa; tensile strength of filamentary boron 3450 to 4830 MPa; amorphous: black or dark brown powd; hardness 11 Mohs [MER89] [KIR78] [CRC93]
Solubility: i H_2O [MER89]
Density, g/cm^3: α: 2.46; β: 2.35; α': 2.31; amorphous: 2.350 [MER89]
Melting Point, °C: 2190 [KIR78]
Boiling Point, °C: 3660 [KIR78]
Thermal Conductivity, W/(m·K): 31.8 (0°C); 27.4 (25°C); 18.8 (100°C) [HO 72]

483
Compound: Boron carbide
Synonyms: norbide
Formula: B_4C
Molecular Formula: CB_4
Molecular Weight: 55.255
CAS RN: 12069-32-8
Properties: hard black shiny cryst, -325 mesh with 99.5% purity; rhomb; hardness 9.3 Mohs; less brittle than most ceramics; does not burn in oxygen flame; used as an abrasive; Knoop hardness ~27 GPa; produced by reducing B_2O_3 with carbon at 1400-2300°C; used in crucible form as a container for molten salts except molten caustic, and as a 99.5% pure sputtering target for producing semiconductor and wear-resistant films [KIR78] [HAW93] [MER89] [CER91]
Solubility: not attacked by hot HF, HNO_3 or chromic acid [MER89]

Density, g/cm³: 2.508-2.512 [MER89]
Melting Point, °C: 2350 [MER89]
Boiling Point, °C: >3500 [MER89]
Reactions: decomposed by molten alkalis at red heat [MER89]

484
Compound: Boron nitride
Formula: BN
Molecular Formula: BN
Molecular Weight: 24.818
CAS RN: 10043-11-5
Properties: white powd, 1 micron or less 99.5% pure; hex, most common form: a = 0.2504 nm, c = 0.6661 nm; fcc: a = 0.3615 nm; hardness: hex like graphite, cub approaches that of diamond; band gap ~7.5 eV at 300 K; dielectric 7.1; used in furnace insulation, and in crucibles for melting aluminum, boron, iron and silicon, also as sputtering target for dielectrics, diffusion masks, passivation layers [KIR81] [HAW93] [MER89] [CER91]
Density, g/cm³: hex: 2.34; fcc: 3.43 [CIC73]
Melting Point, °C: hex: 3000 under N_2 [KIR78]
Boiling Point, °C: sublimes sl below 3000 [MER89]
Reactions: decomposes in vacuum ~2700°C [MER89]
Thermal Conductivity, W/(m·K): hex: 15 [KIR81]
Thermal Expansion Coefficient: 7.51 x 10⁻⁶/°C [KIR81]

485
Compound: Boron oxide
Synonyms: boron anhydride
Formula: B_2O_3
Molecular Formula: B_2O_3
Molecular Weight: 69.620
CAS RN: 1303-86-2
Properties: colorless, brittle, vitreous; 12 mm pieces and smaller (fused) of 99.9995% purity; hygr; heat of solution -75.9 kJ/mol [MER89] [KIR78] [CER91]
Solubility: %w soln, H_2O: 4.72 (20°C), 8.08 (40°C), 27.53 (100°C); in alcohol: 94.4 g/L (25°C) [KIR78]
Density, g/cm³: amorphous: 1.8; cryst: 2.46 [MER89]
Melting Point, °C: cryst: 450 [MER89]
Boiling Point, °C: ~1860 [STR93]

486
Compound: Boron oxide glass
Synonyms: vitreous boric oxide

Formula: B_2O_3
Molecular Formula: B_2O_3
Molecular Weight: 69.620
CAS RN: 1303-86-2
Properties: colorless, glassy solid; enthalpy of formation from elements, 25°C, -1252.2 kJ/mol; heat capacity (25°C) 62.969 J/(mol·K) [KIR78]
Density, g/cm³: 0°C, 1.8766; 18-25°C, 1.844 [KIR78]
Boiling Point, °C: 2316, extrapolated [KIR78]

487
Compound: Boron phosphate
Synonyms: borophosphoric acid
Formula: BPO_4
Molecular Formula: BO_4P
Molecular Weight: 105.782
CAS RN: 13308-51-5
Properties: white not hygr cryst; prepared by reacting boric acid and phosphoric acid up to 1200°C: $B(OH)_3 + H_3PO_4 = BPO_4 + 3H_2O$; used in special glasses [HAW93]
Solubility: s H_2O [HAW93]
Density, g/cm³: 1.873 [HAW93]
Melting Point, °C: vaporizes slowly without decomposition >1200 [KIR78]

488
Compound: Boron phosphide
Formula: BP
Molecular Formula: BP
Molecular Weight: 41.785
CAS RN: 20205-91-8
Properties: maroon powd; -100 mesh at 97.5% purity; refractory; hardness 9.5 Mohs; band gap 2.0 eV (300 K); material with formula $B_{13}P_2$ [12008-82-1] is -325 mesh, 10 micron or less of 99% purity [HAW93] [CER91] [KIR82]
Solubility: reacts with H_2O and acids, releasing toxic fumes [HAW93]
Reactions: ignites at 200°C [CRC94]

489
Compound: Boron silicide
Formula: B_6Si
Molecular Formula: B_6Si
Molecular Weight: 92.952
CAS RN: 12008-29-6
Properties: black cryst; -200 mesh; also material with composition B_4Si [12007-81-7] and 98% purity [CRC94] [CER91] [ALF93]
Density, g/cm³: 2.47 [CRC94]

490

Compound: Boron tribromide
Formula: BBr_3
Molecular Formula: BBr_3
Molecular Weight: 250.523
CAS RN: 10294-33-4
Properties: 99.999% and 99.9+% purity doping grade; colorless fuming liq; sensitive to moisture; critical temp 300°C; enthalpy of vaporization 30.5 kJ/mol; used as a catalyst in the manufacture of diborane [HAW93] [MER89] [KIR78] [STR93] [CER91] [CRC93]
Solubility: decomposed by H_2O, alcohol [MER89]
Density, g/cm^3: 2.698 [MER89]
Melting Point, °C: -46.0 [MER89]
Boiling Point, °C: 90 [ALD94]
Thermal Conductivity, W/(m·K): 0.112 (20°C) [KIR78]

491

Compound: Boron trichloride
Formula: BCl_3
Molecular Formula: BCl_3
Molecular Weight: 117.169
CAS RN: 10294-34-5
Properties: colorless fuming liq or gas; critical temp 178.8°C; critical pressure 3901.0 kPa; enthalpy of vaporization 23.77 kJ/mol; enthalpy of fusion 2.10 kJ/mol; vapor pressure: 0.53 (-80°C), 8.9 (-40°C), 63.5 (0°C), 243 (40°C), 689 (80°C); used as catalyst in organic synthesis, and as VLSI etchant in electronics industry [HAW93] [MER89] [KIR78] [AIR87] [CRC93]
Solubility: decomposed by H_2O, alcohol [MER89]
Density, g/cm^3: 1.35 (12°C), 1.3728 (0°C) [MER89]; 1.434 (0°C) [KIR78]
Melting Point, °C: -107 [MER89]
Boiling Point, °C: 12.5 [MER89]

492

Compound: Boron trifluoride
Formula: BF_3
Molecular Formula: BF_3
Molecular Weight: 67.806
CAS RN: 7637-07-2
Properties: colorless gas, pungent suffocating odor; forms dense white fumes in moist air; triple point -128.4°C at 8.34 kPa; critical temp -12.25°C; critical pressure 4984 kPa; enthalpy of fusion 4.20 kJ/mol; enthalpy of vaporization 19.33 kJ/mol; can be produced by reacting borax, fluorspar and sulfuric acid; used as a catalyst in organic synthesis, and in electronics [HAW93] [KIR78] [MER89] [AIR87] [CRC93]

Solubility: 332g/100g H_2O (0°C) with some hydrolysis [MER89]
Density, g/cm^3: gas: 3.07666 g/L (STP) [KIR78]
Melting Point, °C: -126.8 [CRC93]
Boiling Point, °C: -101 [CRC93]
Reactions: forms the solid complex $HNO_3 \cdot 2BF_3$ with HNO_3 [MER89]

493

Compound: Boron trifluoride etherate
Synonyms: boron fluoride-ether
Formula: $BF_3(CH_3CH_2)_2O$
Molecular Formula: $C_4H_{10}BF_3O$
Molecular Weight: 141.929
CAS RN: 109-63-7
Properties: fuming liq; sensitive to moisture; prepared by vapor phase reaction between BF_3 and ethyl ether; used as catalyst in organic reactions [MER89] [ALF95]
Solubility: hydrolyzed by H_2O [MER89]
Density, g/cm^3: 1.125 [MER89]
Melting Point, °C: -60.4 [MER89]
Boiling Point, °C: 125.7 [MER89]

494

Compound: Boron triiodide
Formula: BI_3
Molecular Formula: BI_3
Molecular Weight: 391.524
CAS RN: 13517-10-7
Properties: needles or cryst with 99.9% purity; unstable; enthalpy of vaporization 40.5 kJ/mol [CRC93] [CER91]
Density, g/cm^3: 3.35 [KIR78]
Melting Point, °C: 49.9 [COT88]
Boiling Point, °C: 210 [KIR78]

495

Compound: Boron trisulfide
Formula: B_2S_3
Molecular Formula: B_2S_3
Molecular Weight: 117.820
CAS RN: 12007-33-9
Properties: white; -200 mesh, 99.9% purity [CRC94] [CER91]
Density, g/cm^3: 1.55 [CRC94]
Melting Point, °C: 310 [AES93]

496

Compound: Bromic acid
Formula: $HBrO_3$

Molecular Formula: $BrHO_3$
Molecular Weight: 128.910
CAS RN: 7789-31-3
Properties: colorless or sl yellowish liq; stable only in very dil aq solutions; oxidizing agent; used in dyes [HAW93]
Solubility: can exist only in aq media [HAW93]
Density, g/cm³: 3.28 [HAW93]
Melting Point, °C: decomposes 100 [CRC94]

497
Compound: Bromine
Formula: Br_2
Molecular Formula: Br_2
Molecular Weight: 159.808 (atomic weight, 79.904)
CAS RN: 7726-95-6
Properties: dark reddish, volatile liq; suffocating odor; vaporizes rapidly at room temp; oxidant; viscosity (30°C) 0.288 mm²/s; surface tension 40.9 mN/m (25°C); electrical resistivity, 6.5 x 10^{+10} ohm cm at 25°C; dielectric constant 3.33 x 10^{+5} Hz (25°C); enthalpy of vaporization 29.97 kJ/mol; enthalpy of fusion 10.57 kJ/mol; electronegativity, 3.0; critical temp 311°C; critical pressure 10.3 atm; used in flame-retardant materials [MER89] [KIR78] [CRC93]
Solubility: 3.35g/100g soln, H_2O (25°C) [KIR78]; forms HOBr with H_2O [MER89]
Density, g/cm³: 3.1055 (25°C); vapor density 7.139 g/L (0°C) [KIR78]
Melting Point, °C: -7.2 [CRC93]
Boiling Point, °C: 59.10 [CRC93]
Reactions: thermal dissociation >600°C [KIR78]
Thermal Conductivity, W/(m·K): 0.123 (25°C) [KIR78]
Thermal Expansion Coefficient: 20°C to 30°C, 0.0011/°C [KIR78]

498
Compound: Bromine azide
Formula: BrN_3
Molecular Formula: BrN_3
Molecular Weight: 121.924
CAS RN: 13973-87-0
Properties: cryst, or red liq; oxidizing agent; used in detonators and other explosive devices [HAW93]
Melting Point, °C: ~45 [HAW93]
Boiling Point, °C: explodes [HAW93]

499
Compound: Bromine chloride

Formula: BrCl
Molecular Formula: BrCl
Molecular Weight: 115.357
CAS RN: 13863-41-7
Properties: reddish yellow liq; formed by reaction of bromide and chlorine, in the vapor or liq states; easily hydrolyzed by water; used as a disinfectant in industry, for wastewater treatment [HAW93] [KIR78]
Solubility: s H_2O with ready hydrolysis, s carbon disulfide and ether [HAW93]
Density, g/cm³: 5.062 g/L [LID94]
Melting Point, °C: -66 [HAW93]
Boiling Point, °C: decomposes evolving Cl_2 at 10 [HAW93]

500
Compound: Bromine dioxide
Formula: BrO_2 or Br_2O_4
Molecular Formula: BrO_2
Molecular Weight: 111.903
CAS RN: 21255-83-4
Properties: yellowish orange solid; obtained by ozonation of Br_2 in Freon 11 at -50°C, and subsequent evaporation [KIR78]
Melting Point, °C: decomposes at 0 [KIR78]

501
Compound: Bromine monofluoride
Formula: BrF
Molecular Formula: BrF
Molecular Weight: 98.902
CAS RN: 13863-59-7
Properties: red to brownish-red; very unstable, rapidly forming bromine and higher fluorides; prepared by reaction of Br_2 and F_2 [KIR78]
Density, g/cm³: 4.34 g/L [LID94]
Melting Point, °C: ~-33 [KIR78]
Boiling Point, °C: ~20 [KIR78]

502
Compound: Bromine oxide
Synonyms: bromine monoxide
Formula: Br_2O
Molecular Formula: Br_2O
Molecular Weight: 175.807
CAS RN: 21308-80-5
Properties: dark brown solid; stable below -40°C; formed by reaction of HgO and Br_2 in CCl_4 in the absence of light; used for bromination reactions [KIR78]
Melting Point, °C: -17.5, decomposes [KIR78]

503
Compound: Bromine pentafluoride
Formula: BrF_5
Molecular Formula: BrF_5
Molecular Weight: 174.896
CAS RN: 7789-30-2
Properties: colorless fuming liq; vapor pressure 7 psia (21.1°C); enthalpy of vaporization 30.6 kJ/mol; enthalpy of fusion 5.66 kJ/mol; specific conductivity (25°C) 9.1×10^{-8} ohm·cm; highly reactive, e.g. reacts with all known elements except the inert gases, nitrogen and oxygen; can be prepared by reacting Br_2 and F_2; used as a fluorinating agent in organic synthesis and as an oxidizing agent in liq rocket fuels [KIR78] [HAW93]
Solubility: explodes on contact with H_2O [MER89]
Density, g/cm³: 2.460 [MER89]
Melting Point, °C: -60.5 [MER89]
Boiling Point, °C: 40.76 [MER89]
Reactions: thermally stable up to 460°C [MER89]

504
Compound: Bromine trifluoride
Formula: BrF_3
Molecular Formula: BrF_3
Molecular Weight: 136.899
CAS RN: 7787-71-5
Properties: colorless liq, if pure; long prisms when solid; fumes in air; very reactive; vapor pressure 0.15 psia (21.1°C); formed by reaction of Br_2 and F_2; enthalpy of vaporization 47.57 kJ/mol; enthalpy of fusion 12.01 kJ/mol; used as a fluorinating agent and as a solvent for fluorides [HAW93] [MER89] [AIR87] [KIR78] [CRC93]
Solubility: violent reaction with H_2O [HAW93]
Density, g/cm³: 2.803 [MER89]
Melting Point, °C: 8.77 [MER89]
Boiling Point, °C: 125.75 [MER89]

505
Compound: Bromochloromethane
Formula: CH_2BrCl
Molecular Formula: CH_2BrCl
Molecular Weight: 129.383
CAS RN: 74-97-5
Properties: liq [ALF95]
Density, g/cm³: 1.991 [ALF95]
Melting Point, °C: -88 [ALF95]
Boiling Point, °C: 68 [ALF95]

506
Compound: Bromosilane
Formula: SiH_3Br
Molecular Formula: BrH_3Si
Molecular Weight: 111.014
CAS RN: 13465-73-1
Properties: enthalpy of vaporization 24.4 kJ/mol; entropy of vaporization 88.3 J/(mol·K) [CIC73] [CRC93]
Melting Point, °C: -94 [CIC73]
Boiling Point, °C: 1.9 [CIC73]

507
Compound: Cacodylic acid
Synonyms: dimethylarsenic acid, hydroxydimethylarsine oxide
Formula: $(CH_3)_2As(O)OH$
Molecular Formula: $C_2H_7AsO_2$
Molecular Weight: 137.998
CAS RN: 75-60-5
Properties: colorless odorless cryst; hygr; preparation: distillation of a mixture of As_2O_3 and CH_3COOK, followed by oxidation of resulting product with HgO; uses: herbicide, dermatologic [MER89] [HAW93]
Solubility: s 0.5 parts H_2O; v s alcohol [MER89]
Melting Point, °C: 195-196 [ALD94]

508
Compound: Cadmium
Formula: Cd
Molecular Formula: Cd
Molecular Weight: 112.411
CAS RN: 7440-43-9
Properties: soft, silvery white, with blue tinge metal; distorted hex; easily cut with knife; slowly oxidized by moist air to CdO; hardness 2.0 Mohs; fusion enthalpy 6.19 kJ/mol; vaporization enthalpy 99.87 kJ/mol; electrical resistivity (22°C) 7.2 μohm·cm; Brinell hardness 16.23 kg/mm²; Poissons' ratio 0.33; used in easily fusible alloys, solder for aluminum, photoelectric cells, and as 99.999% pure sputtering target for dielectric films [KIR78] [CER91] [CRC93]
Solubility: i H_2O; reacts with dil HNO_3, slowly with hot HCl [MER89]
Density, g/cm³: 8.65 [MER89]
Melting Point, °C: 321.07 [CRC93]
Boiling Point, °C: 767 [CRC93]
Thermal Conductivity, W/(m·K): 98 (0°C), 95 (100°C) 89 (300°C) [KIR78]

Thermal Expansion Coefficient: 20°C, 31.3
    microcm/(cm·°C) [KIR78]

509
Compound: Cadmium acetate
Synonyms: acetic acid, cadmium salt
Formula: Cd(CH₃COO)₂
Molecular Formula: C₄H₆CdO₄
Molecular Weight: 230.501
CAS RN: 543-90-8
Properties: colorless cryst; used in ceramics and
    electroplating baths [HAW93]
Solubility: s H₂O, alcohol [HAW93]
Density, g/cm³: 2.34 [MER89]
Melting Point, °C: 255 [MER89]

510
Compound: Cadmium acetate dihydrate
Synonyms: acetic acid, cadmium salt dihydrate
Formula: Cd(CH₃COO)₂·2H₂O
Molecular Formula: C₄H₁₀CdO₆
Molecular Weight: 266.530
CAS RN: 5743-04-4
Properties: white powd; cryst; slight acetic acid
    odor [MER89] [STR93]
Solubility: v s H₂O [MER89]
Density, g/cm³: 2.01 [HAW93]
Reactions: minus 2H₂O ~130°C [MER89]

511
Compound: Cadmium acetylacetonate
Synonyms: 2,4-pentanedione, cadmium(II)
    derivative
Formula: Cd(CH₃COCH=C(O)CH₃)₂
Molecular Formula: C₁₀H₁₄CdO₄
Molecular Weight: 310.630
CAS RN: 14689-45-3
Properties: white powd [STR93]

$$[CH_3-C=CH-C-CH_3]_2 Cd$$

with O– and O substituents above

Melting Point, °C: decomposes [STR93]

512
Compound: Cadmium antimonide
Formula: CdSb
Molecular Formula: CdSb
Molecular Weight: 234.171
CAS RN: 12014-29-8

Properties: 6 mm pieces and smaller of 99.999%
    purity; ortho-rhomb, a = 0.6471 nm, b = 0.8253
    nm; semiconductor; uses: in thermoelectric
    devices; enthalpy of fusion 32.05 kJ/mol;
    enthalpy of evaporation 138 kJ/mol; there is also
    Cd₃Sb₂, 12014-29-8 [CER91] [KIR78]
    [HAW93]
Density, g/cm³: 6.92 [KIR78]
Melting Point, °C: 456 [KIR78]

513
Compound: Cadmium arsenide
Formula: Cd₃As₂
Molecular Formula: As₂Cd₃
Molecular Weight: 487.076
CAS RN: 12006-15-4
Properties: gray tetr cryst, a = 0.8945 nm, c = 1.265
    nm [KIR78]
Density, g/cm³: 6.21 [KIR78]
Melting Point, °C: 721 [KIR78]

514
Compound: Cadmium azide
Formula: Cd(N₃)₂
Molecular Formula: CdN₆
Molecular Weight: 196.451
CAS RN: 14215-29-3
Properties: yellow-white ortho [LID94]
Density, g/cm³: 3.24 [LID94]

515
Compound: Cadmium borotungstate
    octadecahydrate
Formula: Cd₅(BW₁₂O₄₀)·18H₂O
Molecular Formula: BCd₅H₃₆O₅₈W₁₂
Molecular Weight: 3732.386
CAS RN: 1306-26-9
Properties: yellow, heavy cryst; used to separate
    minerals [HAW93]
Solubility: 1250 g/100 mL (19°C) H₂O, yielding a
    yellow to light brown solution [HAW93]
    [CRC77]
Melting Point, °C: 75 [CRC77]

516
Compound: Cadmium bromide
Formula: CdBr₂
Molecular Formula: Br₂Cd
Molecular Weight: 272.219
CAS RN: 7789-42-6

Properties: white powd, -80 mesh 99.9% pure; hygr; hex, a = 0.395 nm, c = 1.867 nm; enthalpy of fusion 20.90 kJ/mol; enthalpy of vaporization 115 kJ/mol; pearly flakes; crystallizes as the monohydrate below 36°C, as tetrahydrate above 36°C; finds use in lithography and photography [HAW93] [MER89] [STR93] [KIR78] [CER91] [CRC93]

Solubility: g/100 g soln, H_2O: 36.0 (0°C), 52.9 (25°C), 61.65 (100°C); Solid phase, $CdBr_2 \cdot 4H_2O$ (0°C, 25°C), $CdBr_2$ (100°C) [KRU93]; s acetone, alcohol and acids [HAW93]

Density, g/cm³: 5.192 [MER89]

Melting Point, °C: 568 [CRC93]

Boiling Point, °C: 1136 [COT88]

517

Compound: Cadmium bromide tetrahydrate

Formula: $CdBr_2 \cdot 4H_2O$

Molecular Formula: $Br_2CdH_8O_4$

Molecular Weight: 344.281

CAS RN: 13464-92-1

Properties: white to yellowish efflorescent cryst [HAW93]

Solubility: 121 g/100mL H_2O (10°C) [CRC94]; s acetone, alcohol, acids [HAW93]

Melting Point, °C: transition 36 [CRC94]

518

Compound: Cadmium carbonate

Synonyms: otavite

Formula: $CdCO_3$

Molecular Formula: $CCdO_3$

Molecular Weight: 172.420

CAS RN: 513-78-0

Properties: -200 mesh with 99.999%, 99.9% and 99% purity; white amorphous powd or rhomb leaflets; a = 0.61306 nm [HAW93] [MER89] [KIR78] [CER91]

Solubility: 2.8 microgram/100g H_2O; s dil acids [KIR78] [MER89]

Density, g/cm³: 4.258 [HAW93]

Melting Point, °C: decomposes 500 [HAW93]

519

Compound: Cadmium chlorate dihydrate

Formula: $Cd(ClO_3)_2 \cdot 2H_2O$

Molecular Formula: $CdCl_2H_4O_8$

Molecular Weight: 315.343

CAS RN: 7790-78-5

Properties: colorless cryst; hygr [HAW93]

Solubility: mol/100 mol soln, H_2O: 25.92 (0°C); Solid phase, $Cd(ClO_3)_2 \cdot 2H_2O$ [KRU93]; s alcohol, acetone [HAW93]; g/100g H_2O: 299 (0°C), 322 (20°C), 455 (60°C) [LAN85]

Density, g/cm³: 2.28 (18°C) [HAW93]

Melting Point, °C: 80 [HAW93]

520

Compound: Cadmium chloride

Formula: $CdCl_2$

Molecular Formula: $CdCl_2$

Molecular Weight: 183.316

CAS RN: 10108-64-2

Properties: -80 mesh of 99.9% purity; white odorless hygr; hex, a = 0.3854 nm, c = 1.746 nm; enthalpy of fusion 48.58 kJ/mol; enthalpy of vaporization at bp 124.3 kJ/mol [MER89] [CRC93]

Solubility: g/100 g soln, H_2O: 47.3 (0°C), 54.65 (25°C), 59.55 (100°C); Solid phase, $CdCl_2 \cdot 2$-$1/2H_2O$ (0°C, 25°C), $CdCl_2 \cdot H_2O$ (100°C) [KRU93]; s acetone [HAW93]

Density, g/cm³: 4.05 [STR93]

Melting Point, °C: 564 [CRC93]

Boiling Point, °C: 960 [CRC93]

521

Compound: Cadmium chloride hemipentahydrate

Formula: $CdCl_2 \cdot 2$-$1/2H_2O$

Molecular Formula: $CdCl_2H_5O_{2.5}$

Molecular Weight: 228.354

CAS RN: 7790-78-5

Properties: white efflorescent granules or rhomb leaflets [MER89] [STR93]

Solubility: g/100g H_2O: 90 (0°C), 113 (20°C), 132 (30°C) [LAN85]; s acetone [HAW93]

Density, g/cm³: 3.327 [HAW93]

Melting Point, °C: 960 [HAW93]

522

Compound: Cadmium chromate

Formula: $CdCrO_4$

Molecular Formula: $CdCrO_4$

Molecular Weight: 228.405

CAS RN: 14312-00-6

Properties: yellow solid; used in catalysts and in pigments [KIR78]

Solubility: i H_2O [KIR78]

Density, g/cm³: 4.5 [LID94]

523
Compound: Cadmium cyanide
Formula: $Cd(CN)_2$
Molecular Formula: C_2CdN_2
Molecular Weight: 164.446
CAS RN: 542-83-6
Properties: cryst or white powd; turns brown if heated in air; used to electroplate cadmium [HAW93] [MER89]
Solubility: 1.71 g/100mL H_2O (15°C) [MER89]
Density, g/cm³: 2.23 [MER89]
Melting Point, °C: decomposes >200 [CRC94]

524
Compound: Cadmium dichromate monohydrate
Formula: $CdCr_2O_7 \cdot H_2O$
Molecular Formula: $CdCr_2H_2O_8$
Molecular Weight: 346.414
CAS RN: 69239-51-6
Properties: orange solid used in metal finishing [KIR78]
Solubility: s H_2O [KIR78]

525
Compound: Cadmium fluoride
Formula: CdF_2
Molecular Formula: CdF_2
Molecular Weight: 150.408
CAS RN: 7790-79-6
Properties: 99.9% pure melted pieces, 3-6 mm; cub cryst, a = 0.53880 nm; enthalpy of fusion 22.60 kJ/mol; enthalpy of vaporization 214 kJ/mol; available as 99.89% pure cryst; used in electronic and optical materials, in the preparation of laser cryst, and as a sputtering target to produce multilayers [HAW93] [MER89] [CER91] [CRC93]
Solubility: mol/kg soln, H_2O: 0.4652 (0°C), 0.291 (25°C), 0.12 (100°C) [KRU93]; s acids, i alkalies [HAW93]
Density, g/cm³: 6.33 [MER89]; 6.64 [STR93]
Melting Point, °C: 1110 [CRC93]
Boiling Point, °C: 1748 [CRC93]

526
Compound: Cadmium hydroxide
Formula: $Cd(OH)_2$
Molecular Formula: CdH_2O_2
Molecular Weight: 146.426
CAS RN: 21041-95-2

Properties: white amorphous powd, or trig and hex cryst, a = 0.3475 nm, c = 0.467 nm; can absorb atm H_2O and CO_2 [HAW93] [MER89] [KIR78]
Solubility: i H_2O; s dil acid, NH_4OH [MER89]; mmol/L soln, H_2O: 0.038 (18°C) [KRU93]
Density, g/cm³: 4.79 [MER89]
Melting Point, °C: 150 decomposes [KIR78]
Reactions: minus $2H_2O$ at 130-200°C [MER89]

527
Compound: Cadmium iodate
Formula: $Cd(IO_3)_2$
Molecular Formula: CdI_2O_6
Molecular Weight: 462.216
CAS RN: 7790-81-0
Properties: fine white powd; uses: oxidizing agent [HAW93]
Solubility: mol/L soln, H_2O: (1.97±0.13) x 10^{-3} (25°C) [KRU93]; s HNO_3, NH_4OH [HAW93]
Density, g/cm³: 6.48 [HAW93]
Melting Point, °C: decomposes [CRC77]

528
Compound: Cadmium iodide
Formula: CdI_2
Molecular Formula: CdI_2
Molecular Weight: 366.220
CAS RN: 7790-80-9
Properties: -40 mesh 99.5% pure; lustrous, flake-like cryst; odorless; two forms: α and β; α-form is hex, a = 0.424 nm, c = 0.684 nm, enthalpy of fusion 15.30 kJ/mol, enthalpy of vaporization 115 kJ/mol; becomes yellow on prolonged exposure to air and light; finds applications in photography and in lubricants [HAW93] [MER89] [CRC93]
Solubility: s H_2O, alcohol, ether, acetone [MER89]; g/100 g soln, H_2O: 44.05 (0°), 46.30 (25°C), 55.55 (100°C); Solid phase, CdI_2 [KRU93]
Density, g/cm³: α: 5.67; β: 5.30 [HAW93]
Melting Point, °C: α:388; β: 404 [HAW93]
Boiling Point, °C: 742 [CRC93]

529
Compound: Cadmium metasilicate
Formula: $CdSiO_3$
Molecular Formula: CdO_3Si
Molecular Weight: 188.495
CAS RN: 13477-19-5
Properties: colorless; monocl, a = 1.504 nm, b = 0.710 nm, c = 0.696 nm [KIR78] [CRC77]
Solubility: v sl s H_2O [CRC77]

Density, g/cm³: 4.928 [KIR78]
Melting Point, °C: 1252 [KIR78]

530
Compound: Cadmium molybdate(VI)
Formula: $CdMoO_4$
Molecular Formula: $CdMoO_4$
Molecular Weight: 272.349
CAS RN: 13972-68-4
Properties: -200 mesh 99.9% pure; yellow cryst; scheelite structure, c/a = 2.174; applications in electronic and optical materials [HAW93] [KIR81]
Solubility: 0.0067 g/100g H_2O, s acids [HAW93] [KIR81]
Density, g/cm³: 5.347 [HAW93]
Melting Point, °C: ~900 decomposes [KIR81]

531
Compound: Cadmium niobate
Formula: $Cd_2Nb_2O_7$
Molecular Formula: $Cd_2Nb_2O_7$
Molecular Weight: 522.631
CAS RN: 12187-14-3
Properties: cub; -200 mesh 99.9% purity [LID94] [CER91]
Density, g/cm³: 6.2 [LID94]
Melting Point, °C: ~1450 [LID94]

532
Compound: Cadmium nitrate
Formula: $Cd(NO_3)_2$
Molecular Formula: CdN_2O_6
Molecular Weight: 236.427
CAS RN: 10325-94-7
Properties: white amorphous pieces, or hygr needles; cub, a = 0.756 nm; used in coloring glass, and in photographic emulsions [HAW93] [KIR78]
Solubility: g/100 g soln, H_2O: 55.1 (0.6°C), 61.3 (25°C), 87.2 (100°C); Solid phase, $Cd(NO_3)_2\cdot4H_2O$ (0°C, 25°C), $Cd(NO_3)_2$ (100°C) [KRU93]; s alcohol and ammonia [HAW93]
Density, g/cm³: 3.6 [LID94]
Melting Point, °C: 350 [HAW93]

533
Compound: Cadmium nitrate tetrahydrate
Formula: $Cd(NO_3)_2\cdot4H_2O$
Molecular Formula: $CdH_8N_2O_{10}$
Molecular Weight: 308.482

CAS RN: 10022-68-1
Properties: colorless; hygr; ortho-rhomb cryst, a = 0.583 nm, b = 2.575 nm, c = 1.099 nm; enthalpy of fusion 32.636 kJ/mol [MER89] [STR93] [KIR78]
Solubility: 132 g/100g H_2O (0°C) [KIR78]
Density, g/cm³: 2.45 [STR93]
Melting Point, °C: 59.5 [MER89]
Boiling Point, °C: 132 [HAW93]

534
Compound: Cadmium oxalate
Synonyms: ethanedioic acid, cadmium salt
Formula: CdC_2O_4
Molecular Formula: C_2CdO_4
Molecular Weight: 200.431
CAS RN: 814-88-0
Properties: colorless [LAN52]
Solubility: mol/kg soln, H_2O: 0.00030 (25°C); equilibrium solid phase $CdC_2O_4\cdot3H_2O$ [KRU93]
Density, g/cm³: 3.32 [HAW93]
Melting Point, °C: 330, decomposes [HAW93]

535
Compound: Cadmium oxalate trihydrate
Synonyms: ethanedioic acid, cadmium salt trihydrate
Formula: $CdC_2O_4\cdot3H_2O$
Molecular Formula: $C_2H_6CdO_7$
Molecular Weight: 254.477
CAS RN: 20712-42-9
Properties: white amorphous powd [HAW93]
Solubility: i H_2O, alcohol; s dil acids, NH_4OH [HAW93]
Melting Point, °C: 340, decomposes [HAW93]

536
Compound: Cadmium oxide
Formula: CdO
Molecular Formula: CdO
Molecular Weight: 128.410
CAS RN: 1306-19-0
Properties: -100 mesh 99.999% pure, -200 mesh 99.9% and 99.5% purity; two forms: (1) colorless amorphous powd, and (2) brown or red cryst; cub cryst, a = 0.46953 nm; enthalpy of fusion 243.5 kJ/mol; used in cadmium plating baths and in ceramics [HAW93] [KIR78] [CER91]
Solubility: 0.00094 g/100g H_2O; s dil acids [MER89] [KIR78]
Density, g/cm³: (1) 6.95; (2) 8.15 [HAW93]

Melting Point, °C: sublimes 1540 [KIR78]
Thermal Conductivity, W/(m·K): 0.7 [CRC93]

537
Compound: Cadmium perchlorate
Formula: $Cd(ClO_4)_2$
Molecular Formula: $CdCl_2O_8$
Molecular Weight: 311.311
CAS RN: 79490-00-9
Solubility: g/100 g soln, H_2O: 58.7 (25°C), 66.9 (100°C); Solid phase, $Cd(ClO_4)_2 \cdot 6H_2O$ [KRU93]

538
Compound: Cadmium perchlorate hexahydrate
Formula: $Cd(ClO_4)_2 \cdot 6H_2O$
Molecular Formula: $CdCl_2H_{12}O_{14}$
Molecular Weight: 419.403
CAS RN: 10326-28-0
Properties: white cryst [STR93]
Solubility: g/100g H_2O: 180 (10°C), 188 (20°C), 272 (100°C) [LAN85]
Density, g/cm³: 2.37 [LID94]

539
Compound: Cadmium phosphide
Formula: CdP_2
Molecular Formula: CdP_2
Molecular Weight: 174.359
CAS RN: 12133-44-7
Properties: -100 mesh of 99.9% purity; there is a Cd_3P_2, 12014-28-7, melting point 700°C, with same properties [CER91] [ALD94]
Density, g/cm³: Cd_3P_2: 5.60 [CRC94]
Melting Point, °C: Cd_3P_2: 700 [CRC94]

540
Compound: Cadmium selenate dihydrate
Formula: $CdSeO_4 \cdot 2H_2O$
Molecular Formula: CdH_4O_6Se
Molecular Weight: 291.399
CAS RN: 10060-09-0
Properties: ortho-rhomb cryst [MER89]
Solubility: g/100 g H_2O: 72.44 (0°C), 61.23 (26°C), 22.01 (98.5°C); Solid phase $CdSeO_4 \cdot 2H_2O$ [KRU93]
Density, g/cm³: 3.632 [MER89]
Melting Point, °C: decomposes 100 [MER89]

541
Compound: Cadmium selenide

Formula: CdSe
Molecular Formula: CdSe
Molecular Weight: 191.371
CAS RN: 1306-24-7
Properties: 3-12 mm pieces (sintered), and -325 mesh 10 microns or less with 99.999% purity; white to brown; cub or hex cryst; hex: a = 0.4309 nm, c = 0.7021 nm, enthalpy of fusion 305.307 kJ/mol; becomes red in sunlight; used as a red pigment, in semiconductors, and as an evaporation material and sputtering target to produce photoconductive films and infrared filters; possible use in the electrofabrication of microdiode arrays [HAW93] [MER89] [CER91] [KLE93] [KRE91]

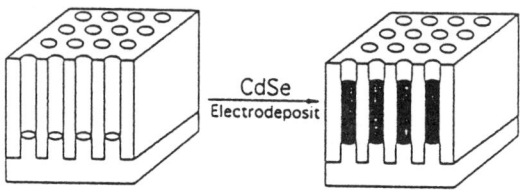

(By permission of the American Chemical Society.)

Solubility: i H_2O [MER89]
Density, g/cm³: α: 5.81 [KIR78]
Melting Point, °C: 1240 [CRC93]

542
Compound: Cadmium selenite
Formula: $CdSeO_3$
Molecular Formula: CdO_3Se
Molecular Weight: 239.369
CAS RN: 13814-59-0
Properties: -80 mesh 99.5% pure [CER91]

543
Compound: Cadmium stearate
Formula: $Cd(OOC_{18}H_{35})_2$
Molecular Formula: $C_{36}H_{70}CdO_4$
Molecular Weight: 679.361
CAS RN: 2223-93-0
Properties: white powd; uses: lubricant and plastics stabilizer [HAW93] [STR93]
Density, g/cm³: 1.21 [KIR78]
Melting Point, °C: 104 [KIR78]

544
Compound: Cadmium succinate
Synonyms: succinic acid, cadmium salt
Formula: $Cd(OOCCH_2CH_2COO)$
Molecular Formula: $C_4H_4CdO_4$
Molecular Weight: 228.484

CAS RN: 141-00-4

Properties: white powd; needles or plates; preparation: reaction of $CdCO_3$ with succinic acid; uses: plant fungicide [MER89] [HAW93]

Solubility: 0.367 g/100 mL (40°C) H_2O; i alcohol [MER89] [HAW93]

545

Compound: Cadmium sulfate

Formula: $CdSO_4$

Molecular Formula: CdO_4S

Molecular Weight: 208.475

CAS RN: 10124-36-4

Properties: colorless odorless cryst; ortho-rhomb, a = 0.4717 nm, b = 0.6559 nm, c = 0.4701 nm; enthalpy of fusion 20.084 kJ/mol; used as a pigment [HAW93] [KIR78]

Solubility: g/100g H_2O: 75.6±0.1 (0°C), 76.7 (25°C), 58.4 (99°C); Solid phase, $CdSO_4 \cdot 8/3H_2O$ (0°C, 25°C), β-$CdSO_4 \cdot H_2O$ (99°C) [KRU93]

Density, g/cm³: 4.691 [KIR78]

Melting Point, °C: 1000 [HAW93]

546

Compound: Cadmium sulfate monohydrate

Formula: $CdSO_4 \cdot H_2O$

Molecular Formula: CdH_2O_5S

Molecular Weight: 226.490

CAS RN: 13477-20-8

Properties: monocl, a = 7.607 nm, b = 7.541 nm, c = 8.186 nm [KIR78]

Density, g/cm³: 3.79 [KIR78]

Melting Point, °C: 105 [KIR78]

547

Compound: Cadmium sulfate octahydrate

Formula: $3CdSO_4 \cdot 8H_2O$

Molecular Formula: $Cd_3H_{16}O_{20}S_3$

Molecular Weight: 769.546

CAS RN: 15244-35-6

Properties: colorless odorless monocl cryst; preparation: reaction of dil sulfuric acid with Cd or CdO; uses: pigment, electrolyte in Weston standard cell [MER89] [HAW93]

Solubility: v s H_2O [MER89]

Density, g/cm³: 3.08 [MER89]

Reactions: minus H_2O above 40°C, forms monohydrate by 80°C; does not become anhydrous if heated further [MER89]

548

Compound: Cadmium sulfide

Synonyms: greenockite

Formula: CdS

Molecular Formula: CdS

Molecular Weight: 144.477

CAS RN: 1306-23-6

Properties: 3-6 mm pieces (highly dense, pressure sintered) and -325 mesh 99.999% pure; light yellow or orange colored cub or hex cryst; hex, a = 0.41348 nm, c = 0.6749 nm; enthalpy of fusion 201.669 kJ/mol; used in pigments and inks, and as evaporation material and pure sputtering target to produce photoconductive films, infrared filters and solar cells [HAW93] [MER89] [KIR78] [CER91]

Solubility: 0.13mg/100g (18°C) [MER89]; mol/L in H_2O: 1.42 x 10^{-10} (25°C); 8.56 x 10^{-10} (100°C) [KRU93]; s acids and ammonia [HAW93]

Density, g/cm³: cub: 4.50; hex: 4.82 [MER89]

Melting Point, °C: 1750 [STR93]; sublimes 980 [MER89]

Reactions: decomposed by warm dil mineral acids to evolve H_2S [MER89]

549

Compound: Cadmium sulfite

Formula: $CdSO_3$

Molecular Formula: CdO_3S

Molecular Weight: 192.475

CAS RN: 13477-23-1

Properties: cryst [LAN52]

Solubility: mol/kg H_2O: 0.00221 (0°C), 0.00207 (90°C) [KRU93]

Melting Point, °C: decomposes [LAN52]

550

Compound: Cadmium tantalate

Formula: $Cd_2Ta_2O_7$

Molecular Formula: $Cd_2O_7Ta_2$

Molecular Weight: 698.714

CAS RN: 12050-35-0

Properties: -200 mesh 99.9% pure [CER91]

551

Compound: Cadmium telluride

Formula: CdTe

Molecular Formula: CdTe

Molecular Weight: 240.011

CAS RN: 1306-25-8

Properties: 3-6 mm pieces (fused) of 99.999% purity; brownish black; cub cryst when prepared by sublimation in H_2 atm, also a hex form; hex, a = 0.457 nm, c = 0.747 nm; oxidizes eventually in moist air; used in phosphors, in semiconductor materials, and as an evaporation material and sputtering target to produce photoconductive films and infrared filters [HAW93] [MER89] [CER91]
Solubility: i H_2O, dil acids; s with decomposition in HNO_3 [MER89]
Density, g/cm³: 6.2 [MER89]
Melting Point, °C: 1041 [MER89]

552
Compound: Cadmium tellurite
Formula: $CdTeO_3$
Molecular Formula: CdO_3Te
Molecular Weight: 288.009
CAS RN: 15851-44-2
Properties: -80 mesh 99% pure [CER91]

553
Compound: Cadmium tetrafluoroborate
Formula: $Cd(BF_4)_2$
Molecular Formula: B_2CdF_8
Molecular Weight: 286.020
CAS RN: 14486-19-2
Properties: colorless liq [STR93]

554
Compound: Cadmium titanate
Formula: $CdTiO_3$
Molecular Formula: CdO_3Ti
Molecular Weight: 208.276
CAS RN: 12014-14-1
Properties: ortho-rhomb [KIR83]
Density, g/cm³: 6.5 [KIR83]

555
Compound: Cadmium tungstate(VI)
Formula: $CdWO_4$
Molecular Formula: CdO_4W
Molecular Weight: 360.249
CAS RN: 7790-85-4
Properties: -325 mesh, 10 microns or less of 99.9% purity; white or yellowish monocl cryst or powd; finds use in fluorescent paint, x-ray screens [HAW93] [MER89] [CER91]
Solubility: i H_2O [MER89]; s NH4OH, alkali cyanides [HAW93]

Density, g/cm³: 8.0 [LID94]

556
Compound: Cadmium vanadate
Formula: CdV_2O_6
Molecular Formula: CdO_6V_2
Molecular Weight: 310.290
CAS RN: 12422-12-7
Properties: -200 mesh [CER91]

557
Compound: Cadmium zirconate
Formula: $CdZrO_3$
Molecular Formula: CdO_3Zr
Molecular Weight: 251.633
CAS RN: 12139-23-0
Properties: -200 mesh 99.5% purity [CER91]

558
Compound: Calcium
Formula: Ca
Molecular Formula: Ca
Molecular Weight: 40.078
CAS RN: 7440-70-2
Properties: lustrous silver-white surface (freshly cut); much harder than sodium, softer than aluminum or magnesium; acquires bluish gray tarnish in moist air; enthalpy of fusion 8.54 kJ/mol; enthalpy of vaporization 161.5 kJ/mol; enthalpy of combustion 634.3 kJ/mol; electrical resistivity (20°C) 3.5 μohm·cm; Brinell hardness 17 [MER89] [CRC93] [KIR78]
Solubility: reacts with H_2O, alcohols, dil acids to evolve H_2 [MER89]
Density, g/cm³: 1.54 [MER89]
Melting Point, °C: 843 [COT88]
Boiling Point, °C: 1440 [MER89]
Reactions: burns in air; contact with alkali hydroxides or carbonates may cause explosion [MER89]
Thermal Conductivity, W/(m·K): 201 (25°C) [ALD94]
Thermal Expansion Coefficient: from 0 to 400°C coefficient is 22.3 x 10⁻⁶ m/(m·K) [KIR78]

559
Compound: Calcium acetate
Formula: $Ca(CH_3COO)_2$
Molecular Formula: $C_4H_6CaO_4$
Molecular Weight: 158.168
CAS RN: 62-54-4

Properties: very hygr; rod shaped cryst [MER89]
Solubility: 26g/100g soln (25°C); 23g/100g soln (100°C) [CIC73]
Density, g/cm³: 1.50 [MER89]
Melting Point, °C: decomposes [ALF93]
Reactions: decomposes >160°C to acetone and CaCO₃ [MER89]

560
Compound: Calcium acetate dihydrate
Formula: Ca(CH₃COO)₂·2H₂O
Molecular Formula: C₄H₁₀CaO₆
Molecular Weight: 194.197
CAS RN: 14977-17-4
Properties: colorless long, transparent needles [CRC94] [MER89]
Solubility: g/100g H₂O: 37.4 (0°C); 34.7 (20°C); 29.7 (100°C) [LAN85]
Reactions: minus H₂O on standing in air to form monohydrate [MER89]

561
Compound: Calcium acetate monohydrate
Formula: Ca(CH₃COO)₂·H₂O
Molecular Formula: C₄H₈CaO₅
Molecular Weight: 176.183
CAS RN: 5743-26-0
Properties: brown, gray or white powd; hygr; sl bitter taste; an odor of acetic acid; decomposes if heated; used in the manufacture of acetone, acetic acid; as a mordant in dyeing textiles [HAW93] [ALD94] [STR93]
Solubility: 48.6/100 mL (0°C), 34.3/100 mL (100°C) H₂O; sl s alcohol [CRC77] [HAW93]
Melting Point, °C: decomposes [CRC77]

562
Compound: Calcium acetylacetonate hydrate
Synonyms: 2,4-pentanedione, calcium derivative
Formula: Ca(CH₃COCH=C(O)CH₃)₂·xH₂O
Molecular Formula: C₁₀H₁₄CaO₄ (anhydrous)
Molecular Weight: 238.297 (anhydrous)
CAS RN: 19372-44-2
Properties: white powd; x~0.5 [ALD94] [STR93]
Melting Point, °C: 175, decomposes [STR93]

563
Compound: Calcium aluminate
Formula: CaO·Al₂O₃
Molecular Formula: Al₂CaO₄
Molecular Weight: 158.039

CAS RN: 12042-68-1
Properties: white, monocl, tricl or rhomb; -200 mesh 99% powd [CER91] [CRC77]
Solubility: decomposed by H₂O [CRC77]
Density, g/cm³: 2.98 [ALF93]
Melting Point, °C: 1600 [ALF93]
Thermal Expansion Coefficient: (volume) 100°C (0.12), 200°C (0.28), 400°C (0.63), 800°C (1.49), 1200°C (2.37) [CLA66]

564
Compound: Calcium aluminate (β)
Formula: 3CaO·Al₂O₃
Molecular Formula: Al₂Ca₃O₆
Molecular Weight: 270.193
CAS RN: 12042-78-3
Properties: white cryst or powd; refractory material, important in cement [HAW93]
Solubility: s acids [HAW93]
Density, g/cm³: 3.038 [HAW93]
Melting Point, °C: 1535, decomposes [HAW93]
Thermal Expansion Coefficient: (volume) 100°C (0.18), 200°C (0.41), 400°C (0.97), 800°C (2.23), 1200°C (3.66) [CLA66]

565
Compound: Calcium aluminum silicate
Synonyms: gehlenite
Formula: Ca₂Al₂SiO₇
Molecular Formula: Al₂Ca₂O₇Si
Molecular Weight: 274.201
CAS RN: 1327-39-5
Properties: colorless, tetr; mineral; used in cement and refractories [MER89] [CRC77]
Density, g/cm³: 3.048 [CRC77]
Melting Point, °C: 1500 [CRC77]
Thermal Expansion Coefficient: (volume) 100°C (0.20), 200°C (0.45), 400°C (0.93), 800°C (1.97), 1200°C (3.07) [CLA66]

566
Compound: Calcium arsenate
Formula: Ca₃(AsO₄)₂
Molecular Formula: As₂Ca₃O₈
Molecular Weight: 398.072
CAS RN: 7778-44-1
Properties: -80 mesh 99% purity; white powd; decomposes if heated; preparation: from calcium chloride and sodium arsenate; used as an insecticide and germicide [HAW93] [CER91]
Solubility: 0.013 g/100mL H₂O (25°C) [CRC94]; s dil acids [MER89]

Density, g/cm³: 3.62 [STR93]
Melting Point, °C: 1455 [CRC94]

571

Compound: Calcium bromate
Formula: $Ca(BrO_3)_2$
Molecular Formula: Br_2CaO_6
Molecular Weight: 295.882
CAS RN: 10102-75-7
Properties: 99% pure powd [ALF93]
Melting Point, °C: 180 [ALF93]

567

Compound: Calcium arsenite
Formula: $CaHAsO_3$
Molecular Formula: $AsCaHO_3$
Molecular Weight: 163.206
CAS RN: 52740-16-6
Properties: white powd; uncertain composition; preparation: passing steam over mixture of CaO and As_2O_3; uses: insecticide, germicide [MER89] [HAW93]

572

Compound: Calcium bromate monohydrate
Formula: $Ca(BrO_3)_2 \cdot H_2O$
Molecular Formula: $Br_2CaH_2O_7$
Molecular Weight: 313.898
CAS RN: 10102-75-7
Properties: white monocl cryst powd; oxidizing agent; used as a maturing agent, a dough conditioner [HAW93] [CRC77]
Solubility: v s H_2O [HAW93]
Density, g/cm³: 3.329 [HAW93]
Reactions: minus H_2O at 180°C [HAW93]

568

Compound: Calcium bis(2,2,6,6-tetramethyl-3,5-heptanedionate)
Formula: $Ca[(CH_3)_3CCOCH=C(O)C(CH_3)_3]_2$
Molecular Formula: $C_{22}H_{38}CaO_4$
Molecular Weight: 406.619
CAS RN: 118448-18-3
Properties: uses: preparation of thin-film semiconductors [ALD94]
Melting Point, °C: 221-224 [ALD94]

573

Compound: Calcium bromide
Formula: $CaBr_2$
Molecular Formula: Br_2Ca
Molecular Weight: 199.886
CAS RN: 7789-41-5
Properties: -80 mesh of 99.5% purity; white odorless, deliq granules, or rhomb cryst; becomes yellow on long exposure to air; sharp saline taste; enthalpy of fusion 29.08 kJ/mol; used in medicine, photography [HAW93] [MER89] [CER91] [STR93] [CRC93]
Solubility: v s H_2O; s alcohol, acetone [HAW93]
Density, g/cm³: 3.353 [MER89]
Melting Point, °C: 742 [CRC94]
Boiling Point, °C: 1815 [CRC94]
Reactions: when strongly heated in air, forms lime and bromine [MER89]

569

Compound: Calcium borate hexahydrate
Formula: $CaB_4O_7 \cdot 6H_2O$
Molecular Formula: $B_4CaH_{12}O_{13}$
Molecular Weight: 303.409
CAS RN: 13701-64-9
Properties: white powd; prepared by fusion of $CaCO_3$ and B_2O_3; used as a flux in metallurgy, in fire-retardant paint [MER89] [STR93]
Solubility: g/100g H_2O: 2.32 (0°C), 2.72 (20°C), 8.70 (100°C) [LAN85]
Melting Point, °C: 986 (anhydrous) [CRC94]

574

Compound: Calcium bromide dihydrate
Formula: $CaBr_2 \cdot 2H_2O$
Molecular Formula: $Br_2CaH_4O_2$
Molecular Weight: 235.917
CAS RN: 22208-73-7
Properties: white cryst powd [ALF93] [STR93]

570

Compound: Calcium boride
Synonyms: calcium hexaboride
Formula: CaB_6
Molecular Formula: B_6Ca
Molecular Weight: 104.944
CAS RN: 12007-99-7
Properties: black cub; -200 mesh 99.5% pure; refractory material [KIR78] [CER91] [CRC94]
Density, g/cm³: 2.3 [ALD94]
Melting Point, °C: 2235 [KIR78]

575

Compound: Calcium bromide hexahydrate

Formula: $CaBr_2 \cdot 6H_2O$
Molecular Formula: $Br_2CaH_{12}O_6$
Molecular Weight: 307.977
CAS RN: 13477-28-6
Properties: white cryst or powd, odorless, sharp saline taste, very deliq; also $CaBr_2 \cdot xH_2O$ [HAW93] [STR93]
Solubility: g/100g H_2O: 125 (0°C), 143 (20°C), 312 (105°C) [LAN85]; s alcohol, acetone [HAW93]
Density, g/cm³: 2.295 [STR93]
Melting Point, °C: 38.2 [STR93]
Boiling Point, °C: 149, decomposes [HAW93]

576
Compound: Calcium carbide
Synonyms: acetylenogen
Formula: CaC_2
Molecular Formula: C_2Ca
Molecular Weight: 64.100
CAS RN: 75-20-7
Properties: 9-40 mm grayish black, irregular lumps or ortho-rhomb cryst; garlic like odor; can be prepared by reacting the metal or CaO with carbon in an electric furnace; used to generate acetylene gas, to reduce copper sulfide to metallic Cu [HAW93] [MER89] [ALF93] [COT88]
Solubility: decomposed by H_2O with evolution of acetylene [MER89]
Density, g/cm³: 2.22 [MER89]
Melting Point, °C: 2300 [MER89]

577
Compound: Calcium carbonate
Synonyms: aragonite
Formula: $CaCO_3$
Molecular Formula: $CCaO_3$
Molecular Weight: 100.087
CAS RN: 471-34-1
Properties: odorless, tasteless powd; ortho-rhomb; formed >30°C [MER89]
Solubility: mol/kg H_2O: 7.76 x 10^{-5} (0.7°C), 6.79 x 10^{-5} (25°C), 3.1 x 10^{-5} (90°C) [KRU93]; solubility data are also found in [PLU82]
Density, g/cm³: 2.83 [MER89]
Melting Point, °C: 825 decomposes [MER89]
Reactions: transforms to calcite ~400°C [KIR78]
Thermal Expansion Coefficient: (volume) 100°C (0.36), 200°C (1.00), 400°C (2.48) [CLA66]

578
Compound: Calcium carbonate

Synonyms: calcite
Formula: $CaCO_3$
Molecular Formula: $CCaO_3$
Molecular Weight: 100.087
CAS RN: 471-34-1
Properties: -325 mesh 99.95% pure, 10 microns; white odorless, tasteless hex cryst or powd; formed below 30°C; occurs naturally as mineral calcite; enthalpy of fusion 53.10 kJ/mol; used as a source of lime [KIR78] [CER91] [HAW93] [CRC93] [MER89]
Solubility: mol/kg H_2O: 6.44 x 10^{-5} (0.2°C), 5.75 x 10^{-5} (25°C), 2.76 x 10^{-5} (89.7°C) [KRU93]; s in acids, evolving CO_2 [HAW93]; solubility data are also found in [PLU82]
Density, g/cm³: 2.930 [STR93]
Reactions: decomposes to CaO ~800°C [MER89]
Thermal Expansion Coefficient: (volume) 100°C (0.105), 200°C (0.285), 400°C (0.765), 600°C (1.395) [CLA66]

579
Compound: Calcium carbonate
Synonyms: vaterite
Formula: $CaCO_3$
Molecular Formula: $CCaO_3$
Molecular Weight: 100.087
CAS RN: 471-34-1
Solubility: mol/kg H_2O: 1.34 x 10^{-4} (0°C), 1.10 x 10^{-4} (25.1°C), 4.48 x 10^{-5} (90°C) [KRU93]; solubility data are also found in [PLU82]

580
Compound: Calcium chlorate dihydrate
Formula: $Ca(ClO_3)_2 \cdot 2H_2O$
Molecular Formula: $CaCl_2H_4O_8$
Molecular Weight: 243.010
CAS RN: 10035-05-9
Properties: white to yellowish cryst; prepared by reaction between chlorine and hot $Ca(OH)_2$ slurry; used in photography and as a powd to control poison ivy by dusting [HAW93]
Solubility: g/100g soln, H_2O: 63.0 (0.5°C), 66.0 (25°C), 78.0 (93°); Solid phase, $Ca(ClO_3)_2 \cdot 2H_2O$ (25°C), $Ca(ClO_3)_2$ (93.0°C) [KRU93]
Density, g/cm³: 2.711 [HAW93]
Melting Point, °C: 340 [HAW93]

581
Compound: Calcium chloride
Synonyms: hydrophilite
Formula: $CaCl_2$

Molecular Formula: $CaCl_2$
Molecular Weight: 110.983
CAS RN: 10043-52-4
Properties: 99.99% pure -325 mesh white powd; cub cryst, granules of fused masses; very hygr; enthalpy of fusion 28.54 kJ/mol; infinite enthalpy of solution -81.82 kJ/mol; used on roads to control ice and dust [HAW93] [MER89] [KIR78] [STR93] [CRC93]
Solubility: g/100g soln, H_2O: 37.3 (0°C), 45.3 (25°C), 61.4 (100°C); Solid phase, $CaCl_2 \cdot 6H_2O$ (0°C, 25°C), $CaCl_2 \cdot 2H_2O$ (100°C) [KRU93]
Density, g/cm^3: 2.152 [MER89]
Melting Point, °C: 772 [MER89]
Boiling Point, °C: 1935 [KIR78]

582
Compound: Calcium chloride dihydrate
Formula: $CaCl_2 \cdot 2H_2O$
Molecular Formula: $CaCl_2H_4O_2$
Molecular Weight: 147.014
CAS RN: 10035-04-8
Properties: hygr granules, flakes or powd; enthalpy of fusion 88 J/g; enthalpy of infinite solution -44.78 kJ/mol [KIR78] [MER89] [ALF93]
Solubility: v s H_2O [MER89]
Density, g/cm^3: 1.85 [KIR78]
Melting Point, °C: 176 [KIR78]

583
Compound: Calcium chloride hexahydrate
Synonyms: antarcticite
Formula: $CaCl_2 \cdot 6H_2O$
Molecular Formula: $CaCl_2H_{12}O_6$
Molecular Weight: 219.074
CAS RN: 7774-34-7
Properties: white deliq trig cryst; enthalpy of fusion 209 J/g; enthalpy of infinite solution 15.77 kJ/mol [KIR78] [MER89]
Solubility: g/100g H_2O: 59.5 (0°C), 74.5 (20°C), 159 (100°C) [LAN85]
Density, g/cm^3: 1.68 [MER89]
Melting Point, °C: 29.9 [KIR78]
Reactions: minus $6H_2O$ at 200°C [MER89]

584
Compound: Calcium chloride monohydrate
Formula: $CaCl_2 \cdot H_2O$
Molecular Formula: $CaCl_2H_2O$
Molecular Weight: 128.998
CAS RN: 13477-29-7

Properties: white deliq cryst, lumps, granules, flakes; enthalpy of fusion 17.28 kJ/mol; heat of infinite solution -52.24 kJ/mol [KIR78] [HAW93]
Solubility: s H_2O, alcohol [HAW93]
Density, g/cm^3: 2.24 [KIR78]
Melting Point, °C: 187 [KIR78]

585
Compound: Calcium chloride tetrahydrate
Formula: $CaCl_2 \cdot 4H_2O$
Molecular Formula: $CaCl_2H_8O_4$
Molecular Weight: 183.045
CAS RN: 25094-02-4
Properties: enthalpy of fusion 29.86 kJ/mol; heat of infinite dilution -10.87 kJ/mol [KIR78]
Density, g/cm^3: 1.83 [KIR78]

586
Compound: Calcium chlorite
Formula: $Ca(ClO_2)_2$
Molecular Formula: $CaCl_2O_4$
Molecular Weight: 174.981
CAS RN: 14674-72-7
Properties: cub white cryst; oxidizing agent [HAW93] [CRC77]
Solubility: decomposed by H_2O [HAW93]
Density, g/cm^3: 2.71 [HAW93]

587
Compound: Calcium chromate
Synonyms: calcium chrome yellow
Formula: $CaCrO_4$
Molecular Formula: $CaCrO_4$
Molecular Weight: 156.072
CAS RN: 10060-08-9
Properties: yellow cryst, monocl or rhomb; -20 mesh 99.9% pure; used as a pigment and as a corrosion inhibitor [STR93] [CER91]
Solubility: sl s H_2O; s dil acids [MER89]

588
Compound: Calcium chromate dihydrate
Formula: $CaCrO_4 \cdot 2H_2O$
Molecular Formula: $CaCrH_4O_6$
Molecular Weight: 192.102
CAS RN: 13765-19-0
Properties: bright yellow powd; used as a pigment, corrosion inhibitor and oxidizing agent [HAW93]

Solubility: g/100g H_2O: 17.3 (0°C), 16.6 (30°C), 16.1 (40°C) [LAN85]
Density, g/cm³: 2.5 [LID94]
Reactions: minus $2H_2O$ at 200°C [HAW93]

589
Compound: Calcium citrate tetrahydrate
Synonyms: tricalcium dicitrate tetrahydrate
Formula:
  $[OOCCH_2C(OH)(COO)CH_2COO]_2Ca_3 \cdot 4H_2O$
Molecular Formula: $C_{12}H_{18}Ca_3O_{18}$
Molecular Weight: 570.498
CAS RN: 5785-44-4
Properties: white odorless needles or powd; can be obtained from citrus fruit; used as a dietary supplement, sequesterant and firming agent in foods, used to prepare citric acid [HAW93] [MER89] [KIR78]
Solubility: 0.088 g/100mL H_2O at 18°C, 0.096 g/100mL at 23°C; 0.0065 g/100mL alcohol at 18°C; i ether [KIR78]
Reactions: minus most of its water at 100°C, all H_2O at 120°C [MER89]

590
Compound: Calcium cyanamide
Synonyms: calcium carbimide
Formula: $N{\equiv}C-N{=}Ca$
Molecular Formula: $CCaN_2$
Molecular Weight: 80.102
CAS RN: 156-62-7
Properties: -12 mesh; colorless cryst or powd; pure material glistens, hex cryst; used as a fertilizer and pesticide, and in manufacturing iron and $Ca(CN)_2$ [HAW93] [MER89] [ALF93]
Solubility: i H_2O, but undergoes hydrolysis releasing acetylene and ammonia [HAW93] [MER89]
Density, g/cm³: 2.29 [MER89]; 1.083 [HAW93]
Melting Point, °C: ~1340 [MER89]
Reactions: sublimes at 1150-1200°C [MER89]

591
Compound: Calcium cyanide
Synonyms: cyanogas
Formula: $Ca(CN)_2$
Molecular Formula: C_2CaN_2
Molecular Weight: 92.113
CAS RN: 592-01-8
Properties: colorless or white rhomb cryst or powd; decomposes in moist air liberating HCN; used as a rodenticide, as a fumigant in greenhouses, and in flour mills [HAW93] [MER89]

Solubility: s H_2O, gradually releasing HCN [MER89]
Melting Point, °C: 640 estimated [KIR78]

592
Compound: Calcium dichromate trihydrate
Formula: $CaCr_2O_7 \cdot 3H_2O$
Molecular Formula: $CaCr_2H_6O_{10}$
Molecular Weight: 310.112
CAS RN: 14307-33-6
Properties: bipyramidal, reddish orange cryst; not hygr, if pure; used as a catalyst in manufacturing $CrCl_3$ and CrO_3, and as a corrosion inhibitor [MER89]
Solubility: v s H_2O; i ether, CCl_4; reacts with alcohol [MER89]
Density, g/cm³: 2.37 [MER89]
Melting Point, °C: decomposes above 100 [MER89]
Reactions: decomposes when heated forming $CaCrO_4$ and CrO_3 [MER89]

593
Compound: Calcium dihydrogen phosphate monohydrate
Formula: $Ca(H_2PO_4)_2 \cdot H_2O$
Molecular Formula: $CaH_6O_9P_2$
Molecular Weight: 252.068
CAS RN: 10031-30-8
Properties: white large shining tricl plates, cryst powd or granules; not hygr [MER89] [STR93]
Solubility: moderately s H_2O; s dil HCl, HNO_3, acetic acid [MER89]
Density, g/cm³: 2.22 [MER89]
Reactions: minus H_2O at 100°C; decomposes at 200°C [MER89]

594
Compound: Calcium ferrocyanide dodecahydrate
Formula: $Ca_2Fe(CN)_6 \cdot 12H_2O$
Molecular Formula: $C_6H_{24}Ca_2FeN_6O_{12}$
Molecular Weight: 508.291
Properties: yellow tricl cryst; decomposes when heated; used to remove metallic impurities from citric, tartaric and other acids [CRC77] [HAW93]
Solubility: 86.8 g/100 mL (25°C), 115 g/100 mL (65°C) H_2O; i alcohol [HAW93] [CRC77]
Density, g/cm³: 1.68 [HAW93]
Melting Point, °C: decomposes [CRC77]

595
Compound: Calcium fluoride
Synonyms: fluorspar, fluorite
Formula: CaF_2
Molecular Formula: CaF_2
Molecular Weight: 78.075
CAS RN: 7789-75-5
Properties: white powd or cub cryst; enthalpy of fusion 29.71 kJ/mol; enthalpy of vaporization 335 kJ/mol; hardness 4 Mohs; electrical conductivity 1.3×10^{-18} (ohm cm)$^{-1}$ at 20°C, 6 x 10^{-5} at 650°C; becomes luminous on heating; pure material prepared from $CaCO_3$ and HF solution; mineral fluorspar is main source of fluorine; 99.95% pure material used in spectroscopy, lasers, and electronics, for sputtering anti-reflection coating on glass [HAW93] [MER89] [CER91] [CRC93]
Solubility: g/L soln, H_2O: 0.013 (0°C), 0.016 (25°C) [KRU93]
Density, g/cm³: 3.18 [MER89]
Melting Point, °C: 1418 [CRC93]
Boiling Point, °C: 2500 [MER89]
Thermal Conductivity, W/(m·K): 10.96 [KIR78]
Thermal Expansion Coefficient: (volume) 100°C (0.47), 200°C (1.12) [CLA66]

596
Compound: Calcium fluorophosphate
Synonyms: fluoroapatite
Formula: $Ca_5(PO_4)_3F$
Molecular Formula: $Ca_5FO_{12}P_3$
Molecular Weight: 504.302
CAS RN: 1306-05-4
Properties: fluoroapatite is a mineral for fluorine; can be prepared from $CaCl_2$ and sodium monofluorophosphate; used as a laser cryst, may have the lowest energy threshold of any cryst at room temp [KIR78] [HAW93]

597
Compound: Calcium fluorophosphate dihydrate
Formula: $CaPO_3F \cdot 2H_2O$
Molecular Formula: $CaFH_4O_5P$
Molecular Weight: 174.079
CAS RN: 37809-19-1
Properties: monocl cryst powd; tendency to form twins; loses fluorine on heating [MER89] [STR93]
Solubility: 0.417 g/100mL soln (27°C); i in most common organic solvents [MER89]

598
Compound: Calcium formate
Synonyms: formic acid, calcium salt
Formula: $Ca(HCOO)_2$
Molecular Formula: $C_2H_2CaO_4$
Molecular Weight: 130.114
CAS RN: 544-17-2
Properties: ortho-rhomb cryst or powd; slight odor similar to acetic acid; obtained from $Ca(OH)_2$ and CO reaction at high temperatures and pressures; used as food preservative, as binder for fine-ore briquettes and in drilling fluids [MER89]
Solubility: g/100g H_2O: 16.15 (0°C), 16.60 (20°C), 17.95 (80°C), 18.40 (100°C) [LAN85]; i alcohol [MER89]
Density, g/cm³: 2.02 [MER89]
Melting Point, °C: 300 [HAW93]

599
Compound: Calcium hexaborate pentahydrate
Synonyms: colemanite
Formula: $Ca_2B_6O_{11} \cdot 5H_2O$
Molecular Formula: $B_6Ca_2H_{10}O_{16}$
Molecular Weight: 411.091
CAS RN: 12291-65-5
Properties: monocl; forms when saturated solutions of inyoite or higher hydrates are heated [KIR78]
Solubility: about 1% in H_2O at 25°C [KIR78]
Density, g/cm³: 2.42 [KIR78]

600
Compound: Calcium hexafluoroacetylacetonate dihydrate
Synonyms: 1,1,1,5,5,5-hexafluoro-2,4-pentanedione, calcium derivative
Formula: $Ca[CF_3COCH=C(O)CF_3]_2 \cdot 2H_2O$
Molecular Formula: $C_{10}H_6CaF_{12}O_6$
Molecular Weight: 490.211
Properties: off-white powd; precursor for metal oxide chemical vapor deposition [STR94] [STR93]
Melting Point, °C: 135-140 [STR93]
Boiling Point, °C: decomposes, 230-240 [STR93]
Reactions: sublimes at 180°C, 0.7 mm Hg [STR94]

601
Compound: Calcium hexafluorosilicate dihydrate
Formula: $CaSiF_6 \cdot 2H_2O$
Molecular Formula: $CaF_6H_4O_2Si$
Molecular Weight: 218.185
CAS RN: 16925-39-6

Properties: colorless, tetr; powd; obtained by reaction of a Ca salt and H_2SiF_6; used as a flotation agent and insecticide [HAW93] [MER89] [CRC77]
Solubility: i cold H_2O, partially decomposed in hot H_2O [MER89]
Density, g/cm³: 2.25 [MER89]

602
Compound: Calcium hydride
Formula: CaH_2
Molecular Formula: CaH_2
Molecular Weight: 42.094
CAS RN: 7789-78-8
Properties: -40 mesh, -20 mesh 98% pure; sensitive to moisture; ortho-rhomb cryst or powd; commercial product is gray; decomposes into Ca and H_2 at 990°C without melting; prepared by heating calcium metal to ~300°C under 1 atm of H_2; used to obtain some metals by reduction of their oxides, and to dry gases and unreactive liq [KIR80] [MER89] [STR93]
Solubility: decomposed by H_2O, with evolution of H_2 [HAW93]
Density, g/cm³: 1.7 [MER89]; 1.90 [KIR80]
Melting Point, °C: 816 under H_2 [STR93]

603
Compound: Calcium hydrogen phosphate
Synonyms: calcium phosphate, dibasic
Formula: $CaHPO_4$
Molecular Formula: $CaHO_4P$
Molecular Weight: 136.057
CAS RN: 7757-93-9
Properties: white, tasteless tricl cryst; prepared by reacting phosphoric acid and milk of lime (calcium hydroxide suspended in water); used as food supplement, in medicine, constituent of a dentrifice and fertilizer [HAW93] [MER89]
Solubility: i H_2O, alcohol [MER89]
Reactions: at red heat, dehydrated to calcium pyrophosphate [MER89]

604
Compound: Calcium hydrogen phosphate dihydrate
Synonyms: brushite
Formula: $CaHPO_4 \cdot 2H_2O$
Molecular Formula: CaH_5O_6P
Molecular Weight: 172.088
CAS RN: 7789-77-7
Properties: monocl [MER89]
Solubility: i H_2O, alcohol; s dil HCl, HNO_3 [MER89]

Density, g/cm³: 2.31 [MER89]
Reactions: minus H_2O <100°C; dehydrates at red heat to calcium pyrophosphate [MER89]

605
Compound: Calcium hydrogen sulfite
Formula: $Ca(HSO_3)_2$
Molecular Formula: $CaH_2O_6S_2$
Molecular Weight: 202.222
CAS RN: 13780-03-5
Properties: solution of calcium sulfite in an aq solution of sulfur dioxide; yellowish liq; used in bleaching textiles [HAW93]
Density, g/cm³: 1.06 [HAW93]

606
Compound: Calcium hydrosulfide hexahydrate
Formula: $Ca(HS)_2 \cdot 6H_2O$
Molecular Formula: $CaH_{14}O_6S_2$
Molecular Weight: 214.317
Properties: colorless transparent cryst; used in the leather industry [HAW93]
Solubility: s H_2O, alcohol [HAW93]
Melting Point, °C: decomposes in air at 15-18 [HAW93]

607
Compound: Calcium hydroxide
Synonyms: portlandite, slaked lime
Formula: $Ca(OH)_2$
Molecular Formula: CaH_2O_2
Molecular Weight: 74.093
CAS RN: 1305-62-0
Properties: cryst or soft, odorless granules or powd; sl bitter, alkaline taste; readily absorbs CO_2 from air; manufactured from lime and water; used in mortar, plasters, cement [HAW93] [MER89]
Solubility: g/100g H_2O: 0.189 (0°C), 0.173 (20°C), 0.076 (100°C) [LAN85]; s acids (caution) [MER89]; s glycerol; i alcohol [HAW93]
Density, g/cm³: 2.08-2.34 [MER89]
Reactions: minus H_2O at 580°C [HAW93]

608
Compound: Calcium hypochlorite
Synonyms: losantin
Formula: $Ca(OCl)_2$
Molecular Formula: $CaCl_2O_2$
Molecular Weight: 142.982
CAS RN: 7778-54-3

Properties: grayish white powd; oxidizing agent; used as an algicide, bactericide, deodorant, disinfectant, as a bleach, and to refine sugar; there is a dihydrate, 22464-76-2 [KIR78] [MER89] [HAW93] [STR93]
Solubility: decomposes in both H_2O and alcohol [HAW93]
Density, g/cm³: 2.35 [HAW93]
Melting Point, °C: 100, decomposes [HAW93]

609
Compound: Calcium hypophosphite
Formula: $Ca(H_2PO_2)_2$
Molecular Formula: $CaH_4O_4P_2$
Molecular Weight: 170.055
CAS RN: 7789-79-9
Properties: white monocl, prismatic cryst or granular powd; used as a corrosion inhibitor and in medicine [MER89] [HAW93]
Solubility: s H_2O, sl s in glycerol; i alcohol [MER89]
Melting Point, °C: decomposes [ALF93]
Reactions: when heated >300°C, evolves phosphine which spontaneously ignites [MER89]

610
Compound: Calcium iodate
Synonyms: lautarite
Formula: $Ca(IO_3)_2$
Molecular Formula: CaI_2O_6
Molecular Weight: 389.883
CAS RN: 7789-80-2
Properties: white powd; monocl prismatic cryst; the hexahydrate is ortho-rhomb; not hygr; obtained by passing Cl_2 into a hot solution of lime containing dissolved iodine; used as a deodorant, in mouthwashes, as a food additive and dough conditioner [MER89] [STR93]
Solubility: s HNO_3; i alcohol [MER89]; g/100 g soln, H_2O: 0.0906 (0°C), 0.306 (25°C), 0.668 (90°C); Solid phase, $Ca(IO_3)_2 \cdot 6H_2O$ (0°C, 25°C), $Ca(IO_3)_2$ (90°C) [KRU93]
Density, g/cm³: 4.519 [MER89]
Melting Point, °C: stable up to 540 [MER89]
Reactions: sensitive to reducing agents [MER89]

611
Compound: Calcium iodide
Formula: CaI_2
Molecular Formula: CaI_2
Molecular Weight: 293.887
CAS RN: 10102-68-8

Properties: -20 mesh 99.5% pure powd; very hygr; hex; becomes yellow and completely insoluble on exposure to air due to liberation of I_2 and absorption of CO_2; enthalpy of fusion 41.80 kJ/mol; finds use as an expectorant [MER89] [CRC93]
Solubility: v s H_2O, methanol, ethanol, acetone; i ether [MER89]; g/100g soln, H_2O: 64.6 (0°C), 68.3 (25°C), 81.0 (100°C) [KRU93]
Density, g/cm³: 4.0 [HAW93]
Melting Point, °C: 779 [CRC93]
Boiling Point, °C: 1100 [MER89]

612
Compound: Calcium iodide hexahydrate
Formula: $CaI_2 \cdot 6H_2O$
Molecular Formula: $CaH_{12}I_2O_6$
Molecular Weight: 401.978
CAS RN: 71626-98-7
Properties: white powd, hex, thick needles, plates or lumps; very hygr; becomes yellow in air; absorbs atm CO_2; used in photography and in medicine; formula also given as $CaI_2 \cdot xH_2O$ [HAW93] [MER89] [STR93]
Solubility: v s H_2O, alcohol [MER89]
Density, g/cm³: 2.55 [HAW93]
Melting Point, °C: 783 [HAW93]
Boiling Point, °C: ~1100 [HAW93]
Reactions: minus $6H_2O$ at 42°C [HAW93]

613
Compound: Calcium molybdate
Synonyms: powellite
Formula: $CaMoO_4$
Molecular Formula: $CaMoO_4$
Molecular Weight: 200.016
CAS RN: 7789-82-4
Properties: -200 mesh 99.9% pure; white cryst powd; can be produced by reacting $CaSO_4$ with sodium molybdate; used in optical and electronic applications, as an alloying agent in iron and steel manufacturing [HAW93] [MER89] [CER91]
Solubility: 0.0050 g/100g H_2O [KIR81]; s conc mineral acids [MER89]; mg/100g soln, H_2O: 0.0022 (0°C), 0.0025 (22°C), 0.0085 (100°C) [KRU93]
Density, g/cm³: 4.38-4.53 [STR93]
Melting Point, °C: 965 decomposes [KIR81]; ~1250 [HAW93]

614
Compound: Calcium nitrate

Formula: $Ca(NO_3)_2$
Molecular Formula: CaN_2O_6
Molecular Weight: 164.087
CAS RN: 10124-37-5
Properties: white deliq granules; oxidizing agent; used in pyrotechnics, explosives and in fertilizers [HAW93] [MER89]
Solubility: v s H_2O, evolves heat; s methanol, ethanol, acetone [MER89]; g/100g soln, H_2O: 50.50 (0°C), 57.98 (25°C), 78.43 (100°C); Solid phase, $Ca(NO_3)_2 \cdot 4H_2O$ (0°C, 25°C), $Ca(NO_3)_2$ (100°C) [KRU93]
Density, g/cm³: 2.36 [HAW93]
Melting Point, °C: 561 [HAW93]

615
Compound: Calcium nitrate tetrahydrate
Formula: $Ca(NO_3)_2 \cdot 4H_2O$
Molecular Formula: $CaH_8N_2O_{10}$
Molecular Weight: 236.149
CAS RN: 13477-34-4
Properties: -4 mesh 99.999% pure; white cryst [STR93] [CER91]
Solubility: g/100g H_2O: 102 (0°C), 129 (20°C), 363 (100°C) [LAN85]; s alcohol, acetone [HAW93]
Density, g/cm³: 1.82 [STR93]
Melting Point, °C: 39.7 [STR93]

616
Compound: Calcium nitride
Formula: Ca_3N_2
Molecular Formula: Ca_3N_2
Molecular Weight: 148.247
CAS RN: 12013-82-0
Properties: 12 mm pieces and smaller, -200 mesh 99% pure; brown cryst; hex: a = 0.3533 nm, c = 0.411 nm; cub: a = 1.138 nm [CIC73] [CER91]
Solubility: s H_2O, releasing ammonia; s dil acids; i absolute alcohol [HAW93]
Density, g/cm³: hex: 2.62; cub: 2.54 [CIC73]
Melting Point, °C: 1195 [HAW93]

617
Compound: Calcium nitrite
Formula: $Ca(NO_2)_2$
Molecular Formula: CaN_2O_4
Molecular Weight: 132.089
CAS RN: 13780-06-8

Properties: white or yellowish deliq, hex cryst; prepared from nitric oxide and a mixture consisting of calcium ferrate and calcium nitrate; used to inhibit corrosion in lubricants and concrete [MER89]
Solubility: sl s alcohol [MER89]; g/100 g soln, H_2O: 38.3 (0°C), 43.0 (18.5°C), 71.2 (91°C); Solid phase $Ca(NO_2)_2 \cdot 4H_2O$ (0°C, 18.5°C), $Ca(NO_2)_2 \cdot 2H_2O$ (91°C) [KRU93]
Density, g/cm³: 2.23 [MER89]

618
Compound: Calcium nitrite monohydrate
Formula: $Ca(NO_2)_2 \cdot H_2O$
Molecular Formula: $CaH_2N_2O_5$
Molecular Weight: 150.104
CAS RN: 13780-06-8
Properties: colorless or yellowish cryst; hygr; used in lubricants as a corrosion inhibitor [HAW93]
Solubility: g$Ca(NO_3)_2 \cdot 4H_2O$/100g H_2O: 63.9 (0°C), 104 (30°C), 178 (100°C) [LAN85]; sl s alcohol [HAW93]
Density, g/cm³: 2.23 (anhydrous, 34°C) [HAW93]
Reactions: minus H_2O at 100°C [HAW93]

619
Compound: Calcium oleate
Synonyms: 9-octadecanoic acid, calcium salt
Formula: $Ca(C_{18}H_{33}O_2)_2$
Molecular Formula: $C_{36}H_{66}CaO_4$
Molecular Weight: 602.996
CAS RN: 142-17-6
Properties: pale yellow transparent solid; slowly absorbs moisture from air to form monohydrate; used as a thickening agent for grease [HAW93] [MER89]
Solubility: 0.04g/100 mL (25°C), 0.03g/100 mL (50°C) H_2O; i alcohol, ether, acetone; s benzene, chloroform [MER89] [CRC77]
Melting Point, °C: decomposes >140 [MER89]

620
Compound: Calcium oxalate
Synonyms: ethanedioic acid, calcium salt
Formula: CaC_2O_4
Molecular Formula: C_2CaO_4
Molecular Weight: 128.098
CAS RN: 563-72-4
Properties: white cryst powd; obtained from calcium formate or calcium cyanamide; used to prepare oxalic acid, glazes and to separate rare earths [HAW93] [MER89]

Solubility: g/L soln, H_2O: 0.0069 (25°C), 0.0142 (95°C); Solid phase, $CaC_2O_4 \cdot H_2O$ [KRU93]; s dil HCl, HNO_3 [HAW93]

Density, g/cm³: 2.2 [HAW93]

Melting Point, °C: decomposes [CRC94]

621

Compound: Calcium oxalate monohydrate

Synonyms: ethanedioic acid, calcium salt monohydrate

Formula: $CaC_2O_4 \cdot H_2O$

Molecular Formula: $C_2H_2CaO_5$

Molecular Weight: 146.113

CAS RN: 5794-28-5

Properties: colorless cub cryst; hygr; uses: ceramic glazes [MER89] [CRC77]

Solubility: i H_2O, acetic acid; s dil HCl, HNO_3 [MER89]

Density, g/cm³: 2.2 [MER89]

Reactions: minus $2H_2O$ at 200°C; when ignited converts into $CaCO_3$ or CaO without appreciable charring [MER89]

622

Compound: Calcium oxide

Synonyms: lime; quicklime

Formula: CaO

Molecular Formula: CaO

Molecular Weight: 56.077

CAS RN: 1305-78-8

Properties: -325 mesh, 10 microns or less, 99.99% and 99.5% pure; cryst, white or grayish lumps, or granular powd; readily absorbs CO_2 and H_2O from air; odorless; enthalpy of fusion 59.00 kJ/mol; produced from limestone; used in pulp and paper, dehairing of hides, in brick, mortar and stucco [HAW93] [MER89] [CRC93]

Solubility: reacts with H_2O to form $Ca(OH)_2$; s acids [MER89]

Density, g/cm³: 3.32-3.35 [MER89]

Melting Point, °C: 2927 [CRC93]

Boiling Point, °C: 2850 [MER89]

Thermal Conductivity, W/(m·K): 8.0 (500°C), 7.8 (1000°C) [KIR80]

Thermal Expansion Coefficient: (volume) 100°C (0.225), 200°C (0.571), 400°C (1.402), 800°C (3.107), 1200°C (5.078) [CLA66]

623

Compound: Calcium palmitate

Synonyms: hexadecanoic acid, calcium salt

Formula: $Ca(C_{16}H_{31}O_2)_2$

Molecular Formula: $C_{32}H_{62}CaO_4$

Molecular Weight: 550.920

CAS RN: 542-42-7

Properties: white or pale yellow powd; used for waterproofing, as a thickener for lubricating oils [HAW93] [MER89]

Solubility: i H_2O, alcohol, ether, acetone; sl s benzene [MER89]

Melting Point, °C: decomposes above 155 [MER89]

624

Compound: Calcium perborate heptahydrate

Formula: $Ca(BO_3)_2 \cdot 7H_2O$

Molecular Formula: $B_2CaH_{14}O_{13}$

Molecular Weight: 283.803

Properties: grayish white lumps or powd; uses: medicine, as a bleach and in tooth powd [HAW93]

Solubility: s H_2O, acids; evolves oxygen [HAW93]

625

Compound: Calcium perchlorate

Synonyms: perchloric acid, calcium salt

Formula: $Ca(ClO_4)_2$

Molecular Formula: $CaCl_2O_8$

Molecular Weight: 238.978

CAS RN: 13477-36-6

Properties: white cryst; oxidizing agent [HAW93]

Solubility: g/100g soln, H_2O: 65.35 (25°C) [KRU93]; s alcohol [HAW93]

Density, g/cm³: 2.651 [HAW93]

Melting Point, °C: decomposes 270 [HAW93]

626

Compound: Calcium perchlorate tetrahydrate

Formula: $Ca(ClO_4)_2 \cdot 4H_2O$

Molecular Formula: $CaCl_2H_8O_{12}$

Molecular Weight: 311.039

CAS RN: 15627-86-8

Properties: white cryst [STR93]

627

Compound: Calcium permanganate

Formula: $Ca(MnO_4)_2$

Molecular Formula: $CaMn_2O_8$

Molecular Weight: 277.949

CAS RN: 10118-76-0

Properties: violet or dark purple, deliq cryst; made by reacting $KMnO_4$ and $CaCl_2$; used in the textile industry, to sterilize water and as a deodorizer [HAW93] [MER89]

Solubility: v s H_2O; decomposed by alcohol
[MER89]
Density, g/cm³: 2.4 [HAW93]

628
Compound: Calcium peroxide
Synonyms: calcium dioxide
Formula: CaO_2
Molecular Formula: CaO_2
Molecular Weight: 72.077
CAS RN: 1305-79-9
Properties: tetr; white or yellowish, odorless, almost
tasteless powd; decomposes in moist air; used to
disinfect seeds, in dentrifices [HAW93]
[MER89] [CRC77]
Solubility: sl s H_2O; s in acids forming H_2O_2
[MER89]
Density, g/cm³: 2.92 [CRC77]
Melting Point, °C: decomposes ~200 [HAW93]

629
Compound: Calcium phenoxide
Synonyms: calcium phenolate, calcium carbolate
Formula: $Ca(OC_6H_5)_2$
Molecular Formula: $C_{12}H_{10}CaO_2$
Molecular Weight: 226.288
CAS RN: 5793-84-0
Properties: reddish powd; decomposes in air; used
as a detergent for lubricating oil, and as an
emulsifier [HAW93] [MER89]
Solubility: sl s H_2O, alcohol [MER89]

630
Compound: Calcium phosphate
Synonyms: whitlockite
Formula: $Ca_3(PO_4)_2$
Molecular Formula: $Ca_3O_8P_2$
Molecular Weight: 310.177
CAS RN: 7758-87-4
Properties: amorphous odorless tasteless white
powd; produced from phosphate rock; used in
ceramics, as a polishing powd [HAW93]
[MER89] [STR93]
Solubility: i H_2O, alcohol, acetic acid; s dil HCl,
HNO_3 [MER89]
Density, g/cm³: 3.14 [MER89]
Melting Point, °C: 1670 [HAW93]

631
Compound: Calcium phosphate hydroxide
Synonyms: Durapatite, hydroxylapatite
Formula: $3Ca_3(PO_4)_2 \cdot Ca(OH)_2$
Molecular Formula: $Ca_{10}H_{26}O_{26}P_6$
Molecular Weight: 748.143
CAS RN: 1306-06-5
Properties: occurs naturally as mineral; hex needles
with rosettes arrangement; preparation from
$Ca(NO_3)_2$ and KH_2PO_4; uses: purification of
DNA, Ca and P supplement, prosthetic aid
[ALD94] [MER89]
Solubility: i H_2O [MER89]
Melting Point, °C: decomposes >1100 [MER89]

632
Compound: Calcium phosphide
Synonyms: photophor
Formula: Ca_3P_2
Molecular Formula: Ca_3P_2
Molecular Weight: 182.182
CAS RN: 1305-99-3
Properties: 0.5 inch pieces and down; reddish
brown cryst powd or gray lumps; decomposed by
moist air, evolving flammable phosphine, which
can ignite spontaneously; obtained by heating
quicklime in phosphorus vapor; used in signal
fires, torpedoes, and pyrotechnics [HAW93]
[MER89] [STR93] [KIR82]
Solubility: decomposes in H_2O to form flammable
phosphine [MER89]; i alcohol, ether [HAW93]
Density, g/cm³: 2.51 (15°C) [HAW93]
Melting Point, °C: ~1600 [MER89]
Reactions: may be heated up to 1250°C in H_2
without decomposition [KIR82]

633
Compound: Calcium phosphite monohydrate
Formula: $CaHPO_3 \cdot H_2O$
Molecular Formula: CaH_3O_4P
Molecular Weight: 138.073
CAS RN: 21056-98-4
Properties: cryst; used in fertilizers and
polymerization catalysts [MER89]
Solubility: sl s H_2O; i alcohol [MER89]
Reactions: minus H_2O at 200°C, decomposes
>300°C [MER89]

634
Compound: Calcium plumbate
Formula: Ca_2PbO_4
Molecular Formula: Ca_2O_4Pb
Molecular Weight: 351.354

Properties: orange to brown cryst powd; used as an oxidizing agent, in safety matches and storage batteries [HAW93]

Solubility: i H_2O, decomposed by hot H_2O; s acids, with decomposition [HAW93]

Density, g/cm³: 5.71 [HAW93]

Melting Point, °C: decomposes [CRC94]

635

Compound: Calcium propionate

Synonyms: propionic acid, calcium salt

Formula: $(C_2H_5COO)_2Ca$

Molecular Formula: $C_6H_{10}CaO_4$

Molecular Weight: 186.221

CAS RN: 4075-81-4

Properties: white powd or monocl cryst; uses: mold-retardant additive for bread, tobacco, pharmaceuticals, anti-fungal agent [HAW93]

Solubility: s H_2O, sl sl methanol, ethanol [MER89]

636

Compound: Calcium pyrophosphate

Synonyms: calcium diphosphate

Formula: $Ca_2P_2O_7$

Molecular Formula: $Ca_2O_7P_2$

Molecular Weight: 254.099

CAS RN: 7790-76-3

Properties: 99.95% pure 6-8 micron; white polymorphous cryst or powd; can be made by igniting $CaHPO_4$; used as a polishing agent in dentrifices, as a mild abrasive to polish metals [HAW93] [MER89] [STR93]

Solubility: i H_2O; s dil HCl, HNO_3 [MER89]

Density, g/cm³: 3.09 [MER89]

Melting Point, °C: 1230 [STR93]

637

Compound: Calcium selenate dihydrate

Formula: $CaSeO_4 \cdot 2H_2O$

Molecular Formula: CaH_4O_6Se

Molecular Weight: 219.066

CAS RN: 7790-74-1

Properties: colorless; monocl powd; used as a general pesticide [HAW93] [CRC77]

Solubility: g/100g H_2O: 9.73 (0°C), 9.22 (20°C), 7.14 (40°C) [LAN85]

Density, g/cm³: 2.7 [HAW93]

638

Compound: Calcium selenide

Formula: CaSe

Molecular Formula: CaSe

Molecular Weight: 119.038

CAS RN: 1305-84-6

Properties: -20 mesh 99.5% pure; white powd; in air may turn red within a few minutes, and light brown in a few hours; obtained by reduction of $CaSeO_4$ with H_2 at 400-500°C; used in electron emitter devices [CER91] [MER89]

Solubility: decomposed by H_2O; releases H_2Se gas and forms red Se in HCl [MER89]

Density, g/cm³: 3.82 [MER89]

639

Compound: Calcium silicate

Synonyms: wollastonite

Formula: β-$CaSiO_3$

Molecular Formula: CaO_3Si

Molecular Weight: 116.162

CAS RN: 1344-95-2

Properties: -200 mesh 99% white powd; used as an absorbent, antacid, as a filler paper and for paper coatings [HAW93] [ALD93]

Solubility: i H_2O [HAW93]

Density, g/cm³: 2.9 [HAW93]

Reactions: transition to pseudowollastonite at 1200°C [ROB78]

640

Compound: Calcium silicide

Formula: $CaSi_2$

Molecular Formula: $CaSi_2$

Molecular Weight: 96.249

CAS RN: 12013-56-8

Properties: 3 mm pieces and smaller -140 mesh of 99.5% purity; powd [STR93] [CER91]

Solubility: i cold H_2O, decomposed by hot H_2O; s acids and alkalies [HAW93]

Density, g/cm³: 2.5 [HAW93]

Melting Point, °C: 1000 [ALF93]

641

Compound: Calcium stannate trihydrate

Formula: $CaSnO_3 \cdot 3H_2O$

Molecular Formula: CaH_6O_6Sn

Molecular Weight: 260.832

CAS RN: 12013-46-6

Properties: white cryst powd; used as an additive for ceramic capacitors, in the production of ceramic colors; anhydrous is -325 mesh, 5 microns or less 99% pure [HAW93] [CER91]

Solubility: i H_2O [HAW93]

Reactions: minus $3H_2O$ ~350°C [HAW93]

642

Compound: Calcium stearate
Synonyms: octadecanoic acid, calcium salt
Formula: $Ca[CH_3(CH_2)_{16}COO]_2$
Molecular Formula: $C_{36}H_{70}CaO_4$
Molecular Weight: 607.028
CAS RN: 1592-23-0
Properties: white granular, fatty powd; used as a water repellent; flattening agent in paints [HAW93] [MER89]
Solubility: i H_2O, ether, chloroform; sl s hot mineral oils [MER89]
Density, g/cm³: 1.12 [KIR78]
Melting Point, °C: 147-149 [MER89]

643

Compound: Calcium succinate trihydrate
Synonyms: butanedioic acid, calcium salt trihydrate
Formula: $Ca(OOCCH_2CH_2COO)\cdot_3H_2O$
Molecular Formula: $C_4H_{10}CaO_7$
Molecular Weight: 210.197
CAS RN: 140-99-8
Properties: needles or granules [MER89]
Solubility: g/100g H_2O: 1.127 (0°C), 1.28 (20°C), 0.66 (100°C) [LAN85]; i alcohol; s dil acids [MER89]

644

Compound: Calcium sulfate
Synonyms: anhydrite
Formula: $CaSO_4$
Molecular Formula: CaO_4S
Molecular Weight: 136.142
CAS RN: 7778-18-9
Properties: ortho-rhomb; various colors; odorless; white with blue, gray or red tinge; hardness 3-3.5 Mohs; enthalpy of fusion 28.03 kJ/mol; used in cement formulations, and as a paper filler [MER89] [CRC93]
Solubility: g/100g soln, H_2O: 0.63 (25°C), 0.151 (100°C); Solid phase, $CaSO_4$ [KRU93]
Density, g/cm³: 2.96 [MER89]
Melting Point, °C: 1450 [CRC93]

645

Compound: Calcium sulfate dihydrate
Synonyms: gypsum
Formula: $CaSO_4\cdot2H_2O$
Molecular Formula: CaH_4O_6S
Molecular Weight: 172.172
CAS RN: 10101-41-4

Properties: monocl; hardness 1.5-2.0 Mohs; lumps or white powd; used in manufacturing portland cement, plaster of Paris, and artificial marble [KIR78] [MER89] [STR93]
Solubility: g/100mL soln, H_2O: 0.1759 (0°C), 0.2080 (25°C), 0.1619 (100°C) [KRU93]
Density, g/cm³: 2.32 [KIR79]
Reactions: minus 1.5H_2O at 128°C, minus 2H_2O at 163°C [KIR78]
Thermal Expansion Coefficient: (volume) 100°C (0.58) [CLA66]

646

Compound: Calcium sulfate hemihydrate
Synonyms: plaster of Paris
Formula: $CaSO_4\cdot1/2-H_2O$
Molecular Formula: $CaHO_{4.5}S$
Molecular Weight: 145.145
CAS RN: 10034-76-1
Properties: odorless and tasteless fine powd; hygr; uses: wall plasters, wallboards and tiles [MER89] [ALD94]
Solubility: sets to hard mass when mixed with H_2O [MER89]; g/100g soln, H_2O: 1.23 (0°C), 0.71 (25°C), 0.189 (100°C); Solid phase, $CaSO_4\cdot1/2H_2O$ [KRU93]
Reactions: minus 1/2H_2O at 163°C [KIR78]

647

Compound: Calcium sulfide
Synonyms: oldhamite
Formula: CaS
Molecular Formula: CaS
Molecular Weight: 72.144
CAS RN: 20548-54-3
Properties: -325 mesh, 10 micron or less, 99.99% and 99% pure; white powd, if pure; else may be yellowish to a pale gray; odor of H_2S in moist air; unpleasant alkaline taste; oxidizes in dry air and decomposes in moist air; can be obtained by reacting $CaCO_3$, H_2S and H_2 at 1000°C [MER89]
Solubility: sl s cold H_2O, more s hot H_2O with partial decomposition [MER89]
Density, g/cm³: 2.59 [MER89]
Melting Point, °C: >2000 [MER89]

648

Compound: Calcium sulfite dihydrate
Formula: $CaSO_3\cdot2H_2O$
Molecular Formula: CaH_4O_5S
Molecular Weight: 156.173
CAS RN: 10257-55-3

Properties: hex; white cryst or powd; slowly oxidizes in air to CaSO$_4$; used in brewing, as a disinfectant in sugar manufacturing [HAW93] [MER89] [CRC77]
Solubility: 0.0043 g/100 mL (18°C), 0.001 g/100 mL (100°C) H$_2$O;sl s alcohol; s in acid solutions with SO$_2$ evolution [MER89] [CRC77]
Reactions: minus 2H$_2$O at 100°C [HAW93]

649
Compound: Calcium tartrate tetrahydrate
Synonyms: 2,3-dihydroxybutanedioic acid, calcium salt tetrahydrate
Formula: CaC$_4$H$_4$O$_6$·4H$_2$O
Molecular Formula: C$_4$H$_{12}$CaO$_{10}$
Molecular Weight: 260.211
CAS RN: 3164-34-9
Properties: rhomb; white cryst; uses: food preservative, antacid [HAW93] [CRC77]

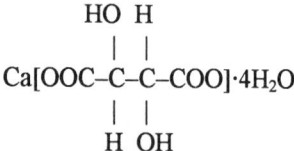

Solubility: g/100g H$_2$O: 0.026 (0°C), 0.034 (20°C), 0.130 (80°C) [LAN85]; s dil HCl, HNO$_3$ [MER89]
Melting Point, °C: decomposes [CRC77]

650
Compound: Calcium telluride
Formula: CaTe
Molecular Formula: CaTe
Molecular Weight: 167.678
CAS RN: 12013-57-9
Properties: cub; -20 mesh 99.5% purity [CER91] [CRC77]
Density, g/cm^3: 4.873 [CRC77]

651
Compound: Calcium thiocyanate tetrahydrate
Formula: Ca(SCN)$_2$·4H$_2$O
Molecular Formula: C$_2$H$_8$CaN$_2$O$_4$S$_2$
Molecular Weight: 228.307
CAS RN: 2092-16-2
Properties: hygr cryst or powd [MER89]
Solubility: v s H$_2$O; s methanol, ethanol, acetone [MER89]
Reactions: decomposes if heated above 160°C [MER89]

652
Compound: Calcium thioglycollate trihydrate
Synonyms: mercaptoacetic acid, calcium salt trihydrate
Formula: Ca(HSCH$_2$COO)$_2$·3H$_2$O
Molecular Formula: C$_4$H$_{12}$CaO$_7$S$_2$
Molecular Weight: 276.345
CAS RN: 814-71-1
Properties: white powd or prismatic rod cryst; odorless or faint mercaptan odor; used in depilatories and in hair wave preparations [HAW93] [MER89]
Solubility: s H$_2$O [MER89]
Reactions: slowly loses H$_2$O above 95°C; darkens at 220°C; partially fuses with decomposition at 280-290°C [MER89]

653
Compound: Calcium thiosulfate hexahydrate
Synonyms: calcium hyposulfite hexahydrate
Formula: CaS$_2$O$_3$·6H$_2$O
Molecular Formula: CaH$_{12}$O$_9$S$_2$
Molecular Weight: 260.300
CAS RN: 10124-41-1
Properties: tricl cryst; when dry, decomposes on standing to form a yellow crust on the surface; spontaneously decomposed at 43-49°C; used to treat dermatitis and jaundice caused by arsphenamine [MER89]
Solubility: 100g/100 mL H$_2$O (3°C), decomposed by hot water; i alcohol [MER89] [CRC77]
Density, g/cm^3: 1.87 [MER89]
Melting Point, °C: decomposes [CRC77]

654
Compound: Calcium titanate
Synonyms: perovskite
Formula: CaTiO$_3$
Molecular Formula: CaO$_3$Ti
Molecular Weight: 135.956
CAS RN: 12049-50-2
Properties: -150, +325 mesh 99% pure; occurs naturally as the mineral perovskite; can be made by heating CaO and TiO$_2$ to 1350°C; used as an additive to BaTiO$_3$ [KIR83] [CER91]
Density, g/cm^3: 4.10 [STR93]
Melting Point, °C: 1975 [STR93]

655
Compound: Calcium tungstate
Synonyms: scheelite
Formula: CaWO$_4$

Molecular Formula: CaO_4W
Molecular Weight: 287.916
CAS RN: 7790-75-2
Properties: -325 mesh, 10 micron or less 99.9% pure; occurs in nature as mineral scheelite; white tetr powd, a = 0.524 nm, c = 1.138 nm; can be prepared by heating tungstic acid and CaO or $CaCO_3$; used in tumor treatment and in luminous paint [KIR83] [STR93] [MER89]
Solubility: i H_2O; decomposed by hot HCl, HNO_3 [MER89]
Density, g/cm³: 6.062 [STR93]
Melting Point, °C: 1620 [STR93]

656
Compound: Calcium vanadate
Formula: CaV_2O_6
Molecular Formula: CaO_6V_2
Molecular Weight: 237.957
CAS RN: 12135-52-3
Properties: -325 mesh, 10 micron or less, 99.9% pure [CER91]

657
Compound: Calcium zirconate
Formula: $CaZrO_3$
Molecular Formula: CaO_3Zr
Molecular Weight: 179.300
CAS RN: 12013-47-7
Properties: colorless, monocl; -100, +200 mesh, other sizes, 99% pure; white powd [STR93] [CER91] [CRC77]
Solubility: s HNO_3 [HAW93]
Density, g/cm³: 4.78 [STR93]
Melting Point, °C: 2550 [STR93]

658
Compound: Californium
Formula: Cf
Molecular Formula: Cf
Molecular Weight: 251
CAS RN: 7440-71-3
Properties: α-form: hex, a = 0.339 nm, c = 1.101 nm; β: fcc, a = 0.494 nm; γ: fcc, a = 0.575 nm; ionic radius of Cf^{+++} is 0.0934 nm, of Cf^{++++} is 0.0851 nm; discovered in 1950; ^{252}Cf is an intense neutron source, 1 g emits 2.4 x 10^{+12} neutrons per sec; has application to neutron activation analysis and for field use in mineral prospecting and oil-well logging, potential use in medical applications [KIR78]

Density, g/cm³: all at 25°C: α: 15.1; β: 13.7; γ: 8.70 [KIR78]
Melting Point, °C: 900 [KIR91]

659
Compound: Carbon
Synonyms: graphite bromide
Formula: C_8Br
Molecular Formula: C_8Br
Molecular Weight: 175.992
CAS RN: 12079-58-2
Properties: -100 mesh 99.9% pure [CER91]

660
Compound: Carbon
Synonyms: graphite oxide
Formula: $C_7O_2(H_2)$
Molecular Formula: $C_7H_2O_2$
Molecular Weight: 118.092
CAS RN: 1399-57-1
Properties: formula also given as $C_4O(OH)$; light yellow flakes or plates; preparation: oxidation of graphite with KNO_3 in nitric and sulfuric acids; uses: rocket propellant mixtures; membranes in the desalination of seawater by reverse osmosis [HAW93] [MER89]

661
Compound: Carbon
Synonyms: graphite
Formula: C
Molecular Formula: C
Molecular Weight: 12.011
CAS RN: 7782-42-5
Properties: hex; soft, slippery feel; steel gray to black color with metallic sheen; electrical resistivity (20°C) 1375 μohm·cm; tensile strength 400-2000 psi; compressive strength ~2000-8000 psi; coefficient of friction 0.1 μ; enthalpy of fusion 104.6 kJ/mol; enthalpy of vaporization 711 kJ/mol; semiconductor: band gap 5.47 eV (300 K); mobility (300 K) cm²/(V s), 1800 electron, 1200 hole; effective mass: 0.2 electron, 0.25 hole [KIR78] [HAW93] [COT88] [ALD94] [CRC93]
Density, g/cm³: 2.0-2.25 [HAW93]
Melting Point, °C: sublimes 3650 [HAW93]
Thermal Conductivity, W/(m·K): 119-165 (25°C) [ALD94]; 13.4 (500°C), 9.9 (1000°C) [KIR80]
Thermal Expansion Coefficient: (linear) to 1000°C is 4.0 x 10^{-6}/°C [KIR80]

662

Compound: Carbon

Synonyms: graphite fluoride

Formula: $(CF_x)_n$

Molecular Formula: x = 0.8 to 1.5

CAS RN: 11113-63-6

Properties: -200 mesh, 99.9% pure; polymer [ALD94] [CER91]

663

Compound: Carbon

Synonyms: diamond

Formula: C

Molecular Formula: C

Molecular Weight: 12.011

CAS RN: 7782-40-3

Properties: cryst modification of carbon; fcc, a = 3.56683 to 3.56725 nm; specific heat 6.184 J/(mol·K); hardness 10 Mohs; resistivity, 20°C, $>10^{+16}$ ohm cm (Type I, most Type IIa), 10 to 10^{+3} (Type IIb); dielectric constant (27°C, 0 to 3 kHz) 5.58; Type I: diamonds containing 0.1% to 0.2% nitrogen; Type IIa, free of nitrogen; Type IIb very pure, generally blue in color; obtained by mining; uses: jewelry, polishing, grinding [KIR78] [MER89]

Density, g/cm³: 3.51524 [KIR78]

Melting Point, °C: >3550 [COT88]

Boiling Point, °C: 4827 [COT88]

Reactions: diamond to graphite transition >1500°C in absence of air [KIR78]

Thermal Conductivity, W/(m·K): 20°C: Type I, 900; Type IIa, 2600 [KIR78]

Thermal Expansion Coefficient: 20°C: 0.8 x 10^{-6}; -100°C: 0.4 x 10^{-6}; 100 to 900°C: (1.5 to 4.8) x 10^{-6} [KIR78]

664

Compound: Carbon (amorphous)

Synonyms: carbon black

Formula: C

Molecular Formula: C

Molecular Weight: 12.011

CAS RN: 7440-44-0

Properties: a quasi graphitic form of carbon of small particle size; [MER89]

665

Compound: Carbon dioxide

Synonyms: carbonic acid anhydride

Formula: CO_2

Molecular Formula: CO_2

Molecular Weight: 44.010

CAS RN: 124-38-9

Properties: colorless, odorless, noncombustible gas; faint acid taste; colorless, odorless volatile liq; white, snow like flakes or cubes in the solid form; critical temp 31.3°C; critical pressure 7.38 MPa; enthalpy of vaporization 25.21 kJ/mol; enthalpy of fusion 9.02 kJ/mol [HAW93] [MER89] [AIR87] [CRC93]

Solubility: mL CO_2/100mL H_2O at 760mm: 171 (0°C), 88 (20°C), 36 (60°C) [MER89]

Density, g/cm³: gas: 1.527 (air = 1) [MER89]

Melting Point, °C: -56.6 (5.2 atm) [MER89]

Boiling Point, °C: sublimes -78.5 [MER89]

666

Compound: Carbon diselenide

Formula: CSe_2

Molecular Formula: CSe_2

Molecular Weight: 169.931

CAS RN: 506-80-9

Properties: light sensitive, golden yellow, strongly refractive liq; odor of rotten radishes; turns brown to black on storage [MER89]

Solubility: i H_2O; miscible with CCl_4, CS_2, toluene [MER89]

Density, g/cm³: 2.6824 [MER89]

Melting Point, °C: -40 to -45 [KIR82]

Boiling Point, °C: 125-126 [MER89]

667

Compound: Carbon disulfide

Formula: CS_2

Molecular Formula: CS_2

Molecular Weight: 76.143

CAS RN: 75-15-0

Properties: clear, colorless or faintly yellow liq; refractive, mobile, flammable; decomposes on standing for a long time; burns with blue flame to CO_2 and SO_2; enthalpy of fusion 4.39 kJ/mol; enthalpy of vaporization at bp 26.74 kJ/mol, 27.51 at 25°C; refractive index 1.6232; flash point -30°C; autoignition temp 100°C; used as a solvent, in the manufacture of viscose rayon, cellophane [HAW93] [CRC93] [MER89] [CIC73]

Solubility: g/100g H_2O: 0.204 (0°C), 0.179 (20°C), 0.111 (40°C) [LAN85]; s alcohol, benzene, ether [HAW93]

Density, g/cm³: liq: 1.2632 (20°C); vapor: 2.67 [MER89]

Melting Point, °C: -111.6 [MER89]

Boiling Point, °C: 46 [COT88]

668

Compound: Carbon fluoride
Formula: C_4F
Molecular Formula: C_4F
Molecular Weight: 67.042
CAS RN: 12774-81-1
Properties: solid, nonconductor formed on carbon anodes when molten KF-HF mixtures are oxidized at carbon electrodes to generate fluorine [HAW93]
Melting Point, °C: decomposes >60 [HAW93]

669

Compound: Carbon fullerenes
Synonyms: (5,6) fullerene
Formula: C_{70}
Molecular Formula: C_{70}
Molecular Weight: 840.770
CAS RN: 115383-22-7
Properties: black powd; the C_{70} fullerene has five cryst structures, depending on the temp; fcc, T>67°C, a = 1.501 nm [DRE93] [STR93] [ALD94]
Solubility: s benzene, toluene [LID94]
Melting Point, °C: >280 [LID94]

670

Compound: Carbon monoxide
Formula: CO
Molecular Formula: CO
Molecular Weight: 28.010
CAS RN: 630-08-0
Properties: highly poisonous, odorless, colorless, tasteless gas; very flammable, burns with bright blue flame; autoignition temp 609°C; enthalpy of fusion 0.83 kJ/mol; enthalpy of vaporization 6.04 kJ/mol; critical temp -140.21°C; critical pressure 34.529 atm; critical density 0.3010 g/cm³; triple point 205.0°C at 115.4 mm Hg; produced by partial oxidation of hydrocarbon gases; used as a reducing agent in metallurgy, e.g. for Ni [CIC73] [CRC93] [MER89] [AIR87]
Solubility: mL/100mL H_2O: 3.3 (0°C), 2.3 (20°C) [MER89]
Density, g/cm³: gas: 0.968 (air = 1.000) [MER89]
Melting Point, °C: -205.0 [MER89]
Boiling Point, °C: -191.5 [MER89]

671

Compound: Carbon oxyselenide
Synonyms: carbonyl selenide
Formula: COSe

Molecular Formula: COSe
Molecular Weight: 106.970
CAS RN: 1603-84-5
Properties: colorless gas; light sensitive; unstable [KIR82] [CRC77]
Solubility: decomposed by H_2O [CRC77]
Density, g/cm³: 4.694 g/L [LID94]
Melting Point, °C: -124.4 [CRC77]
Boiling Point, °C: -21.7 [CRC77]

672

Compound: Carbon oxysulfide
Synonyms: carbonyl sulfide
Formula: COS
Molecular Formula: COS
Molecular Weight: 60.076
CAS RN: 463-58-1
Properties: colorless gas with sulfide odor unless pure; flammable [HAW93] [COT88] [ALD94]
Solubility: mL/100mL H_2O: 133.3 (0°C), 56.1 (20°C), 40.3 (30°C) [LAN85]; slowly decomposes in H_2O [COT88]; s alcohol [HAW93]
Density, g/cm³: 2.636 g/L [LID94]
Melting Point, °C: -138.8 [HAW93]
Boiling Point, °C: -50.2 [HAW93]

673

Compound: Carbon soot
Formula: Cx
CAS RN: 1333-86-4
Properties: black powd; contains 2-20% C_{60}/C_{70}; preparation: from resistive heating of graphite, 5-10% yield; uses: precursor to the fullerenes [STR93] [ALD94]

674

Compound: Carbon suboxide
Synonyms: 1,2-propadiene-1,3-dione
Formula: C_3O_2
Molecular Formula: C_3O_2
Molecular Weight: 68.032
CAS RN: 504-64-3
Properties: colorless, highly refractive liq or colorless gas which burns with a blue sooty flame; odor like acrolein or mustard oil; structure: O=C=C=C=O [MER89]
Solubility: forms malonic acid with H_2O [MER89]
Density, g/cm³: 2.985 g/L [LID94]
Melting Point, °C: -111.3 [MER89]
Boiling Point, °C: 6.8 [MER89]

675
Compound: Carbon sulfide selenide
Formula: CSSe
Molecular Formula: CSSe
Molecular Weight: 123.037
CAS RN: 5951-19-9
Properties: yellow, oily liq; unstable, sensitive to light [KIR82] [CRC77]
Solubility: i H_2O [CRC77]
Density, g/cm^3: 1.9874 [CRC77]
Melting Point, °C: -85 [CRC77]
Boiling Point, °C: 84.5 [CRC77]

676
Compound: Carbon sulfide telluride
Synonyms: carbon sulfotelluride
Formula: CSTe
Molecular Formula: CSTe
Molecular Weight: 171.677
CAS RN: 10340-06-4
Properties: reddish yellow liq; odor of garlic; decomposed by light even at -50°C forming CS_2 and Te [KIR83]
Density, g/cm^3: 2.9 [CRC77]
Melting Point, °C: -54 [CRC77]
Boiling Point, °C: decomposes [CRC77]

677
Compound: Carbon tetrabromide
Synonyms: tetrabromomethane
Formula: CBr$_4$
Molecular Formula: CBr$_4$
Molecular Weight: 331.627
CAS RN: 558-13-4
Properties: colorless cryst; slight decomposition if boiled [HAW93] [COT88]
Solubility: i H_2O; s alcohol, ether and chloroform [HAW93]
Density, g/cm^3: 3.42 [HAW93]
Melting Point, °C: 90.1 [HAW93]
Boiling Point, °C: 189.5 [HAW93]

678
Compound: Carbon tetrachloride
Synonyms: tetrachloromethane
Formula: CCl$_4$
Molecular Formula: CCl$_4$
Molecular Weight: 153.822
CAS RN: 56-23-5

Properties: colorless clear nonflammable heavy liq; sweetish odor; refractive index 1.4607; vapor pressure 91.3 mm Hg (20°C); enthalpy of vaporization 29.82 kJ/mol; enthalpy of fusion 3.28 kJ/mol [CRC93] [MER89]
Solubility: 1 mL dissolves in 2000 mL H_2O [MER89]; miscible with alcohol, ether [HAW93]
Density, g/cm^3: 1.589 [MER89]
Melting Point, °C: -23 [MER89]
Boiling Point, °C: 76.8 [CRC93]

679
Compound: Carbon tetrafluoride
Synonyms: tetrafluoromethane, Freon 14
Formula: CF$_4$
Molecular Formula: CF$_4$
Molecular Weight: 88.003
CAS RN: 75-73-0
Properties: colorless, odorless gas; thermally stable; chemically very inert; critical temp -45.7°C; critical pressure 3.74 MPa; enthalpy of vaporization 11.98 kJ/mol; obtained by reaction of C or CO and F_2; used as a gaseous insulator, and in electronics production [HAW93] [MER89] [AIR87]
Solubility: mL/100mL in H_2O: 0.595 (10°C), 0.490 (20°C), 0.366 (40°C) [LAN85]
Density, g/cm^3: solid (-195°C), 1.98; liq (-183°C), 1.89 [MER89]
Melting Point, °C: -183.6 [MER89]
Boiling Point, °C: -127.8 [MER89]

680
Compound: Carbon tetraiodide
Synonyms: tetraiodomethane
Formula: CI$_4$
Molecular Formula: CI$_4$
Molecular Weight: 519.629
CAS RN: 507-25-5
Properties: red cub cryst; odor of iodine; decomposes to iodine and tetraiodoethylene under influence of heat or light [MER89]
Solubility: sl s H_2O with hydrolysis; s benzene, chloroform [MER89]
Density, g/cm^3: 4.32 [MER89]
Melting Point, °C: 171 [MER89]
Boiling Point, °C: decomposes [COT88]
Reactions: can be sublimed at low pressure [COT88]

681
Compound: Carbonyl bromide

Synonyms: bromophosgene
Formula: COBr$_2$
Molecular Formula: CBr$_2$O
Molecular Weight: 187.818
CAS RN: 593-95-3
Properties: heavy, colorless liq with a strong odor; fumes in air; decomposed by light and heat; used in making cryst-violet type coloring agents, and as a poison gas [HAW93]
Solubility: hydrolyzes in H$_2$O to form CO$_2$ and HBr [COT88]
Density, g/cm^3: 2.5 (~15°C) [HAW93]
Melting Point, °C: 65 [COT88]

682
Compound: Carbonyl chloride
Synonyms: phosgene,carbon oxychloride
Formula: COCl$_2$
Molecular Formula: CCl$_2$O
Molecular Weight: 98.910
CAS RN: 75-44-5
Properties: colorless to light yellow gas; haylike odor in small concentrations; enthalpy of fusion 5.74 kJ/mol; used in organic synsthesis for isocyanates, polyurethane and polycarbonate resins [HAW93] [CRC93]
Solubility: sl s and hydrolyzed in H$_2$O; s benzene, toluene [HAW93]
Density, g/cm^3: 4.34 g/L [LID94]
Melting Point, °C: -127.9 [CRC93]
Boiling Point, °C: 8.2 [HAW93]

683
Compound: Carbonyl fluoride
Synonyms: fluorophosgene
Formula: COF$_2$
Molecular Formula: CF$_2$O
Molecular Weight: 66.007
CAS RN: 353-50-4
Properties: pungent very hygr gas [MER89]
Solubility: instantly hydrolyzed by H$_2$O [MER89]
Density, g/cm^3: solid: (-190°C), 1.388; liq: (-114°C), 1.139 [MER89]
Melting Point, °C: -114.0 [MER89]
Boiling Point, °C: -83.1 [MER89]

684
Compound: Ceric basic nitrate trihydrate
Formula: Ce(OH)(NO$_3$)$_3$·3H$_2$O
Molecular Formula: CeH$_7$N$_3$O$_{13}$
Molecular Weight: 397.183

Properties: long red needle; prepartion: evaporating a solution of ceric hydroxide in nitric acid solution [KIR78] [CRC92]
Solubility: s H$_2$O [CRC92]

685
Compound: Ceric fluoride
Synonyms: cerium(IV) fluoride
Formula: CeF$_4$
Molecular Formula: CeF$_4$
Molecular Weight: 216.109
CAS RN: 10060-10-3
Properties: white powd; hygr; minute cryst; thermally stable below 550°C; monocl, a = 1.2587 nm, c = 0.8041 nm; preparation: reaction of F$_2$ with CeF$_3$; uses: fluorinating agent [STR93] [MER89] [CRC92]
Solubility: i H$_2$O, very slowly hydrolyzed by cold H$_2$O [MER89]
Density, g/cm^3: 4.77 [MER89]
Melting Point, °C: 650, decomposes [STR93]

686
Compound: Ceric hydroxide
Synonyms: cerium(IV) hydroxide
Formula: Ce(OH)$_4$
Molecular Formula: CeH$_4$O$_4$
Molecular Weight: 208.145
CAS RN: 12014-56-1
Properties: addition of NaOH or NH$_4$OH to a solution of Ce^{++++} results in a gelatinous precipitate of CeO$_2$·xH$_2$O (x = 0.5 to 2); yellowish white powd when dried; granular Ce(OH)$_4$ obtained by boiling insoluble Ce^{++++} salt with NaOH [STR93] [KIR78]
Solubility: i H$_2$O, s conc a [HAW93]

687
Compound: Ceric oxide
Synonyms: cerianite
Formula: CeO$_2$
Molecular Formula: CeO$_2$
Molecular Weight: 172.114
CAS RN: 1306-38-3
Properties: brownish white powd or cub cryst, but usually pale yellow; refractory material; a = 0.54110 nm; used in ceramics, as an abrasive for polishing glass, as evaporated material of 99.9% purity in high index film for dielectric beam splitters, interference filters, and in multilayers as anti-reflection coating; can be prepared by calcining cerous oxalate or hydroxide [KIR78] [HAW93] [TAY85] [MER89] [CER91]

Solubility: i H$_2$O; s H$_2$SO$_4$, HNO$_3$; i dil acid [HAW93]
Density, g/cm^3: 7.65 [HAW93]
Melting Point, °C: 2400 [KNA91]
Thermal Expansion Coefficient: from 25°C to: 100°C (0.24), 200°C (0.54), 400°C (1.20), 600°C (1.92), 800°C (2.70), 1000°C (3.51), 1200°C (4.38) [TAY85]

688
Compound: Ceric oxide hydrate
Synonyms: cerium dioxide hydrate
Formula: CeO$_2$·xH$_2$O
Molecular Formula: CeO$_2$ (anhydrous)
Molecular Weight: 172.114 (anhydrous)
CAS RN: 12014-56-1
Properties: white powd; used to produce cerium salts, as an opacifier to impart a yellow color to glasses and enamels [HAW93]
Solubility: i H$_2$O; s conc mineral acids [HAW93]

689
Compound: Ceric sulfate tetrahydrate
Synonyms: sulfuric acid, cerium(IV) salt tetrahydrate
Formula: Ce(SO$_4$)$_2$·4H$_2$O
Molecular Formula: CeH$_8$O$_{12}$S$_2$
Molecular Weight: 404.304
CAS RN: 10294-42-5
Properties: yellow to orange powd or ortho-rhomb cryst; oxidizer [MER89]
Solubility: s small quantity of H$_2$O, but decomposes in excess H$_2$O [MER89]; s dil H$_2$SO$_4$ [HAW93]
Density, g/cm^3: 3.91 [HAW93]
Melting Point, °C: decomposes above 350 [MER89]
Reactions: minus 4H$_2$O at 180-200°C [MER89]

690
Compound: Ceric titanate
Synonyms: cerium(IV) titanate
Formula: CeTiO$_4$
Molecular Formula: CeO$_4$Ti
Molecular Weight: 251.980
CAS RN: 52014-82-1
Properties: -325 mesh 10 microns or less at 99.9% purity [CER91]

691
Compound: Ceric vanadate
Synonyms: cerium(IV) vanadate
Formula: CeVO$_4$

Molecular Formula: CeO$_4$V
Molecular Weight: 255.055
CAS RN: 13597-19-8
Properties: -200 mesh with 99.9% purity [CER91]

692
Compound: Ceric zirconate
Synonyms: cerium(IV) zirconate
Formula: CeZrO$_4$
Molecular Formula: CeO$_4$Zr
Molecular Weight: 295.337
CAS RN: 53169-24-7
Properties: -325 mesh with 99.5% purity; pyrochlore type structure [TAY88a] [CER91]
Thermal Expansion Coefficient: from 25°C to: 100°C (0.24), 200°C (0.57), 400°C (1.23), 600°C (1.92); 800°C (2.58); 1000°C (3.24); 1200°C (3.90) [TAY88a]

693
Compound: Cerium
Formula: Ce
Molecular Formula: Ce
Molecular Weight: 140.115
CAS RN: 7440-45-1
Properties: gray metal; α-Ce, fcc; β-Ce, hex; γ-Ce, fcc; δ-Ce, bcc; for γ: heat capacity 26.96 J/(mol·K); compressibility 4.18 x 10^{-2} GPa; Young's modulus 30 GPa; shear modulus 12 GPa; Poisson's ratio, 0.248; Vicker's hardness 235 MPa; yield strength 91.2 MPa; enthalpy of fusion 5.46 kJ/mol; enthalpy of sublimation 422.6 kJ/mol; electrical reistivity, 20°C, 73 µohm·cm; radius of atom 0.1824 nm; radius of ion 0.1034 nm for Ce^{+++} [KIR82] [CRC93] [ALD94]
Solubility: s dil mineral acids [KIR78]
Density, g/cm^3: hex 6.689, cub 6.773 [CRC92], [KIR78]
Melting Point, °C: 798 [KIR78]
Boiling Point, °C: 3433 [KIR82]
Reactions: reacts vigorously with the halogens >200°C [KIR78]
Thermal Conductivity, W/(m·K): 11.3, 25°C [ALD94]
Thermal Expansion Coefficient: 8.5 x 10^{-6}/°C [KIR78]

694
Compound: Cerium carbide
Formula: CeC$_2$
Molecular Formula: C$_2$Ce

Molecular Weight: 164.137
CAS RN: 12012-32-7
Properties: red, hex; 6 mm pieces and smaller of 99.5% purity [CER91] [CRC92]
Solubility: decomposes in H_2O; s a [CRC92]
Density, g/cm³: 5.23 [CRC92]
Melting Point, °C: 2420 [KNA91]

695
Compound: Cerium dihydride
Formula: CeH_2
Molecular Formula: CeH_2
Molecular Weight: 142.131
CAS RN: 13569-50-1
Properties: black pyrophoric solid; ignites spontaneously in air; can be prepared by reacting cerium and hydrogen at 345°C; used to store H_2 in the system $CeMg_2$ [KIR80] [KIR78]
Solubility: reacts with H_2O at 0°C [KIR80]
Density, g/cm³: 5.45 [LID94]
Melting Point, °C: ignites [CRC94]

696
Compound: Cerium hexaboride
Formula: CeB_6
Molecular Formula: B_6Ce
Molecular Weight: 204.981
CAS RN: 12008-02-5
Properties: refractory material; blue cub; -325 mesh, 10 microns or less at 99.9% purity [CRC92] [KIR78] [CER91]
Solubility: i H_2O; i HCl [CRC92]
Density, g/cm³: 4.87 [LID94]
Melting Point, °C: 2190 [CRC94]

697
Compound: Cerium monosulfide
Formula: CeS
Molecular Formula: CeS
Molecular Weight: 172.181
CAS RN: 12014-82-3
Properties: yellow cub; 3 mm pieces and smaller (fused) 99.9% [CER91] [LID94]
Density, g/cm³: 5.9 [LID94]
Melting Point, °C: 2445 [KNA91]

698
Compound: Cerium nitride
Formula: CeN
Molecular Formula: CeN
Molecular Weight: 154.122

CAS RN: 25764-08-3
Properties: -60 mesh, 99.9% pure; NaCl cryst system; a = 0.501 nm [CIC73] [CER91]
Density, g/cm³: 7.89 [LID94]
Melting Point, °C: 2557 [KNA89]

699
Compound: Cerium oxysulfide
Formula: Ce_2O_2S
Molecular Formula: Ce_2O_2S
Molecular Weight: 344.295
CAS RN: 12442-45-4
Properties: -200 mesh with 99.9% purity [CER91]

700
Compound: Cerium silicide
Formula: $CeSi_2$
Molecular Formula: $CeSi_2$
Molecular Weight: 196.286
CAS RN: 12014-85-6
Properties: 6.35 mm & down pieces, 99.9% purity [CER91] [ALF93]
Solubility: i H_2O [CRC92]
Density, g/cm³: 5.67 [CRC92]
Melting Point, °C: 1620 [LID94]

701
Compound: Cerium stannate
Formula: $CeO_2 \cdot SnO_2$
Molecular Formula: CeO_4Sn
Molecular Weight: 322.822
CAS RN: 53169-23-6
Properties: -325 mesh, 10 micron average reacted product of 99.9% purity; pyrochlore type structure [CER91] [TAY88a]
Thermal Expansion Coefficient: from 25°C to: 100°C (0.21), 200°C (0.48), 400°C (1.05), 600°C (1.62), 800°C (2.16), 1000°C (2.73), 1200°C (3.30) [TAY88a]

702
Compound: Cerium trihydride
Formula: CeH_3
Molecular Formula: CeH_3
Molecular Weight: 143.139
CAS RN: 13864-02-3
Properties: dark black amorphous powd [CRC92]
Solubility: decomposes in H_2O [CRC92]

703

Compound: Cerous acetate hemitrihydrate
Synonyms: cerium(III) acetate hemitrihydrate
Formula: $Ce(CH_3COO)_3 \cdot 1\text{-}1/2H_2O$
Molecular Formula: $C_6H_{12}CeO_{7.5}$
Molecular Weight: 344.272
CAS RN: 537-00-8
Properties: white powd [STR93]
Solubility: g/100mL H_2O: 26.5 (15°C), 16.2 (75°C) [CRC94]
Boiling Point, °C: decomposes [CRC94]
Reactions: minus 1-1/2 H_2O at 115°C [CRC94]

704

Compound: Cerous acetylacetonate hydrate
Synonyms: 2,4-pentanedione, cerium(III) derivative
Formula: $Ce(CH_3COCH=C(O)CH_3)_3 \cdot xH_2O$
Molecular Formula: $C_{15}H_{21}CeO_6$ (anhydrous)
Molecular Weight: 437.443 (anhydrous)
CAS RN: 15653-01-7
Properties: tan powd; hygr [STR93] [ALD94]
Melting Point, °C: 131-132 [CRC94]

705

Compound: Cerous bromide
Synonyms: cerium(III) bromide
Formula: $CeBr_3$
Molecular Formula: Br_3Ce
Molecular Weight: 379.827
CAS RN: 14457-87-5
Properties: orange powd; hygr [STR93]
Melting Point, °C: 730 [KNA91]
Boiling Point, °C: 1457 [KNA91]

706

Compound: Cerous bromide heptahydrate
Formula: $CeBr_3 \cdot 7H_2O$
Molecular Formula: $Br_3CeH_{14}O_7$
Molecular Weight: 505.924
CAS RN: 14457-87-5
Properties: colorless, deliq needles; anhydrous $CeBr_3$ -20 mesh at 99.9% purity [MER89] [CER91]
Solubility: s H_2O, alcohol [MER89]
Melting Point, °C: 732 [MER89]

707

Compound: Cerous carbonate
Synonyms: cerium(III) carbonate
Formula: $Ce_2(CO_3)_3$
Molecular Formula: $C_3Ce_2O_9$

Molecular Weight: 460.258
CAS RN: 537-01-9
Properties: white powd; if a solution of the carbonate in water is boiling, the product can be $Ce(OH)(CO_3)$ [KIR78] [MER83]
Solubility: i H_2O, s mineral acids [HAW93]
Reactions: decomposes at 500°C to CeO_2 with evolution of CO, CO_2 [KIR78]

708

Compound: Cerous carbonate pentahydrate
Synonyms: cerium(III) carbonate pentahydrate
Formula: $Ce_2(CO_3)_3 \cdot 5H_2O$
Molecular Formula: $C_3H_{10}Ce_2O_{14}$
Molecular Weight: 550.334
CAS RN: 72520-94-6
Properties: white cryst; pentahydrate can be obtained by adding an alkali bicarbonate solution to a solution of Ce^{+++} [KIR78] [STR93] [MER89]
Solubility: i H_2O, s dil acid [MER89]

709

Compound: Cerous chloride
Synonyms: cerium(III) chloride
Formula: $CeCl_3$
Molecular Formula: $CeCl_3$
Molecular Weight: 246.473
CAS RN: 7790-86-5
Properties: -20 mesh with 99.9% purity; white very fine powd; can be prepared by dissolving cerium carbonate in HCl [CER91] [KIR78] [STR93] [MER89]
Solubility: s H_2O, alcohol (exothermic) [MER89]
Density, g/cm³: 3.97 [MER89]
Melting Point, °C: 807 [KNA91]
Boiling Point, °C: 1725 (estimated) [KNA91]

710

Compound: Cerous chloride heptahydrate
Synonyms: cerium(III) chloride heptahydrate
Formula: $CeCl_3 \cdot 7H_2O$
Molecular Formula: $CeCl_3H_{14}O_7$
Molecular Weight: 372.580
CAS RN: 18618-55-8
Properties: colorless to yellow deliq ortho-rhomb cryst; can be prepared by evaporating $CeCl_3$ solution, or by saturating conc $CeCl_3$ solution with HCl [KIR78] [MER89]
Solubility: v s H_2O, alcohol [MER89]
Density, g/cm³: 3.92 [ALD94]

Reactions: minus H_2O >90°C, becomes anhydrous by 230°C [MER89]

711
Compound: Cerous chloride hydrate
Synonyms: cerium(III) chloride hydrate
Formula: $CeCl_3 \cdot xH_2O$
Molecular Formula: $CeCl_3$ (anhydrous)
Molecular Weight: 246.473 (anhydrous)
CAS RN: 19423-76-8
Properties: -4 mesh with 99.9% purity; white or off-white cryst [STR93] [CER91]

712
Compound: Cerous fluoride
Synonyms: cerium(III) fluoride
Formula: CeF_3
Molecular Formula: CeF_3
Molecular Weight: 197.110
CAS RN: 7758-88-5
Properties: white powd, or 99.9% pure melted pieces of 3-6 mm; hygr; hex, a = 0.71306 nm, c = 0.72805 nm; melted pieces used as an evaporation material and sputtering target for multilayers and thin film capacitors [STR93] [GME76] [CER91]
Solubility: i H_2O, but slowly hydrolyzed [MER89]
Density, g/cm³: 6.157 [MER89]
Melting Point, °C: 1437 [KNA91]
Boiling Point, °C: 2280 (estimated) [KNA91]

713
Compound: Cerous hydroxide
Synonyms: cerium(III) hydroxide
Formula: $Ce(OH)_3$
Molecular Formula: CeH_3O_3
Molecular Weight: 191.137
CAS RN: 15785-09-8
Properties: white gelatinous precipitate; however, impurities impart yellow, brown or pink coloration; used to produce cerium salts to color glass [HAW93]
Solubility: i H_2O; s acids [HAW93]

714
Compound: Cerous iodide
Synonyms: cerium(III) iodide
Formula: CeI_3
Molecular Formula: CeI_3
Molecular Weight: 520.828
CAS RN: 7790-87-6

Properties: -20 mesh with 99.9% purity; bright yellow, ortho-rhomb; decomposes in moist air [MER89] [CER91]
Solubility: s H_2O [MER89]
Melting Point, °C: 760 [KNA91]
Boiling Point, °C: 1500 (estimated) [KNA91]

715
Compound: Cerous iodide nonahydrate
Synonyms: cerium(III) iodide nonahydrate
Formula: $CeI_3 \cdot 9H_2O$
Molecular Formula: $CeH_{18}I_3O_9$
Molecular Weight: 682.966
CAS RN: 7790-87-6
Properties: white or reddish white cryst [MER89]
Solubility: v s H_2O, solution decomposes with I_2 liberated; s alcohol [MER89]
Melting Point, °C: 752 [CRC92]
Boiling Point, °C: 1400 [CRC92]

716
Compound: Cerous nitrate hexahydrate
Synonyms: cerium(III) nitrate hexahydrate
Formula: $Ce(NO_3)_3 \cdot 6H_2O$
Molecular Formula: $CeH_{12}N_3O_{15}$
Molecular Weight: 434.221
CAS RN: 10294-41-4
Properties: white cryst; deliq; oxidizing agent; used as a catalyst in the hydrolysis of phosphoric acid esters [HAW93]
Solubility: s H_2O, alcohol, acetone [HAW93]
Boiling Point, °C: decomposes 200 [HAW93]
Reactions: minus $3H_2O$ at 150°C [HAW93]

717
Compound: Cerous oxalate nonahydrate
Synonyms: cerium(III) oxalate nonahydrate
Formula: $Ce_2(C_2O_4)_3 \cdot 9H_2O$
Molecular Formula: $C_6H_{18}Ce_2O_{21}$
Molecular Weight: 706.426
CAS RN: 13266-83-6
Properties: white or sl pink, tasteless powd; odorless; decomposes upon heating; used in medicine and in the extraction of cerium metals; can be prepared by precipitation with oxalic acid, from sl acidic cerium solutions [KIR78] [HAW93] [MER89]
Solubility: i H_2O; s H_2SO_4, HCl; i oxalic acid, alkali, ether, alcohol [KIR78]
Melting Point, °C: decomposes [STR93]

718
Compound: Cerous oxide
Synonyms: cerium(III) oxide
Formula: Ce_2O_3
Molecular Formula: Ce_2O_3
Molecular Weight: 328.228
CAS RN: 1345-13-7
Properties: -100 mesh golden green with 99.9% purity; trig; can be prepared by heating powd carbon and CeO_2 at 1250°C in CO atm [KIR78] [CER91]
Solubility: i H_2O; s H_2SO_4; i HCl [CRC92]
Density, g/cm³: 6.86 [CRC92]
Melting Point, °C: 2177 [KNA91]

719
Compound: Cerous perchlorate hexahydrate
Synonyms: cerium(III) perchlorate hexahydrate
Formula: $Ce(ClO_4)_3 \cdot 6H_2O$
Molecular Formula: $CeCl_3H_{12}O_{18}$
Molecular Weight: 546.557
CAS RN: 14017-47-1
Properties: white cryst [STR93]

720
Compound: Cerous phosphate hydrate
Synonyms: monazite
Formula: $CePO_4 \cdot xH_2O$
Molecular Formula: $CePO_4$ (anhydrous)
Molecular Weight: 235.087 (anhydrous)
CAS RN: 13454-71-2
Properties: off-white powd [STR93]

721
Compound: Cerous selenate
Synonyms: cerium(III) selenate
Formula: $Ce_2(SeO_4)_3$
Molecular Formula: $Ce_2O_{12}Se_3$
Molecular Weight: 709.103
Properties: rhomb [CRC92]
Solubility: g/100g H_2O: 39.5 (0°C), 35.2 (20°C), 2.1 (90°C) [LAN85]
Density, g/cm³: 4.456 [CRC92]

722
Compound: Cerous sulfate
Synonyms: cerium(III) sulfate
Formula: $Ce_2(SO_4)_3$
Molecular Formula: $Ce_2O_{12}S_3$
Molecular Weight: 568.421
CAS RN: 13454-94-9

Properties: colorless to green, monocl or rhomb; prepared by heating hydrated salt at 350 to 400°C, or by reducing a solution of ceric sulfate with H_2O_2 solution [KIR78] [CRC92]
Solubility: 10.1 g/100mL H_2O (0°C), 0.25 g/100mL H_2O (100°C) [CRC92]
Density, g/cm³: 3.912 [CRC92]
Melting Point, °C: 630 [HAW93]; 920 [CRC93]

723
Compound: Cerous sulfate octahydrate
Synonyms: cerium(III) sulfate octahydrate
Formula: $Ce_2(SO_4)_3 \cdot 8H_2O$
Molecular Formula: $Ce_2H_{16}O_{20}S_3$
Molecular Weight: 712.543
CAS RN: 10450-59-6
Properties: white; ortho-rhomb, octahedral cryst [MER89] [STR93]
Solubility: g/100g H_2O: 9.43 (20°C), 5.70 (40°C), 4.04 (60°C) [LAN85]
Density, g/cm³: 2.87 [MER89]
Melting Point, °C: 630 [ALD94]
Reactions: minus $8H_2O$ when 250°C is reached [MER89]

724
Compound: Cerous sulfide
Synonyms: cerium(III) sulfide
Formula: Ce_2S_3
Molecular Formula: Ce_2S_3
Molecular Weight: 376.428
CAS RN: 12014-93-6
Properties: red cryst; dark brown powd, or purple; -325 mesh 10 micron or less 99.9% purity [CER91] [CRC92]
Solubility: i H_2O [CRC92]
Density, g/cm³: 5.02 [CRC92]
Melting Point, °C: 2100 [CRC92]

725
Compound: Cerous telluride
Synonyms: cerium(III) telluride
Formula: Ce_2Te_3
Molecular Formula: Ce_2Te_3
Molecular Weight: 663.030
CAS RN: 12014-97-0
Properties: -20 mesh 99.9% purity [CER91]

726
Compound: Cerous tungstate
Synonyms: cerium(III) tungstate

Formula: Ce$_2$(WO$_4$)$_3$
Molecular Formula: Ce$_2$O$_{12}$W$_3$
Molecular Weight: 1023.743
CAS RN: 13454-74-5
Properties: yellow tetr; -200 mesh at 99.9% purity; white monocl powd, a = 1.151 nm, b = 1.172 nm, c = 0.782 nm [KIR83] [STR93] [CRC92]
Density, g/cm^3: 6.77 [KIR83]
Melting Point, °C: 1089 [KIR83]

727
Compound: Cesium
Formula: Cs
Molecular Formula: Cs
Molecular Weight: 132.90543
CAS RN: 7440-46-2
Properties: bcc; atomic radius 0.274 nm; silvery white ductile metal; oxidizes rapidly in moist air, can ignite spontaneously; hardness 0.2 Mohs; electrical resistivity 19 (0°), 36.6 (30°C) μohm·cm; specific heat (20°C) 0.217 J/(g K); enthalpy of fusion 2.087 kJ/mol; enthalpy of vaporization 68.85 KJ/mol [KIR79] [HAW93] [MER89] [ALD94]
Solubility: reacts with H$_2$O to evolve H$_2$; s liq NH$_3$ [MER89]
Density, g/cm^3: solid: (18°C) 1.892; liq: (40°C) 1.827 [KIR79]
Melting Point, °C: 28.44 [LID94]
Boiling Point, °C: 671 [LID94]
Thermal Conductivity, W/(m·K): 35.9 [ALD94]; liq, at mp: 18.4; vapor at bp 0.0046 [KIR78]

728
Compound: Cesium acetate
Synonyms: acetic acid, cesium salt
Formula: CH$_3$COOCs
Molecular Formula: C$_2$H$_3$CsO$_2$
Molecular Weight: 191.950
CAS RN: 3396-11-0
Properties: lump; hygr [STR93]
Solubility: 945.1 g/100 mL (-2.5°C), 1345.5 g/100 mL (88.5°C) [CRC77]
Melting Point, °C: 194 [STR93]

729
Compound: Cesium acetylacetonate
Synonyms: 2,4-pentanedione, cesium derivative
Formula: Cs[CH$_3$COCH=C(O)CH$_3$]
Molecular Formula: C$_5$H$_7$CsO$_2$
Molecular Weight: 232.015
CAS RN: 25937-78-4

Properties: hygr [ALD94]

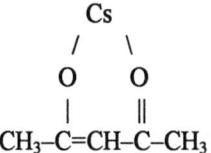

730
Compound: Cesium aluminum sulfate dodecahydrate
Synonyms: cesium alum
Formula: CsAl(SO$_4$)$_2$·12H$_2$O
Molecular Formula: AlCsH$_{24}$O$_{20}$S$_2$
Molecular Weight: 568.198
CAS RN: 7784-17-0
Properties: colorless cub cryst; used in mineral waters [HAW93] [CRC77]
Solubility: g anhydrous/100g H$_2$O: 18.8 (0°C), 0.40 (20°C), 22.7 (100°C) [LAN85]; i alcohol [HAW93]
Density, g/cm^3: 2.0215 [HAW93]
Melting Point, °C: 117 [HAW93]

731
Compound: Cesium amide
Formula: CsNH$_2$
Molecular Formula: CsH$_2$N
Molecular Weight: 148.928
CAS RN: 22205-57-8
Properties: white needles; tetr [CRC77] [CIC73]
Solubility: decomposed by H$_2$O; v s liq NH$_3$ [CIC73] [CRC77]
Density, g/cm^3: 3.44 [CRC77]
Melting Point, °C: 262 [CRC77]

732
Compound: Cesium azide
Formula: CsN$_3$
Molecular Formula: CsN$_3$
Molecular Weight: 174.925
CAS RN: 22750-57-8
Properties: colorless needles; hygr; tetr, a = 0.672 nm, c = 0.804 nm; Cs-N$_3$ bond length, 0.334 nm [CIC73] [CRC77]
Solubility: 224.2 g/100 mL H$_2$O (0°C) [CRC77]
Density, g/cm^3: ~3.5 [LID94]
Melting Point, °C: 310 [CRC77]

733
Compound: Cesium bromate

Formula: $CsBrO_3$
Molecular Formula: $BrCsO_3$
Molecular Weight: 260.807
CAS RN: 13454-75-6
Properties: hex cryst [CRC77]
Solubility: g/100g soln, H_2O: 3.66 (25°C); Solid phase, $CsBrO_3$ [KRU93]; g/100g H_2O: 0.21 (0°C), 5.30 (35°C) [LAN85]
Density, g/cm³: 4.10 [LAN52]
Melting Point, °C: 420 [LAN52]

734
Compound: Cesium bromide
Formula: CsBr
Molecular Formula: BrCs
Molecular Weight: 212.809
CAS RN: 7787-69-1
Properties: white cryst; hygr; used in infrared spectroscopy, scintillation counters [HAW93] [STR93]
Solubility: s alcohol; i acetone [MER89]; g/100g H_2O: 55.24 (25°C) [KRU93]
Density, g/cm³: 4.44 [MER89]
Melting Point, °C: 636 [MER89]
Boiling Point, °C: ~1300 [MER89]

735
Compound: Cesium bromoiodide
Synonyms: cesium dibromoiodide
Formula: $CsIBr_2$
Molecular Formula: Br_2CsI
Molecular Weight: 419.617
CAS RN: 18278-82-5
Properties: rhomb; -8 mesh with 99.9% purity [CRC77] [CER91]
Solubility: 4.61 g/100 mL (20°C) H_2O [CRC77]
Density, g/cm³: 4.25 [CRC77]
Melting Point, °C: 248 [CRC77]
Boiling Point, °C: 330, decomposes [CRC77]

736
Compound: Cesium carbonate
Formula: Cs_2CO_3
Molecular Formula: CCs_2O_3
Molecular Weight: 325.820
CAS RN: 534-17-8
Properties: -20 mesh with 99.996 and 99.9% purity; white powd; very deliq cryst; preparation: addition of CO_2 to a solution of CsOH; uses: catalyst for ethylene oxide polymerization [KIR79] [STR93] [MER89] [CER91]

Solubility: 260.5 g/100mL H_2O (15°C) [CRC94]; s alcohol, ether [MER89]
Density, g/cm³: 4.24 [LID94]
Melting Point, °C: 610, decomposes [STR93]

737
Compound: Cesium chlorate
Formula: $CsClO_3$
Molecular Formula: $ClCsO_3$
Molecular Weight: 216.356
CAS RN: 13763-67-2
Properties: small cryst [CRC77]
Solubility: g/100g H_2O: 2.46 (0°C), 7.6 (25°C), 79.0 (100°C) [KRU93]
Density, g/cm³: 3.57 [CRC77]

738
Compound: Cesium chloride
Formula: CsCl
Molecular Formula: ClCs
Molecular Weight: 168.358
CAS RN: 7647-17-8
Properties: -4 mesh with 99.999% and 99.9% purity; white deliq cub cryst; enthalpy of fusion 15.90 kJ/mol [CRC93] [MER89] [STR93] [CER91]
Solubility: v s H_2O; s alcohol [MER89]; g/100g solution H_2O: 61.7 (0°C), 65.6 (25°C), 73.0 (100°C) [KRU93]; 11.382±0.010 mol/(kg-H_2O) at 25°C [RAR85b]
Density, g/cm³: 3.988 [STR93]
Melting Point, °C: 646 [MER89]
Boiling Point, °C: 1303 [MER89]

739
Compound: Cesium chromate
Formula: Cs_2CrO_4
Molecular Formula: $CrCs_2O_4$
Molecular Weight: 381.805
CAS RN: 56320-90-2
Properties: -20 mesh with 99.9% purity; yellow hex or ortho-rhomb; used in electronics [KIR78] [CER91]
Solubility: 71.4 g/100 mL H_2O (15°C), 95.95 g/100 mL (30°C) [CRC77]
Density, g/cm³: 4.23 [KIR78]

740
Compound: Cesium cyanide
Formula: CsCN
Molecular Formula: CCsN

Molecular Weight: 158.923
CAS RN: 21159-32-0
Properties: white cryst, has odor of HCN; very hygr [KIR78]
Solubility: v s H_2O [KIR78]
Density, g/cm³: 2.93 [CRC77]
Melting Point, °C: 350 [LID94]

741
Compound: Cesium fluoride
Formula: CsF
Molecular Formula: CsF
Molecular Weight: 151.903
CAS RN: 13400-13-0
Properties: -4 mesh of 99.9% purity; enthalpy of fusion 21.70 kJ/mol; hygr white powd; used in optics specialty glasses [HAW93] [STR93] [JAN85] [CER91]
Solubility: 366.5 parts CsF dissolves in 100 parts H_2O (18°C) [KIR79]; s methanol; i dioxane, pyridine [HAW93]
Density, g/cm³: 4.115 [HAW93]
Melting Point, °C: 703 [JAN71]
Boiling Point, °C: 1251 [HAW93]

742
Compound: Cesium fluoroborate
Formula: $CsBF_4$
Molecular Formula: $BCsF_4$
Molecular Weight: 219.710
CAS RN: 18909-69-8
Properties: white; ortho-rhomb below 140°C, a = 0.7647 nm, b = 0.9675 nm, c = 0.5885 nm; cub above 140°C [KIR78]
Solubility: 1.6 g/100mL H_2O (17°C), 30 g/100mL H_2O (100°C) [KIR78]
Density, g/cm³: 3.20 [KIR78]
Melting Point, °C: 555 decomposes [KIR78]

743
Compound: Cesium hexafluorogermanate
Formula: Cs_2GeF_6
Molecular Formula: Cs_2F_6Ge
Molecular Weight: 452.411
Properties: white cryst solid [HAW93]
Solubility: sl s cold H_2O; v s hot H_2O; sl s acids [HAW93]
Density, g/cm³: 4.10 [HAW93]
Melting Point, °C: ~675 [HAW93]

744
Compound: Cesium hydrogen carbonate
Formula: $CsHCO_3$
Molecular Formula: $CHCsO_3$
Molecular Weight: 193.92
CAS RN: 29703-01-3
Properties: rhomb white powd [STR93] [CRC77]
Solubility: 209.3 g/100 mL (15°C) [CRC77]
Reactions: minus $1/2H_2O$ at 175°C [CRC77]

745
Compound: Cesium hydroxide
Formula: CsOH
Molecular Formula: CsHO
Molecular Weight: 149.912
CAS RN: 21351-79-1
Properties: white or yellowish; fused; very deliq cryst mass; readily absorbs atm CO_2; strongest known base; preparation: by the electrolysis of Cs salts; uses: battery electrolyte, catalyst [KIR79] [MER89]
Solubility: s in about 0.25 parts H_2O, with evolution of heat [MER89]; g/100g soln, H_2O: 75.18 (30°C) [KRU93]
Density, g/cm³: 3.68 [MER89]
Melting Point, °C: 272 [MER89]

746
Compound: Cesium hydroxide monohydrate
Formula: $CsOH·H_2O$
Molecular Formula: CsH_3O_2
Molecular Weight: 167.928
CAS RN: 35103-79-8
Properties: -4 mesh with 99.9% purity, contains up to 10% Cs_2CO_3; cryst; contains 15 to 20% H_2O [STR93] [CER91]
Density, g/cm³: 3.675 [STR93]
Melting Point, °C: 272 [ALD94]

747
Compound: Cesium iodate
Formula: $CsIO_3$
Molecular Formula: $CsIO_3$
Molecular Weight: 307.807
CAS RN: 13454-81-4
Properties: white, monocl; -4 mesh with 99.9% purity [CER91] [CRC77]
Solubility: 2.6 g/100 mL (24°C) [CRC77]
Density, g/cm³: 4.85 [CRC77]

748
Compound: Cesium iodide
Formula: CsI
Molecular Formula: CsI
Molecular Weight: 259.809
CAS RN: 7789-17-5
Properties: -20 mesh with 99.9% purity; white; deliq cryst or cryst powd; used as an optical material in infrared spectrophotometers and in scintillation counters [MER89] [STR93] [CER91]
Solubility: s ethanol; sl s methanol; i acetone [MER89]; g/100 g soln, H_2O: 30.6 (0°C), 46.5 (25°C), 70.2 (102.8°); Solid phase, CsI [KRU93]
Density, g/cm³: 4.5 [MER89]
Melting Point, °C: 621 [MER89]
Boiling Point, °C: ~1280 [MER89]

749
Compound: Cesium metavanadate
Formula: $CsVO_3$
Molecular Formula: CsO_3V
Molecular Weight: 231.845
CAS RN: 14644-55-4
Properties: -100 mesh with 99.9% purity [CER91]

750
Compound: Cesium molybdate
Formula: Cs_2MoO_4
Molecular Formula: Cs_2MoO_4
Molecular Weight: 425.749
CAS RN: 13597-64-3
Properties: -200 mesh with 99.9% purity; white [KIR81] [CER91]
Solubility: 67.07 g/100g H_2O (18°C) [KIR81]
Melting Point, °C: 936 [KIR81]

751
Compound: Cesium niobate
Formula: $CsNbO_3$
Molecular Formula: $CsNbO_3$
Molecular Weight: 273.809
CAS RN: 12053-66-6
Properties: -200 mesh with 99.9% purity [CER91]

752
Compound: Cesium nitrate
Formula: $CsNO_3$
Molecular Formula: $CsNO_3$
Molecular Weight: 194.910

CAS RN: 7789-18-6
Properties: -4 mesh with 99.9% purity; white, lustrous hex or cub prisms; preparation: from pollucite (cesium aluminum silicate); uses: preparation of other cesium salts [MER89] [CER91] [HAW93]
Solubility: s acetone [MER89]; g/100g soln, H_2O: 8.54 (0°C), 21.53 (25°C), 66.3 (100°C) [KRU93]
Density, g/cm³: 3.64-3.68 [MER89]
Melting Point, °C: 414 [MER89]
Boiling Point, °C: decomposes >414 [MER89]

753
Compound: Cesium orthovanadate
Formula: Cs_3VO_4
Molecular Formula: Cs_3O_4V
Molecular Weight: 513.656
CAS RN: 34283-69-7
Properties: -100 mesh with 99.9% purity [CER91]

754
Compound: Cesium oxide
Formula: Cs_2O
Molecular Formula: Cs_2O
Molecular Weight: 281.810
CAS RN: 20281-00-9
Properties: 6 mm pieces and smaller with 99% purity; yellow-brown powd; lemon yellow at -80°C, reddish orange cryst at room temp, cherry red >180°C [KIR79] [HAW93] [STR93] [CER91]
Solubility: v s H_2O; s acids [HAW93]
Density, g/cm³: 4.25 [STR93]
Melting Point, °C: 490 [STR93]

755
Compound: Cesium perchlorate
Formula: $CsClO_4$
Molecular Formula: $ClCsO_4$
Molecular Weight: 232.356
CAS RN: 13454-84-7
Properties: -4 mesh with 99.9% purity; white cryst; hygr; oxidizing agent [HAW93] [STR93] [CER91]
Solubility: g/100 g H_2O: 0.8 (0°C), 2.0 (25°C), 30.0 (100°C) [KRU93]
Density, g/cm³: 3.327 [STR93]
Melting Point, °C: 250 [STR93]
Reactions: decomposes to CsCl at 575°C [KIR79]

756
Compound: Cesium pyrovanadate
Formula: $Cs_4V_2O_7$
Molecular Formula: $Cs_4O_7V_2$
Molecular Weight: 745.501
CAS RN: 55343-67-4
Properties: -100 mesh with 99.9% purity [CER91]

757
Compound: Cesium rubidium fullerene
Formula: Cs_2RbC_{60}
Molecular Formula: $C_{60}Cs_2Rb$
Molecular Weight: 1071.939
Properties: fcc; lattice parameter 1.4493 nm; superconductor, T_c 33 K [CEN92] [DRE93]

758
Compound: Cesium sulfate
Formula: Cs_2SO_4
Molecular Formula: Cs_2O_4S
Molecular Weight: 361.875
CAS RN: 10294-54-9
Properties: -20 mesh with 99.9% purity; white hygr cryst; ortho-rhomb or hex prisms; enthalpy of fusion 35.70 kJ/mol [CRC93] [MER89] [STR93] [CER91]
Solubility: v s H_2O; i alcohol, acetone, pyridine [MER89]; g/100g soln, H_2O: 62.6 (0°C), 64.5 (25°C), 68.8 (100°C) [KRU93]
Density, g/cm³: 4.24 [MER89]
Melting Point, °C: 1005 [CRC93]

759
Compound: Cesium sulfide
Formula: Cs_2S
Molecular Formula: Cs_2S
Molecular Weight: 297.877
CAS RN: 12214-16-3
Properties: -100 mesh with 99.9% purity; tetrahydrate: white cryst, hygr [CRC77] [CER91]
Solubility: tetrahydrate: v s H_2O [CRC77]

760
Compound: Cesium superoxide
Formula: CsO_2
Molecular Formula: CsO_2
Molecular Weight: 164.904
CAS RN: 12018-61-0

Properties: bright yellow cryst; oxidizing agent; preparation: by reaction of Cs metal with O_2 at 330°C [KIR79]
Density, g/cm³: 3.77 [LID94]
Melting Point, °C: 432 [LID94]
Reactions: forms Cs_2O_2 on heating at 280 to 360°C [KIR79]

761
Compound: Cesium tantalate
Formula: $CsTaO_3$
Molecular Formula: CsO_3Ta
Molecular Weight: 361.851
CAS RN: 12158-56-4
Properties: -200 mesh with 99.9% purity reacted product [CER91]

762
Compound: Cesium titanate
Formula: Cs_2TiO_3
Molecular Formula: Cs_2O_3Ti
Molecular Weight: 361.676
CAS RN: 51222-65-2
Properties: reacted product, -200 mesh with 99.9% purity [CER91]

763
Compound: Cesium trifluoroacetate
Synonyms: trifluoroacetic acid, cesium salt
Formula: CF_3COOCs
Molecular Formula: $C_2CsF_3O_2$
Molecular Weight: 245.921
CAS RN: 21907-50-6
Properties: hygr; uses: biochemical for detection of DNA-DNA crosslinks, and isolation of proteoglycans [ALD94]
Melting Point, °C: 114-116 [ALD94]

764
Compound: Cesium trioxide
Formula: Cs_2O_3
Molecular Formula: Cs_2O_3
Molecular Weight: 313.809
CAS RN: 12134-22-4
Properties: chocolate brown cryst [HAW93]
Solubility: decomposed by H_2O [HAW93]
Density, g/cm³: 4.25 [HAW93]
Melting Point, °C: 400 [HAW93]

765

Compound: Cesium tungstate
Formula: Cs_2WO_4
Molecular Formula: Cs_2O_4W
Molecular Weight: 513.649
CAS RN: 52350-17-1
Properties: -200 mesh with 99.9% purity [CER91]

766

Compound: Cesium zirconate
Formula: Cs_2ZrO_3
Molecular Formula: Cs_2O_3Zr
Molecular Weight: 405.033
CAS RN: 51222-66-3
Properties: reacted product, -200 mesh with 99.9% purity [CER91]

767

Compound: Chloric acid heptahydrate
Formula: $HClO_3 \cdot 7H_2O$
Molecular Formula: $ClH_{15}O_{10}$
Molecular Weight: 210.566
CAS RN: 7790-93-4
Properties: can occur only in aq solution; oxidizing agent; preparation: reaction between H_2SO_4 and barium chlorate; used as a catalyst in the polymerization of acrylonitrile, as an oxidizing agent [MER89] [HAW93]
Solubility: v s H_2O [CRC77]
Density, g/cm³: 1.282 [CRC77]
Melting Point, °C: <-20 [CRC77]
Reactions: decomposes, 40°C [CRC77]

768

Compound: Chlorine
Formula: Cl_2
Molecular Formula: Cl_2
Molecular Weight: 70.906 (atomic weight 35.4527)
CAS RN: 7782-50-5
Properties: greenish yellow diatomic gas; suffocating odor; oxidizing agent; critical temp 144.0°C; critical pressure 78.525 atm; critical density 573 g/L; critical volume 1.763 L/kg; viscosity of gas 14.0 µPa·s at 20°C; specific volume 0.34 m³/kg (21.1°C); enthalpy of vaporization 20.41 kJ/mol; enthalpy of fusion 6.40 kJ/mol; strongly electronegative [HAW93] [AIR87] [MER89] [KIR78] [CRC93]
Solubility: 0.64 g/100g H_2O [HAW93]; 0.062 mol/L H_2O (25°C) [MER89]
Density, g/cm³: 3.209 g/L (0°C) [KIR78]
Melting Point, °C: -101.5 [CRC93]

Boiling Point, °C: -34.05 [MER89]
Thermal Conductivity, W/(m·K): 0.0089 at 25°C [ALD94]

769

Compound: Chlorine dioxide
Formula: ClO_2
Molecular Formula: ClO_2
Molecular Weight: 67.452
CAS RN: 10049-04-4
Properties: strongly oxidizing, yellow to reddish yellow gas at room temp; unstable in light; reacts violently with organic material; vapor pressure at mp is 1.3 kPa; enthalpy of vaporization at bp 30 kJ/mol [MER89] [CRC93]
Solubility: 3.01 g/L H_2O (25°C) at 34.5mm Hg [MER89]; g/100g H_2O: 2.76 (0°C), 6.00 (10°C), 8.70 (15°C) [LAN85]
Density, g/cm³: 1.62 (11°C); liq: 1.765 (-59°C) [KIR78]
Melting Point, °C: -59 [KIR78]
Boiling Point, °C: 11 [MER89]

770

Compound: Chlorine heptoxide
Formula: Cl_2O_7
Molecular Formula: Cl_2O_7
Molecular Weight: 182.901
CAS RN: 10294-48-1
Properties: colorless, very volatile oily liq; explodes violently upon concussion, or when in contact with iodine or a flame; preparation: dehydration of $HClO_4$ with phosphorus pentoxide; uses: catalyst [KIR78] [MER89]
Solubility: slowly hydrolyzed in H_2O to form $HClO_4$ [MER89]
Density, g/cm³: 1.86 [MER89]
Melting Point, °C: -91.5 [MER89]
Boiling Point, °C: 82 [MER89]
Reactions: decomposes with evolution of Cl_2 and O_2 at 0.2 to 10.7 kPa pressures, and at temperatures from 100 to 120°C [KIR78]

771

Compound: Chlorine monofluoride
Formula: ClF
Molecular Formula: ClF
Molecular Weight: 54.451
CAS RN: 7790-89-8

Properties: colorless gas; sl yellow when liq; enthalpy of vaporization 24 kJ/mol; specific conductivity 1.9 x 10^{-7} ohm cm; destroys glass instantly, attacks quartz readily in presence of moisture; organic matter bursts into flame instantly on contact [MER89] [CRC93]

Solubility: violent reaction with H_2O [MER89]

Density, g/cm³: gas: 2.389 g/L [LID94]; liq (-108°C), 1.67 [MER89]

Melting Point, °C: -155.6 [MER89]

Boiling Point, °C: -101.1 [CRC93]

772

Compound: Chlorine monoxide

Formula: Cl_2O

Molecular Formula: Cl_2O

Molecular Weight: 80.905

CAS RN: 7791-21-1

Properties: yellowish brown gas; disagreeable penetrating odor; explodes on contact with organic matter; decomposes at a moderate rate at room temp; anhydride of hypochlorous acid; Henry's constant at 3.46°C is 14.23 kPa/(molarity); enthalpy of vaporization 25.9 kJ/mol; can be prepared by reacting Cl_2 with HgO; used as an intermediate in manufacture of calcium hypochlorite and in sterilization; reacts with a variety of organic compounds [MER89] [KIR78]

Solubility: 1 volume of water dissolves more than 100 volumes (0°C); saturation solubility is 143.6 g/100g H_2O (-9.4°C) [MER89] [KIR78]

Density, g/cm³: 3.813 g/L [LID94]

Melting Point, °C: -116 [COT88]

Boiling Point, °C: 2.2 [CRC93]

Reactions: forms HClO in water [MER89]

773

Compound: Chlorine pentafluoride

Formula: ClF_5

Molecular Formula: ClF_5

Molecular Weight: 130.445

CAS RN: 13637-63-3

Properties: colorless gas; critical temp 142.6°C; enthalpy of vaporization 22.21 kJ/mol; specific conductivity 1.25 x 10^{-9} ohm·cm [KIR78]

Density, g/cm³: 5.724 g/L [LID94]

Melting Point, °C: -103 [KIR78]

Boiling Point, °C: -13.1 [KIR78]

774

Compound: Chlorine trifluoride

Formula: ClF_3

Molecular Formula: ClF_3

Molecular Weight: 92.450

CAS RN: 7790-91-2

Properties: corrosive colorless gas or pale yellow liq; somewhat sweet, suffocating odor; extremely reactive; critical temp 154.5°C; enthalpy of vaporization 27.53 kJ/mol; specific conductivity 4.9 x 10^{-9} ohm cm; prepared by reaction of F_2 and Cl_2; used as a fluorinating agent for nuclear reactor fuels, rocket igniter and propellant, and pyrolysis inhibitor for fluoropolymers [KIR78] [MER89] [CRC93]

Solubility: violently hydrolyzed by H_2O [MER89]

Density, g/cm³: 1.825 (at bp) [KIR78]

Melting Point, °C: -76.34 [MER89]

Boiling Point, °C: 11.75 [MER89]

775

Compound: Chlorohydridotris(triphenylphosphine) ruthenium(II)

Formula: $[(C_6H_5)_3P]_3Ru(Cl)H$

Molecular Formula: $C_{54}H_{46}ClP_3Ru$

Molecular Weight: 924.403

CAS RN: 55102-19-7

Properties: sensitive to moisture; used as highly active hydrogenation catalyst [ALD94]

Melting Point, °C: 130, decomposes [ALD94]

776

Compound: Chloropentafluoroethane

Synonyms: halocarbon-115

Formula: $C_2Cl_2F_5$

Molecular Formula: $C_2Cl_2F_5$

Molecular Weight: 189.919

CAS RN: 76-15-3

Properties: colorless nonflammable gas with ethereal odor; critical temp 80.0°C; critical pressure 66.4 MPa; enthalpy of vaporization 23.59 kJ/mol; used in electronics [AIR87]

Melting Point, °C: -106.0 [AIR87]

Boiling Point, °C: -39.1 [AIR87]

777

Compound: Chlorosilane

Formula: SiH_3Cl

Molecular Formula: ClH_3Si

Molecular Weight: 66.563

CAS RN: 13465-78-6

Properties: colorless gas; enthalpy of vaporization 21 kJ/mol; entropy of vaporization 82.8 kJ/(mol·K) [CIC73] [CRC93]

Density, g/cm³: 3.033 g/L [CRC93]

Melting Point, °C: -118 [CIC73]
Boiling Point, °C: -30.4 [CIC73]

778

Compound: Chlorosulfonic acid
Formula: $ClSO_3H$
Molecular Formula: $ClHO_3S$
Molecular Weight: 116.525
CAS RN: 7790-94-5
Properties: colorless to light yellow, fuming liq; pungent odor; used in synthetic detergents, pharmaceuticals and pesticides [HAW93]
Solubility: decomposed by H_2O to HCl and H_2SO_4, decomposed by alcohol and acids [HAW93]
Density, g/cm³: 1.76-1.77 [HAW93]
Melting Point, °C: -80 [HAW93]
Boiling Point, °C: 158 [HAW93]

779

Compound: Chromium
Formula: Cr
Molecular Formula: Cr
Molecular Weight: 51.9961
CAS RN: 7440-47-3
Properties: bluish white refractory metal; bcc; a = 0.2844 to 0.2848 nm; enthalpy of fusion 21.00 kJ/mol; enthalpy of vaporization (2680°C) 320.6 kJ/mol; specific heat 23.9 kJ/(mol·K); electrical resistivity (20°C) 0.129 μohm·m; elastic modulus 250 GPa [KIR78] [CRC93]
Solubility: reacts with dil HCl, H_2SO_4 [MER89]
Density, g/cm³: 7.19 (20°C) [KIR78]
Melting Point, °C: 1857 [COT88]
Boiling Point, °C: 2680 [COT88]
Thermal Conductivity, W/(m·K): 93.9 (25°C) [ALD94]
Thermal Expansion Coefficient: linear coefficient at 20°C is 6.2 x 10^{-6} [KIR78]

780

Compound: Chromium antimonide
Formula: CrSb
Molecular Formula: CrSb
Molecular Weight: 173.756
CAS RN: 12053-12-2
Properties: hex cryst; -100 mesh with 99% purity [CER91]
Density, g/cm³: 7.11 [LID94]
Melting Point, °C: 1110 [LID94]

781

Compound: Chromium arsenide
Formula: Cr_2As
Molecular Formula: $AsCr_2$
Molecular Weight: 178.914
CAS RN: 12254-85-2
Properties: tetr cryst; -60 mesh with 99% purity [LID94] [CER91]
Density, g/cm³: 7.04 [LID94]

782

Compound: Chromium boride
Formula: Cr_2B
Molecular Formula: BCr_2
Molecular Weight: 114.803
CAS RN: 12006-80-3
Properties: -325 mesh 10 microns or less with 99.5% purity; borides are generally used for wear-resistant films, and to produce semiconductor films [CER91]

783

Compound: Chromium boride
Formula: Cr_5B_3
Molecular Formula: B_3Cr_5
Molecular Weight: 292.414
CAS RN: 12007-38-4
Properties: tetr cryst; -325 mesh 10 microns or less with 99.5% purity; borides are generally used to provide wear-resistant films, and to produce semiconductor films [LID94] [CER91]
Density, g/cm³: 6.1 [LID94]
Melting Point, °C: 1900 [LID94]

784

Compound: Chromium carbide
Formula: Cr_3C_2
Molecular Formula: C_2Cr_3
Molecular Weight: 180.010
CAS RN: 12012-35-0
Properties: gray powd; ortho-rhomb cryst; microhardness 2700 kg/mm² Hg with 50 g load; used as a 99.5 % pure sputtering target to produce wear-resistant films and semiconductor films; there are two other carbides: Cr_7C_3, 12075-40-0 and $Cr_{23}C_6$, 12105-81-6 [HAW93] [STR93] [CER91]
Density, g/cm³: 6.68 [STR93]
Melting Point, °C: 1890 [STR93]
Boiling Point, °C: 3800 [STR93]
Thermal Expansion Coefficient: 10.3 x 10^{-6}/K [KIR78]

785

Compound: Chromium carbonyl
Synonyms: chromium hexacarbonyl
Formula: $Cr(CO)_6$
Molecular Formula: C_6CrO_6
Molecular Weight: 220.058
CAS RN: 13007-92-6
Properties: white cryst; ortho-rhomb; stable in air; preparation: reaction of Cr salt and CO gas in presence of Grignard reagent; uses: catalyst, gasoline additive [MER89] [KIR78] [DOU83]
Solubility: i H_2O, ether, ethanol, benzene; sl s CCl_4 [KIR78]
Density, g/cm³: 1.77 [KIR78]
Melting Point, °C: 154-155 [STR93]
Boiling Point, °C: 130 decomposes [MER89]
Reactions: sinters at 90°C, explodes at 210°C [MER89]

786

Compound: Chromium diboride
Formula: CrB_2
Molecular Formula: B_2Cr
Molecular Weight: 73.618
CAS RN: 12007-16-8
Properties: -150,+325 mesh with 99.5% purity; refractory material; high melting point, very hard, very high corrosion resistance; used as metallurgical additive, and as a sputtering target to produce films which can be wear-resistant and semiconducting [KIR78] [HAW93] [CER91]
Density, g/cm³: 5.15 [HAW93]
Melting Point, °C: 1850 [HAW93]; 2130 [KIR78]

787

Compound: Chromium disilicide
Formula: $CrSi_2$
Molecular Formula: $CrSi_2$
Molecular Weight: 108.167
CAS RN: 12018-09-6
Properties: gray powd; used as 99.99% and 99.5% pure sputtering targets to fabricate wear-resistant interconnections and gate electrodes in IC devices [STR93] [CER91]
Density, g/cm³: 4.7 [STR93]
Melting Point, °C: 1490 [STR93]

788

Compound: Chromium monoboride
Formula: CrB
Molecular Formula: BCr

Molecular Weight: 62.807
CAS RN: 12006-79-0
Properties: silvery refractory material; powd; used as a sputtering target with 99.9% purity to produce wear-resistant films and semiconductor films [STR93] [CRC77] [KIR78] [CER91]
Solubility: i H_2O [CRC77]
Density, g/cm³: 6.17 [STR93]
Melting Point, °C: 2060 [KIR78]

789

Compound: Chromium nitride
Formula: Cr_2N
Molecular Formula: Cr_2N
Molecular Weight: 117.999
CAS RN: 12053-27-9
Properties: -325 mesh 15 microns or less with 99% purity; hex, a = 0.274 nm, c = 0.445 nm [CER91] [CIC73]
Density, g/cm³: 6.8 [LID94]

790

Compound: Chromium nitride
Formula: CrN
Molecular Formula: CrN
Molecular Weight: 66.003
CAS RN: 24094-93-7
Properties: gray; fcc, a = 0.4150 nm; electrical resistivity 640 μohm·cm; microhardness 1090; not superconductive; can be prepared by reacting NH_3 with chromium halide [KIR81]
Density, g/cm³: 6.14 [KIR81]
Melting Point, °C: decomposes 1282 [COT88]
Thermal Conductivity, W/(m·K): 11.7 [KIR81]

791

Compound: Chromium phosphide
Formula: CrP
Molecular Formula: CrP
Molecular Weight: 82.970
CAS RN: 26342-61-0
Properties: grayish-black cryst; -100 mesh with 99.5% purity [CER91] [CRC77]
Solubility: i H_2O [CRC77]
Density, g/cm³: 5.7 [CRC77]

792

Compound: Chromium selenide
Formula: $CrSe$
Molecular Formula: $CrSe$
Molecular Weight: 130.956

CAS RN: 12053-13-3
Properties: -325 mesh 10 micron or less with 99.5% purity [CER91]
Density, g/cm^3: 6.1 [LID94]

793
Compound: Chromium silicide
Formula: Cr$_3$Si
Molecular Formula: Cr$_3$Si
Molecular Weight: 184.074
CAS RN: 12018-36-9
Properties: cub cryst; -150,+325 mesh; in the form of 99.99% and 99.5% pure material, used as a sputtering target to produce resistant semiconductor films, and to fabricate interconnections and gate electrodes in IC devices [CER91] [LID94]
Density, g/cm^3: 6.4 [LID94]
Thermal Expansion Coefficient: (volume): 100°C (0.191), 200°C (0.468), 400°C (1.077), 800°C (2.516), 1000°C (3.368) [CLA66]

794
Compound: Chromium(II) acetate monohydrate
Synonyms: chromous acetate monohydrate
Formula: Cr(CH$_3$COO)$_2$·H$_2$O
Molecular Formula: C$_4$H$_8$CrO$_5$
Molecular Weight: 188.101
CAS RN: 628-52-4
Properties: deep red powd or monocl cryst; easily oxidized, especially when moist, to chromic acetate; also listed as the dimer, CAS RN 14976-80-8; uses: preparation of other Cr salts, to absorb O$_2$ in gas analyses [MER89] [ALD94]
Solubility: sl s cold H$_2$O, readily s hot H$_2$O [MER89]
Density, g/cm^3: 1.79 [MER89]
Reactions: minus H$_2$O when dried over P$_2$O$_5$ at 100°C [MER89]

795
Compound: Chromium(II) bromide
Synonyms: chromous bromide
Formula: CrBr$_2$
Molecular Formula: Br$_2$Cr
Molecular Weight: 211.804
CAS RN: 10049-25-9
Properties: white; monocl cryst, becomes yellow when heated; stable in dry air, oxidizes in moist air [MER89]
Solubility: s H$_2$O, exothermal, blue soln [MER89]
Density, g/cm^3: 4.236 [MER89]

Melting Point, °C: 842 [MER89]

796
Compound: Chromium(II) chloride
Synonyms: chromous chloride
Formula: CrCl$_2$
Molecular Formula: Cl$_2$Cr
Molecular Weight: 122.901
CAS RN: 10049-05-5
Properties: -80 mesh with 99.9% purity; white cryst; tetr; has strongly reducing aq solution; enthalpy of vaporization 197 kJ/mol; enthalpy of fusion 32.20 kJ/mol; uses: to manufacture Cr metal and Cr compounds, as a polymerization catalyst [ALF95] [CRC93] [KIR78] [CER91]
Solubility: s H$_2$O, blue soln; absorbs O$_2$ [KIR78]
Density, g/cm^3: 2.93 [KIR78]
Melting Point, °C: 815 [KIR78]
Boiling Point, °C: 1300 [CRC93]

797
Compound: Chromium(II) chloride tetrahydrate
Synonyms: chromous chloride tetrahydrate
Formula: Cr(H$_2$O)$_4$Cl$_2$·4H$_2$O
Molecular Formula: Cl$_2$CrH$_{16}$O$_8$
Molecular Weight: 267.023
CAS RN: 13931-94-7
Properties: bright blue, hygr cryst; transforms to isomeric green modification >38°C [MER89]
Solubility: s H$_2$O, oxidized on standing with H$_2$(g) liberated [MER89]
Reactions: minus H$_2$O to form trihydrate at 51°C [MER89]

798
Compound: Chromium(II) fluoride
Synonyms: chromous fluoride
Formula: CrF$_2$
Molecular Formula: CrF$_2$
Molecular Weight: 89.993
CAS RN: 10049-10-2
Properties: hygr bluish-green monocl cryst, with iridescent sheen [MER89] [ALD94]
Solubility: sl s H$_2$O, s boiling HCl [MER89]
Density, g/cm^3: 3.79 [MER89]
Melting Point, °C: 894 [MER89]
Boiling Point, °C: >1300 [CRC77]
Reactions: transforms to Cr$_2$O$_3$ when heated in air [MER89]

799
Compound: Chromium(II) formate monohydrate
Synonyms: chromous formate monohydrate
Formula: $Cr(HOOC)_2 \cdot H_2O$
Molecular Formula: $C_2H_4CrO_5$
Molecular Weight: 160.047
CAS RN: 4493-37-2
Properties: red needles; preparation: reaction between $CrCl_2$ and sodium formate; used in chromium electroplating solutions and as a catalyst for organic reactions [MER89] HAW93]
Solubility: s H_2O to give blue soln [MER89]

800
Compound: Chromium(II) oxalate monohydrate
Synonyms: chromous oxalate monohydrate
Formula: $CrC_2O_4 \cdot H_2O$
Molecular Formula: $C_2H_2CrO_5$
Molecular Weight: 158.031
CAS RN: 814-90-4
Properties: yellow to yellowish green cryst powd; not appreciably oxidized by moist air [MER89]
Solubility: i cold H_2O, s hot H_2O [MER89]
Density, g/cm³: 2.468 [MER89]

801
Compound: Chromium(II) sulfate pentahydrate
Synonyms: chromous sulfate pentahydrate
Formula: $CrSO_4 \cdot 5H_2O$
Molecular Formula: $CrH_{10}O_9S$
Molecular Weight: 238.136
CAS RN: 13825-86-0
Properties: blue cryst; stable in air if dry; water solutions rapidly oxidized by air [MER89]
Solubility: s H_2O; s dil H_2SO_4, decomposed by conc H_2SO_4 [MER89]

802
Compound: Chromium(III) acetate hexahydrate
Synonyms: chromic acetate hexahydrate
Formula: $Cr(CH_3COO)_3 \cdot 6H_2O$
Molecular Formula: $C_6H_{21}CrO_{12}$
Molecular Weight: 337.222
CAS RN: 1066-30-4
Properties: bluish violet needles; solution in water is blue under incident light, red under transmitted light [MER89]
Solubility: readily s H_2O with partial hydrolysis [MER89]

803
Compound: Chromium(III) acetate hydroxide
Synonyms: basic chromic acetate
Formula: $Cr(CH_3COO)_2(OH)$
Molecular Formula: $C_4H_7CrO_5$
Molecular Weight: 187.093
CAS RN: 39430-51-8
Properties: violet powd; commercial material used as a mordant in dyeing, in tanning, and as an oxidation catalyst; formula also written as $Cr_3(CH_3COO)_7(OH)_2$ [ALD94] [MER89]
Solubility: readily s H_2O [MER89]

804
Compound: Chromium(III) acetate monohydrate
Synonyms: chromic acetate monohydrate
Formula: $Cr(CH_3COO)_3.H_2O$
Molecular Formula: $C_6H_{11}CrO_7$
Molecular Weight: 247.145
CAS RN: 1066-30-4
Properties: greenish gray powd or violet plates; used as a mordant for textiles, and to harden emulsions [HAW93] [MER89]
Solubility: sl s H_2O; i alcohol [MER89]

805
Compound: Chromium(III) acetylacetonate
Synonyms: 2,4-pentanedione, chromium(III) derivative
Formula: $Cr(CH_3COCH=C(O)CH_3)_3$
Molecular Formula: $C_{15}H_{21}CrO_6$
Molecular Weight: 349.324
CAS RN: 21679-31-2
Properties: reddish violet cryst; monocl [KIR78]

$$[CH_3-\overset{\overset{\textstyle O-}{|}}{C}=CH-\overset{\overset{\textstyle O}{\|}}{C}-CH_3]_3Cr$$

Solubility: i H_2O; s benzene [KIR78]
Density, g/cm³: 1.34 [KIR78]
Melting Point, °C: 208 [KIR78]
Boiling Point, °C: 345 [KIR78]

806
Compound: Chromium(III) basic sulfate
Synonyms: chromic basic sulfate
Formula: $Cr(OH)SO_4$
Molecular Formula: $CrHO_5S$
Molecular Weight: 165.067
CAS RN: 12336-95-7

Properties: prepared by reduction of sodium chromate in H_2SO_4 solution; used in tanning leather [KIR79]

807
Compound: Chromium(III) bromide
Synonyms: chromic bromide
Formula: $CrBr_3$
Molecular Formula: Br_3Cr
Molecular Weight: 291.708
CAS RN: 10031-25-1
Properties: black cryst; used as a catalyst for polymerization of olefins [HAW93]
Solubility: i cold H_2O, s hot H_2O [HAW93]
Density, g/cm³: 4.25 [CRC77]
Melting Point, °C: 1130 [LID94]

808
Compound: Chromium(III) bromide hexahydrate
Synonyms: chromic bromide hexahydrate
Formula: $CrBr_3 \cdot 6H_2O$
Molecular Formula: $Br_3CrH_{12}O_6$
Molecular Weight: 399.799
CAS RN: 13478-06-3
Properties: two isomeric forms: dibromotetraaquochromium bromide dihydrate: green, deliq cryst; hexaaquochromium tribromide: violet, deliq cryst [MER89]
Solubility: s H_2O; i alcohol, ether [MER89]
Density, g/cm³: 5.4 [ALF95]

809
Compound: Chromium(III) carbonate hydrate
Synonyms: chromic carbonate hydrate
Formula: $Cr_2(CO_3)_3 \cdot xH_2O$
Molecular Formula: $C_3Cr_2O_9$ (anhydrous)
Molecular Weight: 284.019 (anhydrous)
CAS RN: 29689-14-3
Properties: bluish green amorphous powd [MER89]
Solubility: i H_2O; s mineral acids [MER89]

810
Compound: Chromium(III) chloride
Synonyms: chromic chloride
Formula: $CrCl_3$
Molecular Formula: Cl_3Cr
Molecular Weight: 158.354
CAS RN: 10025-73-7
Properties: bright purple plates; hex [KIR78]
Solubility: s H_2O extremely slowly [MER89]
Density, g/cm³: 2.87 [KIR78]

Melting Point, °C: 1152 [MER89]
Boiling Point, °C: dissociates above 1300 [MER89]

811
Compound: Chromium(III) chloride hexahydrate
Synonyms: chromic chloride hexahydrate
Formula: $CrCl_3 \cdot 6H_2O$
Molecular Formula: $Cl_3CrH_{12}O_6$
Molecular Weight: 266.445
CAS RN: 10060-12-5
Properties: bright green cryst: tricl or monocl; violet cryst: rhomb; several known isomers [KIR78] [MER89]
Solubility: s H_2O, gives green or violet soln [KIR78]
Density, g/cm³: 1.835 [KIR78]
Melting Point, °C: tricl/moncl: 95; rhomb: 90 [KIR78]

812
Compound: Chromium(III) fluoride
Synonyms: chromic fluoride
Formula: CrF_3
Molecular Formula: CrF_3
Molecular Weight: 108.991
CAS RN: 7788-97-8
Properties: dark green needles [MER89]
Solubility: i H_2O; s HCl, violet color [MER89]
Density, g/cm³: 3.8 [HAW93]
Melting Point, °C: 1400 [LID94]

813
Compound: Chromium(III) fluoride tetrahydrate
Synonyms: chromic fluoride tetrahydrate
Formula: $CrF_3 \cdot 4H_2O$
Molecular Formula: $CrF_3H_8O_4$
Molecular Weight: 181.052
CAS RN: 123333-98-2
Properties: fine, green cryst [HAW93]
Solubility: i H_2O, alcohol; s HCl [HAW93]

814
Compound: Chromium(III) fluoride trihydrate
Synonyms: chromic fluoride trihydrate
Formula: $CrF_3 \cdot 3H_2O$
Molecular Formula: $CrF_3H_6O_3$
Molecular Weight: 163.037
CAS RN: 16671-27-5
Properties: green cryst from solutions of Cr or $Cr(OH)_3$ in hydrofluoric acid [MER89]
Solubility: sl s H_2O [MER89]
Density, g/cm³: 2.2 [LID94]

815
Compound: Chromium(III) hydroxide trihydrate
Synonyms: chromic hydroxide trihydrate
Formula: $Cr(OH)_3 \cdot 3H_2O$
Molecular Formula: CrH_9O_6
Molecular Weight: 157.063
CAS RN: 1308-14-1
Properties: bluish green powd [MER89]
Solubility: i H_2O; s dil mineral acids, when freshly prepared [MER89]

816
Compound: Chromium(III) iodide
Synonyms: chromic iodide
Formula: CrI_3
Molecular Formula: CrI_3
Molecular Weight: 432.709
CAS RN: 13569-75-0
Properties: black cryst; -60 mesh with 99.5% purity [CER91] [CRC77]
Density, g/cm³: 5.32 [LID94]
Melting Point, °C: decomposes, 500 [LID94]
Reactions: minus I_2 when heated in vacuum at 350°C [CRC77]

817
Compound: Chromium(III) nitrate
Synonyms: chromic nitrate
Formula: $Cr(NO_3)_3$
Molecular Formula: CrN_3O_9
Molecular Weight: 238.011
CAS RN: 13548-38-4
Properties: pale green, extremely deliq powd [MER89]
Solubility: g/100g soln, H_2O: 39.21 (5°C), 44.0±0.8 (25°C); Solid phase, $Cr(NO_3)_3 \cdot 9H_2O$ [KRU93]
Melting Point, °C: decomposes >60 [MER89]

818
Compound: Chromium(III) nitrate nonahydrate
Synonyms: chromic nitrate nonahydrate
Formula: $Cr(NO_3)_3 \cdot 9H_2O$
Molecular Formula: $CrH_{18}N_3O_{18}$
Molecular Weight: 400.148
CAS RN: 7789-02-8
Properties: greenish black, deep violet; rhomb, monocl cryst; hygr [MER89] [STR93]
Solubility: 74% H_2O (25°C) [KIR78]
Density, g/cm³: 1.80 [KIR78]
Melting Point, °C: 66.3 [KIR78]
Boiling Point, °C: decomposes above 100 [MER89]

819
Compound: Chromium(III) oxide
Synonyms: chromic oxide, eskolaite
Formula: Cr_2O_3
Molecular Formula: Cr_2O_3
Molecular Weight: 151.990
CAS RN: 1308-38-9
Properties: -325 mesh 10 micron or less, precipitated with 99.999% purity; green powd or hex cryst; cryst Cr_2O_3 is extremely hard, and will scratch quartz, zircon, topaz; enthalpy of fusion 130.00 kJ/mol; evaporated material of 99.8% purity used for hard, durable light absorption films with a medium refractive index, and as a sputtering target to produce an absorbent brown film with medium index [KIR78] [CRC93] [MER89] [CER91]
Solubility: i H_2O [KIR78]
Density, g/cm³: 5.22 [KIR78]
Melting Point, °C: 2330 [CRC93]
Boiling Point, °C: 4000 [HAW93]
Thermal Expansion Coefficient: (volume) 100°C (0.15), 200°C (0.36), 400°C (0.78), 800°C (1.68), 1200°C (2.61) [CLA66]

820
Compound: Chromium(III) perchlorate
Formula: $Cr(ClO_4)_3$
Molecular Formula: Cl_3CrO_{12}
Molecular Weight: 350.347
CAS RN: 55147-94-9
Properties: greenish blue cryst; there is also a hexahydrate [STR93] [ALD94]
Solubility: g/100g soln, H_2O: 50.99 (0°C), 57.73 (25°C); Solid phase, $Cr(ClO_4)_3 \cdot 9H_2O$ [KRU93]

821
Compound: Chromium(III) phosphate
Synonyms: chromic phosphate
Formula: $CrPO_4$
Molecular Formula: CrO_4P
Molecular Weight: 146.967
CAS RN: 7789-04-0
Properties: grayish brown to black cryst or amorphous solid; partially oxidizes to CrO_3 on heating in air [MER89]
Solubility: i H_2O, acetic acid, HCl, aqua regia [MER89]
Density, g/cm³: 2.94 [MER89]
Melting Point, °C: >1800 [MER89]

822
Compound: Chromium(III) phosphate
  hemiheptahydrate
Synonyms: chromic phosphate hemiheptahydrate,
  Arnaudon's green, Plessy's green
Formula: $CrPO_4 \cdot 3\text{-}1/2H_2O$
Molecular Formula: $CrH_7O_{7.5}P$
Molecular Weight: 210.021
CAS RN: 7789-04-0
Properties: bluish green powd [MER89]; formula
  probably also given as the tetrahydrate, see the
  reference [HAW93]
Solubility: i H_2O; s acids [MER89]
Density, g/cm^3: 2.15 [MER89]

823
Compound: Chromium(III) phosphate hexahydrate
Synonyms: chromic phosphate hexahydrate
Formula: $CrPO_4 \cdot 6H_2O$
Molecular Formula: $CrH_{12}O_{10}P$
Molecular Weight: 255.059
CAS RN: 84359-31-9
Properties: violet cryst; loses water gradually on
  heating, becomes anhydrous after 1 h at 800°C,
  or 3-4 h at 500°C [MER89]
Solubility: i H_2O; sl s acetic acid solutions [MER89]
Density, g/cm^3: 2.121 [MER89]
Melting Point, °C: decomposes >500 [LID94]

824
Compound: Chromium(III) potassium sulfate
  dodecahydrate
Synonyms: chrome alum
Formula: $CrK(SO_4)_2 \cdot 12H_2O$
Molecular Formula: $CrH_{24}KO_{20}S_2$
Molecular Weight: 499.405
CAS RN: 7788-99-0
Properties: dark reddish violet cryst; efflorescent;
  used in tanning, as a textile mordant, in
  ceramics [HAW93]
Solubility: s H_2O [HAW93]
Density, g/cm^3: 1.813 [HAW93]
Melting Point, °C: 89 [HAW93]
Reactions: minus all H_2O at 400°C [MER89]

825
Compound: Chromium(III) sulfate
Synonyms: chromic sulfate
Formula: $Cr_2(SO_4)_3$
Molecular Formula: $Cr_2O_{12}S_3$
Molecular Weight: 392.183
CAS RN: 10101-53-8

Properties: peach colored solid [MER89]
Solubility: i H_2O, acids [MER89]
Density, g/cm^3: 3.012 [MER89]

826
Compound: Chromium(III) sulfate hydrate
Synonyms: chromic sulfate hydrate
Formula: $Cr_2(SO_4)_3 \cdot xH_2O$
Molecular Formula: $Cr_2O_{12}S_3$ (anhydrous)
Molecular Weight: 392.183 (anhydrous)
CAS RN: 15244-38-9
Properties: greenish black powd; amorphous
  [STR93] [KIR78]

827
Compound: Chromium(III) sulfate octadecahydrate
Synonyms: chromic sulfate octadecahydrate
Formula: $Cr_2(SO_4)_3 \cdot 18H_2O$
Molecular Formula: $Cr_2H_{36}O_{30}S_3$
Molecular Weight: 716.458
CAS RN: 10101-53-8
Properties: violet [KIR78]
Solubility: 120 g/100 mL H_2O (20°C) [CRC77]
Density, g/cm^3: 1.7 [CRC77]
Reactions: minus $12H_2O$ on heating [CRC77]

828
Compound: Chromium(III) sulfide
Synonyms: chromic sulfide
Formula: Cr_2S_3
Molecular Formula: Cr_2S_3
Molecular Weight: 200.190
CAS RN: 12018-22-3
Properties: -200 mesh with 99% purity; brownish
  black powd [CER91] [STR93]
Solubility: decomposed in H_2O [CRC77]
Density, g/cm^3: 3.77 [STR93]
Reactions: minus sulfur at 1350°C [CRC77]

829
Compound: Chromium(III) telluride
Synonyms: chromic telluride
Formula: Cr_2Te_3
Molecular Formula: Cr_2Te_3
Molecular Weight: 486.792
CAS RN: 12053-39-3
Properties: hex cryst; -325 mesh 10 microns or less;
  with 99.5% purity [CER91] [LID94]
Density, g/cm^3: 7.0 [LID94]

830
Compound: Chromium(IV) chloride
Formula: CrCl$_4$
Molecular Formula: Cl$_4$Cr
Molecular Weight: 193.807
CAS RN: 15597-88-3
Properties: gas; stable only at high temp [LID94] [KIR78]
Boiling Point, °C: 830 [KIR78]

831
Compound: Chromium(IV) oxide
Synonyms: chromium dioxide
Formula: CrO$_2$
Molecular Formula: CrO$_2$
Molecular Weight: 83.995
CAS RN: 12018-01-8
Properties: dark brown or black powd; tetr; used in magnetic tapes [KIR78]; ferromagnetic; rutile structure [MER89]
Solubility: i H$_2$O, s acids giving Cr^{+++} and dichromate [KIR78]
Density, g/cm^3: 4.98 (calc) [KIR78]
Boiling Point, °C: decomposes to Cr$_2$O$_3$ [KIR78]
Reactions: metastable in air; decomposes to Cr$_2$O$_3$ at many reported temperatures [MER89]

832
Compound: Chromium(VI) morpholine
Synonyms: morpholine chromate
Formula: (OC$_4$H$_8$NH$_2$)$_2$CrO$_4$
Molecular Formula: C$_8$H$_{20}$CrN$_2$O$_6$
Molecular Weight: 292.252
CAS RN: 36969-05-8
Properties: yellow, oily; corrosion inhibitor [KIR78]

833
Compound: Chromium(VI) oxide
Synonyms: chromium trioxide, chromic acid, chromic anhydride
Formula: CrO$_3$
Molecular Formula: CrO$_3$
Molecular Weight: 99.994
CAS RN: 1333-82-0
Properties: ruby red cryst; ortho-rhomb; oxidizing agent; deliq [HAW93] [KIR78]
Solubility: g/100g H$_2$O: 164.8 (0°), 167.2 (20°C), 206.8 (100°C) [LAN85]; s acetic acid, H$_2$SO$_4$, ether [KIR78]
Density, g/cm^3: 2.7 [KIR78]
Melting Point, °C: 197 [KIR78]

Boiling Point, °C: decomposes at 250 to Cr$_2$O$_3$ and O$_2$ [MER89]

834
Compound: Chromyl chloride
Synonyms: chromium(VI) oxychloride
Formula: CrO$_2$Cl$_2$
Molecular Formula: Cl$_2$CrO$_2$
Molecular Weight: 154.900
CAS RN: 14977-61-8
Properties: cherry red liq; enthalpy of vaporization 35.1 kJ/mol [CRC93] [KIR78] [MER89]
Solubility: i H$_2$O, hydrolyzes; s CS$_2$, CCl$_4$ [KIR78]
Density, g/cm^3: 1.91 [KIR78]
Melting Point, °C: -96.5 [KIR78]
Boiling Point, °C: 117 [ALD94]

835
Compound: Cobalt
Formula: Co
Molecular Formula: Co
Molecular Weight: 58.93320
CAS RN: 7440-48-4
Properties: gray, hard, magnetic, ductile, somewhat malleable metal; two allotropic forms, hex and cub, both can exist at room temp; transformation temp 417°C; enthalpy of transformation 14.79 kJ/mol; enthalpy of fusion 16.20 kJ/mol; enthalpy of vaporization 369.86 kJ/mol; electrical resistivity (20°C) 6.24 μohm cm; Young's modulus 211 GPa; Poisson's ratio 0.32; Curie temp 1121°C; stable in air or water at ordinary temp [KIR79] [MER89] [CRC93]
Solubility: s dil HNO$_3$, very slowly attacked by cold H$_2$SO$_4$ or HCl [MER89]
Density, g/cm^3: 8.92 [MER89]
Melting Point, °C: 1495 [CRC93]
Boiling Point, °C: 2927 [CRC93]
Thermal Conductivity, W/(m·K): 100 (25°C) [ALD94]
Thermal Expansion Coefficient: coefficient of thermal expansion 12.5 x 10^{-6}/°C at room temp (hex); 14.2 x 10^{-6}/°C (cub) at 417°C [KIR78]

836
Compound: Cobalt aluminate
Synonyms: Thenard's blue
Formula: CoAl$_2$O$_4$
Molecular Formula: Al$_2$CoO$_4$
Molecular Weight: 176.894
CAS RN: 13820-62-7

Properties: blue cub solid; used in nickel and cobalt alloys [KIR79]
Solubility: i H_2O [KIR79]

837
Compound: Cobalt antimonide
Formula: CoSb
Molecular Formula: CoSb
Molecular Weight: 180.693
CAS RN: 12052-42-5
Properties: hex cryst; 6 mm pieces and smaller with 99.5% purity [CER91] [LID94]
Density, g/cm³: 8.8 [LID94]
Melting Point, °C: 1202 [LID94]

838
Compound: Cobalt arsenic sulfide
Synonyms: cobaltite
Formula: CoAsS
Molecular Formula: AsCoS
Molecular Weight: 165.921
CAS RN: 12254-82-9
Properties: silvery white to gray mineral with metallic luster; 5.5 Mohs hardness; used as an important source of cobalt and in ceramics [HAW93]
Density, g/cm³: 6-6.3 [HAW93]
Melting Point, °C: decomposes [CRC77]

839
Compound: Cobalt arsenide
Formula: CoAs
Molecular Formula: AsCo
Molecular Weight: 133.855
CAS RN: 27016-73-5
Properties: ortho-rhomb cryst; -10 mesh with 99.5% purity [CER91] [LID94]
Density, g/cm³: 8.22 [LID94]
Melting Point, °C: 1180 [LID94]

840
Compound: Cobalt arsenide
Synonyms: skutterudite
Formula: CoAs₃
Molecular Formula: As₃Co
Molecular Weight: 283.698
CAS RN: 12196-91-7
Properties: mineral, cub, hardness 6.0 Mohs; 6 mm pieces and smaller [CER91] [KIR79]
Density, g/cm³: 6.5 [KIR79]
Melting Point, °C: 942 [LID94]

841
Compound: Cobalt arsenide
Formula: CoAs₂
Molecular Formula: As₂Co
Molecular Weight: 208.776
CAS RN: 12044-42-7
Properties: cub mineral, also ortho-rhomb; hardness: cub 6.0 Mohs, ortho-rhomb 5.0 Mohs [KIR79]
Density, g/cm³: cub: 6.5; ortho-rhomb: 7.2 [KIR79]

842
Compound: Cobalt boride
Formula: CoB
Molecular Formula: BCo
Molecular Weight: 69.744
CAS RN: 12006-77-8
Properties: refractory material; cryst; used in ceramics [HAW93]
Solubility: decomposed by H_2O; s HNO_3 [HAW93]
Density, g/cm³: 7.25 [HAW93]
Melting Point, °C: 1460 [KIR78]

843
Compound: Cobalt boride
Formula: Co₂B
Molecular Formula: BCo₂
Molecular Weight: 128.677
CAS RN: 12045-01-1
Properties: refractory material; there is also a Co₃B, 12006-78-9 [KIR78]
Density, g/cm³: 8.1 [LID94]
Melting Point, °C: Co₂B: 1285; Co₃B: 1125, decomposes [KIR78]

844
Compound: Cobalt carbonyl
Synonyms: dicobalt octacarbonyl
Formula: Co₂(CO)₈
Molecular Formula: C₈Co₂O₈
Molecular Weight: 341.949
CAS RN: 10210-68-1
Properties: dark orange cryst; stabilized with 5-10% hexane; sensitive to oxidation; freezing point of stabilized compound is -22.8°C; used as catalyst in Oxo process [HAW93] [STR93]
Solubility: i H_2O; s alcohol, ether, carbon disulfide [HAW93]
Density, g/cm³: 1.78 [HAW93]
Melting Point, °C: 51-52, decomposes [STR93]

845
Compound: Cobalt disilicide
Formula: $CoSi_2$
Molecular Formula: $CoSi_2$
Molecular Weight: 115.104
CAS RN: 12017-12-8
Properties: gray rhomb powd [ALF93] [KIR79]
Solubility: s hot HCl [KIR79]
Density, g/cm³: 5.3 [KIR79]
Melting Point, °C: 1277 [ALF93]

846
Compound: Cobalt disulfide
Formula: CoS_2
Molecular Formula: CoS_2
Molecular Weight: 123.065
CAS RN: 12013-10-4
Properties: black cub; -200 mesh with 99.5% purity [CRC94] [CER91]
Solubility: i H_2O; s HNO_3 [CRC94]
Density, g/cm³: 4.3 [LID94]
Melting Point, °C: decomposes 269 [CRC94]

847
Compound: Cobalt dodecacarbonyl
Synonyms: tetracobalt dodecacarbonyl
Formula: $Co_4(CO)_{12}$
Molecular Formula: $C_{12}Co_4O_{12}$
Molecular Weight: 571.858
CAS RN: 17786-31-1
Properties: black cryst; air sensitive [DOU83] [STR93]
Solubility: sl s H_2O [CRC77]
Density, g/cm³: 2.09 [STR93]
Melting Point, °C: 60, decomposes [STR93]

848
Compound: Cobalt metaborate hydrate
Formula: $Co(BO_2)_2 \cdot xH_2O$ (x=2,3)
Molecular Formula: B_2CoO_4 (x=0)
Molecular Weight: 144.553 (x=0)
Properties: prepartion: precipitation from an aq Co^{++} solution to which borax is added, the product is $CoO \cdot 3B_2O_3 \cdot 10H_2O$; stirring a Co^{++} solution with boric acid results in $CoB_4O_7 \cdot 4H_2O$; uses: as acid catalyst [KIR78]

849
Compound: Cobalt metatitanate
Formula: $CoTiO_3$
Molecular Formula: CoO_3Ti

Molecular Weight: 154.798
CAS RN: 12017-01-5
Properties: green; rhombohedral; -325 mesh 10 microns average or less, with 99.9% purity [CER91] [LID94] [KIR83]
Density, g/cm³: 5.0 [KIR83]

850
Compound: Cobalt molybdate
Formula: $CoMoO_4$
Molecular Formula: CoO_4Mo
Molecular Weight: 218.871
CAS RN: 13762-14-6
Properties: -325 mesh 10 microns or less with 99.9% purity; green powd; three forms: α, β, γ; obtained by reaction of the two oxides; used as a catalyst to desulfurize petroleum [KIR81] [HAW93] [STR93] [CER91]
Density, g/cm³: α: 3.6; β: 4.5; γ: 4.1 [KIR81]
Melting Point, °C: 1040 [KIR81]

851
Compound: Cobalt nitrosocarbonyl
Synonyms: cobalt tricarbonyl nitrosyl
Formula: $Co(NO)(CO)_3$
Molecular Formula: C_3CoNO_4
Molecular Weight: 172.971
CAS RN: 14096-82-3
Properties: dark red liq; air sensitive [STR93]
Solubility: i H_2O [CRC77]
Melting Point, °C: -1.05 [CRC77]
Boiling Point, °C: 50 [STR93]

852
Compound: Cobalt nitrosodicarbonyl
Formula: $Co(NO)(CO)_2$
Molecular Formula: C_2CoNO_3
Molecular Weight: 144.960
CAS RN: 12021-68-0
Properties: cherry red liq [KIR79]
Solubility: i H_2O [KIR79]

853
Compound: Cobalt orthotitanate
Formula: Co_2TiO_4
Molecular Formula: Co_2O_4Ti
Molecular Weight: 229.731
CAS RN: 12017-38-8
Properties: greenish black cub cryst [KIR79]
Solubility: s conc HCl [HAW93]
Density, g/cm³: 5.07-5.12 [HAW93]

854

Compound: Cobalt phosphide
Formula: Co_2P
Molecular Formula: Co_2P
Molecular Weight: 148.840
CAS RN: 12134-02-0
Properties: -100 mesh with 99% purity; gray
    needles [KIR79] [CER91]
Solubility: i H_2O; s HNO_3 [KIR79]
Density, g/cm³: 6.4 [KIR79]
Melting Point, °C: 1386 [KIR79]

855

Compound: Cobalt stearate
Formula: $Co[CH_3(CH_2)_{16}COO]_2$
Molecular Formula: $C_{36}H_{70}CoO_4$
Molecular Weight: 625.883
CAS RN: 1002-88-6
Properties: purple pellets; used as drying agent
    [KIR79] [STR93]
Density, g/cm³: 1.13 [KIR78]
Melting Point, °C: 140 [KIR78]

856

Compound: Cobalt zirconate
Formula: $CoZrO_3$
Molecular Formula: CoO_3Zr
Molecular Weight: 198.155
CAS RN: 39361-25-6
Properties: reacted product, -325 mesh 10 micron or
    less with 99.5% purity [CER91]

857

Compound: Cobalt(II) acetate tetrahydrate
Synonyms: cobaltous acetate tetrahydrate
Formula: $Co(C_2H_3O_2)_2 \cdot 4H_2O$
Molecular Formula: $C_4H_{14}CoO_8$
Molecular Weight: 249.083
CAS RN: 6147-53-1
Properties: pink; monocl, prismatic cryst; used as a
    drying agent for lacquers and varnishes, and as
    a catalyst [KIR79] [MER89] [STR93]
Solubility: s H_2O, alcohol, dil acids [MER89]
Density, g/cm³: 1.705 [MER89]
Reactions: minus $4H_2O$ by 140°C [MER89]

858

Compound: Cobalt(II) acetylacetonate
Synonyms: 2,4-pentanedione, cobalt(II) derivative
Formula: $Co(CH_3COCH=C(O)CH_3)_2$
Molecular Formula: $C_{10}H_{14}CoO_4$

Molecular Weight: 257.152
CAS RN: 14024-48-7
Properties: black monocl; tetramer; used for vapor
    deposition of chromium; hydrate is a pink powd
    [KIR79] [STR93] [COT88]

$$[CH_3–\overset{\overset{\displaystyle O-}{|}}{C}=CH–\overset{\overset{\displaystyle O}{\|}}{C}–CH_3]_2Co$$

Density, g/cm³: 1.43 [KIR79]
Melting Point, °C: 241 [KIR79]

859

Compound: Cobalt(II) arsenate octahydrate
Synonyms: cobaltous arsenate octahydrate
Formula: $Co_3(AsO_4)_2 \cdot 8H_2O$
Molecular Formula: $As_2Co_3H_{16}O_{16}$
Molecular Weight: 598.760
CAS RN: 24719-19-5
Properties: pink to blood red; monocl, fine needles
    [MER89]
Solubility: i H_2O; s dil mineral acids, NH_4OH
    [MER89]
Density, g/cm³: 2.9-3.1 [MER89]
Melting Point, °C: decomposes by 1000 to
    $Co_4As_2O_{11}$ [MER89]
Reactions: minus $8H_2O$ by 400°C [MER89]

860

Compound: Cobalt(II) basic carbonate
Formula: $2CoCO_3 \cdot 3Co(OH)_2 \cdot H_2O$
Molecular Formula: $C_2H_8Co_5O_{13}$
Molecular Weight: 534.744
CAS RN: 7542-09-8
Properties: form of commercial cobalt carbonate;
    reddish violet cryst; used in pigments [HAW93]
    [MER89]
Solubility: i cold H_2O, decomposed by hot H_2O; s
    acids [HAW93]
Melting Point, °C: decomposes [HAW93]

861

Compound: Cobalt(II) bromate hexahydrate
Formula: $Co(BrO_3)_2 \cdot 6H_2O$
Molecular Formula: $Br_2CoH_{12}O_{12}$
Molecular Weight: 422.829
CAS RN: 13476-01-2
Properties: violet cryst [STR93]; red octahedral
    [KIR79]
Solubility: 45.5 g/100mL (17°) H_2O; s NH_4OH
    [CRC94]

Density, g/cm^3: ~2.462 [STR93]

862
Compound: Cobalt(II) bromide
Synonyms: cobaltous bromide
Formula: CoBr$_2$
Molecular Formula: Br$_2$Co
Molecular Weight: 218.741
CAS RN: 7789-43-7
Properties: bright green solid or lustrous green cryst leaflets; hygr; forms hexahydrate in air [MER89]
Solubility: g/100g H$_2$O: 91.9 (0°C), 112 (20°C), 257 (100°C) [LAN85]; s methanol, ethanol, acetone [MER89]
Density, g/cm^3: 4.91 [MER89]
Melting Point, °C: 678 (under HBr and N$_2$) [MER89]

863
Compound: Cobalt(II) bromide hexahydrate
Synonyms: cobaltous bromide hexahydrate
Formula: CoBr$_2$·6H$_2$O
Molecular Formula: Br$_2$CoH$_{12}$O$_6$
Molecular Weight: 326.832
CAS RN: 13762-12-4
Properties: red to reddish purple; deliq, prismatic cryst [MER89] [ALD94]
Solubility: s H$_2$O to give red or blue solution, depending on temp [MER89]
Density, g/cm^3: 2.46 [MER89]
Melting Point, °C: 47-48 [MER89]
Reactions: minus 4H$_2$O at 100°C forming purple dihydrate [MER89]

864
Compound: Cobalt(II) carbonate
Synonyms: spherocobaltite
Formula: CoCO$_3$
Molecular Formula: CCoO$_3$
Molecular Weight: 118.942
CAS RN: 513-79-1
Properties: -325 mesh 10 microns or less with 99.5% purity; pink powd or rhomb cryst; there is a monohydrate, CAS RN 137506-60-6 [MER89] [STR93] [CER91] [ALD94]
Solubility: i H$_2$O, alcohol, methyl acetate [MER89]
Density, g/cm^3: 4.13 [MER89]
Melting Point, °C: decomposes [KIR79]
Reactions: oxidized by air to cobalt(III) carbonate [MER89]

865
Compound: Cobalt(II) chlorate hexahydrate
Synonyms: cobaltous chlorate hexahydrate
Formula: Co(ClO$_3$)$_2$·6H$_2$O
Molecular Formula: Cl$_2$CoH$_{12}$O$_{12}$
Molecular Weight: 333.926
Properties: red, cub, hygr [CRC77]
Solubility: mol/100 mol H$_2$O: 10.75 (0°C), 14.51 (21°C); Solid phase, Co(ClO$_3$)$_2$·6H$_2$O (0°C), Co(ClO$_3$)$_2$·4H$_2$O (21°C) [KRU93]; g/100g H$_2$O: 135 (0°C), 180 (20°C), 316 (60°C) [LAN85]
Density, g/cm^3: 1.92 [CRC77]
Melting Point, °C: 50 [CRC77]
Reactions: decomposes at 100°C [CRC77]

866
Compound: Cobalt(II) chloride
Synonyms: cobaltous chloride
Formula: CoCl$_2$
Molecular Formula: Cl$_2$Co
Molecular Weight: 129.838
CAS RN: 7646-79-9
Properties: -80 mesh with 99.9% purity; pale blue powd; hygr leaflets; turns pink in moist air; enthalpy of fusion 45.00 kJ/mol [CRC93] [MER89] [CER91]
Solubility: s H$_2$O, alcohols, acetone, ether, glycerol, acetone [MER89]; g/100g soln, H$_2$O: 30.3 (0°C), 36.0 (25°C), 51.5 (100°C); Solid phase, CoCl$_2$·6H$_2$O (0°C, 25°C), CoCl$_2$·2H$_2$O (100°C) [KRU93]
Density, g/cm^3: 3.367 [MER89]
Melting Point, °C: 740 [CRC93]
Boiling Point, °C: 1049 [MER89]
Reactions: decomposes if subjected to lengthy heating in air at 400°C, sublimes at 500°C in HCl gas [MER89]

867
Compound: Cobalt(II) chloride dihydrate
Synonyms: cobaltous chloride dihydrate
Formula: CoCl$_2$·2H$_2$O
Molecular Formula: Cl$_2$CoH$_4$O$_2$
Molecular Weight: 165.869
CAS RN: 16544-92-6
Properties: violet or blue cryst [MER89]
Solubility: s H$_2$O [CRC94]
Density, g/cm^3: 2.477 [MER89]

868
Compound: Cobalt(II) chloride hexahydrate
Synonyms: cobaltous chloride hexahydrate

Formula: $CoCl_2 \cdot 6H_2O$
Molecular Formula: $Cl_2CoH_{12}O_6$
Molecular Weight: 237.929
CAS RN: 7791-13-1
Properties: monocl cryst; pink to red, sl deliq [MER89]
Solubility: 76.7 g/100mL cold H_2O, 190.7 g/100mL hot H_2O; s alcohols, acetone, ether, glycerol [MER89]
Density, g/cm^3: 1.924 [MER89]
Melting Point, °C: 87 [MER89]
Reactions: minus $4H_2O$ at 52-56°C forming dihydrate; minus H_2O at 100°C giving monohydrate; loses remaining H_2O at 120-140°C [MER89]

869
Compound: Cobalt(II) chromate
Synonyms: cobaltous chromate
Formula: $CoCrO_4$
Molecular Formula: $CoCrO_4$
Molecular Weight: 174.927
CAS RN: 24613-38-5
Properties: brown or yellowish brown powd, but pure material is gray-black; used in ceramics [HAW93]
Solubility: i H_2O; s mineral acids [HAW93]
Density, g/cm^3: ~4.0 [LID94]
Melting Point, °C: decomposes [CRC77]

870
Compound: Cobalt(II) chromite
Synonyms: cobalt dichromium tetraoxide
Formula: $CoCr_2O_4$
Molecular Formula: $CoCr_2O_4$
Molecular Weight: 226.923
CAS RN: 12016-69-2
Properties: -200 mesh with 99.5% purity; brilliant greenish blue powd; cub spinel structure; used in pigments and catalysts [KIR79] [MER89] [CER91]
Solubility: almost i conc HCl, HNO_3 [MER89]

871
Compound: Cobalt(II) citrate dihydrate
Synonyms: citric acid, cobalt(II) salt dihydrate
Formula: $Co_3(C_6H_5O_7)_2 \cdot 2H_2O$
Molecular Formula: $C_{12}H_{14}Co_3O_{16}$
Molecular Weight: 591.033
CAS RN: 18727-04-3
Properties: rose red [KIR79]
Solubility: 0.8 g/100mL cold H_2O [KIR79]

Reactions: minus $2H_2O$ at 150°C [KIR79]

872
Compound: Cobalt(II) cyanide dihydrate
Synonyms: cobaltous cyanide
Formula: $Co(CN)_2 \cdot 2H_2O$
Molecular Formula: $C_2H_4CoN_2O_2$
Molecular Weight: 146.999
CAS RN: 20427-11-6
Properties: pink to reddish brown powd or needles [MER89]
Solubility: i H_2O, acids, methyl acetate; s alkali cyanides [MER89]
Density, g/cm^3: anhydrous: 1.872 [KIR79]
Melting Point, °C: decomposes 300 [CRC77]
Reactions: minus $2H_2O$ at 280°C [KIR79]

873
Compound: Cobalt(II) cyanide trihydrate
Synonyms: cobaltous cyanide trihydrate
Formula: $Co(CN)_2 \cdot 3H_2O$
Molecular Formula: $C_2H_6CoN_2O_3$
Molecular Weight: 165.014
CAS RN: 20427-11-6
Properties: pink to reddish brown powd or needles [MER89]
Solubility: almost i H_2O, acids; s alkali cyanide solutions [MER89]
Reactions: minus $3H_2O$ at 250°C [CRC77]

874
Compound: Cobalt(II) diiron tetroxide
Synonyms: cobaltous ferrite
Formula: $CoFe_2O_4$
Molecular Formula: $CoFe_2O_4$
Molecular Weight: 234.621
CAS RN: 12052-28-7
Properties: a magnetic ferrite [HAW93]

875
Compound: Cobalt(II) ferricyanide
Synonyms: cobaltous ferricyanide
Formula: $Co_3[Fe(CN)_6]_2$
Molecular Formula: $C_{12}Co_3Fe_2N_{12}$
Molecular Weight: 600.703
CAS RN: 15415-49-3
Properties: red needles [KIR79]
Solubility: i H_2O, HCl; s NH_4OH [KIR79]

876
Compound: Cobalt(II) ferrocyanide hydrate
Synonyms: cobaltous ferrocyanide
Formula: $Co_2[Fe(CN)_6] \cdot xH_2O$
Molecular Formula: $C_6Co_2FeN_6$ (anhydrous)
Molecular Weight: 329.819 (anhydrous)
CAS RN: 4049-81-1
Properties: gray green [KIR79]
Solubility: i H_2O, HCl; s KCN [KIR79]

877
Compound: Cobalt(II) fluoride
Synonyms: cobaltous fluoride
Formula: CoF_2
Molecular Formula: CoF_2
Molecular Weight: 96.930
CAS RN: 10026-17-2
Properties: rosy red tetr cryst; pink powd; forms di-, tri- and tetrahydrates, all soluble in H_2O; enthalpy of fusion 59.00 kJ/mol; can be prepared by reacting $CoCO_3$ with anhydrous HF; finds use in the manufacture of CoF_3 [MER89] [STR93] [KIR78] [CRC93]
Solubility: sl s H_2O; readily s warm mineral acids [MER89]; g/100ml soln, H_2O: 1.415 (25°C); Solid phase, $CoF_2 \cdot 4H_2O$ [KRU93]
Density, g/cm³: 4.46 [HAW93]
Melting Point, °C: 1127 [CRC93]
Boiling Point, °C: volatilizes ~1400 [MER89]

878
Compound: Cobalt(II) fluoride tetrahydrate
Synonyms: cobaltous fluoride tetrahydrate
Formula: $CoF_2 \cdot 4H_2O$
Molecular Formula: $CoF_2H_8O_4$
Molecular Weight: 168.992
CAS RN: 13817-37-3
Properties: red ortho-rhomb; hygr; senstive to moisture [LID94] [ALD94]
Solubility: s H_2O [MER89]
Density, g/cm³: 2.19 [ALD94]
Boiling Point, °C: decomposes [MER89]

879
Compound: Cobalt(II) hexafluorosilicate hexahydrate
Formula: $CoSiF_6 \cdot 6H_2O$
Molecular Formula: $CoF_6H_{12}O_6Si$
Molecular Weight: 309.100
CAS RN: 15415-49-3
Properties: pale red cryst; used in ceramics [HAW93]

Solubility: 118.1 g/100mL cold H_2O [KIR79]
Density, g/cm³: 2.087 [HAW93]

880
Compound: Cobalt(II) hydroxide
Synonyms: cobaltous hydroxide
Formula: $Co(OH)_2$
Molecular Formula: CoH_2O_2
Molecular Weight: 92.948
CAS RN: 21041-93-0
Properties: blue green cryst, or rose-red powd, or microscopic rhomb; red form most stable of two; easily oxidized by air to $Co(OH)_3$; amphoteric [MER89]
Solubility: v sl s H_2O; readily s in acids; almost i alkalies [MER89]; mol/L soln, H_2O: 2×10^{-5} (25°C) [KRU93]
Density, g/cm³: 3.597 [HAW93]
Melting Point, °C: decomposes [KIR79]
Reactions: minus H_2O on heating in vacuum [MER89]

881
Compound: Cobalt(II) iodate
Formula: $Co(IO_3)_2$
Molecular Formula: CoI_2O_6
Molecular Weight: 408.738
CAS RN: 13455-28-2
Properties: black violet needles [KIR79]
Solubility: mol/L soln, H_2O: 0.040 (30°C), 0.031 (100°C); Solid phase, $Co(IO_3)_2$ [KRU93]; g/100g H_2O: 1.02 (20°C), 0.88 (40°C), 0.70 (100°C) [LAN85]
Density, g/cm³: 5.08 [KIR79]
Melting Point, °C: decomposes 200 [KIR79]

882
Compound: Cobalt(II) iodide
Synonyms: cobaltous iodide
Formula: CoI_2
Molecular Formula: CoI_2
Molecular Weight: 312.742
CAS RN: 15238-00-3
Properties: -60 mesh with 99.5% purity; two isomorphous forms: α-CoI_2 is black, graphite like solid; very hygr; blackish green on exposure to air; β-CoI_2 is ochre yellow powd; blackens at 400°C; very hygr; deliq in moist air forming green droplets [MER89] [CER91]

Solubility: g/100g soln in H_2O: 58.0 (0°C), 67.0 (25°C), 80.7 (100°C); Solid phase: $CoI_2 \cdot H_2O$ (green) (0,25°C), $CoI_2 \cdot H_2O$ (yellow) (100°C) [KRU93]

Density, g/cm³: α: 5.584; β: 5.45 [MER89]

Melting Point, °C: α: 515-520 (high vacuum); β: blackens at 400 [MER89]

Reactions: transition β to α at 400°C [MER89]

883

Compound: Cobalt(II) iodide dihydrate

Formula: $CoI_2 \cdot 2H_2O$

Molecular Formula: $CoH_4I_2O_2$

Molecular Weight: 348.773

CAS RN: 13455-29-3

Properties: green cryst; deliq [CRC94] [STR93]

Solubility: 376.2 g/100mL H_2O (45°C) [CRC94]

Melting Point, °C: 100 decomposes [STR93]

884

Compound: Cobalt(II) iodide hexahydrate

Synonyms: cobaltous iodide hexahydrate

Formula: $CoI_2 \cdot 6H_2O$

Molecular Formula: $CoH_{12}I_2O_6$

Molecular Weight: 420.813

CAS RN: 13455-29-3

Properties: dark red; hex prisms; loses iodine when exposed to light and air [MER89]

Solubility: s H_2O to give red solution below 20°C, olive green color 20-40°C [MER89]

Density, g/cm³: 2.90 [MER89]

Reactions: minus $6H_2O$ by 130°C [MER89]

885

Compound: Cobalt(II) linoleate

Formula: $Co(C_{18}H_{31}O_2)_2$

Molecular Formula: $C_{36}H_{62}CoO_4$

Molecular Weight: 617.819

CAS RN: 14666-96-7

Properties: brown amorphous powd; used as a drier for paint and varnish, especially for enamels and white paints [HAW93]

Solubility: i H_2O; s alcohol, ether and acids [HAW93]

886

Compound: Cobalt(II) molybdate monohydrate

Formula: $CoMoO_4 \cdot H_2O$

Molecular Formula: CoH_2MoO_5

Molecular Weight: 236.886

CAS RN: 18601-87-1

Properties: black powd [STR93]

887

Compound: Cobalt(II) nitrate

Synonyms: cobaltous nitrate

Formula: $Co(NO_3)_2$

Molecular Formula: CoN_2O_6

Molecular Weight: 182.942

CAS RN: 10141-05-6

Properties: pale red powd [MER89]

Solubility: g/100g soln, H_2O: 45.66 (0°), 50.5±0.2 (25°C), 77.21 (91°C); Solid phase, $Co(NO_3)_2 \cdot 6H_2O$ (0°C, 25°C), $Co(NO_3)_2 \cdot 3H_2O$ (91°C) [KRU93]

Density, g/cm³: 2.49 [MER89]

Melting Point, °C: decomposes 100-105 [MER898]

888

Compound: Cobalt(II) nitrate hexahydrate

Synonyms: cobaltous nitrate hexahydrate

Formula: $Co(NO_3)_2 \cdot 6H_2O$

Molecular Formula: $CoH_{12}N_2O_{12}$

Molecular Weight: 291.034

CAS RN: 10026-22-9

Properties: red monocl cryst; deliq [MER89]

Solubility: 133.8 g/100mL H_2O (0°C), 0.217 g/100mL H_2O (80°C) [CRC94]; s alcohol, most organic solvents [MER89]

Density, g/cm³: 1.88 [MER89]

Melting Point, °C: 55-56 [STR93]

Reactions: decomposes to oxide >74°C [MER89]

889

Compound: Cobalt(II) nitrite

Formula: $Co(NO_2)_2$

Molecular Formula: CoN_2O_4

Molecular Weight: 150.944

CAS RN: 18488-96-5

Solubility: g/100g soln, H_2O: 0.076 (0°C), 0.49 (25°C) [KRU93]

890

Compound: Cobalt(II) oleate

Formula: $Co(C_{18}H_{33}O_2)_2$

Molecular Formula: $C_{36}H_{66}CoO_4$

Molecular Weight: 621.851

CAS RN: 14666-94-5

Properties: brown amorphous powd; used in driers for paints and varnishes [HAW93]

Solubility: i H_2O; s alcohol, ether [HAW93]

Melting Point, °C: 235 [HAW93]

891
Compound: Cobalt(II) oxalate
Synonyms: cobaltous oxalate
Formula: CoC_2O_4
Molecular Formula: C_2CoO_4
Molecular Weight: 146.953
CAS RN: 814-89-1
Properties: pink powd; readily absorbs moisture from air to form hydrates [MER89] [STR93]
Solubility: i H_2O; s acids, NH_4OH [KIR79]
Density, g/cm³: 3.021 [MER89]
Melting Point, °C: 250 decomposes [KIR79]

892
Compound: Cobalt(II) oxalate dihydrate
Synonyms: cobaltous oxalate dihydrate
Formula: $CoC_2O_4 \cdot 2H_2O$
Molecular Formula: $C_2H_4CoO_6$
Molecular Weight: 182.984
CAS RN: 5965-38-8
Properties: light pink; microcryst powd or needles; decomposes on heating with KOH or Na_2CO_3 aq solutions [MER89]
Solubility: i H_2O; sl s acids; freely s aq ammonia [MER89]; g/L soln, H_2O: 0.0346 (25°C) [KRU93]
Reactions: minus H_2O ~190°C [CRC77]

893
Compound: Cobalt(II) oxide
Synonyms: cobaltous oxide
Formula: CoO
Molecular Formula: CoO
Molecular Weight: 74.932
CAS RN: 1307-96-6
Properties: -325 mesh 10 microns or less with 99.5% purity; powd or cub or hex cryst; color varies from olive green to red depending on particle size; commercial material usually dark gray; readily absorbs O_2, even at room temp [MER89] [CER91]
Solubility: i H_2O; s acids or alkalis [MER89]
Density, g/cm³: 5.7 to 6.7 [MER89]
Melting Point, °C: 1805 [JAN85]

894
Compound: Cobalt(II) perchlorate
Formula: $Co(ClO_4)_2$
Molecular Formula: Cl_2CoO_8
Molecular Weight: 257.833
CAS RN: 13455-31-7

Properties: red needles; used as a chemical reagent, oxidizing agent [HAW93]
Solubility: g/100g H_2O: 100.0 (0°C), 113.4 (26°C); Solid phase, $Co(ClO_4)_2 \cdot 5H_2O$ [KRU93]; i alcohol, acetone [KIR79]
Density, g/cm³: 3.327 [HAW93]

895
Compound: Cobalt(II) perchlorate hexahydrate
Formula: $Co(ClO_4)_2 \cdot 6H_2O$
Molecular Formula: $Cl_2CoH_{12}O_{14}$
Molecular Weight: 365.925
CAS RN: 13478-33-6
Properties: red cryst; hygr [STR93]
Solubility: 225 g/100mL H_2O (18°C) [CRC94]
Melting Point, °C: 1534 decomposes [CRC94]

896
Compound: Cobalt(II) phosphate octahydrate
Synonyms: cobaltous phosphate octahydrate
Formula: $Co_3(PO_4)_2 \cdot 8H_2O$
Molecular Formula: $Co_3H_{16}O_{16}P_2$
Molecular Weight: 510.865
CAS RN: 10294-50-5
Properties: pink to lavender amorphous powd [MER89]
Solubility: i H_2O; s mineral acids [MER89]
Density, g/cm³: 2.77 [MER89]
Reactions: minus $8H_2O$ at 200°C [HAW93]

897
Compound: Cobalt(II) selenate pentahydrate
Formula: $CoSeO_4 \cdot 5H_2O$
Molecular Formula: $CoH_{10}O_9Se$
Molecular Weight: 291.967
CAS RN: 14590-19-3
Properties: ruby red triclinic [KIR79]
Solubility: v s H_2O [KIR79]
Density, g/cm³: 2.512 [KIR79]
Melting Point, °C: decomposes [KIR79]

898
Compound: Cobalt(II) selenide
Formula: CoSe
Molecular Formula: CoSe
Molecular Weight: 137.893
CAS RN: 1307-99-9
Properties: 6 mm pieces and smaller with 99.5% purity; yellow hex [KIR79] [CER91]
Solubility: s HNO_3, aqua regia; i alkali [KIR79]
Density, g/cm³: 7.65 [KIR79]

Melting Point, °C: 1055 [LID94]

899

Compound: Cobalt(II) selenite dihydrate
Formula: $CoSeO_3 \cdot 2H_2O$
Molecular Formula: CoH_4O_5Se
Molecular Weight: 221.922
CAS RN: 19034-13-0
Properties: blue red powd [HAW93] [MER52]
Solubility: i H_2O [HAW93]

900

Compound: Cobalt(II) silicate
Synonyms: cobalt orthosilicate
Formula: Co_2SiO_4
Molecular Formula: Co_2O_4Si
Molecular Weight: 209.950
CAS RN: 12017-08-2
Properties: violet cryst [KIR79]
Solubility: i H_2O; s dil HCl [KIR79]
Density, g/cm³: 4.63 [KIR79]
Melting Point, °C: 1345 [KIR79]

901

Compound: Cobalt(II) stannate
Formula: Co_2SnO_4
Molecular Formula: Co_2O_4Sn
Molecular Weight: 300.574
CAS RN: 12139-93-4
Properties: greenish blue cub; used as paint and varnish drier [KIR79]
Solubility: s alkali [KIR79]
Density, g/cm³: 6.30 [KIR79]

902

Compound: Cobalt(II) sulfate
Synonyms: cobaltous sulfate
Formula: $CoSO_4$
Molecular Formula: CoO_4S
Molecular Weight: 154.997
CAS RN: 10124-43-3
Properties: red to lavender dimorphic, ortho-rhomb cryst; used in ceramics [MER89] [KIR79]
Solubility: dissolves slowly in boiling H_2O [MER89]; g/100g soln, H_2O: 19.7±0.1 (0°C), 27.2±0.1 (25°C), 27.8 (100°C); Solid phase, $CoSO_4 \cdot 7H_2O$ (0°C, 25°C), $CoSO_4 \cdot H_2O$ (100°C) [KRU93]
Density, g/cm³: 3.71 [MER89]
Melting Point, °C: decomposes 1140 [JAN85]

903

Compound: Cobalt(II) sulfate heptahydrate
Synonyms: bieberite
Formula: $CoSO_4 \cdot 7H_2O$
Molecular Formula: $CoH_{14}O_{11}S$
Molecular Weight: 281.103
CAS RN: 10026-24-1
Properties: pink to red monocl, prismatic cryst [MER89]
Solubility: 60.4 g/100mL cold H_2O, 67 g/100mL hot H_2O; sl s methanol, ethanol [MER89] [KIR78]
Density, g/cm³: 1.948 [STR93]
Melting Point, °C: 96.8 [STR93]
Reactions: minus H_2O at 41.5°C; minus $6H_2O$ at 71°C [MER89]

904

Compound: Cobalt(II) sulfate monohydrate
Formula: $CoSO_4 \cdot H_2O$
Molecular Formula: CoH_2O_5S
Molecular Weight: 173.012
CAS RN: 13455-34-0
Properties: red cryst; used as a pigment for porcelain and glazes [KIR78]
Solubility: s H_2O [KIR79]
Density, g/cm³: 3.075 [KIR79]
Melting Point, °C: decomposes [KIR79]

905

Compound: Cobalt(II) sulfide
Synonyms: sycoporite, cobaltous sulfide
Formula: CoS
Molecular Formula: CoS
Molecular Weight: 90.999
CAS RN: 1317-42-6
Properties: -200 mesh with 99.5% purity; exists in two forms; α-CoS: black, amorphous powd; forms Co(OH)S in air; β-CoS: gray powd or reddish-silver octahedral cryst [MER89] [CER91]
Solubility: 0.00038 g/100mL cold H_2O; s acids [MER89] [KIR79]
Density, g/cm³: 5.45 [MER89]
Melting Point, °C: 1182 [LID94]

906

Compound: Cobalt(II) telluride
Formula: CoTe
Molecular Formula: CoTe
Molecular Weight: 186.533
CAS RN: 12017-13-9

Properties: hex cryst; 6 mm pieces and smaller [LID94] [CER91]
Density, g/cm³: ~8.8 [LID94]

907
Compound: Cobalt(II) thiocyanate
Synonyms: cobaltous thiocyanate
Formula: Co(SCN)$_2$
Molecular Formula: C$_2$CoN$_2$S$_2$
Molecular Weight: 175.100
CAS RN: 3017-60-5
Properties: yellowish brown powd [MER89]
Solubility: s H$_2$O, giving rose-colored soln [MER89]; g/100g soln, H$_2$O: 50.7 (25°C) [KRU93]

908
Compound: Cobalt(II) thiocyanate trihydrate
Synonyms: cobaltous thiocyanate trihydrate
Formula: Co(SCN)$_2$·3H$_2$O
Molecular Formula: C$_2$H$_6$CoN$_2$O$_3$S$_2$
Molecular Weight: 229.146
CAS RN: 97126-35-7
Properties: violet to brownish violet rhomb cryst; red in transmitted light; used as an indicator of relative humidity [KIR79] [MER89]
Solubility: s H$_2$O, resulting in blue soln [MER89]
Reactions: minus 3H$_2$O at 105°C [KIR79]

909
Compound: Cobalt(II) tungstate
Formula: CoWO$_4$
Molecular Formula: CoO$_4$W
Molecular Weight: 306.771
CAS RN: 12640-47-0
Properties: -325 mesh 10 microns or less with 99.9% purity; reddish orange powd; used as a pigment and antiknock agent [HAW93] [CER91]
Solubility: i H$_2$O; s hot conc acids [HAW93]
Density, g/cm³: 8.42 [HAW93]

910
Compound: Cobalt(II,III) oxide
Synonyms: cobaltic-cobaltous oxide
Formula: Co$_3$O$_4$
Molecular Formula: Co$_3$O$_4$
Molecular Weight: 240.798
CAS RN: 1308-06-1

Properties: -325 mesh 10 microns or less with 99.5% purity; black or gray octahedral cryst; nonstoichometric; commercial material is black [MER89] [CER91]
Solubility: i H$_2$O; s acids, alkalies [MER89]
Density, g/cm³: 6.11 [MER89]
Melting Point, °C: 947, decomposes [JAN85]
Reactions: minus O$_2$ >900°C forming CoO [MER89]

911
Compound: Cobalt(III) acetate
Synonyms: cobaltic acetate
Formula: Co(CH$_3$COO)$_3$
Molecular Formula: C$_6$H$_9$CoO$_6$
Molecular Weight: 236.066
CAS RN: 917-69-1
Properties: dark green, very hygr powd or green octahedral cryst; used as a catalyst for cumene hydroperoxide [KIR79] [MER89]
Solubility: s H$_2$O with slow hydrolysis, becoming rapid at 60-70°C [MER89]
Melting Point, °C: decomposes 100 [KIR79]
Reactions: becomes black and decomposes on heating to 100°C [MER89]

912
Compound: Cobalt(III) acetylacetonate
Synonyms: 2,4-pentanedione, cobalt(III) derivative
Formula: Co(CH$_3$COCH=C(O)CH$_3$)$_3$
Molecular Formula: C$_{15}$H$_{21}$CoO$_6$
Molecular Weight: 356.261
CAS RN: 21679-46-9
Properties: sensitive to moisture; green cryst [STR93] [ALD94]

$$[CH_3-\underset{\underset{O-}{|}}{C}=CH-\underset{\underset{O}{\|}}{C}-CH_3]_3Co$$

Melting Point, °C: 216 [STR93]
Boiling Point, °C: 340 [STR93]

913
Compound: Cobalt(III) fluoride
Synonyms: cobaltic fluoride, cobalt trifluoride
Formula: CoF$_3$
Molecular Formula: CoF$_3$
Molecular Weight: 115.928
CAS RN: 10026-18-3

Properties: light brown; hex cryst; strong oxidizing agent; hygr and discolors rapidly in moist air; can be obtained by reacting F_2 with $CoCl_2$ or CoF_2 at 150-180°C; used as a fluorinating agent to replace hydrogen by fluorine in halocarbons and hydrocarbons [KIR78] [MER89]

Solubility: reacts with H_2O to evolve O_2 [MER89]

Density, g/cm³: 3.88 [MER89]

Melting Point, °C: 927 [LID94]

Reactions: comparatively stable to heat [MER89]

914

Compound: Cobalt(III) fluoride dihydrate

Formula: $Co_2F_6 \cdot 2H_2O$

Molecular Formula: $Co_2F_6H_4O_2$

Molecular Weight: 267.887

CAS RN: 54496-71-8

Properties: two forms: α is red, rhomb; β is rose powd [KIR79]

Solubility: s H_2O; i alcohol [KIR79]

Density, g/cm³: 2.192 [KIR79]

915

Compound: Cobalt(III) hydroxide trihydrate

Formula: $Co(OH)_3 \cdot 3H_2O$

Molecular Formula: CoH_9O_6

Molecular Weight: 164.001

CAS RN: 1307-86-4

Properties: dark brown or pink powd; uses: catalyst [STR93] [HAW93]

Solubility: i H_2O, alcohol; s cold conc acids [HAW93]

Density, g/cm³: 4.46 [HAW93]

Melting Point, °C: decomposes [STR93]

Reactions: minus water at 100°C [HAW93]

916

Compound: Cobalt(III) nitrate

Formula: $Co(NO_3)_3$

Molecular Formula: CoN_3O_9

Molecular Weight: 244.948

CAS RN: 15520-84-0

Properties: green hygr, reacts vigorously with organic solvents [KIR79]

Solubility: s H_2O [LID94]

Density, g/cm³: ~3.0 [LID94]

917

Compound: Cobalt(III) oxide

Synonyms: cobaltic oxide; cobalt black

Formula: Co_2O_3

Molecular Formula: Co_2O_3

Molecular Weight: 165.864

CAS RN: 1308-04-9

Properties: steel gray or black powd; used as a pigment to color enamels, and in glazing pottery [HAW93]

Solubility: i H_2O; s conc acids [HAW93]

Density, g/cm³: 4.81-5.60 [HAW93]

Melting Point, °C: 895, decomposes [HAW93]

918

Compound: Cobalt(III) oxide hydroxide

Synonyms: cobaltic oxide monohydrate

Formula: $CoO(OH)$

Molecular Formula: $CoHO_2$

Molecular Weight: 91.940

CAS RN: 12016-80-7

Properties: dark brown to black powd; hex; formula also written as $Co_2O_3 \cdot H_2O$ [MER89]

Solubility: i H_2O; s HCl, evolving Cl_2; s HNO_3, H_2SO_4 [MER89]

Reactions: transforms to Co_3O_4 if heated to 148-150°C [MER89]

919

Compound: Cobalt(III) sepulchrate trichloride

Synonyms: 1,3,6,8,10,13,16,19-octaazabicyclo(6,6,6)eicosanecobalt trichloride

Molecular Weight: 451.72

CAS RN: 71963-57-0

Properties: promising sensitizer for water-splitting systems [ALD94] [HOU82]

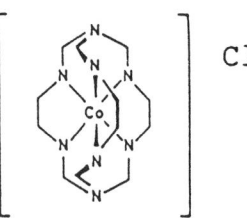

Melting Point, °C: 262, decomposes [ALD94]

920

Compound: Cobalt(III) sulfide

Formula: Co_2S_3

Molecular Formula: Co_2S_3

Molecular Weight: 214.064

CAS RN: 1332-71-4

Properties: black cryst [KIR79]

Solubility: decomposes in acid, aqua regia [KIR79]

Density, g/cm³: 4.8 [KIR79]

921

Compound: Cobaltocene

Synonyms: bis(cyclopentadienyl)cobalt(II)

Formula: (C$_5$H$_5$)$_2$Co

Molecular Formula: C$_{10}$H$_{10}$Co

Molecular Weight: 189.122

CAS RN: 1277-43-6

Properties: purplish-black cryst; sensitive to light, air and heat; forms intercalation compound SnS$_2$[Co(C$_5$H$_5$)$_2$] by reaction with SnS$_2$ [CEN94] [STR93] [ALD94]

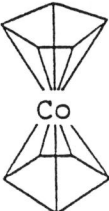

Melting Point, °C: 176-180, decomposes [ALD94]

922

Compound: Cobaltocenium hexafluorophosphate

Synonyms: bis(cyclopentadienyl)cobalt(II) hexafluorophosphate

Formula: (C$_5$H$_5$)$_2$CoPF$_6$

Molecular Formula: C$_{10}$H$_{10}$CoF$_6$P

Molecular Weight: 334.086

CAS RN: 12427-42-8

Properties: yellow cryst; moisture sensitive [STR93] [ALD94]

923

Compound: Copper

Formula: Cu

Molecular Formula: Cu

Molecular Weight: 63.546

CAS RN: 7440-50-8

Properties: reddish, lustrous, ductile metal; fcc; excellent conductor of electricity; becomes dull when exposed to air, coated with green basic carbonate in moist air; hardness 3.0 Mohs; electrical resistivity (20°C) 1.673 μohm·cm; electronegativity 2.43; enthalpy of fusion 13.26 kJ/mol; enthalpy of vaporization 300.4 kJ/mol; ionic radius of Cu^{++} 0.096 nm [KIR78] [CRC93] [MER89] [ALD94]

Solubility: very slowly attacked by cold HCl and dil H$_2$SO$_4$, readily by HNO$_3$ [MER89]

Density, g/cm^3: 8.94 [MER89]

Melting Point, °C: 1084.62 [CRC93]

Boiling Point, °C: 2562 [CRC93]

Thermal Conductivity, W/(m·K): 394 (25°C) [ALD94]

Thermal Expansion Coefficient: linear coefficient of expansion at 20°C is 16.5 x 10^{-6}/°C [KIR78]

924

Compound: Copper arsenide

Formula: Cu$_3$As

Molecular Formula: AsCu$_3$

Molecular Weight: 265.560

CAS RN: 12005-75-3

Properties: 6 mm pieces and smaller with 99% purity [CER91]

925

Compound: Copper citrate hemipentahydrate

Formula: Cu$_2$C$_6$H$_4$O$_7$·2-1/2H$_2$O

Molecular Formula: C$_6$H$_9$Cu$_2$O$_{9.5}$

Molecular Weight: 360.219

CAS RN: 10402-15-0

Properties: green or bluish green; odorless; cryst powd; can be prepared by reacting hot citric acid with copper sulfate solutions; used as an antiseptic and astringent [MER89]

Solubility: sl s H$_2$O; s ammonia, dil acids [MER89]

Reactions: minus 2.5 H$_2$O at 100°C [MER89]

926

Compound: Copper nitride

Formula: Cu$_3$N

Molecular Formula: Cu$_3$N

Molecular Weight: 204.645

CAS RN: 1308-80-1

Properties: dark green powd; -200 mesh with 99.5% purity; cub, a = 0.380 nm [CER91] [CIC73] [CRC77]

Solubility: decomposes in H$_2$O [CRC77]

Density, g/cm^3: 5.84 [CIC73]

Melting Point, °C: 300, decomposes [ALF93]

927

Compound: Copper phosphide

Formula: Cu$_3$P

Molecular Formula: Cu$_3$P

Molecular Weight: 221.612

CAS RN: 12019-57-7

Properties: grayish-black; -100 mesh of 99.5% purity; other phosphides are Cu$_3$P$_2$, 12134-35-9, Cu$_2$P, 12324-28-6, CuP$_2$, 12019-11-3 [KIR82] [CER91]

Solubility: i H$_2$O [CRC77]

Density, g/cm³: 6.4 to 6.8 [CRC77]
Melting Point, °C: decomposes [CRC77]

928
Compound: Copper silicide
Formula: Cu₅Si
Molecular Formula: Cu₅Si
Molecular Weight: 345.816
CAS RN: 12159-07-8
Properties: 6.35 mm & down pieces [ALF93]
Melting Point, °C: 825 [ALF93]

929
Compound: Copper vanadate
Formula: CuV₂O₆
Molecular Formula: CuO₆V₂
Molecular Weight: 261.425
CAS RN: 12789-09-2
Properties: -200 mesh with 99.5% purity [CER91]

930
Compound: Copper zirconate
Formula: CuZrO₃
Molecular Formula: CuO₃Zr
Molecular Weight: 202.768
CAS RN: 70714-64-6
Properties: reacted product of CuO+ZrO₂, -200 mesh with 99.5% purity [CER91]

931
Compound: Copper(I) acetate
Synonyms: acetic acid, copper(I) salt
Formula: Cu(CH₃COO)
Molecular Formula: C₂H₃CuO₂
Molecular Weight: 122.591
CAS RN: 598-54-9
Properties: tan powd; stable when dry, decomposes slowly in H₂O [STR93] [KIR78]
Solubility: rapidly hydrolyzed by H₂O to form yellow Cu₂O [MER89]
Melting Point, °C: volatilizes by heating [MER89]
Boiling Point, °C: decomposes if strongly heated [MER89]

932
Compound: Copper(I) acetylide
Synonyms: cuprous acetylide
Formula: CuC≡CCu
Molecular Formula: C₂Cu₂
Molecular Weight: 151.114

CAS RN: 1117-94-8
Properties: amorphous red powd; unstable; oxidizes in air to Cu₂O, carbon and water; structure CuC≡CCu; can explode when subjected to shock or heated; obtained by reacting acetylene, HC≡CH, with soluble cuprous salt in water; used in detonators and other explosives [HAW93] [KIR78]
Solubility: v sl s H₂O [CRC77]
Melting Point, °C: explodes [CRC77]

933
Compound: Copper(I) azide
Formula: CuN₃
Molecular Formula: CuN₃
Molecular Weight: 105.566
CAS RN: 14336-80-2
Properties: colorless tetr, a = 0.865 nm, c = 0.559 nm; M-N₃ bond length, 0.223 nm; highly explosive [CRC77] [CIC73]
Solubility: 0.00075 g/100 mL H₂O (20°C) [CRC77]
Density, g/cm³: 3.26 [CRC77]

934
Compound: Copper(I) bromide
Synonyms: cuprous bromide
Formula: CuBr
Molecular Formula: BrCu
Molecular Weight: 143.450
CAS RN: 7787-70-4
Properties: -80 mesh with 99.999% and 99% purity; white powd; cub cryst; hygr; turns green to dark blue in sunlight; used as a catalyst for organic reactions [HAW93] [MER89] [STR93] [CER91]
Solubility: mol/dm³ CuBr (mol/dm³ KBr), 24.8°C: 0.00064 (0.0989), 0.00633 (0.3955), 0.01539 (0.5933), 0.0496 (0.9888), 0.334 (1.978), 1.005 (2.966), 1.860 (3.955) [FRI87]
Density, g/cm³: 4.72 [ALD94]
Melting Point, °C: 497 [CRC93]
Boiling Point, °C: 1345 [KIR78]
Reactions: oxidizes slowly in air becoming green [KIR78]
Thermal Conductivity, W/(m·K): 1.25 [CRC93]
Thermal Expansion Coefficient: 15.4 x 10⁻⁶/K [CRC93]

935
Compound: Copper(I) chloride
Synonyms: nantokite, cuprous chloride
Formula: CuCl
Molecular Formula: ClCu

Molecular Weight: 98.999
CAS RN: 7758-89-6
Properties: -80 mesh with 99.999% and 99% purity; white cub cryst; stable to air and light, if dry; turns green in presence of moisture, sensitive to light, becoming brown; enthalpy of fusion 10.20 kJ/mol; used as a catalyst, preservative and fungicide [HAW93] [MER89] [CER91] [CRC93]
Solubility: sl s H_2O; s conc HCl, conc NH_4OH with complex formation [MER89]
Density, g/cm³: 4.14 [MER89]
Melting Point, °C: 430 [CRC93]
Boiling Point, °C: 1490 [STR93]; 1366 [KIR78]
Thermal Conductivity, W/(m·K): 0.84 [CRC93]
Thermal Expansion Coefficient: 12.1 x 10^{-6}/K (300K) [CRC93]

936
Compound: Copper(I) cyanide
Synonyms: cuprous cyanide
Formula: CuCN
Molecular Formula: CCuN
Molecular Weight: 89.564
CAS RN: 544-92-3
Properties: white to cream colored powd; colorless or dark green ortho-rhomb cryst; used in plating copper, antifouling paints and insecticide [HAW93] [MER89]
Solubility: i H_2O, alcohol, cold dil acids; s NH_4OH, sodium and potassium cyanide solutions [HAW93] [MER89]
Density, g/cm³: 1.9 [HAW93]; 2.92 [STR93]
Melting Point, °C: 474 [MER89]
Boiling Point, °C: decomposes [STR93]
Reactions: decomposed by HNO_3 or boiling HCl [MER89]

937
Compound: Copper(I) hydride
Formula: CuH
Molecular Formula: CuH
Molecular Weight: 64.554
CAS RN: 13517-00-5
Properties: light reddish brown; composition can vary from $Cu_{0.6}$ to CuH; can be prepared by precipitation with hypophosphorus acid at 65°C; used as a reducing agent in some organic reactions [KIR78]
Solubility: decomposes in H_2O at 65°C [CRC77]
Density, g/cm³: 6.38 [CRC77]
Melting Point, °C: decomposes >60 [KIR78]

938
Compound: Copper(I) iodide
Synonyms: marshite,cuprous iodide
Formula: CuI
Molecular Formula: CuI
Molecular Weight: 190.450
CAS RN: 7681-65-4
Properties: -80 mesh with 99.999% purity and -60 mesh with 99% purity; pure white or brownish powd; light sensitive; can be prepared by reacting solution of Cu^{++} with I^- to precipitate CuI; has been used in manufacture of photographic emulsions, catalyst, and in cloud seeding to assist rainfall [KIR78] [CER91] [ALD94]
Solubility: i H_2O, dil acids; s aq NH_3 solutions [MER89]
Density, g/cm³: 5.63 [MER89]
Melting Point, °C: 605 [CRC93]
Boiling Point, °C: ~1290 [MER89]
Reactions: decomposed by conc H_2SO_4 and HNO_3 [MER89]
Thermal Conductivity, W/(m·K): 1.68 (25°C) [CRC93]
Thermal Expansion Coefficient: 19.2 x 10^{-6}/K [CRC93]

939
Compound: Copper(I) mercury iodide
Synonyms: cuprous mercuric iodide
Formula: Cu_2HgI_4
Molecular Formula: Cu_2HgI_4
Molecular Weight: 835.300
CAS RN: 13876-85-2
Properties: -60 mesh with 99.5% purity; α-form tetra, deep red; β-form cub, brownish; thermochromic; used for detecting overheating of machine bearings; red color changes to brownish-black at 60-70°C, then back to red when cooled; obtained by precipitation from a solution of K_2HgI_4 and cuprous chloride [KIR81] [MER89] [CER91] [CRC77]
Solubility: i H_2O [KIR81]
Density, g/cm³: α: 6.116; β: 6.102 [CRC77]

940
Compound: Copper(I) oxide
Synonyms: cuprite, cuprous oxide
Formula: Cu_2O
Molecular Formula: Cu_2O
Molecular Weight: 143.091
CAS RN: 1317-39-1

Properties: -200 mesh with 99% purity; cub cryst; color may be yellow, red or brown, depending on method of preparation; brownish red mineral; enthalpy of fusion 64.8 kJ/mol; stable at high temp, can be made by thermal decomposition of CuO above 1030°C; used as a catalyst, fungicide, in purification of helium, and as an antioxidant in lubricants [MER89] [KIR78] [CER91] [JAN85]

Solubility: i H_2O; s NH_4OH, HCl; reacts with dil H_2SO_4, HNO_3 [MER89]

Density, g/cm³: 6.0 [MER89]

Melting Point, °C: 1235 [KIR78]

Boiling Point, °C: decomposes >1800 [KIR78]

941

Compound: Copper(I) selenide

Synonyms: cuprous selenide

Formula: Cu_2Se

Molecular Formula: Cu_2Se

Molecular Weight: 206.052

CAS RN: 20405-64-5

Properties: 6 mm pieces and smaller with 99.5% purity; bluish black to black; tetr or cub cryst; metallic luster; semiconductor [MER89] [CER91]

Solubility: i H_2O; s HCl, evolves H_2Se [MER89]

Density, g/cm³: 6.84 [MER89]

Melting Point, °C: 1113 [MER89]

942

Compound: Copper(I) sulfide

Synonyms: chalcocite, cuprous sulfide

Formula: Cu_2S

Molecular Formula: Cu_2S

Molecular Weight: 159.158

CAS RN: 22205-45-4

Properties: -200 mesh with 99.999% purity; blue to grayish black lustrous powd; ortho-rhomb cryst; mineral chalcocite has hardness 2.5-3 Mohs [KIR78] [MER89] [CER91]

Solubility: i H_2O, acetic acid; sl s HCl [MER89]

Density, g/cm³: 5.6 [MER89]

Melting Point, °C: ~1100 [MER89]

Reactions: decomposed by HNO_3 and conc H_2SO_4 [MER89]

943

Compound: Copper(I) sulfite hemihydrate

Synonyms: Etard's salt; cuprous sulfite hemihydrate

Formula: $Cu_2SO_3 \cdot 1/2H_2O$

Molecular Formula: $Cu_2HO_{3.5}S$

Molecular Weight: 216.164

CAS RN: 13982-53-1

Properties: white to pale yellow; hex cryst; fungicide [MER89]

Solubility: sl s H_2O; s HCl, NH_4OH, alkali solutions [MER89]

944

Compound: Copper(I) sulfite monohydrate

Synonyms: Rogojski's salt; cuprous sulfite monohydrate

Formula: $Cu_2SO_3 \cdot H_2O$

Molecular Formula: $Cu_2H_2O_4S$

Molecular Weight: 225.172

CAS RN: 13982-53-1

Properties: white cryst powd or brick red solid; shown to have an equimolar composition of metallic Cu and Chevreul's salt: $Cu_3O_6S_2$, cupro-cupric sulfate; used as a catalyst, fungicide and in dyeing textiles [HAW93] [MER89]

Solubility: i H_2O; s NH_4OH, decomposes in HCl [HAW93]

Density, g/cm³: 3.83 [HAW93]

Melting Point, °C: decomposes [CRC94]

945

Compound: Copper(I) telluride

Formula: Cu_2Te

Molecular Formula: Cu_2Te

Molecular Weight: 254.692

CAS RN: 12019-52-2

Properties: black hex; 3 mm pieces and smaller with 99.5% purity [CER91] [LID94]

Density, g/cm³: 4.6 [STR93]

946

Compound: Copper(I) thiocyanate

Synonyms: cuprous thiocyanate

Formula: CuSCN

Molecular Formula: CCuNS

Molecular Weight: 121.630

CAS RN: 1111-67-7

Properties: white to yellow amorphous powd; used in manufacturing organic chemicals, antifouling paints and in printing textiles [HAW93] [MER89]

Solubility: i H_2O, dil acids, alcohol, acetone; s NH_4OH, ether [MER89]

Density, g/cm³: 2.843 [HAW93]

Melting Point, °C: 1084 [HAW93]

Reactions: decomposed by conc mineral acids [MER89]

947

Compound: Copper(I,II) sulfite dihydrate

Synonyms: Chevreul's salt

Formula: $Cu_2SO_3 \cdot CuSO_3 \cdot 2H_2O$

Molecular Formula: $Cu_3H_4O_8S_2$

Molecular Weight: 386.797

CAS RN: 13814-81-8

Properties: red; microcryst powd or prismatic cryst [MER89]

Solubility: i H_2O, alcohol; s NH_4OH, HCl [MER89]

Density, g/cm³: 3.57 [CRC77]

Melting Point, °C: decomposes at 200 [CRC77]

948

Compound: Copper(II) acetate

Synonyms: acetic acid, Cu(II) salt

Formula: $Cu(CH_3COO)_2$

Molecular Formula: $C_4H_6CuO_4$

Molecular Weight: 181.636

CAS RN: 142-71-2

Properties: bluish green powd; hygr [STR93]

949

Compound: Copper(II) acetate metaarsenite

Synonyms: Paris green; cupric acetoarsenite

Formula: $Cu(CH_3COO)_2 \cdot 3Cu(AsO_2)_2$

Molecular Formula: $C_4H_6As_6Cu_4O_{16}$

Molecular Weight: 1013.796

CAS RN: 12002-03-8

Properties: emerald green cryst powd; stable to air and light; used as an insecticide, wood preservative and paint pigment [KIR78] [MER89]

Solubility: i H_2O, decomposed by prolonged heating; unstable in acids [MER89]

950

Compound: Copper(II) acetate monohydrate

Synonyms: neutral verdigris; cupric acetate monohydrate

Formula: $Cu(CH_3COO)_2 \cdot H_2O$

Molecular Formula: $C_4H_8CuO_5$

Molecular Weight: 199.651

CAS RN: 6046-93-1

Properties: dark green; monocl cryst; forms dimeric units; effloresces in dry air [KIR78] [MER89]

Solubility: s H_2O, alcohol; sl s ether, glycerol [MER89]

Density, g/cm³: 1.882 [MER89]

Melting Point, °C: 115 [MER89]

Boiling Point, °C: decomposes 240 [MER89]

951

Compound: Copper(II) acetylacetonate

Synonyms: 2,4-pentanedione, copper(II) derivative

Formula: $Cu(CH_3COCH=C(O)CH_3)_2$

Molecular Formula: $C_{10}H_{14}CuO_4$

Molecular Weight: 261.765

CAS RN: 13395-16-9

Properties: cryst blue powd; does not hydrolyze readily [HAW93] [STR93]

$$[CH_3-\overset{\overset{\displaystyle O-}{|}}{C}=CH-\overset{\overset{\displaystyle O}{\|}}{C}-CH_3]_2 Cu$$

Solubility: sl s H_2O, alcohol; s chloroform [HAW93]

Melting Point, °C: 284-285, decomposes [STR93]

Boiling Point, °C: sublimes 78 at 0.05mm Hg pressure [STR93]

952

Compound: Copper(II) arsenate

Formula: $Cu_3(AsO_4)_2$

Molecular Formula: $As_2Cu_3O_8$

Molecular Weight: 468.476

CAS RN: 10103-61-4

Properties: light blue, blue or greenish blue; can have a variable composition; used as an insecticide and fungicide [HAW93]

Solubility: i H_2O, alcohol; s dil acids, NH_4OH [HAW93]

953

Compound: Copper(II) arsenite

Synonyms: Scheele's green

Formula: $CuHAsO_3$ (also, $Cu(AsO_2)_2$)

Molecular Formula: $AsCuHO_3$

Molecular Weight: 187.474

CAS RN: 10290-12-7

Properties: yellowish green powd; used as an insecticide, fungicide, paint pigment and as a wood preservative [HAW93] [MER89]

Solubility: i H_2O, alcohol; s acids, ammonia [MER89]

Melting Point, °C: decomposes [HAW93]

954

Compound: Copper(II) azide

Formula: $Cu(N_3)_2$
Molecular Formula: CuN_6
Molecular Weight: 147.586
CAS RN: 14215-30-6
Properties: brownish-red; ortho-rhomb, a = 0.923 nm, b = 1.323 nm, c = 0.307 nm [CRC77] [CIC73]
Solubility: 0.008 g/100 mL H_2O (20°C) [CRC77]
Density, g/cm³: 2.604 [CRC77]
Reactions: explodes at 215°C [CRC77]

955

Compound: Copper(II) basic acetate
Synonyms: blue verdigris
Formula: $Cu(CH_3COO)_2 \cdot CuO \cdot 6H_2O$
Molecular Formula: $C_4H_{18}Cu_2O_{11}$
Molecular Weight: 369.272
CAS RN: 52503-64-7
Properties: formula is approximate; compounds are in the form of blue cryst or blue to green powd [MER89] [CRC77]
Solubility: sl s H_2O, alcohol; s dil acids, ammonia [MER89]

956

Compound: Copper(II) basic chromate
Synonyms: basic cupric chromate
Formula: $CuCrO_4 \cdot 2Cu(OH)_2$
Molecular Formula: $CrCu_3H_4O_8$
Molecular Weight: 340.646
CAS RN: 12433-14-6
Properties: light chocolate brown powd; used as a mordant in dyeing, as a wood preservative and to treat seeds for fungus [HAW93]
Solubility: i H_2O; s HNO_3 [HAW93]
Reactions: minus water at 260°C [HAW93]

957

Compound: Copper(II) basic nitrite
Formula: $Cu(NO_2)_2 \cdot 3Cu(OH)_2$
Molecular Formula: $Cu_4H_6N_2O_{10}$
Molecular Weight: 448.239
Properties: green powd [HAW93]
Solubility: sl s H_2O; s NH_4OH; decomposes in dil acids [HAW93]
Melting Point, °C: 120, decomposes [HAW93]
Reactions: decomposed by hot H_2O [CRC77]

958

Compound: Copper(II) benzoate dihydrate
Formula: $Cu(C_6H_5COO)_2 \cdot 2H_2O$

Molecular Formula: $C_{14}H_{14}CuO_6$
Molecular Weight: 341.807
Properties: blue cryst odorless powd [HAW93]
Solubility: sl s cold H_2O, alcohol and acids [HAW93]
Reactions: minus $2H_2O$ at 100°C [HAW93]

959

Compound: Copper(II) borate
Synonyms: copper(II) metaborate
Formula: $Cu(BO_2)_2$
Molecular Formula: B_2CuO_4
Molecular Weight: 149.166
CAS RN: 39290-85-2
Properties: amorphous or bluish green cryst powd; can be prepared by adding borax to an aq Cu^{++} sulfate or chloride solution, which precipitates the borate; used as an oil pigment, a dehydrogenation catalyst, wood preservative and fire retardant additive [HAW93] [KIR78]
Solubility: i H_2O; s acids [HAW93]
Density, g/cm³: 3.859 [HAW93]

960

Compound: Copper(II) bromide
Synonyms: cupric bromide
Formula: $CuBr_2$
Molecular Formula: Br_2Cu
Molecular Weight: 223.354
CAS RN: 7789-45-9
Properties: almost black, deliq, monocl cryst or gray powd [MER89] [STR93]
Solubility: s alcohol, acetone, ammonia; i benzene, ether [MER89]; g/100 g soln, H_2O: 55.7 (0°); Solid phase, $CuBr_2$ [KRU93]; g/100g H_2O: 107 (0°C), 126 (20°C), 131 (50°C) [LAN85]
Density, g/cm³: 4.710 [MER89]
Melting Point, °C: 498 [MER89]
Boiling Point, °C: 900 [MER89]

961

Compound: Copper(II) butyrate monohydrate
Synonyms: cupric butyrate monohydrate
Formula: $Cu(CH_3CH_2CH_2COO)_2 \cdot H_2O$
Molecular Formula: $C_8H_{16}CuO_5$
Molecular Weight: 255.758
CAS RN: 540-16-9
Properties: large, dark green; monocl, hex plates; becomes dull and disintegrates after several days exposure to air [MER89]
Solubility: s H_2O, dioxane, benzene; sl s alcohol, chloroform [MER89]

962
Compound: Copper(II) carbonate hydroxide
Synonyms: malacite, Bremen green
Formula: $CuCO_3 \cdot Cu(OH)_2$
Molecular Formula: $CH_2Cu_2O_5$
Molecular Weight: 221.116
CAS RN: 12069-69-1
Properties: green to blue amorphous powd, or dark green monocl cryst; malacite hardness 3.5-4 Mohs [KIR78] [MER89]
Solubility: i H_2O, alcohol; s dil acids, ammonia [MER89]
Density, g/cm³: 4.0 [STR93]
Melting Point, °C: 200, decomposes [STR93]

963
Compound: Copper(II) chlorate hexahydrate
Synonyms: cupric chlorate hexahydrate
Formula: $Cu(ClO_3)_2 \cdot 6H_2O$
Molecular Formula: $Cl_2CuH_{12}O_{12}$
Molecular Weight: 338.539
CAS RN: 14721-21-2
Properties: blue to green; deliq; octahedral cryst [MER89]
Solubility: mol/100 mol H_2O: 11.02 (0.8°C), equilibrium solid phase $Cu(ClO_3)_2 \cdot 4H_2O$ [KRU93]; v s alcohol [MER89]
Melting Point, °C: 65 [MER89]
Boiling Point, °C: decomposes at 100 [MER89]

964
Compound: Copper(II) chloride
Synonyms: cupric chloride
Formula: $CuCl_2$
Molecular Formula: Cl_2Cu
Molecular Weight: 134.451
CAS RN: 7447-39-4
Properties: -80 mesh with 99% purity; yellow to brown; deliq; microcryst powd; usually exists as blue-green dihydrate; enthalpy of fusion 20.40 kJ/mol [CRC93] [MER89] [KIR78] [CER91]
Solubility: s H_2O, alcohol, acetone [MER89]; g/100g soln, H_2O: 40.7±0.2 (0°C), 43.8±0.6 (25°C), 54.6 (100°C); Solid phase, $CuCl_2 \cdot 2H_2O$ [KRU93]
Density, g/cm³: 3.386 [STR93]
Melting Point, °C: 620 [ALD94]
Boiling Point, °C: 993, decomposes to CuCl [KIR78]

965
Compound: Copper(II) chloride dihydrate

Synonyms: eriochaleite, cupric chloride dihydrate
Formula: $CuCl_2 \cdot 2H_2O$
Molecular Formula: $Cl_2CuH_4O_2$
Molecular Weight: 170.482
CAS RN: 10125-13-0
Properties: green to blue powd, or ortho-rhomb cryst; deliq in moist air, effloresces in dry air; used as catalyst in organic synthesis, e.g. chlorination; used as pigment [MER89] [KIR78]
Solubility: 1g dihydrate/1mL H_2O (25°C) [KIR78]; s methanol, ethanol [MER89]
Density, g/cm³: 2.51 [MER89]
Melting Point, °C: ~100 [MER89]
Reactions: minus $2H_2O$ if heated at 120°C in a stream of HCl [KIR78]

966
Compound: Copper(II) chromate
Formula: $CuCrO_4$
Molecular Formula: $CrCuO_4$
Molecular Weight: 179.540
CAS RN: 13548-42-0
Properties: reddish brown; used as a fungicide, to weatherproof textiles, in epoxy adhesives and to preserve wood [KIR78]
Solubility: mol/L soln, H_2O: 0.0020 (25°C); Solid phase, $CuCrO_4 \cdot H_2O$ [KRU93]
Melting Point, °C: decomposes 400 [KIR78]
Reactions: minus O_2 to form blue-black $CuOCr_2O_3$ at 400°C [KIR78]

967
Compound: Copper(II) chromite
Synonyms: cupric chromite
Formula: $CuCr_2O_4$
Molecular Formula: Cr_2CuO_4
Molecular Weight: 231.536
CAS RN: 12018-10-9
Properties: grayish black to black; tetr cryst; used in automobile exhaust catalysts; there is also $2CuO \cdot Cr_2O_3$, CAS RN 12053-18-8 [KIR78] [MER89] [ALD94]
Solubility: i H_2O, dil acids, conc HCl [MER89]
Density, g/cm³: 5.4 [LID94]

968
Compound: Copper(II) cyanide
Synonyms: cupric cyanide
Formula: $Cu(CN)_2$
Molecular Formula: C_2CuN_2
Molecular Weight: 115.581
CAS RN: 14763-77-0

Properties: green powd; used to electroplate copper onto iron [HAW93]
Solubility: i H_2O; s acids and alkalies [HAW93]
Melting Point, °C: decomposes [CRC77]

969
Compound: Copper(II) dichromate dihydrate
Formula: $CuCr_2O_7 \cdot 2H_2O$
Molecular Formula: $Cr_2CuH_4O_9$
Molecular Weight: 315.565
CAS RN: 13675-47-3
Properties: reddish brown; tricl; used in catalysts and as a wood preservative [KIR78]
Solubility: v s H_2O [KIR78]
Density, g/cm³: 2.286 [KIR78]
Reactions: minus $2H_2O$ at 100°C [CRC77]

970
Compound: Copper(II) ferrate
Synonyms: copper diiron tetroxide
Formula: $CuFe_2O_4$
Molecular Formula: $CuFe_2O_4$
Molecular Weight: 239.234
CAS RN: 12018-79-0
Properties: black cryst; spinel structure; magnetic properties; formed by reaction of CuO and Fe_2O_3 at ~900°C, or by coprecipitation and sintering at 900°C; has been used to manufacture piezomagnets and magnetic tapes; also used as a catalyst, e.g. exhaust control [KIR78] [STR93]

971
Compound: Copper(II) ferrocyanide
Synonyms: Hatchett's brown, cupric ferrocyanide
Formula: $Cu_2Fe(CN)_6$
Molecular Formula: $C_6Cu_2FeN_6$
Molecular Weight: 339.043
CAS RN: 13601-13-3
Properties: reddish brown powd or cub cryst; precipitates as gel or colloid [MER89]
Solubility: i H_2O, dil acids; s NH_4OH [MER89]

972
Compound: Copper(II) ferrous sulfide
Synonyms: chalcopyrite
Formula: $CuFeS_2$
Molecular Formula: $CuFeS_2$
Molecular Weight: 183.523
CAS RN: 1308-56-1

Properties: yellow brass colored cryst; tetr; hardness 3.5-4.0; chalcopyrite is an ore; can be synthesized by reacting KFeS with ammonia copper solution; used in semiconductor research [MER89]
Solubility: s HNO_3, aqua regia; i HCl [MER89]
Density, g/cm³: 4.1-4.3 [MER89]
Melting Point, °C: 950 [MER89]
Thermal Expansion Coefficient: (volume) 100°C (0.42) [CLA66]

973
Compound: Copper(II) fluoride
Synonyms: cupric fluoride
Formula: CuF_2
Molecular Formula: CuF_2
Molecular Weight: 101.543
CAS RN: 7789-19-7
Properties: -100 mesh with 99.5% purity; white powd; monocl cryst; hygr, turns blue in moist air, forming dihydrate; enthalpy of fusion 55.23 kJ/mol; prepared by reaction of copper oxide or basic carbonate with anhydrous HF, followed by dehydration in stream of anhydrous HF at high temp; finds use as a catalyst in organic reactions, as a fluorinating agent and in high-density batteries [KIR78] [MER89] [STR93] [CER91] [CRC93] [JAN82]
Solubility: 4.7 g/100mL H_2O (20°C); hydrolyzed to CuFOH in hot H_2O [MER89]; g/100 ml soln, H_2O: 0.075 (25°C); Solid phase, $CuF_2 \cdot 2H_2O$ [KRU93]
Density, g/cm³: 4.23 [STR93]
Melting Point, °C: 836 [CRC93]
Boiling Point, °C: 1678 [JAN85]

974
Compound: Copper(II) fluoride dihydrate
Synonyms: cupric fluoride dihydrate
Formula: $CuF_2 \cdot 2H_2O$
Molecular Formula: $CuF_2H_4O_2$
Molecular Weight: 137.574
CAS RN: 13454-88-1
Properties: blue; monocl cryst; a trihydrate is also listed with density 2.93 [MER89] [ALD94]
Solubility: sl s cold H_2O, hydrolyzed to CuFOH in hot H_2O [MER89]
Density, g/cm³: 2.934 [MER89]
Melting Point, °C: decomposes >130 [MER89]

975
Compound: Copper(II) formate
Synonyms: cupric formate

Formula: $Cu(HCOO)_2$
Molecular Formula: $C_2H_2CuO_4$
Molecular Weight: 153.582
CAS RN: 544-19-4
Properties: three forms of anhydrous formate exist: powd blue, turquoise, or royal blue cryst [MER89]
Solubility: 12.5 g/100 mL H_2O [CRC77]; i most organic solvents [MER89]
Density, g/cm³: 1.831 [STR93]

976
Compound: Copper(II) formate tetrahydrate
Synonyms: cupric formate tetrahydrate
Formula: $Cu(HCOO)_2 \cdot 4H_2O$
Molecular Formula: $C_2H_{10}CuO_8$
Molecular Weight: 225.642
CAS RN: 5893-61-8
Properties: large, light blue; monocl; powd blue modification formed by dehydration over $CaCl_2$, under reduced pressure [MER89]
Solubility: 6.2 g/100 mL H_2O [CRC77]; v sl s alcohol [MER89]
Density, g/cm³: 1.81 [CRC77]
Reactions: minus H_2O at 130°C [CRC77]

977
Compound: Copper(II) gluconate
Formula: $Cu(CH_2OH(CHOH)_4COO)_2$
Molecular Formula: $C_{12}H_{22}CuO_{14}$
Molecular Weight: 453.845
CAS RN: 527-09-3
Properties: forms light blue to bluish green water soluble cryst; used as a feed additive, dietary supplement and mouth deodorant, and to treat arthritis [HAW93] [KIR78] [STR93]
Solubility: s H_2O; i alcohol, ether, acetone [HAW93]

978
Compound: Copper(II) glycinate monohydrate
Synonyms: cupric glycinate hydrate
Formula: $Cu(H_2NCH_2COO)_2 \cdot H_2O$
Molecular Formula: $C_4H_{10}CuN_2O_5$
Molecular Weight: 229.679
CAS RN: 13479-54-4
Properties: deep blue; long rhomb needles [MER89]
Solubility: s H_2O, sl s alcohol [MER89]
Melting Point, °C: chars 213 [MER89]
Boiling Point, °C: decomposes with evolution of gas at 228 [MER89]

979
Compound: Copper(II) hexafluoroacetylacetonate
Synonyms: 1,1,1,5,5,5-hexafluoro-2,4-pentanedione, Cu(II) derivative
Formula: $Cu(CF_3COCHCOCF_3)_2$
Molecular Formula: $C_{10}H_2CuF_{12}O_4$
Molecular Weight: 477.650
CAS RN: 14781-45-4
Properties: blue-green cryst; hygr [STR94]

$$\begin{array}{cc} O- & O \\ | & \| \end{array}$$
$$[CF_3-C=CH-C-CF_3]_2Cu$$

Melting Point, °C: 85-89 [STR93]
Boiling Point, °C: sublimes 70 (0.05 mm Hg) [STR93]
Reactions: decomposes at 220°C [STR93]

980
Compound: Copper(II) hexafluorosilicate tetrahydrate
Synonyms: cupric hexafluorosilicate tetrahydrate
Formula: $CuSiF_6 \cdot 4H_2O$
Molecular Formula: $CuF_6H_8O_4Si$
Molecular Weight: 277.684
CAS RN: 12062-24-7
Properties: blue monocl efflorescent cryst; decomposed by heating [HAW93] [MER89]
Solubility: g anhydrous/100g H_2O: 73.5 (0°C), 81.6 (20°C), 93.2 (75°C) [LAN85]; sl s alcohol [HAW93] [MER89]
Density, g/cm³: 2.56 [MER89]

981
Compound: Copper(II) hexafluroacetylacetonate hydrate
Synonyms: 1,1,1,5,5,5-hexafluoro-2,4-pentanedione, Cu(II) derivative hydrate
Formula: $Cu[CF_3COCH=C(O)CF_3]_2 \cdot xH_2O$
Molecular Formula: $C_{10}H_2CuF_{12}O_4$ (anhydrous)
Molecular Weight: 477.650 (anhydrous)
CAS RN: 14781-45-4
Properties: bluish-green cryst; precursor for metal oxide chemical vapor deposition [STR94] [STR93] [ALD94]
Melting Point, °C: 130-134 [ALD94]

982
Compound: Copper(II) hydroxide
Synonyms: cupric hydroxide
Formula: $Cu(OH)_2$

Molecular Formula: CuH_2O_2
Molecular Weight: 97.561
CAS RN: 20427-59-2
Properties: blue to bluish green gel or light blue cryst powd which decomposes to CuO by sl warming; can be prepared by anodic dissolution of a copper anode in sodium sulfate or trisodium phosphate solution; used as a fungicide and as a constituent of antifouling marine paints [MER89]
Solubility: i H_2O; when freshly precipitated, s conc alkali; s acids, NH_4OH [MER89]; mol/L soln, H_2O: 3 x 10^{-5} (25°C) [KRU93]
Density, g/cm³: 3.37 [MER89]
Reactions: minus H_2O on heating [CRC94]

983
Compound: Copper(II) hydroxy chloride
Formula: $CuCl_2 \cdot 3Cu(OH)_2$
Molecular Formula: $Cl_2Cu_4H_6O_6$
Molecular Weight: 427.133
CAS RN: 16004-08-3
Properties: this basic chloride as well as other compositions can be synthesized by precipitation at controlled pH; other compositions include Cu_2OCl_2, 12167-76-6, and $Cu_4O_3Cl_2$, 12356-86-4; compounds have been used in crop protection, electronics, metallurgy and as catalysts [KIR78]
Solubility: i H_2O [CRC94]
Density, g/cm³: 3.75 [CRC94]
Reactions: minus H_2O at 250°C [CRC94]

984
Compound: Copper(II) lactate dihydrate
Formula: $Cu(C_3H_5O_3)_2 \cdot 2H_2O$
Molecular Formula: $C_6H_{14}CuO_8$
Molecular Weight: 277.718
Properties: greenish blue cryst or granules; fungicide [HAW93]
Solubility: 16.7 g/100 mL H_2O [CRC77]; NH4OH [HAW93]

985
Compound: Copper(II) molybdate
Formula: $CuMoO_4$
Molecular Formula: $CuMoO_4$
Molecular Weight: 223.484
CAS RN: 13767-34-5
Properties: light green cryst; used as a paint pigment, corrosion inhibitor and in protective coatings [HAW93] [KIR81] [MOI86]
Solubility: 0.038 g/100g H_2O [KIR81]

Density, g/cm³: 3.4 [HAW93]
Melting Point, °C: 820 decomposes [KIR81]

986
Compound: Copper(II) nitrate
Synonyms: cupric nitrate
Formula: $Cu(NO_3)_2$
Molecular Formula: CuN_2O_6
Molecular Weight: 187.555
CAS RN: 3251-23-8
Properties: large, bluish green; deliq; ortho-rhomb cryst; can be obtained by sublimation of $Cu(NO_3)_2$ under vacuum from a well mixed mixture of $AgNO_3$ and $CuBr_2$ at 200°C; used as ceramic color, mordant in dyeing, and a catalyst [MER89]
Solubility: s H_2O, ethyl acetate, dioxane; reacts with ether [MER89]; g/100 g soln, H_2O: 45.5 (0°C), 60.1 (25°C), 71.2 (100°C); Solid phase, $Cu(NO_3)_2 \cdot 6H_2O$ (0°C, 25°C), $Cu(NO_3)_2 \cdot 2$-1/2H_2O (100°C) [KRU93]
Melting Point, °C: 255-256 [MER89]

987
Compound: Copper(II) nitrate hexahydrate
Synonyms: cupric nitrate hexahydrate
Formula: $Cu(NO_3)_2 \cdot 6H_2O$
Molecular Formula: $CuH_{12}N_2O_{12}$
Molecular Weight: 295.647
CAS RN: 13478-38-1
Properties: blue; deliq; prismatic cryst [MER89]
Solubility: 243.7 g/100 mL H_2O (0°C) [CRC77]; s alcohol [MER89]
Density, g/cm³: 2.07 [MER89]
Reactions: minus $3H_2O$ at 26.4°C [HAW93]

988
Compound: Copper(II) nitrate trihydrate
Synonyms: cupric nitrate trihydrate, gerhardite
Formula: $Cu(NO_3)_2 \cdot 3H_2O$
Molecular Formula: $CuH_6N_2O_9$
Molecular Weight: 241.602
CAS RN: 10031-43-3
Properties: blue; deliq; rhomb plates; there is a hemipentahydrate with CAS RN 19004-19-4 with same density and melting point as the trihydrate [MER89] [ALD94]
Solubility: 137.8 g/100 mL H_2O (0°C), 1270 g/100 mL H_2O (100°C) [CRC77]; v s alcohol; i ethyl acetate [MER89]
Density, g/cm³: 2.32 [HAW93]
Melting Point, °C: 114.5 [MER89]

Boiling Point, °C: 170, decomposes [HAW93]

989

Compound: Copper(II) oleate

Synonyms: 9-octadecanoic acid, Cu(II) salt

Formula: $Cu(C_{17}H_{33}COO)_2$

Molecular Formula: $C_{36}H_{66}CuO_4$

Molecular Weight: 626.464

CAS RN: 1120-44-1

Properties: blue to green solid; can be prepared by reacting the acid with CuO or basic carbonate of Cu; coalesces mercury drops; improves oil combustion by reducing smoke and fumes of burning oil; used as textile fungicide, used in antifouling marine paints [MER89]

Solubility: i H_2O; sl s alcohol; s ether [MER89]

990

Compound: Copper(II) oxalate

Synonyms: ethanedioic acid, Cu(II) salt

Formula: CuC_2O_4

Molecular Formula: C_2CuO_4

Molecular Weight: 151.566

CAS RN: 814-91-5

Properties: bluish white powd [MER89]

Solubility: i H_2O, alcohol, ether, acetic acid; s NH_4OH [MER89]

Melting Point, °C: decomposes 310 [MER89]

991

Compound: Copper(II) oxalate hemihydrate

Synonyms: cupric oxalate hemihydrate

Formula: $CuC_2O_4 \cdot 1/2H_2O$

Molecular Formula: $C_2HCuO_{4.5}$

Molecular Weight: 160.573

CAS RN: 814-91-5

Properties: bluish white powd; loses any hydrated water by 200°C [MER89]

Solubility: i H_2O, alcohol, ether, acetic acid; s NH_4OH [MER89]

Melting Point, °C: anhydrous decomposes ~300 to copper oxide [HAW93]

992

Compound: Copper(II) oxide

Synonyms: tenorite [1317-92-6], cupric oxide

Formula: CuO

Molecular Formula: CuO

Molecular Weight: 79.545

CAS RN: 1317-38-0

Properties: -20 mesh with 99.999% and -200 mesh with 99.9% purity; black to brownish black amorphous or cryst powd or granules; enthalpy of fusion 11.80 kJ/mol; can be prepared by oxidation of Cu turnings at 800°C; used as a fungicide, herbicide and in cloud seeding, as a heat collector surface in solar energy; reduces tar in tobacco smoke [MER89] [KIR78] [CER91] [CRC93]

Solubility: 0.000320 moles/kg alkaline phosphate: 0.203×10^{-6} mol/kg-H_2O (26.1°C), 0.675×10^{-6} mol/kg-H_2O (94.4°C), 5.241×10^{-6} moles/kg-H_2O (190.0°C), 20.43×10^{-6} m/kg-H_2O (260.0°C) [ZIE92a]

Density, g/cm³: 6.315 [MER89]

Melting Point, °C: 1446 [CRC93]

993

Compound: Copper(II) oxychloride

Synonyms: Brunswick green

Formula: $CuCl_2 \cdot 3CuO \cdot 3\text{-}1/2H_2O$

Molecular Formula: $Cl_2Cu_4H_7O_{6.5}$

Molecular Weight: 436.141

CAS RN: 1332-40-7

Properties: bluish green powd; used as a pigment, pesticide and to control fungus growth in grapevines [HAW93]

Solubility: i H_2O; s acids, ammonia [HAW93]

Reactions: minus $3H_2O$ at 140°C [CRC77]

994

Compound: Copper(II) perchlorate

Synonyms: cupric perchlorate

Formula: $Cu(ClO_4)_2$

Molecular Formula: Cl_2CuO_8

Molecular Weight: 262.446

CAS RN: 13770-18-8

Properties: very pale green; hygr cryst; volatilized by heating; thermally stable to 130°C [MER89]

Solubility: s H_2O, ether, dioxane, ethyl acetate; i benzene, CCl_4 [MER89]; g/100g soln, H_2O: 54.3 (0°C), 59.3 (30°C); Solid phase, $Cu(ClO_4)_2 \cdot 6H_2O$ [KRU93]

Melting Point, °C: ~230-240 [MER89]

Reactions: decomposes to basic perchlorate >130°C [MER89]

995

Compound: Copper(II) perchlorate hexahydrate

Synonyms: cupric perchlorate hexahydrate

Formula: $Cu(ClO_4)_2 \cdot 6H_2O$

Molecular Formula: $Cl_2CuH_{12}O_{14}$

Molecular Weight: 370.538

CAS RN: 10294-46-9
Properties: deep blue; monocl cryst [MER89]
Solubility: v H_2O, methanol, ethanol, acetic acid [MER89]
Density, g/cm³: 2.225 [STR93]
Melting Point, °C: 82 [STR93]
Boiling Point, °C: 120, decomposes [STR93]

996

Compound: Copper(II) phosphate trihydrate
Synonyms: cupric phosphate trihydrate
Formula: $Cu_3(PO_4)_2 \cdot 3H_2O$
Molecular Formula: $Cu_3H_6O_{11}P_2$
Molecular Weight: 434.627
CAS RN: 10031-48-8
Properties: blue or olive green; ortho-rhomb cryst; used in chemical analysis, as a fungicide, catalyst and to inhibit oxidation of metals [HAW93] [MER89]
Solubility: i cold H_2O, sl s hot H_2O; s acids, NH_4OH [MER89]
Reactions: decomposes if heated [MER89]

997

Compound: Copper(II) phthalocyanine
Synonyms: Pigment blue 15C
Molecular Formula: $C_{32}H_{16}CuN_8$
Molecular Weight: 576.079
CAS RN: 147-14-8
Properties: bright blue cryst; both α and β forms, β more stable; used in inks and paints [MER89] [ALD94]
Solubility: s 98% H_2SO_4; i H_2O [MER89]
Melting Point, °C: sublimes ~580 at low pressure of N_2 [MER89]
Reactions: decomposed by hot HNO_3 [MER89]

998

Compound: Copper(II) pyrophosphate hydrate
Synonyms: Unichrome
Formula: $Cu_2P_2O_7 \cdot xH_2O$
Molecular Formula: $Cu_2O_7P_2$ (anhydrous)
Molecular Weight: 301.035 (anhydrous)
CAS RN: 10102-90-6
Properties: used in electroplating copper [HAW93]

999

Compound: Copper(II) selenate pentahydrate
Synonyms: cupric selenate pentahydrate
Formula: $CuSeO_4 \cdot 5H_2O$
Molecular Formula: $CuH_{10}O_9Se$

Molecular Weight: 296.580
CAS RN: 10031-45-5
Properties: light blue; tricl cryst; used as a black colorant for copper [HAW93] [MER89]
Solubility: g/100 g soln H_2O: 10.6±0.2 (0°C), 16.0±0.1 (25°C), Solid phase $CuSeO_4 \cdot 5H_2O$ [KRU93]; s acids, NH_4OH; v sl s acetone; i alcohol [HAW93] [MER89]
Density, g/cm³: 2.56 [MER89]
Reactions: forms monohydrate at 150-200°C, anhydrous by 265°C; decomposes to selenite and basic selenite ~480°C, then to CuO ~700°C [MER89]

1000

Compound: Copper(II) selenide
Synonyms: cupric selenide
Formula: CuSe
Molecular Formula: CuSe
Molecular Weight: 142.506
CAS RN: 1317-41-5
Properties: 6 mm pieces and smaller with 99.5% purity; bluish black to greenish black; prismatic needles or hex plates; decomposes at dull red heat [MER89] [CER91]
Solubility: s HCl, evolves H_2Se; s H_2SO_4, evolves SO_2 [MER89]
Density, g/cm³: 5.99 [STR93]
Melting Point, °C: 550, decomposes [STR93]
Reactions: oxidized to $CuSeO_3$ by HNO_3 [MER89]

1001

Compound: Copper(II) selenite dihydrate
Synonyms: cupric selenite dihydrate
Formula: $CuSeO_3 \cdot 2H_2O$
Molecular Formula: CuH_4O_5Se
Molecular Weight: 226.535
CAS RN: 15168-20-4
Properties: blue; ortho-rhomb or monocl cryst [MER89]
Solubility: i H_2O, H_2SeO_3; s acids, NH_4OH [MER89]
Density, g/cm³: 3.31 [MER89]
Reactions: minus $2H_2O$ by 265°C; decomposes to $CuSeO_3 \cdot CuO$ >460°C; decomposes to CuO >660°C [MER89]

1002

Compound: Copper(II) silicate dihydrate
Synonyms: chrysocolla
Formula: $CuSiO_3 \cdot 2H_2O$
Molecular Formula: CuH_4O_5Si

Molecular Weight: 175.661
CAS RN: 26318-99-0
Properties: green to blue ortho-rhomb; hardness 2.4 Mohs [KIR78]
Density, g/cm³: 2 to 2.24 [CRC77]

1003
Compound: Copper(II) stannate
Formula: CuSnO₃
Molecular Formula: CuO₃Sn
Molecular Weight: 230.254
CAS RN: 12019-07-7
Properties: blue powd [STR93]

1004
Compound: Copper(II) stearate
Synonyms: octadecanoic acid, Cu(II) salt
Formula: Cu[CH₃(CH₂)₁₆COO]₂
Molecular Formula: C₃₆H₇₀CuO₄
Molecular Weight: 630.496
CAS RN: 660-60-6
Properties: pale blue to bluish green; amorphous powd; used in coatings for xerographic plates, and in heat sensitive coatings [MER89] [KIR78]
Solubility: i H₂O, ethanol, ether; s pyridine, dioxane [MER89]
Density, g/cm³: 1.10 [KIR78]
Melting Point, °C: 112 [KIR78]

1005
Compound: Copper(II) sulfate
Synonyms: chalcocyanite
Formula: CuSO₄
Molecular Formula: CuO₄S
Molecular Weight: 159.610
CAS RN: 7758-98-7
Properties: grayish white to greenish white; rhomb cryst or amorphous powd [MER89]
Solubility: i alcohol [MER89]; g/100 g soln, H₂O: 12.7±0.3 (0°C), 18.4±0.2 (25°C), 42.9±0.5 (100°C); Solid phase, CuSO₄·5H₂O (0°C, 25°C), CuSO₄·3H₂O (100°C) [KRU93]
Density, g/cm³: 3.603 [STR93]
Melting Point, °C: 560, decomposes [LID94]

1006
Compound: Copper(II) sulfate pentahydrate
Synonyms: blue vitriol
Formula: CuSO₄·5H₂O
Molecular Formula: CuH₁₀O₉S
Molecular Weight: 249.686

CAS RN: 7758-99-8
Properties: blue; tricl cryst; can be made by reaction of copper with hot, conc H₂SO₄; used as fungicide, source of Cu in animal nutrition; slowly efflorescent in air [KIR78] [MER89]
Solubility: v s H₂O; s methanol; sl s ethanol [MER89]
Density, g/cm³: 2.286 [MER89]
Melting Point, °C: 110, decomposes [STR93]
Reactions: minus 2H₂O at 30°C; minus 2H₂O at 110°C; becomes anhydrous by 250°C [MER89]

1007
Compound: Copper(II) sulfide
Synonyms: covellite, cupric sulfide
Formula: CuS
Molecular Formula: CuS
Molecular Weight: 95.612
CAS RN: 1317-40-4
Properties: -100 mesh with 99.999% and -200 mesh with 99.5% purity; black powd; stable in dry air, oxidized to CuSO₄ in moist air; used in antifouling paints, catalysts; covellite mineral is indigo-blue or darker; hex; hardness 1.5-2 Mohs [KIR78] [MER89] [CER91]
Solubility: i H₂O, alcohol, dil acids and alkalies; s KCN, hot HNO₃ [MER89]
Density, g/cm³: 4.6 [STR93]
Melting Point, °C: 220, decomposes [HAW93]

1008
Compound: Copper(II) tartrate trihydrate
Synonyms: cupric tartrate trihydrate
Formula: CuC₄H₄O₆·3H₂O
Molecular Formula: C₄H₁₀CuO₉
Molecular Weight: 265.664
CAS RN: 815-82-7
Properties: bluish green cryst [STR93]
Solubility: 0.02 g/100 mL H₂O (15°C), 0.14 g/100 mL H₂O (85°C) [CRC77]; s acids, alkali solutions [MER89]
Melting Point, °C: decomposes [CRC77]

1009
Compound: Copper(II) telluride
Formula: CuTe
Molecular Formula: CuTe
Molecular Weight: 191.146
CAS RN: 12019-23-7
Properties: -60 mesh with 99.5% purity; formula is generally Cu₁.₄Te [CER91]
Density, g/cm³: 7.1 [LID94]

1010

Compound: Copper(II) tellurite
Formula: $CuTeO_3$
Molecular Formula: CuO_3Te
Molecular Weight: 239.144
CAS RN: 13812-58-3
Properties: black glassy; -60 mesh with 99% purity [CER91] [CRC77]
Solubility: i H_2O [CRC77]

1011

Compound: Copper(II) tetraammine sulfate monohydrate
Synonyms: tetraaminecopper(II) sulfate monohydrate
Formula: $Cu(NH_3)_4SO_4 \cdot H_2O$
Molecular Formula: $CuH_{14}N_4O_5S$
Molecular Weight: 245.747
CAS RN: 10380-29-7
Properties: dark blue cryst; odor of ammonia; decomposes in air; made by dissolving $CuSO_4$ in water containing ammonia, then by precipitating with ethanol; used in textile printing and as a fungicide [MER89]
Solubility: 18.5 g/100 mL H_2O (21.5°C) [MER89]
Density, g/cm³: 1.810 [ALD94]
Reactions: minus H_2O and $2NH_3$ at 120°C, additional $2NH_3$ at 160°C [MER89]

1012

Compound: Copper(II) tetrafluoroborate
Synonyms: copper(II) fluoroborate
Formula: $Cu(BF_4)_2$
Molecular Formula: B_2CuF_8
Molecular Weight: 237.155
CAS RN: 14735-84-3
Properties: prepared by neutralizing HBF with cupric hydroxide or cupric carbonate, then crystallizing; usually a hydrate; used in electroplating bath formulation [KIR78]

1013

Compound: Copper(II) titanate
Formula: $CuTiO_3$
Molecular Formula: CuO_3Ti
Molecular Weight: 159.411
CAS RN: 12019-08-8
Properties: reacted product -325 mesh 10 microns with 99.5% purity; gray powd [STR93] [CER91]

1014

Compound: Copper(II) trifluoroacetylacetonate
Synonyms: 1,1,1-trifluoro-2,4-pentanedione, Cu(II) derivative
Formula: $Cu(CF_3COCHCOCH_3)_2$
Molecular Formula: $C_{10}H_8CuF_6O_4$
Molecular Weight: 369.707
CAS RN: 14324-82-4
Properties: purple powd [STR93]

$$[CH_3-\overset{\overset{\displaystyle O-}{\displaystyle |}}{C}=CH-\overset{\overset{\displaystyle O}{\displaystyle \|}}{C}-CF_3]_2Cu$$

Melting Point, °C: 194-196 [STR94]
Boiling Point, °C: sublimes 140 (0.1 mm Hg) [STR93]
Reactions: decomposes at 260°C [STR93]

1015

Compound: Copper(II) tungstate
Synonyms: cupric tungstate
Formula: $CuWO_4$
Molecular Formula: CuO_4W
Molecular Weight: 311.384
CAS RN: 13587-35-4
Properties: -200 mesh with 99.5% purity; brown powd [STR93] [CER91]
Density, g/cm³: 7.5 [LID94]

1016

Compound: Copper(II) tungstate dihydrate
Synonyms: cupric tungstate dihydrate
Formula: $CuWO_4 \cdot 2H_2O$
Molecular Formula: CuH_4O_6W
Molecular Weight: 347.414
CAS RN: 13587-35-4
Properties: light green powd; turns brown to grayish yellow by heating, with loss of H_2O; used in semiconductors, nuclear reactors, catalyst for polyester formation [MER89]
Solubility: i H_2O; sl s acetic acid; s NH_4OH, H_3PO_4 [MER89]
Reactions: decomposed by mineral acids [MER89]

1017

Compound: Copper(II) zirconate
Formula: $CuZrO_3$
Molecular Formula: CuO_3Zr
Molecular Weight: 202.768
CAS RN: 70714-64-6

Properties: reacted product -200 mesh with 99.5% purity [CER91]

1018
Compound: Curium (α)
Formula: α-Cm
Molecular Formula: Cm
Molecular Weight: 247
CAS RN: 7440-51-9
Properties: silvery white, hard brittle metal; chemistry of trivalent state similar to that of trivalent lanthanides; α-emitter; hex, a = 0.3496 nm, c = 1.1331 nm; enthalpy of vaporization 1340 kJ/mol; ionic radius of Cm^{+++} is 0.0970 nm; discovered in 1944; used in generating thermoelectric power for remote locations and in space; β-Cm is fcc which is stable <1340°C [HAW93] [MER89] [KIR78]
Density, g/cm³: 13.51 (25°C) [KIR78]
Melting Point, °C: 1345 [KIR91]
Boiling Point, °C: 3110 [KIR91]

1019
Compound: Curium (β)
Formula: β-Cm
Molecular Formula: Cm
Molecular Weight: 247
CAS RN: 7440-51-9
Properties: fcc, a = 0.5039 nm; silvery, hard, brittle metal; oxidizes rapidly in the presence of traces of oxygen; chemistry of trivalent state similar to that of trivalent lanthanides; discovered in 1944; stable <1340°C [KIR78] [MER89]
Density, g/cm³: 12.66 (25°C) [KIR78]
Melting Point, °C: 1350 [MER89]

1020
Compound: Cyanogen
Synonyms: dicyan
Formula: N≡C-C≡N
Molecular Formula: C_2N_2
Molecular Weight: 52.035
CAS RN: 460-19-5
Properties: colorless highly poisonous gas; almond like odor; burns with pink flame with a bluish border; critical pressure 59.6 atm; critical temp 123.3°C; used in organic synthesis, welding and as a rocket propellant [HAW93] [MER89]
Solubility: 1.1-1.3 g/100g H_2O; 26 g/100g alcohol; 5 g/100g ether [CIC73]
Density, g/cm³: liq at b.p: 0.9537 [MER89]; gas: 2.321 g/L [CIC73]

Melting Point, °C: -28 [COT88]
Boiling Point, °C: -21.17 [MER89]
Reactions: slowly hydrolyzed to oxalic acid and ammonia [MER89]

1021
Compound: Cyanogen azide
Synonyms: carbon pernitride
Formula: N=N=N-C≡N
Molecular Formula: CN_4
Molecular Weight: 68.038
CAS RN: 764-05-6
Properties: clear, colorless, oily liq; can detonate on thermal, electrical or mechanical shock; decomposes in acetonitrile solvent; made by suspending NaN_3 in dry acetonitrile, followed by distillation of cyanogen chloride into the cooled suspension; used in organic synthesis for example reacts with alkanes to produce primary alkylcyanamides [MER89]
Thermal Expansion Coefficient: from 25°C to: 100°C (0.18), 200°C (0.48), 400°C (1.11), 600°C (1.77) [TOU77]

1022
Compound: Cyanogen bromide
Synonyms: bromocyanide
Formula: BrC≡N
Molecular Formula: CBrN
Molecular Weight: 105.922
CAS RN: 506-68-3
Properties: white cryst; sensitive to moisture; corrodes most metals; prepared from Br_2 and KCN; used in organic synthesis, as a fumigant and in gold extraction [HAW93] [STR93]
Solubility: decomposed slowly by cold H_2O [HAW93]
Density, g/cm³: 2.015 [STR93]
Melting Point, °C: 52 [STR93]
Boiling Point, °C: 61.4 [STR93]

1023
Compound: Cyanogen chloride
Synonyms: chlorocyanide
Formula: ClC≡N
Molecular Formula: CClN
Molecular Weight: 61.470
CAS RN: 506-77-4
Properties: colorless gas or liq with irritating vapor; prepared from Cl_2 and HCN; used in chemical syntheses [HAW93] [MER89]
Solubility: s H_2O, ether, alcohol [MER89]

Density, g/cm³: 2.697 g/L [LID94]
Melting Point, °C: -6 [MER89]
Boiling Point, °C: 13.8 [MER89]

1024

Compound: Cyanogen fluoride
Synonyms: fluorocyanide
Formula: FC≡N
Molecular Formula: CFN
Molecular Weight: 45.016
CAS RN: 1495-50-7
Properties: colorless gas; made by reaction of AgF and cyanogen iodide; used in tear gas and in organic synthesis [HAW93]
Solubility: i H_2O [HAW93]
Density, g/cm³: 1.975 g/L [LID94]
Melting Point, °C: -82 [LID94]
Boiling Point, °C: -46 [LID94]

1025

Compound: Cyclohexadiene iron tricarbonyl
Formula: $C_6H_8Fe(CO)_3$
Molecular Formula: $C_{11}H_8FeO_3$
Molecular Weight: 220.008
CAS RN: 12152-72-6
Properties: orange-yellow liq; air sensitive [STR93]
Melting Point, °C: 8 [STR93]

1026

Compound: Cyclooctatetraene iron tricarbonyl
Formula: $C_8H_8Fe(CO)_3$
Molecular Formula: $C_{11}H_8FeO_3$
Molecular Weight: 244.029
CAS RN: 12093-05-9
Properties: red-brown cryst; sensitive to air [STR93]
Melting Point, °C: 93-95 [STR93]

1027

Compound: Cyclopentadienylindium(I)
Formula: C_5H_5In
Molecular Formula: C_5H_5In
Molecular Weight: 179.915
CAS RN: 34822-89-4
Properties: off-white cryst; sensitive to air, light and heat [STR93]
Melting Point, °C: sublimes 50 (0.01 mm Hg) [STR93]

1028

Compound: Cyclopentadienyliron dicarbonyl dimer
Formula: $[C_5H_5Fe(CO)_2]_2$
Molecular Formula: $C_{14}H_{10}Fe_2O_4$
Molecular Weight: 353.925
CAS RN: 12154-95-9
Properties: purple-red cryst; air sensitive [STR93]
Melting Point, °C: 194 decomposes [STR93]

1029

Compound: Cyclopentadienylniobium tetrachloride
Formula: $C_5H_5NbCl_4$
Molecular Formula: $C_5H_5Cl_4Nb$
Molecular Weight: 229.812
CAS RN: 33114-15-7
Properties: red-brown powd; sensitive to moisture [STR93]
Melting Point, °C: 180, decomposes [STR93]

1030

Compound: Decaborane(14)
Formula: $B_{10}H_{14}$
Molecular Formula: $B_{10}H_{14}$
Molecular Weight: 122.221
CAS RN: 17702-41-9
Properties: white cryst [STR93]
Solubility: sl s cold H_2O [MER89]
Density, g/cm³: 0.94 [STR93]
Melting Point, °C: 100 [STR93]
Boiling Point, °C: 213 (extrapolated) [COT88]
Reactions: hydrolyzed by hot H_2O [MER89]

1031

Compound: Deuterium
Synonyms: heavy water
Formula: D_2
Molecular Formula: D_2
Molecular Weight: 4.032
CAS RN: 7782-39-0
Properties: colorless odorless gas; flammable; stable, not radioactive; specific volume 6.00 m³/kg at 21.1°C and 101.3 kPa; critical temp -234.75°C; critical pressure 16.432 atm; enthalpy of vaporization 1.23 kJ/mol; enthalpy of fusion 197 J/mol; used extensively at trace levels in measuring rates of chemical reactions [MER89] [AIR87] [KIR78]
Density, g/cm³: gas: 0.00018; liq: 0.169 at -252.7°C [MER89] [ALF93]
Melting Point, °C: -254.6 [ALF93]
Boiling Point, °C: -249.7 [ALF93]

Thermal Conductivity, W/(m·K): 0.126 (-252.7°C) [KIR78]

1032

Compound: Deuterium bromide
Formula: DBr
Molecular Formula: BrD
Molecular Weight: 81.919
CAS RN: 13536-59-9
Properties: corrosive [ALD94]
Density, g/cm³: 1.537 [ALD93]
Boiling Point, °C: 126 [ALD93]

1033

Compound: Deuterium chloride
Formula: DCl
Molecular Formula: ClD
Molecular Weight: 37.468
CAS RN: 7698-05-7
Properties: gas [CRC77]
Solubility: 11.9 cm³/100 mL H₂O (25°C), 8.4 cm³/100 mL H₂O (40°C) [CRC77]
Melting Point, °C: -254.6 [CRC77]
Boiling Point, °C: -249.7 [CRC77]

1034

Compound: Deuterium iodide
Formula: DI
Molecular Formula: DI
Molecular Weight: 128.919
CAS RN: 14104-45-1
Properties: hygr [ALD94]

1035

Compound: Deuterium oxide
Synonyms: water-2$_d$
Formula: D₂O
Molecular Formula: D₂O
Molecular Weight: 20.028
CAS RN: 7789-20-0
Properties: ordinary water contains about one part of D₂O to 6500 parts of H₂O; triple point 3.82°C; critical temp 371.5°C; enthalpy of fusion 6.280 kJ/mol; enthalpy of vaporization 41.493 kJ/mol; dielectric constant (25°C) 78.06; finds use in the study of rates and mechanisms of chemical reactions [MER89] [HAW93]
Density, g/cm³: 1.1056 [HAW93]
Melting Point, °C: 3.81 [MER89]
Boiling Point, °C: 101.42 [MER89]

1036

Compound: Deuterosulfuric acid
Formula: D₂SO₄
Molecular Formula: D₂O₄S
Molecular Weight: 100.094
CAS RN: 13813-19-9
Properties: liq; 96-98%, in D₂O [ALF93]
Density, g/cm³: 1.878 [ALD94]

1037

Compound: Diborane(6)
Synonyms: boroethane
Formula: B₂H₆
Molecular Formula: B₂H₆
Molecular Weight: 27.670
CAS RN: 19287-45-7
Properties: gas; spontaneously flammable in air; critical temp 16.7°C; critical pressure 4.00 MPa; enthalpy of vaporization 14.28 kJ/mol; can be prepared from NaBH₄ and H₃PO₄; used as a catalyst and as a reducing agent [AIR87] [COT88] [CRC93] [MER89]
Solubility: hydrolyzes quickly [COT88]
Density, g/cm³: 1.214 g/L [LID94]
Melting Point, °C: -164.9 [AIR87]
Boiling Point, °C: -87.55 [CRC93]
Reactions: decomposes at red heat to B + H₂ [MER89]

1038

Compound: Dibromosilane
Formula: SiH₂Br₂
Molecular Formula: Br₂H₂Si
Molecular Weight: 189.910
CAS RN: 13768-94-0
Properties: colorless liq; flammable; enthalpy of vaporization 31 kJ/mol; entropy of vaporization 91.21 kJ/(mol·K) [CRC93] [CIC73]
Solubility: decomposed by H₂O [CRC77]
Density, g/cm³: 2.17 (0°C) [CRC77]
Melting Point, °C: -70.1 [CIC73]
Boiling Point, °C: 66 [CIC73]

1039

Compound: Dicarbonylacetylacetonate iridium(I)
Formula: Ir(CO)₂(CH₃COCHCOCH₃)
Molecular Formula: C₇H₇IrO₄
Molecular Weight: 347.350
CAS RN: 14023-80-4
Properties: copper-brown cryst [STR93]

1040
Compound: Dichlorodiamminepalladium(II)
Formula: $Pd(NH_3)_2Cl_2$
Molecular Formula: $Cl_2H_6N_2Pd$
Molecular Weight: 211.386
CAS RN: cis: 15684-18-1; trans: 13782-33-7
Properties: yellow powd [STR93]
Density, g/cm³: trans: 2.50 [ALD94]

1041
Compound: Dichlorodiammineplatinum(II) (cis)
Synonyms: cisplatin
Formula: $Pt(NH_3)_2Cl_2$
Molecular Formula: $Cl_2H_6N_2Pt$
Molecular Weight: 300.046
CAS RN: 15663-27-1
Properties: deep yellow solid; uses: antineoplastic, has antitumour activity [MER89] [ALD94]
Solubility: 0.253 g/L H_2O (25°C) [MER89]
Melting Point, °C: 270, decomposes [MER89]
Reactions: cis → trans change slowly in aq media [MER89]

1042
Compound: Dichlorodiammineplatinum(II)-trans
Formula: $Pt(NH_3)_2Cl_2$
Molecular Formula: $Cl_2H_6N_2Pt$
Molecular Weight: 300.046
CAS RN: 14913-33-8
Melting Point, °C: 340, decomposes [ALD94]

1043
Compound: Dichlorodifluoromethane
Synonyms: halocarbon-12
Formula: CCl_2F_2
Molecular Formula: CCl_2F_2
Molecular Weight: 120.913
CAS RN: 75-71-8
Properties: colorless gas; nonflammable; critical temp 112.04°C; critical pressure 4.14 MPa; enthalpy of vaporization 19.99 kJ/mol; used in electronics industry [AIR87]
Melting Point, °C: -157.8 [AIR87]
Boiling Point, °C: -29.8 [AIR87]

1044
Compound: Dichlorosilane
Formula: SiH_2Cl_2
Molecular Formula: Cl_2H_2Si
Molecular Weight: 101.007
CAS RN: 4109-96-0

Properties: colorless flammable gas; sharp pungent odor; autoignition temp 100°C; critical temp 176.0°C; critical pressure 4.68 MPa; enthalpy of vaporization 25 kJ/mol; entropy of vaporization 89.5 kJ/(mol·K); used in electronics industry [AIR87] [CIC73] [CRC93]
Solubility: hydrolyzes in H_2O [AIR87]
Density, g/cm³: 3.47 (air = 1) [AIR87]
Melting Point, °C: -122 [CIC73]
Boiling Point, °C: 8.3 [CIC73]

1045
Compound: Diethylaluminum chloride
Synonyms: DEAC
Formula: $(C_2H_5)_2AlCl$
Molecular Formula: $C_4H_{10}AlCl$
Molecular Weight: 120.558
CAS RN: 96-10-6
Properties: colorless volatile liq; flammable; sensitive to moisture; prepared from ethyl halide and Al; uses: aldol condensation reagent [ALD94] [MER89]
Density, g/cm³: 0.961 [ALD94]
Melting Point, °C: -50 [ALD94]
Boiling Point, °C: 125-126 at 50 mm Hg [ALD94]
Reactions: can be cleaved by H_2O [MER89]

1046
Compound: Diethylzinc
Synonyms: zinc diethyl
Formula: $(C_2H_5)_2Zn$
Molecular Formula: $C_4H_{10}Zn$
Molecular Weight: 123.513
CAS RN: 557-20-0
Properties: liq; can ignite in air; preparation: reaction of Zn and diethyl iodide; uses: organic synthesis, preservation of archival papers [MER89]
Solubility: miscible with ether, petroleum ether, benzene [MER89]
Density, g/cm³: 1.2065 [MER89]
Melting Point, °C: -28 [ALD94]
Boiling Point, °C: 117 [ALD94]

1047
Compound: Difluorophosphoric acid
Formula: HPO_2F_2
Molecular Formula: F_2HO_2P
Molecular Weight: 101.978
CAS RN: 13779-41-4
Properties: mobile colorless liq; fumes in air [KIR78]

Density, g/cm³: 1.583 [KIR78]
Melting Point, °C: -96.5 or -91.3 [KIR78]
Boiling Point, °C: 107-111 decomposes [KIR78]

1048
Compound: Difluorosilane
Formula: SiH_2F_2
Molecular Formula: F_2H_2Si
Molecular Weight: 68.099
CAS RN: 13824-36-7
Properties: enthalpy of vaporization 16.3 kJ/mol; entropy of vaporization 84 kJ/(mol·K) [CIC73] [CRC93]
Melting Point, °C: -122 [CIC73]
Boiling Point, °C: -77.8 [CRC93]

1049
Compound: Dihydrazine sulfate
Formula: $(N_2H_4)_2·H_2SO_4$
Molecular Formula: $H_{10}N_4O_4S$
Molecular Weight: 162.170
CAS RN: 13464-80-7
Properties: white cryst; deliq [LAN52] [MER89]
Solubility: g/100g H_2O: 221 (30°C), 300 (40°C), 554 (60°C) [LAN85]
Melting Point, °C: ~104 [MER89]
Boiling Point, °C: decomposes [MER89]

1050
Compound: Diiodosilane
Formula: SiH_2I_2
Molecular Formula: H_2I_2Si
Molecular Weight: 283.911
CAS RN: 13760-02-6
Properties: enthalpy of vaporization 36.8 kJ/mol; entropy of vaporization 87.0 kJ/(mol·K) [CIC73]
Melting Point, °C: -1 [CIC73]
Boiling Point, °C: 150 [CIC73]

1051
Compound: Diisobutylaluminum chloride
Formula: $[(CH_3)_2CHCH_2]_2AlCl$
Molecular Formula: $C_8H_{18}AlCl$
Molecular Weight: 176.665
CAS RN: 1779-25-5
Properties: liq; pyrophoric; sensitive to air; uses: transmetallating reagent [ALD94]
Density, g/cm³: 0.905 [ALD94]
Melting Point, °C: -40 [ALD94]
Boiling Point, °C: 152 (10 mm Hg) [ALD94]

1052
Compound: Dimethylaminotrimethyltin
Synonyms: pentamethylstannanamine
Formula: $(CH_3)_3SnN(CH_3)_2$
Molecular Formula: $C_5H_{15}NSn$
Molecular Weight: 207.891
CAS RN: 993-50-0
Properties: liq; uses: a dehydrochlorinating agent [ALD94]
Density, g/cm³: 1.274 [ALD94]
Melting Point, °C: 1 [ALD94]
Boiling Point, °C: 126 [ALD94]

1053
Compound: Dimethylgermanium dichloride
Formula: $(CH_3)_2GeCl_2$
Molecular Formula: $C_2H_6Cl_2Ge$
Molecular Weight: 173.585
CAS RN: 1529-48-2
Properties: liq; flammable [ALD94]
Density, g/cm³: 1.505 [ALD94]
Melting Point, °C: -22 [ALD94]
Boiling Point, °C: 123 [ALD94]

1054
Compound: Dimethylmercury
Synonyms: methylmercury
Formula: $(CH_3)_2Hg$
Molecular Formula: C_2H_6Hg
Molecular Weight: 230.659
CAS RN: 593-74-8
Properties: colorless, volatile liq; flammable; uses: inorganic reagent [MER89]
Solubility: i H_2O; s ether, alcohol [MER89]
Density, g/cm³: 2.961 [ALD94]
Melting Point, °C: -43 [ALD94]
Boiling Point, °C: 93-94 [ALD94]

1055
Compound: Disilane
Formula: Si_2H_6
Molecular Formula: H_6Si_2
Molecular Weight: 62.219
CAS RN: 1590-87-0
Properties: colorless gas; extremely reactive, ignites in air spontaneously; critical temp 150.9°C; critical pressure 5.15 MPa; enthalpy of vaporization 21.21 kJ/mol; critical temp 109°C; made by photolysis of SiH_4 and H_2 mixture; used in electronics industry [AIR87] [CIC73] [COT88]
Solubility: s CS_2, ethanol, benzene [MER89]

Density, g/cm³: liq, at bp: 0.69 [CIC73]
Melting Point, °C: -132.5 [CIC73]
Boiling Point, °C: -14.5 [CIC73]
Reactions: decomposes at 300°C; KOH causes evolution of H_2 [MER89]

1056
Compound: Disulfur decafluoride
Synonyms: sulfur pentafluoride
Formula: S_2F_{10}
Molecular Formula: $F_{10}S_2$
Molecular Weight: 254.116
CAS RN: 5714-22-7
Properties: colorless gas; odor of SO_2; vapor pressure is 561 torr at 20°C [HAW93]
Solubility: i H_2O [HAW93]
Density, g/cm³: 2.08 [HAW93]
Melting Point, °C: -92 [HAW93]
Boiling Point, °C: 29 [HAW93]

1057
Compound: Dysprosium
Formula: Dy
Molecular Formula: Dy
Molecular Weight: 162.50
CAS RN: 7429-91-6
Properties: silver metal; tarnishes in moist air; hex close-packed; forms greenish yellow salts; enthalpy of fusion 10.782 kJ/mol; enthalpy of sublimation 290.4 kJ/mol; radius of atom 0.17743 nm; radius of ion 0.0908 nm for Dy^{+++}; light yellow solutions; electrical reisitivity at 20°C 89 μohm·cm [KIR82] [MER89] [ALD94]
Solubility: reacts slowly with H_2O; s dil acids [HAW93]
Density, g/cm³: 8.55 [KIR82]
Melting Point, °C: 1412 [KIR82]
Boiling Point, °C: 2567 [KIR82]
Thermal Conductivity, W/(m·K): 10.7 (25°C) [ALD94]
Thermal Expansion Coefficient: 9.9×10^{-6}/K [CRC93]

1058
Compound: Dysprosium acetate tetrahydrate
Formula: $Dy(CH_3COO)_3 \cdot 4H_2O$
Molecular Formula: $C_6H_{17}DyO_{10}$
Molecular Weight: 411.695
CAS RN: 15280-55-4
Properties: yellow cryst [ALF93]
Solubility: s H_2O [CRC77]
Melting Point, °C: 120, decomposes [ALF93]

1059
Compound: Dysprosium acetylacetonate
Synonyms: 2,4-pentanedione, dysprosium(III) derivative
Formula: $Dy(CH_3COCH=C(O)CH_3)_3$
Molecular Formula: $C_{15}H_{21}DyO_6$
Molecular Weight: 459.828
CAS RN: 14637-88-8
Properties: powd [STR93]

$$[CH_3-\overset{\overset{O-}{|}}{C}=CH-\overset{\overset{O}{\|}}{C}-CH_3]_3Dy$$

1060
Compound: Dysprosium boride
Formula: DyB_4
Molecular Formula: B_4Dy
Molecular Weight: 205.744
CAS RN: 12310-43-9
Properties: -60 mesh with 99.9% purity; there is a DyB_6 material, -60 mesh and 99.9% purity, 12008-04-7 [CER91]
Density, g/cm³: 6.98 [LID94]
Melting Point, °C: 2500 [LID94]

1061
Compound: Dysprosium bromide
Formula: $DyBr_3$
Molecular Formula: Br_3Dy
Molecular Weight: 402.212
CAS RN: 14456-48-5
Properties: colorless cryst; -20 mesh with 99.9% purity [CER91] [CRC77]
Solubility: s H_2O [CRC77]
Melting Point, °C: 881 [AES93]
Boiling Point, °C: 1480 [CRC77]

1062
Compound: Dysprosium carbonate tetrahydrate
Formula: $Dy_2(CO_3)_3 \cdot 4H_2O$
Molecular Formula: $C_3H_8Dy_2O_{13}$
Molecular Weight: 577.089
CAS RN: 38245-35-1
Properties: white powd; cryst [ALF93]
Solubility: i H_2O [CRC77]
Reactions: minus $3H_2O$ at 15°C [CRC77]

1063
Compound: Dysprosium chloride
Formula: $DyCl_3$

Molecular Formula: Cl_3Dy
Molecular Weight: 268.858
CAS RN: 10025-74-8
Properties: -20 mesh with 99.9% purity; white powd [STR93]; yellow, shining cryst [MER89] [CER91]
Density, g/cm³: 3.67 [MER89]
Melting Point, °C: 680 [MER89]; 718 [STR93]

1064
Compound: Dysprosium chloride hexahydrate
Formula: $DyCl_3 \cdot 6H_2O$
Molecular Formula: $Cl_3DyH_{12}O_6$
Molecular Weight: 376.949
CAS RN: 15059-52-6
Properties: -4 mesh with 99.9% purity; light yellow cryst; hygr [STR93] [CER91]
Melting Point, °C: 718 [ALF93]

1065
Compound: Dysprosium fluoride
Formula: DyF_3
Molecular Formula: DyF_3
Molecular Weight: 219.495
CAS RN: 13569-80-7
Properties: white powd, or 99.9% pure melted pieces of 3-12 mm; hygr; melted pieces used as evaporation material for possible application to multilayers [STR93] [CER91]
Melting Point, °C: 1154 [LID94]
Boiling Point, °C: >2200 [STR93]

1066
Compound: Dysprosium hydride
Formula: DyH_3
Molecular Formula: DyH_3
Molecular Weight: 165.524
CAS RN: 13537-09-2
Properties: hex; -40 mesh with 99.9% purity; lump [ALF93] [CER91] [LID94]
Density, g/cm³: 7.1 [LID94]

1067
Compound: Dysprosium hydroxide
Formula: $Dy(OH)_3$
Molecular Formula: DyH_3O_3
Molecular Weight: 213.522
CAS RN: 1308-85-6
Properties: gelatinous precipitate; forms a blue colloidal solution [MER89]

1068
Compound: Dysprosium iodide
Formula: DyI_3
Molecular Formula: DyI_3
Molecular Weight: 544.213
CAS RN: 15474-63-2
Properties: greenish yellow cryst; -20 mesh with 99.9% purity; yellow powd [STR93] [CER91] [CRC77]
Solubility: s H_2O [CRC77]
Melting Point, °C: 955 [STR93]
Boiling Point, °C: 1320 [STR93]

1069
Compound: Dysprosium nitrate pentahydrate
Formula: $Dy(NO_3)_3 \cdot 5H_2O$
Molecular Formula: $DyH_{10}N_3O_{14}$
Molecular Weight: 438.591
CAS RN: 10031-49-9
Properties: yellow cryst; hygr [STR93]
Solubility: s H_2O [MER89]
Melting Point, °C: at 88.6, melts in waters of hydrate [MER89]

1070
Compound: Dysprosium nitride
Formula: DyN
Molecular Formula: DyN
Molecular Weight: 176.507
CAS RN: 12019-88-4
Properties: -60 mesh with 99.9% purity [CER91]
Density, g/cm³: 9.93 [LID94]

1071
Compound: Dysprosium oxalate decahydrate
Formula: $Dy_2(C_2O_4)_3 \cdot 10H_2O$
Molecular Formula: $C_6H_{20}Dy_2O_{22}$
Molecular Weight: 769.210
CAS RN: 24670-07-3
Properties: white cryst [STR93]
Solubility: i H_2O [CRC77]
Melting Point, °C: 40 [ALF93]
Reactions: minus H_2O at 40°C [CRC77]

1072
Compound: Dysprosium oxide
Synonyms: dysprosia
Formula: Dy_2O_3
Molecular Formula: Dy_2O_3
Molecular Weight: 372.998
CAS RN: 1308-87-8

Properties: white powd; can be prepared by heating the oxalate or sulfate; more magnetic than ferric oxide; sl hygr; absorbs atm H_2O and CO_2; used with nickel in cermets, and as an evaporated film of 99.9% purity it is reactive to radio frequencies [HAW93] [MER89] [CER91]
Solubility: s acids [HAW93]
Density, g/cm³: 7.81 [MER89]
Melting Point, °C: 2330-2350 [STR93]

1073
Compound: Dysprosium perchlorate hydrate
Formula: $Dy(ClO_4)_3 \cdot xH_2O$
Molecular Formula: Cl_3DyO_{12} (anhydrous)
Molecular Weight: 460.851 (anhydrous)
CAS RN: 14692-17-2
Properties: white cryst; hygr [STR93]

1074
Compound: Dysprosium silicide
Formula: $DySi_2$
Molecular Formula: $DySi_2$
Molecular Weight: 218.671
CAS RN: 12133-07-2
Properties: ortho-rhomb; 10 mm & down lump, 6 mm pieces and smaller with 99.9% purity [ALF93] [CER91] [LID94]
Density, g/cm³: 5.2 [LID94]

1075
Compound: Dysprosium sulfate octahydrate
Formula: $Dy_2(SO_4)_3 \cdot 8H_2O$
Molecular Formula: $Dy_2H_{16}O_{20}S_3$
Molecular Weight: 757.313
CAS RN: 10031-50-2
Properties: yellow cryst; can be prepared by dissolving the oxide in H_2SO_4, then precipitating with alcohol; stable in air at 110°C [MER89]
Solubility: s H_2O [HAW93]
Reactions: minus $8H_2O$ at 360°C [MER89]

1076
Compound: Dysprosium sulfide
Formula: Dy_2S_3
Molecular Formula: Dy_2S_3
Molecular Weight: 421.198
CAS RN: 12133-10-7
Properties: reddish brown monocl; -200 mesh with 99.9% purity [CER91] [LID94]
Density, g/cm³: 6.08 [LID94]

1077
Compound: Dysprosium telluride
Formula: Dy_2Te_3
Molecular Formula: Dy_2Te_3
Molecular Weight: 707.800
CAS RN: 12159-43-2
Properties: -20 mesh with 99.9% purity [CER91]

1078
Compound: Einsteinium
Formula: Es
Molecular Formula: Es
Molecular Weight: 252
CAS RN: 7429-92-7
Properties: man made radioisotope; identified by Ghiorso and colleagues at Berkeley in December 1952, as part of debris from first large thermonuclear explosion; chemical properties similar to those of holmium; ionic radius of Es^{+++} is 0.0925 nm; has lowest enthalpy of vaporization of any of the divalent elements; cub, a = 0.575 nm; discovered in 1952; $t_{1/2}$ of ^{253}Es is 20.5 days, $t_{1/2}$ of ^{254}Es is 276 days, $t_{1/2}$ of ^{255}Es is 40 days [HAW93] [KIR78]
Melting Point, °C: 860 [KIR91]

1079
Compound: Erbium
Formula: Er
Molecular Formula: Er
Molecular Weight: 167.26
CAS RN: 7440-52-0
Properties: soft, malleable dark gray metallic solid; hex close-pack cryst; similar to other rare earths; used in nuclear controls, room temp laser; enthalpy of fusion 19.90 kJ/mol; enthalpy of sublimation 317.10 kJ/mol; electrical resistivity at 20°C 86 μohm·cm; radius of atom 0.17566 nm; radius of ion 0.0881 nm, Er^{+++}, pink colored solutions [MER89] [HAW93] [KIR82] [ALD94]
Solubility: i H_2O; s acids [HAW93]
Density, g/cm³: 9.066 [KIR82]
Melting Point, °C: 1529 [KIR82]
Boiling Point, °C: 2868 [KIR82]
Thermal Conductivity, W/(m·K): 14.5 (25°C) [CRC93]
Thermal Expansion Coefficient: 12.2 x 10^{-6}/K [CRC93]

1080
Compound: Erbium acetate tetrahydrate

Formula: $Er(CH_3COO)_3 \cdot 4H_2O$
Molecular Formula: $C_6H_{17}ErO_{10}$
Molecular Weight: 416.455
CAS RN: 15280-57-6
Properties: pink cryst; tricl [STR93] [CRC77]
Density, g/cm³: 2.114 [STR93]

1081
Compound: Erbium acetylacetonate hydrate
Synonyms: 2,4-pentanedione, erbium(III) derivative
Formula: $Er(CH_3COCH=C(O)CH_3)_3 \cdot xH_2O$
Molecular Formula: $C_{15}H_{21}ErO_6$ (anhydrous)
Molecular Weight: 464.588 (anhydrous)
CAS RN: 14553-08-3
Properties: off-white powd [STR93]

1082
Compound: Erbium barium copper oxide
Formula: $ErBa_2Cu_3O_x$
Molecular Formula: $Ba_2Cu_3ErO_x$
Properties: 99.9% and 99.999%, 0.2 micron and 20 micron powd, high T_c superconductor [ALF93]

1083
Compound: Erbium boride
Formula: ErB_4
Molecular Formula: B_4Er
Molecular Weight: 210.504
CAS RN: 12310-44-0
Properties: tetr; -100 mesh with 99.9% purity [CER91] [LID94]
Density, g/cm³: 7.0 [LID94]
Melting Point, °C: 2450 [LID94]

1084
Compound: Erbium bromide
Formula: $ErBr_3$
Molecular Formula: Br_3Er
Molecular Weight: 406.972
CAS RN: 13536-73-7
Properties: -20 mesh with 99.9% purity; powd [ALF93] [CER91]
Melting Point, °C: 923 (under argon) [ALF95]

1085
Compound: Erbium bromide nonahydrate
Formula: $ErBr_3 \cdot 9H_2O$
Molecular Formula: $Br_3ErH_{18}O_9$
Molecular Weight: 569.110
CAS RN: 13536-73-7

Properties: rose cryst; deliq [MER89]
Solubility: s H_2O [CRC77]
Boiling Point, °C: 1460 [CRC77]

1086
Compound: Erbium carbonate hydrate
Formula: $Er_2(CO_3)_3 \cdot xH_2O$
Molecular Formula: $C_3Er_2O_9$ (anhydrous)
Molecular Weight: 514.548 (anhydrous)
CAS RN: 22992-83-2
Properties: pink powd [STR93]

1087
Compound: Erbium chloride
Formula: $ErCl_3$
Molecular Formula: Cl_3Er
Molecular Weight: 273.618
CAS RN: 10138-41-7
Properties: -20 mesh with 99.9% purity; pinkish powd [MER89] [CER91]
Density, g/cm³: 4.1 [MER89]
Melting Point, °C: 774 [STR93]
Boiling Point, °C: 1500 [STR93]

1088
Compound: Erbium chloride hexahydrate
Formula: $ErCl_3 \cdot 6H_2O$
Molecular Formula: $Cl_3ErH_{12}O_6$
Molecular Weight: 381.709
CAS RN: 10025-75-9
Properties: pink; hygr cryst [STR93]
Solubility: s H_2O; s sl alcohol [MER89]
Melting Point, °C: 1500 [ALF93]

1089
Compound: Erbium fluoride
Formula: ErF_3
Molecular Formula: ErF_3
Molecular Weight: 224.255
CAS RN: 13760-83-3
Properties: rose powd, or 99.9% pure melted pieces of 3-6 mm ; hygr; melted pieces used as evaporation material for possible application to multilayers [STR93] [CER91]
Density, g/cm³: 7.814 [STR93]
Melting Point, °C: 1350 [STR93]
Boiling Point, °C: 2200 [STR93]

1090
Compound: Erbium hydride

Formula: ErH$_3$
Molecular Formula: ErH$_3$
Molecular Weight: 170.284
CAS RN: 13550-53-3
Properties: hex; -60 mesh with 99.9% purity; lump [ALF93] [CER91] [LID94]
Density, g/cm^3: ~7.6 [LID94]

1091
Compound: Erbium hydroxide
Formula: Er(OH)$_3$
Molecular Formula: ErH$_3$O$_3$
Molecular Weight: 218.282
CAS RN: 14646-16-3
Properties: pale pink, gelatinous precipitate [MER89]

1092
Compound: Erbium iodide
Formula: ErI$_3$
Molecular Formula: ErI$_3$
Molecular Weight: 547.973
CAS RN: 13813-42-8
Properties: -20 mesh with 99.9% purity; red hygr powd [STR93] CER91]
Density, g/cm^3: ~5.5 [LID94]
Melting Point, °C: 1020 [STR93]
Boiling Point, °C: 1280 [STR93]

1093
Compound: Erbium nitrate pentahydrate
Formula: Er(NO$_3$)$_3$·5H$_2$O
Molecular Formula: ErH$_{10}$N$_3$O$_{14}$
Molecular Weight: 443.351
CAS RN: 10031-51-3
Properties: reddish cryst; deliq [MER89]
Solubility: 5.5223±0.0053 mol/(kg·H$_2$O) at 25°C; reference is uncertain about the number of hydrated waters, and pentahydrate is assumed for this solubility [RAR85b]
Reactions: minus 4H$_2$O at 130°C [MER89]

1094
Compound: Erbium nitride
Formula: ErN
Molecular Formula: ErN
Molecular Weight: 181.267
CAS RN: 12020-21-2
Properties: cub; -60 mesh with 99.9% purity [CER91] [LID94]
Density, g/cm^3: 10.6 [LID94]

1095
Compound: Erbium oxalate decahydrate
Formula: Er$_2$(C$_2$O$_4$)$_3$·10H$_2$O
Molecular Formula: C$_6$H$_{20}$Er$_2$O$_{22}$
Molecular Weight: 778.732
CAS RN: 30618-31-6
Properties: pink cryst [STR93]
Solubility: s H$_2$O, dil acids [HAW93]
Density, g/cm^3: 2.64 [HAW93]
Melting Point, °C: 575, decomposes [HAW93]

1096
Compound: Erbium oxide
Synonyms: erbia
Formula: Er$_2$O$_3$
Molecular Formula: Er$_2$O$_3$
Molecular Weight: 382.518
CAS RN: 12061-16-4
Properties: pinkish powd; changes into cub cryst at 1300°C; readily absorbs atm H$_2$O and CO$_2$; used as a phosphor activator, in infrared absorbing glass, and as an evaporated material of 99.9% purity it is reactive to radio frequencies [HAW93] [MER89] [CER91]
Solubility: 1.28 x 10^{-5} g mol/L H$_2$O (29°C); v s a [MER89]
Density, g/cm^3: 8.64 [MER89]
Melting Point, °C: 2400 [STR93]

1097
Compound: Erbium perchlorate hydrate
Formula: Er(ClO$_4$)$_3$·xH$_2$O
Molecular Formula: Cl$_3$ErO$_{12}$ (anhydrous)
Molecular Weight: 465.611 (anhydrous)
CAS RN: 61565-07-9
Properties: pink cryst; hygr [STR93]

1098
Compound: Erbium silicide
Formula: ErSi$_2$
Molecular Formula: ErSi$_2$
Molecular Weight: 223.431
CAS RN: 12020-28-9
Properties: ortho-rhomb; 10 mm & down lump [ALF93] [LID94]
Density, g/cm^3: 7.26 [LID94]

1099
Compound: Erbium sulfate
Formula: Er$_2$(SO$_4$)$_3$
Molecular Formula: Er$_2$O$_{12}$S$_3$

Molecular Weight: 622.711
CAS RN: 13478-49-4
Properties: powd; hygr; dissociates on heating in H$_2$O with evolution of heat [MER89]
Density, g/cm^3: 3.678 [MER89]
Melting Point, °C: decomposes [LID94]

1100
Compound: Erbium sulfate octahydrate
Formula: Er$_2$(SO$_4$)$_3$·8H$_2$O
Molecular Formula: Er$_2$H$_{16}$O$_{20}$S$_3$
Molecular Weight: 766.833
CAS RN: 10031-52-4
Properties: pink; monocl cryst [MER89]
Solubility: parts/100 parts H$_2$O: 16 (20°C), 6.53 (40°C) [MER89]
Density, g/cm^3: 3.217 [STR93]
Melting Point, °C: decomposes [STR93]
Reactions: minus 8H$_2$O at 400°C [HAW93]

1101
Compound: Erbium sulfide
Formula: Er$_2$S$_3$
Molecular Formula: Er$_2$S$_3$
Molecular Weight: 430.718
CAS RN: 12159-66-9
Properties: reddish brown monocl; -200 mesh with 99.9% purity [CER91] [LID94]
Density, g/cm^3: 6.07 [LID94]
Melting Point, °C: 1730 [LID94]

1102
Compound: Erbium telluride
Formula: Er$_2$Te$_3$
Molecular Formula: Er$_2$Te$_3$
Molecular Weight: 717.320
CAS RN: 12020-39-2
Properties: ortho-rhomb; -20 mesh with 99.9% purity [CER91] [LID94]
Density, g/cm^3: 7.11 [LID94]
Melting Point, °C: 1213 [LID94]

1103
Compound: Ethylenediaminetetraacetic acid dihydrate disodium salt
Synonyms: Edetate disodium
Formula: see under Properties
Molecular Formula: C$_{10}$H$_{14}$N$_2$Na$_2$O$_8$
Molecular Weight: 336.209
CAS RN: 6381-92-6

Properties: Used as a sequestering chelating agent for metals; forms strong chelates with most metals [MER89]

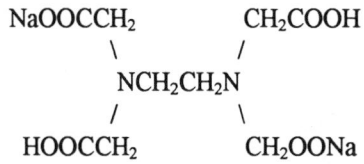

Solubility: g/100g H$_2$O: 10.6 (0°C), 11.1 (20°C), 27.0 (98°C) [LAN85]; pH ~5.3 [MER89]
Melting Point, °C: 252, decomposes [MER89]

1104
Compound: Ethylenediaminetetraacetic acid
Synonyms: Edetic acid, EDTA
Formula: (HOOCCH$_2$)$_2$NCHCH$_2$CH$_2$N(CH$_2$COOH)$_2$
Molecular Formula: C$_{10}$H$_{16}$N$_2$O$_8$
Molecular Weight: 292.246
CAS RN: 60-00-4
Properties: colorless cryst; prepared by addition of NaCN and formaldehyde to basic solution of ethylenediamine, forming the tetrasodium salt; used as food antioxidant, and as chelating agent for pharmaceutics, added to detergents, shampoos, liq soaps [HAW93] [MER89]
Solubility: 0.5 g/L H$_2$O (25°C) [MER89]
Melting Point, °C: 220, decomposes [MER89]
Reactions: decarboxylates when heated to 150°C [MER89]

1105
Compound: Europium
Formula: Eu
Molecular Formula: Eu
Molecular Weight: 151.965
CAS RN: 7440-53-1
Properties: soft, silvery metal; enthalpy of fusion 9.1 kJ/mol; radius of atom 0.20418 nm; radius of Eu^{+++} is 0.0950 nm; bcc, a = 0.4582 nm; enthalpy of vaporization 176 kJ/mol; can be made by reduction of Eu$_2$O$_3$ with La under vacuum, followed by distillation; used as a neutron absorber, in color TV phosphors, and in phosphors for postage-stamp glues for electronic identification of first-class mail [ALD94] [HAW93] [MER89] [RAR85a] [KIR82]
Solubility: reacts with H$_2$O to evolve hydrogen gas; s liq ammonia [HAW93] [MER89]
Density, g/cm^3: 5.234 [KIR82]
Melting Point, °C: 822 [ALD94]
Boiling Point, °C: 1527 [ALD94]

Thermal Conductivity, W/(m·K): 13.9 (25°C)
[ALD94]
Thermal Expansion Coefficient: 35 x 10^{-6}/K
[CRC93]

1106
Compound: Europium boride
Formula: EuB_6
Molecular Formula: B_6Eu
Molecular Weight: 216.831
CAS RN: 12008-05-8
Properties: cub; -60 mesh with 99.9% purity
[LID94] [CER91]
Density, g/cm^3: 4.91 [LID94]
Melting Point, °C: ~2600 [LID94]

1107
Compound: Europium hydride
Formula: EuH_{2-3}
Molecular Formula: EuH_2; EuH_3
Molecular Weight: EuH_2: 153.981; EuH_3: 154.989
CAS RN: 70446-10-5
Properties: -60 mesh with 99.9% purity [CER91]

1108
Compound: Europium nitride
Formula: EuN
Molecular Formula: EuN
Molecular Weight: 165.972
CAS RN: 12020-58-5
Properties: -60 mesh with 99.9% purity [CER91]

1109
Compound: Europium silicide
Formula: $EuSi_2$
Molecular Formula: $EuSi_2$
Molecular Weight: 208.136
CAS RN: 12434-24-1
Properties: tetr; 6 mm pieces and smaller with
99.9% purity, and 10 mm & down lump
[ALF93] [LID94] [CER91]
Density, g/cm^3: 5.46 [LID94]
Melting Point, °C: 1500 [LID94]

1110
Compound: Europium(II) chloride
Synonyms: europous chloride
Formula: $EuCl_2$
Molecular Formula: Cl_2Eu
Molecular Weight: 222.870

CAS RN: 13769-20-5
Properties: white ortho-rhomb cryst; amorphous
powd [LID94] [MER89]
Solubility: s H_2O [MER89]
Density, g/cm^3: 4.9 [LID94]
Melting Point, °C: 731 [LID94]

1111
Compound: Europium(II) fluoride
Synonyms: europous fluoride
Formula: EuF_2
Molecular Formula: EuF_2
Molecular Weight: 189.962
CAS RN: 14077-37-5
Properties: yellow, cub; -200 mesh (precipitated)
with 99.9% purity [CER91] [STR93] [LID94]
Solubility: i H_2O [CRC86]
Density, g/cm^3: 6.495 [CRC86]
Melting Point, °C: 1380 [CRC86]
Boiling Point, °C: >2400 [CRC86]

1112
Compound: Europium(II) iodide
Synonyms: europous iodide
Formula: EuI_2
Molecular Formula: EuI_2
Molecular Weight: 405.774
CAS RN: 22015-35-6
Properties: olive green cryst; -20 mesh with 99.9%
purity [CER91] [CRC86]
Solubility: s H_2O [CRC86]
Density, g/cm^3: 5.50 [CRC86]
Melting Point, °C: 580 [LID94]
Boiling Point, °C: 1580 [CRC86]

1113
Compound: Europium(II) selenide
Formula: EuSe
Molecular Formula: EuSe
Molecular Weight: 230.925
CAS RN: 12020-66-5
Properties: brown cub; -100 mesh with 99.9%
purity [LID94] [CER91]
Density, g/cm^3: 6.45 [LID94]
Melting Point, °C: 2027 [CRC93]
Thermal Conductivity, W/(m·K): 0.24 [CRC93]

1114
Compound: Europium(II) sulfate
Synonyms: europous sulfate
Formula: $EuSO_4$

Molecular Formula: EuO₄S
Molecular Weight: 248.029
CAS RN: 10031-54-6
Properties: ortho-rhomb; colorless cryst [MER89] [CRC86]
Solubility: i H₂O, dil acids [MER89]
Density, g/cm³: 4.989 [CRC86]

1115
Compound: Europium(II) sulfide
Formula: EuS
Molecular Formula: EuS
Molecular Weight: 184.031
CAS RN: 12020-65-4
Properties: cub; -200 mesh with 99.9% purity [LID94] [CER91]
Density, g/cm³: 5.7 [LID94]

1116
Compound: Europium(II) telluride
Formula: EuTe
Molecular Formula: EuTe
Molecular Weight: 279.565
CAS RN: 12020-69-8
Properties: black, cub; -20 mesh with 99.9% purity [LID94] [CER91]
Density, g/cm³: 6.48 [LID94]
Melting Point, °C: 1526 [LID94]

1117
Compound: Europium(III) acetylacetonate
Synonyms: 2,4-pentanedione, Eu(III) derivative
Formula: Eu(CH₃COCH=C(O)CH₃)₃·xH₂O
Molecular Formula: C₁₅H₂₁EuO₆ (anhydrous)
Molecular Weight: 449.293 (anhydrous)
Properties: hygr [ALD94]

$$[CH_3-\overset{\overset{\displaystyle O-}{\displaystyle |}}{C}=CH-\overset{\overset{\displaystyle O}{\displaystyle \|}}{C}-CH_3]_3Eu$$

Melting Point, °C: decomposes 140 [ALD94]

1118
Compound: Europium(III) bromide
Synonyms: europic bromide
Formula: EuBr₃
Molecular Formula: Br₃Eu
Molecular Weight: 391.677
CAS RN: 13759-88-1

Properties: gray cryst; -20 mesh with 99.9% purity [LID94] [CER91]
Solubility: s H₂O [CRC86]
Melting Point, °C: 702 [CRC86]
Boiling Point, °C: decomposes [CRC86]

1119
Compound: Europium(III) carbonate hydrate
Formula: Eu₂(CO₃)₃·xH₂O
Molecular Formula: C₃Eu₂O₉ (anhydrous)
Molecular Weight: 483.958 (anhydrous)
CAS RN: 86546-99-8
Properties: white powd; hygr [STR93] [ALD94]

1120
Compound: Europium(III) chloride
Synonyms: europic chloride
Formula: EuCl₃
Molecular Formula: Cl₃Eu
Molecular Weight: 258.323
CAS RN: 10025-76-0
Properties: -20 mesh with 99.9% purity; yellowish-white powd; hygr [STR93] [CER91]
Density, g/cm³: 4.89 [STR93]
Melting Point, °C: 623 [LID94]
Reactions: yields EuCl₂ by reduction with H₂ at 600°C [MER89]

1121
Compound: Europium(III) chloride hexahydrate
Formula: EuCl₃·6H₂O
Molecular Formula: Cl₃EuH₁₂O₆
Molecular Weight: 366.414
CAS RN: 13759-92-7
Properties: yellow needles; white cryst; hygr [HAW93] [STR93]
Solubility: s H₂O [HAW93]
Density, g/cm³: 4.89 (20°C) [HAW93]
Melting Point, °C: 850 [HAW93]

1122
Compound: Europium(III) fluoride
Formula: EuF₃
Molecular Formula: EuF₃
Molecular Weight: 208.960
CAS RN: 13765-25-8
Properties: white powd, or 99.9% pure melted pieces of 3-6 mm; hygr; pieces used as evaporation material for possible applications in multilayers [STR93] [CER91]
Melting Point, °C: 1390 [STR93]

Boiling Point, °C: 2280 [STR93]

1123
Compound: Europium(III) nitrate hexahydrate
Synonyms: europic nitrate hexahydrate
Formula: $Eu(NO_3)_3 \cdot 6H_2O$
Molecular Formula: $EuH_{12}N_3O_{15}$
Molecular Weight: 446.071
CAS RN: 10031-53-5
Properties: white to pale pink cryst; hygr [HAW93] [STR93]
Solubility: 4.2732±0.0061 mol/kg in H_2O (25°C) [RAR84]
Melting Point, °C: 85 [HAW93]

1124
Compound: Europium(III) nitrate pentahydrate
Formula: $Eu(NO_3)_3 \cdot 5H_2O$
Molecular Formula: $EuH_{10}N_3O_{14}$
Molecular Weight: 428.071
CAS RN: 63026-01-7
Properties: white cryst [STR93]

1125
Compound: Europium(III) oxalate
Formula: $Eu_2(C_2O_4)_3$
Molecular Formula: $C_6Eu_2O_{12}$
Molecular Weight: 567.989
CAS RN: 14175-02-1
Properties: white powd; hygr [HAW93] [ALD94]
Solubility: i H_2O; sl s acids [HAW93]

1126
Compound: Europium(III) oxide
Synonyms: europia
Formula: Eu_2O_3
Molecular Formula: Eu_2O_3
Molecular Weight: 351.928
CAS RN: 1308-96-9
Properties: -325 mesh 5 microns or less of 99.995% and 99.9% purity; pale rose powd; used in red and infrared sensitive phosphors and in nuclear-reactor control rods, as an evaporated material of 99.9% purity it is reactive to radio frequencies [HAW93] [CER91]
Solubility: i H_2O; s acids [HAW93]
Density, g/cm³: 7.42 [MER89]
Melting Point, °C: 2350 [LID94]

1127
Compound: Europium(III) perchlorate hexahydrate
Formula: $Eu(ClO_4)_3 \cdot 6H_2O$
Molecular Formula: $Cl_3EuH_{12}O_{18}$
Molecular Weight: 558.407
CAS RN: 36907-40-1
Properties: off-white cryst; hygr [STR93]

1128
Compound: Europium(III) sulfate octahydrate
Formula: $Eu_2(SO_4)_3 \cdot 8H_2O$
Molecular Formula: $Eu_2H_{16}O_{20}S_3$
Molecular Weight: 736.243
CAS RN: 10031-52-4
Properties: white cryst; prepared by dissolving the oxide in H_2SO_4 [MER89] [STR93]
Solubility: parts/100 parts H_2O: 2.56 (20°C), 1.93 parts (40°C) [MER89]
Density, g/cm³: 4.95 [CRC86]
Reactions: minus $8H_2O$ at 375°C [HAW93]

1129
Compound: Fermium
Formula: Fm
Molecular Formula: Fm
Molecular Weight: 257
CAS RN: 7440-72-4
Properties: chemical properties similar to those of erbium; man made radioactive element; discovered in 1952 in debris from thermonuclear explosion by Ghiorso and colleagues; $t_{1/2}$ of ^{257}Fm is 100 days [KIR78] [HAW93] [MER89]
Melting Point, °C: 1527 [LID94]

1130
Compound: Ferric acetylacetonate
Synonyms: 2,4-pentanedione, iron(III) derivative
Formula: $Fe(CH_3COCH=C(O)CH_3)_3$
Molecular Formula: $C_{15}H_{21}FeO_6$
Molecular Weight: 353.173
CAS RN: 14024-18-1
Properties: reddish orange cryst; resistant to hydrolysis; used as a moderating and combustion catalyst [HAW93] [STR93]

$$[CH_3{-}\overset{\overset{\displaystyle O-}{|}}{C}{=}CH{-}\overset{\overset{\displaystyle O}{\|}}{C}{-}CH_3]_3Fe$$

Solubility: sl s H_2O; s in most organic solvents [HAW93]
Density, g/cm³: 5.24 [STR93]

Melting Point, °C: 184 [STR93]

Melting Point, °C: sublimes, decomposes [STR93]

1131
Compound: Ferric arsenate dihydrate
Formula: $FeAsO_4 \cdot 2H_2O$
Molecular Formula: $AsFeH_4O_6$
Molecular Weight: 230.795
CAS RN: 10102-49-5
Properties: green or brown powd; used as an insecticide [HAW93]
Solubility: i H_2O; s dil mineral acids [HAW93]
Density, g/cm³: 3.18 [HAW93]
Melting Point, °C: decomposes when heated [HAW93]

1132
Compound: Ferric basic acetate
Formula: $FeOH(CH_3COO)_2$
Molecular Formula: $C_4H_7FeO_5$
Molecular Weight: 190.942
CAS RN: 10450-55-2
Properties: brownish red scales, or amorphous powd; uses: mordant in textile dyeing, wood preservative, and in medicine [HAW93] [MER89]
Solubility: i H_2O; s alcohol, acids [MER89]

1133
Compound: Ferric basic arsenite
Formula: $2FeAsO_3 \cdot Fe_2O_3 \cdot 5H_2O$
Molecular Formula: $As_2Fe_4H_{10}O_{14}$
Molecular Weight: 607.294
CAS RN: 63989-69-5
Properties: brownish yellow powd; used in medicine [HAW93]
Solubility: s in acids [HAW93]
Melting Point, °C: decomposes [CRC86]

1134
Compound: Ferric bromide
Synonyms: iron(III) bromide
Formula: $FeBr_3$
Molecular Formula: Br_3Fe
Molecular Weight: 295.557
CAS RN: 10031-26-2
Properties: dark red or black; hex cryst; very hygr; used as a catalyst for brominations [HAW93] [MER89]
Solubility: s H_2O, alcohol, ether, acetic acid [MER89]
Density, g/cm³: 4.5 [LID94]

1135
Compound: Ferric chloride
Synonyms: molysite
Formula: $FeCl_3$
Molecular Formula: Cl_3Fe
Molecular Weight: 162.203
CAS RN: 7705-08-0
Properties: hex; dark leaflets or plates; red by transmitted light, green by reflected light; very hygr; readily absorbs H_2O from air to form hexahydrate; enthalpy of fusion 43.10 kJ/mol; used in water purification, etching agent [HAW93] [MER89] [CRC93]
Solubility: s alcohol, ether, acetone [MER89]; mol/100 mol H_2O: 2.06 (0°C), 7.77 (25°C), 14.88 (100°C); Solid phase, $FeCl_3 \cdot 6H_2O$ (0°C), $FeCl_3 \cdot 3\text{-}1/2H_2O$ (25°C), $FeCl_3$ (100°C) [KRU93]
Density, g/cm³: 2.804 [STR93]
Melting Point, °C: 304 [CRC93]
Boiling Point, °C: ~316 [MER89]

1136
Compound: Ferric chloride hexahydrate
Synonyms: iron(III) chloride hexahydrate
Formula: $FeCl_3 \cdot 6H_2O$
Molecular Formula: $Cl_3FeH_{12}O_6$
Molecular Weight: 270.294
CAS RN: 10025-77-1
Properties: brownish yellow or orange; monocl cryst [MER89]
Solubility: s H_2O, alcohol, acetone, ether [MER89]
Density, g/cm³: 1.82 [MER89]
Melting Point, °C: ~37 [MER89]
Boiling Point, °C: 280-285 [ALD94]

1137
Compound: Ferric chromate
Synonyms: iron(III) chromate
Formula: $Fe_2(CrO_4)_3$
Molecular Formula: $Cr_3Fe_2O_{12}$
Molecular Weight: 459.671
CAS RN: 10294-52-7
Properties: yellow powd; used in metallurgy, ceramics and as a paint pigment [HAW93]
Solubility: i H_2O, alcohol; s acids [HAW93]

1138
Compound: Ferric citrate pentahydrate
Synonyms: iron(III) citrate pentahydrate

Formula: $FeC_6H_5O_7 \cdot 5H_2O$
Molecular Formula: $C_6H_{15}FeO_{12}$
Molecular Weight: 335.023
CAS RN: 28633-45-6
Properties: reddish brown scales; light sensitive; used in medicine and in blueprinting paper [HAW93]
Solubility: s H_2O; i alcohol [HAW93]

1139
Compound: Ferric dichromate
Synonyms: iron(III) dichromate
Formula: $Fe_2(Cr_2O_7)_3$
Molecular Formula: $Cr_6Fe_2O_{21}$
Molecular Weight: 759.654
CAS RN: 10294-53-8
Properties: reddish brown granules; oxidizing action; used in the preparation of paint pigments [HAW93]
Solubility: s H_2O, acids [HAW93]

1140
Compound: Ferric ferrocyanide
Synonyms: Prussian blue
Formula: $Fe_4[Fe(CN)_6]_3$
Molecular Formula: $C_{18}Fe_7N_{18}$
Molecular Weight: 859.234
CAS RN: 14038-43-8
Properties: dark blue powd or lumps [MER89]
Solubility: i H_2O, dil acids; s aq oxalic acid when freshly prepared [MER89]
Density, g/cm³: 1.80 [MER89]

1141
Compound: Ferric fluoride
Synonyms: iron(III) fluoride
Formula: FeF_3
Molecular Formula: F_3Fe
Molecular Weight: 112.840
CAS RN: 7783-50-8
Properties: green hex cryst; a preparation is by reaction of $FeCl_3$ with F_2 or anhydrous HF; uses: ceramics, catalyst [HAW93] [MER89]
Solubility: v sl s H_2O; s dil HF [MER89]; g/100 g soln, H_2O: 5.59 (25°C); Solid phase, $FeF_3 \cdot 3H_2O$ [KRU93]
Density, g/cm³: 3.87 [MER89]
Melting Point, °C: sublimes >1000 [MER89]

1142
Compound: Ferric fluoride trihydrate

Synonyms: iron(III) fluoride trihydrate
Formula: $FeF_3 \cdot 3H_2O$
Molecular Formula: $F_3FeH_6O_3$
Molecular Weight: 166.886
CAS RN: 15469-38-2
Properties: yellowish brown powd; readily prepared from reaction of Fe_2O_3 and aq HF; dehydration produces a mixture of oxyfluorides [KIR78] [STR93]
Density, g/cm³: 2.3 [LID94]

1143
Compound: Ferric hydroxide
Synonyms: iron(III) hyroxide
Formula: $Fe(OH)_3$
Molecular Formula: FeH_3O_3
Molecular Weight: 106.867
CAS RN: 1309-33-7
Properties: brown flocculant precipitate; used in water purification, manufacturing pigments [HAW93]
Solubility: i H_2O, alcohol, ether; s acids [HAW93]
Density, g/cm³: 3.4-3.9 [HAW93]
Reactions: minus water ~500°C [HAW93]

1144
Compound: Ferric hypophosphite
Synonyms: iron(III) hypophosphite
Formula: $Fe(H_2PO_2)_3$
Molecular Formula: $FeH_6O_6P_3$
Molecular Weight: 250.811
CAS RN: 7783-84-8
Properties: white or grayish white powd; odorless and tasteless [HAW93] [MER89]
Solubility: s in 2300 parts cold H_2O, 1200 parts boiling H_2O [MER89]

1145
Compound: Ferric metavanadate
Synonyms: iron(III) vanadate
Formula: $Fe(VO_3)_3$
Molecular Formula: FeO_9V_3
Molecular Weight: 352.665
CAS RN: 65842-03-7
Properties: grayish brown powd; used in metallurgy [HAW93]
Solubility: i H_2O, alcohol; s acids [HAW93]

1146
Compound: Ferric nitrate nonahydrate
Synonyms: iron(III) nitrate nonahydrate

Formula: Fe(NO$_3$)$_3$·9H$_2$O
Molecular Formula: FeH$_{18}$N$_3$O$_{18}$
Molecular Weight: 403.997
CAS RN: 7782-61-8
Properties: pale violet to grayish white cryst; somewhat deliq; used in dyeing, tanning and analytical chemistry [HAW93] [MER89]
Solubility: g/100g H$_2$O: 112.0 (0°C), 137.7 (20°C), 175.0 (40°C) [LAN85]; v s alcohol, acetone [MER89]
Density, g/cm^3: 1.684 [HAW93]
Melting Point, °C: 47.2 [HAW93]
Boiling Point, °C: decomposes below 100 [MER89]

1147
Compound: Ferric oxalate
Synonyms: iron(III) oxalate
Formula: Fe$_2$(C$_2$O$_4$)$_3$
Molecular Formula: C$_6$Fe$_2$O$_{12}$
Molecular Weight: 375.749
CAS RN: 19469-07-9
Properties: pale yellow amorphous odorless powd; used in photographic printing paper; also a hexahydrate [HAW93] [STR93]
Solubility: s H$_2$O, acids; i alkalies [HAW93]
Melting Point, °C: 100 (decomposes) [HAW93]

1148
Compound: Ferric oxide
Synonyms: hematite,maghemite
Formula: hematite: α-Fe$_2$O$_3$; maghemite: γ-Fe$_2$O$_3$
Molecular Formula: Fe$_2$O$_3$
Molecular Weight: 159.688
CAS RN: 1309-37-1
Properties: hematite: reddish brown powd; used in metallurgy, gas purification, as a paint pigment, and as an evaporated material and sputtering target of 99.99% and 99.9% purity for beam splitter and interference layers, and magnetic films [HAW93] [STR93] [CER91]
Solubility: i H$_2$O; s acids [HAW93]
Density, g/cm^3: 5.12-5.24 [HAW93]
Melting Point, °C: 1565 [HAW93]
Thermal Expansion Coefficient: (volume) 100°C (0.202), 200°C (0.485), 400°C (1.175) [CLA66]

1149
Compound: Ferric oxide hydroxide
Synonyms: goethite
Formula: α-FeO(OH)
Molecular Formula: FeHO$_2$
Molecular Weight: 88.852

CAS RN: 20344-49-4
Properties: red to brown powd or cryst [MER89]
Solubility: i H$_2$O; s mineral acids [MER89]
Density, g/cm^3: 3.4-3.9 [MER89]

1150
Compound: Ferric oxide monohydrate
Formula: Fe$_2$O$_3$·H$_2$O
Molecular Formula: Fe$_2$H$_2$O$_4$
Molecular Weight: 177.704
CAS RN: 51274-00-1
Properties: yellow powd [STR93]

1151
Compound: Ferric perchlorate hexahydrate
Synonyms: iron(III) perchlorate hexahydrate
Formula: Fe(ClO$_4$)$_3$·6H$_2$O
Molecular Formula: Cl$_3$FeH$_{12}$O$_{18}$
Molecular Weight: 462.287
CAS RN: 13537-24-1
Properties: yellow cryst; hygr [STR93]

1152
Compound: Ferric perchlorate hydrate
Synonyms: iron(III) perchlorate
Formula: Fe(ClO$_4$)$_3$·xH$_2$O
Molecular Formula: Cl$_3$FeO$_{12}$ (anhydrous)
Molecular Weight: 354.196 (anhydrous)
CAS RN: 14013-71-9
Properties: purple cryst; hygr [STR93]
Solubility: g/100 g soln, H$_2$O: 74.32 (0°C), 79.86 (25°C); Solid phase, Fe(ClO$_4$)$_3$·10H$_2$O [KRU93]

1153
Compound: Ferric phosphate dihydrate
Synonyms: iron(III) phosphate dihydrate
Formula: FePO$_4$·2H$_2$O
Molecular Formula: FeH$_4$O$_6$P
Molecular Weight: 186.847
CAS RN: 10045-86-0
Properties: white, grayish white or light pink; ortho-rhomb or monocl cryst, and amorphous powd; used in fertilizers and as a food and feed additive [HAW93] [MER89]
Solubility: i H$_2$O; s HCl [MER89]
Density, g/cm^3: 2.87 [MER89]
Reactions: minus H$_2$O >140°C [MER89]

1154
Compound: Ferric phosphate hydrate

Synonyms: iron(III) phosphate hydrate
Formula: $FePO_4 \cdot xH_2O$
Molecular Formula: FeO_4P (anhydrous)
Molecular Weight: 150.817 (anhydrous)
CAS RN: 13463-10-0
Properties: white powd [STR93]

1155
Compound: Ferric pyrophosphate nonahydrate
Synonyms: iron(III) pyrpophosphate nonahydrate
Formula: $Fe_4(P_2O_7)_3 \cdot 9H_2O$
Molecular Formula: $Fe_4H_{18}O_{30}P_6$
Molecular Weight: 907.348
CAS RN: 10058-44-3
Properties: yellowish white powd [MER89]
Solubility: i H_2O, acetic acid; s mineral acids
[MER89]

1156
Compound: Ferric sulfate
Synonyms: iron(III) sulfate
Formula: $Fe_2(SO_4)_3$
Molecular Formula: $Fe_2O_{12}S_3$
Molecular Weight: 399.881
CAS RN: 10028-22-5
Properties: grayish white powd, or rhomb cryst;
very hygr [MER89]
Solubility: slowly s H_2O with hydrolysis [MER89]
Density, g/cm³: 3.097 [MER89]
Melting Point, °C: 1178 decomposes [JAN85]

1157
Compound: Ferric sulfate hydrate
Synonyms: iron(III) sulfate hydrate
Formula: $Fe_2(SO_4)_3 \cdot xH_2O$
Molecular Formula: $Fe_2O_{12}S_3$ (anhydrous)
Molecular Weight: 399.881 (anhydrous)
CAS RN: 15244-10-7
Properties: off-white powd [STR93]

1158
Compound: Ferric sulfate nonahydrate
Synonyms: coquimbite
Formula: $Fe_2(SO_4)_3 \cdot 9H_2O$
Molecular Formula: $Fe_2H_{18}O_{21}S_3$
Molecular Weight: 562.018
CAS RN: 13520-56-4
Properties: rhomb yellow cryst or grayish white
powd [HAW93] [CRC86]
Solubility: 440 g/100 mL H_2O [CRC86]
Density, g/cm³: 2.0-2.1 [HAW93]

Melting Point, °C: decomposes at 480 [HAW93]
Reactions: minus $7H_2O$ at 1750°C [CRC86]

1159
Compound: Ferric thiocyanate
Synonyms: iron(III) thiocyanate
Formula: $Fe(SCN)_3$
Molecular Formula: $C_3FeN_3S_3$
Molecular Weight: 230.096
CAS RN: 4119-52-2
Properties: cub red cryst; deliq; decomposes on
heating [MER89] [CRC86]
Solubility: s H_2O, alcohol, ether, acetone, pyridine; i
$CHCl_3$, toluene [MER89]

1160
Compound: Ferric trifluoroacetylacetonate
Synonyms: 1,1,1-trifluoro-2,4-pentanedione, Fe(III)
derivative
Formula: $Fe(CF_3COCH=C(O)CH_3)_3$
Molecular Formula: $C_{15}H_{12}F_9FeO_6$
Molecular Weight: 515.089
CAS RN: 14526-22-8
Properties: cryst [ALF95]

$$[CH_3\text{–}C=CH\text{–}C\text{–}CF_3]_3Fe$$

Melting Point, °C: 110-112 decomposes [ALF95]

1161
Compound: 1,1'-Bis(diphenylphosphino)ferrocene
Formula: $(C_6H_5)_2PC_5H_4FeC_5H_4P(C_6H_5)_2$
Molecular Formula: $C_{34}H_{28}FeP_2$
Molecular Weight: 554.395
CAS RN: 12150-46-8
Properties: yellow-orange cryst [STR93]

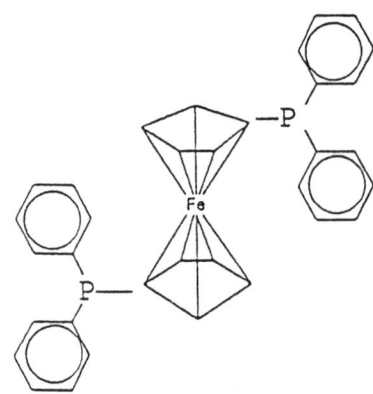

Melting Point, °C: 180 [STR93]

1162
Compound: Ferrocene
Synonyms: dicyclopentadienyliron
Formula: Fe(C$_5$H$_5$)$_2$
Molecular Formula: C$_{10}$H$_{10}$Fe
Molecular Weight: 186.036
CAS RN: 102-54-5
Properties: orange needles; camphor odor; volatile in steam; diamagnetic; thermally stable to >500°C; preparation: reaction of ferric chloride with cyclopentadienylmagnesium bromide in a solvent of diethylether/benzene (T.J.Kealy; P.L.Pauson, Nature 1951, vol. 168, 1039); use: antiknock gasoline additive, catalyst [MER89] [COT88] [ALD94]

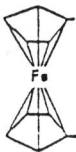

Solubility: s alcohol, ether, benzene; dissolves in dil HNO$_3$ and in conc H$_2$SO$_4$ solutions with deep red solutions [MER89]
Melting Point, °C: 174-176 [ALD94]
Boiling Point, °C: 249 [ALD94]

1163
Compound: Ferrocenium hexafluorophosphate
Molecular Weight: 331.00
CAS RN: 11077-24-0
Properties: hygr; uses: preparation of catalysts for asymmetric epoxidation of olefins and other organics [ALD94]

1164
Compound: Ferrocenium tetrafluoroborate
Molecular Weight: 272.84
CAS RN: 1282-37-7
Properties: hygr; uses: preparation of catalysts for asymmetric epoxidation of olefins and other organics [ALD94]

1165
Compound: Ferrous acetate
Synonyms: iron(II) acetate
Formula: Fe(CH$_3$COO)$_2$
Molecular Formula: C$_4$H$_6$FeO$_4$
Molecular Weight: 173.935
CAS RN: 3094-87-9

Properties: off-white to light brown powd; sensitive to atm oxygen and moisture; used to prepare dark shades of inks and dyes; and as a mordant [KIR82] [STR93]
Density, g/cm^3: 190-200, decomposes [ALD94]

1166
Compound: Ferrous acetate tetrahydrate
Synonyms: iron(II) acetate tetrahydrate
Formula: Fe(CH$_3$COO)$_2$·4H$_2$O
Molecular Formula: C$_4$H$_{14}$FeO$_8$
Molecular Weight: 245.995
CAS RN: 3094-87-9
Properties: greenish cryst, but usually partly brownish due to air oxidation; used in dyeing textiles and leather, in medicine and as a wood preservative [HAW93]
Solubility: s H$_2$O, alcohol [HAW93]
Melting Point, °C: decomposes [CRC86]

1167
Compound: Ferrous acetylacetonate
Synonyms: 2,4-pentanedione, iron(II) derivative
Formula: Fe(CH$_3$COCH=C(O)CH$_3$)$_2$
Molecular Formula: C$_{10}$H$_{14}$FeO$_4$
Molecular Weight: 254.064
CAS RN: 14024-17-0
Properties: powd [STR93]

$$[CH_3-\overset{\overset{\displaystyle O-}{|}}{C}=CH-\overset{\overset{\displaystyle O}{\|}}{C}-CH_3]_2Fe$$

1168
Compound: Ferrous arsenate hexahydrate
Synonyms: iron(II) orthoarsenate hexahydrate
Formula: Fe$_3$(AsO$_4$)$_2$·6H$_2$O
Molecular Formula: As$_2$Fe$_3$H$_{12}$O$_{14}$
Molecular Weight: 553.465
CAS RN: 10102-50-8
Properties: green amorphous powd; used in insecticides [HAW93]
Solubility: i H$_2$O; s acids [HAW93]
Melting Point, °C: decomposes [CRC86]

1169
Compound: Ferrous bromide
Synonyms: iron(II) bromide
Formula: FeBr$_2$
Molecular Formula: Br$_2$Fe
Molecular Weight: 215.653

CAS RN: 7789-46-0

Properties: light yellow to dark brown; hygr cryst; enthalpy of fusion 50.20 kJ/mol [CRC93] [MER89]

Solubility: v s alcohol [MER89]; g/100 g soln, H_2O: 54.6±0.5 (25°C), 64.8 (100°C); Solid phase, $FeBr_2 \cdot 6H_2O$ (25°C), $FeBr_2 \cdot 2H_2O$ (100°C) [KRU93]

Density, g/cm³: 4.636 [STR93]

Melting Point, °C: 684 [ALD94]

Boiling Point, °C: 934 [ALD94]

1170

Compound: Ferrous bromide hexahydrate

Synonyms: iron(II) bromide hexahydrate

Formula: $FeBr_2 \cdot 6H_2O$

Molecular Formula: $Br_2FeH_{12}O_6$

Molecular Weight: 323.744

CAS RN: 13463-12-2

Properties: pale green to bluish green cryst powd; rhomb prisms; very deliq; rapidly oxidized in moist air; used as a polymerization catalyst [HAW93] [MER89]

Solubility: s H_2O, alcohol [HAW93]

Density, g/cm³: 4.636 [HAW93]

Melting Point, °C: 27 [HAW93]

Reactions: minus $2H_2O$ at 49°C; minus $4H_2O$ forming dihydrate at 83°C [MER89]

1171

Compound: Ferrous bromide hydrate

Formula: $FeBr_2 \cdot xH_2O$

Molecular Formula: Br_2Fe (anhydrous)

Molecular Weight: 295.557 (anhydrous)

CAS RN: 13463-12-2

Properties: orange cryst [STR93]

Melting Point, °C: 27 [STR93]

1172

Compound: Ferrous carbonate

Synonyms: siderite

Formula: $FeCO_3$

Molecular Formula: $CFeO_3$

Molecular Weight: 115.854

CAS RN: 563-71-3

Properties: gray, yellow, brown, green, white or brownish red mineral; hardness is 3.5-4 Mohs; obtained as a white precipitate by adding alkaline carbonate solution to a ferrous solution; used as a flame retardant and diet supplement; forms $Fe(HCO_3)_2$, 6013-77-0, in solutions containing CO_2 [KIR81] [HAW93]

Solubility: 0.0067 g/100 mL H_2O (25°C) [CRC86]

Density, g/cm³: 3.8 [CRC86]

Melting Point, °C: decomposes [CRC86]

1173

Compound: Ferrous chloride

Synonyms: iron(II) chloride

Formula: $FeCl_2$

Molecular Formula: Cl_2Fe

Molecular Weight: 126.750

CAS RN: 7758-94-3

Properties: white, sometimes with green tint; rhomb cryst; very hygr; readily oxidized; enthalpy of fusion 43.01 kJ/mol [CRC93] [HAW93] [MER89]

Solubility: v s alcohol, acetone; i ether [MER89]: g/100g soln, H_2O: 33.2 (0°C), 39.4±0.2 (25°C), 48.7 (100°C); Solid phase, $FeCl_2 \cdot 6H_2O$ (0°C), $FeCl_2 \cdot 4H_2O$ (25°C), $FeCl_2 \cdot 2H_2O$ (100°C) [KRU93]

Density, g/cm³: 3.16 [MER89]

Melting Point, °C: 677 [CRC93]

Boiling Point, °C: 1023 [MER89]

Reactions: can sublime in HCl atm at 700°C without decomposition; forms $FeCl_3$ and Fe_2O_3 when heated in air [MER89]

1174

Compound: Ferrous chloride dihydrate

Synonyms: iron(II) chloride dihydrate

Formula: $FeCl_2 \cdot 2H_2O$

Molecular Formula: $Cl_2FeH_4O_2$

Molecular Weight: 162.781

CAS RN: 16399-77-2

Properties: white with pale green tint; monocl cryst [MER89]

Reactions: minus H_2O at 120°C [MER89]

1175

Compound: Ferrous chloride tetrahydrate

Synonyms: iron(II) chloride tetrahydrate

Formula: $FeCl_2 \cdot 4H_2O$

Molecular Formula: $Cl_2FeH_8O_4$

Molecular Weight: 198.812

CAS RN: 13478-10-9

Properties: pale green to bluish green; monocl cryst or cryst powd [MER89]

Solubility: s H_2O, alcohol [MER89]

Density, g/cm³: 1.93 [MER89]

Reactions: minus $2H_2O$ ~105-115°C [MER89]

1176

Compound: Ferrous chromite

Synonyms: chromite

Formula: $FeCr_2O_4$

Molecular Formula: Cr_2FeO_4

Molecular Weight: 223.835

CAS RN: 1308-31-2

Properties: natural oxide of the two metals; iron black to brownish black; hardness 5.5 Mohs; it is the only commercial source of chromium and chromium compounds [HAW93]

Solubility: i H_2O [CRC86]

Density, g/cm³: 3.6 [HAW93]

Thermal Expansion Coefficient: (volume) 100°C (0.09), 200°C (0.21), 400°C (0.48), 800°C (1.26), 1000°C (1.74) [CLA66]

1177

Compound: Ferrous citrate monohydrate

Synonyms: iron(II) citrate monohydrate

Formula: $FeC_6H_6O_7 \cdot H_2O$

Molecular Formula: $C_6H_8FeO_8$

Molecular Weight: 263.970

CAS RN: 23383-11-1

Properties: white, rhomb powd; prepared from iron powd and citric acid; used as supplements to animal diets [KIR81] [MER89] [CRC86]

Solubility: i H_2O, alcohol, acetone [MER89]

Melting Point, °C: decomposes under H_2 at 350 [CRC86]

1178

Compound: Ferrous fluoride

Synonyms: iron(II) fluoride

Formula: FeF_2

Molecular Formula: F_2Fe

Molecular Weight: 93.842

CAS RN: 7789-28-8

Properties: off-white powd; tetr cryst or powd; enthalpy of fusion 52.00 kJ/mol; can be prepared by reacting anhydrous HF with metallic Fe; used in ceramics and as a catalyst [HAW93] [MER89] [CRC93]

Solubility: sl s H_2O; s dil HF; i alcohol, ether [MER89] [HAW93]

Density, g/cm³: 4.09 [MER89]

Melting Point, °C: 1100 [CRC93]

1179

Compound: Ferrous fluoride tetrahydrate

Synonyms: iron(II) fluoride tetrahydrate

Formula: $FeF_2 \cdot 4H_2O$

Molecular Formula: $F_2FeH_8O_4$

Molecular Weight: 165.904

CAS RN: 13940-89-1

Properties: rhomb, colorless cryst; can be prepared by dissolution of Fe or FeF_2 in HF solution; decomposes to Fe_2O_3 if heated in air [KIR78] [CRC86]

Solubility: v sl s H_2O [CRC86]

Density, g/cm³: 2.095 [CRC86]

Melting Point, °C: decomposes [CRC86]

1180

Compound: Ferrous hexafluorosilicate hexahydrate

Synonyms: iron(II) hexafluorosilicate hexahydrate

Formula: $FeSiF_6 \cdot 6H_2O$

Molecular Formula: $F_6FeH_{12}O_6Si$

Molecular Weight: 306.012

CAS RN: 12021-70-4

Properties: colorless, trig [CRC77]

Solubility: g/100g H_2O: 72.1 (0°C), 77.0 (25°C), 100.1 (106°C) [LAN85]

Density, g/cm³: 1.961 [CRC77]

1181

Compound: Ferrous hydroxide

Synonyms: iron(II) hydroxide

Formula: $Fe(OH)_2$

Molecular Formula: FeH_2O_2

Molecular Weight: 89.860

CAS RN: 18624-44-7

Properties: white amorphous powd, or white to pale green hex cryst; oxidizes in air [MER89]

Solubility: 0.00015 g/100mL H_2O (18°C) [CRC86]

Density, g/cm³: 3.4 [CRC86]

Melting Point, °C: decomposes [CRC86]

1182

Compound: Ferrous iodide

Synonyms: iron(II) iodide

Formula: FeI_2

Molecular Formula: FeI_2

Molecular Weight: 309.654

CAS RN: 7783-86-0

Properties: gray powd; large, thin, reddish violet cryst; very hygr; aq solution readily oxidizes in air; enthalpy of fusion 45.00 kJ/mol [STR93] [MER89] [CRC93]

Solubility: s H_2O, alcohol, ether [MER89]

Density, g/cm³: 5.31 [STR93]

Melting Point, °C: 587 [CRC93]

1183
Compound: Ferrous iodide tetrahydrate
Formula: $FeI_2 \cdot 4H_2O$
Molecular Formula: $FeH_8I_2O_4$
Molecular Weight: 381.71
CAS RN: 13492-45-0
Properties: dark violet to black; hygr leaflets; sensitive to light; uses: manufacture of alkali metal iodides, pharmaceutical preparations, as a catalyst [HAW93]
Solubility: s H_2O, alcohol [HAW93]
Density, g/cm³: 2.873 [HAW93]
Melting Point, °C: decomposes at 90-98 [ALF95]

1184
Compound: Ferrous nitrate hexahydrate
Synonyms: iron(II) nitrate hexahydrate
Formula: $Fe(NO_3)_2 \cdot 6H_2O$
Molecular Formula: $FeH_{12}N_2O_{12}$
Molecular Weight: 287.946
CAS RN: 13476-08-9
Properties: green rhomb cryst [CRC86]
Solubility: g/100g H_2O: 113 (0°C), 134 (10°C), 266 (60°C) [LAN85]
Melting Point, °C: decomposes [LAN52]

1185
Compound: Ferrous oxalate dihydrate
Synonyms: iron(II) oxalate dihydrate
Formula: $FeC_2O_4 \cdot 2H_2O$
Molecular Formula: $C_2H_4FeO_6$
Molecular Weight: 179.895
CAS RN: 6047-25-2
Properties: pale yellow, odorless, cryst powd; decomposes at 160°C, evolving carbon monoxide; used as a photographic developer, a pigment for glass, in paints [HAW93] [MER89]
Solubility: sl s H_2O; s mineral acids [MER89]
Density, g/cm³: 2.28 [MER89]
Melting Point, °C: decomposes 150 [LID94]

1186
Compound: Ferrous oxide
Synonyms: wustite
Formula: FeO
Molecular Formula: FeO
Molecular Weight: 71.844
CAS RN: 1345-25-1
Properties: jet black powd; easily oxidized in air; strong base; readily absorbs CO_2; enthalpy of fusion 24.00 kJ/mol; used as a catalyst, glass colorant [HAW93] [MER89] [CRC93]

Solubility: i H_2O, alkalies; s acid [HAW93] [MER89]
Density, g/cm³: 5.7 [HAW93]
Melting Point, °C: 1377 [CRC93]
Thermal Expansion Coefficient: (volume) 100°C (0.30), 200°C (0.63), 400°C (1.32), 600°C (2.10) [CLA66]

1187
Compound: Ferrous perchlorate hexahydrate
Synonyms: iron(II) perchlorate hexahydrate
Formula: $Fe(ClO_4)_2 \cdot 6H_2O$
Molecular Formula: $Cl_2FeH_{12}O_{14}$
Molecular Weight: 362.839
CAS RN: 13520-69-9
Properties: green cryst [AES93]
Solubility: g/100g soln, H_2O: 63.39 (0°C), 67.76 (25°C); Solid phase, $Fe(ClO_4)_2 \cdot 6H_2O$ [KRU93]
Melting Point, °C: decomposes >100 [CRC86]

1188
Compound: Ferrous phosphate octahydrate
Synonyms: vivianite
Formula: $Fe_3(PO_4)_2 \cdot 8H_2O$
Molecular Formula: $Fe_3H_{16}O_{16}P_2$
Molecular Weight: 501.600
CAS RN: 14940-41-1
Properties: grayish blue powd or monocl cryst; hygr; used in ceramics and as a catalyst [HAW93] [MER89]
Solubility: i H_2O; s mineral acids [MER89]
Density, g/cm³: 2.58 [MER89]

1189
Compound: Ferrous phosphide
Formula: Fe_2P
Molecular Formula: Fe_2P
Molecular Weight: 142.664
CAS RN: 1310-43-6
Properties: gray; hex needles; or, bluish gray powd; ferromagnetic; used in the manufacture of iron and steel; there is also an FeP, 26508-33-8, and Fe_3P, 12023-53-9; can be formed by heating phosphorus rock, silica and coke [KIR82] [HAW93] [MER89] [CER91]
Solubility: i H_2O, dil acid, dil alkali; reacts with hot mineral acids [MER89]
Density, g/cm³: 6.85 [MER89]
Melting Point, °C: 1290 [HAW93]

1190
Compound: Ferrous selenide
Synonyms: iron(II) selenide
Formula: FeSe
Molecular Formula: FeSe
Molecular Weight: 134.805
CAS RN: 1310-32-3
Properties: black mass, with metallic luster; stable in air; decomposes on heating in O_2; used in semiconductor technology [MER89] CER91] [HAW93]
Solubility: i H_2O; s HCl evolving H_2Se [MER89]
Density, g/cm^3: 6.78 [MER89]

1191
Compound: Ferrous sulfate
Synonyms: iron(II) sulfate
Formula: FeSO$_4$
Molecular Formula: FeO$_4$S
Molecular Weight: 151.909
CAS RN: 7720-78-7
Properties: white, ortho-rhomb, hygr [LID94]
Solubility: g/100g soln, H_2O: 13.6 (0°C), 22.8 (25°C), 24.0 (100°C); Solid phase, FeSO$_4$·7H$_2$O (0°C, 25°C), FeSO$_4$·H$_2$O (100°C) [KRU93]
Density, g/cm^3: 3.65 [LID94]
Melting Point, °C: decomposes 671 [JAN85]

1192
Compound: Ferrous sulfate heptahydrate
Synonyms: melanterite
Formula: FeSO$_4$·7H$_2$O
Molecular Formula: FeH$_{14}$O$_{11}$S
Molecular Weight: 278.015
CAS RN: 7782-63-0
Properties: off-white powd; bluish green; monocl cryst or granules; efflorescent in dry air; oxidizes in moist air; odorless with saline taste; used as a pigment, in water and sewage treatment, process engraving; minus 7H$_2$O by 300°C [HAW93] [STR93] [MER89]
Solubility: g/100g H_2O: 28.8 (0°C), 48.0 (20°C), 57.8 (100°C) [LAN85]
Density, g/cm^3: 1.897 [MER89]
Melting Point, °C: decomposes ~60 [LID94]
Reactions: minus 3H$_2$O at 56.6°C; minus 6H$_2$O at 65°C [MER89]

1193
Compound: Ferrous sulfate monohydrate
Synonyms: szomolnokite
Formula: FeSO$_4$·H$_2$O

Molecular Formula: FeH$_2$O$_5$S
Molecular Weight: 169.924
CAS RN: 17375-41-6
Properties: white to yellow cryst powd [MER89]
Solubility: s H_2O [MER89]
Density, g/cm^3: 2.970 [CRC86]
Reactions: minus H_2O ~300°C; decomposes at higher temp [MER89]

1194
Compound: Ferrous sulfide
Synonyms: troilite,iron(II) sulfide
Formula: FeS
Molecular Formula: FeS
Molecular Weight: 87.911
CAS RN: 1317-37-9
Properties: when pure: colorless, hex cryst; usually gray to brownish black lumps; oxidized by moist air to S and Fe$_2$O$_3$; enthalpy of fusion 31.50 kJ/mol [MER89] [CRC93]
Solubility: i H_2O; s acids, evolving H$_2$S [MER89]
Density, g/cm^3: 4.84 [MER89]
Melting Point, °C: 1194 [MER89]
Boiling Point, °C: decomposes [HAW93]

1195
Compound: Ferrous thiocyanate trihydrate
Synonyms: iron(II) thiocyanate trihydrate
Formula: Fe(SCN)$_2$·3H$_2$O
Molecular Formula: C$_2$H$_6$FeN$_2$O$_3$S$_2$
Molecular Weight: 226.058
CAS RN: 6010-09-9
Properties: pale green, monocl prisms; rapidly oxidized when exposed to air [MER89]
Solubility: s H_2O, alcohol, ether [MER89]
Reactions: decomposed by heat [MER89]

1196
Compound: Ferrous titanate
Synonyms: ilmenite
Formula: FeTiO$_3$
Molecular Formula: FeO$_3$Ti
Molecular Weight: 151.710
CAS RN: 12168-52-4
Properties: opaque black with almost metallic luster; rhombohedral; occurs naturally as the mineral ilmenite; finds extensive use in manufacturing Ti paint pigments; there is also Fe$_2$TiO$_5$, 12789-64-9, -100 mesh with 99.9% purity [KIR83] [CER91]
Density, g/cm^3: 4.72 [KIR83]
Melting Point, °C: ~1470 [KIR83]

1197
Compound: Fluorine
Formula: F_2
Molecular Formula: F_2
Molecular Weight: 18.9984032 (atomic wt)
CAS RN: 7782-41-4
Properties: pale yellow diatomic gas, condenses to yellowish orange liq at -188°C, solidifies to yellow solid at -220°C; critical temp -129°C; critical pressure 55 atm; most reactive nonmetal; enthalpy of fusion 0.51 kJ/mol; enthalpy of vaporization 6.62 kJ/mol; enthalpy of dissociation 157.7 kJ/mol; reacts vigorously with most oxidizable substances; produced by electrolysis of dil solution of KF in anhydrous HF [KIR78] [MER89] [CRC93]
Density, g/cm³: gas: 1.695 g/L [KIR78]; liq: 1.5127 at bp [MER89]
Melting Point, °C: -219.66 [CRC93]
Boiling Point, °C: -188.11 [CRC93]
Thermal Conductivity, W/(m·K): 24.77 x 10⁻⁷ at 0°C [KIR78]; 0.0277 at 25°C [ALD94]

1198
Compound: Fluorine dioxide
Synonyms: dioxygen difluoride
Formula: F_2O_2
Molecular Formula: F_2O_2
Molecular Weight: 69.996
CAS RN: 7783-44-0
Properties: thermally unstable gas at room temp; pale yellow solid, or yellow liq; enthalpy of vaporization 19.1 kJ/mol; produced by reacting O_2 and F_2 at cryogenic temperatures in an electrical discharge [KIR78] [MER89] [CRC93]
Density, g/cm³: 3.071 g/L [LID94]
Melting Point, °C: -154 [MER89]
Boiling Point, °C: -57 [CRC93]
Reactions: decomposes to F_2 and O_2 at -100°C [MER89]

1199
Compound: Fluorine monoxide
Formula: F_2O
Molecular Formula: F_2O
Molecular Weight: 53.996
CAS RN: 7783-41-7
Properties: colorless gas; yellowish brown when liq; gas may be kept over water unchanged for a month; does not attack glass in the cold; enthalpy of vaporization 11.09 kJ/mol [CRC93] [MER89]
Solubility: 6.8mL gas/100mL H_2O (0°C) [MER89]

Density, g/cm³: 2.369 g/L [LID94]; liq: 1.90 (-224°C) [MER89]
Melting Point, °C: -223.8 [MER89]
Boiling Point, °C: -144.75 [CRC93]

1200
Compound: Fluorine nitrate
Formula: FNO_3
Molecular Formula: FNO_3
Molecular Weight: 81.003
CAS RN: 7789-26-6
Properties: colorless gas; moldy, acrid odor; liq explodes on slight percussion; hydrolyzed by water to OF_2, O_2, HF and HNO_3; burns with alcohol, ether, aniline [MER89]
Solubility: s acetone [MER89]
Density, g/cm³: 3.554 g/L [LID94]
Melting Point, °C: -175 [LID94]
Boiling Point, °C: -46 [LID94]

1201
Compound: Fluorine perchlorate
Synonyms: chlorine tetroxyfluoride
Formula: $FOClO_3$
Molecular Formula: $ClFO_4$
Molecular Weight: 118.449
CAS RN: 10049-03-3
Properties: colorless gas; pungent, acrid odor; readily explodes on contact with solids, or heated [MER89]
Solubility: reacts with H_2O [LID94]
Density, g/cm³: 5.197 g/L [LID94]
Melting Point, °C: -167.3 [MER89]
Boiling Point, °C: -15.9 (755 mm) [MER89]

1202
Compound: Fluoroantimonic acid
Formula: $HF \cdot SbF_5$
Molecular Formula: F_6HSb
Molecular Weight: 236.758
CAS RN: 16950-06-4
Properties: superacid; moisture sensitive [KIR78] [ALD94]

1203
Compound: Fluoroboric acid
Formula: HBF_4
Molecular Formula: BF_4H
Molecular Weight: 87.813
CAS RN: 16872-11-0

Properties: does not exist as a free pure material; colorless strongly acid liq; stable in conc solutions; produced by reacting 70% HF with boric acid, H_3BO_3; used to produce fluoroborates, in electrolytic brightening of aluminum [HAW93] [KIR78]
Solubility: miscible with H_2O, alcohol [HAW93]
Density, g/cm³: ~1.84 [HAW93]
Boiling Point, °C: 130, decomposes [HAW93]

1204
Compound: Fluorosilane
Formula: SiH_3F
Molecular Formula: FH_3Si
Molecular Weight: 50.108
CAS RN: 13537-33-2
Properties: enthalpy of vaporization 18.8 kJ/mol; entropy of vaporization 107.9 kJ/(mol·K) [CIC73]
Boiling Point, °C: -98.6 [CIC73]

1205
Compound: Fluorosulfonic acid
Formula: HSO_3F
Molecular Formula: FHO_3S
Molecular Weight: 100.070
CAS RN: 7789-21-1
Properties: colorless liq; fumes in moist air; stable to 900°C; considerably more acidic than 100 % H_2SO_4; does not attack glass when anhydrous and pure; viscosity 1.56 MPa s; dielectric constant ~120; specific conductance 1.08×10^{-6} (ohm m)$^{-1}$; used as a catalyst in organic synthesis, electropolishing and as a fluorinating agent [HAW93] [MER89] [KIR78]
Solubility: hydrolyzes violently in H_2O; reddish-brown color in acetone [MER89]
Density, g/cm³: 1.726 [MER89]
Melting Point, °C: freezing point -89 [MER89]
Boiling Point, °C: 163 [MER89]

1206
Compound: Fluorotrimethylsilane
Synonyms: trimethylsilyl fluoride
Formula: $(CH_3)_3SiF$
Molecular Formula: C_3H_9FSi
Molecular Weight: 92.158
CAS RN: 420-56-4
Properties: gas; flammable; sensitive to moisture [ALD94]
Density, g/cm³: 0.793 [ALD94]
Melting Point, °C: -74 [ALD94]

Boiling Point, °C: 16 [ALD94]

1207
Compound: Francium
Formula: Fr
Molecular Formula: Fr
Molecular Weight: 223
CAS RN: 7440-73-5
Properties: heaviest of the alkali metal family; may exist only as radioactive isotopes; only natural isotope is ^{223}Fr with a half-life of 21 min; discovered in 1939 by Mll. M. Perey, Curie Inst., Paris; formed from α-decay of actinium [CRC86] [HAW93]
Melting Point, °C: 27 [CRC86]
Boiling Point, °C: 677 [CRC86]

1208
Compound: Fullerene
Synonyms: carbon fullerenes
Formula: C_{60}
Molecular Formula: C_{60}
Molecular Weight: 720.660
CAS RN: 99685-96-8
Properties: fcc, lattice constant 1.417 nm; mean ball diameter 0.683 nm; compressibility 6.9×10^{-12} cm²/dyne; bulk modulus 14 GPa; binding energy per atom 7.40 eV; structural phase transitions -18°C, -108°C; sound velocity vt 2.1×10^5 cm/s, vl 3.6×10^5 cm/s; Debye temp -88°C; static dielectric constant 4.0-4.5; synthesized by ac discharge of graphite electrodes under He at 200 torr [DRE93]

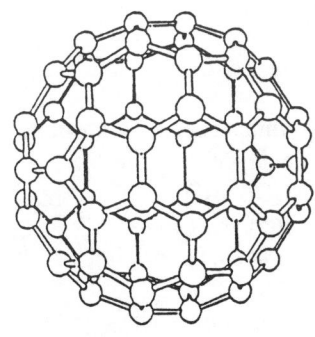

Solubility: s organic solvents [LID94]
Density, g/cm³: 1.72 [DRE93]
Melting Point, °C: >280 [LID94]
Reactions: bromination gives $C_{60}Br_6$, $C_{60}Br_8$ [IUP93]
Thermal Conductivity, W/(m·K): 0.4 (7°C) [DRE93]
Thermal Expansion Coefficient: volume thermal expansion 6.2×10^{-5}/K [DRE93]

1209
Compound: Fullerenes
Formula: C_{60}/C_{70}
Molecular Formula: C_{60}/C_{70}
Molecular Weight: C_{60}: 720.660; C_{70}: 840.777
CAS RN: 131159-39-2
Properties: black powd; contains 10-15% C_{70} [STR93]
Density, g/cm³: 1.6 [STR93]

1210
Compound: Gadolinium
Formula: Gd
Molecular Formula: Gd
Molecular Weight: 157.25
CAS RN: 7440-54-2
Properties: colorless or faintly yellowish metal; tarnishes in moist air; hex close-packed; magnetic, especially at low temperatures; enthalpy of fusion 9.81 kJ/mol; enthalpy of sublimation 397.5 kJ/mol; electrical resistivity (20°C) 126 μohm·cm; radius of atom 0.10813 nm; radius of Gd^{+++} ion 0.0938 nm; solutions are colorless; used in neutron shielding, garnets for microwave filter [KIR82] [MER89] [HAW93] [CRC93] [ALD94]
Solubility: reacts slowly with H_2O; s dil acid [HAW93]
Density, g/cm³: 7.9004 [KIR82]
Melting Point, °C: 1312 [MER89]
Boiling Point, °C: 3273 [KIR82]
Thermal Conductivity, W/(m·K): 10.5 (25°C) [ALD94]
Thermal Expansion Coefficient: 9×10^{-6}/K [CRC93]

1211
Compound: Gadolinium acetate tetrahydrate
Synonyms: acetic acid, Gd(III) salt
Formula: $Gd(CH_3COO)_3 \cdot 4H_2O$
Molecular Formula: $C_6H_{17}GdO_{10}$
Molecular Weight: 406.445
CAS RN: 15280-53-2
Properties: tricl white cryst [STR93] [CRC94]
Solubility: 11.6 g/100 mL H_2O (25°C) [CRC94]
Density, g/cm³: 1.611 [STR93]
Boiling Point, °C: decomposes [ALF95]

1212
Compound: Gadolinium acetylacetonate dihydrate
Synonyms: 2,4-pentanedione, gadolinium(III) derivative

Formula: $Gd(CH_3COCH=C(O)CH_3)_3 \cdot 2H_2O$
Molecular Formula: $C_{15}H_{25}GdO_8$
Molecular Weight: 490.609
CAS RN: 14284-87-8
Properties: off-white powd [STR93]

$$\begin{array}{cc} O- & O \\ | & \| \\ \end{array}$$
$$Ga[CH_3-C=CH-C-CH_3]_3 \cdot 2H_2O$$

Melting Point, °C: 143, decomposes [ALD94]

1213
Compound: Gadolinium boride
Formula: GdB_6
Molecular Formula: B_6Gd
Molecular Weight: 222.116
CAS RN: 12008-06-9
Properties: brownish-black cub; -325 mesh 10 microns or less with 99.9% purity; refractory material [LID94] [KIR78] [CER91]
Density, g/cm³: 5.31 [LID94]
Melting Point, °C: 2100 [KIR78]

1214
Compound: Gadolinium bromide
Formula: $GdBr_3$
Molecular Formula: Br_3Gd
Molecular Weight: 396.962
CAS RN: 13818-75-2
Properties: white hygr cryst; -20 mesh with 99.9% purity [LID94] [CER91]
Melting Point, °C: 770 [LID94]

1215
Compound: Gadolinium chloride
Formula: $GdCl_3$
Molecular Formula: Cl_3Gd
Molecular Weight: 263.608
CAS RN: 10138-52-0
Properties: -20 mesh with 99.9% purity; white; monocl cryst; hygr [STR93] [MER89] [CER91]
Solubility: s H_2O [MER89]
Density, g/cm³: 4.52 (0°C) [MER89]
Melting Point, °C: ~609 [MER89]

1216
Compound: Gadolinium chloride hexahydrate
Formula: $GdCl_3 \cdot 6H_2O$
Molecular Formula: $Cl_3GdH_{12}O_6$
Molecular Weight: 371.654

CAS RN: 13450-84-5

Properties: -4 mesh with 99.9% purity; colorless deliq cryst, obtained from aq solutions; used as a source of Gd metal [MER89] [CER91] [HAW93]

Solubility: s H$_2$O [HAW93]

Density, g/cm^3: 2.424 [MER89]

1217

Compound: Gadolinium fluoride

Formula: GdF$_3$

Molecular Formula: F$_3$Gd

Molecular Weight: 214.245

CAS RN: 13765-26-9

Properties: white powd, or 99.9% pure melted pieces of 3-6 mm; pieces used as evaporation material for possible application to multilayers [STR93] [CER91]

Melting Point, °C: 1231 [LID94]

1218

Compound: Gadolinium gallium garnet

Formula: Gd$_3$Ga$_5$O$_{12}$

Molecular Formula: Ga$_5$Gd$_3$O$_{12}$

Molecular Weight: 1012.358

CAS RN: 12024-36-1

Properties: lump [ALF95]

Density, g/cm^3: 7.09 [ALD94]

1219

Compound: Gadolinium hydride

Formula: GdH$_{2-3}$

Molecular Formula: GdH$_2$; GdH$_3$

Molecular Weight: GdH$_2$: 159.266; GdH$_3$: 160.274

CAS RN: 13572-97-9

Properties: -60 mesh with 99.9% purity [CER91]

1220

Compound: Gadolinium iodide

Formula: GdI$_3$

Molecular Formula: GdI$_3$

Molecular Weight: 537.963

CAS RN: 13572-98-0

Properties: yellow; -20 mesh with 99.9% purity [CER91] [CRC94]

Solubility: s H$_2$O [CRC94]

Melting Point, °C: 926 [AES93]

Boiling Point, °C: 1340 [CRC94]

1221

Compound: Gadolinium nitrate hexahydrate

Formula: Gd(NO$_3$)$_3$·6H$_2$O

Molecular Formula: GdH$_{12}$N$_3$O$_{15}$

Molecular Weight: 451.356

CAS RN: 19598-90-4

Properties: deliq; tricl cryst [MER89]

Solubility: s H$_2$O, alcohol [MER89]

Density, g/cm^3: 2.332 [MER89]

Melting Point, °C: 91 [MER89]

1222

Compound: Gadolinium nitrate pentahydrate

Formula: Gd(NO$_3$)$_3$·5H$_2$O

Molecular Formula: GdH$_{10}$N$_3$O$_{14}$

Molecular Weight: 433.341

CAS RN: 52788-53-1

Properties: hygr white cryst [STR93]

Solubility: i H$_2$O [MER89]

Density, g/cm^3: 2.406 [MER89]

Melting Point, °C: 92 [MER89]

1223

Compound: Gadolinium nitride

Formula: GdN

Molecular Formula: GdN

Molecular Weight: 171.257

CAS RN: 25764-15-2

Properties: -60 mesh with 99.9% purity; NaCl cryst system, a = 0.499 nm [CIC73] [CER91]

Density, g/cm^3: 9.10 [LID94]

1224

Compound: Gadolinium oxalate decahydrate

Formula: Gd$_2$(C$_2$O$_4$)$_3$·10H$_2$O

Molecular Formula: C$_6$H$_{20}$Gd$_2$O$_{22}$

Molecular Weight: 758.712

CAS RN: 22992-15-0

Properties: monocl white powd [STR93] [CRC94]

Solubility: i H$_2$O; s sl acids [HAW93]

Reactions: minus 6H$_2$O at 110°C [HAW93]

1225

Compound: Gadolinium oxide

Synonyms: gadolinia

Formula: Gd$_2$O$_3$

Molecular Formula: Gd$_2$O$_3$

Molecular Weight: 362.498

CAS RN: 12064-62-9

Properties: -325 mesh 5 microns or less with 99.999% purity; white to cream colored powd; hygr; absorbs CO_2 from air; used in neutron shields, in special glasses, and as an evaporated material of 99.9% purity, it is reactive to radio frequencies [HAW93] [MER89] [CER91]
Solubility: i H_2O; s in acids [HAW93]
Density, g/cm³: 7.407 [MER89]
Melting Point, °C: 2310 [STR93]

1226
Compound: Gadolinium perchlorate hydrate
Formula: $Gd(ClO_4)_3 \cdot xH_2O$
Molecular Formula: Cl_3GdO_{12} (anhydrous)
Molecular Weight: 455.601 (anhydrous)
CAS RN: 14017-52-8
Properties: white cryst; hygr; x=6 [ALF95] [STR93]

1227
Compound: Gadolinium silicide
Formula: $GdSi_2$
Molecular Formula: $GdSi_2$
Molecular Weight: 213.421
CAS RN: 12134-75-7
Properties: 10 mm & down lump; 6 mm pieces and smaller with 99.9% purity [ALF93] [CER91]
Density, g/cm³: 5.9 [LID94]

1228
Compound: Gadolinium sulfate
Formula: $Gd_2(SO_4)_3$
Molecular Formula: $Gd_2O_{12}S_3$
Molecular Weight: 602.691
CAS RN: 13450-87-8
Properties: colorless [CRC94]
Solubility: g/100g H_2O: 3.98 (0°C), 2.60 (20°C), 2.32 (40°C) [LAN85]
Density, g/cm³: 4.139 [LAN52]
Melting Point, °C: decomposes, 500 [CRC77]

1229
Compound: Gadolinium sulfate octahydrate
Formula: $Gd_2(SO_4)_3 \cdot 8H_2O$
Molecular Formula: $Gd_2H_{16}O_{20}S_3$
Molecular Weight: 746.813
CAS RN: 13450-87-8
Properties: colorless monocl cryst; used in cryogenic research [MER89] [HAW93]
Solubility: 3.28 g/100mL H_2O (20°C), 2.54 g/100mL H_2O (40°C) [CRC94]
Density, g/cm³: 3.010 [STR93]

Reactions: minus $8H_2O$ at 400°C [MER89]

1230
Compound: Gadolinium sulfide
Formula: Gd_2S_3
Molecular Formula: Gd_2S_3
Molecular Weight: 410.698
CAS RN: 12134-77-9
Properties: yellow, hygr; -200 mesh, 99.9% purity [CER91]
Solubility: decomposed by H_2O [CRC94]
Density, g/cm³: 6.1 [LID94]

1231
Compound: Gadolinium telluride
Formula: Gd_2Te_3
Molecular Formula: Gd_2Te_3
Molecular Weight: 697.300
CAS RN: 12160-99-5
Properties: ortho-rhomb; -20 mesh with 99.9% purity [LID94] [CER91]
Density, g/cm³: 7.7 [LID94]
Melting Point, °C: 1255 [LID94]

1232
Compound: Gadolinium titanate
Formula: $Gd_2Ti_2O_7$
Molecular Formula: $Gd_2O_7Ti_2$
Molecular Weight: 522.230
CAS RN: 12024-89-4
Properties: -100 mesh with 99.9% purity [CER91]

1233
Compound: Gadopentetic acid
Synonyms: diethylenetriaminepentaacetic acid, gadolinium(III) salt
Molecular Formula: $C_{14}H_{20}GdN_3O_{10}$
Molecular Weight: 547.577
CAS RN: 80529-93-7
Properties: uses: metal ion complex for MRI, diagnosis of cerebral tumors [ALD94] [MER89]
Melting Point, °C: 129, decomposes [ALD94]

1234
Compound: Gallium
Formula: Ga
Molecular Formula: Ga
Molecular Weight: 69.723
CAS RN: 7440-55-3

Properties: silvery white liq or grayish metal; has tendency to remain in supercooled state; contracts on melting; ortho-rhomb, a = 0.45198 nm, b = 0.76602 nm, c = 0.45258 nm; electron affinity, 0.18 eV; enthalpy of fusion 5.59 kJ/mol; enthalpy of vaporization 254 kJ/mol; electrical resistivity 15.05 μohm·cm (20°C) for polycrystalline form, 25.79 μohm·cm for liq (30°C); radius of atom 0.138 nm, radius of Ga^{+++} 0.133 nm [CRC93] [KIR78] [MER89] [CIC73]

Solubility: reacts with alkalies to evolve H_2 [MER89]

Density, g/cm³: solid: 5.907; liq: 6.095 [KIR78] [CIC73]

Melting Point, °C: 29.78 [ALD94]

Boiling Point, °C: 2403 [ALD94]

Reactions: attacked by halogens and cold conc HCl [MER89]

Thermal Conductivity, W/(m·K): a axis: 88.4, b axis: 16.0, c axis 40.8 (20°C); liq 28.7 (77°C) [KIR78]

Thermal Expansion Coefficient: cub coefficient/°C: 0-20°C, 5.98 x 10^{-5} (solid); liq 1.2 x 10^{-4} (103°C), 1.03 x 10^{-4} (600°C) [KIR78]

1235

Compound: Gallium acetylacetonate

Synonyms: 2,4-pentanedione, gallium(III) derivative

Formula: $Ga(CH_3COCH=C(O)CH_3)_3$

Molecular Formula: $C_{15}H_{21}GaO_6$

Molecular Weight: 367.051

CAS RN: 14405-43-7

Properties: monocl white powd [STR93] [CRC94]

$$[CH_3-\overset{\overset{\displaystyle O-}{|}}{C}=CH-\overset{\overset{\displaystyle O}{\|}}{C}-CH_3]_3Ga$$

Density, g/cm³: 1.42 [STR93]

Melting Point, °C: 192-194 [STR93]

Boiling Point, °C: 140 (10 mm Hg) sublimes [STR93]

1236

Compound: Gallium antimonide

Formula: GaSb

Molecular Formula: GaSb

Molecular Weight: 191.483

CAS RN: 12064-03-8

Properties: cub; 6mm pieces and smaller with 99.99% purity; band gap, eV, 0.81 (0 K), 0.72 (300 K); mobility (300 K), cm²/(V s), 5000 for electrons, 850 for holes; effective mass 0.042 for electrons, 0.40 for holes; dielectric constant 15.7; enthalpy of fusion 25.10 kJ/mol; used in semiconducting devices; obtained by direct reaction of Ga and Sb at high temp [HAW93] [KIR82] [CER91] [CRC93]

Density, g/cm³: 6.096 [KIR78]

Melting Point, °C: 703 [CRC93]

Thermal Conductivity, W/(m·K): 27 [CRC93]

Thermal Expansion Coefficient: 6.1 x 10^{-6}/K [CRC93]

1237

Compound: Gallium arsenide

Formula: GaAs

Molecular Formula: AsGa

Molecular Weight: 144.645

CAS RN: 1303-00-0

Properties: cub cryst; 3-12 mm pieces of 99.999% purity, 25 mm and down polycrystalline pieces; dark gray, with metallic sheen; hardness 4.5; dielectric constant 13.1; band gap, eV, 1.52 (0 K), 1.42 (300 K); mobility (300 K), cm²/(V s), 8500 electrons and 400 holes; electroluminescent in infrared light; obtained by direct reaction of Ga and As at high temp; used as a semiconductor in light-emitting diodes for telephone dials [HAW93] [MER89] [STR93] [KIR82]

Density, g/cm³: 5.3176 [LID94]

Melting Point, °C: 1238 [MER89]

Thermal Conductivity, W/(m·K): 0.52 [MER89]

Thermal Expansion Coefficient: 5.9 x 10^{-6}/°C [MER89]

1238

Compound: Gallium azide

Synonyms: gallium(III) azide

Formula: $Ga(N_3)_3$

Molecular Formula: GaN_9

Molecular Weight: 195.784

CAS RN: 73157-11-6

Properties: prepared by decomposition of $GaF_3 \cdot NH_3$, 73157-06-9, at ~250°C [KIR78]

1239

Compound: Gallium nitride

Formula: GaN

Molecular Formula: GaN

Molecular Weight: 83.730

CAS RN: 25617-97-4
Properties: gray powd; -100 mesh with 99.9% purity; hex, a = 0.319 nm, c = 0.518 nm; has both semiconductor and electroluminescence properties; bandgap, eV, 3.50 (0 K) and 3.36 (300 K); electron mobility (300 K) 380 cm^2/(V s); effective mass 0.19 electrons and 0.60 holes; dielectric 12.2; can be prepared by reaction of Ga with ammonia at ~1000°C [KIR81] [CIC73] [KIR82] [CER91] [CRC94]
Density, g/cm^3: 6.1 [LID94]
Melting Point, °C: 600 (vacuum) [CIC73]
Boiling Point, °C: decomposes >600 [KIR78]
Thermal Conductivity, W/(m·K): 6.56 [CRC93]

1240

Compound: Gallium phosphide
Formula: GaP
Molecular Formula: GaP
Molecular Weight: 100.697
CAS RN: 12063-98-8
Properties: amber; cub; 6 mm pieces and smaller with 99.999% purity; translucent, amber colored cryst; dielectric constant 11.1; band gap, eV, 2.34 (0 K) and 2.26 (300 K); mobility (300 K), cm^2/(V s), 75 holes and 110 electrons; effective mass 0.82 electrons and 0.60 holes; obtained by direct reaction of Ga and P at high temperatures; used in semiconductor devices; electroluminescent in visible light [HAW93] [MER89] [KIR82] [CER91] [STR93]
Density, g/cm^3: 4.138 [LID94]
Melting Point, °C: 1457 [LID94]
Thermal Conductivity, W/(m·K): 75.2 (25°C) [CRC93]
Thermal Expansion Coefficient: 5.3 x 10^{-6}/K [CRC93]

1241

Compound: Gallium suboxide
Formula: Ga$_2$O
Molecular Formula: Ga$_2$O
Molecular Weight: 155.445
CAS RN: 12024-20-3
Properties: brown powd; obtained by heating Ga$_2$O$_3$ and Ga at 700°C; stable in dry air [MER89]
Solubility: i H$_2$O [CRC94]
Density, g/cm^3: 4.77 [KIR78]
Melting Point, °C: decomposes above 800 [MER89]
Reactions: oxidized to the trivalent state by HNO$_3$ or Br$_2$ [MER89]

1242

Compound: Gallium(II) chloride
Synonyms: gallium dichloride
Formula: GaCl$_2$
Molecular Formula: Cl$_2$Ga
Molecular Weight: 140.628
CAS RN: 24597-12-4
Properties: white; deliq; cryst; can be prepared by heating GaCl$_3$ with Ga [MER89] [CRC94]
Solubility: decomposed by H$_2$O [CRC94]
Density, g/cm^3: 2.74 [LID94]
Melting Point, °C: 172.4 [MER89]; 164 [STR93]
Boiling Point, °C: 535 [STR93]

1243

Compound: Gallium(II) selenide
Formula: GaSe
Molecular Formula: GaSe
Molecular Weight: 148.683
CAS RN: 12024-11-2
Properties: dark red; hex; 6 mm pieces and smaller with 99.999% purity [CER91] [KIR78] [CRC94]
Density, g/cm^3: 5.01 [KIR78]
Melting Point, °C: 960-965 [KIR78]

1244

Compound: Gallium(II) sulfide
Formula: GaS
Molecular Formula: GaS
Molecular Weight: 101.789
CAS RN: 12024-10-1
Properties: 6 mm pieces and smaller with 99.999% purity; hex, lamellar structure; air sensitive [KIR78] [STR93] [CER91]
Density, g/cm^3: 3.86 [STR93]
Melting Point, °C: ~965 [STR93]

1245

Compound: Gallium(II) telluride
Formula: GaTe
Molecular Formula: GaTe
Molecular Weight: 197.323
CAS RN: 12024-14-5
Properties: 6 mm pieces and smaller with 99.999% purity; mono or hex [KIR78] [CER91]
Density, g/cm^3: 5.44 [KIR78]
Melting Point, °C: 825 [KIR78]

1246

Compound: Gallium(III) bromide

Formula: GaBr$_3$
Molecular Formula: Br$_3$Ga
Molecular Weight: 309.435
CAS RN: 13450-88-9
Properties: -8 mesh with 99.999% purity; ortho-rhomb white cryst; enthalpy of vaporization 38.9 kJ/mol; enthalpy of fusion 11.70 kJ/mol; formed by reacting Br$_2$ vapor with Ga in N$_2$ atm; has seven hydrates with 1, 2, 2.5, 3, 4, 6 and 15 H$_2$O [CER91] [STR93] [KIR78] [CRC93]
Density, g/cm^3: 3.69 [STR93]
Melting Point, °C: 121.5 [CRC93]
Boiling Point, °C: 279 [CRC93]

1247
Compound: Gallium(III) chloride
Formula: GaCl$_3$
Molecular Formula: Cl$_3$Ga
Molecular Weight: 176.081
CAS RN: 13450-90-3
Properties: solid ingot in glass with 99.999% purity; tricl, colorless needles; enthalpy of vaporization 23.9 kJ/mol; enthalpy of fusion 10.90 kJ/mol; prepared by reacting Ga metal with Cl$_2$ or HCl in nitrogen atm at 200°C; a trihydrate, 23306-52-7, is known [MER89] [KIR78] [CER91] [CRC93]
Solubility: >800 gGaCl$_3$/L H$_2$O [KIR78]
Density, g/cm^3: 2.47 [STR93]
Melting Point, °C: 78 [ALD94]
Boiling Point, °C: 201 [CRC93]

1248
Compound: Gallium(III) fluoride
Synonyms: gallium trifluoride
Formula: GaF$_3$
Molecular Formula: F$_3$Ga
Molecular Weight: 126.718
CAS RN: 7783-51-9
Properties: trig; -60 mesh with 99.95% purity; white powd, colorless needles; formed by thermal decomposition of ammonium hexafluorogallate in Ar atm [MER89] [KIR78] [CER91]
Solubility: 0.0024 g/100mL H$_2$O (25°C) [MER89]
Density, g/cm^3: 4.47 (heated in F$_2$ atm, 630°C) [MER89]
Melting Point, °C: >1000 [MER89]
Reactions: can be sublimed in N$_2$ atm at 800°C without decomposition [MER89]

1249
Compound: Gallium(III) fluoride trihydrate
Formula: GaF$_3$·3H$_2$O
Molecular Formula: F$_3$GaH$_6$O$_3$
Molecular Weight: 180.764
CAS RN: 22886-66-4
Properties: -60 mesh with 99.5% purity; white cryst; obtained by dissolution of Ga or Ga(OH)$_3$ in HF [KIR78] [STR93] [CER91]
Solubility: more soluble than anhydrous GaF$_3$ in H$_2$O [MER89]
Melting Point, °C: >140 [MER89]
Reactions: thermally decomposed to 16[Ga(OH,F)$_3$]·6H$_2$O at 200°C [KIR78]

1250
Compound: Gallium(III) hydride
Synonyms: gallane
Formula: GaH$_3$
Molecular Formula: GaH$_3$
Molecular Weight: 72.747
CAS RN: 13572-93-5
Properties: viscous liq; can be prepared by reacting (CH$_3$)$_3$N·GaH$_3$, 19528-13-3, with BF$_3$ at -20°C [LID94] [KIR78]
Melting Point, °C: decomposes above -15 [KIR78]

1251
Compound: Gallium(III) hydroxide
Formula: Ga(OH)$_3$
Molecular Formula: GaH$_3$O$_3$
Molecular Weight: 120.745
CAS RN: 12023-99-3
Properties: white; unstable gelatinous precipitate; obtained by adding ammonia to solution of Ga(III) salt [MER89] [KIR78] [CRC94]
Solubility: i H$_2$O [CRC94]
Melting Point, °C: decomposes at 440 [CRC94]

1252
Compound: Gallium(III) iodide
Formula: GaI$_3$
Molecular Formula: GaI$_3$
Molecular Weight: 450.436
CAS RN: 13450-91-4
Properties: -20 mesh with 99.999% purity; monocl; enthalpy of vaporization 56.5 kJ/mol; enthalpy of fusion 16.30 kJ/mol; can be prepared by direct reaction of Ga and I$_2$ [CRC93] [CER91] [KIR78]
Density, g/cm^3: 4.15 [KIR78]
Melting Point, °C: 212 [KIR78]

Boiling Point, °C: sublimes at 340 [ALD94]

1253
Compound: Gallium(III) nitrate
Formula: Ga(NO₃)₃
Molecular Formula: GaN₃O₉
Molecular Weight: 255.738
CAS RN: 13494-90-1
Properties: white; cryst powd [MER89]
Solubility: s warm and cold H₂O, absolute alcohol, ether [MER89]
Melting Point, °C: 110 decomposes [AES93]

1254
Compound: Gallium(III) nitrate hydrate
Formula: Ga(NO₃)₃·xH₂O
Molecular Formula: GaN₃O₉ (anhydrous)
Molecular Weight: 255.738 (anhydrous)
CAS RN: 69365-72-6
Properties: obtained by dissolving Ga metal or the oxide in conc HNO₃ [MER89]
Solubility: s H₂O [CRC94]
Melting Point, °C: decomposes 110 [AES93]
Reactions: decomposes to Ga₂O₃ at 200°C [CRC94]

1255
Compound: Gallium(III) oxide
Formula: Ga₂O₃
Molecular Formula: Ga₂O₃
Molecular Weight: 187.444
CAS RN: 12024-21-4
Properties: α, β, γ, δ, ε forms, β is the most stable; white cryst; α and β obtained by thermal decomposition of salts; γ formed by rapid dehydration of Ge(OH)₃ gels at ~400°C, δ prepared by decomposing Ge(NO₃)₃ at ~250°C, ε formed by briefly heating δ form at ~550°C; used in spectroscopic analysis, and as an evaporated material and sputtering target of 99.999% purity in dielectric films [HAW93] [MER89] [CER91]
Solubility: s hot acid [HAW93]
Density, g/cm³: α: 6.44; β: 5.88 [HAW93]
Melting Point, °C: 1725 [LID94]
Reactions: reacts violently with Mg to give Ga [MER89]

1256
Compound: Gallium(III) oxide hydroxide
Formula: GaOOH
Molecular Formula: GaHO₂

Molecular Weight: 102.730
CAS RN: 20665-52-5
Properties: ortho-rhomb; prepared by oxidation of Ga with H₂O at ~200°C under pressure [KIR78]
Density, g/cm³: 5.23 [LID94]
Reactions: GaOOH → α-Ga₂O₃ at 300-500°C [KIR78]

1257
Compound: Gallium(III) perchlorate hexahydrate
Formula: Ga(ClO₄)₃·6H₂O
Molecular Formula: Cl₃GaH₁₂O₁₈
Molecular Weight: 476.166
CAS RN: 17835-81-3
Properties: cryst [ALF95]
Melting Point, °C: 175 decomposes [ALF95]

1258
Compound: Gallium(III) selenide
Formula: Ga₂Se₃
Molecular Formula: Ga₂Se₃
Molecular Weight: 376.326
CAS RN: 12024-24-7
Properties: monocl; 6 mm pieces and smaller with 99.999% purity [CER91] [KIR78]
Density, g/cm³: 4.95 [KIR78]
Melting Point, °C: 1005-1010 [KIR78]
Thermal Conductivity, W/(m·K): 5 [CRC93]
Thermal Expansion Coefficient: 8.9 x 10⁻⁶/K [CRC93]

1259
Compound: Gallium(III) sulfate
Formula: Ga₂(SO₄)₃
Molecular Formula: Ga₂O₁₂S₃
Molecular Weight: 427.637
CAS RN: 13494-94-2
Properties: white powd; prepared by evaporation of a solution of GaOOH, 20665-52-5, in 50% sulfuric acid, followed by drying at 360°C; crystallizes from aq solution as the octadecahydrate [KIR78] [CRC94]
Solubility: v s H₂O [CRC94]

1260
Compound: Gallium(III) sulfate octadecahydrate
Formula: Ga₂(SO₄)₃·18H₂O
Molecular Formula: Ga₂H₃₆O₃₀S₃
Molecular Weight: 751.912
CAS RN: 13780-42-2

Properties: octahedral cryst; formed by dissolving Ga_2O_3 or $Ga(OH)_3$ in sulfuric acid, and precipitating with ether or alcohol [MER89]
Solubility: s H_2O, 60% alcohol [MER89]
Density, g/cm³: 3.86 [STR93]

1261
Compound: Gallium(III) sulfide
Formula: Ga_2S_3
Molecular Formula: Ga_2S_3
Molecular Weight: 235.644
CAS RN: 12024-22-5
Properties: -100 mesh with 99.95% purity; monocl [CER91] [KIR78]
Density, g/cm³: 3.77 [KIR78]
Melting Point, °C: 1090 [KIR78]

1262
Compound: Gallium(III) telluride
Formula: Ga_2Te_3
Molecular Formula: Ga_2Te_3
Molecular Weight: 522.246
CAS RN: 12024-27-0
Properties: 6 mm pieces and smaller with 99.999% purity; two forms: cub and tetra; tetra, 73623-48-0, is stable only at 400-495°C; the pentavalent telluride Ga_2Te_5, 73623-48-0, is stable only in the range 400-495°C [KIR78] [CER91]
Density, g/cm³: cub: 5.57; tetr: 5.85; Ga_2Te_5: 5.85 [KIR78]
Melting Point, °C: 792 [KIR78]
Thermal Conductivity, W/(m·K): 4.7 [CRC93]

1263
Compound: Germanium
Formula: Ge
Molecular Formula: Ge
Molecular Weight: 72.61
CAS RN: 7440-56-4
Properties: grayish white brittle metalloid; stable to oxidation in air up to 400°C; cub, a = 0.56574 nm; enthalpy of fusion 36.94 kJ/mol; enthalpy of vaporization 334 kJ/mol; Poisson's ratio 0.278; hardness 6 Mohs; resistivity 53,000 μohm·cm (25°C); electronegativity 1.8-1.9; band gap, eV, 0.74 (0 K), 0.66 (300 K); mobility (300 K), $cm^2/(V·s)$, 3900 electron and 1900 holes; used in transistors and semiconductor applications [MER89] [KIR78] [COT88] [CRC93]
Solubility: i H_2O, HCl, dil alkali hydroxides; attacked by aqua regia [MER89]

Density, g/cm³: 5.323 [MER89]
Melting Point, °C: 938.25 [LID94]
Boiling Point, °C: 2830 [KIR78]
Thermal Conductivity, W/(m·K): 60.2 (25°C) [ALD94]
Thermal Expansion Coefficient: $6.1 \times 10^{-6}/°C$ [MER89]

1264
Compound: Germanium nitride
Formula: Ge_3N_4
Molecular Formula: Ge_3N_4
Molecular Weight: 273.857
CAS RN: 12065-36-0
Properties: brownish-white powd; -200 mesh with 99.999% purity; prepared by reacting Ge powd and ammonia at 700-850°C; ortho-rhomb; a = 1.384 nm, b = 0.406 nm, c = 0.818 nm [CIC73] [CER91] [CRC94]
Solubility: i H_2O; does not react with most mineral acids, aqua regia or caustic solutions [KIR78] [CRC94]
Density, g/cm³: 5.25 [CRC94]
Melting Point, °C: 900-1000 decomposes [CIC73]

1265
Compound: Germanium tetrahydride
Synonyms: germane
Formula: GeH_4
Molecular Formula: GeH_4
Molecular Weight: 76.642
CAS RN: 7782-65-2
Properties: colorless gas; spontaneously flammable in air; enthalpy of vaporization 14.06 kJ/mol; can be prepared in small quantity by reaction: $GeCl_4 + 4NaB_4 + 12H_2O = GeH_4(gas) + 4NaCl + 4B(OH)_3 + 12 H_2(gas)$; used to produce highly pure electronic grade germanium by thermal decomposition at ~350°C [KIR80] [CRC93]
Solubility: i H_2O; s liq ammonia, sl s hot HCl [HAW93]
Density, g/cm³: liq: 1.523 at -142°C; gas: 3.43 g/L (0°C) [KIR80]
Melting Point, °C: -165 [HAW93]
Boiling Point, °C: -90 [MER89]

1266
Compound: Germanium(II) chloride
Formula: $GeCl_2$
Molecular Formula: Cl_2Ge
Molecular Weight: 143.515
CAS RN: 10060-11-4

Properties: white powd; unstable; decomposes into polymer subchloride at low temp [MER89] [HAW93]
Solubility: decomposes in H_2O; s ether, benzene; i alcohol and chloroform [HAW93] [MER89]
Melting Point, °C: decomposes [HAW93]

1267

Compound: Germanium(II) fluoride
Formula: GeF_2
Molecular Formula: F_2Ge
Molecular Weight: 110.607
CAS RN: 13940-63-1
Properties: white solid; decomposes above 130°C to form GeF_4(gas), Ge and GeF(gas); deliq in moist air forming Ge(II) hydroxide; can be formed by reduction of GeF_4 with metallic Ge [KIR78]
Solubility: s HF solutions [KIR78]
Melting Point, °C: 110 [LID94]
Boiling Point, °C: decomposes, 130 [LID94]

1268

Compound: Germanium(II) iodide
Formula: GeI_2
Molecular Formula: GeI_2
Molecular Weight: 326.419
CAS RN: 13573-08-5
Properties: -10 mesh with 99.999% purity; yellow powd [STR93] [CER91]
Solubility: s H_2O [CRC94]
Density, g/cm³: 5.73 [STR93]
Melting Point, °C: decomposes 550 [LID94]

1269

Compound: Germanium(II) oxide
Synonyms: germanium monoxide
Formula: GeO
Molecular Formula: GeO
Molecular Weight: 88.609
CAS RN: 20619-16-3
Properties: black solid; stable at room temp; best prepared in a pure form by heating Ge and GeO_2 in oxygen free atm, GeO sublimes above 710°C and is condensed [KIR78] [HAW93]
Solubility: s in about 250 parts cold H_2O, 100 parts boiling H_2O; s acids [MER89]
Melting Point, °C: 710 (sublimes) [HAW93]

1270

Compound: Germanium(II) selenide

Formula: GeSe
Molecular Formula: GeSe
Molecular Weight: 151.570
CAS RN: 12065-10-0
Properties: gray ortho-rhomb, or brown powd; 6 mm pieces and smaller with 99.999% purity [CER91] [LID94]
Density, g/cm³: 5.6 [LID94]
Melting Point, °C: 667 [LID94]

1271

Compound: Germanium(II) sulfide
Formula: GeS
Molecular Formula: GeS
Molecular Weight: 104.676
CAS RN: 12025-32-0
Properties: reddish-yellow, amorphous or rhomb cryst; -20 mesh with 99.95% purity [CER91] [CRC94]
Solubility: 0.24 g/100mL H_2O [CRC94]
Density, g/cm³: amorphous: 3.31; rhomb: 4.01 [CRC94]
Melting Point, °C: 615 [LID94]
Reactions: sublimes at 430°C [CRC94]

1272

Compound: Germanium(II) telluride
Formula: GeTe
Molecular Formula: GeTe
Molecular Weight: 200.210
CAS RN: 12025-39-7
Properties: 6 mm pieces and smaller with 99.999% purity; good semiconductor [HAW93] [CER91]
Density, g/cm³: 6.14 [CRC93]
Melting Point, °C: 725 [HAW93]

1273

Compound: Germanium(IV) bromide
Synonyms: germanium tetrabromide
Formula: $GeBr_4$
Molecular Formula: Br_4Ge
Molecular Weight: 392.226
CAS RN: 13450-92-5
Properties: solid ingot in glass with 99.999% purity; white cryst; enthalpy of vaporization 41.4 kJ/mol; can be prepared readily by reacting Ge with Br_2, or with GeO_2 and HBr solutions [KIR78] [STR93] [CER91] [CRC93]
Density, g/cm³: 3.132 [STR93]
Melting Point, °C: 26.1 [STR93]
Boiling Point, °C: 186.5 [ALD94]

1274
Compound: Germanium(IV) chloride
Synonyms: germanium tetrachloride
Formula: GeCl$_4$
Molecular Formula: Cl$_4$Ge
Molecular Weight: 214.421
CAS RN: 10038-98-9
Properties: 99.9999% purity; colorless liq; fumes in air; appreciably volatile at room temp; refractive index 1.464; enthalpy of vaporization 27.9 kJ/mol; vapor pressure, Pa: 100 (-48°C), 1000 (-20°C), 10,000 (21°C), 10^5 (83°C), 10^6 (190°C); prepared by reacting germanium oxides or germanates with HCl [KIR78] [HAW93] [MER89] [CER91] [CRC93]
Solubility: hydrolyzed in H$_2$O; s benzene, ether, carbon disulfide, alcohol, chloroform [HAW93]
Density, g/cm^3: 1.874 [HAW93]
Melting Point, °C: -49.5 [HAW93]
Boiling Point, °C: 86.55 [CRC93]

1275
Compound: Germanium(IV) ethoxide
Formula: Ge(OC$_2$H$_5$)$_4$
Molecular Formula: C$_8$H$_{20}$GeO$_4$
Molecular Weight: 252.84
CAS RN: 14165-55-0
Properties: liq [ALF95]
Solubility: s alcohol, benzene [ALF95]
Density, g/cm^3: 1.140 [ALF95]
Melting Point, °C: -72 [ALF95]
Boiling Point, °C: 185.5 [ALF95]
Reactions: hydrolyzed by H$_2$O [ALF95]

1276
Compound: Germanium(IV) fluoride
Synonyms: germanium tetrafluoride
Formula: GeF$_4$
Molecular Formula: F$_4$Ge
Molecular Weight: 148.604
CAS RN: 7783-58-6
Properties: 99.99% purity; colorless gas; fumes strongly in air; odor of garlic; thermally stable up to ~1000°C; triple point reported as -50°C and 404.1 kPa; vapor pressure ~100 kPa at -36.5°C; pure GeF$_4$ usually prepared by decomposing BaGeF$_6$ at ~700°C [KIR78] [MER89] [CER91]
Solubility: hydrolyzes to GeO$_2$ and H$_2$GeF$_6$ in H$_2$O [MER89]
Density, g/cm^3: liq: 2.162; solid (-195°C): 3.148 [MER89]

Melting Point, °C: -15 (3032 mm pressure) [MER89]
Boiling Point, °C: sublimes at -36.5 [MER89]
Reactions: corrodes Hg and grease [MER89]

1277
Compound: Germanium(IV) fluoride trihydrate
Synonyms: germanium tetrafluoride trihydrate
Formula: GeF$_4$·3H$_2$O
Molecular Formula: F$_4$GeH$_6$O$_3$
Molecular Weight: 202.650
CAS RN: 7783-58-6
Properties: white cryst; deliq; obtained by slow evaporation of GeO$_2$ in 20% HF [MER89] [CRC94]
Solubility: s H$_2$O [CRC94]
Melting Point, °C: decomposes [CRC94]

1278
Compound: Germanium(IV) iodide
Synonyms: germanium tetraiodide
Formula: GeI$_4$
Molecular Formula: GeI$_4$
Molecular Weight: 580.228
CAS RN: 13450-95-8
Properties: -10 mesh with 99.999% purity; reddish orange cryst; a method of preparation is to react Ge with I$_2$, or GeO$_2$ with HI solutions [KIR78] [STR93] [CER91]
Density, g/cm^3: 4.416 [STR93]
Melting Point, °C: 146 [STR93]
Boiling Point, °C: 350 [STR93]

1279
Compound: Germanium(IV) oxide
Synonyms: germanium dioxide
Formula: GeO$_2$
Molecular Formula: GeO$_2$
Molecular Weight: 104.609
CAS RN: 1310-53-8
Properties: white powd; two forms: hex, tetr amorphous (vitreous); hex can be produced by the hydrolysis of GeCl$_4$ in H$_2$O, or by igniting Ge sulfides; tetr form is insoluble, it can be produced by heating the hex form at 300-900°C; amorphous material formed when the hex or tetr forms are melted and cooled; used in phosphors, transistors and diodes, in infrared transmitting glass; hex form of 99.999% purity is a sputtering target for dielectric film preparation [HAW93] [CER91]

Solubility: (probably hex form) g/100g H_2O: 0.49 (10°C), 0.43 (20°C), 0.61 (40°C) [LAN85]; hex: s HCl, HF, NaOH solutions [KIR78]

Density, g/cm³: hex: 4.228; tetr: 6.239; amorphous: 3.637 [KIR78]

Melting Point, °C: hex: 1116; tetr: 1086 [KIR78]

1280

Compound: Germanium(IV) selenide

Synonyms: germanium diselenide

Formula: $GeSe_2$

Molecular Formula: $GeSe_2$

Molecular Weight: 230.530

CAS RN: 12065-11-1

Properties: 6 mm pieces and smaller with 99.999% purity; orange cryst [STR93] [CER91]

Solubility: i H_2O [CRC94]

Density, g/cm³: 4.56 [STR93]

Melting Point, °C: 707 [STR93]

Boiling Point, °C: decomposes [CRC94]

1281

Compound: Germanium(IV) sulfide

Synonyms: germanium disulfide

Formula: GeS_2

Molecular Formula: GeS_2

Molecular Weight: 136.742

CAS RN: 12025-34-2

Properties: black cryst; can be prepared by reacting GeO_2 and sulfur [KIR78] [STR93]

Solubility: decomposed by H_2O [CRC94]

Density, g/cm³: 2.94 [CRC94]

Melting Point, °C: 530 [STR93]

1282

Compound: Germanium(IV) telluride

Synonyms: germanium ditelluride

Formula: $GeTe_2$

Molecular Formula: $GeTe_2$

Molecular Weight: 327.810

CAS RN: 12260-55-8

Properties: reacted product, 6mm pieces and smaller with 99.999% purity [CER91]

1283

Compound: Gold

Formula: Au

Molecular Formula: Au

Molecular Weight: 196.96654

CAS RN: 7440-57-5

Properties: yellow; soft metal; cub, a = 0.407 nm; electronegativity, 2.88; electrical resistivity (20°C) 2.35 μohm·cm; temp coefficient of electrical resistivity (0-100°C) 0.004; enthalpy of fusion 12.55 kJ/mol; enthalpy of vaporization 324 kJ/mol; specific heat (18°C) 131 J/(kg-°C); hardness 2.5-3.0 Mohs; forms $AuTe_2$ (calaverite), 12006-61-0, by reaction with Te at ~475°C [KIR78] [MER89] [CRC93]

Solubility: s aqua regia, alkali cyanide solutions [MER89]

Density, g/cm³: 19.32 (20°C) [KIR78]

Melting Point, °C: 1064.43 [ALD94]

Boiling Point, °C: 2808 [ALD94]

Reactions: extremely stable: not attacked by acids or air [MER89]

Thermal Conductivity, W/(m·K): 318 (25°C) [ALD94]

Thermal Expansion Coefficient: 100°C: 14.16 x 10^{-6}/K [KIR78]

1284

Compound: Gold(I) bromide

Synonyms: aurous bromide

Formula: AuBr

Molecular Formula: AuBr

Molecular Weight: 276.871

CAS RN: 10294-27-6

Properties: yellowish gray mass [HAW93]

Solubility: i H_2O [HAW93]

Density, g/cm³: 7.9 [CRC94]

Melting Point, °C: decomposes 165 [HAW93]

1285

Compound: Gold(I) carbonyl chloride

Formula: Au(CO)Cl

Molecular Formula: CAuClO

Molecular Weight: 260.430

CAS RN: 50960-82-2

Properties: off-white powd; sensitive to atm oxygen [STR93]

1286

Compound: Gold(I) chloride

Synonyms: aurous chloride

Formula: AuCl

Molecular Formula: AuCl

Molecular Weight: 232.420

CAS RN: 10294-29-8

Properties: yellowish powd [MER89]

Solubility: i H_2O with slow decomposition [MER89]

Density, g/cm³: 7.57 [MER89]

Melting Point, °C: decomposes ~289 [MER89]
Reactions: decomposes to Au and Cl$_2$ at 289°C [MER89]

1287
Compound: Gold(I) cyanide
Synonyms: aurous cyanide
Formula: AuCN
Molecular Formula: CAuN
Molecular Weight: 222.985
CAS RN: 506-65-0
Properties: yellow powd; hex; odorless; iridescent in sunlight; slowly decomposes in presence of moisture [MER89]
Solubility: i H$_2$O, alcohol; dil acid; s NH$_3$, NaCN soln [MER89]
Density, g/cm^3: 7.14 [MER89]
Melting Point, °C: decomposes to Au and CN when ignited [MER89]
Reactions: evolves HCN gas if warmed with HCl [MER89]

1288
Compound: Gold(I) iodide
Synonyms: aurous iodide
Formula: AuI
Molecular Formula: AuI
Molecular Weight: 323.871
CAS RN: 10294-31-2
Properties: yellowish to greenish yellow powd; decomposes slowly at ordinary temp, rapidly at elevated temp [MER89]
Solubility: i H$_2$O; s alkali iodide or cyanide [MER89]
Density, g/cm^3: 8.25 [MER89]
Melting Point, °C: 120, decomposes [STR93]
Reactions: decomposed by warm acids [MER89]

1289
Compound: Gold(I) sulfide
Synonyms: aurous sulfide
Formula: Au$_2$S
Molecular Formula: Au$_2$S
Molecular Weight: 425.999
CAS RN: 1303-60-2
Properties: brownish black powd; forms colloid in H$_2$O when freshly prepared by treatment of acidified KAu(CN)$_2$ solutions with H$_2$S [MER89] [KIR78]
Solubility: i H$_2$O, dil single acids; s aqua regia, alkali cyanide solutions [MER89] [KIR78]
Density, g/cm^3: ~11 [LID94]

Melting Point, °C: decomposes at 240 [CRC94]

1290
Compound: Gold(III) bromide
Synonyms: auric bromide
Formula: AuBr$_3$
Molecular Formula: AuBr$_3$
Molecular Weight: 436.679
CAS RN: 10294-28-7
Properties: brownish orange powd; used to test for alkaloids and for testing spermatic fluid, medicinal uses [HAW93] [STR93]
Solubility: s H$_2$O, alcohol, glycerol [MER89]
Melting Point, °C: ~160, decomposes [MER89]
Reactions: slowly decomposed by alcohol and glycerol [MER89]

1291
Compound: Gold(III) chloride
Synonyms: auric chloride
Formula: AuCl$_3$
Molecular Formula: AuCl$_3$
Molecular Weight: 303.325
CAS RN: 13453-07-1
Properties: -8 mesh with 99% purity; yellow to red cryst; can be made by heating Au and Cl$_2$ at 200°C; exists as dimer Au$_2$Cl$_3$ in both solid and gas phases under Cl$_2$ below 254°C [HAW93] [CER91]
Solubility: s H$_2$O, alcohol, ether [HAW93]
Density, g/cm^3: 4.7 [LID94]
Melting Point, °C: decomposes >160 [LID94]
Reactions: decomposes to AuCl, then Au, in Cl$_2$ atm >254°C [KIR78]

1292
Compound: Gold(III) cyanide trihydrate
Synonyms: cyanoauric acid
Formula: Au(CN)$_3$·3H$_2$O
Molecular Formula: C$_3$H$_6$AuN$_3$O$_3$
Molecular Weight: 329.066
CAS RN: 535-37-5
Properties: colorless; deliq cryst; used as an electrolyte for plating gold [HAW93] [MER89]
Solubility: v s H$_2$O; sl s alcohol, ether [HAW93]
Melting Point, °C: decomposes 50 [MER89]

1293
Compound: Gold(III) hydroxide
Synonyms: auric hydroxide
Formula: Au(OH)$_3$

Molecular Formula: AuH_3O_3
Molecular Weight: 247.989
CAS RN: 1303-52-2
Properties: brown powd; decomposed by sunlight to Au metal; also decomposes on standing; decomposes to Au and O_2 >160°C; precipitates from $AuCl_4^-$ solutions following addition of alkali hydroxides; used in gilding liq, in decorating porcelain [HAW93] [MER89] [KIR78]
Solubility: i H_2O; s in NaCN soln, HCl, conc HNO_3 [MER89]
Melting Point, °C: decomposes ~100 [LID94]
Reactions: $Au(OH)_3 + NH_3 \rightarrow$ gold fulminate (explosive) [MER89]

1294
Compound: Gold(III) iodide
Synonyms: auric iodide
Formula: AuI_3
Molecular Formula: AuI_3
Molecular Weight: 577.680
CAS RN: 13453-24-2
Properties: green powd, unstable, converts to AuI, 10294-31-2 [KIR78] [STR93]
Solubility: i cold H_2O, decomposed by hot H_2O [CRC94]

1295
Compound: Gold(III) oxide
Synonyms: auric oxide, gold trioxide
Formula: Au_2O_3
Molecular Formula: Au_2O_3
Molecular Weight: 441.931
CAS RN: 1303-58-8
Properties: -100 mesh with 99.9% purity; brownish orange powd; slowly decomposed by sunlight; forms when $Au(OH)_3$ is heated at 140°C; used in gold plating [MER89] [HAW93] [CER91] [KIR78]
Solubility: i H_2O; s HCl, conc HNO_3, NaCN solns [MER89]
Melting Point, °C: decomposes ~150 [LID94]
Reactions: evolves O_2 at 110°C; decomposed to metal at 250°C [MER89]

1296
Compound: Gold(III) selenate
Synonyms: auric selenate
Formula: $Au_2(SeO_4)_3$
Molecular Formula: $Au_2O_{12}Se_3$
Molecular Weight: 822.806

CAS RN: 10294-32-3
Properties: small, yellow cryst; decomposes in light [MER89]
Solubility: i H_2O; s H_2SO_4, HNO_3 [MER89]

1297
Compound: Gold(III) selenide
Synonyms: auric selenide
Formula: Au_2Se_3
Molecular Formula: Au_2Se_3
Molecular Weight: 630.813
CAS RN: 1303-62-4
Properties: black amorphous solid [MER89]
Solubility: s aqua regia, alkali cyanides [MER89]
Density, g/cm³: 4.65 [MER89]
Melting Point, °C: decomposed by heat [MER89]

1298
Compound: Gold(III) sulfide
Synonyms: auric sulfide, gold trisulfide
Formula: Au_2S_3
Molecular Formula: Au_2S_3
Molecular Weight: 490.131
CAS RN: 1303-61-3
Properties: black powd [MER89]; not very stable, can be prepared by adding H_2S to ether solution of $AuCl_3$ [KIR78]
Solubility: i H_2O [CRC94]
Density, g/cm³: 8.75 [CRC94]
Reactions: decomposes to metal and sulfur at 200°C [MER89]

1299
Compound: Hafnium
Formula: Hf
Molecular Formula: Hf
Molecular Weight: 178.49
CAS RN: 7440-58-6
Properties: gray, highly lustrous, hard ductile metal; two forms; thermal neutron cross section 115 barns; good corrosion resistance and high strength; electrical resistivity 3.57×10^{-7} ohm·m (0°C), 6.24×10^{-7} (200°C); enthalpy of vaporization 571 kJ/mol; enthalpy of fusion 27.20 kJ/mol; entropy of fusion 54.4 J/(kg·K); used in controls for nuclear reactors, in light bulb filaments, electrodes and special glasses [HAW93] [MER89] [KIR80] [CRC93] [ALD94]
Solubility: s HF; slowly reacts with conc H_2SO_4, aqua regia [KIR80]
Density, g/cm³: 13.28 [KIR80]
Melting Point, °C: 2227 [ALD94]

Boiling Point, °C: 4602 [ALD94]
Reactions: transition α to β at 1777°C [KIR80]
Thermal Conductivity, W/(m·K): 23.0 (25°C)
[ALD94]; 22.3 (50°C), 20.7 (400°C) [KIR80]
Thermal Expansion Coefficient: 5.9 x 10^{-6}/K
[KIR80]

1300
Compound: Hafnium acetylacetonate
Synonyms: 2,4-pentanedione, hafnium(IV)
derivative
Formula: Hf(CH$_3$COCH=C(O)CH$_3$)$_4$
Molecular Formula: C$_{20}$H$_{28}$HfO$_8$
Molecular Weight: 574.927
CAS RN: 17475-67-1
Properties: powd [STR93]

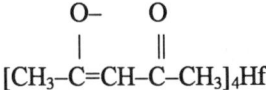

1301
Compound: Hafnium boride
Formula: HfB$_2$
Molecular Formula: B$_2$Hf
Molecular Weight: 200.112
CAS RN: 12007-23-7
Properties: gray hex cryst solid, a = 0.3141 nm, c =
0.3470 nm; can be prepared by heating HfO$_2$ +
C + B$_2$O$_3$; hardness 2900 kgf/mm^2; resistivity
8.8 μohm·cm; used as a refractory material, and
as a sputtering target with 99.5% purity to
produce films which may be wear-resistant and
semiconducting [HAW93] [KIR80] [CER91]
Density, g/cm^3: 10.5 (theoretical 11.2) [KIR80]
Melting Point, °C: 3250 [KIR78]
Reactions: attacked by HF, else highly resistant
[KIR80]
Thermal Expansion Coefficient: 5.7 x 10^{-6} [KIR80]

1302
Compound: Hafnium bromide
Synonyms: hafnium tetrabromide
Formula: HfBr$_4$
Molecular Formula: Br$_4$Hf
Molecular Weight: 498.106
CAS RN: 13777-22-5
Properties: -20 mesh with 99.7% purity; hygr; white
cub, a = 0.095 nm; or tan powd [KIR80]
[STR93] [CER91]
Density, g/cm^3: 4.90 (5.09 theoretical) [KIR80]
Melting Point, °C: 424 (3.34 MPa) [KIR80]

Boiling Point, °C: sublimation point 322 [KIR80]

1303
Compound: Hafnium carbide
Formula: HfC
Molecular Formula: CHf
Molecular Weight: 190.501
CAS RN: 12069-85-1
Properties: dark gray brittle solid; fcc, a = 0.4640
nm; high cross section for absorption of thermal
neutrons; resistivity 8.8 μohm cm; most
refractory binary material known; hardness 2300
kgf/mm^2; used in control rods of nuclear
reactors; can be prepared by heating HfO$_2$ with
lampblack under H$_2$ at 1900-2300°C; used in
crucible form for melting hafnium oxide, other
oxides [KIR80] [HAW93] [CER91]
Density, g/cm^3: 12.2 (theoretical 12.7) [KIR80]
Melting Point, °C: 3950 [KIR80] [CIC73]
Thermal Expansion Coefficient: 6.59 x 10^{-6}/K
[KIR80]

1304
Compound: Hafnium chloride
Synonyms: hafnium tetrachloride
Formula: HfCl$_4$
Molecular Formula: Cl$_4$Hf
Molecular Weight: 320.301
CAS RN: 13499-05-3
Properties: -80 mesh with 99.99% and 99% purity;
white cryst monocl, a = 0.631 nm, b = 0.7407
nm, c = 0.6256 nm; can be obtained by heating
the oxide in Cl$_2$ with carbon >317°C [MER89]
[CER91] [KIR80]
Solubility: hydrolyzed by H$_2$O to HfOCl$_2$ [MER89]
Melting Point, °C: 432 [KIR80]
Boiling Point, °C: sublimation point 317 [KIR80]

1305
Compound: Hafnium fluoride
Synonyms: hafnium tetrafluoride
Formula: HfF$_4$
Molecular Formula: F$_4$Hf
Molecular Weight: 254.484
CAS RN: 13709-52-9

Properties: white, or highly dense pressure sintered pieces of 3-6 mm; monocl, a = 0.957 nm, b = 0.993 nm, c = 0.7730 nm; can be formed by careful thermal decomposition of ammonium fluorohafnate, 16925-24-9; sintered pieces used as evaporation material for possible application to low-index material as a replacement for ThF_4, 99.95% pure material used as a sputtering target to produce anti-reflection coatings on glass [KIR80] [CER91]
Density, g/cm³: 7.1 [LID94]
Melting Point, °C: >968 [KIR80]
Boiling Point, °C: 968 sublimes [KIR80]

1306
Compound: Hafnium hydride
Formula: HfH_2
Molecular Formula: H_2Hf
Molecular Weight: 180.506
CAS RN: 12770-26-2
Properties: -325 mesh 10 microns or less with 99.8% purity; brittle solid, fcc; prepared by reacting Hf and H_2 above 250°C [KIR80] [CER91]
Density, g/cm³: 11.4 [LID94]

1307
Compound: Hafnium iodide
Synonyms: hafnium tetraiodide
Formula: HfI_4
Molecular Formula: HfI_4
Molecular Weight: 686.108
CAS RN: 13777-23-6
Properties: yellowish orange cub, a = 1.176 nm, or red powd; sensitive to moisture [STR93] [KIR80]
Density, g/cm³: 5.6 [LID94]
Melting Point, °C: 449 (3.34 MPa) [KIR80]
Boiling Point, °C: sublimation point 393 [KIR80]

1308
Compound: Hafnium nitride
Formula: HfN
Molecular Formula: HfN
Molecular Weight: 192.497
CAS RN: 25817-87-2

Properties: yellowish brown cryst; fcc, a = 0.4518 nm; most refractory of all known metal nitrides; hardness 1640 kgf/mm²; electrical resistivity 33 μohm·cm; can be prepared by heating Hf in N_2 or NH_3 atm at 1000-1500°C; as a 99.5% pure material, used as a sputtering target to increase electrical stability of diodes, transistors and integrated circuits [KIR80] [KIR81] [HAW93] [CER91]
Density, g/cm³: 13.84 (theoretical) [KIR80]
Melting Point, °C: 3305 [HAW93]
Thermal Conductivity, W/(m·K): 11.1 [KIR81]
Thermal Expansion Coefficient: 6.9 x 10⁻⁶ [KIR80]

1309
Compound: Hafnium oxide
Synonyms: hafnia
Formula: HfO_2
Molecular Formula: HfO_2
Molecular Weight: 210.489
CAS RN: 12055-23-1
Properties: white solid; cub, a = 0.51156 nm, b = 0.51722 nm, c = 0.52948 nm; obtained by ignition of the hydroxide, oxalate or sulfate; hardness 1050 kgf/mm²; resistivity >10⁺⁸ μohm·cm; used as a refractory metal oxide, an evaporated material of 99.9% purity for dielectric coatings, to coat wires for emitters, and as a 99.95% pure sputtering target to provide very hard, adherent film; stabilized with 10-15% CaO [HAW93] [MER89] [KIR80] [CER91]
Solubility: i H_2O [HAW93]
Density, g/cm³: 9.68 [MER89]
Melting Point, °C: 2900 [KIR80]
Reactions: transformation tetr to cub above 2700°C [KIR80]
Thermal Expansion Coefficient: (volume) 100°C (0.144), 200°C (0.319), 400°C (0.696), 800°C (1.499), 1000°C (2.085) [CLA66]

1310
Compound: Hafnium oxychloride octahydrate
Formula: $HfOCl_2 \cdot 8H_2O$
Molecular Formula: $Cl_2H_{16}HfO_9$
Molecular Weight: 409.517
CAS RN: 14456-34-9
Properties: -6 mesh with 99.99% purity; white powd or tetr cryst; produced by addition of $HfCl_4$ to H_2O, or by dissolution of HfO_2 in HCl; when heated, first dissolves in waters of crystallization, then decomposes [KIR80] [STR93] [CER91]
Solubility: s H_2O [KIR80]

Melting Point, °C: decomposes [KIR80]

1311
Compound: Hafnium phosphide
Formula: HfP
Molecular Formula: HfP
Molecular Weight: 209.464
CAS RN: 12325-59-6
Properties: -100 mesh with 99% purity; hex, a = 0.365 nm, c = 1.237 nm [KIR80] [CER91]
Density, g/cm³: 9.78 (theoretical) [KIR80]

1312
Compound: Hafnium selenide
Synonyms: hafnium diselenide
Formula: HfSe$_2$
Molecular Formula: HfSe$_2$
Molecular Weight: 336.410
CAS RN: 12162-21-9
Properties: -325 mesh 10 microns or less with 99.5% purity; dark brown; hex, a = 0.375 nm, b = 0.616 nm; resistivity 20 μohm·cm [CER91] [KIR80]
Density, g/cm³: 7.46 [KIR80]

1313
Compound: Hafnium silicate
Formula: HfSiO$_4$
Molecular Formula: HfO$_4$Si
Molecular Weight: 270.574
CAS RN: 37248-04-7
Properties: zircon cryst structure; a = 0.65725 nm, c = 0.59632 nm [SUB90]
Thermal Expansion Coefficient: 1020°C is 3.1 x 10^{-6}/°C [SUB90]

1314
Compound: Hafnium silicide
Formula: HfSi$_2$
Molecular Formula: HfSi$_2$
Molecular Weight: 234.661
CAS RN: 12401-56-8
Properties: gray powd; rhomb, a = 0.3677 nm, b = 1.455 nm, c = 0.3649 nm; hardness 930 kgf/mm²; as 99.5% pure material, used as sputtering target to produce wear-resistant films, and semiconducting films for use in integrated circuits [KIR80] [STR93] [CER91]
Density, g/cm³: 7.2 (8.03 theoretical) [KIR80]
Melting Point, °C: 1680 [STR93]; 1750 [KIR80]

1315
Compound: Hafnium sulfate
Formula: Hf(SO$_4$)$_2$
Molecular Formula: HfO$_8$S$_2$
Molecular Weight: 370.617
CAS RN: 15823-43-5
Properties: can be prepared by reacting fuming sulfuric acid with HfCl$_4$ [MER89]
Melting Point, °C: decomposes >500 [MER89]

1316
Compound: Hafnium sulfide
Formula: HfS$_2$
Molecular Formula: HfS$_2$
Molecular Weight: 242.622
CAS RN: 18855-94-2
Properties: -200 mesh with 99.9% purity; purple brown; hex, a = 0.364 nm, b = 0.584 nm; resistivity at room temp 1 μohm·cm; can be prepared by reacting the elements at 500°C; used as a solid lubricant [HAW93] [CER91]
Density, g/cm³: 6.03 [KIR80]

1317
Compound: Hafnium telluride
Formula: HfTe$_2$
Molecular Formula: HfTe$_2$
Molecular Weight: 433.690
CAS RN: 39082-23-0
Properties: -325 mesh 10 microns or less with 99.5% purity [CER91]

1318
Compound: Hafnium titanate
Formula: HfTiO$_4$
Molecular Formula: HfO$_4$Ti
Molecular Weight: 290.355
CAS RN: 12055-24-2
Properties: -325 mesh 10 microns or less with 99.5% purity; off-white powd [STR93] [CER91]

1319
Compound: Hafnocene dichloride
Synonyms: bis(cyclopentadienyl)hafnium dichloride
Molecular Weight: 379.59
CAS RN: 12116-66-4
Properties: moisture sensitive; uses: synthesis of many organometallic compounds and early-transition-metal complexes [ALD94]

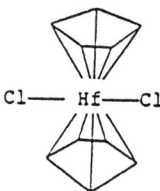

Melting Point, °C: 230-233 [ALD94]

1320
Compound: Helium
Formula: He
Molecular Formula: He
Molecular Weight: 4.002602
CAS RN: 7440-59-7
Properties: inert gas; nonflammable, colorless, odorless, tasteless; critical temp -267.9°C; critical pressure 227.5 kPa; enthalpy of vaporization 81.70 J/mol; enthalpy of fusion 0.0138 kJ/mol; heat capacity (101.32 kPa, 25°C) 20.78 J/(mol·K); sonic velocity (101.32 kPa, 0°C) 973 m/s; viscosity (101.32 kPa, 0°C) 19.86 Pa s, specific volume (21.1°C, 101.3 kPa) 6.04 m³/kg [MER89] [KIR78] [AIR87] [ALD94]
Solubility: 8.61 mL/1000g H_2O (101.32 kPa, 0°C) [KIR78]; Henry's law constants, k x 10^{-4}: 9.856 (104°C), 6.739 (149.4°C), 2.524 (250.6°C), 1.796 (275.1°C) [POT78]
Density, g/cm³: gas (101.3 kPa, 0°C) 0.00017850 [KIR78]
Melting Point, °C: -272.2 (25 atm) [HAW93]
Boiling Point, °C: -268.93 [KIR78]
Thermal Conductivity, W/(m·K): 0.14184 (gas) (101.32 kPa, 0°C) [KIR78]

1321
Compound: Helium-3
Formula: ³He
Molecular Formula: He
Molecular Weight: 3.0160
CAS RN: 14762-55-7
Properties: gas; enthalpy of vaporization 0.0829 kJ/mol; heat capacity (101.32 kPa, 25°C) 20.78 J/(mol·K); viscosity (101.32 kPa, 25°C) ~17.2 Pa·s [KIR78] [CRC93]
Density, g/cm³: 0.0001347 (gas, 101.3 kPa, 0°C) [KIR78]
Boiling Point, °C: -268.93 [CRC93]
Thermal Conductivity, W/(m·K): gas (101.32 kPa, 0°C) ~0.1636 [KIR78]

1322
Compound: Hexaamminecobalt(III) chloride

Formula: Co(NH₃)₆Cl₃
Molecular Formula: Cl₃CoH₁₈N₆
Molecular Weight: 267.474
CAS RN: 10534-89-1
Properties: wine red monocl [KIR79]
Solubility: 5.99 g/100mL cold H_2O; s conc HCl; i alcohol, NH₄OH [KIR79]
Density, g/cm³: 1.71 [KIR79]
Reactions: evolves NH₃ at 215°C [KIR79]

1323
Compound: Hexaammineruthenium(III) chloride
Formula: Ru(NH₃)₆Cl₃
Molecular Formula: Cl₃H₁₈N₆Ru
Molecular Weight: 309.612
CAS RN: 14282-91-8
Properties: off-white powd; there is also a hexaammineruthenium(II) chloride, CAS RN 15305-72-3 [STR93] [ALD94]

1324
Compound: Hexaborane(10)
Formula: B₆H₁₀
Molecular Formula: B₆H₁₀
Molecular Weight: 74.945
CAS RN: 23777-80-2
Properties: colorless liq; slowly decomposes at 25°C [COT88]
Solubility: hydrolyzed in H_2O if heated [COT88]
Density, g/cm³: 0.67 [LID94]
Melting Point, °C: -62.3 [KIR78]
Boiling Point, °C: 108 [KIR78]

1325
Compound: Hexachlorodisilane
Formula: Cl₃SiSiCl₃
Molecular Formula: Cl₆Si₂
Molecular Weight: 268.89
CAS RN: 13465-77-5
Properties: colorless liq; precursor to substituted disilanes [ALD94] [CRC94]
Density, g/cm³: 1.562 [ALD94]
Melting Point, °C: -1 [CRC94]
Boiling Point, °C: 144-145.5 [ALD94]

1326
Compound: Hexafluorophosphoric acid
Formula: HPF₆
Molecular Formula: F₆HP
Molecular Weight: 145.972
CAS RN: 16940-81-1

Properties: clear liq; fumes due to evolving HF; forms a hexahydrate [KIR78] [STR93]
Density, g/cm³: 1.651 [ALD94]

1327
Compound: Holmium
Formula: Ho
Molecular Formula: Ho
Molecular Weight: 164.93032
CAS RN: 7440-60-0
Properties: cryst solid, has metallic luster; hex; forms yellow-green salts; sensitive to air and moisture; electrical resistivity (20°C) 94 μohm·cm; enthalpy of fusion 11.76 kJ/mol; enthalpy of sublimation 300.8 kJ/mol; atom radius 0.17661 nm; radius of Ho^{+++} ion 0.0894 nm, has yellow colored solutions; used as a getter in vacuum tubes, applications to electrochemical and spectrochemical research [HAW93] [MER89] [KIR82] [CRC93] [ALD94]
Solubility: reacts slowly with H_2O, s dil acids [HAW93]
Density, g/cm³: 8.7947 [KIR82]
Melting Point, °C: 1470 [ALD94]
Boiling Point, °C: 2700 [ALD94]
Thermal Conductivity, W/(m·K): 16.2 (25°C) [CRC93]
Thermal Expansion Coefficient: 11.2 x 10^{-6}/K [CRC93]

1328
Compound: Holmium acetate monohydrate
Formula: Ho(CH₃COO)₃·H₂O
Molecular Formula: C₆H₁₁HoO₇
Molecular Weight: 360.079
CAS RN: 25519-09-9
Properties: peach powd [STR93]

1329
Compound: Holmium bromide
Formula: HoBr₃
Molecular Formula: Br₃Ho
Molecular Weight: 404.642
CAS RN: 13825-76-8
Properties: -20 mesh with 99.9% purity; off-white powd; hygr [CER91] [STR93]
Melting Point, °C: 914 [MER89]
Boiling Point, °C: 1470 [STR93]

1330
Compound: Holmium carbonate hydrate

Formula: Ho₂(CO₃)₃·xH₂O
Molecular Formula: C₃Ho₂O₉ (anhydrous)
Molecular Weight: 509.888 (anhydrous)
CAS RN: 38245-34-0
Properties: white powd; hygr [STR93] [ALD94]

1331
Compound: Holmium chloride
Formula: HoCl₃
Molecular Formula: Cl₃Ho
Molecular Weight: 271.288
CAS RN: 10138-62-2
Properties: -20 mesh with 99.9% purity; off-white powd; bright yellow solid; hygr [HAW93] [CER91] [MER89] [STR93]
Solubility: s H_2O [HAW93]
Melting Point, °C: 718 [MER89]
Boiling Point, °C: 1500 [HAW93]

1332
Compound: Holmium chloride hexahydrate
Formula: HoCl₃·6H₂O
Molecular Formula: Cl₃H₁₂HoO₆
Molecular Weight: 379.380
CAS RN: 14914-84-2
Properties: -4 mesh with 99.9% purity; off-white cryst [CER91] [STR93]
Melting Point, °C: 718 [AES93]

1333
Compound: Holmium fluoride
Formula: HoF₃
Molecular Formula: F₃Ho
Molecular Weight: 221.925
CAS RN: 13760-78-6
Properties: bright yellow solid, or 99.9% pure melted pieces of 3-6 mm; hygr; melted pieces used as evaporation material for possible application in multilayers [STR93] [HAW93] [CER91]
Solubility: s H_2O [HAW93]
Density, g/cm³: 7.644 [STR93]
Melting Point, °C: 1143 [STR93]
Boiling Point, °C: >2200 [STR93]

1334
Compound: Holmium hydride
Formula: HoH₂₋₃
Molecular Formula: H₂Ho; H₃Ho
Molecular Weight: H₂Ho: 166.946; H₃Ho: 167.954
CAS RN: 13598-41-9

Properties: -60 mesh with 99.9% purity [CER91]

1335
Compound: Holmium iodide
Formula: HoI_3
Molecular Formula: HoI_3
Molecular Weight: 545.643
CAS RN: 13813-41-7
Properties: -20 mesh with 99.9% purity; light yellow solid [CER91] [MER89]
Melting Point, °C: 1010 [MER89]
Boiling Point, °C: 1300 [CRC94]

1336
Compound: Holmium nitrate pentahydrate
Formula: $Ho(NO_3)_3 \cdot 5H_2O$
Molecular Formula: $H_{10}HoN_3O_{14}$
Molecular Weight: 441.022
CAS RN: 14483-18-2
Properties: orange cryst; hygr [STR93]

1337
Compound: Holmium nitride
Formula: HoN
Molecular Formula: HoN
Molecular Weight: 178.937
CAS RN: 12029-81-1
Properties: cub; -60 mesh with 99.9% purity [LID94] [CER91]
Density, g/cm³: 10.6 [LID94]

1338
Compound: Holmium oxalate decahydrate
Formula: $Ho_2(C_2O_4)_3 \cdot 10H_2O$
Molecular Formula: $C_6H_{20}Ho_2O_{22}$
Molecular Weight: 774.072
CAS RN: 28965-57-3
Properties: off-white powd [STR93]
Reactions: minus H_2O, 40°C [AES93]

1339
Compound: Holmium oxide
Synonyms: holmia
Formula: Ho_2O_3
Molecular Formula: Ho_2O_3
Molecular Weight: 377.859
CAS RN: 12055-62-8

Properties: light yellow solid; sl hygr; used in refractories and as a special catalyst, and as an evaporated film of 99.9% purity, it is reactive to radio frequencies [HAW93] [CER91]
Solubility: s inorganic acids [HAW93]
Density, g/cm³: 8.36 [STR93]
Melting Point, °C: 2415 [LID94]

1340
Compound: Holmium perchlorate hexahydrate
Formula: $Ho(ClO_4)_3 \cdot 6H_2O$
Molecular Formula: $Cl_3H_{12}HoO_{18}$
Molecular Weight: 571.373
CAS RN: 14017-54-0
Properties: cryst; hygr [STR93]

1341
Compound: Holmium silicide
Formula: $HoSi_2$
Molecular Formula: $HoSi_2$
Molecular Weight: 221.101
CAS RN: 12136-24-2
Properties: hex; 10 mm & down lump, 6 mm pieces and smaller with 99.9% purity [ALF93] [LID94] [CER91]
Density, g/cm³: 7.1 [LID94]

1342
Compound: Holmium sulfate octahydrate
Formula: $Ho_2(SO_4)_3 \cdot 8H_2O$
Molecular Formula: $H_{16}Ho_2O_{20}S_3$
Molecular Weight: 762.174
CAS RN: 13473-57-9
Properties: hygr cryst [AES93] [ALD94]
Solubility: g/100g H_2O: 8.18 (20°C), 6.71 (25°C), 4.52 (40°C) [LAN85]

1343
Compound: Holmium sulfide
Formula: Ho_2S_3
Molecular Formula: Ho_2S_3
Molecular Weight: 426.059
CAS RN: 12162-59-3
Properties: mono; -200 mesh with 99.9% purity [LID94] [CER91]
Density, g/cm³: 5.92 [LID94]

1344
Compound: Holmium telluride
Formula: Ho_2Te_3

Molecular Formula: Ho_2Te_3
Molecular Weight: 712.661
CAS RN: 12162-61-7
Properties: -20 mesh with 99.9% purity [CER91]

1345
Compound: Holmium boride
Synonyms: holmium tetraboride
Formula: HoB_4
Molecular Formula: B_4Ho
Molecular Weight: 208.174
CAS RN: 12045-77-1
Properties: -100 mesh of 99.9% purity [CER91]

1346
Compound: Hydrazine
Formula: H_2NNH_2
Molecular Formula: H_4N_2
Molecular Weight: 32.045
CAS RN: 302-01-2
Properties: colorless, oily liq, fuming in air; burns with violet flame; N-N distance 0.146 nm; vapor pressure 14.38 mm (25°C); viscosity 0.009 dyne·s/cm^2 (25°C); surface tension 66.67 dyne/cm (25°C); dielectric constant 58.5 (0°C); enthalpy of vaporization 41.8 kJ/mol; enthalpy of fusion 12.60 kJ/mol [CRC93] [MER89] [CIC73]
Solubility: miscible with H_2O and the following alcohols: methyl, ethyl, propyl, isobutyl [MER89]
Density, g/cm^3: liq: 1.00; solid: 1.146 (-5°C) [CIC73]
Melting Point, °C: 1.4 [CRC93]
Boiling Point, °C: 113.55 [CRC93]
Reactions: good reducing agent [MER89]

1347
Compound: Hydrazine acetate
Formula: $H_2NNH_2 \cdot (CH_3COOH)$
Molecular Formula: $C_2H_8N_2O_2$
Molecular Weight: 92.098
CAS RN: 13255-48-6
Melting Point, °C: 100-102 [ALD94]

1348
Compound: Hydrazine azide
Formula: $N_2H_4 \cdot HN_3$
Molecular Formula: H_5N_5
Molecular Weight: 75.074
CAS RN: 14662-04-5

Properties: white; deliq prism [CRC77]
Solubility: v s H_2O [CRC77]
Melting Point, °C: 75.4 explodes [CIC73]

1349
Compound: Hydrazine dihydrochloride
Formula: $N_2H_4 \cdot 2HCl$
Molecular Formula: $Cl_2H_6N_2$
Molecular Weight: 104.966
CAS RN: 5341-61-7
Properties: white, cryst powd; ortho-rhomb, a = 1.249 nm, b = 2.185 nm, c = 0.441 nm [CIC73] [MER89]
Solubility: 27.2 g/100g H_2O (32°C) [CIC73]; sl s alcohol [HAW93]
Density, g/cm^3: 1.42 [CIC73]
Melting Point, °C: 198 (minus HCl) [CIC73]
Boiling Point, °C: decomposes 200 [CIC73]

1350
Compound: Hydrazine dinitrate
Formula: $N_2H_4 \cdot 2HNO_3$
Molecular Formula: $H_6N_4O_6$
Molecular Weight: 158.071
CAS RN: 13464-98-7
Properties: needles [LAN52]
Solubility: 20.2 g/100g H_2O (35°C) [CIC73]
Melting Point, °C: 104 [LAN52]
Boiling Point, °C: decomposes [LAN52]
Reactions: decomposes quickly at 104°C, slowly at 80°C [CIC73]

1351
Compound: Hydrazine hydrate
Formula: $H_2NNH_2 \cdot xH_2O$
Molecular Formula: H_4N_2 (anhydrous)
Molecular Weight: 32.048 (anhydrous)
CAS RN: 10217-52-4
Properties: x~1.5 [ALD94]
Density, g/cm^3: 1.029 [ALD94]
Melting Point, °C: 95 [ALD94]

1352
Compound: Hydrazine monohydrate
Formula: $N_2H_4 \cdot H_2O$
Molecular Formula: H_6N_2O
Molecular Weight: 50.060
CAS RN: 7803-57-8
Properties: fuming refractive liq; trig, a = 0.4873 nm, c = 1.094 nm [CIC73] [MER89]

Solubility: miscible with H_2O, alcohol; i chloroform, ether [MER89]
Density, g/cm³: 1.0305 [CIC73]
Melting Point, °C: -51.7 [CIC73]
Boiling Point, °C: 118.5 [CIC73]
Reactions: reducing agent [MER89]

1353
Compound: Hydrazine monohydrobromide
Formula: $N_2H_4 \cdot HBr$
Molecular Formula: BrH_5N_2
Molecular Weight: 112.957
CAS RN: 13775-80-9
Properties: white cryst flakes; used in soldering flux; monocl, a = 1.285 nm, b = 0.454 nm, c = 1.194 nm [CIC73] [HAW93]
Solubility: s H_2O, lower alcohols; i most organic solvents [HAW93]
Density, g/cm³: 2.3 [LID94]
Melting Point, °C: 81-87 [HAW93]
Boiling Point, °C: decomposes ~190 [HAW93]

1354
Compound: Hydrazine monohydrochloride
Formula: $N_2H_4 \cdot HCl$
Molecular Formula: ClH_5N_2
Molecular Weight: 68.506
CAS RN: 2644-70-4
Properties: white cryst flakes; ortho-rhomb, a = 1.249 nm, b = 2.185 nm, c = 0.441 nm [CIC73]
Solubility: 37 g/100g H_2O (20°C); i most organic solvents [HAW93]
Density, g/cm³: 1.5 [LID94]
Melting Point, °C: 89 [CIC73]
Boiling Point, °C: decomposes 240 [CIC73]

1355
Compound: Hydrazine monohydroiodide
Formula: $N_2H_4 \cdot HI$
Molecular Formula: H_5IN_2
Molecular Weight: 159.957
CAS RN: 10039-55-1
Properties: colorless prism [CRC77]
Solubility: s H_2O [CIC73]
Melting Point, °C: 125 [CIC73]

1356
Compound: Hydrazine mononitrate
Formula: $N_2H_4 \cdot HNO_3$
Molecular Formula: $H_5N_3O_3$
Molecular Weight: 95.058

CAS RN: 37836-27-4
Properties: colorless needles; explosive; α,β forms; monocl, a = 1.123 nm, b = 1.173 nm, c = 0.517 nm [HAW93] [CIC73] [CRC94]
Solubility: 327 g/100g H_2O (25°C) [CIC73]; g/100g H_2O: 175 (10°C), 266 (20°C), 2127 (60°C) [LAN85]
Melting Point, °C: α: 70.71; β: 62.09 [CIC73]
Boiling Point, °C: sublimes at 140 [CRC94]

1357
Compound: Hydrazine monooxalate
Formula: $2N_2H_4 \cdot H_2C_2O_4$
Molecular Formula: $C_2H_{10}N_4O_4$
Molecular Weight: 154.13
CAS RN: 108249-27-0
Properties: monocl, a = 0.3580 nm., b = 0.3321 nm, c = 0.5097 nm [CIC73]
Solubility: 200 g/100mL H_2O (15°C) [CRC94]
Melting Point, °C: 148 [CRC94]

1358
Compound: Hydrazine perchlorate hemihydrate
Formula: $N_2H_4 \cdot HClO_4 \cdot 1/2H_2O$
Molecular Formula: $ClH_6N_2O_{4.5}$
Molecular Weight: 141.511
CAS RN: 13762-65-7
Properties: solid; used as a rocket propellant [HAW93]
Solubility: decomposes in H_2O; s alcohol; i ether, benzene, chloroform, carbon disulfide [HAW93]
Density, g/cm³: 1.939 [CIC73]
Melting Point, °C: 137 [HAW93]
Boiling Point, °C: 145 [HAW93]
Reactions: can explode [CRC94]

1359
Compound: Hydrazine sulfate
Formula: $N_2H_4 \cdot H_2SO_4$
Molecular Formula: $H_6N_2O_4S$
Molecular Weight: 130.125
CAS RN: 10034-93-2
Properties: glasslike plates or prisms; ortho-rhomb cryst, a = 0.8251 nm, b = 0.9159 nm, c = 0.5532 nm [MER89] [CIC73]
Solubility: 3.415 g/100g H_2O (25°C) [CIC73]; g/100g H_2O: 2.87 (20°C), 4.15 (40°C), 14.39 (80°C) [LAN85]
Density, g/cm³: 1.378 [MER89]
Melting Point, °C: 254 [CIC73]

1360
Compound: Hydrazoic acid
Synonyms: hydrogen azide
Formula: HN_3
Molecular Formula: HN_3
Molecular Weight: 43.028
CAS RN: 7782-79-8
Properties: colorless volatile liq; intolerable pungent odor; highly explosive; enthalpy of vaporization 30.5 kJ/mol [CRC93] [MER89] [HAW93]
Solubility: v s H_2O [HAW93]
Density, g/cm³: 1.092 [CRC94]
Melting Point, °C: -80 [MER89]
Boiling Point, °C: 35.7 [CRC93]

1361
Compound: Hydrofluoric acid, 70%
Formula: HF
Molecular Formula: HF
Molecular Weight: 20.006
CAS RN: 7664-39-3
Properties: colorless, fuming, mobile liq; solid phase is $HF \cdot H_2O$ at freezing point; specific conductivity 0.79 $(ohm \cdot cm)^{-1}$ at 0°C; attacks glass, silica; manufactured by reacting fluorspar with sulfuric acid to form calcium sulfate and HF(gas); used in aluminum production, acidizing oil wells, gasoline production [HAW93] [KIR78]
Density, g/cm³: 1.22 [KIR78]
Melting Point, °C: -69 [KIR78]
Boiling Point, °C: 66.4 [KIR78]

1362
Compound: Hydrogen
Formula: H_2
Molecular Formula: H_2
Molecular Weight: 2.016 (at wt 1.00794)
CAS RN: 1333-74-0
Properties: colorless, odorless, tasteless diatomic gas; flammable or explosive when mixed with air, oxygen, chlorine; critical temp -239.96°C; critical pressure 1315 kPa; critical volume 66.949 cm³/mol; enthalpy of vaporization 0.898 kJ/mol; enthalpy of fusion 0.12 kJ/mol; velocity of sound (0°C) 1246 m/s; viscosity (0°C) 0.00843 mPa s; dielectric constant at bp 1.231, 1.000271 at 0°C [MER89] [CRC93]
Solubility: s in about 50 vols H_2O (0°C) [MER89]
Density, g/cm³: 0.088 g/L [LID94]
Melting Point, °C: -259.34 [CRC93]
Boiling Point, °C: -252.87 [CRC93]

Thermal Conductivity, W/(m·K): liq: at bp: 0.10; at triple point: 0.074; gas: 0.1739 [KIR80]

1363
Compound: Hydrogen bromide
Formula: HBr
Molecular Formula: BrH
Molecular Weight: 80.912
CAS RN: 10035-10-6
Properties: colorless, corrosive, nonflammable gas; fumes in moist air; specific conductance 1.4 x 10^{-10} $(ohm \cdot cm)^{-1}$ at -84°C; dielectric constant 7.33 at -84°C; enthalpy of vaporization (25°C) 12.69 kJ/mol; enthalpy of fusion 2.41 kJ/mol [CRC93] [MER89] [COT88]
Solubility: g/100g H_2O: 221.2 (0°C), 204.0 (15°C), 171.5 (50°C), 130.0 (100°C) [LAN85]; s alcohol [HAW93]
Density, g/cm³: 3.55 g/L [LID94]
Melting Point, °C: -86.81 [CRC93]
Boiling Point, °C: -66.38 [CRC93]

1364
Compound: Hydrogen chloride
Formula: HCl
Molecular Formula: ClH
Molecular Weight: 36.461
CAS RN: 7647-01-0
Properties: colorless gas; fumes in air; suffocating odor; enthalpy of fusion 2.00 kJ/mol; enthalpy of vaporization 16.15 kJ/mol; triple point -114.25°C; critical temp 54.4°C; critical pressure 8.316 MPa; critical volume 0.069 L/mol; critical density 424 g/L; dielectric constant liq: 14.2 (10°C), gas: 1.0046 (25°C); specific conductance 3.5 x 10^{-9} (-85°C) [KIR80] [MER89] [COT88] [CRC93]
Solubility: g/100g H_2O: 82.3 (0°C); 67.3 (30°C); 56.1 (60°C) [MER89]
Density, g/cm³: 1.268 (air = 1.000) [MER89]
Melting Point, °C: -114.18 [CRC93]
Boiling Point, °C: -85 [CRC93]
Thermal Conductivity, W/(m·K): liq: (-154.99°C) 3.35; vapor: (0°C) 1.34 [KIR80]

1365
Compound: Hydrogen cyanide
Synonyms: hydrocyanic acid, prussic acid
Formula: HCN
Molecular Formula: CHN
Molecular Weight: 27.026
CAS RN: 74-90-8

Properties: colorless gas or liq; burns in air with blue flame; weakly acidic solutions; critical pressure 55 atm; critical temp 183.5°C; viscosity at 20°C 0.00201 dyne·s/cm^2; dielectric constant at 15.6°C is 123; enthalpy of fusion 8.41 kJ/mol [CIC73] [MER89] [CRC93]

Solubility: miscible with H_2O, alcohol; sl s ether [MER89]

Density, g/cm^3: gas: 0.941 (air = 1); liq: 0.687 [MER89]

Melting Point, °C: -13.4 [MER89]

Boiling Point, °C: 25.6 [MER89]

1366

Compound: Hydrogen fluoride

Formula: HF

Molecular Formula: FH

Molecular Weight: 20.006

CAS RN: 7664-39-3

Properties: colorless liq or gas; fumes in air; irritating, corrosive and poisonous; vapor pressure 122.9 MPa at 25°C; enthalpy of vaporization 7493 J/mol; enthalpy of fusion 4.58 kJ/mol; critical temp 188°C; critical pressure 6.480 MPa; critical density 0.29 g/cm^3; viscosity 0.25 mPa·s at 0°C; dielectric constant 83.6 at 0°C; specific conductance 1.6 x 10^{-6} (ohm·cm)$^{-1}$ at 0°C [KIR78] [MER89] [COT88] [CRC93]

Solubility: v s H_2O, alcohol; sl s ether, other organic solvents [MER89]

Density, g/cm^3: liq: 0.9576 25°C [KIR78]

Melting Point, °C: -83.36 [CRC93]

Boiling Point, °C: 19.51 [MER89]

1367

Compound: Hydrogen hexachloroiridate(IV) hydrate

Synonyms: chloroiridic acid

Formula: $H_2IrCl_6 \cdot xH_2O$

Molecular Formula: Cl_6H_2Ir (anhydrous)

Molecular Weight: 406.952 (anhydrous)

CAS RN: 110802-84-1

Properties: black cryst; hygr [STR93]

1368

Compound: Hydrogen hexachloroplatinate(IV)

Synonyms: platinic acid

Formula: H_2PtCl_6

Molecular Formula: Cl_6H_2Pt

Molecular Weight: 409.812

CAS RN: 16941-12-1

Properties: color red to brown; liq [KIR82] [ALF95]

Solubility: v s H_2O, alcohol [KIR82]

1369

Compound: Hydrogen hexachloroplatinate(IV) hexahydrate

Synonyms: platinic acid hexahydrate

Formula: $H_2PtCl_6 \cdot 6H_2O$

Molecular Formula: $Cl_6H_{14}O_6Pt$

Molecular Weight: 517.903

CAS RN: 16941-12-1

Properties: brownish yellow; very deliq cryst; sensitive to light [MER89]

Solubility: v s H_2O, alcohol [MER89]

Density, g/cm^3: 2.431 [MER89]

Melting Point, °C: 60 [MER89]

1370

Compound: Hydrogen hexafluorosilicic acid

Formula: H_2SiF_6

Molecular Formula: F_6H_2Si

Molecular Weight: 144.092

CAS RN: 16961-83-4

Properties: aq solution: colorless fuming liq; attacks glass and stoneware; produced as a byproduct of the reaction between H_2SO_4 and phosphate rocks which contain fluorides and silica; used to fluoridate water, to increase the hardness of ceramics, and in electroplating [HAW93]

Density, g/cm^3: 1.22 [ALD94]

1371

Compound: Hydrogen hexahydroxyplatinate(IV)

Formula: $H_2Pt(OH)_6$

Molecular Formula: H_8O_6Pt

Molecular Weight: 299.15

CAS RN: 51850-20-5

Properties: yellow needles; hygr [ALD94]

Reactions: minus $2H_2O$ at 100; minus $3H_2O$ at 120°C [CRC94]

1372

Compound: Hydrogen iodide

Formula: HI

Molecular Formula: HI

Molecular Weight: 127.912

CAS RN: 10034-85-2

Properties: pale yellow or colorless nonflammable gas; fumes in moist air; decomposed by light; critical temp 150°C; critical pressure 8.3 MPa; dielectric constant 3.57 at -45°C; specific conductance 8.5 x 10^{-10} (ohm·cm)$^{-1}$ at -45°C; enthalpy of vaporization 19.76 kJ/mol; enthalpy of fusion 2.87 kJ/mol; produced by reaction of I_2 and hydrazine in the presence of H_2O [KIR81] [MER89] [CRC93] [COT88]

Solubility: g/100g H_2O: 234 (10°C), 900 (0°C); s organic solvents [MER89]

Density, g/cm³: gas: 5.23 g/L (25°C); liq: 2.85 g/cm³ (-4.7°C) [KIR81]

Melting Point, °C: -50.77 [CRC93]

Boiling Point, °C: -35.55 [CRC93]

1373

Compound: Hydrogen peroxide

Formula: H_2O_2

Molecular Formula: H_2O_2

Molecular Weight: 34.015

CAS RN: 7722-84-1

Properties: clear, colorless liq; weakly acidic, dissociation constant 1.78 x 10^{-12}; viscosity 1.245 mPa s (20°C); surface tension, 80.4 mN/m (20°C); enthalpy of dissociation 34.3 kJ/mol; enthalpy of vaporization 51.6 kJ/g at 25°C; enthalpy of fusion 12.50 kJ/mol; specific conductance (25°C) 4 x 10^{-7} ohm·cm [CRC93] [KIR81]

Solubility: miscible with H_2O [KIR81]

Density, g/cm³: 1.443 [KIR81]

Melting Point, °C: -0.43 [CRC93]

Boiling Point, °C: 150.2 [KIR81]

1374

Compound: Hydrogen selenide

Formula: H_2Se

Molecular Formula: H_2Se

Molecular Weight: 80.976

CAS RN: 7783-07-5

Properties: colorless gas with disagreeable odor; flammable; toxic; liquefies at 0°C under 6.6 atm pressure; thermodynamically unstable at room temp, but rate of decomposition is slow; reducing agent; enthalpy of vaporization 19.7 kJ/mol; can be prepared by adding HCl to ferrous selenide [KIR82] [MER89] [CRC93]

Solubility: mL/100mL H_2O: 377 (4°C), 270 (22.5°C) [MER89]; mL/100g H_2O at standard temp, pressure: 386 (0°C), 351 (10°C), 289 (20°C) [LAN85]

Density, g/cm³: 3.553 g/L [LID94]; (-42°C) 2.12 [MER89]

Melting Point, °C: -65.73 [MER89]

Boiling Point, °C: -41.3 [KIR82]

Reactions: reacts directly with most metals to form highly insoluble selenides [MER89]

1375

Compound: Hydrogen sulfide

Formula: H_2S

Molecular Formula: H_2S

Molecular Weight: 34.082

CAS RN: 7783-06-4

Properties: colorless flammable gas with characteristic rotten egg odor; burns in air with blue flame; critical temp 100.5°C; critical pressure 9.02 MPa; enthalpy of vaporization 18.67 kJ/mol; enthalpy of fusion 23.80 kJ/mol [CRC93] [AIR87] [HAW93] [MER89]

Solubility: 1g dissolves in H_2O: 187mL (10°C), 242mL (20°C), 314 (30°C) [MER89]

Density, g/cm³: 1.19 (air = 1) [MER89]

Melting Point, °C: -85.49 [MER89]

Boiling Point, °C: -59.55 [CRC93]

1376

Compound: Hydrogen telluride

Formula: H_2Te

Molecular Formula: H_2Te

Molecular Weight: 129.616

CAS RN: 7783-09-7

Properties: colorless gas with offensive odor, like garlic; one liter weighs 6.234g; liq H_2Te readily decomposed by light; dry gas stable to light, but decomposes in presence of dust; enthalpy of vaporization 19.2 kJ/mol; can be prepared by adding $AlTe_3$ to water in the absence of air; easily oxidized [KIR83] [MER89] [CRC93]

Solubility: s H_2O, unstable solution; s alcohol and alkalies [HAW93]

Density, g/cm³: 5.687 g/L [LID94]; (-12°C) 2.68 [MER89]

Melting Point, °C: -49 [MER89]

Boiling Point, °C: -2 [MER89]

1377

Compound: Hydrogen tetrabromoaurate(III) pentahydrate

Formula: $HAuBr_4 \cdot 5H_2O$

Molecular Formula: $AuBr_4H_{11}O_5$

Molecular Weight: 607.667

CAS RN: 17083-68-0

Properties: dark reddish brown needle shaped cryst, or granular masses; odorless; acidic taste [HAW93]

Solubility: s H₂O, alcohol [HAW93]
Melting Point, °C: 27 [HAW93]

1378
Compound: Hydrogen tetracarbonylferrate(II)
Formula: $H_2Fe(CO)_4$
Molecular Formula: $C_4H_2FeO_4$
Molecular Weight: 169.903
CAS RN: 12002-28-7
Properties: colorless cryst [MER89]
Solubility: s alkalies [MER89]
Melting Point, °C: -70 [MER89]

1379
Compound: Hydrogen tetrachloroaurate(III) hydrate
Formula: $HAuCl_4 \cdot xH_2O$
Molecular Formula: $AuCl_4H$ (anhydrous)
Molecular Weight: 339.785 (anhydrous)
CAS RN: 27988-77-8
Properties: yellowish orange cryst; hygr; sensitive to light [STR93]
Melting Point, °C: decomposes [STR93]

1380
Compound: Hydrogen tetrachloroaurate(III) tetrahydrate
Formula: $HAuCl_4 \cdot 4H_2O$
Molecular Formula: $AuCl_4H_9O_4$
Molecular Weight: 411.847
CAS RN: 16903-35-8
Properties: golden yellow to reddish yellow; very hygr; deliq; monocl cryst; readily affected by sunlight; there is a trihydrate, CAS RN 16961-25-4 [MER89] [ALD94]
Solubility: v s H₂O, alcohol; s ether [MER89]
Density, g/cm³: ~3.9 [MER89]
Reactions: decomposes if strongly heated forming Cl_2, HCl, Au [MER89]

1381
Compound: Hydroxylamine
Formula: H_2NOH
Molecular Formula: H_3NO
Molecular Weight: 33.030
CAS RN: 7803-49-8
Properties: unstable large white flakes or needles; ortho-rhomb, a = 0.729 nm, b = 0.439 nm, c = 0.488 nm; very hygr; vapor pressure 5.3 mm (32°C), 400 mm (99.2°C); dielectric constant 77.63-77.85; used as a reducing agent in photography [CIC73] [MER89]

Solubility: v s H₂O, methanol; decomposed by hot H₂O [MER89]
Density, g/cm³: liq: 1.204 (33°C) [CIC73]
Melting Point, °C: 32.05 [CIC73]
Boiling Point, °C: 56-57 (22 mm) [CIC73]

1382
Compound: Hydroxylamine hydrobromide
Formula: $H_2NOH \cdot HBr$
Molecular Formula: BrH_4NO
Molecular Weight: 113.942
CAS RN: 41591-55-3
Properties: monocl, a = 0.729 nm, b = 0.613 nm, c = 0.804 nm [CIC73]
Density, g/cm³: 2.3514 [CIC73]

1383
Compound: Hydroxylamine hydrochloride
Formula: $H_2NOH \cdot HCl$
Molecular Formula: ClH_4NO
Molecular Weight: 69.491
CAS RN: 5470-11-1
Properties: colorless; monocl, a = 0.695 nm, b = 0.595 nm, c = 0.770 nm [CIC73] [CRC94]
Solubility: 94.4 g/100g H₂O (25°C) [CIC73]
Density, g/cm³: 1.680 [CIC73]
Melting Point, °C: 152 decomposes [CIC73]

1384
Compound: Hydroxylamine perchlorate
Formula: $H_2NOH \cdot HClO_4$
Molecular Formula: ClH_4NO_5
Molecular Weight: 133.489
Properties: ortho-rhomb, a = 0.752 nm, b = 0.714 nm, c = 1.599 nm [CIC73]
Melting Point, °C: 88-89 [CIC73]
Boiling Point, °C: 120 decomposes [CIC73]

1385
Compound: Hydroxylamine sulfate
Formula: $(H_2NOH)_2 \cdot H_2SO_4$
Molecular Formula: $H_8N_2O_6S$
Molecular Weight: 164.139
CAS RN: 10039-54-0
Properties: colorless monocl cryst [CRC94]
Solubility: 32.9 g/100mL H₂O (0°C), 68.5 g/100mL (20°C) [CRC94]
Melting Point, °C: 170 decomposes [CIC73]

1386

Compound: Hypobromous acid

Formula: HOBr

Molecular Formula: BrHO

Molecular Weight: 96.911

CAS RN: 13517-11-8

Properties: stable only in solution, produced by the hydrolysis of bromine chloride; used as a bactericide and a disinfectant [HAW93]

Solubility: s H_2O, decomposed by hot H_2O [CRC94]

1387

Compound: Hypochlorous acid

Formula: HOCl

Molecular Formula: ClHO

Molecular Weight: 52.460

CAS RN: 7790-92-3

Properties: greenish yellow; can exist only in aq solution; very unstable weak acid, which decomposes to HCl and oxygen; used in bleaching textiles and fibers, in water purification, and as an antiseptic [HAW93]

1388

Compound: Hypophosphoric acid

Formula: $H_4O_6P_2$

Molecular Formula: $H_4O_6P_2$

Molecular Weight: 161.976

CAS RN: 7803-60-3

Properties: plate like cryst; available commercially as a water solution; readily forms hydrates; conc solution decomposes; used in baking powd in the form of the sodium salt [HAW93]

Melting Point, °C: 73 decomposes [LID94]

1389

Compound: Hypophosphorous acid

Synonyms: phosphinic acid

Formula: H_3PO_2

Molecular Formula: H_3O_2P

Molecular Weight: 65.997

CAS RN: 6303-21-5

Properties: colorless oily liq, or deliq cryst; sour odor; reducing agent; enthalpy of fusion 9.70 kJ/mol; strong monobasic acid; sold in the form of a solution; can be prepared by heating baryta water with white phosphorus, followed by treatment with H_2SO_4 and filtering; used in electroplating baths [HAW93] [CRC93]

Density, g/cm³: 1.439 [HAW93]

Melting Point, °C: 26.5 [HAW93]

Boiling Point, °C: decomposes at 130 [CRC94]

1390

Compound: Indium

Formula: In

Molecular Formula: In

Molecular Weight: 114.818

CAS RN: 7440-74-6

Properties: soft, white metal with bluish tinge; ductile; quite stable in air; hardness 1.2 Mohs; tetr, a = 0.5979 nm, c = 0.49467 nm; enthalpy of fusion 3.28 kJ/mol; enthalpy of vaporization 55.57 kJ/mol; electrical resistivity (22°C) 8.8 μohm·cm; tensile strength 2.645 MPa; elongation 22%; modulus of elasticity 10.8 GPa; superconductor at 3.38 K; uses: production of bearings, and in solder [MER89] [KIR81] [CRC93]

Solubility: i H_2O; attacked by mineral acids; not attacked by alkalies [MER89]

Density, g/cm³: 7.31 [CIC73]

Melting Point, °C: 156.6 [KIR81]

Boiling Point, °C: 2080 [KIR81]

Thermal Conductivity, W/(m·K): 81.8 (25°C) [ALD94]

Thermal Expansion Coefficient: linear expansion is 25 x 10^{-6}/°C from 0-100°C [KIR81]

1391

Compound: Indium acetate

Formula: $In(CH_3COO)_3$

Molecular Formula: $C_6H_9InO_6$

Molecular Weight: 291.951

CAS RN: 25114-58-3

Properties: white hygr cryst [STR93]

1392

Compound: Indium acetylacetonate

Synonyms: 2,4-pentanedione, indium(III) derivative

Formula: $In(CH_3COCH=C(O)CH_3)_3$

Molecular Formula: $C_{15}H_{21}InO_6$

Molecular Weight: 412.146

CAS RN: 14405-45-9

Properties: off-white powd [STR93]

$$[CH_3-\underset{|}{\overset{O-}{C}}=CH-\overset{O}{\overset{\|}{C}}-CH_3]_3$$

Melting Point, °C: 180-185 [STR93]

1393

Compound: Indium antimonide

Formula: InSb

Molecular Formula: InSb
Molecular Weight: 236.578
CAS RN: 1312-41-0
Properties: 6 mm pieces and smaller (fused) with 99.999% purity; black cryst; semiconductor; band gap, eV, 0.23 (0 K) and 0.17 (300 K); electron mobility 80,000 cm²/(V s) and hole mobility 1250 cm²/(V s); dielectric constant 17.7; effective mass 0.0145 for electrons and 0.40 for holes; enthalpy of fusion 25.50 kJ/mol [KIR82] [MER89] [STR93] [CER91] [CRC93]
Density, g/cm³: 5.7747 [LID94]; liq: 6.48 [MER89]
Melting Point, °C: 525 [CRC93]
Thermal Conductivity, W/(m·K): 16 [CRC93]
Thermal Expansion Coefficient: 4.7 x 10⁻⁶/K [CRC93]

1394
Compound: Indium arsenide
Formula: InAs
Molecular Formula: AsIn
Molecular Weight: 189.740
CAS RN: 1303-11-3
Properties: 6 mm pieces and smaller (fused) with 99.999% purity; semiconductor; metallic appearance; band gap, 0.42 (0 K) and 0.36 (300 K); mobility (300 K), cm²/(V s), 33,000 electrons and 460 holes; dielectric constant 14.6; effective mass 0.023 for electrons and 0.40 for holes [KIR82] [MER89] [CER91]
Solubility: i acids [HAW93]
Density, g/cm³: 5.67 [LID94]
Melting Point, °C: 943 [MER89]

1395
Compound: Indium nitride
Formula: InN
Molecular Formula: InN
Molecular Weight: 128.825
CAS RN: 25617-98-5
Properties: -100 mesh with 99.999% purity; wurtzite system, a = 0.353 nm, c = 0.570 nm; can be prepared by reacting In_2O_3 with ammonia at high temp; has semiconductor and electroluminescence properties [KIR81] [CIC73] [CER91]
Density, g/cm³: 6.89 [CIC73]
Melting Point, °C: decomposes above 600 [KIR81]
Thermal Conductivity, W/(m·K): 55.6 [CRC93]

1396
Compound: Indium phosphide
Formula: InP

Molecular Formula: InP
Molecular Weight: 145.792
CAS RN: 22398-80-7
Properties: black cryst; 6 mm pieces and smaller with 99.999% purity; semiconductor; band gap, eV, 1.42 (0 K) and 1.35 (300 K); mobility (300 K), cm²/(V s), 4600 electrons and 150 holes; dielectric constant 12.4; effective mass 0.077 electrons and 0.64 holes [KIR82] [CER91] [STR93]
Density, g/cm³: 4.81 [LID94]
Melting Point, °C: 1070 [AES93]

1397
Compound: Indium(I) bromide
Formula: InBr
Molecular Formula: BrIn
Molecular Weight: 194.722
CAS RN: 14280-53-6
Properties: can be prepared by reacting In with $InBr_3$ vapor; enthalpy of vaporization 92 kJ/mol [KIR81] [CRC93]
Solubility: reacts in H_2O to form In and $InBr_3$ [KIR81]
Density, g/cm³: 4.96 [CRC94]
Melting Point, °C: 220 [AES93]
Boiling Point, °C: 656 [CRC93]

1398
Compound: Indium(I) chloride
Formula: InCl
Molecular Formula: ClIn
Molecular Weight: 150.271
CAS RN: 13465-10-6
Properties: golden yellow powd; sensitive to atm oxygen and moisture; enthalpy of fusion 17.20 kJ/mol; can be prepared by passing $InCl_3$ vapor over In metal [KIR81] [STR93] [CRC93]
Solubility: decomposes in H_2O to In and $InCl_3$ [KIR81]
Density, g/cm³: 4.19 [STR93]
Melting Point, °C: 225 [CRC93]
Boiling Point, °C: 608 [STR93]

1399
Compound: Indium(I) iodide
Formula: InI
Molecular Formula: IIn
Molecular Weight: 241.722
CAS RN: 13966-94-4

Properties: reddish solid; -8 mesh with 99.999% purity; enthalpy of vaporization 90.8 kJ/mol [CRC94] [CER91]
Density, g/cm³: 5.31 [CRC94]
Melting Point, °C: 351 [AES93]
Boiling Point, °C: 712 [CRC93]

1400
Compound: Indium(II) bromide
Formula: InBr$_2$
Molecular Formula: Br$_2$In
Molecular Weight: 274.626
CAS RN: 21264-43-7
Properties: pale yellow solid; preparation: by reduction of InBr$_3$ with H$_2$/HBr mixture [KIR81] [CRC94]
Solubility: reacts in H$_2$O to form In and InBr$_3$ [KIR81]
Density, g/cm³: 4.22 [CRC94]
Melting Point, °C: 235 [CRC94]
Boiling Point, °C: 632, sublimes [CRC94]

1401
Compound: Indium(II) chloride
Formula: InCl$_2$
Molecular Formula: Cl$_2$In
Molecular Weight: 185.723
CAS RN: 13465-11-7
Properties: colorless rhomb cryst; can be prepared by heating indium to 200°C in HCl, or by reducing InCl$_3$ in H$_2$/HCl mixture below 600°C [KIR81] [CRC94]
Solubility: decomposes in H$_2$O to In and InCl$_3$ [KIR81]
Density, g/cm³: 3.655 [CRC94]
Melting Point, °C: 235 [CRC94]
Boiling Point, °C: 550-570 [CRC94]

1402
Compound: Indium(II) sulfide
Formula: InS
Molecular Formula: InS
Molecular Weight: 146.884
CAS RN: 12030-14-7
Properties: red-brown solid, -100 mesh with 99.999% purity [CER91] [KIR81]
Density, g/cm³: 5.18 [CRC94]
Melting Point, °C: 692 [CRC94]
Boiling Point, °C: sublimes at 850 in vacuum [CRC94]

1403
Compound: Indium(III) bromide
Formula: InBr$_3$
Molecular Formula: Br$_3$In
Molecular Weight: 354.530
CAS RN: 13465-09-3
Properties: -60 mesh with 99.999% purity; white to light yellow powd; hygr [STR93] [CER91]
Density, g/cm³: 4.74 [STR93]
Melting Point, °C: ~436 [STR93]

1404
Compound: Indium(III) chloride
Synonyms: indium trichloride
Formula: InCl$_3$
Molecular Formula: Cl$_3$In
Molecular Weight: 221.176
CAS RN: 10025-82-8
Properties: sublimed flakes with 99.999% purity; tan to yellowish, deliq cryst; can be made by heating indium in presence of excess Cl$_2$; used in electroplating baths; has high vapor pressure [KIR81] [HAW93] [MER89] [CER91]
Solubility: v s H$_2$O [MER89]; s alcohol [HAW93]
Density, g/cm³: 4.0 [MER89]; 3.46 [HAW93]
Melting Point, °C: 586 [MER89]
Boiling Point, °C: sublimes, 600 [CRC94]

1405
Compound: Indium(III) chloride tetrahydrate
Formula: InCl$_3$·4H$_2$O
Molecular Formula: Cl$_3$H$_8$InO$_4$
Molecular Weight: 293.238
CAS RN: 22519-64-8
Properties: white cryst [STR93]

1406
Compound: Indium(III) fluoride
Synonyms: indium trifluoride
Formula: InF$_3$
Molecular Formula: F$_3$In
Molecular Weight: 171.813
CAS RN: 7783-52-0
Properties: -40 mesh with 99.999% purity; white powd; hygr; stable in hot and cold H$_2$O [STR93] [MER89] [CER91]
Solubility: 0.04 g/200mL H$_2$O (25°C); v s dil acids [MER89]
Density, g/cm³: 4.39 [MER89]
Melting Point, °C: 1170 [MER89]
Boiling Point, °C: >1200 [MER89]

1407
Compound: Indium(III) fluoride trihydrate
Synonyms: indium trifluoride trihydrate
Formula: $InF_3 \cdot 3H_2O$
Molecular Formula: $F_3H_6InO_3$
Molecular Weight: 225.859
CAS RN: 14166-78-0
Properties: white cryst [STR93]
Solubility: 8.49 g/100mL H_2O (22°C) [MER89]
Melting Point, °C: 100, decomposes [STR93]

1408
Compound: Indium(III) hydroxide
Formula: $In(OH)_3$
Molecular Formula: H_3InO_3
Molecular Weight: 165.840
CAS RN: 20661-21-6
Properties: yellow-white powd; -100 mesh with 99.999% purity [AES93] [CER91]
Density, g/cm³: 4.4 [LID94]
Reactions: minus H_2O, 150°C [AES93]

1409
Compound: Indium(III) iodide
Formula: InI_3
Molecular Formula: I_3In
Molecular Weight: 495.531
CAS RN: 13510-35-5
Properties: -40 mesh with 99.999% purity; yellow-red solid; hygr [STR93] [CER91]
Density, g/cm³: 4.69 [STR93]
Melting Point, °C: 210 [STR93]

1410
Compound: Indium(III) nitrate trihydrate
Formula: $In(NO_3)_3 \cdot 3H_2O$
Molecular Formula: $H_6InN_3O_{12}$
Molecular Weight: 354.879
CAS RN: 13770-61-1
Properties: -80 mesh with 99.999% purity; can be prepared by dissolving In or the oxide in nitric acid [CER91] [KIR81]
Solubility: v s [KIR81]
Melting Point, °C: decomposes [AES93]
Reactions: minus $2H_2O$ at 100°C [KIR81]

1411
Compound: Indium(III) oxide
Formula: In_2O_3
Molecular Formula: In_2O_3
Molecular Weight: 277.634

CAS RN: 1312-43-2
Properties: white to pale yellow powd in both amorphous and cryst forms; turns red brown if heated; volatilizes at 850°C; can be prepared by heating indium in air, or by calcining indium carbonate; forms In_2O, 12030-22-7 and InO, 12136-26-4, if carefully reduced; used to impart yellow color to glass, as an evaporated film of 99.999% purity provides protective coating for metal mirrors, and as a sputtering target for transparent conductive films in electro-optical displays [KIR82] [MER89]
Solubility: i H_2O; s hot mineral acids [MER89]
Density, g/cm³: 7.179 [HAW93]
Melting Point, °C: ~2000 [LID94]

1412
Compound: Indium(III) perchlorate octahydrate
Formula: $In(ClO_4)_3 \cdot 8H_2O$
Molecular Formula: $Cl_3H_{16}InO_{20}$
Molecular Weight: 557.291
CAS RN: 13465-15-1
Properties: white cryst [STR93]
Melting Point, °C: ~80 [STR93]
Boiling Point, °C: 200, decomposes [STR93]

1413
Compound: Indium(III) phosphate
Formula: $InPO_4$
Molecular Formula: InO_4P
Molecular Weight: 209.789
CAS RN: 14693-82-4
Properties: prepared by adding phosphate solution to an indium solution [KIR81]
Solubility: i H_2O [KIR81]
Density, g/cm³: 4.9 [LID94]

1414
Compound: Indium(III) selenide
Formula: In_2Se_3
Molecular Formula: In_2Se_3
Molecular Weight: 466.516
CAS RN: 12056-07-4
Properties: black cryst; ortho-rhomb; used in the form of 99.999% pure material as a sputtering target for production of semiconductors [MER89] [CER91]
Density, g/cm³: 5.67 [ALD94]
Melting Point, °C: 660 [MER89]

1415
Compound: Indium(III) sulfate
Formula: $In_2(SO_4)_3$
Molecular Formula: $In_2O_{12}S_3$
Molecular Weight: 517.827
CAS RN: 13464-82-9
Properties: -80 mesh with 99.999% purity; grayish deliq powd; decomposed by heat; forms by dissolution of In or In oxides in warm sulfuric acid [KIR81] [HAW93] [CER91]
Solubility: s H_2O [MER89]
Density, g/cm³: 3.438 [HAW93]
Melting Point, °C: decomposes 600 [AES93]

1416
Compound: Indium(III) sulfide
Formula: In_2S_3
Molecular Formula: In_2S_3
Molecular Weight: 325.834
CAS RN: 12030-24-9
Properties: -200 mesh with 99.999% purity; orange powd; a preparation is to heat In and S, or to precipitate with H_2S from weakly acidic solutions [KIR81] [STR93] [CER91]
Density, g/cm³: 4.45 [STR93]
Melting Point, °C: 1050 [STR93]

1417
Compound: Indium(III) telluride
Formula: In_2Te_3
Molecular Formula: In_2Te_3
Molecular Weight: 612.436
CAS RN: 1312-45-4
Properties: 6 mm pieces and smaller with 99.999% purity; black or gray cryst; used in semiconductor technology [HAW93] [STR93] [CER91]
Density, g/cm³: (x-ray) 5.798 [MER89]
Melting Point, °C: 667 [MER89]

1418
Compound: Iodic acid
Formula: HIO_3
Molecular Formula: HIO_3
Molecular Weight: 175.910
CAS RN: 7782-68-5
Properties: colorless rhomb cryst, or white cryst powd; darkens on exposure to light; it is a moderately strong acid; used in analytical chemistry and in medicine [HAW93] [MER89]

Solubility: g/mL H_2O: 269 (20°C), 295 (40°C); i alcohol, ether [MER89]; 310 g/100g H_2O [KIR81]
Density, g/cm³: 4.629 [MER89]
Melting Point, °C: 110, with decomposition [MER89]
Reactions: decomposes to I_2O_5 at 220°C [MER89]

1419
Compound: Iodine
Formula: I_2
Molecular Formula: I_2
Molecular Weight: 253.809 (atomic wt 126.90447)
CAS RN: 7553-56-2
Properties: heavy bluish black scales or plates; rhomb; metallic luster; a = 0.47761 nm, b = 0.72501 nm, c = 0.97711 nm; readily sublimes to a violet, corrosive vapor; dielectric constant (23°C) 10.3; enthalpy of fusion 15.52 kJ/mol; enthalpy of sublimation (113.6°C) 238.24 J/g; enthalpy of vaporization 41.57 kJ/mol; electrical resistivity (25°C) 5.85 x 10^{+6} ohm·cm; used in aniline dyes, antiseptics, feed and food additives [HAW93] [MER89] [KIR81] [CRC93]
Solubility: g/100g H_2O: 0.014 (0°C), 0.029 (20°C), 0.445 (100°C) [LAN85]; s benzene, CCl_4, alcohol, glycerol, ether, CS_2 and alkaline iodide solutions [HAW93]
Density, g/cm³: solid 4.940; liq (120°C) 3.960 [KIR81]
Melting Point, °C: 113.60 [MER89]
Boiling Point, °C: 185.25 [CRC93]
Thermal Conductivity, W/(m·K): 0.421 (24.4°C) [KIR81]
Thermal Expansion Coefficient: cub coefficient is 2.81 x 10^{-4}/°C (0 to 113.6°C) [KIR81]

1420
Compound: Iodine cyanide
Synonyms: cyanogen iodide
Formula: ICN
Molecular Formula: CIN
Molecular Weight: 152.922
CAS RN: 506-78-5
Properties: colorless needles; very pungent odor and acrid taste; used in taxidermists' preservatives [HAW93]
Solubility: s H_2O, alcohol, ether [HAW93]
Density, g/cm³: 1.84 [HAW93]
Melting Point, °C: 146.5 [HAW93]

1421
Compound: Iodine dioxide

Formula: IO_2
Molecular Formula: IO_2
Molecular Weight: 158.903
CAS RN: 13494-92-3
Properties: lemon yellow cryst [KIR82]
Density, g/cm³: 4.2 [KIR81]
Reactions: decomposes at 85°C [KIR81]

1422

Compound: Iodine heptafluoride
Formula: IF_7
Molecular Formula: F_7I
Molecular Weight: 259.893
CAS RN: 16921-96-3
Properties: colorless liq; enthalpy of vaporization 24.7 kJ/mol; specific conductivity 10^{-9} ohm·cm [KIR78] [MER89]
Solubility: s H_2O with some decomposition [MER89]
Density, g/cm³: 2.669 [KIR78]
Melting Point, °C: 6.45 [MER89]
Boiling Point, °C: sublimes 4.77 [MER89]

1423

Compound: Iodine monobromide
Formula: IBr
Molecular Formula: BrI
Molecular Weight: 206.808
CAS RN: 7789-33-5
Properties: brownish black cryst, or very hard solid; uses: organic synthesis [HAW93] [MER89]
Solubility: s H_2O, alcohol, ether, CS_2, glacial acetic acid [MER89]
Density, g/cm³: 4.416 [MER89]
Melting Point, °C: 40 [MER89]
Boiling Point, °C: 116, decomposes [MER89]

1424

Compound: Iodine monochloride
Formula: ICl
Molecular Formula: ClI
Molecular Weight: 162.357
CAS RN: 7790-99-0
Properties: reddish brown oily liq; viscosity at 35°C is 1.21 cSt; electrical conductivity 4.60×10^{-3} at 35°C; enthalpy of vaporization 256.4 kJ/kg; enthalpy of fusion 11.60 kJ/mol; readily supercooled; polar solvent; best prepared by direct reaction between I_2 and liq Cl_2; used in analytical chemistry and in organic synthesis [KIR81] [CRC93] [HAW93]

Solubility: s in H_2O (decomposes), alcohol, dil HCl [HAW93]
Density, g/cm³: 3.24 (liq at 34°C) [HAW93]
Melting Point, °C: 27.38 [CRC93]
Boiling Point, °C: 101, decomposes [HAW93]

1425

Compound: Iodine nonoxide
Formula: I_4O_9
Molecular Formula: I_4O_9
Molecular Weight: 651.613
CAS RN: 73560-00-6
Properties: yellow hygr powd [KIR81]
Reactions: decomposes at 75°C [KIR81]

1426

Compound: Iodine pentafluoride
Formula: IF_5
Molecular Formula: F_5I
Molecular Weight: 221.896
CAS RN: 7783-66-6
Properties: fuming straw colored liq; critical temp 300.7°C; critical pressure 5.16 MPa; enthalpy of vaporization 41.3 kJ/mol; enthalpy of fusion 11.21 kJ/mol; specific conductivity 5.4×10^{-6} ohm·cm; attacks glass; used as a fluorinating and incendiary agent [KIR78] [HAW93] [AIR87] [CRC93]
Solubility: reacts violently with H_2O [HAW93]
Density, g/cm³: 3.19 [MER89]
Melting Point, °C: 9.43 [MER89]
Boiling Point, °C: 100.5 [MER89]

1427

Compound: Iodine pentoxide
Synonyms: iodine(V) oxide
Formula: I_2O_5
Molecular Formula: I_2O_5
Molecular Weight: 333.806
CAS RN: 12029-98-0
Properties: -80 mesh with 99.9% purity; white cryst powd; oxidizing agent, used in organic synthesis; oxidizes CO to CO_2 at ordinary temperatures; can be prepared by dehydration of iodic acid [KIR81] [HAW93] [CER91]
Solubility: s H_2O, HNO_3; i absolute alcohol, ether, CS_2 [HAW93]
Density, g/cm³: 4.980 [KIR81]
Melting Point, °C: 300-350, decomposes [STR93]

1428
Compound: Iodine trichloride
Formula: ICl_3
Molecular Formula: Cl_3I
Molecular Weight: 233.262
CAS RN: 865-44-1
Properties: yellowish orange deliq cryst powd; pungent irritating odor; electrical conductivity 8.60×10^{-3} at 102°C; can be formed by adding iodine to liq chlorine; used to introduce iodine and chlorine in organic synthesis, used as a topical antiseptic [HAW93] [KIR81]
Solubility: decomposes in H_2O; s alcohol, benzene [HAW93]
Density, g/cm³: 3.111 [KIR81]
Melting Point, °C: 33 [HAW93]
Reactions: decomposes at 65°C [KIR81]

1429
Compound: Iridium
Formula: Ir
Molecular Formula: Ir
Molecular Weight: 192.217
CAS RN: 7439-88-5
Properties: silver white metal; most corrosion resistant of the elements; cub, a = 0.384 nm; hardness 6.5 Mohs; Poisson's ratio 0.26; not attacked by acids, including aqua regia; attacked by fluorine, chlorine at red heat; modulus of elasticity is 75,000,000 psi, one of the highest; enthalpy of fusion 41.12 kJ/mol; enthalpy of vaporization, 231.84 kJ/mol; electrical resistivity at 20°C, 4.71 μohm·cm [HAW93] [MER89] [KIR82] [CRC93] [ALD94]
Solubility: s slowly in aqua regia and in fused alkalies [HAW93]
Density, g/cm³: 22.65, highest of any element [MER89]
Melting Point, °C: 2410 [ALD94]
Boiling Point, °C: 4130 [ALD94]
Thermal Conductivity, W/(m·K): 147 (25°C) [ALD94]
Thermal Expansion Coefficient: 6.4×10^{-6}/K [CRC93]

1430
Compound: Iridium carbonyl
Synonyms: tetrairidium dodecacarbonyl
Formula: $Ir_4(CO)_{12}$
Molecular Formula: $C_{12}Ir_4O_{12}$
Molecular Weight: 1104.993
CAS RN: 11065-24-0

Properties: yellow powd; stable in air [DOU83] [STR93]
Melting Point, °C: 230, decomposes [STR93]

1431
Compound: Iridium hexafluoride
Formula: IrF_6
Molecular Formula: F_6Ir
Molecular Weight: 306.207
CAS RN: 7783-75-7
Properties: golden yellow; cub; very hygr; enthalpy of vaporization 36 kJ/mol; enthalpy of fusion 8.40 kJ/mol [MER89] [CRC93]
Solubility: decomposes in H_2O [MER89]
Density, g/cm³: 4.8 [LID94]
Melting Point, °C: 44 [CRC93]
Boiling Point, °C: 53 [MER89]
Reactions: volatilized by slow heating; reduced to IrF_4 by halogens [MER89]

1432
Compound: Iridium pentafluoride
Formula: $(IrF_5)_4$
Molecular Formula: $F_{20}Ir_4$
Molecular Weight: 1148.836
CAS RN: 14568-19-5
Properties: tetramer [KIR82]

1433
Compound: Iridium(I) chlorotricarbonyl
Synonyms: chlorotricarbonyliridium(I)
Formula: $[IrCl(CO)_3]n$
Molecular Formula: C_3ClIrO_3 (n = 1)
Molecular Weight: 311.701 (n = 1)
CAS RN: 32594-40-4
Properties: brown powd [STR93]
Melting Point, °C: 235, decomposes [STR93]

1434
Compound: Iridium(III) acetylacetonate
Synonyms: 2,4-pentanedione, Ir(III) derivative
Formula: $Ir[CH_3COCH=C(O)CH_3]_3$
Molecular Formula: $C_{15}H_{21}IrO_6$
Molecular Weight: 489.550
CAS RN: 15635-87-7
Properties:

$$[CH_3-\underset{|}{\overset{O-}{C}}=CH-\underset{\|}{\overset{O}{C}}-CH_3]_3Ir$$

Melting Point, °C: 269-271 [ALD94]

1435
Compound: Iridium(III) bromide tetrahydrate
Synonyms: iridium tribromide tetrahydrate
Formula: $IrBr_3 \cdot 4H_2O$
Molecular Formula: $Br_3H_8IrO_4$
Molecular Weight: 503.991
CAS RN: 10049-24-8
Properties: olive green, brown or black cryst
   [HAW93]
Solubility: s H_2O; i alcohol [HAW93]
Reactions: minus $3H_2O$, 100°C [AES93]

1436
Compound: Iridium(III) chloride
Synonyms: iridium trichloride
Formula: $IrCl_3$
Molecular Formula: Cl_3Ir
Molecular Weight: 298.575
CAS RN: 10025-83-9
Properties: -8 mesh with 99.5% purity; α-$IrCl_3$:
   brown, monocl cryst; β-$IrCl_3$: red, ortho-rhomb
   cryst [MER89] [CER91]
Solubility: i H_2O, acids, alkalies [MER89]
Density, g/cm³: 5.30 [STR93]
Melting Point, °C: decomposes 763 [MER89]

1437
Compound: Iridium(III) chloride hydrate
Formula: $IrCl_3 \cdot xH_2O$
Molecular Formula: Cl_3Ir (anhydrous)
Molecular Weight: 298.575 (anhydrous)
CAS RN: 14996-61-3
Properties: black cryst [STR93]

1438
Compound: Iridium(III) oxide
Synonyms: iridium trioxide
Formula: Ir_2O_3
Molecular Formula: Ir_2O_3
Molecular Weight: 432.432
CAS RN: 1312-46-5
Properties: bluish black powd [MER89]
Solubility: i H_2O; slowly dissolves in boiling HCl
   [MER89]
Reactions: minus O at 400°C; oxidized to IrO_2 by
   HNO_3 [MER89] [CRC94]

1439
Compound: Iridium(IV) oxide
Formula: IrO_2
Molecular Formula: IrO_2
Molecular Weight: 224.216
CAS RN: 12030-49-8
Properties: -325 mesh 10 microns or less with
   99.9% purity; brown powd [STR93] [CER91]
Density, g/cm³: 11.66 [STR93]
Melting Point, °C: 1100, decomposes [STR93]

1440
Compound: Iron
Formula: Fe
Molecular Formula: Fe
Molecular Weight: 55.845
CAS RN: 7439-89-6
Properties: cub; silver-white or gray; soft, ductile;
   somewhat magnetic; tensile strength 30,000 psi;
   Brinell hardness 82-100; electrical resistivity
   (20°C) 9.71 µohm·cm; magnetic permeability
   88,400 gauss (25°C); tensile strength 245-280
   MPa; yield strength 70-140 MPa; enthalpy of
   fusion 13.81 kJ/mol; enthalpy of vaporization
   340 kJ/mol; reducing agent; oxidizes readily in
   moist air; elongation in 5 cm at 20°C is 40-60%
   [MER89] [KIR81] [CRC93] [ALD94]
Solubility: s HCl, H_2SO_4, dil HNO_3 [HAW93]
Density, g/cm³: 7.86 [MER89]
Melting Point, °C: 1535 [MER89]
Boiling Point, °C: 2861 [LID94]
Reactions: reacts with steam to produce hydrogen
   gas [HAW93]
Thermal Conductivity, W/(m·K): 80.4 (25°C)
   [ALD94]
Thermal Expansion Coefficient: (volume) 100°C
   (0.326), 200°C (0.747), 400°C (1.645), 800°C
   (3.523), 1200°C (2.986) [CLA66]

1441
Compound: Iron antimonide
Formula: Fe_3Sb_2
Molecular Formula: Fe_3Sb_2
Molecular Weight: 411.055
CAS RN: 39356-80-4
Properties: 6 mm pieces and smaller with 99.5%
   purity [CER91]

1442
Compound: Iron arsenide
Formula: FeAs
Molecular Formula: AsFe

Molecular Weight: 130.767
CAS RN: 12044-16-5
Properties: white; 6 mm pieces and smaller with 99.5% purity; there are two other arsenides: Fe_2As, 12005-88-8, and $FeAs_2$, 12006-21-2 [CER91] [CRC94]
Solubility: v sl s H_2O [CRC94]
Density, g/cm^3: 7.83 [CRC94]
Melting Point, °C: 1020 [CRC94]

1443
Compound: Iron boride
Formula: FeB
Molecular Formula: BFe
Molecular Weight: 66.656
CAS RN: 12006-84-7
Properties: gray cryst; -35 mesh with 99% purity; refractory material [KIR78] [CER91] [CRC94]
Density, g/cm^3: 7.15 [ALF93]
Melting Point, °C: ~1550 [KIR78]

1444
Compound: Iron boride
Formula: Fe_2B
Molecular Formula: BFe_2
Molecular Weight: 122.501
CAS RN: 12006-85-8
Properties: -35 mesh with 99% purity; refractory material [KIR78] [CER91]

1445
Compound: Iron disilicide
Formula: $FeSi_2$
Molecular Formula: $FeSi_2$
Molecular Weight: 112.016
CAS RN: 12022-99-0
Properties: tetr, gray powd [LID94] [STR93]
Density, g/cm^3: 4.75 [STR93]
Melting Point, °C: 1220 [STR93]

1446
Compound: Iron disulfide
Synonyms: pyrite
Formula: FeS_2
Molecular Formula: FeS_2
Molecular Weight: 119.967
CAS RN: 1317-66-4
Properties: black cub powd [LID94] [STR93]
Density, g/cm^3: 5.02 [LID94]
Melting Point, °C: 1171 [AES93]

Thermal Expansion Coefficient: (volume) 100°C (0.219), 200°C (0.529), 400°C (1.291) [CLA66]

1447
Compound: Iron dodecacarbonyl
Formula: $Fe_3(CO)_{12}$
Molecular Formula: $C_{12}Fe_3O_{12}$
Molecular Weight: 503.660
CAS RN: 12088-65-2
Properties: stabilized with 5-10% methanol; black cryst; sensitive to air [STR93]
Density, g/cm^3: 2.00 [STR93]
Melting Point, °C: 140 STR93]

1448
Compound: Iron molybdate
Formula: $FeMoO_4$
Molecular Formula: $FeMoO_4$
Molecular Weight: 215.783
CAS RN: 13718-70-2
Properties: dark brown, yellow; monocl [KIR81]
Solubility: 0.0076 g/100g H_2O [KIR78]
Density, g/cm^3: 5.6 [LID94]
Melting Point, °C: 1115 [LID94]

1449
Compound: Iron nonacarbonyl
Formula: $Fe_2(CO)_9$
Molecular Formula: $C_9Fe_2O_9$
Molecular Weight: 363.784
CAS RN: 15321-51-4
Properties: orange yellow cryst; sensitive to air [STR93]
Density, g/cm^3: 2.85 [STR93]
Melting Point, °C: 100, decomposes [STR93]

1450
Compound: iron pentacarbonyl
Formula: $Fe(CO)_5$
Molecular Formula: C_5FeO_5
Molecular Weight: 195.897
CAS RN: 13463-40-6
Properties: colorless to yellow, oily liq; decomposed by light to $Fe_2(CO)_9$ and CO; burns in air; used as a catalyst in organic reactions [HAW93] [MER89]
Solubility: i H_2O; s ether, benzene, acetone, CCl_4 [MER89]
Density, g/cm^3: 1.490 [STR93]
Melting Point, °C: -20 [MER89]
Boiling Point, °C: 103 [MER89]

1451
Compound: Iron silicide
Formula: FeSi
Molecular Formula: FeSi
Molecular Weight: 83.931
CAS RN: 12022-95-6
Properties: -20 mesh gray powd [ALF93]
Density, g/cm³: 6.1 [CRC94]
Melting Point, °C: 1410 [ALF93]

1452
Compound: Iron telluride
Formula: FeTe
Molecular Formula: FeTe
Molecular Weight: 183.445
CAS RN: 12125-63-2
Properties: tetr; 6 mm pieces and smaller with 99.5% purity; there is also a $FeTe_2$, 12023-03-9 [LID94] [CER91]
Density, g/cm³: 6.8 [LID94]
Melting Point, °C: 914 [LID94]

1453
Compound: Iron tungstate
Synonyms: ferberite
Formula: $FeWO_4$
Molecular Formula: FeO_4W
Molecular Weight: 303.683
CAS RN: 13870-24-1
Properties: tetr; -200 mesh will 99.5% purity [CRC94] [CER91]
Density, g/cm³: 6.64 [CRC94]

1454
Compound: Iron zirconate
Formula: $Fe_2O_3 \cdot ZrO_2$
Molecular Formula: Fe_2O_5Zr
Molecular Weight: 282.911
CAS RN: 52933-62-7
Properties: reacted product, -200 mesh with 99.7% purity [CER91]

1455
Compound: Iron(II,III) oxide
Synonyms: magnetite
Formula: Fe_3O_4
Molecular Formula: Fe_3O_4
Molecular Weight: 231.533
CAS RN: 1317-61-9

Properties: black cubes or amorphous powd; dull to metallic luster; magnetic properties; hardness 5.5-6.5; enthalpy of fusion 138.00 kJ/mol; used as 99.5% pure material as a sputtering target for magnetic films [HAW93] [CRC93] [MER89] [CER91]
Solubility: i H_2O; s acids [MER89]
Density, g/cm³: 5.2 [MER89]
Melting Point, °C: 1597 [CRC93]
Thermal Expansion Coefficient: (volume) 100°C (0.212), 200°C (0.513), 400°C (1.328), 800°C (3.24), 1000°C (4.26) [CLA66]

1456
Compound: Krypton
Formula: Kr
Molecular Formula: Kr
Molecular Weight: 83.80
CAS RN: 7439-90-9
Properties: colorless, odorless gas; condenses to colorless liq; critical temp -63.8°C; critical pressure 5.50 MPa; enthalpy of vaporization at bp 9.08 kJ/mol; enthalpy of fusion 1.37 kJ/mol; heat capacity (101.32 kPa, 25°C) 20.95 J/(mol·K); sonic velocity of gas (101.32 kPa, 0°C) 213 m/s; viscosity of gas (101.32 kPa, 25°C) 25.3 [CRC93] [MER89] [KIR78]
Solubility: 59.4 mL/1000g H_2O (20°C) [KIR78]; Henry's law constants, k x 10^{-4}: 3.685 (70.2°C), 4.017 (175.0°C), 3.761 (175.0°C), 2.392 (252.5°C) [POT78]
Density, g/cm³: gas: 101.3 kPa, 0°C, 0.0037493 [KIR78]
Melting Point, °C: -157.36 [CRC93]
Boiling Point, °C: -153.23 [CRC93]
Reactions: forms hydrate with water [MER89]
Thermal Conductivity, W/(m·K): gas at 101.32 kPa and 0°C is 0.874 [KIR78]

1457
Compound: Krypton difluoride
Formula: KrF_2
Molecular Formula: F_2Kr
Molecular Weight: 121.797
CAS RN: 13773-81-4
Properties: colorless solid; decomposes rapidly at room temp; tetr, a = 0.6533 nm, c = 0.5831 nm; vapor pressure at 0°C is 29 mm Hg; thermodynamically unstable, can be stored at -78°C; prepared by electrical discharge of Kr and F_2 at -183°C; used in certain types of electric light bulbs [MER89] [KIR78] [COT88]
Density, g/cm³: 3.24 [KIR78]
Melting Point, °C: decomposes ~25 [KIR78]

1458

Compound: Krypton fluoride hexafluoroantimonate

Formula: KrFSbF$_6$

Molecular Formula: F$_7$KrSb

Molecular Weight: 338.549

CAS RN: 52708-44-8

Properties: white [KIR78]

Melting Point, °C: decomposes 45 [KIR78]

1459

Compound: Krypton fluoride monodecafluoroantimonate

Formula: KrFSb$_2$F$_{11}$

Molecular Formula: F$_{12}$KrSb$_2$

Molecular Weight: 555.301

CAS RN: 39578-36-4

Properties: white [KIR78]

Melting Point, °C: decomposes 50 [KIR78]

1460

Compound: Krypton fluoride monodecafluorotantalate

Formula: KrFTaF$_{11}$

Molecular Formula: F$_{12}$KrTa

Molecular Weight: 492.729

CAS RN: 58815-72-8

Properties: white [KIR78]

Melting Point, °C: decomposes ~-20 [KIR78]

1461

Compound: Krypton trifluoride hexafluoroantimonate

Formula: Kr$_2$F$_3$SbF$_6$

Molecular Formula: F$_9$Kr$_2$Sb

Molecular Weight: 460.346

CAS RN: 52721-22-9

Properties: white [KIR78]

Melting Point, °C: decomposes ~25 [KIR78]

1462

Compound: Lanthanum

Formula: La

Molecular Formula: La

Molecular Weight: 138.9055

CAS RN: 7439-91-0

Properties: white, malleable metal; tarnishes in air; three cryst forms: α-La, hex, stable at ordinary temp; β-La, stable at 350°C, fcc; γ-La, stable above 868°C; enthalpy of fusion is 6.20 kJ/mol; enthalpy of sublimation is 431.0 kJ/mol; electrical resistivity (20°C) 54 μohm·cm; radius of atom is 0.1879 nm; radius of La^{+++} ion is 0.1061 nm; solution is colorless [MER89] [KIR82] [ALD94]

Solubility: slowly decomposes in H$_2$O; readily attacked by mineral acids [MER89]

Density, g/cm^3: α: 6.17; β: 6.19; γ: 5.98 [MER89]; α: 6.1453 [KIR82]

Melting Point, °C: 920 [MER89]

Boiling Point, °C: 3464 [KIR82]

Thermal Conductivity, W/(m·K): 13.4 (25°C) [CRC93]

Thermal Expansion Coefficient: 12.1 x 10^{-6}/K [CRC93]

1463

Compound: Lanthanum acetate hydrate

Formula: La(CH$_3$COO)$_3$·xH$_2$O

Molecular Formula: C$_6$H$_9$LaO$_6$ (anhydrous)

Molecular Weight: 316.039 (anhydrous)

CAS RN: 100587-90-4

Properties: white cryst; x=1.5 [ALF95] [STR93]

Solubility: 16.88 g/100mL H$_2$O (25°C) [CRC94]

1464

Compound: Lanthanum acetylacetonate hydrate

Synonyms: 2,4-pentanedione, lanthanum(III) derivative

Formula: La[CH$_3$COCH=C(O)CH$_3$]$_3$·xH$_2$O

Molecular Formula: C$_{15}$H$_{21}$LaO$_6$ (anhydrous)

Molecular Weight: 436.234 (anhydrous)

CAS RN: 64424-12-0

Properties: hygr white powd [STR93] [ALD94]

1465

Compound: Lanthanum aluminum oxide

Formula: LaAlO$_3$

Molecular Formula: AlLaO$_3$

Molecular Weight: 213.885

CAS RN: 12003-65-5

Properties: white powd [STR93]

1466

Compound: Lanthanum boride

Synonyms: lanthanum hexaboride

Formula: LaB$_6$

Molecular Formula: B$_6$La
Molecular Weight: 203.772
CAS RN: 12008-21-8
Properties: refractory material; black powd; used as a sputtering target with 99.9% and 99.5% purity to produce wear-resistant and semiconducting films, and for preparation of thermionic conductor films [STR93] [CER91]
Density, g/cm^3: 4.76 [LID94]
Melting Point, °C: 2715 [LID94]

1467

Compound: Lanthanum bromate nonahydrate
Formula: La(BrO$_3$)$_3$·9H$_2$O
Molecular Formula: Br$_3$H$_{18}$LaO$_{18}$
Molecular Weight: 644.751
CAS RN: 28958-23-8
Properties: hex [CRC94]
Solubility: g/100g H$_2$O: 98 (0°C), 149 (20°C), 200 (30°C) [LAN85]
Melting Point, °C: 37.5 [CRC94]
Reactions: minus 7H$_2$O at 100°C [CRC94]

1468

Compound: Lanthanum bromide
Formula: LaBr$_3$
Molecular Formula: Br$_3$La
Molecular Weight: 378.618
CAS RN: 13536-79-3
Properties: -20 mesh with 99.9% purity; heptahydrate, 13465-19-5, is white cryst [CER91] [STR93]
Density, g/cm^3: 5.057 [CRC94]
Melting Point, °C: 783 [CRC94]
Boiling Point, °C: 1577 [CRC94]

1469

Compound: Lanthanum carbide
Formula: LaC$_2$
Molecular Formula: C$_2$La
Molecular Weight: 162.928
CAS RN: 12071-15-7
Properties: yellow cryst; 12 mm pieces and smaller of 99.9% purity [CER91] [CRC94]
Solubility: decomposed by H$_2$O [CRC94]
Density, g/cm^3: 5.29 [LID94]
Melting Point, °C: 2360 [LID94]

1470

Compound: Lanthanum carbonate octahydrate
Formula: La$_2$(CO$_3$)$_3$·8H$_2$O

Molecular Formula: C$_3$H$_{16}$La$_2$O$_{17}$
Molecular Weight: 601.961
CAS RN: 6487-39-4
Properties: white; cryst powd [MER89]
Solubility: i H$_2$O; s dil mineral acids [MER89]
Density, g/cm^3: 2.6 [HAW93]

1471

Compound: Lanthanum carbonate pentahydrate
Formula: La$_2$(CO$_3$)$_3$·5H$_2$O
Molecular Formula: C$_3$H$_{10}$La$_2$O$_{14}$
Molecular Weight: 547.915
CAS RN: 54451-24-0
Properties: white powd [STR93]
Density, g/cm^3: 2.6-2.7 [STR93]

1472

Compound: Lanthanum chloride
Formula: LaCl$_3$
Molecular Formula: Cl$_3$La
Molecular Weight: 245.264
CAS RN: 10099-58-8
Properties: white powd; hygr [STR93]
Density, g/cm^3: 3.84 [STR93]
Melting Point, °C: 860 [STR93]
Boiling Point, °C: 1000 [STR93]

1473

Compound: Lanthanum chloride heptahydrate
Formula: LaCl$_3$·7H$_2$O
Molecular Formula: Cl$_3$H$_{14}$LaO$_7$
Molecular Weight: 371.371
CAS RN: 10025-84-0
Properties: white transparent hygr tricl cryst [HAW93] [MER89] [ALD94]
Solubility: s H$_2$O, alcohol [MER89]
Melting Point, °C: 91, decomposes [HAW93]
Reactions: forms LaCl$_3$ by heating in HCl at 850°C [MER89]

1474

Compound: Lanthanum chloride hexahydrate
Formula: LaCl$_3$·6H$_2$O
Molecular Formula: Cl$_3$H$_{12}$LaO$_6$
Molecular Weight: 353.355
CAS RN: 17272-45-6
Properties: white cryst [STR93]
Solubility: 4.6147±0.0056 mol/kg (25°C) [RAR87a]
Melting Point, °C: 91, decomposes [STR93]

1475
Compound: Lanthanum chromite
Formula: $LaCrO_3$
Molecular Formula: $CrLaO_3$
Molecular Weight: 238.900
CAS RN: 12017-94-6
Properties: -200 mesh with 99.9% purity [CER91]

1476
Compound: Lanthanum fluoride
Formula: LaF_3
Molecular Formula: F_3La
Molecular Weight: 195.901
CAS RN: 13709-38-1
Properties: white powd, or 99.9% pure melted
    pieces of 3-6 mm; hygr; used in phosphor lamp
    coatings, lasers, and melted pieces are used as
    evaporation material and sputtering material for
    multilayers [HAW93] [STR93] [CER91]
Solubility: i H_2O, acids [HAW93]
Density, g/cm^3: 5.936 [STR93]
Melting Point, °C: 1493 [STR93]

1477
Compound: Lanthanum hydride
Formula: LaH_3
Molecular Formula: H_3La
Molecular Weight: 141.930
CAS RN: 13864-01-2
Properties: black pyrophoric solid; prepared by
    heating La metal in H_2 to 300°C; spontaneously
    ignites in air; there is also the LaH_2 hydride;
    possible use as hydrogenation catalysts; used to
    store H_2 in the system $LaNi_5$, 12196-72-4
    [KIR80]
Solubility: reacts with H_2O at 0°C [KIR80]
Density, g/cm^3: 5.36 [LID94]

1478
Compound: Lanthanum hydroxide
Formula: $La(OH)_3$
Molecular Formula: H_3LaO_3
Molecular Weight: 189.928
CAS RN: 14507-19-8
Properties: white, amorphous precipitate; absorbs
    CO_2 [MER89]
Melting Point, °C: decomposes [STR93]

1479
Compound: Lanthanum iodide
Formula: LaI_3

Molecular Formula: I_3La
Molecular Weight: 519.619
CAS RN: 13813-22-4
Properties: grayish white powd; hygr [STR93]
Density, g/cm^3: 5.63 [STR93]
Melting Point, °C: 772 [STR93]

1480
Compound: Lanthanum nitrate hexahydrate
Formula: $La(NO_3)_3 \cdot 6H_2O$
Molecular Formula: $H_{12}LaN_3O_{15}$
Molecular Weight: 433.012
CAS RN: 10277-43-7
Properties: white; deliq cryst; used as an antiseptic,
    in gas mantles [HAW93] [MER89]
Solubility: g/100g H_2O: 100 (0°C), 136 (20°C), 247
    (60°C) [LAN85]
Melting Point, °C: ~40 [MER89]

1481
Compound: Lanthanum nitride
Formula: LaN
Molecular Formula: LaN
Molecular Weight: 152.913
CAS RN: 25764-10-7
Properties: cub; -60 mesh with 99.9% purity
    [LID94] [CER91]
Density, g/cm^3: 6.73 [LID94]

1482
Compound: Lanthanum oxalate hydrate
Formula: $La_2(C_2O_4)_3 \cdot xH_2O$
Molecular Formula: $C_6La_2O_{12}$ (anhydrous)
Molecular Weight: 541.869 (anhydrous)
CAS RN: 537-03-1
Properties: white powd; x=10 [ALF95] [HAW93]
    [STR93]
Solubility: i H_2O; s acids [HAW93]

1483
Compound: Lanthanum oxide
Formula: La_2O_3
Molecular Formula: La_2O_3
Molecular Weight: 325.809
CAS RN: 1312-81-8

Properties: -325 mesh 5 microns or less with 99.9999% purity; almost white, amorphous powd; absorbs CO_2; can be prepared by decomposition of the hydroxide, or oxalate at high temp; used in calcium lights, optical glass, in refractories, and as an evaporated material or sputtering target of 99.9% and 99.99% purity in thermistor elements, and for thin film capacitors [HAW93] [MER89] [CER91]

Solubility: i H_2O; s dil mineral acids [MER89]

Density, g/cm³: 6.51 [MER89]

Melting Point, °C: 2315 [HAW93]

Boiling Point, °C: 4200 [HAW93]

1484

Compound: Lanthanum oxysulfide

Formula: La_2O_2S

Molecular Formula: La_2O_2S

Molecular Weight: 341.876

CAS RN: 12031-43-5

Properties: -200 mesh with 99.9% purity [CER91]

1485

Compound: Lanthanum perchlorate hexahydrate

Formula: $La(ClO_4)_3 \cdot 6H_2O$

Molecular Formula: $Cl_3H_{12}LaO_{18}$

Molecular Weight: 545.348

CAS RN: 36907-37-6

Properties: white cryst; hygr [STR93]

1486

Compound: Lanthanum phosphate hydrate

Formula: $LaPO_4 \cdot xH_2O$

Molecular Formula: LaO_4P (anhydrous)

Molecular Weight: 233.827 (anhydrous)

CAS RN: 14913-14-5

Properties: white powd [STR93]

1487

Compound: Lanthanum silicide

Formula: $LaSi_2$

Molecular Formula: $LaSi_2$

Molecular Weight: 195.077

CAS RN: 12056-90-5

Properties: gray tetr; 10 mm & down lump [LID94] [ALF93]

Density, g/cm³: 5.0 [LID94]

1488

Compound: Lanthanum strontium copper oxide

Formula: $La_{1.85}Sr_{0.15}CuO_4$

Molecular Formula: $CuLa_{1.85}O_4Sr_{0.15}$

Molecular Weight: 397.662

CAS RN: 111419-39-7

Properties: superconductor; general formula is $La_{2-x}M_xCuO_4$; for x = 0.15, T_c is 36 K; oriented thin films can be prepared from bulk cuprates or the metal oxides by sputtering or by laser ablation; potential uses of superconductors include low-loss microwave devices such as cavities and wave guides, frictionless bearings and electronic devices [CEN92]

1489

Compound: Lanthanum sulfate nonahydrate

Formula: $La_2(SO_4)_3 \cdot 9H_2O$

Molecular Formula: $H_{18}La_2O_{21}S_3$

Molecular Weight: 728.139

CAS RN: 10294-62-9

Properties: hex prisms; decomposes at white heat; refractive index 1.564 (20°C) [HAW93] [MER89]

Solubility: s H_2O, solubility decreases with increasing temp [MER89]; i alcohol [HAW93]

Density, g/cm³: 2.821 [HAW93]

1490

Compound: Lanthanum sulfate octahydrate

Formula: $La_2(SO_4)_3 \cdot 8H_2O$

Molecular Formula: $H_{16}La_2O_{20}S_3$

Molecular Weight: 710.124

CAS RN: 57804-25-8

Properties: white cryst [STR93]

Solubility: g/100g H_2O: 3.00 (0°C), 2.33 (20°C), 0.66 (100°C) [LAN85]

Density, g/cm³: 2.821 [STR93]

Melting Point, °C: decomposes [STR93]

1491

Compound: Lanthanum sulfide

Formula: La_2S_3

Molecular Formula: La_2S_3

Molecular Weight: 374.009

CAS RN: 12031-49-1

Properties: red powd; there is a LaS_2, 12325-81-4, with -200 mesh of 99.9% purity [STR93] [CER91]

Density, g/cm³: 4.911 [STR93]

Melting Point, °C: 2100-2150 [STR93]

1492
Compound: Lanthanum telluride
Formula: La_2Te_3
Molecular Formula: La_2Te_3
Molecular Weight: 660.62
CAS RN: 12031-53-7
Properties: -20 mesh [ALF95]

1493
Compound: Lanthanum tris(cyclopentadienyl)
Synonyms: tris(cyclopentadienyl)lanthanum
Formula: $(C_5H_5)_3La$
Molecular Formula: $C_{15}H_{15}La$
Molecular Weight: 334.189
CAS RN: 1272-23-7
Properties: white cryst; air and moisture sensitive [STR93]
Melting Point, °C: 295 decomposes [STR93]

1494
Compound: Lawrencium
Formula: Lr
Molecular Formula: Lr
Molecular Weight: 262
CAS RN: 22537-19-5
Properties: has two isotopes, 257 and 256; ^{257}Lr has a half-life of 8 sec; discovered in 1961 by Ghioros and colleagues at Lawrence Berkeley Laboratory; can be synthesized by bombardment of californium with boron ions [KIR78] [HAW93] [CRC94]
Melting Point, °C: 1627 [LID94]

1495
Compound: Lead
Formula: Pb
Molecular Formula: Pb
Molecular Weight: 207.2
CAS RN: 7439-92-1
Properties: bluish white, silvery gray metal; fcc; tarnishes in air; very soft and malleable; cub; electrical resistivity 20.65 μohm·cm (20°C), 27.02 μohm·cm (100°C); velocity of sound 122,700 cm/sec; hardness 1.5 Mohs; Young's modulus 16.5 MPa; surface tension (360°C) 442 mN/m; electronegativity 1.8; enthalpy of fusion 4.77 kJ/mol; enthalpy of vaporization 179.5 kJ/mol [CRC93] [KIR78] [MER89] [COT88]
Solubility: reacts with hot conc HNO_3, boiling conc HCl, H_2SO_4 [MER89]
Density, g/cm^3: 11.35 (20°C); 327°C: 11.00 (solid); 10.67 (liq) [KIR78]

Melting Point, °C: 327.4 [MER89]
Boiling Point, °C: 1749 [LID94]
Reactions: attacked by pure water in presence of O_2; resistant to tap water, HF, brine, solvents [MER89]
Thermal Conductivity, W/(m·K): 35.3 (25°C) [ALD94]
Thermal Expansion Coefficient: linear expansion from 0 to 100°C is 29 x 10^{-6}/K [MER89]

1496
Compound: Lead acetate
Synonyms: lead diacetate
Formula: $Pb(CH_3COO)_2$
Molecular Formula: $C_4H_6O_4Pb$
Molecular Weight: 325.289
CAS RN: 301-04-2
Properties: white cryst; -8 mesh with 99.999% purity; can be prepared by dissolution of PbO (litharge) or $PbCO_3$ in acetic acid; used in the preparation of other Pb salts [KIR78] [CER91]
Solubility: g/100g H_2O: 19.7 (0°C), 55.2 (25°C); equilibrium solid phase, $Pb(CH_3COO)_2\cdot3H_2O$ [KRU93]; g/100mL H_2O: 44.3 (20°C), 221 (50°C) [KIR78]
Density, g/cm^3: 3.25 [KIR78]
Melting Point, °C: 280 [KIR78]
Boiling Point, °C: decomposes >280 [KIR78]

1497
Compound: Lead acetate trihydrate
Formula: $Pb(CH_3COO)_2\cdot3H_2O$
Molecular Formula: $C_4H_{12}O_7Pb$
Molecular Weight: 379.319
CAS RN: 6080-56-4
Properties: colorless cryst or white granules or powd; slowly effloresces; absorbs CO_2 from air; obtained by dissolution of PbO in hot dil acetic acid, then cooling to precipitate crysts; used to prepare basic lead carbonate and lead chromate, used as a mordant in cotton dyes, and a water repellent [KIR78] [MER89]
Solubility: g/100mL H_2O: 45.61 (15°C), 200 (100°C) [KIR78]; sl s in alcohol, s in glycerol [HAW93]
Density, g/cm^3: 2.55 [MER89]
Melting Point, °C: 75 (decomposes 200) [KIR78]
Reactions: minus $3H_2O$ at 75°C [HAW93]

1498
Compound: Lead acetylacetonate
Synonyms: 2,4-pentanedione, lead(II) derivative
Formula: $Pb(CH_3C(O)CH=COCH_3)_2$

Molecular Formula: $C_{10}H_{14}O_4Pb$
Molecular Weight: 405.419
CAS RN: 15282-88-9
Properties: white powd; hygr [STR93]

$$\begin{array}{cc} O- & O \\ | & || \\ \end{array}$$
$$[CH_3-C=CH-C-CH_3]_2Pb$$

1499

Compound: Lead antimonate
Synonyms: naples yellow
Formula: $Pb_3(SbO_4)_2$
Molecular Formula: $O_8Pb_3Sb_2$
Molecular Weight: 993.115
CAS RN: 13510-89-9
Properties: orange yellow powd; formed by a reaction between lead nitrate and potassium antimonate solutions, with subsequent crystallization; used in staining glass, crockery and porcelain [HAW93] [MER89]
Solubility: i H_2O, dil acids [MER89]
Density, g/cm³: 6.58 (20°C) [HAW93]

1500

Compound: Lead antimonide
Formula: PbSb
Molecular Formula: PbSb
Molecular Weight: 328.960
CAS RN: 12266-38-5
Properties: gray powd [STR93]

1501

Compound: Lead arsenate
Synonyms: schultenite
Formula: $Pb_3(AsO_4)_2$
Molecular Formula: $As_2O_8Pb_3$
Molecular Weight: 899.438
CAS RN: 7784-40-9
Properties: white cryst; used in insecticides and herbicides [HAW93]
Solubility: i H_2O; s HNO_3 [HAW93]
Density, g/cm³: 5.8 [HAW93]
Melting Point, °C: 1042, decomposes [HAW93]

1502

Compound: Lead arsenite
Formula: $Pb(AsO_2)_2$
Molecular Formula: As_2O_4Pb
Molecular Weight: 421.041
CAS RN: 10031-13-7

Properties: white powd; used as an insecticide [HAW93]
Solubility: i H_2O; s dil HNO_3 [MER89]
Density, g/cm³: 5.85 [MER89]

1503

Compound: Lead azide
Formula: $Pb(N_3)_2$
Molecular Formula: N_6Pb
Molecular Weight: 291.240
CAS RN: 13424-46-9
Properties: needles or white powd; prepared by reaction of dil solutions of lead nitrate and sodium azide; used as a primary detonating compound for high explosives; α-$Pb(N_3)_2$: ortho, a = 0.663 nm, b = 0.546 nm, c = 1.625 nm; β-$Pb(N_3)_3$: monocl, a = 0.509 nm, b = 0.884 nm, c = 1.751 nm; γ-$Pb(N_3)_2$: a = 0.622 nm, b = 1.051 nm, c = 1.217 nm [MER89] [CIC73] [KIR78]
Solubility: 0.023% H_2O (18°C), 0.09% (70°C) [MER89]; i NH_4OH; v s acetic acid [KIR78]
Density, g/cm³: 4.7 [LID94]
Reactions: explodes at 350°C [MER89]

1504

Compound: Lead basic acetate
Synonyms: lead subacetate
Formula: $2Pb(OH)_2 \cdot Pb(CH_3COO)_2$
Molecular Formula: $C_4H_{10}O_8Pb_3$
Molecular Weight: 807.718
CAS RN: 1335-32-6
Properties: white; heavy powd; absorbs atm CO_2 [MER89]
Solubility: s 16 parts cold, 4 parts boiling H_2O [MER89]
Melting Point, °C: decomposes [LID94]

1505

Compound: Lead basic carbonate
Synonyms: hydrocerussite,white lead
Formula: $2PbCO_3 \cdot Pb(OH)_2$
Molecular Formula: $C_2H_2O_8Pb_3$
Molecular Weight: 775.633
CAS RN: 1319-46-6
Properties: white hex cryst; -100 mesh with 99.9% purity; produced by addition of CO_2 to a solution of lead acetate; used in ceramic glazes and as an exterior paint pigment [HAW93] [CER91] [KIR78]
Solubility: i H_2O; s acids [HAW93]
Density, g/cm³: 6.68 [HAW93]
Melting Point, °C: 400, decomposes [HAW93]

1506
Compound: Lead borate monohydrate
Formula: $Pb(BO_2)_2 \cdot H_2O$
Molecular Formula: $B_2H_2O_5Pb$
Molecular Weight: 310.835
CAS RN: 10124-39-8
Properties: white cryst powd; produced by fusing boric acid with either lead carbonate or litharge; used as a varnish and paint drier, to waterproof paints, in electrically conductive ceramic coatings [HAW93] [KIR78]
Solubility: i H_2O; s dil HNO_3 [MER89]; g/100mL soln, H_2O: 2 (room temp) [KRU93]
Density, g/cm³: 5.6 [HAW93]
Melting Point, °C: 500 [KIR78]
Reactions: minus H_2O at 160°C [HAW93]

1507
Compound: Lead bromate monohydrate
Formula: $Pb(BrO_3)_2 \cdot H_2O$
Molecular Formula: $Br_2H_2O_7Pb$
Molecular Weight: 481.020
CAS RN: 10031-21-7
Properties: monocl colorless cryst [CRC94] [MER89]
Solubility: 1.38 g/100mL H_2O (20°C) [CRC94]
Density, g/cm³: 5.53 [HAW93]
Melting Point, °C: ~180, decomposes [HAW93]

1508
Compound: Lead bromide
Formula: $PbBr_2$
Molecular Formula: Br_2Pb
Molecular Weight: 367.008
CAS RN: 10031-22-8
Properties: white ortho-rhomb cryst; -80 mesh with 99.999% purity; enthalpy of vaporization 133 kJ/mol; enthalpy of fusion 16.44 kJ/mol; obtained from PbO or $PbCO_3$ and HBr; finds used as a photopolymerization catalyst, and in photoduplication processes in the 365 nm region [KIR78] [CER91] [CRC93] [MER89]
Solubility: s in ~200 parts H_2O, 20 parts boiling H_2O; sl s in ammonia; i alcohol [MER89] [KIR78]; g/100g soln, H_2O: 0.4554 (0°C), 0.9744 (25°C), 4.751 (100°C) [KRU93]
Density, g/cm³: 6.66 [HAW93]
Melting Point, °C: 371 [CRC93]
Boiling Point, °C: 892 [CRC93]

1509
Compound: Lead carbonate

Synonyms: cerussite
Formula: $PbCO_3$
Molecular Formula: CO_3Pb
Molecular Weight: 267.209
CAS RN: 598-63-0
Properties: colorless ortho-rhomb cryst; made by adding CO_2 to a cold dil solution of lead acetate; many uses such as a catalyst for organic reactions, in high temp greases, as a photoconductor in electrophotography [KIR78]
Solubility: g/L solution, H_2O: 0.0011 - 0.0017 (20°C) [KRU93]
Density, g/cm³: 6.6 [KIR78]
Melting Point, °C: decomposes ~315 [KIR78]

1510
Compound: Lead chlorate
Formula: $Pb(ClO_3)_2$
Molecular Formula: Cl_2O_6Pb
Molecular Weight: 374.101
CAS RN: 10294-47-0
Properties: colorless; deliq cryst [MER89]
Solubility: s 0.7 parts H_2O; v s alcohol [MER89]; g/100 g soln, H_2O: 151.3 to 440 (18°C) [KRU93]
Density, g/cm³: 3.9 [MER89]
Melting Point, °C: decomposes 230 [MER89]

1511
Compound: Lead chloride
Formula: $PbCl_2$
Molecular Formula: Cl_2Pb
Molecular Weight: 278.105
CAS RN: 7758-95-4
Properties: white ortho-rhomb needles; -80 mesh precipitated agglomerates with 99.999% purity, or 0.8-3.4 mm pieces (fused) with 99.999% purity; readily forms basic chlorides, for example $PbCl_2 \cdot Pb(OH)_2$, 15887-88-4, which is Pattinson's lead white pigment; enthalpy of vaporization 127 kJ/mol; enthalpy of fusion 21.90 kJ/mol; can be prepared by reacting PbO or $PbCO_3$ and HCl; uses: preparation of lead salts [CER91] [HAW93] [KIR78] [CRC93]
Solubility: g/100g soln, H_2O: 0.67 (0°C), 1.06±0.02 (25°C), 3.17±0.07 (100°C) [KRU93]; sl s dil HCl and NH_3; i alcohol [KIR78]
Density, g/cm³: 5.85 [KIR78]
Melting Point, °C: 501 [MER89]
Boiling Point, °C: 951 [CRC93]

1512

Compound: Lead chlorite

Formula: $Pb(ClO_2)_2$

Molecular Formula: Cl_2O_4Pb

Molecular Weight: 342.103

CAS RN: 13453-57-2

Properties: yellow monocl [CRC94]

Solubility: g/100g soln, H_2O: 0.035 (0°C), 0.12 (25°C), 0.41 (100°C); Solid phase, $Pb(ClO_2)_2$ [KRU93]

Melting Point, °C: decomposes 126 [LAN52]

1513

Compound: Lead chromate

Synonyms: crucoite, chrome yellow

Formula: $PbCrO_4$

Molecular Formula: CrO_4Pb

Molecular Weight: 323.194

CAS RN: 7758-97-6

Properties: yellow: ortho-rhomb cryst; orange: monocl; red: tetr [KIR78]

Solubility: s dil HNO_3; i acetic acid [MER89] [KIR78]; g/L soln, H_2O: 0.00017 (25°C) [KRU93]

Density, g/cm³: 6.12 (monocl) [KIR78]

Melting Point, °C: 844 (monocl) [KIR78]

1514

Compound: Lead citrate trihydrate

Synonyms: citric acid, lead salt trihydrate

Formula: $Pb_3(C_6H_5O_7)_2 \cdot 3H_2O$

Molecular Formula: $C_{12}H_{16}O_{17}Pb_3$

Molecular Weight: 1053.849

CAS RN: 512-26-5

Properties: white cryst powd [CRC92] [STR93]

Solubility: s H_2O; v sl s alcohol [CRC92]

1515

Compound: Lead cyanide

Formula: $Pb(CN)_2$

Molecular Formula: C_2N_2Pb

Molecular Weight: 259.235

CAS RN: 592-05-2

Properties: white to yellowish powd; used in metallurgy [HAW93]

Solubility: sl s H_2O; decomposed by acid [HAW93]

1516

Compound: Lead dioxide

Synonyms: plattnerite

Formula: PbO_2

Molecular Formula: O_2Pb

Molecular Weight: 239.199

CAS RN: 1309-60-0

Properties: dark brown powd; oxidizing agent; evolves O_2 if heated; produced by oxidation of Pb_3O_4 with Cl_2 in an alkaline media; used as electrodes, to manufacture dyes, and in lead-storage batteries [MER89] [HAW93]

Solubility: i H_2O; s HCl evolving Cl_2; s hot caustic solutions [MER89]

Density, g/cm³: 9.375 [HAW93]

Melting Point, °C: 290, decomposes [HAW93]

Reactions: if heated evolves O_2 to form Pb_3O_4, then PbO at higher temperatures [MER89]

1517

Compound: Lead fluoride

Formula: PbF_2

Molecular Formula: F_2Pb

Molecular Weight: 245.197

CAS RN: 7783-46-2

Properties: white to colorless cryst, or 99.9% pure melted pieces of 3-6 mm or 1-3 mm; ortho-rhomb or cub; enthalpy of vaporization 160.4 kJ/mol; enthalpy of fusion 14.70 kJ/mol; obtained by reaction of HF and $PbCO_3$ or lead hydroxide; used in electronic and optical applications, pieces used as evaporation material and sputtering material for dielectric interference filter in the ultraviolet, and as a high index film in the ultraviolet [HAW93] [CRC93] [MER89] [KIR78] [CER91]

Solubility: 0.057 g/100mL H_2O (0°C), 0.065 (20°C) [MER89]; g/L soln, H_2O: 0.66 (25°C) [KRU93]; s HNO_3; i acetone, ammonia [KIR78]

Density, g/cm³: ortho-rhomb: 8.445; cub: 7.750 [MER89]

Melting Point, °C: 830 [CRC93]

Boiling Point, °C: 1293 [MER89]

Reactions: transition from ortho-rhomb to cub >316°C [MER89]

1518

Compound: Lead fluoroborate

Synonyms: lead borofluoride

Formula: $Pb(BF_4)_2$

Molecular Formula: B_2F_8Pb

Molecular Weight: 380.809

CAS RN: 13814-96-5

Properties: liq; used in lead electroplating, in tin-lead plating [HAW93]

1519
Compound: Lead fluorosilicate dihydrate
Synonyms: lead hexafluorosilicate dihydrate
Formula: $PbSiF_6 \cdot 2H_2O$
Molecular Formula: $F_6H_4O_2PbSi$
Molecular Weight: 385.307
CAS RN: 1310-03-8
Properties: colorless cryst; used in aq solutions for the electrodeposition of lead [HAW93]
Solubility: g/100g H_2O: 190 (0°C), 222 (20°C), 463 (100°C) [LAN85]
Melting Point, °C: decomposes [HAW93]

1520
Compound: Lead hexafluoroacetylacetonate
Synonyms: 1,1,1,5,5,5-hexafluoro-2,4-pentanedione, lead derivative
Formula: $Pb(CF_3COCHCOCF_3)_2$
Molecular Formula: $C_{10}F_{12}O_4Pb$
Molecular Weight: 621.289
CAS RN: 19648-88-5
Properties: white powd [STR93]

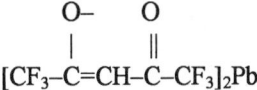

Melting Point, °C: 153-158 [STR93]
Boiling Point, °C: decomposes at 210 [STR93]
Reactions: sublimes at 180°C (0.05 mm Hg) [STR93]

1521
Compound: Lead hydrogen phosphate
Formula: $PbHPO_4$
Molecular Formula: HO_4PPb
Molecular Weight: 303.179
CAS RN: 7446-27-7
Properties: soft, white powd, or fine plate like cryst; used to impart heat resistance and pearlescence to polystyrene and casein plastics [HAW93]
Density, g/cm³: 5.66 [HAW93]
Melting Point, °C: decomposes [HAW93]

1522
Compound: Lead hydroxide
Formula: $Pb(OH)_2$
Molecular Formula: H_2O_2Pb
Molecular Weight: 241.215
CAS RN: 19783-14-3

Properties: white, bulky powd; absorbs atm CO_2; can be prepared by addition of alkali to a solution of lead nitrate; used in Ni-Cd battery electrolytes, as an oxidation catalyst, in electrical insulating paper and in petroleum well plugs [KIR78] [HAW93]
Solubility: 0.0155 g/100mL H_2O (20°C) [KIR78]; s HNO_3 and acetic acid [HAW93]
Density, g/cm³: 7.592 [HAW93]
Melting Point, °C: 145, decomposes [HAW93]

1523
Compound: Lead iodate
Formula: $Pb(IO_3)_2$
Molecular Formula: I_2O_6Pb
Molecular Weight: 557.005
CAS RN: 25659-31-8
Properties: white; -60 mesh with 99.9% purity [CRC94] [CER91]
Solubility: g/L soln, H_2O: 0.0254 (25°C) [KRU93]
Density, g/cm³: 6.5 [LID94]
Melting Point, °C: decomposes at 300 [CRC94]

1524
Compound: Lead iodide
Formula: PbI_2
Molecular Formula: I_2Pb
Molecular Weight: 461.009
CAS RN: 10101-63-0
Properties: yellow hex powd cryst; -100 mesh with 99.999% purity; heavy odorless powd; sensitive to light; enthalpy of vaporization 104 kJ/mol; enthalpy of fusion 23.40 kJ/mol; can be made by mixing a soluble Pb compound with HI, followed by recrystallization from water; used in bronzing, printing, photography and in cloud seeding [HAW93] [CRC93] [MER89] [CER91] [KIR78]
Solubility: 1 gram dissolves in 1350 mL cold H_2O, in 230 mL boiling H_2O [MER89]; g/100 g soln, H_2O: 0.0442 (0°C), 0.0764 (25°C), 0.436 (100°C) [KRU93]
Density, g/cm³: 6.16 [MER89]
Melting Point, °C: 402 [CRC94]
Boiling Point, °C: 954 [CRC94]

1525
Compound: Lead molybdate
Synonyms: wulfenite
Formula: $PbMoO_4$
Molecular Formula: MoO_4Pb
Molecular Weight: 367.138
CAS RN: 10190-55-3

Properties: white; scheelite structure, c/a = 2.23; used in pigments and as an analytical reagent [HAW93] [KIR81]
Solubility: 0.000012 g/100g H_2O; s HNO_3, NaOH when freshly precipitated [HAW93] [MER89] [KIR81]
Density, g/cm³: 6.92 [STR93]
Melting Point, °C: 1060-1070 [STR93]

1526

Compound: Lead niobate
Formula: $PbNb_2O_6$
Molecular Formula: Nb_2O_6Pb
Molecular Weight: 489.009
CAS RN: 12034-88-7
Properties: -200 mesh with 99.9% purity [CER91]
Melting Point, °C: >1200 [LID94]

1527

Compound: Lead nitrate
Formula: $Pb(NO_3)_2$
Molecular Formula: N_2O_6Pb
Molecular Weight: 331.209
CAS RN: 10099-74-8
Properties: colorless cub or monocl cryst; prepared by dissolution of Pb, PbO or $PbCO_3$ in dil HNO_3; -50 mesh with 99.999% purity; hygr; many uses including textile mordant, pyrotechnics and photothermography [KIR78] [MER89] [STR93] [CER91]
Solubility: s alkalies, ammonia; 8.77 g/100mL alcohol (22°C) [KIR78]; g/100 g soln, H_2O: 39.5±0.7 (0°C), 60.0±0.8 (25°C), 130±6 (100°C) [KRU93]
Density, g/cm³: 4.53 [MER89]
Melting Point, °C: 470 [STR93]
Reactions: decomposes with evolution of O_2 and N_2 >205°C [KIR78]

1528

Compound: Lead oxalate
Formula: PbC_2O_4
Molecular Formula: C_2O_4Pb
Molecular Weight: 295.220
CAS RN: 814-93-7
Properties: white, heavy powd [MER89]
Solubility: i H_2O; s dil HNO_3; sl s acetic acid [MER89]; g/L soln, H_2O: 0.0015 (18°C) [KRU93]
Density, g/cm³: 5.28 [MER89]
Melting Point, °C: 300 decomposes [AES93]

1529

Compound: Lead oxide
Synonyms: massicot, litharge
Formula: PbO
Molecular Formula: OPb
Molecular Weight: 223.199
CAS RN: 1317-36-8
Properties: two forms: red to reddish yellow, tetr cryst, stable at ordinary temp; yellow, orthorhomb cryst, stable above 489°C [MER89]
Solubility: i H_2O, alcohol; s acetic acid, dil HNO_3 [MER89]
Density, g/cm³: 9.53 [MER89]
Melting Point, °C: 888 [MER89]
Reactions: transforms slowly to Pb_3O_4 at 300-450°C; reverts to PbO at higher temperatures [MER89]

1530

Compound: Lead oxide
Synonyms: lead sesquioxide
Formula: Pb_2O_3
Molecular Formula: O_3Pb_2
Molecular Weight: 462.398
CAS RN: 1314-27-8
Properties: reddish yellow amorphous powd; obtained from PbO_2 by hydrothermal reaction; used in ceramics, ceramic cements, metallurgy, and in varnishes [HAW93] [MER89] [KIR78]
Solubility: i H_2O; decomposed by conc HCl, H_2SO_4, evolving Cl_2 or O_2 [MER89]; s alkalies [HAW93]
Melting Point, °C: 370 decomposes [KIR78]
Reactions: forms Pb_3O_4 in air at 370°C [MER89]

1531

Compound: Lead perchlorate trihydrate
Formula: $Pb(ClO_4)_2 \cdot 3H_2O$
Molecular Formula: $Cl_2H_6O_{11}Pb$
Molecular Weight: 460.146
CAS RN: 13637-76-8
Properties: white cryst [HAW93]
Solubility: 81.5 g/100g soln in H_2O (25°C), solid phase is $Pb(ClO_4)_2 \cdot 3H_2O$ [KRU93]
Density, g/cm³: 2.6 [HAW93]
Melting Point, °C: 100, decomposes [HAW93]

1532

Compound: Lead phosphate
Formula: $Pb_3(PO_4)_2$
Molecular Formula: $O_8P_2Pb_3$
Molecular Weight: 811.543

CAS RN: 7446-27-7
Properties: white powd; used as a stabilizing agent for plastics [HAW93]
Solubility: g/L H_2O: 1.35 x 10^{-4} (20°C) [KRU93]
Density, g/cm³: 6.9 [MER89]
Melting Point, °C: 1014 [MER89]

1533
Compound: Lead selenate
Formula: $PbSeO_4$
Molecular Formula: O_4PbSe
Molecular Weight: 350.158
CAS RN: 7446-15-3
Properties: ortho-rhomb cryst [MER89]
Solubility: i H_2O; s conc acids [MER89]
Density, g/cm³: 6.37 [MER89]
Melting Point, °C: decomposes [CRC94]

1534
Compound: Lead selenide
Synonyms: clausthalite
Formula: PbSe
Molecular Formula: PbSe
Molecular Weight: 286.160
CAS RN: 12069-00-0
Properties: gray black powd; semiconducting material used in infrared detectors and in thermoelectric devices, and as a 99.999% pure material, to produce photoconductive films [HAW93] [STR93] [CER91]
Solubility: i H_2O; s HNO_3 [HAW93]
Density, g/cm³: 8.10 (15°C) [HAW93]
Melting Point, °C: 1065 [HAW93]
Thermal Conductivity, W/(m·K): 1.7 [CRC93]
Thermal Expansion Coefficient: (volume) 100°C (0.474), 200°C (1.110), 400°C (2.420), 600°C (3.716) [CLA66]

1535
Compound: Lead selenite
Formula: $PbSeO_3$
Molecular Formula: O_3PbSe
Molecular Weight: 334.158
CAS RN: 7488-51-9
Properties: white, monocl; -100 mesh with 99.9% purity; powd [CER91] [MER89] [LID94]
Solubility: v sl s H_2O [MER89]
Density, g/cm³: 7.0 [LID94]
Melting Point, °C: ~500, melts to yellow liq [MER89]
Reactions: decomposes at bright-red heat with formation of selenium oxide [MER89]

1536
Compound: Lead silicate
Synonyms: alamoisite
Formula: $PbSiO_3$
Molecular Formula: O_3PbSi
Molecular Weight: 283.284
CAS RN: 10099-76-0
Properties: white cryst powd; obtained by reacting lead acetate with sodium silicate; used in ceramics and for fireproofing fabrics; formula given also as 1-1/2PbO·SiO_2 for a commercial product [KIR78] [HAW93]
Solubility: i most solvents [HAW93]
Density, g/cm³: 6.50-6.65 [KIR78]
Melting Point, °C: 700-784 [KIR78]

1537
Compound: Lead stearate
Formula: $Pb[CH_3(CH_2)_{16}COO]_2$
Molecular Formula: $C_{36}H_{70}O_4Pb$
Molecular Weight: 774.152
CAS RN: 1072-35-1
Properties: white powd; used as drier in varnishes and lacquer, in high pressure lubricants [HAW93] [MER89]
Solubility: i H_2O; s hot alcohol [MER89]
Density, g/cm³: 1.4 [HAW93]
Melting Point, °C: 100-115 [HAW93]

1538
Compound: Lead sulfate
Synonyms: anglesite
Formula: $PbSO_4$
Molecular Formula: O_4PbS
Molecular Weight: 303.264
CAS RN: 7446-14-2
Properties: white ortho-rhomb, monocl; heavy cryst powd; made by warming PbO, $PbCO_3$ or $Pb(OH)_2$ in H_2SO_4; used in paint pigments and in storage batteries, used to stabilize clay soils [HAW93] [MER89] [KIR78]
Solubility: s in ~2225 parts H_2O; more s dil HCl, HNO_3; s NaOH [MER89]; g/L soln, H_2O: 0.0330 (0°C), 0.0452 (25°C), 0.0574 (50°C) [KRU93]
Density, g/cm³: 6.12-6.39 [HAW93]
Melting Point, °C: 1170 [MER89]

1539
Compound: Lead sulfide
Synonyms: galena
Formula: PbS

Molecular Formula: PbS
Molecular Weight: 239.266
CAS RN: 1314-87-0
Properties: black powd, or silvery cub cryst; enthalpy of fusion 19.00 kJ/mol; photoconductive; can be made by heating Pb in sulfur vapor; used in ceramics, in infrared radiation detection and semiconductors, and in the form of 99.9% or 99.999% pure material as a sputtering target for metallic high reflecting films; galena is natural lead sulfide ore, it has a lead gray color with a metallic luster, a 2.5 Mohs hardness [KIR78] [HAW93] [CER91] [CRC93]
Solubility: 0.01244 g/100mL H_2O (20°C); s HNO_3, hot dil HCl [KIR78] [MER89]
Density, g/cm³: 7.61 [CRC93]
Melting Point, °C: 1113 [CRC93]
Thermal Conductivity, W/(m·K): 2.3 [CRC93]
Thermal Expansion Coefficient: (volume) 100°C (0.490), 200°C (1.099), 400°C (2.402), 600°C (3.878) [CLA66]

1540
Compound: Lead sulfite
Formula: $PbSO_3$
Molecular Formula: O_3PbS
Molecular Weight: 287.264
CAS RN: 7446-10-8
Properties: white powd [HAW93]
Solubility: i H_2O; s HNO_3 [HAW93]
Melting Point, °C: decomposes [HAW93]

1541
Compound: Lead tantalate
Formula: $PbTa_2O_6$
Molecular Formula: O_6PbTa_2
Molecular Weight: 665.092
CAS RN: 12065-68-8
Properties: ortho; -200 mesh with 99.9% purity [LID94] [CER91]
Density, g/cm³: 7.9 [LID94]

1542
Compound: Lead telluride
Synonyms: altaite
Formula: PbTe
Molecular Formula: PbTe
Molecular Weight: 334.800
CAS RN: 1314-91-6

Properties: silver gray cub cryst, 99.999% pure melted pieces of 3-12 mm and 1-3 mm; semiconductor, photoconductor; hardness 3 Mohs; made by melting mixture of Pb and Te; used as an evaporation material and sputtering target for high-index film in infrared filters and infrared detectors [KIR78] [MER89] [CER91]
Solubility: not attacked by HCl, HF, HClO [MER89]
Density, g/cm³: 8.164 [STR93]
Melting Point, °C: 905 [MER89]
Thermal Conductivity, W/(m·K): 2.3 [CRC93]

1543
Compound: Lead tellurite
Formula: $PbTeO_3$
Molecular Formula: O_3PbTe
Molecular Weight: 382.798
CAS RN: 15851-47-5
Properties: -100 mesh with 99.9% purity [CER91]

1544
Compound: Lead tetraacetate
Synonyms: lead(IV) acetate
Formula: $Pb(CH_3COO)_4$
Molecular Formula: $C_8H_{12}O_8Pb$
Molecular Weight: 443.378
CAS RN: 546-67-8
Properties: white to light brown; monocl prisms from glacial acetic acid; readily turns pink; unstable in air, moisture sensitive; hydrolyzes in H_2O to form brown PbO_2 and acetic acid; oxidizing agent; obtained by adding warm glacial water free acetic acid to Pb_3O_4, then cooling; used as a laboratory reagent, as an oxidant in organic synthesis [KIR78] [HAW93] [MER89] [STR93]
Solubility: decomposes in cold H_2O and alcohol; s hot glacial acetic acid, benzene, chloroform [MER89] [KIR78]
Density, g/cm³: 2.228 [MER89]
Melting Point, °C: 175-180 [MER89]
Reactions: reacts with conc halogen acids, HX, to form haloplumbic acids, H_2PbX [MER89]

1545
Compound: Lead tetrafluoride
Formula: PbF_4
Molecular Formula: F_4Pb
Molecular Weight: 283.194
CAS RN: 7783-59-7

Properties: white tetr cryst; readily hydrolyzes and forms PbO_2 in the presence of moisture; can be produced from a reaction of F_2 with PbF_2; used as a very effective fluorinating agent for olefins [KIR78] [MER89]
Density, g/cm³: 6.7 [MER89]
Melting Point, °C: ~600 decomposes [MER89]

1546
Compound: Lead thiocyanate
Formula: $Pb(SCN)_2$
Molecular Formula: $C_2N_2PbS_2$
Molecular Weight: 323.367
CAS RN: 592-87-0
Properties: white or light yellow odorless powd; used as an ingredient of priming mixture for small arms cartridges, safety matches and used in dyeing [HAW93] [MER89]
Solubility: s ~200 parts cold, 50 parts boiling H_2O [MER89]; g/L H_2O: 0.0137 (18°) [KRU93]
Density, g/cm³: 3.82 [MER89]
Melting Point, °C: 190, decomposes [ALD94]

1547
Compound: Lead thiosulfate
Formula: PbS_2O_3
Molecular Formula: O_3PbS_2
Molecular Weight: 319.330
CAS RN: 13478-50-7
Properties: white cryst [HAW93]
Solubility: i H_2O; s acids and sodium thiosulfate solutions [HAW93]
Density, g/cm³: 5.18 [HAW93]
Melting Point, °C: decomposes [HAW93]

1548
Compound: Lead titanate
Synonyms: lead metatitanate
Formula: $PbTiO_3$
Molecular Formula: O_3PbTi
Molecular Weight: 303.065
CAS RN: 12060-00-3
Properties: yellow tetr cryst <490°C, cub >490°C; can be made by calcination of stoichiometric amounts of PbO and TiO_2 at 400°C; also prepared by precipitation from aq solution of lead nitrate, titanium tetrachloride and ammonium hydroxide, followed by calcining at 900°C; has been used as a paint pigment; and in 99.9% purity as a sputtering target for thin film capacitors [KIR78] [KIR83] [STR93] [CER91] [SAF87]

Solubility: i H_2O; decomposed in HCl solution to $PbCl_2$ and TiO_2 [KIR78] [HAW93]
Density, g/cm³: 7.52 [HAW93]

1549
Compound: Lead tungstate
Synonyms: raspite
Formula: $PbWO_4$
Molecular Formula: O_4PbW
Molecular Weight: 455.038
CAS RN: 7759-01-5
Properties: -200 mesh with 99.9% purity; colorless monocl [CER91] [KIR83]
Solubility: 0.03 g/100mL H_2O [CRC94]
Density, g/cm³: 8.46 [KIR83]
Melting Point, °C: 1123 [KIR83]
Reactions: transforms to stolzite ~400°C [LID94]

1550
Compound: Lead tungstate
Synonyms: stolzite
Formula: $PbWO_4$
Molecular Formula: O_4PbW
Molecular Weight: 455.038
CAS RN: 7759-01-5
Properties: white powd; used as a pigment [HAW93]
Solubility: i H_2O, cold HNO_3; s in fixed alkali hydroxides [MER89]
Density, g/cm³: 8.24 [HAW93]
Melting Point, °C: 1130 [HAW93]

1551
Compound: Lead vanadate
Synonyms: lead metavanadate
Formula: $Pb(VO_3)_2$
Molecular Formula: O_6PbV_2
Molecular Weight: 405.079
CAS RN: 10099-79-3
Properties: yellow powd; used to prepare other vanadium compounds and as a pigment [HAW93]
Solubility: i H_2O; decomposed by HNO_3 [MER89]

1552
Compound: Lead zirconate
Formula: $PbZrO_3$
Molecular Formula: O_3PbZr
Molecular Weight: 346.422
CAS RN: 12060-01-4

Properties: colorless cub perovskite >230°, pseudotetr or ortho-rhomb <230°C; can be obtained by heating stoichiometric amounts of lead and zirconium oxides; has high piezoelectric properties; used in high power acoustic transducer, hydrophones, and as a 99.7% pure sputtering target for thin film capacitors [STR93] [CER91] [KIR78]
Solubility: i H_2O, alkalies; s mineral acids [KIR78]
Density, g/cm³: 7.0 [STR93]

1553
Compound: Lead(II,III) oxide
Synonyms: minium, red lead
Formula: Pb_3O_4
Molecular Formula: O_4Pb_3
Molecular Weight: 685.598
CAS RN: 1314-41-6
Properties: bright red, heavy powd; spinel structure; evolves Cl in contact with hot HCl; oxidizing agent; manufactured by heating PbO in air at 450-500°C; used in storage batteries, in the purification of alcohol, as a pigment in corrosion-protective paints [HAW93] [MER89] [KIR78]
Solubility: i H_2O, alcohol; s in excess glacial acetic acid, hot HCl [MER89]
Density, g/cm³: 9.1 [MER89]
Melting Point, °C: 830 (under O_2) [KIR78]
Boiling Point, °C: decomposes 500 [KIR78]

1554
Compound: Lithium
Formula: Li
Molecular Formula: Li
Molecular Weight: 6.941
CAS RN: 7439-93-2
Properties: very soft silver-white metal; bcc from ~-195°C to 180°C, a = 0.350 nm; hardness 0.6 Mohs; specific heat 3.55 J/g; enthalpy of fusion 3.00 kJ/mol; enthalpy of vaporization ~147.8 kJ/mol; electrical resistivity 9.446 μohm·cm; ionic radius 0.060 nm; vapor pressure, kPa: 0.065 (702°C), 0.376 (802°C), 1.61 (902°C), 5.47 (1002°C), 9.4 (1052°C), 12.13 (1077°C); manufactured by electrolysis; used to prepare the amide, nitride and hydride [KIR81] [CRC93]
Solubility: reacts vigorously with H_2O, dil HCl, H_2SO_4, HNO_3 evolving H_2 [MER89]
Density, g/cm³: 0.531 (20°C) [KIR81]
Melting Point, °C: 180.5 [KIR81]
Boiling Point, °C: 1336 [KIR81]
Reactions: transformation to fcc at -133°C; bcc to hex at -199°C [KIR78]

Thermal Conductivity, W/(m·K): 84.7 [CRC93]
Thermal Expansion Coefficient: 46 x 10⁻⁶/K [CRC93]

1555
Compound: Lithium acetate
Formula: CH_3COOLi
Molecular Formula: $C_2H_3LiO_2$
Molecular Weight: 65.986
CAS RN: 546-89-4
Properties: used as an alcoholysis catalyst in the manufacture of alkyl resins [KIR81]
Solubility: g/100g soln, H_2O: 23.76 (0°C), 31.28 (25.8°C), 66.73 (102.8°C); Solid phase, $CH_3COOLi·2H_2O$ (0°C, 25°C), CH_3COOLi (100°C) [KRU93]
Melting Point, °C: 291 [KIR81]

1556
Compound: Lithium acetate dihydrate
Synonyms: acetic acid, lithium salt
Formula: $LiCH_3COO·2H_2O$
Molecular Formula: $C_2H_7LiO_4$
Molecular Weight: 102.017
CAS RN: 6108-17-4
Properties: white powd; rhomb cryst; obtained readily from reaction of acetic acid and Li_2CO_3 or LiOH; used as catalyst in production of polyester, as an anticorrosion agent [KIR81] [STR93] [MER89] [FMC93]
Solubility: $gLiC_2H_3O_2/100g\ H_2O$: 31 (25°C), 66g/100g H_2O (100°C) [KIR81]
Density, g/cm³: 1.3 [STR93]
Melting Point, °C: 57.8; anhydrous, 291 [KIR81]

1557
Compound: Lithium acetylacetonate
Synonyms: 2,4-pentanedione, lithium derivative
Formula: $Li[CH_3COCH=C(O)CH_3]$
Molecular Formula: $C_5H_7LiO_2$
Molecular Weight: 106.051
CAS RN: 18115-70-3
Properties: white powd; hygr [STR93]

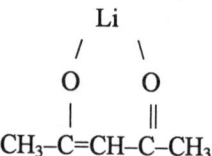

Melting Point, °C: 250, decomposes [ALD94]

1558
Compound: Lithium aluminum deuteride
Formula: $LiAlD_4$
Molecular Formula: AlD_4Li
Molecular Weight: 41.983
CAS RN: 14128-54-2
Properties: white to gray cryst; sensitive to moisture; decomposes >140°C liberating deuterium; used to introduce deuterium atoms into molecules [HAW93]
Solubility: s ether, tetrahydrofuran [HAW93]
Density, g/cm^3: 1.02 [HAW93]
Melting Point, °C: 175, decomposes [ALD94]

1559
Compound: Lithium aluminum hydride
Synonyms: lithium tetrahydridoaluminate
Formula: $LiAlH_4$
Molecular Formula: AlH_4Li
Molecular Weight: 37.955
CAS RN: 16853-85-3
Properties: cryst or gray powd; monocl; stable in dry air, decomposes in moist air; prepared by reaction of LiH with $AlCl_3$; used as a reducing agent for organics to convert esters, aldehydes and ketones to alcohols, used in perfumes, and in pharmaceuticals [HAW93] [MER83]
Solubility: reacts rapidly with H_2O; 30 parts/100 parts ether [MER83]
Density, g/cm^3: 0.917 [STR93]
Melting Point, °C: decomposes, >125 [MER83]
Reactions: slowly evolves H_2 at 120°C [MER83]

1560
Compound: Lithium aluminum silicate
Synonyms: α-spodumene
Formula: $\alpha\text{-}LiAlSi_2O_6$
Molecular Formula: $AlLiO_6Si_2$
Molecular Weight: 186.089
CAS RN: 12068-40-5
Properties: white powd; used in ceramic flux formulations, as a solid electrolyte, and in heat sinks for solar and nuclear applications [FMC93] [STR93]

1561
Compound: Lithium amide
Formula: $LiNH_2$
Molecular Formula: H_2LiN
Molecular Weight: 22.964
CAS RN: 7782-89-0

Properties: gray-to-white powd; tetr; manufactured by reacting LiH and NH_3 gas; used in the pharmaceutical industry to make antihistamines and analgesics [KIR81] [MER89] [FMC93]
Solubility: decomposed by H_2O [MER89]
Density, g/cm^3: 1.178 [ALD94]
Melting Point, °C: 375 [KIR81]
Reactions: transforms to imide >400°C [KIR81]

1562
Compound: Lithium arsenate
Formula: Li_3AsO_4
Molecular Formula: $AsLi_3O_4$
Molecular Weight: 159.743
CAS RN: 13478-14-3
Properties: white powd [HAW93]
Solubility: sl s H_2O; s dil acetic acid [HAW93]
Density, g/cm^3: 3.07 (15°C) [HAW93]

1563
Compound: Lithium azide
Formula: LiN_3
Molecular Formula: LiN_3
Molecular Weight: 48.961
CAS RN: 19597-69-4
Properties: colorless; hygr; body-center rhombohedral [CRC94] [CIC73]
Solubility: g/100g H_2O: 61.3 (0°C), 67.2 (20°C), 86.6 (60°C) [LAN85]
Density, g/cm^3: 1.83 [LID94]
Melting Point, °C: decomposes at 115-298 [CRC94]

1564
Compound: Lithium borate
Synonyms: lithium metaborate
Formula: $LiBO_2$
Molecular Formula: $BLiO_2$
Molecular Weight: 49.751
CAS RN: 13453-69-5
Properties: -80 mesh with 99.9% purity; used in special glass and enamel formulations as a flux, as an electrolyte component for lithium batteries; there is a dihydrate, $LiBO_2 \cdot 2H_2O$ [KIR81] [CER91] [FMC93]
Solubility: g/100g H_2O: 0.9 (0°C), 2.7 (20°C), 5.78 (30°C) [LAN85]
Density, g/cm^3: 2.18 [LID94]
Melting Point, °C: 849 [KIR81]

1565
Compound: Lithium borohydride

Formula: LiBH$_4$
Molecular Formula: BH$_4$Li
Molecular Weight: 21.784
CAS RN: 16949-15-8
Properties: white to gray cryst powd; ortho-rhomb; decomposes in moist air; used as a source of hydrogen; reducing agent for aldehydes, ketones and esters [MER89] [HAW93]
Solubility: s H$_2$O above pH7, ether, tetrahydrofuran, aliphatic amines [MER89]
Density, g/cm^3: 0.66 [MER89]
Melting Point, °C: 268 [MER89]
Boiling Point, °C: decomposes 380 [MER89]
Reactions: reacts with HCl to form hydrogen, diborane and LiCl [MER89]

1566
Compound: Lithium bromate
Formula: LiBrO$_3$
Molecular Formula: BrLiO$_3$
Molecular Weight: 134.843
CAS RN: 13550-28-2
Solubility: g/100g soln, H$_2$O: 61.2 (0°C), 65.4 (25°C), 78.0 (100°C); Solid phase, LiBrO$_3$·H$_2$O (0°C, 25°C), LiBrO$_3$ (100°C) [KRU93]

1567
Compound: Lithium bromide
Formula: LiBr
Molecular Formula: BrLi
Molecular Weight: 86.845
CAS RN: 7550-35-8
Properties: white cub; very deliq; bitter taste; enthalpy of fusion 17.60 kJ/mol manufactured by neutralizing HBr with LiOH or Li$_2$CO$_3$; used in pharmaceuticals, air conditioning, in batteries, LiBr solution is used as a component of refrigrant in absorption air conditioning and as a swelling agent for proteins [CRC93] [HAW93] [FMC93]
Solubility: s alcohol, glycol, ether, amyl alcohol [KIR81] [MER89]; g/100g soln, H$_2$O: 58.4±0.4 (0°C), 63.3±1.8 (25°C), 72.7 (100°C); Solid phase, LiBr·3H$_2$O (0°C), LiBr·2H$_2$O (25°C), LiBr·H$_2$O (100°C) [KRU93]
Density, g/cm^3: 3.464 [LID94]
Melting Point, °C: 552 [CRC93]
Boiling Point, °C: 1310 [KIR81]; 1265 [STR93]

1568
Compound: Lithium bromide monohydrate
Formula: LiBr·H$_2$O

Molecular Formula: BrH$_2$LiO
Molecular Weight: 104.860
CAS RN: 13453-70-8
Properties: white powd; obtained by crystallization from a hot solution of HBr and either LiOH or Li$_2$CO$_3$; can be dried to the anhydrous salt [KIR81] [STR93]

1569
Compound: Lithium carbide
Formula: Li$_2$C$_2$
Molecular Formula: C$_2$Li$_2$
Molecular Weight: 37.904
CAS RN: 1070-75-3
Properties: cryst white powd; decomposes in water, evolves acetylene when dissolved in acid [HAW93]
Density, g/cm^3: 1.65 (18°C) [HAW93]

1570
Compound: Lithium carbonate
Formula: Li$_2$CO$_3$
Molecular Formula: CLi$_2$O$_3$
Molecular Weight: 73.891
CAS RN: 554-13-2
Properties: white, light alkaline powd; enthalpy of fusion 41.00 kJ/mol; produced by reacting hot conc soda ash with LiCl, Li$_2$SO$_4$; used in ceramics and porcelain glazes; Li$_2$CO$_3$ slurry can be dissolved by passing CO$_2$ through the slurry, carbonate reprecipitates if heated [HAW93] [MER89] [CRC93]
Solubility: g/100g H$_2$O: 1.52 (0°C), 1.31 (20°C), 0.71 (100°C); i alcohol [KIR81]; g/100g H$_2$O: 1.54 (0°C), 1.29 (25°C), 0.72 (100°C) [KRU93]
Density, g/cm^3: 2.11 [MER89]
Melting Point, °C: 726 [KIR81]
Boiling Point, °C: 1310, decomposes [STR93]

1571
Compound: Lithium chlorate
Formula: LiClO$_3$
Molecular Formula: ClLiO$_3$
Molecular Weight: 90.392
CAS RN: 13453-71-9
Properties: needle like cryst; deliq; oxidizing agent; decomposes at 270°C; used in air conditioning, in propellants [HAW93]

Solubility: g/100g soln, H_2O: 71.1 (1.5°C), 81.7 (22.1°C), 94.9 (99°C); Solid phase, $LiClO_3\cdot3H_2O$ + $LiClO_3\cdot H_2O$ (1.5°C), $4LiClO_3\cdot H_2O$ (22.1°C), α-$LiClO_3$ + β-$LiClO_3$ (100°C) [KRU93]

Density, g/cm³: 1.119 [HAW93]

Melting Point, °C: 128 [HAW93]

Boiling Point, °C: decomposes [HAW93]

1572

Compound: Lithium chloride

Formula: LiCl

Molecular Formula: ClLi

Molecular Weight: 42.394

CAS RN: 7447-41-8

Properties: white powd; cub; very hygr; sharp saline taste; enthalpy of fusion 19.90 kJ/mol; obtained from reaction of HCl and LiOH or Li_2CO_3 >95°C; used in air conditioning, welding and in soldering flux; component of dry batteries, catalyst for some oxidation reactions, chlorinating agent for steroid substrates [KIR81] [MER89] [STR93] [FMC93]

Solubility: s water, alcohol, acetone, amyl alcohol, pyridine [KIR81] [MER89]; g/100g soln, H_2O: 40.9 (0°C), 45.8 (25°C), 56.2 (100°C); Solid phase, $LiCl\cdot2H_2O$ (0°C), $LiCl\cdot H_2O$ (25°C), LiCl (100°C) [KRU93]

Density, g/cm³: 2.068 [HAW93]

Melting Point, °C: 610 [CRC93]

Boiling Point, °C: 1360 [HAW93]

1573

Compound: Lithium chloride monohydrate

Formula: $LiCl\cdot H_2O$

Molecular Formula: ClH_2LiO

Molecular Weight: 60.409

CAS RN: 16712-20-2

Properties: white cryst; prepared by reacting HCl with LiOH or Li_2CO_3 and crystallizing below 95°C; solution is used for deicing, in fire extinguishers, catalysts and for dehumidifying [FMC93] [KIR81] [STR93]

Solubility: 45.9g/100g saturated solution in water (25°C) [KIR81]

Density, g/cm³: 1.78 [CRC94]

Reactions: minus H_2O at >98°C [CRC94]

1574

Compound: Lithium chromate

Formula: Li_2CrO_4

Molecular Formula: $CrLi_2O_4$

Molecular Weight: 129.876

CAS RN: 14307-35-8

Properties: yellow cryst; clear yellow solution is used as a corrosion inhibitor, and as an additive to industrial batteries [STR93] [FMC93]

Solubility: g/100g soln, H_2O: 47.27 (0.7°C), 48.60 (20°C), 56.82 (100°C); Solid phase, $Li_2CrO_4\cdot2H_2O$ (0.7°C, 20°C), Li_2CrO_4 (100°C) [KRU93]

Melting Point, °C: 495 [AES93]

1575

Compound: Lithium chromate dihydrate

Formula: $Li_2CrO_4\cdot2H_2O$

Molecular Formula: $CrH_4Li_2O_6$

Molecular Weight: 165.906

CAS RN: 7789-01-7

Properties: yellow cryst; ortho-rhomb; deliq powd; oxidizing agent; eutectic in aq solutions is at -60°C; used as a corrosion inhibitor [HAW93] [MER89]

Solubility: 49.6% H_2O (30°C); s methanol, ethanol [KIR78] [MER89]

Density, g/cm³: 2.15 [KIR78]

Reactions: minus $2H_2O$ (74.6°C) to become anhydrous [KIR78]

1576

Compound: Lithium citrate tetrahydrate

Synonyms: citric acid, trilithium salt

Formula:
$LiOOCCH_2C(OH)(COOLi)CH_2COOLi\cdot4H_2O$

Molecular Formula: $C_6H_{13}Li_3O_{11}$

Molecular Weight: 281.985

CAS RN: 6680-58-6

Properties: white granules or cryst powd; deliq in moist air; barely perceptible alkaline taste; used in beverages and pharmaceuticals, as a clay dispersant, in electroplating solutions, and as an buffer for ion chromatography [FMC93] [HAW93] [MER89]

Solubility: 74.5 g/100mL H_2O (25°C), 66.7 g/100mL (100°C) H_2O [CRC94]; sl s alcohol [MER89]

Melting Point, °C: 209.92 [ALD94]

Reactions: minus $4H_2O$ at 105°C becoming anhydrous [MER89]

1577

Compound: Lithium cobaltite

Formula: $LiCoO_2$

Molecular Formula: $CoLiO_2$

Molecular Weight: 97.873

CAS RN: 12190-79-3

Properties: dark gray powd, a = 0.2817 nm, c = 1.4059 nm; can be prepared by reacting Li_2CO_3 or $LiOH \cdot H_2O$ with $CoCO_3$ at 850-900°C for 24 h in air; used in ceramics and as an insertion electrode in Li battery systems; has fluxing properties of lithium oxide, and enhances adhesion similar to cobalt oxide [HAW93] [STR93] [DAH90] [GUM92]

Solubility: i H_2O [HAW93]

1578
Compound: Lithium cyanide
Formula: LiCN
Molecular Formula: CLiN
Molecular Weight: 32.959
CAS RN: 2408-36-8
Properties: colorless to light yellow liq; sensitive to moisture; freezing point 58°C [STR93]
Density, g/cm³: 1.075 (fused) [KIR78]
Melting Point, °C: 160 [KIR78]
Reactions: decomposes to cyanamide and carbon below ~600°C [KIR78]

1579
Compound: Lithium cyclopentadienide
Formula: LiC_5H_5
Molecular Formula: C_5H_5Li
Molecular Weight: 72.036
CAS RN: 16733-97-4
Properties: off-white powd; air and moisture sensitive [STR93] [ALF95]

1580
Compound: Lithium deuteride
Formula: LiD
Molecular Formula: DLi
Molecular Weight: 8.956
CAS RN: 13587-16-1
Properties: off-white powd; sensitive to air and moisture; thermally stable up to its melting point; used in thermonuclear fusion [HAW93] [STR93]
Density, g/cm³: 0.820 [STR93]
Melting Point, °C: ~680 [STR93]

1581
Compound: Lithium dichromate dihydrate
Formula: $Li_2Cr_2O_7 \cdot 2H_2O$
Molecular Formula: $Cr_2H_4Li_2O_9$
Molecular Weight: 265.901
CAS RN: 10022-48-7

Properties: reddish orange cryst powd; deliq; eutectic in aq solutions is at -70°C; used to dehumidify and in refrigeration [HAW93] [KIR78]
Solubility: g/100g soln, H_2O: 62.36 (0.8°C), 65.25 (30°C), 73.55 (100°C); Solid phase $Li2Cr_2O_7 \cdot 2H_2O$ [KRU93]
Density, g/cm³: 2.34 (30°C) [KIR78]
Melting Point, °C: 130 [HAW93]
Reactions: minus $2H_2O$ at 110°C [CRC94]

1582
Compound: Lithium dihydrogen phosphate
Formula: LiH_2PO_4
Molecular Formula: H_2LiO_4P
Molecular Weight: 103.928
CAS RN: 13453-80-0
Properties: white powd; used as a constituent in low-expansion porcelain enamel glazes, and in some laser glasses [STR93] [FMC93]
Solubility: g/100g soln, H_2O: 55.8 (0°C); Solid phase, LiH_2PO_4 [KRU93]
Density, g/cm³: 2.461 [STR93]
Melting Point, °C: >100 [AES93]

1583
Compound: Lithium diisopropylamide
Synonyms: LDA
Formula: $[(CH_3)_2CH]_2NLi$
Molecular Formula: $C_6H_{14}LiN$
Molecular Weight: 107.125
CAS RN: 4111-54-0
Properties: pyrophoric powd; hindered, non-nucleophilic strong base; sensitive to air and moisture; uses: generation of carbanions [ALD94] [MER89]
Melting Point, °C: decomposes [MER89]

1584
Compound: Lithium fluoride
Formula: LiF
Molecular Formula: FLi
Molecular Weight: 25.939
CAS RN: 7789-24-4

Properties: cub cryst, or white fluffy powd; does not form a hydrate; enthalpy of fusion 27.09 kJ/mol; enthalpy of vaporization 147 kJ/mol; index of refraction 1.3915; manufactured by reacting lithium carbonate or lithium hydroxide with dil HF; used as a welding and soldering flux, used in ceramics to reduce firing temperatures and to improve thermal shock resistance; as a 99.9% pure sputtering target for low index, anti-reflection film [HAW93] [MER89] [KIR78] [CER91] [CRC93]

Solubility: 0.133 g/100 soln in H_2O (25°C) [KIR81]; g/100g soln, H_2O: 0.120 (0°C), 0.134±0.008 (25°C) [KRU93]; s acids; i alcohol [HAW93]

Density, g/cm³: 2.640 [MER89]

Melting Point, °C: 848 [KIR81]

Boiling Point, °C: 1673 [CRC93]

Reactions: volatilizes at 1100-1200°C [MER89]

Thermal Expansion Coefficient: (volume) 100°C (0.912), 200°C (2.086), 400°C (4.759) [CLA66]

1585

Compound: Lithium hexafluoroantimonate

Formula: $LiSbF_6$

Molecular Formula: F_6LiSb

Molecular Weight: 242.691

CAS RN: 18424-17-4

Properties: powd; hygr [STR93]

Melting Point, °C: decomposes [ALF95]

1586

Compound: Lithium hexafluoroarsenate

Synonyms: lithium hexafluorarsenate(V)

Formula: $LiAsF_6$

Molecular Formula: AsF_6Li

Molecular Weight: 195.853

CAS RN: 29935-35-1

Properties: white hygr powd; has been used as an electrolyte for organic solvent lithium batteries [KIR78]

1587

Compound: Lithium hexafluorophosphate

Formula: $LiPF_6$

Molecular Formula: F_6LiP

Molecular Weight: 151.905

CAS RN: 21324-40-3

Properties: white to off-white powd; hygr [STR93]

1588

Compound: Lithium hexafluorosilicate

Formula: Li_2SiF_6

Molecular Formula: F_6Li_2Si

Molecular Weight: 155.958

CAS RN: 17347-95-4

Properties: white powd [STR93]

1589

Compound: Lithium hexafluorostannate(IV)

Formula: Li_2SnF_6

Molecular Formula: F_6Li_2Sn

Molecular Weight: 246.582

CAS RN: 17029-16-2

Properties: white powd [STR93]

1590

Compound: Lithium hydride

Formula: LiH

Molecular Formula: HLi

Molecular Weight: 7.949

CAS RN: 7580-67-8

Properties: gray powd; sensitive to moisture; cub; darkens rapidly on exposure to light; very stable thermally, melts without decomposition; enthalpy of fusion 22.59 kJ/mol; can be prepared by adding H_2O to molten lithium at 680-900°C under ~1 atm H_2 pressure; used as a source of hydrogen in military applications and buoyancy devices, and in organic synthesis [KIR80] [CRC93] [KIR81] [MER89]

Solubility: reacts vigorously with H_2O [KIR81]; s ether, i benzene, toluene [HAW93]

Density, g/cm³: 0.78 [KIR81]

Melting Point, °C: 688.7 [CRC93]

Reactions: forms LiOH and H_2 in H_2O; reacts with lower alcohols, carboxylic acids, chlorine and ammonia with evolution of H_2 [MER89]

1591

Compound: Lithium hydrogen carbonate

Synonyms: lithia water

Formula: $LiHCO_3$

Molecular Formula: $CHLiO_3$

Molecular Weight: 67.958

CAS RN: 10377-37-4

Properties: white; prepared by dissolving lithium carbonate in water which contains excess dissolved carbon dioxide; used in medicine and in prepared mineral waters [CRC94] [HAW93]

Solubility: 5.5 g/100mL H_2O (13°C) [CRC94]

1592
Compound: Lithium hydroxide
Formula: LiOH
Molecular Formula: HLiO
Molecular Weight: 23.948
CAS RN: 1310-65-2
Properties: colorless; granular, free flowing powd; tetr; acrid; readily absorbs CO_2 and H_2O from atm; enthlapy of vaporization 188 kJ/mol; enthalpy of fusion 20.88 kJ/mol; can be prepared from Li_2CO_3 and $Ca(OH)_2$; used to manufacture lithium stearate, in storage battery electrolytes, and to absorb CO_2 in space vehicles [HAW93] [CRC93] [MER89] [KIR81]
Solubility: g/100g H_2O: 10.7 (0°C), 11.3 (40°C), 14.8 (100°C); sl s alcohol [KIR81] [MER89]; g/100g soln, H_2O: 12.7 (0°C), 12.9 (25°C), 17.5 (100°C); Solid phase, LiOH·H_2O [KRU93]
Density, g/cm³: 1.45 [LID94]
Melting Point, °C: 471.2 [CRC93]
Boiling Point, °C: 1626 [CRC93]

1593
Compound: Lithium hydroxide monohydrate
Formula: LiOH·H_2O
Molecular Formula; H_3LiO_2
Molecular Weight: 41.964
CAS RN: 1310-66-3
Properties: white powd; hygr; monocl; solid phase in equilibrium with dissolved LiOH from 0 to 100°C; used in manufacturing lithium-based greases, as an additive in alkaline battery electrolyte, dye solubilizer for textiles, and as a heat sink in nuclear reactors [KIR81] [STR93] [KIR81] [FMC93]
Solubility: w/w solubility, H_2O: 10.7% (0°C), 10.9% (20°C), 14.8% (100°C); sl s in alcohol [MER89]
Density, g/cm³: 1.51 [MER89]
Melting Point, °C: 680 [AES93]
Reactions: minus H_2O >100°C [KIR81]

1594
Compound: Lithium hypochlorite
Formula: LiOCl
Molecular Formula: ClLiO
Molecular Weight: 58.393
CAS RN: 13840-33-0
Properties: white granules; oxidant; can ignite organic materials; can be prepared by action of chlorine on a solution of LiOH; used as a bleach and an oxidizing agent, sanitizer for swimming pools, cooling water treatment [HAW93] [KIR81] [FMC93]

1595
Compound: Lithium iodate
Formula: $LiIO_3$
Molecular Formula: $ILiO_3$
Molecular Weight: 181.843
CAS RN: 13765-03-2
Properties: -80 mesh with 99.9% purity; has two forms, α and β; white powd; oxidizing agent [HAW93] [CER91]
Solubility: s H_2O; i alcohol [HAW93]
Density, g/cm³: 4.487 [HAW93]
Melting Point, °C: 50-60 [HAW93]
Reactions: transition from α to β at 50-60° [HAW93]

1596
Compound: Lithium iodide
Formula: LiI
Molecular Formula: ILi
Molecular Weight: 133.845
CAS RN: 10377-51-2
Properties: white powd; hygr; enthalpy of fusion 14.60 kJ/mol; formed by neutralizing HI solutions with LiOH or Li_2CO_3 to obtain the trihydrate followed by careful dehydration in vacuum [KIR81] [STR93] [CRC93]
Solubility: g/100g soln, H_2O: 62.6 (25°C), 71.1 (75°C) [KIR81]; g/100 g H_2O: 149 (0°C), 163±3 (25°C), 476 (99°C); Solid phase, LiI·$3H_2O$ (0°C, 25°C), LiI·H_2O (99°C) [KRU93]
Density, g/cm³: 3.494 [STR93]
Melting Point, °C: 469 [CRC93]
Boiling Point, °C: 1142 [KIR81]; 1180 [STR93]
Reactions: minus iodine when heated in air [KIR81]

1597
Compound: Lithium iodide trihydrate
Formula: LiI·$3H_2O$
Molecular Formula: H_6ILiO_3
Molecular Weight: 187.891
CAS RN: 7790-22-9
Properties: white cryst; extremely hygr; granules of fused masses; becomes yellow due to liberation of I_2 when exposed to atm; used in air conditioning, as a catalyst in acetal formation; there are two other hydrates: LiI·$2H_2O$, 17023-25-5, and LiI·H_2O, 17023-24-4 [KIR81] [HAW93] [MER89]
Solubility: s in about 0.5 parts H_2O or alcohol; v s in amyl alcohol or acetone [MER89]
Density, g/cm³: 3.48 [HAW93]
Melting Point, °C: 73 [MER89]
Boiling Point, °C: 1171 [HAW93]

Reactions: minus $3H_2O$ at 450°C [HAW93]

1598
Compound: Lithium manganate
Formula: $LiMn_2O_3$
Molecular Formula: $LiMn_2O_3$
Molecular Weight: 164.815
CAS RN: 12057-17-9
Properties: monocl, a = 0.4921 nm, b = 0.8526 nm, c = 0.9606 nm; prepared by reacting LiOH and γ-MnO_2 in air at 400°C for several days, or at 700°C for 24 h; used in battery research [ROS91] [RIO92]
Density, g/cm³: 3.90 [ROS91] [RIO92]

1599
Compound: Lithium manganite
Formula: Li_2MnO_3
Molecular Formula: Li_2MnO_3
Molecular Weight: 116.818
CAS RN: 12163-00-7
Properties: reddish brown powd; very highly stable; used as a smelter addition in the manufacture of frit, and as a cathode material for lithium batteries [HAW93] [FMC93]
Solubility: i H_2O [HAW93]

1600
Compound: Lithium metaaluminate
Formula: $LiAlO_2$
Molecular Formula: $AlLiO_2$
Molecular Weight: 65.922
CAS RN: 12003-67-7
Properties: white powd [STR93]
Density, g/cm³: 2.55 [STR93]
Melting Point, °C: >1625 [STR93]

1601
Compound: Lithium metaborate dihydrate
Formula: $LiBO_2 \cdot 2H_2O$
Molecular Formula: BH_4LiO_4
Molecular Weight: 85.782
CAS RN: 15293-74-0
Properties: white cryst powd; used in ceramics as a flux, in welding and brazing [HAW93]
Solubility: s H_2O [HAW93]
Density, g/cm³: 1.8 [STR93]

1602
Compound: Lithium metaphosphate

Formula: $LiPO_3$
Molecular Formula: LiO_3P
Molecular Weight: 85.913
CAS RN: 13762-75-9
Properties: white cryst, or glassy transparent particles; used as a consituent in low-expansion procelain enamel, and in selected laser glasses [FMC93] [AES93]
Solubility: g/100g H_2O: 0.101 (0°C), 0.058 (25°C), 0.048 (40°C) [LAN85]
Melting Point, °C: 656 [AES93]

1603
Compound: Lithium metasilicate
Formula: Li_2SiO_3
Molecular Formula: Li_2O_3Si
Molecular Weight: 89.966
CAS RN: 10102-24-6
Properties: white powd; ortho-rhomb needles; enthalpy of fusion 28.00 kJ/mol; obtained by fusing lithium carbonate and SiO_2; used as a flux in glazes and ceramic enamels [HAW93] [MER89] [FMC93]
Solubility: i cold H_2O, decomposes in boiling H_2O [MER89]
Density, g/cm³: 2.52 [MER89]
Melting Point, °C: 1201 [MER89]

1604
Compound: Lithium molybdate
Synonyms: lithium molybdate(VI)
Formula: Li_2MoO_4
Molecular Formula: Li_2MoO_4
Molecular Weight: 173.820
CAS RN: 13568-40-6
Properties: white cryst; phenacite structure, c/a = 1.153; used in steel coating and in petroleum cracking catalysts [HAW93] [KIR81]
Solubility: g/100 g soln, H_2O: 45.24 (0°C), 44.81 (25°C), 42.50 (98°C); Solid phase, $4Li_2MoO_4 \cdot 3H_2O$ [KRU93]
Density, g/cm³: 2.66 [STR93]
Melting Point, °C: 705 [STR93]

1605
Compound: Lithium niobate
Synonyms: lithium niobate(V)
Formula: $LiNbO_3$
Molecular Formula: $LiNbO_3$
Molecular Weight: 147.845
CAS RN: 12031-63-9

Properties: white powd, also single cryst; can be prepared by hydrolysis of equimolar amounts of lithium ethoxide and niobium ethoxide in absolute alcohol by refluxing at 78.5°C for 24 h, then crystallizing the resulting precipitate by heating 2 h in a stream of oxygen gas at 250-350°C; ferroelectric; used in infrared detectors, in transducers for lasers, and as a sputtering target of 99.9% purity for piezoelectric applications [HAW93] [STR93] [HIR87] [CER91]
Melting Point, °C: 1240 [LID94]

1606
Compound: Lithium nitrate
Formula: $LiNO_3$
Molecular Formula: $LiNO_3$
Molecular Weight: 68.946
CAS RN: 7790-69-4
Properties: white cryst powd; very hygr; enthalpy of fusion 24.90 kJ/mol; can be prepared by reaction of HNO_3 with LiOH or Li_2CO_3, followed by evaporation to dryness and then heating at ~200°C in vacuum; used in ceramics, pyrotechnics, molten salt baths, rocket propellants, refrigerators [HAW93] [CRC93] [MER89] [KIR81] [FMC93]
Solubility: 43 g/100g soln, H_2O (20°C); s alcohol [KIR81] [MER89]; g/100g soln, H_2O: 34.6 (0°C), 45.8 (25°C), 68.0 (90°C); Solid phase, $LiNO_3·3H_2O$ (0°C, 25°C), $LiNO_3$ (90°C) [KRU93]
Density, g/cm³: 2.38 [MER89]
Melting Point, °C: 253 [CRC93]

1607
Compound: Lithium nitride
Formula: Li_3N
Molecular Formula: Li_3N
Molecular Weight: 34.830
CAS RN: 26134-62-3
Properties: reddish brown cryst or freely flowing powd; slowly decomposed by atm moisture; ruby red; hex; a = 0.3658 nm, c = 0.3882 nm; conductivity, 227°C, 0.04 (ohm cm)$^{-1}$; one of most effective solid ionic conductors; can be prepared by direct reaction of Li and nitrogen; used as a nitriding agent in metallurgy [HAW93] [STR93] [CIC73] [KIR81]
Solubility: reacts with H_2O, yielding LiOH and ammonia; i polyethers [HAW93]
Density, g/cm³: 1.27 [LID94]
Melting Point, °C: 813 [KIR81]; 845 [HAW93]

1608
Compound: Lithium nitrite monohydrate
Formula: $LiNO_2·H_2O$
Molecular Formula: H_2LiNO_3
Molecular Weight: 70.962
CAS RN: 13568-33-7
Properties: colorless needles [CRC77]
Solubility: g/100g soln, H_2O: 41.5 (0°C), 50.9 (25°C), 76.4 (99°C); Solid phase, $LiNO_2·H_2O$ (0°C, 25°C), $LiNO_2$ (99°C) [KRU93]
Density, g/cm³: 1.615 [CRC77]
Melting Point, °C: >100 [CRC77]
Boiling Point, °C: decomposes [CRC77]

1609
Compound: Lithium orthosilicate
Formula: Li_4SiO_4
Molecular Formula: Li_4O_4Si
Molecular Weight: 119.848
CAS RN: 13453-84-4
Properties: rhomb; white powd; -100 mesh with 99.% purity; used as a flux in ceramic formulations [FMC93] [CER91] [CRC94]
Density, g/cm³: 2.39 [CRC94]
Melting Point, °C: 1256 [CRC94]

1610
Compound: Lithium oxalate
Formula: $Li_2C_2O_4$
Molecular Formula: $C_2Li_2O_4$
Molecular Weight: 101.902
CAS RN: 30903-87-8
Properties: white cryst powd; used as an anticoagulant in blood analysis [FMC93] [STR93]
Solubility: s in 15 part H_2O [MER89]; g/100g soln, H_2O: 5.87 (25°C) [KRU93]
Density, g/cm³: 2.12 [MER89]
Melting Point, °C: decomposes [STR93]

1611
Compound: Lithium oxide
Synonyms: lithia
Formula: Li_2O
Molecular Formula: Li_2O
Molecular Weight: 29.881
CAS RN: 12057-24-8

Properties: finely divided white powd or crusty material; readily absorbs CO_2 and H_2O from the atm; made by heating LiOH to ~800°C in a vacuum, or by thermal decomposition of lithium peroxide; used in ceramics and special glass formulations, and in lithium thermal batteries [HAW93] [MER89] [KIR81] [FMC93]
Density, g/cm³: 2.013 [MER89]
Melting Point, °C: 1570 [MER89]
Reactions: attacks glass, silica, many metals at elevated temperatures [MER89]

1612
Compound: Lithium perchlorate
Formula: $LiClO_4$
Molecular Formula: $ClLiO_4$
Molecular Weight: 106.392
CAS RN: 7791-03-9
Properties: white powd or ortho-rhomb cryst; hygr; oxidizing agent; enthalpy of fusion 29.00 kJ/mol; prepared from a saturated solution which forms the trihydrate, followed by drying at 300°C; used in solid rocket propellants, as an electrolyte constituent for lithium batteries, and as a catalyst and oxidizing agent [HAW93] [STR93] [KIR79] [FMC93]
Solubility: v s alcohol, acetone, ether, ethyl acetate [MER89]; g/100g soln, H_2O: 29.90 (0°C), 37.48 (25°C), 71.4 (100°C); Solid phase, $LiClO_4·3H_2O$ (0°C, 25°C), $LiClO_4·H_2O$ (100°C) [KRU93]
Density, g/cm³: 2.428 [STR93]
Melting Point, °C: 236 [CRC93]
Boiling Point, °C: 430 [STR93]
Reactions: decomposes rapidly at 450°C to LiCl and O_2 [KIR81]

1613
Compound: Lithium perchlorate trihydrate
Formula: $LiClO_4·3H_2O$
Molecular Formula: ClH_6LiO_7
Molecular Weight: 160.438
CAS RN: 13453-78-6
Properties: white powd; hygr; can be prepared by crystallization from a saturated solution of lithium perchlorate [KIR81] [STR93]
Solubility: 37.5g/100g saturated solution in water (25°C) [KIR81]
Density, g/cm³: 1.841 [STR93]
Melting Point, °C: 95 [KIR79]
Boiling Point, °C: 470 decomposes [KIR79]
Reactions: minus $3H_2O$ at 300°C [KIR81]

1614
Compound: Lithium peroxide
Formula: Li_2O_2
Molecular Formula: Li_2O_2
Molecular Weight: 45.881
CAS RN: 12031-80-0
Properties: light yellow to tan powd; can be made from H_2O_2 and LiOH solution in boiling ethanol; used in fuel cells and as an oxidizing agent [HAW93] [FMC93]
Solubility: solubility in H_2O is 8% (20°C); solubility in acetic acid is 5.6% (20°C); i absolute alcohol (20°C) [HAW93]
Density, g/cm³: 2.14 (20°C) [HAW93]

1615
Compound: Lithium phosphate
Synonyms: lithium phosphate, tribasic
Formula: Li_3PO_4
Molecular Formula: Li_3O_4P
Molecular Weight: 115.794
CAS RN: 10377-52-3
Properties: white powd [STR93]
Solubility: g/100g soln, H_2O: 0.022 (0°C); 0.0297 g/100mL soln, H_2O (25°C); Solid phase, Li_3PO_4 [KRU93]
Density, g/cm³: 2.46 [LID94]
Melting Point, °C: 1205 [STR93]

1616
Compound: Lithium selenate monohydrate
Formula: $Li_2SeO_4·H_2O$
Molecular Formula: $H_2Li_2O_5Se$
Molecular Weight: 174.855
CAS RN: 15593-52-9
Properties: monocl [MER89]
Solubility: v s H_2O [MER89]
Density, g/cm³: 2.565 [MER89]

1617
Compound: Lithium selenite monohydrate
Formula: $Li_2SeO_3·H_2O$
Molecular Formula: $H_2Li_2O_4Se$
Molecular Weight: 158.856
CAS RN: 15593-51-8
Properties: cryst; hygr [MER89]
Solubility: g/100g H_2O: 25.0 (0°C), 21.5 (20°C), 9.9 (100°C) [LAN85]

1618
Compound: Lithium stearate

Synonyms: stearic acid, lithium salt
Formula: $CH_3(CH_2)_{16}COOLi$
Molecular Formula: $C_{18}H_{35}LiO_2$
Molecular Weight: 290.416
CAS RN: 4485-12-5
Properties: white cryst; forms gels in mineral oils; used in cosmetics, plastics, waxes, greases [HAW93]
Solubility: i H_2O, alcohol, ethyl acetate [HAW93]
Density, g/cm³: 1.025 [HAW93]
Melting Point, °C: 220 [HAW93]

1619
Compound: Lithium sulfate
Formula: Li_2SO_4
Molecular Formula: Li_2O_4S
Molecular Weight: 109.946
CAS RN: 10377-48-7
Properties: white hygr cryst; enthalpy of fusion 7.50 kJ/mol; can be obtained from sulfuric acid and LiOH or Li_2CO_3 solutions to form the monohydrate, followed by heating to obtain the anhydrous salt; used in ceramic compositions, as a solubilizer in photographic developers, and as an additive to specialty Portland cements [KIR81] [FMC93]
Solubility: g/100g soln, H_2O: 25.7 (25°C), 23.6 (100°C) [KIR81]; g/100 g soln, H_2O: 25.9±0.5 (0°C), 25.7±0.1 (25°C), 23.5 (100.1°C); Solid phase, $Li_2SO_4 \cdot H_2O$ [KRU93]
Melting Point, °C: 860 [KIR81]

1620
Compound: Lithium sulfate monohydrate
Formula: $Li_2SO_4 \cdot H_2O$
Molecular Formula: $H_2Li_2O_5S$
Molecular Weight: 127.961
CAS RN: 10102-25-7
Properties: colorless cryst; does not form alums; obtained by reacting solution of H_2SO_4 with LiOH or Li_2CO_3; used in ceramics and in pharmaceutical products [HAW93] [KIR81] [STR93]
Solubility: soluble in 2.6 parts H_2O; sl s alcohol [MER89]
Density, g/cm³: 2.06 [MER89]
Melting Point, °C: 130 [HAW93]
Reactions: minus H_2O >100°C [KIR81]

1621
Compound: Lithium sulfide
Formula: Li_2S

Molecular Formula: Li_2S
Molecular Weight: 45.948
CAS RN: 12136-58-2
Properties: off-white powd; sensitive to moisture [STR93]
Density, g/cm³: 1.64 [LID94]
Melting Point, °C: 1372 [LID94]

1622
Compound: Lithium tantalate
Formula: $LiTaO_3$
Molecular Formula: LiO_3Ta
Molecular Weight: 235.887
CAS RN: 12031-66-2
Properties: white powd [STR93]

1623
Compound: Lithium tellurite
Formula: Li_2TeO_3
Molecular Formula: Li_2O_3Te
Molecular Weight: 189.480
CAS RN: 14929-69-2
Properties: -100 mesh with 99.5% purity [CER91]

1624
Compound: Lithium tetraborate
Formula: $Li_2B_4O_7$
Molecular Formula: $B_4Li_2O_7$
Molecular Weight: 169.122
CAS RN: 12007-60-2
Properties: white powd; -100 mesh with 99.9% purity; used as a flux in x-ray fluorescence analysis, in grease formulations, and as an electrolyte component in lithium batteries [FMC93] [CER91] [STR93]
Solubility: 2.89 g/100mL H_2O (20°C), 5.45 g/100mL H_2O (100°C) [CRC94]
Melting Point, °C: 917 [STR93]

1625
Compound: Lithium tetraborate pentahydrate
Formula: $Li_2B_4O_7 \cdot 5H_2O$
Molecular Formula: $B_4H_{10}Li_2O_{12}$
Molecular Weight: 211.200
CAS RN: 1303-94-2
Properties: white cryst powd; used in ceramics, in vacuum spectroscopy, in metal refining and degassing [HAW93]
Solubility: v s H_2O; i alcohol [HAW93]
Reactions: minus $5H_2O$ at 200°C [HAW93]

1626
Compound: Lithium tetrachlorocuprate
Formula: Li_2CuCl_4
Molecular Formula: Cl_4CuLi_2
Molecular Weight: 219.239
CAS RN: 15489-27-7
Properties: 0.1M in THF; freezing point, -17°C; orange liq; sensitive to moisture [STR93]

1627
Compound: Lithium tetracyanoplatinate(II) pentahydrate
Formula: $Li_2Pt(CN)_4 \cdot 5H_2O$
Molecular Formula: $C_4H_{10}Li_2N_4O_5Pt$
Molecular Weight: 403.109
CAS RN: 14402-73-4
Properties: greenish yellow cryst [MER89]
Solubility: sl s H_2O [MER89]

1628
Compound: Lithium tetrafluoroborate
Formula: $LiBF_4$
Molecular Formula: BF_4Li
Molecular Weight: 93.746
CAS RN: 14283-07-9
Properties: white hygr powd; hygr; can be prepared by reacting LiOH with fluoroboric acid [KIR78] [STR93]
Solubility: v s H_2O [KIR78]
Melting Point, °C: decomposes [STR93]

1629
Compound: Lithium thiocyanate hydrate
Formula: $LiSCN \cdot xH_2O$
Molecular Formula: CLiNS (anhydrous)
Molecular Weight: 65.025 (anhydrous)
CAS RN: 123333-85-7
Properties: white, deliq cryst [CRC77]
Solubility: g/100g soln, H_2O: 54.5 (25°C); Solid phase, $LiSCN \cdot 2H_2O$ [KRU93]

1630
Compound: Lithium titanate
Formula: Li_2TiO_3
Molecular Formula: Li_2O_3Ti
Molecular Weight: 109.747
CAS RN: 12031-82-2
Properties: white powd; has strong fluxing properties at low concentrations for use in titanium-bearing enamels [HAW93]
Solubility: i H_2O [HAW93]

Melting Point, °C: 1520-1564 [STR93]

1631
Compound: Lithium tungstate
Formula: Li_2WO_4
Molecular Formula: Li_2O_4W
Molecular Weight: 261.720
CAS RN: 13568-45-1
Properties: trig; white powd [CRC94] [STR93]
Solubility: s H_2O [HAW93]
Density, g/cm³: 3.71 [STR93]
Melting Point, °C: 742 [STR93]

1632
Compound: Lithium vanadate
Formula: $LiVO_3$
Molecular Formula: LiO_3V
Molecular Weight: 105.881
CAS RN: 15060-59-0
Properties: -100 mesh with 99.9% purity; hydrate is yellowish powd [HAW93] [CER91]

1633
Compound: Lithium zirconate
Formula: Li_2ZrO_3
Molecular Formula: Li_2O_3Zr
Molecular Weight: 153.104
CAS RN: 12031-83-3
Properties: white powd; used as a flux in glasses which contain zirconium dioxide [HAW93] [FMC93]

1634
Compound: Lutetium
Synonyms: cassiopeium
Formula: Lu
Molecular Formula: Lu
Molecular Weight: 174.967
CAS RN: 7439-94-3
Properties: silvery white metal; hex; soft and ductile; electrical resistivity (20°C) 54 μohm·cm; enthalpy of fusion 78.03 kJ/mol; enthalpy of sublimation 427.6 kJ/mol; atom radius is 0.17349 nm; ion radius is 0.0850 nm for Lu^{+++}, colorless solutions [HAW93] [KIR82] [ALD94]
Solubility: reacts slowly with H_2O; s in dil acids [HAW93]
Density, g/cm³: 9.840 [KIR82]
Melting Point, °C: 1663 [KIR82]
Boiling Point, °C: 3402 [KIR82]

Thermal Conductivity, W/(m·K): 16.4 at 25°C
[CRC93]
Thermal Expansion Coefficient: 9.9 x 10⁻⁶/K
[CRC93]

1635
Compound: Lutetium acetate hydrate
Formula: $Lu(CH_3COO)_3 \cdot xH_2O$
Molecular Formula: $C_6H_9LuO_6$ (anhydrous)
Molecular Weight: 352.101 (anhydrous)
CAS RN: 18779-08-3
Properties: hygr white cryst [STR93] [ALD94]

1636
Compound: Lutetium boride
Formula: LuB_4
Molecular Formula: B_4Lu
Molecular Weight: 218.211
CAS RN: 12688-52-7
Properties: tetr; -100 mesh of 99.9% purity [LID94]
[CER91]
Density, g/cm³: ~7.0 [LID94]
Melting Point, °C: 2600 [LID94]

1637
Compound: Lutetium bromide
Formula: $LuBr_3$
Molecular Formula: Br_3Lu
Molecular Weight: 414.679
CAS RN: 14456-53-2
Properties: white, hygr cryst; -20 mesh with 99.9%
purity [CER91] [LID94]
Solubility: s H_2O [CRC94]
Density, g/cm³: 1.025 [ALF95]
Melting Point, °C: 1400 [CRC94]

1638
Compound: Lutetium chloride
Formula: $LuCl_3$
Molecular Formula: Cl_3Lu
Molecular Weight: 281.325
CAS RN: 10099-66-8
Properties: colorless cryst [MER89]
Solubility: s H_2O [MER89]
Density, g/cm³: 3.98 [STR93]
Melting Point, °C: 905 [HAW93]
Reactions: sublimes >750°C [MER89]

1639
Compound: Lutetium chloride hexahydrate

Formula: $LuCl_3 \cdot 6H_2O$
Molecular Formula: $Cl_3H_{12}LuO_6$
Molecular Weight: 389.416
CAS RN: 15230-79-2
Properties: -4 mesh with 99.9% purity; white cryst
[CER91] [STR93]
Melting Point, °C: 892 [ALF95]

1640
Compound: Lutetium fluoride
Formula: LuF_3
Molecular Formula: F_3Lu
Molecular Weight: 231.962
CAS RN: 13760-81-1
Properties: ortho; 99.9% pure melted pieces of 3-6
mm; used as an evaporation material for
possible application to multilayers [LID94]
[CER91]
Solubility: i H_2O [HAW93]
Density, g/cm³: 8.3 [LID94]
Melting Point, °C: 1182 [HAW93]
Boiling Point, °C: 2200 [HAW93]

1641
Compound: Lutetium hydride
Formula: LuH_{2-3}
Molecular Formula: H_2Lu; H_3Lu
Molecular Weight: H_2Lu: 176.983; H_3Lu: 177.991
CAS RN: 13598-44-2
Properties: -60 mesh with 99.9% purity; lumps,
under argon [ALF95] [CER91]

1642
Compound: Lutetium iodide
Formula: LuI_3
Molecular Formula: I_3Lu
Molecular Weight: 555.680
CAS RN: 13813-45-1
Properties: powd, under argon; -20 mesh with
99.9% purity [CER91] [ALF95]
Density, g/cm³: ~5.6 [LID94]
Melting Point, °C: 1050 [AES93]

1643
Compound: Lutetium iron oxide
Synonyms: lutetium garnet
Formula: $Lu_3Fe_5O_{12}$
Molecular Formula: $Fe_5Lu_3O_{12}$
Molecular Weight: 996.119
CAS RN: 12023-71-1

Properties: used in 99.9% purity as a sputtering target in the preparation of bubble memory devices [CER91]

1644
Compound: Lutetium nitrate hydrate
Formula: $Lu(NO_3)_3 \cdot xH_2O$
Molecular Formula: LuN_3O_9 (anhydrous)
Molecular Weight: 360.982 (anhydrous)
CAS RN: 10099-67-9
Properties: white cryst [STR93]

1645
Compound: Lutetium nitride
Formula: LuN
Molecular Formula: LuN
Molecular Weight: 188.974
CAS RN: 12125-25-6
Properties: -60 mesh with 99.9% purity [CER91]
Density, g/cm³: 11.6 [LID94]

1646
Compound: Lutetium oxalate hexahydrate
Formula: $Lu_2(C_2O_4)_3 \cdot 6H_2O$
Molecular Formula: $C_6H_{12}Lu_2O_{18}$
Molecular Weight: 722.084
CAS RN: 26677-69-0
Properties: white cryst [ALF95]

1647
Compound: Lutetium oxide
Formula: Lu_2O_3
Molecular Formula: Lu_2O_3
Molecular Weight: 397.932
CAS RN: 12032-20-1
Properties: white powd; cub cryst; absorbs H_2O and CO_2; as an evaporated material of 99.9% purity is reactive to radio frequencies [HAW93] [CER91] [STR93] [MER89]
Density, g/cm³: 9.41 [STR93]
Melting Point, °C: 2487 [STR93]

1648
Compound: Lutetium perchlorate hexahydrate
Formula: $Lu(ClO_4)_3 \cdot 6H_2O$
Molecular Formula: $Cl_3H_{12}LuO_{18}$
Molecular Weight: 581.409
CAS RN: 14646-29-8
Properties: white cryst; hygr [STR93]

1649
Compound: Lutetium silicide
Formula: $LuSi_2$
Molecular Formula: $LuSi_2$
Molecular Weight: 231.138
CAS RN: 12032-13-2
Properties: 6 mm pieces and smaller with 99.9% purity [CER91]

1650
Compound: Lutetium sulfate
Formula: $Lu_2(SO_4)_3$
Molecular Formula: $Lu_2O_{12}S_3$
Molecular Weight: 638.125
CAS RN: 14986-89-1
Properties: white powd [STR93]
Solubility: 0.6260±0.0017 mol/kg in H_2O (25°C) [RAR88]

1651
Compound: Lutetium sulfate octahydrate
Formula: $Lu_2(SO_4)_3 \cdot 8H_2O$
Molecular Formula: $H_{16}Lu_2O_{20}S_3$
Molecular Weight: 782.247
CAS RN: 13473-77-3
Properties: white cryst [STR93]
Solubility: g/100g H_2O: 42.27 (20°C), 16.93 (40°C) [MER89]

1652
Compound: Lutetium sulfide
Formula: Lu_2S_3
Molecular Formula: Lu_2S_3
Molecular Weight: 446.132
CAS RN: 12163-20-1
Properties: gray rhomb cryst; -200 mesh with 99.9% purity [CER91] [LID94]
Density, g/cm³: 6.26 [LID94]
Melting Point, °C: 1750, decomposes [LID94]

1653
Compound: Lutetium telluride
Formula: Lu_2Te_3
Molecular Formula: Lu_2Te_3
Molecular Weight: 732.734
CAS RN: 12163-22-3
Properties: ortho cryst; -20 mesh with 99.9% purity [LID94] [CER91]
Density, g/cm³: 7.8 [LID94]

1654
Compound: Magnesium
Formula: Mg
Molecular Formula: Mg
Molecular Weight: 24.3050
CAS RN: 7439-95-4
Properties: silver-white metal; hex, a = 0.3203 nm, c = 0.5199 nm; slowly oxidizes in moist air; electrical resistivity (20°C) 4.46 μohm· cm; enthalpy of fusion 8.48 kJ/mol; enthalpy of sublimation 150 kJ/mol; enthalpy of combustion 606 kJ/mol; thermal diffusivity (20°C) 0.87 cm^2/sec; Poisson's ratio 0.35; electronegativity 1.56; many uses including ferromagnetic films preparation by diffusion with bismuth [DOU83] [MER89] [KIR81] [CER91]
Solubility: reacts slowly with H$_2$O; evolves H$_2$ with dil acids [MER89]
Density, g/cm^3: 1.738 (20°C) [CIC73]
Melting Point, °C: 649 [CIC73]
Boiling Point, °C: 1105 [CIC73]
Thermal Conductivity, W/(m·K): 155 at 20°C [KIR81]
Thermal Expansion Coefficient: 24.8 x 10^{-6}/K [CRC93]

1655
Compound: Magnesium acetate
Synonyms: cromosan
Formula: Mg(CH$_3$COO)$_2$
Molecular Formula: C$_4$H$_6$MgO$_4$
Molecular Weight: 142.395
CAS RN: 142-72-3
Properties: white; two forms: α ortho-rhomb, a = 1.127 nm, b = 1.501 nm, c = 1.100 nm, obtained by reacting MgO with 13-33% acetic acid in boiling ethyl acetate, and β tricl, a = 1.034 nm, b = 1.29 nm, c = 7.726 nm, obtained from 5-6% acetic acid; odor of acetic acid; used as a dye fixative in textile printing, as a deodorant and disinfectant [KIR81] [HAW93]
Solubility: g/100g soln, H$_2$O: 36.2 (0.1°C), 39.6 (24.9°C); Solid phase, Mg(CH$_3$COO)$_2$·4H$_2$O [KRU93]; s dil alcohol [HAW93]
Density, g/cm^3: α: 1.507, β: 1.502 [KIR81]
Melting Point, °C: 323 decomposes [KIR81]

1656
Compound: Magnesium acetate monohydrate
Formula: Mg(CH$_3$COO)$_2$·H$_2$O
Molecular Formula: C$_4$H$_8$MgO$_5$
Molecular Weight: 160.410
CAS RN: 60582-92-5

Properties: ortho-rhomb, a = 1.175 nm, b = 1.753 nm, c = 0.6662 nm; can be obtained by reacting MgO and acetic acid in isobutyl alcohol which contains some H$_2$O [KIR81]
Density, g/cm^3: 1.553 [KIR81]

1657
Compound: Magnesium acetate tetrahydrate
Synonyms: acetic acid, magnesium salt
Formula: Mg(CH$_3$COO)$_2$·4H$_2$O
Molecular Formula: C$_4$H$_{14}$MgO$_8$
Molecular Weight: 214.455
CAS RN: 16674-78-5
Properties: colorless or white; monocl, a = 0.8550 nm, b = 1.1995 nm, c = 0.4807 nm; deliq cryst; crystallizes from aq solution as only stable phase below 68°C; there is a β phase [KIR81] [MER89]
Solubility: v s H$_2$O, alcohol [MER89]
Density, g/cm^3: 1.45 [LID94]
Melting Point, °C: 80, decomposes [LID94]

1658
Compound: Magnesium acetylacetonate dihydrate
Synonyms: 2,4-pentanedione, magnesium derivative dihydrate
Formula: Mg(CH$_3$COCH=C(O)CH$_3$)$_2$·2H$_2$O
Molecular Formula: C$_{10}$H$_{18}$MgO$_6$
Molecular Weight: 258.554
CAS RN: 68488-07-3
Properties: white powd [STR93]
Melting Point, °C: 265, decomposes [ALD94]

1659
Compound: Magnesium aluminum oxide
Synonyms: spinel
Formula: Mg(AlO$_2$)$_2$
Molecular Formula: Al$_2$MgO$_4$
Molecular Weight: 142.266
CAS RN: 12068-51-8
Properties: 3-12 mm pieces (fused); used as an evaporated ceramic of 99.9% purity to form a high temp dielectric [CER91] [MIT72] [YAM87]
Density, g/cm^3: 3.58 [KIR80]
Melting Point, °C: 2135 [KIR80]
Thermal Conductivity, W/(m·K): 9.1 (500°C), 5.8 (1000°C) [KIR80]
Thermal Expansion Coefficient: (volume) 100°C (0.18), 200°C (0.39), 400°C (0.90), 800°C (2.01), 1200°C (3.24) [CLA66]

1660
Compound: Magnesium aluminum silicate
Synonyms: cordierite
Formula: $Mg_2Al_3(AlSi_5O_{18})$
Molecular Formula: $Al_4Mg_2O_{18}Si_5$
Molecular Weight: 584.953
CAS RN: 61027-88-1
Properties: dielectric constant (26-300°C) 4.00-4.42, hydrothermally prepared material [MOY86]; sol-gel synthesis in [MAE90] and [KAZ90]
Density, g/cm³: when sintered at 1200-1400°C: 2.63-2.35 [KAZ90]
Thermal Expansion Coefficient: 50 to 650°C: 0.3 to 2.5 x 10^{-6}/°C [MOY86]

1661
Compound: Magnesium aluminum zirconate
Synonyms: zirconium spinel
Formula: $MgO \cdot Al_2O_3 \cdot ZrO_2$
Molecular Formula: Al_2MgO_6Zr
Molecular Weight: 265.488
CAS RN: 53169-11-2
Properties: 3-12 sintered pieces powd; used for vacuum deposition [ALF95] [CER91]

1662
Compound: Magnesium amide
Formula: $Mg(NH_2)_2$
Molecular Formula: H_4MgN_2
Molecular Weight: 56.350
CAS RN: 7803-54-5
Properties: white powd or cryst; flammable in air; used as a catalyst for polymerization [HAW93] [MER89]
Solubility: reacts violently with H_2O releasing NH_3 [MER89]
Density, g/cm³: 1.39 [MER89]
Melting Point, °C: decomposes when heated [HAW93]

1663
Compound: Magnesium ammonium phosphate hexahydrate
Synonyms: guanite
Formula: $MgNH_4PO_4 \cdot 6H_2O$
Molecular Formula: $H_{16}MgNO_{10}P$
Molecular Weight: 245.407
CAS RN: 13478-16-5
Properties: white powd; used as a fire retardant for fabrics, and in fertilizer [HAW93]
Solubility: i H_2O, alcohol; s in acids [HAW93]

Density, g/cm³: 1.71 [HAW93]
Melting Point, °C: decomposes to $Mg_2P_2O_7$ [HAW93]

1664
Compound: Magnesium antimonide
Formula: Mg_3Sb_2
Molecular Formula: Mg_3Sb_2
Molecular Weight: 316.435
CAS RN: 12057-75-9
Properties: hex; 6 mm pieces and smaller with 99.5% purity [CER91] [CRC94]
Density, g/cm³: 4.088 [CRC94]
Melting Point, °C: 1245 [LID94]

1665
Compound: Magnesium arsenate hydrate
Formula: $Mg_3(AsO_4)_2 \cdot xH_2O$
Molecular Formula: $As_2Mg_3O_8$ (anhydrous)
Molecular Weight: 350.753 (anhydrous)
CAS RN: 10103-50-1
Properties: white powd; used as an insecticide [HAW93]
Solubility: i H_2O [HAW93]

1666
Compound: Magnesium arsenide
Formula: Mg_3As_2
Molecular Formula: As_2Mg_3
Molecular Weight: 222.758
CAS RN: 12044-49-4
Properties: 6 mm pieces and smaller of 99.5% purity [CER91]
Density, g/cm³: 3.148 [ALF95]
Melting Point, °C: 800 [ALF95]

1667
Compound: Magnesium basic carbonate pentahydrate
Formula: $4MgCO_3 \cdot Mg(OH)_2 \cdot 5H_2O$
Molecular Formula: $C_4H_{12}Mg_5O_{19}$
Molecular Weight: 485.653
CAS RN: 56378-72-4
Properties: white, colorless, bulky powd [MER89]
Solubility: s in ~3300 parts H_2O; more soluble if H_2O contains dissolved CO_2; s dil acids; i alcohol [MER89]
Reactions: converts to MgO ~700°C [MER89]

1668
Compound: Magnesium bis(pentamethylcyclopentadienyl)
Synonyms: bis(pentamethylcyclopentadienyl)magnesium
Formula: $[(CH_3)_5C_5]_2Mg$
Molecular Formula: $C_{20}H_{30}Mg$
Molecular Weight: 294.763
CAS RN: 74507-64-5
Properties: cryst [ALF95]

1669
Compound: Magnesium borate octahydrate
Formula: $Mg(BO_2)_2 \cdot 8H_2O$
Molecular Formula: $B_2H_{16}MgO_{12}$
Molecular Weight: 254.047
CAS RN: 13703-82-7
Properties: white powd [MER89]
Solubility: sl s H_2O [MER89]
Density, g/cm³: 2.30 [CRC94]

1670
Compound: Magnesium boride
Formula: MgB_2
Molecular Formula: B_2Mg
Molecular Weight: 45.927
CAS RN: 12007-25-9
Properties: hex cryst; -100 mesh with 99% purity; refractory material [CER91] [KIR78] [LID94]
Density, g/cm³: 2.57 [LID94]
Melting Point, °C: 800, decomposes [KIR78]

1671
Compound: Magnesium boride
Formula: MgB_6
Molecular Formula: B_6Mg
Molecular Weight: 89.171
CAS RN: 12008-22-9
Properties: refractory material [KIR78]
Melting Point, °C: 1100, decomposes [KIR78]

1672
Compound: Magnesium bromate hexahydrate
Formula: $Mg(BrO_3)_2 \cdot 6H_2O$
Molecular Formula: $Br_2H_{12}MgO_{12}$
Molecular Weight: 388.201
CAS RN: 7789-36-8
Properties: colorless or white cryst; used as an oxidizing agent [HAW93] [MER89]
Solubility: 42 g/100mL H_2O (18°C) [CRC94]
Density, g/cm³: 2.29 [HAW93]

Reactions: minus $6H_2O$ ~200°C; decomposes at higher temp [MER89]

1673
Compound: Magnesium bromide
Formula: $MgBr_2$
Molecular Formula: Br_2Mg
Molecular Weight: 184.113
CAS RN: 7789-48-2
Properties: hex, a = 0.3822 nm, c = 0.6269 nm; off-white powd; hygr; enthalpy of fusion 39.30 kJ/mol; occurs in seawater, brines, the Dead Sea; used in medicine as a sedative and in some dry cell electrolytes for batteries [STR93] [KIR81] [CRC93]
Solubility: g/100g H_2O: 100.6 (25°C), 125.4 (100°C); Solid phase, $MgBr_2 \cdot 6H_2O$ [KRU93]
Density, g/cm³: 3.72 [STR93]
Melting Point, °C: 711 [CRC93]

1674
Compound: Magnesium bromide hexahydrate
Formula: $MgBr_2 \cdot 6H_2O$
Molecular Formula: $Br_2H_{12}MgO_6$
Molecular Weight: 292.204
CAS RN: 13446-53-2
Properties: colorless monocl, a = 1.0286 nm, b = 0.7331 nm, c = 0.6211 nm; very deliq cryst, or white granules; bitter taste; used as a sedative and in organic synthesis [HAW93] [KIR81] [MER89]
Solubility: s 0.3 parts H_2O; s alcohol [MER89]
Density, g/cm³: 2.00 [STR93]
Melting Point, °C: ~165 with decomposition [MER89]

1675
Compound: Magnesium carbonate
Synonyms: magnesite
Formula: $MgCO_3$
Molecular Formula: $CMgO_3$
Molecular Weight: 84.314
CAS RN: 546-93-0
Properties: light, bulky white powd; trig, a = 0.46332 nm, c = 1.5015 nm; magnesite mineral, 13717-00-5, hardness is 3.5-4.5 Mohs; can be prepared in aq systems under high CO_2 pressure; used in heat insulation and inks [HAW93] [KIR81]

Solubility: g $MgCO_3$/100g soln at CO_2 pressure, kPa, 18°C: 3.5 (203), 4.28 (405), 5.90 (1010), 7.49 (1820), 7.49 (5670); at 0°C 8.58 (3445), at 60°C 5.56 (3445); s acids; i alcohol [HAW93] [KIR81]
Density, g/cm³: 3.009 (calculated) [KIR81]
Melting Point, °C: 350, decomposes [HAW93]
Reactions: minus CO_2 at 900°C [CRC94]

1676
Compound: Magnesium carbonate dihydrate
Synonyms: barringtonite
Formula: $MgCO_3 \cdot 2H_2O$
Molecular Formula: CH_4MgO_5
Molecular Weight: 120.345
CAS RN: 5145-48-2
Properties: colorless; tricl, a = 0.9115 nm, b = 0.6202 nm, c = 0.6092 nm [KIR81]
Density, g/cm³: 2.825 (calculated) [KIR81]

1677
Compound: Magnesium carbonate hydroxide tetrahydrate
Synonyms: hydromagnesite
Formula: $4MgCO_3 \cdot Mg(OH)_2 \cdot 4H_2O$
Molecular Formula: $C_4H_{10}Mg_5O_{18}$
Molecular Weight: 467.637
CAS RN: 39409-82-0
Properties: white; monocl, a = 1.011 nm, b = 0.315 nm, c = 0.622 nm; there is a pentahydrate (dypingite) 56378-72-6, and an octahydrate, 75300-49-1 [KIR81]
Density, g/cm³: 2.254 [KIR81]

1678
Compound: Magnesium carbonate hydroxide trihydrate
Synonyms: artinite
Formula: $MgCO_3 \cdot Mg(OH)_2 \cdot 3H_2O$
Molecular Formula: $CH_8Mg_2O_8$
Molecular Weight: 198.680
CAS RN: 12143-96-3
Properties: white; monocl, a = 1.656 nm, b = 0.315 nm, c = 0.622 nm [KIR81]
Density, g/cm³: 2.039 [KIR81]

1679
Compound: Magnesium carbonate pentahydrate
Synonyms: lansfordite
Formula: $MgCO_3 \cdot 5H_2O$
Molecular Formula: $CH_{10}MgO_8$

Molecular Weight: 174.390
CAS RN: 61042-72-6
Properties: white monocl [KIR81]
Solubility: 0.176 g/100mL H_2O (7°C), 0.375 g/100mL H_2O (20°C) [CRC94]
Density, g/cm³: 1.73 (calculated) [KIR81]
Melting Point, °C: decomposes [CRC94]

1680
Compound: Magnesium carbonate trihydrate
Synonyms: nesquehonite
Formula: $MgCO_3.3H_2O$
Molecular Formula: CH_6MgO_6
Molecular Weight: 138.360
CAS RN: 14457-83-1
Properties: colorless to white; monocl, a = 1.2112 nm, b = 0.5365 nm, c = 0.7697 nm [KIR81]
Solubility: 0.179 g/100mL H_2O (16°C) [CRC94]
Density, g/cm³: 1.837 (calculated) [KIR81]
Reactions: minus $3H_2O$ at 100°C [CRC94]

1681
Compound: Magnesium chlorate hexahydrate
Formula: $Mg(ClO_3)_2 \cdot 6H_2O$
Molecular Formula: $Cl_2H_{12}MgO_{12}$
Molecular Weight: 299.298
CAS RN: 10326-21-3
Properties: white; very deliq cryst, or cryst powd; bitter taste; used as a defoliant and dessicant; oxidizing agent [HAW93] [MER89]
Solubility: mol/100 mol H_2O: 10.73 (0°C), 13.52 (25°C), 26.38 (93°C); Solid phase $Mg(ClO_3)_2 \cdot 6H_2O$ (0°C, 25°C), $Mg(ClO_3)_2 \cdot 2H_2O$ (93°C) [KRU93]
Density, g/cm³: 1.80 [MER89]
Melting Point, °C: ~35 [MER89]

1682
Compound: Magnesium chloride
Synonyms: magnogene
Formula: $MgCl_2$
Molecular Formula: Cl_2Mg
Molecular Weight: 95.210
CAS RN: 7786-30-3

Properties: white lustrous, soft highly deliq leaflets; hex, a = 0.3632 nm, c = 1.7795 nm; can be distilled in H_2; attacks fused silica when melted; evolves heat when dissolved in H_2O; can be obtained by dissolution of MgO, $MgCO_3$ or $Mg(OH)_2$ in HCl, followed by cooling and dehydration; enthalpy of fusion 43.10 kJ/mol; used in disinfectants, in fire extinguishers, for fireproofing wood, and in ceramics [HAW93] [CRC93] [MER89] [KIR81]

Solubility: g/100g soln, H_2O: 34.6 (0°C), 35.5 (25°C), 42.3 (100°C); Solid phase, $MgCl_2·6H_2O$ [KRU93]; g/100g alcohol: 3.61 (0°C), 15.89 (60°C) [KIR81]

Density, g/cm³: 2.325 [KIR81]

Melting Point, °C: 714 [CRC93]

Boiling Point, °C: 1412 [CIC73]

Reactions: slow heating releases Cl_2 at 300°C [MER89]

1683

Compound: Magnesium chloride hexahydrate

Synonyms: bischofite

Formula: $MgCl_2·6H_2O$

Molecular Formula: $Cl_2H_{12}MgO_6$

Molecular Weight: 203.301

CAS RN: 7791-18-6

Properties: colorless or white, highly deliq monocl cryst, a = 0.9871 nm, b = 0.7113 nm, c = 0.6097 nm; only stable hydrate from 0 to 100°C; obtained from a solution of MgO, $MgCO_3$ or $Mg(OH)_2$ in HCl [KIR81] [HAW93]

Solubility: 5.8101±0.0017 mol/(kg·H_2O) at 25°C [RAR85b]; 1 g/2mL alcohol [MER89]

Density, g/cm³: 1.56 [MER89]

Melting Point, °C: ~118, decomposes [MER89]

Reactions: minus $2H_2O$ at 95-115°C, minus $4H_2O$ at 135-180°C, minus $5H_2O$ >230°C and decomposes [KIR81]

1684

Compound: Magnesium chromate pentahydrate

Formula: $MgCrO_4·5H_2O$

Molecular Formula: $CrH_{10}MgO_9$

Molecular Weight: 230.375

CAS RN: 16569-85-0

Properties: small yellow cryst; tricl; used as a corrosion inhibitor in the water coolant of gas turbine engines [KIR78] [HAW93]

Solubility: 35.39% H_2O (25°C) [KIR78]

Density, g/cm³: 1.954 [KIR78]

Reactions: transforms to $7H_2O$ (17.2°C) [KIR78]

1685

Compound: Magnesium chromite

Formula: $MgCr_2O_4$

Molecular Formula: Cr_2MgO_4

Molecular Weight: 192.295

CAS RN: 12053-26-8

Properties: brown cub spinel; used as a refractory [KIR78]

Density, g/cm³: 4.415 [KIR78]

1686

Compound: Magnesium citrate pentahydrate

Synonyms: magnesium dibasic citrate

Formula: $MgC_6H_6O_7·5H_2O$

Molecular Formula: $C_6H_{16}MgO_{12}$

Molecular Weight: 304.491

CAS RN: 7779-25-1

Properties: white or sl yellow granules or powd; odorless [MER89]

Solubility: 20 g/100mL H_2O (20°C) [CRC94]

1687

Compound: Magnesium citrate tetradecahydrate

Synonyms: citric acid, magnesium salt tetradecahydrate

Formula: $Mg_3(C_6H_5O_7)_2·14H_2O$

Molecular Formula: $C_{12}H_{38}Mg_3O_{28}$

Molecular Weight: 703.332

CAS RN: 144-23-0

Properties: white, odorless, cryst powd or granules; used as a cathartic [MER89]

Solubility: sl s H_2O; s dil acids [MER89]

1688

Compound: Magnesium dichromate hexahydrate

Formula: $MgCr_2O_7·6H_2O$

Molecular Formula: $Cr_2H_{12}MgO_{13}$

Molecular Weight: 348.384

CAS RN: 16569-85-0

Properties: reddish orange ortho-rhomb; deliq [HAW93]

Solubility: 58.52% H_2O (30°C) [KIR78]

Density, g/cm³: 2.002 [KIR78]

Reactions: minus H_2O at 48.5°C [KIR78]

1689

Compound: Magnesium dititanate

Synonyms: magnesium pyrotitanate

Formula: $MgTi_2O_5$

Molecular Formula: MgO_5Ti_2

Molecular Weight: 176.461

CAS RN: 12032-35-8
Properties: ortho-rhomb; -325 mesh with 99.9% purity [CER91] [KIR83]
Melting Point, °C: 1645 [KIR83]

1690

Compound: Magnesium fluoride
Synonyms: sellaite
Formula: MgF_2
Molecular Formula: F_2Mg
Molecular Weight: 62.302
CAS RN: 7783-40-6
Properties: white powd or cryst, or 99.9% pure melted pieces of 3-6 mm or 0.8-3 mm; enthalpy of fusion 58.2 kJ/mol; enthalpy of vaporization 264 kJ/mol; manufactured by reacting hydrofluoric acid with MgO or $MgCO_3$; hardness 6 Mohs; used in ceramics, melted pieces used as evaporation material and sputtering material for widely used anti-reflection films, low-index film used in multilayers [HAW93] [MER89] [KIR78] [CER91]
Solubility: g/L soln, H_2O: 0.130 (25°C) [KRU93]
Density, g/cm^3: 3.148 [MER89]
Melting Point, °C: 1263 [CIC73]
Boiling Point, °C: 2227 [CIC73]

1691

Compound: Magnesium germanate
Formula: Mg_2GeO_4
Molecular Formula: $GeMg_2O_4$
Molecular Weight: 185.218
CAS RN: 12025-13-7
Properties: white precipitate; -325 mesh 10 microns or less with 99.9% purity [CER91] [CRC94]
Solubility: 0.0016 g/100mL H_2O (25°C) [CRC94]

1692

Compound: Magnesium germanide
Formula: Mg_2Ge
Molecular Formula: $GeMg_2$
Molecular Weight: 121.220
CAS RN: 1310-52-7
Properties: cub cryst; used in semiconductor research [LID94] [MER89]
Density, g/cm^3: 3.09 [LID94]
Melting Point, °C: 1115 [MER89]

1693

Compound: Magnesium hexafluoroacetylacetonate dihydrate
Synonyms: 1,1,1,5,5,5-hexafluoro-2,4-pentanedione, magnesium derivative
Formula: $Mg(CF_3COCH=C(O)CF_3)_2 \cdot 2H_2O$
Molecular Formula: $C_{10}H_6F_{12}MgO_6$
Molecular Weight: 474.440
CAS RN: 19648-85-2
Properties: white powd [STR93] [ALF95]

1694

Compound: Magnesium hexafluorosilicate hexahydrate
Formula: $MgSiF_6 \cdot 6H_2O$
Molecular Formula: $F_6H_{12}MgO_6Si$
Molecular Weight: 274.472
CAS RN: 60950-56-3
Properties: white efflorescent odorless cryst; used to mothproof textile fabrics [MER89] [STR93]
Solubility: anhydrous, g/100g H_2O: 26.3 (0°C), 30.8 (20°C), 44.4 (80°C) [LAN85]
Density, g/cm^3: 1.788 [MER89]
Reactions: minus SiF_4 ~120°C [MER89]

1695

Compound: Magnesium hydride
Formula: MgH_2
Molecular Formula: H_2Mg
Molecular Weight: 26.321
CAS RN: 60616-74-2
Properties: white; nonvolatile mass or tetr cryst; strong reducing agent; readily oxidized; reactivity depends on method of preparation, e.g., if prepared from diethylmagnesium it is very reactive; spontaneously ignites in air forming MgO and H_2O [MER89] [KIR80]
Solubility: reacts violently with H_2O, evolving H_2 [MER89]
Density, g/cm^3: 1.45 [MER89]
Melting Point, °C: decomposes 280 in high vacuum [MER89]

1696

Compound: Magnesium hydrogen phosphate trihydrate
Synonyms: newberyite
Formula: $MgHPO_4 \cdot 3H_2O$
Molecular Formula: H_7MgO_7P
Molecular Weight: 174.331
CAS RN: 7757-86-0

Properties: white, cryst powd; decomposes to
$Mg_2P_2O_7$ when heated; used to fireproof wood,
and as a stabilizer for plastics [HAW93]
Solubility: sl s H_2O; s dil acids [MER89]
Density, g/cm³: 2.13 [MER89]
Melting Point, °C: decomposes at 550-650
[HAW93]

1697

Compound: Magnesium hydroxide
Synonyms: brucite
Formula: $Mg(OH)_2$
Molecular Formula: H_2MgO_2
Molecular Weight: 58.320
CAS RN: 1309-42-8
Properties: white powd; absorbs CO_2 when H_2O is
present; hex, a = 0.3147 nm, c = 0.4769 nm;
hardness 2.5 Mohs; produced from seawater and
brines by precipitation of soluble magnesium
with $Ca(OH)_2$; used in sugar refining, as an
antacid [HAW93] [KIR81] [MER89]
Solubility: mg/L, H_2O: 11.7 (25°C), 4.08 (100°C); s
dil acid [KIR81]; mol/L soln, H_2O: 0.5 x 10^{-4}
(0°C), (2.6±1.5) x 10^{-4} (25°C), 7.2 x 10^{-5}
(100°C) [KRU93]
Density, g/cm³: 2.37 [KIR81]
Melting Point, °C: 350 [KIR81]

1698

Compound: Magnesium iodate tetrahydrate
Formula: $Mg(IO_3)_2 \cdot 4H_2O$
Molecular Formula: $H_8I_2MgO_{10}$
Molecular Weight: 446.171
CAS RN: 7790-32-1
Properties: white, monocl [CRC77]
Solubility: g/100g soln, H_2O: 8.55 (25°C), 13.5
(90°C); Solid phase, $Mg(IO_3)_2 \cdot 4H_2O$ (25°C),
$Mg(IO_3)_2$ (90°C) [KRU93]
Density, g/cm³: 3.3 [CRC77]
Melting Point, °C: 210, decomposes [CRC77]
Reactions: minus $4H_2O$, 210°C [CRC77]

1699

Compound: Magnesium iodide
Formula: MgI_2
Molecular Formula: I_2Mg
Molecular Weight: 278.114
CAS RN: 10377-58-9

Properties: white; hex, a = 0.4148 nm, c = 0.6894
nm; very hygr; decomposes in air evolving I_2;
enthalpy of fusion 29.00 kJ/mol; can be obtained
by heating the hexahydrate in stream of dry H_2
[KIR81] [CRC93]
Solubility: g/100g soln, H_2O: 54.7 (0°C), 59.6
(25°C), 65.2 (100°C); Solid phase, $MgI_2 \cdot 8H_2O$
(0°C, 25°C), $MgI_2 \cdot 6H_2O$ (100°C) [KRU93]
Density, g/cm³: 4.43 [KIR81]
Melting Point, °C: 637 decomposes [KIR81]

1700

Compound: Magnesium iodide hexahydrate
Formula: $MgI_2 \cdot 6H_2O$
Molecular Formula: $H_{12}I_2MgO_6$
Molecular Weight: 386.2005
CAS RN: 75535-11-4
Properties: white; monocl, a = 1.1159 nm, b =
0.7740 nm, c = 0.6323 nm [KIR81]
Density, g/cm³: 2.353 [KIR81]

1701

Compound: Magnesium iodide octahydrate
Formula: $MgI_2 \cdot 8H_2O$
Molecular Formula: $H_{16}I_2MgO_8$
Molecular Weight: 422.236
CAS RN: 7790-31-0
Properties: white; ortho-rhomb, a = 0.9948 nm, b =
1.5652 nm, c = 0.8585 nm; deliq powd;
discolors readily in air and light [MER89]
[KIR81]
Solubility: 81 g/100mL H_2O (20°C), 90.3 g/100mL
H_2O (80°C) [CRC94]
Density, g/cm³: 2.098 [KIR81]
Melting Point, °C: 41 decomposes [KIR81]

1702

Compound: Magnesium metatitanate
Formula: $MgTiO_3$
Molecular Formula: MgO_3Ti
Molecular Weight: 120.183
CAS RN: 12032-30-3
Properties: rhombohedral white powd; used mainly
as an additive for ceramic dielectric materials,
also as a gemstone and a pigment in ultra violet
cured systems [KIR83] [STR93]
Density, g/cm³: 3.36 [STR93]
Melting Point, °C: 1610 [STR93]

1703

Compound: Magnesium molybdate

Synonyms: magnesium molybdate(VI)
Formula: $MgMoO_4$
Molecular Formula: $MgMoO_4$
Molecular Weight: 184.243
CAS RN: 12013-21-7
Properties: -200 mesh with 99.9% purity; white powd; used in electronic and optical applications [CER91] [STR93] [HAW93]
Solubility: g/100g soln, H_2O: 15.90 (25°C), 9.38 (95°C); Solid phase, $MgMoO_4 \cdot 5H_2O$ (25°C), $MgMoO_4 \cdot 2H_2O$ (95°C) [KRU93]
Density, g/cm³: 2.208 [STR93]
Melting Point, °C: ~1060 [HAW93]

1704

Compound: Magnesium niobate
Formula: $MgNb_2O_6$
Molecular Formula: $MgNb_2O_6$
Molecular Weight: 306.114
CAS RN: 12163-26-7
Properties: -200 mesh with 99.9% purity [CER91]

1705

Compound: Magnesium nitrate
Formula: $Mg(NO_3)_2$
Molecular Formula: MgN_2O_6
Molecular Weight: 148.314
CAS RN: 10377-60-3
Properties: white, cub cryst; difficult to isolate in anhydrous form; can be prepared at room temp by dissolution of MgO, $Mg(OH)_2$ or $MgCO_3$ in HNO_3, followed by solvent evaporation and crystallization; finds use as a fertilizer, and in the manufacture of ammonium nitrate [KIR81] [LID94]
Solubility: g/100g soln, H_2O: 38.5 (1.0°C), 42.1 (25°C), 71.7 (100°C); Solid phase, $Mg(NO_3)_2 \cdot 6H_2O$ (1.0°C, 25°C), $Mg(NO_3)_2 \cdot 2H_2O$ (100°C) [KRU93]
Density, g/cm³: ~2.3 [LID94]

1706

Compound: Magnesium nitrate dihydrate
Formula: $Mg(NO_3)_2 \cdot 2H_2O$
Molecular Formula: $H_4MgN_2O_8$
Molecular Weight: 184.345
CAS RN: 15750-45-5
Properties: white cryst; deliq; used in pyrotechnics [HAW93]
Solubility: s H_2O, alcohol [HAW93]
Density, g/cm³: 1.45 [HAW93]
Melting Point, °C: 95-100 [HAW93]

Boiling Point, °C: 330, decomposes [HAW93]

1707

Compound: Magnesium nitrate hexahydrate
Formula: $Mg(NO_3)_2 \cdot 6H_2O$
Molecular Formula: $H_{12}MgN_2O_{12}$
Molecular Weight: 256.406
CAS RN: 13446-18-9
Properties: colorless, clear deliq cryst; monocl [KIR81] [MER89]
Solubility: s in 0.8 parts H_2O; v sl alcohol [MER89]
Density, g/cm³: 1.464 [MER89]
Melting Point, °C: ~95 [MER89]

1708

Compound: Magnesium nitride
Formula: Mg_3N_2
Molecular Formula: Mg_3N_2
Molecular Weight: 100.928
CAS RN: 12057-71-5
Properties: -325 mesh 10 microns or less of 99.6% purity; bcc; a = 0.993 nm [CER91] [CIC73]
Density, g/cm³: 2.71 [ALD94]
Melting Point, °C: 271, decomposes [CIC73]

1709

Compound: Magnesium nitrite trihydrate
Formula: $Mg(NO_2)_2 \cdot 3H_2O$
Molecular Formula: $H_6MgN_2O_7$
Molecular Weight: 170.362
CAS RN: 15070-34-5
Properties: white prism; hygr [CRC77]
Solubility: g/100g soln, H_2O: 47.0 (25.65°C); Solid phase, $Mg(NO_2)_2 \cdot 6H_2O$ [KRU93]
Melting Point, °C: 100, decomposes [CRC77]

1710

Compound: Magnesium orthotitanate
Formula: Mg_2TiO_4
Molecular Formula: Mg_2O_4Ti
Molecular Weight: 160.475
CAS RN: 12032-52-9
Properties: cub [KIR83]
Density, g/cm³: 3.53 [KIR83]
Melting Point, °C: 1840 [KIR83]

1711

Compound: Magnesium oxalate dihydrate
Formula: $MgC_2O_4 \cdot 2H_2O$
Molecular Formula: $C_2H_4MgO_6$

Molecular Weight: 148.355
CAS RN: 6150-88-5
Properties: white powd [MER89]
Solubility: g/L solution in H_2O: 0.38±0.04 (25°C), 0.4 (92°C); Solid phase $MgC_2O_4 \cdot 2H_2O$ [KRU93]
Density, g/cm^3: 2.45 [CRC77]
Melting Point, °C: decomposes 150 [AES93]

1712
Compound: Magnesium oxide
Synonyms: periclase, magnesia
Formula: MgO
Molecular Formula: MgO
Molecular Weight: 40.304
CAS RN: 1309-48-4
Properties: white powd; highly reflective in visible and near UV regions; cub, a = 0.4213 nm; hardness 5.5 Mohs; resistivity 1.3 x 10^{+15} ohm cm (27°C); enthalpy of fusion 78 kJ/mol; used as a refractory material, particularly for steel furnace linings, as a sputtering target of 99.95% and 99.9% purity to prepare high temp dielectrics, and in crucible form to contain melting corrosive salts such as fluorides [MER89] [HAW93] [KIR81] [CER91] [CRC93]
Solubility: v sl s pure H_2O; s dil acids; i alcohol [MER89]
Density, g/cm^3: 3.581 [KIR81]
Melting Point, °C: 2852 [KIR81]
Boiling Point, °C: 3600 [STR93]
Reactions: absorbs CO_2 and H_2O from atm [MER89]
Thermal Conductivity, W/(m·K): 60.0 (27°C), 43.1 (127°C); 13.9 (550°C), 7.0 (1000°C) [KIR80] [KIR81]
Thermal Expansion Coefficient: (volume) 100°C (0.219), 200°C (0.588), 400°C (1.386), 800°C (3.150), 1000°C (4.050) [CLA66]

1713
Compound: Magnesium perborate heptahydrate
Formula: $Mg(BO_3)_2 \cdot 7H_2O$
Molecular Formula: $B_2H_{14}MgO_{13}$
Molecular Weight: 268.030
CAS RN: 14635-87-1
Properties: white powd; decomposes, evolving oxygen; used in driers, in bleaching, and as an antiseptic for tooth powd [HAW93]
Solubility: sl s H_2O [HAW93]

1714
Compound: Magnesium perchlorate

Synonyms: dehydrite
Formula: $Mg(ClO_4)_2$
Molecular Formula: Cl_2MgO_8
Molecular Weight: 223.205
CAS RN: 10034-81-8
Properties: white, very hygr powd; oxidant; evolves heat when dissolving in H_2O; crystallizes from H_2O as the hexahydrate; used as a regenerable drying agent for gases [HAW93] [MER89]
Solubility: mol/100mol H_2O: 0.410 (0°C), 0.448±0.001 (25°C); Solid phase, $Mg(ClO_4)_2 \cdot 6H_2O$ [KRU93]
Density, g/cm^3: 2.21 [STR93]
Melting Point, °C: 251, decomposes [STR93]

1715
Compound: Magnesium perchlorate hexahydrate
Formula: $Mg(ClO_4)_2 \cdot 6H_2O$
Molecular Formula: $Cl_2H_{12}MgO_{14}$
Molecular Weight: 331.297
CAS RN: 13446-19-0
Properties: white cryst; hygr [HAW93]
Solubility: v s H_2O, alcohol [HAW93]
Density, g/cm^3: 1.98 [HAW93]
Melting Point, °C: 185-190 [HAW93]

1716
Compound: Magnesium permanganate hexahydrate
Formula: $Mg(MnO_4)_2 \cdot 6H_2O$
Molecular Formula: $H_{12}MgMn_2O_{14}$
Molecular Weight: 370.268
CAS RN: 10377-62-5
Properties: bluish black cryst; deliq; used as an antiseptic and as a polymerization catalyst [HAW93]
Solubility: s H_2O [HAW93]
Density, g/cm^3: 2.18 [HAW93]
Melting Point, °C: decomposes [HAW93]

1717
Compound: Magnesium peroxide
Synonyms: magnesium dioxide
Formula: MgO_2
Molecular Formula: MgO_2
Molecular Weight: 56.304
CAS RN: 1335-26-8
Properties: white, tasteless, odorless powd; used as a bleaching and oxidizing agent, as an antacid [HAW93] [MER89]
Solubility: i H_2O; s dil acids forming H_2O_2 [MER89]
Melting Point, °C: decomposes >100 [HAW93]

Reactions: gradually decomposed by H_2O evolving O_2 [MER89]

1718
Compound: Magnesium phosphate octahydrate
Synonyms: bobierrite
Formula: $Mg_3(PO_4)_2 \cdot 8H_2O$
Molecular Formula: $H_{16}Mg_3O_{16}P_2$
Molecular Weight: 406.980
CAS RN: 13446-23-6
Properties: soft, bulky, white powd; odorless and tasteless; used as a dentifrice polishing agent, as an antacid [HAW93] [ALF95]
Solubility: i H_2O; s acids [HAW93]
Density, g/cm³: 2.195 [CRC94]
Reactions: minus $5H_2O$ at 150°C [CRC94], minus $8H_2O$ at 400°C [HAW93]

1719
Compound: Magnesium phosphate pentahydrate
Formula: $Mg_3(PO_4)_2 \cdot 5H_2O$
Molecular Formula: $H_{10}Mg_3O_{13}P_2$
Molecular Weight: 352.934
CAS RN: 7757-87-1
Properties: white, cryst powd [MER89]
Solubility: i H_2O; s dil mineral acids [MER89]
Reactions: minus last H_2O ~400°C [MER89]

1720
Compound: Magnesium phosphide
Formula: Mg_3P_2
Molecular Formula: Mg_3P_2
Molecular Weight: 134.863
CAS RN: 12057-74-8
Properties: shiny grayish yellow; stable in dry air; decomposed by moisture; can be prepared directly from magnesium and phosphorus; used with igniting agent such as 1% HNO_3 in sea flares [KIR82]
Solubility: decomposed by H_2O [KIR82]
Density, g/cm³: 2.055 [CRC94]

1721
Compound: Magnesium pyrophosphate
Formula: $Mg_2P_2O_7$
Molecular Formula: $Mg_2O_7P_2$
Molecular Weight: 222.553
CAS RN: 13446-24-7
Properties: colorless; monocl [CRC77]
Density, g/cm³: 2.559 [CRC77]
Melting Point, °C: 1383 [HAW93]

1722
Compound: Magnesium pyrophosphate trihydrate
Formula: $Mg_2P_2O_7 \cdot 3H_2O$
Molecular Formula: $H_6Mg_2O_{10}P_2$
Molecular Weight: 276.600
CAS RN: 10102-34-8
Properties: white powd [MER89]
Solubility: i H_2O; s mineral acids [MER89]
Density, g/cm³: 2.56 [MER89]
Reactions: minus $3H_2O$ at 100°C [MER89]

1723
Compound: Magnesium salicylate tetrahydrate
Formula: $Mg(C_7H_5O_3)_2 \cdot 4H_2O$
Molecular Formula: $C_{14}H_{18}MgO_{10}$
Molecular Weight: 370.596
CAS RN: 18917-95-8
Properties: white, odorless, efflorescent cryst powd; used as an anti-infective in medicine [HAW93] [MER89]
Solubility: s 13 parts H_2O; s alcohol [MER89]

1724
Compound: Magnesium selenate hexahydrate
Formula: $MgSeO_4 \cdot 6H_2O$
Molecular Formula: $H_{12}MgO_{10}Se$
Molecular Weight: 275.354
CAS RN: 14986-91-5
Properties: monocl cryst [MER89]
Solubility: g/100g soln in H_2O: 31.41 (0°C), 35.70 (25°C), 46.50 (99.5°C); Solid phase, $MgSeO_4 \cdot 7H_2O$ (0°C), $MgSeO_4 \cdot 6H_2O$ (25°C), $MgSeO_4 \cdot 4\text{-}1/2H_2O$ (99.5°C) [KRU93]
Density, g/cm³: 1.928 [MER89]

1725
Compound: Magnesium selenide
Formula: MgSe
Molecular Formula: MgSe
Molecular Weight: 103.265
CAS RN: 1313-04-8
Properties: light brown powd; unstable in air [MER89]
Solubility: decomposes in H_2O [MER89]
Density, g/cm³: 4.21 [MER89]

1726
Compound: Magnesium selenite hexahydrate
Formula: $MgSeO_3 \cdot 6H_2O$
Molecular Formula: $H_{12}MgO_9Se$
Molecular Weight: 259.355

CAS RN: 15593-61-0
Properties: colorless ortho-rhomb cryst [LID94] [MER89]
Solubility: i H_2O; s dil acids [MER89]
Density, g/cm³: 2.09 [LID94]
Reactions: loses $5H_2O$ to form monohydrate at 100°C [MER89]

1727

Compound: Magnesium silicate
Synonyms: clinoenstatite
Formula: $MgSiO_3$
Molecular Formula: MgO_3Si
Molecular Weight: 100.389
CAS RN: 1343-88-0
Properties: white monocl cryst [MER89]
Solubility: i H_2O [MER89]
Density, g/cm³: 3.192 [MER89]
Melting Point, °C: decomposes 1557 [MER89]
Thermal Expansion Coefficient: (volume) 100°C (0.19), 200°C (0.42), 400°C (0.96), 800°C (2.28), 1200°C (3.68) [CLA66]

1728

Compound: Magnesium silicate
Synonyms: forsterite
Formula: Mg_2SiO_4
Molecular Formula: Mg_2O_4Si
Molecular Weight: 140.694
CAS RN: 26686-77-1
Properties: white; ortho-rhomb; enthalpy of fusion 71.00 kJ/mol [CRC94]
Density, g/cm³: 3.22 [KIR80]
Melting Point, °C: 1898 [KIR80]
Thermal Conductivity, W/(m·K): 3.1 (500°C), 2.4 (1000°C) [KIR80]
Thermal Expansion Coefficient: linear expansion to 1000°C, 9.5 x 10^{-6}/°C [KIR80]

1729

Compound: Magnesium silicide
Formula: Mg_2Si
Molecular Formula: Mg_2Si
Molecular Weight: 76.696
CAS RN: 22831-39-6
Properties: -20 mesh with 99.5% purity; gray powd; slate blue, cub cryst; used in semiconductor technology [CER91] [HAW93] [STR93] [MER89]
Solubility: decomposed by H_2O, HCl [MER89]
Density, g/cm³: 1.94 [STR93]
Melting Point, °C: 1085 [MER89]; 1102 [ALF93]

1730

Compound: Magnesium stannate trihydrate
Formula: $MgSnO_3 \cdot 3H_2O$
Molecular Formula: H_6MgO_6Sn
Molecular Weight: 245.059
CAS RN: 12032-29-0
Properties: white cryst powd; decomposes ~340°C; used as an additive in ceramic capacitors; anhydrous form -325 mesh 10 microns or less with 99% and 99.9% purity [CER91] [HAW93]
Solubility: s H_2O [HAW93]

1731

Compound: Magnesium stannide
Formula: Mg_2Sn
Molecular Formula: Mg_2Sn
Molecular Weight: 167.320
CAS RN: 1313-08-2
Properties: bluish white metallic compound; resistivity (25°C) 42,000 μohm·cm; used in semiconductors, thermoelectric research [HAW93] [MER89]
Solubility: s H_2O, dil HCl [MER89]
Density, g/cm³: 3.60 [LID94]
Melting Point, °C: 775 [HAW93]

1732

Compound: Magnesium stearate
Formula: $Mg[CH_3(CH_2)_{16}COO]_2$
Molecular Formula: $C_{36}H_{70}MgO_4$
Molecular Weight: 591.255
CAS RN: 557-04-0
Properties: soft, light white powd; odorless, tasteless; nonflammable used in baby dusting powd and as tablet lubricant [HAW93] [MER89]
Solubility: i H_2O; decomposed by dil acids [MER89]
Density, g/cm³: 1.028 [HAW93]
Melting Point, °C: 88.5 [HAW93]

1733

Compound: Magnesium sulfate
Formula: $MgSO_4$
Molecular Formula: MgO_4S
Molecular Weight: 120.369
CAS RN: 7487-88-9

Properties: colorless; ortho-rhomb, a = 0.5182 nm, b = 0.7893 nm, c = 0.6506 nm; occurs widely in minerals; enthalpy of fusion 14.60 kJ/mol; saline bitter taste; prepared by dehydration of its hydrates; used in fireproofing, for warp sizing and loading textile goods [HAW93] [KIR81] [CRC93]

Solubility: g/100g soln, H_2O: 20.5 (0°C), 27.6 (25°C), 42.9 (100°C); Solid phase, $MgSO_4 \cdot 6H_2O$ [KRU93]

Density, g/cm³: 2.66 [STR93]

Melting Point, °C: 1124, decomposes [HAW93]

1734

Compound: Magnesium sulfate heptahydrate

Synonyms: epsomite

Formula: $MgSO_4 \cdot 7H_2O$

Molecular Formula: $H_{14}MgO_{11}S$

Molecular Weight: 246.475

CAS RN: 10034-99-8

Properties: ortho-rhomb, a = 1.186 nm, b = 1.199 nm, c = 0.6858 nm; colorless efflorescent cryst or powd; bitter, saline cooling taste; stable from ~-5 to 48.2°C; there is a hexahydrate, 17830-18-1, stable from 48.2 to 67.5°C [MER89] [KIR81]

Solubility: g/100mL H_2O: 71 (20°C), 91 (40°C); sl s alcohol [MER89]

Density, g/cm³: 1.67 [MER89]

Melting Point, °C: decomposes ~150 [KIR81]

Reactions: loses $6H_2O$ at 150°C, minus H_2O at 200°C [HAW93]

1735

Compound: Magnesium sulfate monohydrate

Synonyms: kieserite

Formula: $MgSO_4 \cdot H_2O$

Molecular Formula: H_2MgO_5S

Molecular Weight: 138.384

CAS RN: 14168-73-1

Properties: colorless cryst; monocl, a = 0.690 nm, b = 0.771 nm, c = 0.754 nm [KIR81]

Solubility: H_2O: 37.1% (67.5°C), 8% (170°C), 0.5% (240°C) [KIR81]

Density, g/cm³: 2.571 [KIR81]

Melting Point, °C: decomposes ~150 [KIR81]

1736

Compound: Magnesium sulfide

Formula: MgS

Molecular Formula: MgS

Molecular Weight: 56.371

CAS RN: 12032-36-9

Properties: powd; sensitive to moisture; reddish brown cryst; used as a source of hydrogen sulfide [STR93] [HAW93]

Solubility: decomposes in H_2O [HAW93]

Density, g/cm³: 2.68 [LID94]

Melting Point, °C: >2000, decomposes [STR93]

1737

Compound: Magnesium sulfite

Formula: $MgSO_3$

Molecular Formula: MgO_3S

Molecular Weight: 104.369

CAS RN: 7757-88-2

Properties: used in systems for flue gas desulfurization in which $Mg(OH)_2$ is the alkaline scrubber [KIR81]

Solubility: g/100g soln, H_2O: 0.338 (0°C), 0.646 (25°C), 0.615 (98°C); Solid phase, $MgSO_3 \cdot 6H_2O$ (0°C, 25°C), $MgSO_3 \cdot 3H_2O$ (98°C) [KRU93]

1738

Compound: Magnesium sulfite hexahydrate

Formula: $MgSO_3 \cdot 6H_2O$

Molecular Formula: $H_{12}MgO_9S$

Molecular Weight: 212.461

CAS RN: 13446-29-2

Properties: white; hex, a = 0.88385 nm, c = 0.9080 nm; gradually oxidizes to sulfate in air; used in the manufacture of paper pulp [HAW93] [MER89] [KIR81]

Solubility: s in ~150 parts H_2O; sl more soluble in hot H_2O [MER89]; i alcohol [HAW93]

Density, g/cm³: 1.725 [HAW93]

Melting Point, °C: 200 decomposes [KIR81]

Reactions: minus all H_2O at 200°C [MER89]

1739

Compound: Magnesium sulfite trihydrate

Formula: $MgSO_3 \cdot 3H_2O$

Molecular Formula: H_6MgO_6S

Molecular Weight: 158.415

CAS RN: 19086-20-5

Properties: colorless; ortho-rhomb, a = 0.939 nm, b = 0.9584 nm, c = 0.5523 nm; used in flue gas desulfurization [KIR81]

Density, g/cm³: 2.117 [KIR81]

1740

Compound: Magnesium tantalate

Formula: $MgTa_2O_6$

Molecular Formula: MgO_6Ta_2
Molecular Weight: 482.197
CAS RN: 12293-61-7
Properties: -200 mesh with 99.9% purity [CER91]

1741
Compound: Magnesium tetrahydrogen phosphate dihydrate
Formula: $MgH_4(PO_4)_2 \cdot 2H_2O$
Molecular Formula: $H_8MgO_{10}P_2$
Molecular Weight: 254.311
CAS RN: 13092-66-5
Properties: white cryst powd; hygr; decomposes when heated to metaphosphate; preparation: reaction between phosphoric acid and magnesium hydroxide; used to fireproof wood, and as a stabilizer for plastics [HAW93]
Solubility: s H_2O, acids; i alcohol [HAW93]

1742
Compound: Magnesium thiocyanate tetrahydrate
Formula: $Mg(SCN)_2 \cdot 4H_2O$
Molecular Formula: $C_2H_8MgN_2O_4S_2$
Molecular Weight: 212.534
CAS RN: 306-61-6
Properties: colorless or white deliq cryst [MER89]
Solubility: v s H_2O, alcohol [MER89]

1743
Compound: Magnesium thiosulfate hexahydrate
Synonyms: magnesium hyposulfite hexahydrate
Formula: $MgS_2O_3 \cdot 6H_2O$
Molecular Formula: $H_{12}MgO_9S_2$
Molecular Weight: 244.527
CAS RN: 10124-53-5
Properties: colorless or white cryst [MER89]
Solubility: g/100g soln, H_2O: 30.69 (0°C), 34.51 (28°C); Solid phase, $MgS_2O_3 \cdot 6H_2O$ [KRU93]; i alcohol [MER89]
Density, g/cm³: 1.82 [MER89]
Melting Point, °C: 1700, decomposes [AES93]
Reactions: minus $3H_2O$ at 170°C [MER89]

1744
Compound: Magnesium trifluoroacetylacetonate dihydrate
Synonyms: 1,1,1-trifluoro-2,4-pentandione, magnesium derivative
Formula: $Mg(CF_3COCH=C(O)CH_3)_2 \cdot 2H_2O$
Molecular Formula: $C_{10}H_{12}F_6MgO_6$
Molecular Weight: 366.497

CAS RN: 53633-79-7
Properties: white powd [STR93]

1745
Compound: Magnesium tungstate
Synonyms: magnesium tungstate(VI)
Formula: $MgWO_4$
Molecular Formula: MgO_4W
Molecular Weight: 272.143
CAS RN: 13573-11-0
Properties: -325 mesh with 99.9% purity, 10 microns or less; white, cryst powd; used in fluorescent x-ray screens; and in luminescent paint [CER91] [HAW93] [MER89]
Solubility: i H_2O and alcohol; s in acids [HAW93]
Density, g/cm³: 5.66 [HAW93]

1746
Compound: Magnesium vanadate
Formula: $2MgO \cdot V_2O_5$
Molecular Formula: $Mg_2O_7V_2$
Molecular Weight: 262.489
CAS RN: 13568-63-3
Properties: tricl, a = 1.3767 nm, b = 0.5414 nm, c = 0.4912 nm; other vanadates are MgV_2O_6, 13573-13-2, MgV_3O_8, 12181-49-6, $Mg_{1.9}V_3O_8$ and $Mg_3V_2O_8$, 13568-68-8; formed by addition of MgO dispersed in oil into the flame zone of utility boilers in order to remove vanadium compounds and thereby to reduce corrosion caused by the presence of vanadium [KIR81] [CER91]
Density, g/cm³: 3.1 [KIR81]

1747
Compound: Magnesium zirconate
Formula: $MgZrO_3$
Molecular Formula: MgO_3Zr
Molecular Weight: 163.527
CAS RN: 12032-31-4
Properties: reacted product, -100 and +200 mesh of 99% purity; powd; used in electronics [CER91] [HAW93]
Density, g/cm³: 4.23 [HAW93]
Melting Point, °C: 2060 [HAW93]

1748
Compound: Magnesium zirconium silicate
Formula: $MgO \cdot ZrO_2 \cdot SiO_2$
Molecular Formula: MgO_5SiZr
Molecular Weight: 223.612

CAS RN: 52110-05-1
Properties: white solid; used in electrical resistor, ceramics, as an opacifier for glazes [HAW93]
Solubility: i H_2O, alkalies; sl s in acids [HAW93]

1749
Compound: Manganese
Formula: Mn
Molecular Formula: Mn
Molecular Weight: 54.93805
CAS RN: 7439-96-5
Properties: steel gray, lustrous, hard, brittle metal; has four allotropes: α-Mn, cub, a = 0.89 nm, stable <710°C; β-Mn, cub, a = 0.63 nm, stable 710-1079°C; γ-Mn, fcc, a = 0.387 nm, stable 1079-1143°C; δ-Mn, bcc, a = 0.309 nm, stable 1143°C to melting; hardness 5.0 Mohs; enthalpy of fusion 12.91 kJ/mol; enthalpy of vaporization 220.9 kJ/mol; electrical resistivity at 20°C for α is 160 μohm·cm [MER89] [KIR81] [CRC93]
Solubility: decomposes slowly in cold H_2O, rapidly in hot H_2O; evolves hydrogen with dil mineral acids, dissolving as Mn++ [MER89]
Density, g/cm³: α: 7.47; β: 7.26; γ: 6.37; δ: 6.28 [MER89]
Melting Point, °C: 1244 [MER89]
Boiling Point, °C: 2095 [MER89]
Thermal Conductivity, W/(m·K): 7.82 (25°C) [CRC93]
Thermal Expansion Coefficient: 100°C: 25.2 x 10^{-6}/°C (α), 43.0 x 10^{-6}/°C (β), 45.2 x 10^{-6}/°C (γ), 41.6 x 10^{-6}/°C (δ) [KIR81]

1750
Compound: Manganese aluminide
Formula: MnAl₃
Molecular Formula: Al₃Mn
Molecular Weight: 135.883
CAS RN: 12253-13-3
Properties: 6 mm pieces and smaller with 99.5% purity [CER91]

1751
Compound: Manganese ammonium sulfate hexahydrate
Formula: $MnSO_4 \cdot (NH_4)_2SO_4 \cdot 6H_2O$
Molecular Formula: $H_{20}MnN_2O_{14}S_2$
Molecular Weight: 391.235
CAS RN: 7785-19-5
Properties: light red cryst [HAW93]
Solubility: s H_2O [HAW93]
Density, g/cm³: 1.83 [HAW93]

1752
Compound: Manganese antimonide
Formula: MnSb
Molecular Formula: MnSb
Molecular Weight: 176.698
CAS RN: 12032-82-5
Properties: hex cryst; 6 mm pieces and smaller with 99.5% purity; there is also an antimonide with the formula Mn₂Sb, 12032-97-2 [CER91] [LID94]
Density, g/cm³: 6.9 [LID94]
Melting Point, °C: 840 [LID94]

1753
Compound: Manganese bis(cyclopentadienyl)
Synonyms: bis(cyclopentadienyl)manganese
Formula: $(C_5H_5)_2Mn$
Molecular Formula: $C_{10}H_{10}Mn$
Molecular Weight: 185.127
CAS RN: 1271-27-8
Properties: sublimed powd, in ampoules [ALF95]
Melting Point, °C: 292 [ALF95]

1754
Compound: Manganese boride
Formula: MnB
Molecular Formula: BMn
Molecular Weight: 65.749
CAS RN: 12045-15-7
Properties: powd; -80 mesh with 99% purity, there is also MnB₂, 12228-50-1, -200 mesh; refractory material [CER91] [KIR81] [CRC94]
Density, g/cm³: 6.2 [CRC94]
Melting Point, °C: 1890 [KIR81]

1755
Compound: Manganese carbide
Formula: Mn₃C
Molecular Formula: CMn₃
Molecular Weight: 176.825
CAS RN: 12266-65-8
Properties: tetr; mixture of Mn₅C₂ and possibly other Mn-C phases, 6 mm pieces and smaller with 99.5% purity; there is also material with formula Mn₂₃C₆, 72266-65-8, -80 mesh with 99.5% purity [CER91] [CRC94]
Density, g/cm³: 6.89 [CRC94]
Melting Point, °C: 1520 [LID94]

1756
Compound: Manganese carbonyl

Synonyms: dimanganese decacarbonyl
Formula: $Mn_2(CO)_{10}$
Molecular Formula: $C_{10}Mn_2O_{10}$
Molecular Weight: 389.980
CAS RN: 10170-69-1
Properties: golden yellow, monocl cryst; stable under CO gas, less stable to air, heat and light in solution; used as an antiknock agent in gasoline [HAW93] [MER89]
Solubility: i H_2O; s organic solvents [MER89]
Density, g/cm^3: 1.75 [HAW93]
Melting Point, °C: 154 [HAW93]
Boiling Point, °C: 80 [ALF95]

1757
Compound: Manganese diboride
Formula: MnB_2
Molecular Formula: B_2Mn
Molecular Weight: 76.560
CAS RN: 12228-50-1
Properties: gray-violet cryst; refractory material [KIR81] [CRC94]
Solubility: decomposed by H_2O [CRC94]
Density, g/cm^3: 5.3 [LID94]
Melting Point, °C: 1827 [LID94]

1758
Compound: Manganese niobate
Formula: $MnNb_2O_6$
Molecular Formula: $MnNb_2O_6$
Molecular Weight: 336.747
CAS RN: 12032-69-8
Properties: -200 mesh with 99.9% purity [CER91]

1759
Compound: Manganese nitride
Formula: MnN
Molecular Formula: MnN
Molecular Weight: 68.945
CAS RN: 36678-21-4
Properties: formed by reaction of Mn and N_2 above 740°C; other nitrides are Mn_6N_5 (64886-63-1), Mn_3N_2 (12033-03-3), Mn_2N (12163-53-0) and Mn_4N (12033-07-7); used in steelmaking as nitrogen containing intermediate alloys [KIR81]

1760
Compound: Manganese pentacarbonyl bromide
Formula: $Mn(CO)_5Br$
Molecular Formula: C_5BrMnO_5
Molecular Weight: 274.894

CAS RN: 14516-54-2
Properties: yellowish orange cryst [STR93]

1761
Compound: Manganese phosphide
Formula: Mn_3P_2
Molecular Formula: Mn_3P_2
Molecular Weight: 226.762
CAS RN: 12397-32-9
Properties: dark gray; mixture of MnP and Mn_2P; -100 mesh with 99% purity [CER91] [CRC94]
Density, g/cm^3: 5.12 [CRC94]
Melting Point, °C: 1095 [AES93]

1762
Compound: Manganese silicate
Synonyms: rhodonite, manganjustite, tephoroite
Formula: $MnSiO_3$
Molecular Formula: MnO_3Si
Molecular Weight: 131.022
CAS RN: 7759-00-4
Properties: red cryst or yellowish red powd; prepared from manganous salts and sodium silicate; used to color glass and pottery [MER89] [HAW93]
Solubility: i H_2O [MER89]
Density, g/cm^3: 3.48 [MER89]
Melting Point, °C: 1323 [HAW93]

1763
Compound: Manganese silicide
Synonyms: manganese disilicide
Formula: $MnSi_2$
Molecular Formula: $MnSi_2$
Molecular Weight: 111.109
CAS RN: 12032-86-9
Properties: gray; possibly $Mn_{15}Si_{26}$; -325 mesh 10 microns or less with 99.5% purity [CER91] [CRC94]
Density, g/cm^3: 5.24 [CRC94]

1764
Compound: Manganese vanadate
Formula: MnV_2O_6
Molecular Formula: MnO_6V_2
Molecular Weight: 252.817
CAS RN: 14986-94-8
Properties: -200 mesh with 99.9% purity [CER91]

1765
Compound: Manganese(II) acetate tetrahydrate
Formula: $Mn(CH_3COO)_2·4H_2O$
Molecular Formula: $C_4H_{14}MnO_8$
Molecular Weight: 245.088
CAS RN: 6156-78-1
Properties: pale red cryst; monocl; used in textile dyeing, as an oxidation catalyst; anhydrous manganese(II) acetate has CAS RN 638-38-0 [ALF93] [HAW93]
Solubility: s H_2O, alcohol [HAW93]
Density, g/cm³: 1.589 [KIR81]
Melting Point, °C: 80 [HAW93]; 180 [AES93]

1766
Compound: Manganese(II) acetylacetonate
Synonyms: 2,4-pentanedione, manganese(II) derivative
Formula: $Mn(CH_3COCH=C(O)CH_3)_2$
Molecular Formula: $C_{10}H_{14}MnO_4$
Molecular Weight: 253.157
CAS RN: 14024-58-9
Properties: tan powd; trimer; hygr [STR93] [COT88]

$$\underset{[CH_3-C=CH-C-CH_3]_2Mn}{\overset{\overset{\displaystyle O-\quad\quad O}{\displaystyle |\quad\quad\; \|}}{}}$$

Melting Point, °C: decomposes at 216 [ALD94]

1767
Compound: Manganese(II) borate octahydrate
Formula: $MnB_4O_7·8H_2O$
Molecular Formula: $B_4H_{16}MnO_{15}$
Molecular Weight: 354.300
CAS RN: 12228-91-0
Properties: brownish-white powd; preparation: by addition of borax to an aq manganese(II) sulfate solution; used in drying varnishes and oils, and in the leather industry [MER89] [KIR81]
Solubility: i H_2O, alcohol; s dil acids [MER89]
Reactions: decomposes on standing in H_2O [MER89]

1768
Compound: Manganese(II) bromide
Formula: $MnBr_2$
Molecular Formula: Br_2Mn
Molecular Weight: 214.746
CAS RN: 13446-03-2

Properties: -80 mesh with 99.5% purity; pink powd; hygr [STR93] [CER91]
Solubility: g/100g soln, H_2O: 56.0 (0°C), 60.2 (25°C), 69.5 (100°C); Solid phase, $MnBr_2·4H_2O$ (0°C, 25°C), $MnBr_2·2H_2O$ (100°C) [KRU93]
Density, g/cm³: 4.385 [STR93]
Melting Point, °C: 698 [LID94]

1769
Compound: Manganese(II) bromide tetrahydrate
Formula: $MnBr_2·4H_2O$
Molecular Formula: $Br_2H_8MnO_4$
Molecular Weight: 286.808
CAS RN: 10031-20-6
Properties: rose red, sl deliq cryst [MER89]
Solubility: s in 0.5 parts H_2O; s alcohol [MER89]
Melting Point, °C: 64, with some decomposition [MER89]

1770
Compound: Manganese(II) carbonate
Synonyms: rhodochrosite
Formula: $MnCO_3$
Molecular Formula: $CMnO_3$
Molecular Weight: 114.947
CAS RN: 598-62-9
Properties: pink solid; trig; gradually turns light brown in air; photoluminescent; hardness 3-4 Mohs; can be made by precipitation of a water soluble Mn(II) salt with an alkali carbonate [HAW93] [KIR81] [MER89]
Solubility: sl s H_2O; s dil acids [KIR81]
Density, g/cm³: 3.125 [KIR81]
Melting Point, °C: decomposes >200 [KIR81]

1771
Compound: Manganese(II) chloride
Synonyms: sacchite
Formula: $MnCl_2$
Molecular Formula: Cl_2Mn
Molecular Weight: 125.843
CAS RN: 7773-01-5
Properties: pink cryst or flakes; trig, hygr; enthalpy of fusion 30.70 kJ/mol; obtained by reaction of Mn, MnO, $Mn(OH)_2$ or $MnCO_3$ and HCl, followed by crystallization and dehydration; used as a chlorination catalyst for organic materials, as a paint dryer, as a dietary supplement [CRC93] [HAW93] [KIR81]

Solubility: s pyridine, ethanol; i ether [KIR78];
  g/100g soln, H_2O: 38.8 (0°C), 43.6 (25°C), 53.5
  (100°C); equilibrium solid phases: $MnCl_2 \cdot 4H_2O$
  (0°C, 25°C), $MnCl_2$ (100°C) [KRU93]
Density, g/cm³: 2.977 [KIR81]
Melting Point, °C: 652 [KIR81]
Boiling Point, °C: 1190 [KIR81]

1772
Compound: Manganese(II) chloride tetrahydrate
Formula: $MnCl_2 \cdot 4H_2O$
Molecular Formula: $Cl_2H_8MnO_4$
Molecular Weight: 197.905
CAS RN: 13446-34-9
Properties: reddish, sl deliq, monocl cryst; there is a
  dihydrate, 20603-88-7 [KIR81] [MER89]
Solubility: s 0.7 parts H_2O; s alcohol; i ether
  [MER89]
Density, g/cm³: 1.913 [HAW93]
Melting Point, °C: 87.5 [HAW93]

1773
Compound: Manganese(II) citrate
Synonyms: manganous citrate
Formula: $Mn_3(C_6H_5O_7)_2$
Molecular Formula: $C_{12}H_{10}Mn_3O_{14}$
Molecular Weight: 543.017
CAS RN: 71799-92-3
Properties: white powd; used as a food and feed
  additive, and as a dietary supplement [HAW93]
Solubility: s H_2O containing dissolved sodium
  citrate [HAW93]

1774
Compound: Manganese(II) dihydrogen phosphate
  dihydrate
Formula: $Mn(H_2PO_4)_2 \cdot 2H_2O$
Molecular Formula: $H_8MnO_{10}P_2$
Molecular Weight: 284.944
CAS RN: 18718-07-5
Properties: almost colorless cryst; deliq [KIR81]
Solubility: s H_2O; i ethanol [KIR81]
Reactions: minus H_2O at 100°C [KIR81]

1775
Compound: Manganese(II) dithionate
Formula: $Mn(SO_3)_2$
Molecular Formula: MnO_6S_2
Molecular Weight: 215.066
CAS RN: 13568-72-4
Properties: tricl cryst [CRC94]

Solubility: s H_2O [HAW93]
Density, g/cm³: 1.76 [HAW93]

1776
Compound: Manganese(II) fluoride
Synonyms: manganous fluoride
Formula: MnF_2
Molecular Formula: F_2Mn
Molecular Weight: 92.935
CAS RN: 7782-64-1
Properties: hygr; -80 mesh with 99.5% purity;
  reddish powd [HAW93] [CER91] [STR93]
Solubility: g/100g soln, H_2O: 0.800 (0°C), 1.00
  (23.5°C), 0.48 (100°C); equilibrium solid phase,
  $MnF_2 \cdot 4H_2O$ (0°C), MnF_2 (25°C, 100°C)
  [KRU93]; s dil HF, conc HCl or HNO_3 [MER89]
Density, g/cm³: 3.98 [MER89]
Melting Point, °C: 856 [MER89]

1777
Compound: Manganese(II) hydrogen phosphate
  trihydrate
Formula: $MnHPO_4 \cdot 3H_2O$
Molecular Formula: H_7MnO_7P
Molecular Weight: 204.959
CAS RN: 10236-39-2
Properties: has two forms: gray, prepared by
  decomposition of $MnCO_3$, and pink, prepared by
  reaction between phosphoric acid and
  $Mn_3(PO_4)_2$ [MER89]
Solubility: pink form: v sl s H_2O; s dil acids; gray
  form: s only in hot conc HCl [MER89]
Reactions: minus H_2O >100°C [CRC94]

1778
Compound: Manganese(II) hydroxide
Synonyms: pyrochroite
Formula: $Mn(OH)_2$
Molecular Formula: H_2MnO_2
Molecular Weight: 88.953
CAS RN: 18933-05-6
Properties: white to pink cryst; hardness is 2.5
  Mohs [HAW93]
Solubility: mol/L soln, H_2O: 3.6 x 10^{-5} (25°C)
  [KRU93]
Density, g/cm³: 3.258 [HAW93]
Melting Point, °C: decomposes [HAW93]

1779
Compound: Manganese(II) hypophosphite
  monohydrate

Formula: $Mn(H_2PO_2)_2 \cdot H_2O$
Molecular Formula: $H_6MnO_5P_2$
Molecular Weight: 202.931
CAS RN: 10043-84-2
Properties: pink, odorless, almost tasteless cryst or powd; used as a food additive and dietary supplement [HAW93] [MER89]
Solubility: 1 g/6.5mL H_2O, 1 g/6 mL boiling H_2O; i alcohol [MER89]
Reactions: evolves phosphine when heated [MER89]

1780
Compound: Manganese(II) iodide
Formula: MnI_2
Molecular Formula: I_2Mn
Molecular Weight: 308.747
CAS RN: 7790-33-2
Properties: red-brown powd; hygr [STR93]
Solubility: s H_2O, with gradual decomposition; s alcohol [HAW93]
Density, g/cm³: 5.01 [HAW93]
Melting Point, °C: 638 [HAW93]

1781
Compound: Manganese(II) iodide tetrahydrate
Formula: $MnI_2 \cdot 4H_2O$
Molecular Formula: $H_8I_2MnO_4$
Molecular Weight: 380.809
CAS RN: 7790-33-2
Properties: rose red cryst; rapidly turns brown when exposed to air and light due to liberation of iodine [MER89]
Solubility: v s H_2O, with gradual decomposition; s alcohol [MER89]
Melting Point, °C: decomposes [CRC94]

1782
Compound: Manganese(II) molybdate
Formula: $MnMoO_4$
Molecular Formula: $MnMoO_4$
Molecular Weight: 214.876
CAS RN: 14013-15-1
Properties: yellow or off-white powd; monocl [KIR81] [STR93]
Density, g/cm³: 4.05 [LID94]

1783
Compound: Manganese(II) nitrate
Formula: $Mn(NO_3)_2$
Molecular Formula: MnN_2O_6

Molecular Weight: 178.948
CAS RN: 10377-66-9
Properties: liq; available in dissolved form; forms pink solutions [ALF95] [STR93]
Solubility: g/100g soln, H_2O: 50.49 (0°C), 61.74 (25°C); solid phase, $Mn(NO_3)_2 \cdot 6H_2O$ [KRU93]

1784
Compound: Manganese(II) nitrate hexahydrate
Formula: $Mn(NO_3)_2 \cdot 6H_2O$
Molecular Formula: $H_{12}MnN_2O_{12}$
Molecular Weight: 287.040
CAS RN: 17141-63-8
Properties: rose colored, deliq, monocl needles; used in ceramics, as a catalyst [HAW93] [MER89]
Solubility: v s H_2O, alcohol [MER89]
Density, g/cm³: 1.8 [MER89]

1785
Compound: Manganese(II) nitrate tetrahydrate
Formula: $Mn(NO_3)_2 \cdot 4H_2O$
Molecular Formula: $H_8MnN_2O_{10}$
Molecular Weight: 251.010
CAS RN: 20694-39-7
Properties: pink, deliq cryst masses below 20°C [MER89]
Solubility: v s H_2O, s alcohol [MER89]
Density, g/cm³: 2.129 [MER89]
Melting Point, °C: 37.1 [MER89]

1786
Compound: Manganese(II) oxalate dihydrate
Formula: $MnC_2O_4 \cdot 2H_2O$
Molecular Formula: $C_2H_4MnO_6$
Molecular Weight: 178.988
CAS RN: 6556-16-7
Properties: white cryst powd; used as a paint and varnish drier [HAW93] [MER89]
Solubility: g/100g soln in H_2O: 0.0198 (0°C), 0.0309±0.0002 (25°C); Solid phase $MnC_2O_4 \cdot 2H_2O$ [KRU93]
Density, g/cm³: 2.453 [HAW93]
Melting Point, °C: decomposes 150 [MER89]
Reactions: minus $2H_2O$ at 100°C [HAW93]

1787
Compound: Manganese(II) oxide
Synonyms: manganosite
Formula: MnO
Molecular Formula: MnO

Molecular Weight: 70.937
CAS RN: 1344-43-0
Properties: grass green powd; cub; enthalpy of fusion 54.40 kJ/mol; produced from MnO_2 ores by roasting in a reducing atm, e.g. methane; used in textile printing, in ceramics, in paints [HAW93] [CRC93]
Solubility: i H_2O; s acids [HAW93]
Density, g/cm³: 5.37 [KIR81]
Melting Point, °C: 1840 [LID94]

1788
Compound: Manganese(II) perchlorate hexahydrate
Formula: $Mn(ClO_4)_2 \cdot 6H_2O$
Molecular Formula: $Cl_2H_{12}MnO_{14}$
Molecular Weight: 361.930
CAS RN: 15364-94-0
Properties: pink cryst; hygr [STR93]
Density, g/cm³: 2.10 [LID94]
Melting Point, °C: 165, decomposes [AES93]

1789
Compound: Manganese(II) phosphate heptahydrate
Formula: $Mn_3(PO_4)_2 \cdot 7H_2O$
Molecular Formula: $H_{14}Mn_3O_{15}P_2$
Molecular Weight: 480.864
CAS RN: 10236-39-2
Properties: reddish white powd; used in conversion coating of steel, aluminum and other metals [HAW93]
Solubility: i H_2O; s mineral acids [HAW93]

1790
Compound: Manganese(II) pyrophosphate
Formula: $Mn_2P_2O_7$
Molecular Formula: $Mn_2O_7P_2$
Molecular Weight: 283.819
CAS RN: 53731-35-4
Properties: white powd [HAW93]
Solubility: i H_2O; s solutions of potassium or sodium pyrophosphate [HAW93]
Density, g/cm³: 3.71 [HAW93]
Melting Point, °C: 1196 [HAW93]

1791
Compound: Manganese(II) pyrophosphate trihydrate
Formula: $Mn_2P_2O_7 \cdot 3H_2O$
Molecular Formula: $H_6Mn_2O_{10}P_2$
Molecular Weight: 337.866
CAS RN: 53731-35-4

Properties: white or nearly white powd [MER89]
Solubility: i H_2O, s in excess of alkali pyrophosphate, acids [MER89]

1792
Compound: Manganese(II) selenide
Formula: MnSe
Molecular Formula: MnSe
Molecular Weight: 133.898
CAS RN: 1313-22-0
Properties: black cryst; -20 mesh with 99.9% purity [CER91] [STR93]
Density, g/cm³: 5.45 [LID94]
Melting Point, °C: 1460 [LID94]

1793
Compound: Manganese(II) sulfate
Formula: $MnSO_4$
Molecular Formula: MnO_4S
Molecular Weight: 151.002
CAS RN: 7785-87-7
Properties: almost white; ortho-rhomb [KIR81]
Solubility: g/100g soln, H_2O: 34.6 (0°C), 39.2 (25°C), 26.1 (100.7°C); Solid phase, $MnSO_4 \cdot 7H_2O$ (0°C), $MnSO_4 \cdot H_2O$ (25°C, 100.7°C) [KRU93]
Density, g/cm³: 3.25 [KIR81]
Melting Point, °C: 700 [HAW93]
Boiling Point, °C: decomposes at 850 [HAW93]

1794
Compound: Manganese(II) sulfate monohydrate
Formula: $MnSO_4 \cdot H_2O$
Molecular Formula: H_2MnO_5S
Molecular Weight: 169.017
CAS RN: 10034-96-5
Properties: pale red, sl efflorescent cryst; used in dyeing [MER89]
Solubility: s in about 1 part cold H_2O, 0.6 parts boiling; i alcohol [MER89]
Density, g/cm³: 2.95 [LID94]
Reactions: minus H_2O at 400-450°C [MER89]

1795
Compound: Manganese(II) sulfate tetrahydrate
Formula: $MnSO_4 \cdot 4H_2O$
Molecular Formula: H_8MnO_8S
Molecular Weight: 223.063
CAS RN: 10101-68-5

Properties: translucent, pale rose red; efflorescent
    prisms; used in fertilizers, as a feed additive, in
    ceramics [HAW93]
Solubility: s H_2O; i alcohol [HAW93]
Density, g/cm³: 2.107 [HAW93]
Melting Point, °C: 30 [HAW93]

1796
Compound: Manganese(II) sulfide
Synonyms: alabandite, manganblende
Formula: MnS
Molecular Formula: MnS
Molecular Weight: 87.004
CAS RN: 18820-29-6
Properties: pink, green or brownish green powd;
    three cryst forms: α: green cub; β: red cub; γ:
    red hex; used as an additive in steel production
    [HAW93] [MER89]
Solubility: i H_2O; s dil acids [MER89]
Density, g/cm³: 3.99 [STR93]
Melting Point, °C: decomposes 1610 [AES93]
Reactions: readily oxidized in moist air to sulfate
    [MER89]

1797
Compound: Manganese(II) telluride
Formula: MnTe
Molecular Formula: MnTe
Molecular Weight: 182.538
CAS RN: 12032-88-1
Properties: hex cryst; possibly contains some
    $MnTe_2$, 12032-89-2; 6 mm pieces and smaller
    with 99.9% purity [LID94] [CER91]
Density, g/cm³: 6.0 [LID94]

1798
Compound: Manganese(II) titanate
Synonyms: pyrophenite
Formula: $MnTiO_3$
Molecular Formula: MnO_3Ti
Molecular Weight: 150.803
CAS RN: 12032-74-5
Properties: -100 mesh with 99.9% purity; red
    hexagonal cryst [LID94] [CER91] [KIR83]
Density, g/cm³: 4.54 [KIR83]
Melting Point, °C: 1360 [KIR83]

1799
Compound: Manganese(II) tungstate
Formula: $MnWO_4$
Molecular Formula: MnO_4W

Molecular Weight: 302.776
CAS RN: 13918-22-4
Properties: -200 mesh with 99.9% purity; off-white
    powd [STR93] [CER91]
Density, g/cm³: 7.2 [LID94]

1800
Compound: Manganese(II) zirconate
Formula: $MnZrO_3$
Molecular Formula: MnO_3Zr
Molecular Weight: 194.160
CAS RN: 70692-94-3
Properties: reacted product; -200 mesh with 99.5%
    purity [CER91]

1801
Compound: Manganese(II,III) oxide
Synonyms: hausmannite
Formula: Mn_3O_4
Molecular Formula: Mn_3O_4
Molecular Weight: 228.812
CAS RN: 1317-35-7
Properties: brownish powd; tetr, a = 0.5762 nm, c =
    0.9470 nm [THA92] [HAW93]
Solubility: i H_2O [KIR81]; s HCl, releasing Cl_2
    [MER89]
Density, g/cm³: 4.876 [HAW93]
Melting Point, °C: 1564 [HAW93]

1802
Compound: Manganese(III) acetate dihydrate
Formula: $Mn(CH_3COO)_3 \cdot 2H_2O$
Molecular Formula: $C_6H_{13}MnO_8$
Molecular Weight: 268.102
CAS RN: 19513-05-4
Properties: brown cryst; mild selective oxidizing
    agent [ALD94] [STR93]

1803
Compound: Manganese(III) acetylacetonate
Synonyms: 2,4-pentanedione, manganese(III)
    derivative
Formula: $Mn(CH_3COCH=C(O)CH_3)_3$
Molecular Formula: $C_{15}H_{21}MnO_6$
Molecular Weight: 352.266
CAS RN: 14284-89-0
Properties: black cryst; hygr [STR93]

$$[CH_3-C=CH-C-CH_3]_3Mn$$

with the substituents: O– and O (one single-bonded, one double-bonded)

Melting Point, °C: decomposes at 160 [ALD94]

1804
Compound: Manganese(III) fluoride
Formula: MnF_3
Molecular Formula: F_3Mn
Molecular Weight: 111.933
CAS RN: 7783-53-1
Properties: red cryst; monocl; hygr; used as a fluorinating agent [HAW93] [KIR81]
Solubility: decomposed by H_2O [KIR81]
Density, g/cm³: 3.54 [KIR81]
Melting Point, °C: decomposes >600 [KIR81]

1805
Compound: Manganese(III) hydroxide
Synonyms: manganite
Formula: γ-MnO(OH)
Molecular Formula: $HMnO_2$
Molecular Weight: 87.945
CAS RN: 1332-63-4
Properties: black solid; monocl [KIR81]
Solubility: i H_2O; disproportionates in dil acid [KIR81]
Density, g/cm³: 4.2-4.4 [KIR81]
Melting Point, °C: decomposes at 250 [KIR81]
Reactions: transformed to Mn_2O_3 at 250°C [KIR81]

1806
Compound: Manganese(III) oxide
Synonyms: braunite
Formula: α-Mn_2O_3
Molecular Formula: Mn_2O_3
Molecular Weight: 157.874
CAS RN: 1317-34-6
Properties: black to brown solid; rhomb, cub; very hard [HAW93] [KIR81]
Solubility: i H_2O [KIR81]; s HCl, evolving Cl_2 [MER89]
Density, g/cm³: 4.50 [HAW93]
Melting Point, °C: 1080, decomposes [HAW93]

1807
Compound: Manganese(IV) oxide
Synonyms: manganese dioxide, pyrolusite
Formula: MnO_2
Molecular Formula: MnO_2
Molecular Weight: 86.937
CAS RN: 1313-13-9

Properties: black cryst or powd; α-MnO_2: a = 0.97876 nm, c = 0.28650 nm; λ-MnO_2: cub, a = 0.8029 nm; used as an oxidizing agent, in dry cell batteries [HAW93] [THA92] [ROS92]
Solubility: i H_2O, HNO_3, cold H_2SO_4; slowly dissolves in HCl, evolving Cl_2; s dil H_2SO_4, dil HNO_3 in presence of H_2O_2 or oxalic acid [MER89]
Density, g/cm³: 5.026 [STR93]; α: 4.21 [ROS92]
Melting Point, °C: 535, decomposes [STR93]
Reactions: minus O, 535°C [CRC94]

1808
Compound: Manganese(IV) telluride
Synonyms: manganese ditelluride
Formula: $MnTe_2$
Molecular Formula: $MnTe_2$
Molecular Weight: 310.14
CAS RN: 12032-89-2
Properties: 6 mm pieces and down [STR93]

1809
Compound: Manganese(VII) oxide
Synonyms: manganese heptoxide
Formula: Mn_2O_7
Molecular Formula: Mn_2O_7
Molecular Weight: 221.872
CAS RN: 12057-92-0
Properties: dark red oil; hygr [KIR81]
Solubility: v s H_2O [KIR81]
Density, g/cm³: 2.396 [KIR81]
Melting Point, °C: 5.9 [KIR81]
Boiling Point, °C: decomposes 55 [KIR81]

1810
Compound: Mendelevium
Formula: Md
Molecular Formula: Md
Molecular Weight: 258
CAS RN: 7440-11-1
Properties: discovered in 1955 by Ghiorso and colleagues at Lawrence Berkeley Laboratory; preparation: bombardment of ^{253}Es with He ions [KIR78]
Melting Point, °C: 827 [LID94]

1811
Compound: Mercury
Synonyms: quicksilver
Formula: Hg
Molecular Formula: Hg

Molecular Weight: 200.59
CAS RN: 7439-97-6
Properties: silvery white, heavy liq metal; surface tension is 480 dynes/cm; enthalpy of vaporization 59.11 kJ/mol; enthalpy of fusion 2.29 kJ/mol; resistivity (20°C) 95.8 μohm·cm; critical density 3.56 g/cm^3; critical pressure 74.2 MPa; critical temp 1677°C; expansion coefficient of liq at 20°C is 182 x 10^{-6}/°C; viscosity at 20°C is 1.55 mPa s [KIR81] [HAW93] [COT88] [CRC93]
Solubility: 20-30 microgram/L in H_2O; s in boiling H_2SO_4, HNO_3 [KIR81] [HAW93]
Density, g/cm^3: 13.534 [MER89]
Melting Point, °C: -38.87 [MER89]
Boiling Point, °C: 356.73 [CRC93]
Thermal Conductivity, W/(m·K): 9.2 (25°C) [KIR81]
Thermal Expansion Coefficient: liq: 100°C (1.458), 200°C (3.307) [CLA66]

1812
Compound: Mercury(I) acetate
Synonyms: mercurous acetate
Formula: $Hg_2(CH_3COO)_2$
Molecular Formula: $C_4H_6Hg_2O_4$
Molecular Weight: 519.370
CAS RN: 631-60-7
Properties: colorless scales; lustrous leaflets or cryst powd; light sensitive, becomes dark; aq solutions decomposed quickly by light and heat; used in medicine as an anti-bacteria agent [HAW93] [MER89]
Solubility: 0.75 g/100mL H_2O (12°C) [CRC94]
Melting Point, °C: decomposes [CRC94]

1813
Compound: Mercury(I) bromide
Synonyms: mercurous bromide
Formula: Hg_2Br_2
Molecular Formula: Br_2Hg_2
Molecular Weight: 560.988
CAS RN: 15385-58-7
Properties: white, odorless tasteless powd; tetr; sensitive to light (darkens); decomposed by hot HCl or alkali bromides; becomes yellow when heated, returns to white color when cooled; prepared by oxidation of Hg with Br_2, or as a precipitate by addition of NaBr to $HgNO_3$ solution [KIR81] [HAW93] [MER89]
Solubility: g/L soln, H_2O: 3.9 x 10^{-5} (25°C) [KRU93]; i alcohol, ether [MER89]; s in fuming HNO_3, hot conc H_2SO_4 [HAW93]
Density, g/cm^3: 7.307 [HAW93]

Melting Point, °C: 405 [HAW93]
Boiling Point, °C: sublimes at 340-350 [HAW93]

1814
Compound: Mercury(I) carbonate
Formula: Hg_2CO_3
Molecular Formula: CHg_2O_3
Molecular Weight: 461.189
CAS RN: 50968-00-8
Properties: yellow powd [LAN52]
Solubility: g/L soln, H_2O: 8.8 x 10^{-9} (25°C) [KRU93]
Melting Point, °C: 130, decomposes [LAN52]

1815
Compound: Mercury(I) chlorate
Synonyms: mercurous chlorate
Formula: $Hg_2(ClO_3)_2$
Molecular Formula: $Cl_2Hg_2O_6$
Molecular Weight: 568.081
CAS RN: 10294-44-7
Properties: white cryst; decomposes ~250°C to O_2, HgO, $HgCl_2$ [MER89]
Solubility: sl s H_2O, hydrolyzed in hot H_2O forming basic salt [MER89]; s alcohol, acetic acid [HAW93]
Density, g/cm^3: 6.409 [HAW93]
Melting Point, °C: 250, decomposes [HAW93]

1816
Compound: Mercury(I) chloride
Synonyms: mercurous chloride, calomel
Formula: Hg_2Cl_2
Molecular Formula: Cl_2Hg_2
Molecular Weight: 472.085
CAS RN: 10112-91-1
Properties: white, odorless, tasteless powd; rhomb cryst; slowly decomposed by sunlight to liq Hg and $HgCl_2$; can be obtained by oxidation of Hg with Cl_2; used as a fungicide, as a reference electrode, in ceramic painting; decomposed by alkalies [HAW93] [MER89]
Solubility: g/100g soln, H_2O: 0.000140 (0.5°C); mol/L soln, H_2O: 7.5 x 10^{-6} (25°C) [KRU93]; i alcohol, ether [MER89]
Density, g/cm^3: 7.15 [MER89]
Melting Point, °C: 302 [HAW93]
Boiling Point, °C: 384 [HAW93]

1817
Compound: Mercury(I) chromate

Synonyms: mercurous chromate
Formula: Hg$_2$CrO$_4$
Molecular Formula: CrHg$_2$O$_4$
Molecular Weight: 517.174
CAS RN: 13444-75-2
Properties: brick red powd; ortho-rhomb cryst; can have a variable composition; used for coloring ceramics green [HAW93] [KIR78]
Solubility: i H$_2$O, alcohol; s conc HNO$_3$ [HAW93]
Melting Point, °C: decomposes [CRC94]

1818
Compound: Mercury(I) fluoride
Synonyms: mercurous fluoride
Formula: Hg$_2$F$_2$
Molecular Formula: F$_2$Hg$_2$
Molecular Weight: 439.177
CAS RN: 13967-25-4
Properties: small, yellow cub cryst; blackens in light; preparation by reaction of Hg$_2$CO$_3$ and HF [MER89]
Solubility: hydrolyzed in H$_2$O to Hg(l), HgO, HF [MER89]
Density, g/cm^3: 8.73 [MER89]
Melting Point, °C: ~240, sublimes [MER89]
Boiling Point, °C: decomposes 570 [MER89]

1819
Compound: Mercury(I) iodate
Formula: Hg$_2$(IO$_3$)$_2$
Molecular Formula: Hg$_2$I$_2$O$_6$
Molecular Weight: 750.985
CAS RN: 13465-35-9
Properties: yellowish [LAN52]
Solubility: g/L soln, H$_2$O: 6.0 x 10^{-7} (25°C) [KRU93]
Melting Point, °C: 250, volatilizes [LAN52]

1820
Compound: Mercury(I) iodide
Synonyms: mercurous iodide
Formula: Hg$_2$I$_2$
Molecular Formula: Hg$_2$I$_2$
Molecular Weight: 654.989
CAS RN: 15385-57-6

Properties: bright yellow, amorphous, odorless, tasteless powd; darkens or becomes greenish in light forming HgI and liq Hg; color changes to dark yellow, orange and orange-red when heated, reverse order when cooled; enthalpy of fusion 27.00 kJ/mol; can be made by precipitation in HgNO$_3$ solution with KI; used as a topical antibacterial agent in medicine [HAW93] [MER89] [KIR81] [CRC93]
Solubility: g/L soln, H$_2$O: 2.0 x 10^{-7} (25°C) [KRU93]; i alcohol, ether [MER89]; s castor oil, ammonia [HAW93]
Density, g/cm^3: 7.65-7.75 [HAW93]
Melting Point, °C: 290, when rapidly heated, decomposes to HgI, Hg(l) [MER89]
Boiling Point, °C: sublimes at 140 [HAW93]

1821
Compound: Mercury(I) nitrate dihydrate
Synonyms: mercurous nitrate
Formula: HgNO$_3$·2H$_2$O
Molecular Formula: H$_4$HgNO$_5$
Molecular Weight: 298.626
CAS RN: 10415-75-5
Properties: short prismatic cryst; effloresces, becoming anhydrous in dry air; light sensitive; used as an analytical reagent [HAW93]
Solubility: sensitive to H$_2$O: s in small quantities of warm H$_2$O, but hydrolyzes in larger amounts [HAW93]
Density, g/cm^3: 4.785 (3.9°C) [HAW93]
Melting Point, °C: 70, decomposes [HAW93]

1822
Compound: Mercury(I) nitrate monohydrate
Formula: HgNO$_3$·H$_2$O
Molecular Formula: H$_2$HgNO$_4$
Molecular Weight: 280.611
CAS RN: 7782-86-7
Properties: white cryst; anhydrous is white monocl which can be prepared by dissolution of Hg in hot dil HNO$_3$ followed by crystallization, some Hg(NO$_3$)$_2$ is also formed [KIR81] [STR93]
Density, g/cm^3: 4.79 [STR93]
Melting Point, °C: 70 [STR93]

1823
Compound: Mercury(I) oxide
Synonyms: mercurous oxide
Formula: Hg$_2$O
Molecular Formula: Hg$_2$O
Molecular Weight: 417.179
CAS RN: 15829-53-5

Properties: black or brownish black powd; light sensitive; no evidence that Hg_2O has been isolated; on adding NaOH solution to $HgNO_3$, x-ray data indicate could be intimate mixture of HgO and Hg [MER89]

Solubility: i H_2O; s HNO_3; reacts with HCl to form calomel, Hg_2Cl_2 [MER89]

Density, g/cm^3: 9.8 [HAW93]

Melting Point, °C: 100, decomposes [HAW93]

1824

Compound: Mercury(I) perchlorate tetrahydrate

Formula: $HgClO_4 \cdot 4H_2O$

Molecular Formula: ClH_8HgO_8

Molecular Weight: 372.102

CAS RN: 65202-12-2

Properties: white cryst [STR93]

Solubility: g/100g soln in H_2O: 282 (0°C), 394 (25°C), 580 (99°C); Solid phase: $Hg_2(ClO_4)_2 \cdot 4H_2O$ (0°C, 25°C), $Hg_2(ClO_4)_2 \cdot 2H_2O$ (99°C) [KRU93]

1825

Compound: Mercury(I) sulfate

Synonyms: mercurous sulfate

Formula: Hg_2SO_4

Molecular Formula: Hg_2O_4S

Molecular Weight: 497.224

CAS RN: 7783-36-0

Properties: white to sl yellow cryst powd; becomes gray and forms Hg and $HgSO_4$ in light; decomposes on heating; forms as a precipitate by addition of dil H_2SO_4 to a solution of $HgNO_3$; used in Clark and Weston standard cells [KIR81] [HAW93] [MER89]

Solubility: 0.05 g/100g H_2O [KIR81]; s dil HNO_3 [MER89]

Density, g/cm^3: 7.56 [MER89]

Melting Point, °C: decomposes [AES93]

1826

Compound: Mercury(I) sulfide

Formula: Hg_2S

Molecular Formula: Hg_2S

Molecular Weight: 433.246

CAS RN: 51595-71-2

Properties: black [LAN52]

Solubility: g/L soln, H_2O: 2.8×10^{-23} (25°C) [KRU93]

Melting Point, °C: decomposes [LAN52]

1827

Compound: Mercury(II) acetate

Synonyms: mercuric acetate

Formula: $Hg(CH_3COO)_2$

Molecular Formula: $C_4H_6HgO_4$

Molecular Weight: 318.680

CAS RN: 1600-27-7

Properties: white cryst or light yellow powd; -60 mesh with 99.9% purity; sensitive to light; prepared by dissolving HgO in warm 20% acetic acid; used in pharmaceuticals and as a catalyst in organic synthesis [HAW93] [STR93] [KIR81] [CER91]

Solubility: s H_2O, alcohol [HAW93]

Density, g/cm^3: 3.25 [HAW93]

Melting Point, °C: 179-182 [ALD94]

1828

Compound: Mercury(II) arsenate

Synonyms: mercuric arsenate

Formula: $HgHAsO_4$

Molecular Formula: $AsHHgO_4$

Molecular Weight: 340.518

CAS RN: 7784-37-4

Properties: yellow powd; used to waterproof paints, and in antifouling paints [HAW93] [MER89]

Solubility: i H_2O; s HCl, HNO_3 [MER89]

1829

Compound: Mercury(II) basic carbonate

Formula: $HgCO_3 \cdot 3HgO$

Molecular Formula: CHg_4O_6

Molecular Weight: 910.367

Properties: brown precipitate formed by adding sodium carbonate solution to $HgCl_2$ solution, forming a slurry; refluxing the slurry decomposes carbonate to red HgO [KIR81]

1830

Compound: Mercury(II) benzoate monohydrate

Formula: $Hg(C_7H_5O_2)_2 \cdot H_2O$

Molecular Formula: $C_{14}H_{12}HgO_5$

Molecular Weight: 460.836

CAS RN: 583-15-3

Properties: white cryst; odorless cryst powd; sensitive to light [MER89] [HAW93]

Solubility: 1.2 g/100mL H_2O (15°C), 2.5 g/100mL H_2O (100°C) [CRC94]

Melting Point, °C: 165 [HAW93]

Reactions: hydrolyzed in boiling H_2O to a basic salt and benzoic acid [MER89]

1831
Compound: Mercury(II) bromide
Formula: $HgBr_2$
Molecular Formula: Br_2Hg
Molecular Weight: 360.398
CAS RN: 7789-47-1
Properties: white rhomb cryst or powd; sensitive to light; enthalpy of vaporization 58.89 kJ/mol; enthalpy of fusion 17.90 kJ/mol; can be prepared by precipitation below 75°C from $Hg(NO_3)_2$ solution with NaBr, followed by drying; used in medicine [HAW93] [MER89] [KIR81] [CRC93]
Solubility: g/100g soln, H_2O: 0.3 (0°C), 0.611±0.002 (25°C), 4.7 (100°C) [KRU93]; v s hot alcohol, methanol, HCl, HBr; sl s chloroform [MER89]
Density, g/cm³: 6.109 [STR93]
Melting Point, °C: 236 [CRC94]
Boiling Point, °C: 322 [CRC93]

1832
Compound: Mercury(II) chloride
Synonyms: corrosive sublimate
Formula: $HgCl_2$
Molecular Formula: Cl_2Hg
Molecular Weight: 271.495
CAS RN: 7487-94-7
Properties: cryst or white granules or powd; odorless; volatilizes unchanged at ~300°C; vapor pressure 13 Pa (100°C), 400 Pa (150°C); enthalpy of vaporization 58.9 kJ/mol; enthalpy of fusion 19.41 kJ/mol; obtained by reaction of Hg with Cl_2; used to coagulate albumin, to manufacture mercury compounds, as a disinfectant [HAW93] [MER89] [KIR81] [JAN71]
Solubility: g/100g soln, H_2O: 3.5 (0°C), 6.8 (25°C), 36.5 (100°C) [KRU93]; 1g dissolves in 3.8mL alcohol, 200mL benzene, 22mL ether, 12mL glycerol, 40mL acetic acid; s other organics [MER89]
Density, g/cm³: 5.44 [HAW93]; vapor: 9.8 [KIR81]
Melting Point, °C: 276 [JAN71]
Boiling Point, °C: 304 [JAN71]

1833
Compound: Mercury(II) chloride ammoniated
Synonyms: ammoniated mercuric chloride
Formula: $Hg(NH_2)Cl$
Molecular Formula: ClH_2HgN
Molecular Weight: 252.066
CAS RN: 10124-48-8
Properties: white lumps, odorless powd; stable in air; darkens when exposed to light; used in medicine as a topical anti-infective [HAW93] [MER89]
Solubility: i H_2O, alcohol; s warm HCl, HNO_3, acetic acid [MER89]
Density, g/cm³: 5.38 [MER89]

1834
Compound: Mercury(II) chromate
Synonyms: mercuric chromate
Formula: $HgCrO_4$
Molecular Formula: $CrHgO_4$
Molecular Weight: 316.584
CAS RN: 13444-75-2
Properties: red ortho-rhomb; used in antifouling formulations [KIR78]
Solubility: sl s H_2O [KIR78]
Density, g/cm³: 6.06 [LID94]
Melting Point, °C: decomposes [CRC94]

1835
Compound: Mercury(II) cyanide
Synonyms: mercuric cyanide
Formula: $Hg(CN)_2$
Molecular Formula: C_2HgN_2
Molecular Weight: 252.625
CAS RN: 592-04-1
Properties: colorless, odorless; tetr cryst; darkens if exposed to light; can be made by reaction of aq slurry of yellow HgO with excess HCN, followed by heating to 95°C and filtration; used as an antiseptic in medicine, in germicidal soaps, and in photography [HAW93] [MER89]
Solubility: g/100g soln, H_2O: 6.31 (0°C), 10.06±0.06 (25°C), 35.05 (101.1°C) [KRU93]; 1g dissolves in 13mL alcohol, 4mL methanol; sl s ether; slowly s glycerol [MER89]
Density, g/cm³: 3.996 [MER89]
Melting Point, °C: decomposes 320 [MER89]

1836
Compound: Mercury(II) dichromate
Synonyms: mercuric dichromate
Formula: $HgCr_2O_7$
Molecular Formula: Cr_2HgO_7
Molecular Weight: 416.578
CAS RN: 7789-10-8
Properties: red, heavy, cryst powd [MER89]
Solubility: i H_2O; s HCl, HNO_3 [MER89]

1837
Compound: Mercury(II) fluoride
Synonyms: mercuric fluoride
Formula: HgF_2
Molecular Formula: F_2Hg
Molecular Weight: 238.587
CAS RN: 7783-39-3
Properties: transparent cryst; white powd or cub cryst; very sensitive to moisture; can be obtained from a reaction of F_2 with $HgCl_2$ or HgO; used in the synthesis of organic fluoride compounds [HAW93] [MER89] [KIR78]
Solubility: turns yellow and hydrolyzes in H_2O after prolonged time [MER89]; moderately s in alcohol [HAW93]
Density, g/cm³: 8.95 [MER89]
Melting Point, °C: 645, decomposes [HAW93]

1838
Compound: Mercury(II) fulminate
Formula: $Hg(CNO)_2$
Molecular Formula: $C_2HgN_2O_2$
Molecular Weight: 284.624
CAS RN: 628-86-4
Properties: gray, cryst powd; explodes readily if dry; used to manufacture caps and detonators for explosives [HAW93]
Solubility: sl s cold H_2O, s hot H_2O; s alcohol, NH_4OH [HAW93]
Density, g/cm³: 4.42 [HAW93]
Melting Point, °C: explodes [HAW93]

1839
Compound: Mercury(II) iodate
Synonyms: mercuric iodate
Formula: $Hg(IO_3)_2$
Molecular Formula: HgI_2O_6
Molecular Weight: 550.395
CAS RN: 7783-32-6
Properties: white powd [MER89]
Solubility: g/100g H_2O: 0.002 (20°C) [MER89]
Melting Point, °C: decomposes 175 [LID94]

1840
Compound: Mercury(II) iodide(α)
Synonyms: mercuric iodide
Formula: α-HgI_2
Molecular Formula: HgI_2
Molecular Weight: 454.399
CAS RN: 7774-29-0

Properties: scarlet red heavy odorless, almost tasteless powd; tetr; sensitive to light; enthalpy of vaporization 59.2 kJ/mol; enthalpy of fusion 18.90 kJ/mol; obtained as a precipitate from a solution of $HgCl_2$ and KI; used to treat skin diseases and as an analytical reagent [KIR81] [CRC93] [MER89]
Solubility: 0.006 g/100g H_2O (25°C); 1g dissolves in: 115mL alcohol, 20mL boiling alcohol, about 120mL ether, about 60mL acetone, 910mL chloroform, 75mL ethyl acetate [MER89]
Density, g/cm³: 6.28 [MER89]
Melting Point, °C: 259 [MER89]
Boiling Point, °C: 354 [CRC93]
Reactions: transition red → yellow at 130°C [MER89]

1841
Compound: Mercury(II) iodide(β)
Synonyms: coccinite
Formula: β-HgI_2
Molecular Formula: HgI_2
Molecular Weight: 454.399
CAS RN: 7774-29-0
Properties: rhomb; turns yellow from red α-form when α-form is heated to 130°C, then reverses color change when cooled; used as an antiseptic in medicine and as Nessler's reagent [HAW93] [STR93] [CRC94]
Solubility: g/L soln, H_2O: 0.050±0.06 (25°C) [KRU93]; s boiling alcohol [HAW93]
Density, g/cm³: 6.094 [CRC94]
Melting Point, °C: 259 (under argon) [STR93]
Boiling Point, °C: 349 [HAW93]

1842
Compound: Mercury(II) nitrate hemihydrate
Synonyms: mercuric nitrate
Formula: $Hg(NO_3)_2 \cdot 1/2H_2O$
Molecular Formula: $HHgN_2O_{6.5}$
Molecular Weight: 333.607
CAS RN: 10045-94-0
Properties: colorless cryst or white deliq powd; decomposes by heating; made by dissolution of Hg in hot conc HNO_3, then cooling to crystallize product; used in the nitration of aromatic organic compounds and in felt manufacture [HAW93]
Solubility: s H_2O, HNO_3; i alcohol [HAW93]
Density, g/cm³: 4.39 [CRC94]
Melting Point, °C: 79 [HAW93]
Boiling Point, °C: decomposes [CRC94]

1843
Compound: Mercury(II) nitrate monohydrate
Synonyms: mercuric nitrate monohydrate
Formula: $Hg(NO_3)_2 \cdot H_2O$
Molecular Formula: $H_2HgN_2O_7$
Molecular Weight: 342.615
CAS RN: 7783-34-8
Properties: white or sl yellow, deliq cryst powd [MER89]
Solubility: s H_2O, decomposes; s dil acids [MER89]
Density, g/cm³: 4.3 [STR93]

1844
Compound: Mercury(II) oleate
Synonyms: mercuric oleate
Formula: $Hg(C_{17}H_{33}COO)_2$
Molecular Formula: $C_{36}H_{66}HgO_4$
Molecular Weight: 763.508
CAS RN: 1191-80-6
Properties: yellowish brown; odor of oleic acid; sensitive to light; used as an antiseptic, and in antifouling paints [HAW93] [MER89]
Solubility: i H_2O; sl s alcohol, ether; v s in fixed oils [MER89]

1845
Compound: Mercury(II) oxalate
Formula: HgC_2O_4
Molecular Formula: C_2HgO_4
Molecular Weight: 288.610
CAS RN: 3444-13-1
Properties: powd [LAN52]
Solubility: g/100g H_2O: 0.0107 (20°C) [KRU93]
Melting Point, °C: decomposes, 165 [LAN52]

1846
Compound: Mercury(II) oxide red
Synonyms: red mercuric oxide
Formula: HgO
Molecular Formula: HgO
Molecular Weight: 216.589
CAS RN: 21908-53-2
Properties: bright red or orange red, odorless cryst; ortho-rhomb; sensitive to light, decomposes to Hg and O_2; obtained either by thermal decomposition of $Hg(NO_3)_2$ or by precipitation from a hot solution of HgCl by Na_2CO_3; used in paint pigments, perfumery and cosmetics, and as an antiseptic [HAW93] [MER89] [KIR81]
Solubility: i H_2O; s dil HCl, HNO_3; i alcohol [MER89]
Density, g/cm³: 11.14 [MER89]

Melting Point, °C: 500, decomposes [STR93]
Reactions: decomposes to Hg(l) and O_2 at 500°C [MER89]

1847
Compound: Mercury(II) oxide yellow
Synonyms: yellow mercuric oxide
Formula: HgO
Molecular Formula: HgO
Molecular Weight: 216.589
CAS RN: 21908-53-2
Properties: yellow or light orange yellow, odorless powd; ortho-rhomb; stable in air; darkens on exposure to light; can be prepared by precipitation from a water soluble mercuric salt by an alkali; used as an antiseptic [KIR81] [HAW93] [MER89]
Solubility: i H_2O; s dil HCl, HNO_3 [MER89]
Density, g/cm³: 11.14 [MER89]
Melting Point, °C: 500, decomposes [STR93]

1848
Compound: Mercury(II) oxycyanide
Synonyms: mercuric oxycyanide
Formula: $Hg(CN)_2 \cdot HgO$
Molecular Formula: $C_2Hg_2N_2O$
Molecular Weight: 469.214
CAS RN: 1335-31-5
Properties: white, ortho-rhomb cryst, or cryst powd; exploded by percussion or if in contact with flame; prepared from excess yellow HgO in a slurry with HCN [MER89] [KIR81]
Solubility: 1 g/80mL cold H_2O, more soluble in hot H_2O [MER89]
Density, g/cm³: 4.44 [MER89]
Melting Point, °C: can explode [CRC94]

1849
Compound: Mercury(II) perchlorate trihydrate
Formula: $Hg(ClO_4)_2 \cdot 3H_2O$
Molecular Formula: $Cl_2H_6HgO_{11}$
Molecular Weight: 453.536
CAS RN: 73491-34-6
Properties: white cryst [STR93]
Density, g/cm³: ~4 [STR93]

1850
Compound: Mercury(II) phosphate
Formula: $Hg_3(PO_4)_2$
Molecular Formula: $Hg_3O_8P_2$
Molecular Weight: 791.713

CAS RN: 7782-66-3
Properties: heavy, white or yellowish powd [HAW93]
Solubility: i H_2O, alcohol; s in acids [HAW93]

1851
Compound: Mercury(II) selenide
Synonyms: tiemannite
Formula: HgSe
Molecular Formula: HgSe
Molecular Weight: 279.550
CAS RN: 20601-83-6
Properties: gray powd; sublimes in a vacuum [HAW93] [STR93]
Solubility: i H_2O [HAW93]
Density, g/cm³: 8.266 [HAW93]
Melting Point, °C: sublimes in vacuum [CRC94]

1852
Compound: Mercury(II) sulfate
Synonyms: mercuric sulfate
Formula: $HgSO_4$
Molecular Formula: HgO_4S
Molecular Weight: 296.654
CAS RN: 7783-35-9
Properties: white, odorless granules or cryst powd; sensitive to light; decomposes at red heat; can be obtained by reacting yellow HgO with H_2SO_4 solution; used as a catalyst for converting acetylene to acetaldehyde, and as a battery electrolyte [HAW93] [MER89] [KIR81]
Solubility: decomposed in H_2O to yellow insoluble basic sulfate and H_2SO_4; s HCl, hot dil H_2SO_4, conc NaCl solution [MER89]
Density, g/cm³: 6.47 [MER89]
Melting Point, °C: decomposes [AES93]

1853
Compound: Mercury(II) sulfide(α)
Synonyms: cinnabar
Formula: α-HgS
Molecular Formula: HgS
Molecular Weight: 232.656
CAS RN: 1344-48-5
Properties: bright scarlet red powd, lumps, hex cryst; sensitive to light (blackens); not attacked by HNO_3 or cold HCl; decomposed by hot conc H_2SO_4; can be prepared by heating black HgS in a conc solution of alkali polysulfide [KIR81] [MER89]
Solubility: i H_2O; s aqua regia, warm HI [MER89]
Density, g/cm³: 8.10 [STR93]

Melting Point, °C: 583 sublimes [STR93]

1854
Compound: Mercury(II) sulfide(β)
Synonyms: metacinnabar
Formula: β-HgS
Molecular Formula: HgS
Molecular Weight: 232.656
CAS RN: 1344-48-5
Properties: black or grayish black, odorless, tasteless, cub; can exist indefinitely in metastable state at room temp; prepared by mixing soluble mercuric salts and sulfides [KIR81] [MER89]
Solubility: i H_2O, alcohol, dil mineral acids [MER89]
Density, g/cm³: 7.70 [LID94]
Melting Point, °C: 583 sublimes [STR93]

1855
Compound: Mercury(II) telluride
Formula: HgTe
Molecular Formula: HgTe
Molecular Weight: 328.190
CAS RN: 12068-90-5
Properties: gray powd; used as a semiconductor in solar cells, in thin-film transistors and in infrared detectors [HAW93] [STR93]
Density, g/cm³: 8.17 [CRC93]
Melting Point, °C: 673 [CRC93]
Thermal Conductivity, W/(m·K): 2 (25°C) [CRC93]

1856
Compound: Mercury(II) tetrathiocyanatocobaltate(II)
Synonyms: cobalt(II) tetrathiocyanatomercurate(II)
Formula: $HgCo(SCN)_4$
Molecular Formula: $C_4CoHgN_4S_4$
Molecular Weight: 491.845
CAS RN: 27685-51-4
Properties: blue cryst; high-purity magnetic susceptibility standard [ALD94] [ALF95]

1857
Compound: Mercury(II) thiocyanate
Synonyms: mercuric thiocyante
Formula: $Hg(SCN)_2$
Molecular Formula: $C_2HgN_2S_2$
Molecular Weight: 316.757
CAS RN: 592-85-8

Properties: odorless powd; radially arranged cryst needles; if heated, swells to considerable volume; sensitive to light [MER89]
Solubility: 0.069 g/100mL H_2O (25°C); more soluble in boiling water, but decomposes; s dil HCl [MER89]
Density, g/cm³: 3.71 [LID94]
Melting Point, °C: 165, decomposes [ALD94]
Reactions: decomposes into Hg, N_2, other products ~165°C [MER89]

1858
Compound: Metaphosphoric acid
Synonyms: phosphoric acid, meta
Formula: $(HPO_3)_n$
Molecular Formula: HO_3P n=1
Molecular Weight: 79.980 n=1
CAS RN: 10343-62-1
Properties: transparent, glass-like solid or soft silky masses; hygr; volatilizes at red heat $(HPO_3)_n$ [MER89]
Solubility: dissolves very slowly in cold H_2O to form H_3PO_4, formation of H_3PO_4 accelerated by boiling; s alcohol [MER89]

1859
Compound: Metavanadic acid
Formula: HVO_3
Molecular Formula: HO_3V
Molecular Weight: 99.948
CAS RN: 13470-24-1
Properties: yellow scales [KIR83]
Solubility: s acids and alkalies [KIR83]

1860
Compound: Methylcyclopentadienylmanganese tricarbonyl
Formula: $C_5H_4CH_3Mn(CO)_3$
Molecular Formula: $C_9H_7MnO_3$
Molecular Weight: 218.091
CAS RN: 12108-13-3
Properties: yellow liq; metallocene derivative [STR93] [ALD94]
Density, g/cm³: 1.38 [ALD94]
Melting Point, °C: -1 [ALD94]
Boiling Point, °C: 232-233 [ALD94]

1861
Compound: Molybdenum
Formula: Mo
Molecular Formula: Mo

Molecular Weight: 95.94
CAS RN: 7439-98-7
Properties: dark gray or black powd, metallic luster or silver-white color; bcc; enthalpy of fusion 37.48 kJ/mol; enthalpy of vaporization 491 kJ/mol; electrical resistivity (0°C) 50 μohm·m, (1000°C) 320; optical reflectivity 46% at 500 nm, 93% at 10,000 nm; has excellent corrosion resistance; used in crucible form to melt reactive metals such as barium, strontium and indium, and evaporate fluorides, chlorides and reactive sulfides [KIR81] [MER89] [CER91] [CRC93]
Solubility: i H_2O, dil acids, conc HCl, alkali hydroxides, fused alkalis; reacts with HNO_3, hot conc H_2SO_4 [MER89]
Density, g/cm³: 10.28 [MER89]
Melting Point, °C: 2622 [MER89]
Boiling Point, °C: 4639 [CRC93]
Reactions: oxidizes rapidly to MoO_3 >600°C [KIR81]
Thermal Conductivity, W/(m·K): 138 (25°C), 122 (500°C), 101 (1000°C), 82 (1500°C) [ALD94] [KIR81]
Thermal Expansion Coefficient: 0-400°C: 0.23%; 0-800°C: 0.46%; 0-1200°C: 0.72% [KIR81]

1862
Compound: Molybdenum acetate dimer
Formula: $[Mo(CH_3COO)_2]_2$
Molecular Formula: $C_8H_{12}Mo_2O_8$
Molecular Weight: 428.058
CAS RN: 14221-06-8
Properties: yellow cryst; sensitive to air [STR93]

1863
Compound: Molybdenum aluminide
Formula: Mo_3Al
Molecular Formula: $AlMo_3$
Molecular Weight: 314.802
CAS RN: 12003-72-4
Properties: -150, +325, -325 mesh with 99.5% purity; compound is a cermet, which can be flame-sprayed [CER91] [HAW93]

1864
Compound: Molybdenum boride
Formula: Mo_2B
Molecular Formula: BMo_2
Molecular Weight: 202.691
CAS RN: 12006-99-4

Properties: tetr; refractory material; one of several borides, other formulas: MoB, MoB_2, Mo_2B_2, Mo_2B_5; used in brazes to join molybdenum and tungsten and tantalum for electronic and corrosion protection applications, and as sputtering material with 99.5% purity to produce wear-resistant and semiconductive films [KIR78] [HAW93] [CER91]
Density, g/cm³: 9.26 [CRC94]
Melting Point, °C: 2280 [KIR78]

1865
Compound: Molybdenum carbide
Formula: MoC
Molecular Formula: CMo
Molecular Weight: 107.951
CAS RN: 12011-97-1
Properties: -325 mesh, 10 microns or less, 95% purity; fcc, a = 0.42810 nm [CIC73] [CER91]
Density, g/cm³: 9.15 [KIR78]
Melting Point, °C: 2577 [CIC73]

1866
Compound: Molybdenum carbide
Synonyms: β-Mo_2C
Formula: Mo_2C
Molecular Formula: CMo_2
Molecular Weight: 203.891
CAS RN: 12069-89-5
Properties: gray powd; ortho-rhomb, a = 0.4733 nm, b = 0.60344 nm, c = 0.52056 nm; in 99.5% purity, used as a sputtering target to produce wear-resistant films and semiconducting films; there is also material with formula MoC, 12011-97-1 [CIC73] [STR93] [CER91]
Density, g/cm³: 9.18 [STR93]
Melting Point, °C: 2687 [STR93]
Thermal Expansion Coefficient: 7.8×10^{-6}/K [KIR78]

1867
Compound: Molybdenum carbonyl
Synonyms: molybdenum hexacarbonyl
Formula: $Mo(CO)_6$
Molecular Formula: C_6MoO_6
Molecular Weight: 264.002
CAS RN: 13939-06-5

Properties: white shiny cryst; ortho-rhomb, a = 1.20 nm, b = 0.64 nm, c = 1.12 nm; vapor pressure ~0.1 mm Hg at 20°C, ~43 mm Hg at 101°C; can be prepared by reacting $MoCl_5$ with zinc dust and CO in ether at high pressure; used in depositing molybdenum, for example to form molybdenum mirror [HAW93] [KIR81]
Solubility: i H_2O; s ceresin, paraffin oil, benzene; sl s ether and other organic solvents [HAW93]
Density, g/cm³: 1.96 [HAW93]
Melting Point, °C: 150-151, decomposes (sublimes) [HAW93]

1868
Compound: Molybdenum disulfide
Synonyms: molybdenite
Formula: MoS_2
Molecular Formula: MoS_2
Molecular Weight: 160.072
CAS RN: 1317-33-5
Properties: lead gray; hex; can be prepared by direct reaction of the elements to form a black and lustrous powd; greasy to the touch with low coefficient of friction; hardness 1-1.5 Mohs; similar to graphite in appearance; principal ore for molybdenum; as a 99% pure material, used as a sputtering target to form lubricant film on bearings and other moving parts [HAW93] [KIR81] [MER89] [CER91] [JAN85]
Solubility: i H_2O, dil acids; s H_2SO_4, conc HNO_3 [HAW93] [MER89]
Density, g/cm³: 5.06 [MER89]; black powd: 4.80 [STR93]
Melting Point, °C: 1750 [JAN85]
Reactions: begins to sublime at 450°C [MER89]

1869
Compound: Molybdenum metaphosphate
Formula: $Mo(PO_3)_6$
Molecular Formula: $MoO_{18}P_6$
Molecular Weight: 569.772
CAS RN: 133863-98-6
Properties: yellow powd [HAW93]
Solubility: i H_2O, most acids; sl s hot aqua regia [HAW93]
Density, g/cm³: 3.28 (0°C) [HAW93]

1870
Compound: Molybdenum mononitride
Formula: MoN
Molecular Formula: MoN
Molecular Weight: 109.947
CAS RN: 12033-19-1

Properties: hex; a = 0.5725 nm, c = 0.5608 nm [CIC73]
Density, g/cm³: 9.20 [LID94]

1871
Compound: Molybdenum nitride
Formula: Mo₂N
Molecular Formula: Mo₂N
Molecular Weight: 205.887
CAS RN: 12033-31-7
Properties: gray; fcc, a = 0.416 nm; microhardness 1700; transition temp 5.0 K [KIR81]
Density, g/cm³: 9.46 [KIR81]
Melting Point, °C: 790 decomposes [KIR81]
Thermal Expansion Coefficient: 6.7 x 10⁻⁶ [KIR81]

1872
Compound: Molybdenum oxytetrafluoride
Formula: MoOF₄
Molecular Formula: F₄MoO
Molecular Weight: 187.933
CAS RN: 14459-59-7
Properties: volatile at moderate temp; there is a tetrahydrate (77727-63-0) and the compound MoO₂F₂ (13824-57-2) [KIR81]
Density, g/cm³: 3.00 [CRC94]
Melting Point, °C: MoOF₄: 98; MoO₂F₂ sublimes at 270 [KIR81]
Boiling Point, °C: 180 [CRC94]

1873
Compound: Molybdenum pentaboride
Formula: Mo₂B₅
Molecular Formula: B₅Mo₂
Molecular Weight: 245.935
CAS RN: 12007-97-5
Properties: refractory material; borides used as a braze to join molybdenum, tungsten and tantalum, and niobium parts for electronic, corrosion and abrasion protection, used as a sputtering target with 99.5% purity to produce films which may be wear-resistant and semiconducting [CER91] [HAW93] [KIR78]
Melting Point, °C: 1600 [HAW93]
Reactions: transformed to MoB₂ at 1600°C [HAW93]

1874
Compound: Molybdenum phosphide
Formula: MoP
Molecular Formula: MoP

Molecular Weight: 126.914
CAS RN: 12163-69-8
Properties: gray-green powd; -200 mesh with 99.5% purity [CER91] [CRC94]
Density, g/cm³: 7.34 [LID94]

1875
Compound: Molybdenum silicide
Synonyms: molybdenum disilicide
Formula: MoSi₂
Molecular Formula: MoSi₂
Molecular Weight: 152.111
CAS RN: 12136-78-6
Properties: a cermet; dark gray cryst powd; has high stress rupture strength; used in electrical resistors, as a high temp protective coating, and in crucible form for melting bismuth, gallium, lead, silicon, silver, tin, zinc and to contain mercury, also as sputtering targets of 99.5 to 99.95% purity to fabricate electrodes for integrated circuits [HAW93] [CER91]
Solubility: s HF and HNO₃; i aqua regia, other acids [HAW93]
Density, g/cm³: 6.31 [HAW93]
Melting Point, °C: 1870-2030 [HAW93]
Thermal Expansion Coefficient: (volume) 100°C (0.166), 200°C (0.400), 400°C (0.899), 800°C (1.960), 1200°C (3.048) [CLA66]

1876
Compound: Molybdenum(II) chloride
Formula: MoCl₂
Molecular Formula: Cl₂Mo
Molecular Weight: 166.845
CAS RN: 13478-17-6
Properties: yellow amorphous solid; -100 mesh with 99.5% purity [CER91] [CRC94]
Density, g/cm³: 3.714 [CRC94]
Melting Point, °C: decomposes 530 [LID94]

1877
Compound: Molybdenum(II) iodide
Formula: MoI₂
Molecular Formula: I₂Mo
Molecular Weight: 349.749
CAS RN: 14055-74-4
Properties: black cryst; sensitive to air and moisture; can be prepared by reduction of MoI₃ (14055-75-5) with Mo, H₂ or hydrocarbon [KIR81] [STR93]
Density, g/cm³: 5.278 [STR93]

1878
Compound: Molybdenum(III) bromide
Formula: $MoBr_3$
Molecular Formula: Br_3Mo
Molecular Weight: 335.652
CAS RN: 13446-57-6
Properties: can be prepared from $MoBr_4$ by
reduction with Mo, H_2 or a hydrocarbon;
another bromide is $MoBr_2$ which can be
prepared by reduction of $MoBr_4$ (13446-56-5);
oxybromides are: $MoOBr_3$ (13596-04-8) and
MoO_2Br_2 (13595-98-7) [KIR81]
Melting Point, °C: decomposes [KIR81]

1879
Compound: Molybdenum(III) chloride
Formula: $MoCl_3$
Molecular Formula: Cl_3Mo
Molecular Weight: 202.298
CAS RN: 13478-18-7
Properties: -100 mesh with 99.5% purity; purple
cryst [STR93] [CER91]
Density, g/cm³: 3.59 [STR93]
Melting Point, °C: disproportionates >410 [KIR81]

1880
Compound: Molybdenum(III) oxide
Synonyms: molybdenum sesquioxide
Formula: Mo_2O_3
Molecular Formula: Mo_2O_3
Molecular Weight: 239.878
CAS RN: 1313-29-7
Properties: known only in the hydrated form
$Mo(OH)_3$, but generally given the formula
Mo_2O_3; grayish-black powd; used as a catalyst
for organic synthesis; feed additive [HAW93]
Solubility: i H_2O; in alkalies; sl s acids [HAW93]

1881
Compound: Molybdenum(III) sulfide
Formula: Mo_2S_3
Molecular Formula: Mo_2S_3
Molecular Weight: 288.078
CAS RN: 12033-33-9
Properties: enthalpy of fusion 129.7 kJ/mol; can be
prepared by reacting Mo and S in a sealed tube,
under vacuum at high temperatures; there is also
MoS_2, 1317-33-5 [CER91] [KIR81] [JAN85]
Density, g/cm³: 5.91 [CRC94]
Melting Point, °C: 1807 [JAN85]
Boiling Point, °C: decomposes 1870 [JAN85]

1882
Compound: Molybdenum(IV) bromide
Formula: $MoBr_4$
Molecular Formula: Br_4Mo
Molecular Weight: 415.556
CAS RN: 13520-59-7
Properties: black needles; deliq; obtained by
bromination of Mo [KIR81] [CRC94]
Solubility: reacts with H_2O [LID94]
Melting Point, °C: decomposes [KIR81]

1883
Compound: Molybdenum(IV) chloride
Synonyms: molybdenum tetrachloride
Formula: $MoCl_4$
Molecular Formula: Cl_4Mo
Molecular Weight: 237.751
CAS RN: 13320-71-3
Properties: black cryst; sensitive to both air and
moisture [STR93]
Solubility: reacts with H_2O [LID94]
Melting Point, °C: decomposes [STR93]

1884
Compound: Molybdenum(IV) oxide
Synonyms: molybdenum dioxide
Formula: MoO_2
Molecular Formula: MoO_2
Molecular Weight: 127.939
CAS RN: 18868-43-4
Properties: tetr; brownish violet or lead gray
nonvolatile powd; can be made by reduction of
MoO_3 with H_2 at 300-400°C, at higher temp Mo
obtained ~500°C [CER91] [HAW93] [KIR81]
Solubility: sl s H_2SO_4; i HCl, HF and alkalies
[HAW93]
Density, g/cm³: 6.47 [STR93]
Melting Point, °C: decomposes [JAN85]

1885
Compound: Molybdenum(IV) selenide
Formula: $MoSe_2$
Molecular Formula: $MoSe_2$
Molecular Weight: 253.860
CAS RN: 12058-18-3
Properties: gray powd; used as a solid lubricant and
lubricant film prepared by sputtering 99.9%
pure material [HAW93] [STR93] [CER91]
Density, g/cm³: 6.0 [STR93]
Melting Point, °C: >1200 [STR93]

1886
Compound: Molybdenum(IV) telluride
Formula: $MoTe_2$
Molecular Formula: $MoTe_2$
Molecular Weight: 351.140
CAS RN: 12058-20-7
Properties: gray hex and 40 micron powd; used as a solid lubricant, and with 99.9% purity as a sputtering target to produce lubricant films [HAW93] [CER91] [LID94]
Density, g/cm³: 7.7 [LID94]

1887
Compound: Molybdenum(V) chloride
Synonyms: molybdenum pentachloride
Formula: $MoCl_5$
Molecular Formula: Cl_5Mo
Molecular Weight: 273.204
CAS RN: 10241-05-1
Properties: greenish black solid, is dark red liq or vapor; hygr; reacts with atm oxygen; enthalpy of vaporization 62.8 kJ/mol; enthalpy of fusion 19.00 kJ/mol; used as a chlorination catalyst for vapor deposition of molybdenum [HAW93] [CRC93]
Solubility: s in dry ether and dry alcohol, in other organic solvents [HAW93]
Density, g/cm³: 2.928 [ALD94]
Melting Point, °C: 194 [CRC93]
Boiling Point, °C: 268 [HAW93]

1888
Compound: Molybdenum(V) oxytrichloride
Formula: $MoOCl_3$
Molecular Formula: Cl_3MoO
Molecular Weight: 218.297
CAS RN: 13814-74-9
Properties: green cryst; can be prepared by refluxing $MoOCl_4$ in benzene; other oxychlorides are: MoOCl (41004-72-2), MoO_2Cl (20770-33-6), $MoOCl_2$ (24989-40-0) sublimes, MoO_2Cl_2 (13637-68-8) mp 184°C in a sealed tube, Mo_2OCl_8 (77727-64-1) [KIR81] [CRC94]
Solubility: reacts with H_2O [LID94]
Melting Point, °C: 302 [KIR81]
Boiling Point, °C: sublimes [LID94]

1889
Compound: Molybdenum(VI) acid monohydrate
Synonyms: molybdic acid monohydrate
Formula: $H_2MoO_4 \cdot H_2O$
Molecular Formula: H_4MoO_5

Molecular Weight: 179.969
CAS RN: 7782-91-4
Properties: white powd [MER89]
Solubility: 0.133 g/100mL H_2O (18°C), 2.568 (70°C) [CRC94]
Density, g/cm³: 3.1 [MER89]
Melting Point, °C: decomposes [CRC94]
Reactions: minus H_2O at 70°C [CRC94]

1890
Compound: Molybdenum(VI) dioxydichloride
Synonyms: molybdenum dichloride dioxide
Formula: MoO_2Cl_2
Molecular Formula: Cl_2MoO_2
Molecular Weight: 198.844
CAS RN: 13637-68-8
Properties: yellowish-orange solid [LID94]
Density, g/cm³: 3.31 [ALD94]
Melting Point, °C: 184 [KIR81]

1891
Compound: Molybdenum(VI) dioxydifluoride
Synonyms: molybdenum dioxide difluoride
Formula: MoO_2F_2
Molecular Formula: F_2MoO_2
Molecular Weight: 165.936
CAS RN: 13824-57-2
Properties: white cryst; hygr [CRC77]
Density, g/cm³: 3.494 [CRC77]
Melting Point, °C: 270 sublimes [KIR81]

1892
Compound: Molybdenum(VI) fluoride
Synonyms: molybdenum hexafluoride
Formula: MoF_6
Molecular Formula: F_6Mo
Molecular Weight: 209.930
CAS RN: 7783-77-9
Properties: volatile, white, cub cryst; very hygr; enthalpy of fusion 4.33 kJ/mol; enthalpy of vaporization 27.7 kJ/mol; evolves blue-white clouds in moist air; can be prepared by reacting molybdenum metal with F_2 at elevated temperatures; other fluorides are: MoF_5, 13819-84-6, mp 64°C, MoF_4, 23412-45-5, MoF_3, 20193-58-2, mp decomposes, MoF_2, 20205-60-1 [KIR81] [MER89] [CRC93]
Solubility: hydrolyzed in H_2O; s anhydrous HF: 1.5 moles/1000g HF [MER89]
Density, g/cm³: 2.30 [ALD94]
Melting Point, °C: 17.5 [MER89]
Boiling Point, °C: 35.0 [MER89]

1893

Compound: Molybdenum(VI) oxide

Synonyms: molybdenum trioxide; molybdic anhydride

Formula: MoO_3

Molecular Formula: MoO_3

Molecular Weight: 143.938

CAS RN: 1313-27-5

Properties: white or sl yellow to sl bluish powd or granules; ortho-rhomb, a = 0.39628 nm, b = 1.3855 nm; c = 0.36964 nm; enthalpy of vaporization 138 kJ/mol; enthalpy of fusion 48.00 kJ/mol; produced by roasting MoS_2; used in 99.99% pure form as a sputtering target for luminescent coatings [KIR81] [MER89] [CER91] [CRC93]

Solubility: g/100g H_2O: 0.134 (20°C), 0.285 (30°C), 1.74 (80°C) [LAN85]; s conc mineral acids, alkali hydroxides; after strongly ignited is v sl s acids [MER89]

Density, g/cm³: 4.696 [MER89]

Melting Point, °C: 801 [CRC93]

Boiling Point, °C: 1155 [MER89]

Reactions: can sublime >795°C [MER89]

1894

Compound: Molybdenum(VI) oxytetrachloride

Formula: $MoOCl_4$

Molecular Formula: Cl_4MoO

Molecular Weight: 253.750

CAS RN: 13814-75-0

Properties: green powd; sensitive to moisture; volatile at ordinary temp [KIR81] [STR93]

Melting Point, °C: 100-101 [STR93]

1895

Compound: Molybdenum(VI) sulfide

Synonyms: molybdenum trisulfide

Formula: MoS_3

Molecular Formula: MoS_3

Molecular Weight: 192.138

CAS RN: 12033-29-3

Properties: brownish black amorphous powd; formed by acidifying a solution of ammonium tetrathiomolybdate [KIR81]

Melting Point, °C: decomposes [CRC94]

1896

Compound: Molybdic silicic acid hydrate

Formula: $H_4SiMo_{12}O_{40} \cdot xH_2O$

Molecular Formula: $H_4Mo_{12}O_{40}Si$ (anhydrous)

Molecular Weight: 1823.373 (anhydrous)

CAS RN: 11089-20-6

Properties: x is commonly 6-8; yellow cryst powd; thermally stable; used in photography, as a catalyst [HAW93]

Solubility: s H_2O, alcohol, acetone; i benzene; decomposes in strongly basic solutions [HAW93]

Density, g/cm³: 2.82 [HAW93]

1897

Compound: Monofluorophosphoric acid

Formula: $(HO)_2P(O)F$

Molecular Formula: FH_2O_3P

Molecular Weight: 99.986

CAS RN: 13537-32-1

Properties: colorless or yellow viscous liq; used in metal cleaners, electrolytic or chemical polishing agents [HAW93] [STR93]

Solubility: miscible with H_2O [HAW93]

Density, g/cm³: 1.818 [HAW93]

1898

Compound: Monoiodosilane

Formula: SiH_3I

Molecular Formula: H_3ISi

Molecular Weight: 158.014

CAS RN: 13598-42-0

Properties: enthalpy of vaporization 28.9 kJ/mol; entropy of vaporization 90.4 kJ/(mol·K) [CIC73]

Melting Point, °C: -57 [CIC73]

Boiling Point, °C: 45.6 [CIC73]

1899

Compound: Neodymium

Formula: Nd

Molecular Formula: Nd

Molecular Weight: 144.24

CAS RN: 7440-00-8

Properties: soft, silver white metal; yellowish in air; hex, at room temp; bcc >868°C; enthalpy of fusion 7.14 kJ/mol; enthalpy of sublimation 327.6 kJ/mol; electrical resistivity (20°C) 64.0 μohm·cm; radius of atom is 0.1821 nm; radius of ion 0.0995 nm for Nd^{+++}, rose colored solution [MER89] [KIR82] [CRC93] [ALD94]

Solubility: slowly reacts with cold H_2O; rapidly with hot H_2O [MER89]

Density, g/cm³: 7.003 [MER89]

Melting Point, °C: 1021 [ALD94]

Boiling Point, °C: 3074 [KIR82]

Thermal Conductivity, W/(m·K): 16.5 (25°C) [CRC93]

Thermal Expansion Coefficient: 9.6×10^{-6}/K
[CRC93]

1900
Compound: Neodymium acetate monohydrate
Synonyms: acetic acid, neodymium salt
   monohydrate
Formula: $Nd(CH_3COO)_3 \cdot H_2O$
Molecular Formula: $C_6H_{11}NdO_7$
Molecular Weight: 339.389
CAS RN: 6192-13-8
Properties: light purple cryst [STR93]
Solubility: 26.2 g/100mL H_2O [CRC94]

1901
Compound: Neodymium boride
Formula: NdB_6
Molecular Formula: B_6Nd
Molecular Weight: 209.106
CAS RN: 12008-23-0
Properties: -325 mesh 10 microns or less with
   99.9% purity; refractory material [KIR78]
   [CER91]
Density, g/cm³: 4.93 [LID94]
Melting Point, °C: 2540 [KIR78]

1902
Compound: Neodymium bromate nonahydrate
Formula: $Nd(BrO_3)_3 \cdot 9H_2O$
Molecular Formula: $Br_3H_{18}NdO_{18}$
Molecular Weight: 690.084
CAS RN: 15162-92-2
Properties: red; hex [CRC77]
Solubility: g/100g H_2O: 43.9 (0°C), 75.6 (20°C),
   116 (40°C) [LAN85]
Melting Point, °C: 66.7 [CRC77]
Reactions: minus $9H_2O$, 150°C [CRC77]

1903
Compound: Neodymium bromide
Formula: $NdBr_3$
Molecular Formula: Br_3Nd
Molecular Weight: 383.952
CAS RN: 13536-80-6
Properties: -20 mesh with 99.9% purity; green
   powd; hygr [STR93] [CER91]
Density, g/cm³: 5.3 [LID94]
Melting Point, °C: 684 [STR93]
Boiling Point, °C: 1540 [STR93]

1904
Compound: Neodymium carbonate hydrate
Formula: $Nd_2(CO_3)_3 \cdot xH_2O$
Molecular Formula: $C_3Nd_2O_9$ (anhydrous)
Molecular Weight: 468.508 (anhydrous)
CAS RN: 38245-38-4
Properties: light purple powd [STR93]
Solubility: i H_2O; s acids [HAW93]

1905
Compound: Neodymium cerium copper oxide
Formula: $Nd_{1.85}Ce_{0.15}CuO_4$
Molecular Formula: $Ce_{0.15}CuNd_{1.85}O_4$
Molecular Weight: 415.405
Properties: superconductor; general formula is
   $Nd_{2-x}Ce_xCuO_4$; material with formula where x =
   0.15 has T_c 20K; cuprate semiconductors are
   metastable at low temperatures; oriented thin
   films can be prepared from bulk cuprates or
   from metal oxides by sputtering or by laser
   ablation; potential uses of superconductors
   include frictionless bearings, microwave and
   electronic devices [CEN92]

1906
Compound: Neodymium chloride
Formula: $NdCl_3$
Molecular Formula: Cl_3Nd
Molecular Weight: 250.598
CAS RN: 10024-93-8
Properties: -20 mesh with 99.9% purity; violet
   powd [ALD94] [STR93] [CER91]
Solubility: g/100g H_2O: 96.7 (10°C), 98.0 (20°C),
   105 (60°C) [LAN85]; v s alcohol; i ether,
   chloroform [HAW93] [MER89]
Density, g/cm³: 4.134 [STR93]
Melting Point, °C: 784 [STR93]
Boiling Point, °C: 1600 [HAW93]

1907
Compound: Neodymium chloride hexahydrate
Formula: $NdCl_3 \cdot 6H_2O$
Molecular Formula: $Cl_3H_{12}NdO_6$
Molecular Weight: 358.689
CAS RN: 13477-89-9
Properties: -4 mesh with 99.9% purity; purple cryst
   [STR93] [CER91]
Solubility: 2.46 parts per 1 part H_2O [MER89]; s
   alcohol [HAW93]
Density, g/cm³: 2.282 [STR93]
Melting Point, °C: 124 [MER89]
Reactions: minus $6H_2O$ at 160°C [HAW93]

1908
Compound: Neodymium fluoride
Formula: NdF$_3$
Molecular Formula: F$_3$Nd
Molecular Weight: 201.235
CAS RN: 13709-42-7
Properties: purple powd, or 99.9% pure sintered tablets; hygr; tablets used as evaporation and sputtering material for multilayers, used with ZnS [STR93] [CER91]
Solubility: i H$_2$O [HAW93]
Density, g/cm^3: 6.506 [STR93]
Melting Point, °C: 1410 [HAW93]
Boiling Point, °C: 2300 [HAW93]

1909
Compound: Neodymium hexafluoroacetylacetonate dihydrate
Synonyms: 1,1,1,5,5,5-hexafluoro-2,4-pentanedione, Nd
Formula: Nd(CF$_3$COCHCOCF$_3$)$_3$·2H$_2$O
Molecular Formula: C$_{15}$H$_7$F$_{18}$NdO$_8$
Molecular Weight: 801.427
CAS RN: 47814-18-6
Properties: cryst [STR93]

1910
Compound: Neodymium hydride
Formula: NdH$_{2-3}$
Molecular Formula: H$_2$Nd; H$_3$Nd
Molecular Weight: NdH$_2$: 146.256; NdH$_3$: 147.264
CAS RN: 13864-04-5
Properties: lumps under argon atm; -60 mesh with 99.9% purity [CER91] [ALF95]

1911
Compound: Neodymium hydroxide
Formula: Nd(OH)$_3$
Molecular Formula: H$_3$NdO$_3$
Molecular Weight: 195.262
CAS RN: 16469-17-3
Properties: bluish or pink precipitate; heating at 300-350°C converts compound to grayish-brown 2Nd$_2$O$_3$·3H$_2$O, further heating converts hydroxide to Nd$_2$O$_3$·H$_2$O [MER89] [ALF95]

1912
Compound: Neodymium iodide
Formula: NdI$_3$
Molecular Formula: I$_3$Nd
Molecular Weight: 524.953

CAS RN: 13813-24-6
Properties: -20 mesh with 99.9% purity; green powd; hygr [STR93] [CER91]
Melting Point, °C: 775 [STR93]

1913
Compound: Neodymium nitrate hexahydrate
Formula: Nd(NO$_3$)$_3$·6H$_2$O
Molecular Formula: H$_{12}$N$_3$NdO$_{15}$
Molecular Weight: 438.346
CAS RN: 14517-29-4
Properties: purple cryst; hygr [STR93]
Solubility: g/100g H$_2$O: 127 (0°C), 142 (20°C), 211 (60°C) [LAN85]

1914
Compound: Neodymium nitride
Formula: NdN
Molecular Formula: NNd
Molecular Weight: 158.247
CAS RN: 25764-11-8
Properties: black powd; -60 mesh with 99.9% purity; NaCl cryst system, a = 0.514 nm [CIC73] [CER91] [CRC94]
Solubility: decomposed by H$_2$O [CRC94]
Density, g/cm^3: 7.69 [LID94]

1915
Compound: Neodymium oxalate decahydrate
Formula: Nd$_2$(C$_2$O$_4$)$_3$·10H$_2$O
Molecular Formula: C$_6$H$_{20}$Nd$_2$O$_{22}$
Molecular Weight: 732.692
CAS RN: 14551-74-7
Properties: rose cryst [STR93]
Solubility: 0.000074 g/100mL H$_2$O (25°C) [CRC94]
Reactions: minus H$_2$O 40-50°C [AES93]

1916
Compound: Neodymium oxide
Synonyms: neodymia
Formula: Nd$_2$O$_3$
Molecular Formula: Nd$_2$O$_3$
Molecular Weight: 336.478
CAS RN: 1313-97-9
Properties: pure material is a blue powd; technical material has a brown color; hygr; absorbs atm CO$_2$; hex; has sl red fluorescence; used in ceramic capacitors, in coloring glass and in television tubes, and as an evaporated material of 99.9% purity, it is possibly reactive to radio frequencies [HAW93] [MER89] [CER91]

Solubility: 5.7×10^{-6} g-mol/L H_2O (29°C); s dil acids [MER89]
Density, g/cm³: 7.24 [STR93]
Melting Point, °C: 2272 [STR93]

1917
Compound: Neodymium perchlorate hexahydrate
Formula: $Nd(ClO_4)_3 \cdot 6H_2O$
Molecular Formula: $Cl_3H_{12}NdO_{18}$
Molecular Weight: 550.683
CAS RN: 17522-69-9
Properties: cryst [ALF95]

1918
Compound: Neodymium phosphate hydrate
Formula: $NdPO_4 \cdot xH_2O$
Molecular Formula: NdO_4P (anhydrous)
Molecular Weight: 239.211 (anhydrous)
CAS RN: 14298-32-9
Properties: powd [ALF95]

1919
Compound: Neodymium silicide
Formula: $NdSi_2$
Molecular Formula: $NdSi_2$
Molecular Weight: 200·411
CAS RN: 12137-04-1
Properties: 10 mm & down lump [ALF93]

1920
Compound: Neodymium sulfate
Formula: $Nd_2(SO_4)_3$
Molecular Formula: $Nd_2O_{12}S_3$
Molecular Weight: 576.671
CAS RN: 101509-27-7
Properties: pinkish needles [MER89]
Solubility: g/100g H_2O: 13.0 (0°C), 7.1 (20°C), 1.2 (90°C) [LAN85]
Melting Point, °C: decomposes 700-800 [MER89]

1921
Compound: Neodymium sulfate octahydrate
Formula: $Nd_2(SO_4)_3 \cdot 8H_2O$
Molecular Formula: $H_{16}Nd_2O_{20}S_3$
Molecular Weight: 720.793
CAS RN: 13477-91-3
Properties: purple cryst [STR93]
Solubility: s cold H_2O, less soluble in hot H_2O [HAW93]
Density, g/cm³: 2.85 [HAW93]

Melting Point, °C: 800, decomposes [HAW93]

1922
Compound: Neodymium sulfide
Formula: Nd_2S_3
Molecular Formula: Nd_2S_3
Molecular Weight: 384.678
CAS RN: 12035-32-4
Properties: -200 mesh with 99.9% purity [CER91]
Density, g/cm³: 5.46 [LID94]
Melting Point, °C: 2207 [LID94]

1923
Compound: Neodymium telluride
Formula: Nd_2Te_3
Molecular Formula: Nd_2Te_3
Molecular Weight: 671.280
CAS RN: 12035-35-7
Properties: -20 mesh with 99.9% purity gray powd [STR93] [CER91]
Density, g/cm³: 7.0 [LID94]
Melting Point, °C: 1377 [LID94]

1924
Compound: Neodymium trifluoroacetylacetonate
Synonyms: 1,1,1-trifluoro-2,4-pentanedione, Nd derivative
Formula: $Nd(CF_3COCH=C(O)CH_3)_3$
Molecular Formula: $C_{15}H_{12}F_9NdO_6$
Molecular Weight: 603.482
CAS RN: 37473-67-9
Properties: blue-pink cryst [STR93]

$$[CF_3\!-\!\underset{\underset{\displaystyle O-}{\mid}}{C}\!=\!CH\!-\!\underset{\underset{\displaystyle O}{\parallel}}{C}\!-\!CH_3]_3Nd$$

Melting Point, °C: 140-142 [STR93]

1925
Compound: Neodymium tris(cyclopentadienyl)
Synonyms: tris(cyclopentadienyl)neodymium
Formula: $Nd(C_5H_5)_3$
Molecular Formula: $C_{15}H_{15}Nd$
Molecular Weight: 339.53
CAS RN: 1273-98-9
Properties: powd; sensitive to air and moisture [STR93]
Melting Point, °C: 417 [STR93]
Boiling Point, °C: 220 sublimes (0.01 mm Hg) [STR93]

1926
Compound: Neon
Formula: Ne
Molecular Formula: Ne
Molecular Weight: 20.1797
CAS RN: 7440-01-9
Properties: inert, odorless gas; critical temp -228.7°C; critical pressure 2.65 MPa; enthalpy of vaporization 1.71 kJ/mol; enthalpy of fusion 0.34 kJ/mol; heat capacity (25°C) 20.79; sonic velocity at 0°C is 433 m/s; viscosity (25°C) 31.73 Pa s [KIR78] [CRC93] [AIR87]
Solubility: 10.5 mL/100g H_2O (20°C, 101.32 kPa) [KIR78]; Henry's law constants, k x 10^{-4}: 13.023 (70.0°C), 12.022 (124.5°C), 9.805 (174.5°C), 7.166 (226.4°C), 4.160 (283.7°C) [POT78]
Density, g/cm³: gas (101.3 kPa, 0°C) 0.0009000 [KIR78]
Melting Point, °C: -248.59 [CRC93]
Boiling Point, °C: -246.08 [CRC93]
Thermal Conductivity, W/(m·K): gas: (101.32 kPa, 0°C) 0.04607 [KIR78]

1927
Compound: Neptunium
Formula: Np
Molecular Formula: Np
Molecular Weight: 237
CAS RN: 7439-99-8
Properties: silvery metal; α: ortho-rhomb, a = 0.4721 nm, b = 0.4888 nm, c = 0.6661 nm, stable from room temp to 280°C; β: tetra, a = 0.4895nm, c = 0.3386 nm, stable from 280 to 577°C; γ: a = 0.3518 nm; bcc, stable from 577 to 637°C; enthalpy of vaporization 418 kJ/mol; enthalpy of fusion 3.20 kJ/mol; ^{237}Np, $t_{1/2}$ = 2.14 x 10^{+6} years, $t_{1/2}$ of ^{236}Np 1.29 x 10^{+6} years; discovered in 1940; produced in kg amounts as a byproduct of plutonium production [MER89] [KIR78] [CRC93]
Solubility: s HCl [HAW93]
Density, g/cm³: 20.45 [KIR78]
Melting Point, °C: 637 [KIR91]
Boiling Point, °C: 3900 [KIR91]
Thermal Conductivity, W/(m·K): 6.3 [CRC93]

1928
Compound: Neptunium(IV) oxide
Synonyms: neptunium dioxide
Formula: NpO_2
Molecular Formula: NpO_2
Molecular Weight: 269
CAS RN: 12035-79-9

Properties: cub; dark olive powd [HAW93] [CRC94]
Density, g/cm³: 11.11 [CRC94]
Melting Point, °C: 2547 [LID94]

1929
Compound: Nickel
Formula: Ni
Molecular Formula: Ni
Molecular Weight: 58.6934
CAS RN: 7440-02-0
Properties: white ferromagnetic metal; fcc, a = 0.35238 nm; decomposes steam at red heat; electrical resistivity (20°C) 6.844 μohm·cm; Curie temp 353°C; Poisson's ratio 0.30; saturation magnetization 0.617 T; residual magnetization 0.300 T; coercive force 239 A/m; initial permeability 0.251 mH/m, maximum permeability 2.51-3.77 mH/m; enthalpy of vaporization 377.5 kJ/mol; enthalpy of fusion 17.48 kJ/mol [KIR81] [MER89] [CRC93] [ALD94]
Solubility: i H_2O; slowly attacked by dil HCl, H_2SO_4, readily by HNO_3 [MER89]
Density, g/cm³: 8.908 [HAW93]
Melting Point, °C: 1453 [ALD94]
Boiling Point, °C: 2732 (extrapolated) [KIR81]
Thermal Conductivity, W/(m·K): 90.0 (25°C), 82.8 (100°C), 63.6 (300°C), 61.9 (500°C) [KIR81] [ALD94]
Thermal Expansion Coefficient: 500°C: 15.2 x 10^{-6}/°C [KIR81]

1930
Compound: Nickel acetate tetrahydrate
Synonyms: acetic acid, nickel(II) salt
Formula: $Ni(CH_3COO)_2\cdot4H_2O$
Molecular Formula: $C_4H_{14}NiO_8$
Molecular Weight: 248.843
CAS RN: 6018-89-9
Properties: green monocl cryst; hygr; used in textiles as a mordant [HAW93]
Solubility: s in 6 parts H_2O; s alcohol [MER89]
Density, g/cm³: 1.744 [MER89]
Melting Point, °C: 250, decomposes [HAW93]

1931
Compound: Nickel acetylacetonate
Synonyms: 2,4-pentanedione, Ni(II) derivative
Formula: $Ni(CH_3COCH=C(O)CH_3)_2$
Molecular Formula: $C_{10}H_{14}NiO_4$
Molecular Weight: 256.912

CAS RN: 3264-82-2
Properties: light green powd; hygr; trimer; there is a hydrate, CAS RN 120156-44-7 [STR93] [ALD94] [COT88]

$$\text{[CH}_3\text{–C=CH–C–CH}_3]_2\text{Ni}$$

with O– (single bond) and O (double bond) substituents above

Melting Point, °C: 238 decomposes [STR93]

1932

Compound: Nickel aluminide
Formula: NiAl$_3$
Molecular Formula: Al$_3$Ni
Molecular Weight: 139.638
CAS RN: 12004-71-6
Properties: -20 mesh with 99.9% purity; there is also NiAl, 12003-78-0; a cermet that can be flame-sprayed [CER91] [HAW93]

1933

Compound: Nickel antimonide
Synonyms: breithauptite
Formula: NiSb
Molecular Formula: NiSb
Molecular Weight: 180.453
CAS RN: 12035-52-8
Properties: reddish hex; 6 mm pieces and smaller with 99.5% purity; there is also Ni$_3$Sb, 12503-49-0 [CER91] [CRC94]
Density, g/cm³: 8.74 [LID94]
Melting Point, °C: 1158 [CRC94]
Boiling Point, °C: decomposes at 1400 [CRC94]

1934

Compound: Nickel arsenate octahydrate
Formula: Ni$_3$(AsO$_4$)$_2$·8H$_2$O
Molecular Formula: As$_2$H$_{16}$Ni$_3$O$_{16}$
Molecular Weight: 598.040
CAS RN: 7784-48-7
Properties: yellowish green powd; formed when an aq solution of arsenic anhydride is reacted with nickel carbonate; used as a catalyst for hardening fats which are used in soap [HAW93] [KIR81]
Solubility: i H$_2$O; s acids [HAW93]
Density, g/cm³: 4.98 [HAW93]
Melting Point, °C: decomposes to NiO and As$_2$O$_5$ [KIR81]

1935

Compound: Nickel arsenide
Synonyms: niccolite
Formula: NiAs
Molecular Formula: AsNi
Molecular Weight: 133.615
CAS RN: 27016-75-7
Properties: hex; 6 mm pieces and smaller of 99.5% purity; there is also NiAs$_2$, 12068-61-0, of the same purity [CER91] [CRC94]
Density, g/cm³: 7.57 [CRC94]
Melting Point, °C: 968 [CRC94]

1936

Compound: Nickel basic carbonate tetrahydrate
Synonyms: zaratite
Formula: NiCO$_3$·2Ni(OH)$_2$·4H$_2$O
Molecular Formula: CH$_{12}$Ni$_3$O$_{11}$
Molecular Weight: 376.179
CAS RN: 3333-67-3
Properties: light green rhomb cryst or brown powd; can be obtained by adding sodium carbonate to a nickel salt solution; used in the manufacture of catalyst, electroplating, ceramic colors and glazes [HAW93] [KIR81]
Solubility: i H$_2$O; s in dil acids and in ammonia [HAW93]
Density, g/cm³: 2.6 [HAW93]

1937

Compound: Nickel bis(cyclopentadienyl)
Synonyms: bis(cyclopentadienyl)nickel
Formula: Ni(C$_5$H$_5$)$_2$
Molecular Formula: C$_{10}$H$_{10}$Ni
Molecular Weight: 188.883
CAS RN: 1271-28-9
Properties: bright green cryst; fairly stable in air [COT88] [ALF95]
Melting Point, °C: 171-173 [ALF95]
Reactions: forms CpNiNO with NO [COT88]

1938

Compound: Nickel boride
Formula: Ni$_2$B
Molecular Formula: BNi$_2$
Molecular Weight: 128.198
CAS RN: 12007-01-1
Properties: -35 mesh with 99% purity; refractory material [CER91] [KIR78]
Density, g/cm³: 7.90 [LID94]
Melting Point, °C: 1230 [KIR78]

1939
Compound: Nickel boride
Formula: NiB
Molecular Formula: BNi
Molecular Weight: 69.504
CAS RN: 12007-00-0
Properties: -35 mesh with 99% purity; refractory material; silver green [CER91] [STR93] [KIR78]
Density, g/cm³: 7.39 [STR93]
Melting Point, °C: 1080 [KIR78]

1940
Compound: Nickel boride
Formula: Ni₃B
Molecular Formula: BNi₃
Molecular Weight: 186.891
CAS RN: 12007-02-2
Properties: -35 mesh with 99% purity; refractory material [KIR78]
Density, g/cm³: 8.17 [LID94]
Melting Point, °C: 1155 [KIR78]

1941
Compound: Nickel bromide
Formula: NiBr₂
Molecular Formula: Br₂Ni
Molecular Weight: 218.501
CAS RN: 13462-88-9
Properties: brownish yellow, or yellow, lustrous scales [HAW93]
Solubility: g/100g soln, H_2O: 53.0 (0°C), 57.3 (25°C), 60.8 (100°C); solid phase, NiBr₂·6H₂O [KRU93]
Density, g/cm³: 5.098 [STR93]
Melting Point, °C: 963 [STR93]

1942
Compound: Nickel bromide trihydrate
Formula: NiBr₂·3H₂O
Molecular Formula: Br₂H₆NiO₃
Molecular Weight: 272.547
CAS RN: 13462-88-9
Properties: yellowish green, very deliq cryst; hexahydrate obtained from a reaction of black NiO and HBr [KIR81] [MER89]
Solubility: 199 g/100mL H_2O (0°C), 316 g/100mL H_2O (100°C) [CRC94]
Melting Point, °C: 300, decomposes [STR93]
Reactions: minus 3H₂O ~200°C [MER89]

1943
Compound: Nickel carbonate
Formula: NiCO₃
Molecular Formula: CNiO₃
Molecular Weight: 118.702
CAS RN: 3333-67-3
Properties: light green rhomb; -325 mesh 10 microns and smaller; can be prepared by adding Na₂CO₃ solution to a nickel salt solution, or by heating Ni powd in NH₃ and CO₂, followed by boiling off NH₃ to obtain purer material; used in manufacture of catalysts, colored glass and certain nickel pigments [KIR81] [CER91]
Solubility: 0.0093 g/100mL H_2O (25°C) [CRC94]
Density, g/cm³: 4.39 [LID94]
Melting Point, °C: decomposes [CRC94]

1944
Compound: Nickel carbonate hydroxide tetrahydrate
Formula: 2NiCO₃·3Ni(OH)₂·4H₂O
Molecular Formula: C₂H₁₄Ni₅O₁₆
Molecular Weight: 587.591
CAS RN: 12244-51-8
Properties: green powd [AES93] [STR93] [ALD94]
Melting Point, °C: decomposes [CRC94]

1945
Compound: Nickel carbonyl
Formula: Ni(CO)₄
Molecular Formula: C₄NiO₄
Molecular Weight: 170.735
CAS RN: 13463-39-3
Properties: colorless, volatile liq; oxidizes in air; explodes ~60°C; vapor pressure, kPa: 19.2 (0°C), 28.7 (10°C), 44.0 (20°C); critical temp 200°C; vapor density is four times greater than air; manufactured by reacting CO with Ni; used to prepare highly pure nickel and in alkaline nickel batteries [KIR81] [HAW93] [MER89]
Solubility: s ~5000 parts air free H_2O; s alcohol, benzene, chloroform, acetone, CCl₄ [MER89]
Density, g/cm³: 1.3185 [HAW93]
Melting Point, °C: -19.3 [MER89]
Boiling Point, °C: 42.6 [KIR81]
Reactions: decomposed by heat to Ni and 4CO [DOU83]

1946
Compound: Nickel chlorate hexahydrate
Formula: Ni(ClO₃)₂·6H₂O
Molecular Formula: Cl₂H₁₂NiO₁₂
Molecular Weight: 333.687

CAS RN: 67952-43-6
Properties: dark red [CRC77]
Solubility: mol/100mol H_2O: 8.88 (0°C), 12.02 (25°C); Solid phase, $Ni(ClO_3)_2 \cdot 6H_2O$ (0°C), $Ni(ClO_3)_2 \cdot 4H_2O$ (25°C) [KRU93]; g/100g H_2O: 111 (0°C), 133 (20°C), 308 (80°C) [LAN85]
Density, g/cm^3: 2.07 [CRC77]
Melting Point, °C: decomposes, 80 [CRC77]

1947

Compound: Nickel chloride
Formula: $NiCl_2$
Molecular Formula: Cl_2Ni
Molecular Weight: 129.598
CAS RN: 7718-54-9
Properties: golden yellow powd; hygr; readily absorbs NH_3; enthalpy of fusion 71.20 kJ/mol [CRC93] [STR93] [MER89]
Solubility: g/100g soln, H_2O: 35.04 (0°C), 39.6 (25°C), 46.7 (100.2°C); Solid phase, $NiCl_2 \cdot 6H_2O$ (0°C, 25°C), $NiCl_2 \cdot 2H_2O$ (100.2°C) [KRU93]
Density, g/cm^3: 3.55 [HAW93]
Melting Point, °C: 1031 [CRC93]
Boiling Point, °C: sublimes [LID94]
Reactions: sublimable in absence of air [MER89]

1948

Compound: Nickel chloride hexahydrate
Formula: $NiCl_2 \cdot 6H_2O$
Molecular Formula: $Cl_2H_{12}NiO_6$
Molecular Weight: 237.689
CAS RN: 7791-20-0
Properties: green cryst; monocl; formed when Ni or NiO is reacted with hot aq HCl solution; used in nickel electroplating [KIR81] [MER89]
Solubility: 4.9208±0.0028 mol/kg, and 4.9172±0.0049 mol/kg, in H_2O (25°C) [RAR87b], [RAR92]; s alcohol [MER89]
Reactions: sublimes at 973°C [STR93]

1949

Compound: Nickel chromate
Formula: $NiCrO_4$
Molecular Formula: $CrNiO_4$
Molecular Weight: 174.687
CAS RN: 14721-18-7
Properties: maroon; used in catalysts; gives dark red aq solutions; there is also nickel chromium oxide, $NiCr_2O_4$, 12018-18-7, -100 mesh with 99% purity [KIR78] [CER91]
Solubility: v sl s H_2O [KIR78]

1950

Compound: Nickel cyanide tetrahydrate
Formula: $Ni(CN)_2 \cdot 4H_2O$
Molecular Formula: $C_2H_8N_2NiO_4$
Molecular Weight: 182.790
CAS RN: 13477-95-7
Properties: apple green plates or powd; prepared by reacting KCN and $NiSO_4$; used in metallurgy and in electroplating; anhydrous $Ni(CN)_2$, 557-19-7 [STR93] [MER89] [KIR81]
Solubility: i H_2O; sl s dil acids; v s alkali cyanides, ammonia, ammonium carbonate [MER89]
Melting Point, °C: decomposes above 200 [KIR81]
Reactions: minus $4H_2O$ at 200°C [HAW93]

1951

Compound: Nickel disilicide
Formula: $NiSi_2$
Molecular Formula: $NiSi_2$
Molecular Weight: 114.864
CAS RN: 12201-89-7
Properties: -80 mesh powd [ALF93]
Density, g/cm^3: 4.83 [LID94]
Melting Point, °C: 1200 [ALF93]

1952

Compound: Nickel fluoride
Formula: NiF_2
Molecular Formula: F_2Ni
Molecular Weight: 96.690
CAS RN: 10028-18-9
Properties: yellowish to green tetr cryst; hygr [MER89] [STR93]
Solubility: g/100g soln, H_2O: 2.50 (25°C), 2.52 (90°C); Solid phase, $NiF_2 \cdot 4H_2O$ [KRU93]; i alcohol, ether [MER89]
Density, g/cm^3: 4.72 [MER89]
Melting Point, °C: ~1100 [KIR78]
Reactions: sublimes under HF >1000°C [MER89]

1953

Compound: Nickel fluoride tetrahydrate
Formula: $NiF_2 \cdot 4H_2O$
Molecular Formula: $F_2H_8NiO_4$
Molecular Weight: 168.752
CAS RN: 13940-83-5
Properties: green powd; can be prepared by reacting $NiCO_3$ with aq HF [KIR78] [STR93]

1954

Compound: Nickel hexafluoroacetylacetonate hydrate

Synonyms: 1,1,1,5,5,5-hexafluoro-2,4-pentanedione, Ni derivative

Formula: $Ni(CF_3COCH=C(O)CF_3)_2 \cdot xH_2O$

Molecular Formula: $C_{10}H_2F_{12}NiO_4$ (anhydrous)

Molecular Weight: 472.798

CAS RN: 14949-69-0

Properties: green cryst [STR94] [ALD94]

$$\text{O}- \quad \text{O}$$
$$| \qquad \|$$
$$[\text{CF}_3-\text{C}=\text{CH}-\text{C}-\text{CF}_3]_2\text{Ni}$$

Melting Point, °C: 211-213 [STR93]

1955

Compound: Nickel hydroxide

Formula: $Ni(OH)_2$

Molecular Formula: H_2NiO_2

Molecular Weight: 92.708

CAS RN: 12054-48-7

Properties: fine green powd; formed when a nickel sulfate solution is treated with sodium hydroxide, then neutralized and filtered; used in the manufacture of Ni-Cd batteries [KIR81] [HAW93]

Solubility: mol/L soln, H_2O: 1×10^{-4} (20°C) [KRU93]; s in acids and in NH_4OH [HAW93]

Density, g/cm³: 4.15 [HAW93]

Melting Point, °C: 230, decomposes [STR93]

Reactions: forms NiO and H_2O on decomposing [MER89]

1956

Compound: Nickel iodate

Formula: $Ni(IO_3)_2$

Molecular Formula: I_2NiO_6

Molecular Weight: 408.498

CAS RN: 13477-98-0

Properties: yellow needles [LAN52]

Solubility: mol/100mol H_2O: 0.023 (0°C), 0.035 (25°C), 0.044 (90°C); Solid phase, $Ni(IO_3)_2 \cdot 2H_2O$ (0°C, 25°C), $Ni(IO_3)_2$ (90°C) [KRU93]

Density, g/cm³: 5.07 [LAN52]

1957

Compound: Nickel iodate tetrahydrate

Formula: $Ni(IO_3)_2 \cdot 4H_2O$

Molecular Formula: $H_8I_2NiO_{10}$

Molecular Weight: 480.560

CAS RN: 13477-99-1

Properties: hex cryst; deliq [CRC94] [LAN52]

Solubility: g/100g H_2O: 0.74 (0°C), 1.09 (20°C), 1.43 (30°C) [LAN85]

Melting Point, °C: decomposes, 100 [LAN52]

1958

Compound: Nickel iodide

Formula: NiI_2

Molecular Formula: I_2Ni

Molecular Weight: 312.502

CAS RN: 13462-90-3

Properties: iron black color; hygr; sublimes in absence of air [MER89] [STR93]

Solubility: g/100g soln, H_2O: 55.4 (0°C), 60.7 (25°C), 65.3 (90°C) [KRU93]

Density, g/cm³: 5.83 [KIR81]

Melting Point, °C: 797 [KIR81]

Boiling Point, °C: sublimes [LID94]

1959

Compound: Nickel iodide hexahydrate

Formula: $NiI_2 \cdot 6H_2O$

Molecular Formula: $H_{12}I_2NiO_6$

Molecular Weight: 420.593

CAS RN: 7790-34-3

Properties: bluish green, very deliq cryst; obtained when nickel carbonate is reacted with hydriodic acid [KIR81]

Solubility: v s H_2O, alcohol [MER89]

Density, g/cm³: 5.83 [KIR81]

Melting Point, °C: 797 [KIR81]

1960

Compound: Nickel molybdate

Formula: $NiMoO_4$

Molecular Formula: $MoNiO_4$

Molecular Weight: 218.631

CAS RN: 14177-55-0

Properties: green; three forms; α monocl, β and γ; obtained by reacting the metal oxides [KIR81]

Density, g/cm³: α: 3.5; β: 4.9 [KIR81]

Melting Point, °C: 970 [KIR78]

1961

Compound: Nickel nitrate

Formula: $Ni(NO_3)_2$

Molecular Formula: N_2NiO_6

Molecular Weight: 182.702

CAS RN: 13138-45-9

Properties: prepared by reacting red fuming nitric acid with nickel nitrate hexahydrate [KIR81]
Solubility: g/100g soln, H_2O: 44.2 (0°C), 50.0 (25°C), 69.2 (99.5°C); Solid phase, $Ni(NO_3)_2 \cdot 6H_2O$ (0°C, 25°C), $Ni(NO_3)_2 \cdot 2H_2O$ (99.5°C) [KRU93]

1962
Compound: Nickel nitrate hexahydrate
Formula: $Ni(NO_3)_2 \cdot 6H_2O$
Molecular Formula: $H_{12}N_2NiO_{12}$
Molecular Weight: 290.794
CAS RN: 13478-00-7
Properties: green monocl deliq cryst; manufactured by several methods for example slowly adding nickel powd to a stirred mixture of nitric acid and water; used as an intermediate in the manufacture of nickel catalysts [KIR81] [MER89]
Solubility: s in 0.4 parts H_2O; s alcohol [MER89]
Density, g/cm³: 2.065 [HAW93]
Melting Point, °C: 56.7 [MER89]
Boiling Point, °C: 137 [MER89]
Reactions: minus H_2O on heating [KIR81]

1963
Compound: Nickel oxalate dihydrate
Formula: $NiC_2O_4 \cdot 2H_2O$
Molecular Formula: $C_2H_4NiO_6$
Molecular Weight: 182.743
CAS RN: 6018-94-6
Properties: light green powd [MER89]
Solubility: g/L soln, H_2O: 0.0118 (25°C); Solid phase is $NiC_2O_4 \cdot 2H_2O$ [KRU93]

1964
Compound: Nickel oxide
Synonyms: bunsenite
Formula: NiO
Molecular Formula: NiO
Molecular Weight: 74.692
CAS RN: 1313-99-1
Properties: green powd, yellow when hot; cub; prepared by reaction of pure nickel powd and water in air at 1000°C; inert and refractory material; there is a black form with 76-77% Ni, whereas green form has 78.5% Ni; black form is microcrystalline and prepared by calcination of the $NiCO_3$ or $Ni(NO_3)_2$ at 600°C, green form prepared by heating a mixture of H_2O and Ni powd in air at 1000°C; used in the manufacture of steels [KIR81] [MER89]

Solubility: in 0.00054 m alkaline phosphate solution: 0.0157×10^{-6} m (26.8°C), 0.0169×10^{-6} m (93.3°C), 0.0102×10^{-6} m (262.2°C) [ZIE89]; s acids [MER89]
Density, g/cm³: 7.45 [KIR81]
Melting Point, °C: 2090 [KIR81]
Reactions: β-$Ni(OH)_2 = NiO + H_2O$ at 195°C [ZIE89]

1965
Compound: Nickel perchlorate hexahydrate
Formula: $Ni(ClO_4)_2 \cdot 6H_2O$
Molecular Formula: $Cl_2Hl_2NiO_{14}$
Molecular Weight: 365.686
CAS RN: 13520-61-1
Properties: green cryst; hygr [STR93]
Solubility: g/100g soln, H_2O: 104.6 (0°C), 112.2 (26°C); Solid phase $Ni(ClO_4)_2 \cdot 5H_2O$ (0°C) [KRU93]
Melting Point, °C: 140 [STR93]

1966
Compound: Nickel phosphate heptahydrate
Formula: $Ni_3(PO_4)_2 \cdot 7H_2O$
Molecular Formula: $H_{14}Ni_3O_{15}P_2$
Molecular Weight: 492.130
CAS RN: 10381-36-9
Properties: light green powd; can be prepared by reaction of nickel with hot dil phosphoric acid solution; used in electroplating [KIR81] [HAW93]
Solubility: i H_2O; s acids and NH_4OH [HAW93]
Melting Point, °C: decomposes on heating [KIR81]

1967
Compound: Nickel phosphate octahydrate
Formula: $Ni_3(PO_4)_2 \cdot 8H_2O$
Molecular Formula: $H_{16}Ni_3O_{16}P_2$
Molecular Weight: 510.145
CAS RN: 10381-36-9
Properties: light green powd [MER89]
Solubility: i H_2O; s acids, ammonia [MER89]
Melting Point, °C: decomposes [ALF95]

1968
Compound: Nickel phosphide
Formula: Ni_2P
Molecular Formula: Ni_2P
Molecular Weight: 148.361
CAS RN: 12035-64-2

Properties: gray cryst; -100 mesh with 99.5% purity [CER91] [CRC94]
Density, g/cm³: 7.33 [LID94]
Melting Point, °C: 1112 [AES93]

1969
Compound: Nickel selenate hexahydrate
Formula: NiSeO₄·6H₂O
Molecular Formula: H₁₂NiO₁₀Se
Molecular Weight: 309.743
CAS RN: 75060-62-5
Properties: green; tetr [CRC77]
Solubility: g/100g H₂O: 27.36 (0°C), 36.20 (21.6°C), 83.99 (100°C); Solid phase, NiSeO₄·6H₂O (0°C, 25°C), NiSeO₄·4H₂O (100°C) [KRU93]
Density, g/cm³: 2.314 [CRC77]

1970
Compound: Nickel selenide
Formula: NiSe
Molecular Formula: NiSe
Molecular Weight: 137.653
CAS RN: 1314-05-2
Properties: cub gray powd [CRC94]
Density, g/cm³: 7.2 [LID94]

1971
Compound: Nickel silicide
Formula: Ni₂Si
Molecular Formula: Ni₂Si
Molecular Weight: 145.473
CAS RN: 12059-14-2
Properties: -20 mesh powd [ALF93]
Density, g/cm³: 7.4 [LID94]
Melting Point, °C: 1255 [LID94]

1972
Compound: Nickel stannate dihydrate
Formula: NiSnO₃·2H₂O
Molecular Formula: H₄NiO₅Sn
Molecular Weight: 261.432
CAS RN: 12035-38-0
Properties: green powd; light colored cryst powd; used as an additive in ceramic capacitors [HAW93]
Reactions: minus 2H₂O ~120°C [HAW93]

1973
Compound: Nickel stearate

Formula: Ni[CH₃(CH₂)₁₆COO]₂
Molecular Formula: C₃₆H₇₀NiO₄
Molecular Weight: 625.643
CAS RN: 2223-95-2
Properties: waxy green solid [STR93]
Density, g/cm³: 1.13 [STR93]
Melting Point, °C: 180 [KIR78]

1974
Compound: Nickel subsulfide
Synonyms: heazlewoodite
Formula: Ni₃S₂
Molecular Formula: Ni₃S₂
Molecular Weight: 240.212
CAS RN: 12035-72-2
Properties: yellowish; naturally occurring nickel mineral [KIR81] [CRC94]
Density, g/cm³: 5.82 [CRC94]
Melting Point, °C: 790 [CRC94]

1975
Compound: Nickel sulfate
Formula: NiSO₄
Molecular Formula: NiO₄S
Molecular Weight: 154.757
CAS RN: 7786-81-4
Properties: greenish yellow cryst; used in manufacture of nickel catalysts; nickel plating [HAW93]
Solubility: g/100g H₂O: 27.6 (0°C), 40.8 (25°C), 78.0 (100°C); Solid phase, NiSO₄·7H₂O (0°C, 25°C), β-NiSO₄·6H₂O (100°C) [KRU93]; i alcohol and ether [HAW93]
Density, g/cm³: 3.68 [HAW93]
Melting Point, °C: 840, minus SO₃ [HAW93]

1976
Compound: Nickel sulfate heptahydrate
Formula: NiSO₄·7H₂O
Molecular Formula: H₁₄NiO₁₁S
Molecular Weight: 280.863
CAS RN: 10101-98-1
Properties: green cryst [HAW93]
Solubility: g/100g H₂O: 26.2 (0°C), 37.7 (20°C), 50.4 (40°C) [LAN85]; s alcohol [HAW93]
Density, g/cm³: 1.98 [HAW93]
Reactions: minus H₂O at 99°C, minus 6H₂O at 103°C [CRC94] [HAW93]

1977
Compound: Nickel sulfate hexahydrate

Formula: NiSO$_4$·6H$_2$O
Molecular Formula: H$_{12}$NiO$_{10}$S
Molecular Weight: 262.848
CAS RN: 10101-97-0
Properties: two phases: α: blue to bluish green tetr cryst; β: green transparent cryst, stable at 40°C; sweet, astringent taste; somewhat efflorescent; produced by adding Ni powd to hot dil H$_2$SO$_4$; used as an electrolyte in metal finishing and in electroless plating [KIR81] [MER89]
Solubility: g/100g H$_2$O: (blue green) 40.1 (20°C), 43.6 (30°C), 47.6 (40°C); (green) 44.4 (20°C), 46.6 (30°C), 76.7 (100°C) [LAN85]; sl s alcohol, more s methanol [MER89]
Density, g/cm^3: 2.07 [STR93]
Melting Point, °C: decomposes ~100 [LID94]
Reactions: transition α to β at 53.3°C; minus 5H$_2$O ~100°C; decomposes to NiO and SO$_3$ >800°C [KIR81] [MER89]

1978
Compound: Nickel sulfide
Synonyms: millerite
Formula: NiS
Molecular Formula: NiS
Molecular Weight: 90.759
CAS RN: 16812-54-7
Properties: black powd; trig, yellow metallic luster; enthalpy of fusion 30.10 kJ/mol; can be formed by fusing Ni powd and molten sulfur; other sulfides include Ni$_2$S, 12137-08-5, Ni$_3$S$_2$, 12035-72-2 (heazlewoodite), NiS$_2$, 12035-51-7, and Ni$_3$S, 12137-12-1 (polydymite) [KIR81] [STR93] [CRC93] [JAN85]
Solubility: 0.00036 g/100mL H$_2$O (18°C) [CRC94]
Density, g/cm^3: 5.3-5.65 [STR93]
Melting Point, °C: 976 [LID94]

1979
Compound: Nickel telluride
Formula: NiTe
Molecular Formula: NiTe
Molecular Weight: 186.293
CAS RN: 12142-88-0
Properties: 6 mm pieces and smaller with 99.9% purity [CER91]

1980
Compound: Nickel tetrafluoroborate hexahydrate
Formula: Ni(BF$_4$)$_2$·6H$_2$O
Molecular Formula: B$_2$F$_8$H$_{12}$NiO$_6$
Molecular Weight: 340.394

CAS RN: 15684-36-3
Properties: green cryst [STR93]
Density, g/cm^3: 1.47 [STR93]

1981
Compound: Nickel thiocyanate
Formula: Ni(SCN)$_2$
Molecular Formula: C$_2$N$_2$NiS$_2$
Molecular Weight: 174.860
CAS RN: 13689-92-4
Properties: green powd [STR93]
Solubility: g/100g soln, H$_2$O: 35.48 (25°C) [KRU93]

1982
Compound: Nickel titanate
Formula: NiTiO$_3$
Molecular Formula: NiO$_3$Ti
Molecular Weight: 154.558
CAS RN: 12035-39-1
Properties: brown powd or canary yellow with rhomb structure; -325 mesh 10 microns or less with 99.9% purity [CER91] [KIR83] [STR93]
Density, g/cm^3: 4.56 [STR93]

1983
Compound: Nickel trifluoroacetylacetonate dihydrate
Synonyms: 1,1,1-trifluoro-2,4-pentandione nickel derivative
Formula: Ni(CF$_3$COCH=C(O)CH$_3$)$_2$·2H$_2$O
Molecular Formula: C$_{10}$H$_{12}$F$_6$NiO$_6$
Molecular Weight: 400.885
CAS RN: 14324-83-5
Properties: powd [ALF95]

$$Ni[CF_3-\overset{\overset{\textstyle O-}{|}}{C}=CH-\overset{\overset{\textstyle O}{\|}}{C}-CH_3]_2 \cdot 2H_2O$$

1984
Compound: Nickel tungstate
Formula: NiWO$_4$
Molecular Formula: NiO$_4$W
Molecular Weight: 306.531
CAS RN: 14177-51-6
Properties: -325 mesh 10 microns or less with 99.9% purity [CER91]

1985
Compound: Nickel vanadate

Formula: NiV_2O_6
Molecular Formula: NiO_6V_2
Molecular Weight: 256.572
CAS RN: 52502-12-2
Properties: -200 mesh with 99.9% purity [CER91]
[ALF95]

1986
Compound: Nickel(II,III) sulfide
Synonyms: polydymite
Formula: Ni_3S_4
Molecular Formula: Ni_3S_4
Molecular Weight: 304.344
CAS RN: 12137-12-1
Properties: gray-black; naturally occurring nickel
mineral [KIR81] [CRC94]
Density, g/cm³: 4.77 [LID94]

1987
Compound: Nickel(III) oxide
Synonyms: nickel sesquioxide
Formula: Ni_2O_3
Molecular Formula: Ni_2O_3
Molecular Weight: 165.385
CAS RN: 1314-06-3
Properties: gray black powd [MER89]
Solubility: i H_2O; v sl s cold acid; dissolves in hot
HCl releasing Cl_2; dissolves in hot H_2SO_4 or
HNO_3 evolving O_2 [MER89]
Reactions: decomposes ~600°C to give NiO and O_2
[MER89]

1988
Compound: Niobium
Formula: Nb
Molecular Formula: Nb
Molecular Weight: 92.90638
CAS RN: 7440-03-1
Properties: steel gray, lustrous metal; pure metal is
ductile and malleable; bcc, lattice constant
0.33004 nm; enthalpy of fusion 30.0 kJ/mol;
enthalpy of vaporization 697 kJ/mol; enthalpy of
combustion 949 kJ/mol; vapor pressure
(2300°C) 22 mPa; evaporation rate (2300°C) 1.9
microgram/(cm² s); electrical resistivity 13-16
µohm·cm; produced from Nb_2O_5 and carbon at
1800-2000°C; used as an anodic film for
rectification [KIR81] [MER89] [CER91]
[CRC93]
Solubility: inert to HCl, HNO_3, aqua regia; attacked
by fusion with alkali hydroxides [MER89]
Density, g/cm³: 8.57 [MER89]

Melting Point, °C: 2477 [CRC93]
Boiling Point, °C: 4944 [CRC93]
Thermal Conductivity, W/(m·K): 52.3 (25°C)
[KIR81]
Thermal Expansion Coefficient: coefficient of
linear expansion 7.1 x 10^{-6}/°C from 18 to 100°C
[KIR81]

1989
Compound: Niobium carbide
Formula: Nb_2C
Molecular Formula: CNb_2
Molecular Weight: 197.824
CAS RN: 12011-99-3
Properties: -325 mesh, 10 microns or less, 99.5%
purity; hex, a = 0.3127 nm, c = 0.4972 nm
[CER91] [KIR81]
Density, g/cm³: 7.80 [KIR81]
Melting Point, °C: 3090 [KIR81]

1990
Compound: Niobium diboride
Formula: NbB_2
Molecular Formula: B_2Nb
Molecular Weight: 114.528
CAS RN: 12007-29-3
Properties: gray powd; hex, a = 0.3089 nm, c =
0.3303 nm; hardness 8+ Mohs; resistivity 65
µohm·cm at 25°C; refractory material; used as a
sputtering target with 99.5% purity to produce
thermionic conductor film [KIR81] [CER91]
[ALF93]
Density, g/cm³: 6.97 [ALD94]
Melting Point, °C: 2900 [KIR78]
Thermal Conductivity, W/(m·K): 17 at 296K
[KIR81]

1991
Compound: Niobium disilicide
Formula: $NbSi_2$
Molecular Formula: $NbSi_2$
Molecular Weight: 149.077
CAS RN: 12034-80-9
Properties: -325 mesh powd; cryst solid; used as a
refractory, and as 99.5 to 99.95% pure material
as a sputtering target in the fabrication of
integrated circuits [HAW93] [ALF93] [CER91]
Density, g/cm³: 5.7 [LID94]
Melting Point, °C: 1950 [HAW93]

1992
Compound: Niobium hydride
Formula: NbH
Molecular Formula: HNb
Molecular Weight: 93.914
CAS RN: 13981-86-7
Properties: gray powd; bcc; sensitive to moisture; reaction of H_2 with Nb at 300-1500°C can result in the formation of $NbH_{0.85}$ [STR93] [KIR81]
Density, g/cm³: 6.6 [STR93]

1993
Compound: Niobium monoboride
Formula: NbB
Molecular Formula: BNb
Molecular Weight: 103.717
CAS RN: 12045-19-1
Properties: gray powd; ortho-rhomb, a = 0.3298 nm, b = 0.872 nm, c = 0.3166 nm; resistivity 64.5 µohm·cm at 25°C; refractory material; commonly prepared by hot-pressing boron with niobium or niobium hydride; used as sputtering target with 99.5% purity to produce wear-resistant and semiconductor films, and can provide neutron absorbing layers on nuclear fuel pellets [KIR78] [KIR81] [CER91] [ALF93]
Density, g/cm³: 7.5 [KIR81]
Melting Point, °C: 2270 [KIR78]

1994
Compound: Niobium nitride
Formula: NbN
Molecular Formula: NNb
Molecular Weight: 106.913
CAS RN: 24621-21-4
Properties: dark gray; fcc, a = 0.4388 nm; hardness, 8+ Mohs; electrical resistivity 78 µohm·cm; transition temp 15.2 K; can be prepared by heating Nb metal in excess N_2 or NH_3 to 700-1100°C; used in the form of 99.5% pure sputtering target for increasing electrical stability of diodes, transistors and integrated circuits [KIR81] [CIC73] [CER91]
Solubility: i HCl, HNO_3, H_2SO_4; attacked by hot caustic, lime or strong alkali, evolving NH_3 [KIR81]
Density, g/cm³: 8.47 [KIR81]
Melting Point, °C: 2575 [STR93]
Thermal Conductivity, W/(m·K): 3.8 [KIR81]
Thermal Expansion Coefficient: 10.1×10^{-6} [KIR81]

1995
Compound: Niobium phosphide
Formula: NbP
Molecular Formula: NbP
Molecular Weight: 123.880
CAS RN: 12034-66-1
Properties: tetr cryst; -200 mesh with 99.5% purity [LID94] [CER91]
Density, g/cm³: 6.5 [LID94]

1996
Compound: Niobium(II) oxide
Formula: NbO
Molecular Formula: NbO
Molecular Weight: 108.905
CAS RN: 12034-57-0
Properties: gray metallic appearance; can be obtained by reduction of Nb_2O_5 in H_2 at 1300-1700°C; -100 mesh with 99.9% purity; cub, a = 0.42108 nm; enthalpy of fusion 85.00 kJ/mol [CRC93] [CER91] [KIR81]
Density, g/cm³: 7.30 [KIR81]
Melting Point, °C: 1937 [CRC93]

1997
Compound: Niobium(IV) carbide
Formula: NbC
Molecular Formula: CNb
Molecular Weight: 104.917
CAS RN: 12069-94-2
Properties: lavender gray powd; -325 mesh, 10 microns or less, 99.9% purity; fcc, a = 0.4471 nm; hardness 9+ Mohs; resistivity 180 µohm·cm maximum; used in special steels, coating graphite for nuclear reactors; as a 99.5% pure material, used as a sputtering target to produce wear-resistant and semiconducting films [HAW93] [KIR81] [CER91]
Solubility: i H_2O, acids; s in a mixture of HNO_3 and HF [HAW93]
Density, g/cm³: 7.82 [STR93]
Melting Point, °C: 3500 [STR93]
Boiling Point, °C: 4300 [KIR81]
Thermal Conductivity, W/(m·K): 14 at 23°C [KIR81]
Thermal Expansion Coefficient: (volume) 100°C (0.141), 200°C (0.329), 400°C (0.740), 800°C (1.626), 1200°C (2.565) [CLA66]

1998
Compound: Niobium(IV) oxide
Synonyms: niobium dioxide

Formula: NbO_2
Molecular Formula: NbO_2
Molecular Weight: 124.905
CAS RN: 12034-59-2
Properties: white powd; -200 mesh with 99.9% purity; tetr, a = 0.371 nm, c = 0.5985 nm; enthalpy of fusion 92.00 kJ/mol; can be prepared by reduction of Nb_2O_5 with H_2 at 800-1300°C [KIR81] [STR93] [CER91] [CRC93]
Density, g/cm³: 5.9 [STR93]
Melting Point, °C: 1902 [CRC93]

1999
Compound: Niobium(IV) selenide
Formula: $NbSe_2$
Molecular Formula: $NbSe_2$
Molecular Weight: 250.826
CAS RN: 12034-77-4
Properties: gray black solid; has a higher electrical conductivity than graphite; used as a lubricant and conductor at high temperatures, and in high vacuum, and as a 99.8% pure material, as a sputtering target to produce electrically conductive lubricant film [HAW93] [CER91]
Density, g/cm³: 6.3 [LID94]
Melting Point, °C: >1316 [HAW93]

2000
Compound: Niobium(IV) sulfide
Formula: NbS_2
Molecular Formula: NbS_2
Molecular Weight: 157.038
CAS RN: 12136-97-9
Properties: black powd; formula also given as $NbS_{1.75}$; as a 99.8% pure material, used as sputtering target to produce lubricant film on bearings and other moving parts [STR93] [CER91]
Density, g/cm³: 4.4 [LID94]

2001
Compound: Niobium(IV) telluride
Formula: $NbTe_2$
Molecular Formula: $NbTe_2$
Molecular Weight: 348.106
CAS RN: 12034-83-2
Properties: hex cryst; -325 mesh with 10 microns average or less; 99.8% pure material used as a sputtering target for lubricant film [CER91] [LID94]
Density, g/cm³: 7.6 [LID94]

2002
Compound: Niobium(V) bromide
Synonyms: niobium pentabromide
Formula: $NbBr_5$
Molecular Formula: Br_5Nb
Molecular Weight: 492.426
CAS RN: 13478-45-0
Properties: yellow powd; ortho-rhomb, a = 0.6127 nm, b = 1.2198 nm, c = 1.855 nm; sensitive to moisture; can be formed by reacting Br_2 and niobium at ~500°C [STR93] [KIR81]
Solubility: s H_2O, alcohol, ethyl bromide [KIR81]
Density, g/cm³: 4.36 [KIR81]
Melting Point, °C: 150 [STR93]; 254 [KIR81]
Boiling Point, °C: 361.6 [STR93]

2003
Compound: Niobium(V) chloride
Synonyms: niobium pentachloride
Formula: $NbCl_5$
Molecular Formula: Cl_5Nb
Molecular Weight: 270.170
CAS RN: 10026-12-7
Properties: yellow, very deliq; monocl, a = 0.1830 nm, b = 1.798 nm, c = 0.5888 nm; melts to a reddish orange liq; decomposes in moist air evolving HCl; enthalpy of vaporization 52.7 kJ/mol; enthalpy of fusion 33.90 kJ/mol; can be obtained by chlorination of Nb metal at 300-350°C [MER89] [KIR81] [CRC93]
Solubility: hydrolyzes in H_2O; s HCl, CCl_4 [MER89] [KIR81]
Density, g/cm³: 2.75 [MER89]
Melting Point, °C: 204.7 [CRC93]
Boiling Point, °C: 254.05 [CRC93]
Reactions: starts to sublime at 125°C [MER89]

2004
Compound: Niobium(V) ethoxide
Formula: $Nb(OC_2H_5)_5$
Molecular Formula: $C_{10}H_{25}NbO_5$
Molecular Weight: 318.212
CAS RN: 3236-82-6
Properties: liq; flammable; moisture sensitive [ALD94]
Solubility: decomposed by H_2O [ALF95]
Density, g/cm³: 1.258 [ALD94]
Melting Point, °C: 6 [ALD94]
Boiling Point, °C: 140-142 (0.1 mm Hg) [ALD94]

2005
Compound: Niobium(V) fluoride

Synonyms: niobium pentafluoride
Formula: NbF_5
Molecular Formula: F_5Nb
Molecular Weight: 187.898
CAS RN: 7783-68-8
Properties: strongly refractive, deliq, colorless monocl cryst; very hygr; lattice parameters a = 0.963 nm, b = 1.443 nm, c = 0.512 nm; enthalpy of vaporization 52.3 kJ/mol; enthalpy of fusion 36.00 kJ/mol; obtained from Nb and F_2 or anhydrous HF at 250-300°C [KIR81] [MER89] [CRC93]
Solubility: hydrolyzes in H_2O; sl s CS_2, $CHCl_3$ [MER89]
Density, g/cm^3: 2.6955 [MER89]; 3.92 [STR93]
Melting Point, °C: 80 [MER89]
Boiling Point, °C: 229 [CRC93]

2006
Compound: Niobium(V) fluorodioxide
Synonyms: niobium dioxide fluoride
Formula: NbO_2F
Molecular Formula: $FNbO_2$
Molecular Weight: 143.903
CAS RN: 15195-33-2
Properties: cub, a = 0.3902 nm; can be prepared by dissolution of Nb_2O_5 in 48% HF, followed by evaporation to dryness and heating to 250°C [KIR81]
Density, g/cm^3: 4.0 [LID94]

2007
Compound: Niobium(V) iodide
Synonyms: niobium pentiodide
Formula: NbI_5
Molecular Formula: I_5Nb
Molecular Weight: 727.428
CAS RN: 13779-92-5
Properties: black powd; monocl, a = 1.058 nm, b = 0.658 nm, c = 1.388 nm; sensitive to moisture; obtained by reaction of excess I_2 with Nb metal in a sealed tube [STR93] [KIR81]
Density, g/cm^3: 5.32 [LID94]
Melting Point, °C: decomposes ~200 [KIR81]
Reactions: decomposes to NbI_4, 13870-21-8, at 206-270°C in vacuum [KIR81]

2008
Compound: Niobium(V) oxide
Synonyms: niobium pentoxide
Formula: Nb_2O_5
Molecular Formula: Nb_2O_5

Molecular Weight: 265.810
CAS RN: 1313-96-8
Properties: white, ortho-rhomb cryst; becomes yellow when heated; α form is monocl, a = 2.116 nm, b = 0.3822 nm, c = 1.935 nm; enthalpy of fusion 104.3 kJ/mol; used as an evaporated material and sputtering target with 99.95% and 99.5% purity to prepare dielectric coatings and multilayers [KIR81] [MER89] [CER91] [CRC93]
Solubility: i H_2O; s HF, hot H_2SO_4 [MER89]
Density, g/cm^3: 4.47 [ALD94]
Melting Point, °C: 1520 [MER89]

2009
Compound: Niobium(V) oxybromide
Formula: $NbOBr_3$
Molecular Formula: Br_3NbO
Molecular Weight: 348.617
CAS RN: 14459-75-7
Properties: yellowish brown solid; hydrolyzes readily in moist air; prepared by reacting Br_2 with a mixture of Nb_2O_5 and carbon at 550°C [KIR81]
Melting Point, °C: sublimes in vacuum at 180 [KIR81]
Boiling Point, °C: ~320 decomposes [KIR81]
Reactions: decomposes in vacuum to $NbBr_5$ and Nb_2O_5 [KIR81]

2010
Compound: Niobium(V) oxychloride
Formula: $NbOCl_3$
Molecular Formula: Cl_3NbO
Molecular Weight: 215.263
CAS RN: 13597-20-1
Properties: white solid; tetr, a = 1.087 nm, c = 0.396 nm; can be prepared by air oxidation of $NbCl_5$ [KIR81]
Density, g/cm^3: 3.72 [KIR81]
Melting Point, °C: sublimes in vacuum ~200 [KIR81]

2011
Compound: Niobocene dichloride
Synonyms: bis(cyclopentadienyl)niobium
Formula: $(C_5H_5)_2NbCl_2$
Molecular Formula: $C_{10}H_{10}Cl_2Nb$
Molecular Weight: 294.00
CAS RN: 12793-14-5

Properties: sensitive to moisture; uses: synthesis of transition metal complexes and organometallic compounds [ALD93]

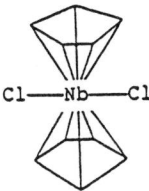

2012

Compound: Nitric acid

Synonyms: aqua fortis

Formula: HNO_3

Molecular Formula: HNO_3

Molecular Weight: 63.013

CAS RN: 7697-37-2

Properties: transparent, colorless or yellowish liq; fuming; hygr; corrosive; attacks almost all metals; yellowish color is due to formation of nitrogen dioxide when exposed to light; strong oxidizing agent; viscosity (25°C) is 0.761 cp; vapor pressure (25°C) is 62 mm Hg; specific conductance 3.77×10^{-2} (ohm·cm)$^{-1}$ at 25°C; enthalpy of vaporization 39.1 kJ/mol at 25°C; enthalpy of fusion 10.50 kJ/mol; [CRC93] [HAW93] [COT88]

Solubility: miscible with H_2O; decomposed by alcohol [HAW93]

Density, g/cm³: 1.504 [HAW93]

Melting Point, °C: -41.6 [CRC93]

Boiling Point, °C: 78, decomposes [HAW93]

2013

Compound: Nitric oxide

Synonyms: nitrogen monoxide

Formula: NO

Molecular Formula: NO

Molecular Weight: 30.006

CAS RN: 10102-43-9

Properties: colorless gas; reacts readily with oxygen at room temp to form NO_2; enthalpy of vaporization 13.83 kJ/mol; enthalpy of fusion 2.30 kJ/mol [CRC93] [HAW93]

Solubility: sl s H_2O [HAW93]

Density, g/cm³: 1.317 g/L [LID94]

Melting Point, °C: -163.6 [COT88]

Boiling Point, °C: -151.8 [COT88]

2014

Compound: Nitrogen

Formula: N_2

Molecular Formula: N_2

Molecular Weight: 28.013 (atomic weight 14.00674)

CAS RN: 7727-37-9

Properties: colorless, tasteless, odorless gas; chemically unreactive; specific volume (21.1°C, 1 atm) 0.86 m³/kg; critical temp -147.1°C; critical pressure 33.5 atm; critical density 0.311 g/cm³; enthalpy of vaporization 5.57 kJ/mol; enthalpy of fusion 0.71 kJ/mol [CRC93] [AIR87] [HAW93] [MER89]

Solubility: sl s H_2O; i alcohol [HAW93]

Density, g/cm³: gas: 1.25046 g/L [MER89]

Melting Point, °C: -210.01 [MER89]

Boiling Point, °C: -195.79 [MER89]

Thermal Conductivity, W/(m·K): 0.02583 (25°C) [ALD94]

2015

Compound: Nitrogen dioxide

Formula: NO_2

Molecular Formula: NO_2

Molecular Weight: 46.006

CAS RN: 10102-44-0

Properties: reddish brown gas; brown liq below 21.15°C; colorless solid at about -11°C; used in the production of nitric acid, as an oxidizing agent [HAW93] [MER89]

Solubility: decomposes in H_2O to HNO_3 and releases NO; s conc H_2SO_4, HNO_3 [MER89]

Density, g/cm³: gas: 2.0198 g/L [LID94]; liq: 1.448 [MER89]

Melting Point, °C: -9.3 [MER89]

Boiling Point, °C: 21.15 [MER89]

2016

Compound: Nitrogen selenide

Synonyms: selenium nitride

Formula: N_4Se_4

Molecular Formula: N_4Se_4

Molecular Weight: 371.867

CAS RN: 12033-88-4

Properties: orange red, amorphous powd or monocl cryst; preparation: by reaction of dry NH_3 and $SeCl_4$; explosive [MER89]

Solubility: i H_2O, ether, absolute alcohol; sl s CS_2, benzene, acetic acid [MER89]

Density, g/cm³: 4.2 [MER89]

Melting Point, °C: explodes [LID94]

2017
Compound: Nitrogen trichloride
Formula: NCl_3
Molecular Formula: Cl_3N
Molecular Weight: 120.365
CAS RN: 10025-85-1
Properties: yellowish, thick, oily, liq; pungent odor; evaporates rapidly in air; very unstable; decomposes in light; explodes when heated to 93°C, or when subjected to a flash of direct sunlight [MER89]
Solubility: i H_2O, decomposes in H_2O after 24 h; s CS_2, phosphorus trichloride, benzene, CCl_4, $CHCl_3$ [MER89]
Density, g/cm³: 1.653 [MER89]
Melting Point, °C: <40 [HAW93]
Boiling Point, °C: <71 [HAW93]

2018
Compound: Nitrogen trifluoride
Formula: NF_3
Molecular Formula: F_3N
Molecular Weight: 71.002
CAS RN: 7783-54-2
Properties: colorless gas; oxidizing agent; moldy odor; critical temp -39.25°C; critical pressure 4.53 MPa; critical volume 123.8 cm³/mol; enthalpy of vaporization 11.59 kJ/mol; enthalpy of fusion 398 J/mol; chemically inert, does not attack glass, mercury; decomposed by electric sparks; used in electronics industry [MER89] [AIR87]
Solubility: 1.4 x 10^{-5} mol/mol H_2O at 25°C [KIR78]
Density, g/cm³: gas: 3.116 g/L [LID94]; liq, at bp: 1.885 [MER89]
Melting Point, °C: -208.5 [MER89]
Boiling Point, °C: -129 [MER89]

2019
Compound: Nitrogen triiodide
Formula: NI_3
Molecular Formula: I_3N
Molecular Weight: 394.720
CAS RN: 13444-85-4
Properties: black cryst; unstable; can explode if touched; more stable if kept wet [HAW93]
Reactions: can explode [CRC94]

2020
Compound: Nitrogen trioxide
Synonyms: dinitrogen trioxide
Formula: N_2O_3

Molecular Formula: N_2O_3
Molecular Weight: 76.011
CAS RN: 10544-73-7
Properties: blue liq; used as an oxidant in special fuel systems [HAW93]
Density, g/cm³: 1.447 (2°C) [HAW93]
Melting Point, °C: -102 [COT88]
Boiling Point, °C: 3.5, decomposes [COT88]

2021
Compound: Nitrogen(V) oxide
Synonyms: dinitrogen pentoxide
Formula: N_2O_5
Molecular Formula: N_2O_5
Molecular Weight: 108.010
CAS RN: 10102-03-1
Properties: colorless, hex cryst [MER89]
Solubility: v s chloroform without appreciable decomposition; less s CCl_4 [MER89]
Density, g/cm³: 2.0 [LID94]
Melting Point, °C: 30 [COT88]
Boiling Point, °C: 47, decomposes [COT88]
Reactions: sublimes 32.4°C [MER89]

2022
Compound: Nitronium hexafluoroantimonate
Formula: NO_2SbF_6
Molecular Formula: F_6NO_2Sb
Molecular Weight: 281.756
CAS RN: 17856-92-7
Properties: white cryst; sensitive to moisture [STR93]

2023
Compound: Nitronium hexafluorophosphate
Formula: NO_2PF_6
Molecular Formula: F_6NO_2P
Molecular Weight: 190.970
CAS RN: 19200-21-6
Properties: white cryst; sensitive to moisture [STR93]

2024
Compound: Nitronium tetrafluoroborate
Formula: NO_2BF_4
Molecular Formula: BF_4NO_2
Molecular Weight: 132.811
CAS RN: 13826-86-3
Properties: white cryst; sensitive to moisture [ALD94]

2025
Compound: Nitrosyl chloride
Formula: NOCl
Molecular Formula: ClNO
Molecular Weight: 65.459
CAS RN: 2696-92-6
Properties: yellowish red liq or yellow gas; forms nitric oxide and chlorine when heated; enthalpy of vaporization 25.78 kJ/mol; used as a catalyst [HAW93] [CRC93]
Solubility: decomposes in H_2O; s fuming H_2SO_4 [HAW93]
Density, g/cm³: gas: 2.872 g/L [LID94]; liq: 1.273 [HAW93]
Melting Point, °C: -61.5 [HAW93]
Boiling Point, °C: -5.55 [CRC93]

2026
Compound: Nitrosyl fluoride
Formula: FNO
Molecular Formula: FNO
Molecular Weight: 49.005
CAS RN: 7789-25-5
Properties: colorless gas, has bluish color if impure; reacts rapidly with glass; used as an oxidizer in rocket propellants and as a reagent for fluorination [MER89]
Solubility: reacts with H_2O to form NO, HNO_3 and HF [MER89]
Density, g/cm³: gas: 2.150 g/L [LID94]; liq: 1.326; solid: 1.719 [MER89]
Melting Point, °C: -132.5 [MER89]
Boiling Point, °C: -59.9 [MER89]

2027
Compound: Nitrosylsulfuric acid
Synonyms: nitrosyl sulfate
Formula: $HOSO_2ONO$
Molecular Formula: HNO_5S
Molecular Weight: 127.078
CAS RN: 7782-78-7
Properties: prisms; oxidizing agent; forms as an intermediate in the Chamber process for fuming sulfuric acid; obtained by reaction of SO_3, nitrogen oxides and H_2O; uses: preparation of cryst diazonium sulfates, bleaching cereal milling products [MER89] [ALD94]
Solubility: decomposes in H_2O; s H_2SO_4 [MER89]
Melting Point, °C: 73.5, decomposes [MER89]
Reactions: decomposes in moist air to form H_2SO_4 and HNO_3 [MER89]

2028
Compound: Nitrous acid
Formula: HNO_2
Molecular Formula: HNO_2
Molecular Weight: 47.014
CAS RN: 7782-77-6
Properties: weak acid; stable only in solution; light blue [HAW93]

2029
Compound: Nitrous oxide
Formula: N_2O
Molecular Formula: N_2O
Molecular Weight: 44.012
CAS RN: 10024-97-2
Properties: colorless gas; sweet taste; critical temp 36.5°C; critical pressure 7.26 MPa; enthalpy of vaporization 16.53 kJ/mol; enthalpy of fusion 6.54 kJ/mol; used in dentistry and medicine as an anesthetic, and in electronics industry [CRC93] [AIR87] [HAW93]
Solubility: sl s H_2O; s alcohol, ether, conc H_2SO_4 [HAW93]
Density, g/cm³: gas: 1.931 g/L [LID94]; liq: 1.22 (-89°C) [HAW93]
Melting Point, °C: -90.8 [HAW93]
Boiling Point, °C: -88.5 [COT88]

2030
Compound: Nitryl chloride
Formula: NO_2Cl
Molecular Formula: $ClNO_2$
Molecular Weight: 81.459
CAS RN: 13444-90-1
Properties: colorless gas; odor of chlorine; yellow tinge to its solutions; enthalpy of vaporization 25.7 kJ/mol [CRC93] [HAW93]
Density, g/cm³: gas: 3.574 g/L [LID94]; liq: 1.33 [HAW93]
Melting Point, °C: -145 [HAW93]
Boiling Point, °C: -14.3 [MER89]

2031
Compound: Nitryl fluoride
Formula: NO_2F
Molecular Formula: FNO_2
Molecular Weight: 65.004
CAS RN: 10022-50-1
Properties: colorless gas; strong oxidizing agent; enthalpy of vaporization 18.05 kJ/mol; used in rocket propellants, and as a fluorinating agent [HAW93] [CRC93]

Solubility: hydrolyzes with HNO_3 and HF as products [HAW93]

Density, g/cm³: gas: 2.852 g/L [LID94]; liq: 1.80 [HAW93]

Melting Point, °C: -165 [HAW93]

Boiling Point, °C: -72.4 [MER89]

2032

Compound: Nobelium

Formula: No

Molecular Formula: No

Molecular Weight: 259

CAS RN: 10028-14-5

Properties: synthetic radioactive element; one of the actinides; has nine very short lived isotopes; discovered in 1958 by Ghiorso and colleagues [HAW93]

Melting Point, °C: 827 [LID94]

2033

Compound: Osmium

Formula: Os

Molecular Formula: Os

Molecular Weight: 190.230

CAS RN: 7440-04-2

Properties: bluish white, lustrous metal; has ten oxidation states, -2 to +8, with higher oxidation states the most stable; closed packed hex, a = 0.27341 nm; stable in cold air; hardness is 7.0 Mohs; electrical resistivity, μohm·cm: 8.12 (0°C), 9.66 (20°C); vapor pressure at mp 1.8 Pa; Young's modulus 558.6 GPa; enthalpy of fusion 57.85 kJ/mol; enthalpy of vaporization 738 kJ/mol [KIR82] [MER89] [ALD94]

Solubility: attacked by aqua regia; barely affected by HCl, H_2SO_4 [MER89]

Density, g/cm³: 22.61 [MER89]

Melting Point, °C: 3045 [ALD94]

Boiling Point, °C: 5027 [ALD94]

Thermal Conductivity, W/(m·K): 87.6 (25°C) [CRC93]

Thermal Expansion Coefficient: at 20°C is 6.1 x 10^{-6}/°C [KIR82]

2034

Compound: Osmium bis(cyclopentadienyl)

Synonyms: bis(cyclopentadienyl)osmium

Formula: $(C_5H_5)_2Os$

Molecular Formula: $C_{10}H_{10}Os$

Molecular Weight: 320.419

CAS RN: 1273-81-0

Properties: cryst [ALF95]

Melting Point, °C: 226-228 [ALF95]

2035

Compound: Osmium carbonyl

Synonyms: triosmium dodecacarbonyl

Formula: $Os_3(CO)_{12}$

Molecular Formula: $C_{12}O_{12}Os_3$

Molecular Weight: 906.815

CAS RN: 15696-40-9

Properties: yellow cryst; stable in air [DOU83] [STR93]

Density, g/cm³: 3.48 [STR93]

Melting Point, °C: 224 [ALD94]

2036

Compound: Osmium(II) chloride

Synonyms: osmium dichloride

Formula: $OsCl_2$

Molecular Formula: Cl_2Os

Molecular Weight: 261.135

CAS RN: 13444-92-3

Properties: dark green needles; hygr; unstable with respect to atm oxygen [HAW93]

Solubility: i H_2O; s alcohol, ether [HAW93]

2037

Compound: Osmium(III) chloride

Synonyms: osmium trichloride

Formula: $OsCl_3$

Molecular Formula: Cl_3Os

Molecular Weight: 296.588

CAS RN: 13444-93-4

Properties: dark gray cub solid [KIR82]

Solubility: i H_2O; s HNO_3 [KIR82]

Melting Point, °C: decomposes >450 [KIR82]

2038

Compound: Osmium(III) chloride hydrate

Formula: $OsCl_3 \cdot xH_2O$

Molecular Formula: Cl_3Os (anhydrous)

Molecular Weight: 296.588 (anhydrous)

CAS RN: 14996-60-2

Properties: black cryst [STR93]

Melting Point, °C: >500, decomposes [STR93]

2039

Compound: Osmium(IV) chloride

Synonyms: osmium tetrachloride

Formula: $OsCl_4$

Molecular Formula: Cl_4Os

Molecular Weight: 332.041
CAS RN: 10026-01-4
Properties: red cryst [MER89]
Solubility: s H_2O resulting in a yellow solution, hydrolysis occurs after standing [MER89]
Density, g/cm³: 4.38 [MER89]
Reactions: sublimes at 450°C [MER89]

2040
Compound: Osmium(IV) oxide
Synonyms: osmium dioxide
Formula: OsO_2
Molecular Formula: O_2Os
Molecular Weight: 222.229
CAS RN: 12036-02-1
Properties: dark bluish black powd, with rutile cryst form [KIR82] [ALD94]
Solubility: i H_2O, acids [KIR82]
Density, g/cm³: 11.4 [KIR82]

2041
Compound: Osmium(VI) fluoride
Synonyms: osmium hexafluoride
Formula: OsF_6
Molecular Formula: F_6Os
Molecular Weight: 304.220
CAS RN: 13768-38-2
Properties: pale yellow, volatile solid; hydrolyzed when exposed to moisture [MER89]
Density, g/cm³: 4.1 [LID94]
Melting Point, °C: 32.1 [MER89]
Boiling Point, °C: 45.9 [MER89]

2042
Compound: Osmium(VIII) oxide
Synonyms: osmium tetroxide
Formula: OsO_4
Molecular Formula: O_4Os
Molecular Weight: 254.228
CAS RN: 20816-12-0
Properties: pale yellow solid; monocl cryst; acrid, chlorine like odor; volatile; vapor pressure 11 mm Hg (27°C); critical temp 405°C; enthalpy of fusion 9.80 kJ/mol; obtained by heating finely divided Os in air or O_2; used as an oxidation agent and catalyst [MER89] [CRC93]
Solubility: g/100g H_2O: 5.26 (0°C), 5.75 (10°C), 6.43 (20°C) [LAN85]; 375 g/100 CCl4; s benzene [MER89]
Density, g/cm³: 5.10 [MER89]
Melting Point, °C: 40.6 [MER89]
Boiling Point, °C: 130.0 [MER89]

Reactions: begins to sublime and distill below the boiling point [MER89]

2043
Compound: Oxalic acid
Synonyms: ethanedioic acid
Formula: $(COOH)_2$
Molecular Formula: $C_2H_2O_4$
Molecular Weight: 90.035
CAS RN: 144-62-7
Properties: colorless odorless hygr solid; has two forms: rhomb α and monocl β; α prepared from sublimation of the dihydrate, β prepared by crystallization from acetic acid; rhomb is thermodynamically stable form at room temp; enthalpy of sublimation 90.58 kJ/mol; enthalpy of solution in water -9.58 kJ/mol; enthalpy of combustion -245.61 kJ/mol; enthalpy of decomposition 826.78 kJ/mol; vapor pressure (P), kPa, (57-107°C) log P = -(4726.95/T) + 11.3478 [KIR81]
Solubility: g/100g H_2O: 3.54 (0°C), 9.52 (20°C), 120 (90°C) [LAN85]; g/100g: 23.7 in ethanol (15.6°C), 1.5 in ethyl ether (25°C) [KIR81]
Density, g/cm³: α: 1.900; β: 1.896 [KIR81]
Melting Point, °C: α: 189.5; β:182 [KIR81]
Reactions: sublimes starting at 100°C, rapidly by 125°C [KIR81]
Thermal Conductivity, W/(m·K): 0.9 at 0°C [KIR81]

2044
Compound: Oxalic acid dihydrate
Formula: $HOOCCOOH \cdot 2H_2O$
Molecular Formula: $C_2H_6O_6$
Molecular Weight: 126.066
CAS RN: 6153-56-6
Properties: transparent, colorless cryst; enthalpy of solution in water -35.5 kJ/mol; crystallizes from water; used to clean automobile radiators [HAW93] [KIR81]
Solubility: s H_2O, ether [KIR81]
Density, g/cm³: 1.653 [KIR81]
Melting Point, °C: 101.5 [HAW93]
Reactions: minus $2H_2O$ at 98-100°C [KIR81]

2045
Compound: Oxalyl chloride
Formula: $ClCOCOCl$
Molecular Formula: $C_2Cl_2O_2$
Molecular Weight: 126.926
CAS RN: 79-37-8

Properties: liq [ALF95]
Density, g/cm³: 1.455 [ALF95]
Boiling Point, °C: 63-64 [ALF95]

2046
Compound: Oxygen
Formula: O_2
Molecular Formula: O_2
Molecular Weight: 31.999 (atomic weight: 15.9994)
CAS RN: 7782-44-7
Properties: colorless, odorless, tasteless gas; diatomic; can be liquefied at -183°C to sl bluish liq; solidifies at -218°C; critical temp -118.95°C; critical pressure 50.14 atm; enthalpy of vaporization 6.820 kJ/mol; enthalpy of fusion 0.44 kJ/mol; specific volume (21.1°C,101.3kPa) 0.75 m³/kg [HAW93] [MER89] [AIR87] [CRC93] [ALD94]
Solubility: one vol gas dissolves in 32 volumes H_2O (20°C), in 7 volumes alcohol (20°C); s other organic liq, usually higher solubility than in H_2O [MER89]
Density, g/cm³: gas: 1.404 g/L [LID94]; liq: 1.14 g/mL [MER89]
Melting Point, °C: -218.79 [CRC93]
Boiling Point, °C: -182.96 [CRC93]
Thermal Conductivity, W/(m·K): 0.02658 (25°C) [ALD94]

2047
Compound: Ozone
Synonyms: triatomic oxygen
Formula: O_3
Molecular Formula: O_3
Molecular Weight: 47.998
CAS RN: 10028-15-6
Properties: blue gas; unstable; pungent odor; oxidizing agent; can be liquefied at -12.1°C; solid is black violet; prepared by silent electric discharge in oxygen; used in drinking water purification, in industrial waste treatment [MER89] [COT88] [DOU83]

$$\begin{array}{ccc} O & & O \\ \| & \leftrightarrow & | \\ O{-}O & & O{=}O \end{array}$$

Density, g/cm³: gas: 2.144; liq: 1.614 [MER89]
Melting Point, °C: -193 [MER89]
Boiling Point, °C: -111.9 [MER89]
Reactions: reacts with most compounds at 25°C [COT88]

2048
Compound: Palladium
Formula: Pd
Molecular Formula: Pd
Molecular Weight: 106.42
CAS RN: 7440-05-3
Properties: silvery white metal; fcc, a = 0.389 nm; also occurs as black powd, spongy compressible mass; hardness, 4.8 Mohs; electrical resistivity 10.0 μohm·cm; Poisson's ratio 0.39; appreciably volatile at high temperatures; absorbs up to 800 times its own volume of $H_2(g)$; enthalpy of fusion 16.74 kJ/mol; enthalpy of vaporization 362 kJ/mol [HAW93] [MER89] [KIR82] [CRC93] [ALD94]
Solubility: s aqua regia, fused alkalies [HAW93]
Density, g/cm³: 12.02 [MER89]
Melting Point, °C: 1555 [MER89]
Boiling Point, °C: 3167 [MER89]
Reactions: forms dihalides at red heat with fluorine, chlorine [MER89]
Thermal Conductivity, W/(m·K): 75.3 (25°C) [KIR82]
Thermal Expansion Coefficient: 11.1 x 10⁻⁶/°C [KIR82]

2049
Compound: Palladium(II) acetate
Synonyms: palladous acetate
Formula: $Pd(CH_3COO)_2$
Molecular Formula: $C_4H_6O_4Pd$
Molecular Weight: 224.510
CAS RN: 3375-31-3
Properties: orange brown cryst; there is a trimer $[Pd(CH_3COO)_2]_3$ [MER89] [AES93]
Solubility: i H_2O; s with decomposition, HCl; s $CHCl_3$, methylene dichloride, acetone, acetonitrile, diethyl ether [MER89]
Melting Point, °C: 205, decomposes [KIR82]

2050
Compound: Palladium(II) acetylacetonate
Synonyms: 2,4-pentanedione, palladium(II) derivative
Formula: $Pd(CH_3C(O)CH=COCH_3)_2$
Molecular Formula: $C_{10}H_{14}O_4Pd$
Molecular Weight: 304.639
CAS RN: 14024-61-4
Properties: yellow cryst [STR93]

$$\begin{array}{cc} O{-} & O \\ | & \| \\ [CH_3{-}C{=}CH{-}C{-}CH_3]_2Pd \end{array}$$

Melting Point, °C: 205 decomposes [STR93]

2051
Compound: Palladium(II) bromide
Formula: PdBr$_2$
Molecular Formula: Br$_2$Pd
Molecular Weight: 266.228
CAS RN: 13444-94-5
Properties: black cryst; hygr [STR93]
Density, g/cm^3: 5.173 [STR93]
Melting Point, °C: decomposes [AES93]

2052
Compound: Palladium(II) chloride
Synonyms: palladous chloride
Formula: PdCl$_2$
Molecular Formula: Cl$_2$Pd
Molecular Weight: 177.325
CAS RN: 7647-10-1
Properties: dark brown powd or cryst; red rhomb; hygr; used in the electroless deposition process [KIR82] [HAW93] [STR93] [MER89]
Solubility: s H$_2$O, HCl, alcohol and acetone [HAW93]
Density, g/cm^3: 4.0 [STR93]
Melting Point, °C: 675, decomposes [HAW93]

2053
Compound: Palladium(II) chloride dihydrate
Formula: PdCl$_2$·2H$_2$O
Molecular Formula: Cl$_2$H$_4$O$_2$Pd
Molecular Weight: 213.356
CAS RN: 7647-10-1
Properties: dark brown cryst; reduced by H$_2$ or CO in solution to the metal [MER89]
Solubility: s H$_2$O, alcohol, acetone [MER89]

2054
Compound: Palladium(II) cyanide
Formula: Pd(CN)$_2$
Molecular Formula: C$_2$N$_2$Pd
Molecular Weight: 158.455
CAS RN: 2035-66-7
Properties: yellow powd [STR93]
Melting Point, °C: decomposes [AES93]

2055
Compound: Palladium(II) fluoride
Formula: PdF$_2$

Molecular Formula: F$_2$Pd
Molecular Weight: 144.417
CAS RN: 13444-96-7
Properties: violet hygr cryst; paramagnetic [LID94] [KIR82]
Solubility: reacts with H$_2$O [LID94]
Density, g/cm^3: 5.76 [LID94]
Melting Point, °C: 952 [LID94]

2056
Compound: Palladium(II) hexafluoroacetylacetonate
Synonyms: 1,1,1,5,5,5-hexafluoro-2,4-pentanedione Pd(II) derivative
Formula: Pd(CF$_3$COCHCOCF$_3$)$_2$
Molecular Formula: C$_{10}$H$_2$F$_{12}$O$_4$Pd
Molecular Weight: 520.524
CAS RN: 64916-48-9
Properties: cryst [ALF95] [ALD94]

$$[CF_3-\underset{\underset{\displaystyle O-}{|}}{C}=CH-\underset{\underset{\displaystyle O}{\|}}{C}-CF_3]_2Pd$$

2057
Compound: Palladium(II) iodide
Formula: PdI$_2$
Molecular Formula: I$_2$Pd
Molecular Weight: 360.229
CAS RN: 7790-38-7
Properties: black powd [HAW93]
Solubility: s KI soln; i H$_2$O, alcohol, ether [HAW93]
Density, g/cm^3: 6.003 [HAW93]
Melting Point, °C: decomposes at 350 [HAW93]

2058
Compound: Palladium(II) nitrate
Synonyms: palladous nitrate
Formula: Pd(NO$_3$)$_2$
Molecular Formula: N$_2$O$_6$Pd
Molecular Weight: 230.429
CAS RN: 10102-05-3
Properties: brown deliq cryst; heating causes decomposition; used as a catalyst [MER89] [HAW93]
Solubility: s H$_2$O giving a turbid solution, may form precipitate; s dil HNO$_3$ [MER89]
Density, g/cm^3: 1.118 [ALD94]

2059
Compound: Palladium(II) oxalate
Formula: Pd(C$_2$O$_4$)$_2$

Molecular Formula: C_4O_8Pd
Molecular Weight: 194.42
CAS RN: 57592-57-1
Properties: powd [ALF95]

2060
Compound: Palladium(II) oxide
Synonyms: palladium monoxide
Formula: PdO
Molecular Formula: OPd
Molecular Weight: 122.419
CAS RN: 1314-08-5
Properties: -20 mesh with 99.95% purity; black green or amber solid [HAW93] [CER91]
Solubility: i H_2O, acids; sl s aqua regia, 48% HBr [MER89]
Density, g/cm^3: 8.70 [STR93]
Melting Point, °C: 870 [ALD94]

2061
Compound: Palladium(II) sulfate dihydrate
Formula: $PdSO_4 \cdot 2H_2O$
Molecular Formula: H_4O_6PdS
Molecular Weight: 238.514
CAS RN: 13566-03-5
Properties: brown cryst [STR93]

2062
Compound: Palladium(II) sulfide
Formula: PdS
Molecular Formula: PdS
Molecular Weight: 138.486
CAS RN: 12125-22-3
Properties: gray, tetr cryst [LID94]
Density, g/cm^3: 6.60 [ALD94]
Melting Point, °C: 950 [LAN52]

2063
Compound: Palladium(II) tetraammine chloride monohydrate
Formula: $Pd(NH_3)_4Cl_2 \cdot H_2O$
Molecular Formula: $Cl_2H_{14}N_4OPd$
Molecular Weight: 263.463
CAS RN: 13933-31-8
Properties: hygr [ALD94]

2064
Compound: Palladium(III) fluoride
Formula: PdF_3
Molecular Formula: F_3Pd

Molecular Weight: 163.415
CAS RN: 12021-58-8
Properties: consists of Pd(II) and Pd(IV) with the formula $Pd(II)[Pd(IV)]F_6$ [KIR82]

2065
Compound: Pentaborane(11)
Synonyms: dihydropentaborane(9)
Formula: B_5H_{11}
Molecular Formula: B_5H_{11}
Molecular Weight: 65.142
CAS RN: 18433-84-6
Properties: unstable liq; prepared from diborane; decomposes when heated or when allowed to stand for long periods of time, producing various products including diborane, tetraborane, hydrogen; spontaneously flammable in air; enthalpy of vaporization 31.8 kJ/mol [MER89] [CRC93]
Solubility: hydrolyzes in H_2O [MER89]
Melting Point, °C: -122 [COT88]
Boiling Point, °C: 63 [MER89]

2066
Compound: Pentaborane(9)
Synonyms: pentaboron nonahydride
Formula: B_5H_9
Molecular Formula: B_5H_9
Molecular Weight: 63.126
CAS RN: 19624-22-7
Properties: liq; vapor pressure (0°C) 66 mm Hg; spontaneously flammable in air; can be prepared from diborane; forms diammine by reaction with NH_3 [MER89]
Solubility: hydrolyzed if heated [COT88]
Density, g/cm^3: 0.61 [MER89]
Melting Point, °C: -46.6 [KIR78]
Boiling Point, °C: 60 [COT88]
Reactions: decomposes slowly at 150°C [MER89]

2067
Compound: Pentamethylcyclopentadienyltantalum tetrachloride
Formula: $[C_5(CH_3)_5]TaCl_4$
Molecular Formula: $C_{10}H_{15}Cl_4Ta$
Molecular Weight: 457.988
CAS RN: 71414-47-6
Properties: orange powd; sensitive to atm moisture and oxygen [STR93]
Melting Point, °C: 220 [STR93]

2068

Compound: Pentamminechlorocobalt(III) chloride
Formula: [Co(NH₃)₅Cl]Cl₂
Molecular Formula: Cl₃CoH₁₅N₅
Molecular Weight: 250.444
CAS RN: 13859-51-3
Properties: brick red cryst [KIR79] [ALD94]
Solubility: 24.87 g/100mL cold H₂O; sl s HCl; i alcohol [KIR79]

2069

Compound: Perchloric acid
Formula: HClO₄
Molecular Formula: ClHO₄
Molecular Weight: 100.459
CAS RN: 7601-90-3
Properties: colorless, fuming liq; hygr; conc acid is unstable, e.g. sensitive to shock; commercial aq acid contains 65-70% HClO₄; some hydrates are: HClO₄·H₂O, 60477-26-1, a colorless oily liq, mp 50°C, bp decomposes; HClO₄·2H₂O, 13445-00-6, colorless, mp -17.5°C, bp 203°C; HClO₄·3H₂O, 35468-32-7, two forms: α, mp -37°C, β, mp -43.2°C [HAW93] [KIR79]
Solubility: s H₂O with evolution of heat [HAW93]
Density, g/cm³: 1.77 [KIR79]
Melting Point, °C: -112 [HAW93]
Boiling Point, °C: 110 (extrapolated) [KIR79]

2070

Compound: Performic acid
Formula: HCOOOH
Molecular Formula: CH₂O₃
Molecular Weight: 62.024
CAS RN: 107-32-4
Properties: colorless liq; solutions are unstable; used in epoxidation and hydroxylation reactions [HAW93]
Solubility: miscible with H₂O, alcohol, ether; s benzene, chloroform [HAW93]

2071

Compound: Periodic acid
Formula: HIO₄·2H₂O
Molecular Formula: H₅IO₆
Molecular Weight: 227.940
CAS RN: 10450-60-9
Properties: white cryst; oxidizing agent; preparation: oxidation of an iodate by Cl₂ in basic solution; uses: quantitative measurement of α,β-dihydroxyorganic compounds, to increase the wet strength of paper [HAW93] [DOU83]

Solubility: s H₂O, alcohol; sl s ether [HAW93]
Melting Point, °C: 122 [HAW93]
Boiling Point, °C: decomposes, 130 [HAW93]
Reactions: minus 2H₂O at 100°C [HAW93]

2072

Compound: Peroxysulfuric acid
Synonyms: Caro's acid
Formula: H₂SO₅
Molecular Formula: H₂O₅S
Molecular Weight: 114.079
CAS RN: 7722-86-3
Properties: white cryst; oxidant; used for testing aniline; in dye manufacturing [HAW93]
Melting Point, °C: 45, decomposing [HAW93]

2073

Compound: Perrhenic acid
Formula: HReO₄
Molecular Formula: HO₄Re
Molecular Weight: 251.213
CAS RN: 13768-11-1
Properties: colorless liq; only exists in solution; strong, very stable acid [HAW93] [STR93]
Solubility: v s H₂O and in organic solvents [HAW93]

2074

Compound: Phenylmercuric acetate
Synonyms: PMA
Formula: CH₃COOHgC₆H₅
Molecular Formula: C₈H₈HgO₂
Molecular Weight: 336.740
CAS RN: 62-38-4
Properties: cryst prisms; preparation: heating mercuric acetate with benzene; uses: herbicide, fungicide [KIR81] [MER89]
Solubility: s ~600 parts H₂O, s CH₃COONH₄ aq solutions, s alcohol, benzene, acetone [MER89]
Melting Point, °C: 150-152 [ALD94]

2075

Compound: Phenylmercuric chloride
Synonyms: chlorophenylmercury
Formula: C₆H₅HgCl
Molecular Formula: C₆H₅ClHg
Molecular Weight: 313.15
CAS RN: 100-56-1
Properties: white leaflets; uses: fungicide [MER89]
Solubility: ~20,000 parts H₂O; s benzene, ether [MER89]

Melting Point, °C: 248-250, decomposes [ALD94]

2076
Compound: Phenylmercuric nitrate, basic
Formula: $C_6H_5HgNO_3 \cdot C_6H_5HgOH$
Molecular Formula: $C_{12}H_{11}Hg_2NO_4$
Molecular Weight: 634.404
CAS RN: 8003-05-2
Properties: pearly scales; preparation: boiling benzene and mercuric acetate, followed by treatment with an alkali nitrate; uses: antimicrobial, fungicide for tree treatment [MER89]
Solubility: s ~1250 parts H_2O; sl s alcohol [MER89]
Melting Point, °C: decomposes, 176-186 [ALD94]

2077
Compound: Phosphine
Formula: PH_3
Molecular Formula: H_3P
Molecular Weight: 33.998
CAS RN: 7803-51-2
Properties: colorless gas; spontaneously flammable in air; decaying fish odor; reacts violently with halogens, O_2; forms phosphonium salts with halogen acids; critical temp 51.6°C; critical pressure 6.53 MPa; enthalpy of vaporization 14.6 kJ/mol; formed from white phosphorus and an aq alkali hydroxide [AIR87] [MER89] [CRC93]
Solubility: 0.26 volumes in H_2O (20°C); i hot H_2O; sl s alcohol, ether, cuprous chloride solutions [HAW93] [MER89]
Density, g/cm³: 1.492 g/L [LID94]
Melting Point, °C: -133 [MER89]
Boiling Point, °C: -87.75 [CRC93]
Reactions: decomposed to H_2 and metal phosphide by hot metal [MER89]

2078
Compound: Phosphomolybdic acid hydrate
Formula: $H_3[P(Mo_3O_{10})_4] \cdot xH_2O$
Molecular Formula: $H_3Mo_{12}O_{40}P$ (anhydrous)
Molecular Weight: 1825.254 (anhydrous)
CAS RN: 11104-88-4
Properties: yellow cryst; oxidizing agent; used as a reagent for alkaloids, as a pigment; imparts water resistance to plastics [HAW93]
Solubility: s in less than 0.4 parts H_2O, alcohol, ether [MER89]
Density, g/cm³: 3.15 [HAW93]
Melting Point, °C: 78-90 [HAW93]

2079
Compound: Phosphonitrilic chloride trimer
Formula: $(PNCl_2)_3$
Molecular Formula: $Cl_6N_3P_3$
Molecular Weight: 347.657
CAS RN: 940-71-6
Properties: white cryst; sensitive to moisture [STR93]
Density, g/cm³: 1.98 [STR93]
Melting Point, °C: 128.8 [STR93]
Boiling Point, °C: 127 (12 mm Hg) [STR93]

2080
Compound: Phosphonium iodide
Formula: PH_4I
Molecular Formula: H_4IP
Molecular Weight: 161.910
CAS RN: 12125-09-6
Properties: large, transparent, colorless cryst; tetr; sublimes at room temp; decomposes to PH_3 and HI when heated, or in presence of alcohol or water; rapid heating causes detonation [MER89]
Solubility: decomposed by H_2O, alcohol, evolving phosphine gas [HAW93]
Density, g/cm³: 2.86 [HAW93]
Melting Point, °C: 18.5 [HAW93]
Boiling Point, °C: 80 [HAW93]
Reactions: sublimes at 61.8°C [HAW93]

2081
Compound: Phosphoric acid
Synonyms: orthophosphoric acid
Formula: H_3PO_4
Molecular Formula: H_3O_4P
Molecular Weight: 97.995
CAS RN: 7664-38-2
Properties: colorless, odorless, sparkling, syrupy liq; or, unstable ortho-rhomb cryst; acid dissociation constants: $K_1 = 7.107 \times 10^{-3}$, $K_2 = 7.99 \times 10^{-8}$, $K_3 = 4.8 \times 10^{-13}$; enthalpy of fusion 13.40 kJ/mol [MER89] [CRC93]
Solubility: s H_2O, alcohol [MER89]
Density, g/cm³: cryst: 1.834 [HAW93]
Melting Point, °C: cryst: 42.35 [MER89]
Boiling Point, °C: 407 [LID94]
Reactions: minus $1/2H_2O$ at 213°C, forming pyrophosphoric acid [HAW93]

2082
Compound: Phosphorous acid
Synonyms: orthophosphorus acid
Formula: H_3PO_3

Molecular Formula: H₃O₃P
Molecular Weight: 81.996
CAS RN: 13598-36-2
Properties: white, very hygr and deliq, cryst mass; slowly oxidized in air to H₃PO₄; enthalpy of fusion 12.80 kJ/mol [MER89] [DOU83]

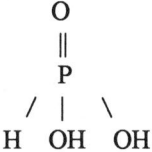

Solubility: v s H₂O, alcohol [MER89]
Density, g/cm³: 1.65; liq: 1.597 [MER89]
Melting Point, °C: 74.4 [CRC93]
Boiling Point, °C: decomposes to PH₃, H₃PO₄ above 180 [MER89]

2083
Compound: Phosphorus (black)
Synonyms: black phosphorus
Formula: P
Molecular Formula: P
Molecular Weight: 30.973762
CAS RN: 7723-14-0
Properties: black solid, resembling graphite; obtained by heating white phosphorus under pressure; ortho-rhomb cryst; stable in air; not spontaneously flammable in air; under high pressure, transformed reversibly to a second rhomb cryst, density 3.56, and cub cryst, density 3.83; conducts electricity [HAW93] [MER89]
Solubility: i organic solvents [MER89]
Density, g/cm³: 2.691 [MER89]

2084
Compound: Phosphorus (red)
Synonyms: red phosphorus
Formula: P
Molecular Formula: P
Molecular Weight: 30.973762
CAS RN: 7723-14-0
Properties: red to violet amorphous powd; 6 mm pieces and smaller with 99.995% purity; obtained from white phosphorus using catalysts by heating at 240°; less active than white phosphorus; has high electrical resistivity [HAW93] [MER89] [CER91]
Solubility: i organic solvents; s PBr₃ [MER89]
Density, g/cm³: 2.34 [MER89]
Melting Point, °C: sublimes 416 [MER89]
Reactions: burns in air to form P₂O₅ at 260°C [MER89]

2085
Compound: Phosphorus (white)
Synonyms: white, yellow phosphorus
Formula: P₄
Molecular Formula: P₄
Molecular Weight: 123.895 (atomic weight: 30.974)
CAS RN: 12185-10-3
Properties: waxlike transparent cryst; darkens when exposed to light; impurities cause yellow color; α-form: exists at room temp; β-form: hex cryst; prepared from α-form at -79.6°C; volatile; sublimes in vacuum at ordinary temperatures; vapor density corresponds to formula P₄; hardness is 0.5 Mohs; electronegativity 2.06; enthalpy of vaporization 12.4 kJ/mol; enthalpy of fusion 0.66 kJ/mol; electrical resistivity (20°C) 10 μohm·cm [CRC93] [ALD94] [HAW93] [MER89] [COT88]
Solubility: 1 part/300,000 parts H₂O; 1 g/400mL absolute alcohol; 1 g/200mL CHCl₂; 1 g/40mL benzene [MER89]
Density, g/cm³: α: 1.83; β: 1.88 [MER89]
Melting Point, °C: 44.1 (0.181 mm) [MER89]
Boiling Point, °C: 277 [CRC93]
Reactions: ignites in moist at ~30°C [MER89]
Thermal Conductivity, W/(m·K): 0.236 (25°C) [ALD94]

2086
Compound: Phosphorus heptasulfide
Formula: P₄S₇
Molecular Formula: P₄S₇
Molecular Weight: 348.357
CAS RN: 12037-82-0
Properties: light yellow cryst [HAW93]
Solubility: sl s carbon disulfide [HAW93]
Density, g/cm³: 2.19 [HAW93]
Melting Point, °C: 310 [HAW93]
Boiling Point, °C: 523 [HAW93]

2087
Compound: Phosphorus nitride
Formula: P₃N₅
Molecular Formula: N₅P₃
Molecular Weight: 162.955
CAS RN: 17739-47-8
Properties: white amorphous solid; stable in air [HAW93]
Solubility: i cold H₂O, decomposed by hot H₂O [HAW93]; s in common organic solvents [HAW93]
Melting Point, °C: 800, decomposes [HAW93]

2088
Compound: Phosphorus oxybromide
Synonyms: phosphoryl bromide
Formula: POBr$_3$
Molecular Formula: Br$_3$OP
Molecular Weight: 286.685
CAS RN: 7789-59-5
Properties: colorless cryst; thin plates with faint orange tint; enthalpy of vaporization 38 kJ/mol; used as an intermediate in chemical processes [HAW93] [MER89] [CRC93]
Solubility: hydrolyzes slowly in H$_2$O to H$_3$PO$_4$ and HBr; s ether, benzene, chloroform, CS$_2$, conc H$_2$SO$_4$ [MER89]
Density, g/cm^3: 2.822 [MER89]
Melting Point, °C: 56 [MER89]
Boiling Point, °C: 191.7 [CRC93]

2089
Compound: Phosphorus oxychloride
Synonyms: phosphoryl chloride
Formula: POCl$_3$
Molecular Formula: Cl$_3$OP
Molecular Weight: 153.331
CAS RN: 10025-87-3
Properties: colorless, clear liq; 99.9 % purity; fumes strongly; liberates heat by reaction with H$_2$O, alcohol; pungent odor; enthalpy of vaporization 34.35 at bp, 38.6 at 25°C; enthalpy of fusion 13.10 kJ/mol; used to manufacture esters for plasticizers and gasoline additives [HAW93] [MER89] [CER91] [CRC93]
Solubility: reacts with H$_2$O, alcohol [HAW93]
Density, g/cm^3: 1.645 [MER89]
Melting Point, °C: 1.25 [ALD94]
Boiling Point, °C: 105.8 [MER89]

2090
Compound: Phosphorus oxyfluoride
Formula: POF$_3$
Molecular Formula: F$_3$OP
Molecular Weight: 103.968
CAS RN: 13478-20-1
Properties: colorless gas; critical temp 73.3°C; critical pressure 4.23 MPa; enthalpy of fusion 14.9 kJ/mol; enthalpy of vaporization 23.2 kJ/mol [KIR78]
Solubility: hydrolyzes [KIR78]
Density, g/cm^3: 4.562 g/L [LID94]
Melting Point, °C: -39.1 [KIR78]
Boiling Point, °C: -39.7 [KIR78]

2091
Compound: Phosphorus triselenide
Formula: P$_2$Se$_3$
Molecular Formula: P$_2$Se$_3$
Molecular Weight: 298.828
CAS RN: 1314-86-9
Properties: dark red mass; heating causes decomposition; decomposed by moist air and in H$_2$O [MER89]
Solubility: s KOH; i CS$_2$ [MER89]
Density, g/cm^3: 1.31 [LID94]
Melting Point, °C: 245 [LID94]
Boiling Point, °C: ~380 [LID94]

2092
Compound: Phosphorus(III) bromide
Synonyms: phosphorus tribromide
Formula: PBr$_3$
Molecular Formula: Br$_3$P
Molecular Weight: 270.686
CAS RN: 7789-60-8
Properties: colorless, fuming liq; 99.9% purity; penetrating odor; enthalpy of vaporization 38.8 kJ/mol [CRC93] [CER91] [HAW93] [MER89]
Solubility: s with decomposition, H$_2$O and alcohol; s acetone, CS$_2$ [MER89]
Density, g/cm^3: 2.852 (15°C) [STR93]
Melting Point, °C: -41.5 [MER89]
Boiling Point, °C: 172.95 [CRC93]

2093
Compound: Phosphorus(III) chloride
Synonyms: phosphorus trichloride
Formula: PCl$_3$
Molecular Formula: Cl$_3$P
Molecular Weight: 137.332
CAS RN: 7719-12-2
Properties: 99.9% purity; colorless, clear, fuming liq; enthalpy of vaporization 30.5 kJ/mol at bp, 32.1 kJ/mol at 25°C; enthalpy of fusion 7.10 kJ/mol [MER89] [CRC93] [CER91]
Solubility: decomposed by H$_2$O, alcohol; s benzene, chloroform, ether, CS$_2$ [MER89]
Density, g/cm^3: 1.574 [MER89]
Melting Point, °C: -112 [CRC93]
Boiling Point, °C: 75.95 [CRC93]

2094
Compound: Phosphorus(III) fluoride
Synonyms: phosphorus trifluoride
Formula: PF$_3$
Molecular Formula: F$_3$P

Molecular Weight: 87.969
CAS RN: 7783-55-3
Properties: colorless gas; sensitive to air and moisture; critical temp -2.05°C; critical pressure 4.33 MPa; enthalpy of vaporization 16.5 kJ/mol [KIR78] [STR93]
Solubility: slowly hydrolyzed by H_2O [MER89]
Density, g/cm³: gas: 3.907 g/L [MER89]
Melting Point, °C: -151.30 [MER89]
Boiling Point, °C: -101.38 [MER89]

2095
Compound: Phosphorus(III) iodide
Synonyms: phosphorus triiodide
Formula: PI_3
Molecular Formula: I_3P
Molecular Weight: 411.687
CAS RN: 13455-01-1
Properties: red or orange cryst; hygr; enthalpy of vaporization 43.9 kJ/mol [HAW93] [CRC93]
Solubility: decomposed by H_2O; s alcohol, CS_2 [HAW93]
Density, g/cm³: 4.18 [STR93]
Melting Point, °C: 61.5 [LID94]
Boiling Point, °C: 227 decomposes [CRC93]

2096
Compound: Phosphorus(III) oxide
Synonyms: phosphorus trioxide
Formula: P_2O_3
Molecular Formula: O_3P_2
Molecular Weight: 109.946
CAS RN: 1314-24-5
Properties: transparent cryst; monocl; or colorless liq; disproportionates to red P and P_2O_4 if heated above 210°C [MER89]
Solubility: slowly forms H_3PO_3 in cold H_2O; reacts violently in hot H_2O forming red P, PH_3 and H_3PO_4 [MER89]
Density, g/cm³: 2.135 [MER89]
Melting Point, °C: 23.8 [MER89]
Boiling Point, °C: 173.1 (under N_2) [MER89]

2097
Compound: Phosphorus(III) sulfide
Synonyms: phosphorus trisulfide
Formula: P_2S_3
Molecular Formula: P_2S_3
Molecular Weight: 158.146
CAS RN: 81129-00-2

Properties: grayish yellow mass; tasteless and odorless; decomposed by atm moisture [HAW93]
Solubility: s alcohol, CS_2, ether [HAW93]
Melting Point, °C: 290 [HAW93]
Boiling Point, °C: 490 [HAW93]

2098
Compound: Phosphorus(V) bromide
Synonyms: phosphorus pentabromide
Formula: PBr_5
Molecular Formula: Br_5P
Molecular Weight: 430.494
CAS RN: 7789-69-7
Properties: yellow cryst; -60 mesh with 99.9% purity; decomposed by H_2O or alcohol; used as a brominating agent [HAW93] [MER89] [CER91]
Solubility: s CS_2, CCl_4 [MER89]
Melting Point, °C: 106, decomposes [HAW93]

2099
Compound: Phosphorus(V) chloride
Synonyms: phosphorus pentachloride
Formula: PCl_5
Molecular Formula: Cl_5P
Molecular Weight: 208.238
CAS RN: 10026-13-8
Properties: white to pale yellow; -60 mesh with 99.9% purity; deliq; fumes; sublimes without melting from 160 to 165°C; irritating odor; used as a chlorinating agent, catalyst and dehydrating agent [HAW93] [MER89] [CER91]
Solubility: hydrolyzed in H_2O to H_3PO_4, HCl; s CS_2, CCl_4 [MER89]
Density, g/cm³: 3.60 [LID94]
Melting Point, °C: 148 (under pressure) [MER89]
Boiling Point, °C: 160 [MER89]

2100
Compound: Phosphorus(V) fluoride
Synonyms: phosphorus pentafluoride
Formula: PF_5
Molecular Formula: F_5P
Molecular Weight: 125.966
CAS RN: 7647-19-0

Properties: colorless gas; nonflammable; fumes strongly in air; high thermal stability; critical temp 144.5°C; critical pressure 3.39 MPa; enthalpy of fusion 12.1 kJ/mol; enthalpy of vaporization 17.2 kJ/mol; can be prepared by reaction of PF_3 with F_2; used as a polymerization catalyst, and in electronics industry [AIR87] [HAW93] [MER89] [CRC93] [KIR78]

Solubility: hydrolyzed in H_2O, eventually to H_3PO_4 [MER89]

Density, g/cm³: gas: 5.527 g/L [LID94]

Melting Point, °C: -93.8 [MER89]

Boiling Point, °C: -84.6 [MER89]

2101

Compound: Phosphorus(V) oxide

Synonyms: phosphorus pentoxide

Formula: P_2O_5

Molecular Formula: O_5P_2

Molecular Weight: 141.945

CAS RN: 1314-56-3

Properties: soft, white powd; -100 mesh with 99.9% purity; several cryst and amorphous forms; very deliq; corrosive; not flammable; readily absorbs H_2O from air; forms H_3PO_4 in water, releasing heat; enthalpy of fusion 27.20 kJ/mol [CRC93] [CER91] [HAW93] [MER89]

Density, g/cm³: 2.39 [STR93]

Melting Point, °C: 420 [CRC93]

Boiling Point, °C: sublimes 360 [MER89]

2102

Compound: Phosphorus(V) selenide

Synonyms: phosphorus pentaselenide

Formula: P_2Se_5

Molecular Formula: P_2Se_5

Molecular Weight: 456.748

CAS RN: 1314-82-5

Properties: amorphous glass; blackish purple solid; decomposes in steam and boiling water [MER89]

Solubility: reacts with CCl_4; i CS_2 [MER89]

2103

Compound: Phosphorus(V) sulfide

Synonyms: phosphorus pentasulfide

Formula: P_2S_5

Molecular Formula: P_2S_5

Molecular Weight: 222.278

CAS RN: 1314-80-3

Properties: light yellow or greenish yellow cryst; H_2S like odor; very hygr; forms P_2O_5 and SO_2 when ignited in air; decomposed by moist air; used in production of lubricating oils, in flotation agents and safety matches [HAW93]

Solubility: s alkali hydroxides, CS_2 [HAW93]

Density, g/cm³: 2.03 [HAW93]

Melting Point, °C: 286-290 [HAW93]

Boiling Point, °C: 515 [HAW93]

2104

Compound: Phosphotungstic acid 24-hydrate

Synonyms: tungstophosphoric acid

Formula: $H_3PW_{12}O_{40} \cdot 24H_2O$

Molecular Formula: $H_{51}O_{64}PW_{12}$

Molecular Weight: 3312.420

CAS RN: 12067-99-1

Properties: H_2O content can vary appreciably; white or sl yellowish green cryst or powd; efflorescent; used in analytical chemistry to detect many organic compounds such as phenols, alkaloids and albumin [MER89]

Solubility: s ~0.5 parts H_2O; s alcohol, ether [MER89]

Melting Point, °C: 89 [HAW93]

2105

Compound: Platinum

Formula: Pt

Molecular Formula: Pt

Molecular Weight: 195.08

CAS RN: 7440-06-4

Properties: silvery gray, lustrous, ductile metal; fcc, a = 0.39231 nm; also has black powd and spongy mass forms; vapor pressure at mp 0.0187 Pa; electrical resistivity, μohm·cm: 10.6 (20°C), 9.85 (0°C); Brinell hardness 97; does not corrode or tarnish; attacked by Cl_2 at high temperatures; Poisson's ratio 0.39; enthalpy of fusion 22.17 kJ/mol; used as a catalyst for chemical, automotive and petroleum industries [KIR82] [HAW93] [MER89] [CRC93]

Solubility: i H_2O, mineral acids; s aqua regia [HAW93] [MER89]

Density, g/cm³: 21.447 (calc) [MER89]

Melting Point, °C: 1768.4 [CRC93]

Boiling Point, °C: 3825 [CRC93]

Thermal Conductivity, W/(m·K): 71.1 (25°C) [KIR82]

Thermal Expansion Coefficient: (volume) 100°C (0.216), 200°C (0.494), 400°C (1.074), 800°C (2.339), 1200°C (3.750) [CLA66]

2106
Compound: Platinum acetylacetonate
Synonyms: 2,4-pentanedione, platinum(II)
   derivative
Formula: $Pt(CH_3C(O)CH=COCH_3)_2$
Molecular Formula: $C_{10}H_{14}O_4Pt$
Molecular Weight: 393.299
CAS RN: 15170-57-7
Properties: pale yellow cryst [STR93]

$$[CH_3-\overset{\underset{|}{O-}}{C}=CH-\overset{\underset{\|}{O}}{C}-CH_3]_2Pt$$

Melting Point, °C: 250-252 [ALD94]

2107
Compound: Platinum hexafluoride
Formula: PtF_6
Molecular Formula: F_6Pt
Molecular Weight: 309.070
CAS RN: 13693-05-5
Properties: dark brick red rhomb solid; strong
   oxidizing agent; there are also PtF_4, 13455-15-7,
   and PtF_5, 13782-84-8 [KIR82]
Density, g/cm³: 3.83 [KIR82]
Melting Point, °C: 61.3 [KIR82]
Boiling Point, °C: 69.14 [KIR82]

2108
Compound: Platinum silicide
Formula: PtSi
Molecular Formula: PtSi
Molecular Weight: 223.166
CAS RN: 12137-83-6
Properties: ortho-rhomb cryst; -100 mesh powd
   [LID94] [ALF93]
Density, g/cm³: 12.4 [LID94]
Melting Point, °C: 1229 [ALF93]

2109
Compound: Platinum(II) bromide
Synonyms: platinum dibromide
Formula: $PtBr_2$
Molecular Formula: Br_2Pt
Molecular Weight: 354.888
CAS RN: 13455-12-4
Properties: brown powd; $PtBr_4$, 13455-11-3, exists
   [KIR82] [STR93]
Density, g/cm³: 6.65 [STR93]
Melting Point, °C: 250, decomposes [STR93]

2110
Compound: Platinum(II) chloride
Synonyms: platinum dichloride
Formula: $PtCl_2$
Molecular Formula: Cl_2Pt
Molecular Weight: 265.985
CAS RN: 10025-65-7
Properties: grayish green to brown powd; hex
   [KIR82] [MER89]
Solubility: i H_2O, alcohol, ether; s HCl [MER89]
Density, g/cm³: 5.87 [MER89]; 6.05 [STR93]
Melting Point, °C: 581, decomposes [STR93]
Reactions: decomposes at red heat yielding platinum
   [HAW93]

2111
Compound: Platinum(II) cyanide
Formula: $Pt(CN)_2$
Molecular Formula: C_2N_2Pt
Molecular Weight: 247.115
CAS RN: 592-06-3
Properties: yellowish green cryst [STR93]

2112
Compound: Platinum(II) hexafluoroacetylacetonate
Synonyms: 1,1,1,5,5,5-hexafluoro-2,4-pentanedione
   Pt(II) derivative
Formula: $Pt(CF_3COCHCOCF_3)_2$
Molecular Formula: $C_{10}H_2F_{12}O_4Pt$
Molecular Weight: 609.185
CAS RN: 65353-51-7
Properties: orange cryst [STR93]

$$[CF_3-\overset{\underset{|}{O-}}{C}=CH-\overset{\underset{\|}{O}}{C}-CF_3]_2Pt$$

Melting Point, °C: sublimes 65 [STR93]

2113
Compound: Platinum(II) iodide
Synonyms: platinum diiodide
Formula: PtI_2
Molecular Formula: I_2Pt
Molecular Weight: 448.889
CAS RN: 7790-39-8
Properties: heavy, black powd [MER89]
Solubility: i H_2O, alkali iodides [MER89]
Density, g/cm³: 6.40 [ALD94]
Melting Point, °C: decomposes 325 [MER89]

2114
Compound: Platinum(II) oxide
Synonyms: platinum monoxide
Formula: PtO
Molecular Formula: OPt
Molecular Weight: 211.079
CAS RN: 12035-82-4
Properties: tetr black cryst [LID94] [KIR82]
Solubility: i H_2O, alcohol; s aqua regia [KIR82]
Density, g/cm³: 14.9 [KIR82]
Melting Point, °C: decomposes 500 [KIR82]

2115
Compound: Platinum(IV) chloride
Synonyms: platinum tetrachloride
Formula: $PtCl_4$
Molecular Formula: Cl_4Pt
Molecular Weight: 336.891
CAS RN: 37773-49-2
Properties: reddish brown cryst; sensitive to moisture [STR93]
Density, g/cm³: 4.303 [STR93]
Melting Point, °C: 370, decomposes [STR93]

2116
Compound: Platinum(IV) chloride pentahydrate
Formula: $PtCl_4 \cdot 5H_2O$
Molecular Formula: $Cl_4H_{10}O_5Pt$
Molecular Weight: 426.967
CAS RN: 13454-96-1
Properties: red cryst [HAW93]
Solubility: s H_2O and alcohol [HAW93]
Density, g/cm³: 2.43 [HAW93]
Reactions: minus $4H_2O$ at 100°C [HAW93]

2117
Compound: Platinum(IV) iodide
Synonyms: platinic iodide
Formula: PtI_4
Molecular Formula: I_4Pt
Molecular Weight: 702.698
CAS RN: 7790-46-7
Properties: brownish black powd; PtI_3, 58782-50-6, exists [KIR82] [MER89]
Solubility: s H_2O [MER89]
Melting Point, °C: decomposes 130 [LID94]

2118
Compound: Platinum(IV) oxide
Synonyms: Adams' catalyst
Formula: PtO_2

Molecular Formula: O_2Pt
Molecular Weight: 227.079
CAS RN: 1314-15-4
Properties: -100 mesh with 99.9% purity; black powd [HAW93] [CER91]
Solubility: s conc acids, s dil KOH solutions [HAW93]
Density, g/cm³: 11.8 [LID94]
Melting Point, °C: 450 [AES93]

2119
Compound: Plutonium
Formula: α-Pu
Molecular Formula: Pu
Molecular Weight: 244
CAS RN: 7440-07-5
Properties: silvery white metal; highly reactive; α form: monocl, a = 0.6183 nm, b = 0.4822 nm, c = 1.0963 nm; ionic radius of Pu^{++++} is 0.0887 nm; stable form from room temp to 115°C; enthalpy of vaporization 333.5 kJ/mol; enthalpy of fusion 2.82 kJ/mol; discovered in 1940-1941; prepared in ton quantities in nuclear reactors; ^{238}Pu produced in kg amounts from ^{237}Np; important fuel for producing power for terrestrial and extraterrestrial applications [MER89] [KIR78] [CRC93]
Density, g/cm³: 19.86 [KIR78]
Melting Point, °C: 646 [KIR91]
Boiling Point, °C: 3235 [KIR91]
Reactions: transitions: α→β, 115°C; β→γ, 185°C; γ→δ, 310°C; δ→δ', 452°C; δ'→ε, 480°C [KIR78]

2120
Compound: Plutonium nitride
Formula: PuN
Molecular Formula: NPu
Molecular Weight: 258
CAS RN: 12033-54-4
Properties: dark gray; fcc, a = 0.4907 nm [KIR81]
Density, g/cm³: 14.4 [KIR81]
Melting Point, °C: 2550 [KIR81]

2121
Compound: Plutonium(III) chloride
Formula: $PuCl_3$
Molecular Formula: Cl_3Pu
Molecular Weight: 350
CAS RN: 13569-62-5
Properties: emerald green; hex, a = 0.7394 nm, c = 0.4243 nm [KIR78]

Density, g/cm³: 5.71 [KIR78]
Melting Point, °C: 760 [KIR91]

2122
Compound: Plutonium(III) fluoride
Formula: PuF$_3$
Molecular Formula: F$_3$Pu
Molecular Weight: 301
CAS RN: 13842-83-6
Properties: purple; hex, a = 0.7092 nm, c = 0.7254 nm [KIR78]
Density, g/cm³: 9.33 [KIR78]
Melting Point, °C: 1425 [KIR91]

2123
Compound: Plutonium(III) iodide
Formula: PuI$_3$
Molecular Formula: I$_3$Pu
Molecular Weight: 625
CAS RN: 13813-46-2
Properties: green; ortho-rhomb, a = 0.4326 nm, b = 1.3962 nm, c = 0.9974 nm [KIR78]
Density, g/cm³: 6.92 [KIR78]

2124
Compound: Plutonium(IV) chloride
Formula: PuCl$_4$
Molecular Formula: Cl$_4$Pu
Molecular Weight: 386
CAS RN: 13536-92-0
Properties: greenish yellow; tetr, a = 0.8377 nm, c = 0.7481 nm [KIR78]
Density, g/cm³: 4.72 [KIR78]

2125
Compound: Plutonium(IV) oxide
Synonyms: plutonium dioxide
Formula: PuO$_2$
Molecular Formula: O$_2$Pu
Molecular Weight: 276
CAS RN: 11116-03-3
Properties: yellowish green to brown; cub, a = 0.53960 nm [KIR78]
Density, g/cm³: 11.46 [KIR78]
Melting Point, °C: 2400 [KIR91]

2126
Compound: Plutonium(VI) hexafluoride
Formula: PuF$_6$
Molecular Formula: F$_6$Pu

Molecular Weight: 358
CAS RN: 13693-06-6
Properties: reddish brown; ortho-rhomb, a = 0.9912 nm, b = 0.8942 nm, c = 0.5206 nm [KIR78]
Density, g/cm³: 5.081 [KIR78]
Melting Point, °C: 52 [KIR91]

2127
Compound: Polonium
Formula: α-Po
Molecular Formula: Po
Molecular Weight: 209
CAS RN: 7440-08-6
Properties: radioactive solid, α-emitter; resistivity at 0°C is 42 μohm cm; resembles Te and Bi in chemical properties; coexists with β-form over temp range 18 to 54°C [MER89]
Density, g/cm³: 9.196 [MER89]
Melting Point, °C: 254 [MER89]
Boiling Point, °C: 962 [MER89]
Thermal Conductivity, W/(m·K): 20 [CRC93]

2128
Compound: Polonium
Formula: β-Po
Molecular Formula: Po
Molecular Weight: 209
CAS RN: 7440-08-6
Properties: radioactive solid; α-emitter; resistivity (0°C) 44 μohm cm; chemical behavior resembles Te, Bi; coexists with α-form over temp range 18 to 54°C; electronegativity 1.76 [MER89] [COT88]
Density, g/cm³: 9.398 [MER89]
Melting Point, °C: 254 [MER89]
Boiling Point, °C: 962 [MER89]

2129
Compound: Polonium(IV) chloride
Synonyms: polonium tetrachloride
Formula: PoCl$_4$
Molecular Formula: Cl$_4$Po
Molecular Weight: 351
CAS RN: 10026-02-5
Properties: hygr; bright yellow cryst; monocl or tricl; hydrolyzed in moist air to form a white solid; vapors are purplish brown, becoming bluish green >500°C [MER89]
Solubility: s H$_2$O, slowly hydrolyzing; v s HCl, thionyl chloride; s ethanol, acetone; decomposed by dil HNO$_3$ [MER89]
Melting Point, °C: ~300 (in chlorine) [MER89]

Boiling Point, °C: 390 [LID94]
Reactions: turns scarlet red at 350°C [MER89]

2130
Compound: Polonium(IV) oxide
Synonyms: polonium dioxide
Formula: PoO_2
Molecular Formula: O_2Po
Molecular Weight: 241
CAS RN: 7446-06-2
Properties: two cryst forms: low temp, yellow fcc; high temp, red tetr [MER89]
Solubility: s phosphoric acid, ammonium carbonate solution [MER89]
Density, g/cm³: 8.9 [LID94]
Melting Point, °C: sublimes 885, color darkens to chocolate brown [MER89]
Reactions: decomposes into elements at 500°C under vacuum [MER89]

2131
Compound: Potassium
Synonyms: kalium
Formula: K
Molecular Formula: K
Molecular Weight: 39.0983
CAS RN: 7440-09-7
Properties: soft, silvery white metal; tarnishes in air; bcc; brittle at low temperatures; reacts vigorously with O_2, H_2O, acids, hydroxides, halogens; enthalpy of fusion 2.32 kJ/mol; enthalpy of vaporization 81.13 kJ/mol; electrical resistivity (20°C) 6.1 μohm·cm; surface tension (100°C) 86 mN/m; viscosity (25°C) 0.258 mPa s; ionic radius 0.133 nm; Pauling electronegativity 0.8 [KIR82] [MER89] [ALD94]
Solubility: s liq ammonia, ethylenediamine, aniline; some metals [MER89]
Density, g/cm³: 0.856 [MER89]
Melting Point, °C: 63.7 [CAB85]
Boiling Point, °C: 760 [CAB85]
Thermal Conductivity, W/(m·K): 102.5 (25°C) [ALD94]

2132
Compound: Potassium acetate
Synonyms: acetic acid, potassium salt
Formula: CH_3COOK
Molecular Formula: $C_2H_3KO_2$
Molecular Weight: 98.143
CAS RN: 127-08-2

Properties: white, lustrous cryst; very deliq; saline taste; usually prepared by reacting potassium carbonate with acetic acid; used as a dehydrating agent and as a textile conditioner [HAW93] [MER89] [KIR82]
Solubility: g/100g, H_2O: 216 (0°C), 256 (20°C), 398 (90°C); Solid phase, CH_3COOK [LAN85]; s alcohol, i ether [HAW93]
Density, g/cm³: 1.57 [MER89]
Melting Point, °C: 292 [MER89]

2133
Compound: Potassium acetylacetonate hemihydrate
Synonyms: 2,4-pentanedione, potassium derivative hemihdyrate
Formula: $K(CH_3COCH=C(O)CH_3)\cdot1/2H_2O$
Molecular Formula: $C_5H_8KO_{2.5}$
Molecular Weight: 147.215
CAS RN: 57402-46-7
Properties: off-white powd; hygr [ALD94] [STR93]
Melting Point, °C: 215 decomposes [STR93]

2134
Compound: Potassium aluminate trihydrate
Formula: $K_2Al_2O_4\cdot3H_2O$
Molecular Formula: $Al_2H_6K_2O_7$
Molecular Weight: 250.202
CAS RN: 12003-63-1
Properties: hard, lustrous cryst; using in dyeing, printing, paper sizing [HAW93] [MER89]
Solubility: v s H_2O, hydrolyzes giving alkaline solution; i alcohol [HAW93] [MER89]

2135
Compound: Potassium aluminum sulfate
Synonyms: burnt alum
Formula: $KAl(SO_4)_2$
Molecular Formula: $AlKO_8S_2$
Molecular Weight: 258.207
CAS RN: 10043-67-1
Properties: white powd; absorbs atm moisture [MER83]
Solubility: g/100g H_2O: 3.00 (0°C), 5.90 (20°C), 109 (90°C) [LAN85]

2136
Compound: Potassium aluminum sulfate dodecahydrate
Synonyms: kalinite
Formula: $KAl(SO_4)_2\cdot12H_2O$
Molecular Formula: $AlH_{24}KO_{20}S_2$

Molecular Weight: 474.391
CAS RN: 7784-24-9
Properties: colorless, trans cryst [MER83]
Solubility: 1 g/7.2mL H_2O, 1 g/0.3mL boiling H_2O; s glyercol [MER83]
Density, g/cm³: 1.725 [MER83]
Melting Point, °C: 92.5 [MER83]
Reactions: minus $12H_2O$ ~200°C [MER83]

2137
Compound: Potassium antimony oxalate trihydrate
Synonyms: antimony potassium oxalate
Formula: $K_3[Sb(OOCCOO)_3]\cdot 3H_2O$
Molecular Formula: $C_6H_6K_3O_{15}Sb$
Molecular Weight: 557.160
CAS RN: 5965-33-3
Properties: cryst powd [MER89]
Solubility: s H_2O [MER89]

2138
Compound: Potassium antimony tartrate hemihydrate
Synonyms: tartar emetic
Formula: $K(SbO)C_4O_6\cdot 1/2H_2O$
Molecular Formula: $C_4HKO_{7.5}Sb$
Molecular Weight: 329.903
CAS RN: 28300-74-5
Properties: transparent cryst; effloresces in air; sweetish metallic taste; used in medicine, in textiles and leather, as an insecticide [HAW93] [MER89]
Solubility: 1 g/12mL H_2O, 1 g/3mL boiling H_2O [MER89]
Density, g/cm³: 2.6 [HAW93]
Reactions: dehydrates at 100°C [HAW93]

2139
Compound: Potassium azide
Formula: KN_3
Molecular Formula: KN_3
Molecular Weight: 81.118
CAS RN: 20762-60-1
Properties: colorless; body-center; tetr, a = 0.6091 nm, c = 0.7056 nm [CIC73] [CRC94]
Solubility: g/100g H_2O: 41.4 (0°C), 50.8 (20°C), 61.0 (40°C), 106 (100°C) [LAN85]
Density, g/cm³: 2.04 [CRC94]
Melting Point, °C: 350 (vacuum) [CRC94]

2140
Compound: Potassium bis(oxalato)platinate(II) dihydrate
Formula: $K_2Pt(C_2O_4)_2\cdot 2H_2O$
Molecular Formula: $C_4H_4K_2O_{10}Pt$
Molecular Weight: 485.346
CAS RN: 14224-64-5
Properties: cryst [ALF95]

2141
Compound: Potassium borohydride
Synonyms: potassium tetrahydroborate
Formula: KBH_4
Molecular Formula: BH_4K
Molecular Weight: 53.941
CAS RN: 13762-51-1
Properties: white cryst powd; non-hygr cryst; supports combustion; decomposes without melting at ~500°C; thermally more stable and less reactive than sodium borohydride; used as a reducing agent for aldehydes, ketones and acid chlorides [MER89]
Solubility: w/w H_2O, 19% (25°C); alkaline solutions stable [MER89]
Density, g/cm³: 1.11 [LID94]
Melting Point, °C: 500 decomposes [KIR80]

2142
Compound: Potassium bromate
Formula: $KBrO_3$
Molecular Formula: $BrKO_3$
Molecular Weight: 167.000
CAS RN: 7758-01-2
Properties: white cryst or granules; oxidizing agent; used as a laboratory reagent, in permanent wave formulations, as a food additive [HAW93] [MER89]
Solubility: g/100g soln, H_2O: 2.96 (0°C), 7.53 (25°C), 33.3 (100°C) [KRU93]; i alcohol [MER89]
Density, g/cm³: 3.27 [MER89]
Melting Point, °C: ~350 [MER89]
Reactions: ~370°C, decomposes with evolution of O_2 [MER89]

2143
Compound: Potassium bromide
Formula: KBr
Molecular Formula: BrK
Molecular Weight: 119.002
CAS RN: 7758-02-3

Properties: colorless cub cryst, or white granules, or powd; hygr; strong, bitter, saline taste; enthalpy of fusion 25.50 kJ/mol; can be prepared by reacting bromine with potassium carbonate; used in photography, engraving and lithography, spectroscopy [HAW93] [CRC93] [STR93] [MER89] [KIR82]

Solubility: g/100g soln, H_2O: 40.61 (25°C); Solid phase, KBr [KRU93]; 1 g/250mL alcohol; 1 g/4.6mL glycerol [MER89]

Density, g/cm³: 2.75 [MER89]

Melting Point, °C: 734 [CRC93]

Boiling Point, °C: 1435 [STR93]

Thermal Expansion Coefficient: (volume) 100°C (0.951), 200°C (2.265), 400°C (5.116) [CLA66]

2144

Compound: Potassium carbonate

Synonyms: salt of tartar, pearl ash

Formula: K_2CO_3

Molecular Formula: CK_2O_3

Molecular Weight: 138.206

CAS RN: 584-08-7

Properties: white monocl; hygr odorless granules or translucent powd; enthalpy of fusion 27.60 kJ/mol; commonly produced by the carbonation of KOH; used in special glasses, e.g. optical and color TV tubes, pigments; general-purpose food additive [HAW93] [MER89] [KIR82] [CRC93]

Solubility: g/100g soln, H_2O: 51.25 (0°C), 52.85 (25°C), 60.90 (100°C); Solid phase, $K_2CO_3 \cdot 1$-1/2H_2O [KRU93]

Density, g/cm³: 2.428 [HAW93]

Melting Point, °C: 891 [MER89]

Boiling Point, °C: decomposes [STR93]

2145

Compound: Potassium carbonate hemitrihydrate

Synonyms: potassium carbonate sesquihydrate

Formula: $K_2CO_3 \cdot 1$-1/2H_2O

Molecular Formula: $CH_3K_2O_{4.5}$

Molecular Weight: 165.229

CAS RN: 6381-79-9

Properties: small, granular cryst; not hygr, if fully hydrated with 1.5 H_2O; formula also given as $2K_2CO_3 \cdot 3H_2O$ [MER89]

Solubility: 129.4 g/100mL H_2O (0°C), 268 g/100mL H_2O (100°C) [LAN52]

Density, g/cm³: 2.043 [CRC94]

Melting Point, °C: 891 [ALD94]

2146

Compound: Potassium chlorate

Synonyms: potcrate

Formula: $KClO_3$

Molecular Formula: $ClKO_3$

Molecular Weight: 122.549

CAS RN: 3811-04-9

Properties: monoclinic lustrous cryst; or white granules, or powd; cooling, saline taste; used as an oxidizing agent, in explosives, in matches, is a source of oxygen; can react violently with organic matter; explodes with H_2SO_4 [MER89] [HAW93]

Solubility: g/100g soln, H_2O: 3.2 (0°C), 7.9 (25°C), 36.0 (100°C) [KRU93]; ~1 g/50mL glycerol; i alcohol [MER89]

Density, g/cm³: 2.32 [MER89]

Melting Point, °C: 368 [MER89]

Boiling Point, °C: decomposes to perchlorate and O_2, >368 [MER89]

2147

Compound: Potassium chloride

Synonyms: sylvite

Formula: KCl

Molecular Formula: ClK

Molecular Weight: 74.551

CAS RN: 7447-40-7

Properties: white cub cryst or powd; -40 mesh with 99.999% purity; strong saline taste; enthalpy of fusion 26.53 kJ/mol; used in fertilizer, pharmaceutical preparations [MER89] [HAW93] [CER91] [CRC93]

Solubility: g/100g soln, H_2O: 21.92 (0°C), 26.4 (25°C), 36.0 (100°C); equilibrium solid phase KCl at 25°C [KRU93]; 1 g/14mL glycerol, 1 g/~250mL alcohol [MER89]; 4.8088±0.0024 mol/(kg·H_2O) at 25°C [RAR85b]

Density, g/cm³: 1.984 [STR93]

Melting Point, °C: 776 [DOU83]

Boiling Point, °C: 1500, sublimes [STR93]

Thermal Expansion Coefficient: (volume) 100°C (0.887), 200°C (1.994), 400°C (4.850), 600°C (8.805) [CLA66]

2148

Compound: Potassium chlorochromate

Synonyms: Peligot's salt

Formula: $KCrO_3Cl$

Molecular Formula: $ClCrKO_3$

Molecular Weight: 174.545

Properties: red or orange cryst; monocl; evolves chlorine when heated; used as an oxidizing agent [HAW93] [KIR78]

Solubility: s H_2O, hydrolyzes [KIR78]

Density, g/cm³: 2.497 [KIR78]

Melting Point, °C: decomposes [KIR78]

2149
Compound: Potassium chromate
Synonyms: tarapacaite
Formula: K_2CrO_4
Molecular Formula: CrK_2O_4
Molecular Weight: 194.191
CAS RN: 7789-00-6
Properties: yellow cryst; ortho-rhomb; used in analytical chemistry as a reagent, and in inks [HAW93] [KIR78]
Solubility: g/100g H_2O: 58.8 (0°C), 65.1 (25°C), 80.1 (100°C) [KRU93]
Density, g/cm³: 2.732 [KIR78]
Melting Point, °C: 971 [KIR78]

2150
Compound: Potassium chromium(III) oxalate trihydrate
Synonyms: potassium tris(oxalato) chromate
Formula: $K_3Cr(C_2O_4)_3 \cdot 3H_2O$
Molecular Formula: $C_6H_6CrK_3O_{15}$
Molecular Weight: 487.396
CAS RN: 15275-09-9
Properties: black-green, monocl; hygr; prepared from reaction of oxalic acid, potassium oxalate and potassium dichromate; used in tanning, and dyeing wool [MER89]
Solubility: s H_2O [MER89]

2151
Compound: Potassium chromium(III) sulfate dodecahydrate
Synonyms: chrome alum
Formula: $CrK(SO_4)_2 \cdot 12H_2O$
Molecular Formula: $CrH_{24}KO_{20}S_2$
Molecular Weight: 499.405
CAS RN: 7789-99-0
Properties: large, reddish violet to black; efflorescent; octahedral cub cryst; ruby red under transmitted light; aq solution is violet when cold, green when hot, color returns to violet on cooling; used in tanning, textile dyeing, ceramics [HAW93] [MER89]
Solubility: 4 parts cold, 2 parts boiling H_2O [MER89]
Density, g/cm³: 1.813 [HAW93]
Melting Point, °C: 89 [HAW93]
Reactions: minus $10H_2O$ at 100°C [HAW93]

2152
Compound: Potassium citrate
Formula: $K_3C_6H_5O_7$
Molecular Formula: $C_6H_5K_3O_7$
Molecular Weight: 306.397
CAS RN: 866-84-2
Properties: used to modify the burning rate of papers [KIR78]
Solubility: 60.91g/100g saturated solution in water (25°C) [MER89]
Melting Point, °C: decomposes at 230 [KIR78]

2153
Compound: Potassium citrate monohydrate
Synonyms: citric acid, tripotassium salt monohydrate
Formula:
$KOOCCH_2C(OH)(COOK)CH_2COOK \cdot H_2O$
Molecular Formula: $C_6H_7K_3O_8$
Molecular Weight: 324.412
CAS RN: 6100-05-6
Properties: colorless or white cryst or powd; deliq; cooling, saline taste, odorless; used as an antacid, as a sequestrant for metals, and buffer in foods [HAW93]
Solubility: g anhydrous/100g H_2O: 153 (10°C), 172 (20°C), 194 (30°C) [LAN85]; 1 g/2.5mL glycerol (dissolves slowly); sl s alcohol [MER89] [HAW93]
Density, g/cm³: 1.98 [HAW93]
Melting Point, °C: 230, decomposes [HAW93]
Reactions: minus H_2O at 180°C [HAW93]

2154
Compound: Potassium cobalt(II) selenate hexahydrate
Formula: $K_2Co(SeO_4)_2 \cdot 6H_2O$
Molecular Formula: $CoH_{12}K_2O_{14}Se_2$
Molecular Weight: 531.137
CAS RN: 28041-86-3
Properties: garnet red; monocl cryst; stable in air [MER89]
Density, g/cm³: 2.514 [MER89]

2155
Compound: Potassium copper(I) cyanide
Formula: $KCu(CN)_2$
Molecular Formula: C_2CuKN_2
Molecular Weight: 150.679
CAS RN: 13682-73-0
Properties: white cryst salt; used in copper plating baths [HAW93]

2156
Compound: Potassium cyanate
Formula: KCNO
Molecular Formula: CKNO
Molecular Weight: 81.115
CAS RN: 590-28-3
Properties: white; cryst powd; used in herbicides, treatment of sickle cell anemia [MER89] [HAW93]
Solubility: s H_2O; v sl s alcohol [MER89]
Density, g/cm³: 2.05 [MER89]
Melting Point, °C: decomposes at 700-900 [HAW93]

2157
Compound: Potassium cyanide
Formula: KCN
Molecular Formula: CKN
Molecular Weight: 65.116
CAS RN: 151-50-8
Properties: white deliq granular powd or fused pieces; odor of hydrogen cyanide; gradually decomposed in moist air; enthalpy of fusion 1470 J/mol; enthalpy of solution 11700 J/mol [MER89] [KIR78]
Solubility: s in: 2 parts cold H_2O, 1 part boiling H_2O, 2 parts glycerol, 100 parts alcohol, 25 parts methanol [MER89]
Density, g/cm³: cub: 1.553 (20°C); ortho-rhomb 1.62 (-60°C) [KIR78]
Melting Point, °C: 634 [MER89]

2158
Compound: Potassium cyanoaurite
Synonyms: potassium dicyanoaurate(I)
Formula: KAu(CN)$_2$
Molecular Formula: C_2AuKN_2
Molecular Weight: 288.104
CAS RN: 13967-50-5
Properties: white cryst powd; used in electrogilding [HAW93]
Solubility: 1 g/7mL H_2O, 1 g/0.5mL boiling H_2O; sl s alcohol; i ether [MER89]
Density, g/cm³: 3.45 [ALD94]

2159
Compound: Potassium dichromate
Synonyms: potassium bichromate
Formula: $K_2Cr_2O_7$
Molecular Formula: $Cr_2K_2O_7$
Molecular Weight: 294.185
CAS RN: 7778-50-9

Properties: bright reddish orange cryst; tricl; not hygr or deliq; decomposes ~500°C; oxidizing agent [KIR78] [MER89]
Solubility: g/100g H_2O: 4.6 (0°C), 15.0 (25°C), 97 (100°C); Solid phase, $K_2Cr_2O_7$ [KRU93]
Density, g/cm³: 2.676 [KIR78]
Melting Point, °C: 398 [KIR78]
Boiling Point, °C: decomposes ~500 [LID94]

2160
Compound: Potassium dihydrogen arsenate
Synonyms: Macquer's salt
Formula: KH$_2$AsO$_4$
Molecular Formula: AsH_2KO_4
Molecular Weight: 180.034
CAS RN: 7784-41-0
Properties: colorless cryst, or white cryst mass, or powd; used in preserving hides, to print textiles, in insecticides [HAW93] [MER89]
Solubility: g/100g soln, H_2O: 15.7 (0°C), 23.6 (25°C), 48.0 (100°C); Solid phase, KH$_2$AsO$_4$·H$_2$O (0°C, 25°C), KH$_2$AsO$_4$ (100°C) [KRU93]; slowly s in 1.6 parts glycerol; i alcohol [MER89]
Density, g/cm³: 2.867 [HAW93]
Melting Point, °C: 288 [HAW93]

2161
Compound: Potassium dihydrogen hypophosphite
Formula: KH$_2$PO$_2$
Molecular Formula: H_2KO_2P
Molecular Weight: 104.087
CAS RN: 7782-87-8
Properties: white cryst or granules; deliq; decomposes when strongly heated in air, evolving PH$_3$; can explode when mixed with oxidizing agents, e.g. chlorates [MER89]
Solubility: 200 g/100mL H_2O (25°C) [CRC94]
Melting Point, °C: decomposes [CRC94]

2162
Compound: Potassium dihydrogen phosphate
Synonyms: potassium phosphate monobasic
Formula: KH$_2$PO$_4$
Molecular Formula: H_2KO_4P
Molecular Weight: 136.085
CAS RN: 7778-77-0
Properties: tetragonal cryst or white granular powd; used in baking powd, in yeast foods, as a buffer and sequestrant for metals [HAW93] [MER89] [KIR82]

Solubility: g/100g soln, H_2O: 12.4 (0°C), 20.0
  (25°C), 45.5 (90°C) [KRU93]; i alcohol
  [MER89]
Density, g/cm³: 2.338 [HAW93]
Melting Point, °C: 253 [HAW93]
Reactions: minus H_2O forming metaphosphate at
  400°C [MER89]

2163
Compound: Potassium dihydrogen phosphite
Synonyms: potassium monobasic phosphite
Formula: KH_2PO_3
Molecular Formula: H_2KO_3P
Molecular Weight: 120.086
Properties: white powd; hygr; slowly oxidized in air
  to the phosphate [HAW93]
Solubility: 220g/100mL H_2O (20°C) [CRC94]
Melting Point, °C: decomposes [CRC94]

2164
Compound: Potassium dithionate
Synonyms: potassium hyposulfate
Formula: $K_2(SO_3)_2$
Molecular Formula: $K_2O_6S_2$
Molecular Weight: 238.325
CAS RN: 13455-20-4
Properties: colorless cryst; used as an analytical
  reagent [HAW93]
Solubility: g/100g H_2O: 2.6 (0°C), 6.6 (20°C), 9.3
  (30°C) [LAN85]; i alcohol [HAW93]
Density, g/cm³: 2.27 [HAW93]
Melting Point, °C: decomposes [CRC94]

2165
Compound: Potassium ferricyanide
Synonyms: potassium hexacyanoferrate(III)
Formula: $K_3Fe(CN)_6$
Molecular Formula: $C_6FeK_3N_6$
Molecular Weight: 329.248
CAS RN: 13746-66-2
Properties: ruby red cryst or powd; lustrous; used to
  temper steel, as an etching liq, in electroplating
  [HAW93] [MER89]
Solubility: g/100g H_2O: 30.2 (0°C), 46 (20°C), 70
  (60°C) [LAN85]; s alcohol; decomposed by
  acids [MER89]
Density, g/cm³: 1.89 [MER89]
Melting Point, °C: decomposes [STR93]

2166
Compound: Potassium ferrocyanide trihydrate

Synonyms: potassium hexacyanoferrate(II)
  trihydrate
Formula: $K_4Fe(CN)_6·3H_2O$
Molecular Formula: $C_6H_6FeK_4N_6O_3$
Molecular Weight: 422.390
CAS RN: 14459-95-1
Properties: lemon yellow; soft, sl efflorescent cryst;
  used in dyeing, in tempering steel [HAW93]
  [MER89]
Solubility: g anhydrous/100g H_2O: 14.3 (0°C), 28.2
  (20°C), 74.2 (100°C) [LAN85]; i alcohol
  [HAW93]
Density, g/cm³: 1.85 [MER89]
Melting Point, °C: 70, decomposes [STR93]
Reactions: begins to give up H_2O at 60°C;
  anhydrous by 100°C [MER89]

2167
Compound: Potassium fluoride
Formula: KF
Molecular Formula: FK
Molecular Weight: 58.096
CAS RN: 7789-23-3
Properties: cub cryst; usually white, deliq powd;
  sharp saline taste; enthalpy of fusion 27.2
  kJ/mol; enthalpy of vaporization 173 kJ/mol;
  commercial preparation is from KOH and HF,
  followed by drying; used to etch glass, as a
  preservative [HAW93] [MER89] [CRC93]
Solubility: g/100g soln, H_2O: 30.90 (0°C), 50.41
  (25°C), 60.01 (80°C); Solid phase, $KF·4H_2O$
  (0°C), $KF·2H_2O$ (25°C), KF (80°C) [KRU93]; s
  a, HF, liq NH_3 [MER89]
Density, g/cm³: 2.481 [MER89]
Melting Point, °C: 858 [CRC93]
Boiling Point, °C: 1505 [MER89]

2168
Compound: Potassium fluoride dihydrate
Formula: $KF·2H_2O$
Molecular Formula: FH_4KO_2
Molecular Weight: 94.127
CAS RN: 13455-21-5
Properties: white monocl cryst; has been used as
  stationary phase in gas chromatography
  [ALD94] [MER89]
Solubility: 349.3 g/100mL H_2O (18°C) [MER89]
Density, g/cm³: 2.454 [STR93]
Melting Point, °C: 41 decomposes [LID94]

2169
Compound: Potassium fullerene

Formula: K_3C_{60}
Molecular Formula: $C_{60}K_3$
Molecular Weight: 837.955
CAS RN: 137232-17-8
Properties: fcc, lattice constant 1.4253 nm; superconductor, Tc 19.3 K; bulk modulus 28 GPa; cohesive energy 24.2 eV; enthalpy of formation 4.9 eV; density of states 25/(eV/C_{60}); electron effective mass 1.3; hole effective mass 1.5, 3.4; potential uses include optical limiter to protect materials from damage due to high light intensities, and for the fabrication of industrial diamonds [DRE93]

2170
Compound: Potassium gold(III) oxide trihydrate
Synonyms: potassium aurate
Formula: $KAuO_2 \cdot 3H_2O$
Molecular Formula: AuH_6KO_5
Molecular Weight: 322.110
Properties: yellow cryst; used to prepare other gold compounds [HAW93]
Solubility: s H_2O, alcohol [HAW93]
Melting Point, °C: decomposes [CRC94]

2171
Compound: Potassium heptafluoroniobate
Formula: K_2NbF_5
Molecular Formula: F_7K_2Nb
Molecular Weight: 304.092
CAS RN: 16924-03-1
Properties: lump [ALF95]

2172
Compound: Potassium heptafluorotantalate
Formula: K_2TaF_7
Molecular Formula: F_7K_2Ta
Molecular Weight: 392.134
CAS RN: 16924-00-8
Properties: colorless rhomb needles when crystallized from solution containing HF, KF and tantalum fluoride [KIR83]
Solubility: hydrolyzes in H_2O [KIR83]
Density, g/cm³: 5.24 [KIR83]
Melting Point, °C: 740 [KIR83]

2173
Compound: Potassium heptaiodobismuthate
Synonyms: bismuth potassium iodide
Formula: K_4BiI_7
Molecular Formula: BiI_7K_4

Molecular Weight: 1253.704
CAS RN: 41944-01-8
Properties: red cryst; used to precipitate vitamins and antibiotics from aq solutions [HAW93] [MER89]
Solubility: decomposes in H_2O; s alkali iodide soln [MER89]

2174
Compound: Potassium hexabromoplatinate(IV)
Formula: K_2PtBr_6
Molecular Formula: Br_6K_2Pt
Molecular Weight: 752.701
CAS RN: 16920-93-7
Properties: reddish brown powd; hygr [STR93]
Density, g/cm³: 4.66 [STR93]
Melting Point, °C: 400, decomposes [STR93]

2175
Compound: Potassium hexachloroiridate(IV)
Formula: K_2IrCl_6
Molecular Formula: Cl_6IrK_2
Molecular Weight: 483.130
CAS RN: 16920-56-2
Properties: black powd; hygr; there is also K_3IrCl_6, CAS RN 14024-41-0 [STR93] [ALD94]
Density, g/cm³: 3.546 [STR93]
Melting Point, °C: decomposes [STR93]

2176
Compound: Potassium hexachloroosmiate(IV)
Formula: K_2OsCl_6
Molecular Formula: Cl_6K_2Os
Molecular Weight: 481.143
CAS RN: 16871-60-6
Properties: dark red to almost black; cub cryst; hygr [STR93] [MER89]
Solubility: v s H_2O; sl s alcohol [MER89]
Density, g/cm³: 3.42 [ALD94]
Melting Point, °C: 600, decomposes [ALD94]

2177
Compound: Potassium hexachloropalladate(IV)
Formula: K_2PdCl_6
Molecular Formula: Cl_6K_2Pd
Molecular Weight: 397.333
CAS RN: 16919-73-6
Properties: red powd; hygr [STR93]
Density, g/cm³: 2.738 [STR93]

2178
Compound: Potassium hexachloroplatinate(IV)
Formula: K_2PtCl_6
Molecular Formula: Cl_6K_2Pt
Molecular Weight: 485.993
CAS RN: 16921-30-5
Properties: small, yellowish orange cryst or powd; used in photography and as a reagent [HAW93]
Solubility: g/100g H_2O: 0.48 (0°C), 0.78 (20°C), 5.03 (100°C) [LAN85]; i alcohol [HAW93]
Density, g/cm³: 3.50 [HAW93]
Melting Point, °C: 250, decomposes [HAW93]

2179
Compound: Potassium hexachlororhenate(IV)
Formula: K_2ReCl_6
Molecular Formula: Cl_6K_2Re
Molecular Weight: 477.120
CAS RN: 16940-97-9
Properties: green powd; hygr [STR93]
Density, g/cm³: 3.34 [STR93]

2180
Compound: Potassium tetracyanocadmium
Synonyms: cadmium potassium cyanide
Formula: $K_2Cd(CN)_4$
Molecular Formula: $C_4CdK_2N_4$
Molecular Weight: 294.679
CAS RN: 14402-75-6
Properties: highly refractive, cub cryst; when heated, melts to colorless liq, solidifying to a gray, cryst mass on cooling [MER89]
Solubility: s 3 parts cold H_2O, 1 part hot H_2O [MER89]
Density, g/cm³: 1.846 [MER89]
Melting Point, °C: ~450 [MER89]

2181
Compound: Potassium hexacyanocobalt(III)
Synonyms: potassium cobalticyanide
Formula: $K_3Co(CN)_6$
Molecular Formula: $C_6CoK_3N_6$
Molecular Weight: 332.334
CAS RN: 13963-58-1
Properties: faintly yellow, monocl cryst when obtained from H_2O; sensitive to light; unstable if stored, generating HCN [MER89]
Solubility: v s H_2O, acetic acid solutions; i alcohol [MER89]
Density, g/cm³: 1.906 [MER89]
Melting Point, °C: decomposes, forming olive green mass [MER89]

2182
Compound: Potassium hexacyanoplatinate(IV)
Formula: $K_2Pt(CN)_6$
Molecular Formula: $C_6K_2N_6Pt$
Molecular Weight: 429.383
CAS RN: 16920-94-8
Properties: white powd [STR93]

2183
Compound: Potassium hexafluorantimonate
Formula: $KSbF_6$
Molecular Formula: F_6KSb
Molecular Weight: 274.846
CAS RN: 16893-92-8
Properties: -6 mesh of 99.9% purity [CER91]

2184
Compound: Potassium hexafluoroarsenate(V)
Formula: $KAsF_6$
Molecular Formula: AsF_6K
Molecular Weight: 228.010
CAS RN: 17029-22-0
Properties: grayish white cryst [STR93]
Melting Point, °C: 400 [STR93]

2185
Compound: Potassium hexafluorogermanate
Formula: K_2GeF_6
Molecular Formula: F_6GeK_2
Molecular Weight: 264.797
CAS RN: 7783-73-5
Properties: white cryst; stable up to 500°C [HAW93]
Solubility: g/100g H_2O: 0.25 (0°C), 0.50 (20°C), 0.96 (40°C) [LAN85]; s alcohol [HAW93]
Melting Point, °C: 730 [CRC94]
Boiling Point, °C: ~835 [CRC94]

2186
Compound: Potassium hexafluoromanganate(IV)
Formula: K_2MnF_6
Molecular Formula: F_6K_2Mn
Molecular Weight: 247.125
CAS RN: 16962-31-5
Properties: golden yellow; hex platelets; turns brown when heated, but returns to original color when cooled [MER89]
Solubility: hydrolyzed in H_2O, precipitates MnO_4 [MER89]
Melting Point, °C: decomposes [CRC94]

2187
Compound: Potassium hexafluoronickelate(IV)
Synonyms: potassium nickel(IV) fluoride
Formula: K_2NiF_6
Molecular Formula: F_6K_2Ni
Molecular Weight: 250.880
CAS RN: 17218-47-2
Properties: powd; can be a source of F_2 because F_2 is evolved when the compound is heated [STR93]
Melting Point, °C: 400 decomposes [STR93]

2188
Compound: Potassium hexafluorophosphate
Formula: KPF_6
Molecular Formula: F_6KP
Molecular Weight: 184.062
CAS RN: 17084-13-8
Properties: white cryst; hygr [ALD94] [STR93]
Solubility: 9.3 g/100mL H_2O (25°C), 20.6 g/100mL H_2O (50°C) [CRC94]
Density, g/cm³: 2.55 [STR93]
Melting Point, °C: ~575 [STR93]
Boiling Point, °C: decomposes [STR93]

2189
Compound: Potassium hexafluorosilicate
Synonyms: hieratite
Formula: K_2SiF_6
Molecular Formula: F_6K_2Si
Molecular Weight: 220.273
CAS RN: 16871-90-2
Properties: white fine powd or cryst; used to manufacture opalescent glass, used in porcelain enamel and as an insecticide [MER89]
Solubility: g/100g H_2O: 0.077 (0°C), 0.151 (20°C), 0.253 (40°C) [LAN85]; hydrolyzes in hot water to KF, HF and silicic acid; i alcohol [MER89]
Density, g/cm³: 2.27 [MER89]
Melting Point, °C: decomposes [STR93]

2190
Compound: Potassium hexafluorotitanate monohydrate
Formula: $K_2TiF_6 \cdot H_2O$
Molecular Formula: $F_6H_2K_2OTi$
Molecular Weight: 258.082
CAS RN: 16919-27-0
Properties: colorless; monocl [CRC77]
Solubility: g/100g H_2O: 0.55 (0°C), 0.91 (10°C), 1.28 (20°C) [LAN85]
Melting Point, °C: 780 [CRC77]

Reactions: minus H_2O, 32°C [CRC77]

2191
Compound: Potassium hexafluorozirconate
Formula: K_2ZrF_6
Molecular Formula: F_6K_2Zr
Molecular Weight: 283.411
CAS RN: 16923-95-8
Properties: colorless monocl cryst [MER89] [CRC94]
Solubility: 0.781 g/100mL H_2O (2°C), 25 g/100mL H_2O (100°C) [CRC94]
Density, g/cm³: 3.48 [CRC94]

2192
Compound: Potassium hexametaphosphite
Formula: $(KPO_3)_6$
Molecular Formula: $K_6O_{18}P_6$
Molecular Weight: 708.420
CAS RN: 7790-53-6
Properties: white powd; hygr [CRC94] [STR93]
Solubility: g/100mL soln, H_2O: 0.0041 (25°C) [KRU93]
Density, g/cm³: 1.207 [CRC94]
Melting Point, °C: 807 [STR93]
Boiling Point, °C: 1320 [CRC94]

2193
Compound: Potassium hexanitritocobalt(III)
Synonyms: Fischer's salt
Formula: $K_3Co(NO_2)_6$
Molecular Formula: $CoK_3N_6O_{12}$
Molecular Weight: 452.261
CAS RN: 66942-97-0
Properties: yellow cryst powd [HAW93]
Solubility: 0.9 g/100mL H_2O (17°C), decomposed by hot H_2O [CRC94]
Melting Point, °C: 200, decomposes [HAW93]

2194
Compound: Potassium hexanitritorhodate(III)
Formula: $K_3Rh(NO_2)_6$
Molecular Formula: $K_3N_6O_{12}Rh$
Molecular Weight: 496.234
CAS RN: 17712-66-2
Properties: white powd [STR93]

2195
Compound: Potassium hexathiocyantoplatinate(IV)
Formula: $K_2Pt(SCN)_6$

Molecular Formula: $C_6K_2N_6PtS_6$
Molecular Weight: 621.779
CAS RN: 17069-38-4
Properties: carmine red cryst [MER89]
Solubility: s H_2O [MER89]

2196
Compound: Potassium hydride
Formula: KH
Molecular Formula: HK
Molecular Weight: 40.106
CAS RN: 7693-26-7
Properties: white needles; slurry of gray powd in oil; sensitive to atm oxygen and moisture [STR93] [CRC94]
Solubility: decomposed by H_2O [CRC94]
Density, g/cm³: 1.47 [CRC94]
Melting Point, °C: decomposes [CRC94]

2197
Compound: Potassium hydrogen arsenite
Formula: $KAsO_2 \cdot HAsO_2$
Molecular Formula: As_2HKO_4
Molecular Weight: 253.947
CAS RN: 10124-50-2
Properties: white hygr powd; gradually decomposed by atm CO_2; used in the manufacture of silver mirrors to reduce silver salts to metallic silver [MER89]
Solubility: s H_2O [MER89]

2198
Compound: Potassium hydrogen carbonate
Synonyms: potassium bicarbonate
Formula: $KHCO_3$
Molecular Formula: $CHKO_3$
Molecular Weight: 100.115
CAS RN: 298-14-6
Properties: monoclinic transparent cryst, white granules, or powd; sl alkaline or salty taste; obtained by addition of CO_2 to a solution of potassium carbonate; used in baking powd, soft drinks, as an antacid [HAW93] [MER89] [KIR82]
Solubility: g/100 g soln, H_2O: 18.6 (0°C), 26.6 (25°C); Solid phase, $KHCO_3$ [KRU93]; i alcohol [MER89]
Density, g/cm³: 2.17 [HAW93]
Melting Point, °C: decomposes 100-120 [HAW93]

2199
Compound: Potassium hydrogen fluoride
Formula: KHF_2
Molecular Formula: F_2HK
Molecular Weight: 78.103
CAS RN: 7789-29-9
Properties: colorless tetr cryst; decomposed by heat; enthalpy of fusion 6.62 kJ/mol; produced by reaction of KOH or K_2CO_3 with HF; used to etch glass, as a flux for silver solders [HAW93] [MER89] [CRC93]
Solubility: g/100mL H_2O: 30.1 (10°C), 39.2 (20°C), 114.0 (80°C); s dil alcohol; i absolute alcohol [MER89]
Density, g/cm³: 2.37 [MER89]
Melting Point, °C: 238.7 [MER89]

2200
Compound: Potassium hydrogen iodate
Synonyms: potassium acid iodate
Formula: $KH(IO_3)_2$
Molecular Formula: HI_2KO_6
Molecular Weight: 389.911
CAS RN: 13455-24-8
Properties: colorless monocl or rhomb cryst; used as an alkalimetric standard [KIR81]
Solubility: 9.15 parts/100 parts H_2O at 50°C [KIR81]

2201
Compound: Potassium hydrogen oxalate hemihydrate
Synonyms: potassium binoxalate, sorrel salt
Formula: $KHC_2O_4 \cdot 1/2H_2O$
Molecular Formula: $C_2H_2KO_{4.5}$
Molecular Weight: 137.133
CAS RN: 127-95-7
Properties: white, odorless cryst; bitter sharp taste; somewhat hygr; used to remove ink stains, in scouring metals, cleaning wood; decomposes when heated [HAW93] [MER89]
Solubility: s 40 parts cold H_2O, 6 parts boiling H_2O [MER89]
Density, g/cm³: 2.088 [HAW93]

2202
Compound: Potassium hydrogen phosphite
Formula: K_2HPO_3
Molecular Formula: HK_2O_3P
Molecular Weight: 158.177
CAS RN: 13492-26-7

Properties: white; deliq powd; slowly oxidized in
air to the phosphate; decomposed by heating
[MER89]
Solubility: v s H_2O; i alcohol [MER89]

2203
Compound: Potassium hydrogen selenite
Synonyms: potassium biselenite
Formula: $KHSeO_3$
Molecular Formula: HKO_3Se
Molecular Weight: 167.064
CAS RN: 7782-70-9
Properties: ortho-rhomb prisms; very deliq;
selenium oxide liberated by heating >100°C
[MER89]
Solubility: s H_2O; sl s alcohol [MER89]
Reactions: slowly loses H_2O at 100°C [MER89]

2204
Compound: Potassium hydrogen sulfate
Synonyms: mercallite, misenite
Formula: $KHSO_4$
Molecular Formula: HKO_4S
Molecular Weight: 136.170
CAS RN: 7646-93-7
Properties: white deliq monoclinic or rhomb cryst;
used as a flux, and in the manufacture of mixed
fertilizers [HAW93] [MER89] [KIR82]
Solubility: g/100g soln, H_2O: 26.6 (0°C), 34.0
(25°C), 54.9 (100°C) [KRU93]
Density, g/cm³: 2.322 [STR93]
Melting Point, °C: 210 [KIR82]
Boiling Point, °C: decomposes [STR93]
Reactions: gives up H_2O at high temp to form
pyrosulfate [MER89]

2205
Compound: Potassium hydrogen sulfide
hemihydrate
Formula: $KHS \cdot 1/2H_2O$
Molecular Formula: $H_2KO_{0.5}S$
Molecular Weight: 81.180
CAS RN: 1310-61-8
Properties: tetr; turns yellow rapidly in air, forming
polysulfides and H_2S; forms dark red liq when
melted [MER89]
Solubility: v s H_2O, alcohol [MER89]
Density, g/cm³: 1.70 [MER89]
Melting Point, °C: 450-510 [MER89]
Reactions: loses H_2O at 175-200°C [MER89]

2206
Compound: Potassium hydrogen sulfite
Synonyms: potassium bisulfite
Formula: $KHSO_3$
Molecular Formula: HKO_3S
Molecular Weight: 120.170
CAS RN: 7773-03-7
Properties: white, cryst powd; odor of sulfur
dioxide; preparation: passing SO_2 through aq
solution of K_2CO_3; uses: antiseptic, bleaching
straw and textiles, reduce various organic
compounds [HAW93]
Solubility: s H_2O; i alcohol [HAW93]
Melting Point, °C: 190, decomposes [HAW93]

2207
Compound: Potassium hydrogen tartrate
Synonyms: tartaric acid, monopotassium salt
Formula: KOOCCH(OH)CH(OH)COOH
Molecular Formula: $C_4H_5KO_6$
Molecular Weight: 188.178
CAS RN: 868-14-4
Properties: colorless cryst, or white, cryst powd;
pleasant acidic taste; used in baking powd,
medicine, as a food additive [HAW93] [MER89]
Solubility: g/100g H_2O: 0.231 (0°C), 0.523 (20°C),
0.762 (30°C) [LAN85]; v s dil mineral acids,
solutions of alkalies [MER89]
Density, g/cm³: 1.984 [HAW93]

2208
Compound: Potassium hydroxide
Formula: KOH
Molecular Formula: HKO
Molecular Weight: 56.105
CAS RN: 1310-58-3
Properties: white or sl yellow lumps, rods, pellets;
rhomb; deliq; absorbs H_2O and CO_2 from air;
enthalpy of fusion 8.60 kJ/mol; manufactured by
electrolysis of KCl solutions; used in soap
manufacture, for bleaching [HAW93] [MER89]
[KIR82]
Solubility: g/100g soln, H_2O: 49.00 (0°C), 54.27
(25°C), 64.59 (100°C); Solid phase, $KOH \cdot 2H_2O$
(0°C, 25°C), $KOH \cdot H_2O$ (100°C) [KRU93]; s 3
parts alcohol, 2.5 parts glycerol [MER89]
Density, g/cm³: 2.044 [STR93]
Melting Point, °C: 406 [LID94]
Boiling Point, °C: 1324 [STR93]

2209
Compound: Potassium iodate

Formula: KIO_3
Molecular Formula: IKO_3
Molecular Weight: 214.001
CAS RN: 7758-05-6
Properties: white, odorless monocl cryst or powd; -80 mesh with 99.9% purity; can be prepared by electrochemical oxidation of KI; forms KIO_2F_2 when reacted with HF [KIR81] [MER89] [CER91]
Solubility: g/100g soln, H_2O: 4.4 (0°C), 8.4 (25°C), 24.4 (100°C) [KRU93]; i alcohol [MER89]
Density, g/cm³: 3.979 [KIR81]
Melting Point, °C: 560, partially decomposes [MER89]

2210

Compound: Potassium iodide
Formula: KI
Molecular Formula: IK
Molecular Weight: 166.003
CAS RN: 7681-11-0
Properties: colorless or white; cub cryst, white granules, or powd; -20 mesh with 99.999% purity; sl deliq in moist air; liberates iodine if exposed to air for a lengthy time; sensitive to light; strong, bitter, saline taste; enthalpy of fusion 24.00 kJ/mol; produced by dissolution of I_2 in KOH solutions; used in infrared transmission instrumentation, as a dietary supplement [HAW93] [MER89] [KIR82] [CER91] [CRC93]
Solubility: g/100g H_2O: 127.5 (0°C), 148 (25°C), 207.2±0.8 (100°C); Solid phase, KI [KRU93]; 1 g/22mL alcohol, 1 g/8mL boiling alcohol, 1 g/51mL absolute alcohol [MER89]
Density, g/cm³: 3.123 [HAW93]
Melting Point, °C: 681 [CRC93]
Boiling Point, °C: 1330 [HAW93]
Thermal Expansion Coefficient: (volume) 100°C (0.90), 200°C (2.04), 400°C (4.92) [CLA66]

2211

Compound: Potassium magnesium chloride sulfate
Synonyms: kainite
Formula: $MgSO_4 \cdot KCl \cdot 3H_2O$
Molecular Formula: ClH_6KMgO_7S
Molecular Weight: 248.966
CAS RN: 1318-72-5
Properties: natural hydrated double salt; white, gray, reddish or colorless; vitreous luster; hardness is 2.5-3; used as a fertilizer [HAW93]
Solubility: 79.56 g/100mL H_2O (18°C) [CRC94]
Density, g/cm³: 2.131 [CRC94]

2212

Compound: Potassium magnesium sulfate
Synonyms: langbeinite
Formula: $K_2SO_4 \cdot 2MgSO_4$
Molecular Formula: $K_2Mg_2O_{12}S_3$
Molecular Weight: 414.998
CAS RN: 13826-56-7
Properties: white, tetr cryst; used in fertilizers [HAW93]
Solubility: g/100g H_2O: 14.0 (0°C), 25.0 (20°C), 63.4 (80°C) [LAN85]
Density, g/cm³: 2.829 [HAW93]
Melting Point, °C: 927 [HAW93]

2213

Compound: Potassium manganate
Formula: K_2MnO_4
Molecular Formula: K_2MnO_4
Molecular Weight: 197.133
CAS RN: 10294-64-1
Properties: dark green cryst; oxidizing agent, liberates Cl_2 from HCl; used in bleaching skins, fibers, oils, as a disinfectant [HAW93] [MER89]
Solubility: s H_2O; s and stable in KOH solutions [MER89]
Melting Point, °C: 190, decomposes [MER89]

2214

Compound: Potassium metaarsenite monohydrate
Formula: $KH(AsO_2)_2 \cdot H_2O$
Molecular Formula: $As_2H_3KO_5$
Molecular Weight: 271.962
CAS RN: 10124-50-2
Properties: white powd; hygr; decomposes slowly in air; used as a reducing agent in silvering mirrors [HAW93]
Solubility: s H_2O, sl s in alcohol [HAW93]

2215

Compound: Potassium molybdate
Formula: K_2MoO_4
Molecular Formula: K_2MoO_4
Molecular Weight: 238.135
CAS RN: 13446-49-6
Properties: white, deliq, microcryst powd; -200 mesh with 99.9% purity; isomorphous with K_2SO_4, K_2CrO_4 [KIR81] [HAW93] [CER91]
Solubility: g/100g soln, H_2O: 64.57±0.1 (25°C), 66.54 (89.96°C); Solid phase, K_2MoO_4 [KRU93]
Density, g/cm³: 2.342 [KIR81]
Melting Point, °C: 919 [HAW93]

2216
Compound: Potassium monohydrogen phosphate
Synonyms: potassium dibasic phosphate
Formula: K_2HPO_4
Molecular Formula: HK_2O_4P
Molecular Weight: 174.176
CAS RN: 7758-11-4
Properties: white somewhat hygr powd; ignited to pyrophosphate [MER89]
Solubility: g/100g soln, H_2O: 45.8±0.3 (0°C), 62.4±0.4 (25°C), 73.8 (99.4°C); Solid phase, $K_2HPO_4 \cdot 6H_2O$ (0°C), $K_2HPO_4 \cdot 3H_2O$ (25°C), K_2HPO_4 (99.4°C) [KRU93]; sl s alcohol [MER89]
Melting Point, °C: decomposes [CRC94]

2217
Compound: Potassium monoxide
Synonyms: potassium oxide
Formula: K_2O
Molecular Formula: K_2O
Molecular Weight: 94.196
CAS RN: 12136-45-7
Properties: gray cryst mass; hygr [CRC94] [HAW93]
Solubility: s H_2O, forming KOH; s alcohol and ether [HAW93]
Density, g/cm³: 2.32 (0°C) [HAW93]
Melting Point, °C: 350, decomposes [HAW93]

2218
Compound: Potassium nickel sulfate hexahydrate
Synonyms: nickel potassium sulfate
Formula: $K_2SO_4 \cdot NiSO_4 \cdot 6H_2O$
Molecular Formula: $H_{12}K_2NiO_{14}S_2$
Molecular Weight: 437.109
CAS RN: 10294-65-2
Properties: blueish green cryst; prepared by crystallization from an aq solution; has limited use as dye mordant and in metal finishing [KIR81] [HAW93]
Solubility: g/100g H_2O: 3.37 (0°C), 5.94 (20°C), 33.4 (100°C) [LAN85]
Density, g/cm³: 2.124 [HAW93]
Melting Point, °C: decomposes <100 [CRC77]

2219
Compound: Potassium niobate
Formula: $KNbO_3$
Molecular Formula: $KNbO_3$
Molecular Weight: 180.002
CAS RN: 12030-85-2

Properties: -100 mesh with 99.9% purity; white powd [STR93] [CER91]

2220
Compound: Potassium niobate hexadecahydrate
Formula: $K_8Nb_6O_{19} \cdot 16H_2O$
Molecular Formula: $H_{32}K_8Nb_6O_{35}$
Molecular Weight: 1462.457
CAS RN: 12502-31-7
Properties: large monocl cryst; forms supersaturated solutions in water [KIR81]
Solubility: saturated soln is 425 g/100g H_2O at room temp, more soluble in hot H_2O [KIR81]

2221
Compound: Potassium nitrate
Synonyms: saltpeter
Formula: KNO_3
Molecular Formula: KNO_3
Molecular Weight: 101.103
CAS RN: 7757-79-1
Properties: colorless transparent prisms, white granular, or rhomb and trig cryst powd; cools when dissolved in H_2O; enthalpy of fusion 10.10 kJ/mol; manufactured by reaction of KCl and HNO_3, with Cl_2 as a byproduct [MER89] [KIR82] [CRC93]
Solubility: g/100g soln, H_2O: 11.7 (0°C), 27.5 (25°C), 71.0 (100°C) [KRU93]; 1 g/620mL alcohol; s glycerol; i absolute alcohol [MER89]
Density, g/cm³: 2.109 [STR93]
Melting Point, °C: 333 [MER89]
Boiling Point, °C: decomposes, 400, evolving oxygen [MER89]

2222
Compound: Potassium nitrite
Formula: KNO_2
Molecular Formula: KNO_2
Molecular Weight: 85.104
CAS RN: 7758-09-0
Properties: white or sl yellow; deliq granules or rods; decomposed by acids, evolving brown fumes of nitrous anhydride [MER89]
Solubility: g/100g soln, H_2O: 73.7 (0°C), 75.75 (25°C), 80.2 (100°C); Solid phase, KNO_2 [KRU93]; sl s alcohol [MER89]
Density, g/cm³: 1.915 [MER89]
Melting Point, °C: 350, decomposes [ALD94]
Boiling Point, °C: explodes at 537 [HAW93]

2223
Compound: Potassium nitroprusside dihydrate
Formula: $K_2[Fe(CN)_5(NO)]\cdot 2H_2O$
Molecular Formula: $C_5H_4FeK_2N_6O_3$
Molecular Weight: 330.167
CAS RN: 14709-57-0
Properties: garnet red; hygr cryst [MER89]
Solubility: 100 g/100mL H_2O (16°C) [CRC94]

2224
Compound: Potassium osmiate dihydrate
Formula: $K_2OsO_4\cdot 2H_2O$
Molecular Formula: $H_4K_2O_6Os$
Molecular Weight: 368.455
CAS RN: 19718-36-6
Properties: purple powd; hygr rhomb cryst; slowly decomposes in aq solution, forming tetroxide [STR93] [MER89] [KIR82]
Solubility: s H_2O; i alcohol, ether [MER89]
Reactions: minus H_2O at 200°C [KIR82]

2225
Compound: Potassium oxalate monohydrate
Formula: $K_2C_2O_4\cdot H_2O$
Molecular Formula: $C_2H_2K_2O_5$
Molecular Weight: 184.232
CAS RN: 6487-48-5
Properties: colorless, odorless cryst; efflorescent in warm, dry air; converted to carbonate by ignition [MER89]
Solubility: g/100g soln, H_2O: 29.32±0.04 (0°C), 27.4 (25°C), 44.5 (100°C); Solid phase $K_2C_2O_4\cdot H_2O$ [KRU93]
Density, g/cm³: 2.13 [MER89]
Melting Point, °C: decomposes when heated [HAW93]
Reactions: minus H_2O ~160°C [MER89]

2226
Compound: Potassium pentaborate octahydrate
Formula: $K_2B_{10}O_{16}\cdot 8H_2O$
Molecular Formula: $B_{10}H_{16}K_2O_{24}$
Molecular Weight: 586.420
CAS RN: 12229-13-9
Properties: ortho-rhomb white powd; enthalpy of dehydration 110.8 kJ/mol from 106.5 to 134°C [STR93] [KIR78]
Solubility: % anhydrous by weight, H_2O: 1.56 (0°C), 3.28 (25°C), 6.88 (50°C), 22.3 (100°C) [KIR78]
Density, g/cm³: 1.74 [KIR78]
Melting Point, °C: 780 [STR93]

2227
Compound: Potassium pentachloronitrosyl iridium(III) hydrate
Formula: $KIr(NO)Cl_5\cdot xH_2O$
Molecular Formula: Cl_5IrKNO (anhydrous)
Molecular Weight: 438.588 (anhydrous)
CAS RN: 22594-86-1
Properties: brown cryst [STR93]

2228
Compound: Potassium pentachlororuthenate(III) hydrate
Formula: $K_2RuCl_5\cdot xH_2O$
Molecular Formula: Cl_5K_2Ru (anhydrous)
Molecular Weight: 356.530 (anhydrous)
CAS RN: 14404-33-2
Properties: brown powd [STR93]

2229
Compound: Potassium perborate monohydrate
Synonyms: potassium peroxyborate
Formula: $2KBO_3\cdot H_2O$
Molecular Formula: $B_2H_2K_2O_7$
Molecular Weight: 213.831
CAS RN: 28876-88-2
Properties: white cryst; formula also given as a hemihydrate [CRC77]
Solubility: g/100g H_2O: 1.25 (0°C) [KRU93]
Melting Point, °C: decomposes, 150 [CRC77]
Reactions: minus O_2, 100°C [CRC77]

2230
Compound: Potassium percarbonate monohydrate
Synonyms: potassium peroxycarbonate
Formula: $K_2C_2O_6\cdot H_2O$
Molecular Formula: $C_2H_2K_2O_7$
Molecular Weight: 216.231
CAS RN: 589-97-9
Properties: white, granular mass; sensitive to light [MER89]
Solubility: 1 part/15 parts cold H_2O; evolves O_2 in hot H_2O [MER89]
Melting Point, °C: 200-300 [HAW93]

2231
Compound: Potassium perchlorate
Formula: $KClO_4$
Molecular Formula: $ClKO_4$
Molecular Weight: 138.549
CAS RN: 7778-74-7

Properties: hygr; colorless cryst, or white powd; oxidizing agent, reacts with organic matter, other oxidizable material; evolves O_2 if heated; can decompose by concussion; used in explosives, photography, pyrotechnics and flares [HAW93] [MER89] [STR93]

Solubility: g/100g soln, H_2O: 0.75 (0°C), 2.03 (25°C), 18.2 (100°C); Solid phase, $KClO_4$ [KRU93]

Density, g/cm³: 2.52 [MER89]

Melting Point, °C: decomposes, 400 [MER89]; 610 [STR93]

2232

Compound: Potassium periodate

Formula: KIO_4

Molecular Formula: IKO_4

Molecular Weight: 230.001

CAS RN: 7790-21-8

Properties: white powd, or colorless, transparent, tetr cryst; strong oxidizing agent in aq solutions, e.g. can oxidize manganese compounds to permanganate [MER89]

Solubility: g/100g soln, H_2O: 0.169 (0.2°C), 0.51 (25°C), 6.83 (97°C) [KRU93]

Density, g/cm³: 3.618 [MER89]

Melting Point, °C: 582 [MER89]

Boiling Point, °C: explodes [HAW93]

2233

Compound: Potassium permanganate

Formula: $KMnO_4$

Molecular Formula: $KMnO_4$

Molecular Weight: 158.034

CAS RN: 7722-64-7

Properties: dark purple cryst; stable in air; decomposes evolving O_2 ~240°C; good oxidizing agent in aq solutions, e.g. oxidizes HCl to Cl_2; astringent taste; odorless; used as an oxidizer, as a disinfectant, bleach, in tanning [HAW93] [MER89]

Solubility: g/100g soln, H_2O: 2.75 (0°C), 7.09 (25°C); Solid phase, $KMnO_4$ [KRU93]

Density, g/cm³: 2.70 [HAW93]

Melting Point, °C: decomposes [HAW93]

2234

Compound: Potassium peroxide

Formula: K_2O_2

Molecular Formula: K_2O_2

Molecular Weight: 110.196

CAS RN: 17014-71-0

Properties: yellow amorphous mass; decomposes in water, evolving oxygen; oxidizing agent; used for bleaching, in oxygen generating gas masks [HAW93]

Solubility: decomposed by H_2O [HAW93]

Melting Point, °C: 490 [HAW93]

2235

Compound: Potassium perrhenate

Formula: $KReO_4$

Molecular Formula: KO_4Re

Molecular Weight: 289.303

CAS RN: 10466-65-6

Properties: white tetrahedral cryst; -40 mesh with 99.9% purity [STR93] [KIR82] [CER91]

Solubility: g/100g H_2O: 0.34 (0°C), 0.99 (20°C), 8.7 (80°C) [LAN85]

Density, g/cm³: 4.887 [STR93]

Melting Point, °C: 555 [STR93]

Boiling Point, °C: ~1365 [STR93]

2236

Compound: Potassium perruthenate

Formula: $KRuO_4$

Molecular Formula: KO_4Ru

Molecular Weight: 204.166

CAS RN: 10378-50-4

Properties: black tetr [KIR82]

Solubility: 1.1 g/100mL H_2O (60°C) [CRC94]

Melting Point, °C: decomposes 400 [KIR82]

2237

Compound: Potassium persulfate

Synonyms: potassium peroxydisulfate

Formula: $K_2S_2O_8$

Molecular Formula: $K_2O_8S_2$

Molecular Weight: 270.324

CAS RN: 7727-21-1

Properties: colorless or white cryst; -100 mesh with 99.9% purity; unstable, gradually evolving O_2 with rate of evolution increasing with temp; strong oxidizing agent in aq solutions [MER89] [CER91]

Solubility: g/100mL soln, H_2O: 1.620 (0°C), 5.840 (25°C) [KRU93]; i alcohol [MER89]

Density, g/cm³: 2.477 [HAW93]

Melting Point, °C: completely decomposed ~100 [MER89]

2238

Compound: Potassium phosphate

Formula: K_3PO_4
Molecular Formula: K_3O_4P
Molecular Weight: 212.266
CAS RN: 7778-53-2
Properties: granular white powd; deliq; ortho-rhomb cryst; used in gasoline purification, to soften water, in liq soaps, and as an emulsifier in foods [HAW93] [MER89]
Solubility: g/100g soln, H_2O: 44.26 (0°C), 51.42 (25°C); Solid phase, $K_3PO_4 \cdot 7H_2O$ [KRU93]; i alcohol [MER89]
Density, g/cm³: 2.564 [MER89]
Melting Point, °C: 1340 [MER89]

2239
Compound: Potassium pyrophosphate trihydrate
Formula: $K_4P_2O_7 \cdot 3H_2O$
Molecular Formula: $H_6K_4O_{10}P_2$
Molecular Weight: 384.374
CAS RN: 7320-34-5
Properties: colorless; deliq granules or cryst; somewhat hygr; used as a builder in soaps and detergents, and sequestering agent; also anhydrous form [HAW93] [MER89] [STR93] [ALD94]
Solubility: v s H_2O; i alcohol [MER89]
Density, g/cm³: 2.33 [HAW93]
Melting Point, °C: 1090 [HAW93]
Reactions: minus $2H_2O$ at 180°C, minus $3H_2O$ ~300°C [CRC94] [HAW93]

2240
Compound: Potassium pyrosulfate
Formula: $K_2S_2O_7$
Molecular Formula: $K_2O_7S_2$
Molecular Weight: 254.325
CAS RN: 7790-62-7
Properties: colorless needles; fused pieces or white cryst powd; used as a laboratory reagent [HAW93] [MER89]
Solubility: s H_2O [MER89]
Density, g/cm³: 2.28 [MER89]
Melting Point, °C: ~325 [MER89]

2241
Compound: Potassium pyrosulfite
Formula: $K_2S_2O_5$
Molecular Formula: $K_2O_5S_2$
Molecular Weight: 222.326
CAS RN: 16731-55-8

Properties: white cryst or cryst powd; odor of sulfur dioxide; SO_2 evolved from acidic solutions; oxidizes to sulfate in air [MER89]
Solubility: g/100g soln, H_2O: 22.1 (0°C), 32.8 (25°C), 55.5 (94.0°C), solid phase $K_2S_2O_5$ [KRU93]
Density, g/cm³: 2.3 [HAW93]
Melting Point, °C: 150-190, decomposes [HAW93]

2242
Compound: Potassium ruthenate(VI)
Formula: K_2RuO_4
Molecular Formula: K_2O_4Ru
Molecular Weight: 243.265
CAS RN: 31111-21-4
Properties: black with green luster; rhomb [KIR82]
Solubility: v s H_2O [KIR82]

2243
Compound: Potassium selenate
Formula: K_2SeO_4
Molecular Formula: K_2O_4Se
Molecular Weight: 221.155
CAS RN: 7790-59-2
Properties: -100 mesh with 99.5% purity; colorless cryst or white powd [MER89] [CER91]
Solubility: g/100g soln, H_2O: 52.7±0.9 (0°C), 53.3±0.3 (25°C), 55.6±0.6 (100°C) [KRU93]
Density, g/cm³: 3.07 [MER89]

2244
Compound: Potassium selenide
Formula: K_2Se
Molecular Formula: K_2Se
Molecular Weight: 157.157
CAS RN: 1312-74-9
Properties: cryst; reddens in air; color changes to brownish black when heated; deliq [MER89]
Solubility: s H_2O; i ammonia [MER89]
Density, g/cm³: 2.29 [LID94]
Melting Point, °C: 800 [LID94]

2245
Compound: Potassium selenite
Formula: K_2SeO_3
Molecular Formula: K_2O_3Se
Molecular Weight: 205.155
CAS RN: 10431-47-7
Properties: -100 mesh with 99.5% purity; white powd; hygr [STR93] [CER91]

Solubility: g/100g soln, H$_2$O: 68.45 (0°C), 68.5 (24.3°C), 68.53 (100.6°C); Solid phase, K$_2$SeO$_3$ (0°C, 100.6°C), K$_2$SeO$_3$·4H$_2$O + K$_2$SeO$_3$ (24.3°C) [KRU93]
Melting Point, °C: 875 [STR93]
Boiling Point, °C: decomposes [STR93]

2246
Compound: Potassium silver cyanide
Synonyms: silver potassium cyanide
Formula: KAg(CN)$_2$
Molecular Formula: C$_2$AgKN$_2$
Molecular Weight: 199.001
CAS RN: 506-61-6
Properties: white cryst; sensitive to light [MER89]
Solubility: s H$_2$O; silver cyanide precipitated from acid solutions [MER89]

2247
Compound: Potassium sodium carbonate hexahydrate
Synonyms: sodium potassium carbonate hexahydrate
Formula: KNaCO$_3$·6H$_2$O
Molecular Formula: CH$_{12}$KNaO$_9$
Molecular Weight: 230.189
CAS RN: 64399-16-2
Properties: colorless cryst; this double salt fuses more readily than the single salts; used as a flux in analysis [HAW93]
Solubility: 185.2 g/100mL H$_2$O (15°C) [CRC94]
Density, g/cm^3: 1.6344 [HAW93]
Melting Point, °C: 135, decomposes [HAW93]
Reactions: minus 6H$_2$O at 100°C [CRC94]

2248
Compound: Potassium stannate trihydrate
Synonyms: potassium hydroxystannate(IV)
Formula: K$_2$SnO$_3$·3H$_2$O
Molecular Formula: H$_6$K$_2$O$_6$Sn
Molecular Weight: 298.951
CAS RN: 12142-33-5
Properties: white to light tan cryst; used to dye and print textiles, in alkaline tin plating baths; anhydrous available as -100 mesh with 99.9% purity [HAW93] [CER91]
Solubility: 110.5 g/100mL H$_2$O at 15°C; i alcohol [MER89] [KIR83]
Density, g/cm^3: 3.197 [MER89]
Melting Point, °C: decomposes 140 [AES93]

2249
Compound: Potassium stannosulfate
Formula: K$_2$Sn(SO$_4$)$_2$
Molecular Formula: K$_2$O$_8$S$_2$Sn
Molecular Weight: 389.034
CAS RN: 27790-37-0
Properties: white cryst; partially decomposed by H$_2$O [MER89]
Solubility: s dil alkaline hydroxide solutions [MER89]

2250
Compound: Potassium stearate
Synonyms: stearic acid, potassium salt
Formula: CH$_3$(CH$_2$)$_{16}$COOK
Molecular Formula: C$_{18}$H$_{35}$KO$_2$
Molecular Weight: 322.573
CAS RN: 593-29-3
Properties: white cryst powd; slight odor of fat; used in softening textiles [HAW93]
Solubility: slowly s cold H$_2$O, more readily s hot H$_2$O, alcohol [MER89]

2251
Compound: Potassium sulfate
Synonyms: arcanite
Formula: K$_2$SO$_4$
Molecular Formula: K$_2$O$_4$S
Molecular Weight: 174.261
CAS RN: 7778-80-5
Properties: colorless or white hard rhomb or hexagonal cryst, granules or powd; bitter saline taste; enthalpy of fusion 36.40 kJ/mol; also used in glass manufacture [MER89] [KIR82] [CRC93]
Solubility: g/100g soln, H$_2$O: 6.9 (0°C), 10.75 (25°C), 19.4 (100°C); Solid phase, K$_2$SO$_4$ [KRU93]; 1 g/75mL glycerol; i alcohol [MER89]
Density, g/cm^3: 2.66 [MER89]
Melting Point, °C: 1069 [CRC93]
Thermal Expansion Coefficient: (volume) 100°C (0.544), 200°C (2.118), 400°C (4.935) [CLA66]

2252
Compound: Potassium sulfide
Formula: K$_2$S
Molecular Formula: K$_2$S
Molecular Weight: 110.263
CAS RN: 1312-73-8

Properties: may contain polysulfides; red or yellowish red; cub cryst or fused plates; discolored in air; very hygr; unstable; may explode when struck, or if rapidly heated; enthalpy of fusion 16.15 kJ/mol [STR93] [MER89] [HAW93] [CRC93]

Solubility: s H_2O, alcohol, glycerol; i ether [HAW93]

Density, g/cm³: 1.74 [MER89]

Melting Point, °C: 948 [CRC93]

2253

Compound: Potassium sulfide pentahydrate

Formula: $K_2S \cdot 5H_2O$

Molecular Formula: $H_{10}K_2O_5S$

Molecular Weight: 200.339

CAS RN: 1312-73-8

Properties: colorless; rhomb; odor of H_2S; turns yellow to yellowish red when exposed to air and light; aq solutions are alkaline and unstable [MER89]

Solubility: v s H_2O, alcohol, glycerol; i ether [MER89]

Melting Point, °C: 60 [MER89]

Reactions: minus $3H_2O$ at 150°C [CRC94]

2254

Compound: Potassium sulfite dihydrate

Formula: $K_2SO_3 \cdot 2H_2O$

Molecular Formula: $H_4K_2O_5S$

Molecular Weight: 194.292

CAS RN: 7790-56-9

Properties: white monoclinic cryst or cryst powd; oxidizes gradually to sulfate in air; used as a photographic developer, as a food and wine preservative [HAW93] [MER89] [KIR82]

Solubility: g/100g soln, H_2O: 47.52 (0°C), 49.01 (25°C), 55.53 (100°C); Solid phase, K_2SO_3 [KRU93]

Melting Point, °C: decomposes [KIR82]

2255

Compound: Potassium superoxide

Synonyms: potassium dioxide

Formula: KO_2

Molecular Formula: KO_2

Molecular Weight: 71.097

CAS RN: 12030-88-5

Properties: yellow powd; sensitive to moisture [STR93]

Solubility: decomposes in H_2O [CRC94]

Density, g/cm³: 2.14 [STR93]

Melting Point, °C: 380 [LID94]

2256

Compound: Potassium tantalate

Formula: $KTaO_3$

Molecular Formula: KO_3Ta

Molecular Weight: 268.044

CAS RN: 12030-91-0

Properties: -100 mesh with 99.9% purity [CER91]

2257

Compound: Potassium tellurate(VI) trihydrate

Formula: $K_2TeO_4 \cdot 3H_2O$

Molecular Formula: $H_6K_2O_7Te$

Molecular Weight: 323.841

CAS RN: 15571-91-2

Properties: white cryst powd [MER89]

Solubility: s in 4 parts H_2O [MER89]

2258

Compound: Potassium tellurite

Formula: K_2TeO_3

Molecular Formula: K_2O_3Te

Molecular Weight: 253.795

CAS RN: 7790-58-1

Properties: white powd; granular; hygr; used in chemical analysis to test for bacteria [HAW93]

Solubility: g/100g H_2O: 8.8 (0°C), 27.5 (20°C), 50.4 (30°C) [LAN85]

Melting Point, °C: 460-470, decomposes [HAW93]

2259

Compound: Potassium tellurite(IV) hydrate

Formula: $K_2TeO_3 \cdot xH_2O$

Molecular Formula: K_2O_3Te (anhydrous)

Molecular Weight: 253.795 (anhydrous)

CAS RN: 123333-66-4

Properties: white hygr powd [ALD94] [STR93]

Melting Point, °C: 465, decomposes [STR93]

2260

Compound: Potassium tetraborate pentahydrate

Formula: $K_2B_4O_7 \cdot 5H_2O$

Molecular Formula: $B_4H_{10}K_2O_{12}$

Molecular Weight: 323.513

CAS RN: 1332-77-0

Properties: white cryst powd [MER89]

Solubility: s 4 parts H_2O; sl s alcohol [MER89]

2261
Compound: Potassium tetraborate tetrahydrate
Formula: $K_2B_4O_7 \cdot 4H_2O$
Molecular Formula: $B_4H_8K_2O_{11}$
Molecular Weight: 305.498
CAS RN: 12045-78-2
Properties: -6 mesh with 99.9% purity; ortho-rhomb white powd; dehydration temp depends on partial pressure of H_2O [CER91] [STR93] [KIR78]
Solubility: % anhydrous by weight, H_2O: 9.02 (10°C), 13.6 (25°C), 24.0 (50°C), 48.4 (100°C) [KIR78]
Density, g/cm³: 1.92 [KIR78]
Melting Point, °C: decomposes [STR93]
Reactions: reversible dehydration from 85 to 111°C [KIR78]

2262
Compound: Potassium tetrabromoaurate(III) dihydrate
Formula: $KAuBr_4 \cdot 2H_2O$
Molecular Formula: $AuBr_4H_4KO_2$
Molecular Weight: 591.712
CAS RN: 14323-32-1
Properties: violet cryst; sensitive to light [MER89]
Solubility: s H_2O, alcohol [MER89]
Melting Point, °C: 120, decomposes [CRC94]

2263
Compound: Potassium tetrabromopalladate(II)
Formula: K_2PdBr_4
Molecular Formula: Br_4K_2Pd
Molecular Weight: 504.233
CAS RN: 13826-93-2
Properties: reddish brown powd; hygr [STR93]

2264
Compound: Potassium tetrabromoplatinate(II)
Formula: K_2PtBr_4
Molecular Formula: Br_4K_2Pt
Molecular Weight: 592.893
CAS RN: 13826-94-3
Properties: red powd [STR93]

2265
Compound: Potassium tetrachloroaurate(III)
Formula: $KAuCl_4$
Molecular Formula: $AuCl_4K$
Molecular Weight: 377.876
CAS RN: 13682-61-6

Properties: yellowish orange cryst [STR93]
Solubility: g/100g H_2O: 38.3 (10°C), 61.8 (20°C), 405 (60°C) [LAN85]
Melting Point, °C: 357, decomposes [AES93]

2266
Compound: Potassium tetrachloroaurate(III) dihydrate
Formula: $KAuCl_4 \cdot 2H_2O$
Molecular Formula: $AuCl_4H_4KO_2$
Molecular Weight: 413.907
CAS RN: 13005-39-5
Properties: yellow monocl cryst; light sensitive [MER89] [HAW93]
Solubility: s H_2O, alcohol, ether [HAW93]

2267
Compound: Potassium tetrachloropalladate(II)
Formula: K_2PdCl_4
Molecular Formula: Cl_4K_2Pd
Molecular Weight: 326.428
CAS RN: 10025-98-6
Properties: brown powd; hygr [STR93]
Density, g/cm³: 2.67 [STR93]
Melting Point, °C: 105, decomposes [ALD94]

2268
Compound: Potassium tetrachloroplatinate(II)
Synonyms: potassium chloroplatinate
Formula: K_2PtCl_4
Molecular Formula: Cl_4K_2Pt
Molecular Weight: 415.088
CAS RN: 10025-99-7
Properties: pink or ruby red; tetr; hygr [MER89] [STR93] [KIR82]
Solubility: s H_2O; i alcohol [MER89] [KIR82]
Density, g/cm³: 3.38 [STR93]
Melting Point, °C: decomposes >500 [KIR82]

2269
Compound: Potassium tetracyanomercurate(II)
Formula: $K_2Hg(CN)_4$
Molecular Formula: $C_4HgK_2N_4$
Molecular Weight: 382.858
CAS RN: 591-89-9
Properties: colorless or white cryst [MER89]
Solubility: s H_2O [MER89]

2270

Compound: Potassium tetracyanonickelate(II) monohydrate

Formula: $K_2Ni(CN)_4 \cdot H_2O$

Molecular Formula: $C_4H_2K_2N_4NiO$

Molecular Weight: 258.976

CAS RN: 14220-17-8

Properties: yellowish orange; cryst powd [MER89]

Solubility: s H_2O [MER89]

Density, g/cm³: 1.875 [CRC94]

Reactions: minus H_2O ~100°C [MER89]

2271

Compound: Potassium tetracyanoplatinate(II)

Formula: $K_2Pt(CN)_4$

Molecular Formula: $C_4K_2N_4Pt$

Molecular Weight: 377.348

CAS RN: 562-76-5

Properties: yellow powd; hygr [STR93]

Solubility: g/100g H_2O: 11.6 (0°C), 33.9 (20°C), 194 (90°C) [LAN85]

2272

Compound: Potassium tetracyanoplatinate(II) trihydrate

Formula: $K_2Pt(CN)_4 \cdot 3H_2O$

Molecular Formula: $C_4H_6K_2N_4O_3Pt$

Molecular Weight: 431.394

CAS RN: 14323-36-5

Properties: almost colorless; rhomb prisms; blue color in direction of principal axis [MER89]

Solubility: s H_2O [MER89]

Density, g/cm³: 2.455 [CRC94]

Melting Point, °C: decomposes 400-600 [CRC94]

Reactions: minus $3H_2O$ at 100°C [CRC94]

2273

Compound: Potassium tetracyanozincate

Formula: $K_2Zn(CN)_4$

Molecular Formula: $C_4K_2N_4Zn$

Molecular Weight: 247.658

CAS RN: 14244-62-3

Properties: cryst powd [MER89]

Solubility: v s H_2O [MER89]

2274

Compound: Potassium tetrafluoroberyllate dihydrate

Synonyms: beryllium potassium fluoride

Formula: $K_2BeF_4 \cdot 2H_2O$

Molecular Formula: $BeF_4H_4K_2O_2$

Molecular Weight: 199.234

CAS RN: 7787-50-0

Properties: brilliant cryst [MER89]

Solubility: s H_2O, conc K_2SO_4 [MER89]

2275

Compound: Potassium tetrafluoroborate

Synonyms: avogadrite

Formula: KBF_4

Molecular Formula: BF_4K

Molecular Weight: 125.903

CAS RN: 14075-53-7

Properties: colorless; ortho-rhomb below 283°C, a = 0.7032 nm, b = 0.8674 nm, c = 0.5496 nm; bipyrimidal or cub cryst; enthalpy of fusion 17.7 kJ/mol; can be prepared as a gelatinous precipitate from fluoroboric acid and KOH or K_2CO_3; used in plating, in flux for soldering [HAW93] [MER89] [KIR78] [JAN71]

Solubility: g/100g H_2O: 0.3 (3°C), 0.448 (20°C), 0.55 (25°C), 1.4 (40°C), 6.27 (100°C); sl s boiling alcohol [MER89]

Density, g/cm³: 2.505 [MER89]

Melting Point, °C: 530 [MER89]

2276

Compound: Potassium tetraiodoaurate(III)

Synonyms: gold potassium iodide

Formula: $KAuI_4$

Molecular Formula: AuI_4K

Molecular Weight: 743.683

CAS RN: 7791-29-9

Properties: black, lustrous cryst; sensitive to light [MER89]

Solubility: s H_2O, decomposes and liberates iodine [MER89]

Melting Point, °C: decomposes at 150 [CRC94]

2277

Compound: Potassium tetraiodocadmium dihydrate

Synonyms: cadmium potassium iodide dihydrate

Formula: $K_2CdI_4 \cdot 2H_2O$

Molecular Formula: $CdH_4I_4K_2O_2$

Molecular Weight: 734.256

CAS RN: 584-10-1

Properties: large, water clear, somewhat distorted octahedra; deliq; turns yellow when aging due to release of I_2 [MER89]

Solubility: 1 part w/w dissolves at 15°C, in: 0.73 parts H_2O, 1.4 parts alcohol; 24.5 parts ether; 4.5 parts of a 1:1 mixture of ether and alcohol; s ethyl acetate [MER89]

Density, g/cm³: 3.359 [MER89]

2278
Compound: Potassium tetraiodomercurate(II)
Formula: K_2HgI_4
Molecular Formula: HgI_4K_2
Molecular Weight: 786.404
CAS RN: 7783-33-7
Properties: sulfur yellow color; cryst; deliq; dihydrate known as Mayers reagent; dihydrate prepared by dissolving stoichiometric amounts of HgI_2 and KI in distilled water which is used as an antiseptic and as a precipitant for alkaloids; in strongly alkaline solutions, known as Nessler's reagent which is used for ammonia detection [KIR81] [MER89]
Solubility: v s H_2O; s alcohol, ether, acetone [MER89]
Density, g/cm³: 4.29 [LID94]
Melting Point, °C: 105 [CRC94]

2279
Compound: Potassium tetranitritoplatinate(II)
Formula: $K_2Pt(NO_2)_4$
Molecular Formula: $K_2N_4O_8Pt$
Molecular Weight: 457.299
CAS RN: 13815-39-9
Properties: white powd [STR93]

2280
Compound: Potassium tetraoxalate dihydrate
Formula: $KHC_2O_4 \cdot H_2C_2O_4 \cdot 2H_2O$
Molecular Formula: $C_4H_7KO_{10}$
Molecular Weight: 254.192
CAS RN: 127-96-8
Properties: colorless or white cryst [MER89]
Solubility: s 60 parts cold H_2O, 12 parts boiling H_2O; sl s alcohol [MER89]
Melting Point, °C: decomposes [CRC94]

2281
Compound: Potassium thioantimonate heminonahydrate
Formula: $K_3SbS_4 \cdot 4\text{-}1/2H_2O$
Molecular Formula: $H_9K_3O_{4.5}S_4Sb$
Molecular Weight: 448.385
CAS RN: 14693-02-8
Properties: colorless to yellowish cryst; formula also given as $2K_3SbS_4 \cdot 4\text{-}1/2H_2O$ [MER89] [CRC94]
Solubility: g anhydrous/100g H_2O: 306 (0°C), 302 (30°C), 381 (80°C) [LAN85]; i alcohol [MER89]

2282
Compound: Potassium thiocarbonate
Formula: K_2CS_3
Molecular Formula: CK_2S_3
Molecular Weight: 186.406
CAS RN: 26750-66-3
Properties: yellowish red; deliq granules or cryst; very hygr; used in medicine, as a soil fumigant [HAW93] [MER89]
Solubility: v s H_2O, giving strongly akaline solution [MER89]
Melting Point, °C: decomposes [CRC94]

2283
Compound: Potassium thiocyanate
Formula: KSCN
Molecular Formula: CKNS
Molecular Weight: 97.182
CAS RN: 333-20-0
Properties: colorless; deliq cryst; temp drops to ~30°C when dissolved in its own weight of H_2O [MER89]
Solubility: g/100g H_2O: 177 (0°C), 239 (25°C), 673.6 (99°C); Solid phase, KSCN [KRU93]; 1 g dissolves in: 0.5mL acetone, 12mL alcohol, 8mL boiling alcohol [MER89]
Density, g/cm³: 1.88 [HAW93]
Melting Point, °C: 173 [HAW93]
Boiling Point, °C: 500, decomposes [HAW93]
Reactions: turns brown, green, blue when fused, white when cooled [HAW93]

2284
Compound: Potassium thiosulfate
Synonyms: potassium hyposulfite
Formula: $K_2S_2O_3$
Molecular Formula: $K_2O_3S_2$
Molecular Weight: 190.327
CAS RN: 10294-66-3
Properties: colorless cryst; hygr [MER89]
Solubility: g/100g H_2O: 96.1 (0°C), 165.0 (25°C), 312.0 (90°C); Solid phase, $K_2S_2O_3 \cdot 2H_2O$ (0°C), $3K_2S_2O_3 \cdot 5H_2O$ (25°C), $K_2S_2O_3$ (90°C) [KRU93]; i alcohol [MER89]
Density, g/cm³: 2.23 [CRC94]

2285
Compound: Potassium titanate
Formula: K_2TiO_3
Molecular Formula: K_2O_3Ti
Molecular Weight: 174.062
CAS RN: 12030-97-6

Properties: white; grayish brown powd; as a fiber, has a high index of refraction, can diffuse and reflect infrared radiation; used in fiber form in rockets, missiles, and in nuclear power as an insulator [HAW93] [STR93]

Solubility: hydrolyzes in H_2O to give a strongly alkaline solution [HAW93]

Density, g/cm^3: 3.1 [LID94]

Melting Point, °C: 1515 [LID94]

2286

Compound: Potassium titanium oxalate dihydrate

Formula: $K_2TiO(C_2O_4)_2 \cdot 2H_2O$

Molecular Formula: $C_4H_4K_2O_{11}Ti$

Molecular Weight: 354.133

CAS RN: 14402-67-6

Properties: colorless lustrous cryst; used as a mordant in cotton and leather dyeing [HAW93]

Solubility: v s H_2O [MER89]

2287

Compound: Potassium triiodide monohydrate

Formula: $KI_3 \cdot H_2O$

Molecular Formula: H_2I_3KO

Molecular Weight: 437.827

CAS RN: 7790-42-3

Properties: dark brown, hygr; monocl prisms; reasonably stable only in the monohydrate form; formula also given as hemihydrate [CRC94] [MER89]

Solubility: s H_2O; s with partial decomposition in alcohol and ether [MER89]

Density, g/cm^3: 3.50 [MER89]

Melting Point, °C: 38 (closed tube) [MER89]

Reactions: decomposes, evolving I_2 with KI residue at 225°C [MER89]

2288

Compound: Potassium triiodozincate

Synonyms: potassium zinc iodide

Formula: $KZnI_3$

Molecular Formula: I_3KZn

Molecular Weight: 485.202

CAS RN: 7790-43-4

Properties: cryst; very hygr [MER89]

Solubility: v s H_2O [MER89]

2289

Compound: Potassium triphosphate

Synonyms: potassium tripolyphosphate

Formula: $K_5P_3O_{10}$

Molecular Formula: $K_5O_{10}P_3$

Molecular Weight: 448.407

CAS RN: 13845-36-8

Properties: white powd; prepared by dehydration of an equimolar mixture of mono-and dipotassium phosphates; used in detergents [KIR82] [STR93]

Density, g/cm^3: 2.54 [STR93]

Melting Point, °C: 620 [STR93]

2290

Compound: Potassium tungstate

Formula: K_2WO_4

Molecular Formula: K_2O_4W

Molecular Weight: 326.035

CAS RN: 7790-60-5

Properties: -100 mesh with 99.5% purity; heavy; deliq; cryst powd [MER89] [CER91]

Solubility: s ~2 parts cold H_2O, ~0.7 parts boiling H_2O; i alcohol [MER89]

Density, g/cm^3: 3.12 [MER89]

Melting Point, °C: 921 [MER89]

2291

Compound: Potassium tungstate dihydrate

Formula: $K_2WO_4 \cdot 2H_2O$

Molecular Formula: $H_4K_2O_6W$

Molecular Weight: 362.065

CAS RN: 7790-60-5

Properties: heavy cryst powd; hygr [ALD94] [HAW93]

Solubility: 51.5 g/100mL H_2O [CRC94]

Density, g/cm^3: 3.1 [HAW93]

Melting Point, °C: 921 [HAW93]

2292

Compound: Potassium uranate

Formula: $K_2U_2O_7$

Molecular Formula: $K_2O_7U_2$

Molecular Weight: 666.251

CAS RN: 7790-63-8

Properties: cub orange powd [LID94] [MER89]

Solubility: i H_2O; s acids [MER89]

Density, g/cm^3: 6.12 [LID94]

2293

Compound: Potassium uranyl nitrate

Formula: $K(UO_2)(NO_3)_3$

Molecular Formula: $KN_3O_{11}U$

Molecular Weight: 495.140

CAS RN: 18078-40-5

Properties: greenish yellow; cryst powd [MER89]

Solubility: s ~1 part H₂O [MER89]

2294
Compound: Potassium uranyl sulfate dihydrate
Formula: $K_2(UO_2)(SO_4)_2 \cdot 2H_2O$
Molecular Formula: $H_4K_2O_{12}S_2U$
Molecular Weight: 576.383
CAS RN: 27709-53-1
Properties: greenish yellow; cryst powd [MER89]
Solubility: v s H₂O [MER89]
Density, g/cm³: 3.363 [CRC94]
Reactions: minus 2H₂O at 120°C [CRC94]

2295
Compound: Potassium vanadate
Synonyms: potassium metavanadate
Formula: KVO_3
Molecular Formula: KO_3V
Molecular Weight: 138.038
CAS RN: 13769-43-2
Properties: colorless cryst; -200 mesh with 99.9% purity; there are also K_3VO_4, 14293-78-8, and $K_4V_2O_7$, 14638-93-8 [CRC94] [CER91]

2296
Compound: Potassium zinc sulfate hexahydrate
Formula: $K_2Zn(SO_4)_2 \cdot 6H_2O$
Molecular Formula: $H_{12}K_2O_{14}S_2Zn$
Molecular Weight: 443.806
CAS RN: 13932-17-7
Properties: cryst [MER89]
Solubility: g/100g H₂O: 13.0 (0°C), 25.9 (20°C), 72.1 (60°C) [LAN85]

2297
Compound: Potassium zirconate
Formula: K_2ZrO_3
Molecular Formula: K_2O_3Zr
Molecular Weight: 217.419
CAS RN: 12030-98-7
Properties: -200 mesh with 99.9% purity [CER91]

2298
Compound: Potassium zirconium sulfate trihydrate
Formula: $K_4Zr(SO_4)_4 \cdot 3H_2O$
Molecular Formula: $H_6K_4O_{19}S_4Zr$
Molecular Weight: 685.918
CAS RN: 53608-79-0
Properties: white cryst powd [HAW93]
Solubility: sl s H₂O [MER89]

2299
Compound: Praseodymium acetate hydrate
Formula: $Pr(CH_3COO)_3 \cdot xH_2O$
Molecular Formula: $C_6H_9O_6Pr$ (anhydrous)
Molecular Weight: 318.041 (anhydrous)
CAS RN: 6192-12-7
Properties: green cryst; hygr [AES93] [STR93]

2300
Compound: Praseodymium acetylacetonate
Synonyms: 2,4-pentanedione, praseodymiun(III) derivative
Formula: $Pr(CH_3COCH=C(O)CH_3)_3$
Molecular Formula: $C_{15}H_{21}O_6Pr$
Molecular Weight: 438.236
CAS RN: 14553-09-4
Properties: powd [STR93]

$$[CH_3-\underset{\underset{O-}{|}}{C}=CH-\underset{\underset{O}{\|}}{C}-CH_3]_3Pr$$

Melting Point, °C: 146 [CRC94]

2301
Compound: Praseodymium barium copper oxide
Formula: $PrBa_2Cu_3O_7$
Molecular Formula: $Ba_2Cu_3O_7Pr$
Molecular Weight: 718.196
CAS RN: 126284-91-1
Properties: a = 0.3878 nm, b = 0.3940 nm, c = 1.1761 nm; can be prepared by heating Pr_6O_{11} which is free of both carbonate and hydroxide with stoichiometric amounts of $BaCO_3$ and CuO, then grinding the reactants and heating the resulting powd at 950°C for 12 h in air; CAS RN 126284-91-1 is for the compound with $PrBa_2CuO_{6.97}$; CAS RN 120309-22-8 refers to compound with $PrBa_2Cu_3O_{7.05}$ [CON87]

2302
Compound: Praseodymium boride
Formula: PrB_6
Molecular Formula: B_6Pr
Molecular Weight: 205.774
CAS RN: 12008-27-4
Properties: -325 mesh with 10 microns average or less [CER91]
Density, g/cm³: 4.84 [LID94]
Melting Point, °C: 2610 [LID94]

2303
Compound: Praseodymium bromate nonahydrate
Formula: $Pr(BrO_3)_3 \cdot 9H_2O$
Molecular Formula: $Br_3H_{18}O_{18}Pr$
Molecular Weight: 686.752
CAS RN: 15162-93-3
Properties: green; hex [CRC77]
Solubility: g/100g H_2O: 55.9 (0°C), 91.8 (20°C), 144 (80°C) [LAN85]
Melting Point, °C: 56.5 [CRC77]
Reactions: minus $7H_2O$, 170°C [CRC77]

2304
Compound: Praseodymium bromide
Formula: $PrBr_3$
Molecular Formula: Br_3Pr
Molecular Weight: 380.620
CAS RN: 13536-53-3
Properties: green cryst; -20 mesh with 99.9% purity [CER91] [CRC94]
Solubility: decomposed by H_2O [CRC94]
Density, g/cm³: 5.28 [LID94]
Melting Point, °C: 691 [AES93]
Boiling Point, °C: 1547 [CRC94]

2305
Compound: Praseodymium carbonate octahydrate
Formula: $Pr_2(CO_3)_3 \cdot 8H_2O$
Molecular Formula: $C_3H_{16}O_{17}Pr_2$
Molecular Weight: 605.965
CAS RN: 14948-62-0
Properties: light green powd [AES93] [STR93]
Reactions: minus $6H_2O$ at 100°C [CRC94]

2306
Compound: Praseodymium chloride
Formula: $PrCl_3$
Molecular Formula: Cl_3Pr
Molecular Weight: 247.266
CAS RN: 10361-79-2
Properties: -20 mesh of 99.9% purity; bluish green powd; hygr [STR93] [CER91]
Solubility: s H_2O, alcohol [MER89]
Density, g/cm³: 4.02 [STR93]
Melting Point, °C: 786 (under argon) [STR93]
Boiling Point, °C: 1700 [STR93]

2307
Compound: Praseodymium chloride heptahydrate
Formula: $PrCl_3 \cdot 7H_2O$
Molecular Formula: $Cl_3H_{14}O_7Pr$

Molecular Weight: 373.373
CAS RN: 10025-90-8
Properties: -4 mesh with 99.9% purity; hygr; green cryst [MER89] [STR93] [CER91]
Solubility: 334 g/100mL H_2O (13°C) [CRC94]
Density, g/cm³: 2.250 [STR93]
Melting Point, °C: 115 [STR93]
Reactions: minus $7H_2O$ at 180-200°C if heated in HCl stream [MER89]

2308
Compound: Praseodymium fluoride
Formula: PrF_3
Molecular Formula: F_3Pr
Molecular Weight: 197.903
CAS RN: 13709-46-1
Properties: hygr light blue powd, and 99.9% pure melted pieces of 3-6 mm; hygr; melted pieces used as evaporation material for infrared multilayers, low and high ends only [STR93] [CER91]
Density, g/cm³: 6.3 [LID94]
Melting Point, °C: 1395 [LID94]

2309
Compound: Praseodymium hexafluoroacetylacetonate
Synonyms: 1,1,1,5,5,5-hexafluoro-2,4-pentanedione, Pr derivative
Formula: $Pr(CF_3COCHCOCF_3)_3$
Molecular Formula: $C_{15}H_3F_{18}O_6Pr$
Molecular Weight: 762.067
CAS RN: 47814-20-0
Properties: light-green powd [STR93]

$$[CF_3-C=CH-C-CF_3]_3Pr$$

2310
Compound: Praseodymium hydride
Formula: PrH_3
Molecular Formula: H_3Pr
Molecular Weight: 143.931
CAS RN: 13864-03-4
Properties: lump under argon atm [ALF95]

2311
Compound: Praseodymium hydroxide
Formula: $Pr(OH)_3$

Molecular Formula: H₃O₃Pr
Molecular Weight: 191.929
CAS RN: 16469-16-2
Properties: gelatinous; pale green; obtained by
reacting hydroxide with Pr salt; purple powd
obtained by reacting Pr carbide with H₂O
[MER89]

2312
Compound: Praseodymium iodide
Formula: PrI₃
Molecular Formula: I₃Pr
Molecular Weight: 521.621
CAS RN: 13813-23-5
Properties: green hygr cryst; -20 mesh with 99.9%
purity [CER91] [CRC94]
Density, g/cm³: ~5.8 [LID94]
Melting Point, °C: 737 [CRC94]

2313
Compound: Praseodymium nitrate hexahydrate
Formula: Pr(NO₃)₃·6H₂O
Molecular Formula: H₁₂N₃O₁₅Pr
Molecular Weight: 435.014
CAS RN: 15878-77-0
Properties: light green cryst; there is a
pentahydrate, 14483-17-1 [AES93] [STR93]
Solubility: 5.0257±0.0087 mol/kg (25°C),
hexahydrate is the stable phase [RAR85b]; g
anhydrous/100g H₂O: 112 (20°C), 162 (30°C),
178 (40°C) [LAN85]

2314
Compound: Praseodymium nitride
Formula: PrN
Molecular Formula: NPr
Molecular Weight: 154.915
CAS RN: 25764-09-4
Properties: -60 mesh with 99.9% purity; NaCl
crystal system, a = 0.515 nm [CER91] [CIC73]
Density, g/cm³: 7.46 [LID94]

2315
Compound: Praseodymium oxalate decahydrate
Formula: Pr₂(C₂O₄)₃·10H₂O
Molecular Formula: C₆H₂₀O₂₂Pr₂
Molecular Weight: 726.027
CAS RN: 24992-60-7
Properties: light green powd [ALF95] [STR93]

2316
Compound: Praseodymium perchlorate hexahydrate
Formula: Pr(ClO₄)₃·6H₂O
Molecular Formula: Cl₃H₁₂O₁₈Pr
Molecular Weight: 547.351
CAS RN: 13498-07-2
Properties: green cryst; hygr [ALF95] [STR93]

2317
Compound: Praseodymium phosphate
Formula: PrPO₄
Molecular Formula: O₄PPr
Molecular Weight: 235.879
CAS RN: 14298-31-8
Properties: green powd [STR93]

2318
Compound: Praseodymium silicide
Formula: PrSi₂
Molecular Formula: PrSi₂
Molecular Weight: 197.079
CAS RN: 12066-83-0
Properties: tetr cryst; 10mm & down lump [LID94]
[ALF93]
Density, g/cm³: 5.46 [LID94]
Melting Point, °C: 1712 [LID94]

2319
Compound: Praseodymium sulfate
Formula: Pr₂(SO₄)₃
Molecular Formula: O₁₂Pr₂S₃
Molecular Weight: 570.006
CAS RN: 10277-44-8
Properties: light green cryst [MER89]
Solubility: 0.1545±0.0019 mol/kg (25°C),
octahydrate is the equilibrium phase [RAR88];
g/100g H₂O: 19.8 (0°C), 12.6 (20°C), 0.91
(100°C) [LAN85]
Density, g/cm³: 3.72 [CRC94]

2320
Compound: Praseodymium sulfate octahydrate
Formula: Pr₂(SO₄)₃·8H₂O
Molecular Formula: H₁₆O₂₀Pr₂S₃
Molecular Weight: 714.128
CAS RN: 13510-41-3
Properties: light green cryst; monocl [CRC94]
[AES93] [STR93]
Solubility: 17.4 g/100mL H₂O (20°C) [CRC94]
Density, g/cm³: 2.827 [CRC94]

2321

Compound: Praseodymium sulfide

Formula: Pr_2S_3

Molecular Formula: Pr_2S_3

Molecular Weight: 378.013

CAS RN: 12038-13-0

Properties: cub cryst; -200 mesh with 99.9% purity [LID94] [CER91]

Density, g/cm³: 5.1 [LID94]

Melting Point, °C: 1765 [LID94]

2322

Compound: Praseodymium telluride

Formula: Pr_2Te_3

Molecular Formula: Pr_2Te_3

Molecular Weight: 664.615

CAS RN: 12038-12-9

Properties: cub cryst; -20 mesh with 99.9% purity; probably also contains $PrTe_3$, as well as other Pr-Te phases [LID94] [CER91]

Density, g/cm³: ~7.0 [LID94]

Melting Point, °C: 1500 [LID94]

2323

Compound: Praseodymium(α)

Formula: α-Pr

Molecular Formula: Pr

Molecular Weight: 140.90765

CAS RN: 7440-10-0

Properties: yellowish metal; hex; tarnishes in moist air, forming oxide film; enthalpy of fusion is 6.912 kJ/mol; enthalpy of sublimation is 355.6 kJ/mol; electrical resistivity 68 μohm·cm; radius of atom is 0.1828 nm; radius of ion is 0.1013 nm, Pr^{+++}; forms green solutions [MER89] [KIR82] [ALD94]

Density, g/cm³: 6.773 [KIR82]

Melting Point, °C: 935 [DOU83]

Reactions: α to β transition at 800°C [MER89]

2324

Compound: Praseodymium(β)

Formula: β-Pr

Molecular Formula: Pr

Molecular Weight: 140.90765

CAS RN: 7440-10-0

Properties: 12 mm pieces and smaller (under oil) of 99.9% purity; yellowish metal; forms oxide film in moist air; bcc; electrical resistivity (20°C) 68 μohm·cm; enthalpy of fusion 6.89 kJ/mol; enthalpy of vaporization 331 kJ/mol [MER89] [CER91] [CRC93] [ALD94]

Density, g/cm³: 6.64 [MER89]

Melting Point, °C: 935 [MER89]

Boiling Point, °C: 3510 [LID94]

Reactions: α to β transition at 800°C [MER89]

Thermal Conductivity, W/(m·K): 12.5 (25°C) [CRC93]

Thermal Expansion Coefficient: 6.7 x 10^{-6}/K [CRC93]

2325

Compound: Praseodymium(III) oxide

Formula: Pr_2O_3

Molecular Formula: O_3Pr_2

Molecular Weight: 329.813

CAS RN: 12036-32-7

Properties: yellow-green amorphous; 3-12 mm pieces with 10 microns average or less with 99.9% purity [CER91] [CRC94]

Solubility: 0.000020 g/100mL H_2O (29°C) [CRC94]

Density, g/cm³: 7.07 [CRC94]

Melting Point, °C: 2300 [LID94]

2326

Compound: Praseodymium(III,IV) oxide

Formula: Pr_6O_{11}

Molecular Formula: $O_{11}Pr_6$

Molecular Weight: 1021.439

CAS RN: 12037-29-5

Properties: black powd, or brown sintered pieces; used as an evaporation material, with interest in its reactivity to radio frequencies [STR93] [CER91]

2327

Compound: Promethium

Formula: Pm

Molecular Formula: Pm

Molecular Weight: 145

CAS RN: 7440-12-2

Properties: silvery metal; enthalpy of sublimation is 348 kJ/mol (estimated); radius of atom is 0.1811 nm; radius of ion is 0.0979 nm; Pr^{+++} has yellow colored solutions [KIR82]

Density, g/cm³: 7.22 [MER89]

Melting Point, °C: 1042 [KIR82]

Boiling Point, °C: ~3000 [KIR82]

Thermal Conductivity, W/(m·K): 15 (25°C) [CRC93]

2328

Compound: 2,2-Bis(ethylferrocenyl)propane

Formula: $(C_2H_5C_5H_3FeC_5H_5C)_2C(CH_3)_2$
Molecular Formula: $C_{27}H_{32}Fe_2$
Molecular Weight: 468.245
CAS RN: 81579-74-0
Properties: red-brown viscous liq [STR93]
Density, g/cm³: 1.275 [STR93]

2329
Compound: Protactinium
Formula: Pa
Molecular Formula: Pa
Molecular Weight: 231.03588
CAS RN: 7440-13-3
Properties: shiny, malleable metal; readily tarnishes in air; α: tetr, a = 0.3929 nm, c = 0.3241 nm, stable from room temp up to 1170°C; β: bcc, a = 0.381 nm, stable from 1170 to 1575°C; first discovered in 1917; enthalpy of fusion 12.34 kJ/mol; can be produced by nuclear reaction $^{230}Th + n \rightarrow {}^{231}Th + \gamma \; {}^{231}Th \rightarrow {}^{231}Pa$, however Pa is also found in natural sources; $t_{1/2}$ ^{231}Pa is 3.25 x 10^{+4} years [KIR78] [MER89] [CRC93]
Density, g/cm³: 15.37 [KIR78]
Melting Point, °C: 1575 [KIR91]
Reactions: reacts with H_2 at 250-300°C forming PaH_3 [MER89]

2330
Compound: Pyrophosphoric acid
Synonyms: diphosphoric acid
Formula: $[P(O)(OH)_2]_2O$
Molecular Formula: $H_4O_7P_2$
Molecular Weight: 177.98
CAS RN: 2466-09-3
Properties: hygr; syrupy liq, glassy cryst; preparation: reaction of H_3PO_4 and $POCl_3$; uses: catalyst, metal treatment, stabilizer for organic peroxides [HAW93] [MER89]
Solubility: 709 g/100 mL H_2O (23°C); s alcohol, ether [MER89]
Melting Point, °C: 61-63 [ALD94]
Reactions: forms phosphoric acid on dissolution in hot H_2O [MER89]

2331
Compound: Radium
Formula: Ra
Molecular Formula: Ra
Molecular Weight: 226
CAS RN: 7440-14-4
Properties: brilliant white metal; bcc; blackens when exposed to air; spontaneously disintegrates, forming radon; compounds closely resemble those of barium; $t_{1/2}$ = 1600 years; enthalpy of sublimation 130 kJ/mol [DOU83] [MER89] [GME77]
Solubility: evolves H_2 in H_2O [CRC94]
Density, g/cm³: 5.5 [MER89]
Melting Point, °C: 700 [MER89]
Boiling Point, °C: 1737 [MER89]

2332
Compound: Radium bromide
Formula: $RaBr_2$
Molecular Formula: Br_2Ra
Molecular Weight: 386
CAS RN: 10031-23-9
Properties: white or sl brownish cryst [MER89]
Solubility: s H_2O [MER89]; s alcohol [HAW93]
Density, g/cm³: 5.79 [MER89]
Melting Point, °C: 728 [MER89]
Boiling Point, °C: sublimes 900 [MER89]

2333
Compound: Radium carbonate
Formula: $RaCO_3$
Molecular Formula: CO_3Ra
Molecular Weight: 286
CAS RN: 7116-98-5
Properties: amorphous radioactive powd; pure material is white, but impurities impart yellow, orange or pink color [HAW93]
Solubility: i H_2O [HAW93]

2334
Compound: Radium chloride
Formula: $RaCl_2$
Molecular Formula: Cl_2Ra
Molecular Weight: 297
CAS RN: 10025-66-8
Properties: radioactive; white or sl brownish cryst [MER89]
Solubility: s H_2O [MER89]; s alcohol [HAW93]
Density, g/cm³: 4.91 [MER89]
Melting Point, °C: 1000 [MER89]

2335
Compound: Radium sulfate
Formula: $RaSO_4$
Molecular Formula: O_4RaS
Molecular Weight: 322

CAS RN: 7446-16-4
Properties: white rhomb cryst when pure, else colored yellow, orange, pink; radioactive [HAW93] [CRC94]
Solubility: i H_2O, acids [HAW93]

2336
Compound: Radon
Formula: Rn
Molecular Formula: Rn
Molecular Weight: 222
CAS RN: 10043-92-2
Properties: colorless inert gas; strongly radioactive; enthalpy of vaporization 18.10 kJ/mol; heat capacity (101.32 kPa,25°C) ~21 J/(mol·K); viscosity (25°C,101.32 kPa) 23.3 Pa·s; ^{222}Rn $t_{1/2}$ = 3.82 days; can be condensed to a colorless, transparent liq [MER89] [HAW93] [KIR78]
Solubility: at 101.32 kPa, 230 mL/L H_2O (20°C) [KIR78]; s organic solvents [MER89]
Density, g/cm^3: gas: (101.32 kPa, 0°C) 9.741 g/L [LID94]; liq: 4.4 [MER89]
Melting Point, °C: -71 [CRC94]
Boiling Point, °C: -61.7 [LID94]

2337
Compound: Rhenium
Formula: Re
Molecular Formula: Re
Molecular Weight: 186.207
CAS RN: 7440-15-5
Properties: 3-6 mm pieces of 99.99% purity; black to silvery gray metal; hex close-packed cryst, a = 0.2760 nm, c = 0.4458 nm; enthalpy of sublimation 791 kJ/mol; enthalpy of fusion 60.43 kJ/mol; electrical resistivity 19.3 μohm·cm at 20°C; tensile strength is 80,000 psi; modulus of elasticity is 0.46 Pa [KIR82] [HAW93] [MER89] [CER91] [CRC93]
Solubility: i HCl; attacked by HNO_3, H_2SO_4 [HAW93]
Density, g/cm^3: 21.02 [MER89]
Melting Point, °C: 3180 [MER89]
Boiling Point, °C: 5596 [CRC93]
Thermal Conductivity, W/(m·K): 47.9 (25°C) [CRC93]
Thermal Expansion Coefficient: 6.2 x 10^{-6}/K [CRC93]

2338
Compound: Rhenium boride
Formula: Re_7B_3

Molecular Formula: B_3Re_7
Molecular Weight: 1335.882
CAS RN: 12355-99-6
Properties: formula is generally Re_7B_3 with possible traces of ReB_2; -100 mesh with 99.5% purity [CER91]

2339
Compound: Rhenium carbonyl
Synonyms: dirhenium pentacarbonyl
Formula: $Re_2(CO)_{10}$
Molecular Formula: $C_{10}O_{10}Re_2$
Molecular Weight: 652.518
CAS RN: 14285-68-8
Properties: yellowish white volatile cryst; can be prepared by reacting Re_2O_7, H_2 and CO under high pressure; used to produce highly pure Re metal [KIR82] [STR93]
Solubility: s in most organic solvents [KIR82]
Density, g/cm^3: 2.87 [STR93]
Melting Point, °C: 170, decomposes [STR93]

2340
Compound: Rhenium pentacarbonyl bromide
Formula: $Re(CO)_5Br$
Molecular Formula: C_5BrO_5Re
Molecular Weight: 406.163
CAS RN: 14220-21-4
Properties: off-white cryst; used to study π-electron complexes [ALD94] [STR93]
Melting Point, °C: 90, sublimes [STR93]

2341
Compound: Rhenium pentacarbonyl chloride
Formula: $Re(CO)_5Cl$
Molecular Formula: C_5ClO_5Re
Molecular Weight: 361.712
CAS RN: 14099-01-5
Properties: octahedral cryst; off-white; used to study π-electron deficient compounds [STR93] [ALD94]

2342
Compound: Rhenium(III) bromide
Synonyms: rhenium tribromide
Formula: $ReBr_3$
Molecular Formula: Br_3Re
Molecular Weight: 425.919
CAS RN: 13569-49-8

Properties: green-black cryst; -100 mesh with 99.9% purity structure consists of Fe_3Br_9 units, linked by bridging [CER91] [COT88] [CRC94]
Density, g/cm³: 6.10 [LID94]
Melting Point, °C: sublimes 500 (vacuum) [CRC94]

2343
Compound: Rhenium(III) chloride
Formula: $ReCl_3$
Molecular Formula: Cl_3Re
Molecular Weight: 292.565
CAS RN: 13569-63-6
Properties: reddish black powd; hygr; sensitive to light; emits green vapor when heated from which the metal can be deposited [HAW93] [STR93]
Solubility: s H_2O and glacial acetic acid [HAW93]
Density, g/cm³: 4.81 [LID94]
Melting Point, °C: decomposes 500 [LID94]

2344
Compound: Rhenium(III) iodide
Synonyms: rhenium triiodide
Formula: ReI_3
Molecular Formula: I_3Re
Molecular Weight: 566.920
CAS RN: 15622-42-1
Properties: black; -80 mesh of 99.9% purity; structure consists of Re_3I_9 molecules linked by bridging; preparation: careful evaporation of $HReO_4$ solution containing excess HI [COT88] [CER91]
Melting Point, °C: decomposes [LID94]

2345
Compound: Rhenium(IV) chloride
Formula: $ReCl_4$
Molecular Formula: Cl_4Re
Molecular Weight: 328.018
CAS RN: 13569-71-6
Properties: greenish black cryst; structure consists of zigzag chains of Fe_2Co_9; -80 mesh; sensitive to moisture; forms when a solution of $HReO_4$ and HCl is carefully evaporated [KIR82] [STR93] [CER91] [COT88]
Density, g/cm³: 4.9 [STR93]
Melting Point, °C: decomposes 300 [LID94]

2346
Compound: Rhenium(IV) fluoride
Formula: ReF_4

Molecular Formula: F_4Re
Molecular Weight: 262.201
CAS RN: 15192-42-4
Properties: dark green; blue [COT88] [CRC94]
Density, g/cm³: 7.49 [LID94]
Melting Point, °C: sublimes >300 [COT88]

2347
Compound: Rhenium(IV) oxide
Formula: ReO_2
Molecular Formula: O_2Re
Molecular Weight: 218.206
CAS RN: 12036-09-8
Properties: black powd; distorted rutile structure; anhydrous available as -100 mesh with 99.95% purity [CER91] [STR93] [COT88]
Density, g/cm³: 11.4 [CRC94]
Melting Point, °C: 1000, decomposes [STR93]

2348
Compound: Rhenium(IV) selenide
Formula: $ReSe_2$
Molecular Formula: $ReSe_2$
Molecular Weight: 344.127
CAS RN: 12038-64-1
Properties: -100 mesh with 99.9% purity; layer structures with considerable Re-Re bonding [COT88] [CER91]

2349
Compound: Rhenium(IV) silicide
Formula: $ReSi_2$
Molecular Formula: $ReSi_2$
Molecular Weight: 242.378
CAS RN: 12038-66-3
Properties: -80 mesh powd of 99.9% purity [CER91] [ALF93]

2350
Compound: Rhenium(IV) sulfide
Formula: ReS_2
Molecular Formula: ReS_2
Molecular Weight: 250.339
CAS RN: 12038-63-0
Properties: tricl cryst; commonly nonstoichiometric; -80 mesh with 99.9% purity; preparation: by heating Re_2S_7 with sulfur in vacuum; uses: effective catalyst for hydrogenation of organic compounds, not poisoned by sulfur compounds, catalyzes reduction of NO to N_2O at 100°C [CER91] [COT88] [LID94]

Density, g/cm³: 7.6 [LID94]

2351
Compound: Rhenium(IV) telluride
Formula: ReTe₂
Molecular Formula: ReTe₂
Molecular Weight: 441.407
CAS RN: 12067-00-4
Properties: well-characterized compound which does not have a layer structure; -60 mesh with 99.9% purity [COT88] [CER91]

2352
Compound: Rhenium(V) bromide
Synonyms: rhenium pentabromide
Formula: ReBr₅
Molecular Formula: Br₅Re
Molecular Weight: 585.727
CAS RN: 30937-53-2
Properties: dark brown; -100 mesh with 99.9% purity [COT88] [CER91]
Melting Point, °C: decomposes 110 [LID94]

2353
Compound: Rhenium(V) chloride
Formula: ReCl₅
Molecular Formula: Cl₅Re
Molecular Weight: 363.471
CAS RN: 39368-69-9
Properties: sensitive to moisture; formula is also given as the dimer Re₂Cl₁₀; dark green to black solid; decomposed by heating; formed by reacting Re with Cl₂ at ~600°C, yielding a dark red-brown liq; starting material for rhenium porphyrin complexes [ALD94] [HAW93] [KIR82] [COT88]
Solubility: decomposes in H₂O; s HCl and alkalies [HAW93]
Density, g/cm³: 4.9 [HAW93]
Melting Point, °C: decomposes [AES93]
Reactions: rapidly hydrolyzed to ReO₄⁻ by aq basic solution [COT88]

2354
Compound: Rhenium(VI) fluoride
Formula: ReF₆
Molecular Formula: F₆Re
Molecular Weight: 300.197
CAS RN: 10049-17-9

Properties: yellow cub cryst; extremely hygr; bluish vapors given off in air; transition point -1.9 or -3.45°C; enthalpy of sublimation 32.657 kJ/mol (above transition), 41.407 kJ/mol (below transition); enthalpy of fusion 4.635 kJ/mol; enthalpy of vaporization 28.75 kJ/mol; entropy of fusion 60.46 J/(mol·K); entropy of vaporization 93.4 J/(mol·K); can be produced by reacting Re powd with F₂ at ~100°C; used in the chemical vapor deposition of Re [KIR78] [MER89]
Solubility: s HNO₃; s 1.75 moles/1000g anhydrous HF [MER89]
Density, g/cm³: 3.58 [KIR78]
Melting Point, °C: 18.5 [MER89]
Boiling Point, °C: 48 [CER91]

2355
Compound: Rhenium(VI) oxide
Synonyms: rhenium trioxide
Formula: ReO₃
Molecular Formula: O₃Re
Molecular Weight: 234.205
CAS RN: 1314-28-9
Properties: -100 mesh with 99.95% purity; red cub cryst; green luster imparted by transmitted light; disproportionates in vacuum to Re₂O₇ and ReO₂ [MER89] [CER91]
Solubility: i H₂O, alkalies, non oxidizing acids [MER89]
Density, g/cm³: 6.9 to 7.4 [MER89]
Melting Point, °C: 400, decomposes [STR93]
Reactions: oxidized to HReO₄ by HNO₃ [MER89]

2356
Compound: Rhenium(VI) oxytetrachloride
Formula: ReOCl₄
Molecular Formula: Cl₄ORe
Molecular Weight: 344.017
CAS RN: 13814-76-1
Properties: orange square pyramid; preparation: by reaction between Re and SO₂Cl₂ at 350°C [COT88] [CRC94] [KIR82]
Melting Point, °C: 29.3 [CRC94]
Boiling Point, °C: 223 [CRC94]

2357
Compound: Rhenium(VI) trioxychloride
Formula: ReO₃Cl
Molecular Formula: ClO₃Re
Molecular Weight: 269.658
CAS RN: 7791-09-5

Properties: clear, colorless liq; can be prepared by chlorination of Re_2O_7 [MER89]

Solubility: hydrolyzed to $HReO_4$ in H_2O; s CCl_4 [MER89]

Density, g/cm³: 3.867 [HAW93]

Melting Point, °C: 4.5 [MER89]

Boiling Point, °C: 128 [MER89]

2358

Compound: Rhenium(VII) oxide

Synonyms: rhenium heptaoxide

Formula: Re_2O_7

Molecular Formula: O_7Re_2

Molecular Weight: 484.410

CAS RN: 1314-68-7

Properties: -6 mesh with 99.99% purity; canary yellow; very deliq cryst; absorbs water readily, forming perrhenic acid, $HReO_4$; enthalpy of fusion 64.20 kJ/mol; structure has infinite array of alternating ReO_4 tetrahedra and ReO_6 octahedra; preparation: reaction with O_2 and heated Re metal [COT88] [CRC93] [MER89] [CER91]

Solubility: v s H_2O, alcohol, ether, ethyl acetate, dioxane, pyridine [MER89]

Density, g/cm³: 6.103 [STR93]

Melting Point, °C: 220 [COT88]

Boiling Point, °C: 360 [ALD94]

Reactions: sublimes at 250°C [MER89]

2359

Compound: Rhenium(VII) sulfide

Formula: Re_2S_7

Molecular Formula: Re_2S_7

Molecular Weight: 596.876

CAS RN: 12038-67-4

Properties: brownish black solid; preparation: precipitation from a saturated solution of 2 to 6M HCl solution of ReO_4^- with H_2S; used as a catalyst [HAW93] [COT88]

Solubility: i H_2O; s alkali sulfide solutions [HAW93]

Density, g/cm³: 4.87 [HAW93]

Melting Point, °C: decomposes to ReS at 600 [HAW93]

2360

Compound: Rhodium

Formula: Rh

Molecular Formula: Rh

Molecular Weight: 102.90550

CAS RN: 7440-16-6

Properties: silvery white, ductile metal; fcc, a = 0.3803 nm; electrical resistivity at 0°C 4.51 μohm·cm; Vicker's hardness 120; Poisson's ratio 0.26; resistant to acids; reacts with aqua regia if finely divided; enthalpy of fusion 26.59 kJ/mol; enthalpy of vaporization 494 kJ/mol [ALD94] [MER89] [KIR82] [CRC93]

Solubility: i acids, aqua regia; s fused $KHSO_4$ [HAW93]

Density, g/cm³: 12.41 [MER89]

Melting Point, °C: 1966 [MER89]

Boiling Point, °C: 3695 [LID94]

Thermal Conductivity, W/(m·K): 150.6 (25°C) [KIR82]

Thermal Expansion Coefficient: 8.3 x 10^{-6}/°C [KIR82]

2361

Compound: Rhodium carbonyl

Synonyms: hexarhodium hexadecacarbonyl

Formula: $Rh_6(CO)_{16}$

Molecular Formula: $C_{16}O_{16}Rh_6$

Molecular Weight: 1065.61

CAS RN: 28407-51-4

Properties: black cryst; cluster structure; used as a catalyst [COT88] [ALD94] [ALF95]

2362

Compound: Rhodium carbonyl chloride

Formula: $[Rh(CO)_2Cl]_2$

Molecular Formula: $C_4Cl_2O_4Rh_2$

Molecular Weight: 388.758

CAS RN: 14523-22-9

Properties: reddish orange cryst, planar structure; stable in dry air; solutions in organic solvents decompose in air; preparation: by passing ethanol saturated CO over $RhCl_3 \cdot 3H_2O$ at ~100°C, followed by sublimation of needles of the carbonyl; uses: homogeneous catalyst [ALD94] [MER89] [COT88]

Solubility: s in most organic solvents; i aliphatic hydrocarbons [MER89]

Melting Point, °C: 124-125 [MER89]

2363

Compound: Rhodium dodecacarbonyl

Formula: $Rh_4(CO)_{12}$

Molecular Formula: $C_{12}O_{12}Rh_4$

Molecular Weight: 747.747

CAS RN: 19584-30-6

Properties: dark red cryst; sensitive to light, atm oxygen and moisture [STR93] [ALD94]

Density, g/cm³: 2.52 [STR93]
Melting Point, °C: 150 [DOU83]

2364
Compound: Rhodium(II) acetate dimer
Formula: $Rh_2(CH_3COO)_4$
Molecular Formula: $C_8H_{12}O_8Rh_2$
Molecular Weight: 441.989
CAS RN: 15956-28-2
Properties: greenish black cryst; preparation: by heating sodium acetate and $RhCl_3 \cdot 3H_2O$ in methanol; uses: homogeneous catalyst [ALD94] [STR93] [COT88]

2365
Compound: Rhodium(III) acetylacetonate
Synonyms: 2,4-pentanedione, rhodium(III) derivative
Formula: $Rh(CH_3COCH=C(O)CH_3)_3$
Molecular Formula: $C_{15}H_{21}O_6Rh$
Molecular Weight: 400.234
CAS RN: 14284-92-5
Properties: yellow cryst [STR93]

$$[CH_3-\underset{|}{\overset{O-}{C}}=CH-\underset{}{\overset{O}{\overset{\|}{C}}}-CH_3]_3Rh$$

Melting Point, °C: 263-264 [ALD94]
Boiling Point, °C: 280 decomposes [STR93]

2366
Compound: Rhodium(III) bromide dihydrate
Synonyms: rhodium tribromide dihydrate
Formula: $RhBr_3 \cdot 2H_2O$
Molecular Formula: $Br_3H_4O_2Rh$
Molecular Weight: 378.649
CAS RN: 15608-29-4
Properties: hygr brownish black cryst [STR93] [ALD94]

2367
Compound: Rhodium(III) chloride
Synonyms: rhodium trichloride
Formula: $RhCl_3$
Molecular Formula: Cl_3Rh
Molecular Weight: 209.264
CAS RN: 10049-07-7
Properties: reddish brown monocl powd [KIR82] [HAW93]

Solubility: i H_2O; s alkali hydroxide or cyanide solutions [MER89]
Density, g/cm³: 5.38 [KIR82]
Melting Point, °C: 450, decomposes [STR93]
Boiling Point, °C: 800, sublimes [HAW93]

2368
Compound: Rhodium(III) chloride hydrate
Synonyms: rhodium trichloride hydrate
Formula: $RhCl_3 \cdot xH_2O$
Molecular Formula: Cl_3Rh (anhydrous)
Molecular Weight: 209.264 (anhydrous)
CAS RN: 20765-98-4
Properties: dark red powd; hygr [STR93]
Melting Point, °C: decomposes 100 [ALD94]

2369
Compound: Rhodium(III) iodide
Synonyms: rhodium triiodide
Formula: RhI_3
Molecular Formula: I_3Rh
Molecular Weight: 483.619
CAS RN: 15492-38-3
Properties: monocl cryst; black powd; hygr [LID94] [STR93]
Density, g/cm³: 6.4 [LID94]

2370
Compound: Rhodium(III) nitrate
Formula: $Rh(NO_3)_3$
Molecular Formula: N_3O_9Rh
Molecular Weight: 288.921
CAS RN: 10139-58-9
Properties: available as amber colored 10% soln [STR93]
Density, g/cm³: soln: 1.410 [ALD94]

2371
Compound: Rhodium(III) nitrate dihydrate
Formula: $Rh(NO_3)_3 \cdot 2H_2O$
Molecular Formula: $H_4N_3O_{11}Rh$
Molecular Weight: 324.951
CAS RN: 13465-43-5
Properties: deliq red cryst [AES93] [CRC94]

2372
Compound: Rhodium(III) oxide
Synonyms: rhodium trioxide
Formula: Rh_2O_3
Molecular Formula: O_3Rh_2

Molecular Weight: 253.809
CAS RN: 12036-35-0
Properties: hex cryst; gray powd [LID94] [STR93]
Density, g/cm³: 8.2 [STR93]
Melting Point, °C: 1100, decomposes [STR93]

2373
Compound: Rhodium(III) oxide pentahydrate
Formula: $Rh_2O_3 \cdot 5H_2O$
Molecular Formula: $H_{10}O_8Rh_2$
Molecular Weight: 343.885
CAS RN: 39373-27-8
Properties: yellow powd [STR93]
Melting Point, °C: decomposes [STR93]

2374
Compound: Rhodium(III) sulfate
Formula: $Rh_2(SO_4)_3$
Molecular Formula: $O_{12}Rh_2S_3$
Molecular Weight: 494.002
CAS RN: 10489-46-0
Properties: reddish yellow solid; ~8% aq soln [ALD94] [KIR82]
Density, g/cm³: soln: 1.217 [ALD94]
Melting Point, °C: >500 decomposes [KIR82]

2375
Compound: Rhodium(IV) oxide dihydrate
Formula: $RhO_2 \cdot 2H_2O$
Molecular Formula: H_4O_4Rh
Molecular Weight: 170.935
CAS RN: 12137-27-8
Properties: olive-green powd [CRC94] [AES93]
Density, g/cm³: 8.20 [CRC94]
Melting Point, °C: decomposes 1100-1150 [CRC94]

2376
Compound: Rubidium
Formula: Rb
Molecular Formula: Rb
Molecular Weight: 85.4678
CAS RN: 7440-17-7

Properties: lustrous, silvery white soft metal; bcc; tarnishes readily in air; evolves H_2 in water; ignites spontaneously in oxygen; reacts with halogens, mercury; ionic radius 0.148 nm; viscosity (39°C) 0.6713 mPa·s; surface tension (39°C) 75 mN/m; electrical resistivity 11.0 μohm·cm; enthalpy of fusion 2.19 kJ/mol; enthalpy of vaporization 75.77 kJ/mol; used in photocells, as a catalyst or catalyst promoter [HAW93] [KIR82] [MER89] [CRC93] [ALD94]
Solubility: s in acids and alcohol [HAW93]
Density, g/cm³: solid: 1.532 [MER89]; liq: 1.472 at 39°C [KIR82]
Melting Point, °C: 38.5 [CAB85]
Boiling Point, °C: 688 [MER89]
Thermal Conductivity, W/(m·K): 58.2 (25°C) [CRC93]; 29.3 for liq [KIR82]

2377
Compound: Rubidium acetate
Synonyms: acetic acid, Rb salt
Formula: CH_3COORb
Molecular Formula: $C_2H_3O_2Rb$
Molecular Weight: 144.513
CAS RN: 563-67-7
Properties: -4 mesh with 99.9% purity; white hygr cryst [CER91] [STR93]
Solubility: 86 g/100mL H_2O (45°C), 89.3 g/100mL (99.4°C) [CRC94]
Melting Point, °C: 246 [STR93]

2378
Compound: Rubidium acetylacetonate
Synonyms: 2,4-pentanedione, rubidium derivative
Formula: $[CH_3COCH=C(O)CH_3]Rb$
Molecular Formula: $C_5H_7O_2Rb$
Molecular Weight: 184.578
CAS RN: 66169-93-5
Properties:

Melting Point, °C: 200, decomposes [ALD94]

2379
Compound: Rubidium aluminum sulfate dodecahydrate
Formula: $RbAl(SO_4)_2 \cdot 12H_2O$
Molecular Formula: $AlH_{24}O_{20}RbS_2$

Molecular Weight: 520.760
CAS RN: 7488-54-2
Properties: colorless cryst [MER83] [HAW93]
Solubility: g/100g H_2O: 0.72 (0°C), 1.50 (20°C), 21.6 (80°C) [LAN85]; i alcohol [MER83]
Density, g/cm^3: 1.867 [HAW93]
Melting Point, °C: dodecahydrate, 99 to 109 [MER83]

2380
Compound: Rubidium azide
Formula: RbN_3
Molecular Formula: N_3Rb
Molecular Weight: 127.488
CAS RN: 22756-36-1
Properties: colorless needles; tetr, a = 0.636 nm, c = 0.741 nm [CRC94] [CIC73]
Solubility: 107 g/100mL H_2O (16°C) [CRC94]
Density, g/cm^3: 2.79 [CRC94]
Melting Point, °C: decomposes ~310 [CRC94]

2381
Compound: Rubidium bromate
Formula: $RbBrO_3$
Molecular Formula: BrO_3Rb
Molecular Weight: 213.370
CAS RN: 13446-70-3
Properties: cub [CRC77]
Solubility: g/100g H_2O: 2.93 (25°C) [KRU93]
Density, g/cm^3: 3.68 [LAN52]
Melting Point, °C: 430 [LAN52]

2382
Compound: Rubidium bromide
Formula: RbBr
Molecular Formula: BrRb
Molecular Weight: 165.372
CAS RN: 7789-39-1
Properties: -4 mesh with 99.9% purity; white, cryst powd; enthalpy of fusion 15.50 kJ/mol [MER89] [CRC93] [CER91]
Solubility: g/100g soln, H_2O: 47.26 (0.5°C), 53.69 (25°C), 67.24 (113.5°C) [KRU93]
Density, g/cm^3: 3.35 [MER89]
Melting Point, °C: 682 [CRC93]
Boiling Point, °C: 1340 [MER89]
Thermal Expansion Coefficient: (volume) 100°C (0.925), 200°C (2.171) [CLA66]

2383
Compound: Rubidium carbonate

Formula: Rb_2CO_3
Molecular Formula: CO_3Rb_2
Molecular Weight: 230.945
CAS RN: 584-09-8
Properties: -20 mesh with 99.9% purity; white monocl cryst; extremely hygr; dissociates above 900°C; used in special glass formulations [LID94] [HAW93] [CER91]
Solubility: 450 g/100mL H_2O at 20°C [KIR82]
Melting Point, °C: 837 [STR93]
Boiling Point, °C: 740 decomposes [KIR82]

2384
Compound: Rubidium chlorate
Formula: $RbClO_3$
Molecular Formula: ClO_3Rb
Molecular Weight: 168.919
CAS RN: 13446-71-4
Properties: trimetric [CRC77]
Solubility: g/100g H_2O: 2.138 (0°C), 5.36 (19.8°C), 62.80 (99°C) [KRU93]
Density, g/cm^3: 3.19 [LAN52]

2385
Compound: Rubidium chloride
Formula: RbCl
Molecular Formula: ClRb
Molecular Weight: 120.921
CAS RN: 7791-11-9
Properties: -4 mesh with 99.9% purity; white, cryst powd; hygr; enthalpy of fusion 18.40 kJ/mol; used in testing for perchloric acid and as a source of Rb metal [HAW93] [CRC93] [CER91]
Solubility: g/100g soln H_2O: 43.5 (0°), 48.4±0.21 (25°), 58.9 (100°); equilibrium solid phase RbCl at 25° [KRU93]; 7.7832±0.0083 mol/(kg·H_2O) at 25°C [RAR85b]
Density, g/cm^3: 2.76 [MER89]
Melting Point, °C: 715 [CRC93]
Boiling Point, °C: 1390 [MER89]
Thermal Expansion Coefficient: (volume) 100°C (0.891), 200°C (2.079) [CLA66]

2386
Compound: Rubidium chromate
Formula: Rb_2CrO_4
Molecular Formula: CrO_4Rb_2
Molecular Weight: 286.930
CAS RN: 13446-72-5
Properties: -20 mesh with 99.9% purity; yellow cryst [STR93] [CER91]

Solubility: g/100g soln, H_2O: 38.27 (0°), 43.265
(25°C) [KRU93]; g/100g H_2O: 62.0 (0°C), 73.6
(20°C), 95.7 (60°C) [LAN85]
Density, g/cm³: 3.518 [STR93]

2387
Compound: Rubidium cobalt(II) sulfate hexahydrate
Formula: $Rb_2Co(SO_4)_2 \cdot 6H_2O$
Molecular Formula: $CoH_{12}O_{14}Rb_2S_2$
Molecular Weight: 530.088
CAS RN: 28038-39-3
Properties: ruby-red, monocl [CRC94]
Solubility: g/100g H_2O: 5.10 (0°C), 10.8 (20°C),
70.1 (100°C) [LAN85]
Density, g/cm³: 2.56 [CRC94]

2388
Compound: Rubidium cyanide
Formula: RbCN
Molecular Formula: CNRb
Molecular Weight: 111.486
CAS RN: 19073-56-4
Properties: cub; white or colorless [LID94] [KIR78]
Density, g/cm³: 2.32 [CRC94]

2389
Compound: Rubidium dichromate
Formula: $Rb_2Cr_2O_7$
Molecular Formula: $Cr_2O_7Rb_2$
Molecular Weight: 386.924
CAS RN: 13446-73-6
Properties: red tricl, or yellow monocl [LAN52]
Solubility: monocl: g/100g H_2O: 5.9 (20°C), 10.0
(30°C), 15.2 (40°C), 32.3 (60°C); tricl: 5.8
(20°C), 9.5 (30°C), 14.8 (40°C), 32.4 (60°C)
[LAN85]
Density, g/cm³: monocl: 3.021; tricl: 3.125 [CRC94]
[LAN52]

2390
Compound: Rubidium fluoride
Formula: RbF
Molecular Formula: FRb
Molecular Weight: 104.466
CAS RN: 13446-74-7
Properties: -4 mesh with 99.9% purity; white cub;
hygr; enthalpy of fusion 17.30 kJ/mol [STR93]
[CRC93] [CER91] [LID94]
Solubility: g/100g soln, H_2O: 75.06 (18°C); solid
phase, $RbF \cdot 1-1/2H_2O$ [KRU93]; i alcohol
[HAW93]

Density, g/cm³: 3.557 [HAW93]
Melting Point, °C: 833 [CRC93]
Boiling Point, °C: 1410 [STR93]

2391
Compound: Rubidium fluoroborate
Synonyms: rubidium borofluoride
Formula: $RbBF_4$
Molecular Formula: BF_4Rb
Molecular Weight: 172.273
CAS RN: 18909-68-7
Properties: ortho-rhomb below 245°C, a = 0.7296
nm, b = 0.9108 nm, c = 0.5636 nm; cub above
245°C [KIR78]
Solubility: 0.6 g/100mL H_2O (17°C) [KIR78]
Density, g/cm³: 2.820 [KIR78]
Melting Point, °C: 612 decomposes [KIR78]

2392
Compound: Rubidium fullerene
Formula: Rb_3C_{60}
Molecular Formula: $C_{60}Rb_3$
Molecular Weight: 977.063
CAS RN: 137926-73-9
Properties: fcc, lattice constant 1.4436 nm; bulk
modulus 22 GPa; density of states 35
states/(eV/C_{60}); superconducting temp 29.4 K
[DRE93]

2393
Compound: Rubidium hexafluorogermanate
Formula: Rb_2GeF_6
Molecular Formula: F_6GeRb_2
Molecular Weight: 357.536
CAS RN: 16962-48-4
Properties: white cryst solid [HAW93]
Solubility: sl s cold H_2O, v s hot H_2O [HAW93]
Melting Point, °C: 696 [HAW93]

2394
Compound: Rubidium hydroxide
Formula: RbOH
Molecular Formula: HORb
Molecular Weight: 102.475
CAS RN: 1310-82-3
Properties: grayish white; deliq; stronger base than
KOH; absorbs atm CO_2; possible use in low
temp storage batteries; there is a hydrate form
[HAW93] [MER89] [STR93] [CER91]
Solubility: g/100g soln, H_2O: 63.39 (30°C)
[KRU93]; s alcohol [MER89]

Density, g/cm³: 3.203 [MER89]
Melting Point, °C: 300 [HAW93]

2395
Compound: Rubidium iodide
Formula: RbI
Molecular Formula: IRb
Molecular Weight: 212.372
CAS RN: 7790-29-6
Properties: -4 mesh with 99.9% purity; white cryst, or cryst powd; discolors if exposed to light, air; enthalpy of fusion 12.50 kJ/mol [CRC93] [MER89] [CER91]
Solubility: g/100g soln, H_2O: 55.50 (0°C), 61.99 (25°C), 73.01 (93°C); solid phase, RbI [KRU93]; s alcohol [MER89]
Density, g/cm³: 3.55 [MER89]
Melting Point, °C: 642 [MER89]
Boiling Point, °C: 1300 [MER89]

2396
Compound: Rubidium iron(III) sulfate dodecahydrate
Formula: $RbFe(SO_4)_2 \cdot 12H_2O$
Molecular Formula: $FeH_{24}O_{20}RbS_2$
Molecular Weight: 549.624
CAS RN: 30622-97-0
Properties: cub [CRC77]
Solubility: g/100g H_2O: 8.0 (10°C), 20 (20°C), 52 (40°C) [LAN85]
Density, g/cm³: 1.91-1.95 [CRC77]
Melting Point, °C: 45-53 [CRC77]

2397
Compound: Rubidium metavanadate
Formula: $RbVO_3$
Molecular Formula: O_3RbV
Molecular Weight: 184.408
CAS RN: 13597-45-0
Properties: -100 mesh with 99.9% purity [CER91]

2398
Compound: Rubidium molybdate
Formula: Rb_2MoO_4
Molecular Formula: MoO_4Rb_2
Molecular Weight: 330.874
CAS RN: 13718-22-4
Properties: -200 mesh with 99.9% purity; white [KIR81] [CER91]
Solubility: g/100g soln, H_2O: 67.88 (18°C) [KRU93]
Melting Point, °C: 958,919 [KIR81]

2399
Compound: Rubidium niobate
Formula: $RbNbO_3$
Molecular Formula: NbO_3Rb
Molecular Weight: 226.372
CAS RN: 12059-51-7
Properties: -200 mesh with 99.9% purity [CER91]

2400
Compound: Rubidium nitrate
Formula: $RbNO_3$
Molecular Formula: NO_3Rb
Molecular Weight: 147.473
CAS RN: 13126-12-0
Properties: -80 mesh with 99.9% purity; white cryst; hygr; enthalpy of fusion 5.60 kJ/mol [STR93] [CRC93] [CER91]
Solubility: g/100g H_2O: 19.5 (0°C), 67.3 (25°C), 452 (100°C) [KRU93]
Density, g/cm³: 3.11 [STR93]
Melting Point, °C: 305 [CRC93]

2401
Compound: Rubidium orthovanadate
Formula: Rb_3VO_4
Molecular Formula: O_4Rb_3V
Molecular Weight: 371.343
CAS RN: 13566-05-7
Properties: -100 mesh with 99.9% purity [CER91]

2402
Compound: Rubidium oxide
Formula: Rb_2O
Molecular Formula: ORb_2
Molecular Weight: 186.935
CAS RN: 18088-11-4
Properties: cub; yellowish brown; sensitive to atm oxygen and moisture [STR93] [CRC94]
Density, g/cm³: 4.0 [LID94]
Melting Point, °C: 400, decomposes [STR93]

2403
Compound: Rubidium perchlorate
Formula: $RbClO_4$
Molecular Formula: ClO_4Rb
Molecular Weight: 184.919
CAS RN: 13510-42-4
Properties: -4 mesh with 99.9% purity; white rhomb cryst; hygr [STR93] [CRC94] [CER91]
Solubility: g/100g H_2O: 1.1 (0°C), 1.8 (25°C), 22 (100°C) [KRU93]

Density, g/cm³: 2.80 [STR93]
Melting Point, °C: 281 [KIR79]
Boiling Point, °C: 606 decomposes [KIR79]
Reactions: transition ortho-rhomb to cub at 551-554 K [KIR79]

2404
Compound: Rubidium permanganate
Formula: $RbMnO_4$
Molecular Formula: MnO_4Rb
Molecular Weight: 204.404
CAS RN: 13465-49-1
Properties: cryst [LAN52]
Solubility: g/100mL soln, H_2O: 1.06 (19°C) [KRU93]
Density, g/cm³: 3.325 [LAN52]
Melting Point, °C: 295, decomposes [LAN52]

2405
Compound: Rubidium pyrovanadate
Formula: $Rb_4V_2O_7$
Molecular Formula: $O_7Rb_4V_2$
Molecular Weight: 555.750
CAS RN: 13597-61-0
Properties: -100 mesh with 99.9% purity [CER91]

2406
Compound: Rubidium selenide
Formula: Rb_2Se
Molecular Formula: Rb_2Se
Molecular Weight: 249.896
CAS RN: 31052-43-4
Properties: white cub cryst; -60 mesh with 99.5% purity [LID94] [CER91]
Density, g/cm³: 3.22 [LID94]
Melting Point, °C: 733 [LID94]

2407
Compound: Rubidium sulfate
Formula: Rb_2SO_4
Molecular Formula: O_4Rb_2S
Molecular Weight: 267.000
CAS RN: 7488-54-2
Properties: -20 mesh with 99.9% purity; white cryst [STR93] [CER91]
Solubility: g/100g H_2O: 36.4 (0°C), 50.8 (25°C), 81.8 (100°C) [KRU93]
Density, g/cm³: 3.613 [STR93]
Melting Point, °C: 1050 [STR93]
Boiling Point, °C: ~1700 [STR93]

2408
Compound: Rubidium sulfide
Formula: Rb_2S
Molecular Formula: Rb_2S
Molecular Weight: 203.002
CAS RN: 31083-74-6
Properties: yellow-white; available as -60 mesh, dry, under argon with 99.9% purity [CER91] [CRC94]
Density, g/cm³: 2.913 [CRC94]
Melting Point, °C: decomposes 530 [CRC94]

2409
Compound: Rubidium tantalate
Formula: $RbTaO_3$
Molecular Formula: O_3RbTa
Molecular Weight: 314.414
CAS RN: 12333-74-3
Properties: -200 mesh with 99.9% purity [CER91]

2410
Compound: Rubidium tetrahydridoborate
Formula: $RbBH_4$
Molecular Formula: BH_4Rb
Molecular Weight: 100.310
CAS RN: 20346-99-0
Properties: powd [ALF95]
Density, g/cm³: 1.92 [ALF95]

2411
Compound: Rubidium titanate
Formula: Rb_2TiO_3
Molecular Formula: O_3Rb_2Ti
Molecular Weight: 266.801
CAS RN: 12137-34-7
Properties: reacted product, -200 mesh with 99.9% purity [CER91]

2412
Compound: Rubidium tungstate
Formula: Rb_2WO_4
Molecular Formula: O_4Rb_2W
Molecular Weight: 418.774
CAS RN: 13597-52-9
Properties: -200 mesh with 99.9% purity [CER91]

2413
Compound: Rubidium zirconate
Formula: Rb_2ZrO_3
Molecular Formula: O_3Rb_2Zr

Molecular Weight: 310.158
CAS RN: 12534-23-5
Properties: -200 mesh with 99.9% purity [CER91]

2414
Compound: Ruthenium
Formula: Ru
Molecular Formula: Ru
Molecular Weight: 101.07
CAS RN: 7440-18-8
Properties: silvery white; lustrous, hard metal; hex close-packed, a = 0.2704 nm, c = 0.4281 nm; enthalpy of sublimation 649.6 kJ/mol; entropy of sublimation 157.8 J/(mol·K); enthalpy of fusion 38.8 kJ/mol; entropy of fusion 15.2 J/(mol·K); vapor pressure at mp is 1.31 Pa; electrical resistivity (20°C) 7.4 μohm·cm; Young's modulus 413.8 GPa; Brinell hardness 220; superconducting transition 0.48 K; used as a hardener for platinum and palladium [HAW93] [MER89] [KIR82] [RAR85]
Solubility: i acids and aqua regia; attacked by conc NaOH, by fused alkalies [HAW93]
Density, g/cm³: 12.45 [MER89]
Melting Point, °C: 2546 [RAR85]
Boiling Point, °C: ~4400 [RAR85]
Thermal Conductivity, W/(m·K): 117 (25°C) [ALD94]
Thermal Expansion Coefficient: 20°C is 9.1 x 10^{-6}/°C [KIR82]

2415
Compound: Ruthenium ammoniated oxychloride
Synonyms: ruthenium red
Formula: [(NH₃)₅Ru-O-Ru(NH₃)₄-ORu(NH₃)₅]Cl₆
Molecular Formula: Cl₆H₄₂N₁₄O₂Ru₃
Molecular Weight: 786.352
CAS RN: 11103-72-3
Properties: brownish red powd; exists as tetrahydrate; preparation: treatment of RuCl₃(aq) with NH₄OH for several days, in air, followed by crystallization [COT88] [MER89]
Solubility: s H₂O, ammonia [MER89]

2416
Compound: Ruthenium dodecacarbonyl
Synonyms: triruthenium dodecacarbonyl
Formula: Ru₃(CO)₁₂
Molecular Formula: C₁₂O₁₂Ru₃
Molecular Weight: 639.335
CAS RN: 15243-33-1

Properties: orange cryst; stable in air [DOU83] [STR93]
Melting Point, °C: 150, decomposes [STR93]

2417
Compound: Ruthenium nitrosyl chloride monohydrate
Formula: Ru(NO)Cl₃·H₂O
Molecular Formula: Cl₃H₂NO₂Ru
Molecular Weight: 255.450
CAS RN: 18902-42-6
Properties: red cryst; hygr [STR93]

2418
Compound: Ruthenium(III) acetylacetonate
Synonyms: 2,4-pentanedione, ruthenium(III) derivative
Formula: Ru(CH₃COCH=C(O)CH₃)₃
Molecular Formula: C₁₅H₂₁O₆Ru
Molecular Weight: 398.398
CAS RN: 14284-93-6
Properties: redish brown cryst [STR93]

$$\underset{[CH_3-C=CH-C-CH_3]_3Ru}{\overset{O-\qquad O}{\overset{|\qquad \|}{}}}$$

Melting Point, °C: 230-235 [STR93]

2419
Compound: Ruthenium(III) bromide
Synonyms: ruthenium tribromide
Formula: RuBr₃
Molecular Formula: Br₃Ru
Molecular Weight: 340.782
CAS RN: 14014-88-1
Properties: hex brown cryst [LID94]
Density, g/cm³: 5.3 [LID94]
Melting Point, °C: decomposes >400 [LID94]

2420
Compound: Ruthenium(III) chloride
Synonyms: ruthenium trichloride
Formula: RuCl₃
Molecular Formula: Cl₃Ru
Molecular Weight: 207.428
CAS RN: 10049-08-8
Properties: α-RuCl₃: black, lustrous cryst; β-RuCl₃: dark brown, fluffy, hex cryst; hygr [MER89] [STR93]

Solubility: i H_2O, decomposed by hot H_2O; sl s
alcohol [HAW93]
Density, g/cm³: 3.11 [HAW93]
Melting Point, °C: >500, decomposes [STR93]

2421
Compound: Ruthenium(III) chloride hydrate
Synonyms: ruthenium trichloride hydrate
Formula: $RuCl_3 \cdot xH_2O$
Molecular Formula: Cl_3Ru (anhydrous)
Molecular Weight: 207.428 (anhydrous)
CAS RN: 14898-67-0
Properties: black powd; hygr [STR93]

2422
Compound: Ruthenium(III) iodide
Formula: RuI_3
Molecular Formula: I_3Ru
Molecular Weight: 481.783
CAS RN: 13896-65-6
Properties: hex black cryst [LID94] [STR93]
Density, g/cm³: 6.0 [LID94]
Melting Point, °C: decomposes 590 [AES93]

2423
Compound: Ruthenium(IV) oxide
Synonyms: ruthenium dioxide
Formula: RuO_2
Molecular Formula: O_2Ru
Molecular Weight: 133.069
CAS RN: 12036-10-1
Properties: -100 mesh with 99.9% purity; dark
grayish black tetr solid [KIR82] [CER91]
Solubility: i H_2O, acids; s fused alkali [KIR82]
Density, g/cm³: 7.0 [KIR82]
Melting Point, °C: decomposes [KIR82]

2424
Compound: Ruthenium(VIII) oxide
Synonyms: ruthenium tetroxide
Formula: RuO_4
Molecular Formula: O_4Ru
Molecular Weight: 165.068
CAS RN: 20427-56-9
Properties: golden yellow; monocl prisms; very
volatile; sublimes at room temp; strong
oxidizing agent, e.g. explosive reaction with
filter paper, alcohol [MER89]
Solubility: 2.03 g/100mL H_2O (20°C); v s CCl_4,
other chlorinated hydrocarbons [MER89]
[KIR82]

Density, g/cm³: 3.29 [KIR82]
Melting Point, °C: 25.4 [MER89]
Boiling Point, °C: 40 [MER89]

2425
Compound: Samarium
Formula: Sm
Molecular Formula: Sm
Molecular Weight: 150.36
CAS RN: 7440-19-9
Properties: yellow metal; tarnishes in air;
rhombhed, room temp; bcc, >917°C; electrical
resistivity (20°C) 91.4 μohm·cm; enthalpy of
fusion 8.62 kJ/mol; enthalpy of sublimation
206.7 kJ/mol; radius of atom is 0.1804 nm;
radius of Sm^{+++} ion 0.0964 nm; Sm^{+++} forms
colorless solutions [MER89] [KIR82]
Density, g/cm³: 7.520 [KIR82]
Melting Point, °C: 1072 [MER89]
Boiling Point, °C: 1794 [KIR82]
Thermal Conductivity, W/(m·K): 13.3 (25°C)
[CRC93]
Thermal Expansion Coefficient: 12.7 x 10^{-6}/K
[CRC93]

2426
Compound: Samarium acetate trihydrate
Formula: $Sm(CH_3COO)_3 \cdot 3H_2O$
Molecular Formula: $C_6H_{15}O_9Sm$
Molecular Weight: 381.540
CAS RN: 17829-86-6
Properties: off-white powd [STR93]
Solubility: 15 g/100 mL H_2O (25°C) [CRC94]
Density, g/cm³: 1.94 [STR93]

2427
Compound: Samarium acetylacetonate
Synonyms: 2,4-pentanedione, samarium(III)
derivative
Formula: $Sm(CH_3COCH=C(O)CH_3)_3$
Molecular Formula: $C_{15}H_{21}O_6Sm$
Molecular Weight: 447.688
CAS RN: 14589-42-5
Properties: white powd [STR93]

$$
\begin{array}{cc}
\text{O--} & \text{O} \\
| & \| \\
\end{array}
$$
$$[CH_3-C=CH-C-CH_3]_3Sm$$

Melting Point, °C: 146 [AES93]

2428
Compound: Samarium boride
Formula: SmB_6
Molecular Formula: B_6Sm
Molecular Weight: 215.226
CAS RN: 12008-29-6
Properties: -325 mesh 10 microns or less with 99.9% purity; refractory material [KIR78] [CER91]
Density, g/cm³: 5.07 [LID94]
Melting Point, °C: 2540 [KIR78]

2429
Compound: Samarium bromate nonahydrate
Formula: $Sm(BrO_3)_3 \cdot 9H_2O$
Molecular Formula: $Br_3H_{18}O_{18}Sm$
Molecular Weight: 696.204
CAS RN: 28958-26-1
Properties: yellow; hex [CRC77]
Solubility: g/100g H_2O: 34.2 (0°C), 62.5 (20°C), 98.5 (40°C) [LAN85]
Melting Point, °C: 75 [CRC77]
Reactions: minus $9H_2O$, 150°C [CRC77]

2430
Compound: Samarium bromide hexahydrate
Formula: $SmBr_3 \cdot 6H_2O$
Molecular Formula: $Br_3H_{12}O_6Sm$
Molecular Weight: 498.164
CAS RN: 13517-12-9
Properties: -20 mesh with 99.9% purity; yellow cryst [ALD94] [CRC94]
Density, g/cm³: 2.971 [ALF95]
Melting Point, °C: >640 [ALF95]

2431
Compound: Samarium carbonate
Formula: $Sm_2(CO_3)_3$
Molecular Formula: $C_3O_9Sm_2$
Molecular Weight: 480.748
CAS RN: 5895-47-6
Properties: white to light yellow powd [STR93]
Melting Point, °C: >500, decomposes [STR93]

2432
Compound: Samarium chloride
Formula: $SmCl_3$
Molecular Formula: Cl_3Sm
Molecular Weight: 256.718
CAS RN: 10361-82-7

Properties: -20 mesh with 99.9% purity; yellowish white powd; hygr [STR93] [CER91]
Solubility: g/100g H_2O: 92.4 (10°C), 93.4 (20°C), 96.9 (40°C) [LAN85]
Density, g/cm³: 4.465 [STR93]
Melting Point, °C: 686 [STR93]

2433
Compound: Samarium chloride hexahydrate
Formula: $SmCl_3 \cdot 6H_2O$
Molecular Formula: $Cl_3H_{12}O_6Sm$
Molecular Weight: 364.809
CAS RN: 13465-55-9
Properties: -4 mesh with 99.9% purity; light yellow cryst [STR93] [CER91]
Density, g/cm³: 2.382 [MER89]
Reactions: minus $5H_2O$ at 110°C [AES93]

2434
Compound: Samarium diiodide
Synonyms: samarium(II) iodide
Formula: SmI_2
Molecular Formula: I_2Sm
Molecular Weight: 404.169
CAS RN: 32248-43-4
Properties: dark brown solid; ampoules under argon, or in liq form; stabilized with Sm powd; sensitive to atm oxygen and moisture; freezing point of liq -17°C in THF solvent [STR93] [ALD94]
Melting Point, °C: solid: 527 [CRC94]
Boiling Point, °C: solid: 1580 [CRC94]

2435
Compound: Samarium fluoride
Formula: SmF_3
Molecular Formula: F_3Sm
Molecular Weight: 207.355
CAS RN: 13765-24-7
Properties: off-white powd, and 99.9% pure melted pieces of 3-6 mm; hygr; pieces used as evaporation material for possible application to multilayers [STR93] [CER91]
Density, g/cm³: 6.928 [STR93]
Melting Point, °C: 1306 [STR93]
Boiling Point, °C: 2323 [STR93]

2436
Compound: Samarium hydride
Formula: SmH_3
Molecular Formula: H_3Sm

Molecular Weight: 153.384
CAS RN: 13598-53-3
Properties: lumps, under argon [AES93]

2437
Compound: Samarium iodide
Formula: SmI_3
Molecular Formula: I_3Sm
Molecular Weight: 531.073
CAS RN: 13813-25-7
Properties: -20 mesh [ALF95]

2438
Compound: Samarium nitrate hexahydrate
Formula: $Sm(NO_3)_3 \cdot 6H_2O$
Molecular Formula: $H_{12}N_3O_{15}Sm$
Molecular Weight: 444.466
CAS RN: 13759-83-6
Properties: yellow cryst; hygr [STR93]
Melting Point, °C: 78 [AES93]

2439
Compound: Samarium oxalate decahydrate
Formula: $Sm_2(C_2O_4)_3 \cdot 10H_2O$
Molecular Formula: $C_6H_{20}O_{22}Sm_2$
Molecular Weight: 744.932
CAS RN: 14175-03-2
Properties: white powd [AES93] [STR93]
Solubility: 0.000054 g/100mL H_2O [CRC94]

2440
Compound: Samarium oxide
Synonyms: samaria, samarium oxide
Formula: Sm_2O_3
Molecular Formula: O_3Sm_2
Molecular Weight: 348.718
CAS RN: 12060-58-1
Properties: yellowish white powd, or sintered pieces; used as an evaporation material because of its reactivity to radio frequencies [MER89] [CER91]
Density, g/cm³: 8.347 [MER89]

2441
Compound: Samarium perchlorate hydrate
Formula: $Sm(ClO_4)_3 \cdot xH_2O$
Molecular Formula: $Cl_3O_{12}Sm$ (anhydrous)
Molecular Weight: 448.711 (anhydrous)
CAS RN: 13569-60-3

Properties: white powd; hygr; x=6 [AES93] [STR93]

2442
Compound: Samarium silicide
Formula: $SmSi_2$
Molecular Formula: Si_2Sm
Molecular Weight: 206.531
CAS RN: 12300-22-0
Properties: 10 mm & down lump [ALF93]

2443
Compound: Samarium sulfate octahydrate
Formula: $Sm_2(SO_4)_3 \cdot 8H_2O$
Molecular Formula: $H_{16}O_{20}S_3Sm_2$
Molecular Weight: 733.033
CAS RN: 13465-58-2
Properties: white monocl cryst; hygr [ALD94] [STR93]
Solubility: 4.4 g/100mL H_2O (25°C), 1.99 g/100mL H_2O (40°C) [CRC94]
Density, g/cm³: 2.93 [STR93]
Reactions: minus $8H_2O$ at 450 [AES93]

2444
Compound: Samarium sulfide
Formula: Sm_2S_3
Molecular Formula: S_3Sm_2
Molecular Weight: 396.918
CAS RN: 12067-22-0
Properties: red powd [STR93]
Density, g/cm³: 5.87 [LID94]
Melting Point, °C: 1720 [LID94]

2445
Compound: Samarium telluride
Formula: Sm_2Te_3
Molecular Formula: Sm_2Te_3
Molecular Weight: 683.520
CAS RN: 12040-00-5
Properties: -20 mesh with 99.9% purity [CER91]
Density, g/cm³: 7.31 [LID94]

2446
Compound: Samarium tris(cyclopentadienyl)
Synonyms: tris(cyclopentadienyl)samarium
Formula: $Sm(C_5H_5)_3$
Molecular Formula: $C_{15}H_{15}Sm$
Molecular Weight: 345.644
CAS RN: 1298-55-1

Properties: orange powd; air and moisture sensitive
[STR93]
Melting Point, °C: 356 [STR93]

2447
Compound: Samarium(II) chloride
Formula: SmCl$_2$
Molecular Formula: Cl$_2$Sm
Molecular Weight: 221.265
CAS RN: 13874-75-4
Properties: dark brown cryst [MER89]
Solubility: decomposes in H$_2$O; i alcohol [MER89]
Density, g/cm^3: 3.687 [MER89]
Melting Point, °C: 740 [CRC94]

2448
Compound: Scandium
Formula: Sc
Molecular Formula: Sc
Molecular Weight: 44.955910
CAS RN: 7440-20-2
Properties: white, silvery metal; electrical resistivity
(20°C) 50.5 μohm·cm; enthalpy of fusion 14.10
kJ/mol; enthalpy of sublimation 377.8 kJ/mol;
radius of atom 0.1640 nm; radius of ion is
0.0732 nm; aq solutions are colorless; α-form:
hex close-packed; existence of β-form
inconclusive; used in catalyst studies of para-to-
ortho hydrogen conversion, in high intensity
lamps, for neutron filters and in ion microprobe
analyzers [KIR82] [CER91] [MER89] [ALD94]
Density, g/cm^3: 2.989 [KIR82]
Melting Point, °C: 1538 [MER89]
Boiling Point, °C: 2836 [KIR82]
Thermal Conductivity, W/(m·K): 15.8 (25°C)
[CRC93]
Thermal Expansion Coefficient: 10.2 x 10^{-6}/K
[CRC93]

2449
Compound: Scandium acetate hydrate
Formula: Sc(CH$_3$COO)$_3$·xH$_2$O
Molecular Formula: C$_6$H$_9$O$_6$Sc (anhydrous)
Molecular Weight: 222.089 (anhydrous)
CAS RN: 3804-23-7
Properties: white cryst [AES93]
Melting Point, °C: decomposes [AES93]

2450
Compound: Scandium boride
Formula: ScB$_2$

Molecular Formula: B$_2$Sc
Molecular Weight: 66.578
CAS RN: 12007-34-0
Properties: -200 mesh with 99.5% purity; refractory
material [KIR78] [CER91]
Density, g/cm^3: 3.17 [LID94]
Melting Point, °C: 2250 [KIR78]

2451
Compound: Scandium bromide
Formula: ScBr$_3$
Molecular Formula: Br$_3$Sc
Molecular Weight: 284.668
CAS RN: 13465-59-3
Properties: -20 mesh with 99.9% purity [CER91]
Density, g/cm^3: 1.914 [CRC94]
Melting Point, °C: sublimes >1000 [CRC94]

2452
Compound: Scandium carbonate hydrate
Formula: Sc$_2$(CO$_3$)$_3$·xH$_2$O
Molecular Formula: C$_3$O$_9$Sc$_2$ (anhydrous)
Molecular Weight: 269.939 (anhydrous)
CAS RN: 17926-77-1
Properties: white powd [STR93]
Melting Point, °C: decomposes >500 [AES93]

2453
Compound: Scandium chloride
Formula: ScCl$_3$
Molecular Formula: Cl$_3$Sc
Molecular Weight: 151.314
CAS RN: 10361-84-9
Properties: -20 mesh with 99.9% purity; white;
deliq [MER89] [CER91]
Solubility: s H$_2$O; i alcohol [MER89]
Density, g/cm^3: 2.39 [STR93]
Melting Point, °C: 960 [MER89]

2454
Compound: Scandium chloride hexahydrate
Formula: ScCl$_3$·6H$_2$O
Molecular Formula: Cl$_3$H$_{12}$O$_6$Sc
Molecular Weight: 259.405
CAS RN: 20662-14-0
Properties: white cryst [STR93]

2455
Compound: Scandium fluoride
Formula: ScF$_3$

Molecular Formula: F_3Sc
Molecular Weight: 101.951
CAS RN: 13709-47-2
Properties: hygr white powd, and 99.9% pure sintered pieces of 3-12 mm; pieces used as evaporation material for possible application in lasers [STR93] [CER91]
Melting Point, °C: 1515 [STR93]

2456
Compound: Scandium nitrate pentahydrate
Formula: $Sc(NO_3)_3 \cdot 5H_2O$
Molecular Formula: $H_{10}N_3O_{14}Sc$
Molecular Weight: 321.047
CAS RN: 13465-60-6
Properties: white cryst [STR93]
Solubility: anhydrous v s H_2O, alcohol [MER89]

2457
Compound: Scandium oxalate pentahydrate
Formula: $Sc_2(C_2O_4)_3 \cdot 5H_2O$
Molecular Formula: $C_6H_{10}O_{17}Sc_2$
Molecular Weight: 444.047
CAS RN: 17926-77-1
Properties: white cryst [STR93]

2458
Compound: Scandium oxide
Synonyms: scandia
Formula: Sc_2O_3
Molecular Formula: O_3Sc_2
Molecular Weight: 137.910
CAS RN: 12060-08-1
Properties: fine white powd, or sintered pieces of 3-12 mm; used as an evaporation material of 99.99% purity to produce fairly hard coating which is very stable, and is useful for antireflection coating on semiconductors with high index [MER89] [CER91]
Solubility: v s hot conc acids [MER89]
Density, g/cm³: 3.864 [MER89]

2459
Compound: Scandium sulfate octahydrate
Formula: $Sc_2(SO_4)_3 \cdot 8H_2O$
Molecular Formula: $H_{16}O_{20}S_3Sc_2$
Molecular Weight: 522.225
CAS RN: 52788-54-2
Properties: white cryst; pentahydrate also exists [STR93] [MER89]

Solubility: pentahydrate: 54.6 g/100mL H_2O (25°C) [MER89]
Density, g/cm³: pentahydrate: 2.519 [MER89]
Reactions: pentahydrate: minus $3H_2O$ to form dihydrate >100°C [MER89]

2460
Compound: Scandium sulfide
Formula: Sc_2S_3
Molecular Formula: S_3Sc_2
Molecular Weight: 186.110
CAS RN: 12166-29-9
Properties: -200 mesh with 99.9% purity [CER91]
Density, g/cm³: 2.91 [LID94]
Melting Point, °C: 1775 [LID94]

2461
Compound: Scandium telluride
Formula: Sc_2Te_3
Molecular Formula: Sc_2Te_3
Molecular Weight: 472.712
CAS RN: 12166-44-8
Properties: black hex cryst; -20 mesh with 99.9% purity (Ta ~1-3%) [CER91] [LID94]
Density, g/cm³: 5.29 [LID94]

2462
Compound: Scandium tris(cyclopentadienyl)
Synonyms: tris(cyclopentadienyl)scandium
Formula: $Sc(C_5H_5)_3$
Molecular Formula: $C_{15}H_{15}Sc$
Molecular Weight: 240.240
CAS RN: 1298-54-0
Properties: powd; sensitive to air and moisture [STR93]
Melting Point, °C: 240 [STR93]
Reactions: sublimes at 200°C (0.05 mm Hg) [STR93]

2463
Compound: Selenic acid
Formula: H_2SeO_4
Molecular Formula: H_2O_4Se
Molecular Weight: 144.974
CAS RN: 7783-08-6
Properties: white hygr solid; easily undercools; oxidizing agent [KIR82] [HAW93]
Solubility: g/100g H_2O: 426 (0°C), 567 (20°C), 1328 (30°C) [LAN85]; decomposed by alcohol [HAW93]
Density, g/cm³: 3.004 [HAW93]

Melting Point, °C: 58 [HAW93]
Boiling Point, °C: 260, decomposes [HAW93]

2464
Compound: Selenium
Formula: Se
Molecular Formula: Se
Molecular Weight: 78.96
CAS RN: 7782-49-2
Properties: gray hex stable at room temp, semiconductor; other forms: monocl red (α and β), amorphous black and red; hex: a = 0.4366 nm, c = 0.4954 nm; trig liq enthalpy of fusion 6.224 J/mol; enthalpy of vaporization 95.48 J/mol; hardness 2.0 Mohs; viscosity 221 mPa·s (220°C) is 70 mPa·s (360°C); rectifies AC voltage to DC; electrical resistivity (0°C) 1.2 μohm·cm; red cryst obtained by evaporation of amorphous red form from CS_2 [MER89] [KIR82] [CRC93] [COT88] [ALD94]
Solubility: i H_2O, alcohol; 2 mg/100mL CS_2; s ether [MER89]
Density, g/cm³: trig: 4.819 (25°C); monocl: 4.4; liq: 3.975 ((200°C) [KIR82]
Melting Point, °C: 217 [ALD94]
Boiling Point, °C: 684.9 [ALD94]
Thermal Conductivity, W/(m·K): 248.1 [KIR82], 4.52 (25°C) [CRC93]; 0.519 for amorphous at 25°C [ALD94] [CRC93]
Thermal Expansion Coefficient: 3.24 x 10^{-5}/°C to 7.5 x 10^{-5}/°C, depending on the form of Se [KIR80]

2465
Compound: Selenium (β)
Formula: β-Se
Molecular Formula: Se
Molecular Weight: 78.96
CAS RN: 7782-49-2
Properties: dark red; monocl, a = 1.285 nm, b = 0.807 nm, c = 0.931 nm; transparent cryst; metastable; prepared by rapid evaporation of solution consisting of Se dissolved in CS_2 [MER89] [KIR82]
Density, g/cm³: 4.39 [LID94]
Melting Point, °C: 221 [COT88]
Boiling Point, °C: 684.8 [COT88]
Reactions: transforms into gray Se if heated [MER89]

2466
Compound: Selenium bromide

Formula: Se_2Br_2
Molecular Formula: Br_2Se_2
Molecular Weight: 317.728
CAS RN: 7789-52-8
Properties: dark red liq; decomposes if heated in moist air [MER89]
Solubility: decomposes in H_2O; s chloroform, ethyl bromide, carbon disulfide [MER89]
Density, g/cm³: 3.604 [MER89]
Boiling Point, °C: 227, decomposes [KIR82]

2467
Compound: Selenium chloride
Formula: Se_2Cl_2
Molecular Formula: Cl_2Se_2
Molecular Weight: 228.825
CAS RN: 10025-68-0
Properties: deep red, oily liq; sensitive to moisture [MER89] [STR93]
Solubility: decomposes in H_2O; s chloroform, benzene, CCl_4, CS_2, fuming sulfuric acid [MER89]
Density, g/cm³: 2.7741 [MER89]
Melting Point, °C: -85 [KIR82]
Boiling Point, °C: 127, decomposes [KIR82]

2468
Compound: Selenium dioxide
Synonyms: selenium(IV) oxide
Formula: SeO_2
Molecular Formula: O_2Se
Molecular Weight: 110.959
CAS RN: 7446-08-4
Properties: off-white powd; has yellowish green vapor; absorbs dry HF, HCl, HBr, HI to form the respective selenium oxyhalide; oxidizing agent, reduced to Se by SO_2, hydrogen, hydrogen sulfide, ammonia [KIR82] [STR93] [MER89]
Solubility: g/100g H_2O: 222 (10°C), 257 (20°C), 440 (60°C) [LAN85]; parts/100 parts: 10.16 methanol (11.8°C); 6.67 93% ethanol (14°C); 4.35 acetone (15.3°C); 1.11 (13.9°C) acetic acid; s H_2SO_4 [MER89]
Density, g/cm³: 3.954 [MER89]
Melting Point, °C: 340 [KIR82]
Boiling Point, °C: 315, sublimes [KIR82]

2469
Compound: Selenium disulfide
Formula: SeS_2
Molecular Formula: S_2Se
Molecular Weight: 143.092

CAS RN: 7488-56-4
Properties: red powd; used in medicine and medicated shampoos [HAW93] [STR93]
Solubility: i H_2O, organic solvents [HAW93]
Melting Point, °C: <100 [STR93]
Boiling Point, °C: decomposes [STR93]

2470
Compound: Selenium hexafluoride
Synonyms: selenium(VI) fluoride
Formula: SeF_6
Molecular Formula: F_6Se
Molecular Weight: 192.950
CAS RN: 7783-79-1
Properties: moisture sensitive; gas; vapor pressure, mm Hg: 651.2 (-48.7°C), 213.1 (-64.8°C), 30.4 (-87.5°C); reacts with NH_3 at 200°C forming Se, N_2 and HF; prepared by reaction of finely divided Se metal in F_2 atm in Cu container; used as an electrical insulator [MER89] [STR93]
Solubility: i H_2O [MER89]
Density, g/cm³: 8.467 g/L [LID94]
Melting Point, °C: -34.6 [KIR82]
Boiling Point, °C: -46.6, sublimes [KIR82]

2471
Compound: Selenium hexasulfide
Formula: Se_2S_6
Molecular Formula: S_6Se_2
Molecular Weight: 350.316
CAS RN: 75926-22-6
Properties: light orange needles when prepared from benzene [MER89]
Solubility: s CS_2; 12 g/L benzene (20°C) [MER89]
Density, g/cm³: 2.44 [MER89]
Melting Point, °C: 121.5 [MER89]

2472
Compound: Selenium monosulfide
Formula: SeS
Molecular Formula: SSe
Molecular Weight: 111.026
CAS RN: 7446-34-6
Properties: brick red [KIR82]
Density, g/cm³: 3.056 [CRC94]
Melting Point, °C: decomposes at 118-119 [CRC94]

2473
Compound: Selenium oxybromide
Formula: $SeOBr_2$
Molecular Formula: Br_2OSe

Molecular Weight: 254.767
CAS RN: 7789-51-7
Properties: reddish yellow solid; decomposed by sulfur and hydrogen sulfide; preparation by reaction of Br_2 with dry mixture of Se and SeO_2; uses: brominating agent [KIR82] [MER89]
Solubility: decomposed by H_2O; s H_2SO_4, CS_2, $CHCl_3$, benzene, toluene, xylene, CCl_4 [MER89]
Density, g/cm³: 3.38 [MER89]
Melting Point, °C: 41 [MER89]
Boiling Point, °C: 217 (740 mm) [MER89]
Reactions: decomposes in air at 50°C [KIR82]

2474
Compound: Selenium oxychloride
Formula: $SeOCl_2$
Molecular Formula: Cl_2OSe
Molecular Weight: 165.864
CAS RN: 7791-23-3
Properties: nearly colorless or yellowish; corrosive liq; fumes in air; excellent solvent for many substances; has high dielectric constant; preparation: Cl_2 reacted with dry mixture of Se and SeO_2; strong chlorinating agent [KIR82] [MER89]
Solubility: decomposed in H_2O to HCl, selenious acid; miscible with CCl_4, $CHCl_3$, CS_2, benzene, toluene [MER89]
Density, g/cm³: 2.44 [MER89]
Melting Point, °C: 10.8 [KIR82]
Boiling Point, °C: 176.4 [STR93]

2475
Compound: Selenium oxydifluoride
Formula: SeO_2F_2
Molecular Formula: F_2O_2Se
Molecular Weight: 148.956
CAS RN: 14984-81-7
Properties: colorless liq; prepared by reaction between selenium oxychloride and silver fluoride [KIR82]
Solubility: reacts with H_2O [KIR82]
Density, g/cm³: 6.536 g/L [LID94]
Melting Point, °C: -99.5 [KIR82]
Boiling Point, °C: -8.4 [KIR82]

2476
Compound: Selenium oxyfluoride
Formula: $SeOF_2$
Molecular Formula: F_2OSe
Molecular Weight: 132.956
CAS RN: 7783-43-9

Properties: colorless liq with pungent odor; reacts with water, glass, silicon and violently with red phosphorus; can be prepared by reacting selenium oxychloride with silver fluoride [KIR82]
Density, g/cm³: 2.8 [MER89]
Melting Point, °C: 15 [KIR82]
Boiling Point, °C: 125 to 126 [KIR82]

2477
Compound: Selenium sulfide
Formula: Se₄S₄
Molecular Formula: S₄Se₄
Molecular Weight: 444.104
CAS RN: 75426-28-2
Properties: red cryst when prepared from benzene [MER89]
Solubility: s benzene, CS₂ [MER89]
Density, g/cm³: 3.20 [MER89]
Melting Point, °C: 113, decomposes [MER89]

2478
Compound: Selenium tetrabromide
Synonyms: selenium(IV) bromide
Formula: SeBr₄
Molecular Formula: Br₄Se
Molecular Weight: 398.576
CAS RN: 7789-65-3
Properties: reddish brown; cryst powd; decomposed in moist air [MER89]
Solubility: decomposed by H₂O; s CS₂, CHCl₃, ethyl bromide [MER89]
Melting Point, °C: 75, decomposes [KIR82]
Reactions: decomposes at 70-80°C [MER89]

2479
Compound: Selenium tetrachloride
Formula: SeCl₄
Molecular Formula: Cl₄Se
Molecular Weight: 220.771
CAS RN: 10026-03-6
Properties: white to pale yellow cryst; sublimes when heated; decomposed by moist air [MER89]
Solubility: decomposed in H₂O; i liq bromine [MER89]
Density, g/cm³: 2.6 [LID94]
Melting Point, °C: 305 [KIR82]
Boiling Point, °C: sublimes [KIR82]

2480
Compound: Selenium tetrafluoride

Formula: SeF₄
Molecular Formula: F₄Se
Molecular Weight: 154.954
CAS RN: 13465-66-2
Properties: colorless liq; fumes in air; strong oxidizing agent; attacks glass, silicon, phosphorus, arsenic, antimony and bismuth; enthalpy of vaporization 47.2 kJ/mol [CRC93] [KIR82] [MER89]
Solubility: reacts violently with H₂O; miscible with ether, ethanol, sulfuric acid; s CHCl₃, CCl₄ [MER89]
Density, g/cm³: 2.75 [MER89]
Melting Point, °C: -13.2 [KIR82]
Boiling Point, °C: 106 [MER89]

2481
Compound: Selenium trioxide
Formula: SeO₃
Molecular Formula: O₃Se
Molecular Weight: 126.958
CAS RN: 13768-86-0
Properties: white cryst; hygr; can be prepared by reacting SO₃ with potassium selenate, or phosphorus pentoxide with selenic acid; strong oxidizing agent; stable in dry air at room temperatures; decomposes, if heated, first to selenium pentoxide, then to the dioxide [KIR82]
Solubility: dissolves in H₂O forming selenic acid [KIR82]
Density, g/cm³: 3.6 [CRC94]
Melting Point, °C: 118 [KIR82]
Boiling Point, °C: decomposes at 180 [CRC94]

2482
Compound: Selenium(α)
Formula: α-Se
Molecular Formula: Se
Molecular Weight: 78.96
CAS RN: 7782-49-2
Properties: dark red; monocl, a = 0.9054 nm, b = 0.9083 nm, c = 1.106 nm; can be prepared by slow evaporation of Se dissolved in CS₂ [MER89] [KIR82]
Density, g/cm³: 4.46 [MER89]
Melting Point, °C: <200 [MER89]
Reactions: transforms to gray Se by heating [MER89]

2483
Compound: Selenous acid
Formula: H₂SeO₃

Molecular Formula: H_2O_3Se
Molecular Weight: 128.974
CAS RN: 7783-00-8
Properties: white cryst; hygr; can be prepared by wet oxidation of selenium, or from an aq solution of SeO_2; oxidizing agent [KIR82] [STR93]
Solubility: g/100g H_2O: 90.1 (0°C), 166.7 (20°C), 385.4 (90°C) [LAN85]; s alcohol; i ammonia [HAW93]
Density, g/cm³: 3.004 [STR93]
Melting Point, °C: 70, decomposes [STR93]

2484
Compound: Silane
Synonyms: silicon tetrahydride
Formula: SiH_4
Molecular Formula: H_4Si
Molecular Weight: 32.118
CAS RN: 7803-62-5
Properties: colorless gas; spontaneously flammable in air; repulsive odor; enthalpy of fusion 667.3 kJ/mol; enthalpy of vaporization 12.48 kJ/mol; triple point -209°C; critical temp -3.5°C; critical pressure 4.84 MPa; prepared by the reaction $4LiH + SiCl_4 = SiH_4(gas) + 4LiCl$; used to prepare semiconducting silicon by thermal decomposition above 600°C [MER89] [HAW93] [CIC73] [AIR87]
Solubility: decomposed by H_2O; i alcohol, benzene [HAW93]
Density, g/cm³: liq: 0.68 at -185°C; gas: 1.44 g/L (20°C) [KIR80]
Melting Point, °C: -200 [HAW93]
Boiling Point, °C: -112 [HAW93]

2485
Compound: Silicon
Formula: Si
Molecular Formula: Si
Molecular Weight: 28.0855
CAS RN: 7440-21-3
Properties: black-gray; cub, needle like cryst or octahedral; burns in F_2, Cl_2 atm; amorphous silicon is dark brown powd; electrical resistivity 100,000 μohm·cm; hardness 7 Mohs; dielectric constant 12; enthalpy of vaporization 359 kJ/mol: band gap, eV, 1.17 (0 K) 1.12 (300 K); mobility (300 K), cm²/(V s), 1500 electrons, 450 holes; enthalpy of fusion 50.21 kJ/mol; uses include index films for infrared filters [COT88] [HAW93] [MER89] [CER91] [CRC93] [ALD94]

Solubility: i H_2O, HNO_3 and HCl; s alkalies and in mixture of HNO_3 and HF acids [HAW93]
Density, g/cm³: 2.33 [MER89]
Melting Point, °C: 1410 [MER89]
Boiling Point, °C: 2355 [HAW93]
Thermal Conductivity, W/(m·K): 149 (25°C) [ALD94]
Thermal Expansion Coefficient: (volume) 100°C (0.066), 200°C (0.171), 400°C (0.398), 800°C (0.875), 1000°C (1.109) [CLA66]

2486
Compound: Silicon acetate
Formula: $Si(CH_3COO)_4$
Molecular Formula: $C_8H_{12}O_8Si$
Molecular Weight: 264.264
CAS RN: 562-90-3
Properties: very hygr white cryst; rapidly hydrolyzed in moist air; reacts violently with H_2O [MER89]
Solubility: s acetone, benzene [MER89]
Melting Point, °C: 110 [MER89]
Boiling Point, °C: 148 (6.0 mm Hg) [MER89]
Reactions: decomposes evolving acetic anhydride at 160-170°C [MER89]

2487
Compound: Silicon boride
Synonyms: silicon tetraboride
Formula: SiB_4
Molecular Formula: B_4Si
Molecular Weight: 71.330
CAS RN: 12007-81-7
Properties: refractory material; grayish black powd; there is also a SiB_6+Si, 12008-29-6, density 2.47 [ALF93] [STR93]
Density, g/cm³: 2.40 [ALD94]
Melting Point, °C: 1870, decomposes [STR93]

2488
Compound: Silicon carbide
Formula: SiC
Molecular Formula: CSi
Molecular Weight: 40.097
CAS RN: 409-21-2

Properties: two forms: β is cub, a = 0.43502 nm; extremely hard; green to bluish black; iridescent; dielectric constant 7.0; hardness 9 Mohs; α is semiconductor: mobility (300 K), cm^2/V s, 400 electrons and 50 holes; band gap, eV, 3.03 (0 K) and 2.996 (300 K); effective mass 0.60 for electrons and 1.00 for holes; electrically conductive; resistant to high temp oxidation; used as an abrasive, and sputtering target [HAW93] [MER89] [CIC73] [CER91] [KIR82]

Solubility: i H_2O, alcohol; s fused alkalies [HAW93]

Density, g/cm^3: 3.16 [LID94]

Melting Point, °C: 2830 [LID94]

Thermal Conductivity, W/(m·K): 22.5 (500°C), 23.7 (1000°C) [KIR80]

Thermal Expansion Coefficient: linear to 1000°C: 5.2 x 10^{-6}/°C [KIR80]

2489

Compound: Silicon decahydride

Synonyms: tetrasilane

Formula: Si_4H_{10}

Molecular Formula: $H_{10}Si_4$

Molecular Weight: 122.421

CAS RN: 7783-29-1

Properties: colorless liq; enthalpy of vaporization 35.56 kJ/mol; vapor pressure at 0°C 9.1 mm Hg; critical temp 249°C [CIC73] [CRC94]

Density, g/cm^3: liq at mp: 0.79 [CIC73]

Melting Point, °C: -84.3 [CIC73]

Boiling Point, °C: 107.4 [CIC73]

2490

Compound: Silicon dioxide

Synonyms: coesite

Formula: SiO_2

Molecular Formula: O_2Si

Molecular Weight: 60.085

CAS RN: 7631-86-9

Properties: cryst from 298.15 K to 1800 K [ROB78]

Thermal Expansion Coefficient: (volume) 100°C (0.059), 200°C (0.145), 400°C (0.345), 800°C (0.849), 1000°C (1.150) [CLA66]

2491

Compound: Silicon dioxide

Synonyms: tridymite

Formula: SiO_2

Molecular Formula: O_2Si

Molecular Weight: 60.085

CAS RN: 7631-86-9

Properties: β form: hex, lattice constant a = 0.503 nm, c = 0.822 nm; α form: rhomb; enthalpy of transition 2903 J/g [CIC73]

Density, g/cm^3: 2.262 (0°C) [CIC73]

Melting Point, °C: 1703 [CIC73]

Boiling Point, °C: 2950 [CIC73]

Reactions: transition α to β at 1470°C [CIC73]

Thermal Expansion Coefficient: (volume) 100°C (0.63), 200°C (2.40), 400°C (3.33), 800°C (3.66), 1200°C (3.60) [CLA66]

2492

Compound: Silicon dioxide

Synonyms: quartz

Formula: SiO_2

Molecular Formula: O_2Si

Molecular Weight: 60.085

CAS RN: 7631-86-9

Properties: two forms: α, trig, a = 0.49127 nm, c = 0.54046 nm, and β, hex, a = 0.501 nm, c = 0.547 nm; transparent cryst or amorphous powd; melts to a glass; enthalpy of fusion of α 8.5 kJ/mol; heat of transition for α to β 10 J/g; hardness of α form 7 Mohs; velocity of sound of α form 5870 m/sec; used in crucible form for melting aluminum, antimony and other metals, as a sputtering target of 99.995% purity for preparing hard durable films with low index [CIC73] [MER89] [CER91]

Solubility: i H_2O, most acids; s HF [MER89]

Density, g/cm^3: α: 2.6507 (0°C); β: 2.533 (600°C) [CIC73]

Melting Point, °C: α: 1423 [JAN85]

Boiling Point, °C: 2950 [CIC73]

Reactions: transition α to β at 573°C [CIC73]

Thermal Conductivity, W/(m·K): 1.6 (500°C), 2.1 (1000°C) [KIR80]

Thermal Expansion Coefficient: (volume) 100°C (0.36), 200°C (0.78), 400°C (1.89), 800°C (4.42), 1000°C (4.29) [CLA66]

2493

Compound: Silicon dioxide

Synonyms: cristobalite

Formula: SiO_2

Molecular Formula: O_2Si

Molecular Weight: 60.085

CAS RN: 14464-46-1

Properties: β form: regular holohedra, α form: tetr; lattice constant 0.711 nm; enthalpy of fusion 9.58 kJ/mol; enthalpy of transition for α to β 50.3 KJ/mol [CIC73] [JAN85]

Density, g/cm^3: 2.21 (0°C) [CIC73]

Melting Point, °C: 1713 [CIC73]
Boiling Point, °C: 2950 [CIC73]
Thermal Expansion Coefficient: (volume) 100°C (0.791), 200°C (1.795), 400°C (6.271), 800°C (6.499), 1200°C (6.651) [CLA66]

2494
Compound: Silicon disulfide
Formula: SiS_2
Molecular Formula: S_2Si
Molecular Weight: 92.218
CAS RN: 13759-10-9
Properties: sublimed cotton like fibrous clumps with 99.9% purity; white, fibrous mass or needles; rhomb, a = 0.560 nm, b = 0.553 nm, c = 0.975 nm; decomposes in moist air to evolve H_2S; burns if ignited by a flame [MER89] [CIC73] [CER91]
Solubility: decomposes in H_2O, alcohol, alkaline solutions; i benzene [MER89]
Density, g/cm³: 2.02 [MER89]
Melting Point, °C: 1092 [CIC73]
Boiling Point, °C: sublimes 1250 [CIC73]

2495
Compound: Silicon monosulfide
Formula: SiS
Molecular Formula: SSi
Molecular Weight: 60.152
CAS RN: 50927-81-6
Properties: yellow or reddish powd; burns in air to SiO_2 and SO_2; very hygr [CIC73]
Solubility: hydrolyzes in H_2O [CIC73]
Density, g/cm³: 1.853 [CIC73]
Melting Point, °C: ~900 [CIC73]

2496
Compound: Silicon monoxide
Synonyms: silicon(II) oxide
Formula: SiO
Molecular Formula: OSi
Molecular Weight: 44.085
CAS RN: 10097-28-6
Properties: brownish black scales when prepared by sublimation; cub, a = 0.64 nm; does not conduct electricity; enthalpy of vaporization 320.6 kJ/mol; dielectric constant 4.9; used as an evaporation material and sputtering target of 99.99% and 99.9% purity to provide protective film for front surface of aluminum mirrors, and for low index layers in infrared filters [CIC73] [MER89] [CER91]

Density, g/cm³: 2.18 [MER89]
Melting Point, °C: >1702 [STR93]
Boiling Point, °C: 1880 [STR93]
Thermal Expansion Coefficient: 4.5×10^{-6} [CIC73]

2497
Compound: Silicon nitride
Formula: Si_3N_4
Molecular Formula: N_4Si_3
Molecular Weight: 140.284
CAS RN: 12033-89-5
Properties: gray, amorphous powd or cryst; hex α, a = 0.77488 nm, c = 0.5618 nm; hex β, a = 0.7608 nm, c = 0.2911 nm; hardness is 9+ Mohs; resistant to oxidation; used in refractory coatings, and in crucible form for melting aluminum, lead, magnesium, tin and as a container for dil acids and caustics, also used as a 99.9% pure (with added small amounts of MgO for strength) sputtering target to provide insulating properties [KIR81] [HAW93] [STR93] [CER91]
Density, g/cm³: hex α: 3.2 [KIR81]
Melting Point, °C: 1900, sublimes [HAW93]
Thermal Conductivity, W/(m·K): 17 [KIR81]
Thermal Expansion Coefficient: 2.5×10^{-6}/K [MER89]

2498
Compound: Silicon octahydride
Synonyms: trisilane
Formula: Si_3H_8
Molecular Formula: H_8Si_3
Molecular Weight: 92.321
CAS RN: 7783-26-8
Properties: colorless liq; enthalpy of vaporization 28.38 kJ/mol; vapor pressure at (0°C) 95 mm Hg; critical temp 189°C; reacts vigorously with CCl_4 and $CHCl_3$ [CIC73] [CRC94] [MER89]
Solubility: decomposed by H_2O [MER89]
Density, g/cm³: liq at mp: 0.725 [CIC73]
Melting Point, °C: -117.4 [CIC73]
Boiling Point, °C: 52.9 [CIC73]

2499
Compound: Silicon tetrabromide
Formula: $SiBr_4$
Molecular Formula: Br_4Si
Molecular Weight: 347.702
CAS RN: 7789-66-4

Properties: colorless fuming liq; becomes yellow in air; enthalpy of vaporization 37.87 kJ/mol; entropy of vaporization 88.7 J/(mol·K); vapor pressure (0°C) 1.8 mm Hg; critical temp 383°C; surface tension (20°C) 16.9 dyne/cm [CIC73] [MER89]

Solubility: decomposed in water to HBr, silicic acid, with evolution of heat [MER89]

Density, g/cm³: 2.772 [STR93]

Melting Point, °C: 5.4 [STR93]

Boiling Point, °C: 153 [ALD94]

Thermal Expansion Coefficient: (20°C) 0.000983 [CIC73]

2500

Compound: Silicon tetrachloride

Formula: $SiCl_4$

Molecular Formula: Cl_4Si

Molecular Weight: 169.897

CAS RN: 10026-04-7

Properties: colorless, clear fuming liq; decomposed in H_2O, evolving heat; enthalpy of vaporization 28.7 kJ/mol; entropy of vaporization 87.9 J/(mol·K); enthalpy of fusion 7.60 kJ/mol; enthalpy of sublimation 38.07 kJ/mol; vapor pressure (0°C) 77 mm Hg; critical temp 233.6°C; critical pressure 36.8 atm; surface tension (20°C) 19.71 dyne/cm; used in electronics industry [CIC73] [CRC93] [MER89] [AIR87]

Solubility: miscible with benzene, ether, $CHCl_3$, petroleum ether [MER89]

Density, g/cm³: 1.5 [LID94]

Melting Point, °C: -68.85 [LID94]

Boiling Point, °C: 57.65 [LID94]

2501

Compound: Silicon tetrafluoride

Synonyms: silicon(IV) fluoride

Formula: SiF_4

Molecular Formula: F_4Si

Molecular Weight: 104.080

CAS RN: 7783-61-1

Properties: colorless gas; odor similar to that of HCl; forms cloud in moist air; critical temp -14.1°C; critical pressure 3.72 MPa; enthalpy of fusion 7.07 kJ/mol; enthalpy of vaporization 18.7 kJ/mol; entropy of vaporization 102.1 J/(mol·K); enthalpy of sublimation 25.9 kJ/mol; vapor pressure at -99.8°C 515 mm Hg; decomposed by H_2O to silicic acid and HF; used in electronics industry [MER89] [CIC73] [AIR87]

Solubility: hydrolyzed by H_2O [AIR87]

Density, g/cm³: gas: 3.57 (15°C) [HAW93]; liq (-80°C): 1.590 [MER89]

Melting Point, °C: -90.2 [MER89]

Boiling Point, °C: -86 [HAW93]

2502

Compound: Silicon tetraiodide

Synonyms: silicon(IV) iodide

Formula: SiI_4

Molecular Formula: I_4Si

Molecular Weight: 535.704

CAS RN: 13465-84-4

Properties: off-white powd; sensitive to moisture; enthalpy of vaporization 50.2 kJ/mol; enthalpy of fusion 19.70 kJ/mol [CRC93] [STR93] [CRC93]

Density, g/cm³: 4.198 [STR93]

Melting Point, °C: 120.5 [CRC93]

Boiling Point, °C: 287.35 [CRC93]

2503

Compound: Silicotungstic acid

Synonyms: tungstosilicic acid

Formula: $H_4SiW_{12}O_{40} \cdot 5H_2O$

Molecular Formula: $H_{14}O_{45}SiW_{12}$

Molecular Weight: 2968.250

CAS RN: 12520-88-6

Properties: white to sl yellow; deliq cryst [MER89]

Solubility: v s H_2O, alcohol [MER89]

2504

Compound: Silver

Formula: Ag

Molecular Formula: Ag

Molecular Weight: 107.8682

CAS RN: 7440-22-4

Properties: white metal; most silver salts are sensitive to light; fcc; electronegativity 2.43; electrical resistivity 1.59 μohm·cm (0°C); temp coefficient of electrical resistivity (0-100°C) 0.0041; Poisson's ratio 0.39 (hard drawn); enthalpy of fusion 11.30 kJ/mol; enthalpy of vaporization 284.34 kJ/mol; used in highly reflective films and in semiconductors [KIR78] [CER91] [MER89] [KIR83]

Solubility: reacts readily with dil HNO_3, hot conc H_2SO_4; s fused alkali hydroxides in air [MER89]

Density, g/cm³: 10.43 [KIR83]

Melting Point, °C: 961.93 [ALD94]

Boiling Point, °C: 2212 [ALD93]

Thermal Conductivity, W/(m·K): 428 (20°C), 356 (450°C) [KIR83]

Thermal Expansion Coefficient: from 0°C to 500°C: 20.61 μm/mK [KIR83]

2505
Compound: Silver acetate
Formula: CH₃COOAg
Molecular Formula: C₂H₃AgO₂
Molecular Weight: 166.913
CAS RN: 563-63-3
Properties: white to sl grayish, lustrous needles or cryst powd; light sensitive; uses: removes and replaces chloride from precious metal complexes [ALD94] [MER89]
Solubility: g CH₃COOAg/100g H₂O: 0.73 (0°C), 1.05 (20°C), 2.59 (80°C) [LAN85]; v s dil HNO₃ [MER89]
Density, g/cm³: 3.26 [MER89]
Melting Point, °C: decomposes [STR93]

2506
Compound: Silver acetylacetonate
Synonyms: 2,3-pentanedione, silver(I) derivative
Formula: Ag(CH₃COCH=C(O)CH₃)
Molecular Formula: C₅H₇AgO₂
Molecular Weight: 206.977
CAS RN: 15525-64-1
Properties: light and moisture sensitive [ALD94]

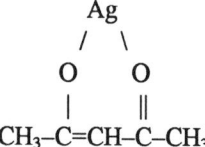

Melting Point, °C: 100, decomposes [ALD94]

2507
Compound: Silver acetylide
Synonyms: silver carbide
Formula: AgC≡CAg
Molecular Formula: C₂Ag₂
Molecular Weight: 239.758
CAS RN: 7659-31-6
Properties: white powd; unstable; prepared by reacting acetylene with silver salts ammoniacal solution; used as a detonator [HAW93] [KIR83]
Reactions: can explode [CRC94]

2508
Compound: Silver azide
Formula: AgN₃

Molecular Formula: AgN₃
Molecular Weight: 149.888
CAS RN: 13863-88-2
Properties: ortho-rhomb, a = 0.559 nm, b = 0.591 nm, c = 0.610 nm; can be prepared by reacting silver nitrate solution with hydrazine or hydrazoic acid; sensitive to shock [KIR83] [CIC73]
Solubility: 0.01 g/100mL H₂O (100°C) [CRC94]
Density, g/cm³: 4.9 [LID94]
Melting Point, °C: ~250 [LID94]
Reactions: violently decomposes when heated [KIR83]

2509
Compound: Silver benzoate
Formula: C₆H₅COOAg
Molecular Formula: C₇H₅AgO₂
Molecular Weight: 228.984
CAS RN: 532-31-0
Properties: powd; sensitive to light [ALD94]
Solubility: s 3 parts cold H₂O [MER89]

2510
Compound: Silver bromate
Formula: AgBrO₃
Molecular Formula: AgBrO₃
Molecular Weight: 235.770
CAS RN: 7783-89-3
Properties: white powd; sensitive to light; decomposes if heated [HAW93]
Solubility: g/100g soln, H₂O: 0.193 (25°C), 1.325 (90°C) [KRU93]
Density, g/cm³: 5.2 [HAW93]

2511
Compound: Silver bromide
Formula: AgBr
Molecular Formula: AgBr
Molecular Weight: 187.772
CAS RN: 7785-23-1
Properties: -20 mesh with 99.999% purity; yellowish powd; darkens when exposed to light; enthalpy of fusion 9.12 kJ/mol; enthalpy of vaporization 198 kJ/mol; used in photographic film, in photochromic glass [HAW93] [MER89] [CER91] [CRC93]
Solubility: 0.135 mg/L H₂O (25°C); i alcohol, most acids; sl s dil ammonia, more soluble in conc ammonia [MER89]
Density, g/cm³: 6.47 [LID94]
Melting Point, °C: 432 [CRC93]

Boiling Point, °C: 1502 [CRC93]

2512
Compound: Silver carbonate
Formula: Ag_2CO_3
Molecular Formula: CAg_2O_3
Molecular Weight: 275.745
CAS RN: 534-16-7
Properties: light yellow powd, when freshly prepared, darkens if permitted to stand; light sensitive; made by adding an alkaline carbonate solution to a conc solution of silver nitrate [KIR83] [MER89]
Solubility: g/1000 g soln, H_2O: 1.15 (25°C) [KRU93]
Density, g/cm³: 6.077 [LID94]
Reactions: ~220 decomposes to Ag_2O and CO_2; at higher temperatures forms Ag [MER89]

2513
Compound: Silver chlorate
Formula: $AgClO_3$
Molecular Formula: $AgClO_3$
Molecular Weight: 191.319
CAS RN: 7783-92-8
Properties: white, tetr cryst; slowly decomposes in presence of light, and darkens; oxidizing agent; used in organic synthesis [MER89]
Solubility: g/100g soln, H_2O: 14.46 (25°C) [KRU93]; g/100g H_2O: 10.4 (10°C), 15.3 (20°C), 26.8 (40°C) [LAN85]; sl s alcohol [MER89]
Density, g/cm³: 4.430 [MER89]
Melting Point, °C: 230 [MER89]
Reactions: decomposes at 270°C to AgCl and evolves O_2 [MER89]

2514
Compound: Silver chloride
Synonyms: ceragyrite
Formula: AgCl
Molecular Formula: AgCl
Molecular Weight: 143.321
CAS RN: 7783-90-6
Properties: white powd; enthalpy of vaporization 199 kJ/mol; enthalpy of fusion 13.20 kJ/mol; sensitive to light, darkens; has several modifications which differ in solubility and in light sensitivity; used in photography, batteries, silver plating [HAW93] [MER89] [CRC93]

Solubility: g/L soln, H_2O: 0.00070 (0°C), 0.00193 (25°C), 0.021 (100°C) [KRU93]; s 250 parts HCl; s in alkali cyanide solutions; i alcohol, dil acids [MER89]
Density, g/cm³: 5.56 [MER89]
Melting Point, °C: 455 [DOU83]
Boiling Point, °C: 1547 [CRC93]

2515
Compound: Silver chlorite
Formula: $AgClO_2$
Molecular Formula: $AgClO_2$
Molecular Weight: 175.320
CAS RN: 7783-91-7
Properties: yellow cryst [CRC77]
Solubility: g/100g soln, H_2O: 0.17 (0°C), 0.44 (18°C), 2.11 (100°C) [KRU93]
Reactions: explodes, 105°C [CRC77]

2516
Compound: Silver chromate
Formula: Ag_2CrO_4
Molecular Formula: Ag_2CrO_4
Molecular Weight: 331.730
CAS RN: 7784-01-2
Properties: dark brownish red to maroon cryst; monocl; formed by reaction of a chromate salt and silver nitrate solution; used as a catalyst to form aldol from alcohol [MER89] [KIR83] [KIR78]
Solubility: g/L soln, H_2O: 0.0142 (0.26°C), 0.028±0.006 (25°C), 0.64 (100°C) [KRU93]; s HNO_3, ammonia [MER89]
Density, g/cm³: 5.625 [KIR78]

2517
Compound: Silver citrate
Synonyms: citric acid, trisilver salt
Formula: $AgOOCCH_2C(OH)(COOAg)CH_2COOAg$
Molecular Formula: $C_6H_5Ag_3O_7$
Molecular Weight: 512.707
CAS RN: 126-45-4
Properties: white heavy cryst powd; light sensitive, causing compound to darken [MER89]
Solubility: s 3500 parts H_2O, more s in boiling H_2O; v s dil HNO_3, ammonia [MER89]

2518
Compound: Silver cyanide
Formula: AgCN
Molecular Formula: CAgN

Molecular Weight: 133.886
CAS RN: 506-64-9
Properties: white or grayish powd; odorless; stable in dry air; darkens in light; can be prepared by mixing stoichiometric amounts of silver nitrate and soluble cyanide; used in silver plating [HAW93] [KIR83] [MER89]
Solubility: 0.000023 g/100mL H_2O (20°C) [CRC94]; s, alcohol, dil acids; s alkali cyanides [MER89]
Density, g/cm³: 3.95 [MER89]
Melting Point, °C: 320 [STR93]
Reactions: evolves HCN from HCl solutions [MER89]

2519
Compound: Silver dichromate
Formula: $Ag_2Cr_2O_7$
Molecular Formula: $Ag_2Cr_2O_7$
Molecular Weight: 431.724
CAS RN: 7784-02-3
Properties: dark red cryst powd [HAW93]
Solubility: 0.0083 g/100mL H_2O (15°C) [CRC94]; s NH4OH, HNO_3 [HAW93]
Density, g/cm³: 4.770 [HAW93]

2520
Compound: Silver diethyldithiocarbamate
Synonyms: diethyldithiocarbamic acid, silver(I) salt
Formula: $(C_2H_5)_2NCS_2Ag$
Molecular Formula: $C_5H_{10}AgNS_2$
Molecular Weight: 256.141
CAS RN: 1470-61-7
Properties: light-sensitive; used as a reagent to detect arsenic [ALD94]
Melting Point, °C: 172-175[ALD94]

2521
Compound: Silver difluoride
Synonyms: Ag(II) fluoride
Formula: AgF_2
Molecular Formula: AgF_2
Molecular Weight: 145.865
CAS RN: 7783-95-1
Properties: white, when pure; usually grayish black or brownish solid; light sensitive; very hygr; reacts violently with H_2O; very strong oxidizing agent, e.g. can evolve ozone from dil acids and oxidizes iodides to iodine; obtained by reaction of F_2 on AgCl at 200°C; fluorinating agent [KIR83] [MER89]
Solubility: decomposed by H_2O [CRC94]

Density, g/cm³: 4.58 [STR93]
Melting Point, °C: 700, decomposes [STR93]

2522
Compound: Silver fluoride
Formula: AgF
Molecular Formula: AgF
Molecular Weight: 126.866
CAS RN: 7775-41-9
Properties: yellow or brownish cryst; cub; very hygr; darkens if exposed to light; forms basic fluoride in moist air; can form several hydrates; used as an antiseptic [HAW93] [MER89]
Solubility: mol/kg H_2O: 6.76 (0°C), 13.97 (25°C), 16.15 (108°C); Solid phase, $AgF \cdot 4H_2O$ (0°C), $AgF \cdot 2H_2O$ (25°C), AgF (108°C) [KRU93]; s HF, NH_3, CH_3CN [MER89]
Density, g/cm³: 5.852 [MER89]
Melting Point, °C: 435 [MER89]
Boiling Point, °C: 1159 [HAW93]

2523
Compound: Silver hexafluoroantimonate(V)
Formula: $AgSbF_6$
Molecular Formula: AgF_6Sb
Molecular Weight: 343.616
CAS RN: 26042-64-8
Properties: -6 mesh with 99.9% purity; white powd; hygr; used as acidic catalyst in epoxide reactions [STR93] [ALD94] [CER91]

2524
Compound: Silver hexafluoroarsenate
Formula: $AgAsF_6$
Molecular Formula: $AgAsF_6$
Molecular Weight: 296.780
CAS RN: 12005-82-2
Properties: hygr powd [STR93]
Melting Point, °C: decomposes [STR93]

2525
Compound: Silver hexafluorophosphate
Formula: $AgPF_6$
Molecular Formula: AgF_6P
Molecular Weight: 252.832
CAS RN: 26042-63-7
Properties: -6 mesh with 99.9% purity; white cryst; hygr; sensitive to light; uses: acidic catalyst to synthesize some sulfides and vinyl fluorides [ALD94] [STR93] [CER91]

2526
Compound: Silver hydrogen fluoride
Formula: $AgHF_2$
Molecular Formula: AgF_2H
Molecular Weight: 146.873
CAS RN: 12249-52-4
Properties: light sensitive; hygr [STR93]
Melting Point, °C: decomposes [STR93]

2527
Compound: Silver iodate
Formula: $AgIO_3$
Molecular Formula: $AgIO_3$
Molecular Weight: 282.770
CAS RN: 7783-97-3
Properties: white, cryst powd; light sensitive [MER89]
Solubility: g/L soln, H_2O: 0.0505 (25°C) [KRU93]; s ~1000 parts 35% HNO_3 (25°C), 2.5 parts 10% ammonia [MER89]
Density, g/cm³: 5.53 [MER89]
Melting Point, °C: >200 [MER89]

2528
Compound: Silver iodide
Synonyms: iodyrite
Formula: AgI
Molecular Formula: AgI
Molecular Weight: 234.772
CAS RN: 7783-96-2
Properties: -20 mesh with 99.999% purity; light yellow powd; darkens when exposed to light; hex or cub cryst; enthalpy of vaporization 143.9 kJ/mol; enthalpy of fusion 9.41 kJ/mol [MER89] [CER91] [CRC93]
Solubility: 0.03 mg/L H_2O; i acids, except HI [MER89]; s KI, KCN, NH_4OH, NaCl solutions [HAW93]
Density, g/cm³: 5.68 [LID94]
Melting Point, °C: 558 [CRC93]
Boiling Point, °C: 1506 [CRC93]
Thermal Expansion Coefficient: -2.5×10^{-6}/K [CRC93]

2529
Compound: Silver lactate monohydrate
Formula: $AgC_3H_5O_3 \cdot H_2O$
Molecular Formula: $C_3H_7AgO_4$
Molecular Weight: 214.955
CAS RN: 128-00-7
Properties: white or sl gray cryst powd; sensitive to light [MER89]

Solubility: s in ~15 parts H_2O; sl s alcohol [MER89]
Melting Point, °C: 212 [CRC94]

2530
Compound: Silver molybdate
Formula: Ag_2MoO_4
Molecular Formula: Ag_2MoO_4
Molecular Weight: 375.674
CAS RN: 13765-74-7
Properties: white, pale yellow (if fused) [KIR81]
Solubility: 3.86 mg/100g soln in H_2O (25°C) [KRU93]
Density, g/cm³: 6.18 [LID94]
Melting Point, °C: 483 [KIR81]

2531
Compound: Silver nitrate
Formula: $AgNO_3$
Molecular Formula: $AgNO_3$
Molecular Weight: 169.873
CAS RN: 7761-88-8
Properties: -10 mesh with 99.999% purity; colorless, transparent rhomb cryst; pure material not light sensitive; enthalpy of fusion 11.50 kJ/mol [CRC93] [MER89] [CER91]
Solubility: g/100g soln, H_2O: 54.8 (0°C), 70.7 (25°C), 88.0 (100°C) [KRU93]; 1g dissolves in: 30mL alcohol, 6.5mL boiling alcohol; 253mL acetone; v s ammonia solution; sl s ether [MER89]
Density, g/cm³: 4.352 [STR93]
Melting Point, °C: 212, forming yellowish liq [MER89]
Boiling Point, °C: decomposes, 440, into Ag, nitrogen, oxygen [MER89]

2532
Compound: Silver nitrite
Formula: $AgNO_2$
Molecular Formula: $AgNO_2$
Molecular Weight: 153.874
CAS RN: 7783-99-5
Properties: pale yellow needles; light sensitive, turning gray; used as a reagent for alcohols, as a standard solution for water analysis [HAW93] [MER89]
Solubility: g/100g soln, H_2O: 0.155 (0°C), 0.4135 (25°C) [KRU93]; i alcohol; decomposed by dil acids [MER89]
Density, g/cm³: 4.453 [STR93]
Melting Point, °C: decomposes, 140 [MER89]

2533
Compound: Silver oxalate
Formula: $Ag_2C_2O_4$
Molecular Formula: $C_2Ag_2O_4$
Molecular Weight: 303.756
CAS RN: 533-51-7
Properties: white, cryst powd [MER89]
Solubility: g/L soln, H_2O: 0.041 (25°C) [KRU93]; s moderately conc HNO_3, ammonia [MER89]
Density, g/cm³: 5.03 [MER89]
Melting Point, °C: decomposes [STR93]
Reactions: can explode [CRC94]

2534
Compound: Silver oxide
Formula: Ag_2O
Molecular Formula: Ag_2O
Molecular Weight: 231.735
CAS RN: 20667-12-3
Properties: brownish black powd; reduced by hydrogen, carbon monoxide, and most metals; light sensitive; used to polish glass, to purify drinking water, as a catalyst [HAW93] [MER89]
Solubility: 0.0013 g/100mL H_2O (20°C), 0.0053 g/100mL H_2O (80°C) [CRC94]; v s dil HNO_3, ammonia; i alcohol [MER89]
Density, g/cm³: 7.2 [LID94]
Melting Point, °C: 300, decomposes [STR93]
Reactions: begins to decompose at ~200°C, rapidly at 250-300°C [MER89]

2535
Compound: Silver perchlorate
Formula: $AgClO_4$
Molecular Formula: $AgClO_4$
Molecular Weight: 207.319
CAS RN: 7783-93-9
Properties: colorless deliq cryst; used in the manufacture of explosives [MER89] [HAW93]
Solubility: g/100g soln, H_2O: 81.7±0.4 (0°C), 84.6±0.1 (25°C), 88.8 (99°C); Solid phase, $AgClO_4 \cdot H_2O$ (0°C, 25°C), $AgClO_4$ (99°C) [KRU93]; s in many organic solvents [MER89]
Density, g/cm³: 2.806 [MER89]
Melting Point, °C: 486, decomposes [MER89]

2536
Compound: Silver perchlorate monohydrate
Formula: $AgClO_4 \cdot H_2O$
Molecular Formula: $AgClH_2O_5$
Molecular Weight: 225.334
CAS RN: 14242-05-8

Properties: white cryst; hygr; stable to 43°C [MER89] [STR93]
Solubility: 84.5g/100g saturated soln of H_2O (25°C) [MER89]

2537
Compound: Silver permanganate
Formula: $AgMnO_4$
Molecular Formula: $AgMnO_4$
Molecular Weight: 226.804
CAS RN: 7783-98-4
Properties: violet; light sensitive; cryst powd; used in gas masks and as an antiseptic [HAW93] [MER89]
Solubility: ~9 g/L H_2O, room temp, more s in hot H_2O [MER89]
Density, g/cm³: 4.49 [MER89]
Melting Point, °C: decomposes [HAW93]

2538
Compound: Silver peroxide
Synonyms: silver(II) oxide
Formula: Ag_2O_2
Molecular Formula: Ag_2O_2
Molecular Weight: 247.735
CAS RN: 1301-96-8
Properties: charcoal gray powd; oxidant; malleable; ortho-rhomb or cub cryst; can be obtained by persulfate oxidation of Ag_2O in alkaline medium at 90°C; strongly oxidizing; used in the manufacture of silver-zinc batteries [HAW93] [MER89] [KIR83]
Solubility: 27 mg/L H_2O, decomposes (25°C); s alkalies, NH_4OH (evolving N_2); s dil acids, evolving O_2 [MER89]
Density, g/cm³: 7.483 [MER89]
Melting Point, °C: decomposes >100 to Ag and O_2 [MER89]

2539
Compound: Silver phosphate
Synonyms: silver phosphate, tribasic
Formula: Ag_3PO_4
Molecular Formula: Ag_3O_4P
Molecular Weight: 418.574
CAS RN: 7784-09-0
Properties: yellow powd; darkened by light; reduced by hydrogen; used in photographic emulsions, in pharmaceuticals [HAW93] [MER89]
Solubility: s 15,550 parts H_2O; sl s dil acids; v s dil HNO_3, ammonia [MER89]
Density, g/cm³: 6.37 [MER89]

Melting Point, °C: 849 [MER89]

2540
Compound: Silver picrate monohydrate
Formula: $AgOC_6H_2(NO_2)_3 \cdot H_2O$
Molecular Formula: $C_6H_4AgN_3O_8$
Molecular Weight: 353.981
CAS RN: 146-84-9
Properties: yellow cryst; used as an antimicrobial agent [HAW93] [MER89]
Solubility: s ~50 parts H_2O; sl s alcohol; i $CHCl_3$, ether [MER89]
Reactions: can explode [HAW93]

2541
Compound: Silver selenate
Formula: Ag_2SeO_4
Molecular Formula: Ag_2O_4Se
Molecular Weight: 358.692
CAS RN: 7784-07-8
Properties: ortho-rhomb cryst; prepared from silver carbonate and sodium selenate [KIR83] [MER89]
Solubility: g/1000g H_2O: 0.870 (25°C), 0.053 (100°C) [KRU93]
Density, g/cm³: 5.72 [MER89]

2542
Compound: Silver selenide
Formula: Ag_2Se
Molecular Formula: Ag_2Se
Molecular Weight: 294.696
CAS RN: 1302-09-6
Properties: gray hex microscopic needles; exists in two forms, transition temp 133°C; oxidized to Ag and selenium oxide when heated in O_2 [MER89]
Solubility: i H_2O [MER89]
Density, g/cm³: 8.216 [MER89]
Melting Point, °C: 880 [MER89]

2543
Compound: Silver selenite
Formula: Ag_2SeO_3
Molecular Formula: Ag_2O_3Se
Molecular Weight: 342.694
CAS RN: 7784-05-6
Properties: needles; decomposes >530°C to Ag, selenium oxide, oxygen [MER89]
Solubility: sl s cold H_2O; v s hot H_2O; s HNO_3 [MER89]

Density, g/cm³: 5.9297 [MER89]
Melting Point, °C: 530, decomposes [MER89]

2544
Compound: Silver subfluoride
Formula: Ag_2F
Molecular Formula: Ag_2F
Molecular Weight: 234.734
CAS RN: 1302-01-8
Properties: hex; bronze colored cryst with green luster; becomes grayish black as a result of prolonged exposure to air; good electrical conductor; quickly hydrolyzes in H_2O precipitating Ag powd; can be prepared by heating a conc solution of AgF with Ag powd [KIR78] [MER89]
Solubility: decomposes in H_2O [KIR78]
Density, g/cm³: 8.57 [MER89]
Melting Point, °C: decomposes to Ag, AgF 100-200 [MER89]

2545
Compound: Silver sulfate
Formula: Ag_2SO_4
Molecular Formula: Ag_2O_4S
Molecular Weight: 311.798
CAS RN: 10294-26-5
Properties: small colorless cryst or cryst powd; light sensitive, slowly darkening; can be made by reaction of metallic Ag and hot H_2SO_4 [KIR84] [MER89]
Solubility: g/100g soln, H_2O: 0.56±0.01 (0°C), 0.834 (25°C), 1.39 (100°C) [KRU93]; s HNO_3, ammonia, conc H_2SO_4 [MER89]
Density, g/cm³: 5.45 [MER89]
Melting Point, °C: 657 [MER89]
Boiling Point, °C: decomposes, 1085 [MER89]

2546
Compound: Silver sulfide
Synonyms: acanthite, argentite
Formula: Ag_2S
Molecular Formula: Ag_2S
Molecular Weight: 247.802
CAS RN: 21548-73-2
Properties: -100 mesh with 99.9% purity; grayish black powd; ortho-rhomb, changes to cub >179°C; enthalpy of fusion 14.10 kJ/mol; used in ceramics [HAW93] [MER89] [CER91]
Solubility: i H_2O; s conc H_2SO_4, HNO_3 [HAW93]
Density, g/cm³: 7.32 [HAW93]
Melting Point, °C: 825 [HAW93]

Boiling Point, °C: decomposes [HAW93]

2547
Compound: Silver telluride
Synonyms: hessite
Formula: Ag_2Te
Molecular Formula: Ag_2Te
Molecular Weight: 343.336
CAS RN: 12002-99-2
Properties: black cryst [STR93]
Density, g/cm³: 8.5 [STR93]
Melting Point, °C: 955 [STR93]

2548
Compound: Silver tetraiodomercurate(II) (α-form)
Synonyms: mercury(II) silver iodide
Formula: Ag_2HgI_4
Molecular Formula: Ag_2HgI_4
Molecular Weight: 923.942
CAS RN: 7784-03-4
Properties: deep yellow; thermochromic powd; becomes blood red at 40-50°C, β-form; can be prepared by precipitation from a solution of $AgNO_3$ and K_2HgI_4 [KIR81] [CRC94] [MER89]
Solubility: i H_2O, dil acids; s in solutions of alkali iodides or cyanides [MER89]
Density, g/cm³: 6.02 [CRC94]
Melting Point, °C: β: decomposes 158 [CRC94]
Reactions: α --→ β transition, 50.7°C [CRC94]

2549
Compound: Silver thiocyanate
Formula: AgSCN
Molecular Formula: CAgNS
Molecular Weight: 165.952
CAS RN: 1701-93-5
Properties: white powd [STR93]
Solubility: g/L soln, H_2O: 0.00018±0.00002 (25°C), 0.0064 (100°C) [KRU93]
Melting Point, °C: decomposes [STR93]

2550
Compound: Silver tungstate
Formula: Ag_2WO_4
Molecular Formula: Ag_2O_4W
Molecular Weight: 463.574
CAS RN: 13465-93-5
Properties: white powd; formula also written as $Ag_8W_4O_{16}$ [ALD94] [STR93]
Solubility: g/L soln, H_2O: 0.235 (25°C) [KRU93]

2551
Compound: Sodium
Synonyms: natrium
Formula: Na
Molecular Formula: Na
Molecular Weight: 22.989768
CAS RN: 7440-23-5
Properties: soft silvery white metal; bcc, a = 0.4282 nm; lustrous, but tarnishes in air; enthalpy of fusion 2.60 kJ/mol; enthalpy of vaporization 97.4 kJ/mol; electrical resistivity (20°C) 4.69 μohm·cm; viscosity at 100°C is 0.680 mPa s; surface tension at 400°C 161 mN/m; decomposes alcohol; burns with a yellow flame; reduces most oxides to elements [MER89] [KIR82] [ALD94]
Solubility: reacts violently with H_2O evolving H_2 and forming NaOH soln; s mercury, liq NH_3 [KIR82] [MER89]
Density, g/cm³: 0.968 at 20°C [MER89]
Melting Point, °C: 97.82 [MER89]
Boiling Point, °C: 881 [MER89]
Thermal Conductivity, W/(m·K): 142 at 25°C [ALD94]
Thermal Expansion Coefficient: 71 x 10^{-6}/K [CRC93]

2552
Compound: Sodium β-aluminum oxide
Formula: $β-Na_2O \cdot 11Al_2O_3$
Molecular Formula: $Al_{22}Na_2O_{34}$
Molecular Weight: 1183.554
CAS RN: 11138-49-1
Properties: hex, a = 0.558 nm, c = 2.245 nm [KIR78]
Density, g/cm³: 3.24 [KIR78]

2553
Compound: Sodium acetate
Synonyms: acetic acid, sodium salt
Formula: CH_3COONa
Molecular Formula: $C_2H_3NaO_2$
Molecular Weight: 82.035
CAS RN: 127-09-3
Properties: white powd; odorless; efflorescent [HAW93]
Solubility: g/100g H_2O: 36.3 (0°C), 50.5 (25°C), 170 (100°C); Solid phase, $CH_3COONa \cdot 3H_2O$ (0°C, 25°C) [KRU93]; s alcohol [MER89]
Density, g/cm³: 1.528 [STR93]
Melting Point, °C: 324 [STR93]

2554

Compound: Sodium acetate trihydrate
Synonyms: acetic acid, sodium salt trihydrate
Formula: NaCH₃COO·3H₂O
Molecular Formula: C₂H₉NaO₅
Molecular Weight: 136.080
CAS RN: 6131-90-4
Properties: transparent cryst or granules; efflorescent in warm air [MER89]
Solubility: 1 g/0.8mL H₂O, 1 g/0.6mL boiling H₂O, 1 g/19mL alcohol [MER89]
Density, g/cm³: 1.45 [MER89]
Melting Point, °C: 58 [MER89]
Reactions: minus 3H₂O 120°C; decomposes at higher temp [MER89]

2555

Compound: Sodium acetylacetonate
Synonyms: 2,4-pentanedione, sodium derivative
Formula: NaCH₃COCH=C(O)CH₃
Molecular Formula: C₅H₇NaO₂
Molecular Weight: 122.100
CAS RN: 15435-71-9
Properties: off-white powd [STR93]

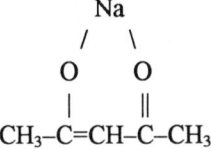

Melting Point, °C: 210 decomposes [STR93]

2556

Compound: Sodium acetylide
Synonyms: sodium carbide
Formula: NaC≡CH
Molecular Formula: C₂HNa
Molecular Weight: 48.020
CAS RN: 1066-26-8
Properties: 18% suspension in xylene with gray color; sensitive to atm oxygen and moisture; freezing point of suspension is 29°C [STR93]

2557

Compound: Sodium aluminate
Formula: NaAlO₂
Molecular Formula: AlNaO₂
Molecular Weight: 81.971
CAS RN: 1302-42-7

Properties: white powd; hygr; can be prepared by fusing sodium carbonate and aluminum acetate in stoichiometric amounts at 800°C; used as a mordant, in water purification, in sizing paper, in cleaning compounds [HAW93] [KIR78]
Solubility: v s H₂O; i alcohol [MER89]
Density, g/cm³: 4.63 [LID94]
Melting Point, °C: 1650 [MER89]

2558

Compound: Sodium aluminum sulfate dodecahydrate
Synonyms: sodium alum
Formula: AlNa(SO₄)₂·12H₂O
Molecular Formula: AlH₂₄NaO₂₀S₂
Molecular Weight: 458.282
CAS RN: 10102-71-3
Properties: colorless cryst [MER83]
Solubility: g anhydrous/100g H₂O: 37.4 (0°C), 39.7 (20°C), 43.8 (40°C) [LAN85]; i alcohol [MER83]
Density, g/cm³: 1.61 [MER83]
Melting Point, °C: ~60 [MER83]

2559

Compound: Sodium amide
Synonyms: sodamide
Formula: NaNH₂
Molecular Formula: H₂NNa
Molecular Weight: 39.013
CAS RN: 7782-92-5
Properties: -40 mesh protected from atm under hexane, with 96% purity; commercial product white to olive green; ortho-rhomb cryst; enthalpy of formation is -118.8 kJ/mol [CIC73] [CER91] [MER89]
Solubility: reacts violently with H₂O, forming NaOH and NH₃ [MER89]; 0.17 g/100g liq NH₃ [CIC73]
Density, g/cm³: 1.39 [CIC73]
Melting Point, °C: 208 [CIC73]
Boiling Point, °C: 400 [HAW93]

2560

Compound: Sodium ammonium hydrogen phosphate tetrahydrate
Synonyms: microcosmic salt
Formula: NaNH₄HPO₄·4H₂O
Molecular Formula: H₁₃NNaO₈P
Molecular Weight: 209.069
CAS RN: 51750-73-3

Properties: odorless; monocl; efflorescent in air, evolving NH_3 [MER89]
Solubility: s about 5 parts cold, 1 part boiling H_2O [MER89]; i alcohol [HAW93]
Density, g/cm³: 1.544 [MER89] [ALD94]
Melting Point, °C: ~80, when rapidly heated [MER89]
Reactions: prolonged heating produces $NaPO_3$ [MER89]

2561
Compound: Sodium antimonate monohydrate
Synonyms: sodium pyroantimonate monohydrate
Formula: $Na_2O \cdot Sb_2O_5 \cdot H_2O$
Molecular Formula: $H_2Na_2O_7Sb_2$
Molecular Weight: 403.512
CAS RN: 33908-66-6
Properties: white, granular powd; formula also given as $NaSb(OH)_6$, -200 mesh with 99.9% purity [MER89] [CER91]
Solubility: sl s H_2O [MER89]

2562
Compound: Sodium arsenate dodecahydrate
Formula: $Na_3AsO_4 \cdot 12H_2O$
Molecular Formula: $AsH_{24}Na_3O_{16}$
Molecular Weight: 424.072
CAS RN: 7778-43-0
Properties: colorless cryst [HAW93]
Solubility: 38.9 g/100mL H_2O (15.5°C) [CRC94]; sl s in alcohol and glycerol; i ether [HAW93]
Density, g/cm³: 1.7539 [HAW93]
Melting Point, °C: 86 [HAW93]

2563
Compound: Sodium arsenite
Synonyms: sodium metaarsenite
Formula: $NaAsO_2$
Molecular Formula: AsO_2Na
Molecular Weight: 129.911
CAS RN: 7784-46-5
Properties: white or grayish white powd; somewhat hygr; absorbs atm CO_2; used in soaps for taxidermists, in insecticides, as a hide preservative [HAW93] [MER89]
Solubility: v s H_2O; sl s alcohol [MER89]
Density, g/cm³: 1.87 [HAW93]

2564
Compound: Sodium azide
Synonyms: smite

Formula: NaN_3
Molecular Formula: N_3Na
Molecular Weight: 65.010
CAS RN: 26628-22-8
Properties: white powd; hex cryst; β-NaN_3: body-center rhomb, with a = 0.5488 nm; decomposes when heated into Na, N_2 [CIC73] [MER89] [STR93]
Solubility: g/100g H_2O: 38.9 (0°C), 40.8 (20°C), 55.3 (100°C) [LAN85]
Density, g/cm³: 1.846 [MER89]
Melting Point, °C: 300, decomposes [STR93]

2565
Compound: Sodium borodeuteride
Formula: $NaBD_4$
Molecular Formula: BD_4Na
Molecular Weight: 41.861
CAS RN: 15681-89-7
Properties: white powd; sensitive to moisture [STR93]
Density, g/cm³: 1.074 [STR93]
Melting Point, °C: ~400 [STR93]

2566
Compound: Sodium borohydride
Formula: $NaBH_4$
Molecular Formula: BH_4Na
Molecular Weight: 37.833
CAS RN: 16940-66-2
Properties: white cub; hygr; stable in dry air up to 300°C, decomposes 400-500°C; strong reducing agent [MER89]
Solubility: w/w H_2O: 55% (25°C), 88.5% (60°C); s liq ammonia, ethylenediamine, other organic solvents [MER89]
Density, g/cm³: 1.074 [MER89]
Melting Point, °C: ~400, decomposes [MER89]

2567
Compound: Sodium bromate
Formula: $NaBrO_3$
Molecular Formula: $BrNaO_3$
Molecular Weight: 150.892
CAS RN: 7789-38-0
Properties: colorless; cub cryst, granules or powd; enthalpy of fusion 28.11 kJ/mol [CRC93] [MER89]
Solubility: g/100g soln, H_2O: 28.29 (25°C), 47.6 (100°C); Solid phase, $NaBrO_3$ [KRU93]
Density, g/cm³: 3.34 [MER89]

Melting Point, °C: 381, decomposes and evolves oxygen [MER89]

2568
Compound: Sodium bromide
Formula: NaBr
Molecular Formula: BrNa
Molecular Weight: 102.894
CAS RN: 7647-15-6
Properties: white cub cryst, a = 0.5977 nm; granules, powd; absorbs moisture from air; enthalpy of fusion 26.11 kJ/mol; preparation: by the addition of stoichometric amount of HBr to NaOH or Na_2CO_3 [CRC93] [KIR82] [MER89]
Solubility: g/100g soln, H_2O: 44.47 (0°C), 48.61 (25°C), 53.8-54.8 (100°C); Solid phase, $NaBr \cdot 2H_2O$ (0°C, 25°C), NaBr (100°C) [KRU93]; s alcohol [HAW93]
Density, g/cm³: 3.203 [STR93]
Melting Point, °C: 747 [CRC94]
Boiling Point, °C: 1390 [STR93]

2569
Compound: Sodium bromide dihydrate
Formula: $NaBr \cdot 2H_2O$
Molecular Formula: BrH_4NaO_2
Molecular Weight: 138.925
CAS RN: 13466-08-5
Properties: white cryst powd [HAW93]
Solubility: 79.5 g/100mL H_2O (0°C), 118.6 g/100mL H_2O (80.5°C) [CRC94]; moderately s in alcohol [HAW93]
Density, g/cm³: 2.176 [HAW93]
Reactions: minus $2H_2O$ at 51°C [CRC94]

2570
Compound: Sodium carbonate
Synonyms: soda ash
Formula: Na_2CO_3
Molecular Formula: CNa_2O_3
Molecular Weight: 105.989
CAS RN: 497-19-8
Properties: -100 mesh with 99.999% purity; hygr powd; gradually absorbs one mole H_2O from air; enthalpy of fusion 29.70 kJ/mol [MER89] [CER91] [CRC93]
Solubility: mol/kg H_2O: 0.66 (0°C), 2.77 (25°C), 4.22 (100°C); Solid phase, $Na_2CO_3 \cdot 10H_2O$ (0°C, 25°C), $Na_2CO_3 \cdot H_2O$ (100°C) [KRU93]; s glycerol; i alcohol [MER89]
Density, g/cm³: 2.54 [LID94]
Melting Point, °C: 858.1 [LID94]

Reactions: minus CO_2 staring at 400°C [MER89]

2571
Compound: Sodium carbonate bicarbonate dihydrate
Synonyms: trona, sodium sesquicarbonate
Formula: $Na_2CO_3 \cdot NaHCO_3 \cdot 2H_2O$
Molecular Formula: $C_2H_5Na_3O_8$
Molecular Weight: 226.026
CAS RN: 497-19-8
Properties: monocl needles; stable in air [MER89]
Solubility: g/100mL H_2O: 13 (0°C), 42 (100°C) [MER89]
Density, g/cm³: 2.112 [MER89]
Melting Point, °C: decomposes [CRC94]

2572
Compound: Sodium carbonate decahydrate
Synonyms: soda, washing soda
Formula: $Na_2CO_3 \cdot 10H_2O$
Molecular Formula: $CH_{20}Na_2O_{13}$
Molecular Weight: 286.142
CAS RN: 6132-02-1
Properties: transparent cryst; effloresces in air [MER89]
Solubility: s 2 parts cold H_2O, 0.25 parts boiling H_2O; glycerol; i alcohol [MER89]
Density, g/cm³: 1.46 [MER89]
Melting Point, °C: 34 [MER89]
Reactions: minus H_2O at 33.5°C [CRC94]

2573
Compound: Sodium carbonate monohydrate
Synonyms: thermonatrite
Formula: $Na_2CO_3 \cdot H_2O$
Molecular Formula: $CH_2Na_2O_4$
Molecular Weight: 124.005
CAS RN: 5968-11-6
Properties: colorless; small cryst or powd; stable under ordinary atm conditions of temp and moisture [MER89]
Solubility: s 3 parts H_2O, 1.8 parts boiling H_2O, 7 parts glycerol; i alcohol [MER89]
Density, g/cm³: 2.25 [MER89]
Melting Point, °C: 109 [HAW93]
Reactions: forms anhydride at 100°C [MER89]

2574
Compound: Sodium carbonate peroxohydrate
Synonyms: sodium percarbonate
Formula: $Na_2CO_3 \cdot 1-1/2H_2O_2$
Molecular Formula: $CH_3Na_2O_6$

Molecular Weight: 157.011

CAS RN: 15630-39-4

Properties: stable, microcryst powd; forms Na_2O_2 and Na_2CO_3 in water; used as a bleaching agent, mild antiseptic and a cleaner for dentures [HAW93]

Solubility: 120 g/kg H_2O (20°C) [HAW93]

2575

Compound: Sodium chlorate

Formula: $NaClO_3$

Molecular Formula: $ClNaO_3$

Molecular Weight: 106.441

CAS RN: 7775-09-9

Properties: cub cryst or white granules; sl hygr; strong oxidizing agent; enthalpy of fusion 22.10 kJ/mol [CRC93] [MER89] [KIR78]

Solubility: g/100g soln, H_2O: 44.3 (0°C), 50.0 (25°C), 66.8 (100°C); Solid phase, $NaClO_3$ [KRU93]; s alcohol [MER89]

Density, g/cm^3: 2.490 [HAW93]

Melting Point, °C: 248-261, decomposes [STR93]

Boiling Point, °C: decomposes 300 [MER89]

Reactions: decomposes at 300°C, with evolution of O_2 [MER89]

2576

Compound: Sodium chloride

Synonyms: halite

Formula: NaCl

Molecular Formula: ClNa

Molecular Weight: 58.443

CAS RN: 7647-14-5

Properties: colorless cub white cryst, granules or powd; hygr; hardness 2.5; enthalpy of fusion 28.16 kJ/mol; enthalpy of solution 3.757 kJ/mol; saturated brine has vapor pressure, kPa: 1.76 (20°C), 2.39 (25°C), 9.26 (50°C), 23.27 (70°C), 52.20 (90°C); bp of saturated brine is 108.7°C; specific gravity of saturated brine is 1.1978 [KIR82] [MER89] [STR93] [CRC93]

Solubility: g/100g soln, H_2O: 35.63 (0°C), 35.92 (25°C), 39.4 (100°C); Solid phase, NaCl [KRU93]; 1 g/10mL glycerol; v sl s alcohol [MER89]; 6.1581±0.0058 mol/(kg·H_2O) at 25°C [RAR85b]

Density, g/cm^3: 2.165 [STR93]

Melting Point, °C: 800.7 [LID94]

Boiling Point, °C: 1465 [LID94]

Thermal Conductivity, W/(m·K): data for aq solutions of NaCl from 20°C to 330°C are found in [OZB80]

Thermal Expansion Coefficient: (volume) 100°C (0.963), 200°C (2.288), 400°C (5.256), 600°C (8.932) [CLA66]

2577

Compound: Sodium chlorite

Formula: $NaClO_2$

Molecular Formula: $ClNaO_2$

Molecular Weight: 90.442

CAS RN: 7758-19-2

Properties: white cryst or powd; sl hygr; strong oxidizing agent; used to improve the taste and odor of potable water, and as a bleaching agent for textiles and wood pulp [HAW93] [MER89]

Solubility: mol/mol soln, H_2O: 6.51 (25°C), 4.1 (60°C); Solid phase, $NaClO_2·3H_2O$ (25°C), $NaClO_2$ (60°C) [KRU93]

Reactions: decomposes at 180-200°C [MER89]

2578

Compound: Sodium chromate

Formula: Na_2CrO_4

Molecular Formula: $CrNa_2O_4$

Molecular Weight: 161.974

CAS RN: 7775-11-3

Properties: yellow cryst; ortho-rhomb [KIR78]

Solubility: g/100 g soln, H_2O: 24.2 (0°C), 45.8 (25°C), 56.1 (100°C); Solid phase, $Na_2CrO_4·10H_2O$ (0°C), $Na_2CrO_4·6H_2O$ (25°C), Na_2CrO_4 (100°C) [KRU93]

Density, g/cm^3: 2.723 [KIR78]

Melting Point, °C: 792 [KIR78]

2579

Compound: Sodium chromate decahydrate

Formula: $Na_2CrO_4·10H_2O$

Molecular Formula: $CrH_{20}Na_2O_{14}$

Molecular Weight: 342.127

CAS RN: 7775-11-3

Properties: yellow, translucent, efflorescent cryst, water content can vary; used in inks, for dyeing, as a paint pigment, wood preservative and corrosion protectant [HAW93] [MER89]

Solubility: s H_2O; sl s in alcohol [HAW93]

Density, g/cm^3: 1.483 [HAW93]

Melting Point, °C: 19.9 [HAW93]

2580

Compound: Sodium chromate tetrahydrate

Formula: $Na_2CrO_4·4H_2O$

Molecular Formula: $CrH_8Na_2O_8$

Molecular Weight: 234.035
CAS RN: 10034-82-9
Properties: yellow; somewhat deliq cryst; used in pigment manufacture, leather tanning and as a corrosion inhibitor [MER89] [HAW93]
Solubility: s ~1 part H_2O; sl s alcohol [MER89]

2581
Compound: Sodium citrate dihydrate
Synonyms: citric acid, trisodium salt dihydrate
Formula:
$NaOOCCH_2C(OH)(COONa)CH_2COONa·2H_2O$
Molecular Formula: $C_6H_9Na_3O_9$
Molecular Weight: 294.101
CAS RN: 6132-04-3
Properties: white cryst, granules or powd; odorless with cool saline taste; used in photography, as a sequestering agent for metals, as an anticoagulant for blood samples, and for emulsifying, acidifying and sequestering food [MER89]
Solubility: g/100mL H_2O: 71 (25°C), 167 (100°C); i alcohol, ether [KIR78]
Reactions: minus $2H_2O$ at 150°C [MER89]

2582
Compound: Sodium citrate pentahydrate
Formula: $Na_3C_6H_5O_7·5H_2O$
Molecular Formula: $C_6H_{15}Na_3O_{12}$
Molecular Weight: 348.147
CAS RN: 6858-44-2
Properties: large, colorless cryst or white granules; effloresces in air [MER89] [KIR78]
Solubility: g/100mL H_2O: 92.6 (25°C), 250 (100°C); sl s alcohol, i ether [KIR78]
Density, g/cm³: 1.857 [CRC94]
Reactions: minus $5H_2O$ at 150°C [CRC94]

2583
Compound: Sodium copper chromate trihydrate
Formula: $Na_2O·4CuO·4CrO_3·3H_2O$
Molecular Formula: $Cr_4Cu_4H_6Na_2O_{20}$
Molecular Weight: 834.184
CAS RN: 68399-60-0
Properties: maroon triclinic; used as an antifouling pigment [KIR78]
Solubility: v sl s H_2O [KIR78]
Density, g/cm³: 3.57 [KIR78]

2584
Compound: Sodium cyanate

Formula: NaOCN
Molecular Formula: CNNaO
Molecular Weight: 65.007
CAS RN: 917-61-3
Properties: colorless needles when prepared from alcohol medium; decomposes in H_2O to urea and Na_2CO_3 [MER89]
Solubility: 0.22 g/100g alcohol (0°C); i ether [MER89]
Density, g/cm³: 1.893 [MER89]
Melting Point, °C: 550 [MER89]

2585
Compound: Sodium cyanide
Synonyms: cyanogran
Formula: NaCN
Molecular Formula: CNNa
Molecular Weight: 49.008
CAS RN: 143-33-9
Properties: white granules or fused pieces; somewhat deliq; enthalpy of fusion 15.4 kJ/mol; enthalpy of vaporization 156.3 kJ/mol; enthalpy of solution 1510 J/mol; vapor pressure, kPa: 0.1013 (800°C), 1.652 (1000°C), 11.9 (1200°C), 41.8 (1360°C); used in electroplating of zinc, copper, brass gold and other metals; forms a dihydrate [KIR78] [MER89]
Solubility: g/100g H_2O: 40.8 (0°C), 58.7 (20°C), 71.2 (30°C) [LAN85]; sl s alcohol [MER89]
Density, g/cm³: cub: 1.60, ortho-rhomb: 1.62-1.624 [KIR78]
Melting Point, °C: 563.7 [STR93]
Boiling Point, °C: 1496 [STR93]

2586
Compound: Sodium cyanoborohydride
Synonyms: sodium cyanotrihydridoborate
Formula: $NaBH_3(CN)$
Molecular Formula: CH_3BNNa
Molecular Weight: 62.843
CAS RN: 25895-60-7
Properties: white powd; hygr; mild reducing agent, e.g. reduces aldehydes, ketones, oximes [MER89] [ALD94]
Solubility: 212 g/100g H_2O (29°C); v s methanol; sl s ethanol [MER89]
Density, g/cm³: 1.199 [MER89]
Melting Point, °C: 240-242 [MER89]

2587
Compound: Sodium deuteride
Formula: NaD

Molecular Formula: DNa
Molecular Weight: 25.005
CAS RN: 15780-28-6
Properties: slurry of gray powd; sensitive to moisture; 20% in oil, 98% isotopic purity [STR93]

2588

Compound: Sodium diacetate
Synonyms: sodium acid acetate
Formula: $CH_3COOH \cdot CH_3COONa$
Molecular Formula: $C_4H_7NaO_4$
Molecular Weight: 142.089
CAS RN: 126-96-5
Properties: white powd; can be used as a source of acetic acid; releases 42.25% of available CH_3COOH; used as a buffer, as a mold inhibitor, food preservative [HAW93] [MER89]
Solubility: s H_2O [MER89]; sl s alcohol; i ether [HAW93]
Melting Point, °C: decomposes >150 [MER89]

2589

Compound: Sodium dichromate dihydrate
Formula: $Na_2Cr_2O_7 \cdot 2H_2O$
Molecular Formula: $Cr_2H_4Na_2O_9$
Molecular Weight: 297.999
CAS RN: 7782-12-0
Properties: reddish orange cryst; monocl; somewhat deliq [KIR78] [MER89] [STR93]
Solubility: g/100g soln in H_2O: 62.17 (0°C), 65.01 (25°C), 80.6 (100°C); Solid phase, $Na_2Cr_2O_7 \cdot 2H_2O$ (0°C, 25°C), $Na_2Cr_2O_7 \cdot 6H_2O$ (100°C) [KRU93]
Density, g/cm³: 2.348 [KIR78]
Melting Point, °C: 84.6 (incongruent) [KIR78]
Reactions: minus $2H_2O$ ~100°C [MER89]

2590

Compound: Sodium dihydrogen phosphate dihydrate
Formula: $NaH_2PO_4 \cdot 2H_2O$
Molecular Formula: H_6NaO_6P
Molecular Weight: 156.008
CAS RN: 7558-80-7
Properties: ortho-rhomb, colorless cryst [MER89]
Solubility: g/100g soln in H_2O: 36.5 (0°C), 48.5 (25°C), 71.0 (100°C); Solid phase, $NaH_2PO_4 \cdot 2H_2O$ (0°C, 25°C), NaH_2PO_4 (100°C) [KRU93]
Density, g/cm³: 1.915 [MER89]
Melting Point, °C: 60 [MER89]

2591

Compound: Sodium dihydrogen phosphate monohydrate
Synonyms: sodium monobasic phosphate
Formula: $NaH_2PO_4 \cdot H_2O$
Molecular Formula: H_4NaO_5P
Molecular Weight: 137.993
CAS RN: 10049-21-5
Properties: white; sl deliq cryst or granules [MER89] [STR93]
Solubility: 59.9g/100mL H_2O (0°C), 427 g/100mL H_2O (100°C) [CRC94]; i alcohol [MER89]
Density, g/cm³: 2.04 [CRC94]
Melting Point, °C: decomposes, 204 [CRC94]
Reactions: minus H_2O at 100°C; forms metaphosphate when ignited [MER89]

2592

Compound: Sodium dihydrogen pyrophosphate
Synonyms: sodium acid pyrophosphate
Formula: $Na_2H_2P_2O_7$
Molecular Formula: $H_2Na_2O_7P_2$
Molecular Weight: 221.939
CAS RN: 7758-16-9
Properties: white; fused masses or powd; used in baking powd [MER89]
Solubility: g/100g H_2O: 4.47 (0°C), 12.0 (20°C), 18.4 (40°C) [LAN85]
Density, g/cm³: ~1.9 [LID94]
Melting Point, °C: decomposes 220 [MER89]

2593

Compound: Sodium dithionate
Synonyms: sodium hyposulfate
Formula: $Na_2(SO_3)_2$
Molecular Formula: $Na_2O_6S_2$
Molecular Weight: 206.108
CAS RN: 7631-94-9
Properties: prepared by reacting MnO_2 with SO_2 to form the product MnS_2O_6, then reacting this product with Na_2CO_3 [MER89]
Solubility: g/100g H_2O: 7.83 (0°C), 17.38 (25°C), 64.74 (100°C); Solid phase, $Na_2S_2O_6 \cdot 2H_2O$ [KRU93]

2594

Compound: Sodium dithionate dihydrate
Formula: $Na_2S_2O_6 \cdot 2H_2O$
Molecular Formula: $H_4Na_2O_8S_2$
Molecular Weight: 242.139
CAS RN: 7631-94-9

Properties: colorless; ortho-rhomb cryst; stable in air [MER89]
Solubility: H₂O, w/w: 6.05% (0°C), 13.39% (20°C), 17.32 (30°C); i alcohol [MER89]
Density, g/cm³: 2.189 [MER89]
Reactions: minus 2H₂O at 110°C; decomposes to Na₂SO₄ + SO₂ at 267°C [MER89]

2595
Compound: Sodium ferricyanide monohydrate
Formula: Na₃Fe(CN)₆·H₂O
Molecular Formula: C₆H₂FeN₆Na₃O
Molecular Weight: 298.935
CAS RN: 14217-21-1
Properties: ruby red deliq cryst; used in the production of pigments, in dyeing and printing [HAW93] [MER89]
Solubility: s 5.5 parts H₂O, 1.5 parts boiling H₂O [MER89]; i alcohol [HAW93]

2596
Compound: Sodium ferrocyanide decahydrate
Synonyms: yellow prussiate of soda
Formula: Na₄Fe(CN)₆·10H₂O
Molecular Formula: C₆H₂₀FeN₆Na₄O₁₀
Molecular Weight: 484.063
CAS RN: 13601-19-9
Properties: pale yellow monocl cryst; somewhat efflorescent; steadily dehydrated >50°C [MER89]
Solubility: H₂O: 10.2% (1°C), 14.7% (17°C), 17.6% (25°C), 28.1% (53°C), 39% (85°C), 39.7% (96.6°C); i most organic solvents [MER89]
Density, g/cm³: 1.458 [HAW93]
Melting Point, °C: decomposes to NaCN, Fe, C, N₂ at 435 [MER89]
Reactions: minus 10H₂O at 81.5°C [MER89]

2597
Compound: Sodium fluoride
Synonyms: villaumite
Formula: NaF
Molecular Formula: FNa
Molecular Weight: 41.988
CAS RN: 7681-49-4

Properties: clear lustrous white powd, or 99.9% pure melted pieces of 3-6 mm; cub or tetr cryst; enthalpy of fusion 33.35 kJ/mol; can be prepared by reacting hydrofluoric acid with soda ash or NaOH; used to fluoridate municipal drinking water, in toothpastes, for cryolite manufacture, as windows in ultraviolet and infrared detectors, in the form of melted pieces used as evaporation material for low index films, and reflection diminishing coatings [MER89] [CER91] [CRC93]
Solubility: g/100g soln, H₂O: 3.53 (0°C), 3.98 (25°C), 4.83 (100°C); Solid phase, NaF (0°C, 25°C) [KRU93]; i alcohol [MER89]
Density, g/cm³: 2.78 [LID94]
Melting Point, °C: 996 [CRC93]
Boiling Point, °C: 1704 [MER89]

2598
Compound: Sodium fluoroborate
Synonyms: sodium tetrafluoroborate
Formula: NaBF₄
Molecular Formula: BF₄Na
Molecular Weight: 109.795
CAS RN: 13755-29-8
Properties: white powd; ortho-rhomb below 240°C, a = 0.68358 nm, b = 0.62619 nm, c = 0.67916 nm; slowly decomposed by heat; can be prepared by reacting NaOH or Na₂CO₃ with fluoroboric acid; used in sand casting of aluminum and magnesium [HAW93] [MER89] [KIR78]
Solubility: g/100mL H₂O: 108 (26°C), 210 (100°C) [MER89]; s alcohol [KIR78]
Density, g/cm³: 2.47 [MER89]
Melting Point, °C: 384 [MER89]

2599
Compound: Sodium fluorophosphate
Formula: Na₂PO₃F
Molecular Formula: FNa₂O₃P
Molecular Weight: 143.939
CAS RN: 10163-15-2
Properties: white powd; used in the preparation of bactericides and fungicides [HAW93] [STR93]
Solubility: s H₂O [HAW93]
Melting Point, °C: ~625 [STR93]

2600
Compound: Sodium fluorosulfonate
Formula: NaSO₃F
Molecular Formula: FNaO₃S
Molecular Weight: 122.052
CAS RN: 14483-63-7

Properties: shiny leaflets; hygr [KIR78]
Solubility: s H_2O, alcohol, acetone; i ether [KIR78]
Melting Point, °C: decomposes [CRC94]

2601
Compound: Sodium formaldehyde sulfoxylate
Formula: $NaHSO_2 \cdot CH_2O \cdot 2H_2O$
Molecular Formula: CH_7NaO_5S
Molecular Weight: 154.120
CAS RN: 149-44-0
Properties: white solid; used as stripping and discharge agent for textiles [HAW93]
Solubility: s H_2O, alcohol [HAW93]
Melting Point, °C: 64 [HAW93]

2602
Compound: Sodium gold cyanide
Synonyms: gold sodium cyanide
Formula: $NaAu(CN)_2$
Molecular Formula: C_2AuN_2Na
Molecular Weight: 271.992
CAS RN: 15280-09-8
Properties: white cryst yellow powd; used for gold plating electronic components [HAW93] [MER89]
Solubility: s H_2O [MER89]

2603
Compound: Sodium gold thiosulfate dihydrate
Synonyms: gold sodium thiosulfate
Formula: $Na_3Au(S_2O_3)_2 \cdot 2H_2O$
Molecular Formula: $AuH_4Na_3O_8S_4$
Molecular Weight: 526.227
CAS RN: 10233-88-2
Properties: white; glistening cryst; needle like or prismatic; darkens slowly when exposed to light; aq solution decomposes on standing, and turns yellow [MER89]
Solubility: 1 gm/2mL H_2O; i alcohol [MER89]
Density, g/cm³: 3.09 [MER89]
Reactions: minus $2H_2O$ at 150-160°C [MER89]

2604
Compound: Sodium hexachloroiridate(III) hydrate
Formula: $Na_3IrCl_6 \cdot xH_2O$
Molecular Formula: Cl_6IrNa_3 (anhydrous)
Molecular Weight: 473.905 (anhydrous)
CAS RN: 123334-23-6
Properties: greenish brown cryst; hygr [STR93] [ALD94]

2605
Compound: Sodium hexachloroiridate(IV) hexahydrate
Formula: $Na_2IrCl_6 \cdot 6H_2O$
Molecular Formula: $Cl_6H_{12}IrNa_2O_6$
Molecular Weight: 559.004
CAS RN: 19567-78-3
Properties: reddish black powd; hygr [STR93]
Solubility: g/100g H_2O: 34.46 (15°C), 56.17 (30°C), 279.3 (80°C) [LAN85]
Melting Point, °C: 600, decomposes [STR93]

2606
Compound: Sodium hexachloroosmiate(IV) hydrate
Formula: $Na_2OsCl_6 \cdot xH_2O$
Molecular Formula: Cl_6Na_2Os (anhydrous)
Molecular Weight: 448.926 (anhydrous)
CAS RN: 1307-81-9
Properties: reddish orange powd; hygr; x<2 [STR93] [ALD94]

2607
Compound: Sodium hexachloropalladate(IV)
Formula: Na_2PdCl_6
Molecular Formula: Cl_6Na_2Pd
Molecular Weight: 365.116
CAS RN: 53823-60-2
Properties: reddish orange cryst; hygr [STR93]

2608
Compound: Sodium hexachloroplatinate(IV)
Formula: Na_2PtCl_6
Molecular Formula: Cl_6Na_2Pt
Molecular Weight: 453.776
CAS RN: 1307-82-0
Properties: yellow cryst; hygr; readily forms hexahydrate in moist air at 25°C when relative humidity >50% [MER89]
Solubility: s H_2O, alcohol [MER89]

2609
Compound: Sodium hexachloroplatinate(IV) hexahydrate
Formula: $Na_2PtCl_6 \cdot 6H_2O$
Molecular Formula: $Cl_6H_{12}Na_2O_6Pt$
Molecular Weight: 561.867
CAS RN: 19583-77-8
Properties: orange powd; hygr [STR93]

2610
Compound: Sodium hexachlororhodate(III) hydrate
Formula: $Na_3RhCl_6 \cdot xH_2O$
Molecular Formula: Cl_6Na_3Rh (anhydrous)
Molecular Weight: 384.591 (anhydrous)
CAS RN: 14972-70-4
Properties: red cryst; hygr [AES93] [STR93]
Melting Point, °C: 900 decomposes [AES93]

2611
Compound: Sodium hexafluoroaluminate
Synonyms: cryolite
Formula: Na_3AlF_6
Molecular Formula: AlF_6Na_3
Molecular Weight: 209.941
CAS RN: 13775-53-6
Properties: white powd or melted pieces; liq vapor pressure 253 Pa (1012°C); monocl, a = 0.546 nm, b = 0.561 nm, c = 0.780 nm; enthalpy of vaporization 225 kJ/mol; hardness 2.5 Mohs; electrical conductivity of liq (1012°C) 2.82 (ohm·cm)$^{-1}$; viscosity of liq 6.7 mPa s (1012°C); enthalpy of transition 8.21 kJ/mol (565°C); used in the molten form to dissolve Al_2O_3 in the manufacture of aluminum [KIR78] [CIC73] [STR93] [CER91]
Solubility: g/100g H_2O: 0.042 (25°C), 0.135 (100°C) [CIC73]
Density, g/cm^3: monocl: 2.97; cub (560°C): 2.77; liq: 2.087 [KIR78]
Melting Point, °C: 1012 [KIR78]
Reactions: transition monocl to cub at 565°C [KIR78]

2612
Compound: Sodium hexafluoroantimonate(V)
Formula: $NaSbF_6$
Molecular Formula: F_6NaSb
Molecular Weight: 258.740
CAS RN: 16925-25-0
Properties: white powd [STR93]
Density, g/cm^3: 3.375 [STR93]

2613
Compound: Sodium hexafluoroarsenate
Formula: $NaAsF_6$
Molecular Formula: AsF_6Na
Molecular Weight: 211.902
CAS RN: 12005-86-6
Properties: white powd [STR93]

2614
Compound: Sodium hexafluoroferrate(III)
Formula: Na_3FeF_6
Molecular Formula: F_6FeNa_3
Molecular Weight: 238.806
CAS RN: 20955-11-7
Properties: off-white to green [STR93]

2615
Compound: Sodium hexafluorogermanate
Formula: Na_2GeF_6
Molecular Formula: F_6GeNa_2
Molecular Weight: 232.580
CAS RN: 36470-39-0
Properties: white powd [STR93]
Solubility: g/100g H_2O: 1.52 (0°C), 2.25 (30°C), 3.36 (80°C) [LAN85]

2616
Compound: Sodium hexafluorophosphate
Formula: $NaPF_6$
Molecular Formula: F_6NaP
Molecular Weight: 167.954
CAS RN: 21324-39-0
Properties: white cryst; hygr [STR93]
Density, g/cm^3: 2.369 [ALD94]

2617
Compound: Sodium hexafluorosilicate
Formula: Na_2SiF_6
Molecular Formula: F_6Na_2Si
Molecular Weight: 188.056
CAS RN: 16893-85-9
Properties: white granular powd; odorless, tasteless, free flowing; prepared from H_2SiF_6 and Na_2CO_3 or NaCl; used in fluoridation, in laundry soaps and for mothproofing woolens [HAW93] [MER89]
Solubility: g/100g H_2O: 4.35 (0°C), 7.2 (20°C), 24.5 (100°C) [LAN85]; i alcohol [MER89]
Density, g/cm^3: 2.679 [STR93]
Melting Point, °C: melts at red heat, decomposing [MER89]

2618
Compound: Sodium hexafluorostannate(IV)
Formula: Na_2SnF_6
Molecular Formula: F_6Na_2Sn
Molecular Weight: 278.680
CAS RN: 16924-51-9
Properties: white powd [STR93]

2619
Compound: Sodium hexafluorotitanate
Formula: Na_2TiF_6
Molecular Formula: F_6Na_2Ti
Molecular Weight: 207.837
CAS RN: 17116-13-1
Properties: white powd [STR93]

2620
Compound: Sodium hexafluorozirconate
Formula: Na_2ZrF_6
Molecular Formula: F_6Na_2Zr
Molecular Weight: 251.194
CAS RN: 16925-26-1
Properties: white powd [STR93]

2621
Compound: Sodium hexametaphosphate
Synonyms: Graham's salt
Formula: $(NaPO_3)_6$
Molecular Formula: $Na_6O_{18}P_6$
Molecular Weight: 611.771
CAS RN: 10124-56-8
Properties: white powd [STR93]

2622
Compound: Sodium hexanitritocobalt(III)
Synonyms: sodium cobaltinitrite
Formula: $Na_3Co(NO_2)_6$
Molecular Formula: $CoN_6Na_3O_{12}$
Molecular Weight: 403.935
CAS RN: 14649-73-1
Properties: yellow to brownish yellow; cryst powd; decomposed by mineral acids [MER89]
Solubility: v s H_2O; sl s alcohol [MER89]

2623
Compound: Sodium hydride
Formula: NaH
Molecular Formula: HNa
Molecular Weight: 23.998
CAS RN: 7646-69-7
Properties: silvery needles; commercial material is grayish white powd; reacts explosively with H_2O, vigorously with lower alcohols; ignites when standing in moist air; manufactured by reaction of H_2 with molten sodium metal which has been dispersed in mineral oil [KIR80] [MER89]
Solubility: s molten NaOH; i liq NH_3 [MER89]
Density, g/cm³: 1.396 [MER89]

Melting Point, °C: 425 with decomposition [MER89]

2624
Compound: Sodium hydrogen arsenate
Formula: Na_2HAsO_4
Molecular Formula: $AsHNa_2O_4$
Molecular Weight: 185.908
CAS RN: 7778-43-0
Properties: powd [MER89]
Solubility: g/100g soln, H_2O: 5.59 (0.1°C), 29.33 (25°C), 66.5 (98.5°C); Solid phase, $Na_2HAsO_4 \cdot 12H_2O$ (0.1°C), $Na_2HAsO_4 \cdot 7H_2O$ (25°C), Na_2HAsO_4 (98.5°C) [KRU93]; sl s alcohol [MER89]

2625
Compound: Sodium hydrogen arsenate heptahydrate
Formula: $Na_2HAsO_4 \cdot 7H_2O$
Molecular Formula: $AsH_{15}Na_2O_{11}$
Molecular Weight: 312.014
CAS RN: 10048-95-0
Properties: odorless cryst; effloresces in warm air; has been used as antimalarial dermatologic [MER89]
Solubility: s in 1.3 parts H_2O; s glycerol; sl s alcohol [MER89]
Density, g/cm³: 1.87 [MER89]
Melting Point, °C: 57 [MER89]
Reactions: minus $5H_2O$ ~50°C, becomes anhydrous at 100°C [MER89]

2626
Compound: Sodium hydrogen carbonate
Synonyms: sodium bicarbonate
Formula: $NaHCO_3$
Molecular Formula: $CHNaO_3$
Molecular Weight: 84.007
CAS RN: 144-55-8
Properties: white cryst powd or granules; readily decomposed by weak acids [MER89]
Solubility: g/100g soln, H_2O: 6.48 (0°C), 9.32 (25°C), 19.1 (100°C); Solid phase, $NaHCO_3$ [KRU93]; i alcohol [MER89]
Density, g/cm³: 2.159 [ALD94]
Reactions: minus CO_2 ~50°C, forms Na_2CO_3 at 100°C [MER89]

2627
Compound: Sodium hydrogen fluoride
Synonyms: sodium bifluoride

Formula: $NaHF_2$
Molecular Formula: F_2HNa
Molecular Weight: 61.995
CAS RN: 1333-83-1
Properties: white cryst powd [MER89]
Solubility: g/100g soln in H_2O: 3.7 (20°C), 16.4 (80°C) [KIR78]
Density, g/cm³: 2.08 [LID94]
Melting Point, °C: decomposes above 160 [KIR78]

2628

Compound: Sodium hydrogen oxalate monohydrate
Formula: $NaHC_2O_4 \cdot H_2O$
Molecular Formula: $C_2H_3NaO_5$
Molecular Weight: 130.033
CAS RN: 1186-49-8
Properties: white powd [STR93]

2629

Compound: Sodium hydrogen phosphate
Formula: Na_2HPO_4
Molecular Formula: HNa_2O_4P
Molecular Weight: 141.959
CAS RN: 7558-79-4
Properties: hygr powd; absorbs from 2 to 7 moles H_2O from atm, depending on humidity and temp [MER89]
Solubility: g/100g H_2O: 1.6 (0°C), 11.8 (25°C), 103.3 (100°C); Solid phase, α-$Na_2HPO_4 \cdot 12H_2O$ (0°C, 25°C), Na_2HPO_4 (100°) [KRU93]
Density, g/cm³: 1.679 [STR93]

2630

Compound: Sodium hydrogen phosphate dodecahydrate
Formula: $Na_2HPO_4 \cdot 12H_2O$
Molecular Formula: $H_{25}Na_2O_{16}P$
Molecular Weight: 358.143
CAS RN: 10039-32-4
Properties: translucent cryst or granules; minus $5H_2O$ at ambient atm conditions [MER89]
Solubility: s 3 parts H_2O; i alcohol [MER89]
Density, g/cm³: ~1.5 [MER89]
Melting Point, °C: 34-35 [MER89]

2631

Compound: Sodium hydrogen phosphate heptahydrate
Formula: $Na_2HPO_4 \cdot 7H_2O$
Molecular Formula: $H_{15}Na_2O_{11}P$
Molecular Weight: 268.066

CAS RN: 7782-85-6
Properties: cryst or granular powd; stable in air [MER89]
Solubility: s 4 parts H_2O, more s in boiling H_2O; i alcohol [MER89]
Density, g/cm³: ~1.7 [MER89]

2632

Compound: Sodium hydrogen phosphite pentahydrate
Formula: $Na_2HPO_3 \cdot 5H_2O$
Molecular Formula: $H_{11}Na_2O_8P$
Molecular Weight: 216.036
CAS RN: 13708-85-5
Properties: white; hygr; cryst powd; used as an antidote to mercuric chloride poisoning [HAW93] [MER89]
Solubility: g anhydrous/100g H_2O: 418 (0°C), 429 (20°C), 566 (30°C) [LAN85]; i alcohol [HAW93]
Melting Point, °C: 53 [HAW93]
Boiling Point, °C: 200-250, decomposes [HAW93]

2633

Compound: Sodium hydrogen sulfate
Synonyms: niter cake
Formula: $NaHSO_4$
Molecular Formula: $HNaO_4S$
Molecular Weight: 120.062
CAS RN: 7681-38-1
Properties: fused hygr; tricl [KIR82] [MER89]
Solubility: s in 2 parts H_2O, 1 part boiling H_2O; decomposes in alcohol to sodium sulfate and H_2SO_4 [MER89]
Density, g/cm³: 2.435 [MER89]
Melting Point, °C: ~315 [MER89]

2634

Compound: Sodium hydrogen sulfate monohydrate
Synonyms: sodium bisulfate monohydrate
Formula: $NaHSO_4 \cdot H_2O$
Molecular Formula: H_3NaO_5S
Molecular Weight: 138.077
CAS RN: 10034-88-5
Properties: cryst; forms pyrosulfate if heated strongly [MER89]
Solubility: s ~0.8 parts H_2O; decomposed by alcohol [MER89]
Density, g/cm³: 2.10 [LID94]

2635
Compound: Sodium hydrogen sulfide
Synonyms: sodium hydrosulfide
Formula: NaHS
Molecular Formula: HNaS
Molecular Weight: 56.064
CAS RN: 16721-80-5
Properties: white to colorless; hydrogen sulfide odor; enthalpy of solution is 15.9 kJ/mol; cub cryst; very hygr; hydrolyzed in moist air to Na_2S and NaOH; color changed by heating in dry air: yellow, then orange as the temp increases [MER89] [KIR83]
Solubility: s H_2O, alcohol, ether [MER89]
Density, g/cm³: 1.79 [MER89]
Melting Point, °C: 350, becomes black liq [MER89]

2636
Compound: Sodium hydrogen sulfide dihydrate
Synonyms: sodium bisulfide
Formula: NaHS·2H₂O
Molecular Formula: H_5NaO_2S
Molecular Weight: 92.095
CAS RN: 16721-80-5
Properties: lemon colored needle or flake forms; used in paper pulping and dyestuff processing [MER89] [HAW93]
Solubility: v s H_2O, alcohol, ether [MER89]
Melting Point, °C: 55 [MER89]

2637
Compound: Sodium hydrogen sulfide trihydrate
Formula: NaHS·3H₂O
Molecular Formula: H_7NaO_3S
Molecular Weight: 110.110
CAS RN: 16721-80-5
Properties: shiny; rhomb [MER89]
Melting Point, °C: 22 [MER89]

2638
Compound: Sodium hydrogen sulfite
Synonyms: sodium bisulfite
Formula: $NaHSO_3$
Molecular Formula: $HNaO_3S$
Molecular Weight: 104.062
CAS RN: 7631-90-5
Properties: white; cryst powd; gradually oxidized to sulfate in air; has SO_2 odor [MER89]
Solubility: s 3.5 parts cold H_2O, 2 parts boiling water, in ~70 parts alcohol [MER89]
Density, g/cm³: 1.48 [MER89]

2639
Compound: Sodium hydrogen tartrate monohydrate
Formula: NaHC₄H₄O₆·H₂O
Molecular Formula: $C_4H_7NaO_7$
Molecular Weight: 190.086
CAS RN: 868-18-8
Properties: white cryst [MER89]
Solubility: s in ~9 parts H_2O, 2 parts boiling H_2O; i alcohol [MER89]

2640
Compound: Sodium hydrosulfite
Synonyms: sodium dithionite
Formula: $Na_2S_2O_4$
Molecular Formula: $Na_2O_4S_2$
Molecular Weight: 174.110
CAS RN: 7775-14-6
Properties: white or grayish white; cryst powd; oxidizes in air to bisulfite and bisulfate, more quickly oxidized in moist atm [MER89]
Solubility: v s H_2O; sl s alcohol [MER89]
Melting Point, °C: 55, decomposes [HAW93]

2641
Compound: Sodium hydroxide
Synonyms: caustic soda; soda lye
Formula: NaOH
Molecular Formula: HNaO
Molecular Weight: 39.997
CAS RN: 1310-73-2
Properties: white, deliq solid; fused solid; absorbs both CO_2 and H_2O from the atm; enthalpy of vaporization 175 kJ/mol; enthalpy of fusion 6.60 kJ/mol [CRC93] [HAW93] [MER89]
Solubility: g/100g soln, H_2O: 29.6 (0°C), 53.3 (25°C), 77.6 (100°C) [KRU93]; 1 g dissolves in: 7.2mL absolute alcohol, 4.2mL methanol; s glycerol [MER89]
Density, g/cm³: 2.13 [MER89]
Melting Point, °C: 323 [CRC93]
Boiling Point, °C: 1390 [STR93]

2642
Compound: Sodium hydroxide monohydrate
Formula: NaOH·H₂O
Molecular Formula: H_3NaO_2
Molecular Weight: 58.013
CAS RN: 12179-02-1
Properties: white powd; hygr [STR93]

2643

Compound: Sodium hypochlorite

Formula: NaClO

Molecular Formula: ClNaO

Molecular Weight: 74.442

CAS RN: 7681-52-9

Properties: very explosive; used to bleach wood pulp and textiles, is a disinfectant for municipal water and sewage, prevents the formation of fungi in oil production [KIR78] [MER89]

Solubility: g/100g soln, H_2O: 22.7 (0°C), 44.0 (24.5°C); Solid phase, NaClO·5H_2O [KRU93]

Density, g/cm³: 1.097 [ALD94]

2644

Compound: Sodium hypochlorite pentahydrate

Formula: NaOCl·5H_2O

Molecular Formula: $ClH_{10}NaO_6$

Molecular Weight: 164.518

CAS RN: 7681-52-9

Properties: pale greenish color; cryst; very unstable; decomposed by atm CO_2; used to bleach paper pulp, textiles [HAW93] [MER89]

Solubility: 29.3 g/100mL H_2O (0°C) [MER89], 94.2 g/100mL H_2O (23°C) [CRC94]

Density, g/cm³: 1.6 [LID94]

Melting Point, °C: 18 [MER89]

2645

Compound: Sodium hypophosphate decahydrate

Formula: $Na_4P_2O_6$·10H_2O

Molecular Formula: $H_{20}Na_4O_{16}P_2$

Molecular Weight: 430.056

CAS RN: 13721-43-2

Properties: colorless or white cryst [MER89]

Solubility: 1.49 g/100mL H_2O (25°C), 5.46 g/100 mL H_2O (60°C) [CRC94]

Density, g/cm³: 1.823 [CRC94]

Melting Point, °C: decomposes [CRC94]

2646

Compound: Sodium hypophosphite monohydrate

Formula: NaH_2PO_2·H_2O

Molecular Formula: H_4NaO_3P

Molecular Weight: 105.994

CAS RN: 123333-67-5

Properties: white, odorless; deliq; evolves flammable phosphine when heated strongly; strong reducing agent, e.g. explodes when triturated with chlorates or other oxidizing agents; used as reducing agent in the electroless plating of nickel onto plastics [MER89]

Solubility: 100g/100g H_2O [KRU93]; s 1 part H_2O, 0.15 parts boiling H_2O; s cold alcohol; i ether [MER89]

Melting Point, °C: decomposes [STR93]

2647

Compound: Sodium iodate

Formula: $NaIO_3$

Molecular Formula: $INaO_3$

Molecular Weight: 197.892

CAS RN: 7681-55-2

Properties: white, cryst powd; oxidizing agent; can be prepared by electrochemical oxidation of NaI; used as an antiseptic, disinfectant, feed additive [HAW93] [MER89] [KIR81]

Solubility: g/100g soln, H_2O: 2.42 (0°C), 8.62 (25°C), 24.8 (100°C); Solid phase, $NaIO_3$·5H_2O (0°C), $NaIO_3$·H_2O (25°C), $NaIO_3$ (100°C) [KRU93]; i alcohol [MER89]

Density, g/cm³: 4.277 [STR93]

Melting Point, °C: decomposes [STR93]

2648

Compound: Sodium iodide

Formula: NaI

Molecular Formula: INa

Molecular Weight: 149.894

CAS RN: 7681-82-5

Properties: white deliq cub cryst or granules; gradually absorbs up to ~5% H_2O from moist atm; iodine slowly evolved, imparting brown coloration; enthalpy of fusion 23.60 kJ/mol; preparation: from a reaction between acidic iodide solution and NaOH or Na_2CO_3; uses: in photography, as a solvent for iodine, in cloud seeding [HAW93] [MER89] [CRC93]

Solubility: g/100g soln, H_2O: 61.54 (0°C), 64.76 (25°C), 75.14 (100°C); Solid phase, NaI·2H_2O (0°C, 25°C), NaI (100°C) [KRU93]; 1 g dissolves in ~2mL alcohol, 1mL glycerol; s acetone [MER89]

Density, g/cm³: 3.667 [KIR82]

Melting Point, °C: 660 [CRC93]

Boiling Point, °C: 1304 [KIR82]

2649

Compound: Sodium iodide dihydrate

Formula: NaI·2H_2O

Molecular Formula: H_4INaO_2

Molecular Weight: 185.925

CAS RN: 13517-06-1

Properties: white, cub cryst or powd [HAW93]

Solubility: 318 g/100mL H_2O (0°C), 1550 g/100mL H_2O (100°C) [CRC94]
Density, g/cm³: 2.448 [HAW93]
Melting Point, °C: 752 [CRC94]

2650
Compound: Sodium metabismuthate
Synonyms: sodium bismuthate
Formula: $NaBiO_3$
Molecular Formula: $BiNaO_3$
Molecular Weight: 279.968
CAS RN: 12232-99-4
Properties: yellow to yellowish brown powd; somewhat hygr; slowly decomposes when stored, decomposition accelerated by moisture and high temp [MER89]
Solubility: i cold H_2O, decomposed in hot H_2O evolving oxygen; evolves chlorine with HCl [MER89]

2651
Compound: Sodium metabisulfite
Synonyms: sodium pyrosulfite
Formula: $Na_2S_2O_5$
Molecular Formula: $Na_2O_5S_2$
Molecular Weight: 190.109
CAS RN: 7681-57-4
Properties: white; cryst or powd; has odor of SO_2; used as a food preservative [HAW93] [MER89]
Solubility: 54 g/100mL H_2O (20°C), 82 g/100mL H_2O (100°C) [CRC94]
Density, g/cm³: 1.48 [ALD94]
Melting Point, °C: decomposes >150 [CRC94]

2652
Compound: Sodium metaborate
Formula: $NaBO_2$
Molecular Formula: $BNaO_2$
Molecular Weight: 65.800
CAS RN: 7775-19-1
Properties: white pieces or powd; enthalpy of fusion 36.20 kJ/mol; used as an herbicide [HAW93] [CRC93] [MER89]
Solubility: g/100g soln, H_2O: 14.10 (0°C), 22.00 (25°C), 55.60 (100°C); Solid phase, $NaBO_2 \cdot 4H_2O$ (0°C, 25°C), $NaBO_2 \cdot 2H_2O$ (100°C) [KRU93]
Density, g/cm³: 2.464 [HAW93]
Melting Point, °C: 966 [CRC93]
Boiling Point, °C: 1434 [HAW93]

2653
Compound: Sodium metaborate dihydrate
Formula: $NaBO_2 \cdot 2H_2O$
Molecular Formula: BH_4NaO_4
Molecular Weight: 101.831
CAS RN: 35585-58-1
Properties: tricl; can be prepared by heating slurry of tetrahydrate above 54°C; stable phase for saturated solutions from 54 to 105°C; a hemihydrate forms at higher temperatures; absorbs atm CO_2 forming borax and sodium carbonate [KIR78]
Solubility: % anhydrous in H_2O: 14.82 (60°C), 19.88 (80°C), 28.22 (100°C); 0.3% in boiling alcohol [KIR78]
Density, g/cm³: 1.91 [KIR78]
Melting Point, °C: 90-95 [KIR78]

2654
Compound: Sodium metaborate tetrahydrate
Formula: $NaBO_2 \cdot 4H_2O$
Molecular Formula: BH_8NaO_6
Molecular Weight: 137.861
CAS RN: 10555-76-7
Properties: tricl; readily obtained by cooling solution of borax and a calculated amount of NaOH; absorbs atm CO_2, forming borax and sodium carbonate; stable phase in saturated solution from 11.5°C to 53.6°C, stable phase above 53.6°C is the dihydrate [KIR78]
Solubility: % anhydrous in H_2O soln: 14.5 (0°C), 21.7 (25°C), 34.1 (50°C); solubility in methanol is 26.4% (40°C) [KIR78]
Density, g/cm³: 1.74 [KIR78]
Melting Point, °C: ~54 melts in waters of hydration [KIR78]
Reactions: minus H_2O 120°C [CRC94]

2655
Compound: Sodium metagermanate
Formula: Na_2GeO_3
Molecular Formula: $GeNa_2O_3$
Molecular Weight: 166.588
CAS RN: 12025-19-3
Properties: white, monocl, deliq [CRC77]
Solubility: g/100g H_2O: 14.4 (0°C), 23.8 (20°C), 116 (80°C) [LAN85]
Density, g/cm³: 3.31 [CRC77]
Melting Point, °C: 1083 [CRC77]

2656
Compound: Sodium metaniobate heptahydrate

Formula: $Na_2Nb_2O_6 \cdot 7H_2O$
Molecular Formula: $H_{14}Na_2Nb_2O_{13}$
Molecular Weight: 453.895
CAS RN: 67211-31-8
Properties: colorless tricl; obtained from niobium pentoxide and sodium hydroxide or sodium carbonate [KIR81]
Solubility: sl s H_2O [HAW93]
Density, g/cm³: 4.512-4.559 [CRC94]
Reactions: minus H_2O 100°C [CRC94]

2657
Compound: Sodium metasilicate
Formula: Na_2SiO_3
Molecular Formula: Na_2O_3Si
Molecular Weight: 122.064
CAS RN: 6834-92-0
Properties: white powd; bead shaped with uniform particle size; hygr; normally in the form of a glass or ortho-rhomb cryst; enthalpy of fusion 52.2 kJ/mol; the nonahydrate $Na_2SiO_3 \cdot 9H_2O$ is ortho-rhomb, efflorescent, melts at 48°C in its waters of cryst, has enthalpy of hydration of -101.04 kJ/mol; normally prepared by fusion of sand (SiO_2) and soda ash (Na_2CO_3); used for cleaning laundry, dairy and metals, for floor cleaning [HAW93] [MER89] [STR93] [OXY93]
Solubility: parts/100 parts soln, H_2O: 15.8 (20°C), 15.6 (35°C), 36.1 (45°C), 46.5 (55°C), 48.3 (60°C), 52.4 (70°C); i alcohol, acids, salt solutions [OXY93] [MER89]
Density, g/cm³: 2.614 [MER89]
Melting Point, °C: 1089 [MER89]

2658
Compound: Sodium metasilicate pentahydrate
Formula: $Na_2SiO_3 \cdot 5H_2O$
Molecular Formula: $H_{10}Na_2O_8Si$
Molecular Weight: 212.140
CAS RN: 13517-24-3
Properties: white powd; uniform particle size; enthalpy of fusion 30.5 kJ/mol; manufactured in brick furnace at 1300°C by fusing sand and soda ash, followed by removal of glass; used in the form of liq solutions for numerous applications such as in coatings, adhesives and cements, gels and catalysts [STR93] [OXY93]
Solubility: parts/100 parts soln in H_2O: 27.5 (20°C), 44.5 (35°C), 62.8 (45°C), 80.9 (55°C), 84.0 (60°C), 91.2 (70°C) [OXY93]
Melting Point, °C: 72.2 [OXY93]

2659
Compound: Sodium metatantalate
Formula: $NaTaO_3$
Molecular Formula: NaO_3Ta
Molecular Weight: 251.936
CAS RN: 12034-15-0
Properties: white powd, -100 mesh of 99.9% purity [CRC91] [STR93]

2660
Compound: Sodium metavanadate
Formula: $NaVO_3$
Molecular Formula: NaO_3V
Molecular Weight: 121.930
CAS RN: 13718-26-8
Properties: -200 mesh with 99.9% purity; colorless, monocl, or pale green cryst powd; used in inks, and in fur dyeing [CER91] [HAW93]
Solubility: 21.1 g/100mL H_2O (25°C), 38.8 g/100mL H_2O (75°C) [CRC94]
Melting Point, °C: 630 [STR93]

2661
Compound: Sodium molybdate
Formula: Na_2MoO_4
Molecular Formula: $MoNa_2O_4$
Molecular Weight: 205.918
CAS RN: 7631-95-0
Properties: white to off-white powd; small, lustrous, cryst plates; prepared by evaporation of aq solution of molybdic oxide and sodium hydroxide to form the dihydrate, followed by heating at 100°C; used in pigments and metal finishing, and to inhibit corrosion in aq media [KIR81] [HAW93] [STR93]
Solubility: g/100g soln, H_2O: 30.62 (0°C), 39.40 (25°C), 45.52 (100°C); Solid phase, $Na_2MoO_4 \cdot 10H_2O$ (0°C), $Na_2MoO_4 \cdot 2H_2O$ (25°C, 100°C) [KRU93]
Density, g/cm³: 3.28 [STR93]
Melting Point, °C: 687 [STR93]

2662
Compound: Sodium molybdate dihydrate
Formula: $Na_2MoO_4 \cdot 2H_2O$
Molecular Formula: $H_4MoNa_2O_6$
Molecular Weight: 241.948
CAS RN: 10102-40-6
Properties: cryst powd [MER89]
Solubility: s 1.7 parts cold H_2O, ~0.7 parts boiling H_2O [MER89]
Density, g/cm³: 3.28 [STR93]

Reactions: minus 2H$_2$O 100°C [CRC94]

2663
Compound: Sodium molybdosilicate hydrate
Formula: Na$_4$SiMo$_{12}$O$_{40}$·xH$_2$O
Molecular Formula: Mo$_{12}$Na$_4$O$_{40}$Si (anhydrous)
Molecular Weight: 1911.301 (anhydrous)
Properties: yellow cryst; used as a catalyst [HAW93]
Solubility: s H$_2$O, acetone, alcohol, ethyl acetate; i ether, benzene, cyclohexane [HAW93]
Density, g/cm^3: 3.44 [HAW93]

2664
Compound: Sodium niobate
Formula: NaNbO$_3$
Molecular Formula: NaNbO$_3$
Molecular Weight: 163.894
CAS RN: 12034-09-2
Properties: -100 mesh with 99.9% purity; white powd [CER91] [STR93]
Density, g/cm^3: 4.55 [LID94]
Melting Point, °C: 1422 [AES93]

2665
Compound: Sodium nitrate
Synonyms: niter
Formula: NaNO$_3$
Molecular Formula: NNaO$_3$
Molecular Weight: 84.995
CAS RN: 7631-99-4
Properties: colorless transparent trigonal cryst, white granules or powd; deliq; enthalpy of fusion 15.00 kJ/mol; preparation: extraction from ore by brine at 70°C, followed by crystallization [KIR82] [MER89] [CRC93]
Solubility: g/100g H$_2$O: 73.0 (0°C), 87.6 (20°C), 180 (100°C) [LAN85]; 1 g dissolves in: 125mL alcohol, 52mL boiling alcohol; 3470mL absolute alcohol; 300mL absolute methanol [MER89]
Density, g/cm^3: 2.261 [STR93]
Melting Point, °C: 307 [CRC93]
Boiling Point, °C: explodes at 537 [HAW93]
Thermal Expansion Coefficient: (volume) 100°C (1.076), 200°C (2.74) [CLA66]

2666
Compound: Sodium nitrite
Formula: NaNO$_2$
Molecular Formula: NNaO$_2$
Molecular Weight: 68.996

CAS RN: 7632-00-0
Properties: white or sl yellow hygr granules, rods or powd; body-centered ortho-rhomb, a = 0.355 nm, b = 0.556 nm, c = 0.557 nm; enthalpy of transition at 158-165°C is 1192 kJ/mol; oxidizing agent; slowly oxidized to the nitrate by atm O$_2$; preparation: by the dissolution of nitrogen oxides in aq alkaline solutions [KIR82] [MER89]
Solubility: g/100g soln, H$_2$O: 41.65 (0°C), 45.92 (25°C), 61.50 (99.9°C); Solid phase, NaNO$_2$ [KRU93]; sl s alcohol; decomposed by weak acids, evolving brown fumes of N$_2$O$_3$ [MER89]
Density, g/cm^3: 2.168 [STR93]
Melting Point, °C: 271 [MER89]
Boiling Point, °C: 320, decomposes [KIR82]
Reactions: forms Na$_2$O and N$_2$ or nitrogen oxide at 320°C [KIR82]

2667
Compound: Sodium nitroferricyanide(III) dihydrate
Synonyms: sodium nitroprusside dihydrate
Formula: Na$_2$[Fe(CN)$_5$NO]·2H$_2$O
Molecular Formula: C$_5$H$_4$FeN$_6$Na$_2$O$_3$
Molecular Weight: 297.950
CAS RN: 13755-38-9
Properties: ruby red; transparent cryst; aq solutions decompose [MER89]
Solubility: s in ~2.3 parts H$_2$O; sl s alcohol [MER89]
Density, g/cm^3: 1.72 [STR93]

2668
Compound: Sodium oleate
Formula: C$_{17}$H$_{33}$COONa
Molecular Formula: C$_{18}$H$_{33}$NaO$_2$
Molecular Weight: 304.449
CAS RN: 143-19-1
Properties: white powd; used in ore flotation, to waterproof textiles [HAW93]
Solubility: s H$_2$O, partially decomposes; s alcohol [HAW93]
Melting Point, °C: 232-235 [CRC94]

2669
Compound: Sodium orthosilicate
Formula: Na$_4$SiO$_4$
Molecular Formula: Na$_4$O$_4$Si
Molecular Weight: 184.043
CAS RN: 13472-30-5

Properties: white powd; used in laundries, for metal cleaning, in heavy duty cleaning [HAW93] [STR93]
Solubility: s H_2O [HAW93]
Melting Point, °C: 1018 [STR93]

2670
Compound: Sodium orthovanadate
Formula: Na_3VO_4
Molecular Formula: Na_3O_4V
Molecular Weight: 183.909
CAS RN: 13721-39-6
Properties: -200 mesh with 99.9% purity; colorless hex prisms [CER91] [KIR83]
Solubility: s H_2O [KIR83]
Melting Point, °C: 850-856 [KIR83]

2671
Compound: Sodium orthovanadate decahydrate
Formula: $Na_3VO_4 \cdot 10H_2O$
Molecular Formula: $H_{20}Na_3O_{14}V$
Molecular Weight: 364.062
CAS RN: 16519-60-1
Properties: white cryst [STR93]
Melting Point, °C: 850-866 [STR93]

2672
Compound: Sodium oxalate
Formula: $Na_2C_2O_4$
Molecular Formula: $C_2Na_2O_4$
Molecular Weight: 134.000
CAS RN: 62-76-0
Properties: white, odorless powd; used in textile and leather finishing, blueprinting [MER89] [HAW93]
Solubility: g/100g soln, H_2O: 2.62 (0°C), 3.48 (25°C), 6.10 (100°C) [KRU93]; i alcohol [MER89]
Density, g/cm³: 2.34 [HAW93]
Melting Point, °C: 250-270, decomposes [HAW93]

2673
Compound: Sodium oxide
Synonyms: sodium monoxide
Formula: Na_2O
Molecular Formula: Na_2O
Molecular Weight: 61.979
CAS RN: 1313-59-3
Properties: white; amorphous pieces or powd; reacts violently with H_2O, forming NaOH; enthalpy of fusion 48.00 kJ/mol [MER89] [CRC93]

Solubility: reacts with H_2O, forming NaOH [HAW93]
Density, g/cm³: 2.27 [MER89]
Melting Point, °C: 1132 [CRC93]
Boiling Point, °C: sublimes at 1274 [HAW93]
Reactions: decomposition begins >400°C to form Na_2O_2 and Na [MER89]

2674
Compound: Sodium paraperiodate
Formula: $Na_3H_2IO_6$
Molecular Formula: $H_2INa_3O_6$
Molecular Weight: 293.885
CAS RN: 13940-38-0
Properties: white cryst solid; used as a selective oxidizing agent for specific carbohydrates and amino acids [HAW93]
Solubility: v sl s H_2O; s in conc NaOH solutions [HAW93]

2675
Compound: Sodium pentaiodobismuthate tetrahydrate
Synonyms: bismuth sodium iodide
Formula: $Na_2BiI_5 \cdot 4H_2O$
Molecular Formula: $BiH_8I_5Na_2O_4$
Molecular Weight: 961.544
CAS RN: 53778-50-0
Properties: odorless, red cryst, astringent taste [MER89]
Solubility: s H_2O, hydrolyzes [MER89]
Melting Point, °C: decomposes 93 [MER89]

2676
Compound: Sodium perborate monohydrate
Formula: $NaBO_3 \cdot H_2O$
Molecular Formula: BH_2NaO_4
Molecular Weight: 99.815
CAS RN: 10332-33-9
Properties: white, amorphous powd; used as a denture cleaner, as a bleaching agent in special detergents [HAW93]
Solubility: v s H_2O, reacting to give H_2O_2 and sodium borate [HAW93]

2677
Compound: Sodium perborate tetrahydrate
Formula: $NaBO_3 \cdot 4H_2O$
Molecular Formula: BH_8NaO_7
Molecular Weight: 153.861
CAS RN: 10486-00-7

Properties: white, odorless, cryst powd; stable when cool and dry, else decomposes evolving O_2 [MER89]

Solubility: s ~40 parts H_2O, soln eventually decomposes in the sequence: $\rightarrow H_2O_2 \rightarrow O_2$ [MER89]

Melting Point, °C: 60, decomposes [ALD94]

2678

Compound: Sodium perchlorate

Formula: $NaClO_4$

Molecular Formula: $ClNaO_4$

Molecular Weight: 122.441

CAS RN: 7601-89-0

Properties: white powd; hygr; used in explosives, jet fuel [HAW93] [STR93]

Solubility: g/100g soln, H_2O: 62.8±0.1 (0°C), 67.7±0.1 (25°C), 76.75 (100°C); Solid phase, $NaClO_4 \cdot H_2O$ (0°C, 25°C), $NaClO_4$ (100°C) [KRU93]

Density, g/cm³: 2.499 [KIR79]

Melting Point, °C: 482 [STR93]

Boiling Point, °C: decomposes [HAW93]

Reactions: transition ortho-rhomb to cub at 577-586K [KIR79]

2679

Compound: Sodium perchlorate monohydrate

Formula: $NaClO_4 \cdot H_2O$

Molecular Formula: ClH_2NaO_5

Molecular Weight: 140.456

CAS RN: 7791-07-3

Properties: white cryst; deliq [MER89]

Solubility: 66 parts in 100 parts H_2O (0°C) [KIR79]

Density, g/cm³: 2.02 [MER89]

Melting Point, °C: decomposes ~130 [MER89]

2680

Compound: Sodium periodate

Synonyms: sodium metaperiodate

Formula: $NaIO_4$

Molecular Formula: $INaO_4$

Molecular Weight: 213.892

CAS RN: 7790-28-5

Properties: white, tetr cryst; oxidant [ALD94] [MER89]

Solubility: g/100g soln, H_2O: 12.62 (25°C); Solid phase, $NaIO_4 \cdot 3H_2O$ [KRU93]; s H_2SO_4, HNO_3, acetic acids [MER89]

Density, g/cm³: 3.865 [MER89]

Melting Point, °C: decomposes ~300 [MER89]

2681

Compound: Sodium periodate trihydrate

Synonyms: sodium metaperiodate

Formula: $NaIO_4 \cdot 3H_2O$

Molecular Formula: H_6INaO_7

Molecular Weight: 267.938

CAS RN: 13472-31-6

Properties: stable phase below 34.5°C; white, efflorescent; trig cryst [MER89] [KIR81]

Solubility: 1 g/8mL H_2O (20°C) [MER89]

Density, g/cm³: 3.219 (18°C) [HAW93]

Melting Point, °C: decomposes 175 [MER89]

2682

Compound: Sodium permanganate trihydrate

Formula: $NaMnO_4 \cdot 3H_2O$

Molecular Formula: H_6MnNaO_7

Molecular Weight: 195.972

CAS RN: 10101-50-5

Properties: purple to reddish black; very hygr; granules; used as an oxidizing agent, disinfectant; there is a monohydrate, CAS RN 79048-36-5 [MER89] [ALD94]

Solubility: v s H_2O; decomposed by alcohol [MER89]

Density, g/cm³: 2.47 [HAW93]

Melting Point, °C: 170, decomposes [HAW93]

2683

Compound: Sodium peroxide

Synonyms: sodium dioxide

Formula: Na_2O_2

Molecular Formula: Na_2O_2

Molecular Weight: 77.979

CAS RN: 1313-60-6

Properties: yellowish white; granular powd; absorbs atm water and CO_2; reacts with dil acids to produce H_2O_2; strong oxidant, e.g. readily reacts with organic matter or other oxidizable materials [MER89]

Solubility: v s H_2O, producing $NaOH$, H_2O_2; H_2O_2 quickly decomposes evolving O_2 [MER89]

Density, g/cm³: 2.805 [STR93]

Melting Point, °C: 460, decomposes [STR93]

2684

Compound: Sodium perrhenate

Formula: $NaReO_4$

Molecular Formula: NaO_4Re

Molecular Weight: 273.195

CAS RN: 13472-33-8

Properties: white powd; hygr [STR93]

Solubility: 100 g/100mL H_2O (20°C) [CRC94]
Density, g/cm³: 5.39 [STR93]
Melting Point, °C: 300 (in oxygen) [STR93]

2685
Compound: Sodium persulfate
Synonyms: sodium peroxydisulfate
Formula: $Na_2S_2O_8$
Molecular Formula: $Na_2O_8S_2$
Molecular Weight: 238.107
CAS RN: 7775-27-1
Properties: white; cryst powd; gradually decomposes if standing, with rate of decomposition accelerated by moisture and high temp; strong oxidizing agent; used as bleaching agent for fats, oils, fabrics [MER89]
Solubility: 549 g/L H_2O (20°C); decomposed by alcohol [MER89]
Density, g/cm³: 2.400 [ALD94]

2686
Compound: Sodium phosphate
Synonyms: trisodium phosphate; sodium orthophosphate
Formula: Na_3PO_4
Molecular Formula: Na_3O_4P
Molecular Weight: 163.940
CAS RN: 7601-54-9
Properties: hygr; -100 mesh powd; uses: photographic developers, clarify sugar, clean boiler scale, soften water, paper manufacturing, laundering, detergents [ALD94] [MER89] [AES93]
Solubility: g/100g H_2O: 5.38 (0°C), 14.53 (25°C), 94.6 (100°C); Solid phase, $Na_3PO_4 \cdot 1/4NaOH \cdot 12H_2O$ (0°C, 25°C), $Na_3PO_4 \cdot 6H_2O$ (100°C) [KRU93]

2687
Compound: Sodium phosphate dodecahydrate
Formula: $Na_3PO_4 \cdot 12H_2O$
Molecular Formula: $H_{24}Na_3O_{16}P$
Molecular Weight: 380.124
CAS RN: 10101-89-0
Properties: colorless or white cryst; used to soften water, as a detergent and metal cleaner [MER89] [HAW93] [STR93]
Solubility: s 3.5 parts H_2O, 1 part boiling H_2O; i alcohol [MER89]
Density, g/cm³: 1.62 [HAW93]
Melting Point, °C: ~75, when heated rapidly [MER89]

Reactions: minus $12H_2O$ at 100°C [HAW93]

2688
Compound: Sodium phosphide
Formula: Na_3P
Molecular Formula: Na_3P
Molecular Weight: 99.943
CAS RN: 12058-85-4
Properties: red solid; decomposes when heated or if immersed in water, evolving phosphine; thermally stable up to 650°C; there are also Na_2P, 12439-14-4, and Na_3P_{11}, 39343-85-6; produced by reacting Na and P, then stored under oil [KIR82] [HAW93]
Solubility: decomposes in H_2O, evolving PH_3 [CRC94] [HAW93]
Melting Point, °C: stable up to 650 [KIR82]

2689
Compound: Sodium phosphomolybdate
Synonyms: sodium 12-molybdophosphate
Formula: $Na_3PO_4 \cdot 12MoO_3$
Molecular Formula: $Mo_{12}Na_3O_{40}P$
Molecular Weight: 1891.199
CAS RN: 1313-30-0
Properties: yellow cryst; used in chemical analysis, neuromicroscopy, imparts water resistance to plastics, adhesives, cements [HAW93] [MER89]
Solubility: v s H_2O [MER89]
Density, g/cm³: 2.83 [HAW93]

2690
Compound: Sodium phosphotungstate
Synonyms: sodium 12-tungstophosphate
Formula: $2Na_2O \cdot P_2O_5 \cdot 12WO_3 \cdot 18H_2O$
Molecular Formula: $H_{36}Na_4O_{61}P_2W_{12}$
Molecular Weight: 3372.236
CAS RN: 51312-42-6
Properties: yellowish white; granular powd; used as a reagent, and in the manufacture of pigments [HAW93] [MER89]
Solubility: v s H_2O, alcohol [HAW93]

2691
Compound: Sodium polyphosphate
Synonyms: sodium polymetaphosphate
Formula: $Na_{(n+2)}P_nO_{(3n+1)}$
CAS RN: 50813-16-6

Properties: clear hygr glass; two most important of the polyphosphates are with n = 2 and with n = 3; preparation: rapid chilling of molten sodium metaphosphate; used in water treatment for sequestering metals [HAW93] [MER89] [ALD94]
Solubility: s H_2O [MER89]
Melting Point, °C: 628 [MER89]

2692
Compound: Sodium potassium tartrate tetrahydrate
Synonyms: Rochelle salt; potassium sodium tartrate
Formula: $NaKC_4H_4O_6 \cdot 4H_2O$
Molecular Formula: $C_4H_{12}KNaO_{10}$
Molecular Weight: 282.221
CAS RN: 304-59-6
Properties: translucent cryst, or white cryst powd; sl efflorescent in warm air; has cool, saline taste; used in baking powd, as a cathartic in medicine, for silvering mirrors [HAW93] [MER89]
Solubility: g anhydrous/100g H_2O: 31.9 (0°C), 67.8 (20°C), 102 (30°C) [LAN85]; sl s alcohol [MER89]
Density, g/cm³: 1.79 [MER89]
Melting Point, °C: 70-80 [MER89]
Boiling Point, °C: decomposition begins at 220 [MER89]
Reactions: minus $3H_2O$ at 100°C; anhydrous 130-140°C [MER89]

2693
Compound: Sodium pyrophosphate
Formula: $Na_4P_2O_7$
Molecular Formula: $Na_4O_7P_2$
Molecular Weight: 265.902
CAS RN: 7722-88-5
Properties: colorless, transparent cryst or white powd; used in water softening, as a metal cleaner [HAW93]
Solubility: g/100g soln, H_2O: 2.236 (0°C), 6.618 (25°C), 31.15 (96°C); Solid phase, $Na_4P_2O_7 \cdot 10H_2O$ (0°C, 25°C), $Na_4P_2O_7$ (100°C) [KRU93]
Density, g/cm³: 2.45 [HAW93]
Melting Point, °C: 880 [HAW93]

2694
Compound: Sodium pyrophosphate decahydrate
Formula: $Na_4P_2O_7 \cdot 10H_2O$
Molecular Formula: $H_{20}Na_4O_{17}P_2$
Molecular Weight: 446.055
CAS RN: 13472-36-1

Properties: white powd [STR93]
Solubility: s H_2O; decomposed by alcohol [HAW93]
Density, g/cm³: 1.815-1.836 [STR93]
Reactions: minus H_2O 94°C [HAW93]

2695
Compound: Sodium pyrovanadate
Formula: $Na_4V_2O_7$
Molecular Formula: $Na_4O_7V_2$
Molecular Weight: 305.838
CAS RN: 13517-26-5
Properties: -200 mesh with 99.9% purity; colorless hex prisms [KIR83] [CER91]
Solubility: s H_2O [KIR83]
Melting Point, °C: 632-654 [KIR83]

2696
Compound: Sodium selenate
Formula: Na_2SeO_4
Molecular Formula: Na_2O_4Se
Molecular Weight: 188.938
CAS RN: 13410-01-0
Properties: -100 mesh with 99.5% purity; white powd; uses: insecticide [MER89] [STR93] [CER91]
Solubility: g/100g soln, H_2O: 11.74 (0°C), 36.91 (25.2°C), 42.14 (100°C); Solid phase, $Na_2SeO_4 \cdot 10H_2O$ (0°C, 25°C), Na_2SeO_4 (100°C) [KRU93]
Density, g/cm³: 3.213 [STR93]

2697
Compound: Sodium selenate decahydrate
Formula: $Na_2SeO_4 \cdot 10H_2O$
Molecular Formula: $H_{20}Na_2O_{14}Se$
Molecular Weight: 369.091
CAS RN: 10102-23-5
Properties: white cryst [MER89]
Solubility: v s H_2O [MER89]
Density, g/cm³: 1.603-1.620 [HAW93]

2698
Compound: Sodium selenide
Formula: Na_2Se
Molecular Formula: Na_2Se
Molecular Weight: 124.940
CAS RN: 1313-85-5

Properties: -60 mesh, dry under argon with 99.9% purity; amorphous cryst: deliq; becomes red if exposed to the atm; decahydrate: needles, becomes red, then brown in air; hexadecahydrate: prisms; decomposes in air to Na_2CO_3, Se and some Na_2Se [MER89] [CER91]

Solubility: decomposes in H_2O [MER89]

Density, g/cm³: 2.625 [MER89]

Melting Point, °C: Na_2Se: >875; hexadecahydrate: 40 [MER89]

2699

Compound: Sodium selenite

Formula: Na_2SeO_3

Molecular Formula: Na_2O_3Se

Molecular Weight: 172.938

CAS RN: 10102-18-8

Properties: -100 mesh with 99.5% purity; white powd; tetr prisms; stable in air [MER89] [STR93] [CER91]

Solubility: g/100g soln, H_2O: 47.28 (24.4°C), 45.3 (103.3°C); Solid phase, $Na_2SeO_3\cdot5H_2O$ (24.4°C), Na_2SeO_3 (103.3°C) [KRU93]; i alcohol [MER89]

Melting Point, °C: decomposes [AES93]

2700

Compound: Sodium selenite pentahydrate

Formula: $Na_2SeO_3\cdot5H_2O$

Molecular Formula: $H_{10}Na_2O_8$

Molecular Weight: 263.014

CAS RN: 26970-82-1

Properties: white acicular cryst; evolves $5H_2O$ in dry air; used in glass manufacturing to control color, in decorating porcelain, and for testing seed germination [HAW93] [MER89] [ALD94]

Solubility: s H_2O; i alcohol [HAW93]

2701

Compound: Sodium silicate

Synonyms: waterglass

Formula: $Na_2O\cdot xSiO_2$

CAS RN: 1344-09-8

Properties: colorless to white; usual compositions: Na_2SiO_3, $Na_6Si_2O_7$, $Na_2Si_3O_7$; contains variable amounts of H_2O, e.g. $Na_2SiO_3\cdot5H_2O$; produced by fusion of sand and soda ash; used as a catalyst and in silica gels, soaps [HAW93] [MER89]

Solubility: v sl s cold H_2O [MER89]

2702

Compound: Sodium stannate trihydrate

Formula: $Na_2SnO_3\cdot3H_2O$

Molecular Formula: $H_6Na_2O_6Sn$

Molecular Weight: 266.734

CAS RN: 12209-98-2

Properties: -100 mesh with 99.9% purity; white or colorless; cryst; decomposed in air and by weak acids; used as a mordant in dyeing, and in ceramics [HAW93] [MER89] [CER91]

Solubility: g/100g H_2O: 46.0 (0°C), 43.7 (20°C), 38.9 (40°C) [LAN85]; i alcohol [MER89]

Reactions: minus $3H_2O$ at 140°C [HAW93]

2703

Compound: Sodium stearate

Synonyms: stearic acid, sodium salt

Formula: $CH_3(CH_2)_{16}COONa$

Molecular Formula: $C_{18}H_{35}NaO_2$

Molecular Weight: 306.465

CAS RN: 822-16-2

Properties: usually contains sodium palmitate; white powd; soapy feel; hydrolyzes in water to given an alkaline solution; used as a waterproofing and gelling agent; in toothpaste and cosmetics [HAW93] [MER89]

Solubility: slowly s cold H_2O, alcohol; v s hot H_2O, alcohol [MER89]

2704

Compound: Sodium sulfate

Synonyms: mirabilite; thenardite

Formula: Na_2SO_4

Molecular Formula: Na_2O_4S

Molecular Weight: 142.044

CAS RN: 7757-82-6

Properties: white odorless powd or ortho-rhomb cryst; enthalpy of fusion is 23.60 kJ/mol; used in the manufacture of kraft paper, paperboard and glass; also used as a filler in synthetic detergents [HAW93] [MER89] [KIR83] [CRC93]

Solubility: g/100g H_2O: 4.5 (0°C), 28.0 (25°C), 42.2 (100°C); Solid phase, $Na_2SO_4\cdot10H_2O$ (0°C, 25°C), Na_2SO_4 (100°C) [KRU93]; i alcohol [MER89]

Density, g/cm³: 2.68 [STR93]

Melting Point, °C: 884 [CRC93]

2705

Compound: Sodium sulfate decahydrate

Synonyms: Glauber's salt, mirabilite

Formula: $Na_2SO_4\cdot10H_2O$

Molecular Formula: $H_{20}Na_2O_{14}S$
Molecular Weight: 322.197
CAS RN: 7727-73-3
Properties: large, transparent monocl cryst or granules; effloresces; enthalpy of crystallization 74.98 J/mol at 25°C; energy storage capacity is more than seven times greater than that of water; used in solar energy to store heat, and in air conditioning [MER89] [HAW93]
Solubility: s in 3.3 parts H_2O (15°C), 1.5 parts (25°C); s glycerol; i alcohol [MER89]
Density, g/cm³: 1.464 (cryst) [HAW93]
Melting Point, °C: 32.4 [MER89]
Reactions: gives up waters of hydration at 100°C [HAW93]

2706
Compound: Sodium sulfate heptahydrate
Formula: $Na_2SO_4 \cdot 7H_2O$
Molecular Formula: $H_{14}Na_2O_{11}S$
Molecular Weight: 268.150
CAS RN: 7727-73-3
Properties: white rhomb; tetr [CRC94] [LAN52]
Solubility: g/100g H_2O: 19.5 (0°C), 30.0 (10°C), 44.1 (20°C) [LAN85]
Reactions: minus $7H_2O$ 24.4°C [CRC94]

2707
Compound: Sodium sulfide
Formula: Na_2S
Molecular Formula: Na_2S
Molecular Weight: 78.046
CAS RN: 1313-82-2
Properties: white cub cryst or granules; very hygr; discolors when exposed to the atm, slowly forming sodium carbonate and sodium thiosulfate; enthalpy of solution is -63.5 kJ/mol; enthalpy of fusion 19.00 kJ/mol; crystallzes from aq solution as the nonahydrate, 1313-84-4 [KIR82] [MER89] [CRC93]
Solubility: g/100g soln, H_2O: 8.8 (0°C), 15.3 (25°C), 60.1 (95°C); Solid phase, $Na_2S \cdot 9H_2O$ (0°C, 25°C), $Na_2S \cdot H_2O$ (95°C) [KRU93]; sl s alcohol; i ether [MER89]
Density, g/cm³: 1.856 [MER89]
Melting Point, °C: 1180 (vacuum) [MER89]

2708
Compound: Sodium sulfide nonahydrate
Formula: $Na_2S \cdot 9H_2O$
Molecular Formula: $H_{18}Na_2O_9S$
Molecular Weight: 240.184

CAS RN: 1313-84-4
Properties: tetr; deliq; cryst; H_2S odor; becomes yellow, then brownish black if subjected to atm exposure; decomposed by acids [MER89]
Solubility: 1 g/0.5mL H_2O (25°C); sl s alcohol; i ether [MER89]
Density, g/cm³: 1.427 [MER89]
Melting Point, °C: 920, decomposes [HAW93]

2709
Compound: Sodium sulfide pentahydrate
Formula: $Na_2S \cdot 5H_2O$
Molecular Formula: $H_{10}Na_2O_5S$
Molecular Weight: 168.122
CAS RN: 1313-83-3
Properties: flat, shiny cryst; flammable; evolves H_2S in acid solutions [MER89] [ALD94]
Solubility: v s H_2O, alcohol; i ether [MER89]
Density, g/cm³: 1.58 [LID94]
Melting Point, °C: 120 [MER89]
Reactions: minus $3H_2O$ 100°C [CRC94], dehydrates at 120°C [MER89]

2710
Compound: Sodium sulfite
Formula: Na_2SO_3
Molecular Formula: Na_2O_3S
Molecular Weight: 126.044
CAS RN: 7757-83-7
Properties: white, small cryst or powd; fairly stable to oxidation; used in paper and dyes industries [HAW93] [MER89]
Solubility: g/100g H_2O: 13.85 (0°C), 30.5±0.4 (25°C), 26.3 (100°C); Solid phase, $Na_2SO_3 \cdot 7H_2O$ (0°C, 25°C), Na_2SO_3 (100°C) [KRU93]
Density, g/cm³: 2.633 [STR93]
Melting Point, °C: decomposes [STR93]

2711
Compound: Sodium sulfite heptahydrate
Formula: $Na_2SO_3 \cdot 7H_2O$
Molecular Formula: $H_{14}Na_2O_{10}S$
Molecular Weight: 252.151
CAS RN: 10102-15-5
Properties: efflorescent cryst; oxidizing in air to sulfate [MER89]
Solubility: s 1.6 parts H_2O, ~30 parts glycerol; sl s alcohol [MER89]
Density, g/cm³: 1.539 [CRC94]
Reactions: minus $7H_2O$ 150°C [CRC94]

2712
Compound: Sodium tartrate dihydrate
Formula: $Na_2C_4H_4O_6 \cdot 2H_2O$
Molecular Formula: $C_4H_8Na_2O_8$
Molecular Weight: 230.083
CAS RN: 868-18-8
Properties: white cryst or granules; used as food additive, as a sequestrant and stabilizer [HAW93] [MER89]
Solubility: s in ~3 parts H_2O, 1.5 parts boiling H_2O; i alcohol [MER89]
Density, g/cm³: 1.794 [HAW93]
Reactions: minus $2H_2O$ at 150°C [HAW93]

2713
Compound: Sodium tellurate(VI)
Formula: Na_2TeO_4
Molecular Formula: Na_2O_4Te
Molecular Weight: 237.578
CAS RN: 10102-83-4
Properties: white powd [MER89]
Solubility: s in 130 parts cold H_2O, 50 parts boiling H_2O [MER89]

2714
Compound: Sodium tellurate(VI) dihydrate
Formula: $Na_2TeO_4 \cdot 2H_2O$
Molecular Formula: $H_4Na_2O_6Te$
Molecular Weight: 273.608
CAS RN: 26006-71-3
Properties: -100 mesh with 99.5% purity; white powd [STR93] [CER91]
Melting Point, °C: decomposes [STR93]

2715
Compound: Sodium tellurite(IV)
Formula: Na_2TeO_3
Molecular Formula: Na_2O_3Te
Molecular Weight: 221.578
CAS RN: 10102-20-2
Properties: -100 mesh with 99.5% purity; white powd; used in bacteriology, and in medicine [HAW93] [CER91]
Solubility: s H_2O [MER89]

2716
Compound: Sodium tetraborate
Synonyms: sodium borate
Formula: $Na_2B_4O_7$
Molecular Formula: $B_4Na_2O_7$
Molecular Weight: 201.220

CAS RN: 1330-43-4
Properties: fused sodium borate; white powd or glassy plates; hygr; becomes opaque in air; partially hydrates in damp air; enthalpy of formation of glass form -3256.6 kJ/mol; has several cryst forms; enthalpy of fusion of cryst form 81.2 kJ/mol; used in the manufacture of glass, enamels and other ceramics [KIR78] [HAW93] [MER89]
Solubility: g/100g soln, H_2O: 1.18 (0°C), 3.13 (25°C), 28.22 (100°C); Solid phase, $Na_2B_4O_7 \cdot 10H_2O$ (0°C, 25°C), $Na_2B_4O_7 \cdot 4H_2O$ (100°C) [KRU93]
Density, g/cm³: 2.367 [STR93]
Melting Point, °C: 741 [STR93]
Boiling Point, °C: 1575 [STR93]

2717
Compound: Sodium tetraborate decahydrate
Synonyms: borax
Formula: $Na_2B_4O_7 \cdot 10H_2O$
Molecular Formula: $B_4H_{20}Na_2O_{17}$
Molecular Weight: 381.373
CAS RN: 1303-96-4
Properties: hard, odorless cryst, granules or powd; effloresces in dry air; monocl; specific heat, 1.611 kJ/(kg·K); cryst habit may be changed by adding various substances, and by altering conditions [MER89] [KIR78]
Solubility: %w anhydrous salt: 1.18 (0°C); 3.13 (25°C); 15.90 (60°C) [KIR78]; 1 g/1mL glycerol; i alcohol [MER89]
Density, g/cm³: 1.73 [MER89]
Melting Point, °C: 75, when heated rapidly [MER89]
Reactions: loses $5H_2O$ at 100°C; loses $9H_2O$ at 150°C, dehydrates at 320°C [MER89]

2718
Compound: Sodium tetraborate pentahydrate
Synonyms: tincalconite
Formula: $Na_2B_4O_7 \cdot 5H_2O$
Molecular Formula: $B_4H_{10}Na_2O_{12}$
Molecular Weight: 291.296
CAS RN: 12045-88-4
Properties: free flowing powd; trig; specific heat 1.32 kJ/(kg·K); enthalpy of formation -4784.4 MJ/ mol; used as a weed killer, and to control fungus growth on citrus fruit [KIR78]
Solubility: %w of anhydrous in H_2O: 16.40 (60°C), 23.38 (80°C), 34.63 (100°C); %w of pentahydrate at 25°C: 16.9% in methanol, 31.1% in ethylene glycol, 10.0% in diethylene glycol [KIR78]

Density, g/cm^3: 1.815 [HAW93]
Reactions: minus H$_2$O at 122 °C [HAW93]

2719
Compound: Sodium tetraborate tetrahydrate
Synonyms: ernite
Formula: Na$_2$B$_4$O$_7$·4H$_2$O
Molecular Formula: B$_4$H$_8$Na$_2$O$_{11}$
Molecular Weight: 273.281
CAS RN: 12045-87-3
Properties: monocl; specific heat ~1.2 kJ/(kg·K); enthalpy of formation -4489.0 kJ/mol; absorbs water to form borax at relative humidities above 70% [KIR78]
Solubility: % anhydrous in H$_2$O: 14.82 (60°C), 17.12 (70°C), 19.88 (80°C), 23.31 (90°C), 28.22 (100°C) [KIR78]
Density, g/cm^3: 1.95 [LID94]

2720
Compound: Sodium tetrabromoaurate(III)
Formula: NaAuBr$_4$
Molecular Formula: AuBr$_4$Na
Molecular Weight: 539.573
CAS RN: 52495-41-7
Properties: reddish black cryst [STR93]

2721
Compound: Sodium tetrachloroaluminate
Formula: NaAlCl$_4$
Molecular Formula: AlCl$_4$Na
Molecular Weight: 191.783
CAS RN: 7784-16-9
Properties: yellow hygr powd; used as a catalyst for organic reactions [CRC94] [HAW93]
Solubility: s H$_2$O [MER89]
Density, g/cm^3: 2.01 [LID94]
Melting Point, °C: 185 [CRC94]

2722
Compound: Sodium tetrachloroaurate(III) dihydrate
Formula: NaAuCl$_4$·2H$_2$O
Molecular Formula: AuCl$_4$H$_4$NaO$_2$
Molecular Weight: 397.799
CAS RN: 13874-02-7
Properties: yellowish orange cryst; rhomb; stable to 100°C; used in photography, in staining fine glass, for decorating porcelain and in medicine [HAW93] [MER89]

Solubility: g anhydrous/100g H$_2$O: 139 (10°C), 151 (20°C), 900 (60°C) [LAN85]; s alcohol, ether [MER89]
Melting Point, °C: 100 decomposes [AES93]

2723
Compound: Sodium tetrachloropalladate(II) trihydrate
Formula: Na$_2$PdCl$_4$·3H$_2$O
Molecular Formula: Cl$_4$H$_6$Na$_2$O$_3$Pd
Molecular Weight: 384.256
CAS RN: 13820-53-6
Properties: reddish brown powd [STR93]

2724
Compound: Sodium tetrafluoroberyllate
Synonyms: beryllium sodium fluoride
Formula: Na$_2$BeF$_4$
Molecular Formula: BeF$_4$Na$_2$
Molecular Weight: 130.986
CAS RN: 13871-27-7
Properties: ortho-rhomb or monocl cryst [MER89]
Solubility: g/100g H$_2$O: 1.33 (0°C), 1.44 (20°C), 2.73 (90°C) [LAN85]
Density, g/cm^3: 2.47 [LID94]
Melting Point, °C: 575 [LID94]

2725
Compound: Sodium tetrasulfide
Formula: Na$_2$S$_4$
Molecular Formula: Na$_2$S$_4$
Molecular Weight: 174.244
CAS RN: 12034-39-8
Properties: yellow, hygr cryst; can be clear, dark red liq; prepared by reacting Na$_2$S with S; used to reduce organic nitro compounds, for the manufacture of sulfur dyes, and in the preparation of metal sulfide finishes [HAW93]
Density, g/cm^3: 1.335 at 15.5°C [KIR83]
Melting Point, °C: cryst: 275 [HAW93]; solidifies at -33 to 10 [KIR83]
Boiling Point, °C: 115 [KIR83]

2726
Compound: Sodium thioantimonate nonahydrate
Synonyms: Schlippe's salt
Formula: Na$_3$SbS$_4$·9H$_2$O
Molecular Formula: H$_{18}$Na$_3$O$_9$S$_4$Sb
Molecular Weight: 481.131
CAS RN: 13776-84-6

Properties: colorless or light yellow large cryst; becomes covered with reddish brown coating of antimony sulfide if exposed to air [MER89]
Solubility: g anhydrous/100g H_2O: 13.4 (0°C), 27.9 (20°C), 88.3 (80°C) [LAN85]; i alcohol; decomposed by weak acids [MER89]
Density, g/cm³: 1.806 [CRC94]
Melting Point, °C: 87 [CRC94]
Boiling Point, °C: decomposes 234 [CRC94]

2727
Compound: Sodium thiocyanate
Formula: NaSCN
Molecular Formula: CNNaS
Molecular Weight: 81.074
CAS RN: 540-72-7
Properties: colorless cryst or white powd; hygr; sensitive to light; used as an analytical reagent, in dyeing and printing textiles [HAW93]
Solubility: g/100g H_2O: 142.6 (25°C), 225.6 (101.4°C); Solid phase, NaSCN·H_2O (25°C), NaSCN (100°C) [KRU93]; s alcohol [HAW93]
Melting Point, °C: 287 [HAW93]

2728
Compound: Sodium thiophosphate dodecahydrate
Formula: $Na_3PO_3S \cdot 12H_2O$
Molecular Formula: $H_{24}Na_3O_{15}PS$
Molecular Weight: 396.191
CAS RN: 51674-17-0
Properties: thin, six sided leaflets when prepared from water solvent; effloresces in dry air [MER89] [ALD94]
Solubility: v s warm H_2O [MER89]
Melting Point, °C: 60 [MER89]

2729
Compound: Sodium thiosulfate
Formula: $Na_2S_2O_3$
Molecular Formula: $Na_2O_3S_2$
Molecular Weight: 158.110
CAS RN: 7772-98-7
Properties: colorless monocl powd; hygr [CRC94] [ALD94] [MER89]
Solubility: s H_2O; i alcohol [MER89]
Density, g/cm³: 1.667 [ALD94]

2730
Compound: Sodium thiosulfate pentahydrate
Synonyms: hypo
Formula: $Na_2S_2O_3 \cdot 5H_2O$

Molecular Formula: $H_{10}Na_2O_8S_2$
Molecular Weight: 248.186
CAS RN: 10102-17-7
Properties: colorless cryst or granules; effloresces in warm dry air; somewhat deliq in moist air [MER89]
Solubility: g/100g H_2O: 50.2 (0°C), 70.1 (20°C), 104 (60°C) [LAN85]; i alcohol [MER89]
Density, g/cm³: 1.729 [STR93]
Melting Point, °C: 48, decomposes [STR93]
Reactions: minus $5H_2O$ at 100°C; decomposes at higher temperatures [MER89]

2731
Compound: Sodium titanate
Formula: $Na_2Ti_3O_7$
Molecular Formula: $Na_2O_7Ti_3$
Molecular Weight: 301.577
CAS RN: 12034-36-5
Properties: -200 mesh with 99.9% purity; white cryst; used in welding [HAW93] [CER91]
Solubility: i H_2O [HAW93]
Density, g/cm³: 3.35-3.50 [STR93]
Melting Point, °C: 1128 [STR93]

2732
Compound: Sodium trimetaphosphate hexahydrate
Synonyms: Knorre's salt
Formula: $(NaPO_3)_3 \cdot 6H_2O$
Molecular Formula: $H_{12}Na_3O_{15}P_3$
Molecular Weight: 413.976
CAS RN: 7785-84-4
Properties: efflorescent; tricl-rhomb prisms [MER89]
Solubility: 1 g/4.5mL H_2O; i alcohol [MER89]
Density, g/cm³: 1.786; anhydrous: 2.49 [MER89]
Melting Point, °C: 53 [MER89]
Reactions: minus H_2O when stored; anhydrous at 100°C [MER89]

2733
Compound: Sodium triphosphate
Synonyms: sodium tripolyphosphate
Formula: $Na_5P_3O_{10}$
Molecular Formula: $Na_5O_{10}P_3$
Molecular Weight: 367.864
CAS RN: 7758-29-4
Properties: white cryst powd; has two cryst forms; sl hygr; granules; used to soften water, as a food additive, and as a sequestering agent [HAW93] [MER89]

Solubility: g/100g soln, H$_2$O: 13.98 (0°C), 12.96 (25°C), 16.50 (70°C); Solid phase, Na$_5$P$_3$O$_{10}$·6H$_2$O [KRU93]
Melting Point, °C: 622 [HAW93]
Reactions: cryst form transition at 417°C [HAW93]

2734
Compound: Sodium tungstate
Formula: Na$_2$WO$_4$
Molecular Formula: Na$_2$O$_4$W
Molecular Weight: 293.818
CAS RN: 13472-45-2
Properties: -200 mesh with 99.9% purity; white rhomb powd [STR93] [KIR83] [CER91]
Solubility: g/100g H$_2$O: 71.5 (0°C), 73.0 (20°C), 97.2 (100°C) [LAN85]
Density, g/cm^3: 4.179 [KIR83]
Melting Point, °C: 698 [KIR83]

2735
Compound: Sodium tungstate dihydrate
Formula: Na$_2$WO$_4$·2H$_2$O
Molecular Formula: H$_4$Na$_2$O$_6$W
Molecular Weight: 329.848
CAS RN: 10213-10-2
Properties: colorless cryst or white rhomb cryst powd; effloresces in dry air [KIR83] [MER89]
Solubility: s in ~1.1 parts H$_2$O; i alcohol [MER89]
Density, g/cm^3: 3.245 [HAW93]
Melting Point, °C: 692 [STR93]
Reactions: minus 2H$_2$O at 100°C [MER89]

2736
Compound: Sodium uranate monohydrate
Synonyms: sodium metauranate
Formula: Na$_2$U$_2$O$_7$·H$_2$O
Molecular Formula: H$_2$Na$_2$O$_8$U$_2$
Molecular Weight: 652.049
CAS RN: 13721-34-1
Properties: yellow powd [MER89]
Solubility: i H$_2$O; s acids [MER89]

2737
Compound: Sodium uranyl carbonate
Formula: 2Na$_2$CO$_3$·UO$_2$CO$_3$
Molecular Formula: C$_3$Na$_4$O$_{11}$U
Molecular Weight: 542.014
CAS RN: 60897-40-7
Properties: yellow; obtained when uranium ores are leached with soda ash at high temperatures and high pressures [KIR83] [CRC94]

Melting Point, °C: decomposes at 400 [KIR83]

2738
Compound: Sodium zirconate
Formula: Na$_2$ZrO$_3$
Molecular Formula: Na$_2$O$_3$Zr
Molecular Weight: 185.202
CAS RN: 12201-48-8
Properties: -200 mesh with 99.5% purity [CER91]

2739
Compound: Stannic bromide
Synonyms: tin(IV) bromide
Formula: SnBr$_4$
Molecular Formula: Br$_4$Sn
Molecular Weight: 438.326
CAS RN: 7789-67-5
Properties: white cryst mass; fumes strongly in air; enthalpy of vaporization 43.5 kJ/mol; enthalpy of fusion 12.00 kJ/mol; used in mineral separations [HAW93] [MER89] [CRC93]
Solubility: v s water, evolving heat; s alcohol [MER89]
Density, g/cm^3: 3.34 [MER89]
Melting Point, °C: 31 [MER89]
Boiling Point, °C: 202 [MER89]

2740
Compound: Stannic chloride
Synonyms: tin(IV) chloride
Formula: SnCl$_4$
Molecular Formula: Cl$_4$Sn
Molecular Weight: 260.521
CAS RN: 7646-78-8
Properties: colorless liq; fumes in air; enthalpy of vaporization 34.9 kJ/mol; enthalpy of fusion 9.20 kJ/mol [MER89] [STR93] [CRC93]
Solubility: s H$_2$O, evolving heat; s alcohol, CCl$_4$, benzene, toluene, acetone, kerosene, gasoline [MER89]
Density, g/cm^3: 2.2788 [HAW93]
Melting Point, °C: -33 [MER89]
Boiling Point, °C: 114 [DOU83]

2741
Compound: Stannic chloride pentahydrate
Synonyms: tin(IV) chloride pentahydrate
Formula: SnCl$_4$·5H$_2$O
Molecular Formula: Cl$_4$H$_{10}$O$_5$Sn
Molecular Weight: 350.697
CAS RN: 10026-06-9

Properties: white or slighlty yellow cryst, or fused
  small lumps; slight odor of HCl [MER89]
Solubility: v s H_2O, alcohol [MER89]
Density, g/cm^3: 2.04 [KIR83]
Melting Point, °C: ~56 decomposes [KIR83]

2742
Compound: Stannic chromate
Formula: $Sn(CrO_4)_2$
Molecular Formula: Cr_2O_8Sn
Molecular Weight: 350.697
CAS RN: 38455-77-5
Properties: brownish yellow, cryst powd; used in
  coloring porcelain [HAW93] [MER89]
Solubility: s H_2O [MER89]
Melting Point, °C: decomposed by heating
  [MER89]

2743
Compound: Stannic fluoride
Synonyms: tin(IV) fluoride
Formula: SnF_4
Molecular Formula: F_4Sn
Molecular Weight: 194.704
CAS RN: 7783-62-2
Properties: snow white, tetr cryst; very hygr; can be
  prepared by reacting F_2 with many stannous or
  stannic compounds [KIR78] [MER89]
Solubility: hydrolyzes in H_2O [MER89]
Density, g/cm^3: 4.78 [MER89]
Melting Point, °C: sublimes 705 [STR93]

2744
Compound: Stannic iodide
Synonyms: tin(IV) iodide
Formula: SnI_4
Molecular Formula: I_4Sn
Molecular Weight: 626.328
CAS RN: 7790-47-8
Properties: -6 mesh with 99.999% purity; yellow to
  reddish cryst; hydrolyzes in H_2O; enthalpy of
  vaporization 56.9 kJ/mol [CRC93] [MER89]
  [CER91]
Solubility: s alcohol, benzene, chloroform, ether,
  carbon disulfide [MER89]
Density, g/cm^3: 4.473 [STR93]
Melting Point, °C: 144.5 [STR93]
Boiling Point, °C: 364.5 [CRC93]

2745
Compound: Stannic oxide

Synonyms: cassiterite
Formula: SnO_2
Molecular Formula: O_2Sn
Molecular Weight: 150.709
CAS RN: 18282-10-5
Properties: white or sl gray powd, or 3-12 mm
  sintered pieces of 99.9% purity; manufactured
  by blowing hot air over molten tin or by
  calcining the hydrated oxide; sintered pieces
  used as an evaporation material for anti static
  film and transparent heating elements, 99.9%
  pure material used as a sputtering target for
  transparent conductive films and in varistors
  [KIR83] [MER89] [CER91]
Solubility: i H_2O, alcohol, cold acids; slowly
  dissolves in hot, conc KOH or NaOH solutions
  [MER89]
Density, g/cm^3: 6.95 [MER89]
Melting Point, °C: 1630 [STR93]
Reactions: sublimes at 1800-1900°C [HAW93]

2746
Compound: Stannic selenide
Synonyms: tin diselenide
Formula: $SnSe_2$
Molecular Formula: Se_2Sn
Molecular Weight: 276.630
CAS RN: 20770-09-6
Properties: reddish brown cryst [MER89]
Solubility: s in alkali, conc acids; decomposed by
  HNO_3 [MER89]
Density, g/cm^3: 4.85 [MER89]
Melting Point, °C: 650 [MER89]

2747
Compound: Stannic selenite
Synonyms: tin selenite
Formula: $Sn(SeO_3)_2$
Molecular Formula: O_6Se_2Sn
Molecular Weight: 372.626
CAS RN: 7446-25-5
Properties: cryst powd [MER89]
Solubility: i H_2O; s in excess warm HCl [MER89]

2748
Compound: Stannic sulfide
Synonyms: mosaic gold; tin disulfide
Formula: SnS_2
Molecular Formula: S_2Sn
Molecular Weight: 182.842
CAS RN: 1315-01-1

Properties: yellow to brown powd; golden leaflets, metallic luster; has been used as a pigment [MER89] [HAW93]
Solubility: i H_2O, dil mineral acids; s aqua regia, alkali hydroxide solutions [MER89]
Density, g/cm³: 4.5 [MER89]
Melting Point, °C: decomposes at 600 [HAW93]

2749
Compound: Stannous acetate
Synonyms: tin(II) acetate
Formula: $Sn(CH_3COO)_2$
Molecular Formula: $C_4H_6O_4Sn$
Molecular Weight: 236.800
CAS RN: 638-39-1
Properties: white, ortho-rhomb cryst; decomposed by water; used as a reducing agent [HAW93] [MER89]
Solubility: s dil HCl [MER89]
Density, g/cm³: 2.31 [MER89]
Melting Point, °C: 182.5-183 [MER89]
Boiling Point, °C: sublimes, 155 (0.1 mm Hg) [STR93]

2750
Compound: Stannous bromide
Synonyms: tin(II) bromide
Formula: $SnBr_2$
Molecular Formula: Br_2Sn
Molecular Weight: 278.518
CAS RN: 10031-24-0
Properties: yellowish powd; oxidizes in air, turning brown; sensitive to moisture; enthalpy of vaporization 102 kJ/mol [CRC93] [MER89] [STR93] [HAW93]
Solubility: s in a small amount of H_2O, decomposed by a larger volume; s alcohol, ether, acetone [MER89]
Density, g/cm³: 5.12 [MER89]
Melting Point, °C: 215 [MER89]
Boiling Point, °C: 639 [CRC93]

2751
Compound: Stannous chloride
Synonyms: tin(II) chloride
Formula: $SnCl_2$
Molecular Formula: Cl_2Sn
Molecular Weight: 189.615
CAS RN: 7772-99-8

Properties: white; ortho-rhomb cryst; mass of flakes; can absorb atm oxygen; enthalpy of vaporization 86.8 kJ/mol; enthalpy of fusion 12.80 kJ/mol; used as a reducing agent [HAW93] [MER89] [CRC93]
Solubility: s H_2O, ethanol, acetone, ether, methyl acetate [MER89]
Density, g/cm³: 3.95 [MER89]
Melting Point, °C: 246 [DOU83]
Boiling Point, °C: 606 [DOU83]

2752
Compound: Stannous chloride dihydrate
Synonyms: tin(II) chloride dihydrate
Formula: $SnCl_2 \cdot 2H_2O$
Molecular Formula: $Cl_2H_4O_2Sn$
Molecular Weight: 225.646
CAS RN: 10025-69-1
Properties: white cryst; absorbs atm O_2, forming insoluble oxychloride; reducing agent [MER89] [STR93]
Solubility: s in less than its own weight of H_2O; forms insoluble material with more water; v s dil, conc HCl; s alcohol, NaOH soln, glacial acetic acid [MER89]
Density, g/cm³: 2.71 [MER89]
Melting Point, °C: 37-38 [MER89]
Boiling Point, °C: 652 [ALD94]
Reactions: decomposes when strongly heated [MER89]

2753
Compound: Stannous fluoride
Synonyms: tin(II) fluoride
Formula: SnF_2
Molecular Formula: F_2Sn
Molecular Weight: 156.707
CAS RN: 7783-47-3
Properties: white powd; hygr; monocl; forms an oxyfluoride in air; can be prepared by reacting SnO with aq HF; used in dental preparations [KIR78] [MER89] [STR93]
Solubility: g/100g in H_2O: 31 (0°), 78.5 (106°C); i alcohol, ether, chloroform [HAW93] [KIR83]
Density, g/cm³: 4.57 [MER89]
Melting Point, °C: 219 [STR93]
Boiling Point, °C: 850 [STR93]

2754
Compound: Stannous fluoroborate
Formula: $Sn(BF)_2$
Molecular Formula: B_2F_2Sn

Molecular Weight: 178.329

Properties: only available as a solution, e.g. 47% w; prepared by dissolution of SnO in fluoroboric acid; used in tin and tin-lead plating baths [KIR83]

2755

Compound: Stannous fluorophosphate

Synonyms: tin(II) fluorophosphate

Formula: $SnPO_3F$

Molecular Formula: FO_3PSn

Molecular Weight: 216.680

CAS RN: 52262-58-5

Properties: white powd [STR93]

2756

Compound: Stannous iodide

Synonyms: tin(II) iodide

Formula: SnI_2

Molecular Formula: I_2Sn

Molecular Weight: 372.519

CAS RN: 10294-70-9

Properties: reddish orange powd; moisture sensitive; enthalpy of vaporization 105 kJ/mol [CRC93] [STR93]

Solubility: g/100g H_2O: 0.99 (20°C), 1.42 (40°C), 4.20 (100°C) [LAN85]; decomposes in H_2O; s alkali chloride or iodide solutions, benzene, $CHCl_3$ [MER89]

Density, g/cm³: 5.28 [MER89]

Melting Point, °C: 320 [MER89]

Boiling Point, °C: 714 [CRC93]

2757

Compound: Stannous oxalate

Synonyms: tin(II) oxalate

Formula: SnC_2O_4

Molecular Formula: C_2O_4Sn

Molecular Weight: 206.730

CAS RN: 814-94-8

Properties: heavy, white cryst powd; used in dyeing and printing textiles [HAW93] [MER89]

Solubility: i H_2O; s dil HCl [MER89]

Density, g/cm³: 3.56 [MER89]

Melting Point, °C: 280, decomposes [STR93]

2758

Compound: Stannous oxide

Synonyms: tin(II) oxide

Formula: SnO

Molecular Formula: OSn

Molecular Weight: 134.709

CAS RN: 21651-19-4

Properties: brownish black powd; unstable in air [HAW93]

Solubility: i H_2O, alcohol; s acids, conc NaOH, KOH solutions [MER89]

Density, g/cm³: 6.45 [MER89]

Melting Point, °C: 1080 (600 mm Hg), decomposes [HAW93]

2759

Compound: Stannous pyrophosphate

Synonyms: tin(II) pyrophosphate

Formula: $Sn_2P_2O_7$

Molecular Formula: $O_7P_2Sn_2$

Molecular Weight: 411.363

CAS RN: 15578-26-4

Properties: white free flowing cryst; prepared from stannous chloride and sodium pyrophosphate; used in toothpastes [HAW93]

Solubility: i H_2O; s conc acid [MER89]

Density, g/cm³: 4.009 [MER89]

Melting Point, °C: decomposes above 400 [KIR83]

2760

Compound: Stannous selenide

Synonyms: tin(II) selenide

Formula: SnSe

Molecular Formula: SeSn

Molecular Weight: 197.670

CAS RN: 1315-06-6

Properties: 3-12 mm pieces with 99.999% purity; steel gray prisms [MER89] [CER91]

Solubility: i H_2O; s aqua regia, alkali sulfide and selenide solutions [MER89]

Density, g/cm³: 6.18 [MER89]

Melting Point, °C: 861 [MER89]

2761

Compound: Stannous stearate

Synonyms: tin(II) stearate

Formula: $Sn[CH_3(CH_2)_{16}COO]_2$

Molecular Formula: $C_{36}H_{70}O_4Sn$

Molecular Weight: 685.660

CAS RN: 7637-13-0

Properties: off-white powd [STR93]

2762

Compound: Stannous sulfate

Synonyms: tin(II) sulfate

Formula: $SnSO_4$

Molecular Formula: O₄SSn
Molecular Weight: 214.774
CAS RN: 7488-55-3
Properties: snow white ortho-rhomb cryst; can be prepared by reacting tin with excess sulfuric acid at 100°C for several days; principal use is in tin plating baths [KIR83] [MER89]
Solubility: 330 g/L H₂O at H₂O, hydrolyzes with precipitation of basic salt; s dil H₂SO₄ [MER89] [KIR83]
Melting Point, °C: decomposes at 378 to SnO₂ and SO₂ [MER89]

2763
Compound: Stannous sulfide
Synonyms: tin(II) sulfide
Formula: SnS
Molecular Formula: SSn
Molecular Weight: 150.776
CAS RN: 1314-95-0
Properties: dark gray cryst, or black amorphous powd; has been used as a pigment [KIR83] [MER89]
Solubility: i H₂O, alkali hydroxides; s conc HCl, hot conc H₂SO₄ [MER89]
Density, g/cm³: 5.08 [MER89]
Melting Point, °C: 880 [HAW93]
Boiling Point, °C: 1230 [HAW93]

2764
Compound: Stannous tartrate
Synonyms: tin(II) tartrate
Formula: SnC₄H₄O₆
Molecular Formula: C₄H₄O₆Sn
Molecular Weight: 266.782
CAS RN: 815-85-0
Properties: heavy, white cryst powd; used in dyeing and printing fabrics [HAW93]
Solubility: s H₂O, dil HCl [MER89]

2765
Compound: Stannous telluride
Synonyms: tin(II) telluride
Formula: SnTe
Molecular Formula: SnTe
Molecular Weight: 246.310
CAS RN: 12040-02-7
Properties: gray cryst; 3-12 mm pieces with 99.999% and 99.8% purity [CER91]
Density, g/cm³: 6.45 [CRC93]
Melting Point, °C: 807 (max) [CRC93]
Thermal Conductivity, W/(m·K): 9.1 [CRC93]

2766
Compound: Strontium
Formula: Sr
Molecular Formula: Sr
Molecular Weight: 87.62
CAS RN: 7440-24-6
Properties: silvery white metal; fcc; active metal, e.g. forms oxide film in air; electrical resistivity (20°C) 23 μohm·cm; enthalpy of fusion 7.43 kJ/mol; enthalpy of vaporization 136.9 kJ/mol; electronegativity 1.10 [CIC73] [MER89] [CRC93] [ALD94]
Solubility: reacts quickly with H₂O; s alcohol [HAW93]
Density, g/cm³: 2.63 [CIC73]
Melting Point, °C: 769 [CRC94]
Boiling Point, °C: 1384 [CRC94]
Thermal Conductivity, W/(m·K): 35.3 (25°C) [ALD94]
Thermal Expansion Coefficient: 22.5 x 10⁻⁶/K [CRC93]

2767
Compound: Strontium acetate
Formula: Sr(CH₃COO)₂
Molecular Formula: C₄H₆O₄Sr
Molecular Weight: 205.710
CAS RN: 543-94-2
Properties: white powd [STR93]
Solubility: g/100g H₂O: 36.93 (0.05°C), 40.19 (25°C), 36.36 (97°C); Solid phase, Sr(CH₃COO)₂·4H₂O (0°C), Sr(CH₃COO)₂·1/2H₂O (25°C, 97°C) [KRU93]
Density, g/cm³: 2.099 [STR93]
Melting Point, °C: decomposes [CRC94]

2768
Compound: Strontium acetate hemihydrate
Formula: Sr(CH₃COO)₂·1/2H₂O
Molecular Formula: C₄H₇O₄.₅Sr
Molecular Weight: 214.717
CAS RN: 543-94-2
Properties: white, cryst powd; ignites to SrCO₃ [ALF95] [MER89]
Solubility: s in 2.5 parts H₂O; sl s alcohol [MER89]
Reactions: minus 1/2H₂O at 150°C [MER89]

2769
Compound: Strontium acetylacetonate
Synonyms: 2,4-pentanedione, strontium derivative
Formula: Sr(CH₃COCH=C(O)CH₃)₂
Molecular Formula: C₁₀H₁₄O₄Sr

Molecular Weight: 285.839
CAS RN: 12193-47-4
Properties: white powd [STR93]

$$[CH_3-C=CH-\overset{\overset{\displaystyle O}{\|}}{C}-CH_3]_2Sr$$
$$\overset{\displaystyle O-}{\underset{\displaystyle |}{}}$$

Melting Point, °C: 220 decomposes [STR93]

2770
Compound: Strontium aluminate
Formula: SrAl$_2$O$_4$
Molecular Formula: Al$_2$O$_4$Sr
Molecular Weight: 205.581
CAS RN: 12004-37-4
Properties: -100 mesh with 99.5% purity [CER91]

2771
Compound: Strontium bromate monohydrate
Formula: Sr(BrO$_3$)$_2$·H$_2$O
Molecular Formula: Br$_2$H$_2$O$_7$Sr
Molecular Weight: 361.440
CAS RN: 14519-18-7
Properties: lustrous powd; white to sl yellow cryst; hygr [HAW93]
Solubility: g/100g soln H$_2$O: 18.32 (0°C), 27.25 (25°C), 41.00 (104°C); Solid phases: Sr(BrO$_3$)$_2$·H$_2$O (0°C,25°C), Sr(BrO$_3$)$_2$ (104°C) [KRU93]
Density, g/cm^3: 3.773 [HAW93]
Melting Point, °C: decomposes at 240 [HAW93]
Reactions: minus H$_2$O at 120°C [HAW93]

2772
Compound: Strontium bromide
Formula: SrBr$_2$
Molecular Formula: Br$_2$Sr
Molecular Weight: 247.428
CAS RN: 10476-81-0
Properties: white powd; -20 mesh with 99.5% purity; enthalpy of fusion 10.12 kJ/mol [CRC94] [CER91]
Solubility: g/100g H$_2$O: 85.2 (0°C), 107.0 (25°C), 222.5 (100°C) [KRU93]
Density, g/cm^3: 4.216 [STR93]
Melting Point, °C: 657 [CRC93]

2773
Compound: Strontium bromide hexahydrate
Formula: SrBr$_2$·6H$_2$O

Molecular Formula: Br$_2$H$_{12}$O$_6$Sr
Molecular Weight: 355.519
CAS RN: 10476-81-0
Properties: colorless; deliq; cryst or white granules; used as a sedative in medicine [HAW93] [MER89]
Solubility: s in 0.35 parts H$_2$O; s alcohol; i ether [MER89]
Density, g/cm^3: 2.386 [CRC94]
Reactions: minus 4H$_2$O at 89°C, minus 6H$_2$O by 180°C [HAW93]

2774
Compound: Strontium carbide
Formula: SrC$_2$
Molecular Formula: C$_2$Sr
Molecular Weight: 111.642
CAS RN: 12071-29-3
Properties: black tetr; -8 mesh with 99% purity [CER91] [CRC94]
Solubility: decomposed by H$_2$O [CRC94]
Density, g/cm^3: 3.2 [CRC94]
Melting Point, °C: >1700 [CRC94]

2775
Compound: Strontium carbonate
Synonyms: strontianite
Formula: SrCO$_3$
Molecular Formula: CO$_3$Sr
Molecular Weight: 147.629
CAS RN: 1633-05-2
Properties: -20 mesh with 99.999% purity, <30 ppm Ba; white powd; hygr [STR93] [CER91]
Solubility: g/L H$_2$O: 0.00082 (8.8°C), 0.0109 (24°C) [KRU93]; s dil acids [MER89]
Density, g/cm^3: 3.70 [STR93]
Melting Point, °C: 1497 [STR93]
Reactions: minus CO$_2$ at 1340 [HAW93]
Thermal Expansion Coefficient: (volume) 100°C (0.541), 200°C (1.168), 400°C (2.473), 800°C (5.726) [CLA66]

2776
Compound: Strontium chlorate
Formula: Sr(ClO$_3$)$_2$
Molecular Formula: Cl$_2$O$_6$Sr
Molecular Weight: 254.521
CAS RN: 7791-10-8
Properties: colorless or white cryst powd [HAW93] [MER89]

Solubility: g/100g soln, H$_2$O: 61.40 (0°C), 63.78 (25°C), 67.08 (95°C); Solid phase, Sr(ClO$_3$)$_2$·3H$_2$O (0°C), Sr(ClO$_3$)$_2$ (25°C, 95°C) [KRU93]; sl s alcohol [MER89]
Density, g/cm^3: 3.15 [MER89]
Melting Point, °C: 120, decomposes [HAW93]

2777
Compound: Strontium chloride
Formula: SrCl$_2$
Molecular Formula: Cl$_2$Sr
Molecular Weight: 158.525
CAS RN: 10476-85-4
Properties: -40 mesh with 99.5% purity; white powd; hygr; enthalpy of fusion 16.20 kJ/mol [STR93] [CRC93] [CER91]
Solubility: g/100g H$_2$O: 43.5 (0°C), 55.8 (25°C), 100.8 (100°C); Solid phase, SrCl$_2$·6H$_2$O (0°C, 25°C), SrCl$_2$·2H$_2$O (100°C) [KRU93]
Density, g/cm^3: 3.052 [STR93]
Melting Point, °C: 874 [CRC93]
Boiling Point, °C: 1250 [STR93]

2778
Compound: Strontium chloride hexahydrate
Formula: SrCl$_2$·6H$_2$O
Molecular Formula: Cl$_2$H$_{12}$O$_6$Sr
Molecular Weight: 266.617
CAS RN: 10025-70-4
Properties: colorless cryst; white granules; effloresces in air; deliq in presence of moisture [MER89]
Solubility: 3.5195±0.0026 mol/(kg·H$_2$O) at 25°C [RAR85b]; s alcohol [MER89]
Density, g/cm^3: 1.96 [MER89]
Melting Point, °C: 115 [STR93]
Reactions: minus 5H$_2$O at 100°C; minus 6H$_2$O at 150°C [MER89]

2779
Compound: Strontium chromate
Formula: SrCrO$_4$
Molecular Formula: CrO$_4$Sr
Molecular Weight: 203.614
CAS RN: 7789-06-2
Properties: -80 mesh with 99% purity; light yellow; monocl; has been used in metal coatings to protect the metal from corrosion [HAW93] [KIR78] [CER91]
Solubility: s dil acids [KIR78]; g/L soln, H$_2$O: 0.91 (25°C), 0.43 (100°C) [KRU93]
Density, g/cm^3: 3.895 [KIR78]

Melting Point, °C: decomposes [KIR78]

2780
Compound: Strontium dithionate tetrahydrate
Formula: Sr(SO$_3$)$_2$·4H$_2$O
Molecular Formula: H$_8$O$_{10}$S$_2$Sr
Molecular Weight: 319.809
CAS RN: 13845-16-4
Properties: trig [CRC77]
Solubility: g/100g soln, H$_2$O: 4.51 (0°C), 10.8 (20°C) [KRU93]
Density, g/cm^3: 2.373 [CRC77]
Reactions: minus 4H$_2$O, 78°C [CRC77]

2781
Compound: Strontium ferrite
Synonyms: strontium dodecairon nonadecaoxide
Formula: SrFe$_{12}$O$_{19}$
Molecular Formula: Fe$_{12}$O$_{19}$Sr
Molecular Weight: 1061.773
CAS RN: 12023-91-5
Properties: powd; -325 mesh with 99.5% purity [CER91] [ALF93]

2782
Compound: Strontium fluoride
Formula: SrF$_2$
Molecular Formula: F$_2$Sr
Molecular Weight: 125.617
CAS RN: 7783-48-4
Properties: white powd, and 99.9% pure melted pieces of 3-6 mm; hygr; enthalpy of fusion 29.70 kJ/mol; pieces used as evaporation and sputtering material for infrared transparent films [STR93] [CER91] [CRC93]
Solubility: g/L soln, H$_2$O: 0.1135 (0.26°C), 0.21±0.13 (25°C) [KRU93]; s dil acids; decomposed by strong acids [MER89]
Density, g/cm^3: 4.24 [MER89]
Melting Point, °C: 1477 [CRC93]
Boiling Point, °C: 2489 [STR93]
Reactions: oxidized to SrO >1000°C [MER89]

2783
Compound: Strontium hexaboride
Synonyms: strontium boride
Formula: SrB$_6$
Molecular Formula: B$_6$Sr
Molecular Weight: 152.486
CAS RN: 12046-54-7

Properties: -325 mesh 10 microns or less with 99.5% purity; refractory material [KIR78] [CER91]
Density, g/cm³: 3.39 [ALD94]
Melting Point, °C: 2235 [KIR78]

2784
Compound: Strontium hydride
Formula: SrH₂
Molecular Formula: H₂Sr
Molecular Weight: 89.636
CAS RN: 13598-33-9
Properties: -60 mesh with 99.5% purity; resembles CaH₂ in both properties and reactivity [KIR80] [CER91]
Density, g/cm³: 3.72 [KIR80]

2785
Compound: Strontium hydroxide
Formula: Sr(OH)₂
Molecular Formula: H₂O₂Sr
Molecular Weight: 121.635
CAS RN: 18480-07-4
Properties: colorless deliq cryst; absorbs H₂O from air; enthalpy of fusion 21.00 kJ/mol [HAW93] [CRC93]
Solubility: g/100g soln in H₂O: 0.90 (0°C), 2.16±0.41 (25°C), 47.71 (100°C); Solid phase, Sr(OH)₂·8H₂O [KRU93]
Density, g/cm³: 3.625 [HAW93]
Melting Point, °C: 512 [JAN71]
Reactions: minus H₂O at 710°C [CRC94]

2786
Compound: Strontium hydroxide octahydrate
Formula: Sr(OH)₂·8H₂O
Molecular Formula: H₁₈O₁₀Sr
Molecular Weight: 265.757
CAS RN: 1311-10-0
Properties: colorless; deliq cryst or white powd; forms SrCO₃ by reaction with atm CO₂ [MER89]
Solubility: s in 50 parts H₂O; 2.1 parts boiling H₂O [MER89]
Density, g/cm³: 1.90 [STR93]
Reactions: minus some H₂O ~100°C [MER89]

2787
Compound: Strontium iodate
Formula: Sr(IO₃)₂
Molecular Formula: I₂O₆Sr

Molecular Weight: 437.425
CAS RN: 13470-01-4
Properties: tricl; -80 mesh with 99.5% purity [CER91] [CRC94]
Solubility: g/100g soln, H₂O: 0.098 (0°C), 0.165 (25°C), 0.350 (100°C); Solid phase, Sr(IO₃)₂ [KRU93]
Density, g/cm³: 5.045 [CRC94]

2788
Compound: Strontium iodide
Formula: SrI₂
Molecular Formula: I₂Sr
Molecular Weight: 341.429
CAS RN: 10476-86-5
Properties: -80 mesh with 99.5% purity; white cryst; enthalpy of fusion 19.70 kJ/mol [CER91] [HAW93] [CRC93]
Solubility: g/100g H₂O: 165.3 (0°C), 181.2 (25°C), 383.1 (100°C); Solid phase, SrI₂·6H₂O (0°C, 25°C), SrI₂·2H₂O (100°C) [KRU93]
Density, g/cm³: 4.549 [HAW93]
Melting Point, °C: 515 [CRC94]
Boiling Point, °C: decomposes [HAW93]

2789
Compound: Strontium iodide hexahydrate
Formula: SrI₂·6H₂O
Molecular Formula: H₁₂I₂O₆Sr
Molecular Weight: 449.520
CAS RN: 10476-86-5
Properties: colorless to yellowish; deliq; sensitive to light and atm O₂ with partial oxidation freeing I₂ [MER89]
Solubility: s in 0.2 parts H₂O; s alcohol [MER89]
Density, g/cm³: 2.67 [HAW93]
Melting Point, °C: ~120, when rapidly heated [MER89]

2790
Compound: Strontium lactate trihydrate
Formula: Sr(C₃H₅O₃)₂·3H₂O
Molecular Formula: C₆H₁₆O₉Sr
Molecular Weight: 319.808
CAS RN: 29870-99-3
Properties: white, odorless, granular powd [MER89]
Solubility: s in 3 parts H₂O, 0.5 parts boiling H₂O; sl s alcohol [MER89]
Reactions: minus 3H₂O at 120°C [MER89]

2791
Compound: Strontium molybdate(VI)
Formula: $SrMoO_4$
Molecular Formula: MoO_4Sr
Molecular Weight: 247.558
CAS RN: 13470-04-7
Properties: -200 mesh with 99.9% purity; white cryst powd; scheelite structure, c/a = 2.23; used as an anticorrosion pigment, used in electronic and optical applications, in solid state lasers [HAW93] [KIR81] [CER91]
Solubility: ~0.003 g/100g H_2O [KIR81]
Density, g/cm³: 4.662 [KIR81]
Melting Point, °C: ~1040 [KIR81]

2792
Compound: Strontium niobate
Formula: $SrNb_2O_6$
Molecular Formula: Nb_2O_6Sr
Molecular Weight: 369.429
CAS RN: 12034-89-8
Properties: monocl cryst; -200 mesh with 99.9% purity [LID94] [CER91]
Density, g/cm³: 5.11 [LID94]
Melting Point, °C: 1225 [LID94]

2793
Compound: Strontium nitrate
Formula: $Sr(NO_3)_2$
Molecular Formula: N_2O_6Sr
Molecular Weight: 211.629
CAS RN: 10042-76-9
Properties: -8 mesh with 99.995% purity; white cryst; used in pyrotechnics, in marine signals, matches, and railroad flares [HAW93] [STR93] [CER91]
Solubility: g/100g soln, H_2O: 28.2 (0.1°C), 40.7 (20°C), 51.2 (105°C); Solid phase, $Sr(NO_3)_2 \cdot 4H_2O$ (0.1°C, 20°C), $Sr(NO_3)_2$ (105°C) [KRU93]; sl s alcohol, acetone [MER89]
Density, g/cm³: 2.99 [MER89]
Melting Point, °C: 570 [MER89]

2794
Compound: Strontium nitride
Formula: Sr_3N_2
Molecular Formula: N_2Sr_3
Molecular Weight: 290.873
CAS RN: 12033-82-8
Properties: -60 mesh with 99.5% purity [CER91]
Melting Point, °C: 1030 [CIC73]

2795
Compound: Strontium nitrite
Formula: $Sr(NO_2)_2$
Molecular Formula: N_2O_4Sr
Molecular Weight: 179.631
CAS RN: 13470-06-9
Properties: white or yellowish powd; hygr needles [HAW93]
Solubility: g/100g soln, H_2O: 43.1 (35°C), 58.1 (98°C); Solid phase, $Sr(NO_2)_2 \cdot H_2O$ [KRU93]; i alcohol [HAW93]
Density, g/cm³: 2.8 [HAW93]
Melting Point, °C: decomposes at 240 [HAW93]

2796
Compound: Strontium oxalate
Formula: SrC_2O_4
Molecular Formula: C_2O_4Sr
Molecular Weight: 175.640
CAS RN: 814-95-9
Properties: white powd [STR93]
Solubility: g/L soln, H_2O: 0.057 (0°C), 0.077 (20°C) [KRU93]

2797
Compound: Strontium oxalate monohydrate
Formula: $SrC_2O_4 \cdot H_2O$
Molecular Formula: $C_2H_2O_5Sr$
Molecular Weight: 193.655
CAS RN: 814-95-9
Properties: white, odorless; cryst powd; used in pyrotechnics, tanning, catalyst manufacturing [HAW93] [MER89]
Solubility: s in 20,000 parts H_2O; 1900 parts 3.5% acetic acid; s dil HCl, HNO_3 [MER89]
Reactions: minus H_2O at 150°C [HAW93]

2798
Compound: Strontium oxide
Synonyms: strontia
Formula: SrO
Molecular Formula: OSr
Molecular Weight: 103.619
CAS RN: 1314-11-0
Properties: white to grayish white; reacts with water, forming $Sr(OH)_2$, with evolution of heat; enthalpy of fusion 75.00 kJ/mol [MER89] [CRC93]
Solubility: g/100g H_2O: 1.03 (30°C), 1.05 (40°C), 12.15 (100°C) [LAN85]; reacts with H_2O to form the hydroxide [HAW93]
Density, g/cm³: 4.7 [MER89]

Melting Point, °C: 2430 [AES93]
Boiling Point, °C: ~3000 [HAW93]

2799
Compound: Strontium perchlorate
Formula: $Sr(ClO_4)_2$
Molecular Formula: Cl_2O_8Sr
Molecular Weight: 286.520
CAS RN: 13450-97-0
Properties: colorless cryst [HAW93]
Solubility: mol/kg H_2O: 8.16 (0°C), 10.67 (25°C), 12.70 (40°C); Solid phase, $Sr(ClO_4)_2\cdot4H_2O$ (0°C), $Sr(ClO_4)_2\cdot2H_2O$ (25°C), $3Sr(ClO_4)_2\cdot2H_2O$ (above 40°C) [KRU93]; s alcohol [HAW93]

2800
Compound: Strontium perchlorate hexahydrate
Formula: $Sr(ClO_4)_2\cdot6H_2O$
Molecular Formula: $Cl_2H_{12}O_{14}Sr$
Molecular Weight: 394.612
CAS RN: 13450-97-0
Properties: white cryst; hygr [STR93]
Melting Point, °C: <100 [STR93]

2801
Compound: Strontium permanganate trihydrate
Formula: $Sr(MnO_4)_2\cdot3H_2O$
Molecular Formula: $H_6Mn_2O_{11}Sr$
Molecular Weight: 379.537
CAS RN: 14446-13-0
Properties: purple cub [CRC94]
Solubility: g/100g soln, H_2O: 2.5 (0°C) [KRU93]
Density, g/cm³: 2.75 [CRC94]
Melting Point, °C: decomposes 175 [CRC94]

2802
Compound: Strontium peroxide
Formula: SrO_2
Molecular Formula: O_2Sr
Molecular Weight: 119.619
CAS RN: 1314-18-7
Properties: white powd; unstable if standing, decomposes in air under ambient conditions; decomposed in water, evolving O_2 [MER89]
Solubility: decomposed to H_2O_2 by dil acids [MER89]
Density, g/cm³: 4.56 [HAW93]
Melting Point, °C: 215, decomposes [HAW93]

2803
Compound: Strontium peroxide octahydrate
Formula: $SrO_2\cdot8H_2O$
Molecular Formula: $H_{16}O_{10}Sr$
Molecular Weight: 263.741
CAS RN: 1314-18-7
Properties: white powd [HAW93]
Solubility: sl s cold H_2O, decomposed by hot H_2O; s NH_4Cl solutions and alcohol [HAW93]
Density, g/cm³: 1.951 [HAW93]
Melting Point, °C: decomposes [CRC94]
Reactions: minus $8H_2O$ at 100°C [HAW93]

2804
Compound: Strontium salicylate dihydrate
Formula: $Sr(C_7H_5O_3)_2\cdot2H_2O$
Molecular Formula: $C_{14}H_{14}O_8Sr$
Molecular Weight: 397.880
CAS RN: 6160-38-9
Properties: white cryst or powd; odorless; light sensitive; heat causes decomposition; used in manufacturing of pharmaceuticals and fine chemicals [HAW93]
Solubility: s H_2O and alcohol [HAW93]
Melting Point, °C: decomposes [HAW93]

2805
Compound: Strontium selenate
Formula: $SrSeO_4$
Molecular Formula: O_4SeSr
Molecular Weight: 230.578
CAS RN: 7446-21-1
Properties: ortho-rhomb cryst; prepared by heating strontium carbonate with selenium or selenium oxide [MER89]
Solubility: i H_2O; s HCl [MER89]
Density, g/cm³: 4.25 [MER89]

2806
Compound: Strontium selenide
Formula: SrSe
Molecular Formula: SeSr
Molecular Weight: 166.580
CAS RN: 1315-07-7
Properties: white cub; -20 mesh with 99.5% purity [CER91] [CRC94]
Density, g/cm³: 4.38 [CRC94]
Melting Point, °C: 1600 [AES93]

2807
Compound: Strontium silicide

Formula: $SrSi_2$
Molecular Formula: Si_2Sr
Molecular Weight: 143.791
CAS RN: 12138-28-2
Properties: silver-gray cub cryst; 10 mm & down lump [LID94] [ALF93]
Density, g/cm³: 3.35 [LID94]
Melting Point, °C: 1100 [LID94]

2808
Compound: Strontium stannate
Formula: $SrSnO_3$
Molecular Formula: O_3SnSr
Molecular Weight: 254.328
CAS RN: 12143-34-9
Properties: -200 mesh with 99.5% purity [CER91]

2809
Compound: Strontium sulfate
Synonyms: celestite
Formula: $SrSO_4$
Molecular Formula: O_4SSr
Molecular Weight: 183.684
CAS RN: 7759-02-6
Properties: hygr white cryst or precipitate; used in pyrotechnics, in ceramics and glass [HAW93] [STR93]
Solubility: g/100mL soln, H_2O: 0.0121 (5°C), 0.0135 (25°C), 0.0113 (95°C); Solid phase, $SrSO_4$ [KRU93]; s HCl, HNO_3 [MER89]
Density, g/cm³: 3.96 [MER89]
Melting Point, °C: 1600 [STR93]

2810
Compound: Strontium sulfide
Formula: SrS
Molecular Formula: SSr
Molecular Weight: 119.686
CAS RN: 1314-96-1
Properties: -200 mesh with 99.9% purity; gray powd; has odor of H_2S in moist atm [MER89] [CER91]
Solubility: sl s H_2O; s acids, decomposes [MER89]
Density, g/cm³: 3.70 [MER89]
Melting Point, °C: >2000 [STR93]

2811
Compound: Strontium tantalate
Formula: $SrTa_2O_6$
Molecular Formula: O_6SrTa_2
Molecular Weight: 545.512

CAS RN: 12065-74-6
Properties: reacted product; -200 mesh with 99.9% purity [CER91]

2812
Compound: Strontium tartrate tetrahydrate
Formula: $SrC_4H_4O_6 \cdot 4H_2O$
Molecular Formula: $C_4H_{12}O_{10}Sr$
Molecular Weight: 307.753
CAS RN: 6100-96-5
Properties: monocl white cryst; used in pyrotechnics [HAW93] [CRC94]
Solubility: sl s H_2O [HAW93]
Density, g/cm³: 1.966 [HAW93]

2813
Compound: Strontium thiosulfate pentahydrate
Synonyms: strontium hyposulfite
Formula: $SrS_2O_3 \cdot 5H_2O$
Molecular Formula: $H_{10}O_8S_2Sr$
Molecular Weight: 289.826
Properties: monocl fine needles [CRC94] [HAW93]
Solubility: g/100g solution H_2O: 8.78 (0°C), 21.10 (27.5°C), 26.80 (40°C); Solid phase: $SrS_2O_3 \cdot 3H_2O$ [KRU93]; i alcohol [HAW93]
Density, g/cm³: 2.17 [HAW93]
Reactions: minus $4H_2O$ at 100°C [HAW93]

2814
Compound: Strontium titanate
Formula: $SrTiO_3$
Molecular Formula: O_3SrTi
Molecular Weight: 183.485
CAS RN: 12060-59-2
Properties: -200 mesh of 99.9% purity; white powd; cub; hardness 6-6.5 Mohs; has properties of refractive index, dispersion and optical transmission which are comparable to diamond; highly pure material can be obtained by calcining the double strontium titanate; used in electronics and in electrical insulation; as 99.9% pure material, used as sputtering target for thin film capacitors [HAW93] [KIR83] [CER91]
Solubility: i H_2O and in most solvents [HAW93]
Density, g/cm³: 4.81 [HAW93]
Melting Point, °C: 2060 [HAW93]

2815
Compound: Strontium tungstate
Formula: $SrWO_4$
Molecular Formula: O_4SrW

Molecular Weight: 335.458
CAS RN: 13451-05-3
Properties: -200 mesh with 99.9% purity; white tetr, a = 0.540 nm, c = 1.109 nm [KIR83] [CER91]
Solubility: 0.14 g/100mL H_2O (15°C) [CRC94]
Density, g/cm^3: 6.187 [KIR83]
Melting Point, °C: decomposes [CRC94]

2816
Compound: Strontium vanadate
Formula: SrV_2O_6
Molecular Formula: O_6SrV_2
Molecular Weight: 285.499
CAS RN: 12435-86-8
Properties: -200 mesh with 99.9% purity [CER91]

2817
Compound: Strontium zirconate
Formula: $SrZrO_3$
Molecular Formula: O_3SrZr
Molecular Weight: 226.842
CAS RN: 12036-39-4
Properties: white powd; used in electronics, and as the 99% pure material, is used as a sputtering target for thin film capacitors [HAW93] [STR93] [CER91]
Melting Point, °C: 2600 [HAW93]

2818
Compound: Sulfur chloride
Synonyms: sulfur monochloride
Formula: S_2Cl_2
Molecular Formula: Cl_2S_2
Molecular Weight: 135.037
CAS RN: 10025-67-9
Properties: light amber to yellowish red; fuming, oily liq; penetrating odor; dielectric constant, 4.9 (22°C); dipole moment, 1.60 [MER89]
Solubility: s alcohol, benzene, ether, CS_2, CCl_4, oils; decomposed in water, forming sulfur, HCl, SO_2, H_2S, sulfite, thiosulfate [MER89]
Density, g/cm^3: 1.6885 [MER89]
Melting Point, °C: -77 [MER89]
Boiling Point, °C: 138 [MER89]

2819
Compound: Sulfur dioxide
Formula: SO_2
Molecular Formula: O_2S
Molecular Weight: 64.065
CAS RN: 7446-09-5

Properties: colorless gas; not flammable; mild reducing agent, e.g. bleaches vegetable colors; vapor pressure is 3.2 atms at 20°C; critical temp 157.5°C; critical pressure 7.87 MPa; enthalpy of vaporization 24.94 kJ/mol [CRC93] [AIR87] [HAW93] [MER89]
Solubility: % H_2O: 17.7 (0°C), 11.9 (15°C), 8.5 (25°C), 6.4 (35°C); % other solvents: 25, alcohol; 32, methanol [MER89]
Density, g/cm^3: liq: 1.5 [MER89]
Melting Point, °C: -72 [MER89]
Boiling Point, °C: -10.0 [CRC93]

2820
Compound: Sulfur hexafluoride
Formula: SF_6
Molecular Formula: F_6S
Molecular Weight: 146.056
CAS RN: 2551-62-4
Properties: colorless, odorless gas; very stable, e.g. to electrical discharge (in transformer oil); does not attack glass; enthalpy of sublimation 23.59 kJ/mol; triple point -50.52°C; enthalpy of vaporization 9.642 kJ/mol; enthalpy of fusion 5.02 kJ/mol; dielectric constant of gas 1.00204; viscosity of gas 0.01576 mPa s; critical temp 45.55°C; critical pressure 3.759 MPa; critical density 0.737 g/cm^3 [KIR78] [MER89] [CRC93]
Solubility: sl s H_2O; s alcohol [MER89]
Density, g/cm^3: gas: 6.5 g/L; liq: 1.67 [HAW93]
Melting Point, °C: -50.8 [MER89]
Boiling Point, °C: sublimes -63.8 [MER89]
Thermal Conductivity, W/(m·K): liq is 0.0583; gas is 0.01415 [KIR78]

2821
Compound: Sulfur tetrafluoride
Formula: SF_4
Molecular Formula: F_4S
Molecular Weight: 108.060
CAS RN: 7783-60-0
Properties: colorless gas; stable to 600°C; reacts violently with H_2O; decomposed by conc H_2SO_4; critical temp 90.9°C; enthalpy of vaporization 26.44 kJ/mol; used in electronics industry [AIR87] [MER89] [CRC93]
Solubility: decomposes in H_2O [HAW93]
Density, g/cm^3: liq: 1.95 (-78°C); solid: 2.349 (-18.3°C) [MER89]
Melting Point, °C: -121.0 [MER89]
Boiling Point, °C: -40.5 [CRC93]
Reactions: attacks glass, but not quartz [MER89]

2822
Compound: Sulfur trioxide N,N-dimethylformamide complex
Formula: HCON(CH$_3$)$_2$·SO$_3$
Molecular Formula: C$_3$H$_7$NO$_4$S
Molecular Weight: 153.159
CAS RN: 29584-42-7
Properties: corrosive; uses: mild sulfating agent [ALD94]
Melting Point, °C: 115-158 [ALD94]

2823
Compound: Sulfur trioxide(α)
Formula: α-SO$_3$
Molecular Formula: O$_3$S
Molecular Weight: 80.064
CAS RN: 7446-11-9
Properties: solid; needles; vapor pressure (25°C) 73 mm; stable form is α; enthalpy of vaporization 43.14 kJ/mol at 25°C; enthalpy of fusion 8.60 kJ/mol [CRC93] [MER89] [HAW93]
Solubility: reacts vigorously with H$_2$O to form H$_2$SO$_4$ [MER89]
Density, g/cm^3: 1.97 [CRC94]
Melting Point, °C: 16.8 [CRC94]
Boiling Point, °C: 45 [CRC94]

2824
Compound: Sulfur trioxide(β)
Formula: β-SO$_3$
Molecular Formula: O$_3$S
Molecular Weight: 80.064
CAS RN: 7446-11-9
Properties: dimer; needles; metastable; vapor pressure, 25°C, 344 mm [MER89] [CRC94]

```
        O
        ‖
        S
      /   \
    O       O
```

Solubility: decomposed by H$_2$O [CRC94]
Melting Point, °C: 32.5 [MER89]

2825
Compound: Sulfur trioxide(γ)
Formula: γ-SO$_3$
Molecular Formula: O$_3$S
Molecular Weight: 80.064
CAS RN: 7446-11-9

Properties: form can be icy mass or liq; metastable [MER89] [DOU83]

Solubility: reacts violently with H$_2$O, forming H$_2$SO$_4$ [MER89]
Density, g/cm^3: liq: 1.9224 [MER89]
Melting Point, °C: 16.8 [MER89]
Boiling Point, °C: 44.8 [MER89]

2826
Compound: Sulfur(α)
Synonyms: brimstone
Formula: α-S
Molecular Formula: S
Molecular Weight: 32.066
CAS RN: 7704-34-9
Properties: yellow cryst; ortho-rhomb; stable form at usual temperatures; electrical resistivity (20°C) 2 x 10^{+23} μohm·cm; electronegativity 2.44; enthalpy of vaporization 45 kJ/mol; enthalpy of fusion 1.72 kJ/mol [CRC93] [MER89] [COT88]
Solubility: i H$_2$O; sl s alcohol and ether; s CS$_2$, CCl$_4$, benzene [HAW93]
Density, g/cm^3: 2.06 [MER89]
Melting Point, °C: 112.8 [ALD94]
Boiling Point, °C: 444.674 [ALD94]
Reactions: forms monocl S at 94.5°C [MER89]
Thermal Conductivity, W/(m·K): 0.205 (25°C) [ALD94]

2827
Compound: Sulfur(β)
Formula: β-S
Molecular Formula: S
Molecular Weight: 32.066
CAS RN: 7704-34-9
Properties: monocl; light yellow; opaque; brittle; stable form from 94.5 to 120°C; transforms if left standing to ortho-rhomb at a slow rate [MER89]
Solubility: i H$_2$O; sl s alcohol, ether; s 1 g/2mL CS$_2$; s benzene, toluene, acetone [MER89]
Density, g/cm^3: 1.957 [ALD94]
Melting Point, °C: 119.0 [ALD94]
Boiling Point, °C: 444.674 [ALD94]

2828
Compound: Sulfur(γ)
Synonyms: mother-of-pearl sulfur
Formula: γ-S
Molecular Formula: S
Molecular Weight: 32.066
CAS RN: 7704-34-9
Properties: yellow, amorphous [CRC77]
Solubility: i H_2O [CRC77]
Density, g/cm^3: 1.92 [CRC77]
Melting Point, °C: 106.8 [MER89]
Boiling Point, °C: 444.6 [CRC77]

2829
Compound: Sulfuric acid
Synonyms: oil of vitriol
Formula: H_2SO_4
Molecular Formula: H_2O_4S
Molecular Weight: 98.080
CAS RN: 7664-93-9
Properties: clear, colorless, oily liq; absorbs moisture from atm; can char organic materials, e.g. sugar; miscible with water, evolving heat; enthalpy of fusion 10.71 kJ/mol; specific conductance 1.044 x 10^{-2} at 25°C; dielectric constant 110 at 20°C [MER89] [COT88] [CRC93]
Solubility: miscible with H_2O [HAW93]
Density, g/cm^3: ~1.84 [MER89]
Melting Point, °C: 10.31 [CRC93]
Boiling Point, °C: ~290 [MER89]
Reactions: decomposes to SO_3 and H_2O at 340°C [MER89]

2830
Compound: Sulfuric acid fuming
Synonyms: oleum
Formula: $H_2SO_4 + SO_3$
Molecular Formula: $H_2S_2O_7$
Molecular Weight: 178.144
CAS RN: 8014-95-7
Properties: commercial acid contains up to 30% SO_3; colorless, or sl colored, viscous liq; choking fumes of SO_3 [MER89]

2831
Compound: Sulfurous acid
Formula: H_2SO_3
Molecular Formula: H_2O_3S
Molecular Weight: 82.080
CAS RN: 7782-99-2
Properties: solution of sulfur dioxide in water; colorless; clear liq; odor of SO_2; gradually oxidized to sulfate by atm O_2; mild reducing agent; e.g. dental bleach [MER89]
Solubility: s H_2O [HAW93]
Density, g/cm^3: ~1.03 [MER89]

2832
Compound: Sulfuryl chloride
Formula: SO_2Cl_2
Molecular Formula: Cl_2O_2S
Molecular Weight: 134.970
CAS RN: 7791-25-5
Properties: colorless liq; pungent odor; turns yellow slowly when standing due to slight dissociation into Cl_2 and SO_2; violent reaction with alkalies; enthalpy of vaporization 31.4 kJ/mol [CRC93] [MER89]
Solubility: slowly decomposed in H_2O, forming H_2SO_4 and HCl [MER89]; miscible with benzene, toluene, ether, acetic acid, other organic materials [MER89]
Density, g/cm^3: 1.664 [MER89]
Melting Point, °C: -54.1, -46 [MER89]
Boiling Point, °C: 69.3 [MER89]
Reactions: forms $SO_2Cl_2 \cdot 15H_2O$ with icy cold H_2O [MER89]

2833
Compound: Sulfuryl fluoride
Formula: SO_2F_2
Molecular Formula: F_2O_2S
Molecular Weight: 102.062
CAS RN: 2699-79-8
Properties: colorless odorless gas; not very reactive; stable to 400°C; not hydrolyzed in H_2O, but hydrolyzes in NaOH solutions; used in insecticides, fumigants [MER89] [HAW93]
Solubility: mL SO_2F_2/100mL solvent: 4-5, H_2O; 24-27, alcohol; 210-220, toluene; 136-138, CCl_4 [MER89]
Density, g/cm^3: gas: 4.55 g/L; liq; 1.7 [CRC94]
Melting Point, °C: -135.8 [MER89]
Boiling Point, °C: -55.4 [MER89]

2834
Compound: Tantalum
Formula: Ta
Molecular Formula: Ta
Molecular Weight: 180.9479
CAS RN: 7440-25-7

Properties: gray, very hard, malleable metal; bcc, a = 0.33026 nm; electrical resistivity (18°C) 12.4 μohm·cm; Poisson's ratio 0.35; Young's modulus at room temp 186; slowly reacts with fused alkalies; reacts with F_2, Cl_2, O_2 when heated; absorbs H_2 at high temp; enthalpy of vaporization 732.8 kJ/mol; enthalpy of fusion 36.57 kJ/mol; used to contain molten metals such as sodium [KIR83] [ALD94] [MER89] [CER91] [CRC93] [CAB93]

Solubility: very resistant to attack by acids except HF, resistant to alkali solutions [KIR83]

Density, g/cm³: 16.69 [MER89]

Melting Point, °C: 2996 [ALD94]

Boiling Point, °C: 5429 [MER89]

Thermal Conductivity, W/(m·K): 54.4 at 20°C [KIR83]

Thermal Expansion Coefficient: 8 x 10^{-6}/°C over the temp range 20-1500°C [HAW93]

2835

Compound: Tantalum aluminide

Formula: $TaAl_3$

Molecular Formula: Al_3Ta

Molecular Weight: 261.859

CAS RN: 12004-76-1

Properties: -80 mesh with 99.5% purity; gray refractory powd; oxidizes slowly in air above 500°C; formed by adding Al metal to potassium fluorotantalate at ~1000°C [KIR83] [CER91]

Solubility: i acids and alkalies [KIR83]

Density, g/cm³: 7.02 [KIR83]

Melting Point, °C: ~1400 [KIR83]

2836

Compound: Tantalum boride

Formula: TaB

Molecular Formula: BTa

Molecular Weight: 191.759

CAS RN: 12007-07-7

Properties: -325 mesh with 99.5% purity; refractory material; used as a sputtering target of 99.5% purity to produce wear-resistant and semiconductive films, and other uses [KIR78] [CER91]

Density, g/cm³: 14.2 [LID94]

Melting Point, °C: 2040 [KIR78]

2837

Compound: Tantalum carbide

Formula: Ta_2C

Molecular Formula: CTa_2

Molecular Weight: 373.907

CAS RN: 12070-07-4

Properties: hex, refractory; -325 mesh, 10 microns or less, 99.5% purity; hex, a = 0.3106 nm [CER91] [CIC73]

Density, g/cm³: 15.1 [LID94]

Melting Point, °C: 3327 [LID94]

2838

Compound: Tantalum diboride

Synonyms: tantalum boride

Formula: TaB_2

Molecular Formula: B_2Ta

Molecular Weight: 202.570

CAS RN: 12077-35-1

Properties: gray metallic powd; hardness >8 mohs; can be formed by heating tantalum and boron in vacuum at ~1800°C; 99.5% pure material used as a sputtering target to produce wear-resistant films and semiconductor films, and other applications [KIR83] [CER91]

Solubility: i acids and alkalies [KIR83]

Density, g/cm³: 11.15 [KIR83]

Melting Point, °C: ~3000 [KIR83]

2839

Compound: Tantalum disulfide

Formula: TaS_2

Molecular Formula: S_2Ta

Molecular Weight: 245.080

CAS RN: 12143-72-5

Properties: black powd or cryst; used as a solid lubricant, also as a 99% pure material used as a sputtering target to form lubricant film on bearings and other moving parts; there is TaS compound [HAW93] [CER91] [STR93]

Solubility: i H_2O [HAW93]

Density, g/cm³: 6.86 [LID94]

Melting Point, °C: for TaS: >1300 [STR93]

2840

Compound: Tantalum ethoxide

Formula: $Ta(OC_2H_5)_5$

Molecular Formula: $C_{10}H_{25}O_5Ta$

Molecular Weight: 406.254

CAS RN: 6074-84-6

Properties: yellow liq; moisture sensitive; 99.999% purity, <100 ppm Nb [CER91] [STR93]

Density, g/cm³: 1.566 [ALD94]

Melting Point, °C: 21 [CER91]

Boiling Point, °C: 145 at 0.1 mm Hg [CER91]

2841
Compound: Tantalum hydride
Formula: TaH
Molecular Formula: HTa
Molecular Weight: 181.956
CAS RN: 13981-95-8
Properties: gray, brittle with metallic luster; can form when H_2 is absorbed by Ta at 450°C; hydrogen is released when TaH is heated above 800°C; material is a superconductor [KIR83]
Solubility: resistant to attack by acids [KIR83]
Density, g/cm³: 15.1 [KIR83]

2842
Compound: Tantalum monocarbide
Formula: TaC
Molecular Formula: CTa
Molecular Weight: 192.959
CAS RN: 12070-06-3
Properties: cryst, very fine golden brown powd; fcc, a = 0.44555 nm; refractory material; resistivity is 30 μohm·cm at room temp; hardness 9-10 Mohs prepared by reaction of tantalum powd and carbon black at ~1900°C in an inert atm; used in crucible form for melting zirconium oxide and similar oxides with high melting points, and as a sputtering target [CER91] [KIR83] [CIC73]
Solubility: i acids except mixture of HF and HNO_3 [KIR83]
Density, g/cm³: 13.9 [KIR83]
Melting Point, °C: 3880 [KIR83]
Boiling Point, °C: 5500 [HAW93]
Thermal Conductivity, W/(m·K): 22 [KIR78]
Thermal Expansion Coefficient: 6.29 x 10⁻⁶/K [KIR78]

2843
Compound: Tantalum nitride(δ)
Formula: δ-TaN
Molecular Formula: NTa
Molecular Weight: 194.955
CAS RN: 12033-62-4
Properties: yellowish gray; fcc, a = 0.4336 nm; microhardness 3200; transition temp 17.8 K; used as a 99.5% pure sputtering target to increase electrical stability of diodes, transistors and integrated circuits [KIR81] [CER91]
Solubility: i acids; decomposed by KOH with evolution of NH_3 [KIR83]
Density, g/cm³: 15.6 [KIR81]
Melting Point, °C: 2950 [KIR81]

2844
Compound: Tantalum nitride(ε)
Formula: ε-TaN
Molecular Formula: NTa
Molecular Weight: 194.955
CAS RN: 12033-62-4
Properties: brown, bronze or black cryst; hex, a = 0.5191 nm, c = 0.2906 nm; electrical resistivity 128 μohm·m; hardness 1100 microhardness; transition temp 1.8 K; forms when tantalum is heated in pure nitrogen ~1100°C [KIR81] [KIR83] [HAW93] [CIC73]
Solubility: i H_2O; sl s in aqua regia, HNO_3 and HF [HAW93]
Density, g/cm³: 13.8 [KIR83]
Melting Point, °C: 2800 [KIR83]
Reactions: evolves N_2 if heated to 2000°C [KIR83]
Thermal Conductivity, W/(m·K): 9.54 [KIR81]

2845
Compound: Tantalum pentabromide
Synonyms: tantalum(V) bromide
Formula: TaBr₅
Molecular Formula: Br₅Ta
Molecular Weight: 580.468
CAS RN: 13451-11-1
Properties: yellow cryst powd; sensitive to moisture; enthalpy of vaporization 62.3 kJ/mol; enthalpy of fusion 45.60 kJ/mol; can be prepared by heating tantalum metal in pure bromine gas above 300°C [STR93] [CRC93]
Density, g/cm³: 4.67 [STR93]
Melting Point, °C: 240 [ALD94]
Boiling Point, °C: 349 [CRC93]

2846
Compound: Tantalum pentachloride
Synonyms: tantalum(V) chloride
Formula: TaCl₅
Molecular Formula: Cl₅Ta
Molecular Weight: 358.212
CAS RN: 7721-01-9
Properties: resublimed yellow cryst powd; monocl; decomposed in moist atm; enthalpy of fusion 35.10 kJ/mol; enthalpy of vaporization 54.8 kJ/mol; formed by reaction of Cl_2 with tantalum at 200°C; used in the chlorination of organic materials [HAW93] [MER89] [CRC93]
Solubility: decomposed in H_2O; s absolute alcohol [MER89]
Density, g/cm³: 3.68 [MER89]
Melting Point, °C: 216 [CRC94]
Boiling Point, °C: 242 [CRC94]

2847
Compound: Tantalum pentafluoride
Synonyms: tantalum(V) fluoride
Formula: TaF_5
Molecular Formula: F_5Ta
Molecular Weight: 275.940
CAS RN: 7783-71-3
Properties: off-white deliq powd; enthalpy of vaporization is 56.9 kJ/mol; slowly etches glass; can be produced by fluorination of Ta metal; used as a catalyst in organic reactions [MER89] [STR93] [HAW93] [CRC93] [KIR78]
Solubility: s H_2O, ether, conc HNO_3; sl s hot CS_2, CCl_4 [MER89]
Density, g/cm³: 4.74 [MER89]
Melting Point, °C: 96.8 [KIR83]
Boiling Point, °C: 229.5 [MER89]

2848
Compound: Tantalum pentaiodide
Synonyms: tantalum(V) iodide
Formula: TaI_5
Molecular Formula: I_5Ta
Molecular Weight: 815.470
CAS RN: 14693-81-3
Properties: hex black powd; sensitive to moisture [LID94] [STR93]
Density, g/cm³: 5.80 [LID94]
Melting Point, °C: 496 [KIR83]
Boiling Point, °C: 543 [KIR83]
Reactions: minus iodine above 1000°C [KIR83]

2849
Compound: Tantalum pentoxide
Synonyms: tantalum(V) oxide
Formula: Ta_2O_5
Molecular Formula: O_5Ta_2
Molecular Weight: 441.893
CAS RN: 1314-61-0
Properties: white microcryst rhomb powd; decomposed by fusing with $KHSO_4$ or KOH; forms potassium tantalate when fused with KOH; enthalpy of fusion 120.00 kJ/mol; can be prepared by igniting Ta in air; used in optical glass, lasers and as a dielectric material, also as an evaporation material and sputtering target of 99.95% purity for dielectric films and multilayers [HAW93] [MER89] [KIR83] [CER91] [CRC93]
Solubility: i H_2O, alcohol, mineral acids; s HF [MER89]
Density, g/cm³: 8.2 [KIR83]
Melting Point, °C: 1800 [KIR83]

Reactions: reacts with carbon at 1900°C to form the carbide [KIR83]

2850
Compound: Tantalum pentoxide hydrate
Synonyms: tantalic acid
Formula: $Ta_2O_5 \cdot xH_2O$
Molecular Formula: O_5Ta_2 (anhydrous)
Molecular Weight: 441.893 (anhydrous)
CAS RN: 75397-94-3
Properties: white insoluble precipitate formed by leaching a potassium pyrosulfate fusion of tantalum in H_2O; tantalic acid forms organic complexes with tannic, oxalic, tartaric, citric and pyrogallic acids; used in analytical chemistry [KIR83]

2851
Compound: Tantalum phosphide
Formula: TaP
Molecular Formula: PTa
Molecular Weight: 211.922
CAS RN: 12037-63-7
Properties: -100 mesh with 99.5% purity [CER91]

2852
Compound: Tantalum selenide
Formula: $TaSe_2$
Molecular Formula: Se_2Ta
Molecular Weight: 338.868
CAS RN: 12039-55-3
Properties: -325 mesh, 10 microns or less, 99.8% pure material used as a sputtering target to produce lubricant films [CER91]

2853
Compound: Tantalum silicide
Formula: $TaSi_2$
Molecular Formula: Si_2Ta
Molecular Weight: 237.119
CAS RN: 12039-79-1
Properties: gray powd; used in the form of a 99.5 to 99.95% pure material as a sputtering target in the fabrication of integrated circuits; there is also the compound Ta_5Si_3, 12067-56-0 [STR93] [CER91]
Density, g/cm³: 9.14 [STR93]
Melting Point, °C: 2200 [STR93]

2854
Compound: Tantalum telluride
Formula: $TaTe_2$
Molecular Formula: $TaTe_2$
Molecular Weight: 436.148
CAS RN: 12067-66-2
Properties: -325 mesh 10 microns or less with 99.8% purity [CER91]

2855
Compound: Tantalum tetroxide
Formula: Ta_2O_4
Molecular Formula: O_4Ta_2
Molecular Weight: 425.894
CAS RN: 12035-90-4
Properties: dark gray powd; probably forms when the pentoxide is partially reduced by carbon at 1900°C [KIR83] [CRC94]
Reactions: oxidizes on heating [CRC94]

2856
Compound: Tantalum trisilicide
Formula: Ta_5Si_3
Molecular Formula: Si_3Ta_5
Molecular Weight: 988.996
CAS RN: 12067-56-0
Properties: -325 mesh powd; used as a 99.5 or 99.95% material as a sputtering target in the fabrication of integrated circuits [ALF93] [CER91]

2857
Compound: Technetium
Formula: Tc
Molecular Formula: Tc
Molecular Weight: 98
CAS RN: 7440-26-8
Properties: closed-packed hex, a = 0.2741 nm, c = 0.4399 nm; enthalpy of sublimation 650 kJ/mol; enthalpy of vaporization ~577 kJ/mol; enthalpy of fusion 33.29 kJ/mol; slowly tarnishes in moist air; when obtained from H_2 reduction of ammonium pertechnate, has silvery gray color, and a spongy mass; resembles rhenium in chemical behavior; Debye constant 455 K; used as a metallurgical tracer, in nuclear medicine, and to protect against corrosion [HAW93] [MER89] [RAR83] [CRC93]
Density, g/cm³: 11.5 [HAW93]
Melting Point, °C: 2167 [RAR83]
Boiling Point, °C: ~4600 [RAR83]
Thermal Conductivity, W/(m·K): 50.6 [CRC93]

Thermal Expansion Coefficient: a-axis: 7.04 x 10^{-6}/K; c-axis: 7.06 x 10^{-6}/K [RAR83]

2858
Compound: Technetium dioxide
Formula: TcO_2
Molecular Formula: O_2Tc
Molecular Weight: 130
CAS RN: 12036-16-7
Properties: there is a monohydrate, CAS RN 42861-23-4, and a dihydrate, CAS RN 60003-95-4 [ERI92]
Solubility: H_2O: $TcO_2 \cdot nH_2O = TcO(OH)_2(aq) + (n-1)H_2O$, log K = -8.16; $TcO_2 \cdot n2H_2O + H_2O = Tc(OH)_3^- + (n-1)H_2O + H^+$, log K = -19.20; reference also contains predominance diagram [ERI92]

2859
Compound: Telluric acid
Synonyms: orthotelluric acid
Formula: H_6TeO_6
Molecular Formula: H_6O_6Te
Molecular Weight: 229.644
CAS RN: 7803-68-1
Properties: -40 mesh with 99.5% purity; white solid; monocl, cub; very weak acid, $K_1 = 2$ x 10^{-8}, $K_2 = 1$ x 10^{-11} [MER89] [CER91]
Solubility: g H_2TeO_6/100g H_2O: 16.2 (0°C), 41.6 (20°C), 155 (100°C) [LAN85]; tends to polymerize; s dil HNO_3 [MER89]
Density, g/cm³: monocl: 3.068; cub: 3.163 [MER89]
Melting Point, °C: 136 [HAW93]
Boiling Point, °C: 160, decomposes [STR93]

2860
Compound: Tellurium
Formula: Te
Molecular Formula: Te
Molecular Weight: 127.60
CAS RN: 13494-80-9
Properties: grayish white, lustrous, brittle; rhomb cryst; hardness, 2.3 Mohs; Poisson's ratio 0.33 at 30°C; enthalpy of fusion 17.87 kJ/mol; enthalpy of vaporization 114.1 kJ/mol; electrical resistivity (20°C) (5.8-33) x 10^{+3} µohm·cm; modulus of elasticity 4140 MPa; viscosity at mp 1.8-1.95 mPa s; electronegativity 2.01; burns with greenish blue flame; p-type semiconductor; used in thin film devices as blocking contact [HAW93] [MER89] [CRC93] [KIR83] [COT88] [CER91] [ALD94]
Solubility: i H_2O, benzene, CS_2 [MER89]

Density, g/cm³: 6.24 (cryst) [KIR83]
Melting Point, °C: 449.8 [MER89]
Boiling Point, °C: 989.8 [ALD94]
Reactions: reacts with HNO_3, conc H_2SO_4, KOH, forming red solution [MER89]
Thermal Conductivity, W/(m·K): 1.97 to 3.38 (25°C) [ALD94]

2861
Compound: Tellurium decafluoride
Formula: Te_2F_{10}
Molecular Formula: $F_{10}Te_2$
Molecular Weight: 445.184
CAS RN: 53214-07-6
Properties: volatile, colorless liq; stable [KIR83]
Melting Point, °C: -33.7 [KIR83]
Boiling Point, °C: 59 [KIR83]

2862
Compound: Tellurium dibromide
Synonyms: tellurous bromide
Formula: $TeBr_2$
Molecular Formula: Br_2Te_2
Molecular Weight: 287.408
CAS RN: 7789-54-0
Properties: greenish black cryst mass, or gray to black needles; very hygr; has a violet vapor [HAW93]
Solubility: decomposed by H_2O; s ether; sl s chloroform [MER89]
Melting Point, °C: 210 [HAW93]
Boiling Point, °C: 339 [HAW93]

2863
Compound: Tellurium dichloride
Synonyms: tellurous chloride
Formula: $TeCl_2$
Molecular Formula: Cl_2Te
Molecular Weight: 198.505
CAS RN: 10025-71-5
Properties: black amorphous solid; hygr; melts to a black liq; purple vapor; disproportionates in ether, dioxane [MER89] [KIR83]
Solubility: decomposed by H_2O; i CCl_4 [MER89] [HAW93]
Density, g/cm³: 6.9 [HAW93]
Melting Point, °C: 208 [MER89]
Boiling Point, °C: 328 [MER89]

2864
Compound: Tellurium dioxide

Synonyms: tellurite
Formula: TeO_2
Molecular Formula: O_2Te
Molecular Weight: 159.599
CAS RN: 7446-07-3
Properties: white cryst; tetr and ortho-rhomb; yellow when heated; made by dissolving Te in strong HNO_3 to form $2TeO_2 \cdot HNO_3$ at 400-430°C, then decomposing this product [KIR83] [MER89]
Solubility: s H_2O ~1:150,000; s NaOH, HCl solutions [MER89]
Density, g/cm³: tetr: 5.75; ortho-rhomb: 6.04 [MER89]
Melting Point, °C: 733, forming yellow liq [MER89]
Boiling Point, °C: 1245 [HAW93]

2865
Compound: Tellurium disulfide
Formula: TeS_2
Molecular Formula: S_2Te
Molecular Weight: 191.732
CAS RN: 7446-35-7
Properties: red powd; eventually turns to brown amorphous powd; fuses to a gray, lustrous mass [HAW93]
Solubility: i H_2O, acids; s in alkali sulfides [HAW93]

2866
Compound: Tellurium hexafluoride
Formula: TeF_6
Molecular Formula: F_6Te
Molecular Weight: 241.590
CAS RN: 7783-80-4
Properties: colorless gas; does not attack glass when pure [MER89]
Solubility: slowly mixes with H_2O, hydrolyzing to telluric acid [MER89]
Density, g/cm³: solid (-191°C): 4.006; liq (-10°C): 2.499 [MER89]
Melting Point, °C: -37.6 [MER89]
Boiling Point, °C: 35.5 [CRC94]
Reactions: reduced by Te to TeF_4 [KIR83]

2867
Compound: Tellurium nitrate
Synonyms: basic tellurium nitrate
Formula: $TeO_2 \cdot NO_3$
Molecular Formula: NO_5Te
Molecular Weight: 221.604

Properties: prepared by dissolution of Te in HNO$_3$ [KIR83]

Melting Point, °C: decomposes from 190-300 [KIR83]

2868

Compound: Tellurium nitride

Formula: Te$_3$N$_4$

Molecular Formula: N$_4$Te$_3$

Molecular Weight: 438.827

CAS RN: 12164-01-1

Properties: citron-yellow colored solid; unstable, and can detonate readily if struck or heated [KIR83]

Solubility: could explode on contact with H$_2$O [KIR83]

2869

Compound: Tellurium sulfate

Synonyms: basic tellurium sulfate

Formula: 2TeO$_2$·SO$_3$

Molecular Formula: O$_7$STe$_2$

Molecular Weight: 399.262

CAS RN: 12068-84-8

Properties: prepared by dissolution of TeO$_2$ solution in H$_2$SO$_4$, followed by slow evaporation to form the compound; stable up to 440-500°C [KIR83]

Solubility: slowly hydrolyzed by cold H$_2$O, rapidly by hot H$_2$O [KIR83]

2870

Compound: Tellurium tetrabromide

Formula: TeBr$_4$

Molecular Formula: Br$_4$Te

Molecular Weight: 447.216

CAS RN: 10031-27-3

Properties: -4 mesh with 99.9% purity; orange cryst, turning red when hot [MER89] [CER91]

Solubility: s in small volume of H$_2$O, hydrolyzes in larger volume [MER89]

Density, g/cm^3: 4.3 [MER89]

Melting Point, °C: ~380 [MER89]

Boiling Point, °C: 420, decomposes to TeBr$_2$ and Br$_2$ [HAW93]

2871

Compound: Tellurium tetrachloride

Formula: TeCl$_4$

Molecular Formula: Cl$_4$Te

Molecular Weight: 269.411

CAS RN: 10026-07-0

Properties: white, cryst sold; very hygr; decomposed by water to TeO$_2$ and HCl; melts to a yellow liq, dark red at higher temperatures; enthalpy of vaporization 77 kJ/mol [MER89] [CRC93]

Solubility: s in absolute alcohol, toluene [MER89]

Density, g/cm^3: 3.01 [MER89]

Melting Point, °C: 225 [MER89]

Boiling Point, °C: ~390 [KIR83]

2872

Compound: Tellurium tetrafluoride

Formula: TeF$_4$

Molecular Formula: F$_4$Te

Molecular Weight: 203.594

CAS RN: 15192-26-4

Properties: white hygr needles; attacks glass, silica, and copper at 200°C, does not attack platinum below 300°C [KIR83]

Solubility: readily hydrolyzed [KIR83]

Melting Point, °C: decomposes to TaF$_6$ at 194 [KIR83]

2873

Compound: Tellurium tetraiodide

Formula: TeI$_4$

Molecular Formula: I$_4$Te

Molecular Weight: 635.218

CAS RN: 7790-48-9

Properties: -4 mesh with 99.9% purity; grayish black volatile cryst; stable in moist air [MER89] [KIR83] [CER91]

Solubility: hydrolyzed by H$_2$O forming TeO$_2$, HI; s HI; sl s acetone [MER89]

Density, g/cm^3: 5.05 [MER89]

Melting Point, °C: 280 [MER89]

Reactions: I$_2$ evolved by heating [MER89]

2874

Compound: Tellurium trioxide

Formula: TeO$_3$

Molecular Formula: O$_3$Te

Molecular Weight: 175.598

CAS RN: 13451-18-8

Properties: -60 mesh with 99.9% purity; two forms: yellowish orange α, and grayish β; α is a strong oxidant, e.g. reacting vigorously with Al, Sn, C, P and S [KIR83] [CER91]

Density, g/cm^3: α: 5.07; β: 6.21 [KIR83]

Melting Point, °C: decomposes [CRC94]

2875
Compound: Tellurous acid
Formula: H_2TeO_3
Molecular Formula: H_2O_3Te
Molecular Weight: 177.614
CAS RN: 10049-23-7
Properties: unstable white cryst, or cryst powd; dehydrates readily to TeO_2 [KIR83] [MER89]
Solubility: sl s H_2O; s in dil acids, alkalies [MER89]
Density, g/cm³: 3.05 [HAW93]
Melting Point, °C: 40, decomposes [HAW93]

2876
Compound: Terbium
Formula: Tb
Molecular Formula: Tb
Molecular Weight: 158.92534
CAS RN: 7440-27-9
Properties: silvery gray metal; easily oxidized by atm O_2; hex close-packed; electrical resistivity (20°C) 116 μohm·cm; enthalpy of fusion is 10.80 kJ/mol; enthalpy of sublimation 288.7 kJ/mol; radius of atom is 0.17833 nm; radius of Tb^{+++} ion is 0.0923 nm; forms colorless solutions; used as a phosphor activator [HAW93] [MER89] [KIR82] [ALD94]
Solubility: slowly reacts with H_2O; s dil acids [HAW93]
Density, g/cm³: 8.27 [MER89]
Melting Point, °C: 1356 [MER89]
Boiling Point, °C: 3230 [ALD94]
Thermal Conductivity, W/(m·K): 11.1 (25°C) [CRC93]
Thermal Expansion Coefficient: 10.3×10^{-6}/K [CRC93]

2877
Compound: Terbium acetate hydrate
Formula: $Tb(CH_3COO)_3 \cdot xH_2O$
Molecular Formula: $C_6H_9O_6Tb$ (anhydrous)
Molecular Weight: 336.059 (anhydrous)
CAS RN: 100587-92-6
Properties: hygr [ALD94]

2878
Compound: Terbium acetylacetonate trihydrate
Synonyms: 2,4-pentanedione, terbium(III) derivative
Formula: $Tb(CH_3COCH=C(O)CH_3)_3 \cdot 3H_2O$
Molecular Formula: $C_{15}H_{27}O_9Tb$
Molecular Weight: 510.299
CAS RN: 14284-95-8
Properties: white powd; hygr [STR93]

Melting Point, °C: 168-170 [STR93]

2879
Compound: Terbium bromide
Formula: $TbBr_3$
Molecular Formula: Br_3Tb
Molecular Weight: 398.637
CAS RN: 14456-47-4
Properties: hex, silvery gray; -20 mesh with 99.9% purity [CER91] [CRC94]
Melting Point, °C: 827 [AES93]
Boiling Point, °C: 1490 [CRC94]

2880
Compound: Terbium carbonate hydrate
Formula: $Tb_2(CO_3)_3 \cdot xH_2O$
Molecular Formula: $C_3O_9Tb_2$ (anhydrous)
Molecular Weight: 497.878 (anhydrous)
CAS RN: 100587-96-0
Properties: white powd [STR93]

2881
Compound: Terbium chloride
Formula: $TbCl_3$
Molecular Formula: Cl_3Tb
Molecular Weight: 265.283
CAS RN: 10042-88-3
Properties: -20 mesh with 99.9% purity; off-white powd; hygr [STR93] [CER91]
Solubility: s H_2O [MER89]
Density, g/cm³: 4.35 [MER89]
Melting Point, °C: 588 [MER89]

2882
Compound: Terbium chloride hexahydrate
Formula: $TbCl_3 \cdot 6H_2O$
Molecular Formula: $Cl_3H_{12}O_6Tb$
Molecular Weight: 373.374
CAS RN: 13798-24-8
Properties: -4 mesh with 99.9% purity; transparent, colorless prismatic cryst; very hygr [HAW93] [CER91]
Solubility: v s H_2O; forms supersaturated solutions [MER89]
Density, g/cm³: 4.35 [HAW93]
Melting Point, °C: 588 (anhydrous) [HAW93]
Reactions: minus $6H_2O$ 180-200°C (in HCl gas stream) [MER89]

2883
Compound: Terbium fluoride
Formula: TbF_3
Molecular Formula: F_3Tb
Molecular Weight: 215.920
CAS RN: 13708-63-9
Properties: white powd, and 99.9% pure melted pieces of 3-12 mm; hygr; pieces used as evaporation material for possible application to multilayers [STR93] [CER91]
Melting Point, °C: 1172 [STR93]

2884
Compound: Terbium hydride
Formula: TbH_{2-3}
Molecular Formula: H_2Tb; H_3Tb
Molecular Weight: TbH_2: 160.941; TbH_3: 161.949
CAS RN: 13598-54-4
Properties: -60 mesh with 99.9% purity [CER91]

2885
Compound: Terbium iodide
Formula: TbI_3
Molecular Formula: I_3Tb
Molecular Weight: 539.638
CAS RN: 13813-40-6
Properties: -20 mesh with 99.9% purity [CER91]
Density, g/cm^3: ~5.2 [LID94]
Melting Point, °C: 946 [AES93]
Boiling Point, °C: >1300 [CRC94]

2886
Compound: Terbium nitrate hexahydrate
Formula: $Tb(NO_3)_3 \cdot 6H_2O$
Molecular Formula: $H_{12}N_3O_{15}Tb$
Molecular Weight: 453.031
CAS RN: 13451-19-9
Properties: colorless; monocl cryst or white powd [HAW93] [MER89]
Solubility: s H_2O [HAW93]
Melting Point, °C: 893 [AES93]

2887
Compound: Terbium nitride
Formula: TbN
Molecular Formula: NTb
Molecular Weight: 172.932
CAS RN: 12033-64-6
Properties: cub cryst; -40 mesh with 99.9% purity [LID94] [CER91]
Density, g/cm^3: 9.55 [LID94]

2888
Compound: Terbium oxalate hydrate
Formula: $Tb_2(C_2O_4)_3 \cdot xH_2O$
Molecular Formula: $C_6O_{12}Tb_2$ (anhydrous)
Molecular Weight: 581.909 (anhydrous)
CAS RN: 24670-06-2
Properties: white powd, x=10 [CRC94] [STR93]
Density, g/cm^3: x=10: 2.60 [STR93]
Reactions: minus H_2O at 40°C [CRC94]

2889
Compound: Terbium perchlorate hexahydrate
Formula: $Tb(ClO_4)_3 \cdot 6H_2O$
Molecular Formula: $Cl_3H_{12}O_{18}Tb$
Molecular Weight: 565.367
CAS RN: 14014-09-6
Properties: white cryst; hygr [STR93]

2890
Compound: Terbium silicide
Formula: $TbSi_2$
Molecular Formula: Si_2Tb
Molecular Weight: 215.096
CAS RN: 12039-80-4
Properties: ortho-rhomb cryst; 10 mm & down lump [LID94] [ALF93]
Density, g/cm^3: 6.66 [LID94]

2891
Compound: Terbium sulfate octahydrate
Formula: $Tb_2(SO_4)_3 \cdot 8H_2O$
Molecular Formula: $H_{16}O_{20}S_3Tb_2$
Molecular Weight: 750.164
CAS RN: 13842-67-6
Properties: white cryst [STR93]
Solubility: s H_2O [HAW93]
Reactions: minus $8H_2O$ at 360°C [HAW93]

2892
Compound: Terbium sulfide
Formula: Tb_2S_3
Molecular Formula: S_3Tb_2
Molecular Weight: 414.049
CAS RN: 12138-11-3
Properties: cub cryst; -200 mesh with 99.9% purity [LID94] [CER91]
Density, g/cm^3: 6.35 [LID94]

2893
Compound: Terbium(III,IV) oxide

Formula: Tb$_4$O$_7$
Molecular Formula: O$_7$Tb$_4$
Molecular Weight: 747.697
CAS RN: 12037-01-3
Properties: dark brown powd, or sintered pieces 3-12 mm; sl hygr; absorbs atm CO$_2$; used as an evaporation material of 99.9% purity; possibly reactive to radio frequencies [HAW93] [CER91]
Solubility: s dil acids [HAW93]
Reactions: minus O$_2$ on heating [CRC94]

2894
Compound: Tetraborane(10)
Formula: B$_4$H$_{10}$
Molecular Formula: B$_4$H$_{10}$
Molecular Weight: 53.323
CAS RN: 18283-93-7
Properties: gas with disagreeable odor; vapor pressure, mm Hg: (0°C) 388, (6°C) 580; enthalpy of vaporization 27.1 kJ/mol; decomposes in a few hours at room temp, more rapidly at 100°C; spontaneously flammable in air, unless pure [MER89] [COT88] [CRC93]
Solubility: hydrolyzes in H$_2$O to boric acid with evolution of hydrogen [MER89]
Density, g/cm^3: 2.34 g/L [LID94]
Melting Point, °C: -120 [KIR78]
Boiling Point, °C: 18 [KIR78]

2895
Compound: Tetraethyl lead
Synonyms: TEL
Formula: (C$_2$H$_5$)$_4$Pb
Molecular Formula: C$_8$H$_{20}$Pb
Molecular Weight: 323.447
CAS RN: 78-00-2
Properties: colorless liq; burns with orange-colored flame; obtained by reacting PbCl$_2$ and zinc-ethyl; uses: formerly used as gasoline additive [ALD94] [MER89]
Solubility: i H$_2$O; s benzene, petroleum ether, gasoline [MER89]
Density, g/cm^3: 1.653 [ALD94]
Melting Point, °C: -136 [ALD94]
Boiling Point, °C: 84-85 at 15 mm Hg [ALD94]

2896
Compound: Tetraethyl silane
Formula: Si(C$_2$H$_5$)$_4$
Molecular Formula: C$_8$H$_{20}$Si
Molecular Weight: 144.332
CAS RN: 631-36-7

Properties: hygr [ALD94]
Density, g/cm^3: 0.766 [ALD94]
Melting Point, °C: -82.5 [ALD94]
Boiling Point, °C: 153-154 [ALD94]

2897
Compound: Tetraethylammonium bromide
Synonyms: TEAB
Formula: (C$_2$H$_5$)$_4$NBr
Molecular Formula: C$_8$H$_{20}$BrN
Molecular Weight: 210.158
CAS RN: 71-91-0
Properties: hygr cryst; prepared from triethylamine and ethyl bromide; used as ganglion blocking agent [MER89] [ALD94]
Solubility: v s H$_2$O, alcohol, chloroform [MER89]
Melting Point, °C: 285, decomposes [ALD94]
Thermal Expansion Coefficient: from 25°C to 100°C (0.18), 200°C (0.42), 400°C (0.90), 600°C (1.38), 800°C (1.86), 1000°C (2.34), 1200°C (2.82) [TAY91b]

2898
Compound: Tetraethylammonium chloride
Synonyms: T.E.A. chloride
Formula: (C$_2$H$_5$)$_4$NCl
Molecular Formula: C$_8$H$_{20}$ClN
Molecular Weight: 165.706
CAS RN: 56-34-8
Properties: deliq cryst, tetrahydrate is monocl; ganglion blocking agent [MER89]
Solubility: s H$_2$O [MER89]
Density, g/cm^3: 1.0801 [MER89]
Melting Point, °C: 37.5 (tetrahydrate) [MER89]

2899
Compound: Tetraethylorthosilicate
Synonyms: ethyl silicate
Formula: Si(OC$_2$H$_5$)$_4$
Molecular Formula: C$_8$H$_{20}$O$_4$Si
Molecular Weight: 208.329
CAS RN: 78-10-4
Properties: colorless, flammable liq; flash point 52°C; prepared by reaction of absolute ethanol with SiCl$_4$; sensitive to moisture; used in the sol-gel preparation of zircon, in weatherproofing and hardening stone [MER89] [ALD94]
Solubility: miscible with ethanol; reacts with H$_2$O [MER89]
Density, g/cm^3: 0.934 [ALD94]
Melting Point, °C: -77 [MER89]
Boiling Point, °C: 165-166 [MER89]

2900
Compound: Tetramethylgermane
Formula: $(CH_3)_4Ge$
Molecular Formula: $C_4H_{12}Ge$
Molecular Weight: 132.749
CAS RN: 865-52-1
Properties: liq; flammable [ALD94]
Density, g/cm³: 0.978 [ALD94]
Melting Point, °C: -88 [ALD94]
Boiling Point, °C: 43-44 [ALD94]

2901
Compound: Tetramethyltin
Formula: $(CH_3)_4Sn$
Molecular Formula: $C_4H_{12}Sn$
Molecular Weight: 178.849
CAS RN: 594-27-4
Properties: liq; flammable [ALD94]
Density, g/cm³: 1.291 [ALD94]
Melting Point, °C: -54 [ALD94]
Boiling Point, °C: 74-75 [ALD94]

2902
Compound: Tetrapropylammonium
perruthenate(VII)
Synonyms: TPAP
Formula: $(CH_3CH_2CH_2)_4NRuO_4$
Molecular Formula: $C_{12}H_{28}NO_4Ru$
Molecular Weight: 351.428
CAS RN: 114615-82-6
Properties: sold; mild catalytic oxidant [ALD94]
Melting Point, °C: 165, decomposes [ALD94]
Reactions: explodes when heated [ALD94]

2903
Compound: Thallium
Formula: Tl
Molecular Formula: Tl
Molecular Weight: 204.3833
CAS RN: 7440-28-0
Properties: α-Tl: hex; β-Tl: cub; bluish white, very
soft; enthalpy of fusion 4.14 kJ/mol;
vaporization enthalpy 165 kJ/mol; Brinell
hardness is 2; resistivity 16.6 μohm·cm;
oxidizes in air, forming oxide film on surface;
used in low melting alloys as electrode to
measure the dissolved oxygen content of waters;
photoelectric [CIC73] [HAW93] [ALD94]
[MER89] [KIR83] [CRC93]
Solubility: i H_2O; reacts with HNO_3, H_2SO_4
[MER89]
Density, g/cm³: α: 11.85; β: 11.86-11.87 [CIC73]

Melting Point, °C: 303.5 [MER89]
Boiling Point, °C: 1553 [COT88]
Reactions: volatilization begins at 174°C [MER89]
Thermal Conductivity, W/(m·K): 46.1 (25°C)
[CRC93]
Thermal Expansion Coefficient: (volume) x
10^{-6}/°C: 90 (20°C) α, 124 β, 140 liq [CIC73]

2904
Compound: Thallium barium calcium copper oxide
Formula: $Tl_4Ba_3Ca_3Cu_4O_{13}$
Molecular Formula: $Ba_3Ca_3Cu_4O_{13}Tl_4$
Molecular Weight: 1808.924
CAS RN: 119000-19-0
Properties: superconductor; 99.99% and 99.9%
purity 20 micron powd [ALF93]

2905
Compound: Thallium barium calcium copper oxide
Formula: $Tl_2Ba_2Ca_2Cu_3O_{10}$
Molecular Formula: $Ba_2Ca_2Cu_3O_{10}Tl_2$
Molecular Weight: 1114.209
CAS RN: 127241-75-2
Properties: superconductor (2223 phase); 99.999%
and 99.9% purity, 20 micron powd; T_c 115-127
K; for material with formula $Tl_2Ba_2Ca_3Cu_4O_{12}$,
T_c is 113-119 K [CEN92] [ALF93] [ASM93]

2906
Compound: Thallium barium calcium copper oxide
Formula: $Tl_2Ba_2CaCu_2O_8$
Molecular Formula: $Ba_2CaCu_2O_8Tl_2$
Molecular Weight: 1018.664
CAS RN: 125720-69-1
Properties: superconductor; 20 micron powd,
99.999% purity and 99.9%; T_c 95-110 K
[ALF93] [ASM93]

2907
Compound: Thallium(I) acetate
Synonyms: thallous acetate
Formula: CH_3COOTl
Molecular Formula: $C_2H_3O_2Tl$
Molecular Weight: 263.428
CAS RN: 563-68-8
Properties: -4 mesh with 99.9% purity; white cryst;
deliq; used in ore flotation separation; there is a
hemitrihydrate, 2570-63-0 [HAW93] [MER89]
[CER91]
Solubility: s H_2O, alcohol [MER89]
Density, g/cm³: 3.68 [HAW93]

Melting Point, °C: 131 [STR93]

2908

Compound: Thallium(I) acetylacetonate

Synonyms: 2,4-pentanedione, thallium(I) derivative

Formula: TlCH$_3$COCH=C(O)CH$_3$

Molecular Formula: C$_5$H$_7$O$_2$Tl

Molecular Weight: 303.493

CAS RN: 25955-51-5

Properties: white powd [STR93]

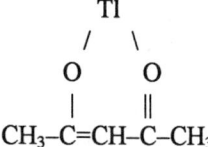

2909

Compound: Thallium(I) azide

Formula: TlN$_3$

Molecular Formula: N$_3$Tl

Molecular Weight: 246.403

CAS RN: 13847-66-0

Properties: yellow; body-center tetr, a = 0.623 nm, c = 0.675 nm [CIC73] [CRC94]

Solubility: g/100g H$_2$O: 0.171 (0°C), 0.236 (10°C), 0.364 (20°C) [LAN85]

Melting Point, °C: 330 (vacuum) [CRC94]

2910

Compound: Thallium(I) bromide

Synonyms: thallous bromide

Formula: TlBr

Molecular Formula: BrTl

Molecular Weight: 284.287

CAS RN: 7789-40-4

Properties: -20 mesh with 99.999% purity; pale yellow; cryst powd; enthalpy of vaporization 99.56 kJ/mol; enthalpy of fusion 25.10 kJ/mol; used in a mixture with TlI in infrared transmitters [HAW93] [MER89] [CER91] [CRC93]

Solubility: g/100g H$_2$O: 0.022 (0°C), 0.048 (20°C), 0.177 (60°C) [LAN85]

Density, g/cm^3: 7.557 [HAW93]

Melting Point, °C: 460 [CRC93]

Boiling Point, °C: 819 [CRC93]

2911

Compound: Thallium(I) carbonate

Synonyms: thallous carbonate

Formula: Tl$_2$CO$_3$

Molecular Formula: CO$_3$Tl$_2$

Molecular Weight: 468.776

CAS RN: 6533-73-9

Properties: shiny white monocl cryst; highly refractive; melts to a dark gray mass; enthalpy of fusion 18.40 kJ/mol; used in testing for carbon disulfide and in artificial diamonds [HAW93] [KIR83] [CRC93]

Solubility: g/100g H$_2$O: 5.3 (20°C), 12.2 (60°C), 27.2 (100°C) [LAN85]; i alcohol [MER89]

Density, g/cm^3: 7.11 [STR93]

Melting Point, °C: 272 [MER89]

2912

Compound: Thallium(I) chlorate

Formula: TlClO$_3$

Molecular Formula: ClO$_3$Tl

Molecular Weight: 287.834

CAS RN: 13453-30-0

Properties: needles [LAN52]

Solubility: g/100g H$_2$O: 2.00 (0°C), 3.92 (20°C), 57.3 (100°C) [LAN85]

Density, g/cm^3: 5.047 [LAN52]

2913

Compound: Thallium(I) chloride

Synonyms: thallous chloride

Formula: TlCl

Molecular Formula: ClTl

Molecular Weight: 239.836

CAS RN: 7791-12-0

Properties: white cryst powd, and 99.9% pure melted pieces of 3-12 mm; turns violet when exposed to light; enthalpy of vaporization 102.2 kJ/mol; enthalpy of fusion 17.80 kJ/mol; used as a chlorination catalyst and in suntan lamp monitors, and as an evaporation material to deposit long wavelength coatings up to >50 μm, has index of ~1.90 at 20 μm [HAW93] [CER91] [CRC93]

Solubility: g/100g H$_2$O: 0.21 (0°C), 0.33 (20°C), 1.80 (100°C) [LAN85]; i alcohol [MER89]

Density, g/cm^3: 7.004 [HAW93]

Melting Point, °C: 430 [MER89]

Boiling Point, °C: 807 [CRC93]

2914

Compound: Thallium(I) cyanide

Synonyms: thallous cyanide

Formula: TlCN

Molecular Formula: CNTl

Molecular Weight: 230.401
CAS RN: 13453-34-4
Properties: white; hex platelets [MER89]
Solubility: 16.8 g/100mL H_2O, s a, alcohol [MER89]
Density, g/cm³: 6.523 [MER89]
Melting Point, °C: decomposes [CRC94]

2915
Compound: Thallium(I) ethoxide
Formula: $TlOC_2H_5$
Molecular Formula: C_2H_5OTl
Molecular Weight: 249.444
CAS RN: 20398-06-5
Properties: cloudy, dense liq; sensitive to moisture [STR93]
Density, g/cm³: 3.493 (20°C) [STR93]
Melting Point, °C: -3 [STR93]
Boiling Point, °C: 130, decomposes [STR93]

2916
Compound: Thallium(I) fluoride
Synonyms: thallous fluoride
Formula: TlF
Molecular Formula: FTl
Molecular Weight: 223.381
CAS RN: 7789-27-7
Properties: white powd; ortho-rhomb; hard, shiny cryst; can deliq, e.g. if breathed upon, however reverts to anhydrous form in dry air; not typically hygr; enthalpy of fusion 14.00 kJ/mol [MER89] [STR93] [CRC93]
Solubility: 78.6 g/100g H_2O at 15°C; s alcohols, HF [KIR83]
Density, g/cm³: 8.36 [MER89]
Melting Point, °C: 322 [MER89]
Boiling Point, °C: 655 [STR93]

2917
Compound: Thallium(I) formate
Formula: HCOOTl
Molecular Formula: CHO_2Tl
Molecular Weight: 249.401
CAS RN: 992-98-3
Properties: colorless needles or white powd; hygr [STR93] [KIR83]
Solubility: 500 g/100g H_2O at 10°C; s methanol [KIR83]
Density, g/cm³: 4.967 [STR93]
Melting Point, °C: 101 [STR93]

2918
Compound: Thallium(I) hexafluoroacetylacetonate
Synonyms: 1,1,1,5,5,5-hexafluoro-2,4-pentanedione, Tl(I) derivative
Formula: $TlCF_3COCHCOCF_3$
Molecular Formula: $C_5HF_6O_2Tl$
Molecular Weight: 411.435
CAS RN: 15444-43-6
Properties: yellow cryst [STR93]

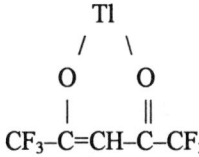

Melting Point, °C: 126-128 [STR93]
Boiling Point, °C: decomposes [STR93]
Reactions: sublimes at 140°C (0.1 mm Hg) [STR94]

2919
Compound: Thallium(I) hexafluorophosphate
Formula: $TlPF_6$
Molecular Formula: F_6PTl
Molecular Weight: 349.347
CAS RN: 60969-19-9
Properties: white cryst [STR93]

2920
Compound: Thallium(I) hydroxide
Synonyms: thallous hydroxide
Formula: TlOH
Molecular Formula: HOTl
Molecular Weight: 221.390
CAS RN: 1310-83-4
Properties: yellow needles; gives strongly alkaline solution when dissolved in H_2O, which turns tumeric paper brown color [MER89]
Solubility: g/100g H_2O: 25.4 (0°C), 35.0 (20°C), 150 (100°C) [LAN85]
Density, g/cm³: 7.44 [KIR83]
Melting Point, °C: 139 decomposes [KIR83]

2921
Compound: Thallium(I) iodide
Synonyms: thallous iodide
Formula: α-TlI
Molecular Formula: ITl
Molecular Weight: 331.287
CAS RN: 7790-30-9

Properties: yellow; rhomb cryst; turns red at 170°C; enthalpy of vaporization 104.7 kJ/mol; enthalpy of fusion 13.10 kJ/mol; used mixed with TlBr in infrared transmitters [HAW93] [MER89] [CRC93] [KIR83]
Solubility: g/100g H$_2$O: 0.002 (0°C), 0.006 (20°C), 0.120 (100°C) [LAN85]
Density, g/cm^3: 7.29 [KIR83]
Melting Point, °C: 440 [MER89]
Boiling Point, °C: 824 [MER89]

2922
Compound: Thallium(I) molybdate
Formula: Tl$_2$MoO$_4$
Molecular Formula: MoO$_4$Tl$_2$
Molecular Weight: 568.705
CAS RN: 34128-09-1
Properties: white when precipitated, yellow during fusion; cub, a = 0.926 nm [KIR81]
Solubility: i H$_2$O [KIR81]
Melting Point, °C: red heat [KIR81]

2923
Compound: Thallium(I) nitrate
Synonyms: thallous nitrate
Formula: TlNO$_3$
Molecular Formula: NO$_3$Tl
Molecular Weight: 266.388
CAS RN: 10102-45-1
Properties: white cryst; enthalpy of fusion 9.60 kJ/mol; used in analysis and as a pyrotechnic (green fire) [HAW93] [MER89] [CRC93]
Solubility: g/100g H$_2$O: 3.90 (0°C), 9.55 (20°C), 414 (100°C) [LAN85]; i alcohol [MER89]
Density, g/cm^3: 5.55 [MER89]
Melting Point, °C: 206 [MER89]
Boiling Point, °C: decomposes, 450 [MER89]

2924
Compound: Thallium(I) nitrite
Formula: TlNO$_2$
Molecular Formula: NO$_2$Tl
Molecular Weight: 250.389
CAS RN: 13824-63-6
Properties: yellow cryst [CRC77]
Solubility: g/100g H$_2$O: 17.9 (0°C), 40.3 (20°C), 750 (90°C) [LAN85]
Melting Point, °C: 182 [CRC77]

2925
Compound: Thallium(I) oxalate

Formula: Tl$_2$C$_2$O$_4$
Molecular Formula: C$_2$O$_4$Tl$_2$
Molecular Weight: 496.786
CAS RN: 30737-24-7
Properties: white powd [STR93]
Solubility: 1.48 g/100mL H$_2$O (15°C), 9.02 g/100mL H$_2$O (100°C) [CRC94]
Density, g/cm^3: 6.31 [STR93]

2926
Compound: Thallium(I) oxide
Synonyms: thallous oxide
Formula: Tl$_2$O
Molecular Formula: OTl$_2$
Molecular Weight: 424.766
CAS RN: 1314-12-1
Properties: black powd; gradually oxides to Tl$_2$O$_3$ in air; used in artificial gems, to test for ozone, in optical glass [MER89]
Solubility: s H$_2$O, hydrolyzes to hydroxide; s alcohol [MER89]
Density, g/cm^3: 9.52 (16°C) [HAW93]
Melting Point, °C: 300 [HAW93]
Boiling Point, °C: 1080 [HAW93]

2927
Compound: Thallium(I) perchlorate
Formula: TlClO$_4$
Molecular Formula: ClO$_4$Tl
Molecular Weight: 303.834
CAS RN: 13453-40-2
Properties: colorless rhomb [CRC94] [LAN52]
Solubility: g/100g H$_2$O: 6.00 (0°C), 13.1 (20°C), 81.5 (80°C) [LAN85]
Density, g/cm^3: 4.89 [LAN52]
Melting Point, °C: 501 [LAN52]
Boiling Point, °C: decomposes [LAN52]

2928
Compound: Thallium(I) picrate
Formula: TlOC$_6$H$_2$(NO$_2$)$_3$
Molecular Formula: C$_6$H$_2$N$_3$O$_7$Tl
Molecular Weight: 432.481
CAS RN: 23293-27-8
Properties: red monocl or yellow tricl [CRC77]
Solubility: g/100g H$_2$O: 0.135 (0°C), 0.40 (20°C), 1.73 (60°C) [LAN85]
Density, g/cm^3: red: 3.164; yellow: 2.993 [CRC77]
Reactions: explodes, 723-725°C [CRC77]

2929
Compound: Thallium(I) selenate
Synonyms: thallous selenate
Formula: Tl_2SeO_4
Molecular Formula: O_4SeTl_2
Molecular Weight: 551.725
CAS RN: 7446-22-2
Properties: ortho-rhomb cryst [MER89]
Solubility: g/100g H_2O: 2.13 (9.3°C), 2.4 (12°C), 10.86 (100°C); i alcohol, ether [MER89]
Density, g/cm³: 6.875 [MER89]
Melting Point, °C: >400 [MER89]

2930
Compound: Thallium(I) selenide
Synonyms: thallous selenide
Formula: Tl_2Se
Molecular Formula: $SeTl_2$
Molecular Weight: 487.727
CAS RN: 15572-25-5
Properties: dark gray plates, with metallic luster [MER89]
Solubility: i H_2O, acids [MER89]
Density, g/cm³: 9.05 [CRC94]
Melting Point, °C: 340 [MER89]

2931
Compound: Thallium(I) sulfate
Synonyms: thallous sulfate
Formula: Tl_2SO_4
Molecular Formula: O_4STl_2
Molecular Weight: 504.831
CAS RN: 7446-18-6
Properties: white, rhomb prisms; enthalpy of fusion 23.00 kJ/mol; used to analyze for iodine in the presence of chlorine, in ozonometry and as a pesticide [HAW93] [MER89] [CRC93]
Solubility: g/100mL H_2O: 2.70 (0°C), 4.87 (20°C), 18.45 (100°C) [MER89]
Density, g/cm³: 6.77 [MER89]
Melting Point, °C: 632 [MER89]
Boiling Point, °C: decomposes [STR93]

2932
Compound: Thallium(I) sulfide
Synonyms: thallous sulfide
Formula: Tl_2S
Molecular Formula: STl_2
Molecular Weight: 440.833
CAS RN: 1314-97-2

Properties: -20 mesh with 99.9% purity; bluish black cryst powd; enthalpy of vaporization 154 kJ/mol; enthalpy of fusion 12.00 kJ/mol; used in infrared photocells [HAW93] [MER89] [CER91] [CRC93]
Solubility: sl s H_2O, alkali hydroxides, sulfides, cyanides; s in mineral acids [MER89]
Density, g/cm³: 8.39 [MER89]
Melting Point, °C: 448.5 [MER89]
Boiling Point, °C: 1367 [CRC93]

2933
Compound: Thallium(I) trifluoroacetylacetonate
Synonyms: 1,1,1-trifluoro-2,4-pentanedione, thallium(I) derivative
Formula: $TlCF_3COCH=C(O)CH_3$
Molecular Formula: $C_5H_4F_3O_2Tl$
Molecular Weight: 357.464
CAS RN: 54412-40-7
Properties:

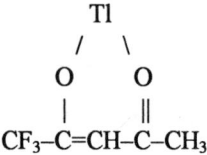

Melting Point, °C: 110 decomposes [ALD94]

2934
Compound: Thallium(III) acetate
Synonyms: acetic acid, thallium(III) salt
Formula: $Tl(CH_3COO)_3$
Molecular Formula: $C_6H_9O_6Tl$
Molecular Weight: 381.517
CAS RN: 2570-63-0
Properties: light sensitive; uses: together with bromine, selective electrophilic aromatic brominator [ALD94]
Melting Point, °C: 182 decomposes [ALD94]

2935
Compound: Thallium(III) bromide
Synonyms: thallic bromide
Formula: $TlBr_3$
Molecular Formula: Br_3Tl
Molecular Weight: 444.095
CAS RN: 13701-90-1
Properties: yellow [KIR83]
Solubility: s H_2O, alcohols [KIR83]

2936
Compound: Thallium(III) chloride
Synonyms: thallic chloride
Formula: $TlCl_3$
Molecular Formula: Cl_3Tl
Molecular Weight: 310.741
CAS RN: 13453-32-2
Properties: hexagonal plate [KIR83]
Solubility: v s H_2O, alcohols, ether [KIR83]
Density, g/cm³: 4.7 [LID94]
Melting Point, °C: 155 [KIR83]

2937
Compound: Thallium(III) chloride hydrate
Synonyms: thallic chloride hydrate
Formula: $TlCl_3 \cdot xH_2O$
Molecular Formula: Cl_3Tl (anhydrous)
Molecular Weight: 310.741 (anhydrous)
CAS RN: 13453-33-3
Properties: white cryst [STR93]
Melting Point, °C: 37 [STR93]

2938
Compound: Thallium(III) fluoride
Synonyms: thallic fluoride
Formula: TlF_3
Molecular Formula: F_3Tl
Molecular Weight: 261.378
CAS RN: 7783-57-5
Properties: olive green ortho-rhomb cryst; very
    sensitive to moisture; quickly decomposed by
    water; decomposed if heated in air [MER89]
    [KIR83]
Solubility: i H_2O [KIR83]
Density, g/cm³: 8.65 [MER89]
Melting Point, °C: decomposes 550 in air [MER89]

2939
Compound: Thallium(III) nitrate
Synonyms: thallic nitrate
Formula: $Tl(NO_3)_3$
Molecular Formula: N_3O_9Tl
Molecular Weight: 390.398
CAS RN: 13746-98-0
Properties: colorless; there is a trihydrate, CAS RN
    13453-38-8 [KIR83] [ALD94]
Solubility: decomposes in H_2O [KIR83]
Melting Point, °C: 102-105 (trihydrate) [ALD94]

2940
Compound: Thallium(III) oxide

Synonyms: thallic oxide
Formula: Tl_2O_3
Molecular Formula: O_3Tl_2
Molecular Weight: 456.765
CAS RN: 1314-32-5
Properties: brown powd; oxidant, e.g. reacts with
    HCl evolving Cl_2, and with H_2SO_4 evolving O_2
    [MER89]
Solubility: i H_2O [MER89]
Density, g/cm³: 10.11 [KIR83]
Melting Point, °C: ~717 [STR93]

2941
Compound: Thallium(III) perchlorate hexahydrate
Formula: $Tl(ClO_4)_3 \cdot 6H_2O$
Molecular Formula: $Cl_3H_{12}O_{18}Tl$
Molecular Weight: 610.825
CAS RN: 15596-83-5
Properties: white cryst; hygr [STR93]

2942
Compound: Thallium(III) trifluoroacetate
Synonyms: trifluoroacetic acid, thallium(III) salt
Formula: $(CF_3COO)_3Tl$
Molecular Weight: 543.42
CAS RN: 23586-53-0
Properties: hygr; oxidizing agent; uses: organic
    sulfide bond formation [ALD94]
Melting Point, °C: 213, decomposes [ALD94]

2943
Compound: Thionyl bromide
Formula: $SOBr_2$
Molecular Formula: Br_2OS
Molecular Weight: 207.873
CAS RN: 507-16-4
Properties: orange-yellow liq; slowly decomposes
    on standing; prepared by reaction of $SOCl_2$ and
    HBr [MER89]
Solubility: hydrolyzed by H_2O; miscible with
    benzene, chloroform, CCl_4 [MER89]
Density, g/cm³: 2.688 [MER89]
Melting Point, °C: -52 [MER89]
Boiling Point, °C: 138 [MER89]

2944
Compound: Thionyl chloride
Synonyms: sulfurous oxychloride
Formula: $SOCl_2$
Molecular Formula: Cl_2OS
Molecular Weight: 118.970

CAS RN: 7719-09-7

Properties: pale yellow to red liq; suffocating odor; sensitive to moisture; enthalpy of vaporization 31.7 kJ/mol at bp, 31 kJ/mol at 25°C; produced by oxidation of SCl_2 by SO_3; used in pesticides, engineering plastics [HAW93] [STR93] [MER89]

Solubility: decomposes in H_2O; s benzene, CCl_4 [HAW93]

Density, g/cm³: 1.638 [HAW93]

Melting Point, °C: -105 [HAW93]

Boiling Point, °C: 79 [ALD94]

Reactions: decomposes at 140°C [HAW93]

2945

Compound: Thionyl fluoride

Formula: SOF_2

Molecular Formula: F_2OS

Molecular Weight: 86.062

CAS RN: 7783-42-8

Properties: colorless gas with suffocating odor; does not attack glass; obtained from reaction between SbF_3 and $SOCl_2$ in presence of SbF_5 [MER89]

Solubility: hydrolyzed by H_2O; s ether, benzene [MER89]

Density, g/cm³: gas: 3.84 g/L [CRC94]; liq: 1.780 (-100°C) [MER89]

Melting Point, °C: -129.5 [MER89]

Boiling Point, °C: -43.8 [MER89]

2946

Compound: Thiophosphoryl chloride

Synonyms: phosphorus sulfochloride

Formula: $PSCl_3$

Molecular Formula: Cl_3PS

Molecular Weight: 169.398

CAS RN: 3982-91-0

Properties: fuming liq; crystallizes to α-form at -40.8°C, to β-form at -36.2°C [MER89] [ALD94]

Solubility: hydrolyzes in H_2O, forming H_3PO_4, HCl, H_2S; hydrolyzes rapidly in alkaline solutions; s benzene, CCl_4, CS_2, chloroform [MER89]

Density, g/cm³: 1.635 [STR93]

Melting Point, °C: -35 [STR93]

Boiling Point, °C: 125 [STR93]

2947

Compound: Thorium

Formula: Th

Molecular Formula: Th

Molecular Weight: 232.0381

CAS RN: 7440-29-1

Properties: soft; grayish white, lustrous metal; somewhat ductile; α: fcc up to 1400°C, a = 0.5086 nm; β: bcc, a = 0.411 nm, stable 1400 to 1750°C; enthalpy of vaporization 564 kJ/mol; enthalpy of fusion 13.81 kJ/mol; electrical resistivity 14 μohm·cm; Poisson's ratio 0.27; thermal diffusivity 0.28 cm²/s at 200°C; $t_{1/2}$ ^{232}Th is 1.41 x 10^{+10} years; ionic radius of Th^{++++} is 0.0972 nm [KIR78] [KIR83] [MER89] [CRC93]

Solubility: s acids; i H_2O, alkalies [HAW93]

Density, g/cm³: α: 11.724 [KIR91]

Melting Point, °C: 1750 [KIR91]

Boiling Point, °C: 3800 [ALD94]

Reactions: phase change from fcc to bcc ~1345°C [KIR83]

Thermal Conductivity, W/(m·K): 54.0 (25°C) [CRC93]

Thermal Expansion Coefficient: 11.0 x 10^{-6}/K [CRC93]

2948

Compound: Thorium acetylacetonate

Synonyms: 2,4-pentanedione, thorium(IV) derivative

Formula: $Th(CH_3COCH=C(O)CH_3)_4$

Molecular Formula: $C_{20}H_{28}O_8Th$

Molecular Weight: 628.475

CAS RN: 102192-40-5

Properties: cryst powd [HAW93]

$$[CH_3-\overset{\overset{\displaystyle O-}{|}}{C}=CH-\overset{\overset{\displaystyle O}{\|}}{C}-CH_3]_4Th$$

Solubility: sl s H_2O, not readily hydrolyzed [HAW93]

2949

Compound: Thorium bromide

Formula: $ThBr_4$

Molecular Formula: Br_4Th

Molecular Weight: 551.654

CAS RN: 13453-49-1

Properties: colorless hygr; -8 mesh with 99.5% purity; can be prepared by reacting Th with Br_2; light sensitive and easily hydrolyzed [KIR83] [CER91] [CRC94]

Density, g/cm³: 5.67 [CRC94]

Melting Point, °C: sublimes 610 [CRC94]

2950
Compound: Thorium carbide
Formula: ThC
Molecular Formula: CTh
Molecular Weight: 244.049
CAS RN: 12012-16-7
Properties: -40 mesh with 99.5% purity; fcc, a = 0.5346 nm; prepared by reacting stoichiometric amounts of Th and C; reactive, e.g. burns in air to form ThO_2; there is a ThC_2, 12071-31-7 [KIR83] [CIC73] [CER91]
Solubility: readily hydrolyzes in H_2O evolving methane [KIR83]
Melting Point, °C: 2655 [KIR83]

2951
Compound: Thorium chloride
Formula: $ThCl_4$
Molecular Formula: Cl_4Th
Molecular Weight: 373.849
CAS RN: 10026-08-1
Properties: tetr; grayish white powd; hygr lustrous needles; enthalpy of vaporization 146.4 kJ/mol; enthalpy of fusion 40.20 kJ/mol; can be prepared by reacting Th with Cl_2; used in incandescent lamps [HAW93] [CRC93] [STR93] [KIR83] [COT88]
Solubility: s H_2O, alcohol [MER89]
Density, g/cm³: 4.59 [MER89]
Melting Point, °C: 770 [MER89]
Boiling Point, °C: 921 [MER89]

2952
Compound: Thorium dicarbide
Formula: ThC_2
Molecular Formula: C_2Th
Molecular Weight: 256.060
CAS RN: 12071-31-7
Properties: yellow solid; α form: monocl, a = 1.0555 nm, b = 0.8233 nm, c = 0.4201 nm; β form: tetr; γ form: cub, a = 0.5808 nm; decomposed by water; forms by heating ThO_2 and excess carbon; used as a nuclear fuel [HAW93] [CIC73] [KIR83]
Solubility: decomposed in H_2O, with evolution of ethane [KIR83]
Density, g/cm³: 8.96 (18°C) [HAW93]
Melting Point, °C: 2630-2680 [HAW93]
Boiling Point, °C: ~5000 [HAW93]
Reactions: α to β transition at 1427°C, β to γ at 1497°C [CIC73]

2953
Compound: Thorium fluoride
Formula: ThF_4
Molecular Formula: F_4Th
Molecular Weight: 308.032
CAS RN: 13709-59-6
Properties: white powd; hygr; reacts with atm moisture to form $ThOF_2$ at temperatures >500°C; enthalpy of vaporization 258 kJ/mol; can be prepared by reaction of Th and F_2; used to produce thorium metal and in high temp ceramics; as 99.99% pure material, used as a sputtering target to produce low-index film with no absorption in visible and ultraviolet [HAW93] [STR93] [KIR83] [CER91] [CRC93]
Density, g/cm³: 6.32 [STR93]
Melting Point, °C: 1068 [KIR91]

2954
Compound: Thorium hexaboride
Formula: ThB_6
Molecular Formula: B_6Th
Molecular Weight: 296.904
CAS RN: 12229-63-9
Properties: -100 mesh with 99.8% purity, there is a ThB_4, 12007-83-9; refractory material [KIR78]
Density, g/cm³: 6.4 [CRC94]
Melting Point, °C: 2195; tetraboride 2500 [KIR78]

2955
Compound: Thorium hexafluoroacetylacetonate
Synonyms: 1,1,1,5,5,5-hexafluoro-2,4-pentanedione, thorium derivative
Formula: $Th(CF_3COCHCOCF_3)_4$
Molecular Formula: $C_{20}H_4F_{24}O_8Th$
Molecular Weight: 1060.247
CAS RN: 18865-75-3
Properties: white powd [STR93]

$$[CF_3-\overset{\underset{|}{O-}}{C}=CH-\overset{\underset{\|}{O}}{C}-CF_3]_4Th$$

Melting Point, °C: 100-101 [STR93]

2956
Compound: Thorium hydride
Formula: ThH_2
Molecular Formula: H_2Th
Molecular Weight: 234.054
CAS RN: 16689-88-6

Properties: tetr cryst; here are also ThH$_3$, 40004-84-0, -60 mesh with 99.5% purity, and ThH$_{3.75}$ (Th$_4$H$_{15}$), 12055-07-1; the third hydride exhibits superconductivity [LID94] [KIR83] [CER91]
Density, g/cm^3: 9.5 [LID94]

2957
Compound: Thorium hydroxide
Formula: Th(OH)$_4$
Molecular Formula: H$_4$O$_4$Th
Molecular Weight: 300.068
CAS RN: 13825-36-0
Properties: prepared by addition of alkali to a solution of Th^{++++} salt, yielding a gelatinous precipitate which is subsequently dehydrated; absorbs CO$_2$ to form ThOCO$_3$ [KIR83]
Melting Point, °C: decomposes [CRC94]
Reactions: minus water >470°C to form ThO$_2$ [KIR83]

2958
Compound: Thorium iodide
Formula: ThI$_4$
Molecular Formula: I$_4$Th
Molecular Weight: 739.656
CAS RN: 7790-49-0
Properties: pale yellow cryst; obtained from a reaction between Th and I$_2$; decomposed by light or heat [MER89] [KIR83]
Melting Point, °C: 556 [KIR91]
Boiling Point, °C: 837 [MER89]

2959
Compound: Thorium nitrate
Formula: Th(NO$_3$)$_4$
Molecular Formula: N$_4$O$_{12}$Th
Molecular Weight: 480.058
CAS RN: 13823-29-5
Properties: plates, deliq; obtained when thorium hydroxide is dissolved in a nitric acid solution [KIR83] [CRC94]
Solubility: g/100g H$_2$O: 186 (0°C), 187 (10°C), 191 (20°C) [LAN85]; additional solubility data are in [SIE94]
Melting Point, °C: decomposes 500 [CRC94]

2960
Compound: Thorium nitrate tetrahydrate
Formula: Th(NO$_3$)$_4$·4H$_2$O
Molecular Formula: H$_8$N$_4$O$_{16}$Th
Molecular Weight: 552.119

CAS RN: 33088-16-3
Properties: white cryst; sl deliq; used in thoriated tungsten filaments, and as a reagent for fluoride determination [HAW93] [MER89]
Solubility: v s H$_2$O, alcohol [MER89]
Melting Point, °C: 500, decomposes [HAW93]

2961
Compound: Thorium nitride
Formula: ThN
Molecular Formula: NTh
Molecular Weight: 246.045
CAS RN: 12033-65-7
Properties: gray refractory solid; fcc, a = 0.5159 nm; electrical resistivity 20 μohm·cm; microhardness 600; there is also the compound Th$_3$N$_4$, CAS RN 12033-90-8, -100 mesh with 99.5% purity [CER91] [KIR81]
Solubility: slowly hydrolyzed by H$_2$O [COT88]
Density, g/cm^3: 11.9 [KIR81]
Melting Point, °C: 2820 [KIR81]

2962
Compound: Thorium orthosilicate
Synonyms: thorite
Formula: ThSiO$_4$
Molecular Formula: O$_4$SiTh
Molecular Weight: 324.122
CAS RN: 51184-23-7
Properties: black to orange; zircon structure, a = 0.71328 nm, c = 0.63188 nm; hardness is 4.5-5 [HAW93] [SUB90]
Density, g/cm^3: 4.4-5.2 [HAW93]
Thermal Expansion Coefficient: (25 to 500°C) 2.5 x 10^{-6}/°C [SUB90]

2963
Compound: Thorium oxalate dihydrate
Formula: Th(C$_2$O$_4$)$_2$·2H$_2$O
Molecular Formula: C$_4$H$_4$O$_{10}$Th
Molecular Weight: 444.108
Properties: white powd; there is a hexahydrate, white cryst, which is precipitated from up to 2M HNO$_3$; used in ceramics [HAW93] [COT88]
Solubility: i H$_2$O and most acids [HAW93]
Density, g/cm^3: anhydrous: 4.637 (16°C) [HAW93]
Melting Point, °C: decomposes >300-400 to ThO$_2$ [HAW93]

2964
Compound: Thorium oxide

Synonyms: thoria,thorianite
Formula: ThO_2
Molecular Formula: O_2Th
Molecular Weight: 264.037
CAS RN: 1314-20-1
Properties: white heavy cryst cub powd, or 3-12 mm sintered pieces; hardness 6.5 Mohs; used in ceramics, gas mantles, crucibles, thoriated tungsten filaments and in crucible form for melting hafnium, iridium, iron, manganese, silicon, thorium, titanium, uranium, zirconium; used as an evaporation material and sputtering target of 99.99% and 99.9% purity for highly durable beam splitter [HAW93] [MER89] [KIR83] [CER91]
Solubility: i H_2O, alkalies [MER89]
Density, g/cm³: 9.86 [STR93]; 10.01 [KIR80]
Melting Point, °C: ~3050 [KIR91]
Boiling Point, °C: 4400 [HAW93]
Thermal Conductivity, W/(m·K): 5.1 (500°C), 3.0 (1000°C) [KIR80]
Thermal Expansion Coefficient: (volume) 100°C (0.234), 200°C (0.517), 400°C (1.100), 800°C (2.249), 1000°C (2.833) [CLA66]

2965
Compound: Thorium oxyfluoride
Formula: $ThOF_2$
Molecular Formula: F_2OTh
Molecular Weight: 286.034
CAS RN: 13597-30-3
Properties: 3-6 mm pieces (sintered) with 99.9% purity [CER91]

2966
Compound: Thorium perchlorate
Formula: $Th(ClO_4)_4$
Molecular Formula: $Cl_4O_{16}Th$
Molecular Weight: 629.839
CAS RN: 16045-17-3
Properties: obtained by dissolution of thorium hydroxide in perchloric acid solution; the tetrahydrate can be crystallized from an acidic solution which is then dehydrated to $Th(ClO_4)_4$ at ~280°C [KIR83]
Solubility: v s H_2O [KIR83]
Melting Point, °C: ~355 decomposes to ThO_2 [KIR83]

2967
Compound: Thorium selenide
Formula: $ThSe_2$
Molecular Formula: Se_2Th

Molecular Weight: 389.958
CAS RN: 60763-24-8
Properties: ortho-rhomb cryst; -80 mesh with 99.5 purity [LID94] [CER91]
Density, g/cm³: 805 [LID94]

2968
Compound: Thorium silicide
Formula: $ThSi_2$
Molecular Formula: Si_2Th
Molecular Weight: 288.209
CAS RN: 12067-54-8
Properties: black tetr; -80 mesh with 99.5% purity [CER91] [CRC94]
Density, g/cm³: 7.96 [CRC94]

2969
Compound: Thorium sulfate nonahydrate
Formula: $Th(SO_4)_2 \cdot 9H_2O$
Molecular Formula: $H_{18}O_{17}S_2Th$
Molecular Weight: 586.303
CAS RN: 10381-37-0
Properties: colorless or white; monocl cryst; decomposes when heated strongly [MER89]
Solubility: g/100g H_2O: 0.74 (0°C), 1.38 (20°C), 3.00 (40°C) [LAN85]
Density, g/cm³: 2.8 [MER89]
Reactions: minus $9H_2O$ at 400°C [CRC94]

2970
Compound: Thorium sulfate octahydrate
Formula: $Th(SO_4)_2 \cdot 8H_2O$
Molecular Formula: $H_{16}O_{16}S_2Th$
Molecular Weight: 568.287
CAS RN: 10381-37-0
Properties: monocl white cryst powd [HAW93] [CRC94]
Solubility: sl s water; s in ice water [HAW93]
Density, g/cm³: 2.8 [HAW93]
Reactions: minus $4H_2O$ at 42°C, minus $8H_2O$ at 400°C [HAW93]

2971
Compound: Thorium sulfate tetrahydrate
Formula: $Th(SO_4)_2 \cdot 4H_2O$
Molecular Formula: $H_8O_{12}S_2Th$
Molecular Weight: 496.227
CAS RN: 10381-37-0
Properties: white needles [CRC77]
Solubility: g/100g H_2O: 4.04 (40°C), 1.63 (60°C) [LAN85]

Reactions: minus 4H$_2$O, 400°C [CRC77]

2972
Compound: Thorium sulfide
Formula: ThS$_2$
Molecular Formula: S$_2$Th
Molecular Weight: 296.170
CAS RN: 12138-07-7
Properties: dark brown cryst; begins to decompose above 1500°C; used as a solid lubricant [HAW93] [KIR83]
Solubility: i H$_2$O; s acids [HAW93] [COT88]
Density, g/cm^3: 7.30 [HAW93]
Melting Point, °C: 1875-1975 (in vacuum) [HAW93]

2973
Compound: Thorium tetracyanoplatinate(II) hexadecahydrate
Formula: Th[Pt(CN)$_4$]$_2$·16H$_2$O
Molecular Formula: C$_8$H$_{32}$N$_8$O$_{16}$Pt$_2$Th
Molecular Weight: 1118.594
CAS RN: 14481-33-5
Properties: yellow cryst [MER89]
Solubility: sl s H$_2$O [MER89]

2974
Compound: Thulium
Formula: Tm
Molecular Formula: Tm
Molecular Weight: 168.9342
CAS RN: 7440-30-4
Properties: silvery white metal; easily worked; hex close-packed; electrical resistivity (20°C) 90 μohm·cm; enthalpy of fusion 16.84 kJ/mol; enthalpy of sublimation 232.2 kJ/mol; radius of atom 0.17462 nm; radius of Tm^{+++} ion 0.0870 nm; forms light green colored solutions [MER89] [KIR82] [ALD94]
Solubility: slowly reacts with H$_2$O; s in dil acids [HAW93]
Density, g/cm^3: 9.32 [KIR82]
Melting Point, °C: 1545 [KIR82]
Boiling Point, °C: 1950 [KIR82]
Thermal Conductivity, W/(m·K): 16.9 (25°C) [CRC93]
Thermal Expansion Coefficient: 13.3 x 10^{-6}/K [CRC93]

2975
Compound: Thulium acetate monohydrate

Formula: Tm(CH$_3$COO)$_3$·H$_2$O
Molecular Formula: C$_6$H$_{11}$O$_7$Tm
Molecular Weight: 364.083
CAS RN: 39156-80-4
Properties: white powd [STR93]

2976
Compound: Thulium acetylacetonate trihydrate
Synonyms: 2,4-pentanedione, thulium(III) derivative
Formula: Tm(CH$_3$COCH=C(O)CH$_3$)$_3$·3H$_2$O
Molecular Formula: C$_{15}$H$_{27}$O$_9$Tm
Molecular Weight: 520.308
CAS RN: 14589-44-7
Properties: white powd [STR93]

2977
Compound: Thulium bromide
Formula: TmBr$_3$
Molecular Formula: Br$_3$Tm
Molecular Weight: 408.646
CAS RN: 14456-51-0
Properties: -20 mesh of 99.9% purity [CER91]
Melting Point, °C: 952 [AES93]
Boiling Point, °C: 1440 [CRC94]

2978
Compound: Thulium chloride
Formula: TmCl$_3$
Molecular Formula: Cl$_3$Tm
Molecular Weight: 275.292
CAS RN: 13537-18-3
Properties: -20 mesh with 99.9% purity; off-white powd; hygr [STR93] [CER91]
Melting Point, °C: 821 [STR93]

2979
Compound: Thulium chloride heptahydrate
Formula: TmCl$_3$·7H$_2$O
Molecular Formula: Cl$_3$H$_{14}$O$_7$Tm
Molecular Weight: 401.399
CAS RN: 13778-39-7
Properties: -4 mesh with 99.9% purity; light green cryst; deliq; there is a hexahydrate, CAS RN 1331-74-4 [HAW93] [STR93] [CER91] [ALD94]
Solubility: s H$_2$O, alcohol [MER89]
Melting Point, °C: 824 [HAW93]
Boiling Point, °C: 1440 [HAW93]

2980
Compound: Thulium fluoride
Formula: TmF_3
Molecular Formula: F_3Tm
Molecular Weight: 225.929
CAS RN: 13760-79-7
Properties: off-white powd; hygr [STR93]
Melting Point, °C: 1158 [STR93]
Boiling Point, °C: >2200 [CRC94]

2981
Compound: Thulium hydroxide
Formula: $Tm(OH)_3$
Molecular Formula: H_3O_3Tm
Molecular Weight: 219.956
CAS RN: 1311-33-7
Properties: white precipitate [MER89]

2982
Compound: Thulium iodide
Formula: TmI_3
Molecular Formula: I_3Tm
Molecular Weight: 549.553
CAS RN: 13813-43-9
Properties: yellow cryst; -20 mesh with 99.9% purity [CRC94] [CER91]
Melting Point, °C: 1015 [CRC94]
Boiling Point, °C: 1260 [CRC94]

2983
Compound: Thulium nitrate hexahydrate
Formula: $Tm(NO_3)_3 \cdot 6H_2O$
Molecular Formula: $H_{12}N_3O_{15}Tm$
Molecular Weight: 463.040
CAS RN: 36548-87-5
Properties: off-white cryst [STR93]

2984
Compound: Thulium oxalate hexahydrate
Formula: $Tm_2(C_2O_4)_3 \cdot 6H_2O$
Molecular Formula: $C_6H_{12}O_{18}Tm_2$
Molecular Weight: 710.018
CAS RN: 26677-68-9
Properties: cryst; greenish white precipitate [STR93] [MER89]
Solubility: s aq solutions of alkali oxalates, forming double oxalates [MER89]
Reactions: minus H_2O at 50°C [HAW93]

2985
Compound: Thulium oxide
Synonyms: thulia
Formula: Tm_2O_3
Molecular Formula: O_3Tm_2
Molecular Weight: 385.866
CAS RN: 12036-44-1
Properties: dense white powd, with greenish tinge, or 3-12 mm sintered pieces; sl hygr; absorbs atm H_2O and CO_2; has reddish incandescence when heated, which changes to yellow and then white if heating is prolonged; used as an evaporation material of 99.9% purity to deposit films possibly reactive to radio frequencies [HAW93] [CER91]
Solubility: slow dissolves in strong acids [HAW93]
Density, g/cm³: 8.6 [HAW93]

2986
Compound: Thulium silicide
Formula: $TmSi_2$
Molecular Formula: Si_2Tm
Molecular Weight: 225.105
CAS RN: 12039-84-8
Properties: 10 mm & down lump [ALF93]

2987
Compound: Thulium sulfate octahydrate
Formula: $Tm_2(SO_4)_3 \cdot 8H_2O$
Molecular Formula: $H_{16}O_{20}S_3Tm_2$
Molecular Weight: 770.181
CAS RN: 13778-40-0
Properties: white cryst; obtained from an aq solution of $TlCl_3$ and H_2SO_4 by precipitating with alcohol [MER89] [STR93]

2988
Compound: Thulium sulfide
Formula: Tm_2S_3
Molecular Formula: S_3Tm_2
Molecular Weight: 434.001
CAS RN: 12166-30-2
Properties: -200 mesh with 99.9% purity [CER91]

2989
Compound: Tin (gray)
Synonyms: gray tin
Formula: Sn
Molecular Formula: Sn
Molecular Weight: 118.710
CAS RN: 7440-31-5

Properties: amorphous; unstable, brittle; formed by white tin at -40°C, but slowly reverts back to white form >20°C [MER89]

Density, g/cm³: 5.77 [KIR83]

Reactions: gray to white transformation at 13.2°C [KIR83]

2990

Compound: Tin (white)

Synonyms: white tin

Formula: Sn

Molecular Formula: Sn

Molecular Weight: 118.710

CAS RN: 7440-31-5

Properties: almost silvery-white, lustrous, very malleable; easily powdered; brittle at 200°C; has thin oxide film; enthalpy of fusion 7.03 kJ/mol; enthalpy of vaporization 296.4 kJ/mol; electrical resistivity (0°C) 11.0 μohm·cm, (100°C) 15.5 μohm·cm; Brinell hardness at 20°C is 3.9; tensile strength at 15°C is 14.5 MPa; electronegativity 1.8-1.9; uses include cryogenic switching devices; band gap 0.082 eV (0 K) [MER89] [KIR83] [KIR82] [COT88] [CER91] [CRC93]

Solubility: reacts slowly with cold dil HCl, dil HNO₃, hot dil H₂SO₄; readily with conc HCl, aqua regia [MER89]

Density, g/cm³: 7.31 [MER89]

Melting Point, °C: 231.9 [KIR83]

Boiling Point, °C: 2270 [ALD94]

Reactions: crumbles to gray amorphous powd at -40°C ('gray tin') [MER89]

Thermal Conductivity, W/(m·K): 66.6 (25°C) [CRC93]

Thermal Expansion Coefficient: (volume) 100°C (0.54), 200°C (1.27) [CLA66]

2991

Compound: Tin hydride

Synonyms: stannane

Formula: SnH₄

Molecular Formula: H₄Sn

Molecular Weight: 122.742

CAS RN: 2406-52-2

Properties: colorless poisonous gas; decomposes rapidly at room temp; enthalpy of vaporization 19.05 kJ/mol [KIR80] [CRC93]

Melting Point, °C: -150 [KIR80]

Boiling Point, °C: -51.8 [CRC93]

2992

Compound: Tin monophosphide

Formula: SnP

Molecular Formula: PSn

Molecular Weight: 149.684

CAS RN: 25324-56-5

Properties: white powd; -100 mesh of 99.5% purity [CER91] [AES93]

Density, g/cm³: 6.56 [CRC94]

Melting Point, °C: decomposes [AES93]

2993

Compound: Tin triphosphide

Formula: Sn₄P₃

Molecular Formula: P₃Sn₄

Molecular Weight: 567.761

CAS RN: 12286-33-8

Properties: white cryst; other phosphides are Sn₂P₃ [53095-87-7] and SnP₃ [37367-13-8] [KIR82] [CRC94]

Density, g/cm³: 5.181 [CRC94]

Melting Point, °C: decomposes <480 [CRC94]

2994

Compound: Titanic acid

Synonyms: orthotitanic acid

Formula: Ti(OH)₄

Molecular Formula: H₄O₄Ti

Molecular Weight: 115.897

CAS RN: 20338-08-3

Properties: white powd; variable water content; can be obtained as a precipitate by adding NaOH solution to a solution of a Ti(IV) salt; used as a mordant [HAW93] [KIR83]

Solubility: i H₂O; s dil HCl [KIR83]

2995

Compound: Titanium

Formula: Ti

Molecular Formula: Ti

Molecular Weight: 47.867

CAS RN: 7440-32-6

Properties: dark gray lustrous metal; two phases: α-form, hex, stable below 882.5°C; β-form, bcc, stable above 882.5°C; brittle when cold, else ductile; Vickers hardness is 80-100; Poisson's ratio ~0.41; enthalpy of fusion 14.15 kJ/mol; enthalpy of vaporization 425 kJ/mol; electrical resistivity 42.0 μohm·cm at 20°C [HAW93] [MER89] [KIR83] [CRC93] [ALD94]

Solubility: i H₂O [HAW93]

Density, g/cm³: α: 4.506; β: 4.400 (885°C) [KIR83] [MER89]

Melting Point, °C: 1660 [ALD94]

Boiling Point, °C: 3277 [MER89]

Reactions: reacts with: F_2 (150°C); Cl_2 (300°C); Br_2 (360°C) [MER89]

Thermal Conductivity, W/(m·K): 21.9 at 25°C [KIR83]

Thermal Expansion Coefficient: (volume) 100°C (0.240), 200°C (0.567), 400°C (1.316), 600°C (2.095) [CLA66]

2996

Compound: Titanium boride

Formula: TiB_2

Molecular Formula: B_2Ti

Molecular Weight: 69.489

CAS RN: 12045-63-5

Properties: gray cryst refractory material; hex, a = 0.3028 nm, c = 0.3228 nm; hardness 9+ Mohs; electrical resistivity 28.4 μohm·cm (20°C); superconducting at 1.26 K; can be prepared by direct reaction of Ti with B at 2000°C; used as a high temp electrical conductor, as a cermet component, in crucibles to melt metals such as aluminum and tin, and as a 99.5% pure sputtering target to form films which increase cutting tool life [HAW93] [KIR78] [KIR83] [CER91]

Density, g/cm³: 4.50 [STR93]

Melting Point, °C: 2900 [STR93]

2997

Compound: Titanium carbide

Formula: TiC

Molecular Formula: CTi

Molecular Weight: 59.878

CAS RN: 12070-08-5

Properties: gray cub, a = 0.4328 nm; extremely hard; resistivity at room temp 60 μohm·cm; enthalpy of fusion 71 kJ/mol; hardness 9-10 Mohs; superconducting at 1.1 K; made by reaction of titanium with carbon at high temperatures; used as an additive in cutting tools, in crucible form for melting metals such as bismuth, zinc and cadmium, and as 99.5% pure sputtering target to prepare wear-resistant semiconducting films [HAW93] [STR93] [CER91] [JAN71]

Solubility: i H_2O; s nitric acid and aqua regia [HAW93]

Density, g/cm³: 4.93 [STR93]

Melting Point, °C: ~3140 [STR93]

Boiling Point, °C: 4820 [STR93]

Thermal Conductivity, W/(m·K): 21 [KIR78]

Thermal Expansion Coefficient: (volume) 100°C (0.125), 200°C (0.326), 400°C (0.771), 800°C (1.736), 1200°C (2.869) [CLA66]

2998

Compound: Titanium dibromide

Formula: $TiBr_2$

Molecular Formula: Br_2Ti

Molecular Weight: 207.675

CAS RN: 13783-04-5

Properties: black powd; strong reducing agent; ignites spontaneously in air; can be made by disporportionation of $TiBr_4$ at 400°C [KIR83]

Solubility: reacts with H_2O evolving hydrogen and forming Ti^{+++} [KIR83]

Density, g/cm³: 4.31 [KIR83]

Melting Point, °C: decomposes >500 [CRC94]

2999

Compound: Titanium dichloride

Formula: $TiCl_2$

Molecular Formula: Cl_2Ti

Molecular Weight: 118.772

CAS RN: 10049-06-6

Properties: black hex cryst; burns in air if heated; enthalpy of vaporization 232 kJ/mol; can be made by heating $TiCl_3$ in vacuum at 475°C [MER89] [KIR83] [CRC93]

Solubility: decomposed by H_2O evolving H_2 and forming $TiCl_3$ [KIR83]; s alcohol; i chloroform, ether, CS_2 [MER89]

Density, g/cm³: 3.13 [MER89]

Melting Point, °C: 1035 [MER89]

Boiling Point, °C: 1500 [KIR83]

3000

Compound: Titanium diiodide

Synonyms: titanium(II) iodide

Formula: TiI_2

Molecular Formula: I_2Ti

Molecular Weight: 301.676

CAS RN: 13783-07-8

Properties: hygr black powd, hex; can be prepared by reacting Ti and I_2 at 440°C [KIR82] [CRC94]

Solubility: reacts quickly with H_2O, evolving hydrogen and forming TiI_3 solution [KIR83]

Density, g/cm³: 4.99 [CRC94]

Melting Point, °C: 600 [CRC94]

Boiling Point, °C: 1000 [CRC94]

3001
Compound: Titanium dioxide
Synonyms: anatase
Formula: TiO_2
Molecular Formula: O_2Ti
Molecular Weight: 79.866
CAS RN: 1317-70-0
Properties: white; tetr, a = 0.3758 nm, c = 0.9514 nm; hardness 5.5-6 Mohs; enthalpy of transition to rutile 12.6 kJ/mol; can be prepared by hydrolysis of titanium tetraethoxide solutions in ethanol solution to yield amorphous hydrous powd, followed by hydrothermal treatment at 200 to 282°C for 5 h and refluxing [KIR83] [OGU88]
Density, g/cm³: 3.9 [KIR83]
Reactions: transforms to rutile ~700°C [KIR83]

3002
Compound: Titanium dioxide
Synonyms: brookite
Formula: TiO_2
Molecular Formula: O_2Ti
Molecular Weight: 79.866
CAS RN: 13463-67-7
Properties: ortho-rhomb, a = 0.9166 nm, b = 0.5436 nm, c = 0.5135 nm; hardness 5.5-6 Mohs; produced by heating amorphous TiO_2 with NaOH or KOH for several days at 200-600°C [KIR83]
Density, g/cm³: 4.0 [KIR83]

3003
Compound: Titanium dioxide
Synonyms: rutile
Formula: TiO_2
Molecular Formula: O_2Ti
Molecular Weight: 79.866
CAS RN: 1317-80-2
Properties: white powd, or 99.9% pure gold sintered tablets; tetr, a = 0.533 nm, c = 0.6645 nm; thermally stable form of TiO_2; hardness 7-7.5 Mohs; dielectric constant 114; used as a white pigment in paints, paper, rubber, etc., as an opacifying agent, and with 99.99% and 99.9% purity as a sputtering target to prepare high index films, and multilayer interference filters [HAW93] [MER89] [CER91] [KIR83]
Solubility: in 0.00058 mol/kg alkaline phosphate solutions: in units of 10^{-6} mol/kg-H_2O: 0.00125 (20°C), 0.00109 (120.5°C), 0.00244 (218.3°C), 0.0232 (287.2°C) [ZIE93]; s hot conc H_2SO_4, HF [MER89]
Density, g/cm³: 4.23 [MER89]

Melting Point, °C: 1855 [MER89]
Boiling Point, °C: 2500-3000 [STR93]
Thermal Conductivity, W/(m·K): 3.8 (500°C), 3.3 (1000°C) [KIR80]
Thermal Expansion Coefficient: (volume) 100°C (0.182), 200°C (0.434), 400°C (0.968), 800°C (2.063), 1000°C (2.861) [CLA66]

3004
Compound: Titanium diselenide
Synonyms: Ti(IV) selenide
Formula: $TiSe_2$
Molecular Formula: Se_2Ti
Molecular Weight: 205.800
CAS RN: 12067-45-7
Properties: -325 mesh white powd [AES93]

3005
Compound: Titanium disulfide
Synonyms: titanium(IV) sulfide
Formula: TiS_2
Molecular Formula: S_2Ti
Molecular Weight: 111.999
CAS RN: 12039-13-3
Properties: yellowish brown powd; hex; sensitive to moisture, forming H_2S and TiO_2; decomposed by steam; is obtained as a product of the reaction of H_2S and $TiCl_4$ at 600°C; used as a solid lubricant [HAW93] [STR93] [KIR83]
Solubility: stable to HCl, s cold or hot H_2SO_4; decomposed by hot NaOH [KIR83]
Density, g/cm³: 3.22 [STR93]

3006
Compound: Titanium ditelluride
Synonyms: Ti(IV) telluride
Formula: $TiTe_2$
Molecular Formula: Te_2Ti
Molecular Weight: 303.080
CAS RN: 12067-15-3
Properties: -325 mesh black powd [AES93]

3007
Compound: Titanium hydride
Formula: TiH_2
Molecular Formula: H_2Ti
Molecular Weight: 49.883
CAS RN: 7704-98-5

Properties: grayish black metallic powd; stable in air; dissociates at 450°C; can produce 448 mL H_2/g; stable at room temp; burns quietly when ignited, but violent reaction if oxidizing agents are present; industrial preparation by reaction of titanium sponge with H_2 at 200-600°C, then cooling in H_2; used as a source for Ti powd, as a getter in electronic tubes and to seal metals [KIR80] [MER89]
Solubility: i H_2O [KIR80]
Density, g/cm³: 3.9 [STR93]
Melting Point, °C: 400, decomposes [STR93]
Reactions: dissociates from 300 to 600°C [KIR80]

3008
Compound: Titanium isopropoxide
Synonyms: titanium isopropylate
Formula: Ti[OCH(CH_3)_2]_4
Molecular Formula: $C_{12}H_{28}O_4Ti$
Molecular Weight: 284.232
CAS RN: 546-68-9
Properties: colorless liq; fumes in air; used as a polymerization catalyst; used to prepare barium titanate, aluminum titanate and TiO_2-CeO_2 coatings by the sol-gel process [MER89] [MAK90] [RIT86] [YAM89] [STR93]
Solubility: decomposes rapidly in H_2O; s absolute ethanol, ether, benzene, chloroform [MER89]
Density, g/cm³: 0.955 [STR93]
Melting Point, °C: ~20 [MER89]
Boiling Point, °C: 220 [MER89]

3009
Compound: Titanium monosulfide
Formula: TiS
Molecular Formula: STi
Molecular Weight: 79.933
CAS RN: 12039-07-5
Properties: hex, dark brown solid; can be obtained by direct reaction of Ti and S [KIR83]
Solubility: attacked by conc HCl and HNO_3, but not by alkalies [KIR83]
Density, g/cm³: 4.05 [KIR83]

3010
Compound: Titanium monoxide
Synonyms: titanium(II) oxide
Formula: TiO
Molecular Formula: OTi
Molecular Weight: 63.866
CAS RN: 12137-20-1

Properties: bronze pellets; fcc; weakly basic oxide with no important industrial uses [HAW93] [STR93] [KIR83]
Solubility: s hot 40% HF and 30% H_2O_2 [KIR83]
Density, g/cm³: 4.95 [STR93]
Melting Point, °C: 1700 [STR93]
Boiling Point, °C: >3000 [STR93]

3011
Compound: Titanium nitride
Formula: TiN
Molecular Formula: NTi
Molecular Weight: 61.874
CAS RN: 25583-20-4
Properties: bronze powd; fcc, a = 0.4246 nm; hardness, 8-9 Mohs; electrical resistivity 21.7 μohm·cm (20°C); transition temp 4.8 K; can be prepared by reaction between finely divided Ti and nitrogen at 1000-1400°C; used in cermets, rectifiers and in crucible form for melting metals such as aluminum, bismuth, cadmium, lead, steel, tin; used as 99.5% pure sputtering target to increase life of cutting tools [HAW93] [CIC73] [KIR81] [KIR83] [CER91]
Solubility: attacked by boiling aqua regia, decomposed by boiling alkalies evolving ammonia; otherwise highly stable [KIR83]
Density, g/cm³: 5.22 [STR93]
Melting Point, °C: 2930 [STR93]
Thermal Conductivity, W/(m·K): 29.1 [KIR81]
Thermal Expansion Coefficient: 9.35 x 10⁻⁶/°C [KIR81]

3012
Compound: Titanium oxalate decahydrate
Formula: $Ti_2(C_2O_4)_3 \cdot 10H_2O$
Molecular Formula: $C_6H_{20}O_{22}Ti_2$
Molecular Weight: 539.946
Properties: yellow prisms; prepared from oxalic acid and $TiCl_3$ [HAW93]
Solubility: s H_2O; i alcohol, ether [HAW93]

3013
Compound: Titanium oxysulfate
Formula: $TiOSO_4$
Molecular Formula: O_5STi
Molecular Weight: 159.930
CAS RN: 13825-75-6
Properties: white or sl yellow powd; decomposed by water [MER89]

3014
Compound: Titanium phosphide
Formula: TiP
Molecular Formula: PTi
Molecular Weight: 78.841
CAS RN: 12037-65-9
Properties: hard, gray metallic powd with hex structure; prepared by heating phosphine with TiCl$_4$; stable to 1400°C; finds use as a catalyst in organic reactions [KIR83]
Solubility: not attacked by common acids [KIR83]
Density, g/cm^3: 4.08 [KIR83]

3015
Compound: Titanium silicide
Synonyms: titanium disilicide
Formula: TiSi$_2$
Molecular Formula: Si$_2$Ti
Molecular Weight: 104.051
CAS RN: 12039-83-7
Properties: ortho-rhomb black powd, a = 0.8236 nm, b = 0.4773 nm, c = 0.8523 nm; hardness 4.5 Mohs; resistivity 123 μohm·cm; can be prepared by reaction of the elements; used in special alloy applications, as a flame resistant coating material, also as 99.5 or 99.9% pure material, as a sputtering target in the fabrication of integrated circuits [HAW93] [STR93] [CER91]
Solubility: s HF, resistant to mineral acids and alkali solutions [KIR83]
Density, g/cm^3: 4.39 [STR93]
Melting Point, °C: 1540 [STR93]

3016
Compound: Titanium sulfate
Synonyms: titanous sulfate
Formula: Ti$_2$(SO$_4$)$_3$
Molecular Formula: O$_{12}$S$_3$Ti$_2$
Molecular Weight: 383.925
CAS RN: 10343-61-0
Properties: green, cryst powd; produced by reduction of Ti(IV) in sulfuric acid solution; used as a reducing agent in the textile industry [HAW93] [MER89] [KIR83]
Solubility: i H$_2$O, alcohol, conc H$_2$SO$_4$; s dil HCl, dil H$_2$SO$_4$ both giving violet solutions [MER89]

3017
Compound: Titanium tetrabromide
Synonyms: titanium(IV) bromide
Formula: TiBr$_4$

Molecular Formula: Br$_4$Ti
Molecular Weight: 367.483
CAS RN: 7789-68-6
Properties: amber yellow or orange cub; very hygr; enthalpy of vaporization 44.37 kJ/mol; enthalpy of fusion 12.90 kJ/mol; can be made by reaction between TiCl$_4$ and HBr [MER89] [KIR83] [CRC93]
Solubility: dissolves with hydrolysis in H$_2$O [KIR83]
Density, g/cm^3: 3.25 [MER89]
Melting Point, °C: 39 [CRC93]
Boiling Point, °C: 230 [CRC93]

3018
Compound: Titanium tetrachloride
Formula: TiCl$_4$
Molecular Formula: Cl$_4$Ti
Molecular Weight: 189.678
CAS RN: 7550-45-0
Properties: pale yellow or colorless liq; absorbs atm moisture and emits dense white cloud; vapor pressure (20°C) 1.33 kPa; enthalpy of vaporization 36.2 kJ/mol; enthalpy of fusion 9.97 kJ/mol; critical temp 358°C; viscosity 0.079 mPa s; dielectric constant (20°C) 2.79; manufactured by the chlorination of titanium compounds, for example rutile [MER89] [STR93] [KIR83] [CRC93]
Solubility: s cold H$_2$O, alcohol; decomposed in hot H$_2$O [MER89]; s dil HCl [HAW93]
Density, g/cm^3: 1.726 [MER89]
Melting Point, °C: -45 [ALD94]
Boiling Point, °C: 136.4 [ALD94]

3019
Compound: Titanium tetrafluoride
Formula: TiF$_4$
Molecular Formula: F$_4$Ti
Molecular Weight: 123.861
CAS RN: 7783-63-3
Properties: powd or white mass; very hygr; can be produced by the reaction between F$_2$ and TiCl$_4$ at 250°C [MER89]
Solubility: hydrolyzes in H$_2$O; s alcohol, pyridine [MER89]
Density, g/cm^3: 2.798 [MER89]
Boiling Point, °C: sublimes, 284 [MER89]

3020
Compound: Titanium tetraiodide
Formula: TiI$_4$
Molecular Formula: I$_4$Ti

Molecular Weight: 555.485
CAS RN: 7720-83-4
Properties: red powd; sensitive to moisture; enthalpy of vaporization 58.4 kJ/mol; enthalpy of fusion 19.80 kJ/mol; prepared by reacting Ti and I_2 under controlled conditions; used extensively as a catalyst in organic reactions [KIR83] [STR93] [CRC93]
Solubility: dissolves and hydrolyzes in H_2O [KIR82]
Density, g/cm³: 4.3 [STR93]
Melting Point, °C: 150 [CRC93]
Boiling Point, °C: 377 [CRC93]

3021
Compound: Titanium tribromide
Formula: $TiBr_3$
Molecular Formula: Br_3Ti
Molecular Weight: 287.579
CAS RN: 13135-31-4
Properties: bluish black cryst powd; hex plates or needles; disproportinates at 400°C to di- and tetrabromides [KIR83]
Solubility: s H_2O resulting in dark violet solution [KIR83]
Melting Point, °C: 115, hexahydrate [CRC94]

3022
Compound: Titanium trichloride
Formula: $TiCl_3$
Molecular Formula: Cl_3Ti
Molecular Weight: 154.225
CAS RN: 7705-07-9
Properties: dark reddish violet cryst; deliq; unstable, decomposes above 500°C; strong reducing agent; reaction between hydrogen and $TiCl_4$ produces α-$TiCl_3$ (violet); β (brown) and γ (violet) forms are produced by reacting $TiCl_4$ with aluminum alkyls; a τ (violet) form is made by grinding the α or γ forms; enthalpy of vaporization 124 kJ/mol; used extensively as a catalyst for the polymerization of hydrocarbons [KIR83] [MER89] [CRC93]
Solubility: s H_2O (exothermic), alcohol [MER89]
Density, g/cm³: 2.640 [STR93]
Melting Point, °C: 440, decomposes [STR93]
Boiling Point, °C: 960 [CRC93]

3023
Compound: Titanium trifluoride
Formula: TiF_3
Molecular Formula: F_3Ti
Molecular Weight: 104.862

CAS RN: 13470-08-1
Properties: violet powd; sensitive to moisture; can be obtained by dissolving Ti metal in aq HF [STR93]
Solubility: i H_2O, dil acids and alkalies [KIR82]
Density, g/cm³: 3.40 [STR93]
Reactions: disproportionates >950°C [CER91]

3024
Compound: Titanium trioxide
Formula: Ti_2O_3
Molecular Formula: O_3Ti_2
Molecular Weight: 143.732
CAS RN: 1344-54-3
Properties: violet sintered tablets; hexagonal; oxidizes to TiO_2; sintered material of 99.9% purity used as an evaporation material for interference films, thin film resistors and capacitors [KIR83] [CER91]
Solubility: 40% hot HF [KIR83]
Density, g/cm³: 4.486 [KIR83]
Melting Point, °C: 1900 [KIR83]

3025
Compound: Titanium trisilicide
Formula: Ti_5Si_3
Molecular Formula: Si_3Ti_5
Molecular Weight: 323.657
CAS RN: 12067-57-1
Properties: -325 mesh gray powd [AES93]
Melting Point, °C: 2130 [AES93]

3026
Compound: Titanium trisulfide
Synonyms: titanium sesquisulfide
Formula: Ti_2S_3
Molecular Formula: S_3Ti_2
Molecular Weight: 191.932
CAS RN: 12039-16-6
Properties: black cryst hex solid; can be prepared by direct reaction of Ti and S at 800°C [KIR83]
Density, g/cm³: 3.52 [KIR83]

3027
Compound: Titanium(IV) oxide acetylacetonate
Synonyms: 2,4-pentanedione, Ti(IV) derivative
Formula: $[CH_3COCH=C(O)CH_3]_2TiO$
Molecular Formula: $C_{10}H_{14}O_5Ti$
Molecular Weight: 262.098
CAS RN: 14024-64-7

Properties: cryst powd; hydrolysis resistant; preparation: reaction of titanium oxychloride with acetylacetone and sodium carbonate; uses: cross-linking agent for cellulosic fibers [HAW93]

$$\text{O–} \quad \text{O}$$
$$| \qquad \|$$
$$[CH_3\text{–}C\text{=}CH\text{–}C\text{–}CH_3]_2OTi$$

Solubility: sl s H_2O [HAW93]
Melting Point, °C: 200, decomposes [ALD94]

3028
Compound: Titanocene dichloride
Synonyms: bis(cyclopentadienyl)titanium dichloride
Formula: $(C_5H_5)_2TiCl_2$
Molecular Formula: $C_{10}H_{10}Cl_2Ti$
Molecular Weight: 248.975
CAS RN: 1271-19-8
Properties: bright red acidular cryst from toluene; sensitive to moisture; uses: synthesis of many transition metal complexes and organometallic compounds, catalyst [MER89] [ALF95]

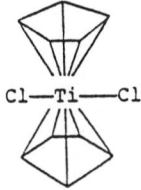

Solubility: sl s H_2O [MER89]
Density, g/cm³: 1.6 [ALD94]
Melting Point, °C: 289-290 [ALF95]
Reactions: with H_2O forms Cp_2TiOH+ [COT88]

3029
Compound: Tribromosilane
Formula: $SiHBr_3$
Molecular Formula: Br_3HSi
Molecular Weight: 268.806
CAS RN: 7789-57-3
Properties: enthalpy of vaporization 34.8 kJ/mol; entropy of vaporization 87.9 kJ/(mol·K) [CRC93] [CIC73]
Melting Point, °C: -73 [CIC73]
Boiling Point, °C: 109 [CRC93]

3030
Compound: Trichlorosilane
Synonyms: silicochloroform
Formula: $SiHCl_3$

Molecular Formula: Cl_3HSi
Molecular Weight: 135.452
CAS RN: 10025-78-2
Properties: colorless volatile mobile liq; supports combustion; enthalpy of vaporization 25.7 kJ/mol at 25°C; entropy of vaporization 82.8 kJ/(mol·K); used in organic synthesis [CIC73] [CRC93] [MER89] [STR93]
Solubility: decomposed by H_2O [MER89]
Density, g/cm³: (25°C) 1.3313 [MER89]
Melting Point, °C: -118 [CIC73]
Boiling Point, °C: 33 [CRC93]

3031
Compound: Triethylphosphine
Formula: $(C_2H_5)_3P$
Molecular Formula: $C_6H_{15}P$
Molecular Weight: 118.16
CAS RN: 554-70-1
Properties: colorless liq; preparation from white phosphorus, ethylene and hydrogen under pressure; used in organic synthesis [MER89] [ALD94]
Solubility: i H_2O; miscible with alcohol, ether [MER89]
Density, g/cm³: 0.80 [MER89]
Melting Point, °C: -17 [ALD94]
Boiling Point, °C: 127-128 [MER89]

3032
Compound: Trifluoromethane
Synonyms: halocarbon-23
Formula: CHF_3
Molecular Formula: CHF_3
Molecular Weight: 70.013
CAS RN: 75-46-7
Properties: colorless gas with ethereal odor; critical temp 25.9°C; critical pressure 4.84 MPa; enthalpy of vaporization 170.5 kJ/mol; used in electronics industry [AIR87]
Melting Point, °C: -155.2 [AIR87]
Boiling Point, °C: -82.2 [AIR87]

3033
Compound: Trifluorosilane
Synonyms: silicofluoroform
Formula: $SiHF_3$
Molecular Formula: F_3HSi
Molecular Weight: 86.089
CAS RN: 13465-71-9

Properties: colorless gas; enthalpy of vaporization 16.1 kJ/mol; entropy of vaporization 87.9 kJ/(mol·K) [CIC73] [CRC94]
Density, g/cm³: 1.86 (0°C) [CRC94]
Melting Point, °C: -131 [CIC73]
Boiling Point, °C: -94.4 [CIC73]

3034
Compound: Triiodosilane
Synonyms: silicoiodoform
Formula: SiHI₃
Molecular Formula: HI₃Si
Molecular Weight: 409.807
CAS RN: 13465-72-0
Properties: colorless liq; enthalpy of vaporization 62.8 kJ/mol; entropy of vaporization 159 kJ/(mol·K) [CIC73] [CRC94]
Density, g/cm³: 3.314 (20°C) [CRC94]
Melting Point, °C: 8 [CIC73]
Boiling Point, °C: 220 decomposes [CIC73]

3035
Compound: Tris(ethylenediamine) cobalt(III) chloride trihydrate
Formula: Co[H₂NCH₂CH₂NH₂]₃Cl₃·3H₂O
Molecular Formula: C₆H₂₈Cl₃CoN₆O₂
Molecular Weight: 399.629
CAS RN: 14883-80-8
Properties: brown prisms [KIR79]
Solubility: v s H₂O [KIR79]
Density, g/cm³: 1.542 [KIR79]
Melting Point, °C: 275, decomposes [ALD94]
Reactions: minus 3H₂O at 100°C [KIR79]

3036
Compound: Tris(triphenylphosphine)rhodium(I) chloride
Synonyms: Wilkinson's catalyst
Formula: Rh[P(C₆H₅)₃]₃
Molecular Formula: C₅₄H₄₅ClP₃Rh
Molecular Weight: 925.231
CAS RN: 14694-95-2
Properties: burgundy-red cryst when prepared in ethanol by reaction between excess triphenylphosphine and RhCl₃·3H₂O; used as a homogeneous catalyst [MER89] [ALD94]
Solubility: ~20 g/L (25°C) chloroform [MER89]
Melting Point, °C: 157-158 [MER89]

3037
Compound: Tritium

Formula: T₂
Molecular Formula: T₂
Molecular Weight: 6.032
CAS RN: 10028-17-8
Properties: gas with $t_{1/2}$ of 12.26 years; low β-emitter; critical temp -232.56 °C; critical pressure 18.317 atm; enthalpy of sublimation 1640 J/mol; enthalpy of vaporization 1390 J/mol; used in hydrogen bomb and in radioactive tracers [MER89] [KIR78]
Density, g/cm³: liq: 45.35 mol/L [KIR78]
Melting Point, °C: -254.54 [MER89]
Boiling Point, °C: -248.12 [MER89]

3038
Compound: Tritium dioxide
Formula: T₂O
Molecular Formula: OT₂
Molecular Weight: 22.032
CAS RN: 14940-65-9
Properties: liq; triple point 4.49°C; liq vapor pressure at 25°C 2.64 kPa; enthalpy of vaporization ~45.81 kJ/mol; ionization constant ~6 x 10⁻¹⁶; temp of maximum density 13.4°C [KIR78]
Density, g/cm³: 1.2138 [KIR78]
Boiling Point, °C: 101.51 [KIR78]

3039
Compound: Tungsten
Synonyms: wolfram
Formula: W
Molecular Formula: W
Molecular Weight: 183.84
CAS RN: 7440-33-7
Properties: steel gray to tin white metal; bcc, a = 0.316524 nm; enthalpy of fusion 52.31 kJ/mol; enthalpy of sublimation (25°C) 859.8 kJ/mol; electrical resistivity (20°C) 5.5 μohm·cm; stable in dry air, unless heated to red heat, then forms WO₃; used in crucible form for growing single cryst from reactive melts, and to evaporate metals and compounds for thin films [MER89] [KIR83] [CER91] [CRC93]
Solubility: s slowly in fused KOH or Na₂CO₃ in the presence of air [MER89]
Density, g/cm³: 19.254 [KIR83]
Melting Point, °C: 3410 [ALD94]
Boiling Point, °C: 5660 [ALD94]
Thermal Conductivity, W/(m·K): 146 (227°C), 118 (727°C), 100 (1727°C), 90 (3127°C) [KIR83]; 173 (25°C) [ALD94]

Thermal Expansion Coefficient: (volume) 100°C
(0.095), 200°C (0.228), 400°C (0.513), 800°C
(1.104), 1200 (1.774) [CLA66]

3040
Compound: Tungsten boride
Formula: WB
Molecular Formula: BW
Molecular Weight: 194.651
CAS RN: 12007-09-9
Properties: refractory material; black powd; in -325
mesh 99.5% pure form, used as a sputtering
target for producing wear-resistant and
semiconducting films, and other applications
[KIR78] [STR93] [CER91]
Density, g/cm³: 15.2 [LID94]
Melting Point, °C: 2660 [KIR78]

3041
Compound: Tungsten boride
Synonyms: ditungsten boride
Formula: W_2B
Molecular Formula: BW_2
Molecular Weight: 378.491
CAS RN: 12007-10-2
Properties: refractory material; black powd; forms
when tungsten and boron are hot pressed;
extremely hard and has almost metallic
electrical conductivity; -325 mesh in 99.5% pure
form used as a sputtering target for fabricating
wear-resistant and semiconductor films [KIR83]
[STR93] [CER91]
Density, g/cm³: 16.0 [LID94]
Melting Point, °C: 2670 [KIR78]

3042
Compound: Tungsten carbide
Formula: W_2C
Molecular Formula: CW_2
Molecular Weight: 379.691
CAS RN: 12070-13-2
Properties: black hex, a = 0.29982 nm, c = 0.4722
nm; can be prepared by heating tungsten and
carbon in the presence of hydrogen at high
temperatures; hardness approaches that of
diamond; brittle; used in hard metals, and as a
99.5% pure sputtering target to produce wear-
resistant films and semiconductor films [KIR83]
[CIC73] [CER91] [CRC94]
Density, g/cm³: 17.15 [KIR83]
Melting Point, °C: ~2800 [KIR83]

3043
Compound: Tungsten carbide
Formula: WC
Molecular Formula: CW
Molecular Weight: 195.851
CAS RN: 12070-12-1
Properties: gray powd; hex, a = 0.29063 nm, c =
0.28386 nm; hardness 9+ Mohs; can be
prepared by heating tungsten and carbon at high
temperatures, sometimes in the presence of
hydrogen; used in dies and cutting tools, wear
resistant parts, electrical resistors, as an
abrasive, in crucible form used to melt copper,
tin, bismuth and cobalt, and as a 99.5% pure
sputtering target to prepare wear-resistant
semiconductor films [HAW93] [CIC73]
[STR93] [KIR83] [CER91] [GEI92]
Solubility: i H_2O; attacked by a mixture of HNO_3
and HF acids [HAW93]
Density, g/cm³: 15.63 [STR93]
Melting Point, °C: ~2870 [STR93]
Boiling Point, °C: 6000 [STR93]
Thermal Conductivity, W/(m·K): 121 [KIR78]
Thermal Expansion Coefficient: (volume) 100°C
(0.104), 200°C (0.236), 400°C (0.507), 800°C
(1.103), 1200°C (1.768) [CLA66]

3044
Compound: Tungsten carbonyl
Synonyms: tungsten hexacarbonyl
Formula: $W(CO)_6$
Molecular Formula: C_6O_6W
Molecular Weight: 351.902
CAS RN: 14040-11-0
Properties: white, volatile, highly refractive cryst;
very stable; vapor pressure is 13.3 Pa (20°C) and
160 Pa (67°C); may be prepared by reducing
WCl_6 with Al in anhydrous ether under 10 MPa
of CO at 70°C [KIR83] [HAW93]
Solubility: i H_2O; s in organic solvents [HAW93]
Density, g/cm³: 2.65 [HAW93]
Melting Point, °C: 169-170, decomposes [STR93]

3045
Compound: Tungsten dibromide
Formula: WBr_2
Molecular Formula: Br_2W
Molecular Weight: 343.648
CAS RN: 13470-10-5
Properties: black powd; prepared by the reduction
of WBr_5 with H_2 [KIR83]
Melting Point, °C: decomposes at 400 [KIR83]

3046
Compound: Tungsten dichloride
Formula: WCl$_2$
Molecular Formula: Cl$_2$W
Molecular Weight: 254.745
CAS RN: 13470-12-7
Properties: gray amorphous powd; prepared by reduction of WCl$_6$ by Al in molten NaAlCl$_6$ [KIR83] [CRC94]
Density, g/cm^3: 5.436 [CRC94]

3047
Compound: Tungsten diiodide
Formula: WI$_2$
Molecular Formula: I$_2$W
Molecular Weight: 437.649
CAS RN: 13470-17-2
Properties: brown powd [KIR83]
Density, g/cm^3: 6.79 [KIR83]
Melting Point, °C: decomposes [CRC94]

3048
Compound: Tungsten dinitride
Formula: WN$_2$
Molecular Formula: N$_2$W
Molecular Weight: 211.853
CAS RN: 60922-26-1
Properties: brown; hex, a = 0.2893 nm, c = 0.2826 nm [CIC73] [CRC94]
Density, g/cm^3: 7.7 [LID94]
Melting Point, °C: decomposes 600 [CIC73]

3049
Compound: Tungsten dioxide
Synonyms: tungsten(IV) oxide
Formula: WO$_2$
Molecular Formula: O$_2$W
Molecular Weight: 215.839
CAS RN: 12036-22-5
Properties: brown powd, may become purple on standing; formed when WO$_3$ is reduced by H$_2$ at 575-600°C [KIR83] [STR93]
Density, g/cm^3: 12.11 [STR93]
Melting Point, °C: 1500-1600 [STR93]

3050
Compound: Tungsten dioxydibromide
Formula: WO$_2$Br$_2$
Molecular Formula: Br$_2$O$_2$W
Molecular Weight: 375.647
CAS RN: 13520-75-7

Properties: light red cryst that forms when a mixture of bromine and oxygen is passed over tungsten at 300°C [KIR83]
Melting Point, °C: decomposes [CRC94]

3051
Compound: Tungsten diselenide
Formula: WSe$_2$
Molecular Formula: Se$_2$W
Molecular Weight: 341.760
CAS RN: 12067-46-8
Properties: -325 mesh black powd; dry, solid lubricant with exceptional stability at high temperatures, and in high vacuum; used in the form of 99.8% pure material as a sputtering target to produce lubricant films [HAW93] [CER91] [AES93]

3052
Compound: Tungsten disilicide
Formula: WSi$_2$
Molecular Formula: Si$_2$W
Molecular Weight: 240.011
CAS RN: 12039-88-2
Properties: bluish gray tetr powd, a = 0.3212 nm, c = 0.7880 nm; attacked by fluorine, chlorine and fused alkalies; can be used in high temp thermocouples in combination with MoSi$_2$, in an oxidizing atm; as a 99.95% and 99.5% pure material, used as a sputtering target in the fabrication of integrated circuits [KIR83] [STR93] [CER91]
Solubility: i H$_2$O [KIR83]
Density, g/cm^3: 9.4 [STR93]
Melting Point, °C: 2165 [STR93]
Thermal Expansion Coefficient: (volume) 100°C (0.164), 200°C (0.364), 400°C (0.868), 800°C (1.886), 1200°C (3.047) [CLA66]

3053
Compound: Tungsten disulfide
Synonyms: tungstenite
Formula: WS$_2$
Molecular Formula: S$_2$W
Molecular Weight: 247.972
CAS RN: 12138-09-9

Properties: soft, grayish black powd which is relatively inert and unreactive; exists in mineral form, or can be prepared by heating tungsten powd with sulfur at 900°C; used as a solid lubricant, and as a 99.8% pure material as a sputtering target for lubricant films on bearings and other moving parts [HAW93] [STR93] [KIR83] [CER91]
Solubility: i H_2O, HCl and alkali [KIR83]
Density, g/cm^3: 7.5 [STR93]
Melting Point, °C: 1250, decomposes [STR93]

3054
Compound: Tungsten hexabromide
Formula: WBr$_6$
Molecular Formula: Br$_6$W
Molecular Weight: 663.264
CAS RN: 13701-86-5
Properties: bluish black cryst; made by reaction of BBr$_3$ with WCl$_6$ [KIR83]
Density, g/cm^3: 6.9 [CRC94]
Melting Point, °C: 232 [KIR83]

3055
Compound: Tungsten hexachloride
Synonyms: tungsten(VI) chloride
Formula: WCl$_6$
Molecular Formula: Cl$_6$W
Molecular Weight: 396.556
CAS RN: 13283-01-7
Properties: purple hex cryst; sensitive to moisture; vapor pressure is 43 mm Hg (215°C); enthalpy of vaporization 52.7 kJ/mol; enthalpy of fusion 6.60 kJ/mol; decomposed by moist air and water; can be obtained by direct reaction of Cl_2 and W at 600°C [HAW93] [STR93] [KIR83] [CRC93]
Solubility: s organic solvents such as ethanol [HAW93]
Density, g/cm^3: 3.52 [HAW93]
Melting Point, °C: 275 [CRC93]
Boiling Point, °C: 346.75 [CRC93]

3056
Compound: Tungsten hexafluoride
Synonyms: tungsten(VI) fluoride
Formula: WF$_6$
Molecular Formula: F$_6$W
Molecular Weight: 297.830
CAS RN: 7783-82-6

Properties: colorless gas or pale yellow liq when condensed; critical temp 169.8°C; critical pressure 4.27 MPa; triple point 2.0°C at 55.1 kPa; transition point -8.4°C; enthalpy of vaporization 27 kJ/mol; enthalpy of fusion 4.10 kJ/mol; vapor pressure (17°C) 100.49 kPa; produced by reacting tungsten powd with gaseous fluorine above 350°C; used in electronics industry [KIR78] [MER89] [AIR87] [CRC93]
Solubility: decomposes in H_2O; s anhydrous HF: 3.14 moles/100g [HAW93] [MER89]
Density, g/cm^3: gas: 12.9g/L [STR93]; liq: 3.441 (15°C) [KIR78]
Melting Point, °C: 2.3 [MER89]
Boiling Point, °C: 17.5 [ALD94]

3057
Compound: Tungsten nitride
Formula: W$_2$N
Molecular Formula: NW$_2$
Molecular Weight: 381.687
CAS RN: 12033-72-6
Properties: gray; fcc, a = 0.412 nm; can be prepared by heating tungsten in ammonia; there is also a WN phase, 12058-38-7 [KIR83] [KIR81]
Density, g/cm^3: 17.7 [KIR81]
Melting Point, °C: decomposes [KIR81]

3058
Compound: Tungsten oxychloride
Synonyms: tungsten(VI) tetrachloride monoxide
Formula: WOCl$_4$
Molecular Formula: Cl$_4$OW
Molecular Weight: 341.650
CAS RN: 13520-78-0
Properties: dark red, acicular cryst; enthalpy of vaporization 67.8 kJ/mol; enthalpy of fusion 45.00 kJ/mol; obtained by refluxing SOCl$_2$ with tungsten trioxide; used in incandescent lamps [HAW93] [CRC93]
Solubility: decomposed by H_2O; s CS$_2$, benzene [KIR83] [HAW93]
Density, g/cm^3: 11.92 [HAW93]
Melting Point, °C: 211 [KIR83]
Boiling Point, °C: 227.5 [CRC93]

3059
Compound: Tungsten oxydichloride
Synonyms: tungsten(VI) dichloride dioxide
Formula: WO$_2$Cl$_2$
Molecular Formula: Cl$_2$O$_2$W
Molecular Weight: 286.744

CAS RN: 13520-76-8

Properties: pale yellow cryst solid; obtained by reacting CCl_4 and WO_2 at 250°C [KIR83]

Solubility: i cold H_2O, partially decomposed by hot H_2O; i alkaline solutions [KIR83]

Melting Point, °C: 266 [KIR83]

3060

Compound: Tungsten oxydiiodide

Formula: WOI_2

Molecular Formula: I_2OW

Molecular Weight: 453.648

CAS RN: 14447-89-3

Properties: obtained by heating tungsten, tungsten trioxide and iodine in a 500-700°C temp gradient for 36 h [KIR83]

3061

Compound: Tungsten oxytetrabromide

Formula: $WOBr_4$

Molecular Formula: Br_4OW

Molecular Weight: 519.455

CAS RN: 13520-77-9

Properties: black deliq needles; made by reacting CBr_4 and WO_2 at 250°C [KIR83]

Melting Point, °C: 277 [KIR83]

Boiling Point, °C: 327 [KIR83]

3062

Compound: Tungsten oxytetrafluoride

Formula: WOF_4

Molecular Formula: F_4OW

Molecular Weight: 275.833

CAS RN: 13520-79-1

Properties: colorless plates; can be prepared by reacting W metal with an O_2-F_2 mixture at high temperatures; hygr [KIR83]

Solubility: decomposes to tungstic acid in H_2O [KIR83]

Melting Point, °C: 110 [CRC94]

Boiling Point, °C: 187 [CRC94]

3063

Compound: Tungsten oxytrichloride

Formula: $WOCl_3$

Molecular Formula: Cl_3OW

Molecular Weight: 306.197

CAS RN: 14249-98-0

Properties: green solid; can be obtained by reduction of $WOCl_4$ with Al in a sealed system at 100-140°C [KIR83]

3064

Compound: Tungsten pentaboride

Formula: W_2B_5

Molecular Formula: B_5W_2

Molecular Weight: 421.735

CAS RN: 12007-98-6

Properties: black powd, -325 mesh; refractory material [KIR78] [STR93]

Melting Point, °C: 2365 [KIR78]

3065

Compound: Tungsten pentabromide

Synonyms: tungsten(V) bromide

Formula: WBr_5

Molecular Formula: Br_5W

Molecular Weight: 583.360

CAS RN: 13470-11-6

Properties: brownish violet cryst; high moisture sensitivity; prepared by reaction of bromine vapor on metallic tungsten at 450-500°C [KIR83]

Melting Point, °C: 276 [KIR83]

Boiling Point, °C: 333 [KIR83]

3066

Compound: Tungsten pentachloride

Formula: WCl_5

Molecular Formula: Cl_5W

Molecular Weight: 361.104

CAS RN: 13470-14-9

Properties: black cryst deliq solid; can be prepared by reducing WCl_6 with red phosphorus [KIR83]

Solubility: decomposes in H_2O to a blue oxide; v sl s CS_2 [KIR83]

Density, g/cm³: 3.875 [CRC94]

Melting Point, °C: 243 [KIR83]

Boiling Point, °C: 275.6 [KIR83]

3067

Compound: Tungsten telluride

Synonyms: tungsten(IV) telluride

Formula: WTe_2

Molecular Formula: Te_2W

Molecular Weight: 439.040

CAS RN: 12067-26-4

Properties: gray powd; used in the form of a -325 mesh, 99.8% pure material as a sputtering target to form lubricant film [STR93] [CER91]

3068

Compound: Tungsten tetrabromide

Synonyms: tungsten(IV) bromide
Formula: WBr$_4$
Molecular Formula: Br$_4$W
Molecular Weight: 503.456
CAS RN: 12045-94-2
Properties: black ortho-rhomb cryst; prepared by reduction of WBr$_5$ with Al [KIR83]

3069
Compound: Tungsten tetrachloride
Synonyms: tungsten(IV) chloride
Formula: WCl$_4$
Molecular Formula: Cl$_4$W
Molecular Weight: 325.651
CAS RN: 13470-13-8
Properties: gray powd; sensitive to moisture; decomposes if heated; diamagnetic; obtained by reduction of WCl$_6$ with Al [KIR83] [STR93]
Density, g/cm^3: 4.624 [STR93]
Melting Point, °C: decomposes [CRC94]

3070
Compound: Tungsten tetraiodide
Formula: WI$_4$
Molecular Formula: I$_4$W
Molecular Weight: 691.458
CAS RN: 14055-84-6
Properties: black powd; decomposed by air; prepared by reacting conc hydriodic acid and WCl$_6$ at 100°C [KIR83]
Density, g/cm^3: 5.2 [CRC94]
Melting Point, °C: decomposes [CRC94]

3071
Compound: Tungsten tribromide
Formula: WBr$_3$
Molecular Formula: Br$_3$W
Molecular Weight: 423.552
CAS RN: 15163-24-3
Properties: black powd; thermally unstable; prepared by reaction of bromine and WBr$_2$ at 50°C in a sealed system [KIR83]
Solubility: i H$_2$O [KIR83]

3072
Compound: Tungsten triiodide
Formula: WI$_3$
Molecular Formula: I$_3$W
Molecular Weight: 564.553
CAS RN: 15513-69-6

Properties: can be prepared by reacting iodine and W(CO)$_6$ in a sealed system at 120°C [KIR83]

3073
Compound: Tungsten trioxide
Synonyms: tungsten(VI) oxide
Formula: WO$_3$
Molecular Formula: O$_3$W
Molecular Weight: 231.838
CAS RN: 1314-35-8
Properties: canary yellow heavy powd which becomes dark orange when heated and reverts to original color when cooled, also greenish yellow 3-12 mm sintered pieces; enthalpy of fusion 73.00 kJ/mol; can be prepared from tungstic acid; used as starting material to produce tungsten powd, sintered pieces used as evaporation material and sputtering target for shadow casting in electron microscopy [KIR83] [MER89] [CER91] [CRC93]
Solubility: i H$_2$O; s caustic alkalies; v sl s in acids [MER89]
Density, g/cm^3: 7.16 [STR93]
Melting Point, °C: 1472 [CRC93]
Reactions: phase change from pseudorhomb to tetr above 700°C [KIR83]

3074
Compound: Tungsten trisilicide
Formula: W$_5$Si$_3$
Molecular Formula: Si$_3$W$_5$
Molecular Weight: 1003.457
CAS RN: 12039-95-1
Properties: bluish gray; very hard solid; attacked by fused alkalies and mixtures of nitric and hydrofluoric acids [HAW93]
Solubility: i H$_2$O [HAW93]
Density, g/cm^3: 9.4 [HAW93]
Melting Point, °C: >900 [HAW93]

3075
Compound: Tungsten trisulfide
Formula: WS$_3$
Molecular Formula: S$_3$W
Molecular Weight: 280.038
CAS RN: 12125-19-8
Properties: chocolate brown powd; can be obtained from an alkali metal thiotungstate by treating with HCl [KIR83]
Solubility: sl s cold H$_2$O, forms colloid in hot H$_2$O; s alkali carbonate and hydroxide solutions [KIR83]

3076
Compound: Tungstic acid
Formula: H_2WO_4
Molecular Formula: H_2O_4W
Molecular Weight: 249.854
CAS RN: 7783-03-1
Properties: amorphous yellow powd; prepared by precipitation from hot tungstate solutions with strong acids, followed by boiling in an acidic medium; used in textile and plastics industries [KIR83] [STR93]
Solubility: i H_2O, acids; s alkalies [KIR83]
Density, g/cm³: 5.5 [STR93]
Reactions: minus H_2O at 100°C [CRC94]

3077
Compound: Tungstophosphoric acid hydrate
Synonyms: 12-tungstophosphate
Formula: $H_3PO_4 \cdot 12WO_3 \cdot xH_2O$
Molecular Formula: $H_3O_{40}PW_{12}$ (anhydrous)
Molecular Weight: 2880.053 (anhydrous)
CAS RN: 12501-23-4
Properties: yellowish white solid; hygr [HAW93] [STR93]
Solubility: s H_2O, acetone, ether [HAW93]
Melting Point, °C: 89, for x=24 [HAW93]

3078
Compound: Uranium
Formula: U
Molecular Formula: U
Molecular Weight: 238.0289
CAS RN: 7440-61-1
Properties: silvery white, lustrous; black powd, when obtained by reduction of UF_4; three forms, α-form: ortho-rhomb, a = 0.2854 nm, b = 0.5869 nm, c = 0.4956 nm; β: tetr, a = 1.0763 nm, c = 0.5652 nm; γ: bcc, a = 0.3524 nm; resistivity 29 μohm·cm; enthalpy of fusion 9.14 kJ/mol; enthalpy of sublimation 1062.73 kJ/mol; flammable in air forming U_3O_8; $t_{1/2}$ ^{238}U is 4.47 x 10^{+9} years; ionic radius of U^{++++} 0.0918 nm [MER89] [KIR78] [KIR83] [CRC93]
Density, g/cm³: α: 19.07; β: 18.11; γ: 18.06 [KIR83]
Melting Point, °C: 1132 [KIR91]
Boiling Point, °C: 3818 [ALD94]
Reactions: transition α to β at 667.7°C, β to γ at 774.8°C [MER89]
Thermal Conductivity, W/(m·K): 25.1 (36°C), 26.3 (100°C), 29.7 (200°C), 31.4 (300°C), 32.6 (400°C) [KIR83]

Thermal Expansion Coefficient: 13.9 x 10^{-6}/K [CRC93]

3079
Compound: Uranium diboride
Formula: UB_2
Molecular Formula: B_2U
Molecular Weight: 259.651
CAS RN: 12007-36-2
Properties: hex; -8 mesh with 99.5% purity; refractory material [KIR78] [CER91] [CRC94]
Density, g/cm³: 12.7 [CRC94]
Melting Point, °C: 2385 [KIR78]

3080
Compound: Uranium dicarbide
Formula: UC_2
Molecular Formula: C_2U
Molecular Weight: 262.051
CAS RN: 12071-33-9
Properties: gray tetr cryst, a = 0.35241 nm, c = 0.59962 nm; used in the form of pellets or microspheres to fuel nuclear reactors [HAW93] [CIC73]
Solubility: decomposes in H_2O; sl s alcohol [HAW93]
Density, g/cm³: 11.28 [HAW93]
Melting Point, °C: 2350 [HAW93]
Boiling Point, °C: 4370 [HAW93]
Reactions: transition tetr to cub at 1765°C [CIC73]

3081
Compound: Uranium dioxide
Synonyms: uraninite
Formula: UO_2
Molecular Formula: O_2U
Molecular Weight: 270.028
CAS RN: 1344-57-6
Properties: -100 mesh; brown to black powd; cub cryst; widely used to manufacture fuel pellets for power reactors [KIR83] [MER89] [STR94]
Solubility: i H_2O, dil acids; s conc acids [MER89]
Density, g/cm³: 10.97 [MER89]
Melting Point, °C: 2865 [MER89]
Thermal Conductivity, W/(m·K): 5.1 (500°C), 3.4 (1000°C) [KIR80]
Thermal Expansion Coefficient: (volume) 100°C (0.199), 200°C (0.468), 400°C (1.045), 800°C (2.638), 1000°C (3.115) [CLA66]

3082
Compound: Uranium hexafluoride
Formula: UF_6
Molecular Formula: F_6U
Molecular Weight: 352.019
CAS RN: 7783-81-5
Properties: white; volatile; monocl solid; reacts vigorously with H_2O, forming mainly UO_2F_2 and HF; enthalpy of vaporization at 64.01°C 28.899 kJ/mol; enthalpy of fusion 19.19 kJ/mol; enthalpy of sublimation 48.095 kJ/mol; triple point 64.052°C at 151 kPa; can be prepared by direct reaction of uranium metal and fluorine [KIR83] [MER89] [CRC93]
Solubility: s liq Cl_2, Br_2; gives dark red fuming solution with nitrobenzene; s CCl_4, CH_3Cl [MER89]
Density, g/cm³: solid: 5.09; liq: 3.595 [MER89]
Melting Point, °C: 64 [KIR91]

3083
Compound: Uranium monocarbide
Formula: UC
Molecular Formula: CU
Molecular Weight: 250.040
CAS RN: 12070-09-6
Properties: gray with metallic appearance; fcc, a = 0.49605 nm; can be prepared by arc melting stoichiometric amounts of the elements in an inert atm; reacts with oxygen [CIC73]
Density, g/cm³: 13.63 [KIR78]
Melting Point, °C: 2790 [CIC73]
Thermal Conductivity, W/(m·K): 25 [KIR78]
Thermal Expansion Coefficient: 9.1×10^{-6}/K [KIR78]

3084
Compound: Uranium mononitride
Formula: UN
Molecular Formula: NU
Molecular Weight: 252.036
CAS RN: 25658-43-9
Properties: dark gray; fcc, a = 0.4890 nm; electrical resistivity 176 μohm·cm; Knoop hardness 580; only stable uranium nitride above 1300°C; formed by reacting uranium and nitrogen; there are two other nitrides, $UN_{1.5}$, 12033-85-1, and $UN_{1.75}$, 12266-20-5 [KIR83] [KIR81]
Density, g/cm³: 14.4 [KIR81]
Melting Point, °C: 2800 [CIC73]
Thermal Conductivity, W/(m·K): 15.5 [KIR81]
Thermal Expansion Coefficient: 8.0×10^{-6} [KIR81]

3085
Compound: Uranium pentabromide
Formula: UBr_5
Molecular Formula: Br_5U
Molecular Weight: 637.549
CAS RN: 13775-16-1
Properties: dark brown hygr unstable solid; formed when UBr_4 is extracted with liq Br_2 at 55°C, followed by recrystallization from liq Br_2 [KIR83]

3086
Compound: Uranium pentachloride
Formula: UCl_5
Molecular Formula: Cl_5U
Molecular Weight: 415.293
CAS RN: 13470-21-8
Properties: reddish brown cryst, with metallic luster; can be formed by reaction of UO_3 with CCl_4 or Cl_2 [KIR83]
Solubility: s liq Cl_2 [KIR83]
Density, g/cm³: 3.81 [CRC94]
Melting Point, °C: decomposes 300 [CRC94]

3087
Compound: Uranium pentafluoride
Formula: UF_5
Molecular Formula: F_5U
Molecular Weight: 333.021
CAS RN: 13775-07-0
Properties: two forms: α-UF_5 is grayish white and can be obtained by reduction of UF_6 with HBr, β-UF_5 is yellowish white and is obtained by reacting UF_6 with UF_4 at 150-200°C; both forms are hygr; have blue solutions in anhydrous HF [KIR83]
Density, g/cm³: 5.81 [LID94]
Melting Point, °C: 348 [LID94]

3088
Compound: Uranium tetraboride
Formula: UB_4
Molecular Formula: B_4U
Molecular Weight: 281.273
CAS RN: 12007-84-0
Properties: brown; -8 mesh with 99.% purity and -60 mesh with 99.7% purity; refractory material [KIR78] [CRC94] [CER91]
Density, g/cm³: 5.35 [CRC94]
Melting Point, °C: 2495 [KIR78]

3089
Compound: Uranium tetrabromide
Formula: UBr_4
Molecular Formula: Br_4U
Molecular Weight: 557.645
CAS RN: 13470-20-7
Properties: dark brown, very hygr cryst; prepared by heating uranium turnings in a nitrogen gas stream saturated with bromine vapor; can be purified by vacuum distillation in a similar nitrogen stream [KIR83]
Density, g/cm³: 5.35 [CRC94]
Melting Point, °C: 519 [KIR91]
Boiling Point, °C: 765 [KIR83]

3090
Compound: Uranium tetrachloride
Formula: UCl_4
Molecular Formula: Cl_4U
Molecular Weight: 379.840
CAS RN: 10026-10-5
Properties: dark green octahedral cryst; oxidizes in air; decomposes in water; enthalpy of fusion 45.00 kJ/mol; prepared by the reaction $UO_3 + 3$ $CCl_3CCl=CCl_2 \rightarrow UCl_4 + Cl_2 + 3\ CCl_2=CClOCl$ [KIR83] [CRC93] [MER89]
Solubility: v s H_2O, with decomposition [MER89]
Density, g/cm³: 4.725 [MER89]
Melting Point, °C: 590 [KIR91]
Boiling Point, °C: 791 [MER89]

3091
Compound: Uranium tetrafluoride
Formula: UF_4
Molecular Formula: F_4U
Molecular Weight: 314.023
CAS RN: 10049-14-6
Properties: monocl green cryst; reacts when heated with atm O_2, forming U_3O_8; prepared by reacting UO_2 with excess gaseous HF at ~550°C; used to produce both uranium metal and UF_6 [MER89] [KIR83]
Solubility: i H_2O; s conc acids and alkalies, but decomposes [MER89]
Density, g/cm³: 6.70 [HAW93]
Melting Point, °C: 960 [KIR91]

3092
Compound: Uranium tetraiodide
Formula: UI_4
Molecular Formula: I_4U
Molecular Weight: 745.647

CAS RN: 13470-22-9
Properties: black lustrous cryst; can be made by reaction of iodine and uranium metal; similar properties for UI_3 and for preparation of UI_3, 13775-18-8 [KIR83]
Density, g/cm³: 5.6 [CRC94]
Melting Point, °C: 506 [CRC94]
Boiling Point, °C: 729 [CRC94]

3093
Compound: Uranium tribromide
Formula: UBr_3
Molecular Formula: Br_3U
Molecular Weight: 477.741
CAS RN: 13470-19-4
Properties: reddish brown cryst; prepared by reacting UH_3 and HBr, or directly from the two elements; black cryst obtained when purified by gas phase transport [KIR83]
Density, g/cm³: 6.53 [CRC94]
Melting Point, °C: 730 [KIR91]

3094
Compound: Uranium tricarbide
Formula: U_2C_3
Molecular Formula: C_3U_2
Molecular Weight: 512.091
CAS RN: 12612-73-6
Properties: -60 mesh; gray with metallic appearance; bcc, a = 0.80889 nm; can be prepared by arc welding stoichiometric amounts of the two elements [CIC73] [CER91]
Density, g/cm³: 12.7 [LID94]
Melting Point, °C: decomposes 1727 [CIC73]

3095
Compound: Uranium trichloride
Formula: UCl_3
Molecular Formula: Cl_3U
Molecular Weight: 344.387
CAS RN: 10025-93-1
Properties: dark purple cryst; somewhat hygr; obtained by reacting UH_3 with HCl at 250-300°C; used in molten salt electrolytes to refine uranium metal [KIR83] [MER89]
Solubility: v s H_2O, evolves H_2, solution changes color from purple to green due to oxidation of U^{+++} [MER89]
Density, g/cm³: 5.51 [MER89]
Melting Point, °C: 835 [KIR91]

3096
Compound: Uranium trifluoride
Formula: UF_3
Molecular Formula: F_3U
Molecular Weight: 295.024
CAS RN: 13775-06-9
Properties: black mass containing small deep purple cryst; has been used as a component in molten salt systems [KIR83]
Solubility: i H_2O; s HNO_3, hot H_2SO_4, hot $HClO_4$ [KIR83]
Density, g/cm³: 8.9 [LID94]
Melting Point, °C: >1140, decomposes [KIR91]

3097
Compound: Uranium trihydride
Formula: UH_3
Molecular Formula: H_3U
Molecular Weight: 241.053
CAS RN: 13598-56-6
Properties: -100 mesh; brownish gray to black powd; conducts electricity; prepared by heating uranium metal in a hydrogen atm at 150-200°C; used to prepare finely divided uranium by decomposition reaction [HAW93] [CER91]
Density, g/cm³: 10.92 [HAW93]

3098
Compound: Uranium trinitride
Formula: U_2N_3
Molecular Formula: N_3U_2
Molecular Weight: 518.078
CAS RN: 12033-83-9
Properties: bcc, a = 1.0678 nm [CIC73]
Density, g/cm³: 11.24 [CIC73]
Melting Point, °C: decomposes [CIC73]

3099
Compound: Uranium trioxide
Synonyms: uranium(VI) oxide
Formula: UO_3
Molecular Formula: O_3U
Molecular Weight: 286.027
CAS RN: 1344-58-7
Properties: -100 mesh; has six forms: α is hex brown, β is orange monocl, γ is bright yellow rhomb, δ is red cub, ϵ is brick red tricl, η is rhomb; UO_3 can be obtained by thermal decomposition of uranyl compounds, e.g carbonates, oxalates nitrates [KIR83] [CER91]
Solubility: i H_2O; s acids [MER89]
Density, g/cm³: 7.29 [MER89]

Melting Point, °C: 650, decomposes [KIR91]

3100
Compound: Uranium(IV) sulfate octahydrate
Formula: $U(SO_4)_2 \cdot 8H_2O$
Molecular Formula: $H_{16}O_{16}S_2U$
Molecular Weight: 574.278
CAS RN: 19086-22-7
Solubility: g/100g H_2O: 11.9 (20°C), 17.9 (30°C), 29.2 (40°C), 55.8 (60°C) [LAN85]
Melting Point, °C: decomposes 90 [CRC94]

3101
Compound: Uranium(IV) sulfate tetrahydrate
Formula: $U(SO_4)_2 \cdot 4H_2O$
Molecular Formula: $H_8O_{12}S_2U$
Molecular Weight: 502.218
CAS RN: 13470-23-0
Properties: green, rhomb [LAN52]
Solubility: g/100g H_2O: 10.1 (30°C), 9.0 (40°C), 7.7 (60°C) [LAN85]
Melting Point, °C: decomposes, 90 [CRC77]
Reactions: minus $4H_2O$ at 300 °C [LAN52]

3102
Compound: Uranium(V,VI) oxide
Formula: U_3O_8
Molecular Formula: O_8U_3
Molecular Weight: 842.082
CAS RN: 1344-59-8
Properties: greenish black powd [STR93]
Density, g/cm³: 8.30 [STR93]
Melting Point, °C: 1150, decomposes to UO_2 [CRC94] [KIR91]

3103
Compound: Uranyl acetate dihydrate
Formula: $UO_2(CH_3COO)_2 \cdot 2H_2O$
Molecular Formula: $C_4H_{10}O_8U$
Molecular Weight: 424.147
CAS RN: 6159-44-0
Properties: yellow; cryst powd; slight odor of acetic acid; used as bacterial oxidation activator, and in copying inks [MER89] [HAW93]
Solubility: s in 10 parts H_2O; sl s alcohol [MER89]
Density, g/cm³: 2.89 [MER89]
Reactions: minus $2H_2O$ at 110°C [CRC94]

3104
Compound: Uranyl acetylacetonate

Synonyms: 2,4-pentanedione, uranyl derivative
Formula: $UO_2(CH_3COCH=C(O)CH_3)_2$
Molecular Formula: $C_{10}H_{14}O_6U$
Molecular Weight: 468.248
CAS RN: 18039-69-5
Properties: cryst [AES93]

$$UO_2[CH_3-\overset{\underset{\textstyle |}{\textstyle O-}}{C}=CH-\overset{\underset{\textstyle \|}{\textstyle O}}{C}-CH_3]_2$$

3105
Compound: Uranyl carbonate
Synonyms: rutherfordine
Formula: UO_2CO_3
Molecular Formula: CO_5U
Molecular Weight: 330.037
CAS RN: 12202-79-8
Properties: naturally occurring mineral [KIR83]

3106
Compound: Uranyl chloride
Formula: UO_2Cl_2
Molecular Formula: Cl_2O_2U
Molecular Weight: 340.933
CAS RN: 7791-26-6
Properties: bright yellow cryst; ortho-rhomb; very hygr; very volatile >775°C [MER89]
Solubility: v s H_2O; s acetone, alcohol; i benzene [MER89]
Melting Point, °C: 578 [CRC94]
Boiling Point, °C: decomposes [CRC94]

3107
Compound: Uranyl chloride trihydrate
Formula: $UO_2Cl_2 \cdot 3H_2O$
Molecular Formula: $Cl_2H_6O_5U$
Molecular Weight: 394.979
CAS RN: 13867-67-9
Properties: yellow powd; hygr [STR93]

3108
Compound: Uranyl hydrogen phosphate tetrahydrate
Formula: $UO_2HPO_4 \cdot 4H_2O$
Molecular Formula: $H_9O_{10}PU$
Molecular Weight: 438.068
CAS RN: 18433-48-2
Properties: yellow; microcryst powd [MER89]
Solubility: i H_2O; s acids [MER89]

3109
Compound: Uranyl nitrate hexahydrate
Formula: $UO_2(NO_3)_2 \cdot 6H_2O$
Molecular Formula: $H_{12}N_2O_{14}U$
Molecular Weight: 502.129
CAS RN: 13520-83-7
Properties: yellow ortho-rhomb cryst; hygr; greenish luster by reflected light; can be prepared by heating a solution of uranyl nitrate to 188°C, then cooling to room temp [MER89] [STR93] [KIR83]
Solubility: g/100g H_2O: 98 (0°C), 122 (20°C), 474 (100°C) [LAN85]; v s alcohol, ether [MER89]; data are in [SIE94]
Density, g/cm³: 2.807 [MER89]
Melting Point, °C: 60.2 [STR93]
Boiling Point, °C: 118 [HAW93]

3110
Compound: Uranyl oxalate trihydrate
Formula: $UO_2C_2O_4 \cdot 3H_2O$
Molecular Formula: $C_2H_6O_9U$
Molecular Weight: 412.094
CAS RN: 22429-50-1
Properties: yellow cryst [CRC77]
Solubility: g anhydrous/100g H_2O: 0.45 (10°C), 0.50 (20°C), 3.16 (100°C) [LAN85]
Reactions: minus H_2O, 110°C [CRC94]

3111
Compound: Uranyl sulfate monohydrate
Formula: $UO_2SO_4 \cdot H_2O$
Molecular Formula: H_2O_7SU
Molecular Weight: 384.107
CAS RN: 19415-82-8
Properties: stable to 600°C; used to leach uranyl sulfate at this temp from sulfates of iron and aluminum [KIR83]

3112
Compound: Uranyl sulfate trihydrate
Formula: $UO_2SO_4 \cdot 3H_2O$
Molecular Formula: H_6O_9SU
Molecular Weight: 420.138
CAS RN: 12384-63-3
Properties: lemon yellow; cryst mass; property of stability at 600°C can be used in order to leach uranium from iron and aluminum at 600°C [KIR83] [MER89]
Solubility: s in ~5 parts H_2O, 25 parts alcohol [MER89]
Density, g/cm³: 3.28 [MER89]

Melting Point, °C: decomposes 100 [CRC94]

3113
Compound: Vanadium
Formula: V
Molecular Formula: V
Molecular Weight: 50.9415
CAS RN: 7440-62-2
Properties: light gray or white, lustrous powd, or fused hard lumps; bcc, a = 0.3026 nm; enthalpy of fusion 21.50 kJ/mol; enthalpy of evaporation 458.6 kJ/mol; stable to moist air under typical conditions; electrical resistivity 24.2 μohm·cm at 20°C; Poisson's ratio 0.36; superconductivity transition 5.13 K; inert towards hot or cold HCl, cold H_2SO_4; uses include film resistors [MER89] [KIR83] [CER91] [CRC93]
Solubility: i H_2O; reacts with hot H_2SO_4, HF, HNO_3, aqua regia [MER89]
Density, g/cm^3: 6.11 [MER89]
Melting Point, °C: 1890 [ALD94]
Boiling Point, °C: 3380 [ALD94]
Thermal Conductivity, W/(m·K): 30.7 at 25°C [ALD94]; 31 at 100°C [KIR83]
Thermal Expansion Coefficient: 8.3 x 10^{-6}/°C (23-100°C) [KIR83]

3114
Compound: Vanadium bis(cyclopentadienyl) dichloride
Synonyms: bis(cyclopentadienyl)vanadium dichloride
Formula: V(C$_5$H$_5$)$_2$Cl$_2$
Molecular Formula: C$_{10}$H$_{10}$Cl$_2$V
Molecular Weight: 252.036
CAS RN: 12086-48-6
Properties: cryst [ALF95]
Density, g/cm^3: 1.6 [ALF95]
Melting Point, °C: 250 decomposes [ALF95]

3115
Compound: Vanadium carbide
Formula: VC
Molecular Formula: CV
Molecular Weight: 62.953
CAS RN: 12070-10-9

Properties: black cub cryst, a = 0.41355 nm; hardness 2800 kg/mm^2; resistivity at room temp is 150 μohm·cm; used in alloys for cutting tools, and in 99.5% pure form as a sputtering target for producing wear-resistant films and semiconductor films [HAW93] [KIR83] [CIC73] [CER91]
Solubility: i H_2O; s HNO_3 with decomposition [KIR83]
Density, g/cm^3: 5.77 [KIR83]
Melting Point, °C: 2810 [KIR83]
Boiling Point, °C: 3900 [KIR83]
Thermal Expansion Coefficient: 7.2 x 10^{-6}/K [KIR78]

3116
Compound: Vanadium carbonyl
Synonyms: vanadium hexacarbonyl
Formula: V(CO)$_6$
Molecular Formula: C$_6$O$_6$V
Molecular Weight: 219.004
CAS RN: 20644-87-5
Properties: bluish green cryst; sensitive to atm O$_2$, pyromorphic; paramagnetic [MER89] [HAW93]
Melting Point, °C: 60-70, decomposes without melting [HAW93]
Boiling Point, °C: sublimes, 50 (15 mm Hg) [HAW93]

3117
Compound: Vanadium diboride
Formula: VB$_2$
Molecular Formula: B$_2$V
Molecular Weight: 72.564
CAS RN: 12007-37-3
Properties: refractory material; used as a 99.5% pure sputtering target to produce wear-resistant and semiconductive films [KIR78] [CER91]
Density, g/cm^3: 5.100 [ALD94]
Melting Point, °C: 2450 [KIR78]

3118
Compound: Vanadium dibromide
Formula: VBr$_2$
Molecular Formula: Br$_2$V
Molecular Weight: 210.750
CAS RN: 14890-41-6
Properties: hex brownish orange cryst [LID94] [KIR83]
Density, g/cm^3: 4.58 [LID94]

3119
Compound: Vanadium dichloride
Formula: VCl_2
Molecular Formula: Cl_2V
Molecular Weight: 121.847
CAS RN: 10580-52-6
Properties: apple green; hex plates; strong reducing agent; preparation: heating VCl_3 in N_2 atm, followed by sublimation in N_2 atm; used to purify HCl by removing arsenic [HAW93]
Solubility: decomposed in hot H_2O; s alcohol, ether [HAW93]
Density, g/cm³: 3.23 (18°C) [HAW93]

3120
Compound: Vanadium diiodide
Synonyms: vanadium(II) iodide
Formula: VI_2
Molecular Formula: I_2V
Molecular Weight: 304.751
CAS RN: 15513-84-5
Properties: red hex [CRC94] [KIR83]
Density, g/cm³: 5.44 [CRC94]
Melting Point, °C: sublimes 750-800 [CRC94]

3121
Compound: Vanadium dioxide
Synonyms: vanadium(IV) oxide
Formula: V_2O_4
Molecular Formula: O_4V_2
Molecular Weight: 165.881
CAS RN: 12036-21-4
Properties: formula also VO_2; bluish black powd; slowly oxidizes in air; can be prepared by reacting V_2O_5 at its melting point with reductants such as sugar or oxalic acid; used as a high temp catalyst [HAW93] [KIR83]
Solubility: i H_2O; s acids and alkalies [HAW93]
Density, g/cm³: 4.339 [STR93]
Melting Point, °C: 1967 [STR93]

3122
Compound: Vanadium disilicide
Formula: VSi_2
Molecular Formula: Si_2V
Molecular Weight: 336.817
CAS RN: 12039-87-1
Properties: metallic prisms; -325 mesh; as a 99.5% pure material, used as a sputtering target in the fabrication of integrated circuits, and as an electrochemical cathode [ALD94] [KIR83] [ALF93] [CER91]

Solubility: s HF [KIR83]
Density, g/cm³: 4.42 [KIR83]

3123
Compound: Vanadium gallide
Formula: V_3Ga
Molecular Formula: GaV_3
Molecular Weight: 222.548
Properties: -100 mesh of 99.5% purity; superconducting material [KIR83] [CER91]

3124
Compound: Vanadium monoboride
Formula: VB
Molecular Formula: BV
Molecular Weight: 61.753
CAS RN: 12045-27-1
Properties: refractory material; in the form of a 99.5% pure material used as a sputtering target to produce semiconductor and wear-resistant films [KIR78] [CER91]
Melting Point, °C: 2250 [KIR78]

3125
Compound: Vanadium monocarbide
Synonyms: divanadium carbide
Formula: V_2C
Molecular Formula: CV_2
Molecular Weight: 113.894
CAS RN: 12012-17-8
Properties: hex, a = 0.41655 nm, b = 0.29020 nm, c = 0.4577 nm [CIC73]
Melting Point, °C: 2167 [CIC73]

3126
Compound: Vanadium monosilicide
Formula: V_3Si
Molecular Formula: SiV_3
Molecular Weight: 180.911
CAS RN: 12039-76-8
Properties: cub cryst; superconducting; -100 mesh; as a 99.5% pure sputtering target, used to produce resistant and semiconducting films in the fabrication of integrated circuits [LID94] [ALF93] [CER91]
Density, g/cm³: 5.70 [LID94]
Melting Point, °C: 1935 [ALF93]

3127
Compound: Vanadium monoxide

Synonyms: vanadium(II) oxide
Formula: VO
Molecular Formula: OV
Molecular Weight: 66.941
CAS RN: 12035-98-2
Properties: -80 mesh powd; light green cryst; enthalpy of fusion 63.00 kJ/mol [CRC93] [KIR83] [STR93]
Solubility: s acids [KIR83]
Density, g/cm³: 5.758 [KIR83]
Melting Point, °C: 1790 [CRC93]

3128
Compound: Vanadium nitride
Formula: VN
Molecular Formula: NV
Molecular Weight: 64.949
CAS RN: 24646-85-3
Properties: black powd; fcc, a = 0.4140 nm; hardness 9-10 Mohs; electrical resistivity 85 μohm·cm; transition temp 7.5 K; used as a 99.5% pure sputtering target to produce films [KIR81] [CIC73] [STR93] [CER91]
Solubility: i H₂O; s aqua regia [HAW93]
Density, g/cm³: 6.13 [STR93]
Melting Point, °C: 2320 [STR93]
Thermal Conductivity, W/(m·K): 11.3 [KIR81]
Thermal Expansion Coefficient: 8.1 x 10⁻⁶ [KIR81]

3129
Compound: Vanadium oxytrichloride
Synonyms: vanadium(V) oxytrichloride
Formula: VOCl₃
Molecular Formula: Cl₃OV
Molecular Weight: 173.299
CAS RN: 7727-18-6
Properties: yellow liq; evolves red fumes in moist atm; can be used as a nonionizing solvent dissolving most nonmetals; hydrolyzes in moisture; enthalpy of vaporization 36.78 kJ/mol [CRC93] [HAW93] [MER89]
Solubility: decomposes in H₂O to vanadic acid and HCl; s methanol, ether, acetone, acids [KIR83] [MER89]
Density, g/cm³: 1.829 [STR93]
Melting Point, °C: -77 [MER89]
Boiling Point, °C: 126-127 [ALD94]

3130
Compound: Vanadium oxytrifluoride
Synonyms: vanadium(V) oxytrifluoride
Formula: VOF₃

Molecular Formula: F₃OV
Molecular Weight: 123.936
CAS RN: 13709-31-4
Properties: yellowish orange powd; sensitive to moisture [STR93]
Density, g/cm³: 2.459 [STR93]
Melting Point, °C: 300 [STR93]
Boiling Point, °C: 480 [STR93]

3131
Compound: Vanadium pentafluoride
Synonyms: vanadium(V) fluoride
Formula: VF₅
Molecular Formula: F₅V
Molecular Weight: 145.934
CAS RN: 7783-72-4
Properties: liq; etches glass slowly at room temp; appreciable vapor pressure at room temp; enthalpy of vaporization 44.52 kJ/mol; enthalpy of fusion 49.96 kJ/mol [CRC93] [MER89]
Solubility: hydrolyzed in H₂O, dil alkali; s anhydrous HF [MER89]
Density, g/cm³: 2.502 [MER89]
Melting Point, °C: 19.5 [MER89]

3132
Compound: Vanadium pentasulfide
Synonyms: vanadium(V) sulfide
Formula: V₂S₅
Molecular Formula: S₅V₂
Molecular Weight: 262.213
CAS RN: 12138-17-9
Properties: greenish black powd; decomposes when heated [HAW93]
Density, g/cm³: 3.0 [CRC94]
Melting Point, °C: decomposes [CRC94]

3133
Compound: Vanadium pentoxide
Synonyms: vanadium(V) oxide
Formula: V₂O₅
Molecular Formula: O₅V₂
Molecular Weight: 181.880
CAS RN: 1314-62-1
Properties: yellow to rust brown ortho-rhomb cryst; reversibly evolves O₂, 700-1125°C; enthalpy of fusion 64.50 kJ/mol [CRC93] [MER89]
Solubility: ~1 g/125mL H₂O; s conc acids, forming red to yellow solutions [MER89]
Density, g/cm³: 3.357 [STR93]
Melting Point, °C: 690 [ALD94]
Boiling Point, °C: 1750, decomposes [HAW93]

3134
Compound: Vanadium sulfide
Formula: V_2S_2
Molecular Formula: S_2V_2
Molecular Weight: 166.015
CAS RN: 12138-08-8
Properties: black; formula also VS; used as a solid lubricant, and as an electrode in lithium based batteries [HAW93] [CRC94]
Density, g/cm³: 4.2 [CRC94]
Melting Point, °C: decomposes [CRC94]

3135
Compound: Vanadium tetrachloride
Synonyms: vanadium(IV) chloride
Formula: VCl_4
Molecular Formula: Cl_4V
Molecular Weight: 192.753
CAS RN: 7632-51-1
Properties: red liq; decomposes slowly to VCl_3 and Cl_2 below 63°C; enthalpy of vaporization 41.4 kJ/mol (bp), 42.5 kJ/mol (25°C); enthalpy of fusion 2.30 kJ/mol [CRC93] [HAW93]
Solubility: s alcohol and ether [HAW93]
Density, g/cm³: 1.816 [HAW93]
Melting Point, °C: -28 [ALD94]
Boiling Point, °C: 154 [ALD94]

3136
Compound: Vanadium tetrafluoride
Synonyms: vanadium(IV) fluoride
Formula: VF_4
Molecular Formula: F_4V
Molecular Weight: 126.936
CAS RN: 10049-16-8
Properties: bright lime green powd; very hygr; disproportionates in vacuum to VF_3 and VF_5 at 100-120°C [MER89]
Solubility: v s H_2O imparting blue color [MER89]
Density, g/cm³: 3.15 [MER89]
Melting Point, °C: 325, decomposes [STR93]

3137
Compound: Vanadium tribromide
Synonyms: vanadium(III) bromide
Formula: VBr_3
Molecular Formula: Br_3V
Molecular Weight: 290.654
CAS RN: 13470-26-3
Properties: -20 mesh; black powd; sensitive to moisture [STR93]
Density, g/cm³: 4.0 [STR93]

3138
Compound: Vanadium trichloride
Synonyms: vanadium(III) chloride
Formula: VCl_3
Molecular Formula: Cl_3V
Molecular Weight: 157.300
CAS RN: 7718-98-1
Properties: purple powd; sensitive to moisture; decomposes if heated; used to prepare organovanadium compounds [HAW93] [STR93]
Solubility: decomposes in H_2O; s alcohol and ether [HAW93]
Density, g/cm³: 3.00 [STR93]
Melting Point, °C: decomposes [KIR83]

3139
Compound: Vanadium trifluoride
Synonyms: vanadium(III) fluoride
Formula: VF_3
Molecular Formula: F_3V
Molecular Weight: 107.937
CAS RN: 10049-12-4
Properties: greenish yellow powd; sublimes at bright red heat [MER89]
Solubility: i H_2O, alcohol [MER89]
Density, g/cm³: 3.363 [MER89]
Melting Point, °C: 1406, decomposes [STR93]
Boiling Point, °C: sublimes, 800 [STR93]

3140
Compound: Vanadium trifluoride trihydrate
Formula: $VF_3 \cdot 3H_2O$
Molecular Formula: $F_3H_6O_3V$
Molecular Weight: 161.983
CAS RN: 10049-12-4
Properties: dark green; rhomb cryst [MER89]
Solubility: sl s H_2O [MER89]
Reactions: minus H_2O at 100°C [MER89]

3141
Compound: Vanadium trioxide
Synonyms: vanadium(III) oxide
Formula: V_2O_3
Molecular Formula: O_3V_2
Molecular Weight: 149.881
CAS RN: 1314-34-7
Properties: black powd; gradually forms indigo blue cryst, V_2O_4, in air; used as a catalyst to convert ethylene to ethanol [HAW93] [MER89]
Solubility: i H_2O; s with difficulty in acids [MER89]
Density, g/cm³: 4.87 [MER89]
Melting Point, °C: 1940 [MER89]

3142
Compound: Vanadium trisulfate
Synonyms: vanadium(III) sulfate
Formula: $V_2(SO_4)_3$
Molecular Formula: $O_{12}S_3V_2$
Molecular Weight: 390.074
CAS RN: 13701-70-7
Properties: lemon yellow powd; decomposes to $VOSO_4$ and SO_2, when heated ~410°C in vacuum; stable in dry air; strong reducing agent [MER89]
Solubility: very slowly dissolves in H_2O at room temp; s HNO_3 [MER89]
Melting Point, °C: decomposes ~400 [MER89]

3143
Compound: Vanadium trisulfide
Synonyms: vanadium(III) sulfide
Formula: V_2S_3
Molecular Formula: S_3V_2
Molecular Weight: 198.081
CAS RN: 1315-03-3
Properties: -325 mesh; greenish black powd; heating causes decomposition [MER89] [STR93]
Solubility: i H_2O, cold HCl, dil H_2SO_4; s hot HCl, hot dil H_2SO_4, HNO_3 [MER89]
Density, g/cm³: 4.7 [MER89]
Melting Point, °C: decomposes >600 [CRC94]

3144
Compound: Vanadium(II) sulfate heptahydrate
Formula: $VSO_4 \cdot 7H_2O$
Molecular Formula: $H_{14}O_{11}SV$
Molecular Weight: 273.112
CAS RN: 36907-42-3
Properties: violet monocl [CRC94] [KIR83]
Reactions: decomposes on heating in air [CRC94]

3145
Compound: Vanadium(III) acetylacetonate
Synonyms: 2,4-pentanedione, vanadium(III) derivative
Formula: $V(CH_3COCH=C(O)CH_3)_3$
Molecular Formula: $C_{15}H_{21}O_6V$
Molecular Weight: 348.270
CAS RN: 13476-99-8
Properties: brown cryst; sensitive to air [KIR83] [STR93]

$$[CH_3\text{–}C\text{=}CH\text{–}C\text{–}CH_3]_3V$$
with O– and O (the O– attached by single bond, O attached by double bond)

Solubility: s methanol, acetone, benzene, chloroform [KIR83]
Density, g/cm³: 0.9-1.2 [KIR83]
Melting Point, °C: 178-190 [KIR83]
Boiling Point, °C: sublimes 170 (0.05 mm Hg) [STR93]

3146
Compound: Vanadocene
Synonyms: bis(cyclopentadienyl)vanadium
Formula: $V(C_5H_5)_2$
Molecular Formula: $C_{10}H_{10}V$
Molecular Weight: 181.131
CAS RN: 1277-47-0
Properties: purple cryst; air and moisture sensitive [STR93]

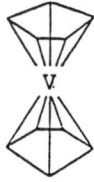

Reactions: sublimes at 200°C (0.1 mm Hg) [STR93]

3147
Compound: Vanadyl bromide
Formula: VOBr
Molecular Formula: BrOV
Molecular Weight: 146.845
CAS RN: 13520-88-2
Properties: violet [KIR83]
Density, g/cm³: 4.0 [CRC94]
Melting Point, °C: decomposes 482 [CRC94]

3148
Compound: Vanadyl chloride
Formula: VOCl
Molecular Formula: ClOV
Molecular Weight: 102.394
CAS RN: 13520-87-1
Properties: yellow-brown powd [CRC94] [KIR83]
Density, g/cm³: 2.824 [CRC94]

3149
Compound: Vanadyl dibromide

Formula: VOBr₂
Formula: VOBr$_2$
Molecular Formula: Br$_2$OV
Molecular Weight: 226.749
CAS RN: 13520-89-3
Properties: yellowish brown powd; deliq [CRC94] [KIR83]
Melting Point, °C: decomposes 180 [CRC94]

3150
Compound: Vanadyl dichloride
Formula: VOCl$_2$
Molecular Formula: Cl$_2$OV
Molecular Weight: 137.846
CAS RN: 10213-09-9
Properties: green cryst; very deliq; disproportionates at 384°C to VOCl and VOCl$_3$ [MER89]
Solubility: slowly decomposed in water; s absolute alcohol, glacial acetic acid [MER89]
Density, g/cm³: 2.88 [MER89]

3151
Compound: Vanadyl difluoride
Formula: VOF$_2$
Molecular Formula: F$_2$OV
Molecular Weight: 104.938
CAS RN: 13814-83-0
Properties: yellow [KIR83]
Density, g/cm³: 3.396 [CRC94]
Melting Point, °C: decomposes [CRC94]

3152
Compound: Vanadyl selenite monohydrate
Formula: VOSeO$_3$·H$_2$O
Molecular Formula: H$_2$O$_5$SeV
Molecular Weight: 211.915
CAS RN: 133578-89-9
Properties: green plates; tricl, a = 0.5969 nm, b = 0.6155 nm, c = 0.6349 nm; has magnetic properties; can be prepared by heating H$_2$SeO$_3$ and V$_2$O$_5$ in an autoclave at 200°C for ~48 h; material selectively intercalates alcohols [HUA91]
Density, g/cm³: 3.506 [HUA91]
Reactions: minus H$_2$O at 240-280 °C [HUA91]

3153
Compound: Vanadyl sulfate dihydrate
Formula: VOSO$_4$·2H$_2$O
Molecular Formula: H$_4$O$_7$SV
Molecular Weight: 199.036

CAS RN: 27774-13-6
Properties: blue cryst powd; used as a mordant, catalyst reducing agent, colorant in glasses and ceramics; there is a trihydrate, CAS RN 12210-47-8 [HAW93] [MER89] [ALD94]
Solubility: s H$_2$O [MER89]

3154
Compound: Vanadyl tribromide
Formula: VOBr$_3$
Molecular Formula: Br$_3$OV
Molecular Weight: 306.653
CAS RN: 13520-90-6
Properties: deep red liq [KIR83]
Density, g/cm³: 2.933 [CRC94]
Boiling Point, °C: 130 [CRC94]
Reactions: decomposes 180°C [CRC94]

3155
Compound: Vitreous silica
Formula: SiO$_2$
Molecular Formula: O$_2$Si
Molecular Weight: 60.085
CAS RN: 60676-86-0
Properties: hardness 5.5 Mohs; velocity of sound 5730 m/sec [CIC73]
Density, g/cm³: 2.1957 at 0°C [CIC73]
Melting Point, °C: ~1500 [CIC73]
Boiling Point, °C: 2950 [CIC73]
Thermal Conductivity, W/(m·K): 1.423 100°C [CIC73]

3156
Compound: Water
Synonyms: hydrogen oxide
Formula: H$_2$O
Molecular Formula: H$_2$O
Molecular Weight: 18.015
CAS RN: 7732-18-5
Properties: colorless, odorless, tasteless liq; dielectric constant 78.54; viscosity 1.005 cp (20°C); vapor pressure 760 mm Hg (100°C); triple point 273.16 K at 4.6 mm Hg; surface tension 73 dyne/cm (20°C); enthalpy of fusion (ice) 6.008 kJ/mol; enthalpy of vaporization 40.65 kJ/mol; critical temp 374.2°C [DOU83] [HAW93] [MER89]
Density, g/cm³: 0.9970 [LID94]
Melting Point, °C: 0 [MER89]
Boiling Point, °C: 100 [MER89]
Thermal Conductivity, W/(m·K): values from 20°C to 330°C are found in [OZB80]

3157
Compound: Xenon
Formula: Xe
Molecular Formula: Xe
Molecular Weight: 131.29
CAS RN: 7440-63-3
Properties: colorless, odorless gas; enthalpy of vaporization 12.62 kJ/mol; enthalpy of fusion 1.81 kJ/mol; critical temp 16.6°C; critical pressure 5.84 MPa; sonic velocity (101.32 kPa, 0°C) 168 m/s; viscosity (101.32 kPa, 25°C) 23.1 Pa·s; dielectric constant 1.0012 at 25°C and 1 atm; used in flash lamps, lasers, anesthesia [HAW93] [KIR78] [AIR87] [CRC93]
Solubility: 101.32 kPa: 108.1 mL/1000g H_2O (20°C) [KIR78]; Henry's law constants, k x 10^{-4}: 2.558 (70.3°C), 2.586 (125.5°C), 2.485 (175.7), 2.048 (225.1°C), 1.308 (284.2°C) [POT78]
Density, g/cm³: gas: 101.3 kPa, 0°C, 0.0058971 [KIR78]
Melting Point, °C: -111.75 [CRC93]
Boiling Point, °C: -108.05 [CRC93]
Thermal Conductivity, W/(m·K): gas (101.32 kPa, 0°C) 0.00565 [ALD94]

3158
Compound: Xenon difluoride
Synonyms: xenon fluoride
Formula: XeF_2
Molecular Formula: F_2Xe
Molecular Weight: 169.287
CAS RN: 13709-36-9
Properties: white stable cryst; powerful oxidizing agent; enthalpy of sublimation 55.73 kJ/mol; tetr, a = 0.4315 nm, c = 0.6990 nm; obtained from F_2 and Xe under high pressure [DOU83] [KIR78]
Solubility: 25 g/L H_2O (0°C) [MER89]; hydrolyzes to Xe + O_2; v s liq HF [COT88]
Density, g/cm³: 4.32 [KIR78]
Melting Point, °C: 129.03 [KIR78]
Boiling Point, °C: sublimes without decomposition [MER89]

3159
Compound: Xenon dioxydifluoride
Formula: XeO_2F_2
Molecular Formula: F_2O_2Xe
Molecular Weight: 201.286
CAS RN: 13875-06-4
Properties: colorless; ortho-rhomb, a = 0.6443 nm, b = 0.6288 nm, c = 0.8312 nm [KIR78]
Density, g/cm³: 4.10 [KIR78]

Melting Point, °C: 30.8 explodes [KIR78]

3160
Compound: Xenon fluoride hexafluoroantimonate
Formula: XeF_3SbF_6
Molecular Formula: F_9SbXe
Molecular Weight: 424.036
CAS RN: 39797-63-2
Properties: yellowish green; monocl, a = 0.5394 nm, b = 1.5559 nm, c = 0.5394 nm [KIR78]
Density, g/cm³: 3.92 [KIR78]
Melting Point, °C: 109-113 [KIR78]

3161
Compound: Xenon fluoride hexafluoroarsenate
Formula: $Xe_2F_3AsF_6$
Molecular Formula: AsF_9Xe_2
Molecular Weight: 377.198
CAS RN: 50432-32-1
Properties: yellowish green; monocl, a = 1.5443 nm, b = 0.8678 nm, c = 2.0888 nm [KIR78]
Density, g/cm³: 3.62 [KIR78]
Melting Point, °C: 99 [KIR78]

3162
Compound: Xenon fluoride hexafluororuthenate
Formula: $XeFRuF_6$
Molecular Formula: F_7RuXe
Molecular Weight: 365.349
CAS RN: 22527-13-5
Properties: yellowish green; monocl, a = 0.7991 nm, b = 1.1086 nm, c = 0.7250 nm [KIR78]
Density, g/cm³: 3.78 [KIR78]
Melting Point, °C: 110-111 [KIR78]

3163
Compound: Xenon fluoride monodecafluoroantimonate
Formula: $XeFSb_2F_{11}$
Molecular Formula: $F_{12}Sb_2Xe$
Molecular Weight: 602.785
CAS RN: 15364-10-0
Properties: yellow; monocl, a = 0.807 nm, b = 0.955 nm, c = 0.733 nm [KIR78]
Density, g/cm³: 3.69 [KIR78]
Melting Point, °C: 63 [KIR78]

3164
Compound: Xenon hexafluoride
Formula: XeF_6

Molecular Formula: F$_6$Xe
Molecular Weight: 245.280
CAS RN: 13693-09-9
Properties: colorless solid; has greenish yellow vapor; monocl, a = 0.933 nm, b = 1.096 nm, c = 0.895 nm; enthalpy of sublimation 59.12 kJ/mol; very strong oxidizing agent; preparation: reaction between Xe and excess F$_2$ at 250°C [DOU83] [MER89] [KIR78]
Solubility: hydrolyzed in H$_2$O, forming XeOF$_4$ and XeO$_3$ [MER89]
Density, g/cm^3: 3.56 [KIR78]
Melting Point, °C: 49.48 [KIR78]
Boiling Point, °C: 75.57 [MER89]

3165
Compound: Xenon oxydifluoride
Formula: XeOF$_2$
Molecular Formula: F$_2$OXe
Molecular Weight: 185.286
CAS RN: 13780-64-8
Properties: yellow; formed as an unstable product of the partial hydrolysis of XeF$_4$ [KIR78]
Melting Point, °C: explodes ~0 [KIR78]

3166
Compound: Xenon oxytetrafluoride
Formula: XeOF$_4$
Molecular Formula: F$_4$OXe
Molecular Weight: 223.283
CAS RN: 13774-85-1
Properties: colorless volatile stable liq at 25°C [KIR78]
Melting Point, °C: -46.2 [KIR78]

3167
Compound: Xenon pentafluoride hexafluoroarsenate
Formula: XeF$_5$AsF$_6$
Molecular Formula: AsF$_{11}$Xe
Molecular Weight: 415.194
CAS RN: 20328-94-3
Properties: white; monocl, a = 0.5886 nm, b = 1.6564 nm, c = 0.8051 nm [KIR78]
Density, g/cm^3: 3.51 [KIR78]
Melting Point, °C: 130.5 [KIR78]

3168
Compound: Xenon pentafluoride hexafluororuthenate
Formula: XeF$_5$RuF$_6$
Molecular Formula: F$_{11}$RuXe

Molecular Weight: 441.342
CAS RN: 39796-98-0
Properties: green; ortho-rhomb, a = 1.6771 nm, b = 0.8206 nm, c = 0.5617 nm [KIR78]
Density, g/cm^3: 3.79 [KIR78]
Melting Point, °C: 152 [KIR78]

3169
Compound: Xenon tetrafluoride
Formula: XeF$_4$
Molecular Formula: F$_4$Xe
Molecular Weight: 207.284
CAS RN: 13709-61-0
Properties: colorless cryst; readily prepared by mixing fluorine and xenon; enthalpy of sublimation 60.92 kJ/mol; monocl, a = 0.5050 nm, b = 0.5922 nm, c = 0.5771 nm [KIR78]
Solubility: reacts violently with H$_2$O, forming Xe, O$_2$, HF and XeO$_3$ [DOU83]
Density, g/cm^3: 4.04 [KIR78]
Melting Point, °C: 117.10 [KIR78]

3170
Compound: Xenon tetroxide
Formula: XeO$_4$
Molecular Formula: O$_4$Xe
Molecular Weight: 195.288
CAS RN: 12340-14-6
Properties: yellow solid; unstable, can explode [KIR78]
Melting Point, °C: decomposes <0 [KIR78]

3171
Compound: Xenon trifluoride monodecafluoroantimonate
Formula: XeF$_3$Sb$_2$F$_{11}$
Molecular Formula: F$_{14}$Sb$_2$Xe
Molecular Weight: 640.776
CAS RN: 35718-37-7
Properties: yellowish green; tricl, a = 0.8237 nm, b = 0.9984 nm, c = 0.8004 nm [KIR78]
Density, g/cm^3: 3.98 [KIR78]
Melting Point, °C: 81-83 [KIR78]

3172
Compound: Xenon trioxide
Formula: XeO$_3$
Molecular Formula: O$_3$Xe
Molecular Weight: 179.288
CAS RN: 13776-58-4

Properties: colorless solid; hygr; strongly explosive; ortho-rhomb, a = 0.6163 nm, b = 0.8115 nm, c = 0.5234 nm [MER89] [KIR78]
Solubility: s H_2O [KIR78]
Density, g/cm³: 4.55 [KIR78]
Melting Point, °C: explodes ~25 [KIR78]
Reactions: forms $HXeO_4^-$ in basic solutions [DOU83]

3173
Compound: Ytterbium
Formula: Yb
Molecular Formula: Yb
Molecular Weight: 173.04
CAS RN: 7440-64-4
Properties: silvery, ductile metal; fcc, room temp; bcc >798°C; enthalpy of fusion 7.657 kJ/mol; enthalpy of sublimation 152.1 kJ/mol; atom radius 0.19392 nm; ion radius 0.0858 nm, Yb^{+++}, colorless; electrical resistivity (20°C) 28 μohm·cm [MER89] [KIR82] [ALD94]
Solubility: slowly reacts with H_2O; s dil acids and ammonia [HAW93]
Density, g/cm³: fcc: 6.9654 [KIR82]; bcc: 6.54 [MER89]
Melting Point, °C: 819 [KIR82]
Boiling Point, °C: 1196 [KIR82]
Thermal Conductivity, W/(m·K): 34.9 (25°C) [CRC93]
Thermal Expansion Coefficient: 26.3 x 10⁻⁶/K [CRC93]

3174
Compound: Ytterbium acetate tetrahydrate
Formula: $Yb(CH_3COO)_3 \cdot 4H_2O$
Molecular Formula: $C_6H_{17}O_{10}Yb$
Molecular Weight: 422.235
CAS RN: 15280-58-7
Properties: white hygr powd; cryst aggregates [AES93] [STR93] [ALD94]
Density, g/cm³: 2.09 [STR93]
Reactions: minus $4H_2O$ at 100°C [CRC94]

3175
Compound: Ytterbium acetylacetonate
Synonyms: 2,4-pentanedione, ytterbium(III) derivative
Formula: $Yb(CH_3COCH=C(O)CH_3)_3$
Molecular Formula: $C_{15}H_{21}O_6Yb$
Molecular Weight: 470.368
CAS RN: 14284-98-1
Properties: powd [STR93]

$$\underset{[CH_3-C=CH-C-CH_3]_3Yb}{\overset{\displaystyle O- \qquad O}{\overset{\displaystyle | \qquad \parallel}{}}}$$

3176
Compound: Ytterbium bromide hydrate
Formula: $YbBr_3 \cdot xH_2O$
Molecular Formula: Br_3Yb (anhydrous)
Molecular Weight: 412.752 (anhydrous)
CAS RN: 15163-03-8
Properties: white cryst; anhydrous $YbBr_3$, 13759-89-2, -20 mesh with 99.9% purity [STR93] [CER91]
Melting Point, °C: 956 [AES93] (anhydrous)

3177
Compound: Ytterbium carbonate hydrate
Formula: $Yb_2(CO_3)_3 \cdot xH_2O$
Molecular Formula: $C_3O_9Yb_2$ (anhydrous)
Molecular Weight: 526.112 (anhydrous)
Properties: cryst [AES93]
Melting Point, °C: decomposes [AES93]

3178
Compound: Ytterbium chloride
Formula: $YbCl_3$
Molecular Formula: Cl_3Yb
Molecular Weight: 279.398
CAS RN: 10361-91-8
Properties: -20 mesh with 99.9% purity; white powd; hygr [STR93] [CER91]
Melting Point, °C: 875 [LID94]

3179
Compound: Ytterbium chloride hexahydrate
Formula: $YbCl_3 \cdot 6H_2O$
Molecular Formula: $Cl_3H_{12}O_6Yb$
Molecular Weight: 387.489
CAS RN: 10035-01-5
Properties: -4 mesh with 99.9% purity; green cryst; hygr [HAW93] [CER91]
Solubility: v s H_2O [HAW93]
Density, g/cm³: 2.575 [MER89]
Melting Point, °C: 865 [HAW93]
Reactions: minus $6H_2O$ at 180°C [HAW93]

3180
Compound: Ytterbium fluoride

Formula: YbF₃
Molecular Formula: F₃Yb
Molecular Weight: 230.035
CAS RN: 13760-80-0
Properties: 3-12 mm fused pieces with 99.9% purity; white powd; hygr [STR93] [CER91]
Solubility: i H₂O [HAW93]
Density, g/cm³: 8.168 [STR93]
Melting Point, °C: 1157 [STR93]
Boiling Point, °C: 2200 [STR93]

3181
Compound: Ytterbium hydride
Formula: YbH₃
Molecular Formula: H₃Yb
Molecular Weight: 176.064
CAS RN: 32997-62-9
Properties: in the form of lumps, in ampoule under Ar [AES93]

3182
Compound: Ytterbium nitrate pentahydrate
Formula: Yb(NO₃)₃·5H₂O
Molecular Formula: H₁₀N₃O₁₄Yb
Molecular Weight: 449.131
CAS RN: 35725-34-9
Properties: white cryst [STR93] [ALD94]

3183
Compound: Ytterbium oxalate decahydrate
Formula: Yb₂(C₂O₄)₃·10H₂O
Molecular Formula: C₆H₂₀O₂₂Yb₂
Molecular Weight: 790.292
CAS RN: 51373-68-3
Properties: white cryst [STR93]
Solubility: 0.0001 g/100mL H₂O [CRC94]
Density, g/cm³: 2.644 [CRC94]

3184
Compound: Ytterbium oxide
Synonyms: ytterbia
Formula: Yb₂O₃
Molecular Formula: O₃Yb₂
Molecular Weight: 394.078
CAS RN: 1314-37-0

Properties: if pure, colorless mass; brownish or yellowish if thulia is present, also 3-12 mm sintered pieces of 99.9% purity; sl hygr; absorbs atm H₂O and NH₃; used in special alloys, dielectric ceramics and special glasses, also sintered pieces used as evaporation material to form film with reactivity to radio frequencies [HAW93] [MER89] [CER91]
Solubility: s dil acids [MER89]
Density, g/cm³: 9.2 [HAW93]
Melting Point, °C: 2346 [HAW93]

3185
Compound: Ytterbium perchlorate
Formula: Yb(ClO₄)₃
Molecular Formula: Cl₃O₁₂Yb
Molecular Weight: 471.390
CAS RN: 13498-08-3
Properties: off-white cryst [STR93]

3186
Compound: Ytterbium silicide
Formula: YbSi₂
Molecular Formula: Si₂Yb
Molecular Weight: 229.211
CAS RN: 12039-89-3
Properties: hex cryst; 10 mm & down lump [LID94] [ALF93]
Density, g/cm³: 7.54 [LID94]

3187
Compound: Ytterbium sulfate
Formula: Yb₂(SO₄)₃
Molecular Formula: O₁₂S₃Yb₂
Molecular Weight: 634.271
CAS RN: 10034-98-7
Properties: colorless cryst [CRC77]
Solubility: g/100g H₂O: 44.2 (0°C), 22.2 (30°C), 4.7 (100°C) [LAN85]
Density, g/cm³: 3.793 [LAN52]
Melting Point, °C: decomposes 900 [LAN52]

3188
Compound: Ytterbium sulfate octahydrate
Formula: Yb₂(SO₄)₃·8H₂O
Molecular Formula: H₁₆O₂₀S₃Yb₂
Molecular Weight: 778.393
CAS RN: 10034-98-7
Properties: lustrous, colorless cryst [MER89]
Solubility: s H₂O, solubility decreases with temp increase [MER89]

Density, g/cm³: 3.286 [STR93]

3189
Compound: Yttrium
Formula: Y
Molecular Formula: Y
Molecular Weight: 88.90585
CAS RN: 7440-65-5
Properties: white; silvery metal; hex close-packed cryst; enthalpy of fusion 11.43 kJ/mol; enthalpy of sublimation 424.7 kJ/mol; radius of atom 0.1801 nm; ion radius of Y^{+++} 0.0893 nm aq solutions are colorless; electrical resistivity (20°C) 57 μohm·cm [KIR82] [ALD94]
Solubility: decomposes in cold H_2O, more rapidly in hot H_2O [MER89]; s dil acids and KOH solutions [HAW93]
Density, g/cm³: 4.4689 [KIR82]
Melting Point, °C: 1522 [KIR82]
Boiling Point, °C: 3338 [KIR82]
Thermal Conductivity, W/(m·K): 17.2 (25°C) [CRC93]
Thermal Expansion Coefficient: 10.6 x 10^{-6}/K [CRC93]

3190
Compound: Yttrium acetate hydrate
Formula: $Y(CH_3COO)_3 \cdot xH_2O$
Molecular Formula: $C_6H_9O_6Y$ (anhydrous)
Molecular Weight: 266.039 (anhydrous)
CAS RN: 23363-14-6
Properties: white cryst; x=4, Molecular Weight 338.10 [AES93] [STR93]
Melting Point, °C: decomposes [AES93]

3191
Compound: Yttrium acetylacetonate trihydrate
Synonyms: 2,4-pentanedione, yttrium(III) derivative
Formula: $Y(CH_3COCH=C(O)CH_3)_3 \cdot 3H2O$
Molecular Formula: $C_{15}H_{27}O_9Y$
Molecular Weight: 440.280
CAS RN: 15554-47-9
Properties: yellowish white cryst [STR93]
Melting Point, °C: 138-140 [STR93]

3192
Compound: Yttrium aluminum oxide
Synonyms: YAG
Formula: $Y_3Al_5O_{12}$
Molecular Formula: $Al_5O_{12}Y_3$
Molecular Weight: 593.619

CAS RN: 12005-21-9
Properties: green cub cryst; 3-12 mm fused pieces; used in the form of a 99.99% pure material as a sputtering target to prepare bubble memory devices [CER91] [LID94]
Density, g/cm³: ~4.5 [LID94]

3193
Compound: Yttrium antimonide
Formula: YSb
Molecular Formula: SbY
Molecular Weight: 210.666
CAS RN: 12186-97-9
Properties: cub cryst; high purity semiconductor [HAW93] [LID94]
Density, g/cm³: 5.97 [LID94]
Melting Point, °C: 2310 [LID94]

3194
Compound: Yttrium arsenide
Formula: YAs
Molecular Formula: AsY
Molecular Weight: 163.828
CAS RN: 12255-48-0
Properties: cub cryst; high purity semiconductor [LID94] [HAW93]
Density, g/cm³: 5.59 [LID94]

3195
Compound: Yttrium barium copper oxide
Synonyms: supercon N-124, O-124
Formula: $YBa_2Cu_4O_8$
Molecular Formula: $Ba_2Cu_4O_8Y$
Molecular Weight: 745.739
CAS RN: 107539-20-8
Properties: N-124: 0.2 micron powd; wet processed from ACS grade nitrates; O-124: 20 micron powd; dry processed from ACS grade oxides; T_c is 81 K [STR93] [ASM93]

3196
Compound: Yttrium barium copper oxide
Synonyms: supercon N-123
Formula: $YBa_2Cu_3O_7$
Molecular Formula: $Ba_2Cu_3O_7Y$
Molecular Weight: 666.194
CAS RN: 107539-20-8

Properties: ortho-rhomb; superconductor; 0.2 micron powd; wet processed from 99.999% nitrates; a hard grayish black sintered material prepared by reacting carbonate and hydroxide free Y_2O_3 with stoichiometric amounts of $BaCO_3$ and CuO in air at 950°C for 12 h; sensitive to atm CO_2 and H_2O; a = 0.3835 nm, b = 0.3884 nm, c = 1.1681 nm; transition temp, T_c, is 92 K; for $YBa_2Cu_{3.5}O_{7.5}$, T_c is 94 K [STR93] [CON87] [CEN92] [ASM93] [ALF93]

3197

Compound: Yttrium barium copper oxide
Synonyms: supercon O-123
Formula: $YBa_2Cu_3O_7$
Molecular Formula: $Ba_2Cu_3O_7Y$
Molecular Weight: 666.194
CAS RN: 107539-20-8
Properties: 20 micron powd; dry processed from ACS grade oxides; T_c is 93 K for $YBa_2Cu_3O_7$; interplanar spacing 0.58 nm; effective mass to free electron mass ratio 2-2.5; coherence length parallel and perpendicular to conduction plane 1.5 nm and 0.15-3 nm; mean free path of charge carrier above T_c is 10.0 nm; penetration depth 145.0 nm; carrier density $3.1 \times 10^{+21}$ per cm^3; for x = 7-δ, T_c is 92 K [CEN92] [STR93] [ASM93]
Reactions: metastable at low temperatures [CEN92]

3198

Compound: Yttrium boride
Synonyms: yttrium hexaboride
Formula: YB_6
Molecular Formula: B_6Y
Molecular Weight: 153.772
CAS RN: 12008-32-1
Properties: -325 mesh 10 microns or less with 99.9% purity; refractory material [CER91] [KIR78]
Density, g/cm³: 3.2 [LID94]
Melting Point, °C: 2600, decomposes [KIR78]

3199

Compound: Yttrium bromide
Formula: YBr_3
Molecular Formula: Br_3Y
Molecular Weight: 328.618
CAS RN: 13469-98-2
Properties: deliq; -20 mesh with 99.9% purity [CRC94] [CER91]
Solubility: g/100g H_2O: 63.9 (0°C), 75.1 (20°C), 123 (90°C) [LAN85]

Melting Point, °C: 904 [HAW93]

3200

Compound: Yttrium bromide nonahydrate
Formula: $YBr_3 \cdot 9H_2O$
Molecular Formula: $Br_3H_{18}O_9Y$
Molecular Weight: 490.756
CAS RN: 13469-98-2
Properties: colorless cryst [HAW93]
Solubility: s H_2O and alcohol; i ether [HAW93]

3201

Compound: Yttrium carbide
Formula: YC_2
Molecular Formula: C_2Y
Molecular Weight: 112.928
CAS RN: 12071-35-1
Properties: yellow; 12 mm pieces and smaller of 99.5% purity [CER91] [CRC94]
Density, g/cm³: 4.13 [CRC94]
Melting Point, °C: ~2400 [LID94]

3202

Compound: Yttrium carbonate trihydrate
Formula: $Y_2(CO_3)_3 \cdot 3H_2O$
Molecular Formula: $C_3H_6O_{12}Y_2$
Molecular Weight: 411.886
CAS RN: 5970-44-5
Properties: white to reddish white powd [MER89] [STR93]
Solubility: i H_2O; s dil mineral acids [MER89]

3203

Compound: Yttrium chloride
Formula: YCl_3
Molecular Formula: Cl_3Y
Molecular Weight: 195.264
CAS RN: 10361-92-9
Properties: -20 mesh with 99.9% purity; white powd; hygr [STR93] [CER91]
Solubility: g/100g H_2O: 77.3 (0°C), 78.8 (20°C), 80.8 (40°C) [LAN85]
Density, g/cm³: 2.67 [STR93]
Melting Point, °C: 721 [STR93]
Boiling Point, °C: 1507 [CRC94]

3204

Compound: Yttrium chloride hexahydrate
Formula: $YCl_3 \cdot 6H_2O$
Molecular Formula: $Cl_3H_{12}O_6Y$

Molecular Weight: 303.355
CAS RN: 10025-94-2
Properties: -4 mesh with 99.9% purity; colorless cryst; deliq [MER89] [CER91]
Solubility: 217 g/100mL H_2O (20°C), 235 g/100mL H_2O (50°C) [CRC94]
Density, g/cm³: 2.18 [STR93]
Reactions: minus $6H_2O$ by heating in HCl stream [MER89]

3205
Compound: Yttrium fluoride
Formula: YF_3
Molecular Formula: F_3Y
Molecular Weight: 145.901
CAS RN: 13709-49-4
Properties: white powd; hygr; in the form of a 99.9% pure material, is used as a sputtering target for multilayers [STR93] [CER91]
Density, g/cm³: 4.01 [STR93]
Melting Point, °C: 1152 [STR93]

3206
Compound: Yttrium hexafluoroacetylacetonate
Synonyms: 1,1,1,5,5,5-hexafluoro-2,4-pentanedione, yttrium derivative
Formula: $Y(CF_3COCHCOCF_3)_3$
Molecular Formula: $C_{15}H_3F_{18}O_6Y$
Molecular Weight: 710.062
CAS RN: 18911-76-7
Properties: white cryst [STR93]

$$\begin{matrix} O- & & O \\ | & & \| \\ \end{matrix}$$
$$[CF_3-C=CH-C-CF_3]_3Y$$

Melting Point, °C: 166-170 [STR93]
Boiling Point, °C: decomposes 240 [STR93]
Reactions: sublimes at 100°C (0.2 mm Hg) [STR93]

3207
Compound: Yttrium hydride
Formula: YH_3
Molecular Formula: H_3Y
Molecular Weight: 91.929
CAS RN: 13598-57-7
Properties: -60 mesh of 99.9% purity and lumps [CER91] [AES93]

3208
Compound: Yttrium hydroxide

Formula: $Y(OH)_3$
Molecular Formula: H_3O_3Y
Molecular Weight: 139.928
CAS RN: 16469-22-0
Properties: white; gelatinous precipitate; dries to a white powd, which absorbs atm CO_2 [MER89]
Reactions: decomposes on heating [CRC94]

3209
Compound: Yttrium iodide
Formula: YI_3
Molecular Formula: I_3Y
Molecular Weight: 469.619
CAS RN: 13470-38-7
Properties: white deliq flakes [CRC94] [AES93]
Melting Point, °C: 1004 [AES93]

3210
Compound: Yttrium iron oxide
Synonyms: yttrium garnet
Formula: $Y_3Fe_5O_{12}$
Molecular Formula: $Fe_5O_{12}Y_3$
Molecular Weight: 737.936
CAS RN: 12063-56-8
Properties: used as 99.99% and 99.9% pure sputtering target to prepare ferromagnetic films, and in bubble memory devices [CER91]

3211
Compound: Yttrium nitrate hexahydrate
Formula: $Y(NO_3)_3 \cdot 6H_2O$
Molecular Formula: $H_{12}N_3O_{15}Y$
Molecular Weight: 383.012
CAS RN: 13494-98-9
Properties: white deliq cryst [MER89] [STR93]
Solubility: g anhydrous/100g H_2O: 93.1 (0°C), 123 (20°C), 200 (60°C) [LAN85]; partially decomposed in water to basic nitrates [MER89]; 5.2759±0.0009 mol/(kg·H_2O) at 25°C [RAR85b]
Density, g/cm³: 2.68 [CRC94]
Reactions: minus $3H_2O$ at 100°C [CRC94]

3212
Compound: Yttrium oxalate nonahydrate
Formula: $Y_2(C_2O_4)_3 \cdot 9H_2O$
Molecular Formula: $C_6H_{18}O_{21}Y_2$
Molecular Weight: 604.008
CAS RN: 13266-82-5
Properties: white cryst [STR93]
Solubility: 0.0001 g/100mL H_2O [CRC94]
Melting Point, °C: decomposes [AES93]

3213
Compound: Yttrium oxide
Synonyms: yttria
Formula: Y_2O_3
Molecular Formula: O_3Y_2
Molecular Weight: 225.810
CAS RN: 1314-36-9
Properties: white powd, or sintered tablets and pieces of 99.9% purity; bcc; readily absorbs atm CO_2; enthalpy of fusion 105.00 kJ/mol; used in crucible form for experimental, proprietary melting, also sintered pieces used as evaporation material for hard film dielectric coating and thin film capacitors, and as 99.999%, 99.99%, 99.9% pure sputtering target for preparing hard films, dielectric coatings, and thin film capacitor [CER91] [MER89] [CRC93]
Solubility: s dil acids [MER89]
Density, g/cm³: 5.03 [MER89]
Melting Point, °C: 2439 [CRC93]

3214
Compound: Yttrium perchlorate hexahydrate
Formula: $Y(ClO_4)_3 \cdot 6H_2O$
Molecular Formula: $Cl_3H_{12}O_{18}Y$
Molecular Weight: 495.348
CAS RN: 14017-56-2
Properties: white cryst [STR93]

3215
Compound: Yttrium phosphide
Formula: YP
Molecular Formula: PY
Molecular Weight: 119.880
CAS RN: 12294-01-8
Properties: cub cryst; high purity semiconductor [LID94] [HAW93]
Density, g/cm³: ~4.4 [LID94]

3216
Compound: Yttrium sulfate octahydrate
Formula: $Y_2(SO_4)_3 \cdot 8H_2O$
Molecular Formula: $H_{16}O_{20}S_3Y_2$
Molecular Weight: 610.125
CAS RN: 7446-33-5
Properties: small, reddish white; monocl cryst [MER89] [HAW93]
Solubility: g/100g H_2O: 8.05 (0°C), 7.30 (20°C), 2.2 (90°C) [LAN85]; s conc H_2SO_4 reacting to produce $Y(HSO_4)_3$ [MER89]
Density, g/cm³: 2.558 [STR93]
Reactions: minus $8H_2O$ at 120°C [HAW93]

3217
Compound: Yttrium sulfide
Formula: Y_2S_3
Molecular Formula: S_3Y_2
Molecular Weight: 274.010
CAS RN: 12039-19-9
Properties: -200 mesh with 99.9% purity; yellow powd [STR93] [CER91]
Density, g/cm³: 3.87 [LID94]
Melting Point, °C: 1925 [LID94]

3218
Compound: Yttrium vanadate
Formula: YVO_3
Molecular Formula: O_3VY
Molecular Weight: 187.845
CAS RN: 12143-39-4
Properties: white cryst; 99.9% pure, doped with Eu_2O_3; used in phosphorescent coating on special currency papers, and as red phosphor in television tubes [HAW93] [CER91]

3219
Compound: Zinc
Formula: Zn
Molecular Formula: Zn
Molecular Weight: 65.39
CAS RN: 7440-66-6
Properties: lustrous bluish white; hex closed-packed metal; reacts with atm moisture producing surface of basic zinc carbonate; hardness 2.5 Mohs; electrical resistivity, 20°C, 5.8 µohm·cm; malleable when heated to 100-150°C; brittle at 210°C; enthalpy of fusion 7.387 kJ/mol; enthalpy of vaporization 114.8 kJ/mol; uses include evaporating metal for metallized paper and capacitor dielectric films [KIR84] [MER89] [CER91] [ALD94]
Solubility: i H_2O; s HCl, HNO_3, alkaline hydroxides [MER89]
Density, g/cm³: 7.14 [MER89]
Melting Point, °C: 419.5 [MER89]
Boiling Point, °C: 908 [MER89]
Thermal Conductivity, W/(m·K): 113.0 (18°C), 96.0 (419.5°C); liq: 60.7 (419.5°C), 56.5 (750°C) [KIR84]
Thermal Expansion Coefficient: (volume) 100°C (0.717), 200°C (1.656), 400°C (3.699) [CLA66]

3220
Compound: Zinc acetate
Formula: $Zn(CH_3COO)_2$

Molecular Formula: $C_4H_6O_4Zn$
Molecular Weight: 183.479
CAS RN: 557-34-6
Properties: prepared from zinc nitrate and acetic anhydride; used to preserve wood, as a mordant in dyeing, and a blood test reagent [MER89]
Density, g/cm³: 1.84 [ALD94]

3221
Compound: Zinc acetate dihydrate
Formula: $Zn(CH_3COO)_2 \cdot 2H_2O$
Molecular Formula: $C_4H_{10}O_6Zn$
Molecular Weight: 219.509
CAS RN: 5970-45-6
Properties: white powd; pearly luster; somewhat efflorescent; monocl cryst; used as wood preservative, mordant, antiseptic and a catalyst; anhydrous form: 99.99% pure powd; CAS RN 557-34-6 [AES93] [KIR84] [MER89] [STR93] [HAW93]
Solubility: g/100g H_2O: 40 (25°C), 67 (100°C); 3 g/100g alcohol at 25°C [KIR84]
Density, g/cm³: 1.735 [MER89]
Melting Point, °C: 237 [MER89]
Reactions: minus $2H_2O$ at 100°C [HAW93]

3222
Compound: Zinc acetylacetonate hydrate
Synonyms: 2,4-pentanedione, zinc(II) derivative
Formula: $Zn(CH_3COCH=C(O)CH_3)_2 \cdot xH_2O$
Molecular Formula: $C_{10}H_{14}O_4Zn$ (anhydrous)
Molecular Weight: 263.609 (anhydrous)
CAS RN: 108503-47-5
Properties: cryst solid; trimer; hydrate is a white powd; used as a catalyst in the synthesis of long chain alcohols and aldehydes, and as a textile weighting agent [HAW93] [STR93] [COT88]
Solubility: decomposed by H_2O; s benzene, acetone [HAW93]
Melting Point, °C: 138 [HAW93]
Boiling Point, °C: sublimes [HAW93]

3223
Compound: Zinc ammonium chloride
Formula: $ZnCl_2 \cdot 2NH_4Cl$
Molecular Formula: $Cl_4H_8N_2Zn$
Molecular Weight: 243.278
CAS RN: 52628-25-8
Properties: white powd or cryst; used in galvanizing, as a flux for solder, and in adhesives [KIR84] [HAW93]

Solubility: g/100g H_2O: 66 (0°C), 69 (30°C) [KIR84]
Density, g/cm³: 1.88 [KIR84]
Melting Point, °C: 150 (decomposes) [KIR84]

3224
Compound: Zinc antimonide
Formula: ZnSb
Molecular Formula: SbZn
Molecular Weight: 187.150
CAS RN: 12039-35-9
Properties: silvery white cryst, 99.5% pure melted pieces of 6 mm and smaller; used in thermoelectric devices, and thermionic studies [HAW93] [CER91]
Solubility: decomposed by H_2O [HAW93]
Density, g/cm³: 6.33 [HAW93]
Melting Point, °C: 570 [HAW93]

3225
Compound: Zinc arsenate octahydrate
Synonyms: koettigite
Formula: $Zn_3(AsO_4)_2 \cdot 8H_2O$
Molecular Formula: $As_2H_{16}O_{16}Zn_3$
Molecular Weight: 618.130
CAS RN: 13464-44-4
Properties: white powd; used as an insecticide and wood preservative [HAW93]
Solubility: i H_2O; s in acids and alkalies [HAW93]
Density, g/cm³: 3.31 (15°C) [HAW93]
Reactions: minus H_2O at 100°C [HAW93]

3226
Compound: Zinc arsenide
Formula: Zn_3As_2
Molecular Formula: As_2Zn_3
Molecular Weight: 346.013
CAS RN: 12006-40-5
Properties: gray tetr; -20 mesh with 99% purity, for arsine generation; 99.9999% purity electronic doping grade [CER91] [CRC94]
Density, g/cm³: 5.528 [ALD94]
Melting Point, °C: 1015 [AES93]

3227
Compound: Zinc arsenite
Synonyms: zinc metaarsenite
Formula: $Zn(AsO_2)_2$
Molecular Formula: As_2O_4Zn
Molecular Weight: 279.231
CAS RN: 10326-24-6

Properties: colorless powd; used as insecticide and wood preservative [HAW93]

Solubility: i H_2O; s acids [HAW93]

3228

Compound: Zinc borate

Synonyms: zinc diborate

Formula: $3ZnO \cdot 2B_2O_3$

Molecular Formula: $B_4O_9Zn_3$

Molecular Weight: 383.409

CAS RN: 27043-84-1

Properties: white, amorphous powd; used in medicine, for fireproofing textiles, as an inhibitor of fungus and mildew, and as a ceramic flux; most common borate is $2ZnO \cdot 3B_2O_3 \cdot 7H_2O$, x-ray structure is given as $Zn[B_2O_3(OH)_5] \cdot H_2O$; this hydrate can be prepared by adding borax to a solution of a soluble zinc salt; there are other hydrated borates, for example $Zn_3B_4O_9 \cdot 5H_2O$, CAS RN 12536-65-1, -325 mesh white powd [AES93] [HAW93] [KIR78]

Solubility: g/100g H_2O: 0.007 (25°C); s dil acids [KIR84]

Density, g/cm³: 3.64 [HAW93]

Melting Point, °C: 980 [HAW93]

3229

Compound: Zinc borate hemiheptahydrate

Formula: $2ZnO \cdot 3B_2O_3 \cdot 3\text{-}1/2H_2O$

Molecular Formula: $B_6H_7O_{14.5}Zn_2$

Molecular Weight: 434.690

CAS RN: 12513-27-8

Properties: white tricl or powd; has about 20% hydrate water; used as a fire retardant material [KIR84] [HAW93] [CRC94]

Solubility: i H_2O [KIR84]

Density, g/cm³: 4.22 [KIR84]

Melting Point, °C: 980 [KIR84]

3230

Compound: Zinc borate pentahydrate

Formula: $3ZnO \cdot 2B_2O_3 \cdot 5H_2O$

Molecular Formula: $B_4H_{10}O_{14}Zn_3$

Molecular Weight: 473.487

CAS RN: 12536-65-1

Properties: -325 mesh white powd; formula also given as a dihydrate; used for fireproofing, in ceramics and fungicides [KIR84] [AES93]

Solubility: 0.007 g/100g H_2O at 25°C; sl s HCl [KIR84]

Density, g/cm³: 3.64 [KIR84]

3231

Compound: Zinc bromate hexahydrate

Formula: $Zn(BrO_3)_2 \cdot 6H_2O$

Molecular Formula: $Br_2H_{12}O_{12}Zn$

Molecular Weight: 429.286

CAS RN: 13517-27-6

Properties: white solid; deliq; oxidizing agent [HAW93]

Solubility: v s H_2O [HAW93]

Density, g/cm³: 2.566 [HAW93]

Melting Point, °C: 100 [HAW93]

Reactions: minus $6H_2O$ at 200°C [CRC94]

3232

Compound: Zinc bromide

Formula: $ZnBr_2$

Molecular Formula: Br_2Zn

Molecular Weight: 225.198

CAS RN: 7699-45-8

Properties: white granular powd; very hygr; enthalpy of vaporization 118 kJ/mol; enthalpy of fusion 16.70 kJ/mol; used in photographic emulsions, to manufacture rayon [MER89] [HAW93] [CRC93]

Solubility: mol/100mol soln, H_2O: 31.1 (0°C), 37.6 (25°C), 53.8 (100°C); Solid phase, $ZnBr_2 \cdot 2H_2O$ (0°C, 25°C), $ZnBr_2$ (100°C) [KRU93]; 1 gm/0.5mL 90% alcohol; s ether, alkali hydroxide solutions [MER89]

Density, g/cm³: 4.5 [LID94]

Melting Point, °C: 394 [CRC93]

Boiling Point, °C: 650 [CRC94]

3233

Compound: Zinc caprylate

Formula: $Zn(C_8H_{15}O_2)_2$

Molecular Formula: $C_{16}H_{30}O_4Zn$

Molecular Weight: 351.802

CAS RN: 557-09-5

Properties: lustrous scales; decomposes in moist atm forming caprylic acid; used as a fungicide [HAW93]

Solubility: sl s boiling H_2O; s boiling alcohol [HAW93]

Melting Point, °C: 136 [HAW93]

3234

Compound: Zinc carbonate

Synonyms: smithsonite

Formula: $ZnCO_3$

Molecular Formula: CO_3Zn

Molecular Weight: 125.399

CAS RN: 3486-35-9
Properties: white, cryst powd; rhombohedr cryst; used in ceramics, as fireproofing filler, in cosmetics and lotions [HAW93] [MER89]
Solubility: mol/L soln, H_2O: 1.64×10^{-4} (solubility is given at $Pco_2 = 0.00032$ bar) [KRU93]; s dil acids, alkalies, NH_3 solutions [MER89]
Density, g/cm³: 4.42-4.45 [HAW93]
Melting Point, °C: decomposes [KIR84]
Reactions: evolves CO_2 at 300°C [HAW93]

3235
Compound: Zinc carbonate hydroxide
Formula: $3Zn(OH)_2 \cdot 2ZnCO_3$
Molecular Formula: $C_2H_6O_{12}Zn_5$
Molecular Weight: 549.013
CAS RN: 3486-35-9
Properties: white powd [STR93]
Density, g/cm³: 4.398 [STR93]

3236
Compound: Zinc chlorate
Formula: $Zn(ClO_3)_2$
Molecular Formula: Cl_2O_6Zn
Molecular Weight: 232.291
CAS RN: 10361-95-2
Properties: colorless to yellowish cryst; deliq; oxidizing agent [HAW93]
Solubility: mol/100mol H_2O: 59.19 (0°C), 66.52 (18°C), 75.44 (55°C); Solid phase, $Zn(ClO_3)_2 \cdot 6H_2O$ (0°C), $Zn(ClO_3)_2 \cdot 4H_2O$ (18°C, 55°C) [KRU93]; s alcohol, glycerol, ether [HAW93]
Density, g/cm³: 2.15 [HAW93]
Melting Point, °C: decomposes at 60 [HAW93]

3237
Compound: Zinc chloride
Formula: $ZnCl_2$
Molecular Formula: Cl_2Zn
Molecular Weight: 136.295
CAS RN: 7646-85-7
Properties: white granules; very deliq; enthalpy of vaporization 126 kJ/mol; used as a catalyst, dehydrating agent in organic synthesis, as an antiseptic, and for fireproofing [HAW93] [MER89] [CRC93]
Solubility: g/100g H_2O: 342 (0°C), 432 (25°), 615 (100°C); Solid phase, $ZnCl_2 \cdot H_2O$ (0°C), $ZnCl_2$ (25°C, 100°C) [KRU93]; 1 g soluble in: 0.25mL 2% HCl, 1.3mL alcohol, 2mL glycerol; v s acetone [MER89]

Density, g/cm³: 2.91 [STR93]
Melting Point, °C: 283 [STR93]
Boiling Point, °C: 732 [STR93]

3238
Compound: Zinc chromate heptahydrate
Formula: $ZnCrO_4 \cdot 7H_2O$
Molecular Formula: $CrH_{14}O_{11}Zn$
Molecular Weight: 307.490
CAS RN: 13530-65-9
Properties: yellow solid; used as a pigment [HAW93]
Density, g/cm³: 3.4 [CRC94] (anhydrous)

3239
Compound: Zinc chromite
Formula: $ZnCr_2O_4$
Molecular Formula: Cr_2O_4Zn
Molecular Weight: 233.380
CAS RN: 12018-19-8
Properties: green cub spinel; pellets; used in catalysts [KIR78] [STR93]
Density, g/cm³: 5.3 [CRC94]

3240
Compound: Zinc citrate dihydrate
Synonyms: citric acid, zinc salt dihydrate
Formula: $Zn_3(C_6H_5O_7)_2 \cdot 2H_2O$
Molecular Formula: $C_{12}H_{14}O_{16}Zn_3$
Molecular Weight: 610.403
CAS RN: 546-46-3
Properties: colorless powd; can be made from zinc carbonate and citric acid; used in toothpaste and as a mouthwash [MER89]
Solubility: sl s H_2O; s dil mineral acids and alkali hydroxides [MER89]

3241
Compound: Zinc cyanide
Formula: $Zn(CN)_2$
Molecular Formula: C_2N_2Zn
Molecular Weight: 117.425
CAS RN: 557-21-1
Properties: white powd; readily decomposed by dil mineral acids, evolving HCN; has been used in plating, and as an insecticide [MER89]
Solubility: mol/L soln, H_2O: 4.2×10^{-5} (room temp) [KRU93]; s dil mineral acids, evolving HCN; i alcohol [HAW93]
Density, g/cm³: 1.852 [HAW93]
Melting Point, °C: 800, decomposes [HAW93]

3242
Compound: Zinc dichromate trihydrate
Formula: $ZnCr_2O_7 \cdot 3H_2O$
Molecular Formula: $Cr_2H_6O_{10}Zn$
Molecular Weight: 335.424
CAS RN: 7789-12-0
Properties: yellowish orange powd; preparation: reaction between chromic acid and $Zn(OH)_2$; used as a pigment [HAW93]
Solubility: s hot H_2O, acids; i alcohol and ether [HAW93]

3243
Compound: Zinc dimethyldithiocarbamate
Synonyms: dimethyldithiocarbamic acid, zinc salt
Formula: $Zn[(CH_3)_2NCS_2]_2$
Molecular Formula: $C_6H_{12}N_2S_4Zn$
Molecular Weight: 305.828
CAS RN: 137-30-4
Properties: cryst when obtained by reaction from hot chloroform + alcohol; used as accelerator for rubber vulcanization [MER89]
Solubility: i H_2O [MER89]
Density, g/cm³: 1.66
Melting Point, °C: 250-252 [ALD94]

3244
Compound: Zinc fluoride
Formula: ZnF_2
Molecular Formula: F_2Zn
Molecular Weight: 103.387
CAS RN: 7783-49-5
Properties: hygr; tetr needles, or white cryst; enthalpy of vaporization 190.1 kJ/mol; finds use as ceramic glaze, and as a wood preservative [HAW93] [MER89] [CRC93] [STR93]
Solubility: g/100mL soln, H_2O: 1.516 (25°C); Solid phase, $ZnF_2 \cdot 4H_2O$ [KRU93]; sl s HF solutions; s HCl, HNO_3, NH_4OH [MER89]; i alcohol [HAW93]
Density, g/cm³: 4.95 [STR93]
Melting Point, °C: 872 [MER89]
Boiling Point, °C: 1500 [MER89]

3245
Compound: Zinc fluoride tetrahydrate
Formula: $ZnF_2 \cdot 4H_2O$
Molecular Formula: $F_2H_8O_4Zn$
Molecular Weight: 175.449
CAS RN: 13986-18-0
Properties: white cryst; rhombohedr [MER89] [STR93]

Solubility: mol/L soln, H_2O: 0.151±0.004 (25°C) [KRU93]
Density, g/cm³: 2.255 [STR93]
Reactions: releases $4H_2O$ at 100°C [MER89]

3246
Compound: Zinc fluoroborate hexahydrate
Formula: $Zn(BF_4)_2 \cdot 6H_2O$
Molecular Formula: $B_2F_8H_{12}O_6Zn$
Molecular Weight: 347.090
CAS RN: 27860-83-9
Properties: hex cryst; used in zinc plating, for bonding and in the textile industry as a resin cure [LID94] [KIR84]
Solubility: >100 g/100g H_2O at 25°C; s alcohol [KIR84]
Density, g/cm³: 2.12 [LID94]
Reactions: minus H_2O at 60°C [KIR84]

3247
Compound: Zinc formaldehyde sulfoxylate
Formula: $Zn(HSO_2 \cdot CH_2O)_2$
Molecular Formula: $C_2H_6O_6S_2Zn$
Molecular Weight: 255.588
CAS RN: 24887-06-7
Properties: rhomb prisms; used as a stripping and discharging agent in the textile industry [HAW93]
Solubility: v s H_2O; i alcohol; decomposed by acids [HAW93]
Melting Point, °C: 90 decomposes [KIR84]

3248
Compound: Zinc formate
Synonyms: formic acid, zinc salt
Formula: $Zn(CHOO)_2$
Molecular Formula: $C_2H_2O_4Zn$
Molecular Weight: 155.426
CAS RN: 557-41-5
Properties: colorless; readily forms dihydrate; can be obtained from reaction between zinc carbonate and formic acid [MER89] [CRC94]
Solubility: g/100g H_2O: 3.70 (0°C), 5.20 (20°C), 38.0 (100°C) [LAN85]
Density, g/cm³: 2.368 [CRC94]
Reactions: decomposes on heating [CRC94]

3249
Compound: Zinc formate dihydrate
Formula: $Zn(CHO_2)_2 \cdot 2H_2O$
Molecular Formula: $C_2H_6O_6Zn$

Molecular Weight: 191.456
CAS RN: 5970-62-7
Properties: white cryst; has been used as a catalyst for the production of methanol, and as a waterproofing agent in the textile industry [HAW93]
Solubility: 5.2g anhydrous/100g H_2O (20°C); i alcohol [MER89]
Density, g/cm³: 2.207 [MER89]
Reactions: minus $2H_2O$ at 140°C [HAW93]

3250
Compound: Zinc hexafluoroacetylacetonate dihydrate
Synonyms: 1,1,1,5,5,5-hexafluoro-2,4-pentanedione, zinc derivative
Formula: $Zn(CF_3COCHCOCF_3)_2 \cdot 2H_2O$
Molecular Formula: $C_{10}H_6F_{12}O_6Zn$
Molecular Weight: 515.525
CAS RN: 16743-33-2
Properties: white powd [STR93]

$$[CF_3-\overset{\overset{\displaystyle O-}{|}}{C}=CH-\overset{\overset{\displaystyle O}{\|}}{C}-CF_3]_2Zn$$

Melting Point, °C: 157-158 [STR93]

3251
Compound: Zinc hexafluorosilicate hexahydrate
Formula: $ZnSiF_6 \cdot 6H_2O$
Molecular Formula: $F_6H_{12}O_6SiZn$
Molecular Weight: 315.557
CAS RN: 16871-71-9
Properties: white cryst; uses: harden concrete, preservative [HAW93] [MER89]
Solubility: s H_2O [MER89]
Density, g/cm³: 2.104 [CRC94]
Reactions: decomposes at 100°C [CRC94]

3252
Compound: Zinc hydroxide
Formula: $Zn(OH)_2$
Molecular Formula: H_2O_2Zn
Molecular Weight: 99.405
CAS RN: 20427-58-1
Properties: colorless cryst; used as an absorbent in surgical dressings and in rubber compounding [HAW93]
Solubility: sl s H_2O [HAW93]
Density, g/cm³: 3.053 [HAW93]
Melting Point, °C: 125, decomposes [HAW93]

3253
Compound: Zinc hypophosphite monohydrate
Formula: $Zn(H_2PO_2)_2 \cdot H_2O$
Molecular Formula: $H_6O_5P_2Zn$
Molecular Weight: 213.383
CAS RN: 7783-14-4
Properties: white cryst; hygr [HAW93]
Solubility: s H_2O and alkalies [HAW93]

3254
Compound: Zinc iodate
Formula: $Zn(IO_3)_2$
Molecular Formula: I_2O_6Zn
Molecular Weight: 415.195
CAS RN: 7790-37-6
Properties: white, cryst powd [MER89]
Solubility: g/100g soln, H_2O: 0.622 (25°C) [KRU93]; s in 77 parts hot H_2O [MER89]
Density, g/cm³: 5.063 [CRC94]
Reactions: decomposes on heating [CRC94]

3255
Compound: Zinc iodide
Formula: ZnI_2
Molecular Formula: I_2Zn
Molecular Weight: 319.199
CAS RN: 10139-47-6
Properties: white or almost white; hygr cryst, granular powd; sensitive to air and light, turning brown due to iodine; used as topical antiseptic [HAW93] [MER89]
Solubility: g/100g soln, H_2O: 81.11 (0°C), 81.20 (18°C), 83.62 (100°C); Solid phase, ZnI_2 [KRU93]; 1 g/2mL glycerol; v s alcohol, ether [MER89]
Density, g/cm³: 4.74 [MER89]
Melting Point, °C: ~446 [MER89]
Boiling Point, °C: ~625, decomposing [MER89]

3256
Compound: Zinc laurate
Synonyms: lauric acid, zinc salt
Formula: $Zn(C_{12}H_{23}O_2)_2$
Molecular Formula: $C_{24}H_{46}O_4Zn$
Molecular Weight: 464.017
CAS RN: 2452-01-9
Properties: white powd; used in paints and varnishes, in rubber compounding [HAW93]
Solubility: 0.01 g/100mL H_2O (15°C), 0.019 g/100mL H_2O (100°C) [CRC94]
Melting Point, °C: 128 [HAW93]

3257

Compound: Zinc molybdate
Formula: $ZnMoO_4$
Molecular Formula: MoO_4Zn
Molecular Weight: 225.328
CAS RN: 13767-32-3
Properties: white tetr; can be prepared from the two metal oxides [KIR81]
Solubility: 0.5 g/100g H_2O [KIR81]
Density, g/cm³: 4.3 [LID94]
Melting Point, °C: >700 [KIR81]

3258

Compound: Zinc nitrate hexahydrate
Formula: $Zn(NO_3)_2 \cdot 6H_2O$
Molecular Formula: $H_{12}N_2O_{12}Zn$
Molecular Weight: 297.491
CAS RN: 10196-18-6
Properties: colorless cryst or lumps; used as catalyst, and in latex coagulation [HAW93]
Solubility: g/100g soln in H_2O: 48.3 (0.4°C), 56.1 (25.1°C), 90.0 (70.7°C); Solid phase: $Zn(NO_3)_2 \cdot 6H_2O$ (0°C,25.1°C), $Zn(NO_3)_2 \cdot H_2O$ (70.7°C) [KRU93]
Density, g/cm³: 2.065 [MER89]
Melting Point, °C: 36.4 [STR93]
Reactions: minus $6H_2O$ between 105 and 131°C [HAW93]

3259

Compound: Zinc nitride
Formula: Zn_3N_2
Molecular Formula: N_2Zn_3
Molecular Weight: 224.183
CAS RN: 1313-49-1
Properties: -200 mesh with 99.9% purity; bluish gray, cryst material; a = 0.972 nm [CER91] [MER89] [CIC73]
Density, g/cm³: 6.22 [CRC94]

3260

Compound: Zinc nitrite
Formula: $Zn(NO_2)_2$
Molecular Formula: N_2O_4Zn
Molecular Weight: 157.401
CAS RN: 10102-02-0
Properties: rapidly hydrolyzes in water; prepared by reaction of sodium nitrite and zinc sulfate in alcohol [MER89]

3261

Compound: Zinc oleate
Formula: $Zn(C_{18}H_{33}O_2)_2$
Molecular Formula: $C_{36}H_{66}O_4Zn$
Molecular Weight: 628.308
CAS RN: 557-07-3
Properties: white to tan, greasy powd; used in paints, resins and varnishes [HAW93]
Solubility: i H_2O; s alcohol, ether, CS_2, benzene, petroleum ether [MER89]
Melting Point, °C: 70 [HAW93]

3262

Compound: Zinc oxalate dihydrate
Formula: $ZnC_2O_4 \cdot 2H_2O$
Molecular Formula: $C_2H_4O_6Zn$
Molecular Weight: 189.440
CAS RN: 547-68-2
Properties: white powd; used in organic synthesis [HAW93]
Solubility: g/L soln, H_2O: 0.018 (0°C), 0.0256 (25°C) [KRU93]; s dil mineral acids, ammonia solutions [MER89]
Density, g/cm³: 2.562 [HAW93]
Melting Point, °C: 100, decomposes [HAW93]

3263

Compound: Zinc oxide
Synonyms: zincite
Formula: ZnO
Molecular Formula: OZn
Molecular Weight: 81.389
CAS RN: 1314-13-2
Properties: white or yellowish white powd, hex cryst or 99.9% sintered tablets; sublimes at normal pressure; absorbs atm CO_2; has greatest UV absorption of all commercial pigments; enthalpy of fusion 52.3 kJ/mol; used in ointments, pigments, as UV absorber in plastics, in ceramics and sintered tablets used as an evaporation material for sensors, also as 99.999% and 99.9% pure sputtering target for dielectric, varistors and gas sensors [HAW93] [MER89] [CRC93]
Solubility: 0.00042 g/100g H_2O at 18°C; s dil acetic acid, dil mineral acids, ammonia solutions, alkali hydroxides [MER89] [KIR84]; see also [ZIE92b]
Density, g/cm³: 5.607 [MER89]
Melting Point, °C: ~1975 sublimes [KIR84]
Thermal Conductivity, W/(m·K): 25.2 [KIR84]
Thermal Expansion Coefficient: 4.0×10^{-6}/°C [KIR84]

3264
Compound: Zinc perchlorate hexahydrate
Formula: $Zn(ClO_4)_2 \cdot 6H_2O$
Molecular Formula: $Cl_2H_{12}O_{14}Zn$
Molecular Weight: 372.382
CAS RN: 10025-64-6
Properties: white rhomb deliq cryst [CRC94]
[STR93]
Solubility: mol/kg H_2O: 3.97 (0°C), 4.30 (25°C),
4.74 (50°C); Solid phase: $Zn(ClO_4)_2 \cdot 7H_2O$
(0°C), $Zn(ClO_4)_2 \cdot 6H_2O$ (25°C,50°C) [KRU93]
Density, g/cm³: 2.252 [STR93]
Melting Point, °C: 106 [MER89]
Boiling Point, °C: 200, decomposes [STR93]

3265
Compound: Zinc permanganate hexahydrate
Formula: $Zn(MnO_4)_2 \cdot 6H_2O$
Molecular Formula: $H_{12}Mn_2O_{14}Zn$
Molecular Weight: 411.353
CAS RN: 23414-72-4
Properties: brownish violet or almost black; deliq
cryst; light sensitive, decomposes; used as a
curative for rubber and elastomers [MER89]
[HAW93]
Solubility: s 3 parts H_2O; decomposed by alcohol
[MER89]
Density, g/cm³: 2.47 [CRC94]
Reactions: minus $5H_2O$ at 100°C [CRC94]

3266
Compound: Zinc peroxide
Formula: ZnO_2
Molecular Formula: O_2Zn
Molecular Weight: 97.389
CAS RN: 1314-22-3
Properties: white powd; decomposes rapidly at
temperatures >150°C; used as curative for
rubber and elastomers, and in high temp
oxidation [HAW93]
Solubility: i H_2O, and decomposed; decomposed by
acids, alcohol and acetone [HAW93]
Density, g/cm³: 1.57 [KIR84]
Melting Point, °C: 212, explodes [KIR84]

3267
Compound: Zinc phosphate
Synonyms: zinc orthophosphate
Formula: $Zn_3(PO_4)_2$
Molecular Formula: $O_8P_2Zn_3$
Molecular Weight: 385.513
CAS RN: 7779-90-0

Properties: colorless rhomb powd [CRC94]
[AES93]
Density, g/cm³: 3.99 [CRC94]
Melting Point, °C: 900 [CRC94]

3268
Compound: Zinc phosphate tetrahydrate
Formula: $Zn_3(PO_4)_2 \cdot 4H_2O$
Molecular Formula: $H_8O_{12}P_2Zn_3$
Molecular Weight: 458.174
CAS RN: 7543-51-3
Properties: white powd; used in metal coatings and
dental cement; also a dihydrate and an
anhydrous form [AE93] [MER89]
Solubility: i H_2O, alcohol; s dil mineral acids, acetic
acid, ammonia solutions, alkali hydroxide
solutions [MER89]
Density, g/cm³: 3.04 [CRC94]

3269
Compound: Zinc phosphide
Formula: Zn_3P_2
Molecular Formula: P_2Zn_3
Molecular Weight: 258.118
CAS RN: 1314-84-7
Properties: dark gray; tetr cryst; lustrous or dull
powd; reducing properties, e.g. vigorous
reaction with conc H_2SO_4, HNO_3, other
oxidizing agents; used as a rodenticide; there is
a ZnP_2, 12037-79-5 [CER91] [MER89]
[HAW93]
Solubility: decomposed by H_2O; i alcohol; s
benzene, CS_2 [MER89] [HAW93]
Density, g/cm³: 4.55 [MER89]
Melting Point, °C: 420 [MER89]
Boiling Point, °C: 1100 [MER89]
Reactions: reacts with HCl and H_2SO_4 evolving PH_3
[MER89]

3270
Compound: Zinc propionate
Synonyms: propionic acid, zinc salt
Formula: $Zn(CH_3CH_2COO)_2$
Molecular Formula: $C_6H_{10}O_4Zn$
Molecular Weight: 211.533
CAS RN: 557-28-8
Properties: plate or tablet form; sensitive to
moisture; prepared by dissolution of ZnO in dil
propionic acid; used as a fungicide [MER89]
Solubility: solubility in H_2O, w/w%: 32 (15°C)
[MER89]

Reactions: evolves propionic acid in moist air [MER89]

3271
Compound: Zinc pyrophosphate
Formula: $Zn_2P_2O_7$
Molecular Formula: $O_7P_2Zn_2$
Molecular Weight: 304.723
CAS RN: 7446-26-6
Properties: white, cryst powd; used as a pigment [HAW93] [MER89]
Solubility: i H_2O; s dil mineral acids [MER89]
Density, g/cm³: 3.75 [MER89]

3272
Compound: Zinc salicylate trihydrate
Formula: $Zn(C_7H_5O_3)_2 \cdot 3H_2O$
Molecular Formula: $C_{14}H_{16}O_9Zn$
Molecular Weight: 393.666
CAS RN: 16283-36-6
Properties: white needles or cryst powd; used as an antiseptic [HAW93] [MER89]
Solubility: 5 g/100mL H_2O (20°C) [CRC94], s alcohol [MER89]

3273
Compound: Zinc selenate pentahydrate
Formula: $ZnSeO_4 \cdot 5H_2O$
Molecular Formula: $H_{10}O_9SeZn$
Molecular Weight: 298.424
CAS RN: 13597-54-1
Properties: white tricl cryst [CRC94] [MER89]
Solubility: g/100g H_2O: 49.37 (0°C), 60.87 (22°C), 46.22 (98.5°C); Solid phase: $ZnSeO_4 \cdot 6H_2O$ (0°C,22°C), $ZnSeO_4$ (98.5°C) [KRU93]
Density, g/cm³: 2.591 [MER89]
Melting Point, °C: decomposes >50 [MER89]

3274
Compound: Zinc selenide
Synonyms: stilleite
Formula: ZnSe
Molecular Formula: SeZn
Molecular Weight: 144.350
CAS RN: 1315-09-9
Properties: yellowish to reddish powd, and 99.999% pure melted pieces of 3-6 mm and 1-3 mm; cub cryst; decomposes in air; used in infrared instruments, and as an evaporation material for deposition of multilayers, laser mirrors, photoconductive and infrared films and filters, also used as a sputtering target [HAW93] [MER89] [CER91]
Solubility: i H_2O; s dil HNO_3 [MER89]
Density, g/cm³: 5.42 [CRC93]
Melting Point, °C: 1517 [CRC93]
Thermal Conductivity, W/(m·K): 14 (25°C) [CRC93]
Thermal Expansion Coefficient: (volume) 100°C (0.151), 200°C (0.380), 400°C (0.902), 800°C (2.059) [CLA66]

3275
Compound: Zinc selenite
Formula: $ZnSeO_3$
Molecular Formula: O_3SeZn
Molecular Weight: 192.348
CAS RN: 13597-46-1
Properties: white powd [AES93]

3276
Compound: Zinc silicate
Synonyms: zinc orthosilicate
Formula: Zn_2SiO_4
Molecular Formula: O_4SiZn_2
Molecular Weight: 222.864
CAS RN: 13597-65-4
Properties: white powd; occurs as mineral willemite; can be prepared by heating ZnO and SiO_2 ~1200°C; used in television screens [MER89]
Solubility: i H_2O, dil acids [MER89]
Density, g/cm³: 4.103 [HAW93]
Melting Point, °C: 1509 [HAW93]

3277
Compound: Zinc stearate
Formula: $Zn[CH_3(CH_2)_{16}COO]_2$
Molecular Formula: $C_{36}H_{70}O_4Zn$
Molecular Weight: 632.340
CAS RN: 557-05-1
Properties: white; hydrophobic; fine, soft, bulky powd; slight odor; used in cosmetics, lacquers, and ointments [MER89] [HAW93]
Solubility: i H_2O, alcohol, ether; s benzene; decomposed by dil acids [MER89]
Density, g/cm³: 1.095 [HAW93]

Melting Point, °C: 130 [HAW93]

3278
Compound: Zinc sulfate
Synonyms: zinkosite
Formula: ZnSO$_4$
Molecular Formula: O$_4$SZn
Molecular Weight: 161.454
CAS RN: 7733-02-0
Properties: used in zinc plating, and as a mordant [KIR84]
Solubility: g/100g H$_2$O: 28.58 (0°C), 36.67 (25°C), 37.7 (100°C); Solid phase, ZnSO$_4$·7H$_2$O (0°C, 25°C), ZnSO$_4$·H$_2$O (100°C) [KRU93]
Density, g/cm^3: 3.54 [KIR84]
Melting Point, °C: 680 decomposes [KIR84]

3279
Compound: Zinc sulfate heptahydrate
Synonyms: goslarite
Formula: ZnSO$_4$·7H$_2$O
Molecular Formula: H$_{14}$O$_{11}$SZn
Molecular Weight: 287.560
CAS RN: 7446-20-0
Properties: colorless; cryst, granules, or powd; effloresces in dry atm; used in rayon manufacture, in animal feeds, and as a wood preservative [MER89] [HAW93]
Solubility: 1 g soluble in: 0.6mL H$_2$O, 2.5mL glycerol; i alcohol [MER89]
Density, g/cm^3: 1.97 [MER89]
Melting Point, °C: 100 [MER89]
Reactions: minus 7H$_2$O at 280°C [MER89]

3280
Compound: Zinc sulfate hexahydrate
Formula: ZnSO$_4$·6H$_2$O
Molecular Formula: H$_{12}$O$_{10}$SZn
Molecular Weight: 269.545
CAS RN: 13986-24-8
Properties: colorless monocl; formed from heptahydrate >39°C by loss of H$_2$O [KIR84] [CRC94]
Solubility: 47.7% in H$_2$O at 70°C [KIR84]
Density, g/cm^3: 2.072 [CRC94]
Reactions: minus 5H$_2$O at 70°C, minus 6H$_2$O at 238°C [KIR84] [CRC94]

3281
Compound: Zinc sulfate monohydrate
Formula: ZnSO$_4$·H$_2$O

Molecular Formula: H$_2$O$_5$SZn
Molecular Weight: 179.469
CAS RN: 7446-19-7
Properties: white powd or granules; used in rayon manufacture, agricultural sprays, dyestuffs [MER89] [STR93]
Solubility: 101 g/100g H$_2$O at 70°C, 87 g/100g H$_2$O at 105°C; i alcohol [MER89] [KIR84]
Density, g/cm^3: 3.28 [KIR84]
Melting Point, °C: 238 decomposes [KIR84]
Reactions: minus H$_2$O >238°C [MER89]

3282
Compound: Zinc sulfide(α)
Synonyms: wurtzite
Formula: α-ZnS
Molecular Formula: SZn
Molecular Weight: 97.456
CAS RN: 1314-98-3
Properties: hex; white to grayish white or yellowish powd, and 99.99% pure highly dense pressure sintered cubes and pieces of 9 mm and 3-12 mm; can slowly oxidize in air; used in phosphors, as a white pigment and in dental materials, and as an evaporation and sputtering material for multilayers, high index film used in nonabsorbing beam splitters [KIR84] [MER89] [CER91]
Solubility: 0.0007 g/100g H$_2$O at 18°C; i alcohol; s dil mineral acids [MER89] [KIR84]
Density, g/cm^3: 4.087 [MER89]
Melting Point, °C: 1700 [STR93]
Thermal Conductivity, W/(m·K): 25.1 [CRC93]
Thermal Expansion Coefficient: (volume) 100°C (0.163), 200°C (0.395), 400°C (0.919), 800°C (2.146), 1000°C (2.839) [CLA66]

3283
Compound: Zinc sulfide(β)
Synonyms: sphalerite
Formula: β-ZnS
Molecular Formula: SZn
Molecular Weight: 97.456
CAS RN: 1314-98-3
Properties: cub; white to grayish white or yellow powd; can slowly oxidize in air to sulfate [MER89]
Solubility: i H$_2$O, alcohol, s dil mineral acids [MER89]
Density, g/cm^3: 4.102 [MER89]
Melting Point, °C: 1700 [STR93]
Thermal Expansion Coefficient: (volume) 100°C (0.156), 200°C (0.386), 400°C (0.898), 800°C (1.996), 1000°C (2.559) [CLA66]

3284
Compound: Zinc sulfite dihydrate
Formula: $ZnSO_3 \cdot 2H_2O$
Molecular Formula: H_4O_5SZn
Molecular Weight: 181.485
CAS RN: 7488-52-0
Properties: white, cryst powd; absorbs atm oxygen, and is oxidized to sulfate; used as a preservative for anatomical specimens [HAW93]
Solubility: 0.01733 mol/kg H_2O (25°C) [KRU93]; decomposed by hot H_2O; s H_2SO_3 [HAW93]
Melting Point, °C: 200, decomposes [HAW93]
Reactions: minus $2H_2O$ at 100°C [HAW93]

3285
Compound: Zinc tartrate dihydrate
Formula: $Zn(C_4H_4O_6)_2 \cdot 2H_2O$
Molecular Formula: $C_8H_{12}O_{14}Zn$
Molecular Weight: 397.565
CAS RN: 551-64-4
Properties: white cryst powd [CRC94] [MER89]
Solubility: g anhydrous/100g H_2O: 0.022 (20°C), 0.041 (30°C), 0.060 (40°C), 0.059 (80°C) [LAN85]

3286
Compound: Zinc telluride
Formula: ZnTe
Molecular Formula: TeZn
Molecular Weight: 192.990
CAS RN: 1315-11-3
Properties: gray or brownish red powd, or 99.999% pure melted pieces 3-12 mm; cub ruby red cryst; stable in dry air; used in semiconductor work, and as an evaporation material and sputtering target to prepare thermionic power generators [HAW93] [MER89] [CER91]
Solubility: i H_2O, prolonged contact with H_2O or dil HCl evolves H_2, H_2Te gases [MER89]
Density, g/cm³: 5.9 [LID94]
Melting Point, °C: 1239 [MER89]
Thermal Conductivity, W/(m·K): 10.8 [CRC93]
Thermal Expansion Coefficient: 6.6×10^{-6}/K [CRC93]

3287
Compound: Zinc thiocyanate
Formula: $Zn(SCN)_2$
Molecular Formula: $C_2N_2S_2Zn$
Molecular Weight: 181.557
CAS RN: 557-42-6

Properties: white cryst; deliq; used as a swelling agent for cellulose esters, and to assist dyeing [HAW93] [MER89]
Solubility: mol/L soln, H_2O: 0.144 (18°C) [KRU93]; s alcohol [MER89]

3288
Compound: Zinc titanate
Formula: $ZnTiO_3$
Molecular Formula: O_3TiZn
Molecular Weight: 161.255
CAS RN: 12036-43-0
Properties: -200 mesh with 99.9% purity; off-white powd with spinel structure [KIR83] [STR93] [CER91]
Density, g/cm³: 5.12 [KIR83]

3289
Compound: Zinc valerate dihydrate
Synonyms: valeric acid, zinc salt dihydrate
Formula: $Zn(CH_3(CH_2)_3COO)_2 \cdot 2H_2O$
Molecular Formula: $C_{10}H_{22}O_6Zn$
Molecular Weight: 303.671
CAS RN: 556-38-7
Properties: lustrous scales or powd; odorous; gradually decomposes in air [MER89]
Solubility: 1 g soluble in: 70mL H_2O, 22mL alcohol; decomposed by acid [MER89]

3290
Compound: Zirconium
Formula: α-Zr; β-Zr
Molecular Formula: Zr
Molecular Weight: 91.224
CAS RN: 7440-67-7
Properties: bluish black amorphous powd, or hard, shiny, ductile metal; α: hex, a = 0.3231 nm, c = 0.5146 nm; β-Zr: bcc; enthalpy of fusion 21.0 kJ/mol; enthalpy of vaporization 573.2 kJ/mol; Brinell hardness 90-130; electrical resistivity (5°C) 43.74 μohm·cm; Poisson's ratio 0.33; resistant to corrosion by water and steam; uses include foundry sand, and evaporated film for depositing interference layers [KIR84] [MER89] [CER91] [CRC93]
Solubility: i H_2O, cold acids; s hot, very conc acids [HAW93]
Density, g/cm³: 6.5107 [KIR84]
Melting Point, °C: 1852 [KIR84]
Boiling Point, °C: 4409 [LID94]
Reactions: transition $\alpha \rightarrow \beta$ at 863°C; heat of transition 3.89 kJ/mol [KIR84]

Thermal Conductivity, W/(m·K): 22.6 (25°C) [ALD94]

Thermal Expansion Coefficient: single cryst, perpendicular to c-axis, 1 + 5.145 x 10^{-T}, T = (273.15 + °C) [KIR84]

3291

Compound: Zirconium acetylacetonate

Synonyms: 2,4-pentanedione, zirconium(IV) derivative

Formula: $Zr(CH_3COCH=C(O)CH_3)_4$

Molecular Formula: $C_{20}H_{28}O_8Zr$

Molecular Weight: 487.661

CAS RN: 17501-44-9

Properties: white powd; hygr [ALD94] [STR93]

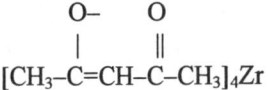

Melting Point, °C: 171-173 [STR93]

3292

Compound: Zirconium aluminide

Formula: $ZrAl_3$

Molecular Formula: Al_3Zr

Molecular Weight: 172.169

CAS RN: 12004-83-0

Properties: 6 mm pieces and smaller of 99.5% purity [CER91]

3293

Compound: Zirconium boride

Synonyms: zirconium diboride

Formula: ZrB_2

Molecular Formula: B_2Zr

Molecular Weight: 112.846

CAS RN: 12045-64-6

Properties: gray metallic cryst or powd; Mohs hardness 8; electrical resistivity 9.2 μohm·cm; excellent thermal shock resistance; hex; considered to have best oxidation resistance of all refractory 'hard metals'; used as a hot pressed crucible for melting metals such as aluminum, bismuth, brass, manganese and tin, and as a container for acidic and basic slags and for cryolite, and as a 99.5% pure sputtering target, make films to increase cutting tool life [HAW93] [KIR84] [CER91]

Density, g/cm³: 6.085 [HAW93]

Melting Point, °C: 3245 [KIR84]

3294

Compound: Zirconium bromide

Formula: $ZrBr_4$

Molecular Formula: Br_4Zr

Molecular Weight: 410.840

CAS RN: 13777-25-8

Properties: -40 mesh with 99.8% purity; off-white powd; cub, a = 1.095 nm [KIR84] [STR93] [CER91]

Density, g/cm³: 3.98 [LID94]

Melting Point, °C: 450 [KIR84]

Reactions: sublimes 357°C [KIR84]

3295

Compound: Zirconium carbide

Formula: ZrC

Molecular Formula: CZr

Molecular Weight: 103.235

CAS RN: 12020-14-3

Properties: dark gray brittle solid; cub, a = 0.46983 nm; it is not a fixed stoichiometric compound; hard, high melting carbide used in UC-fueled reactors; enthalpy of fusion 79.4 kJ/mol; hardness 8+ Mohs; used as a hot pressed crucible to melt metals such as bismuth, cadmium, lead, tin and high melting point oxides such as ZrO_2, and as a 99.5% pure sputtering target to produce wear-resistant and other films [KIR84] [HAW93] [CER91] [JAN71]

Solubility: s HF solutions which contain NO_3^- or peroxide [KIR84]

Density, g/cm³: 6.73 [STR93]

Melting Point, °C: 3540 [STR93]

Boiling Point, °C: 5100 [STR93]

Reactions: forms tetrahalides with halogens above 250°C [KIR84]

Thermal Expansion Coefficient: (volume) 100°C (0.141), 200°C (0.326), 400°C (0.711), 800°C (1.509), 1200°C (2.344) [CLA66]

3296

Compound: Zirconium carbonate basic hydrate

Formula: $3ZrO_2 \cdot CO_2 \cdot xH_2O$

Molecular Formula: CO_8Zr_3 (anhydrous)

Molecular Weight: 413.678 (anhydrous)

CAS RN: 12671-00-0

Properties: white powd [STR93]

Solubility: i H_2O; s acids [HAW93]

3297

Compound: Zirconium chloride

Formula: $ZrCl_4$
Molecular Formula: Cl_4Zr
Molecular Weight: 233.035
CAS RN: 10026-11-6
Properties: lustrous white; monocl cryst, a = 0.6361 nm, b = 0.7404 nm, c = 0.6256 nm; quickly hygr, fuming in moist air, reacting to emit HCl vapor; enthalpy of fusion 50.00 kJ/mol [CRC93] [KIR84]
Solubility: instantly hydrolyzed in H_2O to $ZrOCl_2 \cdot 8H_2O$, and HCl; s alcohol, ether [KIR84], [MER89]
Density, g/cm³: 2.803 [STR93]
Melting Point, °C: 437 [AES93]
Reactions: sublimes 331°C [KIR84]

3298
Compound: Zirconium cyclopentadienyl trichloride
Synonyms: cyclopentadienylzirconium trichloride
Formula: $Zr(C_5H_5)Cl_3$
Molecular Formula: $C_5H_5Cl_3Zr$
Molecular Weight: 262.677
CAS RN: 34767-44-7
Properties: off-white powd; sensitive to moisture [STR93]
Melting Point, °C: 237, decomposes [STR93]

3299
Compound: Zirconium fluoride
Formula: ZrF_4
Molecular Formula: F_4Zr
Molecular Weight: 167.218
CAS RN: 7783-64-4
Properties: white; monocl, a = 0.957 nm, b = 0.993 nm, c = 0.773 nm; enthalpy of fusion 64.20 kJ/mol; component of molten salts for use in nuclear reactors [KIR84] [HAW93] [CRC93]
Solubility: 1.32 g/100mL H_2O (20°C); v s HF acid [MER89]
Density, g/cm³: 4.43 [KIR84]
Melting Point, °C: 932 [KIR84]
Reactions: sublimes 903°C [KIR84]

3300
Compound: Zirconium hexafluoroacetylacetonate
Synonyms: 1,1,1,5,5,5-hexafluoro-2,4-pentanedione Zr derivative
Formula: $Zr(CF_3COCHCOCF_3)_4$
Molecular Formula: $C_{20}H_4F_{24}O_8Zr$
Molecular Weight: 919.433
CAS RN: 19530-02-0
Properties: white to off-white cryst; hygr [STR93]

$$[CF_3-\overset{\overset{\displaystyle O-}{\displaystyle |}}{C}=CH-\overset{\overset{\displaystyle O}{\displaystyle \|}}{C}-CF_3]_4Zr$$

Melting Point, °C: 41-43 [STR93]
Boiling Point, °C: 225 [STR93]

3301
Compound: Zirconium hydride
Formula: ZrH_2
Molecular Formula: H_2Zr
Molecular Weight: 93.240
CAS RN: 7704-99-6
Properties: grayish black metallic powd; strong reducing agent; absorbs and desorbs hydrogen reversibly; enthalpy of formation 167.4 kJ/mol; does not have fixed stoichiometry; most commercial hydride powd contains δ and ε phases with compositions $ZrH_{1.5}$ to $ZrH_{1.7}$ and $ZrH_{1.8}$ to ZrH_2, respectively; stable in air [MER89] [KIR84] [HAW93] [KIR80]
Solubility: no reaction with H_2O [MER89]
Density, g/cm³: 5.6 [HAW93]
Reactions: dissociates between 300 to 700°C [KIR80]

3302
Compound: Zirconium hydroxide
Formula: $Zr(OH)_4$
Molecular Formula: H_4O_4Zr
Molecular Weight: 159.254
CAS RN: 14475-63-9
Properties: white, bulky, amorphous powd [MER89]
Solubility: i H_2O and alkalies; s mineral acids [MER89] [HAW93]
Density, g/cm³: 3.25 [MER89]
Reactions: minus $2H_2O$ at 500°C [CRC94]

3303
Compound: Zirconium iodide
Synonyms: zirconium tetraiodide
Formula: ZrI_4
Molecular Formula: I_4Zr
Molecular Weight: 598.842
CAS RN: 13986-26-0
Properties: orange cryst; fumes heavily in air; cub, a = 1.179 nm [KIR84] [MER89]
Solubility: dissolves in H_2O, evolving steam [MER89]
Density, g/cm³: 4.85 [KIR78]

Melting Point, °C: 499 (high pressure) [MER89]
Boiling Point, °C: sublimes 431 [MER89]

3304
Compound: Zirconium nitrate pentahydrate
Formula: $Zr(NO_3)_4 \cdot 5H_2O$
Molecular Formula: $H_{10}N_4O_{17}Zr$
Molecular Weight: 429.320
CAS RN: 13746-89-9
Properties: white, very hygr cryst [MER89]
Solubility: v s H_2O; s alcohol [MER89]
Melting Point, °C: decomposes 100 [HAW93]

3305
Compound: Zirconium nitride
Formula: ZrN
Molecular Formula: NZr
Molecular Weight: 105.231
CAS RN: 25658-42-8
Properties: brittle, yellow solid; fcc, a = 0.4577 nm; electrical resistivity 21 µohm·cm; hardness 8+ Mohs; transition temp -264°C; can be obtained by reacting NH_3 with a zirconium halide; used as a hot pressed crucible for melting metals such as aluminum, bismuth, cadmium, lead, steel, tin and as a container for acid and basic slags and cryolite, also used as a 99.5% pure sputtering target to produce films [KIR81] [CIC73] [KIR84] [CER91]
Solubility: s conc HF, slowly in hot conc H_2SO_4 [KIR84]; sl s dil HCl and H_2SO_4 [HAW93]
Density, g/cm³: 7.09 [STR93]
Melting Point, °C: 2980 [CIC73]
Reactions: oxidizes to ZrO_2 above 700°C in air [KIR84]
Thermal Conductivity, W/(m·K): 10.9 [KIR81]
Thermal Expansion Coefficient: 7.24×10^{-6} [KIR81]

3306
Compound: Zirconium oxide
Synonyms: baddeleyite
Formula: ZrO_2
Molecular Formula: O_2Zr
Molecular Weight: 123.223
CAS RN: 12036-23-6
Properties: baddeleyite mineral: white, heavy, amorphous powd, or monocl cryst: stable up to ~1100°C, transition to tetr form as temp increases to 1200°C; hardness 6.5 Mohs; enthalpy of fusion 87.00 kJ/mol [CRC93] [MER89] [KIR84]

Solubility: i H_2O; sl s HCl, HNO_3; s in heated mixture of 2 part H_2SO_4, 1 part H_2O [MER89]
Density, g/cm³: 5.85 [MER89]
Melting Point, °C: 2710 [KIR84]
Boiling Point, °C: 4300 [MER89]
Thermal Expansion Coefficient: (volume) 100°C (0.209), 200°C (0.423), 400°C (0.854), 800°C (1.688), 1000°C (2.267) [CLA66]

3307
Compound: Zirconium oxide
Synonyms: zirconia
Formula: ZrO_2
Molecular Formula: O_2Zr
Molecular Weight: 123.223
CAS RN: 1314-23-4
Properties: heavy, white amorphous powd, or 99.7% pure 3-12 mm white sintered pieces; most heat resistant of commercial refractories; four phases: monocl, tetr, ortho-rhomb, cub; monocl stable up to ~1100°C; generally stable to most reagents; used as a crucible for melting metals such as hafnium, iridium, platinum, iron, plutonium, zirconium and vanadium; as 99.7% pure sputtering target produces adherent high index coating and dielectric coating [HAW93] [KIR84] [CER91]
Solubility: i H_2O and most acids and alkalies at room temp [HAW93]; slowly dissolves in hot conc HF, hot conc H_2SO_4 [KIR84]
Density, g/cm³: 5.68 [LID94]
Melting Point, °C: 2710 [KIR84]
Reactions: transforms from monocl to tetr at 1200°C; tetr to monocl between 1000 and 850°C; tetr to cub transition at 2370°C [KIR84]
Thermal Conductivity, W/(m·K): 2.1 (500°C), 2.3 (1000°C) [KIR80]

3308
Compound: Zirconium oxide yttria stabilized
Formula: ZrO_2 (10-15% Y_2O_3)
Molecular Formula: O_2Zr
Molecular Weight: 123.223 (ZrO_2)
CAS RN: 64417-98-7
Properties: -150, +325 mesh with 99% purity; powd [STR93] [CER91]

3309
Compound: Zirconium phosphate trihydrate
Formula: $ZrO(H_2PO_4)_2 \cdot 3H_2O$
Molecular Formula: $H_{10}O_{12}P_2Zr$
Molecular Weight: 355.243
CAS RN: 59129-80-5

Properties: white, dense, amorphous powd; decomposes when heated; extensively hydrolyzed in basic solutions [HAW93]
Solubility: i H_2O; s acids [HAW93]

3310
Compound: Zirconium phosphide
Formula: ZrP_2
Molecular Formula: P_2Zr
Molecular Weight: 153.172
CAS RN: 12037-80-8
Properties: gray, brittle; solubility of phosphorus in zirconium metal is low, ~50 ppm; at higher concentrations phosphorus collects as separate globules at the metal grain boundaries, possibly in the form of Zr_3P; other phosphides include ZrP_3, $ZrP_{0.6}$ [KIR84] [CRC94]
Density, g/cm³: 4.77 [CRC94]

3311
Compound: Zirconium pyrophosphate
Formula: ZrP_2O_7
Molecular Formula: O_7P_2Zr
Molecular Weight: 265.165
CAS RN: 13565-97-4
Properties: white solid; used as a refractory, phosphor and olefin polymerization catalyst [HAW93]
Solubility: i H_2O and dil acids; s HF [HAW93]
Melting Point, °C: decomposes 1550 [HAW93]
Thermal Expansion Coefficient: 5×10^{-6} at 1000°C [HAW93]

3312
Compound: Zirconium selenide
Formula: $ZrSe_2$
Molecular Formula: Se_2Zr
Molecular Weight: 249.144
CAS RN: 12166-47-1
Properties: -325 mesh powd [ALF95]

3313
Compound: Zirconium silicate
Synonyms: zircon
Formula: $ZrSiO_4$
Molecular Formula: O_4SiZr
Molecular Weight: 183.308
CAS RN: 10101-52-7

Properties: white powd; also mineral; colorless, unless contains impurities, can then be brown, gray, red; hardness 7.5; tetr cryst, a = 0.6607 nm, c = 0.5982 nm; dissociates >1540°C into ZrO_2 and SiO_2; highly stable; can be prepared by reacting ZrO_2 with SiO_2 in an arc furnace; sol-gel synthesis by reacting $ZrO(NO_3)_2 \cdot xH_2O$ with tetraethylorthosilicate; used in refractories, ceramics, glazes, and as a gem stone [KIR84] [HAW93] [SUB90] [SCH88]
Solubility: very inert to most solvents [MER89]; i acids [HAW93]
Density, g/cm³: 4.56 [STR93]
Melting Point, °C: 2550 [STR93]
Reactions: transforms to scheelite type structure at 900°C and 12GPa [KIR84]
Thermal Conductivity, W/(m·K): 4.3 (500°C), 4.1 (1000°C) [KIR80]
Thermal Expansion Coefficient: (1000°C) 4.1×10^{-6}/°C [SUB90]

3314
Compound: Zirconium silicide
Formula: $ZrSi_2$
Molecular Formula: Si_2Zr
Molecular Weight: 147.396
CAS RN: 12039-90-6
Properties: gray solid powd; in the form of a 99.5% pure material, used as a sputtering target to produce resistant semiconducting films in the fabrication of integrated circuits [ALF93] [HAW93] [CER91]
Solubility: i H_2O, aqua regia; s HF [HAW93]
Density, g/cm³: 4.88 (22°C) [HAW93]
Melting Point, °C: 1620 [LID94]

3315
Compound: Zirconium sulfate tetrahydrate
Formula: $Zr(SO_4)_2 \cdot 4H_2O$
Molecular Formula: $H_8O_{12}S_2Zr$
Molecular Weight: 355.413
CAS RN: 14644-61-2
Properties: white cryst [STR93]
Solubility: 52.5 g/100g soln, H_2O (18°C); deposits solid with elementary composition $4ZrO_2 \cdot 3SO_3 \cdot 15H_2O$, on standing at ambient temp [MER89]
Density, g/cm³: 3.22 [CRC94]
Reactions: minus $3H_2O$ at 100°C; minus $4H_2O$ at 380°C [MER89]

3316
Compound: Zirconium sulfide

Formula: ZrS_2
Molecular Formula: S_2Zr
Molecular Weight: 155.356
CAS RN: 12039-15-5
Properties: reddish brown powd; hex; semiconductor; has layered structure of stacked sandwiches each containing single sheets of metal cations between two sheets of anions [KIR84] [STR93]
Solubility: i H_2O [HAW93]
Density, g/cm³: 3.82 [LID94]
Melting Point, °C: 1480 [LID94]

3317
Compound: Zirconium telluride
Formula: $ZrTe_2$
Molecular Formula: Te_2Zr
Molecular Weight: 346.424
CAS RN: 32321-65-6
Properties: -325 mesh powd [ALF95]

3318
Compound: Zirconium tungstate
Formula: $Zr(WO_4)_2$
Molecular Formula: O_8W_2Zr
Molecular Weight: 586.899
CAS RN: 16853-74-0
Properties: -200 mesh with 99.7% purity; light green powd [STR93] [CER91]

3319
Compound: Zirconocene dichloride
Synonyms: bis(cyclopentadienyl)zirconium chloride
Formula: $Zr(C_5H_5)_2Cl_2$
Molecular Formula: $C_{10}H_{10}Cl_2Zr$
Molecular Weight: 292.339
CAS RN: 1291-32-3
Properties: sensitive to moisture; uses: synthesis of metal complexes and organometallic compounds [ALD94]

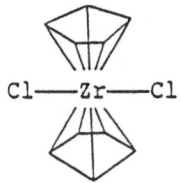

Melting Point, °C: 242-245 [ALD93]

3320
Compound: Zirconyl acetate hydroxide
Formula: $Zr(CH_3COO)_2(OH)_2$
Molecular Formula: $C_4H_8O_6Zr$
Molecular Weight: 243.328
CAS RN: 14311-93-4
Properties: 22% soln; solidifies when heated under reduced pressure; stable at room temp; resinous in appearance; amorphous; highly polymerized material; formula is an approximation [HAW93] [MER89]
Solubility: s H_2O, hydrolyzes when heated [MER89]
Density, g/cm³: 1.46 [MER89]
Melting Point, °C: -7 [HAW93]

3321
Compound: Zirconyl basic nitrate
Formula: $ZrO(OH)NO_3$
Molecular Formula: HNO_5Zr
Molecular Weight: 186.236
CAS RN: 71965-17-8
Properties: aq solution [HAW93]
Density, g/cm³: aq solution, 1.35 (25°C) [HAW93]

3322
Compound: Zirconyl chloride hydrate
Formula: $ZrOCl_2 \cdot xH_2O$
Molecular Formula: Cl_2OZr (anhydrous)
Molecular Weight: 178.129 (anhydrous)
CAS RN: 15461-27-5
Properties: white hygr powd [ALD94] [STR93]
Density, g/cm³: 1.910 [STR93]

3323
Compound: Zirconyl chloride octahydrate
Formula: $ZrOCl_2 \cdot 8H_2O$
Molecular Formula: $Cl_2H_{16}O_9Zr$
Molecular Weight: 322.251
CAS RN: 13520-92-8
Properties: tetr cryst when prepared from water [MER89]
Solubility: v s H_2O, alcohol [MER89]
Density, g/cm³: 1.91 [MER89]
Melting Point, °C: 400, decomposes [STR93]
Reactions: minus $6H_2O$ at 150°C, minus $8H_2O$ at 210°C [CRC94]

3324
Compound: Zirconyl hydroxychloride hydrate
Formula: $ZrO(OH)Cl \cdot xH_2O$
Molecular Formula: $ClHO_2Zr$ (anhydrous)

Molecular Weight: 159.683 (anhydrous)
CAS RN: 10119-31-0
Properties: colorless or sl amber liq (aq solutions); soluble glass formed when evaporated; reacts with alkalies to form hydrous zirconia; used in pharmaceuticals, deodorants, and as a water repellent for textiles [HAW93]

3325
Compound: Zirconyl nitrate hydrate
Formula: $ZrO(NO_3)_2 \cdot xH_2O$
Molecular Formula: N_2O_7Zr (anhydrous)
Molecular Weight: 231.233 (anhydrous)
CAS RN: 14985-18-3
Properties: -8 mesh soft cryst sl wet with HNO_3; white powd [STR93] [CER91]

3326
Compound: Zirconyl perchlorate octahydrate
Synonyms: zirconium diperchlorate oxide
Formula: $ZrO(ClO_4)_2 \cdot 8H_2O$
Molecular Formula: $Cl_2H_{16}O_{17}Zr$
Molecular Weight: 450.246
CAS RN: 12205-73-1
Properties: white cryst [STR93]

CAS Registry Number Index

56-23-5	678: Carbon tetrachloride	298-14-6	2198: Potassium hydrogen carbonate	544-19-4	975: Copper(II) formate
56-34-8	2898: Tetraethyl-ammonium chloride	301-04-2	1496: Lead acetate	544-60-5	200: Ammonium oleate
60-00-4	1104: Ethylenediamine-tetraacetic acid	302-01-2	1346: Hydrazine	544-92-3	936: Copper(I) cyanide
		304-59-6	2692: Sodium potassium tartrate tetrahydrate	546-46-3	3240: Zinc citrate dihydrate
62-38-4	2074: Phenylmercuric acetate	306-61-6	1742: Magnesium thiocyanate tetrahydrate	546-67-8	1544: Lead tetraacetate
62-54-4	559: Calcium acetate			546-68-9	3008: Titanium isopropoxide
62-76-0	2672: Sodium oxalate	333-20-0	2283: Potassium thiocyanate	546-89-4	1555: Lithium acetate
71-91-0	2897: Tetraethyl-ammonium bromide	353-50-4	683: Carbonyl fluoride	546-93-0	1675: Magnesium carbonate
74-90-8	1365: Hydrogen cyanide	409-21-2	2488: Silicon carbide	547-68-2	3262: Zinc oxalate dihydrate
74-97-5	505: Bromochloromethane	420-56-4	1206: Fluorotrimethyl-silane	551-64-4	3285: Zinc tartrate dihydrate
75-15-0	667: Carbon disulfide				
75-20-7	576: Calcium carbide	460-19-5	1020: Cyanogen	554-13-2	1570: Lithium carbonate
75-44-5	682: Carbonyl chloride	463-58-1	672: Carbon oxysulfide	554-70-1	3031: Triethylphosphine
75-46-7	3032: Trifluoromethane	471-34-1	577: Calcium carbonate	555-31-7	52: Aluminum isopropoxide
75-60-5	507: Cacodylic acid	471-34-1	578: Calcium carbonate		
75-71-8	1043: Dichlorodifluoro-methane	471-34-1	579: Calcium carbonate	555-35-1	56: Aluminum monopalmitate
		497-19-8	2570: Sodium carbonate	555-35-1	69: Aluminum palmitate
75-73-0	679: Carbon tetrafluoride	497-19-8	2571: Sodium carbonate bicarbonate dihydrate	555-75-9	37: Aluminum ethoxide
76-15-3	776: Chloropentafluoro-ethane	504-64-3	674: Carbon suboxide	556-38-7	3289: Zinc valerate dihydrate
78-00-2	2895: Tetraethyl lead	506-61-6	2246: Potassium silver cyanide	557-04-0	1732: Magnesium stearate
78-10-4	2899: Tetraethylortho-silicate	506-64-9	2518: Silver cyanide	557-05-1	3277: Zinc stearate
79-37-8	2045: Oxalyl chloride	506-65-0	1287: Gold(I) cyanide	557-07-3	3261: Zinc oleate
96-10-6	1045: Diethylaluminum chloride	506-66-1	415: Beryllium carbide	557-09-5	3233: Zinc caprylate
		506-68-3	1022: Cyanogen bromide	557-20-0	1046: Diethylzinc
100-56-1	2075: Phenylmercuric chloride	506-77-4	1023: Cyanogen chloride	557-21-1	3241: Zinc cyanide
		506-78-5	1420: Iodine cyanide	557-28-8	3270: Zinc propionate
102-54-5	1162: Ferrocene	506-80-9	666: Carbon diselenide	557-34-6	3220: Zinc acetate
107-32-4	2070: Performic acid	506-87-6	122: Ammonium carbonate	557-41-5	3248: Zinc formate
109-63-7	493: Boron trifluoride etherate	507-16-4	2943: Thionyl bromide	557-42-6	3287: Zinc thiocyanate
124-38-9	665: Carbon dioxide	507-25-5	680: Carbon tetraiodide	558-13-4	677: Carbon tetrabromide
126-45-4	2517: Silver citrate	512-26-5	1514: Lead citrate trihydrate	562-76-5	2271: Potassium tetracyanoplatinate(II)
126-96-5	2588: Sodium diacetate	513-74-6	140: Ammonium dithiocarbamate	562-90-3	2486: Silicon acetate
127-08-2	2132: Potassium acetate			563-63-3	2505: Silver acetate
127-09-3	2553: Sodium acetate	513-77-9	319: Barium carbonate	563-67-7	2377: Rubidium acetate
127-95-7	2201: Potassium hydrogen oxalate hemihydrate	513-78-0	518: Cadmium carbonate	563-68-8	2907: Thallium(I) acetate
		513-79-1	864: Cobalt(II) carbonate	563-71-3	1172: Ferrous carbonate
127-96-8	2280: Potassium tetraoxalate dihydrate	516-02-9	360: Barium oxalate	563-72-4	620: Calcium oxalate
128-00-7	2529: Silver lactate monohydrate	527-09-3	977: Copper(II) gluconate	583-15-3	1830: Mercury(II) benzoate monohydrate
		528-94-9	217: Ammonium salicylate	584-08-7	2144: Potassium carbonate
131-74-8	215: Ammonium picrate	532-31-0	2509: Silver benzoate		
137-30-4	3243: Zinc dimethyldi-thiocarbamate	533-51-7	2533: Silver oxalate	584-09-8	2383: Rubidium carbonate
		534-16-7	2512: Silver carbonate	584-10-1	2277: Potassium tetraiodocadmium dihydrate
139-12-8	17: Aluminum acetate	534-17-8	736: Cesium carbonate		
140-99-8	643: Calcium succinate trihydrate	535-37-5	1292: Gold(III) cyanide trihydrate	589-97-9	2230: Potassium percarbonate monohydrate
141-00-4	544: Cadmium succinate	537-00-8	703: Cerous acetate hemitrihydrate		
142-03-0	33: Aluminum diacetate			590-28-3	2156: Potassium cyanate
142-17-6	619: Calcium oleate	537-01-9	707: Cerous carbonate	591-89-9	2269: Potassium tetracyanomercurate(II)
142-71-2	948: Copper(II) acetate	537-03-1	1482: Lanthanum oxalate hydrate		
142-72-3	1655: Magnesium acetate			592-01-8	591: Calcium cyanide
143-19-1	2668: Sodium oleate	538-17-0	89: Aluminum thiocyanate	592-04-1	1835: Mercury(II) cyanide
143-33-9	2585: Sodium cyanide	540-16-9	961: Copper(II) butyrate monohydrate	592-05-2	1515: Lead cyanide
144-23-0	1687: Magnesium citrate tetradecahydrate			592-06-3	2111: Platinum(II) cyanide
		540-69-2	150: Ammonium formate		
144-55-8	2626: Sodium hydrogen carbonate	540-72-7	2727: Sodium thiocyanate	592-85-8	1857: Mercury(II) thiocyanate
		542-42-7	623: Calcium palmitate		
144-62-7	2043: Oxalic acid	542-62-1	326: Barium cyanide	592-87-0	1546: Lead thiocyanate
146-84-9	2540: Silver picrate monohydrate	542-83-6	523: Cadmium cyanide	593-26-0	203: Ammonium palmitate
		543-80-6	302: Barium acetate		
147-14-8	997: Copper(II) phthalocyanine	543-81-3	407: Beryllium acetate	593-29-3	2250: Potassium stearate
		543-90-8	509: Cadmium acetate	593-74-8	1054: Dimethylmercury
149-44-0	2601: Sodium formaldehyde sulfoxylate	543-94-2	2767: Strontium acetate	593-95-3	681: Carbonyl bromide
		543-94-2	2768: Strontium acetate hemihydrate	594-27-4	2901: Tetramethyltin
151-50-8	2157: Potassium cyanide				
156-62-7	590: Calcium cyanamide	544-17-2	598: Calcium formate	598-54-9	931: Copper(I) acetate

598-62-9	1770: Manganese(II) carbonate	1070-75-3	1569: Lithium carbide
598-63-0	1509: Lead carbonate	1072-35-1	1537: Lead stearate
628-52-4	794: Chromium(II) acetate monohydrate	1111-67-7	946: Copper(I) thiocyanate
628-86-4	1838: Mercury(II) fulminate	1111-71-3	419: Beryllium formate
630-08-0	670: Carbon monoxide	1111-78-0	121: Ammonium carbamate
631-36-7	2896: Tetraethyl silane	1113-38-8	201: Ammonium oxalate
631-60-7	1812: Mercury(I) acetate	1117-94-8	932: Copper(I) acetylide
631-61-8	112: Ammonium acetate	1120-44-1	989: Copper(II) oleate
631-61-8	170: Ammonium hydrogen acetate	1185-57-5	142: Ammonium ferric citrate
637-12-7	35: Aluminum distearate	1186-49-8	2628: Sodium hydrogen oxalate monohydrate
637-12-7	48: Aluminum hydroxystearate	1191-80-6	1844: Mercury(II) oleate
637-12-7	91: Aluminum tristearate	1271-19-8	3028: Titanocene dichloride
638-39-1	2749: Stannous acetate	1271-27-8	1753: Manganese bis(cyclopentadienyl)
660-60-6	1004: Copper(II) stearate		
685-83-6	435: Bis(diethylamino)-chlorophosphine	1271-28-9	1937: Nickel bis(cyclopentadienyl)
688-37-9	60: Aluminum oleate	1271-55-2	1: Acetylferrocene
764-05-6	1021: Cyanogen azide	1272-23-7	1493: Lanthanum tris(cyclopentadienyl)
813-93-4	444: Bismuth citrate		
814-71-1	652: Calcium thioglycollate trihydrate	1273-81-0	2034: Osmium bis(cyclopentadienyl)
814-87-9	61: Aluminum oxalate monohydrate	1273-98-9	1925: Neodymium tris(cyclopentadienyl)
814-88-0	534: Cadmium oxalate	1277-43-6	921: Cobaltocene
814-89-1	891: Cobalt(II) oxalate	1277-47-0	3146: Vanadocene
814-90-4	800: Chromium(II) oxalate monohydrate	1282-37-7	1164: Ferrocenium tetrafluoroborate
814-91-5	990: Copper(II) oxalate	1287-13-4	434: Bis(cyclopenta-dienyl)ruthenium
814-91-5	991: Copper(II) oxalate hemihydrate	1291-32-3	3319: Zirconocene dichloride
814-93-7	1528: Lead oxalate	1298-54-0	2462: Scandium tris(cyclopentadienyl)
814-94-8	2757: Stannous oxalate		
814-95-9	2796: Strontium oxalate	1298-55-1	2446: Samarium tris(cyclopentadienyl)
814-95-9	2797: Strontium oxalate monohydrate	1299-86-1	26: Aluminum carbide
815-78-1	86: Aluminum tartrate	1301-96-8	2538: Silver peroxide
815-82-7	1008: Copper(II) tartrate trihydrate	1302-01-8	2544: Silver subfluoride
815-85-0	2764: Stannous tartrate	1302-09-6	2542: Silver selenide
822-16-2	2703: Sodium stearate	1302-42-7	2557: Sodium aluminate
865-44-1	1428: Iodine trichloride	1302-52-9	410: Beryllium aluminum silicate
865-52-1	2900: Tetramethyl-germane	1302-74-5	63: Aluminum oxide (α)
866-84-2	2152: Potassium citrate	1302-76-7	77: Aluminum silicate
868-14-4	2207: Potassium hydrogen tartrate	1302-81-4	85: Aluminum sulfide
868-18-8	2639: Sodium hydrogen tartrate monohydrate	1302-82-5	76: Aluminum selenide
		1302-93-8	80: Aluminum silicate
868-18-8	2712: Sodium tartrate dihydrate	1303-00-0	1237: Gallium arsenide
917-61-3	2584: Sodium cyanate	1303-11-3	1394: Indium arsenide
917-69-1	911: Cobalt(III) acetate	1303-28-2	294: Arsenic(V) oxide
940-71-6	2079: Phosphonitrilic chloride trimer	1303-32-8	282: Arsenic(II) sulfide
		1303-33-9	290: Arsenic(III) sulfide
992-98-3	2917: Thallium(I) formate	1303-34-0	296: Arsenic(V) sulfide
993-50-0	1052: Dimethylaminotri-methyltin	1303-35-1	280: Arsenic hemiselenide
		1303-36-2	289: Arsenic(III) selenide
		1303-37-3	295: Arsenic(V) selenide
1002-88-6	855: Cobalt stearate	1303-52-2	1293: Gold(III) hydroxide
1002-89-7	221: Ammonium stearate	1303-58-8	1295: Gold(III) oxide
1066-26-8	2556: Sodium acetylide	1303-60-2	1289: Gold(I) sulfide
1066-30-4	802: Chromium(III) acetate hexahydrate	1303-61-3	1298: Gold(III) sulfide
		1303-62-4	1297: Gold(III) selenide
1066-30-4	804: Chromium(III) acetate monohydrate	1303-86-2	485: Boron oxide
		1303-86-2	486: Boron oxide glass
1066-33-7	173: Ammonium hydrogen carbonate	1303-94-2	1625: Lithium tetraborate pentahydrate
1068-22-0	199: Ammonium O,O-diethyldithiophosphate	1303-96-4	2717: Sodium tetraborate decahydrate
		1304-28-5	362: Barium oxide
1304-29-6	366: Barium peroxide		
1304-39-8	370: Barium selenide		
1304-40-1	375: Barium silicide		
1304-54-7	425: Beryllium nitride		
1304-56-9	427: Beryllium oxide		
1304-76-3	456: Bismuth oxide		
1304-82-1	473: Bismuth telluride		
1304-85-4	449: Bismuth hydroxide nitrate oxide		
1305-62-0	607: Calcium hydroxide		
1305-78-8	622: Calcium oxide		
1305-79-9	628: Calcium peroxide		
1305-84-6	638: Calcium selenide		
1305-99-3	632: Calcium phosphide		
1306-05-4	596: Calcium fluorophosphate		
1306-06-5	631: Calcium phosphate hydroxide		
1306-19-0	536: Cadmium oxide		
1306-23-6	548: Cadmium sulfide		
1306-24-7	541: Cadmium selenide		
1306-25-8	551: Cadmium telluride		
1306-26-9	515: Cadmium borotungstate octadecahydrate		
1306-38-3	687: Ceric oxide		
1307-81-9	2606: Sodium hexachloroosmiate(IV) hydrate		
1307-82-0	2608: Sodium hexachloroplatinate(IV)		
1307-86-4	915: Cobalt(III) hydroxide trihydrate		
1307-96-6	893: Cobalt(II) oxide		
1307-99-9	898: Cobalt(II) selenide		
1308-04-9	917: Cobalt(III) oxide		
1308-06-1	910: Cobalt(II,III) oxide		
1308-14-1	815: Chromium(III) hydroxide trihydrate		
1308-31-2	1176: Ferrous chromite		
1308-38-9	819: Chromium(III) oxide		
1308-56-1	972: Copper(II) ferrous sulfide		
1308-80-1	926: Copper nitride		
1308-85-6	1067: Dysprosium hydroxide		
1308-87-8	1072: Dysprosium oxide		
1308-96-9	1126: Europium(III) oxide		
1309-33-7	1143: Ferric hydroxide		
1309-37-1	1148: Ferric oxide		
1309-42-8	1697: Magnesium hydroxide		
1309-48-4	1712: Magnesium oxide		
1309-60-0	1516: Lead dioxide		
1309-64-4	258: Antimony(III) oxide		
1310-03-8	1519: Lead fluorosilicate dihydrate		
1310-32-3	1190: Ferrous selenide		
1310-43-6	1189: Ferrous phosphide		
1310-52-7	1692: Magnesium germanide		
1310-53-3	1279: Germanium(IV) oxide		
1310-58-3	2208: Potassium hydroxide		
1310-61-8	2205: Potassium hydrogen sulfide hemihydrate		
1310-65-2	1592: Lithium hydroxide		
1310-66-3	1593: Lithium hydroxide monohydrate		
1310-73-2	2641: Sodium hydroxide		
1310-82-3	2394: Rubidium hydroxide		

CAS RN	Entry	CAS RN	Entry	CAS RN	Entry
1310-83-4	2920: Thallium(I) hydroxide	1314-68-7	2358: Rhenium(VII) oxide	1333-83-1	2627: Sodium hydrogen fluoride
1311-10-0	2786: Strontium hydroxide octahydrate	1314-80-3	2103: Phosphorus(V) sulfide	1333-86-4	673: Carbon soot
1311-33-7	2981: Thulium hydroxide	1314-82-5	2102: Phosphorus(V) selenide	1335-26-8	1717: Magnesium peroxide
1311-90-6	214: Ammonium phosphotungstate dihydrate	1314-84-7	3269: Zinc phosphide	1335-31-5	1848: Mercury(II) oxycyanide
1311-93-9	242: Ammonium tungstate pentahydrate	1314-86-9	2091: Phosphorus triselenide	1335-32-6	1504: Lead basic acetate
1312-41-0	1393: Indium antimonide	1314-87-0	1539: Lead sulfide	1336-21-6	185: Ammonium hydroxide
1312-43-2	1411: Indium(III) oxide	1314-91-6	1542: Lead telluride	1341-49-7	175: Ammonium hydrogen fluoride
1312-45-4	1417: Indium(III) telluride	1314-95-0	2763: Stannous sulfide	1343-88-0	1727: Magnesium silicate
1312-46-5	1438: Iridium(III) oxide	1314-96-1	2810: Strontium sulfide	1344-09-8	2701: Sodium silicate
1312-73-8	2252: Potassium sulfide	1314-97-2	2932: Thallium(I) sulfide	1344-28-1	62: Aluminum oxide
1312-73-8	2253: Potassium sulfide pentahydrate	1314-98-3	3282: Zinc sulfide(α)	1344-28-1	64: Aluminum oxide (γ)
1312-74-9	2244: Potassium selenide	1314-98-3	3283: Zinc sulfide(β)	1344-28-1	65: Aluminum oxide (δ)
1312-81-8	1483: Lanthanum oxide	1315-01-1	2748: Stannic sulfide	1344-28-1	66: Aluminum oxide (κ)
1313-04-8	1725: Magnesium selenide	1315-03-3	3143: Vanadium trisulfide	1344-43-0	1787: Manganese(II) oxide
1313-08-2	1731: Magnesium stannide	1315-04-4	274: Antimony(V) sulfide	1344-48-5	1853: Mercury(II) sulfide(α)
1313-13-9	1807: Manganese(IV) oxide	1315-05-5	262: Antimony(III) selenide	1344-48-5	1854: Mercury(II) sulfide(β)
1313-22-0	1792: Manganese(II) selenide	1315-06-6	2760: Stannous selenide	1344-54-3	3024: Titanium trioxide
1313-27-5	1893: Molybdenum(VI) oxide	1315-07-7	2806: Strontium selenide	1344-57-6	3081: Uranium dioxide
1313-29-7	1880: Molybdenum(III) oxide	1315-09-9	3274: Zinc selenide	1344-58-7	3099: Uranium trioxide
1313-30-0	2689: Sodium phosphomolybdate	1315-11-3	3286: Zinc telluride	1344-59-8	3102: Uranium(V,VI) oxide
1313-49-1	3259: Zinc nitride	1317-33-5	1868: Molybdenum disulfide	1344-95-2	639: Calcium silicate
1313-59-3	2673: Sodium oxide	1317-34-6	1806: Manganese(III) oxide	1345-04-6	264: Antimony(III) sulfide
1313-60-6	2683: Sodium peroxide	1317-35-7	1801: Manganese(II,III) oxide	1345-07-9	472: Bismuth sulfide
1313-82-2	2707: Sodium sulfide	1317-36-8	1529: Lead oxide	1345-13-7	718: Cerous oxide
1313-83-3	2709: Sodium sulfide pentahydrate	1317-37-9	1194: Ferrous sulfide	1345-25-1	1186: Ferrous oxide
1313-84-4	2708: Sodium sulfide nonahydrate	1317-38-0	992: Copper(II) oxide	1399-57-1	660: Carbon
1313-85-5	2698: Sodium selenide	1317-39-1	940: Copper(I) oxide	1470-61-7	2520: Silver diethyldithiocarbamate
1313-96-8	2008: Niobium(V) oxide	1317-40-4	1007: Copper(II) sulfide	1495-50-7	1024: Cyanogen fluoride
1313-97-9	1916: Neodymium oxide	1317-41-5	1000: Copper(II) selenide	1529-48-2	1053: Dimethylgermanium dichloride
1313-99-1	1964: Nickel oxide	1317-42-6	905: Cobalt(II) sulfide	1590-87-0	1055: Disilane
1314-05-2	1970: Nickel selenide	1317-61-9	1455: Iron(II,III) oxide	1592-23-0	642: Calcium stearate
1314-06-3	1987: Nickel(III) oxide	1317-66-4	1446: Iron disulfide	1600-27-7	1827: Mercury(II) acetate
1314-08-5	2060: Palladium(II) oxide	1317-70-0	3001: Titanium dioxide	1603-84-5	671: Carbon oxyselenide
1314-11-0	2798: Strontium oxide	1317-80-2	3003: Titanium dioxide	1633-05-2	2775: Strontium carbonate
1314-12-1	2926: Thallium(I) oxide	1317-98-2	257: Antimony(III) oxide	1701-93-5	2549: Silver thiocyanate
1314-13-2	3263: Zinc oxide	1318-23-6	67: Aluminum oxyhydroxide (α)	1762-95-4	239: Ammonium thiocyanate
1314-15-4	2118: Platinum(IV) oxide	1318-72-5	2211: Potassium magnesium chloride sulfate	1779-25-5	1051: Diisobutylaluminum chloride
1314-18-7	2802: Strontium peroxide	1319-46-6	1505: Lead basic carbonate	1863-63-4	117: Ammonium benzoate
1314-18-7	2803: Strontium peroxide octahydrate	1327-39-5	565: Calcium aluminum silicate	2035-66-7	2054: Palladium(II) cyanide
1314-20-1	2964: Thorium oxide	1327-41-9	47: Aluminum hydroxychloride	2092-16-2	651: Calcium thiocyanate tetrahydrate
1314-22-3	3266: Zinc peroxide	1327-50-0	265: Antimony(III) telluride	2092-17-3	390: Barium thiocyanate
1314-23-4	3307: Zirconium oxide	1327-53-3	287: Arsenic(III) oxide	2223-93-0	543: Cadmium stearate
1314-24-5	2096: Phosphorus(III) oxide	1327-53-3	288: Arsenic(III) oxide	2223-95-2	1973: Nickel stearate
1314-27-8	1530: Lead oxide	1330-43-4	2716: Sodium tetraborate	2406-52-2	2991: Tin hydride
1314-28-9	2355: Rhenium(VI) oxide	1332-40-7	993: Copper(II) oxychloride	2408-36-8	1578: Lithium cyanide
1314-32-5	2940: Thallium(III) oxide	1332-52-1	411: Beryllium basic acetate	2452-01-9	3256: Zinc laurate
1314-34-7	3141: Vanadium trioxide	1332-58-7	82: Aluminum silicate dihydrate	2457-01-4	301: Barium 2-ethylhexanoate
1314-35-8	3073: Tungsten trioxide	1332-63-4	1805: Manganese(III) hydroxide	2466-09-3	2330: Pyrophosphoric acid
1314-36-9	3213: Yttrium oxide	1332-71-4	920: Cobalt(III) sulfide	2551-62-4	2820: Sulfur hexafluoride
1314-37-0	3184: Ytterbium oxide	1332-77-0	2260: Potassium tetraborate pentahydrate	2570-63-0	2934: Thallium(III) acetate
1314-41-6	1553: Lead(II,III) oxide	1332-81-6	266: Antimony(IV) oxide	2644-70-4	1354: Hydrazine monohydrochloride
1314-56-3	2101: Phosphorus(V) oxide	1332-81-6	267: Antimony(IV) oxide	2696-92-6	2025: Nitrosyl chloride
1314-60-9	271: Antimony(V) oxide	1333-74-0	1362: Hydrogen	2699-79-8	2833: Sulfuryl fluoride
1314-61-0	2849: Tantalum pentoxide	1333-82-0	833: Chromium(VI) oxide	3012-65-5	174: Ammonium hydrogen citrate
1314-62-1	3133: Vanadium pentoxide				

3017-60-5	907: Cobalt(II) thiocyanate	
3094-87-9	1165: Ferrous acetate	
3094-87-9	1166: Ferrous acetate tetrahydrate	
3095-65-6	183: Ammonium hydrogen tartrate	
3164-29-2	226: Ammonium tartrate	
3164-34-9	649: Calcium tartrate tetrahydrate	
3236-82-6	2004: Niobium(V) ethoxide	
3251-23-8	986: Copper(II) nitrate	
3264-82-2	1931: Nickel acetylacetonate	
3333-67-3	1936: Nickel basic carbonate tetrahydrate	
3333-67-3	1943: Nickel carbonate	
3375-31-3	2049: Palladium(II) acetate	
3396-11-0	728: Cesium acetate	
3444-13-1	1845: Mercury(II) oxalate	
3458-72-8	131: Ammonium citrate tribasic	
3486-35-9	3234: Zinc carbonate	
3486-35-9	3235: Zinc carbonate hydroxide	
3643-76-3	250: Antimony(III) acetate	
3804-23-7	2449: Scandium acetate hydrate	
3811-04-9	2146: Potassium chlorate	
3982-91-0	2946: Thiophosphoryl chloride	
4049-81-1	876: Cobalt(II) ferrocyanide hydrate	
4075-81-4	635: Calcium propionate	
4109-96-0	1044: Dichlorosilane	
4111-54-0	1583: Lithium diisopropylamide	
4119-52-2	1159: Ferric thiocyanate	
4485-12-5	1618: Lithium stearate	
4493-37-2	799: Chromium(II) formate monohydrate	
5142-76-7	470: Bismuth subacetate	
5145-48-2	1676: Magnesium carbonate dihydrate	
5341-61-7	1349: Hydrazine dihydrochloride	
5470-11-1	1383: Hydroxylamine hydrochloride	
5714-22-7	1056: Disulfur decafluoride	
5743-04-4	510: Cadmium acetate dihydrate	
5743-26-0	561: Calcium acetate monohydrate	
5785-44-4	589: Calcium citrate tetrahydrate	
5793-84-0	629: Calcium phenoxide	
5794-28-5	621: Calcium oxalate monohydrate	
5892-10-4	439: Bismuth basic carbonate hemihydrate	
5893-61-8	976: Copper(II) formate tetrahydrate	
5895-47-6	2431: Samarium carbonate	
5908-64-5	303: Barium acetate monohydrate	
5908-81-6	386: Barium tartrate	
5951-19-9	675: Carbon sulfide selenide	
5965-33-3	2137: Potassium antimony	

	oxalate trihydrate	
5965-38-8	892: Cobalt(II) oxalate dihydrate	
5968-11-6	2573: Sodium carbonate monohydrate	
5970-44-5	3202: Yttrium carbonate trihydrate	
5970-45-6	3221: Zinc acetate dihydrate	
5970-62-7	3249: Zinc formate dihydrate	
5972-71-4	118: Ammonium bimalate	
5972-72-5	177: Ammonium hydrogen oxalate monohydrate	
5972-76-9	120: Ammonium caprylate	
6009-70-7	202: Ammonium oxalate monohydrate	
6010-09-9	1195: Ferrous thiocyanate trihydrate	
6018-89-9	1930: Nickel acetate tetrahydrate	
6018-94-6	1963: Nickel oxalate dihydrate	
6046-93-1	950: Copper(II) acetate monohydrate	
6047-25-2	1185: Ferrous oxalate dihydrate	
6074-84-6	2840: Tantalum ethoxide	
6080-56-4	1497: Lead acetate trihydrate	
6100-05-6	2153: Potassium citrate monohydrate	
6100-96-5	2812: Strontium tartrate tetrahydrate	
6108-17-4	1556: Lithium acetate dihydrate	
6131-90-4	2554: Sodium acetate trihydrate	
6132-02-1	2572: Sodium carbonate decahydrate	
6132-04-3	2581: Sodium citrate dihydrate	
6147-53-1	857: Cobalt(II) acetate tetrahydrate	
6150-88-5	1711: Magnesium oxalate dihydrate	
6153-56-6	2044: Oxalic acid dihydrate	
6156-78-1	1765: Manganese(II) acetate tetrahydrate	
6159-44-0	3103: Uranyl acetate dihydrate	
6160-38-9	2804: Strontium salicylate dihydrate	
6192-12-7	2299: Praseodymium acetate hydrate	
6192-13-8	1900: Neodymium acetate monohydrate	
6211-24-1	328: Barium diphenylamine-4-sulfonate	
6303-21-5	1389: Hypophosphorous acid	
6381-79-9	2145: Potassium carbonate hemitrihydrate	
6381-92-6	1103: Ethylenediamine-tetraacetic acid dihydrate disodium salt	
6484-52-2	196: Ammonium nitrate	
6487-39-4	1470: Lanthanum carbonate octahydrate	

6487-48-5	2225: Potassium oxalate monohydrate	
6533-73-9	2911: Thallium(I) carbonate	
6556-16-7	1786: Manganese(II) oxalate dihydrate	
6569-51-3	480: Borazole	
6680-58-6	1576: Lithium citrate tetrahydrate	
6834-92-0	2657: Sodium metasilicate	
6858-44-2	2582: Sodium citrate pentahydrate	
6865-35-6	378: Barium stearate	
7047-84-9	57: Aluminum monostearate	
7116-98-5	2333: Radium carbonate	
7320-34-5	2239: Potassium pyrophosphate trihydrate	
7429-90-5	16: Aluminum	
7429-91-6	1057: Dysprosium	
7429-92-7	1078: Einsteinium	
7439-88-5	1429: Iridium	
7439-89-6	1440: Iron	
7439-90-9	1456: Krypton	
7439-91-0	1462: Lanthanum	
7439-92-1	1495: Lead	
7439-93-2	1554: Lithium	
7439-94-3	1634: Lutetium	
7439-95-4	1654: Magnesium	
7439-96-5	1749: Manganese	
7439-97-6	1811: Mercury	
7439-98-7	1861: Molybdenum	
7439-99-8	1927: Neptunium	
7440-00-8	1899: Neodymium	
7440-01-9	1926: Neon	
7440-02-0	1929: Nickel	
7440-03-1	1988: Niobium	
7440-04-2	2033: Osmium	
7440-05-3	2048: Palladium	
7440-06-4	2105: Platinum	
7440-07-5	2119: Plutonium	
7440-08-6	2127: Polonium	
7440-08-6	2128: Polonium	
7440-09-7	2131: Potassium	
7440-10-0	2323: Praseodymium(α)	
7440-10-0	2324: Praseodymium(β)	
7440-11-1	1810: Mendelevium	
7440-12-2	2327: Promethium	
7440-13-3	2329: Protactinium	
7440-14-4	2331: Radium	
7440-15-5	2337: Rhenium	
7440-16-6	2360: Rhodium	
7440-17-7	2376: Rubidium	
7440-18-8	2414: Ruthenium	
7440-19-9	2425: Samarium	
7440-20-2	2448: Scandium	
7440-21-3	2485: Silicon	
7440-22-4	2504: Silver	
7440-23-5	2551: Sodium	
7440-24-6	2766: Strontium	
7440-25-7	2834: Tantalum	
7440-26-8	2857: Technetium	
7440-27-9	2876: Terbium	
7440-28-0	2903: Thallium	
7440-29-1	2947: Thorium	
7440-30-4	2974: Thulium	
7440-31-5	2989: Tin (gray)	
7440-31-5	2990: Tin (white)	
7440-32-6	2995: Titanium	
7440-33-7	3039: Tungsten	
7440-34-8	2: Actinium	
7440-35-9	95: Americium	

7440-36-0	246: Antimony
7440-37-1	275: Argon
7440-38-2	277: Arsenic (α)
7440-38-2	278: Arsenic (β)
7440-39-3	300: Barium
7440-40-6	404: Berkelium (α)
7440-40-6	405: Berkelium (β)
7440-41-7	406: Beryllium
7440-42-8	482: Boron
7440-43-9	508: Cadmium
7440-44-0	664: Carbon (amorphous)
7440-45-1	693: Cerium
7440-46-2	727: Cesium
7440-47-3	779: Chromium
7440-48-4	835: Cobalt
7440-50-8	923: Copper
7440-51-9	1018: Curium (α)
7440-51-9	1019: Curium (β)
7440-52-0	1079: Erbium
7440-53-1	1105: Europium
7440-54-2	1210: Gadolinium
7440-55-3	1234: Gallium
7440-56-4	1263: Germanium
7440-57-5	1283: Gold
7440-58-6	1299: Hafnium
7440-59-7	1320: Helium
7440-60-0	1327: Holmium
7440-61-1	3078: Uranium
7440-62-2	3113: Vanadium
7440-63-3	3157: Xenon
7440-64-4	3173: Ytterbium
7440-65-5	3189: Yttrium
7440-66-6	3219: Zinc
7440-67-7	3290: Zirconium
7440-68-8	299: Astatine
7440-69-9	436: Bismuth
7440-70-2	558: Calcium
7440-71-3	658: Californium
7440-72-4	1129: Fermium
7440-73-5	1207: Francium
7440-74-6	1390: Indium
7446-06-2	2130: Polonium(IV) oxide
7446-07-3	2864: Tellurium dioxide
7446-08-4	2468: Selenium dioxide
7446-09-5	2819: Sulfur dioxide
7446-10-8	1540: Lead sulfite
7446-11-9	2823: Sulfur trioxide(α)
7446-11-9	2824: Sulfur trioxide(β)
7446-11-9	2825: Sulfur trioxide(γ)
7446-14-2	1538: Lead sulfate
7446-15-3	1533: Lead selenate
7446-16-4	2335: Radium sulfate
7446-18-6	2931: Thallium(I) sulfate
7446-19-7	3281: Zinc sulfate monohydrate
7446-20-0	3279: Zinc sulfate heptahydrate
7446-21-1	2805: Strontium selenate
7446-22-2	2929: Thallium(I) selenate
7446-25-5	2747: Stannic selenite
7446-26-6	3271: Zinc pyrophosphate
7446-27-7	1521: Lead hydrogen phosphate
7446-27-7	1532: Lead phosphate
7446-32-4	263: Antimony(III) sulfate
7446-33-5	3216: Yttrium sulfate octahydrate
7446-34-6	2472: Selenium monosulfide
7446-35-7	2865: Tellurium disulfide
7446-70-0	29: Aluminum chloride
7447-39-4	964: Copper(II) chloride

7447-40-7	2147: Potassium chloride
7447-41-8	1572: Lithium chloride
7487-88-9	1733: Magnesium sulfate
7487-94-7	1832: Mercury(II) chloride
7488-51-9	1535: Lead selenite
7488-52-0	3284: Zinc sulfite dihydrate
7488-54-2	2379: Rubidium aluminum sulfate dodecahydrate
7488-54-2	2407: Rubidium sulfate
7488-55-3	2762: Stannous sulfate
7488-56-4	2469: Selenium disulfide
7542-09-8	860: Cobalt(II) basic carbonate
7543-51-3	3268: Zinc phosphate tetrahydrate
7550-35-8	1567: Lithium bromide
7550-45-0	3018: Titanium tetrachloride
7553-56-2	1419: Iodine
7558-79-4	2629: Sodium hydrogen phosphate
7558-80-7	2590: Sodium dihydrogen phosphate dihydrate
7580-67-8	1590: Lithium hydride
7601-54-9	2686: Sodium phosphate
7601-89-0	2678: Sodium perchlorate
7601-90-3	2069: Perchloric acid
7631-86-9	2490: Silicon dioxide
7631-86-9	2491: Silicon dioxide
7631-86-9	2492: Silicon dioxide
7631-90-5	2638: Sodium hydrogen sulfite
7631-94-9	2593: Sodium dithionate
7631-94-9	2594: Sodium dithionate dihydrate
7631-95-0	2661: Sodium molybdate
7631-99-4	2665: Sodium nitrate
7632-00-0	2666: Sodium nitrite
7632-51-1	3135: Vanadium tetrachloride
7637-07-2	492: Boron trifluoride
7637-13-0	2761: Stannous stearate
7646-69-7	2623: Sodium hydride
7646-78-8	2740: Stannic chloride
7646-79-9	866: Cobalt(II) chloride
7646-85-7	3237: Zinc chloride
7646-93-7	2204: Potassium hydrogen sulfate
7647-01-0	1364: Hydrogen chloride
7647-10-1	2052: Palladium(II) chloride
7647-10-1	2053: Palladium(II) chloride dihydrate
7647-14-5	2576: Sodium chloride
7647-15-6	2568: Sodium bromide
7647-17-8	738: Cesium chloride
7647-18-9	268: Antimony(V) chloride
7647-19-0	2100: Phosphorus(V) fluoride
7659-31-6	2507: Silver acetylide
7664-38-2	2081: Phosphoric acid
7664-39-3	1361: Hydrofluoric acid, 70%
7664-39-3	1366: Hydrogen fluoride
7664-41-7	110: Ammonia
7664-93-9	2829: Sulfuric acid
7681-11-0	2210: Potassium iodide
7681-38-1	2633: Sodium hydrogen sulfate

7681-49-4	2597: Sodium fluoride
7681-52-9	2643: Sodium hypochlorite
7681-52-9	2644: Sodium hypochlorite pentahydrate
7681-55-2	2647: Sodium iodate
7681-57-4	2651: Sodium metabisulfite
7681-65-4	938: Copper(I) iodide
7681-82-5	2648: Sodium iodide
7693-26-7	2196: Potassium hydride
7697-37-2	2012: Nitric acid
7698-05-7	1033: Deuterium chloride
7699-45-8	3232: Zinc bromide
7704-34-9	2826: Sulfur(α)
7704-34-9	2827: Sulfur(β)
7704-34-9	2828: Sulfur(γ)
7704-98-5	3007: Titanium hydride
7704-99-6	3301: Zirconium hydride
7705-07-9	3022: Titanium trichloride
7705-08-0	1135: Ferric chloride
7718-54-9	1947: Nickel chloride
7718-98-1	3138: Vanadium trichloride
7719-09-7	2944: Thionyl chloride
7719-12-2	2093: Phosphorus(III) chloride
7720-78-7	1191: Ferrous sulfate
7720-83-4	3020: Titanium tetraiodide
7721-01-9	2846: Tantalum pentachloride
7722-64-7	2233: Potassium permanganate
7722-76-1	138: Ammonium dihydrogen phosphate
7722-84-1	1373: Hydrogen peroxide
7722-86-3	2072: Peroxysulfuric acid
7722-88-5	2693: Sodium pyrophosphate
7723-14-0	2083: Phosphorus (black)
7723-14-0	2084: Phosphorus (red)
7726-95-6	497: Bromine
7727-15-3	24: Aluminum bromide
7727-18-6	3129: Vanadium oxytrichloride
7727-21-1	2237: Potassium persulfate
7727-37-9	2014: Nitrogen
7727-43-7	381: Barium sulfate
7727-54-0	210: Ammonium peroxydisulfate
7727-73-3	2705: Sodium sulfate decahydrate
7727-73-3	2706: Sodium sulfate heptahydrate
7732-18-5	3156: Water
7733-02-0	3278: Zinc sulfate
7757-79-1	2221: Potassium nitrate
7757-82-6	2704: Sodium sulfate
7757-83-7	2710: Sodium sulfite
7757-86-0	1696: Magnesium hydrogen phosphate trihydrate
7757-87-1	1719: Magnesium phosphate pentahydrate
7757-88-2	1737: Magnesium sulfite
7757-93-9	603: Calcium hydrogen phosphate
7758-01-2	2142: Potassium bromate
7758-02-3	2143: Potassium bromide
7758-05-6	2209: Potassium iodate

CAS RN	Compound
7758-09-0	2222: Potassium nitrite
7758-11-4	2216: Potassium monohydrogen phosphate
7758-16-9	2592: Sodium dihydrogen pyrophosphate
7758-19-2	2577: Sodium chlorite
7758-29-4	2733: Sodium triphosphate
7758-87-4	630: Calcium phosphate
7758-88-5	712: Cerous fluoride
7758-89-6	935: Copper(I) chloride
7758-94-3	1173: Ferrous chloride
7758-95-4	1511: Lead chloride
7758-97-6	1513: Lead chromate
7758-98-7	1005: Copper(II) sulfate
7758-99-8	1006: Copper(II) sulfate pentahydrate
7759-00-4	1762: Manganese silicate
7759-01-5	1549: Lead tungstate
7759-01-5	1550: Lead tungstate
7759-02-6	2809: Strontium sulfate
7761-88-8	2531: Silver nitrate
7772-98-7	2729: Sodium thiosulfate
7772-99-8	2751: Stannous chloride
7773-01-5	1771: Manganese(II) chloride
7773-03-7	2206: Potassium hydrogen sulfite
7773-06-0	222: Ammonium sulfamate
7774-29-0	1840: Mercury(II) iodide(α)
7774-29-0	1841: Mercury(II) iodide(β)
7774-34-7	583: Calcium chloride hexahydrate
7775-09-9	2575: Sodium chlorate
7775-11-3	2578: Sodium chromate
7775-11-3	2579: Sodium chromate decahydrate
7775-14-6	2640: Sodium hydrosulfite
7775-19-1	2652: Sodium metaborate
7775-27-1	2685: Sodium persulfate
7775-41-9	2522: Silver fluoride
7778-18-9	644: Calcium sulfate
7778-39-4	292: Arsenic(V) acid hemihydrate
7778-43-0	2562: Sodium arsenate dodecahydrate
7778-43-0	2624: Sodium hydrogen arsenate
7778-44-1	566: Calcium arsenate
7778-50-9	2159: Potassium dichromate
7778-53-2	2238: Potassium phosphate
7778-54-3	608: Calcium hypochlorite
7778-74-7	2231: Potassium perchlorate
7778-77-0	2162: Potassium dihydrogen phosphate
7778-80-5	2251: Potassium sulfate
7779-25-1	1686: Magnesium citrate pentahydrate
7779-90-0	3267: Zinc phosphate
7782-12-0	2589: Sodium dichromate dihydrate
7782-39-0	1031: Deuterium
7782-40-3	663: Carbon
7782-41-4	1197: Fluorine
7782-42-5	661: Carbon
7782-44-7	2046: Oxygen
7782-49-2	2464: Selenium
7782-49-2	2465: Selenium (β)
7782-49-2	2482: Selenium(α)
7782-50-5	768: Chlorine
7782-61-8	1146: Ferric nitrate nonahydrate
7782-63-0	1192: Ferrous sulfate heptahydrate
7782-64-1	1776: Manganese(II) fluoride
7782-65-2	1265: Germanium tetrahydride
7782-66-3	1850: Mercury(II) phosphate
7782-68-5	1418: Iodic acid
7782-70-9	2203: Potassium hydrogen selenite
7782-77-6	2028: Nitrous acid
7782-78-7	2027: Nitrosylsulfuric acid
7782-79-8	1360: Hydrazoic acid
7782-85-6	2631: Sodium hydrogen phosphate heptahydrate
7782-86-7	1822: Mercury(I) nitrate monohydrate
7782-87-8	2161: Potassium dihydrogen hypophosphite
7782-89-0	1561: Lithium amide
7782-91-4	1889: Molybdenum(VI) acid monohydrate
7782-92-5	2559: Sodium amide
7782-99-2	2831: Sulfurous acid
7783-00-8	2483: Selenous acid
7783-03-1	3076: Tungstic acid
7783-06-4	1375: Hydrogen sulfide
7783-07-5	1374: Hydrogen selenide
7783-08-6	2463: Selenic acid
7783-09-7	1376: Hydrogen telluride
7783-11-1	225: Ammonium sulfite monohydrate
7783-14-4	3253: Zinc hypophosphite monohydrate
7783-18-8	240: Ammonium thiosulfate
7783-19-9	219: Ammonium selenite
7783-20-2	223: Ammonium sulfate
7783-21-3	218: Ammonium selenate
7783-22-4	243: Ammonium uranate(VI)
7783-26-8	2498: Silicon octahydride
7783-28-0	178: Ammonium hydrogen phosphate
7783-29-1	2489: Silicon decahydride
7783-32-6	1839: Mercury(II) iodate
7783-33-7	2278: Potassium tetraiodomercurate(II)
7783-34-8	1843: Mercury(II) nitrate monohydrate
7783-35-9	1852: Mercury(II) sulfate
7783-36-0	1825: Mercury(I) sulfate
7783-39-3	1837: Mercury(II) fluoride
7783-40-6	1690: Magnesium fluoride
7783-41-7	1199: Fluorine monoxide
7783-42-8	2945: Thionyl fluoride
7783-43-9	2476: Selenium oxyfluoride
7783-44-0	1198: Fluorine dioxide
7783-46-2	1517: Lead fluoride
7783-47-3	2753: Stannous fluoride
7783-48-4	2782: Strontium fluoride
7783-49-5	3244: Zinc fluoride
7783-50-8	1141: Ferric fluoride
7783-51-9	1248: Gallium(III) fluoride
7783-52-0	1406: Indium(III) fluoride
7783-53-1	1804: Manganese(III) fluoride
7783-54-2	2018: Nitrogen trifluoride
7783-55-3	2094: Phosphorus(III) fluoride
7783-56-4	253: Antimony(III) fluoride
7783-57-5	2938: Thallium(III) fluoride
7783-58-6	1276: Germanium(IV) fluoride
7783-58-6	1277: Germanium(IV) fluoride trihydrate
7783-59-7	1545: Lead tetrafluoride
7783-60-0	2821: Sulfur tetrafluoride
7783-61-1	2501: Silicon tetrafluoride
7783-62-2	2743: Stannic fluoride
7783-63-3	3019: Titanium tetrafluoride
7783-63-3	3123: Vanadium gallide
7783-64-4	3299: Zirconium fluoride
7783-66-6	1426: Iodine pentafluoride
7783-68-8	2005: Niobium(V) fluoride
7783-70-2	270: Antimony(V) fluoride
7783-71-3	2847: Tantalum pentafluoride
7783-72-4	3131: Vanadium pentafluoride
7783-73-5	2185: Potassium hexafluorogermanate
7783-75-7	1431: Iridium hexafluoride
7783-77-9	1892: Molybdenum(VI) fluoride
7783-79-1	2470: Selenium hexafluoride
7783-80-4	2866: Tellurium hexafluoride
7783-81-5	3082: Uranium hexafluoride
7783-82-6	3056: Tungsten hexafluoride
7783-84-8	1144: Ferric hypophosphite
7783-86-0	1182: Ferrous iodide
7783-89-3	2510: Silver bromate
7783-90-6	2514: Silver chloride
7783-91-7	2515: Silver chlorite
7783-92-8	2513: Silver chlorate
7783-93-9	2535: Silver perchlorate
7783-95-1	2521: Silver difluoride
7783-96-2	2528: Silver iodide
7783-97-3	2527: Silver iodate
7783-98-4	2537: Silver permanganate
7783-99-5	2532: Silver nitrite
7784-01-2	2516: Silver chromate
7784-02-3	2519: Silver dichromate
7784-03-4	2548: Silver tetraiodo-mercurate(II) (α-form)
7784-05-6	2543: Silver selenite
7784-07-8	2541: Silver selenate
7784-09-0	2539: Silver phosphate
7784-11-4	25: Aluminum bromide hexahydrate
7784-13-6	30: Aluminum chloride hexahydrate
7784-14-7	229: Ammonium

	tetrachloroaluminate
7784-16-9	2721: Sodium
	tetrachloroaluminate
7784-17-0	730: Cesium aluminum
	sulfate dodecahydrate
7784-18-1	38: Aluminum fluoride
7784-19-2	164: Ammonium
	hexafluoroaluminate
7784-21-6	42: Aluminum hydride
7784-22-7	49: Aluminum
	hypophosphite
7784-23-8	50: Aluminum iodide
7784-24-9	2136: Potassium
	aluminum sulfate
	dodecahydrate
7784-25-0	113: Ammonium
	aluminum sulfate
7784-26-1	114: Ammonium
	aluminum sulfate
	dodecahydrate
7784-27-2	58: Aluminum nitrate
	nonahydrate
7784-30-7	72: Aluminum phosphate
7784-30-7	73: Aluminum phosphate
	dihydrate
7784-31-8	84: Aluminum sulfate
	octadecahydrate
7784-33-0	283: Arsenic(III) bromide
7784-34-1	284: Arsenic(III) chloride
7784-35-2	285: Arsenic(III) fluoride
7784-36-3	293: Arsenic(V) fluoride
7784-37-4	1828: Mercury(II)
	arsenate
7784-40-9	1501: Lead arsenate
7784-41-0	2160: Potassium
	dihydrogen arsenate
7784-42-1	298: Arsine
7784-44-3	171: Ammonium
	hydrogen arsenate
7784-45-4	286: Arsenic(III) iodide
7784-46-5	2563: Sodium arsenite
7784-48-7	1934: Nickel arsenate
	octahydrate
7785-19-5	1751: Manganese
	ammonium sulfate
	hexahydrate
7785-20-8	195: Ammonium nickel
	sulfate hexahydrate
7785-23-1	2511: Silver bromide
7785-84-4	2732: Sodium
	trimetaphosphate
	hexahydrate
7785-87-7	1793: Manganese(II)
	sulfate
7786-30-3	1682: Magnesium chloride
7786-81-4	1975: Nickel sulfate
7787-32-8	332: Barium fluoride
7787-33-9	347: Barium iodide
	dihydrate
7787-34-0	345: Barium iodate
	monohydrate
7787-35-1	349: Barium
	manganate(VI)
7787-36-2	365: Barium
	permanganate
7787-37-3	354: Barium molybdate
7787-38-4	359: Barium nitrite
	monohydrate
7787-39-5	384: Barium sulfite
7787-40-8	392: Barium thiosulfate
	monohydrate
7787-41-9	369: Barium selenate
7787-42-0	397: Barium tungstate

7787-46-4	414: Beryllium bromide
7787-47-5	417: Beryllium chloride
7787-48-6	428: Beryllium
	perchlorate tetrahydrate
7787-49-7	418: Beryllium fluoride
7787-50-0	2274: Potassium
	tetrafluoroberyllate
	dihydrate
7787-52-2	420: Beryllium hydride
7787-53-3	423: Beryllium iodide
7787-56-6	432: Beryllium sulfate
	tetrahydrate
7787-57-7	457: Bismuth oxybromide
7787-58-8	441: Bismuth bromide
7787-59-9	458: Bismuth oxychloride
7787-60-2	442: Bismuth chloride
7787-61-3	445: Bismuth fluoride
7787-62-4	462: Bismuth
	pentafluoride
7787-63-5	459: Bismuth oxyiodide
7787-64-6	450: Bismuth iodide
7787-68-0	471: Bismuth sulfate
7787-69-1	734: Cesium bromide
7787-70-4	934: Copper(I) bromide
7787-71-5	504: Bromine trifluoride
7788-97-8	812: Chromium(III)
	fluoride
7788-98-9	129: Ammonium
	chromate(VI)
7788-99-0	824: Chromium(III)
	potassium sulfate
	dodecahydrate
7789-00-6	2149: Potassium chromate
7789-01-7	1575: Lithium chromate
	dihydrate
7789-02-8	818: Chromium(III)
	nitrate nonahydrate
7789-04-0	821: Chromium(III)
	phosphate
7789-04-0	822: Chromium(III)
	phosphate
	hemiheptahydrate
7789-06-2	2779: Strontium chromate
7789-08-4	141: Ammonium ferric
	chromate
7789-09-5	136: Ammonium
	dichromate(VI)
7789-10-8	1836: Mercury(II)
	dichromate
7789-12-0	3242: Zinc dichromate
	trihydrate
7789-17-5	748: Cesium iodide
7789-18-6	752: Cesium nitrate
7789-19-7	973: Copper(II) fluoride
7789-20-0	1035: Deuterium oxide
7789-21-1	1205: Fluorosulfonic acid
7789-23-3	2167: Potassium fluoride
7789-24-4	1584: Lithium fluoride
7789-25-5	2026: Nitrosyl fluoride
7789-26-6	1200: Fluorine nitrate
7789-27-7	2916: Thallium(I) fluoride
7789-28-8	1178: Ferrous fluoride
7789-29-9	2199: Potassium hydrogen
	fluoride
7789-30-2	503: Bromine
	pentafluoride
7789-31-3	496: Bromic acid
7789-33-5	1423: Iodine
	monobromide
7789-36-8	1672: Magnesium
	bromate hexahydrate
7789-38-0	2567: Sodium bromate
7789-39-1	2382: Rubidium bromide

7789-40-4	2910: Thallium(I)
	bromide
7789-41-5	573: Calcium bromide
7789-42-6	516: Cadmium bromide
7789-43-7	862: Cobalt(II) bromide
7789-45-9	960: Copper(II) bromide
7789-46-0	1169: Ferrous bromide
7789-47-1	1831: Mercury(II)
	bromide
7789-48-2	1673: Magnesium
	bromide
7789-51-7	2473: Selenium
	oxybromide
7789-52-8	2466: Selenium bromide
7789-54-0	2862: Tellurium
	dibromide
7789-57-3	3029: Tribromosilane
7789-59-5	2088: Phosphorus
	oxybromide
7789-60-8	2092: Phosphorus(III)
	bromide
7789-61-9	251: Antimony(III)
	bromide
7789-65-3	2478: Selenium
	tetrabromide
7789-66-4	2499: Silicon tetrabromide
7789-67-5	2739: Stannic bromide
7789-68-6	3017: Titanium
	tetrabromide
7789-69-7	2098: Phosphorus(V)
	bromide
7789-75-5	595: Calcium fluoride
7789-77-7	604: Calcium hydrogen
	phosphate dihydrate
7789-78-8	602: Calcium hydride
7789-79-9	609: Calcium
	hypophosphite
7789-80-2	610: Calcium iodate
7789-82-4	613: Calcium molybdate
7789-99-0	2151: Potassium
	chromium(III) sulfate
	dodecahydrate
7790-21-8	2232: Potassium periodate
7790-22-9	1597: Lithium iodide
	trihydrate
7790-28-5	2680: Sodium periodate
7790-29-6	2395: Rubidium iodide
7790-30-9	2921: Thallium(I) iodide
7790-31-0	1701: Magnesium iodide
	octahydrate
7790-32-1	1698: Magnesium iodate
	tetrahydrate
7790-33-2	1780: Manganese(II)
	iodide
7790-33-2	1781: Manganese(II)
	iodide tetrahydrate
7790-34-3	1959: Nickel iodide
	hexahydrate
7790-37-6	3254: Zinc iodate
7790-38-7	2057: Palladium(II) iodide
7790-39-8	2113: Platinum(II) iodide
7790-42-3	2287: Potassium triiodide
	monohydrate
7790-43-4	2288: Potassium
	triiodozincate
7790-44-5	255: Antimony(III) iodide
7790-46-7	2117: Platinum(IV) iodide
7790-47-8	2744: Stannic iodide
7790-48-9	2873: Tellurium
	tetraiodide
7790-49-0	2958: Thorium iodide
7790-53-6	2192: Potassium
	hexametaphosphite

7790-56-9	2254: Potassium sulfite dihydrate	7803-55-6	192: Ammonium metavanadate	10026-01-4	2039: Osmium(IV) chloride
7790-58-1	2258: Potassium tellurite	7803-57-8	1352: Hydrazine monohydrate	10026-02-5	2129: Polonium(IV) chloride
7790-59-2	2243: Potassium selenate				
7790-60-5	2291: Potassium tungstate dihydrate	7803-60-3	1388: Hypophosphoric acid	10026-03-6	2479: Selenium tetrachloride
7790-60-5	2290: Potassium tungstate	7803-62-5	2484: Silane	10026-04-7	2500: Silicon tetrachloride
7790-62-7	2240: Potassium pyrosulfate	7803-63-6	180: Ammonium hydrogen sulfate	10026-06-9	2741: Stannic chloride pentahydrate
7790-63-8	2292: Potassium uranate	7803-65-8	186: Ammonium hypophosphite	10026-07-0	2871: Tellurium tetrachloride
7790-69-4	1606: Lithium nitrate				
7790-74-1	637: Calcium selenate dihydrate	7803-68-1	2859: Telluric acid	10026-08-1	2951: Thorium chloride
		7883-28-0	212: Ammonium phosphate dibasic	10026-10-5	3090: Uranium tetrachloride
7790-75-2	655: Calcium tungstate				
7790-76-3	636: Calcium pyrophosphate	8003-05-2	2076: Phenylmercuric nitrate, basic	10026-11-6	3297: Zirconium chloride
7790-78-5	519: Cadmium chlorate dihydrate	8014-95-7	2830: Sulfuric acid fuming	10026-12-7	2003: Niobium(V) chloride
7790-78-5	521: Cadmium chloride hemipentahydrate	9080-17-5	216: Ammonium polysulfide	10026-13-8	2099: Phosphorus(V) chloride
7790-79-6	525: Cadmium fluoride	10022-31-8	356: Barium nitrate	10026-17-2	877: Cobalt(II) fluoride
7790-80-9	528: Cadmium iodide	10022-47-6	130: Ammonium chromic sulfate dodecahydrate	10026-18-3	913: Cobalt(III) fluoride
7790-81-0	527: Cadmium iodate			10026-22-9	888: Cobalt(II) nitrate hexahydrate
7790-85-4	555: Cadmium tungstate(VI)	10022-48-7	1581: Lithium dichromate dihydrate	10026-24-1	903: Cobalt(II) sulfate heptahydrate
7790-86-5	709: Cerous chloride	10022-50-1	2031: Nitryl fluoride		
7790-87-6	714: Cerous iodide	10022-68-1	533: Cadmium nitrate tetrahydrate	10028-14-5	2032: Nobelium
7790-87-6	715: Cerous iodide nonahydrate			10028-15-6	2047: Ozone
		10024-93-8	1906: Neodymium chloride	10028-17-8	3037: Tritium
7790-89-8	771: Chlorine monofluoride			10028-18-9	1952: Nickel fluoride
		10024-97-2	2029: Nitrous oxide	10028-22-5	1156: Ferric sulfate
7790-91-2	774: Chlorine trifluoride	10025-64-6	3264: Zinc perchlorate hexahydrate	10031-13-7	1502: Lead arsenite
7790-92-3	1387: Hypochlorous acid			10031-16-0	327: Barium dichromate dihydrate
7790-93-4	767: Chloric acid heptahydrate	10025-65-7	2110: Platinum(II) chloride		
				10031-20-6	1769: Manganese(II) bromide tetrahydrate
7790-94-5	778: Chlorosulfonic acid	10025-66-8	2334: Radium chloride		
7790-98-9	208: Ammonium perchlorate	10025-67-9	2818: Sulfur chloride	10031-21-7	1507: Lead bromate monohydrate
		10025-68-0	2467: Selenium chloride		
7790-99-0	1424: Iodine monochloride	10025-69-1	2752: Stannous chloride dihydrate	10031-22-8	1508: Lead bromide
				10031-23-9	2332: Radium bromide
7791-03-9	1612: Lithium perchlorate	10025-70-4	2778: Strontium chloride hexahydrate	10031-24-0	2750: Stannous bromide
7791-07-3	2679: Sodium perchlorate monohydrate			10031-25-1	807: Chromium(III) bromide
		10025-71-5	2863: Tellurium dichloride		
7791-08-4	273: Antimony(V) oxychloride			10031-26-2	1134: Ferric bromide
		10025-73-7	810: Chromium(III) chloride	10031-27-3	2870: Tellurium tetrabromide
7791-09-5	2357: Rhenium(VI) trioxychloride				
		10025-74-8	1063: Dysprosium chloride	10031-30-8	593: Calcium dihydrogen phosphate monohydrate
7791-10-8	2776: Strontium chlorate				
7791-11-9	2385: Rubidium chloride	10025-75-9	1088: Erbium chloride hexahydrate	10031-43-3	988: Copper(II) nitrate trihydrate
7791-12-0	2913: Thallium(I) chloride			10031-45-5	999: Copper(II) selenate pentahydrate
		10025-76-0	1120: Europium(III) chloride		
7791-13-1	868: Cobalt(II) chloride hexahydrate	10025-77-1	1136: Ferric chloride hexahydrate	10031-48-8	996: Copper(II) phosphate trihydrate
7791-16-4	269: Antimony(V) dichlorotrifluoride			10031-49-9	1069: Dysprosium nitrate pentahydrate
		10025-78-2	3030: Trichlorosilane		
7791-18-6	1683: Magnesium chloride hexahydrate	10025-82-8	1404: Indium(III) chloride	10031-50-2	1075: Dysprosium sulfate octahydrate
		10025-83-9	1436: Iridium(III) chloride		
7791-20-0	1948: Nickel chloride hexahydrate	10025-84-0	1473: Lanthanum chloride heptahydrate	10031-51-3	1093: Erbium nitrate pentahydrate
				10031-52-4	1100: Erbium sulfate octahydrate
7791-21-1	772: Chlorine monoxide	10025-85-1	2017: Nitrogen trichloride		
7791-23-3	2474: Selenium oxychloride	10025-87-3	2089: Phosphorus oxychloride	10031-52-4	1128: Europium(III) sulfate octahydrate
				10031-53-5	1123: Europium(III) nitrate hexahydrate
7791-25-5	2832: Sulfuryl chloride	10025-90-8	2307: Praseodymium chloride heptahydrate		
7791-26-6	3106: Uranyl chloride			10031-54-6	1114: Europium(II) sulfate
7791-28-8	316: Barium bromide dihydrate	10025-91-9	252: Antimony(III) chloride		
		10025-93-1	3095: Uranium trichloride	10034-76-1	646: Calcium sulfate hemihydrate
7791-29-9	2276: Potassium tetraiodoaurate(III)	10025-94-2	3204: Yttrium chloride hexahydrate		
				10034-81-8	1714: Magnesium perchlorate
7803-49-8	1381: Hydroxylamine	10025-98-6	2267: Potassium tetrachloropalladate(II)		
7803-51-2	2077: Phosphine			10034-82-9	2580: Sodium chromate tetrahydrate
7803-52-3	254: Antimony(III) hydride	10025-99-7	2268: Potassium tetrachloroplatinate(II)		
7803-54-5	1662: Magnesium amide				

10034-85-2	1372: Hydrogen iodide	
10034-88-5	2634: Sodium hydrogen sulfate monohydrate	
10034-93-2	1359: Hydrazine sulfate	
10034-96-5	1794: Manganese(II) sulfate monohydrate	
10034-98-7	3187: Ytterbium sulfate	
10034-98-7	3188: Ytterbium sulfate octahydrate	
10034-99-8	1734: Magnesium sulfate heptahydrate	
10035-01-5	3179: Ytterbium chloride hexahydrate	
10035-04-8	582: Calcium chloride dihydrate	
10035-05-9	580: Calcium chlorate dihydrate	
10035-06-0	454: Bismuth nitrate pentahydrate	
10035-10-6	1363: Hydrogen bromide	
10038-98-9	1274: Germanium(IV) chloride	
10039-31-3	429: Beryllium selenate tetrahydrate	
10039-32-4	2630: Sodium hydrogen phosphate dodecahydrate	
10039-54-0	1385: Hydroxylamine sulfate	
10039-55-1	1355: Hydrazine monohydroiodide	
10042-76-9	2793: Strontium nitrate	
10042-88-3	2881: Terbium chloride	
10043-01-3	83: Aluminum sulfate	
10043-11-5	484: Boron nitride	
10043-35-3	481: Boric acid	
10043-52-4	581: Calcium chloride	
10043-67-1	2135: Potassium aluminum sulfate	
10043-84-2	1779: Manganese(II) hypophosphite monohydrate	
10043-92-2	2336: Radon	
10045-86-0	1153: Ferric phosphate dihydrate	
10045-89-3	146: Ammonium ferrous sulfate hexahydrate	
10045-94-0	1842: Mercury(II) nitrate hemihydrate	
10048-95-0	2625: Sodium hydrogen arsenate heptahydrate	
10048-98-3	337: Barium hydrogen phosphate	
10048-99-4	389: Barium tetraiodo-mercurate(II)	
10049-01-1	463: Bismuth phosphate	
10049-03-3	1201: Fluorine perchlorate	
10049-04-4	769: Chlorine dioxide	
10049-05-5	796: Chromium(II) chloride	
10049-06-6	2999: Titanium dichloride	
10049-07-7	2367: Rhodium(III) chloride	
10049-08-8	2420: Ruthenium(III) chloride	
10049-10-2	798: Chromium(II) fluoride	
10049-12-4	3139: Vanadium trifluoride	
10049-12-4	3140: Vanadium trifluoride trihydrate	
10049-14-6	3091: Uranium tetrafluoride	

10049-16-8	3136: Vanadium tetrafluoride	
10049-17-9	2354: Rhenium(VI) fluoride	
10049-21-5	2591: Sodium dihydrogen phosphate monohydrate	
10049-23-7	2875: Tellurous acid	
10049-24-8	1435: Iridium(III) bromide tetrahydrate	
10049-25-9	795: Chromium(II) bromide	
10058-44-3	1155: Ferric pyrophosphate nonahydrate	
10060-08-9	587: Calcium chromate	
10060-09-0	540: Cadmium selenate dihydrate	
10060-10-3	685: Ceric fluoride	
10060-11-4	1266: Germanium(II) chloride	
10060-12-5	811: Chromium(III) chloride hexahydrate	
10060-13-6	134: Ammonium copper(II) chloride dihydrate	
10090-53-6	51: Aluminum iodide hexahydrate	
10097-28-6	2496: Silicon monoxide	
10099-58-8	1472: Lanthanum chloride	
10099-66-8	1638: Lutetium chloride	
10099-67-9	1644: Lutetium nitrate hydrate	
10099-74-8	1527: Lead nitrate	
10099-76-0	1536: Lead silicate	
10099-79-3	1551: Lead vanadate	
10101-41-4	645: Calcium sulfate dihydrate	
10101-50-5	2682: Sodium permanganate trihydrate	
10101-52-7	3313: Zirconium silicate	
10101-53-8	825: Chromium(III) sulfate	
10101-53-8	827: Chromium(III) sulfate octadecahydrate	
10101-63-0	1524: Lead iodide	
10101-68-5	1795: Manganese(II) sulfate tetrahydrate	
10101-89-0	2687: Sodium phosphate dodecahydrate	
10101-97-0	1977: Nickel sulfate hexahydrate	
10101-98-1	1976: Nickel sulfate heptahydrate	
10102-02-0	3260: Zinc nitrite	
10102-03-1	2021: Nitrogen(V) oxide	
10102-05-3	2058: Palladium(II) nitrate	
10102-15-5	2711: Sodium sulfite heptahydrate	
10102-17-7	2730: Sodium thiosulfate pentahydrate	
10102-18-8	2699: Sodium selenite	
10102-20-2	2715: Sodium tellurite(IV)	
10102-23-5	2697: Sodium selenate decahydrate	
10102-24-6	1603: Lithium metasilicate	
10102-25-7	1620: Lithium sulfate monohydrate	
10102-34-8	1722: Magnesium pyrophosphate trihydrate	
10102-40-6	2662: Sodium molybdate	

	dihydrate	
10102-43-9	2013: Nitric oxide	
10102-44-0	2015: Nitrogen dioxide	
10102-45-1	2923: Thallium(I) nitrate	
10102-49-5	1131: Ferric arsenate dihydrate	
10102-50-8	1168: Ferrous arsenate hexahydrate	
10102-68-8	611: Calcium iodide	
10102-71-3	2558: Sodium aluminum sulfate dodecahydrate	
10102-75-7	571: Calcium bromate	
10102-75-7	572: Calcium bromate monohydrate	
10102-83-4	2713: Sodium tellurate(VI)	
10102-90-6	998: Copper(II) pyrophosphate hydrate	
10103-50-1	1665: Magnesium arsenate hydrate	
10103-61-4	952: Copper(II) arsenate	
10108-64-2	520: Cadmium chloride	
10112-91-1	1816: Mercury(I) chloride	
10118-76-0	627: Calcium permanganate	
10119-31-0	3324: Zirconyl hydroxychloride hydrate	
10124-36-4	545: Cadmium sulfate	
10124-37-5	614: Calcium nitrate	
10124-39-8	1506: Lead borate monohydrate	
10124-41-1	653: Calcium thiosulfate hexahydrate	
10124-43-3	902: Cobalt(II) sulfate	
10124-48-8	1833: Mercury(II) chloride ammoniated	
10124-50-2	2197: Potassium hydrogen arsenite	
10124-50-2	2214: Potassium metaarsenite monohydrate	
10124-53-5	1743: Magnesium thiosulfate hexahydrate	
10124-56-8	2621: Sodium hexametaphosphate	
10125-13-0	965: Copper(II) chloride dihydrate	
10135-84-9	172: Ammonium hydrogen borate trihydrate	
10138-04-2	144: Ammonium ferric sulfate dodecahydrate	
10138-41-7	1087: Erbium chloride	
10138-52-0	1215: Gadolinium chloride	
10138-62-2	1331: Holmium chloride	
10139-47-6	3255: Zinc iodide	
10139-58-9	2370: Rhodium(III) nitrate	
10141-05-6	887: Cobalt(II) nitrate	
10163-15-2	2599: Sodium fluorophosphate	
10170-69-1	1756: Manganese carbonyl	
10190-55-3	1525: Lead molybdate	
10192-29-7	127: Ammonium chlorate	
10192-30-0	182: Ammonium hydrogen sulfite	
10196-18-6	3258: Zinc nitrate hexahydrate	
10210-64-7	408: Beryllium acetylacetonate	

10210-68-1	844: Cobalt carbonyl		acid	10567-69-8	344: Barium iodate
10213-09-9	3150: Vanadyl dichloride	10361-29-2	220: Ammonium	10580-03-7	241: Ammonium titanium
10213-10-2	2735: Sodium tungstate		sesquicarbonate		oxalate monohydrate
	dihydrate	10361-37-2	322: Barium chloride	10580-52-6	3119: Vanadium
10217-52-4	1351: Hydrazine hydrate	10361-43-0	448: Bismuth hydroxide		dichloride
10233-88-2	2603: Sodium gold	10361-46-3	460: Bismuth oxynitrate	11065-24-0	1430: Iridium carbonyl
	thiosulfate dihydrate	10361-79-2	2306: Praseodymium	11077-24-0	1163: Ferrocenium
10236-39-2	1777: Manganese(II)		chloride		hexafluorophosphate
	hydrogen phosphate	10361-82-7	2432: Samarium chloride	11089-20-6	1896: Molybdic silicic
	trihydrate	10361-84-9	2453: Scandium chloride		acid hydrate
10236-39-2	1789: Manganese(II)	10361-91-8	3178: Ytterbium chloride	11103-72-3	2415: Ruthenium
	phosphate heptahydrate	10361-92-9	3203: Yttrium chloride		ammoniated
10241-05-1	1887: Molybdenum(V)	10361-95-2	3236: Zinc chlorate		oxychloride
	chloride	10377-37-4	1591: Lithium hydrogen	11104-88-4	2078: Phosphomolybdic
10257-55-3	648: Calcium sulfite		carbonate		acid hydrate
	dihydrate	10377-48-7	1619: Lithium sulfate	11113-63-6	662: Carbon
10277-43-7	1480: Lanthanum nitrate	10377-51-2	1596: Lithium iodide	11116-03-3	2125: Plutonium(IV)
	hexahydrate	10377-52-3	1615: Lithium phosphate		oxide
10277-44-8	2319: Praseodymium	10377-58-9	1699: Magnesium iodide	11121-16-7	21: Aluminum borate
	sulfate	10377-60-3	1705: Magnesium nitrate	11126-81-1	23: Aluminum bromate
10290-12-7	953: Copper(II) arsenite	10377-62-5	1716: Magnesium		nonahydrate
10294-26-5	2545: Silver sulfate		permanganate	11138-11-7	330: Barium ferrite
10294-27-6	1284: Gold(I) bromide		hexahydrate	11138-49-1	2552: Sodium β-
10294-28-7	1290: Gold(III) bromide	10377-66-9	1783: Manganese(II)		aluminum oxide
10294-29-8	1286: Gold(I) chloride		nitrate	12002-03-8	949: Copper(II) acetate
10294-31-2	1288: Gold(I) iodide	10378-47-9	126: Ammonium		metaarsenite
10294-32-3	1296: Gold(III) selenate		cerium(IV) sulfate	12002-28-7	1378: Hydrogen
10294-33-4	490: Boron tribromide		dihydrate		tetracarbonylferrate(II)
10294-34-5	491: Boron trichloride	10378-50-4	2236: Potassium	12002-61-8	10: Actinium oxide
10294-38-9	321: Barium chlorate		perruthenate	12002-99-2	2547: Silver telluride
	monohydrate	10380-29-7	1011: Copper(II)	12003-63-1	2134: Potassium
10294-39-0	364: Barium perchlorate		tetraammine sulfate		aluminate trihydrate
	trihydrate		monohydrate	12003-65-5	1465: Lanthanum
10294-40-3	324: Barium chromate	10380-31-1	398: Barium uranium		aluminum oxide
10294-41-4	716: Cerous nitrate		oxide	12003-67-7	1600: Lithium
	hexahydrate	10381-36-9	1966: Nickel phosphate		metaaluminate
10294-42-5	689: Ceric sulfate		heptahydrate	12003-72-4	1863: Molybdenum
	tetrahydrate	10381-36-9	1967: Nickel phosphate		aluminide
10294-44-7	1815: Mercury(I) chlorate		octahydrate	12004-04-5	305: Barium aluminate
10294-46-9	995: Copper(II)	10381-37-0	2969: Thorium sulfate	12004-06-7	409: Beryllium aluminate
	perchlorate hexahydrate		nonahydrate	12004-29-4	74: Aluminum phosphate
10294-47-0	1510: Lead chlorate	10381-37-0	2970: Thorium sulfate		trihydroxide
10294-48-1	770: Chlorine heptoxide		octahydrate	12004-37-4	2770: Strontium
10294-50-5	896: Cobalt(II) phosphate	10381-37-0	2971: Thorium sulfate		aluminate
	octahydrate		tetrahydrate	12004-39-6	90: Aluminum titanate
10294-52-7	1137: Ferric chromate	10402-15-0	925: Copper citrate	12004-50-1	94: Aluminum zirconium
10294-53-8	1139: Ferric dichromate		hemipentahydrate	12004-71-6	1932: Nickel aluminide
10294-54-9	758: Cesium sulfate	10415-75-5	1821: Mercury(I) nitrate	12004-76-1	2835: Tantalum aluminide
10294-62-9	1489: Lanthanum sulfate		dihydrate	12004-83-0	3292: Zirconium
	nonahydrate	10431-47-7	2245: Potassium selenite		aluminide
10294-64-1	2213: Potassium	10450-55-2	1132: Ferric basic acetate	12005-21-9	3192: Yttrium aluminum
	manganate	10450-59-6	723: Cerous sulfate		oxide
10294-65-2	2218: Potassium nickel		octahydrate	12005-67-3	109: Americium(IV) oxide
	sulfate hexahydrate	10450-60-9	2071: Periodic acid	12005-75-3	924: Copper arsenide
10294-66-3	2284: Potassium	10466-65-6	2235: Potassium	12005-82-2	2524: Silver
	thiosulfate		perrhenate		hexafluoroarsenate
10294-70-9	2756: Stannous iodide	10476-81-0	2772: Strontium bromide	12005-86-6	2613: Sodium
10325-94-7	532: Cadmium nitrate	10476-81-0	2773: Strontium bromide		hexafluoroarsenate
10326-21-3	1681: Magnesium chloride		hexahydrate	12006-15-4	513: Cadmium arsenide
	hexahydrate	10476-85-4	2777: Strontium chloride	12006-40-5	3226: Zinc arsenide
10326-24-6	3227: Zinc arsenite	10476-86-5	2788: Strontium iodide	12006-77-8	842: Cobalt boride
10326-26-8	314: Barium bromate	10476-86-5	2789: Strontium iodide	12006-79-0	788: Chromium
	monohydrate		hexahydrate		monoboride
10326-27-9	323: Barium chloride	10486-00-7	2677: Sodium perborate	12006-80-3	782: Chromium boride
	dihydrate		tetrahydrate	12006-84-7	1443: Iron boride
10326-28-0	538: Cadmium perchlorate	10489-46-0	2374: Rhodium(III)	12006-85-8	1444: Iron boride
	hexahydrate		sulfate	12006-99-4	1864: Molybdenum boride
10332-33-9	2676: Sodium perborate	10534-89-1	1322: Hexaammine-	12007-00-0	1939: Nickel boride
	monohydrate		cobalt(III) chloride	12007-01-1	1938: Nickel boride
10340-06-4	676: Carbon sulfide	10544-73-7	2020: Nitrogen trioxide	12007-02-2	1940: Nickel boride
	telluride	10553-31-8	315: Barium bromide	12007-07-7	2836: Tantalum boride
10343-61-0	3016: Titanium sulfate	10555-76-7	2654: Sodium metaborate	12007-09-9	3040: Tungsten boride
10343-62-1	1858: Metaphosphoric		tetrahydrate	12007-10-2	3041: Tungsten boride

12007-16-8	786: Chromium diboride	
12007-23-7	1301: Hafnium boride	
12007-25-9	1670: Magnesium boride	
12007-29-3	1990: Niobium diboride	
12007-33-9	495: Boron trisulfide	
12007-34-0	2450: Scandium boride	
12007-36-2	3079: Uranium diboride	
12007-37-3	3117: Vanadium diboride	
12007-38-4	783: Chromium boride	
12007-60-2	1624: Lithium tetraborate	
12007-81-7	2487: Silicon boride	
12007-84-0	3088: Uranium tetraboride	
12007-97-5	1873: Molybdenum pentaboride	
12007-98-6	3064: Tungsten pentaboride	
12007-99-7	570: Calcium boride	
12008-02-5	696: Cerium hexaboride	
12008-05-8	1106: Europium boride	
12008-06-9	1213: Gadolinium boride	
12008-21-8	1466: Lanthanum boride	
12008-22-9	1671: Magnesium boride	
12008-23-0	1901: Neodymium boride	
12008-27-4	2302: Praseodymium boride	
12008-29-6	489: Boron silicide	
12008-29-6	2428: Samarium boride	
12008-32-1	3198: Yttrium boride	
12009-14-2	355: Barium niobate	
12009-18-6	377: Barium stannate trihydrate	
12009-21-1	401: Barium zirconate	
12009-27-5	393: Barium titanate	
12009-31-3	394: Barium titanate	
12009-36-8	387: Barium telluride	
12011-97-1	1865: Molybdenum carbide	
12011-99-3	1989: Niobium carbide	
12012-16-7	2950: Thorium carbide	
12012-17-8	3125: Vanadium monocarbide	
12012-32-7	694: Cerium carbide	
12012-35-0	784: Chromium carbide	
12013-10-4	846: Cobalt disulfide	
12013-21-7	1703: Magnesium molybdate	
12013-46-6	641: Calcium stannate trihydrate	
12013-47-7	657: Calcium zirconate	
12013-56-8	640: Calcium silicide	
12013-57-9	650: Calcium telluride	
12013-82-0	616: Calcium nitride	
12014-14-1	554: Cadmium titanate	
12014-29-8	512: Cadmium antimonide	
12014-56-1	686: Ceric hydroxide	
12014-56-1	688: Ceric oxide hydrate	
12014-82-3	697: Cerium monosulfide	
12014-85-6	700: Cerium silicide	
12014-93-6	724: Cerous sulfide	
12014-97-0	725: Cerous telluride	
12016-69-2	870: Cobalt(II) chromite	
12016-80-7	918: Cobalt(III) oxide hydroxide	
12017-01-5	849: Cobalt metatitanate	
12017-08-2	900: Cobalt(II) silicate	
12017-12-8	845: Cobalt disilicide	
12017-13-9	906: Cobalt(II) telluride	
12017-38-8	853: Cobalt orthotitanate	
12017-94-6	1475: Lanthanum chromite	
12018-01-8	831: Chromium(IV) oxide	
12018-09-6	787: Chromium disilicide	
12018-10-9	967: Copper(II) chromite	

12018-19-8	3239: Zinc chromite	
12018-22-3	828: Chromium(III) sulfide	
12018-36-9	793: Chromium silicide	
12018-61-0	760: Cesium superoxide	
12018-79-0	970: Copper(II) ferrate	
12019-07-7	1003: Copper(II) stannate	
12019-08-8	1013: Copper(II) titanate	
12019-23-7	1009: Copper(II) telluride	
12019-52-2	945: Copper(I) telluride	
12019-57-7	927: Copper phosphide	
12019-88-4	1070: Dysprosium nitride	
12020-14-3	3295: Zirconium carbide	
12020-21-2	1094: Erbium nitride	
12020-28-9	1098: Erbium silicide	
12020-39-2	1102: Erbium telluride	
12020-58-5	1108: Europium nitride	
12020-65-4	1115: Europium(II) sulfide	
12020-66-5	1113: Europium(II) selenide	
12020-69-8	1116: Europium(II) telluride	
12021-58-8	2064: Palladium(III) fluoride	
12021-68-0	852: Cobalt nitrosodicarbonyl	
12021-70-4	1180: Ferrous hexafluorosilicate hexahydrate	
12022-02-5	152: Ammonium heptafluorotantalate	
12022-95-6	1451: Iron silicide	
12022-99-0	1445: Iron disilicide	
12023-71-1	1643: Lutetium iron oxide	
12023-91-5	2781: Strontium ferrite	
12023-99-3	1251: Gallium(III) hydroxide	
12024-10-1	1244: Gallium(II) sulfide	
12024-11-2	1243: Gallium(II) selenide	
12024-14-5	1245: Gallium(II) telluride	
12024-20-3	1241: Gallium suboxide	
12024-21-4	1255: Gallium(III) oxide	
12024-22-5	1261: Gallium(III) sulfide	
12024-24-7	1258: Gallium(III) selenide	
12024-27-0	1262: Gallium(III) telluride	
12024-36-1	1218: Gadolinium gallium garnet	
12024-89-4	1232: Gadolinium titanate	
12025-13-7	1691: Magnesium germanate	
12025-19-3	2655: Sodium metagermanate	
12025-32-0	1271: Germanium(II) sulfide	
12025-34-2	1281: Germanium(IV) sulfide	
12025-39-7	1272: Germanium(II) telluride	
12026-66-3	111: Ammonium 12-molybdophosphate hydrate	
12027-06-4	188: Ammonium iodide	
12028-48-7	191: Ammonium metatungstate hexahydrate	
12029-81-1	1337: Holmium nitride	
12029-98-0	1427: Iodine pentoxide	
12030-14-7	1402: Indium(II) sulfide	
12030-24-9	1416: Indium(III) sulfide	

12030-49-8	1439: Iridium(IV) oxide	
12030-85-2	2219: Potassium niobate	
12030-88-5	2255: Potassium superoxide	
12030-91-0	2256: Potassium tantalate	
12030-97-6	2285: Potassium titanate	
12030-98-7	2297: Potassium zirconate	
12031-43-5	1484: Lanthanum oxysulfide	
12031-49-1	1491: Lanthanum sulfide	
12031-53-7	1492: Lanthanum telluride	
12031-63-9	1605: Lithium niobate	
12031-66-2	1622: Lithium tantalate	
12031-80-0	1614: Lithium peroxide	
12031-82-2	1630: Lithium titanate	
12031-83-3	1633: Lithium zirconate	
12032-13-2	1649: Lutetium silicide	
12032-20-1	1647: Lutetium oxide	
12032-29-0	1730: Magnesium stannate trihydrate	
12032-30-3	1702: Magnesium metatitanate	
12032-31-4	1747: Magnesium zirconate	
12032-35-8	1689: Magnesium dititanate	
12032-36-9	1736: Magnesium sulfide	
12032-52-9	1710: Magnesium orthotitanate	
12032-69-8	1758: Manganese niobate	
12032-74-5	1798: Manganese(II) titanate	
12032-82-5	1752: Manganese antimonide	
12032-86-9	1763: Manganese silicide	
12032-88-1	1797: Manganese(II) telluride	
12032-89-2	1808: Manganese(IV) telluride	
12033-19-1	1870: Molybdenum mononitride	
12033-29-3	1895: Molybdenum(VI) sulfide	
12033-31-7	1871: Molybdenum nitride	
12033-33-9	1881: Molybdenum(III) sulfide	
12033-54-4	2120: Plutonium nitride	
12033-62-4	2843: Tantalum nitride(δ)	
12033-62-4	2844: Tantalum nitride(ε)	
12033-64-6	2887: Terbium nitride	
12033-65-7	2961: Thorium nitride	
12033-72-6	3057: Tungsten nitride	
12033-82-8	2794: Strontium nitride	
12033-83-9	3098: Uranium trinitride	
12033-88-4	2016: Nitrogen selenide	
12033-89-5	2497: Silicon nitride	
12034-09-2	2664: Sodium niobate	
12034-15-0	2659: Sodium metatantalate	
12034-36-5	2731: Sodium titanate	
12034-39-8	2725: Sodium tetrasulfide	
12034-57-0	1996: Niobium(II) oxide	
12034-59-2	1998: Niobium(IV) oxide	
12034-66-1	1995: Niobium phosphide	
12034-77-4	1999: Niobium(IV) selenide	
12034-80-9	1991: Niobium disilicide	
12034-83-2	2001: Niobium(IV) telluride	
12034-88-7	1526: Lead niobate	
12034-89-8	2792: Strontium niobate	
12035-32-4	1922: Neodymium sulfide	
12035-35-7	1923: Neodymium	

	telluride	12041-54-2	36: Aluminum dodecaboride	12059-14-2	1971: Nickel silicide
12035-38-0	1972: Nickel stannate dihydrate			12059-51-7	2399: Rubidium niobate
12035-39-1	1982: Nickel titanate	12042-68-1	563: Calcium aluminate	12060-00-3	1548: Lead titanate
12035-52-8	1933: Nickel antimonide	12042-78-3	564: Calcium aluminate (β)	12060-01-4	1552: Lead zirconate
12035-64-2	1968: Nickel phosphide			12060-08-1	2458: Scandium oxide
12035-72-2	1974: Nickel subsulfide	12043-29-7	87: Aluminum telluride	12060-58-1	2440: Samarium oxide
12035-79-9	1928: Neptunium(IV) oxide	12044-16-5	1442: Iron arsenide	12060-59-2	2814: Strontium titanate
		12044-42-7	841: Cobalt arsenide	12061-16-4	1096: Erbium oxide
12035-82-4	2114: Platinum(II) oxide	12044-49-4	1666: Magnesium arsenide	12062-24-7	980: Copper(II) hexafluorosilicate tetrahydrate
12035-90-4	2855: Tantalum tetroxide				
12035-98-2	3127: Vanadium monoxide	12044-54-1	291: Arsenic(III) telluride		
		12045-01-1	843: Cobalt boride	12063-56-8	3210: Yttrium iron oxide
12036-02-1	2040: Osmium(IV) oxide	12045-15-7	1754: Manganese boride	12063-98-8	1240: Gallium phosphide
12036-09-8	2347: Rhenium(IV) oxide	12045-19-1	1993: Niobium monoboride	12064-03-8	1236: Gallium antimonide
12036-10-1	2423: Ruthenium(IV) oxide			12064-62-9	1225: Gadolinium oxide
		12045-27-1	3124: Vanadium monoboride	12065-10-0	1270: Germanium(II) selenide
12036-16-7	2858: Technetium dioxide				
12036-21-4	3121: Vanadium dioxide	12045-63-5	2996: Titanium boride	12065-11-1	1280: Germanium(IV) selenide
12036-22-5	3049: Tungsten dioxide	12045-64-6	3293: Zirconium boride		
12036-23-6	3306: Zirconium oxide	12045-77-1	1345: Homium boride	12065-36-0	1264: Germanium nitride
12036-32-7	2325: Praseodymium(III) oxide	12045-78-2	2261: Potassium tetraborate tetrahydrate	12065-68-8	1541: Lead tantalate
				12065-74-6	2811: Strontium tantalate
12036-35-0	2372: Rhodium(III) oxide	12045-87-3	2719: Sodium tetraborate tetrahydrate	12066-83-0	2318: Praseodymium silicide
12036-39-4	2817: Strontium zirconate				
12036-43-0	3288: Zinc titanate	12045-88-4	2718: Sodium tetraborate pentahydrate	12067-00-4	2351: Rhenium(IV) telluride
12036-44-1	2985: Thulium oxide				
12036-46-3	261: Antimony(III) phosphate	12045-94-2	3068: Tungsten tetrabromide	12067-15-3	3006: Titanium ditelluride
				12067-22-0	2444: Samarium sulfide
12037-01-3	2893: Terbium(III,IV) oxide	12046-08-1	333: Barium hexaboride	12067-26-4	3067: Tungsten telluride
		12046-54-7	2783: Strontium hexaboride	12067-45-7	3004: Titanium diselenide
12037-29-5	2326: Praseo-dymium(III,IV) oxide			12067-46-8	3051: Tungsten diselenide
		12047-25-5	348: Barium lead oxide	12067-54-8	2968: Thorium silicide
12037-63-7	2851: Tantalum phosphide	12047-27-7	395: Barium titanate	12067-56-0	2856: Tantalum trisilicide
12037-65-9	3014: Titanium phosphide	12047-34-6	385: Barium tantalate	12067-57-1	3025: Titanium trisilicide
12037-80-8	3310: Zirconium phosphide	12047-79-9	357: Barium nitride	12067-66-2	2854: Tantalum telluride
		12048-50-9	474: Bismuth tetroxide	12067-99-1	2104: Phosphotungstic acid 24-hydrate
12037-82-0	2086: Phosphorus heptasulfide	12048-51-0	476: Bismuth titanate		
		12049-50-2	654: Calcium titanate	12068-40-5	1560: Lithium aluminum silicate
12038-12-9	2322: Praseodymium telluride	12050-35-0	550: Cadmium tantalate		
		12052-28-7	874: Cobalt(II) diiron tetroxide	12068-51-8	1659: Magnesium aluminum oxide
12038-13-0	2321: Praseodymium sulfide				
		12052-42-5	837: Cobalt antimonide	12068-69-8	464: Bismuth selenide
12038-63-0	2350: Rhenium(IV) sulfide	12053-12-2	780: Chromium antimonide	12068-84-8	2869: Tellurium sulfate
				12068-90-5	1855: Mercury(II) telluride
12038-64-1	2348: Rhenium(IV) selenide	12053-13-3	792: Chromium selenide		
		12053-26-8	1685: Magnesium chromite	12069-00-0	1534: Lead selenide
12038-66-3	2349: Rhenium(IV) silicide			12069-32-8	483: Boron carbide
		12053-27-9	789: Chromium nitride	12069-69-1	962: Copper(II) carbonate hydroxide
12038-67-4	2359: Rhenium(VII) sulfide	12053-39-3	829: Chromium(III) telluride		
				12069-85-1	1303: Hafnium carbide
12039-07-5	3009: Titanium monosulfide	12053-66-6	751: Cesium niobate	12069-89-5	1866: Molybdenum carbide
		12054-48-7	1955: Nickel hydroxide		
12039-13-3	3005: Titanium disulfide	12054-85-2	193: Ammonium molybdate tetrahydrate	12069-94-2	1997: Niobium(IV) carbide
12039-15-5	3316: Zirconium sulfide				
12039-16-6	3026: Titanium trisulfide	12055-23-1	1309: Hafnium oxide	12070-06-3	2842: Tantalum monocarbide
12039-19-9	3217: Yttrium sulfide	12055-24-2	1318: Hafnium titanate		
12039-35-9	3224: Zinc antimonide	12055-62-8	1339: Holmium oxide	12070-07-4	2837: Tantalum carbide
12039-55-3	2852: Tantalum selenide	12056-07-4	1414: Indium(III) selenide	12070-08-5	2997: Titanium carbide
12039-76-8	3126: Vanadium monosilicide	12056-90-5	1487: Lanthanum silicide	12070-09-6	3083: Uranium monocarbide
		12057-17-9	1598: Lithium manganate		
12039-79-1	2853: Tantalum silicide	12057-24-8	1611: Lithium oxide	12070-10-9	3115: Vanadium carbide
12039-80-4	2890: Terbium silicide	12057-71-5	1708: Magnesium nitride	12070-12-1	3043: Tungsten carbide
12039-83-7	3015: Titanium silicide	12057-74-8	1720: Magnesium phosphide	12070-13-2	3042: Tungsten carbide
12039-84-8	2986: Thulium silicide			12071-15-7	1469: Lanthanum carbide
12039-87-1	3122: Vanadium disilicide	12057-75-9	1664: Magnesium antimonide	12071-29-3	2774: Strontium carbide
12039-88-2	3052: Tungsten disilicide			12071-31-7	2952: Thorium dicarbide
12039-89-3	3186: Ytterbium silicide	12057-92-0	1809: Manganese(VII) oxide	12071-33-9	3080: Uranium dicarbide
12039-90-6	3314: Zirconium silicide			12071-35-1	3201: Yttrium carbide
12039-95-1	3074: Tungsten trisilicide	12058-18-3	1885: Molybdenum(IV) selenide	12077-35-1	2838: Tantalum diboride
12040-00-5	2445: Samarium telluride			12079-58-2	659: Carbon
12040-02-7	2765: Stannous telluride	12058-20-7	1886: Molybdenum(IV) telluride	12084-29-6	304: Barium acetylacetonate octahydrate
12041-50-8	34: Aluminum diboride	12058-85-4	2688: Sodium phosphide		

12086-48-6	3114: Vanadium bis(cyclopentadienyl) dichloride	
12088-65-2	1447: Iron dodecacarbonyl	
12093-05-9	1026: Cyclooctatetraene iron tricarbonyl	
12108-13-3	1860: Methylcyclopentadienylmanganese tricarbonyl	
12116-66-4	1319: Hafnocene dichloride	
12124-97-9	119: Ammonium bromide	
12124-99-1	181: Ammonium hydrogen sulfide	
12125-01-8	147: Ammonium fluoride	
12125-02-9	128: Ammonium chloride	
12125-08-5	158: Ammonium hexachloroosmiate(IV)	
12125-09-6	2080: Phosphonium iodide	
12125-19-8	3075: Tungsten trisulfide	
12125-22-3	2062: Palladium(II) sulfide	
12125-25-6	1645: Lutetium nitride	
12125-63-2	1452: Iron telluride	
12133-07-2	1074: Dysprosium silicide	
12133-10-7	1076: Dysprosium sulfide	
12133-44-7	539: Cadmium phosphide	
12134-02-0	854: Cobalt phosphide	
12134-22-4	764: Cesium trioxide	
12134-75-7	1227: Gadolinium silicide	
12134-77-9	1230: Gadolinium sulfide	
12135-52-3	656: Calcium vanadate	
12135-76-1	224: Ammonium sulfide	
12136-24-2	1341: Holmium silicide	
12136-45-7	2217: Potassium monoxide	
12136-58-2	1621: Lithium sulfide	
12136-78-6	1875: Molybdenum silicide	
12136-97-9	2000: Niobium(IV) sulfide	
12137-04-1	1919: Neodymium silicide	
12137-12-1	1986: Nickel(II,III) sulfide	
12137-20-1	3010: Titanium monoxide	
12137-27-8	2375: Rhodium(IV) oxide dihydrate	
12137-34-7	2411: Rubidium titanate	
12137-83-6	2108: Platinum silicide	
12138-07-7	2972: Thorium sulfide	
12138-08-8	3134: Vanadium sulfide	
12138-09-9	3053: Tungsten disulfide	
12138-11-3	2892: Terbium sulfide	
12138-17-9	3132: Vanadium pentasulfide	
12138-28-2	2807: Strontium silicide	
12139-23-0	557: Cadmium zirconate	
12139-93-4	901: Cobalt(II) stannate	
12141-45-6	78: Aluminum silicate	
12141-46-7	81: Aluminum silicate	
12142-33-5	2248: Potassium stannate trihydrate	
12142-88-0	1979: Nickel telluride	
12143-34-9	2808: Strontium stannate	
12143-39-4	3218: Yttrium vanadate	
12143-72-5	2839: Tantalum disulfide	
12143-96-3	1678: Magnesium carbonate hydroxide trihydrate	
12150-46-8	1161: 1,1'-Bis(diphenylphosphino)ferrocene	
12152-72-6	1025: Cyclohexadiene iron tricarbonyl	

12154-95-9	1028: Cyclopentadienyliron dicarbonyl dimer
12158-56-4	761: Cesium tantalate
12159-07-8	928: Copper silicide
12159-43-2	1077: Dysprosium telluride
12159-66-9	1101: Erbium sulfide
12160-99-5	1231: Gadolinium telluride
12162-21-9	1312: Hafnium selenide
12162-59-3	1343: Holmium sulfide
12162-61-7	1344: Holmium telluride
12163-00-7	1599: Lithium manganite
12163-20-1	1652: Lutetium sulfide
12163-22-3	1653: Lutetium telluride
12163-26-7	1704: Magnesium niobate
12163-69-8	1874: Molybdenum phosphide
12164-01-1	2868: Tellurium nitride
12164-94-2	116: Ammonium azide
12166-29-9	2460: Scandium sulfide
12166-30-2	2988: Thulium sulfide
12166-44-8	2461: Scandium telluride
12166-47-1	3312: Zirconium selenide
12168-52-4	1196: Ferrous titanate
12179-02-1	2642: Sodium hydroxide monohydrate
12183-80-1	79: Aluminum silicate
12185-10-3	2085: Phosphorus (white)
12186-97-9	3193: Yttrium antimonide
12187-14-3	531: Cadmium niobate
12190-79-3	1577: Lithium cobaltite
12193-47-4	2769: Strontium acetylacetonate
12196-91-7	840: Cobalt arsenide
12201-48-8	2738: Sodium zirconate
12201-89-7	1951: Nickel disilicide
12202-79-8	3105: Uranyl carbonate
12205-73-1	3326: Zirconyl perchlorate octahydrate
12209-98-2	2702: Sodium stannate trihydrate
12211-52-8	135: Ammonium cyanide
12214-16-3	759: Cesium sulfide
12228-50-1	1757: Manganese diboride
12228-86-3	184: Ammonium hydrogen tetraborate dihydrate
12228-87-4	228: Ammonium tetraborate tetrahydrate
12228-91-0	1767: Manganese(II) borate octahydrate
12229-12-8	204: Ammonium pentaborate tetrahydrate
12229-13-9	2226: Potassium pentaborate octahydrate
12229-63-9	2954: Thorium hexaboride
12230-71-6	342: Barium hydroxide octahydrate
12230-74-9	339: Barium hydrosulfide tetrahydrate
12232-99-4	2650: Sodium metabismuthate
12233-34-0	475: Bismuth titanate
12233-56-6	446: Bismuth germanium oxide
12244-51-8	1944: Nickel carbonate hydroxide tetrahydrate
12249-30-8	7: Actinium hydroxide
12249-52-4	2526: Silver hydrogen fluoride
12253-13-3	1750: Manganese aluminide

12254-64-7	103: Americium oxide(α)
12254-64-7	104: Americium oxide(β)
12254-82-9	838: Cobalt arsenic sulfide
12254-85-2	781: Chromium arsenide
12255-36-6	247: Antimony arsenide
12255-48-0	3194: Yttrium arsenide
12255-50-4	309: Barium arsenide
12260-55-8	1282: Germanium(IV) telluride
12266-38-5	1500: Lead antimonide
12266-65-8	1755: Manganese carbide
12279-90-2	279: Arsenic disulfide
12286-33-8	2993: Tin triphosphide
12291-65-5	599: Calcium hexaborate pentahydrate
12293-61-7	1740: Magnesium tantalate
12294-01-8	3215: Yttrium phosphide
12300-22-0	2442: Samarium silicide
12310-43-9	1060: Dysprosium boride
12310-44-0	1083: Erbium boride
12323-03-4	376: Barium sodium niobium oxide
12323-19-2	438: Bismuth antimonide
12325-59-6	1311: Hafnium phosphide
12333-74-3	2409: Rubidium tantalate
12336-95-7	806: Chromium(III) basic sulfate
12338-09-9	465: Bismuth stannate
12340-14-6	3170: Xenon tetroxide
12345-14-1	325: Barium chromate(V)
12355-99-6	2338: Rhenium boride
12384-63-3	3112: Uranyl sulfate trihydrate
12397-32-9	1761: Manganese phosphide
12401-56-8	1314: Hafnium silicide
12412-52-1	259: Antimony(III) oxide
12422-12-7	556: Cadmium vanadate
12427-42-8	922: Cobaltocenium hexafluorophosphate
12433-14-6	956: Copper(II) basic chromate
12434-24-1	1109: Europium silicide
12435-86-8	2816: Strontium vanadate
12442-45-4	699: Cerium oxysulfide
12446-46-7	107: Americium sulfide
12501-23-4	3077: Tungstophosphoric acid hydrate
12502-31-7	2220: Potassium niobate hexadecahydrate
12513-27-8	3229: Zinc borate hemiheptahydrate
12520-88-6	2503: Silicotungstic acid
12534-23-5	2413: Rubidium zirconate
12536-52-6	412: Beryllium borides
12536-65-1	3230: Zinc borate pentahydrate
12612-73-6	3094: Uranium tricarbide
12616-24-9	245: Ammonium zirconyl carbonate dihydrate
12640-47-0	909: Cobalt(II) tungstate
12650-28-1	373: Barium silicate
12671-00-0	3296: Zirconium carbonate basic hydrate
12672-79-6	307: Barium aluminide
12688-52-7	1636: Lutetium boride
12712-36-6	272: Antimony(V) oxide hydrate
12752-71-0	44: Aluminum hydroxide (β')
12770-26-2	1306: Hafnium hydride
12774-81-1	668: Carbon fluoride

CAS RN	Compound
12777-45-6	466: Bismuth stannate pentahydrate
12785-50-1	312: Barium bismuth oxide
12789-09-2	929: Copper vanadate
12793-14-5	2011: Niobocene dichloride
13005-39-5	2266: Potassium tetrachloroaurate(III) dihydrate
13007-92-6	785: Chromium carbonyl
13092-66-5	1741: Magnesium tetrahydrogen phosphate dihydrate
13126-12-0	2400: Rubidium nitrate
13135-31-4	3021: Titanium tribromide
13138-45-9	1961: Nickel nitrate
13255-26-0	353: Barium metasilicate
13255-48-6	1347: Hydrazine acetate
13266-82-5	3212: Yttrium oxalate nonahydrate
13266-83-6	717: Cerous oxalate nonahydrate
13268-42-3	143: Ammonium ferric oxalate trihydrate
13283-01-7	3055: Tungsten hexachloride
13308-51-5	487: Boron phosphate
13320-71-3	1883: Molybdenum(IV) chloride
13327-32-7	421: Beryllium hydroxide(α)
13327-32-7	422: Beryllium hydroxide(β)
13395-16-9	951: Copper(II) acetylacetonate
13400-13-0	741: Cesium fluoride
13410-01-0	2696: Sodium selenate
13424-46-9	1503: Lead azide
13444-75-2	1817: Mercury(I) chromate
13444-75-2	1834: Mercury(II) chromate
13444-85-4	2019: Nitrogen triiodide
13444-90-1	2030: Nitryl chloride
13444-92-3	2036: Osmium(II) chloride
13444-93-4	2037: Osmium(III) chloride
13444-94-5	2051: Palladium(II) bromide
13444-96-7	2055: Palladium(II) fluoride
13446-03-2	1768: Manganese(II) bromide
13446-08-7	149: Ammonium fluorosulfonate
13446-09-8	187: Ammonium iodate
13446-10-1	209: Ammonium permanganate
13446-18-9	1707: Magnesium nitrate hexahydrate
13446-19-0	1715: Magnesium perchlorate hexahydrate
13446-23-6	1718: Magnesium phosphate octahydrate
13446-24-7	1721: Magnesium pyrophosphate
13446-29-2	1738: Magnesium sulfite hexahydrate
13446-34-9	1772: Manganese(II) chloride tetrahydrate
13446-48-5	197: Ammonium nitrite
13446-49-6	2215: Potassium molybdate
13446-53-2	1674: Magnesium bromide hexahydrate
13446-57-6	1878: Molybdenum(III) bromide
13446-70-3	2381: Rubidium bromate
13446-71-4	2384: Rubidium chlorate
13446-72-5	2386: Rubidium chromate
13446-73-6	2389: Rubidium dichromate
13446-74-7	2390: Rubidium fluoride
13450-84-5	1216: Gadolinium chloride hexahydrate
13450-87-8	1228: Gadolinium sulfate
13450-87-8	1229: Gadolinium sulfate octahydrate
13450-88-9	1246: Gallium(III) bromide
13450-90-3	1247: Gallium(III) chloride
13450-91-4	1252: Gallium(III) iodide
13450-92-5	1273: Germanium(IV) bromide
13450-95-8	1278: Germanium(IV) iodide
13450-97-0	2799: Strontium perchlorate
13450-97-0	2800: Strontium perchlorate hexahydrate
13451-05-3	2815: Strontium tungstate
13451-11-1	2845: Tantalum pentabromide
13451-18-8	2874: Tellurium trioxide
13451-19-9	2886: Terbium nitrate hexahydrate
13453-06-0	227: Ammonium tellurate
13453-07-1	1291: Gold(III) chloride
13453-24-2	1294: Gold(III) iodide
13453-30-0	2912: Thallium(I) chlorate
13453-32-2	2936: Thallium(III) chloride
13453-33-3	2937: Thallium(III) chloride hydrate
13453-34-4	2914: Thallium(I) cyanide
13453-40-2	2927: Thallium(I) perchlorate
13453-49-1	2949: Thorium bromide
13453-57-2	1512: Lead chlorite
13453-69-5	1564: Lithium borate
13453-70-8	1568: Lithium bromide monohydrate
13453-71-9	1571: Lithium chloride
13453-78-6	1613: Lithium perchlorate trihydrate
13453-80-0	1582: Lithium dihydrogen phosphate
13453-84-4	1609: Lithium orthosilicate
13454-71-2	720: Cerous phosphate hydrate
13454-74-5	726: Cerous tungstate
13454-75-6	733: Cesium bromate
13454-81-4	747: Cesium iodate
13454-84-7	755: Cesium perchlorate
13454-88-1	974: Copper(II) fluoride dihydrate
13454-94-9	722: Cerous sulfate
13454-96-1	2116: Platinum(IV) chloride pentahydrate
13455-01-1	2095: Phosphorus(III) iodide
13455-12-4	2109: Platinum(II) bromide
13455-20-4	2164: Potassium dithionate
13455-21-5	2168: Potassium fluoride dihydrate
13455-24-8	2200: Potassium hydrogen iodate
13455-28-2	881: Cobalt(II) iodate
13455-29-3	883: Cobalt(II) iodide dihydrate
13455-29-3	884: Cobalt(II) iodide hexahydrate
13455-31-7	894: Cobalt(II) perchlorate
13455-34-0	904: Cobalt(II) sulfate monohydrate
13462-88-9	1941: Nickel bromide
13462-88-9	1942: Nickel bromide trihydrate
13462-90-3	1958: Nickel iodide
13462-93-6	115: Ammonium arsenate hydrate
13462-93-6	137: Ammonium dihydrogen arsenate
13463-10-0	1154: Ferric phosphate hydrate
13463-12-2	1170: Ferrous bromide hexahydrate
13463-12-2	1171: Ferrous bromide hydrate
13463-22-4	361: Barium oxalate monohydrate
13463-39-3	1945: Nickel carbonyl
13463-40-6	1450: Iron pentacarbonyl
13463-67-7	3002: Titanium dioxide
13464-44-4	3225: Zinc arsenate octahydrate
13464-46-5	98: Americium chloride
13464-58-9	297: Arsenious acid
13464-80-7	1049: Dihydrazine sulfate
13464-82-9	1415: Indium(III) sulfate
13464-92-1	517: Cadmium bromide tetrahydrate
13464-98-7	1350: Hydrazine dinitrate
13465-09-3	1403: Indium(III) bromide
13465-10-6	1398: Indium(I) chloride
13465-11-7	1401: Indium(II) chloride
13465-15-1	1412: Indium(III) perchlorate octahydrate
13465-35-9	1819: Mercury(I) iodate
13465-43-5	2371: Rhodium(III) nitrate dihydrate
13465-49-1	2404: Rubidium permanganate
13465-55-9	2433: Samarium chloride hexahydrate
13465-58-2	2443: Samarium sulfate octahydrate
13465-59-3	2451: Scandium bromide
13465-60-6	2456: Scandium nitrate pentahydrate
13465-66-2	2480: Selenium tetrafluoride
13465-71-9	3033: Trifluorosilane
13465-72-0	3034: Triiodosilane
13465-73-1	506: Bromosilane
13465-77-5	1325: Hexachlorodisilane
13465-78-6	777: Chlorosilane
13465-84-4	2502: Silicon tetraiodide
13465-93-5	2550: Silver tungstate
13465-94-6	358: Barium nitrite
13465-95-7	363: Barium perchlorate
13466-08-5	2569: Sodium bromide

	dihydrate	
13466-20-1	352: Barium metaphosphate	
13466-21-2	368: Barium pyrophosphate	
13469-98-2	3199: Yttrium bromide	
13469-98-2	3200: Yttrium bromide nonahydrate	
13470-01-4	2787: Strontium iodate	
13470-04-7	2791: Strontium molybdate(VI)	
13470-06-9	2795: Strontium nitrite	
13470-08-1	3023: Titanium trifluoride	
13470-10-5	3045: Tungsten dibromide	
13470-11-6	3065: Tungsten pentabromide	
13470-12-7	3046: Tungsten dichloride	
13470-13-8	3069: Tungsten tetrachloride	
13470-14-9	3066: Tungsten pentachloride	
13470-17-2	3047: Tungsten diiodide	
13470-19-4	3093: Uranium tribromide	
13470-20-7	3089: Uranium tetrabromide	
13470-21-8	3086: Uranium pentachloride	
13470-22-9	3092: Uranium tetraiodide	
13470-23-0	3101: Uranium(IV) sulfate tetrahydrate	
13470-24-1	1859: Metavanadic acid	
13470-26-3	3137: Vanadium tribromide	
13470-38-7	3209: Yttrium iodide	
13472-30-5	2669: Sodium orthosilicate	
13472-31-6	2681: Sodium periodate trihydrate	
13472-33-8	2684: Sodium perrhenate	
13472-36-1	2694: Sodium pyrophosphate decahydrate	
13472-45-2	2734: Sodium tungstate	
13473-57-9	1342: Holmium sulfate octahydrate	
13473-77-3	1651: Lutetium sulfate octahydrate	
13476-01-2	861: Cobalt(II) bromate hexahydrate	
13476-08-9	1184: Ferrous nitrate hexahydrate	
13476-99-8	3145: Vanadium(III) acetylacetonate	
13477-00-4	320: Barium chlorate	
13477-09-3	336: Barium hydride	
13477-19-5	529: Cadmium metasilicate	
13477-20-8	546: Cadmium sulfate monohydrate	
13477-23-1	549: Cadmium sulfite	
13477-28-6	575: Calcium bromide hexahydrate	
13477-29-7	584: Calcium chloride monohydrate	
13477-34-4	615: Calcium nitrate tetrahydrate	
13477-36-6	625: Calcium perchlorate	
13477-89-9	1907: Neodymium chloride hexahydrate	
13477-91-3	1921: Neodymium sulfate octahydrate	
13477-95-7	1950: Nickel cyanide tetrahydrate	

13477-98-0	1956: Nickel iodate	
13477-99-1	1957: Nickel iodate tetrahydrate	
13478-00-7	1962: Nickel nitrate hexahydrate	
13478-06-3	808: Chromium(III) bromide hexahydrate	
13478-10-9	1175: Ferrous chloride tetrahydrate	
13478-14-3	1562: Lithium arsenate	
13478-16-5	1663: Magnesium ammonium phosphate hexahydrate	
13478-17-6	1876: Molybdenum(II) chloride	
13478-18-7	1879: Molybdenum(III) chloride	
13478-20-1	2090: Phosphorus oxyfluoride	
13478-33-6	895: Cobalt(II) perchlorate hexahydrate	
13478-38-1	987: Copper(II) nitrate hexahydrate	
13478-45-0	2002: Niobium(V) bromide	
13478-49-4	1099: Erbium sulfate	
13478-50-7	1547: Lead thiosulfate	
13479-54-4	978: Copper(II) glycinate monohydrate	
13492-26-7	2202: Potassium hydrogen phosphite	
13492-45-0	1183: Ferrous iodide tetrahydrate	
13494-80-9	2860: Tellurium	
13494-90-1	1253: Gallium(III) nitrate	
13494-92-3	1421: Iodine dioxide	
13494-94-2	1259: Gallium(III) sulfate	
13494-98-9	3211: Yttrium nitrate hexahydrate	
13498-07-2	2316: Praseodymium perchlorate hexahydrate	
13498-08-3	3185: Ytterbium perchlorate	
13499-05-3	1304: Hafnium chloride	
13510-35-5	1409: Indium(III) iodide	
13510-41-3	2320: Praseodymium sulfate octahydrate	
13510-42-4	2403: Rubidium perchlorate	
13510-49-1	430: Beryllium sulfate	
13510-89-9	1499: Lead antimonate	
13517-00-5	937: Copper(I) hydride	
13517-06-1	2649: Sodium iodide dihydrate	
13517-10-7	494: Boron triiodide	
13517-11-8	1386: Hypobromous acid	
13517-12-9	2430: Samarium bromide hexahydrate	
13517-24-3	2658: Sodium metasilicate pentahydrate	
13517-26-5	2695: Sodium pyrovanadate	
13517-27-6	3231: Zinc bromate hexahydrate	
13520-56-4	1158: Ferric sulfate nonahydrate	
13520-59-7	1882: Molybdenum(IV) bromide	
13520-61-1	1965: Nickel perchlorate hexahydrate	
13520-69-9	1187: Ferrous perchlorate hexahydrate	
13520-75-7	3050: Tungsten	

	dioxydibromide	
13520-76-8	3059: Tungsten oxydichloride	
13520-77-9	3061: Tungsten oxytetrabromide	
13520-78-0	3058: Tungsten oxychloride	
13520-79-1	3062: Tungsten oxytetrafluoride	
13520-83-7	3109: Uranyl nitrate hexahydrate	
13520-87-1	3148: Vanadyl chloride	
13520-88-2	3147: Vanadyl bromide	
13520-89-3	3149: Vanadyl dibromide	
13520-90-6	3154: Vanadyl tribromide	
13520-92-8	3323: Zirconyl chloride octahydrate	
13530-65-9	3238: Zinc chromate heptahydrate	
13536-53-3	2304: Praseodymium bromide	
13536-59-9	1032: Deuterium bromide	
13536-73-7	1085: Erbium bromide nonahydrate	
13536-73-7	1084: Erbium bromide	
13536-79-3	1468: Lanthanum bromide	
13536-80-6	1903: Neodymium bromide	
13536-92-0	2124: Plutonium(IV) chloride	
13537-09-2	1066: Dysprosium hydride	
13537-18-3	2978: Thulium chloride	
13537-24-1	1151: Ferric perchlorate hexahydrate	
13537-32-1	1897: Monofluoro-phosphoric acid	
13537-33-2	1204: Fluorosilane	
13548-38-4	817: Chromium(III) nitrate	
13548-42-0	966: Copper(II) chromate	
13550-28-2	1566: Lithium bromate	
13550-53-3	1090: Erbium hydride	
13565-96-3	453: Bismuth molybdenum oxide	
13565-97-4	3311: Zirconium pyrophosphate	
13566-03-5	2061: Palladium(II) sulfate dihydrate	
13566-05-7	2401: Rubidium orthovanadate	
13568-33-7	1608: Lithium nitrite monohydrate	
13568-40-6	1604: Lithium molybdate	
13568-45-1	1631: Lithium tungstate	
13568-63-3	1746: Magnesium vanadate	
13568-72-4	1775: Manganese(II) dithionate	
13569-49-8	2342: Rhenium(III) bromide	
13569-50-1	695: Cerium dihydride	
13569-60-3	2441: Samarium perchlorate hydrate	
13569-62-5	2121: Plutonium(III) chloride	
13569-63-6	2343: Rhenium(III) chloride	
13569-71-6	2345: Rhenium(IV) chloride	
13569-75-0	816: Chromium(III) iodide	
13569-80-7	1065: Dysprosium fluoride	

CAS RN	Compound
13572-93-5	1250: Gallium(III) hydride
13572-97-9	1219: Gadolinium hydride
13572-98-0	1220: Gadolinium iodide
13573-08-5	1268: Germanium(II) iodide
13573-11-0	1745: Magnesium tungstate
13573-16-5	235: Ammonium tetrathiocyanodi-ammonochromate(III) monohydrate
13586-38-4	133: Ammonium cobalt(II) sulfate hexahydrate
13587-16-1	1580: Lithium deuteride
13587-35-4	1016: Copper(II) tungstate dihydrate
13587-35-4	1015: Copper(II) tungstate
13595-87-4	477: Bismuth tungstate
13597-19-8	691: Ceric vanadate
13597-20-1	2010: Niobium(V) oxychloride
13597-30-3	2965: Thorium oxyfluoride
13597-45-0	2397: Rubidium metavanadate
13597-46-1	3275: Zinc selenite
13597-52-9	2412: Rubidium tungstate
13597-54-1	3273: Zinc selenate pentahydrate
13597-61-0	2405: Rubidium pyrovanadate
13597-64-3	750: Cesium molybdate
13597-65-4	3276: Zinc silicate
13597-99-4	424: Beryllium nitrate trihydrate
13598-22-6	433: Beryllium sulfide
13598-33-9	2784: Strontium hydride
13598-36-2	2082: Phosphorous acid
13598-41-9	1334: Holmium hydride
13598-42-0	1898: Monoiodosilane
13598-44-2	1641: Lutetium hydride
13598-53-3	2436: Samarium hydride
13598-54-4	2884: Terbium hydride
13598-56-6	3097: Uranium trihydride
13598-57-7	3207: Yttrium hydride
13598-65-7	211: Ammonium perrhenate
13600-89-0	234: Ammonium tetranitrodiammine-cobaltate(III)
13601-13-3	971: Copper(II) ferrocyanide
13601-19-9	2596: Sodium ferrocyanide decahydrate
13637-63-3	773: Chlorine pentafluoride
13637-68-8	1890: Molybdenum(VI) dioxydichloride
13637-76-8	1531: Lead perchlorate trihydrate
13675-47-3	969: Copper(II) dichromate dihydrate
13682-61-6	2265: Potassium tetrachloroaurate(III)
13682-73-0	2155: Potassium copper(I) cyanide
13689-92-4	1981: Nickel thiocyanate
13693-05-5	2107: Platinum hexafluoride
13693-06-6	2126: Plutonium(VI) hexafluoride
13693-09-9	3164: Xenon hexafluoride
13701-64-9	569: Calcium borate hexahydrate
13701-70-7	3142: Vanadium trisulfate
13701-86-5	3054: Tungsten hexabromide
13701-90-1	2935: Thallium(III) bromide
13703-82-7	1669: Magnesium borate octahydrate
13708-63-9	2883: Terbium fluoride
13708-80-0	99: Americium fluoride
13708-85-5	2632: Sodium hydrogen phosphite pentahydrate
13709-31-4	3130: Vanadium oxytrifluoride
13709-36-9	3158: Xenon difluoride
13709-38-1	1476: Lanthanum fluoride
13709-42-7	1908: Neodymium fluoride
13709-46-1	2308: Praseodymium fluoride
13709-47-2	2455: Scandium fluoride
13709-49-4	3205: Yttrium fluoride
13709-52-9	1305: Hafnium fluoride
13709-59-6	2953: Thorium fluoride
13709-61-0	3169: Xenon tetrafluoride
13718-22-4	2398: Rubidium molybdate
13718-26-8	2660: Sodium metavanadate
13718-50-8	346: Barium iodide
13718-59-7	371: Barium selenite
13718-70-2	1448: Iron molybdate
13721-34-1	2736: Sodium uranate monohydrate
13721-39-6	2670: Sodium orthovanadate
13721-43-2	2645: Sodium hypophosphate decahydrate
13746-66-2	2165: Potassium ferricyanide
13746-89-9	3304: Zirconium nitrate pentahydrate
13746-98-0	2939: Thallium(III) nitrate
13755-29-8	2598: Sodium fluoroborate
13755-32-3	388: Barium tetracyanoplatinate(II) tetrahydrate
13755-38-9	2667: Sodium nitroferricyanide(III) dihydrate
13759-10-9	2494: Silicon disulfide
13759-83-6	2438: Samarium nitrate hexahydrate
13759-88-1	1118: Europium(III) bromide
13759-92-7	1121: Europium(III) chloride hexahydrate
13760-02-6	1050: Diiodosilane
13760-78-6	1333: Holmium fluoride
13760-79-7	2980: Thulium fluoride
13760-80-0	3180: Ytterbium fluoride
13760-81-1	1640: Lutetium fluoride
13760-83-3	1089: Erbium fluoride
13762-12-4	863: Cobalt(II) bromide hexahydrate
13762-14-6	850: Cobalt molybdate
13762-51-1	2141: Potassium borohydride
13762-65-7	1358: Hydrazine perchlorate hemihydrate
13762-75-9	1602: Lithium metaphosphate
13763-67-2	737: Cesium chlorate
13765-03-2	1595: Lithium iodate
13765-19-0	588: Calcium chromate dihydrate
13765-24-7	2435: Samarium fluoride
13765-25-8	1122: Europium(III) fluoride
13765-26-9	1217: Gadolinium fluoride
13765-74-7	2530: Silver molybdate
13767-32-3	3257: Zinc molybdate
13767-34-5	985: Copper(II) molybdate
13768-11-1	2073: Perrhenic acid
13768-38-2	2041: Osmium(VI) fluoride
13768-86-0	2481: Selenium trioxide
13768-94-0	1038: Dibromosilane
13769-20-5	1110: Europium(II) chloride
13769-43-2	2295: Potassium vanadate
13770-18-8	994: Copper(II) perchlorate
13770-56-4	281: Arsenic(II) iodide
13770-61-1	1410: Indium(III) nitrate trihydrate
13773-81-4	1457: Krypton difluoride
13774-24-8	100: Americium hydride
13774-85-1	3166: Xenon oxytetrafluoride
13775-06-9	3096: Uranium trifluoride
13775-07-0	3087: Uranium pentafluoride
13775-16-1	3085: Uranium pentabromide
13775-53-6	2611: Sodium hexafluoroaluminate
13775-80-9	1353: Hydrazine monohydrobromide
13776-58-4	3172: Xenon trioxide
13776-84-6	2726: Sodium thioantimonate nonahydrate
13777-22-5	1302: Hafnium bromide
13777-23-6	1307: Hafnium iodide
13777-25-8	3294: Zirconium bromide
13778-39-7	2979: Thulium chloride heptahydrate
13778-40-0	2987: Thulium sulfate octahydrate
13779-41-4	1047: Difluorophosphoric acid
13779-92-5	2007: Niobium(V) iodide
13780-03-5	605: Calcium hydrogen sulfite
13780-06-8	617: Calcium nitrite
13780-06-8	618: Calcium nitrite monohydrate
13780-42-2	1260: Gallium(III) sulfate octadecahydrate
13780-64-8	3165: Xenon oxydifluoride
13783-04-5	2998: Titanium dibromide
13783-07-8	3000: Titanium diiodide
13798-24-8	2882: Terbium chloride hexahydrate
13812-58-3	1010: Copper(II) tellurite
13813-19-9	1036: Deuterosulfuric acid
13813-22-4	1479: Lanthanum iodide
13813-23-5	2312: Praseodymium

CAS RN	Compound
14221-06-8	1862: Molybdenum acetate dimer
14221-48-8	145: Ammonium ferricyanide trihydrate
14224-64-5	2140: Potassium bis(oxalato)platinate(II)
14242-05-8	2536: Silver perchlorate monohydrate
14244-62-3	2273: Potassium tetracyanozincate
14249-98-0	3063: Tungsten oxytrichloride
14280-53-6	1397: Indium(I) bromide
14282-91-8	1323: Hexaammine-ruthenium(III) chloride
14283-07-9	1628: Lithium tetrafluoroborate
14284-87-8	1212: Gadolinium acetylacetonate dihydrate
14284-89-0	1803: Manganese(III) acetylacetonate
14284-92-5	2365: Rhodium(III) acetylacetonate
14284-93-6	2418: Ruthenium(III) acetylacetonate
14284-95-8	2878: Terbium acetylacetonate trihydrate
14284-98-1	3175: Ytterbium acetylacetonate
14285-68-8	2339: Rhenium carbonyl
14298-31-8	2317: Praseodymium phosphate
14298-32-9	1918: Neodymium phosphate hydrate
14307-33-6	592: Calcium dichromate trihydrate
14307-35-8	1574: Lithium chromate
14311-93-4	3320: Zirconyl acetate hydroxide
14312-00-6	522: Cadmium chromate
14323-32-1	2262: Potassium tetrabromoaurate(III) dihydrate
14323-36-5	2272: Potassium tetracyanoplatinate(II) trihydrate
14324-82-4	1014: Copper(II) trifluoroacetylacetonate
14324-83-5	1983: Nickel trifluoroacetylacetonate dihydrate
14336-80-2	933: Copper(I) azide
14402-67-6	2286: Potassium titanium oxalate dihydrate
14402-70-1	198: Ammonium nitroferricyanide
14402-73-4	1627: Lithium tetracyanoplatinate(II) pentahydrate
14402-75-6	2180: Potassium hexacyanocadmium
14404-33-2	2228: Potassium pentachloro-ruthenate(III) hydrate
14405-43-7	1235: Gallium acetylacetonate
14405-45-9	1392: Indium acetylacetonate
14446-13-0	2801: Strontium permanganate trihydrate
14447-89-3	3060: Tungsten oxydiiodide
14452-39-2	70: Aluminum perchlorate
14456-34-9	1310: Hafnium oxychloride octahydrate
14456-47-4	2879: Terbium bromide
14456-48-5	1061: Dysprosium bromide
14456-51-0	2977: Thulium bromide
14456-53-2	1637: Lutetium bromide
14457-83-1	1680: Magnesium carbonate trihydrate
14457-84-2	68: Aluminum oxyhydroxide (β)
14457-87-5	706: Cerous bromide heptahydrate
14457-87-5	705: Cerous bromide
14459-59-7	1872: Molybdenum oxytetrafluoride
14459-75-7	2009: Niobium(V) oxybromide
14459-95-1	2166: Potassium ferrocyanide trihydrate
14464-46-1	2493: Silicon dioxide
14475-63-9	3302: Zirconium hydroxide
14481-29-9	163: Ammonium hexacyanoferrate(II) monohydrate
14481-33-5	2973: Thorium tetracyanoplatinate(II) hexahydrate
14483-18-2	1336: Holmium nitrate pentahydrate
14483-63-7	2600: Sodium fluorosulfonate
14486-19-2	553: Cadmium tetrafluoroborate
14507-19-8	1478: Lanthanum hydroxide
14516-54-2	1760: Manganese pentacarbonyl bromide
14517-29-4	1913: Neodymium nitrate hexahydrate
14519-18-7	2771: Strontium bromate monohydrate
14523-22-9	2362: Rhodium carbonyl chloride
14526-22-8	1160: Ferric trifluoroacetylacetonate
14551-74-7	1915: Neodymium oxalate decahydrate
14553-08-3	1081: Erbium acetylacetonate hydrate
14553-09-4	2300: Praseodymium acetylacetonate
14568-19-5	1432: Iridium pentafluoride
14589-42-5	2427: Samarium acetylacetonate
14589-44-7	2976: Thulium acetylacetonate trihydrate
14590-13-7	132: Ammonium cobalt(II) phosphate monohydrate
14590-19-3	897: Cobalt(II) selenate pentahydrate
14635-87-1	1713: Magnesium perborate heptahydrate
14637-88-8	1059: Dysprosium acetylacetonate
14639-94-2	165: Ammonium hexafluorogallate
14639-98-6	207: Ammonium pentachlorozincate
14644-55-4	749: Cesium metavanadate
14644-61-2	3315: Zirconium sulfate tetrahydrate
14646-16-3	1091: Erbium hydroxide
14646-29-8	1648: Lutetium perchlorate hexahydrate
14649-73-1	2622: Sodium hexanitritocobalt(III)
14662-04-5	1348: Hydrazine azide
14666-94-5	890: Cobalt(II) oleate
14666-96-7	885: Cobalt(II) linoleate
14674-72-7	586: Calcium chlorite
14689-45-3	511: Cadmium acetylacetonate
14692-17-2	1073: Dysprosium perchlorate hydrate
14693-02-8	2281: Potassium thioantimonate heminonahydrate
14693-56-2	238: Ammonium tetrathiovanadate(IV)
14693-81-3	2848: Tantalum pentaiodide
14693-82-4	1413: Indium(III) phosphate
14694-95-2	3036: Tris(triphenylphos-phine)rhodium(I) chloride
14709-57-0	2223: Potassium nitroprusside dihydrate
14721-18-7	1949: Nickel chromate
14721-21-2	963: Copper(II) chlorate hexahydrate
14735-84-3	1012: Copper(II) tetrafluoroborate
14762-49-3	45: Aluminum hydroxide(α)
14762-55-7	1321: Helium-3
14763-77-0	968: Copper(II) cyanide
14781-45-4	981: Copper(II) hexafluroacetyl-acetonate hydrate
14781-45-4	979: Copper(II) hexafluoroacetyl-acetonate
14871-56-8	380: Barium strontium tungsten oxide
14871-79-5	343: Barium hypophosphite monohydrate
14871-82-0	374: Barium silicate
14883-80-8	3035: Tris(ethylene-diammine) cobalt(III) chloride trihydrate
14890-41-6	3118: Vanadium dibromide
14898-67-0	2421: Ruthenium(III) chloride hydrate
14913-14-5	1486: Lanthanum phosphate hydrate
14913-33-8	1042: Dichlorodiammine-platinum(II)-trans
14914-84-2	1332: Holmium chloride hexahydrate
14929-69-2	1623: Lithium tellurite
14933-38-1	96: Americium bromide
14940-41-1	1188: Ferrous phosphate octahydrate
14940-65-9	3038: Tritium dioxide
14948-62-0	2305: Praseodymium

	carbonate octahydrate	15280-57-6	1080: Erbium acetate
14949-69-0	1954: Nickel hexafluoro-acetylacetonate hydrate		tetrahydrate
14972-70-4	2610: Sodium hexachlororhodate(III) hydrate	15280-58-7	3174: Ytterbium acetate tetrahydrate
14972-90-8	233: Ammonium tetrafluoro-antimonate(III)	15282-88-9	1498: Lead acetylacetonate
14977-17-4	560: Calcium acetate dihydrate	15293-74-0	1601: Lithium metaborate dihydrate
14977-61-8	834: Chromyl chloride	15318-60-2	123: Ammonium cerium(III) nitrate tetrahydrate
14984-81-7	2475: Selenium oxydifluoride		
14985-18-3	3325: Zirconyl nitrate hydrate	15321-51-4	1449: Iron nonacarbonyl
14986-89-1	1650: Lutetium sulfate	15336-18-2	161: Ammonium hexachlororhodate(III) monohydrate
14986-91-5	1724: Magnesium selenate hexahydrate		
14986-94-8	1764: Manganese vanadate	15364-10-0	3163: Xenon fluoride monodecafluoro-antimonate
14996-60-2	2038: Osmium(III) chloride hydrate		
14996-61-3	1437: Iridium(III) chloride hydrate	15364-94-0	1788: Manganese(II) perchlorate hexahydrate
15059-52-6	1064: Dysprosium chloride hexahydrate	15385-57-6	1820: Mercury(I) iodide
		15385-58-7	1813: Mercury(I) bromide
15060-55-6	236: Ammonium tetrathiomolybdate	15415-49-3	879: Cobalt(II) hexafluorosilicate hexahydrate
15060-59-0	1632: Lithium vanadate		
15070-34-5	1709: Magnesium nitrite trihydrate	15415-49-3	875: Cobalt(II) ferricyanide
15098-87-0	40: Aluminum fluoride trihydrate	15435-71-9	2555: Sodium acetylacetonate
15123-80-5	55: Aluminum molybdate	15444-43-6	2918: Thallium(I) hexafluoroacetyl-acetonate
15123-82-7	92: Aluminum tungstate		
15162-92-2	1902: Neodymium bromate nonahydrate	15461-27-5	3322: Zirconyl chloride hydrate
15162-93-3	2303: Praseodymium bromate nonahydrate	15469-38-2	1142: Ferric fluoride trihydrate
15163-03-8	3176: Ytterbium bromide hydrate	15474-63-2	1068: Dysprosium iodide
15163-24-3	3071: Tungsten tribromide	15477-33-5	28: Aluminum chlorate nonahydrate
15168-20-4	1001: Copper(II) selenite dihydrate	15477-33-5	27: Aluminum chlorate
15170-57-7	2106: Platinum acetylacetonate	15489-27-7	1626: Lithium tetrachlorocuprate
15192-26-4	2872: Tellurium tetrafluoride	15491-35-7	396: Barium titanium silicate
15192-42-4	2346: Rhenium(IV) fluoride	15492-38-3	2369: Rhodium(III) iodide
15195-33-2	2006: Niobium(V) fluorodioxide	15513-69-6	3072: Tungsten triiodide
		15513-84-5	3120: Vanadium diiodide
15230-79-2	1639: Lutetium chloride hexahydrate	15520-84-0	916: Cobalt(III) nitrate
		15525-64-1	2506: Silver acetylacetonate
15238-00-3	882: Cobalt(II) iodide		
15243-33-1	2416: Ruthenium dodecacarbonyl	15552-14-4	317: Barium calcium tungstate
15244-10-7	1157: Ferric sulfate hydrate	15554-47-9	3191: Yttrium acetylacetonate trihydrate
15244-35-6	547: Cadmium sulfate octahydrate		
15244-38-9	826: Chromium(III) sulfate hydrate	15571-91-2	2257: Potassium tellurate(VI) trihydrate
15275-09-9	2150: Potassium chromium(III) oxalate trihydrate	15572-25-5	2930: Thallium(I) selenide
		15578-26-4	2759: Stannous pyrophosphate
15280-09-8	2602: Sodium gold cyanide	15593-51-8	1617: Lithium selenite monohydrate
15280-53-2	1211: Gadolinium acetate tetrahydrate	15593-52-9	1616: Lithium selenate monohydrate
15280-55-4	1058: Dysprosium acetate tetrahydrate	15593-61-0	1726: Magnesium selenite hexahydrate
		15596-83-5	2941: Thallium(III) perchlorate hexahydrate
		15597-88-3	830: Chromium(IV) chloride
		15608-29-4	2366: Rhodium(III) bromide dihydrate
		15622-42-1	2344: Rhenium(III) iodide

15627-86-8	626: Calcium perchlorate tetrahydrate
15630-39-4	2574: Sodium carbonate peroxohydrate
15635-87-7	1434: Iridium(III) acetylacetonate
15653-01-7	704: Cerous acetylacetonate hydrate
15663-27-1	1041: Dichlorodiammine-platinum(II) (cis)
15681-89-7	2565: Sodium borodeuteride
cis: 15684-18-1; trans: 13782-33-7	1040: Dichlorodiammine-palladium(II)
15684-36-3	1980: Nickel tetrafluoroborate hexahydrate
15696-40-9	2035: Osmium carbonyl
15750-45-5	1706: Magnesium nitrate dihydrate
15752-05-3	155: Ammonium hexachloroiridate(III)
15771-43-4	426: Beryllium oxalate trihydrate
15780-28-6	2587: Sodium deuteride
15785-09-8	713: Cerous hydroxide
15823-43-5	1315: Hafnium sulfate
15829-53-5	1823: Mercury(I) oxide
15851-44-2	552: Cadmium tellurite
15851-47-5	1543: Lead tellurite
15878-77-0	2313: Praseodymium nitrate hexahydrate
15947-41-8	108: Americium(IV) fluoride
15956-28-2	2364: Rhodium(II) acetate dimer
16004-08-3	983: Copper(II) hydroxy chloride
16045-17-3	2966: Thorium perchlorate
16122-03-5	194: Ammonium nickel chloride hexahydrate
16283-36-6	3272: Zinc salicylate trihydrate
16399-77-2	1174: Ferrous chloride dihydrate
16469-16-2	2311: Praseodymium hydroxide
16469-17-3	1911: Neodymium hydroxide
16469-22-0	3208: Yttrium hydroxide
16519-60-1	2671: Sodium orthovanadate decahydrate
16544-92-6	867: Cobalt(II) chloride dihydrate
16569-85-0	1688: Magnesium dichromate hexahydrate
16569-85-0	1684: Magnesium chromate pentahydrate
16671-27-5	814: Chromium(III) fluoride trihydrate
16674-78-5	1657: Magnesium acetate tetrahydrate
16689-88-6	2956: Thorium hydride
16712-20-2	1573: Lithium chloride monohydrate
16721-80-5	2635: Sodium hydrogen sulfide
16721-80-5	2636: Sodium hydrogen sulfide dihydrate
16721-80-5	2637: Sodium hydrogen

19372-44-2	562: Calcium acetylacetonate hydrate		chloride hydrate
19415-82-8	3111: Uranyl sulfate monohydrate	20770-09-6	2746: Stannic selenide
19423-76-8	711: Cerous chloride hydrate	20816-12-0	2042: Osmium(VIII) oxide
19469-07-9	1147: Ferric oxalate	20859-73-8	75: Aluminum phosphide
19513-05-4	1802: Manganese(III) acetate dihydrate	20955-11-7	2614: Sodium hexafluoroferrate(III)
19530-02-0	3300: Zirconium hexa-fluoroacetylacetonate	21041-93-0	880: Cobalt(II) hydroxide
		21041-95-2	526: Cadmium hydroxide
19567-78-3	2605: Sodium hexachloroiridate(IV) hexahydrate	21056-98-4	633: Calcium phosphite monohydrate
		21109-95-5	382: Barium sulfide
19583-77-8	2609: Sodium hexachloroplatinate(IV) hexahydrate	21109-95-5	383: Barium sulfide
		21159-32-0	740: Cesium cyanide
		21255-83-4	500: Bromine dioxide
19584-30-6	2363: Rhodium dodecacarbonyl	21264-43-7	1400: Indium(II) bromide
		21308-80-5	502: Bromine oxide
19597-69-4	1563: Lithium azide	21324-39-0	2616: Sodium hexafluorophosphate
19598-90-4	1221: Gadolinium nitrate hexahydrate	21324-40-3	1587: Lithium hexafluorophosphate
19624-22-7	2066: Pentaborane(9)	21351-79-1	745: Cesium hydroxide
19648-85-2	1693: Magnesium hexa-fluoroacetylacetonate dihydrate	21548-73-2	2546: Silver sulfide
		21645-51-2	43: Aluminum hydroxide
		21651-19-4	2758: Stannous oxide
19648-88-5	1520: Lead hexafluoro-acetylacetonate	21679-31-2	805: Chromium(III) acetylacetonate
19718-36-6	2224: Potassium osmiate dihydrate	21679-46-9	912: Cobalt(III) acetylacetonate
19783-14-3	1522: Lead hydroxide	21907-50-6	763: Cesium trifluoroacetate
20205-91-8	488: Boron phosphide	21908-53-2	1846: Mercury(II) oxide red
20257-20-9	46: Aluminum hydroxide(β)		
		21908-53-2	1847: Mercury(II) oxide yellow
20281-00-9	754: Cesium oxide		
20328-94-3	3167: Xenon pentafluoride hexafluoroarsenate	21995-38-0	124: Ammonium cerium(III) sulfate tetrahydrate
20328-96-5	256: Antimony(III) nitrate		
20338-08-3	2994: Titanic acid	22015-35-6	1112: Europium(II) iodide
20344-49-4	1149: Ferric oxide hydroxide	22205-45-4	942: Copper(I) sulfide
		22205-57-8	731: Cesium amide
20346-99-0	2410: Rubidium tetrahydridoborate	22208-73-7	574: Calcium bromide dihydrate
20398-06-5	2915: Thallium(I) ethoxide	22306-37-2	437: Bismuth acetate
		22326-55-2	341: Barium hydroxide monohydrate
20405-64-5	941: Copper(I) selenide		
20427-11-6	872: Cobalt(II) cyanide dihydrate	22398-80-7	1396: Indium phosphide
		22429-50-1	3110: Uranyl oxalate trihydrate
20427-11-6	873: Cobalt(II) cyanide trihydrate		
		22519-64-8	1405: Indium(III) chloride tetrahydrate
20427-56-9	2424: Ruthenium(VIII) oxide		
		22527-13-5	3162: Xenon fluoride hexafluororuthenate
20427-58-1	3252: Zinc hydroxide		
20427-59-2	982: Copper(II) hydroxide	22537-19-5	1494: Lawrencium
20548-54-3	647: Calcium sulfide	22594-86-1	2227: Potassium pentachloronitrosyl iridium(III) hydrate
20601-83-6	1851: Mercury(II) selenide		
		22750-57-8	732: Cesium azide
20619-16-3	1269: Germanium(II) oxide	22756-36-1	2380: Rubidium azide
		22831-39-6	1729: Magnesium silicide
20644-87-5	3116: Vanadium carbonyl	22831-42-1	20: Aluminum arsenide
20661-21-6	1408: Indium(III) hydroxide	22886-66-4	1249: Gallium(III) fluoride trihydrate
20662-14-0	2454: Scandium chloride hexahydrate		
		22986-54-5	4: Actinium chloride
20665-52-5	1256: Gallium(III) oxide hydroxide	22992-15-0	1224: Gadolinium oxalate decahydrate
20667-12-3	2534: Silver oxide	22992-83-2	1086: Erbium carbonate hydrate
20694-39-7	1785: Manganese(II) nitrate tetrahydrate		
		23293-27-8	2928: Thallium(I) picrate
20712-42-9	535: Cadmium oxalate trihydrate	23323-79-7	101: Americium hydroxide
20762-60-1	2139: Potassium azide	23363-14-6	3190: Yttrium acetate hydrate
20765-98-4	2368: Rhodium(III)		
23383-11-1	1177: Ferrous citrate monohydrate		
23414-72-4	3265: Zinc permanganate hexahydrate		
23436-05-7	350: Barium metaborate dihydrate		
23586-53-0	2942: Thallium(III) trifluoroacetate		
23777-80-2	1324: Hexaborane(10)		
24094-93-7	790: Chromium nitride		
24304-00-5	59: Aluminum nitride		
24597-12-4	1242: Gallium(II) chloride		
24598-62-7	153: Ammonium hexabromoosmiate(IV)		
24613-38-5	869: Cobalt(II) chromate		
24621-21-4	1994: Niobium nitride		
24646-85-3	3128: Vanadium nitride		
24670-06-2	2888: Terbium oxalate hydrate		
24670-07-3	1071: Dysprosium oxalate decahydrate		
24719-19-5	859: Cobalt(II) arsenate octahydrate		
24887-06-7	3247: Zinc formaldehyde sulfoxylate		
24992-60-7	2315: Praseodymium oxalate decahydrate		
25094-02-4	585: Calcium chloride tetrahydrate		
25114-58-3	1391: Indium acetate		
25152-52-7	19: Aluminum antimonide		
25324-56-5	2992: Tin monophosphide		
25417-81-6	338: Barium hydrosulfide		
25519-09-9	1328: Holmium acetate monohydrate		
25583-20-4	3011: Titanium nitride		
25617-97-4	1239: Gallium nitride		
25617-98-5	1395: Indium nitride		
25658-42-8	3305: Zirconium nitride		
25658-43-9	3084: Uranium mononitride		
25659-31-8	1523: Lead iodate		
25764-08-3	698: Cerium nitride		
25764-09-4	2314: Praseodymium nitride		
25764-10-7	1481: Lanthanum nitride		
25764-11-8	1914: Neodymium nitride		
25764-15-2	1223: Gadolinium nitride		
25817-87-2	1308: Hafnium nitride		
25895-60-7	2586: Sodium cyanoborohydride		
25937-78-4	729: Cesium acetylacetonate		
25955-51-5	2908: Thallium(I) acetylacetonate		
26006-71-3	2714: Sodium tellurate(VI) dihydrate		
26042-63-7	2525: Silver hexafluorophosphate		
26042-64-8	2523: Silver hexafluoro-antimonate(V)		
26124-86-7	351: Barium metaborate monohydrate		
26134-62-3	1607: Lithium nitride		
26318-99-0	1002: Copper(II) silicate dihydrate		
26342-61-0	791: Chromium phosphide		
26628-22-8	2564: Sodium azide		
26677-68-9	2984: Thulium oxalate hexahydrate		
26677-69-0	1646: Lutetium oxalate hexahydrate		
26686-77-1	1728: Magnesium silicate		

26750-66-3 2282: Potassium thiocarbonate
26970-82-1 2700: Sodium selenite pentahydrate
27016-73-5 839: Cobalt arsenide
27016-75-7 1935: Nickel arsenide
27043-84-1 3228: Zinc borate
27133-66-0 367: Barium potassium chromate
27546-07-2 139: Ammonium dimolybdate
27685-51-4 1856: Mercury(II) tetra-thiocyanatocobaltate(II)
27709-53-1 2294: Potassium uranyl sulfate dihydrate
27774-13-6 3153: Vanadyl sulfate dihydrate
27790-37-0 2249: Potassium stannosulfate
27860-83-9 3246: Zinc fluoroborate hexahydrate
27988-77-8 1379: Hydrogen tetrachloroaurate(III) hydrate
28038-39-3 2387: Rubidium cobalt(II) sulfate hexahydrate
28041-86-3 2154: Potassium cobalt(II) selenate hexahydrate
28300-74-5 2138: Potassium antimony tartrate hemihydrate
28407-51-4 2361: Rhodium carbonyl
28633-45-6 1138: Ferric citrate pentahydrate
28876-88-2 2229: Potassium perborate monohydrate
28958-23-8 1467: Lanthanum bromate nonahydrate
28958-26-1 2429: Samarium bromate nonahydrate
28965-57-3 1338: Holmium oxalate decahydrate
29584-42-7 2822: Sulfur trioxide N,N-dimethylformamide complex
29689-14-3 809: Chromium(III) carbonate hydrate
29703-01-3 744: Cesium hydrogen carbonate
29796-57-4 156: Ammonium hexachloroiridate(III) monohydrate
29870-99-3 2790: Strontium lactate trihydrate
29935-35-1 1586: Lithium hexafluoroarsenate
30618-31-6 1095: Erbium oxalate decahydrate
30622-97-0 2396: Rubidium iron(III) sulfate dodecahydrate
30737-24-7 2925: Thallium(I) oxalate
30903-87-8 1610: Lithium oxalate
30937-53-2 2352: Rhenium(V) bromide
31052-43-4 2406: Rubidium selenide
31083-74-6 2408: Rubidium sulfide
31111-21-4 2242: Potassium ruthenate(VI)
31142-56-0 32: Aluminum citrate
32248-43-4 2434: Samarium diiodide
32287-65-3 39: Aluminum fluoride monohydrate
32321-65-6 3317: Zirconium telluride
32594-40-4 1433: Iridium(I) chlorotricarbonyl

32823-06-6 54: Aluminum metaphosphate
32997-62-9 3181: Ytterbium hydride
33088-16-3 2960: Thorium nitrate tetrahydrate
33114-15-7 1029: Cyclopentadienyl-niobium tetrachloride
33445-15-7 190: Ammonium mercuric chloride dihydrate
33689-80-4 5: Actinium fluoride
33689-81-5 3: Actinium bromide
33689-82-6 8: Actinium iodide
33908-66-6 2561: Sodium antimonate monohydrate
34128-09-1 2922: Thallium(I) molybdate
34283-69-7 753: Cesium orthovanadate
34767-44-7 3298: Zirconium cyclopentadienyl trichloride
34822-89-4 1027: Cyclopentadienyl-indium(I)
35103-79-8 746: Cesium hydroxide monohydrate
35585-58-1 2653: Sodium metaborate dihydrate
35718-37-7 3171: Xenon trifluoride monodecafluoro-antimonate
35725-34-9 3182: Ytterbium nitrate pentahydrate
36470-39-0 2615: Sodium hexafluorogermanate
36548-87-5 2983: Thulium nitrate hexahydrate
36678-21-4 1759: Manganese nitride
36907-37-6 1485: Lanthanum perchlorate hexahydrate
36907-40-1 1127: Europium(III) perchlorate hexahydrate
36907-42-3 3144: Vanadium(II) sulfate heptahydrate
36969-05-8 832: Chromium(VI) morpholine
37185-09-4 379: Barium strontium niobium oxide
37248-04-7 1313: Hafnium silicate
37265-86-4 400: Barium yttrium tungsten oxide
37306-42-6 479: Bismuth zirconate
37473-67-9 1924: Neodymium trifluoroacetylacetonate
37541-72-3 176: Ammonium hydrogen oxalate hemihydrate
37773-49-2 2115: Platinum(IV) chloride
37809-19-1 597: Calcium fluorophosphate dihydrate
37836-27-4 1356: Hydrazine mononitrate
37961-19-6 105: Americium oxychloride
38245-34-0 1330: Holmium carbonate hydrate
38245-35-1 1062: Dysprosium carbonate tetrahydrate
38245-38-4 1904: Neodymium carbonate hydrate
38455-77-5 2742: Stannic chromate

39082-23-0 1317: Hafnium telluride
39156-80-4 2975: Thulium acetate monohydrate
39290-85-2 959: Copper(II) borate
39356-80-4 1441: Iron antimonide
39361-25-6 856: Cobalt zirconate
39368-69-9 2353: Rhenium(V) chloride
39373-27-8 2373: Rhodium(III) oxide pentahydrate
39409-82-0 1677: Magnesium carbonate hydroxide tetrahydrate
39416-30-3 399: Barium vanadate
39430-51-8 803: Chromium(III) acetate hydroxide
39483-74-4 443: Bismuth chloride monohydrate
39578-36-4 1459: Krypton fluoride monodecafluoro-antimonate
39733-35-2 189: Ammonium magnesium chloride hexahydrate
39796-98-0 3168: Xenon pentafluoride hexafluororuthenate
39797-63-2 3160: Xenon fluoride hexafluoroantimonate
41591-55-3 1382: Hydroxylamine hydrobromide
41944-01-8 2173: Potassium heptaiodobismuthate
42739-38-8 244: Ammonium valerate
44584-78-3 461: Bismuth oxyperchlorate monohydrate
47814-18-6 1909: Neodymium hexafluoroacetyl-acetonate dihydrate
47814-20-0 2309: Praseodymium hexafluoroacetyl-acetonate
49848-24-0 13: Actinium oxyfluoride
49848-29-5 12: Actinium oxychloride
49848-33-1 11: Actinium oxybromide
50432-32-1 3161: Xenon fluoride hexafluoroarsenate
50647-18-2 15: Actinium sulfide
50813-16-6 2691: Sodium polyphosphate
50813-65-5 318: Barium carbide
50927-81-6 2495: Silicon monosulfide
50960-82-2 1285: Gold(I) carbonyl chloride
50968-00-8 1814: Mercury(I) carbonate
51184-23-7 2962: Thorium orthosilicate
51222-65-2 762: Cesium titanate
51222-66-3 766: Cesium zirconate
51274-00-1 1150: Ferric oxide monohydrate
51312-42-6 2690: Sodium phosphotungstate
51373-68-3 3183: Ytterbium oxalate decahydrate
51503-61-8 179: Ammonium hydrogen phosphite monohydrate
51595-71-2 1826: Mercury(I) sulfide
51674-17-0 2728: Sodium thiophosphate dodecahydrate

51750-73-3	2560: Sodium ammonium hydrogen phosphate tetrahydrate
51850-20-5	1371: Hydrogen hexa-hydroxyplatinate(IV)
51898-99-8	452: Bismuth molybdate
52014-82-1	690: Ceric titanate
52110-05-1	1748: Magnesium zirconium silicate
52262-58-5	2755: Stannous fluorophosphate
52350-17-1	765: Cesium tungstate
52495-41-7	2720: Sodium tetrabromoaurate(III)
52502-12-2	1985: Nickel vanadate
52503-64-7	955: Copper(II) basic acetate
52628-25-8	3223: Zinc ammonium chloride
52708-44-8	1458: Krypton fluoride hexafluoroantimonate
52721-22-9	1461: Krypton trifluoride hexafluoroantimonate
52740-16-6	567: Calcium arsenite
52788-53-1	1222: Gadolinium nitrate pentahydrate
52788-54-2	2459: Scandium sulfate octahydrate
52933-62-7	1454: Iron zirconate
52951-38-9	455: Bismuth oleate
53120-23-3	249: Antimony phosphide
53169-11-2	1661: Magnesium aluminum zirconate
53169-23-6	701: Cerium stannate
53169-24-7	692: Ceric zirconate
53214-07-6	2861: Tellurium decafluoride
53608-79-0	2298: Potassium zirconium sulfate trihydrate
53633-79-7	1744: Magnesium trifluoroacetylacetonate dihydrate
53731-35-4	1790: Manganese(II) pyrophosphate
53731-35-4	1791: Manganese(II) pyrophosphate trihydrate
53778-50-0	2675: Sodium pentaiodobismuthate tetrahydrate
53823-60-2	2607: Sodium hexachloro-palladate(IV)
54412-40-7	2933: Thallium(I) trifluoroacetylacetonate
54451-24-0	1471: Lanthanum carbonate pentahydrate
54496-71-8	914: Cobalt(III) fluoride dihydrate
54723-94-3	213: Ammonium phosphomolybdate
55102-19-7	775: Chlorohydridotris-(triphenylphosphine)-ruthenium(II)
55147-94-9	820: Chromium(III) perchlorate
55343-67-4	756: Cesium pyrovanadate
55576-04-0	308: Barium antimonide
56320-90-2	739: Cesium chromate
56378-72-4	1667: Magnesium basic carbonate pentahydrate
56617-31-3	276: Argon fluoride
57402-46-7	2133: Potassium

	acetylacetonate hemihydrate
57592-57-1	2059: Palladium(II) oxalate
57804-25-8	1490: Lanthanum sulfate octahydrate
57921-51-4	31: Aluminum chromate
58500-12-2	88: Aluminum tellurite
58815-72-8	1460: Krypton fluoride monodecafluoro-tantalate
59129-80-5	3309: Zirconium phosphate trihydrate
59393-06-5	451: Bismuth iron molybdenum oxide
60582-92-5	1656: Magnesium acetate monohydrate
60616-74-2	1695: Magnesium hydride
60676-86-0	3155: Vitreous silica
60763-24-8	2967: Thorium selenide
60883-64-9	416: Beryllium carbonate tetrahydrate
60897-40-7	2737: Sodium uranyl carbonate
60922-26-1	3048: Tungsten dinitride
60936-81-4	6: Actinium hydride
60950-56-3	1694: Magnesium hexafluorosilicate hexahydrate
60969-19-9	2919: Thallium(I) hexafluorophosphate
61027-88-1	1660: Magnesium aluminum silicate
61042-72-6	1679: Magnesium carbonate pentahydrate
61565-07-9	1097: Erbium perchlorate hydrate
63026-01-7	1124: Europium(III) nitrate pentahydrate
63771-33-5	205: Ammonium pentachlororhodate(III) monohydrate
63989-69-5	1133: Ferric basic arsenite
64399-16-2	2247: Potassium sodium carbonate hexahydrate
64417-98-7	3308: Zirconium oxide yttria stabilized
64424-12-0	1464: Lanthanum acetylacetonate hydrate
64916-48-9	2056: Palladium(II) hexa-fluoroacetylacetonate
65202-12-2	1824: Mercury(I) perchlorate tetrahydrate
65277-48-7	260: Antimony(III) perchlorate trihydrate
65353-51-7	2112: Platinum(II) hexa-fluoroacetylacetonate
65842-03-7	1145: Ferric metavanadate
66169-93-5	2378: Rubidium acetylacetonate
66942-97-0	2193: Potassium hexanitritocobalt(III)
67211-31-8	2656: Sodium metaniobate heptahydrate
67952-43-6	1946: Nickel chlorate hexahydrate
68016-36-4	391: Barium thiocyanate trihydrate
68133-88-0	206: Ammonium penta-chlororuthenate(III) monohydrate
68399-60-0	2583: Sodium copper

	chromate trihydrate
68488-07-3	1658: Magnesium acetylacetonate dihydrate
69239-51-6	524: Cadmium dichromate monohydrate
69365-72-6	1254: Gallium(III) nitrate hydrate
70446-10-5	1107: Europium hydride
70692-94-3	1800: Manganese(II) zirconate
70692-95-4	93: Aluminum zirconate
70714-64-6	930: Copper zirconate
70714-64-6	1017: Copper(II) zirconate
71414-47-6	2067: Pentamethylcyclo-pentadienyltantalum tetrachloride
71626-98-7	612: Calcium iodide hexahydrate
71799-92-3	1773: Manganese(II) citrate
71963-57-0	919: Cobalt(III) sepulchrate trichloride
71965-17-8	3321: Zirconyl basic nitrate
72520-94-6	708: Cerous carbonate pentahydrate
73157-11-6	1238: Gallium azide
73491-34-6	1849: Mercury(II) perchlorate trihydrate
73560-00-6	1425: Iodine nonoxide
74507-64-5	1668: Magnesium bis(pentamethylcyclo-pentadienyl)
75060-62-5	1969: Nickel selenate hexahydrate
75397-94-3	2850: Tantalum pentoxide hydrate
75426-28-2	2477: Selenium sulfide
75535-11-4	1700: Magnesium iodide hexahydrate
75926-22-6	2471: Selenium hexasulfide
79490-00-9	537: Cadmium perchlorate
80529-93-7	1233: Gadopentetic acid
81029-06-3	71: Aluminum perchlorate nonahydrate
81129-00-2	2097: Phosphorus(III) sulfide
81579-74-0	2328: 2,2-Bis(ethylferro-cenyl)propane
84359-31-9	823: Chromium(III) phosphate hexahydrate
86546-99-8	1119: Europium(III) carbonate hydrate
97126-35-7	908: Cobalt(II) thiocyanate trihydrate
99685-96-8	1208: Fullerene
100587-90-4	1463: Lanthanum acetate hydrate
100587-92-6	2877: Terbium acetate hydrate
100587-96-0	2880: Terbium carbonate hydrate
101509-27-7	1920: Neodymium sulfate
102192-40-5	2948: Thorium acetylacetonate
107539-20-8	3195: Yttrium barium copper oxide
107539-20-8	3196: Yttrium barium copper oxide
107539-20-8	3197: Yttrium barium copper oxide

108249-27-0   1357: Hydrazine
                monooxalate
108503-47-5   3222: Zinc acetylacetonate
                hydrate
110802-84-1   1367: Hydrogen
                hexachloroiridate(IV)
                hydrate
111419-39-7   1488: Lanthanum
                strontium copper oxide
114615-82-6   2902: Tetrapropyl-
                ammonium
                perruthenate(VII)
114901-61-0   467: Bismuth strontium
                calcium copper oxide
                (1112)
114901-61-0   468: Bismuth strontium
                calcium copper oxide
                (2212)
114901-61-0   469: Bismuth strontium
                calcium copper oxide
                (2223)
115383-22-7   669: Carbon fullerenes
118448-18-3   568: Calcium bis(2,2,6,6-
                tetramethyl-3,5-
                heptanedionate)
119000-19-0   2904: Thallium barium
                calcium copper oxide
123333-66-4   2259: Potassium
                tellurite(IV) hydrate
123333-67-5   2646: Sodium
                hypophosphite
                monohydrate
123333-85-7   1629: Lithium thiocyanate
                hydrate
123333-98-2   813: Chromium(III)
                fluoride tetrahydrate
123334-23-6   2604: Sodium
                hexachloroiridate(III)
                hydrate
125720-69-1   2906: Thallium barium
                calcium copper oxide
126284-91-1   2301: Praseodymium
                barium copper oxide
127241-75-2   2905: Thallium barium
                calcium copper oxide
131159-39-2   1209: Fullerenes
133578-89-9   3152: Vanadyl selenite
                monohydrate
133863-98-6   1869: Molybdenum
                metaphosphate
137232-17-8   2169: Potassium fullerene
137926-73-9   2392: Rubidium fullerene

Molecular Formula Index

MOLECULAR FORMULA INDEX

$Al_2H_3O_7P$	nonahydrate 74: Aluminum phosphate trihydroxide
$Al_2H_4O_9Si_2$	82: Aluminum silicate dihydrate
$Al_2H_6K_2O_7$	2134: Potassium aluminate trihydrate
$Al_2H_{36}O_{30}S_3$	84: Aluminum sulfate octadecahydrate
Al_2MgO_4	1659: Magnesium aluminum oxide
Al_2MgO_6Zr	1661: Magnesium aluminum zirconate
$Al_2Mo_3O_{12}$	55: Aluminum molybdate
Al_2O_3	62: Aluminum oxide 63: Aluminum oxide (α) 64: Aluminum oxide (γ) 65: Aluminum oxide (δ) 66: Aluminum oxide (κ)
Al_2O_4Sr	2770: Strontium aluminate
Al_2O_5Si	78: Aluminum silicate 79: Aluminum silicate 81: Aluminum silicate
Al_2O_5Ti	90: Aluminum titanate
$Al_2O_7Si_2$	77: Aluminum silicate
$Al_2O_9Te_3$	88: Aluminum tellurite
$Al_2O_9Zr_3$	93: Aluminum zirconate
$Al_2O_{12}S_3$	83: Aluminum sulfate
$Al_2O_{12}W_3$	92: Aluminum tungstate
Al_2S_3	85: Aluminum sulfide
Al_2Se_3	76: Aluminum selenide
Al_2Te_3	87: Aluminum telluride
Al_2Zr	94: Aluminum zirconium
Al_3Mn	1750: Manganese aluminide
Al_3Ni	1932: Nickel aluminide
Al_3Ta	2835: Tantalum aluminide
Al_3Zr	3292: Zirconium aluminide
$Al_4B_2O_9$	21: Aluminum borate
Al_4Ba	307: Barium aluminide
$Al_4Mg_2O_{18}Si_5$	1660: Magnesium aluminum silicate
$Al_5O_{12}Y_3$	3192: Yttrium aluminum oxide
$Al_6O_{13}Si_2$	80: Aluminum

	silicate
$Al_{22}Na_2O_{34}$	2552: Sodium β-aluminum oxide
Am	95: Americium
$AmBr_3$	96: Americium bromide
AmClO	105: Americium oxychloride
$AmCl_3$	98: Americium chloride
AmF_3	99: Americium fluoride
AmF_4	108: Americium(IV) fluoride
AmH_3	100: Americium hydride
AmH_3O_3	101: Americium hydroxide
AmI_3	102: Americium iodide
AmO_2	109: Americium(IV) oxide
AmO_4P	106: Americium phosphate
Am_2O_3	103: Americium oxide(α) 104: Americium oxide(β)
Am_2S_3	107: Americium sulfide
Ar	275: Argon
ArF	276: Argon fluoride
As	277: Arsenic (α) 278: Arsenic (β)
$AsBr_3$	283: Arsenic(III) bromide
$AsCaHO_3$	567: Calcium arsenite
$AsCl_3$	284: Arsenic(III) chloride
AsCo	839: Cobalt arsenide
AsCoS	838: Cobalt arsenic sulfide
$AsCr_2$	781: Chromium arsenide
$AsCuHO_3$	953: Copper(II) arsenite
$AsCu_3$	924: Copper arsenide
AsF_3	285: Arsenic(III) fluoride
AsF_5	293: Arsenic(V) fluoride
AsF_6K	2184: Potassium hexafluoro-arsenate(V)
AsF_6Li	1586: Lithium hexa-fluoroarsenate
AsF_6Na	2613: Sodium hexa-fluoroarsenate
AsF_9Xe_2	3161: Xenon fluoride hexa-fluoroarsenate
$AsF_{11}Xe$	3167: Xenon pentafluoride hexafluoro-arsenate
AsFe	1442: Iron arsenide
$AsFeH_4O_6$	1131: Ferric arsenate dihydrate
AsGa	1237: Gallium arsenide
$AsHHgO_4$	1828: Mercury(II)

$AsHNa_2O_4$	arsenate 2624: Sodium hydrogen arsenate
AsH_2KO_4	2160: Potassium dihydrogen arsenate
AsH_3	298: Arsine
AsH_3O_3	297: Arsenious acid
$AsH_4O_{4.5}$	292: Arsenic(V) acid hemihydrate
AsH_6NO_4	137: Ammonium dihydrogen arsenate
$AsH_9N_2O_4$	171: Ammonium hydrogen arsenate
$AsH_{12}N_3O_4$ (anhydrous)	115: Ammonium arsenate hydrate
$AsH_{15}Na_2O_{11}$	2625: Sodium hydrogen arsenate heptahydrate
$AsH_{24}Na_3O_{16}$	2562: Sodium arsenate dodecahydrate
AsI_2	281: Arsenic(II) iodide
AsI_3	286: Arsenic(III) iodide
AsIn	1394: Indium arsenide
$AsLi_3O_4$	1562: Lithium arsenate
AsNi	1935: Nickel arsenide
AsO_2Na	2563: Sodium arsenite
$AsSb_3$	247: Antimony arsenide
AsY	3194: Yttrium arsenide
As_2Ba_3	309: Barium arsenide
$As_2Ca_3O_8$	566: Calcium arsenate
As_2Cd_3	513: Cadmium arsenide
As_2Co	841: Cobalt arsenide
$As_2Co_3H_{16}O_{16}$	859: Cobalt(II) arsenate octahydrate
$As_2Cu_3O_8$	952: Copper(II) arsenate
$As_2Fe_3H_{12}O_{14}$	1168: Ferrous arsenate hexahydrate
$As_2Fe_4H_{10}O_{14}$	1133: Ferric basic arsenite
As_2HKO_4	2197: Potassium hydrogen arsenite
$As_2H_3KO_5$	2214: Potassium metaarsenite monohydrate
$As_2H_{16}Ni_3O_{16}$	1934: Nickel arsenate octahydrate
$As_2H_{16}O_{16}Zn_3$	3225: Zinc arsenate octahydrate
As_2Mg_3	1666: Magnesium arsenide
$As_2Mg_3O_8$ (anhydrous)	1665: Magnesium arsenate hydrate
As_2O_3	287: Arsenic(III) oxide 288: Arsenic(III)

B_2Mn	1757: Manganese diboride
B_2Nb	1990: Niobium diboride
B_2O_3	485: Boron oxide
	486: Boron oxide glass
B_2S_3	495: Boron trisulfide
B_2Sc	2450: Scandium boride
B_2Ta	2838: Tantalum diboride
B_2Ti	2996: Titanium boride
B_2U	3079: Uranium diboride
B_2V	3117: Vanadium diboride
B_2Zr	3293: Zirconium boride
B_3Cr_5	783: Chromium boride
$B_3H_6N_3$	480: Borazole
B_3Re_7	2338: Rhenium boride
$B_4CaH_{12}O_{13}$	569: Calcium borate hexahydrate
B_4Dy	1060: Dysprosium boride
B_4Er	1083: Erbium boride
$B_4H_8K_2O_{11}$	2261: Potassium tetraborate tetrahydrate
$B_4H_8Na_2O_{11}$	2719: Sodium tetraborate tetrahydrate
$B_4H_9NO_9$	184: Ammonium hydrogen tetraborate dihydrate
B_4H_{10}	2894: Tetraborane (10)
$B_4H_{10}K_2O_{12}$	2260: Potassium tetraborate pentahydrate
$B_4H_{10}Li_2O_{12}$	1625: Lithium tetraborate pentahydrate
$B_4H_{10}Na_2O_{12}$	2718: Sodium tetraborate pentahydrate
$B_4H_{10}O_{14}Zn_3$	3230: Zinc borate pentahydrate
$B_4H_{11}NO_{10}$	172: Ammonium hydrogen borate trihydrate
$B_4H_{16}MnO_{15}$	1767: Manganese (II) borate octahydrate
$B_4H_{16}N_2O_{11}$	228: Ammonium tetraborate tetrahydrate
$B_4H_{20}Na_2O_{17}$	2717: Sodium tetraborate decahydrate
B_4Ho	1345: Homium boride
$B_4Li_2O_7$	1624: Lithium tetraborate
B_4Lu	1636: Lutetium boride
$B_4Na_2O_7$	2716: Sodium tetraborate

$B_4O_9Zn_3$	3228: Zinc borate
B_4Si	2487: Silicon boride
B_4U	3088: Uranium tetraboride
B_5H_9	2066: Pentaborane (9)
B_5H_{11}	2065: Pentaborane (11)
$B_5H_{12}NO_{12}$	204: Ammonium pentaborate tetrahydrate
B_5Mo_2	1873: Molybdenum pentaboride
B_5W_2	3064: Tungsten pentaboride
B_6Ba	333: Barium hexaboride
B_6Ca	570: Calcium boride
$B_6Ca_2H_{10}O_{16}$	599: Calcium hexaborate pentahydrate
B_6Ce	696: Cerium hexaboride
B_6Eu	1106: Europium boride
B_6Gd	1213: Gadolinium boride
$B_6H_7O_{14.5}Zn_2$	3229: Zinc borate hemiheptahydrate
B_6H_{10}	1324: Hexaborane (10)
B_6La	1466: Lanthanum boride
B_6Mg	1671: Magnesium boride
B_6Nd	1901: Neodymium boride
B_6Pr	2302: Praseodymium boride
B_6Si	489: Boron silicide
B_6Sm	2428: Samarium boride
B_6Sr	2783: Strontium hexaboride
B_6Th	2954: Thorium hexaboride
B_6Y	3198: Yttrium boride
$B_{10}H_{14}$	1030: Decaborane (14)
$B_{10}H_{16}K_2O_{24}$	2226: Potassium pentaborate octahydrate
Ba	300: Barium
$BaBiO_3$	312: Barium bismuth oxide
$BaBr_2$	315: Barium bromide
$BaBr_2H_2O_7$	314: Barium bromate monohydrate
$BaBr_2H_4O_2$	316: Barium bromide dihydrate
$BaBr_2O_6$	313: Barium bromate
$BaCl_2$	322: Barium chloride
$BaCl_2H_2O_7$	321: Barium chlorate monohydrate
$BaCl_2H_4O_2$	323: Barium chloride dihydrate
$BaCl_2H_6O_{11}$	364: Barium

	perchlorate trihydrate
$BaCl_2O_6$	320: Barium chlorate
$BaCl_2O_8$	363: Barium perchlorate
$BaCrO_4$	324: Barium chromate
$BaCr_2H_4O_9$	327: Barium dichromate dihydrate
$BaCr_2K_2O_8$	367: Barium potassium chromate
BaF_2	332: Barium fluoride
BaF_6Ge	334: Barium hexa-fluorogermanate
BaF_6Si	335: Barium hexafluorosilicate
$BaFe_{12}O_{19}$	330: Barium ferrite
$BaHO_4P$	337: Barium hydrogen phosphate
BaH_2	336: Barium hydride
$BaH_2I_2O_7$	345: Barium iodate monohydrate
$BaH_2N_2O_5$	359: Barium nitrite monohydrate
BaH_2O_2	340: Barium hydroxide
$BaH_2O_4S_2$	392: Barium thiosulfate monohydrate
BaH_2S_2	338: Barium hydrosulfide
$BaH_4I_2O_2$	347: Barium iodide dihydrate
BaH_4O_3	341: Barium hydroxide monohydrate
$BaH_4O_8S_2$	329: Barium dithionate dihydrate
$BaH_6O_5P_2$	343: Barium hypophosphite monohydrate
BaH_6O_6Sn	377: Barium stannate trihydrate
$BaH_{10}O_4S_2$	339: Barium hydrosulfide tetrahydrate
$BaH_{18}O_{10}$	342: Barium hydroxide octahydrate
$BaHgI_4$	389: Barium tetra-iodomercurate(II)
BaI_2	346: Barium iodide
BaI_2O_6	344: Barium iodate
$BaMnO_4$	349: Barium manganate(VI)
$BaMn_2O_8$	365: Barium permanganate
$BaMoO_4$	354: Barium molybdate
BaN_2O_4	358: Barium nitrite
BaN_2O_6	356: Barium nitrate
BaN_6	310: Barium azide
$BaNb_2O_6$	355: Barium niobate
$BaNb_4O_{12}Sr$	379: Barium strontium niobium oxide
BaO	362: Barium oxide

BaO_2	366: Barium peroxide	$Ba_3Cr_2O_8$	325: Barium chromate(V)		hydroxide
BaO_3Pb	348: Barium lead oxide	Ba_3N_2	357: Barium nitride	$BiH_8I_5Na_2O_4$	2675: Sodium penta-iodobismuthate tetrahydrate
BaO_3S	384: Barium sulfite	$Ba_3O_8V_2$	399: Barium vanadate	$BiH_{10}N_3O_{14}$	454: Bismuth nitrate pentahydrate
BaO_3Se	371: Barium selenite	$Ba_3O_9WY_3$	400: Barium yttrium tungsten oxide	$BiIO$	459: Bismuth oxyiodide
BaO_3Si	353: Barium metasilicate	Ba_3Sb_2	308: Barium antimonide	BiI_3	450: Bismuth iodide
BaO_3Ti	395: Barium titanate	$Ba_5O_{21}Si_8$	372: Barium silicate	BiI_7K_4	2173: Potassium heptaiodo-bismuthate
BaO_3Zr	401: Barium zirconate	Be	406: Beryllium		
BaO_4S	381: Barium sulfate	$BeBr_2$	414: Beryllium bromide	$BiNO_4$	460: Bismuth oxynitrate
BaO_4Se	369: Barium selenate	$BeCl_2$	417: Beryllium chloride	$BiNaO_3$	2650: Sodium metabismuthate
BaO_4W	397: Barium tungstate	$BeCl_2H_8O_{12}$	428: Beryllium perchlorate tetrahydrate	BiO_4P	463: Bismuth phosphate
BaO_5Si_2	373: Barium silicate			$BiSb$	438: Bismuth antimonide
BaO_5Ti_2	393: Barium titanate	BeF_2	418: Beryllium fluoride	$Bi_2CaCu_2O_8Sr_2$	468: Bismuth strontium calcium copper oxide (2212)
BaO_6P_2	352: Barium metaphosphate	$BeF_4H_4K_2O_2$	2274: Potassium tetrafluoroberyl-late dihydrate		
BaO_6Ta_2	385: Barium tantalate	BeF_4Na_2	2724: Sodium tetra-fluoroberyllate	$Bi_2Ca_2Cu_3O_{10}Sr_2$	469: Bismuth strontium calcium copper oxide (2223)
BaO_7SiZr	403: Barium zirconium silicate	BeH_2	420: Beryllium hydride		
BaO_7U_2	398: Barium uranium oxide	BeH_2O_2	421: Beryllium hydroxide(α) 422: Beryllium hydroxide(β)	$Bi_2Cr_2O_9$	440: Bismuth basic dichromate
BaO_9Si_3Ti	396: Barium titanium silicate	BeH_4O_6S	431: Beryllium sulfate dihydrate	$Bi_2H_{10}O_{14}Sn_3$	466: Bismuth stannate pentahydrate
BaO_9Ti_4	394: Barium titanate	$BeH_6N_2O_9$	424: Beryllium nitrate trihydrate	Bi_2MoO_6	453: Bismuth molybdenum oxide
BaS	382: Barium sulfide 383: Barium sulfide	BeH_8O_8S	432: Beryllium sulfate tetrahydrate	$Bi_2Mo_3O_{12}$	452: Bismuth molybdate
$BaSe$	370: Barium selenide	BeH_8SeO_8	429: Beryllium selenate tetrahydrate	Bi_2O_3	456: Bismuth oxide
$BaSi_2$	375: Barium silicide			Bi_2O_4	474: Bismuth tetroxide
$BaTe$	387: Barium telluride	BeI_2	423: Beryllium iodide	$Bi_2O_7Sn_2$	465: Bismuth stannate
$Ba_{0.5}O_{12}P_3Zr_2$	402: Barium zirconium phosphate	BeO	427: Beryllium oxide	$Bi_2O_7Ti_2$	476: Bismuth titanate
$Ba_2CaCu_2O_8Tl_2$	2906: Thallium barium calcium copper oxide	BeO_4S	430: Beryllium sulfate	$Bi_2O_8V_2$	478: Bismuth vanadate
		BeS	433: Beryllium sulfide	$Bi_2O_{11}Ti_4$	475: Bismuth titanate
Ba_2CaO_6W	317: Barium calcium tungstate	Be_3N_2	425: Beryllium nitride	$Bi_2O_{12}S_3$	471: Bismuth sulfate
$Ba_2Ca_2Cu_3O_{10}Tl_2$	2905: Thallium barium calcium copper oxide	Bi	436: Bismuth	$Bi_2O_{12}W_3$	477: Bismuth tungstate
		$BiBrO$	457: Bismuth oxybromide	Bi_2S_3	472: Bismuth sulfide
Ba_2Cu_3ErOx	1082: Erbium barium copper oxide	$BiBr_3$	441: Bismuth bromide	Bi_2Se_3	464: Bismuth selenide
$Ba_2Cu_3O_7Pr$	2301: Praseo-dymium barium copper oxide	$BiClH_2O_6$	461: Bismuth oxyperchlorate monohydrate	Bi_2Te_3	473: Bismuth telluride
$Ba_2Cu_3O_7Y$	3196: Yttrium barium copper oxide 3197: Yttrium barium copper oxide	$BiClO$	458: Bismuth oxychloride	$Bi_3FeMo_2O_{12}$	451: Bismuth iron molybdenum oxide
		$BiCl_3$	442: Bismuth chloride	$Bi_4Ge_3O_{12}$	446: Bismuth germanium oxide
$Ba_2Cu_4O_8Y$	3195: Yttrium barium copper oxide	$BiCl_3H_2O$	443: Bismuth chloride monohydrate	$Bi_4O_{12}Zr_3$	479: Bismuth zirconate
$Ba_2NaNb_5O_{15}$	376: Barium sodium niobium oxide	BiF_3	445: Bismuth fluoride	$Bi_5H_9N_4O_{22}$	449: Bismuth hydroxide nitrate oxide
Ba_2O_6SrW	380: Barium strontium tungsten oxide	BiF_5	462: Bismuth pentafluoride	Bk	404: Berkelium (α) 405: Berkelium (β)
$Ba_2O_7P_2$	368: Barium pyrophosphate	BiH_3	447: Bismuth hydride	$BrCl$	499: Bromine chloride
$Ba_2O_8Si_3$	374: Barium silicate	BiH_3O_3	448: Bismuth	$BrCs$	734: Cesium bromide
$Ba_3Ca_3Cu_4O_{13}Tl_4$	2904: Thallium barium calcium copper oxide				

BrCsO$_3$	733: Cesium bromate
BrCu	934: Copper(I) bromide
BrD	1032: Deuterium bromide
BrF	501: Bromine monofluoride
BrF$_3$	504: Bromine trifluoride
BrF$_5$	503: Bromine pentafluoride
BrH	1363: Hydrogen bromide
BrHO	1386: Hypobromous acid
BrHO$_3$	496: Bromic acid
BrH$_2$LiO	1568: Lithium bromide monohydrate
BrH$_3$Si	506: Bromosilane
BrH$_4$N	119: Ammonium bromide
BrH$_4$NO	1382: Hydroxyl-amine hydrobromide
BrH$_4$NaO$_2$	2569: Sodium bromide dihydrate
BrH$_5$N$_2$	1353: Hydrazine monohydro-bromide
BrI	1423: Iodine monobromide
BrIn	1397: Indium(I) bromide
BrK	2143: Potassium bromide
BrKO$_3$	2142: Potassium bromate
BrLi	1567: Lithium bromide
BrLiO$_3$	1566: Lithium bromate
BrN$_3$	498: Bromine azide
BrNa	2568: Sodium bromide
BrNaO$_3$	2567: Sodium bromate
BrOV	3147: Vanadyl bromide
BrO$_2$	500: Bromine dioxide
BrO$_3$Rb	2381: Rubidium bromate
BrRb	2382: Rubidium bromide
BrTl	2910: Thallium(I) bromide
Br$_2$	497: Bromine
Br$_2$Ca	573: Calcium bromide
Br$_2$CaH$_2$O$_7$	572: Calcium bromate monohydrate
Br$_2$CaH$_4$O$_2$	574: Calcium bromide dihydrate
Br$_2$CaH$_{12}$O$_6$	575: Calcium bromide hexahydrate
Br$_2$CaO$_6$	571: Calcium bromate
Br$_2$Cd	516: Cadmium bromide

Br$_2$CdH$_8$O$_4$	517: Cadmium bromide tetrahydrate
Br$_2$Co	862: Cobalt(II) bromide
Br$_2$CoH$_{12}$O$_6$	863: Cobalt(II) bromide hexahydrate
Br$_2$CoH$_{12}$O$_{12}$	861: Cobalt(II) bromate hexahydrate
Br$_2$Cr	795: Chromium(II) bromide
Br$_2$CsI	735: Cesium bromoiodide
Br$_2$Cu	960: Copper(II) bromide
Br$_2$Fe	1169: Ferrous bromide
Br$_2$Fe (anhydrous)	1171: Ferrous bromide hydrate
Br$_2$FeH$_{12}$O$_6$	1170: Ferrous bromide hexahydrate
Br$_2$H$_2$O$_7$Pb	1507: Lead bromate monohydrate
Br$_2$H$_2$O$_7$Sr	2771: Strontium bromate monohydrate
Br$_2$H$_2$Si	1038: Dibromo-silane
Br$_2$H$_6$NiO$_3$	1942: Nickel bromide trihydrate
Br$_2$H$_8$MnO$_4$	1769: Manganese (II) bromide tetrahydrate
Br$_2$H$_{12}$MgO$_6$	1674: Magnesium bromide hexahydrate
Br$_2$H$_{12}$MgO$_{12}$	1672: Magnesium bromate hexahydrate
Br$_2$H$_{12}$O$_6$Sr	2773: Strontium bromide hexahydrate
Br$_2$H$_{12}$O$_{12}$Zn	3231: Zinc bromate hexahydrate
Br$_2$Hg	1831: Mercury(II) bromide
Br$_2$Hg$_2$	1813: Mercury(I) bromide
Br$_2$In	1400: Indium(II) bromide
Br$_2$Mg	1673: Magnesium bromide
Br$_2$Mn	1768: Manganese (II) bromide
Br$_2$Ni	1941: Nickel bromide
Br$_2$O	502: Bromine oxide
Br$_2$OS	2943: Thionyl bromide
Br$_2$OSe	2473: Selenium oxybromide
Br$_2$OV	3149: Vanadyl dibromide
Br$_2$O$_2$W	3050: Tungsten dioxydibromide
Br$_2$Pb	1508: Lead bromide
Br$_2$Pd	2051: Palladium(II) bromide
Br$_2$Pt	2109: Platinum(II)

	bromide
Br$_2$Ra	2332: Radium bromide
Br$_2$Se$_2$	2466: Selenium bromide
Br$_2$Sn	2750: Stannous bromide
Br$_2$Sr	2772: Strontium bromide
Br$_2$Te$_2$	2862: Tellurium dibromide
Br$_2$Ti	2998: Titanium dibromide
Br$_2$V	3118: Vanadium dibromide
Br$_2$W	3045: Tungsten dibromide
Br$_2$Zn	3232: Zinc bromide
Br$_3$Ce	705: Cerous bromide
Br$_3$CeH$_{14}$O$_7$	706: Cerous bromide heptahydrate
Br$_3$Cr	807: Chromium(III) bromide
Br$_3$CrH$_{12}$O$_6$	808: Chromium(III) bromide hexahydrate
Br$_3$Dy	1061: Dysprosium bromide
Br$_3$Er	1084: Erbium bromide
Br$_3$ErH$_{18}$O$_9$	1085: Erbium bromide nonahydrate
Br$_3$Eu	1118: Europium(III) bromide
Br$_3$Fe	1134: Ferric bromide
Br$_3$Ga	1246: Gallium(III) bromide
Br$_3$Gd	1214: Gadolinium bromide
Br$_3$HSi	3029: Tribromo-silane
Br$_3$H$_4$O$_2$Rh	2366: Rhodium(III) bromide dihydrate
Br$_3$H$_8$IrO$_4$	1435: Iridium(III) bromide tetrahydrate
Br$_3$H$_{12}$O$_6$Sm	2430: Samarium bromide hexahydrate
Br$_3$H$_{18}$LaO$_{18}$	1467: Lanthanum bromate nonahydrate
Br$_3$H$_{18}$NdO$_{18}$	1902: Neodymium bromate nonahydrate
Br$_3$H$_{18}$O$_{18}$Pr	2303: Praseo-dymium bromate nonahydrate
Br$_3$H$_{18}$O$_{18}$Sm	2429: Samarium bromate nonahydrate
Br$_3$H$_{18}$O$_9$Y	3200: Yttrium bromide nonahydrate
Br$_3$Ho	1329: Holmium bromide
Br$_3$In	1403: Indium(III) bromide
Br$_3$La	1468: Lanthanum

CH₈Mg₂O₈	carbonate tetrahydrate 1678: Magnesium carbonate hydroxide trihydrate	CNiO₃	carbide 1943: Nickel carbonate	C₂CdN₂	oxalate 523: Cadmium cyanide
CH₈N₂O₃	122: Ammonium carbonate	CO	670: Carbon monoxide	C₂CdO₄	534: Cadmium oxalate
CH₁₀MgO₈	1679: Magnesium carbonate pentahydrate	COS	672: Carbon oxysulfide	C₂Ce	694: Cerium carbide
		COSe	671: Carbon oxyselenide	C₂Cl₂F₅	776: Chloropenta-fluoroethane
CH₁₂KNaO₉	2247: Potassium sodium carbonate hexahydrate	CO₂	665: Carbon dioxide	C₂Cl₂O₂	2045: Oxalyl chloride
		CO₃Pb	1509: Lead carbonate		
CH₁₂Ni₃O₁₁	1936: Nickel basic carbonate tetrahydrate	CO₃Ra	2333: Radium carbonate	C₂CoNO₃	852: Cobalt nitrosodicarbonyl
		CO₃Rb₂	2383: Rubidium carbonate	C₂CoN₂S₂	907: Cobalt(II) thiocyanate
CH₂₀Na₂O₁₃	2572: Sodium carbonate decahydrate	CO₃Sr	2775: Strontium carbonate	C₂CoO₄	891: Cobalt(II) oxalate
CHf	1303: Hafnium carbide	CO₃Tl₂	2911: Thallium(I) carbonate	C₂Cr₃	784: Chromium carbide
CHg₂O₃	1814: Mercury(I) carbonate	CO₃Zn	3234: Zinc carbonate	C₂CsF₃O₂	763: Cesium trifluoroacetate
CHg₄O₆	1829: Mercury(II) basic carbonate	CO₅U	3105: Uranyl carbonate	C₂Cu₂	932: Copper(I) acetylide
CIN	1420: Iodine cyanide	CO₈Zr₃ (anhydrous)	3296: Zirconium carbonate basic hydrate	C₂CuKN₂	2155: Potassium copper(I) cyanide
CI₄	680: Carbon tetraiodide	CSSe	675: Carbon sulfide selenide	C₂CuN₂	968: Copper(II) cyanide
CKN	2157: Potassium cyanide	CSTe	676: Carbon sulfide telluride	C₂CuO₄	990: Copper(II) oxalate
CKNO	2156: Potassium cyanate	CS₂	667: Carbon disulfide	C₂HCuO₄.₅	991: Copper(II) oxalate hemihydrate
CKNS	2283: Potassium thiocyanate	CSe₂	666: Carbon diselenide	C₂HNa	2556: Sodium acetylide
CK₂O₃	2144: Potassium carbonate	CSi	2488: Silicon carbide	C₂H₂BaO₅	361: Barium oxalate monohydrate
CK₂S₃	2282: Potassium thiocarbonate	CTa	2842: Tantalum monocarbide	C₂H₂BeO₄	419: Beryllium formate
CLiN	1578: Lithium cyanide	CTa₂	2837: Tantalum carbide	C₂H₂CaO₄	598: Calcium formate
CLiNS (anhydrous)	1629: Lithium thiocyanate hydrate	CTh	2950: Thorium carbide	C₂H₂CaO₅	621: Calcium oxalate monohydrate
CLi₂O₃	1570: Lithium carbonate	CTi	2997: Titanium carbide	C₂H₂CrO₅	800: Chromium(II) oxalate monohydrate
CMgO₃	1675: Magnesium carbonate	CU	3083: Uranium monocarbide		
		CV	3115: Vanadium carbide	C₂H₂CuO₄	975: Copper(II) formate
CMnO₃	1770: Manganese(II) carbonate	CV₂	3125: Vanadium monocarbide	C₂H₂KO₄.₅	2201: Potassium hydrogen oxalate hemihydrate
CMn₃	1755: Manganese carbide	CW	3043: Tungsten carbide		
CMo	1865: Molybdenum carbide	CW₂	3042: Tungsten carbide	C₂H₂K₂O₅	2225: Potassium oxalate monohydrate
CMo₂	1866: Molybdenum carbide	CZr	3295: Zirconium carbide	C₂H₂K₂O₇	2230: Potassium percarbonate monohydrate
CNNa	2585: Sodium cyanide	C₂AgKN₂	2246: Potassium silver cyanide	C₂H₂O₄	2043: Oxalic acid
		C₂Ag₂	2507: Silver acetylide	C₂H₂O₄Zn	3248: Zinc formate
CNNaO	2584: Sodium cyanate	C₂Ag₂O₄	2533: Silver oxalate	C₂H₂O₅Sr	2797: Strontium oxalate monohydrate
CNNaS	2727: Sodium thiocyanate	C₂AuKN₂	2158: Potassium cyanoaurite		
CNRb	2388: Rubidium cyanide	C₂AuN₂Na	2602: Sodium gold cyanide	C₂H₂O₈Pb₃	1505: Lead basic carbonate
CNTl	2914: Thallium(I) cyanide	C₂Ba	318: Barium carbide	C₂H₃AgO₂	2505: Silver acetate
		C₂BaN₂	326: Barium cyanide	C₂H₃BiO₃	470: Bismuth subacetate
CN₄	1021: Cyanogen azide	C₂BaN₂S₂	390: Barium thiocyanate	C₂H₃CsO₂	728: Cesium acetate
CNa₂O₃	2570: Sodium carbonate	C₂BaO₄	360: Barium oxalate	C₂H₃CuO₂	931: Copper(I) acetate
CNb	1997: Niobium(IV) carbide	C₂Ca	576: Calcium carbide	C₂H₃KO₂	2132: Potassium acetate
CNb₂	1989: Niobium	C₂CaN₂	591: Calcium cyanide	C₂H₃LiO₂	1555: Lithium acetate
		C₂CaO₄	620: Calcium		

$C_2H_3NaO_2$	2553: Sodium acetate	
$C_2H_3NaO_5$	2628: Sodium hydrogen oxalate monohydrate	
$C_2H_3O_2Rb$	2377: Rubidium acetate	
$C_2H_3O_2Tl$	2907: Thallium(I) acetate	
$C_2H_4CoN_2O_2$	872: Cobalt(II) cyanide dihydrate	
$C_2H_4CoO_6$	892: Cobalt(II) oxalate dihydrate	
$C_2H_4CrO_5$	799: Chromium(II) formate monohydrate	
$C_2H_4FeO_6$	1185: Ferrous oxalate dihydrate	
$C_2H_4MgO_6$	1711: Magnesium oxalate dihydrate	
$C_2H_4MnO_6$	1786: Manganese (II) oxalate dihydrate	
$C_2H_4NiO_6$	1963: Nickel oxalate dihydrate	
$C_2H_4O_6Zn$	3262: Zinc oxalate dihydrate	
$C_2H_5Na_3O_8$	2571: Sodium carbonate bicarbonate dihydrate	
C_2H_5OTl	2915: Thallium(I) ethoxide	
$C_2H_6BaN_2O_3S_2$	391: Barium thiocyanate trihydrate	
$C_2H_6BeO_7$	426: Beryllium oxalate trihydrate	
$C_2H_6CdO_7$	535: Cadmium oxalate trihydrate	
$C_2H_6Cl_2Ge$	1053: Dimethyl-germanium dichloride	
$C_2H_6CoN_2O_3$	873: Cobalt(II) cyanide trihydrate	
$C_2H_6CoN_2O_3S_2$	908: Cobalt(II) thiocyanate trihydrate	
$C_2H_6FeN_2O_3S_2$	1195: Ferrous thiocyanate trihydrate	
C_2H_6Hg	1054: Dimethyl-mercury	
$C_2H_6NO_{4.5}$	176: Ammonium hydrogen oxalate hemihydrate	
$C_2H_6O_6$	2044: Oxalic acid dihydrate	
$C_2H_6O_6S_2Zn$	3247: Zinc formaldehyde sulfoxylate	
$C_2H_6O_6Zn$	3249: Zinc formate dihydrate	
$C_2H_6O_9U$	3110: Uranyl oxalate trihydrate	
$C_2H_6O_{12}Zn_5$	3235: Zinc carbonate hydroxide	
$C_2H_7AsO_2$	507: Cacodylic acid	
$C_2H_7LiO_4$	1556: Lithium acetate dihydrate	
$C_2H_7NO_2$	112: Ammonium acetate	

$C_2H_7NO_5$	177: Ammonium hydrogen oxalate monohydrate	
$C_2H_8CaN_2O_4S_2$	651: Calcium thiocyanate tetrahydrate	
$C_2H_8Co_5O_{13}$	860: Cobalt(II) basic carbonate	
$C_2H_8MgN_2O_4S_2$	1742: Magnesium thiocyanate tetrahydrate	
$C_2H_8N_2NiO_4$	1950: Nickel cyanide tetrahydrate	
$C_2H_8N_2O_2$	1347: Hydrazine acetate	
$C_2H_8N_2O_4$	201: Ammonium oxalate	
$C_2H_9NaO_5$	2554: Sodium acetate trihydrate	
$C_2H_{10}CuO_8$	976: Copper(II) formate tetrahydrate	
$C_2H_{10}N_2O_5$	202: Ammonium oxalate monohydrate	
$C_2H_{10}N_4O_4$	1357: Hydrazine monooxalate	
$C_2H_{11}N_3O_5$	220: Ammonium sesquicarbonate	
$C_2H_{14}Ni_5O_{16}$	1944: Nickel carbonate hydroxide tetrahydrate	
C_2HgN_2	1835: Mercury(II) cyanide	
$C_2HgN_2O_2$	1838: Mercury(II) fulminate	
$C_2HgN_2S_2$	1857: Mercury(II) thiocyanate	
C_2HgO_4	1845: Mercury(II) oxalate	
$C_2Hg_2N_2O$	1848: Mercury(II) oxycyanide	
C_2La	1469: Lanthanum carbide	
C_2Li_2	1569: Lithium carbide	
$C_2Li_2O_4$	1610: Lithium oxalate	
C_2N_2	1020: Cyanogen	
$C_2N_2NiS_2$	1981: Nickel thiocyanate	
C_2N_2Pb	1515: Lead cyanide	
$C_2N_2PbS_2$	1546: Lead thiocyanate	
C_2N_2Pd	2054: Palladium(II) cyanide	
C_2N_2Pt	2111: Platinum(II) cyanide	
$C_2N_2S_2Zn$	3287: Zinc thiocyanate	
C_2N_2Zn	3241: Zinc cyanide	
$C_2Na_2O_4$	2672: Sodium oxalate	
C_2O_4Pb	1528: Lead oxalate	
C_2O_4Sn	2757: Stannous oxalate	
C_2O_4Sr	2796: Strontium oxalate	
$C_2O_4Tl_2$	2925: Thallium(I) oxalate	
C_2Sr	2774: Strontium	

		carbide
C_2Th	2952: Thorium dicarbide	
C_2U	3080: Uranium dicarbide	
C_2Y	3201: Yttrium carbide	
$C_3AlN_3S_3$	89: Aluminum thiocyanate	
C_3Al_4	26: Aluminum carbide	
$C_3Ce_2O_9$	707: Cerous carbonate	
C_3ClIrO_3 (n = 1)	1433: Iridium(I) chlorotricarbonyl	
C_3CoNO_4	851: Cobalt nitrosocarbonyl	
$C_3Cr_2O_9$ (anhydrous)	809: Chromium(III) carbonate hydrate	
$C_3Er_2O_9$ (anhydrous)	1086: Erbium carbonate hydrate	
$C_3Eu_2O_9$ (anhydrous)	1119: Europium(III) carbonate hydrate	
$C_3FeN_3S_3$	1159: Ferric thiocyanate	
$C_3H_4Am_2O_{11}$	97: Americium carbonate dihydrate	
$C_3H_6AuN_3O_3$	1292: Gold(III) cyanide trihydrate	
$C_3H_6O_{12}Y_2$	3202: Yttrium carbonate trihydrate	
$C_3H_7AgO_4$	2529: Silver lactate monohydrate	
$C_3H_7NO_4S$	2822: Sulfur trioxide N,N-di-methylformamide complex	
$C_3H_8Dy_2O_{13}$	1062: Dysprosium carbonate tetrahydrate	
C_3H_9FSi	1206: Fluorotri-methylsilane	
$C_3H_{10}Ce_2O_{14}$	708: Cerous carbonate pentahydrate	
$C_3H_{10}La_2O_{14}$	1471: Lanthanum carbonate pentahydrate	
$C_3H_{16}La_2O_{17}$	1470: Lanthanum carbonate octahydrate	
$C_3H_{16}O_{17}Pr_2$	2305: Praseo-dymium carbonate octahydrate	
$C_3H_{17}N_3O_{12}Zr$	245: Ammonium zirconyl carbonate dihydrate	
$C_3Ho_2O_9$ (anhydrous)	1330: Holmium carbonate hydrate	
$C_3Na_4O_{11}U$	2737: Sodium uranyl carbonate	
$C_3Nd_2O_9$ (anhydrous)	1904: Neodymium carbonate hydrate	
C_3O_2	674: Carbon suboxide	
$C_3O_9Sc_2$ (anhydrous)	2452: Scandium carbonate hydrate	
$C_3O_9Sm_2$	2431: Samarium carbonate	
$C_3O_9Tb_2$	2880: Terbium	

(anhydrous)	carbonate hydrate	$C_4H_7CrO_5$	803: Chromium(III)
$C_3O_9Yb_2$	3177: Ytterbium		acetate hydroxide
(anhydrous)	carbonate hydrate	$C_4H_7FeO_5$	1132: Ferric basic
C_3U_2	3094: Uranium		acetate
	tricarbide	$C_4H_7KO_{10}$	2280: Potassium
$C_4CdK_2N_4$	2180: Potassium		tetraoxalate
	hexacyano-		dihydrate
	cadmium	$C_4H_7NaO_4$	2588: Sodium
$C_4Cl_2O_4Rh_2$	2362: Rhodium		diacetate
	carbonyl chloride	$C_4H_7NaO_7$	2639: Sodium
$C_4CoHgN_4S_4$	1856: Mercury(II)		hydrogen tartrate
	tetrathiocyanato-		monohydrate
	cobaltate(II)	$C_4H_7O_{4.5}Sr$	2768: Strontium
C_4F	668: Carbon fluoride		acetate
$C_4HKO_{7.5}Sb$	2138: Potassium		hemihydrate
	antimony tartrate	$C_4H_8BaN_4O_4Pt$	388: Barium tetra-
	hemihydrate		cyanoplatinate(II)
$C_4H_2FeO_4$	1378: Hydrogen		tetrahydrate
	tetracarbonyl-	$C_4H_8BaO_5$	303: Barium acetate
	ferrate(II)		monohydrate
$C_4H_2K_2N_4NiO$	2270: Potassium	$C_4H_8CaO_5$	561: Calcium
	tetracyano-		acetate
	nickelate(II)		monohydrate
	monohydrate	$C_4H_8CrO_5$	794: Chromium(II)
$C_4H_4BaO_6$	386: Barium tartrate		acetate
$C_4H_4CdO_4$	544: Cadmium		monohydrate
	succinate	$C_4H_8CuO_5$	950: Copper(II)
$C_4H_4K_2O_{10}Pt$	2140: Potassium		acetate
	bis(oxalato)-		monohydrate
	platinate(II)	$C_4H_8MgO_5$	1656: Magnesium
$C_4H_4K_2O_{11}Ti$	2286: Potassium		acetate
	titanium oxalate		monohydrate
	dihydrate	$C_4H_8Na_2O_8$	2712: Sodium
$C_4H_4O_6Sn$	2764: Stannous		tartrate dihydrate
	tartrate	$C_4H_8O_6Zr$	3320: Zirconyl
$C_4H_4O_{10}Th$	2963: Thorium		acetate hydroxide
	oxalate dihydrate	$C_4H_9NO_5$	118: Ammonium
$C_4H_5KO_6$	2207: Potassium		bimalate
	hydrogen tartrate	$C_4H_9NO_6$	183: Ammonium
$C_4H_6As_6Cu_4O_{16}$	949: Copper(II)		hydrogen tartrate
	acetate	$C_4H_{10}AlCl$	1045: Diethyl-
	metaarsenite		aluminum
$C_4H_6BaO_4$	302: Barium acetate		chloride
$C_4H_6BeO_4$	407: Beryllium	$C_4H_{10}BF_3O$	493: Boron
	acetate		trifluoride
$C_4H_6CaO_4$	559: Calcium		etherate
	acetate	$C_4H_{10}CaO_6$	560: Calcium
$C_4H_6CdO_4$	509: Cadmium		acetate dihydrate
	acetate	$C_4H_{10}CaO_7$	643: Calcium
$C_4H_6CuO_4$	948: Copper(II)		succinate
	acetate		trihydrate
$C_4H_6FeO_4$	1165: Ferrous	$C_4H_{10}CdO_6$	510: Cadmium
	acetate		acetate dihydrate
$C_4H_6HgO_4$	1827: Mercury(II)	$C_4H_{10}CuN_2O_5$	978: Copper(II)
	acetate		glycinate
$C_4H_6Hg_2O_4$	1812: Mercury(I)		monohydrate
	acetate	$C_4H_{10}CuO_9$	1008: Copper(II)
$C_4H_6K_2N_4O_3Pt$	2272: Potassium		tartrate trihydrate
	tetracyano-	$C_4H_{10}Li_2N_4O_5Pt$	1627: Lithium tetra-
	platinate(II)		cyanoplatinate(II)
	trihydrate		pentahydrate
$C_4H_6MgO_4$	1655: Magnesium	$C_4H_{10}Mg_5O_{18}$	1677: Magnesium
	acetate		carbonate
$C_4H_6O_4Pb$	1496: Lead acetate		hydroxide
$C_4H_6O_4Pd$	2049: Palladium(II)		tetrahydrate
	acetate	$C_4H_{10}N_2O_{10}Ti$	241: Ammonium
$C_4H_6O_4Sn$	2749: Stannous		titanium oxalate
	acetate		monohydrate
$C_4H_6O_4Sr$	2767: Strontium	$C_4H_{10}O_6Zn$	3221: Zinc acetate
	acetate		dihydrate
$C_4H_6O_4Zn$	3220: Zinc acetate	$C_4H_{10}O_8Pb_3$	1504: Lead basic
$C_4H_7AlO_5$	33: Aluminum		acetate
	diacetate	$C_4H_{10}O_8U$	3103: Uranyl acetate

	dihydrate		
$C_4H_{10}Zn$	1046: Diethylzinc		
$C_4H_{11}NO_4$	170: Ammonium		
	hydrogen acetate		
$C_4H_{12}CaO_7S_2$	652: Calcium		
	thioglycollate		
	trihydrate		
$C_4H_{12}CaO_{10}$	649: Calcium		
	tartrate		
	tetrahydrate		
$C_4H_{12}CrN_7OS_4$	235: Ammonium		
	tetrathiocyanodi-		
	ammonochromate		
	(III) monohydrate		
$C_4H_{12}Ge$	2900: Tetramethyl-		
	germane		
$C_4H_{12}KNaO_{10}$	2692: Sodium		
	potassium tartrate		
	tetrahydrate		
$C_4H_{12}Mg_5O_{19}$	1667: Magnesium		
	basic carbonate		
	pentahydrate		
$C_4H_{12}N_2O_6$	226: Ammonium		
	tartrate		
$C_4H_{12}O_7Pb$	1497: Lead acetate		
	trihydrate		
$C_4H_{12}O_{10}Sr$	2812: Strontium		
	tartrate		
	tetrahydrate		
$C_4H_{12}Sn$	2901: Tetra-		
	methyltin		
$C_4H_{14}CoO_8$	857: Cobalt(II)		
	acetate		
	tetrahydrate		
$C_4H_{14}FeO_8$	1166: Ferrous		
	acetate		
	tetrahydrate		
$C_4H_{14}MgO_8$	1657: Magnesium		
	acetate		
	tetrahydrate		
$C_4H_{14}MnO_8$	1765: Manganese		
	(II) acetate		
	tetrahydrate		
$C_4H_{14}NO_2PS_2$	199: Ammonium		
	O,O-diethyldithio-		
	phosphate		
$C_4H_{14}NiO_8$	1930: Nickel acetate		
	tetrahydrate		
$C_4H_{18}Cu_2O_{11}$	955: Copper(II)		
	basic acetate		
$C_4HgK_2N_4$	2269: Potassium		
	tetracyano-		
	mercurate(II)		
$C_4K_2N_4Pt$	2271: Potassium		
	tetracyano-		
	platinate(II)		
$C_4K_2N_4Zn$	2273: Potassium		
	tetracyanozincate		
C_4NiO_4	1945: Nickel		
	carbonyl		
C_4O_8Pd	2059: Palladium(II)		
	oxalate		
C_5BrMnO_5	1760: Manganese		
	pentacarbonyl		
	bromide		
C_5BrO_5Re	2340: Rhenium		
	pentacarbonyl		
	bromide		
C_5ClO_5Re	2341: Rhenium		
	pentacarbonyl		
	chloride		
C_5FeO_5	1450: Iron		
	pentacarbonyl		

$C_5HF_6O_2Tl$ 2918: Thallium(I) hexafluoroacetyl-acetonate

$C_5H_4F_3O_2Tl$ 2933: Thallium(I) trifluoroacetyl-acetonate

$C_5H_4FeK_2N_6O_3$ 2223: Potassium nitroprusside dihydrate

$C_5H_4FeN_6Na_2O_3$ 2667: Sodium nitroferricyanide (III) dihydrate

$C_5H_5Cl_3Zr$ 3298: Zirconium cyclopentadienyl trichloride

$C_5H_5Cl_4Nb$ 1029: Cyclopenta-dienylniobium tetrachloride

C_5H_5In 1027: Cyclopenta-dienylindium(I)

C_5H_5Li 1579: Lithium cyclopentadienide

$C_5H_7AgO_2$ 2506: Silver acetylacetonate

$C_5H_7CsO_2$ 729: Cesium acetylacetonate

$C_5H_7LiO_2$ 1557: Lithium acetylacetonate

$C_5H_7NaO_2$ 2555: Sodium acetylacetonate

$C_5H_7O_2Rb$ 2378: Rubidium acetylacetonate

$C_5H_7O_2Tl$ 2908: Thallium(I) acetylacetonate

$C_5H_8FeN_8O$ 198: Ammonium nitroferricyanide

$C_5H_8KO_{2.5}$ 2133: Potassium acetylacetonate hemihydrate

$C_5H_{10}AgNS_2$ 2520: Silver diethyldithio-carbamate

$C_5H_{13}NO_2$ 244: Ammonium valerate

$C_5H_{15}NSn$ 1052: Dimethyl-aminotrimethyltin

$C_6CoK_3N_6$ 2181: Potassium hexacyanocobalt (III)

$C_6Co_2FeN_6$ (anhydrous)   876: Cobalt(II) ferrocyanide hydrate

C_6CrO_6 785: Chromium carbonyl

$C_6Cu_2FeN_6$ 971: Copper(II) ferrocyanide

$C_6Eu_2O_{12}$ 1125: Europium(III) oxalate

$C_6FeK_3N_6$ 2165: Potassium ferricyanide

$C_6Fe_2O_{12}$ 1147: Ferric oxalate

$C_6H_2Al_2O_{13}$ 61: Aluminum oxalate monohydrate

$C_6H_2FeN_6Na_3O$ 2595: Sodium ferricyanide monohydrate

$C_6H_2N_3O_7Tl$ 2928: Thallium(I) picrate

$C_6H_4AgN_3O_8$ 2540: Silver picrate monohydrate

$C_6H_5Ag_3O_7$ 2517: Silver citrate

$C_6H_5AlO_7$ 32: Aluminum citrate

$C_6H_5BiO_7$ 444: Bismuth citrate

C_6H_5ClHg 2075: Phenyl-mercuric chloride

$C_6H_5K_3O_7$ 2152: Potassium citrate

$C_6H_6CrK_3O_{15}$ 2150: Potassium chromium(III) oxalate trihydrate

$C_6H_6FeK_4N_6O_3$ 2166: Potassium ferrocyanide trihydrate

$C_6H_6K_3O_{15}Sb$ 2137: Potassium antimony oxalate trihydrate

$C_6H_6N_4O_7$ 215: Ammonium picrate

$C_6H_7K_3O_8$ 2153: Potassium citrate monohydrate

$C_6H_8FeO_8$ 1177: Ferrous citrate monohydrate

$C_6H_8GeN_2O_{12}$ (anhydrous)   151: Ammonium germanium oxalate hydrate

$C_6H_9AlO_6$ 17: Aluminum acetate

$C_6H_9BiO_6$ 437: Bismuth acetate

$C_6H_9CoO_6$ 911: Cobalt(III) acetate

$C_6H_9Cu_2O_{9.5}$ 925: Copper citrate hemipentahydrate

$C_6H_9InO_6$ 1391: Indium acetate

$C_6H_9LaO_6$ (anhydrous)   1463: Lanthanum acetate hydrate

$C_6H_9LuO_6$ (anhydrous)   1635: Lutetium acetate hydrate

$C_6H_9Na_3O_9$ 2581: Sodium citrate dihydrate

$C_6H_9O_6Pr$ (anhydrous)   2299: Praseo-dymium acetate hydrate

$C_6H_9O_6Sb$ 250: Antimony(III) acetate

$C_6H_9O_6Sc$ (anhydrous)   2449: Scandium acetate hydrate

$C_6H_9O_6Tb$ (anhydrous)   2877: Terbium acetate hydrate

$C_6H_9O_6Tl$ 2934: Thallium(III) acetate

$C_6H_9O_6Y$ (anhydrous)   3190: Yttrium acetate hydrate

$C_6H_{10}CaO_4$ 635: Calcium propionate

$C_6H_{10}O_4Zn$ 3270: Zinc propionate

$C_6H_{10}O_{17}Sc_2$ 2457: Scandium oxalate pentahydrate

$C_6H_{11}CrO_7$ 804: Chromium(III) acetate monohydrate

$C_6H_{11}HoO_7$ 1328: Holmium acetate monohydrate

$C_6H_{11}NdO_7$ 1900: Neodymium acetate monohydrate

$C_6H_{11}O_7Tm$ 2975: Thulium acetate monohydrate

$C_6H_{12}Ba_2FeN_6O_6$ 331: Barium ferrocyanide hexahydrate

$C_6H_{12}CeO_{7.5}$ 703: Cerous acetate hemitrihydrate

$C_6H_{12}Lu_2O_{18}$ 1646: Lutetium oxalate hexahydrate

$C_6H_{12}N_2S_4Zn$ 3243: Zinc dimethyl-dithiocarbamate

$C_6H_{12}O_{18}Tm_2$ 2984: Thulium oxalate hexahydrate

$C_6H_{13}Li_3O_{11}$ 1576: Lithium citrate tetrahydrate

$C_6H_{13}MnO_8$ 1802: Manganese (III) acetate dihydrate

$C_6H_{14}CuO_8$ 984: Copper(II) lactate dihydrate

$C_6H_{14}LiN$ 1583: Lithium diisopropylamide

$C_6H_{14}N_2O_7$ 174: Ammonium hydrogen citrate

$C_6H_{15}AlO_3$ 37: Aluminum ethoxide

$C_6H_{15}FeO_{12}$ 1138: Ferric citrate pentahydrate

$C_6H_{15}Na_3O_{12}$ 2582: Sodium citrate pentahydrate

$C_6H_{15}O_9Sm$ 2426: Samarium acetate trihydrate

$C_6H_{15}P$ 3031: Triethylphos-phine

$C_6H_{16}MgO_{12}$ 1686: Magnesium citrate pentahydrate

$C_6H_{16}O_9Sr$ 2790: Strontium lactate trihydrate

$C_6H_{17}DyO_{10}$ 1058: Dysprosium acetate tetrahydrate

$C_6H_{17}ErO_{10}$ 1080: Erbium acetate tetrahydrate

$C_6H_{17}GdO_{10}$ 1211: Gadolinium acetate tetrahydrate

$C_6H_{17}N_3O_7$ 131: Ammonium citrate tribasic

$C_6H_{17}O_{10}Yb$ 3174: Ytterbium acetate tetrahydrate

$C_6H_{18}Ce_2O_{21}$ 717: Cerous oxalate nonahydrate

$C_6H_{18}FeN_3O_{15}$ 143: Ammonium ferric oxalate trihydrate

$C_6H_{18}FeN_9O_3$ 145: Ammonium ferricyanide trihydrate

$C_6H_{18}FeN_{10}O$ 163: Ammonium hexacyanoferrate (II) monohydrate

$C_6H_{18}O_{21}Y_2$ 3212: Yttrium oxalate nonahydrate

$C_6H_{20}Ac_2O_{22}$ 9: Actinium oxalate decahydrate

$C_6H_{20}Dy_2O_{22}$ 1071: Dysprosium

	oxalate decahydrate		tonate dihydrate
$C_6H_{20}Er_2O_{22}$	1095: Erbium oxalate decahydrate	$C_{10}H_8CuF_6O_4$	1014: Copper(II) trifluoroacetylacetonate
$C_6H_{20}FeN_6Na_4O_{10}$	2596: Sodium ferrocyanide decahydrate	$C_8H_{12}O_8Pb$	1544: Lead tetraacetate
$C_6H_{20}Gd_2O_{22}$	1224: Gadolinium oxalate decahydrate	$C_8H_{12}O_8Rh_2$	2364: Rhodium(II) acetate dimer
		$C_8H_{12}O_8Si$	2486: Silicon acetate
		$C_8H_{12}O_{14}Zn$	3285: Zinc tartrate dihydrate
$C_6H_{20}Ho_2O_{22}$	1338: Holmium oxalate decahydrate	$C_8H_{16}CuO_5$	961: Copper(II) butyrate monohydrate
$C_6H_{20}Nd_2O_{22}$	1915: Neodymium oxalate decahydrate	$C_8H_{18}AlCl$	1051: Diisobutylaluminum chloride
$C_6H_{20}O_{22}Pr_2$	2315: Praseodymium oxalate decahydrate	$C_8H_{19}NO_2$	120: Ammonium caprylate
$C_6H_{20}O_{22}Sm_2$	2439: Samarium oxalate decahydrate	$C_8H_{20}BrN$	2897: Tetraethylammonium bromide
$C_6H_{20}O_{22}Ti_2$	3012: Titanium oxalate decahydrate	$C_8H_{20}ClN$	2898: Tetraethylammonium chloride
$C_6H_{20}O_{22}Yb_2$	3183: Ytterbium oxalate decahydrate	$C_8H_{20}ClN_2P$	435: Bis(diethylamino)chlorophosphine
$C_6H_{21}CrO_{12}$	802: Chromium(III) acetate hexahydrate	$C_8H_{20}CrN_2O_6$	832: Chromium(VI) morpholine
		$C_8H_{20}GeO_4$	1275: Germanium (IV) ethoxide
$C_6H_{24}Ca_2FeN_6O_{12}$	594: Calcium ferrocyanide dodecahydrate	$C_8H_{20}O_4Si$	2899: Tetraethylorthosilicate
$C_6H_{28}Cl_3CoN_6O_2$	3035: Tris(ethylenediammine) cobalt(III) chloride trihydrate	$C_8H_{20}Pb$	2895: Tetraethyl lead
		$C_8H_{20}Si$	2896: Tetraethyl silane
$C_6K_2N_6Pt$	2182: Potassium hexacyanoplatinate(IV)	$C_9Fe_2O_9$	1449: Iron nonacarbonyl
$C_6K_2N_6PtS_6$	2195: Potassium hexathiocyantoplatinate(IV)	$C_9H_7MnO_3$	1860: Methylcyclopentadienylmanganese tricarbonyl
$C_6La_2O_{12}$ (anhydrous)	1482: Lanthanum oxalate hydrate	$C_9H_{15}AlO_9$	53: Aluminum lactate
C_6MoO_6	1867: Molybdenum carbonyl	$C_9H_{21}AlO_3$	52: Aluminum isopropoxide
C_6O_6V	3116: Vanadium carbonyl	$C_{10}F_{12}O_4Pb$	1520: Lead hexafluoroacetylacetonate
C_6O_6W	3044: Tungsten carbonyl	$C_{10}H_2CuF_{12}O_4$	979: Copper(II) hexafluoroacetylacetonate
$C_6O_{12}Tb_2$ (anhydrous)	2888: Terbium oxalate hydrate	$C_{10}H_2CuF_{12}O_4$ (anhydrous)	981: Copper(II) hexafluroacetylacetonate hydrate
$C_7H_2O_2$	660: Carbon		
$C_7H_5AgO_2$	2509: Silver benzoate	$C_{10}H_2F_{12}NiO_4$ (anhydrous)	1954: Nickel hexafluoroacetylacetonate hydrate
$C_7H_7IrO_4$	1039: Dicarbonylacetylacetonate iridium(I)	$C_{10}H_2F_{12}O_4Pd$	2056: Palladium(II) hexafluoroacetylacetonate
$C_7H_9NO_2$	117: Ammonium benzoate	$C_{10}H_2F_{12}O_4Pt$	2112: Platinum(II) hexafluoroacetylacetonate
$C_7H_9NO_3$	217: Ammonium salicylate	$C_{10}H_6CaF_{12}O_6$	600: Calcium hexafluoroacetylacetonate dihydrate
C_8Br	659: Carbon		
$C_8Co_2O_8$	844: Cobalt carbonyl	$C_{10}H_6F_{12}MgO_6$	1693: Magnesium hexafluoroacetylacetonate dihydrate
$C_8H_8HgO_2$	2074: Phenylmercuric acetate		
$C_8H_{12}Mo_2O_8$	1862: Molybdenum acetate dimer	$C_{10}H_6F_{12}O_6Zn$	3250: Zinc hexafluoroacetylace-
$C_8H_{12}N_8O_6Pt_2Th$	2973: Thorium		

	tetracyanoplatinate(II) hexahydrate	$C_{10}H_8CuF_6O_4$	1014: Copper(II) trifluoroacetylacetonate
		$C_{10}H_{10}Cl_2Nb$	2011: Niobocene dichloride
		$C_{10}H_{10}Cl_2Ti$	3028: Titanocene dichloride
		$C_{10}H_{10}Cl_2V$	3114: Vanadium bis(cyclopentadienyl) dichloride
		$C_{10}H_{10}Cl_2Zr$	3319: Zirconocene dichloride
		$C_{10}H_{10}Co$	921: Cobaltocene
		$C_{10}H_{10}CoF_6P$	922: Cobaltocenium hexafluorophosphate
		$C_{10}H_{10}Fe$	1162: Ferrocene
		$C_{10}H_{10}Mn$	1753: Manganese bis(cyclopentadienyl)
		$C_{10}H_{10}Ni$	1937: Nickel bis(cyclopentadienyl)
		$C_{10}H_{10}Os$	2034: Osmium bis(cyclopentadienyl)
		$C_{10}H_{10}Ru$	434: Bis(cyclopentadienyl)ruthenium
		$C_{10}H_{10}V$	3146: Vanadocene
		$C_{10}H_{12}F_6MgO_6$	1744: Magnesium trifluoroacetylacetonate dihydrate
		$C_{10}H_{12}F_6NiO_6$	1983: Nickel trifluoroacetylacetonate dihydrate
		$C_{10}H_{14}BeO_4$	408: Beryllium acetylacetonate
		$C_{10}H_{14}CaO_4$ (anhydrous)	562: Calcium acetylacetonate hydrate
		$C_{10}H_{14}CdO_4$	511: Cadmium acetylacetonate
		$C_{10}H_{14}CoO_4$	858: Cobalt(II) acetylacetonate
		$C_{10}H_{14}CuO_4$	951: Copper(II) acetylacetonate
		$C_{10}H_{14}FeO_4$	1167: Ferrous acetylacetonate
		$C_{10}H_{14}MnO_4$	1766: Manganese (II) acetylacetonate
		$C_{10}H_{14}N_2Na_2O_8$	1103: Ethylenediaminetetraacetic acid dihydrate disodium salt
		$C_{10}H_{14}NiO_4$	1931: Nickel acetylacetonate
		$C_{10}H_{14}O_4Pb$	1498: Lead acetylacetonate
		$C_{10}H_{14}O_4Pd$	2050: Palladium(II) acetylacetonate
		$C_{10}H_{14}O_4Pt$	2106: Platinum acetylacetonate
		$C_{10}H_{14}O_4Sr$	2769: Strontium acetylacetonate
		$C_{10}H_{14}O_4Zn$ (anhydrous)	3222: Zinc acetylacetonate hydrate
		$C_{10}H_{14}O_5Ti$	3027: Titanium(IV) oxide acetylacetonate
		$C_{10}H_{14}O_6U$	3104: Uranyl acetylacetonate

ClLiO$_4$ 1612: Lithium perchlorate

ClNO   2025: Nitrosyl chloride

ClNO$_2$ 2030: Nitryl chloride

ClNa   2576: Sodium chloride

ClNaO   2643: Sodium hypochlorite

ClNaO$_2$ 2577: Sodium chlorite

ClNaO$_3$ 2575: Sodium chlorate

ClNaO$_4$ 2678: Sodium perchlorate

ClOSb   273: Antimony(V) oxychloride

ClOV   3148: Vanadyl chloride

ClO$_2$ 769: Chlorine dioxide

ClO$_3$Rb   2384: Rubidium chlorate

ClO$_3$Re   2357: Rhenium(VI) trioxychloride

ClO$_3$Tl   2912: Thallium(I) chlorate

ClO$_4$Rb   2403: Rubidium perchlorate

ClO$_4$Tl   2927: Thallium(I) perchlorate

ClRb   2385: Rubidium chloride

ClTl   2913: Thallium(I) chloride

Cl$_2$ 768: Chlorine

Cl$_2$Co   866: Cobalt(II) chloride

Cl$_2$CoH$_4$O$_2$ 867: Cobalt(II) chloride dihydrate

Cl$_2$CoH$_{12}$O$_6$ 868: Cobalt(II) chloride hexahydrate

Cl$_2$CoH$_{12}$O$_{12}$ 865: Cobalt(II) chlorate hexahydrate

Cl$_2$CoH$_{12}$O$_{14}$ 895: Cobalt(II) perchlorate hexahydrate

Cl$_2$CoO$_8$ 894: Cobalt(II) perchlorate

Cl$_2$Cr   796: Chromium(II) chloride

Cl$_2$CrH$_{16}$O$_8$ 797: Chromium(II) chloride tetrahydrate

Cl$_2$CrO$_2$ 834: Chromyl chloride

Cl$_2$Cu   964: Copper(II) chloride

Cl$_2$CuH$_4$O$_2$ 965: Copper(II) chloride dihydrate

Cl$_2$CuH$_{12}$O$_{12}$ 963: Copper(II) chlorate hexahydrate

Cl$_2$CuH$_{12}$O$_{14}$ 995: Copper(II) perchlorate hexahydrate

Cl$_2$CuO$_8$ 994: Copper(II) perchlorate

Cl$_2$Cu$_4$H$_6$O$_6$ 983: Copper(II) hydroxy chloride

Cl$_2$Cu$_4$H$_7$O$_{6.5}$ 993: Copper(II)

Cl$_2$Eu   oxychloride

Cl$_2$Eu   1110: Europium(II) chloride

Cl$_2$F$_3$Sb   269: Antimony(V) dichlorotrifluoride

Cl$_2$Fe   1173: Ferrous chloride

Cl$_2$FeH$_4$O$_2$ 1174: Ferrous chloride dihydrate

Cl$_2$FeH$_8$O$_4$ 1175: Ferrous chloride tetrahydrate

Cl$_2$FeH$_{12}$O$_{14}$ 1187: Ferrous perchlorate hexahydrate

Cl$_2$Ga   1242: Gallium(II) chloride

Cl$_2$Ge   1266: Germanium (II) chloride

Cl$_2$H$_2$Si   1044: Dichlorosilane

Cl$_2$H$_4$O$_2$Pd   2053: Palladium(II) chloride dihydrate

Cl$_2$H$_4$O$_2$Sn   2752: Stannous chloride dihydrate

Cl$_2$H$_6$HgO$_{11}$ 1849: Mercury(II) perchlorate trihydrate

Cl$_2$H$_6$N$_2$ 1349: Hydrazine dihydrochloride

Cl$_2$H$_6$N$_2$Pd   1040: Dichlorodiam-minepalladium(II)

Cl$_2$H$_6$N$_2$Pt   1041: Dichlorodiam-mineplatinum(II) (cis)

1042: Dichlorodiam-mineplatinum(II)-trans

Cl$_2$H$_6$O$_5$U   3107: Uranyl chloride trihydrate

Cl$_2$H$_6$O$_{11}$Pb   1531: Lead perchlorate trihydrate

Cl$_2$H$_8$MnO$_4$ 1772: Manganese(II) chloride tetrahydrate

Cl$_2$H$_{12}$MgO$_6$ 1683: Magnesium chloride hexahydrate

Cl$_2$H$_{12}$MgO$_{12}$ 1681: Magnesium chlorate hexahydrate

Cl$_2$H$_{12}$MgO$_{14}$ 1715: Magnesium perchlorate hexahydrate

Cl$_2$H$_{12}$MnO$_{14}$ 1788: Manganese(II) perchlorate hexahydrate

Cl$_2$H$_{12}$NiO$_6$ 1948: Nickel chloride hexahydrate

Cl$_2$H$_{12}$NiO$_{12}$ 1946: Nickel chlorate hexahydrate

Cl$_2$H$_{12}$O$_6$Sr   2778: Strontium chloride hexahydrate

Cl$_2$H$_{12}$O$_{14}$Sr   2800: Strontium perchlorate hexahydrate

Cl$_2$H$_{12}$O$_{14}$Zn   3264: Zinc perchlorate hexahydrate

Cl$_2$H$_{14}$N$_4$OPd   2063: Palladium(II)

tetraammine chloride monohydrate

Cl$_2$H$_{16}$HfO$_9$ 1310: Hafnium oxychloride octahydrate

Cl$_2$H$_{16}$O$_9$Zr   3323: Zirconyl chloride octahydrate

Cl$_2$H$_{16}$O$_{17}$Zr   3326: Zirconyl perchlorate octahydrate

Cl$_2$Hg   1832: Mercury(II) chloride

Cl$_2$Hg$_2$ 1816: Mercury(I) chloride

Cl$_2$Hg$_2$O$_6$ 1815: Mercury(I) chlorate

Cl$_2$HI$_2$NiO$_{14}$ 1965: Nickel perchlorate hexahydrate

Cl$_2$In   1401: Indium(II) chloride

Cl$_2$Mg   1682: Magnesium chloride

Cl$_2$MgO$_8$ 1714: Magnesium perchlorate

Cl$_2$Mn   1771: Manganese(II) chloride

Cl$_2$Mo   1876: Molybdenum (II) chloride

Cl$_2$MoO$_2$ 1890: Molybdenum (VI) dioxydichloride

Cl$_2$Ni   1947: Nickel chloride

Cl$_2$O   772: Chlorine monoxide

Cl$_2$OS   2944: Thionyl chloride

Cl$_2$OSe   2474: Selenium oxychloride

Cl$_2$OV   3150: Vanadyl dichloride

Cl$_2$OZr (anhydrous)   3322: Zirconyl chloride hydrate

Cl$_2$O$_2$S   2832: Sulfuryl chloride

Cl$_2$O$_2$U   3106: Uranyl chloride

Cl$_2$O$_2$W   3059: Tungsten oxydichloride

Cl$_2$O$_4$Pb   1512: Lead chlorite

Cl$_2$O$_6$Pb   1510: Lead chlorate

Cl$_2$O$_6$Sr   2776: Strontium chlorate

Cl$_2$O$_6$Zn   3236: Zinc chlorate

Cl$_2$O$_7$ 770: Chlorine heptoxide

Cl$_2$O$_8$Sr   2799: Strontium perchlorate

Cl$_2$Os   2036: Osmium(II) chloride

Cl$_2$Pb   1511: Lead chloride

Cl$_2$Pd   2052: Palladium(II) chloride

Cl$_2$Pt   2110: Platinum(II) chloride

Cl$_2$Ra   2334: Radium chloride

Cl$_2$S$_2$ 2818: Sulfur chloride

Cl$_2$Se$_2$ 2467: Selenium chloride

Cl_2Sm	2447: Samarium(II) chloride
Cl_2Sn	2751: Stannous chloride
Cl_2Sr	2777: Strontium chloride
Cl_2Te	2863: Tellurium dichloride
Cl_2Ti	2999: Titanium dichloride
Cl_2V	3119: Vanadium dichloride
Cl_2W	3046: Tungsten dichloride
Cl_2Zn	3237: Zinc chloride
$Cl_3CoH_{15}N_5$	2068: Pentammine-chlorocobalt(III) chloride
$Cl_3CoH_{18}N_6$	1322: Hexaammine-cobalt(III) chloride
Cl_3Cr	810: Chromium(III) chloride
$Cl_3CrH_{12}O_6$	811: Chromium(III) chloride hexahydrate
Cl_3CrO_{12}	820: Chromium(III) perchlorate
Cl_3Dy	1063: Dysprosium chloride
$Cl_3DyH_{12}O_6$	1064: Dysprosium chloride hexahydrate
Cl_3DyO_{12} (anhydrous)	1073: Dysprosium perchlorate hydrate
Cl_3Er	1087: Erbium chloride
$Cl_3ErH_{12}O_6$	1088: Erbium chloride hexahydrate
Cl_3ErO_{12} (anhydrous)	1097: Erbium perchlorate hydrate
Cl_3Eu	1120: Europium(III) chloride
$Cl_3EuH_{12}O_6$	1121: Europium(III) chloride hexahydrate
$Cl_3EuH_{12}O_{18}$	1127: Europium(III) perchlorate hexahydrate
Cl_3Fe	1135: Ferric chloride
$Cl_3FeH_{12}O_6$	1136: Ferric chloride hexahydrate
$Cl_3FeH_{12}O_{18}$	1151: Ferric perchlorate hexahydrate
Cl_3FeO_{12} (anhydrous)	1152: Ferric perchlorate hydrate
Cl_3Ga	1247: Gallium(III) chloride
$Cl_3GaH_{12}O_{18}$	1257: Gallium(III) perchlorate hexahydrate
Cl_3Gd	1215: Gadolinium chloride
$Cl_3GdH_{12}O_6$	1216: Gadolinium chloride hexahydrate
Cl_3GdO_{12} (anhydrous)	1226: Gadolinium perchlorate hydrate

Cl_3HSi	3030: Trichlorosilane
$Cl_3H_2NO_2Ru$	2417: Ruthenium nitrosyl chloride monohydrate
$Cl_3H_6O_{15}Sb$	260: Antimony(III) perchlorate trihydrate
$Cl_3H_8InO_4$	1405: Indium(III) chloride tetrahydrate
$Cl_3H_{12}HoO_6$	1332: Holmium chloride hexahydrate
$Cl_3H_{12}HoO_{18}$	1340: Holmium perchlorate hexahydrate
$Cl_3H_{12}LaO_6$	1474: Lanthanum chloride hexahydrate
$Cl_3H_{12}LaO_{18}$	1485: Lanthanum perchlorate hexahydrate
$Cl_3H_{12}LuO_6$	1639: Lutetium chloride hexahydrate
$Cl_3H_{12}LuO_{18}$	1648: Lutetium perchlorate hexahydrate
$Cl_3H_{12}NdO_6$	1907: Neodymium chloride hexahydrate
$Cl_3H_{12}NdO_{18}$	1917: Neodymium perchlorate hexahydrate
$Cl_3H_{12}O_6Sc$	2454: Scandium chloride hexahydrate
$Cl_3H_{12}O_6Sm$	2433: Samarium chloride hexahydrate
$Cl_3H_{12}O_6Tb$	2882: Terbium chloride hexahydrate
$Cl_3H_{12}O_6Y$	3204: Yttrium chloride hexahydrate
$Cl_3H_{12}O_6Yb$	3179: Ytterbium chloride hexahydrate
$Cl_3H_{12}O_{18}Pr$	2316: Praseodymium perchlorate hexahydrate
$Cl_3H_{12}O_{18}Tb$	2889: Terbium perchlorate hexahydrate
$Cl_3H_{12}O_{18}Tl$	2941: Thallium(III) perchlorate hexahydrate
$Cl_3H_{12}O_{18}Y$	3214: Yttrium perchlorate hexahydrate
$Cl_3H_{14}LaO_7$	1473: Lanthanum chloride heptahydrate
$Cl_3H_{14}O_7Pr$	2307: Praseodymium chloride heptahydrate
$Cl_3H_{14}O_7Tm$	2979: Thulium chloride heptahydrate
$Cl_3H_{16}InO_{20}$	1412: Indium(III) perchlorate octahydrate

$Cl_3H_{16}MgNO_6$	189: Ammonium magnesium chloride hexahydrate
$Cl_3H_{16}NNiO_6$	194: Ammonium nickel chloride hexahydrate
$Cl_3H_{18}N_6Ru$	1323: Hexaammine-ruthenium(III) chloride
Cl_3Ho	1331: Holmium chloride
Cl_3I	1428: Iodine trichloride
Cl_3In	1404: Indium(III) chloride
Cl_3Ir	1436: Iridium(III) chloride
Cl_3Ir (anhydrous)	1437: Iridium(III) chloride hydrate
Cl_3La	1472: Lanthanum chloride
Cl_3Lu	1638: Lutetium chloride
Cl_3Mo	1879: Molybdenum (III) chloride
Cl_3MoO	1888: Molybdenum (V) oxytrichloride
Cl_3N	2017: Nitrogen trichloride
Cl_3NbO	2010: Niobium(V) oxychloride
Cl_3Nd	1906: Neodymium chloride
Cl_3OP	2089: Phosphorus oxychloride
Cl_3OV	3129: Vanadium oxytrichloride
Cl_3OW	3063: Tungsten oxytrichloride
$Cl_3O_{12}Sm$ (anhydrous)	2441: Samarium perchlorate hydrate
$Cl_3O_{12}Yb$	3185: Ytterbium perchlorate
Cl_3Os	2037: Osmium(III) chloride
Cl_3Os (anhydrous)	2038: Osmium(III) chloride hydrate
Cl_3P	2093: Phosphorus (III) chloride
Cl_3PS	2946: Thiophos-phoryl chloride
Cl_3Pr	2306: Praseodymium chloride
Cl_3Pu	2121: Plutonium(III) chloride
Cl_3Re	2343: Rhenium(III) chloride
Cl_3Rh	2367: Rhodium(III) chloride
Cl_3Rh (anhydrous)	2368: Rhodium(III) chloride hydrate
Cl_3Ru	2420: Ruthenium (III) chloride
Cl_3Ru (anhydrous)	2421: Ruthenium (III) chloride hydrate
Cl_3Sb	252: Antimony(III) chloride
Cl_3Sc	2453: Scandium chloride
Cl_3Sm	2432: Samarium

	chloride	
Cl_3Tb	2881: Terbium chloride	
Cl_3Ti	3022: Titanium trichloride	
Cl_3Tl	2936: Thallium(III) chloride	
Cl_3Tl (anhydrous)	2937: Thallium(III) chloride hydrate	
Cl_3Tm	2978: Thulium chloride	
Cl_3U	3095: Uranium trichloride	
Cl_3V	3138: Vanadium trichloride	
Cl_3Y	3203: Yttrium chloride	
Cl_3Yb	3178: Ytterbium chloride	
Cl_4Cr	830: Chromium(IV) chloride	
$Cl_4CuH_{12}N_2O_2$	134: Ammonium copper(II) chloride dihydrate	
Cl_4CuLi_2	1626: Lithium tetrachlorocuprate	
Cl_4Ge	1274: Germanium (IV) chloride	
$Cl_4H_6Na_2O_3Pd$	2723: Sodium tetra-chloropalladate (II) trihydrate	
$Cl_4H_8N_2Pd$	231: Ammonium tetrachloro-palladate(II)	
$Cl_4H_8N_2Pt$	232: Ammonium tetrachloro-platinate(II)	
$Cl_4H_8N_2Zn$	3223: Zinc ammonium chloride	
$Cl_4H_{10}O_5Pt$	2116: Platinum(IV) chloride pentahydrate	
$Cl_4H_{10}O_5Sn$	2741: Stannic chloride pentahydrate	
$Cl_4H_{12}HgN_2O_2$	190: Ammonium mercuric chloride dihydrate	
Cl_4Hf	1304: Hafnium chloride	
Cl_4K_2Pd	2267: Potassium tetrachloro-palladate (II)	
Cl_4K_2Pt	2268: Potassium tetrachloro-platinate(II)	
Cl_4Mo	1883: Molybdenum (IV) chloride	
Cl_4MoO	1894: Molybdenum (VI) oxytetra-chloride	
Cl_4ORe	2356: Rhenium(VI) oxytetrachloride	
Cl_4OW	3058: Tungsten oxychloride	
$Cl_4O_{16}Th$	2966: Thorium perchlorate	
Cl_4Os	2039: Osmium(IV) chloride	
Cl_4Po	2129: Polonium(IV) chloride	
Cl_4Pt	2115: Platinum(IV)	

	chloride	
Cl_4Pu	2124: Plutonium(IV) chloride	
Cl_4Re	2345: Rhenium(IV) chloride	
Cl_4Se	2479: Selenium tetrachloride	
Cl_4Si	2500: Silicon tetrachloride	
Cl_4Sn	2740: Stannic chloride	
Cl_4Te	2871: Tellurium tetrachloride	
Cl_4Th	2951: Thorium chloride	
Cl_4Ti	3018: Titanium tetrachloride	
Cl_4U	3090: Uranium tetrachloride	
Cl_4V	3135: Vanadium tetrachloride	
Cl_4W	3069: Tungsten tetrachloride	
Cl_4Zr	3297: Zirconium chloride	
$Cl_5H_{10}N_2ORh$	205: Ammonium pentachlororhodate (III) monohydrate	
$Cl_5H_{10}N_2ORu$	206: Ammonium pentachloro-ruthenate(III) monohydrate	
$Cl_5H_{12}N_3Zn$	207: Ammonium pentachlorozincate	
Cl_5IrKNO (anhydrous)	2227: Potassium pentachloronitrosyl iridium(III) hydrate	
Cl_5K_2Ru (anhydrous)	2228: Potassium pentachloro-ruthenate(III) hydrate	
Cl_5Mo	1887: Molybdenum (V) chloride	
Cl_5Nb	2003: Niobium(V) chloride	
Cl_5P	2099: Phosphorus(V) chloride	
Cl_5Re	2353: Rhenium(V) chloride	
Cl_5Sb	268: Antimony(V) chloride	
Cl_5Ta	2846: Tantalum pentachloride	
Cl_5U	3086: Uranium pentachloride	
Cl_5W	3066: Tungsten pentachloride	
Cl_6H_2Ir (anhydrous)	1367: Hydrogen hexachloroiridate (IV) hydrate	
Cl_6H_2Pt	1368: Hydrogen hexachloro-platinate(IV)	
$Cl_6H_8IrN_2$	157: Ammonium hexachloroiridate (IV)	
$Cl_6H_8N_2Os$	158: Ammonium hexachloroosmiate (IV)	
$Cl_6H_8N_2Pd$	159: Ammonium hexachloro-palladate(IV)	

$Cl_6H_8N_2Pt$	160: Ammonium hexachloro-platinate(IV)	
$Cl_6H_8N_2Ru$	162: Ammonium hexachloro-ruthenate(IV)	
$Cl_6H_{12}IrN_3$	155: Ammonium hexachloroiridate (III)	
$Cl_6H_{12}IrNa_2O_6$	2605: Sodium hexachloroiridate (IV) hexahydrate	
$Cl_6H_{12}Na_2O_6Pt$	2609: Sodium hexa-chloroplatinate (IV) hexahydrate	
$Cl_6H_{14}IrN_3O$	156: Ammonium hexachloroiridate (III) monohydrate	
$Cl_6H_{14}N_3ORh$	161: Ammonium hexachlororhodate (III) monohydrate	
$Cl_6H_{14}O_6Pt$	1369: Hydrogen hexachloro-platinate(IV) hexahydrate	
$Cl_6H_{42}N_{14}O_2Ru_3$	2415: Ruthenium ammoniated oxychloride	
Cl_6IrK_2	2175: Potassium hexachloroiridate (IV)	
Cl_6IrNa_3 (anhydrous)	2604: Sodium hexa-chloroiridate(III) hydrate	
Cl_6K_2Os	2176: Potassium hexachloroosmiate (IV)	
Cl_6K_2Pd	2177: Potassium hexachloro-palladate(IV)	
Cl_6K_2Pt	2178: Potassium hexachloro-platinate(IV)	
Cl_6K_2Re	2179: Potassium hexachlororhenate (IV)	
$Cl_6N_3P_3$	2079: Phosphonitrilic chloride trimer	
Cl_6Na_2Os (anhydrous)	2606: Sodium hexa-chloroosmiate(IV) hydrate	
Cl_6Na_2Pd	2607: Sodium hexa-chloropalladate (IV)	
Cl_6Na_2Pt	2608: Sodium hexa-chloroplatinate (IV)	
Cl_6Na_3Rh (anhydrous)	2610: Sodium hexa-chlororhodate(III) hydrate	
Cl_6Si_2	1325: Hexachloro-disilane	
Cl_6W	3055: Tungsten hexachloride	
Cm	1018: Curium (α)
	1019: Curium (β)	
Co	835: Cobalt	
$CoCrO_4$	869: Cobalt(II) chromate	
$CoCr_2O_4$	870: Cobalt(II) chromite	
CoF_2	877: Cobalt(II) fluoride	

$CoF_2H_8O_4$ 878: Cobalt(II) fluoride tetrahydrate

CoF_3 913: Cobalt(III) fluoride

$CoF_6H_{12}O_6Si$ 879: Cobalt(II) hexafluorosilicate hexahydrate

$CoFe_2O_4$ 874: Cobalt(II) diiron tetroxide

$CoHO_2$ 918: Cobalt(III) oxide hydroxide

CoH_2MoO_5 886: Cobalt(II) molybdate monohydrate

CoH_2O_2 880: Cobalt(II) hydroxide

CoH_2O_5S 904: Cobalt(II) sulfate monohydrate

$CoH_4I_2O_2$ 883: Cobalt(II) iodide dihydrate

CoH_4O_5Se 899: Cobalt(II) selenite dihydrate

CoH_6NO_5P 132: Ammonium cobalt(II) phosphate monohydrate

CoH_9O_6 915: Cobalt(III) hydroxide trihydrate

$CoH_{10}N_7O_8$ 234: Ammonium tetranitrodiam- minecobaltate(III)

$CoH_{10}O_9Se$ 897: Cobalt(II) selenate pentahydrate

$CoH_{12}I_2O_6$ 884: Cobalt(II) iodide hexahydrate

$CoH_{12}K_2O_{14}Se_2$ 2154: Potassium cobalt(II) selenate hexahydrate

$CoH_{12}N_2O_{12}$ 888: Cobalt(II) nitrate hexahydrate

$CoH_{12}O_{14}Rb_2S_2$ 2387: Rubidium cobalt(II) sulfate hexahydrate

$CoH_{14}O_{11}S$ 903: Cobalt(II) sulfate heptahydrate

$CoH_{20}N_2O_{14}S_2$ 133: Ammonium cobalt(II) sulfate hexahydrate

CoI_2 882: Cobalt(II) iodide

CoI_2O_6 881: Cobalt(II) iodate

$CoK_3N_6O_{12}$ 2193: Potassium hexanitritocobalt (III)

$CoLiO_2$ 1577: Lithium cobaltite

CoN_2O_4 889: Cobalt(II) nitrite

CoN_2O_6 887: Cobalt(II) nitrate

CoN_3O_9 916: Cobalt(III) nitrate

$CoN_6Na_3O_{12}$ 2622: Sodium hexa- nitritocobalt(III)

CoO 893: Cobalt(II) oxide

CoO_3Ti 849: Cobalt metatitanate

CoO_3Zr 856: Cobalt zirconate

CoO_4Mo 850: Cobalt molybdate

CoO_4S 902: Cobalt(II) sulfate

CoO_4W 909: Cobalt(II) tungstate

CoS 905: Cobalt(II) sulfide

CoS_2 846: Cobalt disulfide

$CoSb$ 837: Cobalt antimonide

$CoSe$ 898: Cobalt(II) selenide

$CoSi_2$ 845: Cobalt disilicide

$CoTe$ 906: Cobalt(II) telluride

$Co_2F_6H_4O_2$ 914: Cobalt(III) fluoride dihydrate

Co_2O_3 917: Cobalt(III) oxide

Co_2O_4Si 900: Cobalt(II) silicate

Co_2O_4Sn 901: Cobalt(II) stannate

Co_2O_4Ti 853: Cobalt orthotitanate

Co_2P 854: Cobalt phosphide

Co_2S_3 920: Cobalt(III) sulfide

$Co_3H_{16}O_{16}P_2$ 896: Cobalt(II) phosphate octahydrate

Co_3O_4 910: Cobalt(II,III) oxide

Cr 779: Chromium

$CrCs_2O_4$ 739: Cesium chromate

$CrCuO_4$ 966: Copper(II) chromate

$CrCu_3H_4O_8$ 956: Copper(II) basic chromate

CrF_2 798: Chromium(II) fluoride

CrF_3 812: Chromium(III) fluoride

$CrF_3H_6O_3$ 814: Chromium(III) fluoride trihydrate

$CrF_3H_8O_4$ 813: Chromium(III) fluoride tetrahydrate

$CrHO_5S$ 806: Chromium(III) basic sulfate

$CrH_4Li_2O_6$ 1575: Lithium chromate dihydrate

$CrH_7O_{7.5}P$ 822: Chromium(III) phosphate hemiheptahydrate

$CrH_8N_2O_4$ 129: Ammonium chromate(VI)

$CrH_8Na_2O_8$ 2580: Sodium chromate tetrahydrate

CrH_9O_6 815: Chromium(III) hydroxide trihydrate

$CrH_{10}MgO_9$ 1684: Magnesium chromate pentahydrate

$CrH_{10}O_9S$ 801: Chromium(II) sulfate pentahydrate

$CrH_{12}O_{10}P$ 823: Chromium(III) phosphate hexahydrate

$CrH_{14}O_{11}Zn$ 3238: Zinc chromate heptahydrate

$CrH_{18}N_3O_{18}$ 818: Chromium(III) nitrate nonahydrate

$CrH_{20}Na_2O_{14}$ 2579: Sodium chromate decahydrate

$CrH_{24}KO_{20}S_2$ 824: Chromium(III) potassium sulfate dodecahydrate

2151: Potassium chromium(III) sulfate dodecahydrate

$CrH_{28}NO_{20}S_2$ 130: Ammonium chromic sulfate dodecahydrate

$CrHgO_4$ 1834: Mercury(II) chromate

$CrHg_2O_4$ 1817: Mercury(I) chromate

CrI_3 816: Chromium(III) iodide

CrK_2O_4 2149: Potassium chromate

$CrLaO_3$ 1475: Lanthanum chromite

$CrLi_2O_4$ 1574: Lithium chromate

CrN 790: Chromium nitride

CrN_3O_9 817: Chromium(III) nitrate

$CrNa_2O_4$ 2578: Sodium chromate

$CrNiO_4$ 1949: Nickel chromate

CrO_2 831: Chromium(IV) oxide

CrO_3 833: Chromium(VI) oxide

CrO_4P 821: Chromium(III) phosphate

CrO_4Pb 1513: Lead chromate

CrO_4Rb_2 2386: Rubidium chromate

CrO_4Sr 2779: Strontium chromate

CrP 791: Chromium phosphide

$CrSb$ 780: Chromium antimonide

$CrSe$ 792: Chromium selenide

$CrSi_2$ 787: Chromium disilicide

$Cr_2CuH_4O_9$ 969: Copper(II) dichromate dihydrate

Cr_2CuO_4 967: Copper(II) chromite

$Cr_2FeH_4NO_8$ 141: Ammonium ferric chromate

Cr_2FeO_4 1176: Ferrous chromite

$Cr_2H_4Li_2O_9$ 1581: Lithium dichromate dihydrate

$Cr_2H_4Na_2O_9$	2589: Sodium dichromate dihydrate
$Cr_2H_6O_{10}Zn$	3242: Zinc dichromate trihydrate
$Cr_2H_8N_2O_7$	136: Ammonium dichromate(VI)
$Cr_2H_{12}MgO_{13}$	1688: Magnesium dichromate hexahydrate
$Cr_2H_{36}O_{30}S_3$	827: Chromium(III) sulfate octadecahydrate
Cr_2HgO_7	1836: Mercury(II) dichromate
$Cr_2K_2O_7$	2159: Potassium dichromate
Cr_2MgO_4	1685: Magnesium chromite
Cr_2N	789: Chromium nitride
Cr_2O_3	819: Chromium(III) oxide
Cr_2O_4Zn	3239: Zinc chromite
$Cr_2O_7Rb_2$	2389: Rubidium dichromate
Cr_2O_8Sn	2742: Stannic chromate
$Cr_2O_{12}S_3$	825: Chromium(III) sulfate
$Cr_2O_{12}S_3$ (anhydrous)	826: Chromium(III) sulfate hydrate
Cr_2S_3	828: Chromium(III) sulfide
Cr_2Te_3	829: Chromium(III) telluride
$Cr_3Fe_2O_{12}$	1137: Ferric chromate
Cr_3Si	793: Chromium silicide
$Cr_4Cu_4H_6Na_2O_{20}$	2583: Sodium copper chromate trihydrate
$Cr_6Fe_2O_{21}$	1139: Ferric dichromate
Cs	727: Cesium
CsF	741: Cesium fluoride
$CsHO$	745: Cesium hydroxide
CsH_2N	731: Cesium amide
CsH_3O_2	746: Cesium hydroxide monohydrate
CsI	748: Cesium iodide
$CsIO_3$	747: Cesium iodate
$CsNO_3$	752: Cesium nitrate
CsN_3	732: Cesium azide
$CsNbO_3$	751: Cesium niobate
CsO_2	760: Cesium superoxide
CsO_3Ta	761: Cesium tantalate
CsO_3V	749: Cesium metavanadate
Cs_2F_6Ge	743: Cesium hexa-fluorogermanate
Cs_2MoO_4	750: Cesium molybdate
Cs_2O	754: Cesium oxide
Cs_2O_3	764: Cesium trioxide
Cs_2O_3Ti	762: Cesium titanate
Cs_2O_3Zr	766: Cesium zirconate
Cs_2O_4S	758: Cesium sulfate
Cs_2O_4W	765: Cesium tungstate
Cs_2S	759: Cesium sulfide
Cs_3O_4V	753: Cesium orthovanadate
$Cs_4O_7V_2$	756: Cesium pyrovanadate
Cu	923: Copper
CuF_2	973: Copper(II) fluoride
$CuF_2H_4O_2$	974: Copper(II) fluoride dihydrate
$CuF_6H_8O_4Si$	980: Copper(II) hexafluorosilicate tetrahydrate
$CuFeS_2$	972: Copper(II) ferrous sulfide
$CuFe_2O_4$	970: Copper(II) ferrate
CuH	937: Copper(I) hydride
CuH_2O_2	982: Copper(II) hydroxide
CuH_4O_5Se	1001: Copper(II) selenite dihydrate
CuH_4O_5Si	1002: Copper(II) silicate dihydrate
CuH_4O_6W	1016: Copper(II) tungstate dihydrate
$CuH_6N_2O_9$	988: Copper(II) nitrate trihydrate
$CuH_{10}O_9S$	1006: Copper(II) sulfate pentahydrate
$CuH_{10}O_9Se$	999: Copper(II) selenate pentahydrate
$CuH_{12}N_2O_{12}$	987: Copper(II) nitrate hexahydrate
$CuH_{14}N_4O_5S$	1011: Copper(II) tetraammine sulfate monohydrate
CuI	938: Copper(I) iodide
$CuLa_{1.85}O_4Sr_{0.15}$	1488: Lanthanum strontium copper oxide
$CuMoO_4$	985: Copper(II) molybdate
CuN_2O_6	986: Copper(II) nitrate
CuN_3	933: Copper(I) azide
CuN_6	954: Copper(II) azide
CuO	992: Copper(II) oxide
CuO_3Sn	1003: Copper(II) stannate
CuO_3Te	1010: Copper(II) tellurite
CuO_3Ti	1013: Copper(II) titanate
CuO_3Zr	930: Copper zirconate
	1017: Copper(II) zirconate
CuO_4S	1005: Copper(II) sulfate
CuO_4W	1015: Copper(II) tungstate
CuO_6V_2	929: Copper vanadate
CuS	1007: Copper(II) sulfide
$CuSe$	1000: Copper(II) selenide
$CuTe$	1009: Copper(II) telluride
$Cu_2HO_{3.5}S$	943: Copper(I) sulfite hemihydrate
$Cu_2H_2O_4S$	944: Copper(I) sulfite monohydrate
Cu_2HgI_4	939: Copper(I) mercury iodide
Cu_2O	940: Copper(I) oxide
$Cu_2O_7P_2$ (anhydrous)	998: Copper(II) pyrophosphate hydrate
Cu_2S	942: Copper(I) sulfide
Cu_2Se	941: Copper(I) selenide
Cu_2Te	945: Copper(I) telluride
$Cu_3H_4O_8S_2$	947: Copper(I,II) sulfite dihydrate
$Cu_3H_6O_{11}P_2$	996: Copper(II) phosphate trihydrate
Cu_3N	926: Copper nitride
Cu_3P	927: Copper phosphide
$Cu_4H_6N_2O_{10}$	957: Copper(II) basic nitrite
Cu_5Si	928: Copper silicide
DI	1034: Deuterium iodide
DLi	1580: Lithium deuteride
DNa	2587: Sodium deuteride
D_2	1031: Deuterium
D_2O	1035: Deuterium oxide
D_2O_4S	1036: Deutero-sulfuric acid
Dy	1057: Dysprosium
DyF_3	1065: Dysprosium fluoride
DyH_3	1066: Dysprosium hydride
DyH_3O_3	1067: Dysprosium hydroxide
$DyH_{10}N_3O_{14}$	1069: Dysprosium nitrate pentahydrate
DyI_3	1068: Dysprosium iodide
DyN	1070: Dysprosium nitride
$DySi_2$	1074: Dysprosium silicide
$Dy_2H_{16}O_{20}S_3$	1075: Dysprosium sulfate octahydrate
Dy_2O_3	1072: Dysprosium oxide
Dy_2S_3	1076: Dysprosium sulfide
Dy_2Te_3	1077: Dysprosium telluride
Er	1079: Erbium
ErF_3	1089: Erbium fluoride

ErH_3	1090: Erbium hydride	FNa_2O_3P	2599: Sodium fluorophosphate		difluoride
ErH_3O_3	1091: Erbium hydroxide	$FNbO_2$	2006: Niobium(V) fluorodioxide	F_2Zn	3244: Zinc fluoride
$ErH_{10}N_3O_{14}$	1093: Erbium nitrate pentahydrate	FO_3PSn	2755: Stannous fluorophosphate	F_3Fe	1141: Ferric fluoride
				$F_3FeH_6O_3$	1142: Ferric fluoride trihydrate
ErI_3	1092: Erbium iodide	FRb	2390: Rubidium fluoride	F_3Ga	1248: Gallium(III) fluoride
ErN	1094: Erbium nitride	FTl	2916: Thallium(I) fluoride	$F_3GaH_6O_3$	1249: Gallium(III) fluoride trihydrate
$ErSi_2$	1098: Erbium silicide				
$Er_2H_{16}O_{20}S_3$	1100: Erbium sulfate octahydrate	F_2	1197: Fluorine	F_3Gd	1217: Gadolinium fluoride
Er_2O_3	1096: Erbium oxide	F_2Fe	1178: Ferrous fluoride	F_3HSi	3033: Trifluorosilane
$Er_2O_{12}S_3$	1099: Erbium sulfate	$F_2FeH_8O_4$	1179: Ferrous fluoride tetrahydrate	$F_3H_6InO_3$	1407: Indium(III) fluoride trihydrate
Er_2S_3	1101: Erbium sulfide				
Er_2Te_3	1102: Erbium telluride	F_2Ge	1267: Germanium (II) fluoride	$F_3H_6O_3V$	3140: Vanadium trifluoride trihydrate
Es	1078: Einsteinium	F_2HK	2199: Potassium hydrogen fluoride		
Eu	1105: Europium			F_3Ho	1333: Holmium fluoride
EuF_2	1111: Europium(II) fluoride	F_2HNa	2627: Sodium hydrogen fluoride	F_3In	1406: Indium(III) fluoride
EuF_3	1122: Europium(III) fluoride	F_2HO_2P	1047: Difluorophos- phoric acid	F_3La	1476: Lanthanum fluoride
EuH_2; EuH_3	1107: Europium hydride	F_2H_2Si	1048: Difluorosilane	F_3Lu	1640: Lutetium fluoride
$EuH_{10}N_3O_{14}$	1124: Europium(III) nitrate pentahydrate	F_2H_5N	175: Ammonium hydrogen fluoride	F_3Mn	1804: Manganese (III) fluoride
		$F_2H_8NiO_4$	1953: Nickel fluoride tetrahydrate	F_3N	2018: Nitrogen trifluoride
$EuH_{12}N_3O_{15}$	1123: Europium(III) nitrate hexahydrate	$F_2H_8O_4Zn$	3245: Zinc fluoride tetrahydrate	F_3Nd	1908: Neodymium fluoride
EuI_2	1112: Europium(II) iodide	F_2Hg	1837: Mercury(II) fluoride	F_3OP	2090: Phosphorus oxyfluoride
EuN	1108: Europium nitride	F_2Hg_2	1818: Mercury(I) fluoride	F_3OV	3130: Vanadium oxytrifluoride
EuO_4S	1114: Europium(II) sulfate	F_2Kr	1457: Krypton difluoride	F_3P	2094: Phosphorus (III) fluoride
EuS	1115: Europium(II) sulfide	F_2Mg	1690: Magnesium fluoride	F_3Pd	2064: Palladium(III) fluoride
$EuSe$	1113: Europium(II) selenide	F_2Mn	1776: Manganese(II) fluoride	F_3Pr	2308: Praseodymium fluoride
$EuSi_2$	1109: Europium silicide	F_2MoO_2	1891: Molybdenum (VI) dioxydi- fluoride	F_3Pu	2122: Plutonium(III) fluoride
$EuTe$	1116: Europium(II) telluride			F_3Sb	253: Antimony(III) fluoride
$Eu_2H_{16}O_{20}S_3$	1128: Europium(III) sulfate octahydrate	F_2Ni	1952: Nickel fluoride		
Eu_2O_3	1126: Europium(III) oxide	F_2O	1199: Fluorine monoxide	F_3Sc	2455: Scandium fluoride
FH	1366: Hydrogen fluoride	F_2OS	2945: Thionyl fluoride	F_3Sm	2435: Samarium fluoride
FHO_3S	1205: Fluorosulfonic acid	F_2OSe	2476: Selenium oxyfluoride	F_3Tb	2883: Terbium fluoride
FH_2O_3P	1897: Monofluoro- phosphoric acid	F_2OTh	2965: Thorium oxyfluoride	F_3Ti	3023: Titanium trifluoride
FH_3Si	1204: Fluorosilane	F_2OV	3151: Vanadyl difluoride	F_3Tl	2938: Thallium(III) fluoride
FH_4KO_2	2168: Potassium fluoride dihydrate	F_2OXe	3165: Xenon oxydifluoride	F_3Tm	2980: Thulium fluoride
FH_4N	147: Ammonium fluoride	F_2O_2	1198: Fluorine dioxide	F_3U	3096: Uranium trifluoride
FH_4NO_3S	149: Ammonium fluorosulfonate	F_2O_2S	2833: Sulfuryl fluoride	F_3V	3139: Vanadium trifluoride
FK	2167: Potassium fluoride	F_2O_2Se	2475: Selenium oxydifluoride	F_3Y	3205: Yttrium fluoride
FLi	1584: Lithium fluoride	F_2O_2Xe	3159: Xenon dioxydifluoride	F_3Yb	3180: Ytterbium fluoride
FNO	2026: Nitrosyl fluoride	F_2Pb	1517: Lead fluoride	F_4Ge	1276: Germanium (IV) fluoride
FNO_2	2031: Nitryl fluoride	F_2Pd	2055: Palladium(II) fluoride	$F_4GeH_6O_3$	1277: Germanium (IV) fluoride trihydrate
FNO_3	1200: Fluorine nitrate	F_2Sn	2753: Stannous fluoride		
FNa	2597: Sodium fluoride	F_2Sr	2782: Strontium fluoride	F_4H_4NSb	233: Ammonium tetrafluoro- antimonate(III)
$FNaO_3S$	2600: Sodium fluorosulfonate	F_2Xe	3158: Xenon		

FeH_3O_3 — monohydrate / 1143: Ferric hydroxide

FeH_4O_6P — 1153: Ferric phosphate dihydrate

$FeH_6O_6P_3$ — 1144: Ferric hypophosphite

$FeH_8I_2O_4$ — 1183: Ferrous iodide tetrahydrate

$FeH_{12}N_2O_{12}$ — 1184: Ferrous nitrate hexahydrate

$FeH_{14}O_{11}S$ — 1192: Ferrous sulfate heptahydrate

$FeH_{18}N_3O_{18}$ — 1146: Ferric nitrate nonahydrate

$FeH_{20}N_2O_{14}S_2$ — 146: Ammonium ferrous sulfate hexahydrate

$FeH_{24}O_{20}RbS_2$ — 2396: Rubidium iron(III) sulfate dodecahydrate

$FeH_{28}NO_{20}S_2$ — 144: Ammonium ferric sulfate dodecahydrate

FeI_2 — 1182: Ferrous iodide

$FeMoO_4$ — 1448: Iron molybdate

FeO — 1186: Ferrous oxide

FeO_3Ti — 1196: Ferrous titanate

FeO_4P (anhydrous) — 1154: Ferric phosphate hydrate

FeO_4S — 1191: Ferrous sulfate

FeO_4W — 1453: Iron tungstate

FeO_9V_3 — 1145: Ferric metavanadate

FeS — 1194: Ferrous sulfide

FeS_2 — 1446: Iron disulfide

$FeSe$ — 1190: Ferrous selenide

$FeSi$ — 1451: Iron silicide

$FeSi_2$ — 1445: Iron disilicide

$FeTe$ — 1452: Iron telluride

$Fe_2H_2O_4$ — 1150: Ferric oxide monohydrate

$Fe_2H_{18}O_{21}S_3$ — 1158: Ferric sulfate nonahydrate

Fe_2O_3 — 1148: Ferric oxide

Fe_2O_5Zr — 1454: Iron zirconate

$Fe_2O_{12}S_3$ — 1156: Ferric sulfate

$Fe_2O_{12}S_3$ (anhydrous) — 1157: Ferric sulfate hydrate

Fe_2P — 1189: Ferrous phosphide

$Fe_3H_{16}O_{16}P_2$ — 1188: Ferrous phosphate octahydrate

Fe_3O_4 — 1455: Iron(II,III) oxide

Fe_3Sb_2 — 1441: Iron antimonide

$Fe_4H_{18}O_{30}P_6$ — 1155: Ferric pyrophosphate nonahydrate

$Fe_5Lu_3O_{12}$ — 1643: Lutetium iron oxide

$Fe_5O_{12}Y_3$ — 3210: Yttrium iron oxide

$Fe_{12}O_{19}Sr$ — 2781: Strontium ferrite

Fm — 1129: Fermium

Fr — 1207: Francium

Ga — 1234: Gallium

$GaHO_2$ — 1256: Gallium(III) oxide hydroxide

GaH_3 — 1250: Gallium(III) hydride

GaH_3O_3 — 1251: Gallium(III) hydroxide

GaI_3 — 1252: Gallium(III) iodide

GaN — 1239: Gallium nitride

GaN_3O_9 — 1253: Gallium(III) nitrate

GaN_3O_9 (anhydrous) — 1254: Gallium(III) nitrate hydrate

GaN_9 — 1238: Gallium azide

GaP — 1240: Gallium phosphide

GaS — 1244: Gallium(II) sulfide

$GaSb$ — 1236: Gallium antimonide

$GaSe$ — 1243: Gallium(II) selenide

$GaTe$ — 1245: Gallium(II) telluride

GaV_3 — 3123: Vanadium gallide

$Ga_2H_{36}O_{30}S_3$ — 1260: Gallium(III) sulfate octadecahydrate

Ga_2O — 1241: Gallium suboxide

Ga_2O_3 — 1255: Gallium(III) oxide

$Ga_2O_{12}S_3$ — 1259: Gallium(III) sulfate

Ga_2S_3 — 1261: Gallium(III) sulfide

Ga_2Se_3 — 1258: Gallium(III) selenide

Ga_2Te_3 — 1262: Gallium(III) telluride

$Ga_5Gd_3O_{12}$ — 1218: Gadolinium gallium garnet

Gd — 1210: Gadolinium

GdH_2; GdH_3 — 1219: Gadolinium hydride

$GdH_{10}N_3O_{14}$ — 1222: Gadolinium nitrate pentahydrate

$GdH_{12}N_3O_{15}$ — 1221: Gadolinium nitrate hexahydrate

GdI_3 — 1220: Gadolinium iodide

GdN — 1223: Gadolinium nitride

$GdSi_2$ — 1227: Gadolinium silicide

$Gd_2H_{16}O_{20}S_3$ — 1229: Gadolinium sulfate octahydrate

Gd_2O_3 — 1225: Gadolinium oxide

$Gd_2O_7Ti_2$ — 1232: Gadolinium titanate

$Gd_2O_{12}S_3$ — 1228: Gadolinium sulfate

Gd_2S_3 — 1230: Gadolinium sulfide

Gd_2Te_3 — 1231: Gadolinium telluride

Ge — 1263: Germanium

GeH_4 — 1265: Germanium tetrahydride

GeI_2 — 1268: Germanium(II) iodide

GeI_4 — 1278: Germanium(IV) iodide

$GeMg_2$ — 1692: Magnesium germanide

$GeMg_2O_4$ — 1691: Magnesium germanate

$GeNa_2O_3$ — 2655: Sodium metagermanate

GeO — 1269: Germanium(II) oxide

GeO_2 — 1279: Germanium(IV) oxide

GeS — 1271: Germanium(II) sulfide

GeS_2 — 1281: Germanium(IV) sulfide

$GeSe$ — 1270: Germanium(II) selenide

$GeSe_2$ — 1280: Germanium(IV) selenide

$GeTe$ — 1272: Germanium(II) telluride

$GeTe_2$ — 1282: Germanium(IV) telluride

Ge_3N_4 — 1264: Germanium nitride

HF — 1361: Hydrofluoric acid, 70%

$HHgN_2O_{6.5}$ — 1842: Mercury(II) nitrate hemihydrate

HI — 1372: Hydrogen iodide

HIO_3 — 1418: Iodic acid

HI_2KO_6 — 2200: Potassium hydrogen iodate

HI_3Si — 3034: Triiodosilane

HK — 2196: Potassium hydride

HKO — 2208: Potassium hydroxide

HKO_3S — 2206: Potassium hydrogen sulfite

HKO_3Se — 2203: Potassium hydrogen selenite

HKO_4S — 2204: Potassium hydrogen sulfate

HK_2O_3P — 2202: Potassium hydrogen phosphite

HK_2O_4P — 2216: Potassium monohydrogen phosphate

HLi — 1590: Lithium hydride

$HLiO$ — 1592: Lithium hydroxide

$HMnO_2$ — 1805: Manganese(III) hydroxide

HNO_2 — 2028: Nitrous acid

HNO_3 — 2012: Nitric acid

HNO_5S — 2027: Nitrosylsulfuric acid

HNO_5Zr — 3321: Zirconyl basic nitrate

HN_3 — 1360: Hydrazoic acid

HNa — 2623: Sodium hydride

$HNaO$ — 2641: Sodium hydroxide

HNaO₃S	2638: Sodium hydrogen sulfite
HNaO₄S	2633: Sodium hydrogen sulfate
HNaS	2635: Sodium hydrogen sulfide
HNa₂O₄P	2629: Sodium hydrogen phosphate
HNb	1992: Niobium hydride
HORb	2394: Rubidium hydroxide
HOTl	2920: Thallium(I) hydroxide
HO₃P n=1	1858: Metaphosphoric acid
HO₃V	1859: Metavanadic acid
HO₄PPb	1521: Lead hydrogen phosphate
HO₄Re	2073: Perrhenic acid
HTa	2841: Tantalum hydride
H₂	1362: Hydrogen
H₂Hf	1306: Hafnium hydride
H₂HgNO₄	1822: Mercury(I) nitrate monohydrate
H₂HgN₂O₇	1843: Mercury(II) nitrate monohydrate
H₂Ho; H₃Ho	1334: Holmium hydride
H₂INa₃O₆	2674: Sodium paraperiodate
H₂I₂Si	1050: Diiodosilane
H₂I₃KO	2287: Potassium triiodide monohydrate
H₂KO₀.₅S	2205: Potassium hydrogen sulfide hemihydrate
H₂KO₂P	2161: Potassium dihydrogen hypophosphite
H₂KO₃P	2163: Potassium dihydrogen phosphite
H₂KO₄P	2162: Potassium dihydrogen phosphate
H₂LiN	1561: Lithium amide
H₂LiNO₃	1608: Lithium nitrite monohydrate
H₂LiO₄P	1582: Lithium dihydrogen phosphate
H₂Li₂O₄Se	1617: Lithium selenite monohydrate
H₂Li₂O₅S	1620: Lithium sulfate monohydrate
H₂Li₂O₅Se	1616: Lithium selenate monohydrate
H₂Lu; H₃Lu	1641: Lutetium hydride
H₂Mg	1695: Magnesium hydride

H₂MgO₂	1697: Magnesium hydroxide
H₂MgO₅S	1735: Magnesium sulfate monohydrate
H₂MnO₂	1778: Manganese(II) hydroxide
H₂MnO₅S	1794: Manganese(II) sulfate monohydrate
H₂NNa	2559: Sodium amide
H₂Na₂O₇P₂	2592: Sodium dihydrogen pyrophosphate
H₂Na₂O₇Sb₂	2561: Sodium antimonate monohydrate
H₂Na₂O₈U₂	2736: Sodium uranate monohydrate
H₂Nd; H₃Nd	1910: Neodymium hydride
H₂NiO₂	1955: Nickel hydroxide
H₂O	3156: Water
H₂O₂	1373: Hydrogen peroxide
H₂O₂Pb	1522: Lead hydroxide
H₂O₂Sr	2785: Strontium hydroxide
H₂O₂Zn	3252: Zinc hydroxide
H₂O₃S	2831: Sulfurous acid
H₂O₃Se	2483: Selenous acid
H₂O₃Te	2875: Tellurous acid
H₂O₄S	2829: Sulfuric acid
H₂O₄Se	2463: Selenic acid
H₂O₄W	3076: Tungstic acid
H₂O₅S	2072: Peroxysulfuric acid
H₂O₅SZn	3281: Zinc sulfate monohydrate
H₂O₅SeV	3152: Vanadyl selenite monohydrate
H₂O₇SU	3111: Uranyl sulfate monohydrate
H₂S	1375: Hydrogen sulfide
H₂S₂O₇	2830: Sulfuric acid fuming
H₂Se	1374: Hydrogen selenide
H₂Sr	2784: Strontium hydride
H₂Tb; H₃Tb	2884: Terbium hydride
H₂Te	1376: Hydrogen telluride
H₂Th	2956: Thorium hydride
H₂Ti	3007: Titanium hydride
H₂Zr	3301: Zirconium hydride
H₃ISi	1898: Monoiodosilane
H₃InO₃	1408: Indium(III) hydroxide
H₃La	1477: Lanthanum hydride
H₃LaO₃	1478: Lanthanum hydroxide

H₃LiO₂	1593: Lithium hydroxide monohydrate
H₃Mo₁₂O₄₀P (anhydrous)	2078: Phosphomolybdic acid hydrate
H₃N	110: Ammonia
H₃NO	1381: Hydroxylamine
H₃NaO₂	2642: Sodium hydroxide monohydrate
H₃NaO₅S	2634: Sodium hydrogen sulfate monohydrate
H₃NdO₃	1911: Neodymium hydroxide
H₃O₂P	1389: Hypophosphorous acid
H₃O₃P	2082: Phosphorous acid
H₃O₃Pr	2311: Praseodymium hydroxide
H₃O₃Tm	2981: Thulium hydroxide
H₃O₃Y	3208: Yttrium hydroxide
H₃O₄P	2081: Phosphoric acid
H₃O₄₀PW₁₂ (anhydrous)	3077: Tungstophosphoric acid hydrate
H₃P	2077: Phosphine
H₃Pr	2310: Praseodymium hydride
H₃Sb	254: Antimony(III) hydride
H₃Sm	2436: Samarium hydride
H₃U	3097: Uranium trihydride
H₃Y	3207: Yttrium hydride
H₃Yb	3181: Ytterbium hydride
H₄HgNO₅	1821: Mercury(I) nitrate dihydrate
H₄IN	188: Ammonium iodide
H₄INO₃	187: Ammonium iodate
H₄INaO₂	2649: Sodium iodide dihydrate
H₄IP	2080: Phosphonium iodide
H₄K₂O₅S	2254: Potassium sulfite dihydrate
H₄K₂O₆Os	2224: Potassium osmiate dihydrate
H₄K₂O₆W	2291: Potassium tungstate dihydrate
H₄K₂O₁₂S₂U	2294: Potassium uranyl sulfate dihydrate
H₄MgN₂	1662: Magnesium amide
H₄MgN₂O₈	1706: Magnesium nitrate dihydrate
H₄MnNO₄	209: Ammonium permanganate
H₄MoNa₂O₆	2662: Sodium

	molybdate dihydrate	H₅N₅	1348: Hydrazine azide
H₄MoO₅	1889: Molybdenum (VI) acid monohydrate	H₅NaO₂S	2636: Sodium hydrogen sulfide dihydrate
H₄Mo₁₂O₄₀Si (anhydrous)	1896: Molybdic silicic acid hydrate	H₆ILiO₃	1597: Lithium iodide trihydrate
H₄NO₃V	192: Ammonium metavanadate	H₆INaO₇	2681: Sodium periodate trihydrate
H₄NO₄Re	211: Ammonium perrhenate	H₆InN₃O₁₂	1410: Indium(III) nitrate trihydrate
H₄N₂	1346: Hydrazine	H₆K₂O₆Sn	2248: Potassium stannate trihydrate
H₄N₂ (anhydrous)	1351: Hydrazine hydrate		
H₄N₂O₂	197: Ammonium nitrite	H₆K₂O₇Te	2257: Potassium tellurate(VI) trihydrate
H₄N₂O₃	196: Ammonium nitrate	H₆K₄O₁₀P₂	2239: Potassium pyrophosphate trihydrate
H₄N₃O₁₁Rh	2371: Rhodium(III) nitrate dihydrate	H₆K₄O₁₉S₄Zr	2298: Potassium zirconium sulfate trihydrate
H₄N₄	116: Ammonium azide	H₆MgN₂O₇	1709: Magnesium nitrite trihydrate
H₄NaO₃P	2646: Sodium hypophosphite monohydrate	H₆MgO₆S	1739: Magnesium sulfite trihydrate
H₄NaO₅P	2591: Sodium dihydrogen phosphate monohydrate	H₆MgO₆Sn	1730: Magnesium stannate trihydrate
H₄Na₂O₆Te	2714: Sodium tellurate(VI) dihydrate	H₆Mg₂O₁₀P₂	1722: Magnesium pyrophosphate trihydrate
H₄Na₂O₆W	2735: Sodium tungstate dihydrate	H₆MnNaO₇	2682: Sodium permanganate trihydrate
H₄Na₂O₈S₂	2594: Sodium dithionate dihydrate	H₆MnO₅P₂	1779: Manganese (II) hypo-phosphite monohydrate
H₄NiO₅Sn	1972: Nickel stannate dihydrate	H₆Mn₂O₁₀P₂	1791: Manganese (II) pyrophos-phate trihydrate
H₄O₄Rh	2375: Rhodium(IV) oxide dihydrate	H₆Mn₂O₁₁Sr	2801: Strontium permanganate trihydrate
H₄O₄Th	2957: Thorium hydroxide	H₆NO₂P	186: Ammonium hypophosphite
H₄O₄Ti	2994: Titanic acid	H₆NO₄P	138: Ammonium dihydrogen phosphate
H₄O₄Zr	3302: Zirconium hydroxide		
H₄O₅SZn	3284: Zinc sulfite dihydrate	H₆N₂O	1352: Hydrazine monohydrate
H₄O₆P₂	1388: Hypophos-phoric acid	H₆N₂O₃S	222: Ammonium sulfamate
H₄O₆PdS	2061: Palladium(II) sulfate dihydrate	H₆N₂O₄S	1359: Hydrazine sulfate
H₄O₇P₂	2330: Pyrophos-phoric acid	H₆N₄O₆	1350: Hydrazine dinitrate
H₄O₇SV	3153: Vanadyl sulfate dihydrate	H₆NaO₆P	2590: Sodium dihydrogen phosphate dihydrate
H₄Si	2484: Silane		
H₄Sn	2991: Tin hydride	H₆Na₂O₆Sn	2702: Sodium stannate trihydrate
H₅IN₂	1355: Hydrazine monohydroiodide		
H₅IO₆	2071: Periodic acid	H₆O₅P₂Zn	3253: Zinc hypophosphite monohydrate
H₅NO	185: Ammonium hydroxide	H₆O₆Te	2859: Telluric acid
H₅NO₃S	182: Ammonium hydrogen sulfite	H₆O₉SU	3112: Uranyl sulfate trihydrate
H₅NO₄S	180: Ammonium hydrogen sulfate		
H₅NS	181: Ammonium hydrogen sulfide		
H₅N₃O₃	1356: Hydrazine mononitrate		

H₆Si₂	1055: Disilane		
H₇MgO₇P	1696: Magnesium hydrogen phosphate trihydrate		
H₇MnO₇P	1777: Manganese (II) hydrogen phosphate trihydrate		
H₇NaO₃S	2637: Sodium hydrogen sulfide trihydrate		
H₈I₂MgO₁₀	1698: Magnesium iodate tetrahydrate		
H₈I₂MnO₄	1781: Manganese (II) iodide tetrahydrate		
H₈I₂NiO₁₀	1957: Nickel iodate tetrahydrate		
H₈MgO₁₀P₂	1741: Magnesium tetrahydrogen phosphate dihydrate		
H₈MnN₂O₁₀	1785: Manganese (II) nitrate tetrahydrate		
H₈MnO₈S	1795: Manganese (II) sulfate tetrahydrate		
H₈MnO₁₀P₂	1774: Manganese (II) dihydrogen phosphate dihydrate		
H₈MoN₂S₄	236: Ammonium tetrathio-molybdate		
H₈Mo₂N₂O₇	139: Ammonium dimolybdate		
H₈N₂O₃S₂	240: Ammonium thiosulfate		
H₈N₂O₃Se	219: Ammonium selenite		
H₈N₂O₄S	223: Ammonium sulfate		
H₈N₂O₄Se	218: Ammonium selenate		
H₈N₂O₄Te	227: Ammonium tellurate		
H₈N₂O₆S	1385: Hydroxyl-amine sulfate		
H₈N₂O₇U₂	243: Ammonium uranate(VI)		
H₈N₂O₈S₂	210: Ammonium peroxydisulfate		
H₈N₂S	224: Ammonium sulfide		
H₈N₂S₄W	237: Ammonium tetrathiotungstate		
H₈N₄O₁₆Th	2960: Thorium nitrate tetrahydrate		
H₈O₆Pt	1371: Hydrogen hexahydroxy-platinate(IV)		
H₈O₁₀S₂Sr	2780: Strontium dithionate tetrahydrate		
H₈O₁₂P₂Zn₃	3268: Zinc phosphate tetrahydrate		
H₈O₁₂S₂Th	2971: Thorium sulfate		

Formula	Entry
$H_8O_{12}S_2U$	tetrahydrate 3101: Uranium(IV) sulfate tetrahydrate
$H_8O_{12}S_2Zr$	3315: Zirconium sulfate tetrahydrate
H_8Si_3	2498: Silicon octahydride
$H_9K_3O_{4.5}S_4Sb$	2281: Potassium thioantimonate heminonahydrate
$H_9N_2O_4P$	178: Ammonium hydrogen phosphate 212: Ammonium phosphate dibasic
$H_9O_{10}PU$	3108: Uranyl hydrogen phosphate tetrahydrate
$H_{10}HoN_3O_{14}$	1336: Holmium nitrate pentahydrate
$H_{10}K_2O_5S$	2253: Potassium sulfide pentahydrate
$H_{10}Mg_3O_{13}P_2$	1719: Magnesium phosphate pentahydrate
$H_{10}N_2O_4S$	225: Ammonium sulfite monohydrate
$H_{10}N_3O_{14}Sc$	2456: Scandium nitrate pentahydrate
$H_{10}N_3O_{14}Yb$	3182: Ytterbium nitrate pentahydrate
$H_{10}N_4O_4S$	1049: Dihydrazine sulfate
$H_{10}N_4O_{17}Zr$	3304: Zirconium nitrate pentahydrate
$H_{10}Na_2O_5S$	2709: Sodium sulfide pentahydrate
$H_{10}Na_2O_8$	2700: Sodium selenite pentahydrate
$H_{10}Na_2O_8S_2$	2730: Sodium thiosulfate pentahydrate
$H_{10}Na_2O_8Si$	2658: Sodium metasilicate pentahydrate
$H_{10}O_8Rh_2$	2373: Rhodium(III) oxide pentahydrate
$H_{10}O_8S_2Sr$	2813: Strontium thiosulfate pentahydrate
$H_{10}O_9SeZn$	3273: Zinc selenate pentahydrate
$H_{10}O_{12}P_2Zr$	3309: Zirconium phosphate trihydrate
$H_{10}Si_4$	2489: Silicon decahydride
$H_{11}N_2O_4P$	179: Ammonium hydrogen phosphite monohydrate
$H_{11}Na_2O_8P$	2632: Sodium hydrogen

Formula	Entry
	phosphite pentahydrate
$H_{12}I_2MgO_6$	1700: Magnesium iodide hexahydrate
$H_{12}I_2NiO_6$	1959: Nickel iodide hexahydrate
$H_{12}I_2O_6Sr$	2789: Strontium iodide hexahydrate
$H_{12}K_2NiO_{14}S_2$	2218: Potassium nickel sulfate hexahydrate
$H_{12}K_2O_{14}S_2Zn$	2296: Potassium zinc sulfate hexahydrate
$H_{12}LaN_3O_{15}$	1480: Lanthanum nitrate hexahydrate
$H_{12}MgMn_2O_{14}$	1716: Magnesium permanganate hexahydrate
$H_{12}MgN_2O_{12}$	1707: Magnesium nitrate hexahydrate
$H_{12}MgO_9S$	1738: Magnesium sulfite hexahydrate
$H_{12}MgO_9S_2$	1743: Magnesium thiosulfate hexahydrate
$H_{12}MgO_9Se$	1726: Magnesium selenite hexahydrate
$H_{12}MgO_{10}Se$	1724: Magnesium selenate hexahydrate
$H_{12}MnN_2O_{12}$	1784: Manganese(II) nitrate hexahydrate
$H_{12}Mn_2O_{14}Zn$	3265: Zinc permanganate hexahydrate
$H_{12}Mo_{12}N_3O_{40}P$	213: Ammonium phosphomolybdate
$H_{12}Mo_{12}N_3O_{40}P$ (anhydrous)	111: Ammonium 12-molybdophosphate hydrate
$H_{12}N_2NiO_{12}$	1962: Nickel nitrate hexahydrate
$H_{12}N_2O_{12}Zn$	3258: Zinc nitrate hexahydrate
$H_{12}N_2O_{14}U$	3109: Uranyl nitrate hexahydrate
$H_{12}N_3NdO_{15}$	1913: Neodymium nitrate hexahydrate
$H_{12}N_3O_{15}Pr$	2313: Praseodymium nitrate hexahydrate
$H_{12}N_3O_{15}Sm$	2438: Samarium nitrate hexahydrate
$H_{12}N_3O_{15}Tb$	2886: Terbium nitrate hexahydrate
$H_{12}N_3O_{15}Tm$	2983: Thulium nitrate hexahydrate
$H_{12}N_3O_{15}Y$	3211: Yttrium nitrate hexahydrate
$H_{12}N_3S_4V$	238: Ammonium tetrathiovandate (IV)
$H_{12}Na_3O_{15}P_3$	2732: Sodium trimetaphosphate hexahydrate
$H_{12}NiO_{10}S$	1977: Nickel sulfate hexahydrate
$H_{12}NiO_{10}Se$	1969: Nickel selenate hexahydrate
$H_{12}O_{10}SZn$	3280: Zinc sulfate hexahydrate
$H_{13}NNaO_8P$	2560: Sodium

Formula	Entry
	ammonium hydrogen phosphate tetrahydrate
$H_{14}MgO_{11}S$	1734: Magnesium sulfate heptahydrate
$H_{14}Mn_3O_{15}P_2$	1789: Manganese(II) phosphate heptahydrate
$H_{14}Na_2Nb_2O_{13}$	2656: Sodium metaniobate heptahydrate
$H_{14}Na_2O_{10}S$	2711: Sodium sulfite heptahydrate
$H_{14}Na_2O_{11}S$	2706: Sodium sulfate heptahydrate
$H_{14}NiO_{11}S$	1976: Nickel sulfate heptahydrate
$H_{14}Ni_3O_{15}P_2$	1966: Nickel phosphate heptahydrate
$H_{14}O_{11}SV$	3144: Vanadium(II) sulfate heptahydrate
$H_{14}O_{11}SZn$	3279: Zinc sulfate heptahydrate
$H_{14}O_{45}SiW_{12}$	2503: Silicotungstic acid
$H_{15}Na_2O_{11}P$	2631: Sodium hydrogen phosphate heptahydrate
$H_{16}Ho_2O_{20}S_3$	1342: Holmium sulfate octahydrate
$H_{16}I_2MgO_8$	1701: Magnesium iodide octahydrate
$H_{16}La_2O_{20}S_3$	1490: Lanthanum sulfate octahydrate
$H_{16}Lu_2O_{20}S_3$	1651: Lutetium sulfate octahydrate
$H_{16}MgNO_{10}P$	1663: Magnesium ammonium phosphate hexahydrate
$H_{16}Mg_3O_{16}P_2$	1718: Magnesium phosphate octahydrate
$H_{16}N_3O_{42}PW_{12}$	214: Ammonium phosphotungstate dihydrate
$H_{16}Nd_2O_{20}S_3$	1921: Neodymium sulfate octahydrate
$H_{16}Ni_3O_{16}P_2$	1967: Nickel phosphate octahydrate
$H_{16}O_{10}Sr$	2803: Strontium peroxide octahydrate
$H_{16}O_{16}S_2Th$	2970: Thorium sulfate octahydrate
$H_{16}O_{16}S_2U$	3100: Uranium(IV) sulfate octahydrate
$H_{16}O_{20}Pr_2S_3$	2320: Praseodymium sulfate octahydrate
$H_{16}O_{20}S_3Sc_2$	2459: Scandium sulfate octahydrate
$H_{16}O_{20}S_3Sm_2$	2443: Samarium sulfate octahydrate
$H_{16}O_{20}S_3Tb_2$	2891: Terbium sulfate octahydrate
$H_{16}O_{20}S_3Tm_2$	2987: Thulium sulfate octahydrate

	iodide	KNbO₃	uranyl nitrate	Kr	phosphite
I₃Ru	2422: Ruthenium (III) iodide		2219: Potassium niobate	La	1456: Krypton
I₃Sb	255: Antimony(III) iodide	KO₂	2255: Potassium superoxide	LaN	1462: Lanthanum 1481: Lanthanum nitride
I₃Sm	2437: Samarium iodide	KO₃Ta	2256: Potassium tantalate	LaO₄P (anhydrous)	1486: Lanthanum phosphate hydrate
I₃Tb	2885: Terbium iodide	KO₃V	2295: Potassium vanadate	LaSi₂	1487: Lanthanum silicide
I₃Tm	2982: Thulium iodide	KO₄Re	2235: Potassium perrhenate	La₂O₂S	1484: Lanthanum oxysulfide
I₃W	3072: Tungsten triiodide	KO₄Ru	2236: Potassium perruthenate	La₂O₃	1483: Lanthanum oxide
I₃Y	3209: Yttrium iodide	K₂Mg₂O₁₂S₃	2212: Potassium magnesium sulfate	La₂S₃	1491: Lanthanum sulfide
I₄O₉	1425: Iodine nonoxide	K₂MnO₄	2213: Potassium manganate	La₂Te₃	1492: Lanthanum telluride
I₄Pt	2117: Platinum(IV) iodide	K₂MoO₄	2215: Potassium molybdate	Li	1554: Lithium
I₄Si	2502: Silicon tetraiodide	K₂N₄O₈Pt	2279: Potassium tetranitrito-platinate(II)	LiMn₂O₃	1598: Lithium manganate
I₄Sn	2744: Stannic iodide			LiNO₃	1606: Lithium nitrate
I₄Te	2873: Tellurium tetraiodide	K₂O	2217: Potassium monoxide	LiN₃	1563: Lithium azide
I₄Th	2958: Thorium iodide	K₂O₂	2234: Potassium peroxide	LiNbO₃	1605: Lithium niobate
I₄Ti	3020: Titanium tetraiodide	K₂O₃S₂	2284: Potassium thiosulfate	LiO₃P	1602: Lithium metaphosphate
I₄U	3092: Uranium tetraiodide	K₂O₃Se	2245: Potassium selenite	LiO₃Ta	1622: Lithium tantalate
I₄W	3070: Tungsten tetraiodide	K₂O₃Te	2258: Potassium tellurite	LiO₃V	1632: Lithium vanadate
I₄Zr	3303: Zirconium iodide	K₂O₃Te (anhydrous)	2259: Potassium tellurite(IV) hydrate	Li₂MnO₃	1599: Lithium manganite
I₅Nb	2007: Niobium(V) iodide	K₂O₃Ti	2285: Potassium titanate	Li₂MoO₄	1604: Lithium molybdate
I₅Ta	2848: Tantalum pentaiodide	K₂O₃Zr	2297: Potassium zirconate	Li₂O	1611: Lithium oxide
In	1390: Indium	K₂O₄Ru	2242: Potassium ruthenate(VI)	Li₂O₂	1614: Lithium peroxide
InN	1395: Indium nitride	K₂O₄S	2251: Potassium sulfate	Li₂O₃Si	1603: Lithium metasilicate
InO₄P	1413: Indium(III) phosphate	K₂O₄Se	2243: Potassium selenate	Li₂O₃Te	1623: Lithium tellurite
InP	1396: Indium phosphide	K₂O₄W	2290: Potassium tungstate	Li₂O₃Ti	1630: Lithium titanate
InS	1402: Indium(II) sulfide	K₂O₅S₂	2241: Potassium pyrosulfite	Li₂O₃Zr	1633: Lithium zirconate
InSb	1393: Indium antimonide	K₂O₆S₂	2164: Potassium dithionate	Li₂O₄S	1619: Lithium sulfate
In₂O₃	1411: Indium(III) oxide	K₂O₇S₂	2240: Potassium pyrosulfate	Li₂O₄W	1631: Lithium tungstate
In₂O₁₂S₃	1415: Indium(III) sulfate	K₂O₇U₂	2292: Potassium uranate	Li₂S	1621: Lithium sulfide
In₂S₃	1416: Indium(III) sulfide	K₂O₈S₂	2237: Potassium persulfate	Li₃N	1607: Lithium nitride
In₂Se₃	1414: Indium(III) selenide	K₂O₈S₂Sn	2249: Potassium stannosulfate	Li₃O₄P	1615: Lithium phosphate
In₂Te₃	1417: Indium(III) telluride	K₂S	2252: Potassium sulfide	Li₄O₄Si	1609: Lithium orthosilicate
Ir	1429: Iridium	K₂Se	2244: Potassium selenide	Lr	1494: Lawrencium
IrO₂	1439: Iridium(IV) oxide			Lu	1634: Lutetium
Ir₂O₃	1438: Iridium(III) oxide	K₃N₆O₁₂Rh	2194: Potassium hexanitritorhodate (III)	LuN	1645: Lutetium nitride
K	2131: Potassium	K₃O₄P	2238: Potassium phosphate	LuN₃O₉ (anhydrous)	1644: Lutetium nitrate hydrate
KMnO₄	2233: Potassium permanganate	K₅O₁₀P₃	2289: Potassium triphosphate	LuSi₂	1649: Lutetium silicide
KNO₂	2222: Potassium nitrite	K₆O₁₈P₆	2192: Potassium hexameta-	Lu₂O₃	1647: Lutetium oxide
KNO₃	2221: Potassium nitrate			Lu₂O₁₂S₃	1650: Lutetium sulfate
KN₃	2139: Potassium azide			Lu₂S₃	1652: Lutetium sulfide
KN₃O₁₁U	2293: Potassium			Lu₂Te₃	1653: Lutetium

	telluride	MnO₄S	permanganate	Mo₁₂Na₃O₄₀	
Md	1810: Mendelevium		1793: Manganese (II) sulfate		
Mg	1654: Magnesium	MnO₄W	1799: Manganese (II) tungstate	Mo₁₂Na₄O₄₀ (anhydrous	
MgMoO₄	1703: Magnesium molybdate	MnO₆S₂	1775: Manganese (II) dithionate	NNaO₂	
MgN₂O₆	1705: Magnesium nitrate	MnO₆V₂	1764: Manganese vanadate	NNaO₃	
MgNb₂O₆	1704: Magnesium niobate	MnS	1796: Manganese (II) sulfide	NNb	
MgO	1712: Magnesium oxide	MnSb	1752: Manganese antimonide	NNd	
MgO₂	1717: Magnesium peroxide	MnSe	1792: Manganese (II) selenide	NO	
MgO₃S	1737: Magnesium sulfite	MnSi₂	1763: Manganese silicide	NO₂	
MgO₃Si	1727: Magnesium silicate	MnTe	1797: Manganese (II) telluride	NO₂Tl	
MgO₃Ti	1702: Magnesium metatitanate	MnTe₂	1808: Manganese (IV) telluride	NO₃Rb	
MgO₃Zr	1747: Magnesium zirconate	Mn₂O₃	1806: Manganese (III) oxide	NO₃Tl	
MgO₄S	1733: Magnesium sulfate	Mn₂O₇	1809: Manganese (VII) oxide	NO₅Te	
MgO₄W	1745: Magnesium tungstate	Mn₂O₇P₂	1790: Manganese (II) pyrophos-phate	NPr	
MgO₅SiZr	1748: Magnesium zirconium silicate	Mn₃O₄	1801: Manganese (II,III) oxide	NPu	
MgO₅Ti₂	1689: Magnesium dititanate	Mn₃P₂	1761: Manganese phosphide	NTa	
MgO₆Ta₂	1740: Magnesium tantalate	Mo	1861: Molybdenum	NTb	
MgS	1736: Magnesium sulfide	MoN	1870: Molybdenum mononitride	NTh	
MgSe	1725: Magnesium selenide	MoNa₂O₄	2661: Sodium molybdate	NTi	
Mg₂O₄Si	1728: Magnesium silicate	MoNiO₄	1960: Nickel molybdate	NU	
Mg₂O₄Ti	1710: Magnesium orthotitanate	MoO₂	1884: Molybdenum (IV) oxide	NV	
Mg₂O₇P₂	1721: Magnesium pyrophosphate	MoO₃	1893: Molybdenum (VI) oxide	NW₂	
Mg₂O₇V₂	1746: Magnesium vanadate	MoO₄Pb	1525: Lead molybdate	NZr	
Mg₂Si	1729: Magnesium silicide	MoO₄Rb₂	2398: Rubidium molybdate	N₂	
Mg₂Sn	1731: Magnesium stannide	MoO₄Sr	2791: Strontium molybdate(VI)	N₂NiO₆	
Mg₃N₂	1708: Magnesium nitride	MoO₄Tl₂	2922: Thallium(I) molybdate	N₂O	
Mg₃P₂	1720: Magnesium phosphide	MoO₄Zn	3257: Zinc molybdate	N₂O₃	
Mg₃Sb₂	1664: Magnesium antimonide	MoO₁₈P₆	1869: Molybdenum metaphosphate	N₂O₄Sr	
Mn	1749: Manganese	MoP	1874: Molybdenum phosphide	N₂O₄Zn	
MnMoO₄	1782: Manganese (II) molybdate	MoS₂	1868: Molybdenum disulfide	N₂O₅	
MnN	1759: Manganese nitride	MoS₃	1895: Molybdenum (VI) sulfide	N₂O₆Pb	
MnN₂O₆	1783: Manganese (II) nitrate	MoSe₂	1885: Molybdenum (IV) selenide	N₂O₆Pd	
MnNb₂O₆	1758: Manganese niobate	MoSi₂	1875: Molybdenum silicide	N₂O₆Sr	
MnO	1787: Manganese (II) oxide	MoTe₂	1886: Molybdenum (IV) telluride	N₂O₇Z₁ (anhy	
MnO₂	1807: Manganese (IV) oxide	Mo₂N	1871: Molybdenum nitride	N₂Sr₃	
MnO₃Si	1762: Manganese silicate	Mo₂O₃	1880: Molybdenum (III) oxide	N₂W	
MnO₃Ti	1798: Manganese (II) titanate	Mo₂S₃	1881: Molybdenum (III) sulfide	N₂Zn₃	
MnO₃Zr	1800: Manganese (II) zirconate			N₃Na	
MnO₄Rb	2404: Rubidium			N₃O₉R	
				N₃O₉S	

Formula	Entry
	oxide
	3307: Zirconium oxide
	3308: Zirconium oxide yttria stabilized
O_3	2047: Ozone
O_3P_2	2096: Phosphorus (III) oxide
O_3PbS	1540: Lead sulfite
O_3PbS_2	1547: Lead thiosulfate
O_3PbSe	1535: Lead selenite
O_3PbSi	1536: Lead silicate
O_3PbTe	1543: Lead tellurite
O_3PbTi	1548: Lead titanate
O_3PbZr	1552: Lead zirconate
O_3Pb_2	1530: Lead oxide
O_3Pr_2	2325: Praseo-dymium(III) oxide
O_3RbTa	2409: Rubidium tantalate
O_3RbV	2397: Rubidium metavanadate
O_3Rb_2Ti	2411: Rubidium titanate
O_3Rb_2Zr	2413: Rubidium zirconate
O_3Re	2355: Rhenium(VI) oxide
O_3Rh_2	2372: Rhodium(III) oxide
O_3S	2823: Sulfur trioxide(α)
	2824: Sulfur trioxide(β)
	2825: Sulfur trioxide(γ)
O_3Sb_2	257: Antimony(III) oxide
	258: Antimony(III) oxide
	259: Antimony(III) oxide
O_3Sc_2	2458: Scandium oxide
O_3Se	2481: Selenium trioxide
O_3SeZn	3275: Zinc selenite
O_3Sm_2	2440: Samarium oxide
O_3SnSr	2808: Strontium stannate
O_3SrTi	2814: Strontium titanate
O_3SrZr	2817: Strontium zirconate
O_3Te	2874: Tellurium trioxide
O_3Ti_2	3024: Titanium trioxide
O_3TiZn	3288: Zinc titanate
O_3Tl_2	2940: Thallium(III) oxide
O_3Tm_2	2985: Thulium oxide
O_3U	3099: Uranium trioxide
O_3VY	3218: Yttrium vanadate
O_3V_2	3141: Vanadium

Formula	Entry
	trioxide
O_3W	3073: Tungsten trioxide
O_3Xe	3172: Xenon trioxide
O_3Y_2	3213: Yttrium oxide
O_3Yb_2	3184: Ytterbium oxide
O_4Os	2042: Osmium (VIII) oxide
O_4PPr	2317: Praseo-dymium phosphate
O_4PSb	261: Antimony(III) phosphate
O_4PbS	1538: Lead sulfate
O_4PbSe	1533: Lead selenate
O_4PbW	1549: Lead tungstate
	1550: Lead tungstate
O_4Pb_3	1553: Lead(II,III) oxide
O_4RaS	2335: Radium sulfate
O_4Rb_2S	2407: Rubidium sulfate
O_4Rb_2W	2412: Rubidium tungstate
O_4Rb_3V	2401: Rubidium orthovanadate
O_4Ru	2424: Ruthenium (VIII) oxide
O_4SSn	2762: Stannous sulfate
O_4SSr	2809: Strontium sulfate
O_4STl_2	2931: Thallium(I) sulfate
O_4SZn	3278: Zinc sulfate
O_4Sb_2	266: Antimony(IV) oxide
	267: Antimony(IV) oxide
O_4SeSr	2805: Strontium selenate
O_4SeTl_2	2929: Thallium(I) selenate
O_4SiTh	2962: Thorium orthosilicate
O_4SiZn_2	3276: Zinc silicate
O_4SiZr	3313: Zirconium silicate
O_4SrW	2815: Strontium tungstate
O_4Ta_2	2855: Tantalum tetroxide
O_4V_2	3121: Vanadium dioxide
O_4Xe	3170: Xenon tetroxide
O_5P_2	2101: Phosphorus (V) oxide
O_5STi	3013: Titanium oxysulfate
O_5Sb_2	271: Antimony(V) oxide
O_5Sb_2 (anhydrous)	272: Antimony(V) oxide hydrate
O_5Ta_2	2849: Tantalum pentoxide
O_5Ta_2 (anhydrous)	2850: Tantalum pentoxide hydrate

Formula	Entry
O_5V_2	3133: Vanadium pentoxide
O_6PbTa_2	1541: Lead tantalate
O_6PbV_2	1551: Lead vanadate
O_6Se_2Sn	2747: Stannic selenite
O_6SrTa_2	2811: Strontium tantalate
O_6SrV_2	2816: Strontium vanadate
$O_7P_2Sn_2$	2759: Stannous pyrophosphate
$O_7P_2Zn_2$	3271: Zinc pyrophosphate
O_7P_2Zr	3311: Zirconium pyrophosphate
$O_7Rb_4V_2$	2405: Rubidium pyrovanadate
O_7Re_2	2358: Rhenium(VII) oxide
O_7STe_2	2869: Tellurium sulfate
O_7Tb_4	2893: Terbium (III,IV) oxide
$O_8P_2Pb_3$	1532: Lead phosphate
$O_8P_2Zn_3$	3267: Zinc phosphate
$O_8Pb_3Sb_2$	1499: Lead antimonate
O_8U_3	3102: Uranium (V,VI) oxide
O_8W_2Zr	3318: Zirconium tungstate
$O_{11}Pr_6$	2326: Praseo-dymium(III,IV) oxide
$O_{12}Pr_2S_3$	2319: Praseo-dymium sulfate
$O_{12}Rh_2S_3$	2374: Rhodium(III) sulfate
$O_{12}S_3Sb_2$	263: Antimony(III) sulfate
$O_{12}S_3Ti_2$	3016: Titanium sulfate
$O_{12}S_3V_2$	3142: Vanadium trisulfate
$O_{12}S_3Yb_2$	3187: Ytterbium sulfate
Os	2033: Osmium
P	2083: Phosphorus (black)
	2084: Phosphorus (red)
PSb	249: Antimony phosphide
PSn	2992: Tin monophosphide
PTa	2851: Tantalum phosphide
PTi	3014: Titanium phosphide
PY	3215: Yttrium phosphide
P_2S_3	2097: Phosphorus (III) sulfide
P_2S_5	2103: Phosphorus (V) sulfide
P_2Se_3	2091: Phosphorus triselenide
P_2Se_5	2102: Phosphorus (V) selenide

Formula	Entry
P₂Zn₃	3269: Zinc phosphide
P₂Zr	3310: Zirconium phosphide
P₃Sn₄	2993: Tin triphosphide
P₄	2085: Phosphorus (white)
P₄S₇	2086: Phosphorus heptasulfide
Pa	2329: Protactinium
Pb	1495: Lead
PbS	1539: Lead sulfide
PbSb	1500: Lead antimonide
PbSe	1534: Lead selenide
PbTe	1542: Lead telluride
Pd	2048: Palladium
PdS	2062: Palladium(II) sulfide
Pm	2327: Promethium
Po	2127: Polonium
	2128: Polonium
Pr	2323: Praseodymium(α)
	2324: Praseodymium(β)
PrSi₂	2318: Praseodymium silicide
Pr₂S₃	2321: Praseodymium sulfide
Pr₂Te₃	2322: Praseodymium telluride
Pt	2105: Platinum
PtSi	2108: Platinum silicide
Pu	2119: Plutonium
Ra	2331: Radium
Rb	2376: Rubidium
Rb₂S	2408: Rubidium sulfide
Rb₂Se	2406: Rubidium selenide
Re	2337: Rhenium
ReS₂	2350: Rhenium(IV) sulfide
ReSe₂	2348: Rhenium(IV) selenide
ReSi₂	2349: Rhenium(IV) silicide
ReTe₂	2351: Rhenium(IV) telluride
Re₂S₇	2359: Rhenium(VII) sulfide
Rh	2360: Rhodium
Rn	2336: Radon
Ru	2414: Ruthenium
S	2826: Sulfur(α)
	2827: Sulfur(β)
	2828: Sulfur(γ)
SSe	2472: Selenium monosulfide
SSi	2495: Silicon monosulfide
SSn	2763: Stannous sulfide
SSr	2810: Strontium sulfide
STi	3009: Titanium monosulfide
STl₂	2932: Thallium(I) sulfide
SZn	3282: Zinc sulfide(α)
	3283: Zinc sulfide(β)
S₂Se	2469: Selenium disulfide
S₂Si	2494: Silicon disulfide
S₂Sn	2748: Stannic sulfide
S₂Ta	2839: Tantalum disulfide
S₂Te	2865: Tellurium disulfide
S₂Th	2972: Thorium sulfide
S₂Ti	3005: Titanium disulfide
S₂V₂	3134: Vanadium sulfide
S₂W	3053: Tungsten disulfide
S₂Zr	3316: Zirconium sulfide
S₃Sb₂	264: Antimony(III) sulfide
S₃Sc₂	2460: Scandium sulfide
S₃Sm₂	2444: Samarium sulfide
S₃Tb₂	2892: Terbium sulfide
S₃Ti₂	3026: Titanium trisulfide
S₃Tm₂	2988: Thulium sulfide
S₃V₂	3143: Vanadium trisulfide
S₃W	3075: Tungsten trisulfide
S₃Y₂	3217: Yttrium sulfide
S₄Se₄	2477: Selenium sulfide
S₅Sb₂	274: Antimony(V) sulfide
S₅V₂	3132: Vanadium pentasulfide
S₆Se₂	2471: Selenium hexasulfide
Sb	246: Antimony
SbY	3193: Yttrium antimonide
SbZn	3224: Zinc antimonide
Sb₂Se₃	262: Antimony(III) selenide
Sb₂Te₃	265: Antimony(III) telluride
Sc	2448: Scandium
Sc₂Te₃	2461: Scandium telluride
Se	2464: Selenium
	2465: Selenium (β)
	2482: Selenium(α)
SeSn	2760: Stannous selenide
SeSr	2806: Strontium selenide
SeTl₂	2930: Thallium(I) selenide
SeZn	3274: Zinc selenide
Se₂Sn	2746: Stannic selenide
Se₂Ta	2852: Tantalum selenide
Se₂Th	2967: Thorium selenide
Se₂Ti	3004: Titanium diselenide
Se₂W	3051: Tungsten diselenide
Se₂Zr	3312: Zirconium selenide
Si	2485: Silicon
SiV₃	3126: Vanadium monosilicide
Si₂Sm	2442: Samarium silicide
Si₂Sr	2807: Strontium silicide
Si₂Ta	2853: Tantalum silicide
Si₂Tb	2890: Terbium silicide
Si₂Th	2968: Thorium silicide
Si₂Ti	3015: Titanium silicide
Si₂Tm	2986: Thulium silicide
Si₂V	3122: Vanadium disilicide
Si₂W	3052: Tungsten disilicide
Si₂Yb	3186: Ytterbium silicide
Si₂Zr	3314: Zirconium silicide
Si₃Ta₅	2856: Tantalum trisilicide
Si₃Ti₅	3025: Titanium trisilicide
Si₃W₅	3074: Tungsten trisilicide
Sm	2425: Samarium
Sm₂Te₃	2445: Samarium telluride
Sn	2989: Tin (gray)
	2990: Tin (white)
SnTe	2765: Stannous telluride
Sr	2766: Strontium
T₂	3037: Tritium
Ta	2834: Tantalum
TaTe₂	2854: Tantalum telluride
Tb	2876: Terbium
Tc	2857: Technetium
Te	2860: Tellurium
TeZn	3286: Zinc telluride
Te₂Ti	3006: Titanium ditelluride
Te₂W	3067: Tungsten telluride
Te₂Zr	3317: Zirconium telluride
Th	2947: Thorium
Ti	2995: Titanium
Tl	2903: Thallium
Tm	2974: Thulium
U	3078: Uranium
V	3113: Vanadium
W	3039: Tungsten
x = 0.8 to 1.5	662: Carbon
Xe	3157: Xenon
Y	3189: Yttrium

Name/Synonym Index

NAME/SYNONYM INDEX

Chromium(III) fluoride trihydrate	814
Chromium(III) hydroxide trihydrate	815
Chromium(III) iodide	816
Chromium(III) nitrate	817
Chromium(III) nitrate nonahydrate	818
Chromium(III) oxide	819
Chromium(III) perchlorate	820
Chromium(III) phosphate	821
Chromium(III) phosphate hemiheptahydrate	822
Chromium(III) phosphate hexahydrate	823
Chromium(III) potassium sulfate dodecahydrate	824
Chromium(III) sulfate	825
Chromium(III) sulfate hydrate	826
Chromium(III) sulfate octadecahydrate	827
Chromium(III) sulfide	828
Chromium(III) telluride	829
Chromium(IV) chloride	830
Chromium(IV) oxide	831
Chromium(VI) morpholine	832
Chromium(VI) oxide	833
Chromium(VI) oxychloride	834
Chromous acetate monohydrate	794
Chromous bromide	795
Chromous chloride	796
Chromous chloride tetrahydrate	797
Chromous fluoride	798
Chromous formate monohydrate	799
Chromous oxalate monohydrate	800
Chromous sulfate pentahydrate	801
Chromyl chloride	834
Chrysoberyl	409
Chrysocolla	1002
Cinnabar	1853
Cinnabar	1854
Cisplatin	1041
Citric acid, cobalt(II) salt dihydrate	871
Citric acid, lead salt trihydrate	1514
Citric acid, magnesium salt tetradecahydrate	1687
Citric acid, trilithium salt	1576
Citric acid, tripotassium salt monohydrate	2153
Citric acid, trisilver salt	2517
Citric acid, trisodium salt dihydrate	2581
Citric acid, zinc salt dihydrate	3240
Claudetite	287
Claudetite	288
Clausthalite	1534
Clinoenstatite	1727
Cobalt	835
Cobalt aluminate	836
Cobalt antimonide	837
Cobalt arsenic sulfide	838
Cobalt arsenide	839
Cobalt arsenide	840
Cobalt arsenide	841
Cobalt black	917
Cobalt boride	842
Cobalt boride	843
Cobalt carbonyl	844
Cobalt dichromium tetraoxide	870
Cobalt disilicide	845
Cobalt disulfide	846
Cobalt dodecacarbonyl	847

Cobalt metaborate hydrate	848
Cobalt metatitanate	849
Cobalt molybdate	850
Cobalt nitrosocarbonyl	851
Cobalt nitrosodicarbonyl	852
Cobalt orthosilicate	900
Cobalt orthotitanate	853
Cobalt phosphide	854
Cobalt stearate	855
Cobalt tricarbonyl nitrosyl	851
Cobalt trifluoride	913
Cobalt zirconate	856
Cobalt(II) acetate tetrahydrate	857
Cobalt(II) acetylacetonate	858
Cobalt(II) arsenate octahydrate	859
Cobalt(II) basic carbonate	860
Cobalt(II) bromate hexahydrate	861
Cobalt(II) bromide	862
Cobalt(II) bromide hexahydrate	863
Cobalt(II) carbonate	864
Cobalt(II) chlorate hexahydrate	865
Cobalt(II) chloride	866
Cobalt(II) chloride dihydrate	867
Cobalt(II) chloride hexahydrate	868
Cobalt(II) chromate	869
Cobalt(II) chromite	870
Cobalt(II) citrate dihydrate	871
Cobalt(II) cyanide dihydrate	872
Cobalt(II) cyanide trihydrate	873
Cobalt(II) diiron tetroxide	874
Cobalt(II) ferricyanide	875
Cobalt(II) ferrocyanide hydrate	876
Cobalt(II) fluoride	877
Cobalt(II) fluoride tetrahydrate	878
Cobalt(II) hexafluorosilicate hexahydrate	879
Cobalt(II) hydroxide	880
Cobalt(II) iodate	881
Cobalt(II) iodide	882
Cobalt(II) iodide dihydrate	883
Cobalt(II) iodide hexahydrate	884
Cobalt(II) linoleate	885
Cobalt(II) molybdate monohydrate	886
Cobalt(II) nitrate	887
Cobalt(II) nitrate hexahydrate	888
Cobalt(II) nitrite	889
Cobalt(II) oleate	890
Cobalt(II) oxalate	891
Cobalt(II) oxalate dihydrate	892
Cobalt(II) oxide	893
Cobalt(II) perchlorate	894
Cobalt(II) perchlorate hexahydrate	895
Cobalt(II) phosphate octahydrate	896
Cobalt(II) selenate pentahydrate	897
Cobalt(II) selenide	898
Cobalt(II) selenite dihydrate	899
Cobalt(II) silicate	900
Cobalt(II) stannate	901
Cobalt(II) sulfate	902
Cobalt(II) sulfate heptahydrate	903
Cobalt(II) sulfate monohydrate	904
Cobalt(II) sulfide	905
Cobalt(II) telluride	906
Cobalt(II) tetrathiocyanatomercurate(II)	1856
Cobalt(II) thiocyanate	907
Cobalt(II) thiocyanate trihydrate	908
Cobalt(II) tungstate	909
Cobalt(II,III) oxide	910
Cobalt(III) acetate	911

Cobalt(III) acetylacetonate	912
Cobalt(III) fluoride	913
Cobalt(III) fluoride dihydrate	914
Cobalt(III) hydroxide trihydrate	915
Cobalt(III) nitrate	916
Cobalt(III) oxide	917
Cobalt(III) oxide hydroxide	918
Cobalt(III) sepulchrate trichloride	919
Cobalt(III) sulfide	920
Cobaltic acetate	911
Cobaltic fluoride	913
Cobaltic oxide	917
Cobaltic oxide monohydrate	918
Cobaltic-cobaltous oxide	910
Cobaltite	838
Cobaltocene	921
Cobaltocenium hexafluorophosphate	922
Cobaltous acetate tetrahydrate	857
Cobaltous ammonium phosphate	132
Cobaltous ammonium sulfate	133
Cobaltous arsenate octahydrate	859
Cobaltous bromide	862
Cobaltous bromide hexahydrate	863
Cobaltous chlorate hexahydrate	865
Cobaltous chloride	866
Cobaltous chloride dihydrate	867
Cobaltous chloride hexahydrate	868
Cobaltous chromate	869
Cobaltous cyanide	872
Cobaltous cyanide	873
Cobaltous ferricyanide	875
Cobaltous ferricyanide	879
Cobaltous ferrite	874
Cobaltous ferrocyanide	876
Cobaltous fluoride	877
Cobaltous fluoride tetrahydrate	878
Cobaltous hydroxide	880
Cobaltous iodide	882
Cobaltous iodide hexahydrate	883
Cobaltous iodide hexahydrate	884
Cobaltous nitrate	887
Cobaltous nitrate hexahydrate	888
Cobaltous oxalate	891
Cobaltous oxalate dihydrate	892
Cobaltous oxide	893
Cobaltous phosphate octahydrate	896
Cobaltous sulfate	902
Cobaltous sulfide	905
Cobaltous thiocyanate	907
Cobaltous thiocyanate trihydrate	908
Colemanite	599
Copper	923
Copper arsenide	924
Copper citrate hemipentahydrate	925
Copper diiron tetroxide	970
Copper nitride	926
Copper phosphide	927
Copper silicide	928
Copper vanadate	929
Copper zirconate	930
Copper(I) acetate	931
Copper(I) acetylide	932
Copper(I) azide	933
Copper(I) bromide	934
Copper(I) chloride	935
Copper(I) cyanide	936
Copper(I) hydride	937
Copper(I) iodide	938
Copper(I) mercury iodide	939
Copper(I) oxide	940

hexahydrate	
Holmium silicide	1341
Holmium sulfate octahydrate	1342
Holmium sulfide	1343
Holmium telluride	1344
Holmium tetraboride	1345
Homium boride	1345
Hydrazine	1346
Hydrazine acetate	1347
Hydrazine azide	1348
Hydrazine dihydrochloride	1349
Hydrazine dinitrate	1350
Hydrazine hydrate	1351
Hydrazine monohydrate	1352
Hydrazine monohydrobromide	1353
Hydrazine monohydrochloride	1354
Hydrazine monohydroiodide	1355
Hydrazine mononitrate	1356
Hydrazine monooxalate	1357
Hydrazine perchlorate	1358
hemihydrate	
Hydrazine sulfate	1359
Hydrazoic acid	1360
Hydrocerussite	1505
Hydrocyanic acid	1365
Hydrofluoric acid, 70%	1361
Hydrogen	1362
Hydrogen azide	1360
Hydrogen bromide	1363
Hydrogen chloride	1364
Hydrogen cyanide	1365
Hydrogen fluoride	1366
Hydrogen hexachloroiridate(IV)	1367
hydrate	
Hydrogen	1368
hexachloroplatinate(IV)	
Hydrogen hexachloroplatinate	1369
(IV) hexahydrate	
Hydrogen hexafluorosilicic acid	1370
Hydrogen	1371
hexahydroxyplatinate(IV)	
Hydrogen iodide	1372
Hydrogen oxide	3156
Hydrogen peroxide	1373
Hydrogen selenide	1374
Hydrogen sulfide	1375
Hydrogen telluride	1376
Hydrogen tetrabromoaurate(III)	1377
pentahydrate	
Hydrogen	1378
tetracarbonylferrate(II)	
Hydrogen tetrachloroaurate(III)	1379
hydrate	
Hydrogen tetrachloroaurate(III)	1380
tetrahydrate	
Hydromagnesite	1677
Hydrophilite	581
Hydroxybutanedioic acid,	118
monoammonium salt	
Hydroxydimethylarsine oxide	507
Hydroxylamine	1381
Hydroxylamine hydrobromide	1382
Hydroxylamine hydrochloride	1383
Hydroxylamine perchlorate	1384
Hydroxylamine sulfate	1385
Hydroxylapatite	631
Hypo	2730
Hypobromous acid	1386
Hypochlorous acid	1387
Hypophosphoric acid	1388
Hypophosphorous acid	1389
Ilmenite	1196
Indium	1390

Indium acetate	1391
Indium acetylacetonate	1392
Indium antimonide	1393
Indium arsenide	1394
Indium nitride	1395
Indium phosphide	1396
Indium trichloride	1404
Indium trifluoride	1406
Indium trifluoride trihydrate	1407
Indium(I) bromide	1397
Indium(I) chloride	1398
Indium(I) iodide	1399
Indium(II) bromide	1400
Indium(II) chloride	1401
Indium(II) sulfide	1402
Indium(III) bromide	1403
Indium(III) chloride	1404
Indium(III) chloride tetrahydrate	1405
Indium(III) fluoride	1406
Indium(III) fluoride trihydrate	1407
Indium(III) hydroxide	1408
Indium(III) iodide	1409
Indium(III) nitrate trihydrate	1410
Indium(III) oxide	1411
Indium(III) perchlorate	1412
octahydrate	
Indium(III) phosphate	1413
Indium(III) selenide	1414
Indium(III) sulfate	1415
Indium(III) sulfide	1416
Indium(III) telluride	1417
Iodic acid	1418
Iodine	1419
Iodine cyanide	1420
Iodine dioxide	1421
Iodine heptafluoride	1422
Iodine monobromide	1423
Iodine monochloride	1424
Iodine nonoxide	1425
Iodine pentafluoride	1426
Iodine pentoxide	1427
Iodine trichloride	1428
Iodine(V) oxide	1427
Iodyrite	2528
Iridium	1429
Iridium carbonyl	1430
Iridium hexafluoride	1431
Iridium pentafluoride	1432
Iridium tribromide tetrahydrate	1435
Iridium trichloride	1436
Iridium trioxide	1438
Iridium(I) chlorotricarbonyl	1433
Iridium(III) acetylacetonate	1434
Iridium(III) bromide	1435
tetrahydrate	
Iridium(III) chloride	1436
Iridium(III) chloride hydrate	1437
Iridium(III) oxide	1438
Iridium(IV) oxide	1439
Iron	1440
Iron antimonide	1441
Iron arsenide	1442
Iron boride	1443
Iron boride	1444
Iron disilicide	1445
Iron disulfide	1446
Iron dodecacarbonyl	1447
Iron molybdate	1448
Iron nonacarbonyl	1449
iron pentacarbonyl	1450
Iron silicide	1451
Iron telluride	1452
Iron tungstate	1453

Iron zirconate	1454
Iron(II) acetate tetrahydrate	1165
Iron(II) acetate tetrahydrate	1166
Iron(II) bromide	1169
Iron(II) bromide hexahydrate	1170
Iron(II) bromide hexahydrate	1171
Iron(II) chloride	1173
Iron(II) chloride dihydrate	1174
Iron(II) chloride tetrahydrate	1175
Iron(II) citrate monohydrate	1177
Iron(II) fluoride	1178
Iron(II) fluoride tetrahydrate	1179
Iron(II) hexafluorosilicate	1180
hexahydrate	
Iron(II) hydroxide	1181
Iron(II) iodide	1182
Iron(II) nitrate hexahydrate	1184
Iron(II) orthoarsenate	1168
hexahydrate	
Iron(II) oxalate dihydrate	1185
Iron(II) perchlorate hexahydrate	1187
Iron(II) selenide	1190
Iron(II) sulfate	1191
Iron(II) sulfide	1194
Iron(II) thiocyanate trihydrate	1195
Iron(II,III) oxide	1455
Iron(III) bromide	1134
Iron(III) chloride hexahydrate	1136
Iron(III) chromate	1137
Iron(III) citrate pentahydrate	1138
Iron(III) dichromate	1139
Iron(III) fluoride	1141
Iron(III) fluoride trihydrate	1142
Iron(III) hypophosphite	1144
Iron(III) hyroxide	1143
Iron(III) nitrate nonahydrate	1146
Iron(III) oxalate	1147
Iron(III) perchlorate	1152
Iron(III) perchlorate	1151
hexahydrate	
Iron(III) phosphate dihydrate	1153
Iron(III) phosphate hydrate	1154
Iron(III) pyrpophosphate	1155
nonahydrate	
Iron(III) sulfate	1156
Iron(III) sulfate hydrate	1157
Iron(III) thiocyanate	1159
Iron(III) vanadate	1145
Jeremejevite	21
Kainite	2211
Kalinite	2136
Kalium	2131
Kaolin	82
Kieserite	1735
Knorre's salt	2732
Koettigite	3225
Krypton	1456
Krypton difluoride	1457
Krypton fluoride	1458
hexafluoroantimonate	
Krypton fluoride	1459
monodecafluoroantimonate	
Krypton fluoride	1460
monodecafluorotantalate	
Krypton trifluoride	1461
hexafluoroantimonate	
Kyanite	81
Langbeinite	2212
Lansfordite	1679
Lanthanum	1462
Lanthanum acetate hydrate	1463
Lanthanum acetylacetonate	1464
hydrate	

Manganese(II) telluride	1797
Manganese(II) titanate	1798
Manganese(II) tungstate	1799
Manganese(II) zirconate	1800
Manganese(II,III) oxide	1801
Manganese(III) acetate dihydrate	1802
Manganese(III) acetylacetonate	1803
Manganese(III) fluoride	1804
Manganese(III) hydroxide	1805
Manganese(III) oxide	1806
Manganese(IV) oxide	1807
Manganese(IV) telluride	1808
Manganese(VII) oxide	1809
Manganite	1805
Manganjustite	1762
Manganosite	1787
Manganous citrate	1773
Manganous fluoride	1776
Marshite	938
Mascagnite	223
Massicot	1529
Melanterite	1192
Mendelevium	1810
Mercallite	2204
Mercaptoacetic acid, calcium salt trihydrate	652
Mercuric acetate	1827
Mercuric ammonium chloride	190
Mercuric arsenate	1828
Mercuric barium iodide	389
Mercuric chromate	1817
Mercuric chromate	1834
Mercuric cyanide	1835
Mercuric dichromate	1836
Mercuric fluoride	1837
Mercuric iodate	1839
Mercuric iodide	1840
Mercuric iodide	1841
Mercuric nitrate	1842
Mercuric nitrate monohydrate	1843
Mercuric oleate	1844
Mercuric oxycyanide	1848
Mercuric sulfate	1852
Mercuric thiocyante	1857
Mercurous acetate	1812
Mercurous bromide	1813
Mercurous chlorate	1815
Mercurous chloride, calomel	1816
Mercurous fluoride	1818
Mercurous iodide	1820
Mercurous nitrate	1821
Mercurous oxide	1823
Mercurous sulfate	1825
Mercury	1811
Mercury(I) acetate	1812
Mercury(I) bromide	1813
Mercury(I) carbonate	1814
Mercury(I) chlorate	1815
Mercury(I) chloride	1816
Mercury(I) chromate	1817
Mercury(I) fluoride	1818
Mercury(I) iodate	1819
Mercury(I) iodide	1820
Mercury(I) nitrate dihydrate	1821
Mercury(I) nitrate monohydrate	1822
Mercury(I) oxide	1823
Mercury(I) perchlorate tetrahydrate	1824
Mercury(I) sulfate	1825
Mercury(I) sulfide	1826
Mercury(II) acetate	1827
Mercury(II) arsenate	1828

Mercury(II) basic carbonate	1829
Mercury(II) benzoate monohydrate	1830
Mercury(II) bromide	1831
Mercury(II) chloride	1832
Mercury(II) chloride ammoniated	1833
Mercury(II) chromate	1834
Mercury(II) cyanide	1835
Mercury(II) dichromate	1836
Mercury(II) fluoride	1837
Mercury(II) fulminate	1838
Mercury(II) iodate	1839
Mercury(II) iodide(α)	1840
Mercury(II) iodide(β)	1841
Mercury(II) nitrate hemihydrate	1842
Mercury(II) nitrate monohydrate	1843
Mercury(II) oleate	1844
Mercury(II) oxalate	1845
Mercury(II) oxide red	1846
Mercury(II) oxide yellow	1847
Mercury(II) oxycyanide	1848
Mercury(II) perchlorate trihydrate	1849
Mercury(II) phosphate	1850
Mercury(II) selenide	1851
Mercury(II) silver iodide	2548
Mercury(II) sulfate	1852
Mercury(II) sulfide(α)	1853
Mercury(II) sulfide(β)	1854
Mercury(II) telluride	1855
Mercury(II) tetrathiocyanatocobaltate(II)	1856
Mercury(II) thiocyanate	1857
Metakaolinite	77
Metaphosphoric acid	1858
Metavanadic acid	1859
Methylcyclopentadienyl-manganese tricarbonyl	1860
Methylmercury	1054
Microcosmic salt	2560
Millerite	1978
Minium	1553
Mirabilite	2704
Mirabilite	2705
Misenite	2204
β-Mo$_2$C	1866
Mohr's salt	146
Molybdenite	1868
Molybdenum	1861
Molybdenum acetate dimer	1862
Molybdenum aluminide	1863
Molybdenum boride	1864
Molybdenum carbide	1865
Molybdenum carbide	1866
Molybdenum carbonyl	1867
Molybdenum dichloride dioxide	1890
Molybdenum dioxide	1884
Molybdenum dioxide difluoride	1891
Molybdenum disilicide	1875
Molybdenum disulfide	1868
Molybdenum hexacarbonyl	1867
Molybdenum hexafluoride	1892
Molybdenum metaphosphate	1869
Molybdenum mononitride	1870
Molybdenum nitride	1871
Molybdenum oxytetrafluoride	1872
Molybdenum pentaboride	1873
Molybdenum pentachloride	1887
Molybdenum phosphide	1874
Molybdenum sesquioxide	1880
Molybdenum silicide	1875
Molybdenum tetrachloride	1883

Molybdenum trioxide	1893
Molybdenum trisulfide	1895
Molybdenum(II) chloride	1876
Molybdenum(II) iodide	1877
Molybdenum(III) bromide	1878
Molybdenum(III) chloride	1879
Molybdenum(III) oxide	1880
Molybdenum(III) sulfide	1881
Molybdenum(IV) bromide	1882
Molybdenum(IV) chloride	1883
Molybdenum(IV) oxide	1884
Molybdenum(IV) selenide	1885
Molybdenum(IV) telluride	1886
Molybdenum(V) chloride	1887
Molybdenum(V) oxytrichloride	1888
Molybdenum(VI) acid monohydrate	1889
Molybdenum(VI) dioxydichloride	1890
Molybdenum(VI) dioxydifluoride	1891
Molybdenum(VI) fluoride	1892
Molybdenum(VI) oxide	1893
Molybdenum(VI) oxytetrachloride	1894
Molybdenum(VI) sulfide	1895
Molybdic acid monohydrate	1889
Molybdic anhydride	1893
Molybdic silicic acid hydrate	1896
Molysite	1135
Monazite	720
Monobarium silicate	353
Monofluorophosphoric acid	1897
Monoiodosilane	1898
Morpholine chromate	832
Mosaic gold	2748
Mullite	80
Nantokite	935
Naples yellow	1499
Native aluminum oxide, bauxite	62
Native aluminum oxide, bauxite	64
Native aluminum oxide, bauxite	65
Native aluminum oxide, bauxite	66
Natrium	2551
Neodymia	1916
Neodymium	1899
Neodymium acetate monohydrate	1900
Neodymium boride	1901
Neodymium bromate nonahydrate	1902
Neodymium bromide	1903
Neodymium carbonate hydrate	1904
Neodymium cerium copper oxide	1905
Neodymium chloride	1906
Neodymium chloride hexahydrate	1907
Neodymium fluoride	1908
Neodymium hexafluoroacetyl-acetonate dihydrate	1909
Neodymium hydride	1910
Neodymium hydroxide	1911
Neodymium iodide	1912
Neodymium nitrate hexahydrate	1913
Neodymium nitride	1914
Neodymium oxalate decahydrate	1915
Neodymium oxide	1916
Neodymium perchlorate hexahydrate	1917
Neodymium phosphate hydrate	1918
Neodymium silicide	1919

Titanium diiodide	3000	Tungsten	3039	Uranium tricarbide	3094
Titanium dioxide	3001	Tungsten boride	3040	Uranium trichloride	3095
Titanium dioxide	3002	Tungsten boride	3041	Uranium trifluoride	3096
Titanium dioxide	3003	Tungsten carbide	3042	Uranium trihydride	3097
Titanium diselenide	3004	Tungsten carbide	3043	Uranium trinitride	3098
Titanium disilicide	3015	Tungsten carbonyl	3044	Uranium trioxide	3099
Titanium disulfide	3005	Tungsten dibromide	3045	Uranium(IV) sulfate octahydrate	3100
Titanium ditelluride	3006	Tungsten dichloride	3046	Uranium(IV) sulfate	3101
Titanium hydride	3007	Tungsten diiodide	3047	tetrahydrate	
Titanium isopropoxide	3008	Tungsten dinitride	3048	Uranium(V,VI) oxide	3102
Titanium isopropylate	3008	Tungsten dioxide	3049	Uranium(VI) oxide	3099
Titanium monosulfide	3009	Tungsten dioxydibromide	3050	Uranyl acetate dihydrate	3103
Titanium monoxide	3010	Tungsten diselenide	3051	Uranyl acetylacetonate	3104
Titanium nitride	3011	Tungsten disilicide	3052	Uranyl carbonate	3105
Titanium oxalate decahydrate	3012	Tungsten disulfide	3053	Uranyl chloride	3106
Titanium oxysulfate	3013	Tungsten hexabromide	3054	Uranyl chloride trihydrate	3107
Titanium phosphide	3014	Tungsten hexacarbonyl	3044	Uranyl hydrogen phosphate	3108
Titanium sesquisulfide	3026	Tungsten hexachloride	3055	tetrahydrate	
Titanium silicide	3015	Tungsten hexafluoride	3056	Uranyl nitrate hexahydrate	3109
Titanium sulfate	3016	Tungsten nitride	3057	Uranyl oxalate trihydrate	3110
Titanium tetrabromide	3017	Tungsten oxychloride	3058	Uranyl sulfate monohydrate	3111
Titanium tetrachloride	3018	Tungsten oxydichloride	3059	Uranyl sulfate trihydrate	3112
Titanium tetrafluoride	3019	Tungsten oxydiiodide	3060	Valentinite	257
Titanium tetraiodide	3020	Tungsten oxytetrabromide	3061	Valeric acid	3289
Titanium tribromide	3021	Tungsten oxytetrafluoride	3062	Vanadium	3113
Titanium trichloride	3022	Tungsten oxytrichloride	3063	Vanadium bis(cyclopentadienyl)	3114
Titanium trifluoride	3023	Tungsten pentaboride	3064	dichloride	
Titanium trioxide	3024	Tungsten pentabromide	3065	Vanadium carbide	3115
Titanium trisilicide	3025	Tungsten pentachloride	3066	Vanadium carbonyl	3116
Titanium trisulfide	3026	Tungsten telluride	3067	Vanadium diboride	3117
Titanium(II) iodide	3000	Tungsten tetrabromide	3068	Vanadium dibromide	3118
Titanium(II) oxide	3010	Tungsten tetrachloride	3069	Vanadium dichloride	3119
Titanium(IV) bromide	3017	Tungsten tetraiodide	3070	Vanadium diiodide	3120
Titanium(IV) oxide	3027	Tungsten tribromide	3071	Vanadium dioxide	3121
acetylacetonate		Tungsten triiodide	3072	Vanadium disilicide	3122
Titanium(IV) sulfide	3005	Tungsten trioxide	3073	Vanadium gallide	3123
Titanocene dichloride	3028	Tungsten trisilicide	3074	Vanadium hexacarbonyl	3116
Titanous sulfate	3016	Tungsten trisulfide	3075	Vanadium monoboride	3124
TPAP	2902	Tungsten(IV) bromide	3068	Vanadium monocarbide	3125
Triatomic oxygen	2047	Tungsten(IV) chloride	3069	Vanadium monosilicide	3126
Tribromosilane	3029	Tungsten(IV) oxide	3049	Vanadium monoxide	3127
Tricalcium dicitrate tetrahydrate	589	Tungsten(IV) telluride	3067	Vanadium nitride	3128
Trichlorosilane	3030	Tungsten(V) bromide	3065	Vanadium oxytrichloride	3129
Tridymite	2490	Tungsten(VI) chloride	3055	Vanadium oxytrifluoride	3130
Tridymite	2491	Tungsten(VI) dichloride dioxide	3059	Vanadium pentafluoride	3131
Tridymite	2492	Tungsten(VI) fluoride	3056	Vanadium pentasulfide	3132
Triethylphosphine	3031	Tungsten(VI) oxide	3073	Vanadium pentoxide	3133
Trifluoroacetic acid, cesium salt	763	Tungsten(VI) tetrachloride	3058	Vanadium sulfide	3134
Trifluoroacetic acid,	2942	monoxide		Vanadium tetrachloride	3135
thallium(III) salt		Tungstenite	3053	Vanadium tetrafluoride	3136
Trifluoromethane	3032	Tungstic acid	3076	Vanadium tribromide	3137
Trifluorosilane	3033	Tungstophosphate	3077	Vanadium trichloride	3138
Triiodosilane	3034	Tungstophosphoric acid	2104	Vanadium trifluoride	3139
Trimethylsilyl fluoride	1206	Tungstophosphoric acid hydrate	3077	Vanadium trifluoride trihydrate	3140
Triosmium dodecacarbonyl	2035	Tungstosilicic acid	2503	Vanadium trioxide	3141
Triruthenium dodecacarbonyl	2416	Unichrome	998	Vanadium trisulfate	3142
Tris(cyclopentadienyl)-	1493	Uraninite	3081	Vanadium trisulfide	3143
lanthanum		Uranium	3078	Vanadium(II) iodide	3120
Tris(cyclopentadienyl)-	1925	Uranium diboride	3079	Vanadium(II) oxide	3127
neodymium		Uranium dicarbide	3080	Vanadium(II) sulfate	3144
Tris(cyclopentadienyl)samarium	2446	Uranium dioxide	3081	heptahydrate	
Tris(cyclopentadienyl)scandium	2462	Uranium hexafluoride	3082	Vanadium(III) acetylacetonate	3145
Tris(ethylenediamine)	3035	Uranium monocarbide	3083	Vanadium(III) bromide	3137
cobalt(III) chloride trihydrate		Uranium mononitride	3084	Vanadium(III) chloride	3138
Tris(triphenylphosphine)-	3036	Uranium pentabromide	3085	Vanadium(III) fluoride	3140
rhodium(I) chloride		Uranium pentachloride	3086	Vanadium(III) fluoride	3139
Trisilane	2498	Uranium pentafluoride	3087	Vanadium(III) oxide	3141
Trisodium phosphate	2686	Uranium tetraboride	3088	Vanadium(III) sulfate	3142
Tritium	3037	Uranium tetrabromide	3089	Vanadium(III) sulfide	3143
Tritium dioxide	3038	Uranium tetrachloride	3090	Vanadium(IV) chloride	3135
Troilite	1194	Uranium tetrafluoride	3091	Vanadium(IV) fluoride	3136
Trona	2570	Uranium tetraiodide	3092	Vanadium(IV) oxide	3121
Trona	2571	Uranium tribromide	3093	Vanadium(V) fluoride	3131